TRANSPORTATION AND TRAFFIC THEORY 2007

Related books

HENSHER & BUTTON	Handbooks in Transport 6-Volume Set
MAHMASSANI	Transportation and Traffic Theory: Flow, Dynamics and Human Interaction (ISTTT16)
TAYLOR	Transportation and Traffic Theory in the 21st Century (ISTTT15)
AXHAUSEN	Moving Through Nets
DOHERTY & LEE-GOSSELIN	Integrated Land-Use and Transport Models
STOPHER & STECHER	Transport Survey Methods
TANIGUCHI & THOMPSON	Recent Advances in City Logistics
TIMMERMANS	Progress in Activity-Based Analysis
GOULIAS	Transport Science and Technology
MACARIO	Competition and Ownership in Land Passenger Transport

Related Journals

Transportation Research Part A: Policy and Practice
Editor: Phil Goodwin

Transportation Research Part B: Methodological
Editor: Fred Mannering

Transportation Research Part C: Emerging Technologies
Editor: Markos Papageorgiou

Transportation Research Part D: Transport and Environment
Editor: Ken Button

Transportation Research Part E: Logistics and Transportation Review
Editor: Wayne Talley

Transportation Research Part F: Traffic Psychology and Behaviour
Editors: Talib Rothengatter and John Groeger

Transport Policy
Editors: Moshe Ben-Akiva, Yoshitsugu Hayashi, John Preston & Peter Bonsall

TRANSPORTATION AND TRAFFIC THEORY 2007

Papers selected for presentation at ISTTT17, a peer reviewed series since 1959

edited by

RICHARD E. ALLSOP
University College London, UK

MICHAEL G. H. BELL
Imperial College London, UK

BENJAMIN G. HEYDECKER
University College London, UK

ELSEVIER

Amsterdam • Boston • Heidelberg • London • New York • Oxford
Paris • San Diego • San Francisco • Singapore • Sydney • Tokyo

Elsevier
Linacre House, Jordan Hill, Oxford OX2 8DP, UK
Radarweg 29, PO Box 211, 1000 AE Amsterdam, The Netherlands

First edition 2007

British Library Cataloguing in Publication Data
A catalogue record for this book is available from the British Library

Library of Congress Cataloging-in-Publication Data
A catalog record for this book is available from the Library of Congress

ISBN: 978-0-08-045375-0

For information on all Elsevier publications
visit our website at books.elsevier.com

Printed and bound in The Netherlands

07 08 09 10 11 10 9 8 7 6 5 4 3 2 1

Contents

Preface

The series of International Symposia on Transportation and Traffic Theory provides the main regular opportunity for those concerned with theoretical approaches to the analysis of transport and traffic systems for the benefit of society to meet and exchange their latest thinking on advances in understanding and application. The symposia also provide a unique opening for those who are interested in contributing to or gaining a deeper understanding of this field. The ultimate aim of the symposia is to contribute, though a more complete understanding of transport and traffic phenomena, to more effective approaches to the planning, design and management of transport systems. However, the focus of the symposia is on the underlying science, leading to papers on the more fundamental aspects of transport and traffic, especially where the existing theory is felt to be underdeveloped, inadequate, or in need of correction. In these cases, the formulation, development and testing of new theories as well as the critique, refinement, and extension of existing ones has led to the generation of new perspectives.

After the first symposium that was held in 1959 at the General Motors Research Laboratories in Michigan, the symposia in this series have been held normally every 3 years. However, the International Advisory Committee has taken the view that the international flow of excellent material is now sufficiently great to justify an increase in the frequency to every 2 years from 2005. The level of interest and support for the present symposium has validated this decision. The second symposium took place in London in 1963, and after a total of 7 symposia in north America, 2 in each of Japan and Australia, and symposia in Germany, The Netherlands, France and Israel, we were glad to welcome colleagues from around the world to London for the 17th ISTTT, 23-25 July 2007.

Attention at the Symposia has tracked the themes of interest to leading international researchers in the subject area. By providing a forum at which new insights have been shared and hitherto unsuspected links between different lines of analysis have been explored, the Symposia have facilitated significant advances. The meeting of minds at typically two or three Symposia has engendered ground-breaking developments that have contributed strongly to innovation in practice, often stimulating continuing research in several countries. Some of these themes have been pursued since the early symposia, whilst others have emerged during the series.

The contributions to the present symposium address a wide range of topics on the planning, modelling, management and operation of transport systems. The processes and phenomena that are considered are often both complicated and subtle, reflecting the intricate operation of systems such as transport that are managed to some extent but that are used by many individuals each making decisions on a personal basis. The technical challenges that arise from this are substantial. Within this scope, the contributions range from developments of established topics to the exploratory application of novel methodologies. Several contributions address aspects of traffic flow modelling, including representation of following, merging and lane changing behaviour that can affect the operation of road systems profoundly. Methodological developments presented in this area include analysis of advanced model formulations and solution methods. Investigation of individual behaviour on and operation of networks has given rise to contributions on traffic assignment, both static and dynamic, considering both behaviour and management of travellers. Design and management methods including pricing are topical, and together with issues such as attitude to uncertainty, risk and reliability is given due emphasis. Management of both public access and private operations is considered with reference to logistic systems, public transport and emergency vehicles. The importance of individual decision making and response pervades the contributions, and is the focus of a group that considers aspects of travel demand.

This volume contains contributions selected for presentation at the 17th International Symposium on Transportation and Traffic Theory from the many offers of extended abstracts and subsequently invited submissions of full papers. All the chapters have therefore been subjected to a rigorous three-stage peer review process. In accordance with past practice, the papers were presented wholly in plenary sessions, leading to active discussion across the whole field by participants ranging from young entrants to those approaching retirement. Each of the 77 full papers that were submitted was reviewed by between two and four independent referees of international standing, leading to the selection of the 36 chapters presented in this volume. The result of this is that these proceedings rank with the most highly esteemed international refereed journals. A further 4 papers were identified as having similarly high quality and relevance, but could not be included because of limitations of space and time; the authors of these papers have been recommended to proceed with journal publication in the usual way. The high quality of the resulting papers, notwithstanding the reduction of the interval between successive symposia from the previously usual 3 years to 2, is a testament to the vigour and enthusiasm of researchers in this field, stimulating their activity at the highest levels.

These papers have been prepared by 86 different authors working in 12 countries, with representatives from academia, government and industry. The predominance of multi-authored papers shows that the forefront of research is now being developed by colleagues working together in a collaborative scientific style that is supported by worldwide communication systems. This mode of research is emerging rapidly and becoming established as a norm.

In working on the papers presented at the symposium and preparing this volume for publication, we have called on, and therefore wish to record our thanks to, many colleagues who have given generously of their time, skills and intellectual energy, often against demanding deadlines. Notable amongst these colleagues are, of course, the authors themselves. Alongside them we wish to thank the referees who advised so effectively on content and style of presentation, supporting the authors in their original work and the editors in making what turned out to be difficult choices between so many excellent papers. Members of the local organising committee supported us in planning and arranging the symposium and its sessions through their scientific and organisational experience. We wish to thank members of the International Advisory Committee of the ISTTT who provided regular encouragement and advice on all aspects of the symposium. We also wish to thank Panagiotis Angeloudis, Jackie Sime and Dr Jan-Dirk Schmöcker for their skill and dedication to organisational detail and DLR in Berlin for designing the original website.

Finally, we wish to thank each of the range of organisations whose financial support has made it possible for us to host the 17th ISTTT in London.

Benjamin Heydecker, Michael Bell and Richard Allsop

March 2007

Local Organising Committee

Richard Allsop	University College London
David Bayliss	Halcrow Group
Michael Bell	Imperial College London
Malachy Carey	Queens University, Belfast
Phil Davies	Transport for London
Brian Smith	Rees Jeffreys Road Fund
Benjamin Heydecker	University College London
Neil Hoose	Bittern Consulting
Geoff Hyman	Department for Transport
Frank Kelly	Department for Transport
Rod Kimber	Transport Research Foundation
Terry Mulroy	Institution of Civil Engineers
Tony Ridley	RAC Foundation
Dennis Robertson	
Sally Scarlett	Association for European Transport
John Smart	Institution of Highways and Transportation
Mike Smith	University of York
Derek Turner	Highways Agency
David Watling	University of Leeds
Pilo Willumsen	Steer Davies Gleave

Sponsors

EPSRC	The Engineering and Physical Science Research Council is the main UK government agency for funding research and training in engineering and the physical sciences, investing in a broad range of subjects – from mathematics to materials science, and from information technology to structural engineering. www.epsrc.ac.uk
RJRF	The Rees Jeffreys Road Fund was established in 1950 by the late William Rees Jeffreys to support research and scholarship on highway engineering, vehicle engineering and town planning to improve the safety and pleasantness of road transport. www.reesjeffreys.co.uk
TRF	The Transport Research Foundation group of companies is a non-profit distributing company, limited by guarantee, and established for the impartial furtherance of transport and related research, consultancy, and expert advice. www.transportresearchfoundation.co.uk
TMS Consultancy	TMS is one of the UK's leading independent Road Safety and Traffic Management consultancies providing specialist consultancy, research and training services within this field. www.tmsconsultancy.co.uk
Ove Arup	Ove Arup & Partners Ltd is a professional consultancy practice providing engineering design, planning and project management services in all areas of the built environment. www.arup.com
TPi	Transportation Planning (International) Ltd (TPi) is a leading international independent transportation planning and traffic engineering consultancy providing services to public and private sector clients. www.tpi-online.co.uk
PTV	PTV AG, founded in 1979 and now with over 650 employees worldwide, provides software, consulting and research for travel, traffic and transportation planning. www.english.ptv.de

International Advisory Committee

Honorary Members

Contributors

Zain Adam	George Butler Associates Inc Lenexa, Kansas, USA
Soyoung Ahn	Arizona State University USA
Takashi Akamatsu	Graduate School of Information Sciences Tohoku University, Japan
Jeff X Ban	Institute of Transportation Studies University of California, Berkeley, USA
Michael Bell	Centre for Transport Studies Imperial College London, UK
Giuseppe Bellei	Dipartimento di Idraulica Trasporti e Strade Università "La Sapienza", Rome, Italy
Moshe Ben-Akiva	Department of Civil and Environmental Engineering Massachusetts Institute of Technology, USA
Piet Bovy	Delft University of Technology The Netherlands
Werner Brilon	Ruhr-University Bochum Germany
Michael Cassidy	University of California, Berkeley USA
Avishai Ceder	Transportation Research Institute Technion – Israel Institute of Technology, Haifa, Israel
Anthony Chen	Department of Civil and Environmental Engineering Utah State University, USA
Estelle Chevallier	Laboratoire d'Ingénierie Circulation Transport (ENTPE/INRETS), France
Yu-Chiun Chiou	Institute of Traffic and Transportation National Chiao Tung University, Taiwan
Charisma Choudhury	Department of Civil and Environmental Engineering Massachusetts Institute of Technology, USA
Andy H F Chow	Centre for Transport Studies University College London, UK
Richard Clegg	Department of Electronics and Electrical Engineering University College London, UK
Richard Connors	Institute for Transport Studies University of Leeds, UK
François Dion	Department of Civil and Environmental Engineering Michigan State University, USA
Niamph Dundon	Department of Mathematics University of Massachusetts, Amherst, USA
Terry Friesz	Penn State University USA

Guido Gentile	Dipartimento di Idraulica Trasporti e Strade Università "La Sapienza", Rome, Italy
Bas Groothedde	Kirkman Company, Baarn The Netherlands
Xiaolei Guo	Department of Civil Engineering The Hong Kong University of Science and Technology China
Habib Haj Salem	INRETS/GRETIA France
Timothy Hau	The University of Hong Kong China
Julia Hinkel	Rostock University Germany
H W Ho	The University of Hong Kong China
Serge Hoogendoorn	Transport and Planning Department Delft University of Technology, The Netherlands
Hai-Jun Huang	School of Management Beijing University of Aeronautics and Astronautics, China
Hironori Kato	Department of Civil Engineering The University of Tokyo, Japan
Sirisak Kongsomsaksakul	Department of Civil and Environmental Engineering Utah State University, USA
Reinhart Kühne	German Aerospace Center Transportation Studies Group, Berlin, Germany
Fumitaka Kurauchi	Department of Urban Management Kyoto University, Japan
Changhyun Kwon	Penn State University USA
William H K Lam	Department of Civil and Structural Engineering The Hong Kong Polytechnic University, China
Lawrence Lan	Institute of Traffic and Transportation National Chiao Tung University, Taiwan
Jorge Laval	Laboratoire d'Ingénierie Circulation Transport (ENTPE/INRETS), France
Siriphong Lawphongpanich	Department of Industrial and Systems Engineering University of Florida, USA
Jean-Patrick Lebacque	INRETS/GRETIA France
Ludovic Leclercq	Laboratoire d'Ingénierie Circulation Transport (ENTPE/INRETS), France
Gunwoo Lee	Department of Civil and Environmental Engineering Massachusetts Institute of Technology, USA
Ming Lee	Department of Civil and Environmental Engineering Utah State University, USA

David Levinson	University of Minnesota USA
Zhi-Chun Li	Department of Civil and Structural Engineering The Hong Kong Polytechnic University, China
Henry X Liu	Department of Civil Engineering University of Minnesota, USA
Yu Liu	Department of Civil and Environmental Engineering Michigan State University, USA
Hong Lo	Department of Civil Engineering The Hong Kong University of Science and Technology China
Reinhard Mahnke	Rostock University Germany
Salim Mammar	INRETS/GRETIA France
Lorenzo Meschini	Dipartimento di Idraulica Trasporti e Strade Università "La Sapienza", Rome, Italy
Reetabrata Mookherjee	Oracle Retail USA
Richard Mounce	Department of Mathematics University of York, USA
Takeshi Nagae	Graduate School of Science and Technology Kobe University, Japan
Yu Nie	University of California, Davis USA
Liam O'Brien	Centre for Transport Research Trinity College Dublin, Eire
Margaret O'Mahony	Centre for Transport Research Trinity College Dublin, Eire
Saskia Ossen	Transport and Planning Department Delft University of Technology, The Netherlands
Natale Papola	Dipartimento di Idraulica Trasporti e Strade Università "La Sapienza", Rome, Italy
Anita Rao	Department of Civil and Environmental Engineering Massachusetts Institute of Technology, USA
Will Recker	Department of Civil Engineering University of California, Irvine, USA
Marco Schreuder	Traffic Research Centre Ministry of Transport Public Works and Water Management The Netherlands
Wei Shen	University of California, Davis USA
Mike Smith	Department of Mathematics University of York, UK
Alexandros Sopasakis	Department of Mathematics University of Massachusetts, Amherst, USA

W Y Szeto	Centre for Transport Research Trinity College Dublin, Eire
Michael Taylor	Transport Systems Centre University of South Australia, Australia
Fitsum Teklu	Institute for Transport Studies University of Leeds, UK
Qiong Tian	School of Management Beijing University of Aeronautics and Astronautics, China
Tomer Toledo	Technion – Israel Institute of Technology Haifa, Israel
Ming-Te Wang	Institute of Transportation Ministry of Transportation and Communications, Taiwan
David Watling	Institute for Transport Studies University of Leeds, UK
C K Wong	Department of Building and Construction City University of Hong Kong, China
K I Wong	Department of Transportation Technology and Management National Chiao Tung University, Taiwan
S C Wong	Department of Civil Engineering The University of Hong Kong, China
Feng Xiao	Department of Civil Engineering The Hong Kong University of Science and Technology China
Feng Xie	University of Minnesota USA
Wuping Xin	Civil Engineering Department University of Minnesota, USA
Fan Yang	ESRI Inc Redlands, California, USA
Hai Yang	Department of Civil Engineering The Hong Kong University of Science and Technology China
Yafeng Yin	Department of Civil and Coastal Engineering University of Florida, USA
Kun Zhang	Transport Systems Centre University of South Australia, Australia
Lei Zhang	Dept of Civil Construction and Environmental Engineering Oregon State University, USA
Michael Zhang	University of California, Davis USA
Zhong Zhou	Department of Civil and Environmental Engineering Utah State University, USA
Shanjiang Zhu	University of Minnesota USA

Transportation and Traffic Theory 2007
Edited by R.E. Allsop, M.G.H. Bell and B.G. Heydecker

1

A COMPUTABLE THEORY OF DYNAMIC CONGESTION PRICING

Terry L Friesz, Penn State University, USA
Changhyun Kwon, Penn State University, USA
Reetabrata Mookherjee, Oracle Retail, USA

SUMMARY

In this paper we present a theory of dynamic congestion pricing for the day-to-day as well as the within-day time scales. The equilibrium design problem emphasized herein takes the form of an MPEC, which we call the Dynamic Optimal Toll Problem with Equilibrium Constraints, or DOTPEC. The DOPTEC formulation we employ recalls an important earlier result that allows the equilibrium design problem to be stated as a single level problem, a result which is surprisingly little known. The DOPTEC maintains the usual design objective of minimizing the system travel cost by appropriate toll pricing. We describe how an infinite dimensional mathematical programming perspective may be employed to create an algorithm for the DOTPEC. A numerical example is provided.

INTRODUCTION

The advent of new commitments by municipal, state and federal governments to construct and operate roadways whose tolls may be set dynamically has brought into sharp focus the need for a computable theory of dynamic tolls. Moreover, it is clear from the policy debates that surround the issue of dynamic tolls that pure economic efficiency is not the sole or even the most prominent objective of any dynamic toll mechanism that will be implemented. Rather, equity considerations as well as preferential treatment for certain categories of commuters must be addressed by such a mechanism. Accordingly, we introduce in this paper the dynamic user equilibrium optimal toll problem and discuss two plausible algorithms for its solution; we also provide detailed numerical results that document the performance of the two algorithms.

The dynamic user equilibrium optimal toll problem should not be considered a simple dynamic extension of the traditional congestion pricing paradigm associated with static user equilibrium and usually accredited to Beckmann *et al.* (1956). Rather, the dynamic user equilibrium optimal toll problem is most closely related to the equilibrium network design problem which is now widely recognized to be a specific instance of a mathematical program with equilibrium constraints (MPEC). In fact it will be convenient to refer to the dynamic user equilibrium optimal toll problem as the dynamic optimal toll problem with equilibrium constraints or DOTPEC, where it is understood that the equilibrium of interest is a dynamic user equilibrium.

The relevant background literature for the DOTPEC includes a paper by Friesz *et al.* (2002) who discuss a version of the DOTPEC but for the day-to-day time scale rather than the dual (within-day as well as day-to-day) time scale formulation emphasized in this paper. Also pertinent are the paper by Friesz *et al.* (1996) which discusses dynamic disequilibrium network design and the review by Liu (2004) which considers multi-period efficient tolls. Although the DOTPEC is not the same as the problem of determining efficient tolls including the latter's multiperiod generalization, the exact nature of the differences and similarities is not known and has never been studied. To study the DOTPEC, it is necessary to employ some form of dynamic user equilibrium model. We elect the formulation due to Friesz *et al.* (2001), Friesz and Mookherjee (2006) and its varieties analyzed by Ban *et al.* (2005) and others. The dynamic efficient toll formulation will be constructed by direct analogy to the static efficient toll problem formulation of Hearn and Yildrim (2002).

The main focus of this paper is the formulation and solution of the DOTPEC. To this end, again using the DUE formulation reported in Friesz *et al.* (2001) and Friesz and Mookherjee (2006), we will form a Stackelberg game that envisions a central authority minimizing social costs through its control of link tolls subject to DUE constraints with the potential for additional side constraints for equity and other policy considerations. Also, since we will allow multiple target arrival times of the users, the within-day scale model, we show how to easily extend the formulation to include the day-to-day evolution of demand. Of course there are several ways such a model may be formulated. The dual-time scale formulation we shall emphasize is based on our prior work on differential variational inequalities and equilibrium network design and follows the qualitative theory conjectured (but not analyzed) by Friesz *et al.* (1996).

Central to the study of the DOTPEC in this paper is the dynamic generalization of a result due to Tan *et al.* (1979) and reprised by Friesz and Shah (2001) showing that a system of inequalities expressing the relationship of average effective delay to minimum delay is equivalent to a static user equilibrium. This system of inequalities allows one to state the equilibrium network design problem as a single level mathematical program. Extension of this result to a dynamic setting allows us in this paper to state the DOTPEC as an equivalent, non-hierarchical optimal control problem. We consider two principal methods for solving this optimal control problem: (1) descent in Hilbert space without time discretization, and (2) a finite dimensional approximation solved as a nonlinear program. In both approaches we employ an implicit fixed point scheme like that in Friesz and Mookherjee (2006) for dealing with time shifts in differential variational inequalities. In an example provided near the end of this paper, we numerically study a small network and determine its optimal dynamic tolls.

NOTATION AND MODEL FORMULATION

In this section we purposely repeat key portions of the time-lagged DUE formulation given in Friesz et al. (2001), because of its key role in this manuscript. The reader familiar with the notation and time-shifted DUE model presented in Friesz et al. (2001) may skip this section of the present paper.

Dynamic, Delay Operators and Constraints

The network of interest will form a directed graph $G(N,A)$, where N denotes the set of nodes and A denotes the set of arcs; the respective cardinalities of these sets are $|N|$ and $|A|$. An arbitrary path $p \in P$ of the network is

$$p \equiv \{a_1, a_2, ..., a_i, ..., a_{m(p)}\},$$

where P is the set of all paths and $m(p)$ is the number of arcs of p. We also let t_e denote the time at which flow exists an arc, while t_d is the time of departure from the origin of the same flow. The exit time function $\tau_{a_i}^p$ therefore obeys

$$t_e = \tau_{a_i}^p(t_d)$$

The relevant arc dynamics are

$$\frac{dx_{a_i}^p(t)}{dt} = g_{a_{i-1}}^p(t) - g_{a_i}^p(t) \qquad \forall p \in P, \quad i \in \{1,2,...,m(p)\} \tag{1}$$

$$x_{a_i}^p(t) = x_{a_{i,0}}^p \qquad \forall p \in P, \quad i \in \{1,2,...,m(p)\} \tag{2}$$

where $x_{a_i}^p$ is the traffic volume of arc a_i contributed by path p, $g_{a_i}^p$ is flow exiting arc a_i and $g_{a_{i-1}}^p$ is flow entering arc a_i of path $p \in P$. Also, $g_{a_0}^p$ is the flow exiting the origin of path p; by convention we call this the flow of path p and use the symbolic name $h_p = g_{a_0}^p$.

Furthermore

$$\delta_{a_i p} = \begin{cases} 1 & \text{if } a_i \in p \\ 0 & \text{if } a_i \notin p \end{cases}$$

so that $x_a(t) = \sum_{p \in P} \delta_{ap} \, x_a^p(t) \qquad \forall a \in A$ is the total arc volume.

Arc unit delay is $D_a(x_a)$ for each arc $a \in A$. That is, arc delay depends on the number of vehicles in front of a vehicle as it enters an arc. Of course total path traversal time is

$$D_p(t) = \sum_{i=1}^{m(p)} \left[\tau_{a_i}^p(t) - \tau_{a_{i-1}}^p(t) \right] = \tau_{a_{m(p)}}^p(t) - t \qquad \forall p \in P$$

It is expedient to introduce the following recursive relationships that must hold in light of the above development:

$$\begin{aligned} \tau_{a_1}^p(t) &= t + D_{a_1}\left[x_{a_1}(t)\right] \quad \forall p \in P \\ \tau_{a_i}^p(t) &= \tau_{a_{i-1}}^p(t) + D_{a_i}\left[x_{a_i}(\tau_{a_{i-1}}^p(t))\right] \quad \forall p \in P, \quad i \in \{2,3,\ldots,m(p)\} \end{aligned}$$

from which we have the nested path delay operators first proposed by Friesz *et al.* (1993):

$$D_p(t,x) \equiv \sum_{i=1}^{m(p)} \delta_{a_i p} \Phi_{a_i}(t,x) \quad \forall p \in P,$$

where $x = (x_{a_i}^p : p \in P, i \in \{1,2,\ldots,m(p)\}$

and

$$\begin{aligned} \Phi_{a_1}(t,x) &= D_{a_1}(x_{a_1}(t)) \\ \Phi_{a_2}(t,x) &= D_{a_2}(x_{a_2}(t + \Phi_{a_1})) \\ \Phi_{a_3}(t,x) &= D_{a_3}(x_{a_3}(t + \Phi_{a_1} + \Phi_{a_2})) \\ &\vdots \\ \Phi_{a_i}(t,x) &= D_{a_i}(x_{a_i}(t + \Phi_{a_1} + \cdots + \Phi_{a_{i-1}})) \\ &= D_{a_i}\left(x_{a_i}\left(t + \sum_{j=1}^{i-1} \Phi_{a_j}\right)\right). \end{aligned}$$

To ensure realistic behaviour, we employ asymmetric early/late arrival penalties

$$F\left[t + D_p(t,x) - t_A\right]$$

where t_A is the desired arrival time and

$$\begin{aligned} t + D_p(t,x) > t_A &\Rightarrow \quad F(t + D_p(t,x) - t_A) = \chi^L(x,t) > 0 \\ t + D_p(t,x) < t_A &\Rightarrow \quad F(t + D_p(t,x) - t_A) = \chi^E(x,t) > 0 \\ t + D_p(t,x) = t_A &\Rightarrow \quad F(t + D_p(t,x) - t_A) = 0 \end{aligned}$$

while $\qquad \chi^L(t,x) > \chi^E(t,x)$.

Let us further denote arc tolls by y_a for each arc $a \in A$. We assume that users pay any toll imposed on an arc at the entrance of the arc. Then the path tolls y_p for each path $p \in P$ are

$$y_p(t) = \sum_{i=1}^{m(p)} \delta_{a_i p} y_{a_i}\left(t + \sum_{j=1}^{i-1} \Phi_{a_j}(t,x)\right) \quad \forall p \in P,$$

where $\Phi_{a_0}(t,x) = 0$. If the tolls are paid when users exit arcs, then the path toll becomes

$$y_p(t) = \sum_{i=1}^{m(p)} \delta_{a_i p} y_{a_i}\left(t + \sum_{j=1}^{i} \Phi_{a_j}(t,x)\right) \quad \forall p \in P.$$

We now combine the actual path delays and arrival penalties to obtain the *effective delay operators*

$$\Psi_p(t,x) = D_p(t,x) + F\left(t + D_p(x,t) - T_A\right) \qquad \forall p \in P \tag{3}$$

Since the volume which enters and exits an arc should conserve flow, we must have

$$\int_0^t g_{a_{i-1}}^p(t)\,dt = \int_{D_{a_i}(x_{a_i}(0))}^{t + D_{a_i}(x_{a_i}(t))} g_{a_i}^p(t)\,dt \quad \forall p \in P, i \in [1, m(p)], \tag{4}$$

where $g_{a_0}^p(t) = h_p(t)$. Differentiating both sides of (4) with respect to time t and using the chain rule, we have

$$h_p(t) = g_{a_1}^p(t + D_{a_1}(x_{a_1}(t)))(1 + D'_{a_1}(x_{a_1}(t))\dot{x}_{a_1}) \qquad \forall p \in P \tag{5}$$

$$g_{a_{i-1}}^p(t) = g_{a_i}^p(t + D_{a_i}(x_{a_i}(t)))(1 + D'_{a_i}(x_{a_i}(t))\dot{x}_{a_i}) \qquad \forall p \in P, \quad i \in [2, m(p)]. \tag{6}$$

These are *proper flow progression constraints* derived in a fashion that makes them completely *consistent with the chosen dynamics and point queue model of arc delay*. These constraints involve a state-dependent time lag $D_{a_i}(x_{a_i}(t))$ but make no explicit reference to the exit time functions. These flow propagation constraints describe the expansion and contraction of vehicle platoons; they were presented by Friesz *et al.* (1995). Astarita (1995, 1996) independently proposed flow propagation constraints that may be readily placed in the above form.

The final constraints to consider are those of flow conservation and non-negativity:

$$\sum_{p \in P_{ij}} \int_0^{t_f} h_p(t)\,dt = Q_{ij} \quad \forall (i,j) \in W \tag{7}$$

$$h_p \geq 0 \quad \forall (i,j) \in P_{ij} \tag{8}$$

$$g_{a_i}^p \geq 0 \quad \forall p \in P, i \in [1, m(p)] \tag{9}$$

$$x_{a_i}^p \geq 0 \quad \forall p \in P, i \in [1, m(p)], \tag{10}$$

where W is the set of origin-destination pairs, P_{ij} is the set of paths connecting origin-destination pair (i,j), $t_f > t_0$, and $t_f - t_0$ defines the planning horizon. Furthermore, Q_{ij} is the travel demand (a volume) for the period $[t_0, t_f]$. In what follows h will denote the vector of all path flows, g the vector of all arc exit flows. Finally, we denote the set of all feasible exit flow vectors (h,g) by Ω; that is

$$\Omega \equiv \{(h,g):(1),(2),(5),(6),(7),(8),(9),(10) \text{ are satisfied}\}. \tag{11}$$

Dynamic User Equilibrium

Given the effective unit travel delay Ψ_p for path p, the infinite dimensional variational inequality formulation for dynamic network user equilibrium itself is: find $(g^*,h^*)\in\Omega$ such that

$$\left\langle \Psi(t,x(h^*,g^*)),(h-h^*)\right\rangle = \sum_{p\in P}\int_0^{t_f}\Psi_p\left[t,x(h^*,g^*)\right]\cdot\left[h_p(t)-h_p^*(t)\right]dt \geq 0 \tag{12}$$

for all $(h,g)\in\Omega$, where Ψ denotes the vector of effective path delay operators. Friesz *et al.* (2001) show all solutions of (12) are dynamic user equilibria[1]. In particular the solutions of (12) obey

$$\Psi_p(t,x(g^*,h^*)) > \mu_{ij} \Rightarrow h_p^*(t)=0 \tag{13}$$

$$h_p^*(t)>0 \Rightarrow \Psi_p(t,x(g^*,h^*)) = \mu_{ij} \tag{14}$$

for $p\in P_{ij}$ where μ_{ij} is the lower bound on achievable costs for any ij -traveler, given by

$$\mu_p = ess\inf\{\Theta_p(t,x): t\in[t_0,t_f]\}\geq 0$$

and

$$\mu_{ij} = \min\{\mu_p : p\in P_{ij}\}\geq 0 .$$

We call a flow pattern satisfying (13) and (14) a *dynamic user equilibrium*. The behavior described by (13) and (14) is readily recognized to be a type of Cournot-Nash non-cooperative equilibrium. It is important to note that these conditions do not describe a stationary state, but rather a time varying flow pattern that is a Cournot-Nash equilibrium (or user equilibrium) at each instant of time.

THE DYNAMIC EFFICIENT TOLL PROBLEM (DETP)

Hearn and Yildrim (2002) studied the efficient toll in the static setting with the traveling cost which is linear in the traffic flow. The objective of the efficient toll is to make the user equilibrium traffic flow equivalent to the system optimum by appropriate congestion pricing. To study the dynamic efficient toll problem (DETP), we introduce the notion of a *tolled effective delay operator*:

$$\Theta_p(t,x,y_p) = D_p(t,x) + F\{t + D_p(x,t) - T_A\} + y_p(t) \qquad \forall p\in P,$$

---

[1] Although we have purposely suppressed the functional analysis subtleties of the formulation, it should be noted that (12) involves an inner product in a Hilbert space, namely $\left(L^2[0,T]\right)^{|P|}$.

where y_p denotes the toll for path p. Of course we have the relationship

$$\Theta_p(t,x,y_p) = \Psi_p(t,x) + y_p(t). \tag{15}$$

To make the toll meaningful, we enforce the efficient toll non-negative:

$$y_p(t) \geq 0 \quad \forall t \in [t_0, t_f], p \in P.$$

Analysis of the System Optimum

The dynamic system optimum (DSO) is achieved by solving

$$\min J_1 = \int_0^{t_f} \sum_{p \in P} e^{-rt} \Psi_p(t,x)\, h_p(t)\, dt$$

subject to

$$\frac{dx_{a_i}^p(t)}{dt} = g_{a_{i-1}}^p(t) - g_{a_i}^p(t) \qquad \forall p \in P, \quad i \in [1, m(p)] \tag{16}$$

$$x_{a_i}^p(t) = x_{a_{i,0}}^p \qquad \forall p \in P, \quad i \in [1, m(p)]$$

$$g_{a_{i-1}}^p(t) = g_{a_i}^p(t + D_{a_i}(x_{a_i}(t)))(1 + D_{a_i}^{'}(x_{a_i}(t))\dot{x}_{a_i}) \qquad \forall p \in P, \quad i \in [1, m(p)] \tag{17}$$

$$\sum_{p \in P_{ij}} \int_0^{t_f} h_p(t)\, dt = Q_{ij} \quad \forall (i,j) \in W \tag{18}$$

$$x \geq 0 \qquad g \geq 0 \qquad h \geq 0, \tag{19}$$

where we have used the convention

$$g_{a_0}^p = h_p.$$

It will be convenient to employ the following shorthand for shifted variables:

$$\overline{g}_{a_i}^p \equiv g_{a_i}^p(t + D_{a_i}(x_{a_i}(t))) \quad \forall p \in P, \quad i \in 0, m(p)].$$

Penaltizing (17) we obtain

$$J_1 = \int_0^{t_f} \left\{ \sum_{p \in P} e^{-rt} \Psi_p(t,x) h_p(t) + \sum_{p \in P} \sum_{i=1}^{m(p)} \frac{\mu_{a_i}^p}{2} \left[g_{a_{i-1}}^p(t) - \overline{g}_{a_i}^p(t)(1 + D_{a_i}^{'}(x_{a_i}(t))\dot{x}_{a_i}) \right]^2 \right\} dt, \tag{20}$$

where $\mu_{a_i}^p$ is the penalty coefficient. Let us then define the set of feasible controls

$$\Lambda \equiv \left\{ (h,g) : \sum_{p \in P_{ij}} \int_0^{t_f} h_p(t) dt = Q_{ij} \ \forall (i,j) \in W, h \geq 0, \quad g \geq 0 \right\}. \tag{21}$$

Optimal control problem (20) and (21) is an instance of the time-shifted optimal control problem analyzed in Friesz *et al.* (2001). We also employ the following notation for the state vector and control vector, respectively:

$$x = \left(x_{a_i}^p : p \in P, i \in [1, m(p)] \right)$$

$$g = \left(g_{a_i}^p : p \in P, i \in [1, m(p)] \right).$$

The DSO Hamiltonian is

$$H_1(t,x,h,g,\lambda;\mu) \equiv \sum_{p \in P} e^{-rt} \Psi_p(t,x) h_p(t) + \sum_{p \in P} \sum_{i=1}^{m(p)} \frac{\mu_{a_i}^p}{2} \left\{ g_{a_{i-1}}^p(t) - \bar{g}_{a_i}^p(t)(1 + D'_{a_i}(x_{a_i}(t))\dot{x}_{a_i}) \right\}^2$$
$$+ \sum_{p \in P} \sum_{i=1}^{m(p)} \lambda_{a_i}^p \left(g_{a_{i-1}}^p(t) - g_{a_i}^p(t) \right).$$

Let us introduce the vector

$$F(t,x,h,g,\lambda;\mu) = \left(F_{a_i}^p(t,x,h,g,\lambda;\mu) : p \in P, i \in [0, m(p)] \right),$$

where
$$F_{a_0}^p(t,x,h,g,\lambda;\mu) = \frac{\partial H_1(t,x,h,g,\lambda;\mu)}{\partial h_p} \quad \forall p \in P \tag{22}$$

$$F_{a_i}^p(t,x,h,g,\lambda;\mu) = \begin{cases} \dfrac{\partial H_1(t,x,h,g,\lambda;\mu)}{\partial g_{a_i}^p} & \text{if } t \in [t_0, D_{a_i}(x(t_0))] \\[2ex] \dfrac{\partial H_1(t,x,h,g,\lambda;\mu)}{\partial g_{a_i}^p} + \left[\dfrac{\partial H_1(t,x,h,g,\lambda;\mu)}{\partial \bar{g}_{a_i}^p} \dfrac{1}{1 + D'_{a_i}(x_{a_i}(t))\dot{x}_{a_i}} \right]_{s_{a_i}(t)} & \text{if } t \in [D_{a_i}(x(t_0)), t_f + D_{a_i}(x(t_f))] \end{cases}$$
$$\forall p \in P, \quad i \in [1, m(p)] \tag{23}$$

and each $s_{a_i}(t)$ is a solution of the fixed point problem $s_{a_i}(t) = \arg[s = t - D_{a_i}(x(s))]$. We may write (22) and (23) in detail as

$$F_{a_0}^p(t,x,h,g,\lambda;\mu) = e^{-rt} \left[\Psi_p(t,x) + \frac{\partial \Psi_p(t,x)}{\partial h_p} h_p \right]$$
$$+ \mu_{a_1}^p \left[g_{a_0}^p(t) - \bar{g}_{a_1}^p(t)(1 + D'_{a_1}(x_{a_1}(t))\dot{x}_{a_1}) \right] + \lambda_{a_1}^p \quad \forall p \in P \tag{24}$$

$$F_{a_i}^p(t,x,h,g,\lambda;\mu)=\begin{cases}\mu_{a_{i+1}}^p\left\{g_{a_i}^p(t)-\bar{g}_{a_{i+1}}^p(t)(1+D_{a_{i+1}}'(x_{a_{i+1}}(t))\dot{x}_{a_{i+1}})\right\}-\lambda_{a_i}^p+\lambda_{a_{i+1}}^p \\ \qquad \text{if } t\in\left[t_0,D_{a_i}(x(t_0))\right] \\ \mu_{a_{i+1}}^p\left\{g_{a_i}^p(t)-\bar{g}_{a_{i+1}}^p(t)(1+D_{a_{i+1}}'(x_{a_{i+1}}(t))\dot{x}_{a_{i+1}})\right\}-\lambda_{a_i}^p+\lambda_{a_{i+1}}^p \\ \qquad -\left[\mu_{a_i}^p\left\{g_{a_{i-1}}^p(t)-\bar{g}_{a_i}^p(t)(1+D_{a_i}'(x_{a_i}(t))\dot{x}_{a_i})\right\}\right]_{s_{a_i}(t)} \\ \qquad \text{if } t\in\left[D_{a_i}(x(t_0)),t_f+D_{a_i}(x(t_f))\right]\end{cases} \tag{25}$$

$$\forall p\in P,\quad i\in[1,m(p)-1]$$

$$F_{a_i}^p(t,x,h,g,\lambda;\mu)=\begin{cases}-\lambda_{a_i}^p \\ \qquad \text{if } t\in\left[t_0,D_{a_i}(x(t_0))\right] \\ -\lambda_{a_i}^p-\left[\mu_{a_i}^p\left\{g_{a_{i-1}}^p(t)-\bar{g}_{a_i}^p(t)(1+D_{a_i}'(x_{a_i}(t))\dot{x}_{a_i})\right\}\right]_{s_{a_i}(t)} \\ \qquad \text{if } t\in\left[D_{a_i}(x(t_0)),t_f+D_{a_i}(x(t_f))\right]\end{cases} \tag{26}$$

$$\forall p\in P,\quad i=m(p).$$

Then a necessary condition for $(h^S,g^S)\in\Lambda$ to be the system optimum is

$$0\le\sum_{p\in P}\sum_{i=0}^{m(p)}F_{a_i}^p(t,x^S,h^S,g^S,\lambda^S;\mu)\left(g_{a_i}^p-g_{a_i}^{pS}\right)\qquad\forall(h,g)\in\Lambda \tag{27}$$

for each time instant $t\in\left[t_{0,}\ \sup_{a_i\in A}\{t_f+D_{a_i}(x(t_f))\}\right]$, together with the state dynamics (16) and the following adjoint equations and boundary conditions

$$-\frac{d\lambda_{a_i}^{p,S}}{dt}=\frac{\partial H_1^S}{\partial x_{a_i}^p}=e^{-rt}\frac{\partial\Psi_p(t,x^S)}{\partial x_{a_i}^p}\qquad\forall p\in P,\quad i\in[1,m(p)]$$

$$\lambda_{a_i}^{p,S}(t_f)=0\qquad\forall p\in P,\quad i\in[1,m(p)],$$

where the superscript S denotes a trajectory corresponding to a system optimum.

Analysis of the User Equilibrium in the Presence of Tolls

However, a dynamic tolled user equilibrium must obey

$$\sum_{p\in P}\int_0^{t_f}e^{-rt}\left\{\Theta_p\left[t,x(h^U),y_p^U\right]\right\}\left[h_p(t)-h_p^U(t)\right]dt\ge 0\quad\text{for all }(h,g)\in\Lambda, \tag{28}$$

where the state dynamics as well as all other state and control constraints are identical to those introduced above for DSO. In particular, the set of feasible controls Λ referred to in (28) remains unchanged. We formulate an optimal control problem[2] from the above dynamic user equilibrium variational inequality problem; its objective is

$$\min J_2 = \sum_{p \in P} \int_0^{t_f} e^{-rt} \Theta_p \left[t, x(h^U), y_p^U \right] h_p(t) dt$$

with the same constraints introduced previously. As previously done for the system optimum problem, we penaltize the flow propagation constraints to obtain the modified criterion

$$J_2 = \sum_{p \in P} \int_0^{t_f} \left\{ e^{-rt} \Theta_p \left[t, x(h^U), y_p^U \right] h_p(t) + \sum_{p \in P} \sum_{i=1}^{m(p)} \frac{\mu_{a_i}^p}{2} \left[g_{a_{i-1}}^p(t) - \bar{g}_{a_i}^p(t)(1 + D_{a_i}'(x_{a_i}(t)) \dot{x}_{a_i}) \right]^2 \right\} dt \quad (29)$$

Then we have another standard form time-shifted optimal control problem, although it is subtly but importantly different than that for DSO. In particular, the Hamiltonian now becomes

$$H_2(t, x, h, g, \lambda; \mu) \equiv \sum_{p \in P} e^{-rt} \Theta_p \left[t, x(h^U), y_p^U \right] h_p(t) + \sum_{p \in P} \sum_{i=1}^{m(p)} \frac{\mu_{a_i}^p}{2} \left\{ g_{a_{i-1}}^p(t) - \bar{g}_{a_i}^p(t)(1 + D_{a_i}'(x_{a_i}(t)) \dot{x}_{a_i}) \right\}^2$$
$$+ \sum_{p \in P} \sum_{i=1}^{m(p)} \lambda_{a_i}^p \left(g_{a_{i-1}}^p(t) - g_{a_i}^p(t) \right)$$

An analysis of necessary conditions similar to that for DSO is now possible. The key difference is that the counterpart of (24) must in the user equilibrium case be written as follows:

$$G_{a_0}^p(t, x, h, g, \lambda; \mu) = e^{-rt} \Theta_p \left[t, x(h^U), y_p^U \right]$$
$$+ \mu_{a_1}^p \left[g_{a_0}^p(t) - \bar{g}_{a_1}^p(t)(1 + D_{a_1}'(x_{a_1}(t)) \dot{x}_{a_1}) \right] + \lambda_{a_1}^p \quad \forall p \in P \quad (30)$$

$$G_{a_i}^p(t, x, h, g, \lambda; \mu) = F_{a_i}^p(t, x, h, g, \lambda; \mu) \qquad \forall p \in P, \quad i \in [1, m(p)]. \quad (31)$$

Then a necessary condition for $(h^S, g^S) \in \Lambda$ to be a dynamic user equilibrium (DUE) is

$$0 \le \sum_{p \in P} \sum_{i=0}^{m(p)} G_{a_i}^p \left(t, x^U, h^U, g^U, \lambda^U; \mu \right) \left(g_{a_i}^p - g_{a_i}^{pU} \right) \qquad g \in \Lambda \quad (32)$$

for each time instant $t \in \left[t_0, \sup_{a_i \in A} \{ t_f + D_{a_i}(x(t_f)) \} \right]$, together with the state dynamics (16) and the following adjoint equations and boundary conditions:

$$-\frac{d\lambda_{a_i}^{p,U}}{dt} = \frac{\partial H_2^U}{\partial x_{a_i}^p} = e^{-rt} \frac{\partial \Theta_p \left[t, x(h^U), y_p^U \right]}{\partial x_{a_i}^p} \qquad \forall p \in P, \quad i \in [1, m(p)]$$

$$\lambda_{a_i}^{p,U}(t_f) = 0 \qquad \forall p \in P, \quad i \in [1, m(p)],$$

---

[2] This may not be used for numerical computation as its statement depends on knowledge of the dynamic user equilibrium being sought. However, it may be employed for qualitative analyses like those which follow.

where the superscript U denotes a trajectory corresponding to a dynamic user equilibrium in the presence of tolls.

Characterizing Efficient Tolls

It is the purpose of efficient tolls to make the criteria J_1 and J_2 identical along solution trajectories for which flow propagation and other constraints are satisfied, for then the system optimal total costs are identical to the tolled user optimal total costs. Furthermore, the vectors of path flows (departure rates) obey

$$h^U(t) = h^S(t). \tag{33}$$

There are as well identical arc exit flows and identical arc volumes. Therefore, along solution trajectories

$$\lambda_{a_1}^{p,S} = \frac{\partial J_1}{\partial x_{a_1}^{p,S}} = \frac{\partial J_2}{\partial x_{a_1}^{p,U}} = \lambda_{a_1}^{p,U}. \tag{34}$$

With (34) in mind and upon comparing (27) and (32), we find

$$e^{-rt}\left\{\Psi_p\left(t,x^S\right) + \frac{\partial \Psi_p\left(t,x^S\right)}{\partial h_p}h_p^S\right\} = e^{-rt}\left\{\Theta_p\left[t,x\left(h^U\right),y_p^U\right]\right\}$$

$$= e^{-rt}\left\{\Psi_p\left(t,x^U\right) + y_p^U(t)\right\}.$$

The only toll constraint is non-negativity; hence applying the projection after the expression for $y_p{}^U(t)$ is derived with non-negativity relaxed will give an exact expression:

$$y_p^U(t) = \left[\frac{\partial \Psi_p\left(t,x^S\right)}{\partial h_p}h_p^S\right]^+ \qquad \forall t \in \left[t_0, t_f\right], \tag{35}$$

where $[\cdot]^+$ is the elementary projection operator defined by

$$[v]^+ = \begin{cases} v & \text{if } v \geq 0 \\ 0 & \text{if } v < 0. \end{cases}$$

This result is completely analogous to that for an efficiently tolled static user equilibrium.

THE DYNAMIC OPTIMAL TOLL PROBLEM WITH EQUILIBRIUM CONSTRAINTS (DOTPEC)

We now introduce the dynamic optimal toll problem with equilibrium constraints (DOTPEC). The DOTPEC is a type of dynamic network design problem for which a central authority seeks to minimize congestion in a transport network, whose flows obey a dynamic network user equilibrium, by dynamically adjusting tolls. In particular the central authority seeks to solve the optimal control problem

$$\min J = \int_0^{t_f} \sum_{p \in P} \Psi_p(t,x) h_p(t) dt \tag{36}$$

subject to

$$\sum_{p\in P}\int_0^{t_f}\Theta_p\left[t,x(h,g),y_p\right](w_p-h_p)dt\geq 0\ \forall (w,g)\in\Lambda \tag{37}$$

$$\frac{dx_{a_i}^p(t)}{dt}=g_{a_{i-1}}^p(t)-g_{a_i}^p(t)\qquad \forall p\in P, i\in[1,m(p)] \tag{38}$$

$$x_{a_i}^p(t)=x_{a_{i,0}}^p\qquad \forall p\in P,\quad i\in[1,m(p)] \tag{39}$$

$$h_p(t)=g_{a_1}^p(t+D_{a_1}(x_{a_1}(t)))(1+D_{a_1}^{'}(x_{a_1}(t))\dot{x}_{a_1})\quad \forall p\in P \tag{40}$$

$$g_{a_{i-1}}^p(t)=g_{a_i}^p(t+D_{a_i}(x_{a_i}(t)))(1+D_{a_i}^{'}(x_{a_i}(t))\dot{x}_{a_i})\quad \forall p\in P, i\in[2,m(p)] \tag{41}$$

$$\sum_{p\in P_{ij}}\int_0^{t_f}h_p(t)dt=Q_{ij}\quad \forall (i,j)\in W \tag{42}$$

$$x_{a_i}^p\geq 0\qquad g_{a_i}^p\geq 0\qquad h_p\geq 0\qquad \forall p\in P, i\in[1,m(p)], \tag{43}$$

where Λ is the set of feasible controls (exit flows) defined previously. In the DUE constraints (37), we have introduced the notion of an *effective delay operator in the presence of tolls*, by which is meant

$$\Theta_p(t,x,y_p)=D_p(t,x)+F\{t+D_p(x,t)-T_A\}+y_p(t)\qquad \forall p\in P,$$

where y_p denotes the toll for path p. Of course we have the relationship

$$\Theta_p(t,x,y_p)=\Psi_p(t,x)+y_p(t), \tag{44}$$

where we recall from Friesz, Bernstein, Suo and Tobin (2001) that

$$y_p(t)=\sum_{i=1}^{m(p)}\delta_{a_i p}y_{a_i}\left(t+\Phi_{a_{i-1}}(t,x)\right)\quad \forall p\in P.$$

The variational-inequality constrained optimization problem (35) through (42) is a bi-level problem that is intrinsically difficult to solve. Note in particular that, even for a single instant of time, the number of constraints of the type (37) is uncountable.

In this paper, to numerically solve specific instances of (36)-(43), we may exploit the following alternative to expressing the underlying DUE problem as an infinite dimensional variation inequality:

Theorem 1 *Given that the effective travel delay for path* p *is* $\Theta_p[t,x(t),y_p(t)]$, *a nonnegative path flow vector* $h\geq 0$ *is a user equilibrium if and only if the conditions*

$$\Theta_p\geq\frac{\sum_{p\in P_{ij}}\int_0^{t_f}\Theta_p[t,x(t),y_p(t)]h_p(t)dt}{\sum_{p\in P_{ij}}\int_0^{t_f}h_p(t)dt}=\mu_{ij}\qquad \forall p\in P_{ij},\quad (i,j)\in W \tag{45}$$

are satisfied.

Proof : The dynamic user equilibrium condition stated in (13) and (14) can be modeled as an equivalent complementarity problem, that is

$$\left[\Theta_p\left(t,x^*\right)-\mu_{ij}\right]h_p^*(t)=0, \qquad \Theta_p\left(t,x^*\right)-\mu_{ij}\geq 0, \qquad h_p^*(t)\geq 0 \tag{46}$$

for all $t\in\left[t_0,t_f\right], p\in P_{ij},(i,j)\in W$. To show necessity we integrate the complementarity condition in (46) over the time horizon and summing for all paths, and obtain

$$\sum_{p\in P_{ij}}\int_0^{t_f}\left[\Theta_p\left(t,x^*\right)-\mu_{ij}\right]h_p^*(t)dt=0 \qquad \forall(i,j)\in W$$

or

$$\sum_{p\in P_{ij}}\int_0^{t_f}\Theta_p\left(t,x^*\right)h_p^*(t)dt=\mu_{ij}\sum_{p\in P_{ij}}\int_0^{t_f}h_p^*(t)dt \qquad \forall(i,j)\in W\,. \tag{47}$$

To show sufficiency we re-state (45) as

$$\Theta_p-\mu_{ij}\geq\frac{\sum_{p\in P_{ij}}\int_0^{t_f}\Theta_p\left[t,x(t),y_p(t)\right]h_p(t)dt}{\sum_{p\in P_{ij}}\int_0^{t_f}h_p(t)dt} \qquad \forall p\in P_{ij}, \quad (i,j)\in W \tag{48}$$

and multiply both sides by path flow to obtain

$$0=\left[\Theta_p\left(t,x^*\right)-\mu_{ij}\right]h_p^*(t)\geq\left[\frac{\sum_{p\in P_{ij}}\int_0^{t_f}\Theta_p\left[t,x(t),y_p(t)\right]h_p(t)dt}{\sum_{p\in P_{ij}}\int_0^{t_f}h_p(t)dt}\right]h_p^*(t) \qquad \forall p\in P_{ij}, \quad (i,j)\in W \tag{49}$$

from which (46) follows immediately. ■

By virtue of Theorem 1, we may replace the DUE constraint (37) by the equality and inequality constraints (45) to obtain the following equivalent form of the DOTPEC:

$$\min J=\int_0^{t_f}\sum_{p\in P}\Psi_p(t,x)h_p(t)dt \tag{50}$$

subject to

$$\mu_{ij}=\frac{\sum_{p\in P_{ij}}\int_0^{t_f}\Theta_p\left[t,x(t),y_p(t)\right]h_p(t)dt}{\sum_{p\in P_{ij}}\int_0^{t_f}h_p(t)dt} \qquad \forall(i,j)\in W \tag{51}$$

$$\Theta_p\geq\mu_{ij} \qquad \forall p\in P_{ij}, \quad (i,j)\in W \tag{52}$$

$$\frac{dx_{a_i}^p(t)}{dt}=g_{a_{i-1}}^p(t)-g_{a_i}^p(t) \qquad \forall p\in P, i\in[1,m(p)] \tag{53}$$

$$x_{a_i}^p(t)=x_{a_{i,0}}^p \qquad \forall p\in P, \quad i\in[1,m(p)] \tag{54}$$

$$h_p(t) = g_{a_1}^P (t + D_{a_1}(x_{a_1}(t)))(1 + D'_{a_1}(x_{a_1}(t))\dot{x}_{a_1}) \quad \forall p \in P \tag{55}$$

$$g_{a_{i-1}}^P(t) = g_{a_i}^P (t + D_{a_i}(x_{a_i}(t)))(1 + D'_{a_i}(x_{a_i}(t))\dot{x}_{a_i}) \quad \forall p \in P, i \in [2, m(p)] \tag{56}$$

$$\sum_{p \in P_{ij}} \int_0^{t_f} h_p(t)dt = Q_{ij} \quad \forall (i,j) \in W \tag{57}$$

$$x_{a_i}^P \geq 0 \qquad g_{a_i}^P \geq 0 \qquad h_p \geq 0 \qquad \forall p \in P, i \in [1, m(p)] \,. \tag{58}$$

Note that the above formulation is an infinite dimensional mathematical program with inequality and equality constraints in standard form, and that the number of constraints for any given instant of time is countable.

MULTIPLE TIME SCALES

We have investigated the within-day behavior of road network users so far. In this section we describe a day-to-day adjust process that sets daily travel demand. Our perspective is very simple: if today commuters experiences a level of congestion above a threshold representing the budget or tolerance for congestion of the typical commuter, travel demand will be less tomorrow and more workers will elect to stay at home (telecommute). To operationalize this idea, we take the perspective of evolutionary game theory to describe the day-to-day demand learning process in terms of the *moving average* of congestion and difference equations.

Let $\tau \in \Upsilon \equiv \{1,2,...,L\}$ be one typical discrete day within the planning horizon, and take the length of each day to be Δ, while the continuous clock time t within each day is presented by $t \in [(\tau-1)\Delta, \tau\Delta]$ for all $\tau \in \{1,2,...,L\}$. The entire planning horizon spans L consecutive days. As noted above, we assume the travel demand for each day changes based on the moving average of congestion experienced over previous days. In fact we postulate that the travel demands Q_{ij}^τ for day τ between a given OD pair $(i,j) \in W$ are determined by the following system of difference equations:

$$Q_{ij}^{\tau+1} = \left[Q_{ij}^\tau - s_{ij}^\tau \left\{ \frac{\sum_{p \in P_{ij}} \sum_{j=0}^{\tau-1} \int_{j \cdot \Delta}^{(j+1)\cdot\Delta} \Psi_p \left[t, x(h^*, g^*)\right] dt}{|P_{ij}| \cdot \tau \cdot \Delta} - \chi_{ij} \right\} \right]^+ \quad \forall \tau \in [1, L-1] \tag{59}$$

with boundary condition

$$Q_{ij}^1 = \tilde{Q}_{ij}\,, \tag{60}$$

where $\tilde{Q}_{ij} \in \Re_+$ is the fixed travel demand for the OD pair $(i,j) \in W$ for the first day and χ_{ij} is the representative threshold. The operator $[x]^+$ is shorthand from $\max[0,x]$. The parameter s_{ij}^τ is related to the rate of change of inter-day travel demand.

ALGORITHMS FOR SOLVING THE DOTPEC

In this section, we provide two different algorithms for solving the DOTPEC: (1) descent in Hilbert space without time discretization, and (2) a finite dimensional discrete time approximation solved as a nonlinear program.

The Implicit Fixed Point Perspective

In both approaches, state-dependent time shifts must and can be accommodated using an implicit fixed point perspective, as innovated for the dynamic user equilibrium by Friesz and Mookherjee (2006). More specifically, in such an approach, one employs control and state information from a previous iteration to approximate current time shifted functions. This perspective may be summarized as follows:

1. Articulate the current approximate states (volumes) and controls (arc exit rates) by spline or other curve fitting techniques as continuous functions of time.

2. Using the aforementioned continuous functions of time, express time shifted controls as pure functions of time, while leaving unshifted controls as decision functions to be updated within the current iteration.

3. Update the states and controls, then repeat Step 2 and Step 3 until the control controls converge to a suitable approximate solution.

Descent in Hilbert Space

To articulate what is meant by descent in Hilbert space, it is much easier to study an abstract problem rather than the DOTPEC because of the notational complexity of the underlying DUE problem. To that end, let us consider an abstract optimal control problem with mixed state-control constraints involving state-dependent time shifts from the point of view of infinite dimensional mathematical programming:

$$\min J = \int_0^{t_f} F(x,u,u_D,t)dt \tag{61}$$

subject to

$$x(u,u_D,t) \in \Lambda = \left\{ x : \frac{dx}{dt} = f(x,u,u_D,t), x(0) = 0, G(x,u,u_D,t) = 0, x \geq 0 \right\} \in (H^1[t_0,t_f])^n$$

where

$$\begin{aligned}
u &\in U \subseteq (L^2[t_0,t_f])^m \\
u_D &\equiv u(t+D(x)) : (H^1[t_0,t_f])^n \times \Re^1_+ \to (L^2[t_0,t_f])^m \\
f &: (H^1[t_0,t_f])^n \times (L^2[t_0,t_f])^{2m} \times \Re^1_+ \to (L^2[t_0,t_f])^m \\
F &: (H^1[t_0,t_f])^n \times (L^2[t_0,t_f])^{2m} \times \Re^1_+ \to (L^2[t_0,t_f])^m \\
G &: (H^1[t_0,t_f])^n \times (L^2[t_0,t_f])^{2m} \times \Re^1_+ \to (L^2[t_0,t_f])^m .
\end{aligned}$$

In the above, $(L^2[t_0,t_f])^m$ is the m-fold product of the space of square integrable functions $L^2[t_0,t_f]$ and $(H^1[t_0,t_f])^n$ is the n-fold product of the Sobolev space $H^1[t_0,t_f]$ for the real interval $[t_0,t_f]\subset\Re^1_+$. In applying descent in Hilbert space to this problem, it is convenient to use quadratic-loss penalty functions and a logarithmic barrier function to create the unconstrained program:

$$\min J_1 = \int_0^{t_f} F(x,u,u_D,t)dt + \frac{1}{2}\int_0^{t_f}\sum_i \eta_i (G_i(x,u,u_D,t))^2 dt + \frac{1}{2}\int_0^{t_f}\sum_i \rho_i \min(0,x_i)^2 dt \qquad (62)$$

where it is understood that x denotes the operator

$$x(u,u_D,t)\in \Lambda_1 = \left\{x : \frac{dx}{dt} = f(x,u,u_D,t), x(0)=x_0\right\} \in (H^1[t_0,t_f])^n,$$

and η_i and ρ_i are penalty and barrier multipliers to be adjusted from iteration to iteration.

The resulting problem can be solved using a continuous time steepest descent method. For the penalized criterion (57), the algorithm can be stated as following:

Step 0. Initialization. Pick $u^0(t)\in U$ and set $k=1$.

Step 1. Finding state variables. Solve the state dynamics

$$\begin{aligned}
\frac{dx}{dt} &= f(x,u^{k-1},u_D^{k-1},t) \\
x(0) &= x_0
\end{aligned}$$

and call the solution $x^k(t)$, using curve fitting to create an approximation to $x^k(t)$ when necessary.

Step 2. Finding adjoint variables. Solve the adjoint dynamics

$$\begin{aligned}
-\frac{d\lambda}{dt} &= \left[\nabla_x H(x,u^{k-1},u_D^{k-1},\lambda,t)\right]_{x=x^k} \\
\lambda(t_f) &= 0
\end{aligned}$$

where the Hamiltonian is given by

$$H(x,u,u_D,\lambda,t) = F(x,u,u_D,t) + \frac{1}{2}\sum_i \rho_i \min(0,x_i)^2 + \frac{1}{2}\sum_i \eta_i (G_i(x,u,u_D,t))^2 + \lambda^T f(x,u,u_D,t)$$

Call the solution $\lambda^k(t)$, using curve fitting to create an approximation to $\lambda^k(t)$ when necessary.

Step 3. Finding the gradient. Determine

$$\nabla_u J^k \equiv \left[\nabla_u H(x^k, u, u_D^{k-1}, \lambda^k, t)\right]_{u=u^k}$$

Step 4. Updating the current control. For a suitably small step size

$$\theta_k \in \Re^1_{++}$$

determine

$$u^k(t) = u^{k-1}(t) - \theta_k \nabla_u J^k$$

Step 5. Stopping Test. For $\varepsilon \in \Re^1_{++}$, a preset tolerance, stop if

$$\| u^{k+1} - u^k \| < \varepsilon$$

and declare

$$u^* \approx u^{k+1}$$

Otherwise set $k = k+1$ and go to Step1.

Discrete-time Approximation of DOTPEC

The optimal control problem (45)-(53) may be given the following discrete time approximation:

$$\min J = \sum_{k=0}^{N} \sum_{p \in P} \phi(k) \Psi_p \left[t_k, x(t_k)\right] h_p(t_k) \Delta$$

subject to

$$\mu_{ij} = \frac{\sum_{p \in P_{ij}} \sum_{k=0}^{N} \phi(k) \Theta_p \left[t_k, x(t_k), y_p(t_k)\right] h_p(t_k) \Delta}{\sum_{p \in P_{ij}} \sum_{k=0}^{N} \phi(k) h_p(t_k) \Delta} \qquad \forall (i,j) \in W$$

$$\Theta_p(t_k) \geq \mu_{ij} \qquad \forall k \in [0, N], \quad p \in P_{ij}, \quad (i,j) \in W$$

$$x_{a_i}^p(t_{k+1}) = x_{a_i}^p(t_k) + \Delta\left[g_{a_{i-1}}^p(t_k) - g_{a_i}^p(t_k)\right]$$

$$\forall k \in [0, N-1], \quad p \in P, \quad i \in [1, m(p)]$$

$$x_{a_i}^p(t_0) = x_{a_{i,0}}^p \qquad \forall p \in P, \quad i \in [1, m(p)]$$

$$x(t_k) \geq 0 \qquad \forall k \in [0, N]$$

$$h_p(t_k) = g_{a_1}^p(t_k + D_{a_1}(x_{a_1}(t_k)))(1 + D'_{a_1}(x_{a_1}(t_k))\dot{x}_{a_1}) \quad \forall k \in [0, N], p \in P$$

$$g_{a_{i-1}}^p(t_k) = g_{a_i}^p(t_k + D_{a_i}(x_{a_i}(t_k)))(1 + D'_{a_i}(x_{a_i}(t_k))\dot{x}_{a_i})$$
$$\forall k \in [0, N], p \in P, i \in [2, m(p)]$$

$$\sum_{p \in P_{ij}} \sum_{k=0}^{N} \phi(k) h_p(t_k) \Delta = Q_{ij} \quad \forall (i, j) \in W$$

$$y_a(t_k) \geq 0 \qquad \forall a \in A, \quad k \in [0, N]$$

$$x(t_k) \geq 0 \qquad g(t_k) \geq 0 \qquad h(t_k) \geq 0 \qquad \forall k \in [0, N],$$

where k takes non-negative integer values, Δ is the discrete time step that divides the time interval $[t_0, t_f]$ into N equal segments, $\phi(k)$ is the coefficient which arises from a trapezoidal approximation of integrals, that is

$$\phi(k) = \begin{cases} 0.5 & \text{if } k = 0 \text{ and } N \\ 1 & \text{otherwise} \end{cases}$$

and

$$t_k = k\Delta.$$

One advantage of time discretization is that we can now completely eliminate state variables (arc volumes) from the problem by noting that

$$x_{a_i}^p(t_{k+1}) = x_{a_{i,0}}^p + \sum_{r=0}^{k} \Delta \left[g_{a_{i-1}}^p(t_r) - g_{a_i}^p(t_r) \right] \quad \forall k \in [0, N-1], \quad p \in P, \quad i \in [1, m(p)].$$

As a consequence, one obtains a finite dimensional mathematical program, which may be solved by conventional algorithms developed for such problems. We employ GAMS/MINOS for the numerical example of the next section.

NUMERICAL EXAMPLE

In what follows, we consider a 3 arc, 3 node network shown in Figure 1. The arc labels and arc delay functions for this network are summarized in Table 1.

There are 2 paths connecting the single OD pair formed by nodes 1 and 3, namely:

$$P_{13} = \{p_1, p_2\}, \qquad p_1 = \{a_1, a_2\}, \qquad p_2 = \{a_1, a_3\}.$$

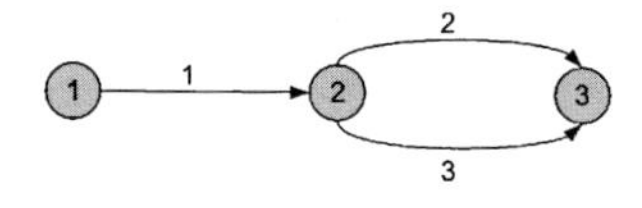

Figure 1. 3-arc 3-node traffic network.

Table 1. Arc labels and delay functions.

Arc name	From node	To node	Arc delay, $D_a(x_a(t))$
a_1	1	2	$2+\frac{1}{100}x_{a_1}$
a_2	2	3	$1+\frac{1}{150}x_{a_2}$
a_3	2	3	$3+\frac{1}{100}x_{a_3}$

The controls (path flows and arc exit flows) and states (path-specific arc traffic volumes) associated with the network are presented in Table 2.

Table 2. Control and state variables.

Path	Path flow	Arc exit flow	Traffic volume of arc
p_1	h_{p_1}	$g_{a_1}^{p_1}, g_{a_2}^{p_1}$	$x_{a_1}^{p_1}, x_{a_2}^{p_1}$
p_2	h_{p_2}	$g_{a_1}^{p_2}, g_{a_3}^{p_2}$	$x_{a_1}^{p_2}, x_{a_3}^{p_2}$

We consider three-day toll planning in which each day is 24 hours, hence, $\Delta = 24$ and $L = 14$ (two weeks). We assume there is the initial travel demand $\tilde{Q} = 150$ units from node 1 (origin) to node 3 (destination). The threshold for travel cost is $\chi = 20000$ and the inter-day rate of change in travel demand is $s_{13} = 0.7$. The desired arrival time for each day is $t_A = 12$, and we employ the symmetric early/late arrival penalty

$$F\left[t + D_p(x,t) - t_A\right] = 5\left[t + D_p(x,t) - t_A\right]^2 .$$

Further, without any loss of generality, we take

$$x_{a_i}^{p}(0) = 0 \quad \forall i \in \left[1, m(p)\right], p \in P .$$

In what follows we forgo the detailed symbolic statement of this example, and, instead, provide numerical results in graphical form.

DOTPEC Computation Based on Time Discretization and GAMS/MINOS

Path flows and arc exit flows for paths p_1 and p_2 are presented in Figures 2 and 3, while path flows and tolls for each arc are given in Figures 4, 5 and 6, for three days from the computed fourteen-day results. We see that tolls tend to be proportional to the path flows. When, for path p_1, we compare the effective path delays (including tolls) with path flows (origin departure rates) by plotting both for the same time scale, Figure 7 is obtained. This figure shows that departure rate peaks when the associated effective path delay achieves a local minimum, thereby demonstrating that a dynamic user equilibrium has been found. Similar comparisons are made for paths p_2 in Figure 8. The daily changes of travel demand from the origin to destination according to the difference equation (54) are given in Figure 9.

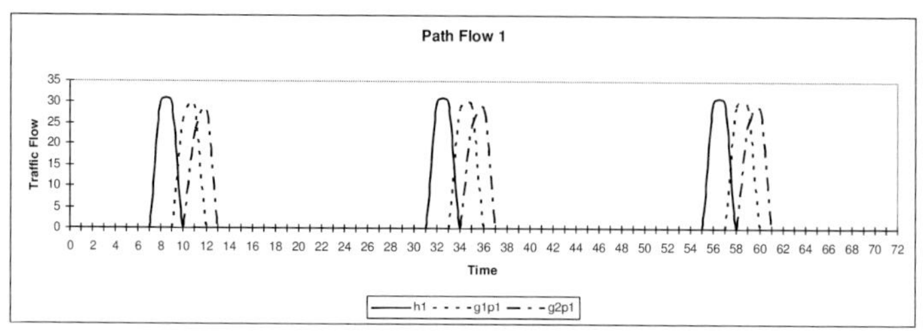

Figure 2. Path and arc exit flows for path p_1

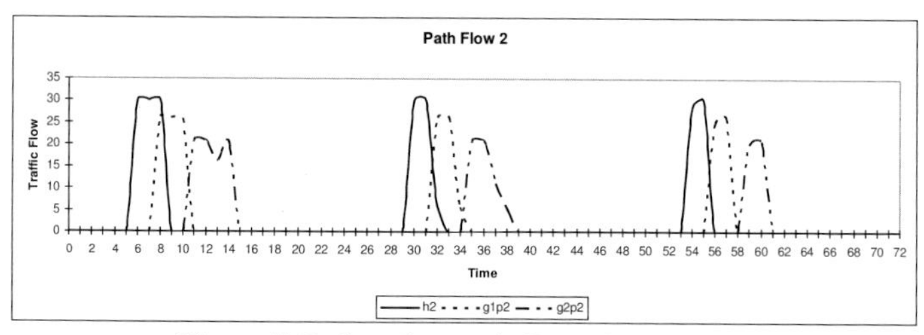

Figure 3. Path and arc exit flows for path p_2.

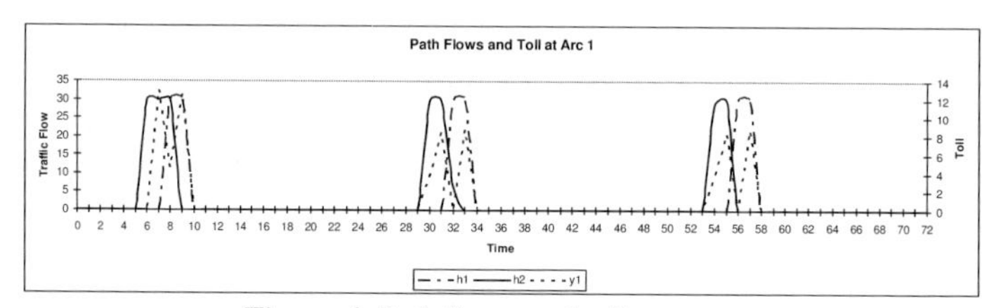

Figure 4. Path flows and toll at arc a_1.

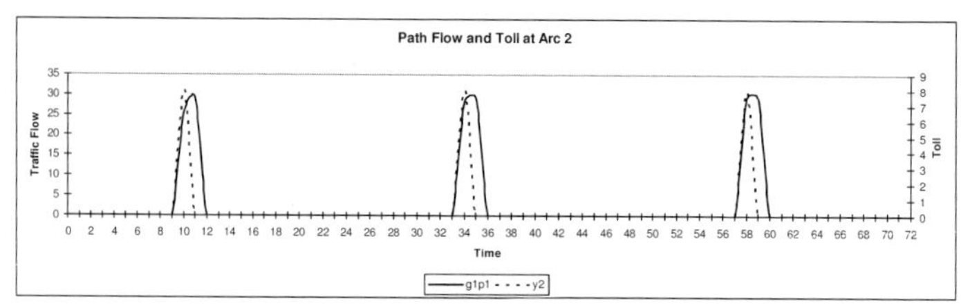

Figure 5. Path flow and toll at arc a_2.

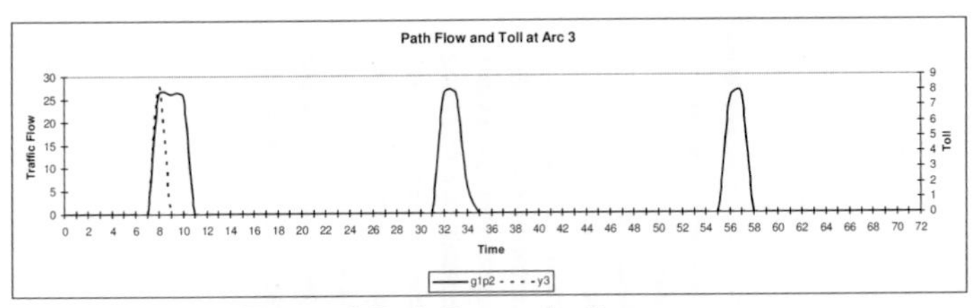

Figure 6. Path flow and toll at arc a_3.

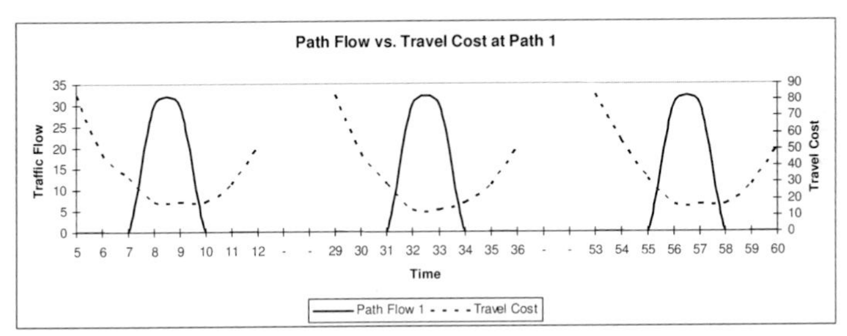

Figure 7. Comparison of path flow and associated unit travel costs for path p_1.

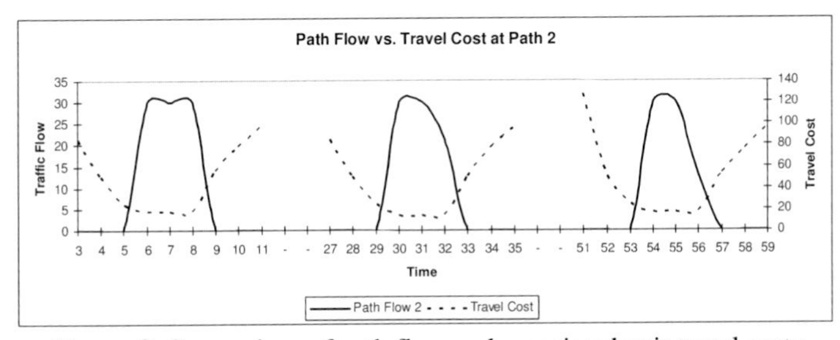

Figure 8. Comparison of path flow and associated unit travel costs for path p_2.

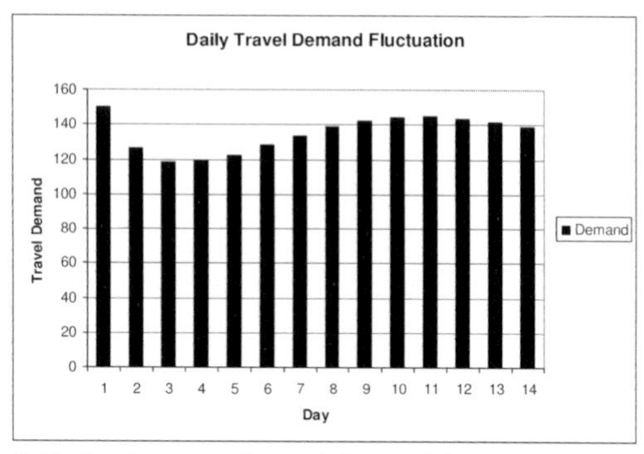

Figure 9. Daily changes of travel demand from the origin (node 1) to the destination (node 3)

DOTPEC Computation based on Descent in Hilbert Space

The same numerical example was also solved by descent in Hilbert space, a continuous-time numerical scheme described in a previous section. While employing the implicit fixed point approach, we penalize the flow propagation constraints, the travel demand constraint, and the DUE conditions which are converted to a set of inequality constraints. We present the path tolls in Figures 10 and 11. As in the previous section we again show the resulting flows are a dynamic user equilibrium by plotting the travel cost and departure flow on the same time axis in Figures 12 and 13.

Comparison of Tolls

To compare, the tolls by DETP and DOTPEC with two algorithms of choice, we suggest a computational scheme for DETP. Recall that the decision rule for the dynamic efficient toll is:

$$y_p^U(t) = \left[\frac{\partial \Psi_p(t, x^S)}{\partial h_p} h_p^S\right]^+ \qquad \forall t \in [t_0, t_f].$$

Note that the partial derivative of $\Psi_p(t, x^S)$ with respect to the path flow h_p is not zero, since the state variable x is an implicit function of the control h_p as the relationship is expressed in the state dynamics. Further we cannot calculate the derivative directly due to the nested delay operator appears in $\Psi_p(\cdot,\cdot)$. However, from the numerical study of the dynamic system optimum traffic assignment, it is known that the controls are zero or singular. When the departure rate is nonzero, it as well as the states obtained from it are smooth and the delay operator is differentiable, although the derivative $\dfrac{\partial \Psi_p(t, x^S)}{\partial h_p}$ does not exist at the time moments where there are kinks in the controls. The derivative is numerically approximated as:

$$\frac{\partial \Psi_p\left[t, x\left(h^*, g^*\right)\right]}{\partial h_p} \cong \frac{\Psi_p\left[t, x(h+\delta, g)\right] - \Psi_p\left[t, x(h, g)\right]}{\delta}.$$

A numerical comparison of the tolls found from the DETP with those from the DOTPEC is given in Figure Figures 14 and 15. We see that the efficient toll has a more spike-like behavior than that for the DOTPEC. It is also interesting to note that the total congestion cost for the DETP is $(26.43, 38.85)$ while the total congestion cost for the DOTPEC is $(38.30, 46.85)$ by discrete approximation and $(43.09, 45.13)$ by descent in Hilbert spaces for paths (p_1, p_2).

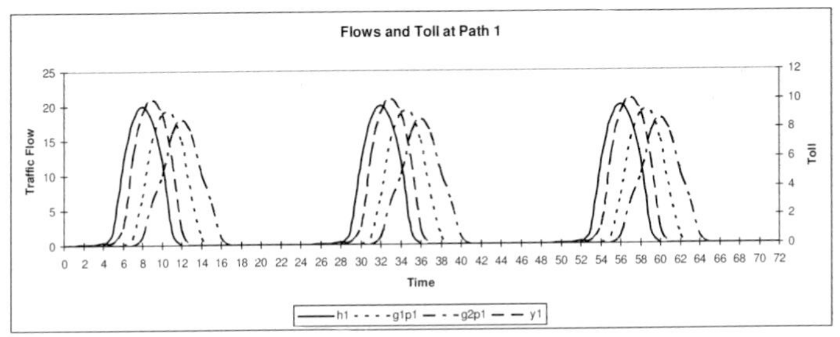

Figure 10. Path flows and toll at path p_1.

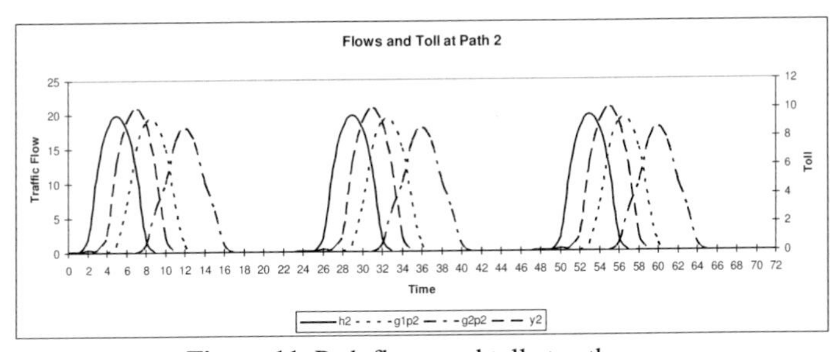

Figure 11. Path flows and toll at path p_2.

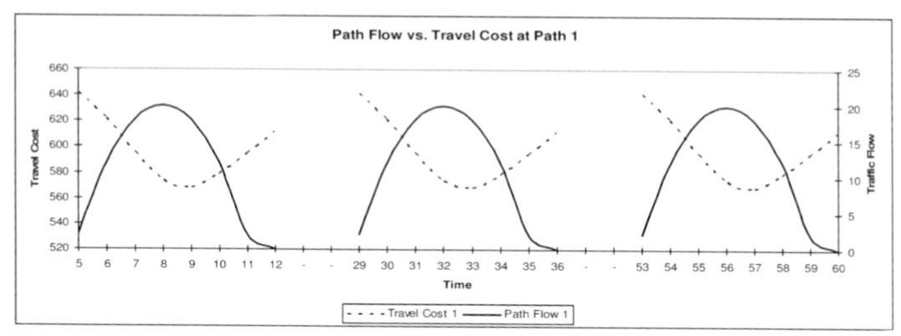

Figure 12. Comparison of path flows and associated unit travel costs for path p_1.

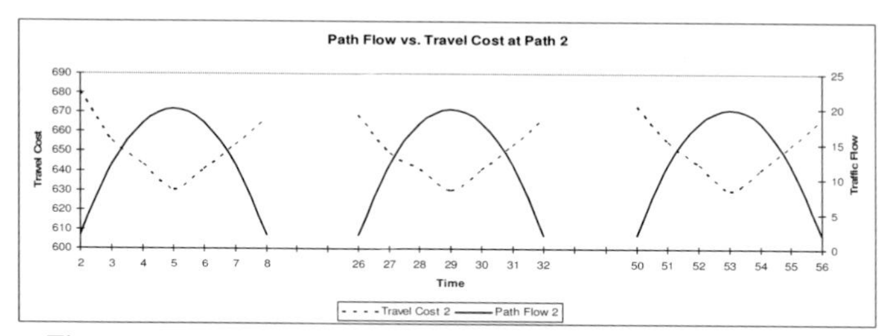

Figure 13. Comparison of path flow and associated unit travel costs for path p_2.

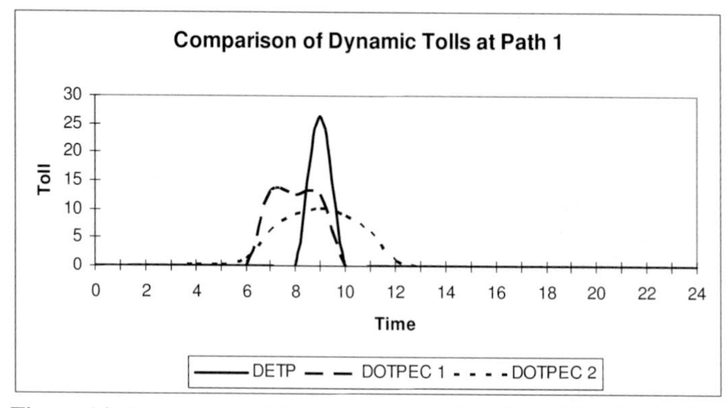

Figure 14. Comparison of Dynamic Tolls by DEPT, DOTPEC solved by discret time approximation (DOTPEC 1), and DOTPEC solved by descent in Hilbert spaces (DOTPEC 2) for path p_1.

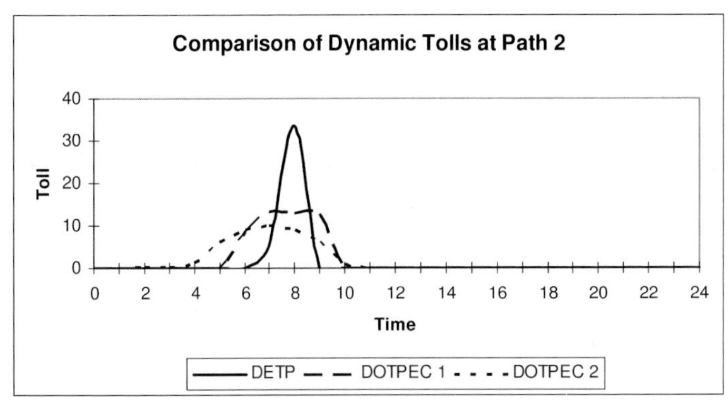

Figure 15. Comparison of Dynamic Tolls by DEPT, DOTPEC solved by discrete time approximation (DOTPEC 1), and DOTPEC solved by descent in Hilbert spaces (DOTPEC 2) for path p_2.

CONCLUDING REMARKS

We have presented a mathematical formulation of the DOTPEC and have shown how it may be directly solved using the notion of descent in Hilbert space for a small illustrative problem. We have also computed solutions using the more familiar approach of time discretization combined with off-the-shelf nonlinear programming software. Clearly, in-depth testing and comparison of these solution methods is required before one can be recommended over the other.

We have not explored in this manuscript the difficult theoretical questions of algorithm convergence, existence of solutions to the dynamic efficient toll and the DOTPEC problems, the Braess paradox and the price of anarchy. These topics are being addressed in a separate manuscript still in preparation. Given that serious efforts are already under way to implement versions of the optimal dynamic toll problem in the U.S. and elsewhere, our initial focus on computation seems fully justified.

We close by commenting that analytical DUE models — in our opinion — are far and away the best starting point for studies of the theoretical aspects of dynamic efficient tolls and dynamic congestion pricing. In particular, we have shown in this paper that an intuitive generalization to a dynamic setting of the efficient static toll rule is correct — something that could not be established in such a definitive way with a simulation model.

REFERENCES

Astarita, V (1995). Flow propagation description in dynamic network loading models. *Proceedings of the IV International Conference on Application of Advanced Technologies in Transportation Engineering* (Y J Stephanedes and F Filippi, eds) 599-603.

Astarita, V (1996). A Continuous Time Link Based Model for Dynamic Network Loading Based on Travel Time Function. *Proceedings of the 13th International Symposium on Transportation and Traffic Theory*, (J-B Lesort, ed), 79-102.

Ban, J, H Liu, M Ferris and B Ran (2005). A link based quasi-vi formulation and solution algorithm for dynamic user equilibria. In: *INFORMS 2005*, San Francisco, CA USA.

Beckmann, M, C B McGuire and C B Winsten (1956). *Studies in the Economics of Transportation*. Yale University Press.

Friesz, T L, D Bernstein and N Kydes (2002). Congestion pricing in disequilibrium. *Networks and Spatial Economics,* **4**, 181-202.

Friesz, T L, D Bernstein, Z Suo and R Tobin (2001). Dynamic network user equilibrium with state-dependent time lags. *Networks and Spatial Economics*, **1**, 319-347.

Friesz, T L and S Shah (2001). An overview of nontraditional formulations of static and dynamic equilibrium network design. *Transportation Research*, **35B**, 5-21.

Friesz, T L, R Tobin, D Bernstein and Z Suo (1995). Proper propagation constraints which obviate exit functions in dynamic traffic assignment. *INFORMS Spring National Meeting*, Los Angeles, April 23-26.

Friesz, T L, D Bernstein, T Smith, R Tobin and B Wie (1993). A variational inequality formulation of the dynamic network user equilibrium problem. *Operations Research*, **41**, 179-191.

Friesz, T L, D Bernstein and R Stough (1996). Dynamic systems, variational inequalities and control theoretic models for predicting urban networks. *Transportation Science*, **30**(1), 14-31.

Friesz, T L and R Mookherjee (2006). Solving the dynamic network user equilibrium with state-dependent time shifts. *Transportation Research*, **40B**, 207-229.

Hearn, D W and M B Yildirim (2002). A Toll Pricing Framework for Traffic Assignment Problems with Elastic Demand, In: *Current Trends in Transportation and Network Analysis—Papers in Honor of Michael Florian* (M Gendreau and P Marcotte, eds), 135-145. Kluwer Academic Publishers.

Liu, L N (2004). Multi-period congestion pricing models and efficient tolls in urban road. *Review of Network Economics*, **3**, 381-391.

Tan, H-N, S Gershwin and M Athans (1979). Hybrid optimization in urban traffic networks. *LIDS Technical Report*, MIT, Cambridge.

Transportation and Traffic Theory 2007
Edited by R.E. Allsop, M.G.H. Bell and B.G. Heydecker

2

BOUNDING THE INEFFICIENCY OF TOLL COMPETITION AMONG CONGESTED ROADS

Feng Xiao, Hai Yang and Xiaolei Guo, Department of Civil Engineering, The Hong Kong University of Science and Technology, Clear Water Bay, Kowloon, Hong Kong, China

INTRODUCTION

Oligopolistic competition in an economic market is known to be inefficient and has been studied extensively in various contexts. Bounding such inefficiency was also considered recently by a few researchers (Cowling and Mueller, 1978; Anderson and Renault, 2003; Johari and Tsitsiklis, 2005; Guo and Yang, 2005). Similarly, competitions in traffic network can also be inefficient. A typical instance that has been studied is the selfish-routing problem, which may cause efficiency loss in comparison with a centrally designed optimal solution, because trip-makers choose their route selfishly with the aim of minimizing their individual generalized travel cost (Roughgarden, 2002 and 2005; Roughgarde and Tardos, 2002; Correa et al, 2004). For selfish-routing problem, bounds are usually established in the spirit of "price of anarchy" determined by looking for the worst possible ratio between the total cost incurred by players in an equilibrium situation and in an outcome of minimum-possible total cost or system optimum. When oligopolistic competition and selfish-routing both come into play in the context of private toll roads, the following issues are of great interest: what are the properties of the inefficiency bounds for the combined problems where private firms control substitutable traffic infrastructures and compete with each other in an oligopolistic market, while taking into account the fact that trip-makers follow the principle of selfish-routing? Can a general expression of the inefficiency for such problems be established? Is the inefficiency bounded? For the combined problem, the inefficiency bound, defined as the worst possible ratio between the total realized social welfares in an outcome of social optimum and in an equilibrium situation, has to be studied by considering the behaviors of both competitive firms and selfish routing users.

A number of studies have been made on the network road pricing games (see Yang and Huang, 2005, for a recent review). Insightful results of toll competition have been obtained for the case of simple networks with parallel links (De Vany and Saving, 1980; De Palma, 1992; De Palma and Lindsey, 2002). In a general network context, Yang and Woo (2000) studied toll road competition using a game-theoretic approach. Wang et al. (2004) examined the strategic interactions and market equilibria of a bilateral monopoly (a private road operator and a private bus service provider) on a private highway. Yang et al. (2006) applied a bilevel variational inequality approach to formulating and solving the toll road competition problem with traffic equilibrium constraints.

Attention is now being paid to the efficiency loss of toll road competition and its bound, but so far only certain specific problems are investigated under restrictive assumptions. Engel et al. (2005) studied toll competition among private asymmetric roads subject to congestion. They found that competition yields tolls that are higher than optimal and that traffic flows are inefficiently small. They also pointed out that increases in competition improve the efficiency. Their limited results are given based on the restrictive assumption that demand grows at the same rate as capacity when the number of roads is increased by network replication. Acemoglu and Ozdaglar (2005) provided an analysis of a similar situation. In contrast with most existing results in the economics literature where greater competition tends to improve the allocation of resources, they found that increasing competition can increase inefficiency. Their observations are made in the case of fixed demand when minimizing the total cost and they found a tight bound of 6/5 on the inefficiency in pure strategy oligopoly equilibria. Ozdaglar (2006) also considered elastic demand function but only for the concave case. A tight bound of 3/2 is given by utilizing a strong assumption that the first-order derivative of demand function is also concave. Hayrapetyan et al. (2005) examined a similar problem with concave demand and linear delay function. They showed that in a network with parallel links, concave demand and linear delays, the inefficiency can be bounded by 5.064. They claimed that this bound is also held even when delays are relaxed to be convex. In particular, when delay is exclusively a congestion effect (without fixed cost in the delay function), they found that the bound can be improved to 3.125. Unsatisfactorily, their discussion is based on a truncated demand curve. Though they stated that the price of anarchy of a truncated demand curve has not decreased in comparison with differentiable demand curve, their bounds are inevitably loose. Xiao et al. (2007) studied the inefficiency of the oligopolistic equilibria of toll road competition. In their model they considered a one-shot game where the road capacity and level of toll charge are determined simultaneously by each firm subject to the resulting traffic flow being in equilibrium. An important property they obtained is that at both oligopolistic equilibria and social optimum, the volume-capacity ratio of each road remains unchanged and is only determined by the road's own unit construction cost. As a result, the travel times are identical for all competitive roads under the assumption that link travel time function is homogeneous of degree zero. With this property, the inefficiency bounds can be developed in a quite simplified manner.

This study is intended to investigate the inefficiency of the equilibria for a situation where two or more profit-maximizing private firms operate multiple toll roads in a parallel road network. It is a continuous development along the line of Xiao et al. (2007), nevertheless, it should be pointed out that the current study focuses on toll competition only, all road capacities are predetermined. In this case, the basic property of identical link travel time at equilibria, as identified in Xiao et al. (2007), no longer holds. Surprisingly, without such a property, the development of inefficiency bounds becomes much more complicated, requiring a totally different methodology.

The inefficiency of oligopolistic toll competition is examined in the absence of any regulatory authority. To derive the inefficiency bound, we technically avoid the interplay of link travel time functions by instead using the market share (percentage of total traffic flow) of each road, which is an equilibrium outcome arising from given toll charges and link travel time functions. The final bounds turn out to be mainly dependent upon the property of the demand function. Furthermore, for the symmetric case where all roads have identical travel time function, we provide analytical expression of the bounds in terms of the number of the parallel roads only.

The rest of the paper is structured as follows. After presenting the model and the results of sensitivity analysis, we establish the necessary conditions for oligopolistic market equilibria and social optimum for further examination. We establish the bounds of inefficiency by considering the worst cases for asymmetric competitive roads, and classify and develop specific bounds by exploring the properties of the various demand functions. We also develop analytical expression of the inefficiency bounds for the specific, simplified case of symmetric roads.

PRELIMINARIES

Description of Traffic Network and Sensitivity Analysis

We consider n roads that join two locations and define N as the set of roads. Each road has a differentiable travel time function, $t_i(q_i)$, where q_i is the traffic flow on road $i \in N$. We assume that $t_i > 0$, $\mathrm{d}t_i/\mathrm{d}q_i > 0$ and $\mathrm{d}^2 t_i/\mathrm{d}q_i^2 > 0$ for any $i \in N$. Suppose, for simplicity, that each firm operates a single toll road consisting of a single link of this parallel network. For each road $i \in N$, a level of toll charge, τ_i, is chosen by the firm operating it. The vectors of tolls and traffic flows are arranged by column vectors $\tau = (\tau_i : i \in N)^{\mathrm{T}}$, and $q = (q_i : i \in N)^{\mathrm{T}}$. Furthermore, we define $B(Q)$ as the marginal benefit of an additional trip when Q trips are already made (it is called marginal benefit function or inverse demand function). It is further assumed that $B' = \mathrm{d}B(Q)/\mathrm{d}Q < 0$, where

$$Q = \sum_{i=1}^{n} q_i \qquad (1)$$

With the above notation, we consider Nash equilibrium among the multiple private firms operating parallel toll roads on the network. To begin with, the following assumption is made and used throughout the whole study.

Assumption 1.

(a) *The travel time function, $t_i(\cdot)$ is convex, strictly increasing and continuously differentiable for $i \in N$;*

(b) *The inverse demand function, $B(\cdot)$ is strictly decreasing and continuously differentiable for $Q \geq 0$;*

(c) *At Nash equilibrium, all firms are active, namely, $q_i^* > 0$ and $\tau_i^* > 0$ for $i \in N$.*

Note that here no assumption is made on the concavity or convexity of $B(Q)$; the travel time function, $t_i(q_i)$ can have a "fixed cost" (i.e. $t_i(0) = t_i^0 \geq 0$, the free flow travel time of road i); part (c) of Assumption 1 means that only actively participating firms are taken into account.

The cost of making one trip on private toll road i has two components. First, the toll τ_i charged by firm $i \in N$; second, the travel time $t_i(q_i)$ when q_i trips are already on the road. Thus, the generalized travel cost faced by a trip-maker is $\tau_i + \beta t_i(q_i)$ (we only consider homogeneous travelers with identical value of time, represented by β). At equilibrium traffic flow q_i is determined by

$$B(Q) = \tau_i + \beta t_i(q_i),\ \forall i \in N \tag{2}$$

So for any two roads $j \neq i,\ i, j \in N$, we have

$$\tau_i + \beta t_i(q_i) = \tau_j + \beta t_j(q_j) \tag{3}$$

Equation (2) implicitly defines q as a function of τ. Differentiating (2) and (3) with respect to τ_i yields

$$B' \sum_{j=1}^{n} \frac{\partial q_j}{\partial \tau_i} = 1 + \beta t_i' \frac{\partial q_i}{\partial \tau_i},\ \forall i \in N \tag{4}$$

$$\beta t_j' \frac{\partial q_j}{\partial \tau_i} = 1 + \beta t_i' \frac{\partial q_i}{\partial \tau_i},\ \forall i, j \in N,\ j \neq i \tag{5}$$

Thus from eqns.(4) and (5) we obtain

$$\frac{\partial q_i}{\partial \tau_i} = \frac{1 - \frac{B'}{\beta} \sum_{j \neq i} \frac{1}{t_j'}}{B' t_i' \sum_{j \neq i} \frac{1}{t_j'} + B' - \beta t_i'},\ \forall i \in N \tag{6}$$

$$\frac{\partial q_j}{\partial \tau_i} = \frac{1}{\beta t_j'} \frac{B'}{B' t_i' \sum_{l \neq i} \frac{1}{t_l'} + B' - \beta t_i'},\ \forall i, j \in N,\ j \neq i \tag{7}$$

Since $B'(Q) < 0$ and $t_i'(q_i) > 0$, we have $\partial q_i / \partial \tau_i < 0$ and $\partial q_j / \partial \tau_i > 0$. This result is expected in the sense that raising toll charge on one road will decrease the traffic flow on that road and correspondingly increase the traffic flows on other roads.

Oligopolistic Market

In an oligopolistic market, each firm $i \in N$ tries to maximize its own profit given by

$$\pi_i = \tau_i q_i(\tau) \tag{8}$$

From definition, if q^* is a Nash equilibrium solution, then for each firm $i \in N$, the following equation must hold

$$\tau_i^* = \arg\max_{\tau_i \geq 0} \tau_i q_i(\tau),\ \ \forall i \in N \tag{9}$$

Thus, if the Nash equilibrium has a solution q^* [1], it must satisfy

$$\left(q_i^* + \tau_i^* \frac{\partial q_i}{\partial \tau_i}\right)\tau_i^* = 0, \quad \forall i \in N \tag{10}$$

$$q_i^* + \tau_i^* \frac{\partial q_i}{\partial \tau_i} \le 0, \quad \forall i \in N \tag{11}$$

By Assumption 1, all roads are active at Nash equilibrium, namely, $q_i^* > 0, \tau_i^* > 0$ for $i \in N$, then the Nash equilibrium conditions (10)-(11) can be simplified into

$$q_i^* + \tau_i^* \frac{\partial q_i}{\partial \tau_i} = 0, \quad \forall i \in N \tag{12}$$

Substituting (2) and (6) into (12) yields

$$B\left(Q^*\right) - \beta t_i\left(q_i^*\right) - \beta q_i^* t_i'\left(q_i^*\right) + \frac{1}{1 - B'\sum_{j \ne i} \frac{1}{\beta t_j'}} q_i^* B'\left(Q^*\right) = 0, \quad \forall i \in N \tag{13}$$

where $Q^* = \sum_{i=1}^{n} q_i^*$ is the total realized traffic flow at Nash equilibrium. Here for convenience we use a parameter ω_i to denote the following term

$$\omega_i = \frac{1}{1 - B'\sum_{j \ne i} \frac{1}{\beta t_j'}}, \quad \forall i \in N \tag{14}$$

Obviously, the value of parameter ω_i depends on the properties of road travel time and inverse demand functions; $\omega_i \in (0,1]$ ($\omega = 1$ when there is only one private firm). With the above definition, eqn.(13) can be rewritten as

$$B\left(Q^*\right) - \beta t_i\left(q_i^*\right) - \beta q_i^* t_i'\left(q_i^*\right) + \omega_i q_i^* B'\left(Q^*\right) = 0, \quad \forall i \in N \tag{15}$$

At equilibrium, each active road has a certain traffic share. Let s_i denote the ith road's equilibrium traffic share defined below

$$s_i = \frac{q_i^*}{Q^*}, \quad \forall i \in N \tag{16}$$

Clearly, $0 < s_i \le 1, \ \forall i \in N$.

---

[1] For the case of concave inverse demand function, a sufficient condition for the existence of Nash equilibrium is provided

$$\frac{1}{\sum_{j \ne i} \frac{1}{t_j'}} - \frac{1}{B'} \ge \frac{1}{t_i'} \sqrt[3]{\sum_{j \ne i} \frac{t_j''}{t_i''}}$$

See Engel et al. (2004) for the proof.

Social Optimum

It is well known that social welfare is defined as the sum of consumers' and producers' surplus, or equivalently total trip-makers' benefits minus total travel cost, which can be mathematically formulated as

$$W = \int_0^Q B(v)\,\mathrm{d}v - \beta\sum_{i=1}^{n} q_i t_i(q_i) \tag{17}$$

Then the social optimization (SO) problem is given by

$$\max_{q\geq 0} \int_0^Q B(v)\mathrm{d}v - \beta\sum_{i=1}^{n} q_i t_i(q_i) \tag{18}$$

Let $\bar{q} = (\bar{q}_i : i \in N)$ be the optimal solution to (18), then the following first-order optimality conditions hold

$$\left[B(\bar{Q}) - \beta t_i(\bar{q}_i) - \beta\bar{q}_i t_i'(\bar{q}_i)\right]\bar{q}_i = 0, \quad \forall i \in N \tag{19}$$

$$B(\bar{Q}) - \beta t_i(\bar{q}_i) - \beta\bar{q}_i t_i'(\bar{q}_i) \leq 0, \quad \forall i \in N \tag{20}$$

where $\bar{Q} = \sum_{i=1}^{n} \bar{q}_i$ is the socially optimal demand.

Let $\bar{W}$ and W^* respectively denote the maximum social welfare and the social welfare at Nash equilibrium

$$\bar{W} = \int_0^{\bar{Q}} B(v)\mathrm{d}v - \beta\sum_{i=1}^{n} \bar{q}_i t_i(\bar{q}_i) \tag{21}$$

$$W^* = \int_0^{Q^*} B(v)\mathrm{d}v - \beta\sum_{i=1}^{n} q_i^* t_i(q_i^*) \tag{22}$$

We can define the following ratio to measure the inefficiency of Nash equilibrium

$$\rho = \frac{\bar{W}}{W^*} \tag{23}$$

Obviously, $\rho \geq 1$, and a larger value of ρ implies a more inefficient Nash equilibrium. Hereafter, our task is to establish an upper bound of this inefficiency ratio ρ.

Interpretation of Parameter ω_i and Property of the Game

In a classic Cournot model, each firm's strategy space is taken as the output. Thus a Cournot equilibrium can be regarded as a set of outputs in which each firm is choosing its profit-maximizing output level given its beliefs about the other firm's choice, and each firm's beliefs about the other firm's choice are actually correct (Varian, 1992). Correspondingly, for the model discussed here, if each firm can control or choose its own road flow, then it is actually a Cournot model. The first-order optimality condition in terms of road flows can be written as

$$B(Q^*) - \beta t_i(q_i^*) - \beta q_i^* t_i'(q_i^*) + q_i^* B'(Q^*) = 0 \quad \text{for } i \in N \tag{24}$$

But for the pricing game on a parallel traffic network, the strategy space of each firm is its direct control variable of toll charge, rather than its output of traffic flow. Even if its belief about the other firms' choice of tolls are correct, the firm can not simply choose a "proper" toll by only examining its own road, because the entire network flow pattern will vary if toll charge on any road changes. The network pricing game cannot be regarded as a classic Bertrand model either, since the total travel cost on each road is not burdened by the firm but the trip-makers using the road.

After the above clarification, we now look at how ω_i, defined in (14), plays an important role in determining the property of the network pricing game, which can be seen from the first-order optimality condition (15). When $\omega_i \to 1$, eqn.(15) reduces to eqn.(24), implying that the game becomes a Cournot game. As a result, it makes no difference whether the strategy space is traffic flow or toll charge. It is clear from (14) that $\omega_i \to 1$ if and only if $t_j' \to \infty$ for all $j \in N, j \neq i$, given that B' is bounded. This just corresponds to the situation when the marginal travel costs on all other roads are very high (i.e., all other roads are highly congested). In this case, a small change of toll charge on road i can hardly affect the traffic flows on other roads, or traffic flows on other roads can be taken as given when firm i sets up its toll charge for maximizing profit. Consequently, profit maximization of firm i with respect to either toll τ_i or flow q_i will give rise to an identical result.

On the other hand, when $t_j' \to 0$ for some $j \in N, j \neq i$ (i.e., there is no congestion on one of the roads), we have $\omega_i \to 0$. By comparing (15) and (19), we find that the toll charged on road i at oligopoly equilibria turns out to be a marginal-cost pricing toll. This can be explained as follows: at equilibrium, a small change in the toll charge of road i will not inflict the generalized total travel cost of road j, since road j is not congested. Thus, to keep the generalized total travel cost of road i unchanged and equal to the generalized total travel cost of road j, the change in the toll charge of road i must lead to the same but opposite change in its travel time, namely, $\Delta\tau = -\Delta t$ (without loss of generality, we consider unit value of time). At the profit-maximizing point, $\tau \cdot q = (\tau + \Delta\tau)(q + \Delta q)$, leading to

$-\tau\Delta q = q\Delta\tau$ after dropping the second-order term. From $\Delta\tau = -\Delta t$, we obtain $\tau = q(\Delta t/\Delta q)$, the marginal-cost pricing toll. In this case, we have the same result as Bertrand model of oligopoly, since the oligopolistic competition in the Bertrand model also leads to marginal-cost pricing.

BOUND EXPRESSIONS FOR ASYMMETRIC ROADS

General Bound Expression

Definition 1. *Let* $\omega_m s_m = \max\{\omega_i s_i, i \in N\}$ *and define a function* $\gamma(\cdot)$ *as*

$$\gamma(x): B(x) + \omega_m s_m x B'(x) = B(\gamma(x)x) \tag{25}$$

or equivalently,

$$\gamma(x) = \frac{B^{-1}\left(B(x) + \omega_m s_m x B'(x)\right)}{x} \tag{26}$$

This function is introduced to facilitate our subsequent analysis. As observed from a geometric interpretation of $\gamma(x)$ in Figure 1, $1 < \gamma(x) \leq 1 + \omega_m s_m$ if $B(\cdot)$ is concave, and $\gamma(x) \geq 1 + \omega_m s_m$ if $B(\cdot)$ is convex (see Appendix A1 for a proof). Of course, for a linear function of $B(\cdot)$, being both convex and concave, we have $\gamma(x) = 1 + \omega_m s_m$. Clearly, $\gamma(x) \to 1$ when $\omega_m s_m \to 0$.

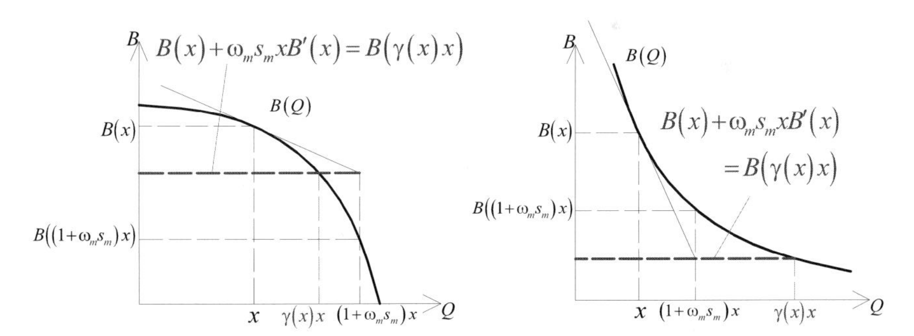

a. $1 < \gamma(x) \leq 1 + \omega_m s_m$ for concave $B(\cdot)$ b. $\gamma(x) \geq 1 + \omega_m s_m$ for convex $B(\cdot)$

Figure 1. Graphical illustration of $\gamma(\cdot)$

When $B(\cdot)$ is concave, $\gamma(\cdot)$ is well upper-bounded by 2, because $\gamma(\cdot)$ has a upper-bound of $1+\omega_m s_m$ and $0<\omega_m s_m \leq 1$. However, from Figure 1(b) we see that, for a convex $B(\cdot)$, $\gamma(\cdot)$ could go to infinity when the tangent line of the demand curve becomes a vertical line (i.e., the demand is highly inelastic in the vicinity of the tangent point). In this extreme case, an upper-bound of $\gamma(\cdot)$ for a convex demand curve can not be simply determined. Nevertheless, given a meaningful specific inverse demand function with a well defined slope anywhere, it is not difficult to find an upper-bound for $\gamma(\cdot)$ expressed in terms of the parameters contained in the inverse demand function. For example, if $B(\cdot)$ takes the widely used form of $B(Q)=aQ^{-b}$, $a>0$, $0<b<1$, we have $\gamma(x)=(1-b\omega_m s_m)^{-1/b}$ and thus, $1<\gamma(x)\leq(1-b)^{-1/b}$; if $B(\cdot)$ takes the form of $B(Q)=B_0-aQ^b$, $a>0$, $0<b<1$, we have $\gamma(x)=(1+b\omega_m s_m)^{1/b}$ and $1<\gamma(x)\leq(1+b)^{1/b}$; and if $B(\cdot)$ takes the exponential form of $B(Q)=-b^{-1}\ln(Q/a)$, $a>0$, $b>0$, we have $\gamma(x)=e^{\omega_m s_m}$ and $1<\gamma(x)\leq e$.

Lemma 1. *Let* $\varepsilon=\bar{Q}/Q^*$*, then it holds that* $1<\varepsilon\leq\gamma(Q^*)$.

Proof: See Appendix A2 for a proof. ■

Definition 2. The following function $\theta(\cdot)$ is introduced

$$\theta(x)=\frac{\frac{1}{2}x^2\left(-B'(x)\right)}{\int_0^x B(v)\,\mathrm{d}v-xB(x)} \tag{27}$$

Figure 2 provides a geometric interpretation of $\theta(x)$. In both cases of Figure 2, the numerator, $x^2\left(-B'(x)\right)/2$, is the area of the triangle with its hypotenuse tangent to the inverse demand curve, and the denominator, $\int_0^x B(v)\,\mathrm{d}v-xB(x)$, is the shaded area. Thus, $\theta(x)$ is the ratio of the two areas, and it can be easily confirmed that $\theta(x)\geq 1$ if $B(\cdot)$ is concave, and $\theta(x)\leq 1$ if $B(\cdot)$ is convex. Of course, for linear $B(\cdot)$, being both convex and concave, we have $\theta(x)=1$.

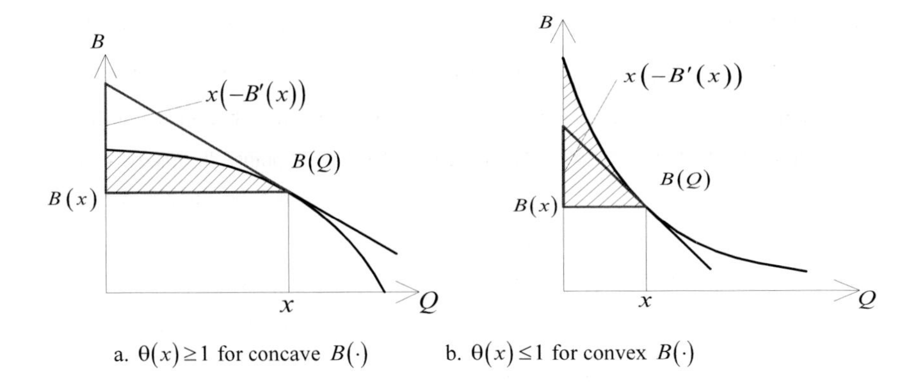

a. $\theta(x) \geq 1$ for concave $B(\cdot)$ b. $\theta(x) \leq 1$ for convex $B(\cdot)$

Figure 2. Geometric meaning of $\theta(x)$

In opposite to $\gamma(\cdot)$, $\theta(\cdot)$ is well upper-bounded by 1 when $B(\cdot)$ is convex. But $\theta(\cdot)$ could go to infinity when $B(\cdot)$ is concave and the shaded area is infinitely small (i.e., the demand is perfectly elastic). Nonetheless, like the function $\gamma(\cdot)$, an upper bound of $\theta(\cdot)$ can be easily established for a given specific concave demand function. For example, if $B(\cdot)$ takes the form of $B(Q) = B_0 - aQ^b$, $a > 0$, $b > 1$, $B_0 > 0$, we have $\theta(x) = (1+b)/2$. If $b = 1$, $\theta(x) = 1$, as expected for a linear $B(\cdot)$.

Lemma 2. *The social welfare W^* at Nash equilibrium satisfies*

$$W^* \geq \left(-B'\left(Q^*\right)\right)\left[\frac{1}{2\theta\left(Q^*\right)}\left(Q^*\right)^2 + \sum_{i=1}^{n} \omega_i \left(q_i^*\right)^2\right] \tag{28}$$

Proof: See Appendix A3 for a proof. ■

Lemma 3. *The following inequality holds*

$$\bar{W} - W^* \leq \left(-B'\left(Q^*\right)\right)\sum_{i=1}^{n} \omega_i \left(\bar{q}_i - q_i^*\right) q_i^* \tag{29}$$

Proof: See Appendix A4 for a proof. ■

Combining Lemma 2 and 3 we readily obtain

$$\rho = \frac{\overline{W}}{W^*} = 1 + \frac{\overline{W} - W^*}{W^*} \leq 1 + \frac{\left(-B'(Q^*)\right)\sum_{i=1}^{n}\omega_i\left(\overline{q}_i - q_i^*\right)q_i^*}{\left(-B'(Q^*)\right)\left[\frac{1}{2\theta(Q^*)}\left(Q^*\right)^2 + \sum_{i=1}^{n}\omega_i\left(q_i^*\right)^2\right]}$$
$$= \frac{\left(Q^*\right)^2 + 2\theta\left(Q^*\right)\sum_{i=1}^{n}\omega_i\overline{q}_i q_i^*}{\left(Q^*\right)^2 + 2\theta\left(Q^*\right)\sum_{i=1}^{n}\omega_i\left(q_i^*\right)^2} \tag{30}$$

By applying the market share s_i defined by (16), eqn.(30) can also be written as follows

$$\rho \leq \frac{1 + 2\theta\left(Q^*\right)\sum_{i=1}^{n}\omega_i s_i \frac{\overline{q}_i}{Q^*}}{1 + 2\theta\left(Q^*\right)\sum_{i=1}^{n}\omega_i s_i^2} \tag{31}$$

Now we try to simplify the right-hand side of inequality (31) and eliminate parameter ω_i. First note that $\omega_m s_m = \max\left\{\omega_i s_i, i \in N\right\}$, we have

$$\rho \leq \frac{1 + 2\theta\left(Q^*\right)\omega_m s_m \sum_{i=1}^{n}\frac{\overline{q}_i}{Q^*}}{1 + 2\theta\left(Q^*\right)\left(\omega_m s_m^2 + \sum_{i=1,i\neq m}^{n}\omega_i s_i^2\right)} \tag{32}$$

$$= \frac{1 + 2\theta\left(Q^*\right)\omega_m s_m \varepsilon}{1 + 2\theta\left(Q^*\right)\left(\omega_m s_m^2 + \sum_{i=1,i\neq m}^{n}\omega_i s_i^2\right)} \tag{33}$$

$$\leq \frac{1 + 2\theta\left(Q^*\right)\omega_m s_m \varepsilon}{1 + 2\theta\left(Q^*\right)\omega_m s_m^2} \tag{34}$$

where (33) is from the definition of ε (see Lemma 1); and by dropping the term $\sum_{i=1,i\neq m}^{n}\omega_i s_i^2$, we obtain (34).

Because $\varepsilon > 1 \geq s_m$, we can further relax the upper-bound of ρ by increasing ω_m. We know that $\omega_m \leq 1$, so we just replace ω_m by 1. Furthermore, from Lemma 1, ε is upper-bounded by $\gamma\left(Q^*\right)$. Therefore, we obtain

$$\rho \leq \frac{1+2\theta(Q^*)\gamma(Q^*)s_m}{1+2\theta(Q^*)s_m^2} \tag{35}$$

From inequality (35), we have the following theorem.

Theorem 1. *For the case of perfect competition, namely* $s_i \to 0, \forall i \in N$, *it holds that* $\rho \to 1$.

Proof: Under perfect competition, it follows that $s_m \to 0$. From Definition 1, when $s_m \to 0$, $\gamma(Q^*) \to 1$. For a given continuous and differentiable inverse demand function, $\theta(Q^*)$ is finite. Thus, the right-hand side of (35) approaches 1. Since $\rho > 1$, it follows immediately that $\rho \to 1$. ∎

Taking the market share s_m of road *m*, as an independent variable in inequality (35), we consider the following maximization problem for obtaining the bound

$$\max_{s_m} \frac{1+2\theta(Q^*)\gamma(Q^*)s_m}{1+2\theta(Q^*)s_m^2} \tag{36}$$

Solving this problem yields

$$\rho \leq \frac{\sqrt{1+2\theta(Q^*)\gamma(Q^*)^2}+1}{2} \tag{37}$$

where "=" can hold only if

$$s_m = \frac{\sqrt{1+2\theta(Q^*)\gamma(Q^*)^2}-1}{2\theta(Q^*)\gamma(Q^*)} \tag{38}$$

To further simplify the above result, we define the following parameters $\hat{\gamma}$ and $\hat{\theta}$

$$\hat{\gamma} = \sup\ \gamma(x) \tag{39}$$

$$\hat{\theta} = \sup\ \theta(x) \tag{40}$$

where $\gamma(\cdot)$ and $\theta(\cdot)$ are two functions defined by (25) and (27), respectively. Therefore, we readily obtain Theorem 2 by setting $\gamma(Q^*) = \hat{\gamma}$ and $\theta(Q^*) = \hat{\theta}$ in (37).

Theorem 2. *With Assumption 1, for any general demand function the inefficiency ratio* ρ *is bounded by*

$$\rho \leq \frac{\sqrt{1+2\hat{\theta}\hat{\gamma}^2}+1}{2} \tag{41}$$

Concave Inverse Demand Function

Lemma 4. *If* $B(\cdot)$ *is concave, the following inequality holds*

$$\bar{W} - W^* \leq \left(-B'\left(Q^*\right)\right)\left[\sum_{i=1}^{n}\omega_i\left(\bar{q}_i - q_i^*\right)q_i^* - \frac{1}{2}\left(\bar{Q}-Q^*\right)^2\right] \tag{42}$$

Proof: With Assumption 1, if $B(\cdot)$ is concave, we have

$$\int_{Q^*}^{\bar{Q}} B(v)\mathrm{d}v \leq \left(\bar{Q}-Q^*\right)B\left(Q^*\right) - \frac{1}{2}\left(-B'\left(Q^*\right)\right)\left(\bar{Q}-Q^*\right)^2 \tag{43}$$

Figure 3 graphically compares the areas represented by the two sides of inequality (43), and the proof follows the same line as in the proof of Lemma 3. Let (43) take the place of (91) in the proof of Lemma 3, then the inequality (29) of Lemma 3 is simply replaced by (42). Therefore, Lemma 4 is true. ■

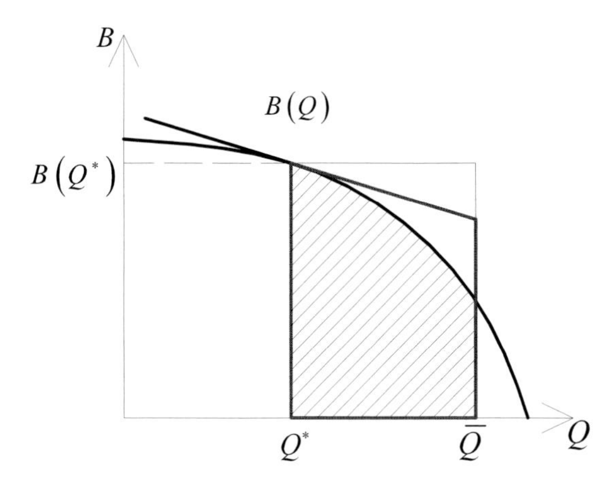

Figure 3. Graphical illustration for Inequality (43)

Lemma 4 for the case of concave inverse demand functions is the counterpart of Lemma 3 for the general inverse demand function. Combining Lemma 2 and Lemma 4, we obtain

$$\rho = 1 + \frac{\bar{W} - W^*}{W^*} \le \frac{\left(Q^*\right)^2 + 2\theta\left(Q^*\right)\left[\sum_{i=1}^{n} \omega_i \bar{q} q_i^* - \frac{1}{2}(\varepsilon - 1)^2 \left(Q^*\right)^2\right]}{\left(Q^*\right)^2 + 2\theta\left(Q^*\right)\sum_{i=1}^{n} \omega_i \left(q_i^*\right)^2} \tag{44}$$

where $\varepsilon = \bar{Q}/Q^*$ satisfies $1 < \varepsilon \le \gamma\left(Q^*\right)$ as shown in Lemma 1. Again, like previous manipulation, we have

$$\rho \le \frac{1 + \theta\left(Q^*\right)\left[2 s_m \varepsilon - (\varepsilon - 1)^2\right]}{1 + 2\theta\left(Q^*\right) s_m^2} \tag{45}$$

From previous analysis of $\gamma\left(Q^*\right)$ and Lemma 1 we know that $\varepsilon \le \gamma\left(Q^*\right) \le 1 + \omega_m s_m \le 1 + s_m$ when $B(\cdot)$ is concave. The right-hand side of inequality (45) is increasing with ε when $\varepsilon \le 1 + s_m$. Thus we can replace ε with $\gamma\left(Q^*\right)$ in (45) and obtain

$$\rho \le \frac{1 + \theta\left(Q^*\right)\left[2 s_m \gamma\left(Q^*\right) - \left(\gamma\left(Q^*\right) - 1\right)^2\right]}{1 + 2\theta\left(Q^*\right) s_m^2} \tag{46}$$

Theorem 3. *With Assumption 1 and the assumption that the inverse demand function $B(\cdot)$ is concave, it holds that*

$$\rho \le \frac{3 + \sqrt{1 + 8\hat{\theta}}}{4} \tag{47}$$

Proof: First, we take $\gamma\left(Q^*\right)$ as a decision variable and solve the following maximization problem

$$\max_{\gamma\left(Q^*\right)} \frac{1 + \theta\left(Q^*\right)\left[2 s_m \gamma\left(Q^*\right) - \left(\gamma\left(Q^*\right) - 1\right)^2\right]}{1 + 2\theta\left(Q^*\right) s_m^2} \tag{48}$$

We readily obtain

$$\rho \le \frac{1 + \theta\left(Q^*\right)\left(2 + s_m\right) s_m}{1 + 2\theta\left(Q^*\right) s_m^2} \tag{49}$$

where "=" can hold only if $\omega_m = 1$ and $\gamma(Q^*) = 1 + s_m$. From our early discussion, we know this is true only if the inverse demand function is linear or $\theta(Q^*) = 1$.

Second, maximizing the right-hand side of inequality (49) with respect to s_m yields

$$\rho \le \frac{3 + \sqrt{1 + 8\theta(Q^*)}}{4} \tag{50}$$

where "=" can hold only if $s_m = \left(\sqrt{1 + 8\theta(Q^*)} - 1\right) / 4\theta(Q^*)$. Thus we readily obtain (47) by setting $\theta(Q^*) = \hat{\theta}$. This completes the proof. ■

From eqn.(47), when $\hat{\theta} = 1$, namely the inverse demand function is linear, the inefficiency ratio is bounded by $3/2$.

Convex Inverse Demand Function

Now we move to the discussion of the case of convex inverse demand function, including the demand function of constant elasticity.

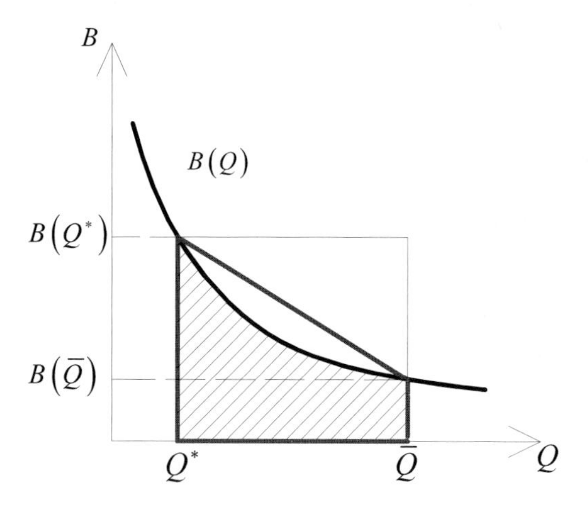

Figure 4. Graphical illustration for Inequality (52)

Lemma 5. *With Assumption 1, and assume that* $B(\cdot)$ *is convex, it holds that*

$$\bar{W} - W^* \le \left(-B'(Q^*)\right)\sum_{i=1}^{n}\omega_i\left(\bar{q}_i - q_i^*\right)q_i^* - \frac{1}{2}\left(\bar{Q} - Q^*\right)\left(B(Q^*) - B(\bar{Q})\right) \tag{51}$$

Proof: With Assumption 1, if $B(\cdot)$ is convex, we have

$$\int_{Q^*}^{\bar{Q}} B(v)\,\mathrm{d}v \le \left(\bar{Q} - Q^*\right)B(Q^*) - \frac{1}{2}\left(\bar{Q} - Q^*\right)\left(B(Q^*) - B(\bar{Q})\right) \tag{52}$$

A geometric interpretation of this inequality is given in Figure 4, where the two sides of (52) are represented by the relevant areas. Then the proof follows the same line as in the proof of Lemma 3. Let (52) take the place of (91) in the proof of Lemma 3, then the inequality (42) of Lemma 4 simply becomes (51). This completes the proof. ■

Lemma 5 for the convex inverse demand function is the counterpart of Lemma 3 for the general inverse demand function. Combining Lemma 2 and Lemma 5 and substituting (28) into (51) yields

$$\rho \le \frac{\left(Q^*\right)^2 + 2\theta(Q^*)\left[\sum_{i=1}^{n}\omega_i\bar{q}q_i^* - \frac{(\varepsilon-1)Q^*\left(B(Q^*) - B(\bar{Q})\right)}{2\left(-B'(Q^*)\right)}\right]}{\left(Q^*\right)^2 + 2\theta(Q^*)\sum_{i=1}^{n}\omega_i\left(q_i^*\right)^2} \tag{53}$$

where $\varepsilon = \bar{Q}/Q^*$ satisfies $1 < \varepsilon \le \gamma(Q^*)$ from Lemma 1. Again, like what we have done earlier, after dropping the term $\sum_{i=1,i\ne m}^{n}\omega_i s_i^2$ in the denominator, we have

$$\rho \le \frac{1 + \theta(Q^*)\left[2\omega_m s_m \varepsilon - (\varepsilon - 1)h(Q^*, \varepsilon)\right]}{1 + 2\theta(Q^*)\omega_m s_m^2} \tag{54}$$

where h is a function defined by

$$h(x, \varepsilon) = \frac{B(x) - B(\varepsilon x)}{-B'(x)x}, \quad 1 < \varepsilon \le \gamma(x) \tag{55}$$

From the definition (55) of $h(x,\varepsilon)$ and the definition (25) of $\gamma(\cdot)$, we have the following relationship

$$h\left(x, \gamma(x)\right) = \omega_m s_m \tag{56}$$

The term $2\omega_m s_m \varepsilon - (\varepsilon - 1)h(Q^*, \varepsilon)$ in (54) has the following property

Lemma 6. *With Assumption 1, if* $B(\cdot)$ *is convex, for any* $\omega_m s_m$, then the term $2\omega_m s_m \varepsilon - (\varepsilon - 1) h(Q^*, \varepsilon)$ *increases with* ε.

Proof: See Appendix A5 for a proof. ∎

With Lemma 1 and Lemma 6, setting $\varepsilon = \gamma(Q^*)$ in the right-hand side of (54) and making use of (56) lead to

$$\rho \le \frac{1 + \theta(Q^*)(1 + \gamma(Q^*))\omega_m s_m}{1 + 2\theta(Q^*)\omega_m s_m^2} \tag{57}$$

Since $\rho > 1$, we have $\theta(Q^*)(1+\gamma(Q^*))s_m > 2\theta(Q^*)s_m^2$. We can just relax ω_m to 1. By setting $\gamma(Q^*) = \hat{\gamma}$ and $\theta(Q^*) = \hat{\theta}$, we obtain

$$\rho \le \frac{1 + \hat{\theta}(1+\hat{\gamma})s_m}{1 + 2\hat{\theta}s_m^2} \tag{58}$$

Clearly, eqn.(58) gives the same result as eqn.(49) for linear inverse demand functions ($\hat{\theta} = 1$ and $\hat{\gamma} = 1 + \omega_m s_m \le 1 + s_m$), which is an expected "coincidence", because a linear function is both convex and concave.

Taking the right-hand side of inequality (58) as a maximization problem with respect to s_m, we readily obtain the following theorem.

Theorem 4. *With Assumption 1, and assume that the inverse demand function* $B(\cdot)$ *is convex, it holds that*

$$\rho \le \frac{2 + \sqrt{4 + 2\hat{\theta}(1+\hat{\gamma})^2}}{4} \tag{59}$$

where "=" can hold only if

$$s_m = \frac{\sqrt{4 + 2\hat{\theta}(1+\hat{\gamma})^2} - 2}{2\hat{\theta}(1+\hat{\gamma})} \tag{60}$$

Theorem 4 bounds the inefficiency of oligopoly with convex inverse demand function. Since parameter $\hat{\gamma}$ is determined by $\omega_m s_m$ and the function form of $B(\cdot)$, Theorem 4 is not straightforwardly tractable without information on the specific inverse demand function.

We conclude this section by providing the inefficiency ratios for some widely used inverse demand functions.

1) *Exponential demand function with* $B(Q) = -b^{-1}\ln(Q/a)$

$B' = -1/bQ < 0,\ B'' = 1/bQ^2 > 0$, the conditions for strictly decreasing and convex demand are satisfied. From Definitions 1 and 2, we have $\gamma(\cdot) = e^{\omega_m s_m} \le e,\ \theta(\cdot) = 1/2$. Substituting them into eqn.(59) yields

$$\rho \le \frac{2+\sqrt{4+(1+e)^2}}{4} \approx 1.56 \tag{61}$$

2) *Polynomial demand function with* $B(Q) = B_0 - aQ^b$ where $a > 0, b > 1,\ B_0 > 0$

Then $B' = -abQ^{b-1} < 0$, $B'' = -ab(b-1)Q^{b-2} < 0$, the conditions for strictly decreasing and concave demand are satisfied. From Definitions 1 and 2, we have $\theta(\cdot) = (1+b)/2$. Substituting $\theta(\cdot) = (1+b)/2$ into eqn.(50) yields

$$\rho \le \frac{3+\sqrt{5+4b}}{4} \tag{62}$$

Especially, when $b = 1$, representing the linear inverse demand function case, we have $\rho \le 1.5$.

SPECIFIC BOUND FOR SYMMETRIC ROADS

The term "symmetric roads" here refers to the case with an identical travel time function $t(\cdot)$ for all parallel roads. As we have proved in Lemma 1, $\bar{Q} > Q^*$. Thus for the symmetric road case we have

$$\bar{q}_i = \frac{\bar{Q}}{n} > \frac{Q^*}{n} = q_i^*,\quad \forall i \in N \tag{63}$$

After relaxing each ω_i to 1 in eqn.(30), we obtain

$$\rho \leq \frac{(Q^*)^2 + 2\theta(Q^*)\sum_{i=1}^{n} \bar{q}_i q_i^*}{(Q^*)^2 + 2\theta(Q^*)\sum_{i=1}^{n} (q_i^*)^2} \tag{64}$$

Substituting (63) into (64) leads to

$$\rho \leq \frac{n + 2\theta(Q^*)\gamma(Q^*)}{n + 2\theta(Q^*)} \tag{65}$$

Since $\gamma(Q^*) > 1$, the right-hand side of inequality (65) is increasing with both $\theta(Q^*)$ and $\gamma(Q^*)$. Substituting $\hat{\theta}$ and $\hat{\gamma}$ for $\theta(Q^*)$ and $\gamma(Q^*)$, we finally obtain

$$\rho \leq \frac{n + 2\hat{\theta}\hat{\gamma}}{n + 2\hat{\theta}} \tag{66}$$

For symmetric roads, it holds that

$$s_m = \frac{1}{n} \tag{67}$$

For concave inverse demand function, from Part (a) of Lemma 4 and (67) we have

$$\gamma(Q^*) \leq 1 + \omega_m s_m \leq 1 + \frac{1}{n} \tag{68}$$

Substituting (67) into (44) and relaxing each ω_i to 1, we have

$$\rho \leq \frac{n + \theta(Q^*)\left[2\varepsilon - n(\varepsilon - 1)^2\right]}{n + 2\theta(Q^*)} \tag{69}$$

When $\varepsilon \leq \gamma(Q^*) \leq 1 + 1/n$, the right-hand side of inequality (69) increases with both ε and $\theta(Q^*)$. Thus we obtain the following largest upper-bound by setting $\varepsilon = 1 + 1/n$ and $\theta(Q^*) = \hat{\theta}$

$$\rho \leq \frac{n + \hat{\theta}\left(2 + \frac{1}{n}\right)}{n + 2\hat{\theta}} \tag{70}$$

When $\hat{\theta} \to \infty$, namely the demand is perfectly elastic when $Q < Q^*$, the right-hand side of inequality (70) reaches an upper-bound of $1 + 1/(2n)$. Thus, if the inverse demand function is concave, we can conclude that

$$\rho \leq 1 + \frac{1}{2n} \tag{71}$$

Here we show that the bound $1+1/(2n)$ is tight by providing a numerical example. Consider an extreme situation that the inverse demand function $B(Q)=1$ for $0 \le Q \le 1$ and $B(Q)=2-Q$ for $1 \le Q \le 2$. There is only one road ($n=1$) and the road has a zero travel time. Then the firm would obtain a maximal profit of 1 by charging a toll of 1 and the social welfare realized is 1. Otherwise, the optimal solution can gain a social welfare of $3/2$ by charging 0. Thus, the inefficiency ratio is $3/2$ and just equal to the upper-bound we calculated from (71). Obviously, it is not difficult to slightly modify this example to ensure Assumption 1(a) and (b).

For convex inverse demand functions, substituting (63) and (56) into (53) yields

$$\rho \le \frac{n+\hat{\theta}(1+\hat{\gamma})}{n+2\hat{\theta}} \tag{72}$$

When $\hat{\theta} \to 1$, the right-hand side of the inequality reaches the upper-bound of $(n+1+\hat{\gamma})/(n+2)$. Thus for any convex inverse demand function, we conclude that

$$\rho \le \frac{n+1+\hat{\gamma}}{n+2} \tag{73}$$

It is worth noting that the bound is tight when the inverse demand function is linear, where $\hat{\gamma} = 1+\omega_m s_m \le 1+1/n$. Thus for the special case of linear inverse demand function

$$\rho \le \frac{(n+1)^2}{n(n+2)} \tag{74}$$

Clearly eqn.(74) is consistent with eqn.(70) when $\hat{\theta}=1$, because a linear inverse demand function is both concave and convex. Especially, for a monopoly market ($n=1$), the inefficiency ratio reaches its maximum of $4/3$, and for a duopoly ($n=2$), the inefficiency ratio reduces to $9/8$.

CONCLUSIONS

Distinguished from previous works, we examined the inefficiency of toll competition with general inverse demand functions. We probe into more details of the inverse demand function and establish more precise bounds under specific situations. We summarize the main results obtained from our study into Tables 1 and 2. In the tables, n is the number of firms or parallel roads; $\hat{\gamma}$ and $\hat{\theta}$ are the upper-bounds of the two parameters defined in Definitions

1 and 2, respectively, and they can be calculated if a specific demand function is given. Here we note that for asymmetric travel cost functions, the bounds established in the above table may not be tight, An obvious observation for the proof is that when $n=1$, the travel cost function is both asymmetric and symmetric, the tight bound obtained in symmetric case is $3/2$ for a concave demand function, while from the result in asymmetric case, the bound could go to infinity if $\hat{\theta} \to \infty$. In the future, we expect new methods to address this disunity and expand the results above by obtaining tighter bounds for the case of asymmetric travel cost functions.

Table 1. General expression of the upper bound of the inefficiency ratio for toll competition

Inverse Demand function		General	Concave	Convex
Asymmetric travel cost functions		$\frac{\sqrt{1+2\hat{\theta}\hat{\gamma}^2}+1}{2}$	$\frac{3+\sqrt{1+8\hat{\theta}}}{4}$	$\frac{2+\sqrt{4+2\hat{\theta}(1+\hat{\gamma})^2}}{4}$
Symmetric Travel cost function	With $\hat{\theta}$	$\frac{n+2\hat{\theta}\hat{\gamma}}{n+2\hat{\theta}}$	$\frac{n+\hat{\theta}\left(2+\frac{1}{n}\right)}{n+2\hat{\theta}}$	$\frac{n+\hat{\theta}(1+\hat{\gamma})}{n+2\hat{\theta}}$
	Without $\hat{\theta}$	$\hat{\gamma}$	$1+\frac{1}{2n}$	$\frac{n+1+\hat{\gamma}}{n+2}$

Table 2. Upper bound of the inefficiency ratio for some specific demand functions

Inverse Demand function	Linear $b-aQ$ $(a>0, b>0)$	Exponential $-b^{-1}\ln(Q/a)$ $(a>0, b>0)$	Polynomial B_0-aQ^b $(a>0, b>1, B_0>0)$	Constant elasticity aQ^{-b} $(a>0, 0<b<1)$
Asymmetric travel cost functions	$\frac{3}{2}$	1.56	$\frac{3+\sqrt{5+4b}}{4}$	$\frac{2+\sqrt{4+(1-b)\left(\left(1+(1-b)^{-\frac{1}{b}}\right)\right)^2}}{4}$
Symmetric Travel cost function	$\frac{(n+1)^2}{n(n+2)}$	$1+\frac{e-1}{2n+2}$	$1+\frac{b+1}{2n(n+b+1)}$	$\frac{2n+(1-b)\left(1+(1-b)^{-\frac{1}{b}}\right)}{2(n+1-b)}$

ACKNOWLEDGEMENTS

The research described here was supported by a grant from the Research Grants Council of the Hong Kong Special Administrative Region, China (Project No. HKUST6215/06E).

REFERENCES

Acemoglu, D. and Ozdaglar, A. (2005). Competition and efficiency in congested markets. *NBER Working Papers 11201, National Bureau of Economic Research, Inc.*

Anderson, S. P. and Renault, R. (2003). Efficiency and surplus bounds in Cournot competition. *Journal of Economic Theory* **113**, 253-264.

Cowling, K. and Mueller, D. (1978). The social costs of monopoly power. *Economic Journal* **88**, 727-748.

Correa, J.R., Schulz, A.S. and Stier-Moses, N.E. (2004). Selfish routing in capcitated networks. *Mathematics of Operations Research* 961-976.

De Vany, A. and Saving, T. (1980). Competition and highway pricing for stochastic traffic. *Journal of Business* **53**, 45-60.

De Palma, A. (1992). A game-theoretic approach to the analysis of simple congested networks. *American Economic Review* **82**, 494-500.

De Palma, A. and Lindsey, R. (2002). Private roads, competition, and incentives to adopt time based congestion tolling. *Journal of Urban Economics* **52**, 217-241.

Engel, E., Fischer, R. and Galetovic, A. (2005). Toll competition among congested roads. *Topics in Economic Analysis & Policy* **4**(1), Article 4.

Guo, X. and Yang, H. (2005). The price of anarchy of Cournot Oligopoly. In: Internet and Network Economics", *Lecture Notes in Computer Scie*nce **3828**, 246-257.

Hayrapetyan, A., Tardos, E. and Wexler, T. (2005). A network pricing game for selfish traffic. *PODC'05, July 17-20, 2005, Las Vegas, Nevada, USA.*

Johari, R. and Tsitsiklis, J. N. (2005). Efficiency loss in Cournot games. *Technical Report LIDS-P-2639*, Laboratory for Information and Decision Systems, MIT.

Ozdaglar, A. (2006). Price competition with elastic traffic. Working paper. Department of Electrical Engineering and Computer Science, Massachusetts Institute of Technology

Roughgarden, T. (2002). The Price of Anarchy is Independent of the Network Topology. *Journal of Computer and System Sciences (special STOC 2002 issue)*, **67**(2), 341-364.

Roughgarden, T. (2005). Selfish Routing and the Price of Anarchy. *The MIT Press, Cambridge.*

Roughgarden, T. and Tardos, E. (2002). How bad is selfish routing? *Journal of the ACM* **49**, 236-259.

Varian, H.R. (1992). Microeconomic analysis. *W.W.Norton & Company, New york.*

Wang, J.Y.T., Yang, H. and Verhoef, E.T. (2004). Strategic interactions of bilateral monopoly on a private highway. *Networks and Spatial Economics* **4**, 203-235.

Xiao, F., Yang, H. and Han, D.R. (2007). Competition and Efficiency of Private Toll Roads. *Transportation research* 41B, 292-308.

Yang, H. and Huang, H.J. (2005). *Mathematical and Economic Theory of Road Pricing.* Elsevier, Oxford.

Yang, H. and Woo, K. K. (2000). Competition and equilibria of private toll roads in a traffic network. *Transportation Research Record* **1733**, 15-22.

Yang, H., Xiao, F. and Huang, H.J. (2006). Private road competition and equilibrium with traffic equilibrium constraints (working paper).

APPENDIX A

A1. Properties of $\gamma(\cdot)$

If $B(\cdot)$ is concave, we have

$$B(x)+\omega_m s_m x B'(x) \ge B(x+\omega_m s_m x) \tag{75}$$

which is illustrated in Figure 1 (a). From the definition of $\gamma(\cdot)$, inequality (75) gives

$$B(\gamma(x)x) \ge B((1+\omega_m s_m)x) \tag{76}$$

Since $B(\cdot)$ is strictly decreasing, inequality (76) simply implies that $\gamma(x) \le 1+\omega_m s_m$. For a convex $B(\cdot)$, the proof is similar. ■

A2. Proof of Lemma 1

(a) We first show that $\varepsilon > 1$. If it holds that $\bar{q}_i > q_i^*$ for all $i \in N$, then $\bar{Q} > Q^*$ and $\varepsilon > 1$. Otherwise, without loss of generality, suppose $q_j^* > \bar{q}_j$, with eqn.(20) we have

$$B(\bar{Q}) \le \beta t_j(\bar{q}_j)+\beta \bar{q}_j t_j'(\bar{q}_j) \tag{77}$$

$$\le \beta t_j(q_j^*)+\beta q_j^* t_j'(q_j^*) \tag{78}$$

$$= B(Q^*)+\omega_j q_j^* B'(Q^*) < B(Q^*) \tag{79}$$

where (78) is from the assumption that travel time function is convex and strictly increasing and (79) is obtained from the Nash equilibrium condition (15). Because $B(\cdot)$ is strictly decreasing, $B(\bar{Q}) < B(Q^*)$ implies $\bar{Q} > Q^*$ and $\varepsilon > 1$.

(b) We prove that $\varepsilon \le \gamma(Q^*)$. As we have proved, $\bar{Q} > Q^*$, then there at least exists one road j on which $\bar{q}_j > q_j^*$. Thus we have

$$B(\bar{Q}) = \beta t_j(\bar{q}_j)+\beta \bar{q}_j t_j'(\bar{q}_j) \tag{80}$$

$$\ge \beta t_j(q_j^*)+\beta q_j^* t_j'(q_j^*) \tag{81}$$

$$= B(Q^*)+\omega_j q_j^* B'(Q^*) \tag{82}$$

$$\ge B(Q^*)+\omega_m s_m Q^* B'(Q^*) \tag{83}$$

$$= B(\gamma(Q^*)Q^*) \tag{84}$$

where (81) is from the assumption that travel time function is convex and strictly increasing; (82) is from Nash equilibrium condition (15); Since $\omega_m s_m = \max\{\omega_i s_i, i \in N\}$, $\omega_j q_j^* \le \omega_m q_m^* = \omega_m s_m Q^*$, we obtain (83); and eqn. (84) is from Definition 1. Because $B(\cdot)$ is strictly decreasing, $B(\bar{Q}) \ge B(\gamma(Q^*)Q^*)$ implies $\bar{Q} < \gamma(Q^*)Q^*$, and thus $\varepsilon \le \gamma(Q^*)$. ■

A3. Proof of Lemma 2

From eqn.(15) we have

$$\beta t_i(q_i^*) = B(Q^*) - \beta q_i^* t_i'(q_i^*) + \omega_i q_i^* B'(Q^*) \le B(Q^*) + \omega_i q_i^* B'(Q^*) \tag{85}$$

Thus

$$W^* = \int_0^{Q^*} B(v)\mathrm{d}v - \beta \sum_{i=1}^n q_i^* t_i(q_i^*) \tag{86}$$

$$\ge \int_0^{Q^*} B(v)\mathrm{d}v - \sum_{i=1}^n q_i^* \left(B(Q^*) + \omega_i q_i^* B'(Q^*)\right) \tag{87}$$

$$= \int_0^{Q^*} B(v)\mathrm{d}v - Q^* B(Q^*) + \left(-B'(Q^*)\right) \sum_{i=1}^n \omega_i (q_i^*)^2 \tag{88}$$

$$= \left(-B'(Q^*)\right)\left[\frac{1}{2\theta(Q^*)}(Q^*)^2 + \sum_{i=1}^n \omega_i (q_i^*)^2\right] \tag{89}$$

where (87) is from (85) and (89) is from Definition 2. ■

A4. Proof of Lemma 3

From eqns.(21)-(22) we have

$$\bar{W} - W^* = \int_{Q^*}^{\bar{Q}} B(v)\mathrm{d}v - \beta \sum_{i=1}^n \left(\bar{q}_i t_i(\bar{q}_i) - q_i^* t_i(q_i^*)\right) \tag{90}$$

Because $B(\cdot)$ is strictly decreasing and from Lemma 1, $\bar{Q} > Q^*$, it follows that

$$\int_{Q^*}^{\bar{Q}} B(v)\mathrm{d}v \le (\bar{Q} - Q^*) B(Q^*) \tag{91}$$

Moreover, because $t_i(\cdot)$ is convex, $q_i t_i(q_i)$ is also convex for $q_i \ge 0$, thus we have

$$\bar{q}_i t_i(\bar{q}_i) - q_i^* t_i(q_i^*) \ge (\bar{q}_i - q_i^*)\left(t_i(q_i^*) + q_i^* t_i'(q_i^*)\right), \quad i \in N \tag{92}$$

Substituting (91) and (92) into (90) yields

$$\bar{W}-W^{*}\leq\left(\bar{Q}-Q^{*}\right)B\left(Q^{*}\right)-\beta\sum_{i=1}^{n}\left(\bar{q}_{i}-q_{i}^{*}\right)\left(t_{i}\left(q_{i}^{*}\right)+q_{i}^{*}t_{i}'\left(q_{i}^{*}\right)\right) \tag{93}$$

$$=\sum_{i=1}^{n}\left(\bar{q}_{i}-q_{i}^{*}\right)\left(B\left(Q^{*}\right)-\beta t_{i}\left(q_{i}^{*}\right)-\beta q_{i}^{*}t_{i}'\left(q_{i}^{*}\right)\right) \tag{94}$$

$$=\left(-B'\left(Q^{*}\right)\right)\sum_{i=1}^{n}\omega_{i}\left(\bar{q}_{i}-q_{i}^{*}\right)q_{i}^{*} \tag{95}$$

where (95) is from the Nash equilibrium condition (15). ■

A5. Proof of Lemma 6

For any given $\omega_m s_m$, we simply rewrite $h(x,\varepsilon)$ as $h(\varepsilon)$, and let $u(\varepsilon)=2\omega_m s_m\varepsilon-(\varepsilon-1)h(\varepsilon)$, $1<\varepsilon\leq\gamma(x)$. Then we have

$$u'(\varepsilon)=2\omega_m s_m-h(\varepsilon)-(\varepsilon-1)h'(\varepsilon),\ 1<\varepsilon\leq\gamma(x) \tag{96}$$

where $h'(\varepsilon)=B'(\varepsilon x)/B'(\varepsilon)$, $1<\varepsilon\leq\gamma(x)$. With Assumption 1, if $B(\cdot)$ is convex, we have

$$B'(\varepsilon x)\geq\frac{B(x)-B(\varepsilon x)}{-(\varepsilon-1)x},\ 1<\varepsilon\leq\gamma(x) \tag{97}$$

Figure 5 graphically compares the two lines' slopes represented by the two sides of inequality (97). Multiplying both sides of (97) by $(\varepsilon-1)/B'(x)$ (a negative term) leads to

$$(\varepsilon-1)h'(\varepsilon)\leq h(\varepsilon),\ 1<\varepsilon\leq\gamma(x) \tag{98}$$

With (98), relaxing $(\varepsilon-1)h'(\varepsilon)$ to $h(\varepsilon)$ in (96) leads to

$$u'(\varepsilon)\geq 2\left(\omega_m s_m-h(\varepsilon)\right),\ 1<\varepsilon\leq\gamma(x) \tag{99}$$

With Assumption 1, it is obvious that $h(\varepsilon)$ increases with ε for $1<\varepsilon\leq\gamma(x)$, and gives $h(\gamma(x))=\omega_m s_m$, thus $h(\varepsilon)<\omega_m s_m$ for $1<\varepsilon<\gamma(x)$. Then it follows immediately from (99) that $u'(\varepsilon)>0$ for $1<\varepsilon<\gamma(x)$ and $u'(\gamma(x))\geq 0$. Therefore, Lemma 6 is true. ■

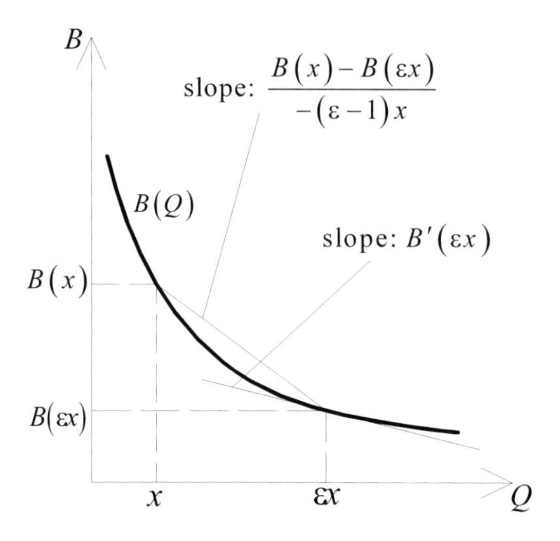

Figure 5. Graphical illustration for Inequality (97)

Transportation and Traffic Theory 2007
Edited by R.E. Allsop, M.G.H. Bell and B.G. Heydecker

3

TIME-DIFFERENTIAL PRICING OF ROAD TOLLS AND PARKING CHARGES IN A TRANSPORT NETWORK WITH ELASTIC DEMAND

Zhi-Chun Li, Department of Civil & Structural Engineering, The Hong Kong Polytechnic University, China and School of Management, Fudan University, Shanghai 200433, China
Hai-Jun Huang, School of Management, Beijing University of Aeronautics and Astronautics, Beijing, China
William H. K. Lam, Department of Civil & Structural Engineering, The Hong Kong Polytechnic University, China
S. C. Wong, Department of Civil Engineering, The University of Hong Kong, China

SUMMARY

This paper investigates competitive equilibrium between road pricing and parking charging when the responses of road users and the equity issue of road pricing are taken into consideration. To determine the equilibrium, this paper presents a multiobjective bilevel programming model with an equity constraint. The lower level is a time-dependent network equilibrium problem with elastic demand that simultaneously considers road users' choice of departure time, route, and parking location, and the upper level is a multiobjective program that maximizes the social welfare of the system with an equity constraint and the total revenue of the car park operator. A penalty function approach that is embedded by a multiobjective simulated annealing method is developed to solve the problem. A numerical example is used to demonstrate the effectiveness of the methodology.

INTRODUCTION

Road and parking congestion is an increasingly serious problem in most densely populated cities around the world due to the dramatic increase in the number of vehicles. Such congestion is mainly attributed to an imbalance between demand and supply (Lam et al., 1999; Tong et al., 2004). A shortage of supply exacerbates traffic congestion on the roads and increases the time that is spent searching for parking, and also causes more illegal roadside parking (Axhausen and Polak, 1991; Young et al., 1991).

A possible solution to the growing road congestion and parking shortage problems in urban areas may be to expand the capacity of the transportation infrastructure. Unfortunately, this can result in the inefficient allocation of resources because it creates excess capacity at most times of day except for the morning and evening peak commuting periods. The expansion of capacity may also induce new traffic demand and thus cause further congestion during peak periods. Furthermore, the resources that are available for the expansion of infrastructure capacity remain limited in most urban areas. Road pricing has been widely suggested as a viable alternative to infrastructure expansion, because it can be used to manage traffic demand and alleviate traffic congestion by changing the route choice of road users (Yang and Lam, 1996; Yang and Bell, 1997; Verhoef, 2002; Lo and Szeto, 2005; Ho et al., 2005) and their choice of departure time (Arnott et al., 1990; Yang and Huang, 1997; Yang and Meng, 1998; Wie and Tobin, 1998). Recent rapid developments in information and communication technologies have further aided and supported the practical implementation of road pricing schemes. Some well-known successful examples of electronic road pricing include the schemes in California, Singapore, and, more recently, London (Yang and Huang, 2005; Wong et al., 2005).

However, the road pricing option comes with social and even political problems, such as the issues of equity. Typically, road pricing is often controversial because of the social equity issue (Button and Verhoef, 1998) and spatial equity issue (Yang and Zhang, 2002). The social and spatial equity issues are concerned about the distributions of costs and benefits across different socioeconomic groups and different areas in a transportation network. Recently, Szeto and Lo (2006) introduced the impacts of road pricing across generations, i.e., the intergeneration equity issue. In the forthcoming discussions, we focus on the spatial equity issue arising from road pricing.

Moreover, it has been observed that the recent pricing practice in London has lessened the traffic demand into and out of central London, which has served to reduce congestion on the roads to the target level that was set by the Department for Transport, but has caused the owners of parking facilities to lose revenue because of the decrease in parking demand. There is therefore an incentive for car park operators to reduce parking charges for attracting customers back into the system. Were this to occur, it would put pressure on the relevant authorities to further increase the toll level to maintain the effectiveness of the road pricing scheme. Naturally, this leads us to ask to the question of where the equilibrium between road pricing and parking charging lies in a dynamic transportation system when the responses of road users and the equity issue are taken into consideration.

To answer this intriguing question, it is necessary to analyze the tradeoff among the players that are involved in the system. In a dynamic transportation system there are three types of players, each of which has a different set of objectives: the road authority, the owners of the parking facilities, and road users. The authority manipulates the differential toll level to maximize the social welfare of the system while taking into consideration the resulting equity issue, where the social welfare of the system is measured by the total benefit to road users minus the total social cost. The car park operator, which represents all of the owners of the parking facilities, determines the parking charge level on a time dependent basis to maximize total revenue. The road users make decisions about their departure time, route and parking location to minimize their individual travel disutility. The two most important decision variables in the system are road tolls and parking charges, both of which significantly influence the objectives of the players. Obviously, each player must consider the responses of the other two players when making a decision.

This paper provides a methodology to explore the tradeoff between the decision variables of road toll and parking charge levels that explicitly considers the spatial equity issue of road pricing. A multiobjective bilevel programming model that incorporates an equity constraint is proposed to optimize the time-differential road tolls and parking charges at all times of day. In the proposed bilevel model, the upper level is a two-player game between the road authority and the car park operator in which time-varying road tolls and parking charges are the decision variables. The lower level is a dynamic or time-dependent network equilibrium problem with elastic demand that governs the choices of the road users regarding their departure time, route, and parking location for a given combination of road tolls and parking charges. A penalty function approach in conjunction with a multiobjective simulated annealing (MOSA) algorithm is developed to solve the proposed model. As a result, the Pareto efficient solutions for the road tolls and parking charges can be obtained by time of the day in a network with various equity levels. We also compare the efficiency of different pricing schemes and explore the impact of equity on network performance.

The remainder of this paper is organized as follows. In the following section, we describe some of the basic considerations in developing the bilevel model. For a given road toll and parking charge pattern, a time-dependent network equilibrium formulation with elastic demand is presented to model a road user's travel and parking choice behavior, and is solved by a projection method. We then formulate the proposed bilevel programming model for the pricing design problem before and after the introduction of road pricing and develop some solution algorithms for the problem. A numerical example is used to illustrate the application of the proposed model. Conclusions are drawn in the final section, and some recommendations for future research are given.

BASIC CONSIDERATIONS

Assumptions

To facilitate the presentation of the essential ideas in this paper, the following assumptions are adopted.

A1 The entire study horizon [0, *T*] is discretized into equally spaced intervals that are sequentially numbered $t \in T = \{0,1,\cdots,\bar{T}\}$, where δ is the length of an interval with $\bar{T}\delta = T$.

A2 The function of the demand elasticity is used to capture the responses of road users to various levels of road tolls and parking charges. The responses include switching to alternative modes, such as the bus or subway, and not making the journey at all.

A3 A road user's travel disutility consists of the travel cost components of travel time from origin to parking location, the time delay of searching for parking, the parking charge at the parking location, the walking time from the parking location to the destination, the schedule delay cost of early or late arrival at the destination, and the cost of road tolls, if any.

A4 All of the users of the road system are assumed to have sufficient and perfect information about the conditions of the roads and parking facilities, and can thus make travel choices about departure time, route, and parking location in a dynamic or time-dependent user-optimal manner (Lam et al., 2006; Tong et al., 2004).

Notation

The following notation is used throughout the paper unless otherwise specified.

I_{rs}	The set of all feasible parking locations that serve users between OD pair (*r*, *s*).
P_{ri}	The set of all routes between origin *r* and parking location *i*.

$T_{ri,p}(t)$ The travel time of road users who leave origin r during interval t to travel to parking location i along route p.

$d_i(t)$ The time delay of searching for an available parking space at location i during interval t.

$z_i(t)$ The parking charge for users who arrive at location i during interval t.

w_{is} The walking time from location i to destination s.

$\Theta_{rs}(t)$ The schedule delay cost of early or late arrival at destination s for users who depart from origin r during interval t.

$f_{rs,ip}(t)$ The departure flow rate during interval t on route p between OD pair (r, s) via location i.

$c_a(k)$ The travel time on link a during interval k.

$u_a(k)$ The inflow to link a during interval k.

$D_i(t)$ The parking accumulation at location i during interval t.

$h_i(t)$ The hourly parking fee at location i during interval t.

$\lambda_{ri,p}(t)$ The road tolls on route p between origin r and parking location i during interval t.

$\varphi_{rs,ip}(t,\mathbf{x})$ The travel disutility of road users who leave origin r during interval t for destination s via route p and park their cars at location i.

$\mathbf{x}$ The vector pair $(\boldsymbol{\lambda},\mathbf{h})$ of road tolls and parking charges with $\boldsymbol{\lambda}=\left(\lambda_{ri,p}(t):\forall r,i,p,t\right)$ and $\mathbf{h}=\left(h_i(t):\forall i,t\right)$.

$\overline{\varphi}_{rs}^{\min}(\mathbf{h})$ The minimal travel disutility between OD pair (r, s) before the introduction of road pricing.

$\varphi_{rs}^{\min}(\mathbf{x})$ The minimal travel disutility between OD pair (r, s) after the introduction of road pricing.

Q_{rs} The total demand between OD pair (r, s) over the entire study horizon.

Travel disutility

Consider a transportation network $G = (N, A)$, where N is the set of all nodes that includes origin nodes, intersection nodes, destination nodes, and parking locations, and A is the set of all directed links that includes road links, parking access links from the road network to the parking location, and walkways or walk links between the parking location and the final destination. Let R denote the set of origin nodes, r represent a single origin node $r \in R \subset N$, S denote the set of destination nodes, and s represent the destination node $s \in S \subset N$.

According to assumption A3, the travel disutility (in terms of equivalent units of time) of road users who leave origin r during interval t for destination s via route p and park their cars at location i for a given combination of road tolls and parking charges can be expressed as

$$\varphi_{rs,ip}(t,\mathbf{x}) = \alpha_1 T_{ri,p}(t) + \alpha_2 d_i\left(t + T_{ri,p}(t)\right)$$
$$+\alpha_3 z_i\left(t + T_{ri,p}(t)\right) + \alpha_4 w_{is} + \alpha_5 \Theta_{rs}(t) + \alpha_6 \lambda_{ri,p}(t), \quad \forall r,s,i,p,t, \quad (1)$$

where the coefficient (α) is used to convert different quantities to units of time.

The route travel time $T_{ri,p}(t)$ in Equation (1) can be represented by the sum of all of the link travel times along this route (Chen, 1998).

$$T_{ri,p}(t) = \sum_a \sum_{k \geq t} c_a(k)\delta^{ri}_{apt}(k), \quad \forall r,i,p,t, \quad (2)$$

where the indicator variable $\delta^{ri}_{apt}(k)$ equals 1 if the flow on route p departing from origin r during interval t to travel to parking location i arrives at link a during interval k, and 0 otherwise.

We now define the link travel time function $c_a(k)$ in Equation (2). Thus far, there is no universally acceptable form of time-dependent or dynamic link travel time function. However, it is generally agreed that the link travel time function can be expressed in terms of the inflow, exit flow, and number of vehicles on the link. Chen and Hsueh (1998) further showed that the exit flow and the number of vehicles on a link can be represented by the inflow. Hence, the travel time $c_a(k)$ on link a during interval k can be expressed as a function of the inflow to that link by interval k, i.e.,

$$c_a(k) = f\left(u_a(1), u_a(2), \cdots, u_a(k)\right), \quad \forall a,k. \quad (3)$$

However, it should be pointed out that such travel time function may possibly violate the first-in-first-out (FIFO) discipline (Daganzo, 1995). In order to satisfy the FIFO condition, it is required that the rate of change in link travel time for each link in the network is greater than −1 (Ran and Boyce, 1996).

The inflow to link a, $u_a(k)$, during interval k in Equation (3) is

$$u_a(k) = \sum_{ri} \sum_p \sum_t u^{ri}_{apt}(k), \quad \forall a,k, \quad (4)$$

where $u_{apt}^{ri}(k)$ is the inflow to link a during interval k that departs from origin r along route p toward parking location i during interval t. This link inflow can be represented by the route inflow using the indicator variable $\delta_{apt}^{ri}(k)$ as follows.

$$u_{apt}^{ri}(k) = \delta_{apt}^{ri}(k) \sum_{s} f_{rs,ip}(t), \quad \forall a, p, r, i, t, k. \tag{5}$$

The time delay of searching for an available parking space can be computed by the following Bureau of Public Roads (BPR) function (Lam et al., 1999, 2006).

$$d_i\left(t + T_{ri,p}(t)\right) = d_i^0 + 0.31\left(D_i(t + T_{ri,p}(t))/C_i\right)^{4.03}, \quad \forall r, i, p, t, \tag{6}$$

where d_i^0 is the free-flow parking access time at parking location i and C_i is the capacity of parking location i. The parameters in Equation (6) are obtained from the calibration results that were based on the survey data in Hong Kong (Lam et al., 1999).

We now derive the parking accumulation formula in Equation (6) through an approach that is similar to that of Lam et al. (2006). The cumulative arrivals $U_{rs,ip}(t)$ of road users at location i along a specific route p between OD pair (r, s) by the beginning of interval t is given by

$$U_{rs,ip}(t) = \sum_{\xi=1}^{t-1} \sum_{k \mid k + T_{ri,p}(k) = \xi} f_{rs,ip}(k), \quad \forall r, s, i, p, t. \tag{7}$$

For a given OD pair (r, s), the cumulative arrivals $U_{rs,i}(t)$ of road users at parking location i along all routes by interval t can then be computed by

$$U_{rs,i}(t) = \sum_{p} U_{rs,ip}(t), \quad \forall r, s, i, t. \tag{8}$$

Therefore, the total cumulative arrivals $U_i(t)$ of road users from all OD pairs at parking location i by interval t is given by

$$U_i(t) = \sum_{rs} U_{rs,i}(t), \quad \forall i, t. \tag{9}$$

Conversely, road users who arrive at parking location i along route p before interval $(t$-$l)$ and park for the duration l will already have left parking location i before interval t, and thus for any OD pair (r, s) the cumulative departures $V_{rs,ip}(t)$ of road users from location i along route p by the beginning of interval t is

$$V_{rs,ip}(t) = \sum_{\xi=1}^{t-1-l} \sum_{k \mid k + T_{ri,p}(k) = \xi} f_{rs,ip}(k), \quad \forall r, s, i, p, t. \tag{10}$$

Hence, for a given OD pair (r, s), the cumulative departures $V_{rs,i}(t)$ of road users from parking location i by interval t is

$$V_{rs,i}(t) = \sum_{p} V_{rs,ip}(t), \quad \forall r, s, i, t, \tag{11}$$

and thus the total cumulative departures $V_i(t)$ of road users for all OD pairs from parking location i by interval t is

$$V_i(t) = \sum_{rs} V_{rs,i}(t), \quad \forall i, t \ . \tag{12}$$

Therefore, the parking accumulation $D_i(t)$ at location i during interval t equals $U_i(t)$ minus $V_i(t)$, that is,

$$D_i(t) = U_i(t) - V_i(t), \quad \forall i, t \ . \tag{13}$$

The parking charge in Equation (1), which is dependent on the desired parking duration l and the arrival time interval at the parking location, can be defined as

$$z_i(t + T_{ri,p}) = l \times h_i(t + T_{ri,p}), \quad \forall r, i, p, t, \tag{14}$$

where $h_i(t + T_{ri,p})$ is the hourly parking fee during interval $t + T_{ri,p}$.

The walking time is computed by

$$w_{is} = \Gamma(i, s)/\omega, \quad \forall i, s, \tag{15}$$

where $\Gamma(i, s)$ is the walking distance from location i to destination s and ω is the average walking speed of users (km/h).

The schedule delay costs of early or late arrival at destination s can be defined as follows (Yang and Meng, 1998; Huang and Lam, 2002).

$$\Theta_{rs}(t) = \begin{cases} \tau\left(t_s^* - \Delta_s - \left(t + T_{ri,p}(t) + d_i\left(t + T_{ri,p}(t)\right) + w_{is}\right)\right), \text{if } t_s^* - \Delta_s > t + T_{ri,p}(t) + d_i\left(t + T_{ri,p}(t)\right) + w_{is} \\ \gamma\left(t + T_{ri,p}(t) + d_i\left(t + T_{ri,p}(t)\right) + w_{is} - t_s^* - \Delta_s\right), \text{if } t_s^* + \Delta_s < t + T_{ri,p}(t) + d_i\left(t + T_{ri,p}(t)\right) + w_{is} \\ 0, \text{otherwise} \end{cases} \tag{16}$$

where $[t_s^* - \Delta_s, t_s^* + \Delta_s]$ is the desired arrival time window of users at destination s without any schedule delay penalty and τ (γ) is the unit cost of arriving early (late) at destination s.

The road toll $\lambda_{ri,p}(t)$ on route p between origin r and parking location i during interval t can be represented by the sum of all of the link tolls along this route as follows.

$$\lambda_{ri,p}(t) = \sum_{a \in \bar{A}} \sum_{k \geq t} \lambda_a(k) \delta_{apt}^{ri}(k), \quad \forall r, i, p, t, \tag{17}$$

where $\lambda_a(k)$ is the road toll on link a during interval k and $\bar{A}$ is the set of toll links.

Time-dependent network equilibrium conditions

For a given combination of road tolls and parking charges, we consider the decision of road users about whether to travel or not, and, if they do decide to travel, then their choice of departure time, route, and parking location. As in Yang and Meng (1998) and Szeto and Lo (2004), we assume that the total demand Q_{rs} between OD pair (r, s) is a continuous and monotonically decreasing function $D_{rs}(\cdot)$ of the OD travel disutility $\varphi_{rs}^{\min}(\mathbf{x})$, that is,

$$Q_{rs} = D_{rs}\left(\varphi_{rs}^{\min}(\mathbf{x})\right), \quad \forall r, s. \tag{18}$$

According to assumption A4, all of the travel choice decisions of road users adhere to the deterministic dynamic or time-dependent user equilibrium condition (Ran and Boyce, 1996; Chen, 1998; Lam et al., 2006) that for each OD pair only the combination of departure time, travel route, and parking location with the minimum travel disutility is actually used. This condition can mathematically be expressed as

$$\varphi_{rs,ip}(t, \mathbf{x}) \begin{cases} = \varphi_{rs}^{\min}(\mathbf{x}), & \text{if } f_{rs,ip}(t) > 0 \\ \geq \varphi_{rs}^{\min}(\mathbf{x}), & \text{if } f_{rs,ip}(t) = 0 \end{cases}, \quad \forall r, s, i, p, t. \tag{19}$$

Equation (19) states that no road user would be better off by unilaterally changing the choice of departure time, travel route, or parking location.

Definition 1. For a given vector pair $\mathbf{x} = (\boldsymbol{\lambda}, \mathbf{h})$ of road tolls and parking charges, a flow pattern $\mathbf{y}^* = \left(\mathbf{f}^*, \mathbf{Q}^*\right)$ is a time-dependent user equilibrium if it satisfies both conditions (18) and (19).

TIME-DEPENDENT NETWORK EQUILIBRIUM FORMULATION

Following the work of Wie et al. (1995) and Chen (1998) on dynamic network models, the deterministic time-dependent network equilibrium problem that is given in Definition 1 can be expressed as a finite-dimensional variational inequality (VI) problem as follows.

Theorem 1. For a given vector pair $\mathbf{x}=(\boldsymbol{\lambda},\mathbf{h})$ of road tolls and parking charges, a time-dependent flow pattern $\mathbf{y}^*=\left(\mathbf{f}^*,\mathbf{Q}^*\right)$ on the road network reaches the user equilibrium state if and only if it satisfies the following VI condition.

$$\sum_{rs}\sum_{i}\sum_{p}\sum_{t}\varphi_{rs,ip}(t,\mathbf{x})\left(f_{rs,ip}(t)-f^*_{rs,ip}(t)\right)+\sum_{rs}D^{-1}_{rs}(Q^*_{rs})\left(Q_{rs}-Q^*_{rs}\right)\geq 0, \tag{20}$$

subject to

$$\sum_{i}\sum_{p}\sum_{t}f_{rs,ip}(t)=Q_{rs},\quad \forall r,s, \tag{21}$$

$$f_{rs,ip}(t)\geq 0,\quad \forall r,s,i,p,t, \tag{22}$$

$$Q_{rs}\geq 0,\quad \forall r,s, \tag{23}$$

where Equation (21) is the OD demand conservation constraint and Equations (22) and (23) are the usual non-negativity constraints on route flow and OD demand, respectively.

Remark 1. The first term on the left-hand side of VI (20) is equivalent to the equilibrium formulation for the departure time, route, and parking location choices, and the second term serves to derive the elastic demand expression (18). Although the proof of Theorem 1 is omitted here, the readers can refer to the studies of Wie et al. (1995), Chen (1998) and Lam et al. (2006) for further details. As all of the functions in VI (20) are continuous and the feasible set is closed because the OD demand is bounded, there is at least a solution to the VI problem (20)-(23) according to Brouwer's fixed point theorem. However, the route travel time is essentially non-linear and non-convex because of the non-linearity and non-convexity of flow propagation constraint (5). It means that the VI problem (20)-(23) is non-convex and multiple local solutions may exist (Chen, 1998).

Note that the VI problem (20)-(23) is formulated in terms of the route flows, and thus a route-based heuristic solution algorithm that requires the explicit enumeration of the routes is developed to solve the VI problem. The solution algorithm is proposed as follows.

Step 0. Initialization. Set iteration n = 1. Choose an initial OD demand pattern $\left\{Q^{(n)}_{rs}\right\}$.

Step 1. Dynamic or time-dependent traffic assignment. Assign the OD demand $\left\{Q^{(n)}_{rs}\right\}$ to the network by employing the existing route-based dynamic or time-dependent traffic assignment methods, and obtain the temporal and spatial link flow distribution.

Step 2. Moving. Compute the minimal OD travel disutility and the auxiliary OD demand pattern $\left\{G^{(n)}_{rs}\right\}$ in terms of Equation (18).

Step 3. Updating. Update the OD demand pattern using the method of successive averages (MSA).

Step 4. Convergence Check. If a certain convergence criterion is satisfied, then stop. Otherwise, set $n = n+1$ and return to Step 1.

Remark 2. In Step 0, the initial OD demand can be determined based on an empty network. In Step 1, the dynamic or time-dependent traffic assignment problem with fixed OD demand can be solved by using various state-of-the-art route-based solution algorithms, such as the disaggregate simplicial decomposition method, the gradient projection method (Chen et al., 1999), the gap-function method (Lo, 1999), or the route/time swapping method (Huang and Lam, 2002). An attractive advantage of the route-flow solutions that are generated by the route-based algorithms is its capability of providing more detailed information about the travel pattern, which is not available with the link-flow solutions. The route-flow solutions have become increasingly important in many real applications, such as route guidance and information systems (Mahmassani and Peeta, 1993). However, the route-based algorithms also have some disadvantages, such as the need of explicit route enumeration. This greatly prohibits the application of the route-based algorithms to realistic large-scale networks. Fortunately, some effective methods, such as the column generation procedure that was recently proposed by Lim and Heydecker (2005), have been developed to generate a set of reasonable routes, in which only the routes that are most utilized are contained. Lim and Heydecker (2005) showed that the number of reasonable routes in a medium-scale or even large-scale network is small after dropping unreasonable routes, which yields encouraging results in the application of the route-based algorithms to a general network. In this study, the GP method in conjunction with the column generation procedure is adopted. In Step 4, the convergence criterion can be set as below (Sheffi, 1985).

$$G = \frac{1}{|RS|} \sum_{rs} \left(\left| \frac{\varphi_{rs}^{\min(n)} - D_{rs}^{-1(n)}(Q_{rs}^{(n)})}{\varphi_{rs}^{\min(n)}} \right| + \left| \frac{\varphi_{rs}^{\min(n)} - \varphi_{rs}^{\min(n-1)}}{\varphi_{rs}^{\min(n)}} \right| \right) < \varepsilon, \tag{24}$$

where $|RS|$ is the cardinality of the OD pairs in the network and ε is a small positive number.

PRICING DESIGN MODEL

In the foregoing sections, we formulate the dynamic or time-dependent network equilibrium problem with elastic demand and develop a solution algorithm for the problem with a given parking charge and road toll pattern $\mathbf{x} = (\boldsymbol{\lambda}, \mathbf{h})$. Yang and Zhang (2002) recently showed that road pricing can create an equity concern, in the sense that a change in the OD travel disutility of road users traveling between various OD pairs may be significantly different before and after the introduction of road tolls at some links. Therefore, to consider the equity issue explicitly, we formulate a pricing design model. For ease of presentation, the before- and after-implementation scenarios are labeled as "Scenario B (Before)" and "Scenario A (After)" respectively.

Pricing design model before the implementation of road tolls: Scenario B

Before the implementation of road tolls there are two types of players in the system, the car park operator and the road users, and one decision variable, the parking charge level. The car park operator attempts to maximize the total revenue by determining the time-varying parking charge level as follows.

$$\max_{\mathbf{h}} Z_0(\mathbf{h}) = \sum_{rs}\sum_{i}\sum_{p}\sum_{t} z_i(t) f_{rs,ip}(t) \ , \tag{25}$$

subject to

$$h_i^{\min}(t) \le h_i(t) \le h_i^{\max}(t), \ \ \forall i,t \, , \tag{26}$$

where $h_i^{\min}(t)$ and $h_i^{\max}(t)$ are the lower and upper bounds, respectively, of the parking charges at location i during interval t. The flow pattern $\mathbf{y} = (\mathbf{f}, \mathbf{Q})$ in Equation (25) is determined by the solution to the lower-level time-dependent network equilibrium problem with elastic demand, which is formulated in the foregoing section as VI problem (20)-(23).

It should be pointed out that the car park operator is mainly concerned with the revenues from parking charges, and does not care about the fairness of the charging level because parking charges are market oriented. Therefore, the equity constraint is not considered in the parking pricing design model.

Pricing design model after the implementation of road tolls: Scenario A

There are three players in Scenario A: the road authority, the car park operator, and the road users. The road tolls and parking charges are the two most important decision variables in designing an efficient transport system in this scenario, as both of them significantly influence the objectives of the players by jointly governing the social welfare of the system, the total revenue of the car park operator, and the temporal and spatial flow distribution and realized OD demand of the road users. Moreover, it has been demonstrated that the introduction of road tolls creates an equity issue that may cause the road pricing policy to be rejected by the public (Yang and Zhang, 2002). The equity issue must therefore be considered in the determination of the road toll level. Consequently, the pricing design problem after the implementation of road tolls aims to achieve an equilibrium between road pricing and parking charging while taking into account the responses of road users and the equity issue. It can be regarded as a two-player game between the road authority and the car park operator, in which the road authority aims to maximize the social welfare of the system by manipulating the road toll level while considering the equity issue, and the car park operator aims to maximize the total revenue by determining the time-varying parking charge level.

Similar to Meng and Yang (2002) and Yang and Zhang (2002), we deal with the equity issue by incorporating an equity constraint into the road toll design model. Equity is measured as the relative change in OD travel disutility before and after the introduction of road tolls. The road toll design problem for the road authority is thus formulated as follows.

$$\max_{\boldsymbol{\lambda}} Z_1(\boldsymbol{\lambda},\mathbf{h}) = \sum_{rs} \int_0^{Q_{rs}} D_{rs}^{-1}(w)\mathrm{d}w - \sum_{rs}\sum_{i}\sum_{p}\sum_{t}\left(\varphi_{rs,ip}(t,\mathbf{x}) - \lambda_{ri,p}(t)\right) f_{rs,ip}(t), \tag{27}$$

subject to

$$\max_{rs}\left\{\frac{\varphi_{rs}^{\min}(\mathbf{x})}{\overline{\varphi}_{rs}^{\min}(\mathbf{h})}\right\} \le \beta, \tag{28}$$

$$\lambda_a^{\min}(t) \le \lambda_a(t) \le \lambda_a^{\max}(t), \quad \forall a \in \bar{A}, t, \tag{29}$$

where expression (28) is the equity constraint, the parameter β is a given appropriate positive constant that measures the degree of equity of the benefit distribution, where a smaller β value implies a more equitable pricing design and vice versa. $\lambda_a^{\min}(t)$ and $\lambda_a^{\max}(t)$ are the lower and upper bounds, respectively, of the road toll on link a during interval t. The interrelation between route toll $\lambda_{ri,p}(t)$ and link toll $\lambda_a(t)$ is determined by Equation (17).

The parking pricing design model for the car park operator is formulated as below.

$$\max_{\mathbf{h}} Z_2(\boldsymbol{\lambda},\mathbf{h}) = \sum_{rs}\sum_{i}\sum_{p}\sum_{t} z_i(t) f_{rs,ip}(t), \tag{30}$$

subject to

$$h_i^{\min}(t) \le h_i(t) \le h_i^{\max}(t), \quad \forall i,t. \tag{31}$$

The pricing design model in Scenario A can be formulated as a multiobjective programming model, as follows.

$$\max_{\boldsymbol{\lambda},\mathbf{h}} Z(\boldsymbol{\lambda},\mathbf{h}) = \begin{pmatrix} Z_1(\boldsymbol{\lambda},\mathbf{h}) \\ Z_2(\boldsymbol{\lambda},\mathbf{h}) \end{pmatrix}$$

$$= \begin{pmatrix} \sum_{rs} \int_0^{Q_{rs}} D_{rs}^{-1}(w)\mathrm{d}w - \sum_{rs}\sum_{i}\sum_{p}\sum_{t}\left(\varphi_{rs,ip}(t,\mathbf{x}) - \lambda_{ri,p}(t)\right) f_{rs,ip}(t) \\ \sum_{rs}\sum_{i}\sum_{p}\sum_{t} z_i(t) f_{rs,ip}(t) \end{pmatrix}, \tag{32}$$

subject to (28)-(29) and (31), where $\varphi_{rs}^{\min}(\mathbf{x})$ and $\mathbf{y} = (\mathbf{f}, \mathbf{Q})$ are the equilibrium OD disutility and flow pattern, respectively, that are obtained by solving the lower-level time-dependent network equilibrium problem (20)-(23).

It should be pointed out that the pricing design models (25)-(26) for Scenario B and (32) for Scenario A with the time-dependent network equilibrium problem (20)-(23) as a constraint actually constitute two bilevel problems. As the lower-level equilibrium constraints (20)-(23) are non-convex, the pricing design models (25)-(26) and (32) are also non-convex. Therefore, it is very difficult to solve the two bilevel programming problems. Fortunately, the recently developed simulated annealing (SA) based algorithms can be used to solve the two bilevel programming problems. In the following section, we develop two SA-based solution algorithms for the proposed pricing design models (25)-(26) and (32).

SOLUTION ALGORITHMS

In this section, we first develop a SA-based solution algorithm for the pricing design model (25)-(26) in Scenario B, and then generate a multiobjective simulated annealing (MOSA) algorithm to find the set of potential Pareto efficient solutions for the proposed multiobjective bilevel programming problem (32) in Scenario A.

Solution to Scenario B

SA algorithms have been successfully applied to solve single-objective continuous equilibrium network design problems (Friesz et al., 1992; Huang and Bell, 1998; Yang and Zhang, 2002). They have the ability to obtain a globally optimal solution without the need of stipulating any details about mathematical structure, such as the shape of the feasible region and the derivatives of the objective functions.

The SA solution method was inspired by the thermodynamic process of the cooling (annealing) of solids (Eglese, 1990). In this process, a solid is heated until it melts, and the temperature of the solid is then slowly decreased in accordance with an annealing schedule until it reaches the lowest energy state, or ground state. If the initial temperature is not high enough or the temperature is decreased too rapidly, then the solid at the ground state will have many defects or imperfections. This annealing process is echoed in the search for a strategy. At the start of the search almost any move is accepted, regardless of whether it is good or bad, which allows the exploration of the solution space. Gradually, the probability of accepting non-improving moves (or neighboring solutions) decreases until in the end only improving moves are accepted. This ability to initially accept non-improving moves avoids the solution becoming trapped at a poor local optimum. The flowchart of the SA-based solution algorithm for the pricing design model (25)-(26) is shown in Figure 1, and the step-by-step procedure of the solution algorithm is given as follows.

Step 0. Initialization.

a) Choose the three classic parameters for the SA procedure: initial temperature T_0, cooling factor $\sigma(<1)$, and length of temperature step in the cooling schedule N_{step}.

b) Choose the two stopping criteria: the final temperature T_{stop} and the maximum number of iterations without improvement N_{stop}.

c) Start with a randomly generated initial solution vector $\mathbf{h}^{(0)}$ and solve the lower-level time-dependent network equilibrium problem (20)-(23) with $\mathbf{h}^{(0)}$. Compute the objective function $Z_0(\mathbf{h}^{(0)})$. Let $V(\mathbf{h}^{(0)})$ be the neighborhood of feasible solutions in the vicinity of $\mathbf{h}^{(0)}$. Set the counter $N_{count} = n = m = 0$.

Step 1.Generation of neighboring solutions. Set a random perturbation and generate a new solution vector $\hat{\mathbf{h}}$ in the neighborhood $V(\mathbf{h}^{(m)})$ of the current solution vector $\mathbf{h}^{(m)}$, solve the lower-level time-dependent network equilibrium problem (20)-(23) with $\hat{\mathbf{h}}$, compute the objective function $Z_0(\hat{\mathbf{h}})$, and then determine $\Delta Z_0 = Z_0(\hat{\mathbf{h}}) - Z_0(\mathbf{h}^{(m)})$.

Step 2. Metropolis' rule. If $\Delta Z_0 \leq 0$, then accept the new solution, let $\mathbf{h}^{(m+1)} = \hat{\mathbf{h}}$, $m = m+1$, and go to Step 3. Otherwise, accept the new solution with a certain probability $p = \exp(-\Delta Z_0 / T_n)$: if $p >$ random [0,1), then $\mathbf{h}^{(m+1)} = \hat{\mathbf{h}}$, $m = m+1$ and go to Step 3. Otherwise, let $\mathbf{h}^{(m+1)} = \mathbf{h}^{(m)}$, $N_{count} = N_{count} + 1$ and go to Step 3.

Step 3. Cooling schedules. If $m < N_{step}$, then go to Step 1. Otherwise, set $n = n+1, T_n = \sigma T_{n-1}$, $m = 0$, and go to Step 4.

Step 4. Termination check. If $N_{count} > N_{stop}$ or $T_n < T_{stop}$, then terminate. Otherwise, go to Step 1.

Remark 3. In Step 1, the new solution vector $\hat{\mathbf{h}}$ can be generated in the neighborhood $V(\mathbf{h}^{(m)})$ of the current solution vector $\mathbf{h}^{(m)}$ by various methods, such as the method of Hooke and Jeeves that was used in Yang and Zhang (2002) or the method of random and independent choice from the empirically determined interval $[-\sqrt{3}, \sqrt{3}]$ as in Friesz et al. (1992) and Huang and Bell (1998). If any entry in $\hat{\mathbf{h}}$ exceeds its upper or lower bound, then it is set at the respective bound.

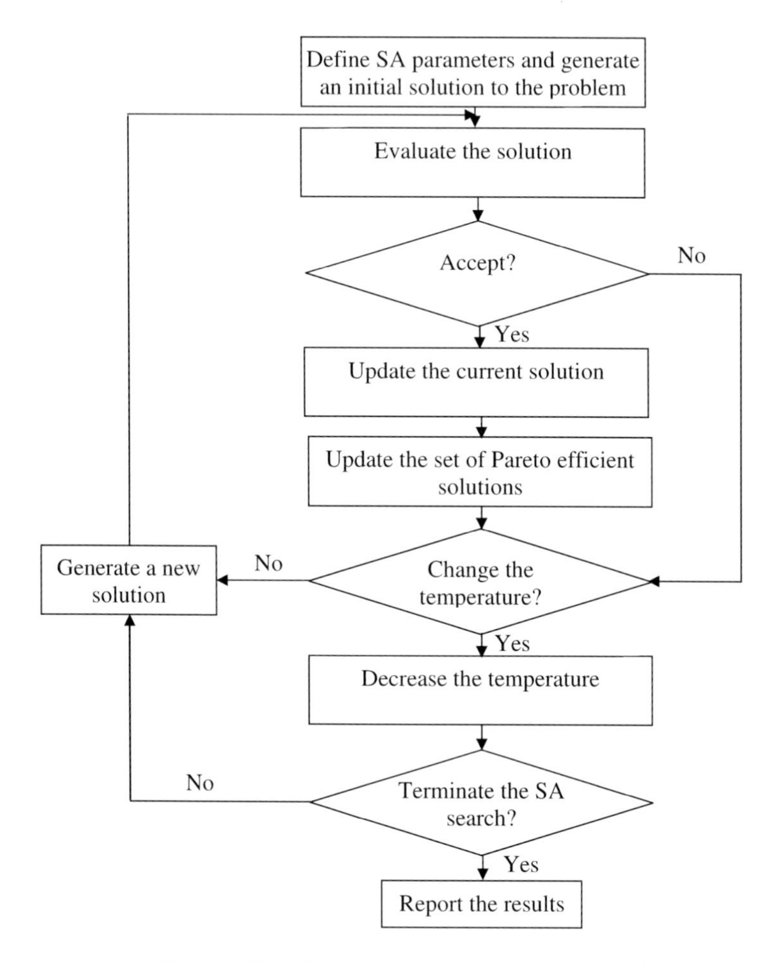

Figure 1 Flowchart of the proposed SA-based algorithm

Solution to Scenario A

It is known that the objectives of the different players in a multiobjective scenario are usually in conflict with each other, and thus a solution may perform well in terms of one criterion but give an adverse result in terms of another criterion. This implies that there is no well-defined optimal solution and the decision makers have to choose a compromise solution from the set of available Pareto efficient (or non-dominated) solutions.

There are various approaches to procure the set of Pareto efficient solutions for multiobjective programming problems, such as the hierarchical approach, utility function approach, simultaneous approach, goal programming approach, and interactive approach (Miettinen, 1999), but these traditional solution methods are difficult to apply for the non-convex multiobjective programming problem as proposed in this paper. However, the SA-based solution algorithms are appropriate for solving such multiobjective programming problems, and Friesz et al. (1993) and Meng and Yang (2002) employed this technique to solve multiobjective static equilibrium network design problems. The MOSA approach that was presented by Ulungu et al. (1999) has been successfully employed to solve multiobjective combinatorial optimization problems, such as the Knapsack problem, assignment problem, and production scheduling problem (Teghem et al., 2000; Suman, 2004; Mansouri, 2006). Mansouri (2006) recently showed that the MOSA algorithm performs very well with multiobjective problems, and outperforms the multiobjective genetic algorithm (MOGA) in terms of solution quality. In this paper, the MOSA approach of Ulungu et al. (1999) is adopted to find the set of potential Pareto efficient solutions for the proposed multiobjective bilevel programming problem (32) with equilibrium constraints (20)-(23). Note that the nonlinear implicit constraint (28) can be incorporated into objective function (32) by using the penalty function approach, which means that we need only to solve the resultant multiobjective programming problem with bound constraints using the MOSA algorithm. The step-by-step procedure of the penalty function method in conjunction with the MOSA algorithm is given as follows.

Step 0. Initialization. Set an initial penalty multiplier ρ_0 and a scale parameter θ. Set $k = 0$.

Step 1. The MOSA algorithm for solving the multiobjective programming problem with bound constraints.

Step 1.0. Set $l = 1$.

Step 1.1. Generation of the weight set for the scalarizing function. Generate a wide diversified set L of uniform random weight vectors $\mathbf{w}^{(l)} = \left(w_k^{(l)} : k = 1,2\right), \forall l \in L$ with $w_k^{(l)} \geq 0$ and $\sum_{k=1}^{2} w_k^{(l)} = 1, \forall l \in L$ for the scalarizing function

$$S(Z, \mathbf{w}^{(l)}) = w_1^{(l)} \left(Z_1(\boldsymbol{\lambda}, \mathbf{h}) + \rho_k \varphi(\boldsymbol{\lambda}, \mathbf{h})\right) + w_2^{(l)} Z_2(\boldsymbol{\lambda}, \mathbf{h}), \tag{33}$$

where

$$\varphi(\boldsymbol{\lambda}, \mathbf{h}) = \max\left\{ \max_{rs} \left\{ \frac{\varphi_{rs}^{\min}(\mathbf{x})}{\varphi_{rs}^{\min}(\mathbf{h})} \right\} - \beta, 0 \right\}. \tag{34}$$

Step 1.2. Use the proposed SA-based algorithm for Scenario B to solve the transformed bilevel programming problem with bound constraints (29) and (31), in which the objective function is the scalarizing function (33) and the decision variable is $\mathbf{x} = (\boldsymbol{\lambda}, \mathbf{h})$. Denote $PE(\mathbf{w}^{(l)})$ as the set of Pareto efficient solutions up to the current iteration.

Step 1.3. Termination check for the weight set. If Steps 1.1-1.2 has been carried out for each weight vector $\mathbf{w}^{(l)}$ of set L, then go to Step 1.4. Otherwise, set l = l+1 and repeat Steps 1.1-1.2 for the next weight.

Step 1.4. Filtering operations. As the solutions that are generated with a certain weight may be dominated by solutions that were generated with other weights, it is necessary to filter set $\cup_{l=1}^{|L|} PE(\mathbf{w}^{(l)})$ to obtain a good approximation of the set of Pareto efficient solutions. This operation can be performed by making pairwise comparisons of all of the solutions in the sets $PE(\mathbf{w}^{(l)})$ and removing the dominated solutions.

Step 2. Termination check. If $\rho_k \varphi(\boldsymbol{\lambda}, \mathbf{h}) < \varepsilon$, then stop. Otherwise, set $\rho_{k+1} = \theta \rho_k$, k = k+1, and go to Step 1.

Remark 4. In Step 1.1, the scalarizing function is used to project the multidimensional objective space into a mono-dimensional space. Different scalarizing functions lead to different projection paradigms. However, the impact of using different scalarizing functions is small because the method is stochastic (Ulungu et al., 1999; Teghem et al., 2000). In this paper, the well-known and easiest scalarizing function, the weighted-sum type, is adopted, but other scalarizing functions could also be chosen, such as the weighted-Chebyshev type. The set of weights is uniformly generated as $w_k^{(l)} = \{0, 1/r, 2/r, \cdots, (r-1)/r, 1\}$, where r is a positive integer.

NUMERICAL STUDY

In this section, a numerical example is given to illustrate the application of the proposed model and solution algorithm. We employ the proposed algorithm to determine the set of potential Pareto efficient solutions for road tolls and parking charges for the proposed multiobjective bilevel programming problem, which we hope will provide some interesting insights for decision makers. We also compare the efficiency of different pricing schemes and investigate the effect of the equity parameter on OD disutility and demand level.

Experimental settings

The test network, which is shown in Figure 2, consists of two OD pairs (1-3 and 1-4), six routes, four nodes, six links, and three parking locations (A, B, and C). The six travel routes are shown in Table 1. The central node 3 represents the city center area. As the city center is very congested, a toll cordon is set around this area, which is depicted by the bold dashed line in Figure 2. Road users traversing the city center area are subject to a cordon toll. Road users traveling between OD pair (1, 3) must select car park A or B to park their cars on reaching the city center area via Routes 1 or 2. Road users traveling between OD pair (1, 4) can reach their final destination by Routes 4 and 5 that traverse the city center area, which involves paying the cordon toll, or by bypassing the area on Routes 3 or 6, which avoids paying the cordon toll.

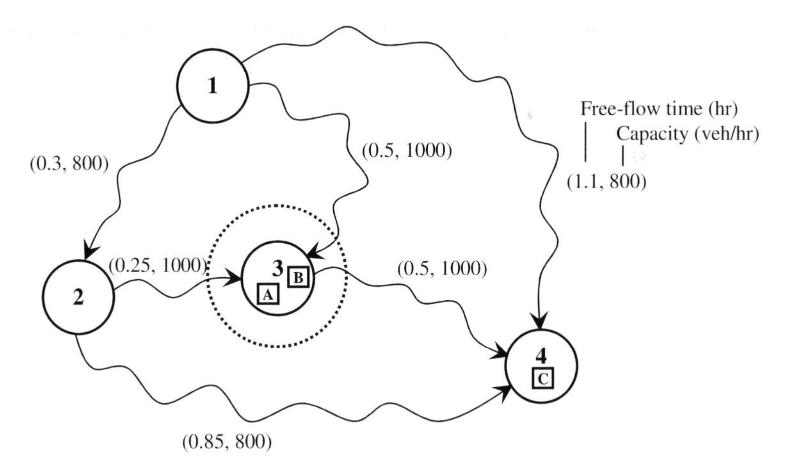

Figure 2 Example network

Table 1 Routes in the example network

OD pair	Route No.	Node sequence
(1,3)	1	1-3
	2	1-2-3
(1,4)	3	1-4
	4	1-3-4
	5	1-2-3-4
	6	1-2-4

For locations A, B, and C, the free-flow parking access times and parking capacities are all 0.1 hour and 1,100 vehicles, respectively. It is assumed that the walking distances from locations A, B, and C to the final destination are 0.5 km, 0.75 km, and 0.75 km, respectively. The average parking duration of the road users is four hours. The study period runs from 06:00 to 14:00 and is divided into 16 intervals of 30 minutes each. For each time interval, the following BPR-type link travel time function is adopted.

$$c_a(t) = c_a^0\left(1.0 + 0.15\left(u_a(t)/C_a\right)^4\right), \quad \forall a, \tag{35}$$

where c_a^0 and C_a are the free-flow travel time and capacity of link *a*, respectively. The values of these parameters are given in Figure 2. The average walking speed of users is assumed to be given and fixed at ω = 5.0 km/h. The parameters in the travel disutility function are $\alpha_1 = 1.0$, $\alpha_2 = 1.4$, $\alpha_3 = 0.15$, $\alpha_4 = 1.8$, $\alpha_5 = 0.15$, $\alpha_6 = 0.15$, $t^* = 9.0$, $\Delta = 0.25$, $\tau = 6.9$, and $\gamma = 15.21$.

The negative exponential demand functions are adopted, and are specified as

$$Q_{rs} = \bar{q}_{rs} \exp(-\rho_{rs}\varphi_{rs}^{\min}(\mathbf{x})), \quad \forall r, s, \tag{36}$$

where ρ_{rs} is a positive parameter of the elastic demand function and $\bar{q}_{rs}$ is the potential demand between OD pair (*r*, *s*). In this numerical example, these parameters are set as $\bar{q}_{13} = 2500$, $\bar{q}_{14} = 1500$, $\rho_{13} = 0.1$, and $\rho_{14} = 0.08$.

After many trial and error experiments to test the influence of the SA parameters, the following values are chosen.

$$T_0 = 100,\ \sigma = 0.9,\ N_{step} = 500,\ T_{stop} = 1.0, \text{ and } N_{stop} = 2500.$$

The proposed solution algorithm is coded in Language C and run on a DELL INSPIRON/510m computer.

Analysis of the numerical results

We first investigate the convergence of the proposed solution algorithms. Figure 3 plots the value of the weighted objective function with the weight $(w_1, w_2) = (0.5, 0.5)$ and the equity parameters $\beta = 1.10$ against the CPU time that is required by the proposed MOSA algorithm.

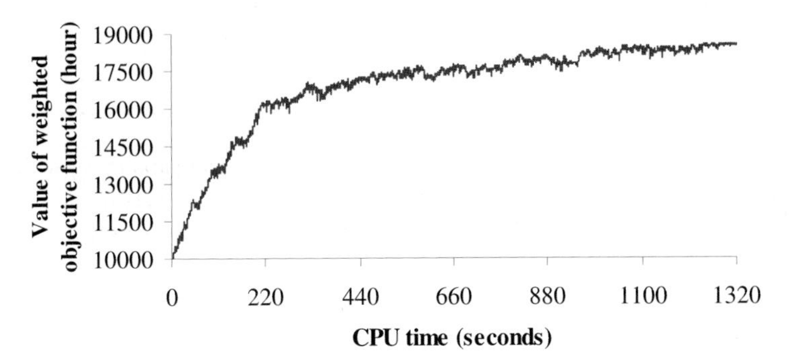

Figure 3 Convergence of the proposed MOSA algorithm

Figure 3 shows that the value of the weighted objective function increases sharply at the beginning of the iterations and then tends to stabilize after about 1,300 seconds. A small oscillation can be observed along the solution path. This is because the SA algorithm allows the acceptance of non-improving neighboring solutions with a certain probability during the course of the solution.

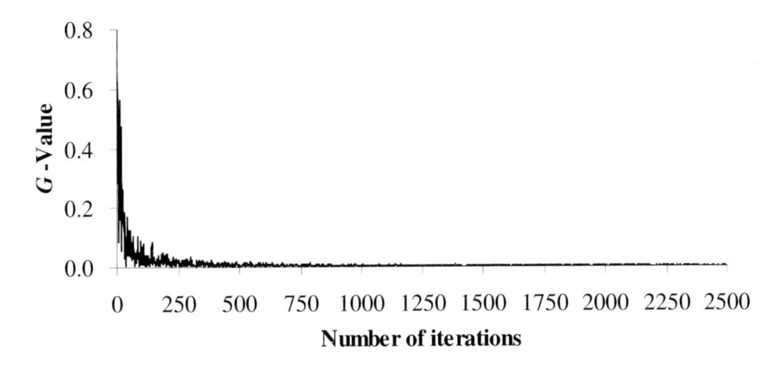

Figure 4 Typical convergence of the algorithm for the lower-level problem

Table 2 Departure flow pattern for the journey to parking locations A and B along different routes and the corresponding disutility (Disu.) for OD pair (1, 3) (hours)

Departure time	A via Route 1		A via Route 2		B via Route 1		B via Route 2	
	Flow	Disu.	Flow	Disu.	Flow	Disu.	Flow	Disu.
06:00-06:30	0.0	2.87	0.0	2.89	0.0	2.93	0.0	2.95
06:30-07:00	0.0	2.64	0.0	2.67	0.0	2.70	0.0	2.72
07:00-07:30	**305.6**	**2.50**	0.0	2.52	0.0	2.55	0.0	2.57
07:30-08:00	**486.8**	**2.50**	0.0	2.51	0.0	2.51	0.0	2.53
08:00-08:30	0.0	2.60	0.0	2.61	**431.2**	**2.50**	0.0	2.51
08:30-09:00	0.0	2.64	0.0	2.73	**291.3**	**2.50**	0.0	2.59
09:00-09:30	0.0	2.94	0.0	3.05	0.0	2.93	0.0	3.04
09:30-10:00	0.0	3.18	0.0	3.29	0.0	3.22	0.0	3.33
10:00-10:30	0.0	3.70	0.0	3.81	0.0	3.76	0.0	3.87
10:30-11:00	0.0	4.29	0.0	4.40	0.0	4.34	0.0	4.44

Figure 4 illustrates the typical convergence pattern of the proposed solution algorithm for the lower-level time-dependent network equilibrium problem. The G-value (see Equation (24)) dramatically oscillates at the beginning of iterations, and then gradually decreases as the number of iterations increases. This is because the MSA method with a pre-specified step size is used in the proposed algorithm for solving the lower-level problem. After 2,500 iterations, the G-value becomes less than 0.0001. Tables 2 and 3 list the resultant departure flow patterns for the journey to various parking locations along different routes and the corresponding disutilities for OD pairs (1, 3) and (1, 4), respectively. The routes and parking locations that are used in all of the intervals have the minimum disutility, whereas the unused routes and parking locations have an equal or greater disutility. Hence, the deterministic time-dependent network equilibrium condition that is given in Definition 1 has been fulfilled.

Table 3 Departure flow pattern for the journey to parking location C along different routes and the corresponding disutility (Disu.) for OD pair (1, 4) (hours)

Departure time	Route 3		Route 4		Route 5		Route 6	
	Flow	Disu.	Flow	Disu.	Flow	Disu.	Flow	Disu.
06:00-06:30	0.0	2.64	0.0	2.90	0.0	2.93	0.0	2.66
06:30-07:00	0.0	2.43	0.0	2.74	0.0	2.76	0.0	2.45
07:00-07:30	**296.7**	**2.27**	0.0	2.61	0.0	2.63	**63.2**	**2.27**
07:30-08:00	**330.8**	**2.27**	0.0	2.67	0.0	2.67	**352.4**	**2.27**
08:00-08:30	0.0	3.16	0.0	3.46	0.0	3.48	0.0	3.27
08:30-09:00	0.0	3.50	0.0	3.65	0.0	3.74	0.0	3.61
09:00-09:30	0.0	3.85	0.0	3.96	0.0	4.07	0.0	3.96
09:30-10:00	0.0	4.33	0.0	4.43	0.0	4.56	0.0	4.44
10:00-10:30	0.0	4.92	0.0	5.02	0.0	5.12	0.0	5.03
10:30-11:00	0.0	5.51	0.0	5.60	0.0	5.71	0.0	5.62

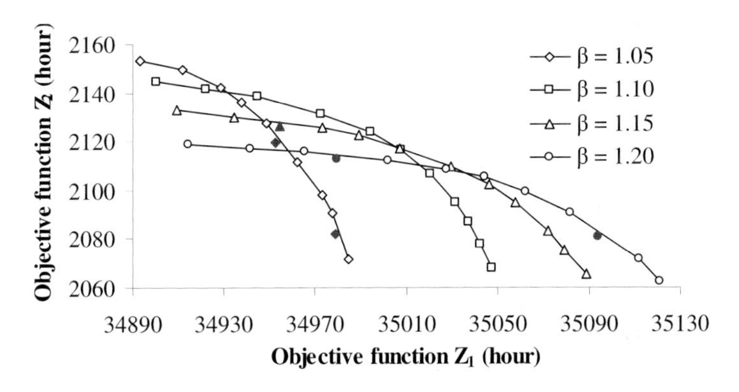

Figure 5 Frontier of the Pareto efficient solutions in the objective space (Z_1, Z_2) for various values of the equity parameter β

After some trial-and-error tests, the number of weight vectors L in the MOSA algorithm is fixed to 13 (r = 12). The resultant sets of potential Pareto efficient solutions for the four values of the equity parameter β (β=1.05, 1.10, 1.15 and 1.20) are plotted in the objective space, as shown in Figure 5. It can be observed from this figure that the number of solutions within the frontier of the potential Pareto efficient solutions for all β values is generally less than the number of weights. This is because the solutions that are generated with a certain weight may be dominated by solutions that are generated with other weights. The dominated solutions are removed from the set of potential Pareto efficient solutions by the filtering operations in the proposed solution algorithm.

It can also be seen in Figure 5 that some solutions (represented by the non-shaded points) are supported (solutions that correspond to the optimization of the scalarizing function (33) for some values of weight vector $\mathbf{w}$) and some solutions (represented by the darkened points) are non-supported (Ulungu et al., 1999). The sets of potential Pareto efficient solutions for the various degrees of equity in Figure 5 present some interesting information on the tradeoffs between the total social welfare (objective function Z_1) and the total parking revenue decreases (objective function Z_2) for road authorities and car park operators.

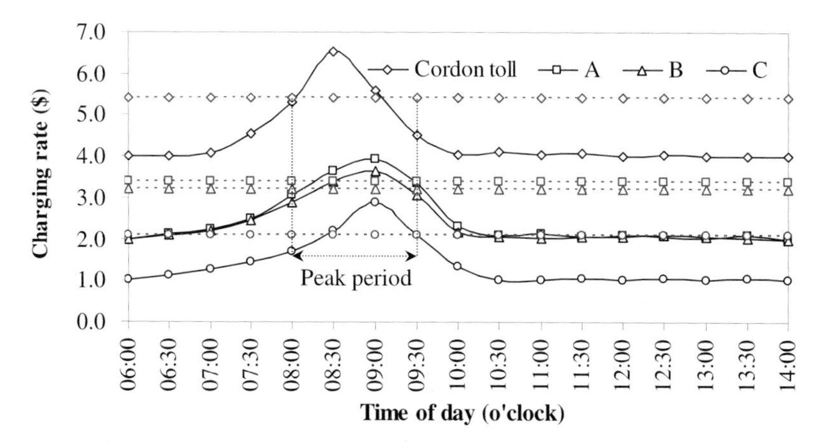

Figure 6 Resultant pricing of road tolls and parking charges at various times of day

As an illustration, Figure 6 displays the time-differential pricing (represented by the solid lines) of road tolls and parking charges that is associated with the weight $(w_1, w_2) = (0.5, 0.5)$ and the equity parameter $\beta = 1.10$. It can be seen that the road toll and parking charge levels during the morning peak period (08:00 to 09:30) are high, but are close to their respective lower bounds during the off-peak periods. It can also be seen that the parking charging level at location A during the peak period is a few higher than that at location B. This is because the walking distance from location A to final destination is 0.25 km nearer than that from location B to the final destination. To compare the efficiency of the different pricing schemes in terms of the total social welfare and total parking revenue, we also compute the optimal uniform pricing for the road tolls and parking charges, represented by the dotted lines in Figure 6.

Table 4 Results generated by the different pricing schemes

Equity parameter		$\beta = 1.05$		$\beta = 1.10$		$\beta = 1.15$		Scenario B
Pricing scheme		S1	S2	S1	S2	S1	S2	S3
Objective function (hour)	Z_1	34948.5	34902.1	34994.3	34963.9	35006.8	34979.4	34812.6
	Z_2	2127.4	2101.5	2123.6	2092.7	2117.1	2078.6	2234.5

Note: S1 and S2 represent the time-differential and uniform pricing schemes for the road tolls and parking charges, respectively. S3 represents a no-toll pricing scheme.

Table 4 shows the values of the objective functions Z_1 and Z_2 for the time-differential (S1) and uniform pricing (S2) schemes for three values of the equity parameter β and the weight $(w_1, w_2) = (0.5, 0.5)$, together with the objective function that is associated with the situation in which there is no road toll pricing scheme (S3). It is noted that as compared with Schemes S1 and S2, Scheme S3 results in the largest Z_2 value and the smallest Z_1 value. This is because Scheme S3 is to maximize the objective function Z_2 at the cost of the reduction of the objective function Z_1. It is also noted that the values of the objective functions Z_1 and Z_2 that are associated with Scheme S1 for each equity parameter are larger than those that are associated with Scheme S2. In other words, the solution that is generated by Scheme S2 is dominated by that which is generated by Scheme S1. This can be explained by examining the route inflow distributions of the two pricing schemes and of Scheme S3, as shown in Tables 5 and 6.

Table 5 shows the route inflow pattern for OD pair (1, 3) the destination of which is located within the road pricing area. It can be seen that compared to Scheme S3, road users traveling between OD pair (1, 3) under Scheme S1 depart from the origin earlier to avoid a high cordon toll when they cross the cordon. This adjustment of departure time will greatly reduce the road congestion level during the peak period. However, road users traveling between OD pair (1, 3) in Scheme S2 choose not to change the departure pattern that they followed before the implementation of road pricing. This implies that Scheme S1 can drive travel demand into a more reasonable and efficient distribution over the network because it provides a more flexible temporal and spatial alternative to road users. Therefore, Scheme S1 is generally superior to Scheme S2.

Table 5 Route inflow pattern for OD pair (1, 3)

Departure time	Route 1			Route 2		
	S1	S2	S3	S1	S2	S3
06:00-06:30	0.0	0.0	0.0	0.0	0.0	0.0
06:30-07:00	0.0	0.0	0.0	0.0	0.0	0.0
07:00-07:30	**305.6**	0.0	0.0	0.0	0.0	0.0
07:30-08:00	**486.8**	0.0	**37. 2**	0.0	0.0	**146.6**
08:00-08:30	**431.2**	**581.3**	**586.3**	0.0	**471.4**	**477.9**
08:30-09:00	**292.0**	**451.7**	**376.4**	0.0	**36.7**	0.0
09:00-09:30	0.0	0.0	0.0	0.0	0.0	0.0
09:30-10:00	0.0	0.0	0.0	0.0	0.0	0.0
10:00-10:30	0.0	0.0	0.0	0.0	0.0	0.0
10:30-11:00	0.0	0.0	0.0	0.0	0.0	0.0

Table 6 Route inflow pattern for OD pair (1, 4)

Departure time	Route 3			Route 4			Route 5			Route 6		
	S1	S2	S3	S1	S2	S3	S1	S2	S3	S1	S2	S3
06:00-06:30	0.0	0.0	0.0	**0.0**	**0.0**	0.0	**0.0**	**0.0**	0.0	0.0	0.0	0.0
06:30-07:00	0.0	0.0	0.0	**0.0**	**0.0**	0.0	**0.0**	**0.0**	0.0	0.0	0.0	0.0
07:00-07:30	296.8	156.3	0.0	**0.0**	**0.0**	328.2	**0.0**	**0.0**	0.0	63. 2	0.0	0.0
07:30-08:00	330.7	390.6	239.7	**0.0**	**0.0**	425.3	**0.0**	**0.0**	59.7	352.4	488.4	0.0
08:00-08:30	0.0	0.0	0.0	**0.0**	**0.0**	0.0	**0.0**	**0.0**	0.0	0.0	0.0	0.0
08:30-09:00	0.0	0.0	0.0	**0.0**	**0.0**	0.0	**0.0**	**0.0**	0.0	0.0	0.0	0.0
09:00-09:30	0.0	0.0	0.0	**0.0**	**0.0**	0.0	**0.0**	**0.0**	0.0	0.0	0.0	0.0
09:30-10:00	0.0	0.0	0.0	**0.0**	**0.0**	0.0	**0.0**	**0.0**	0.0	0.0	0.0	0.0
10:00-10:30	0.0	0.0	0.0	**0.0**	**0.0**	0.0	**0.0**	**0.0**	0.0	0.0	0.0	0.0
10:30-11:00	0.0	0.0	0.0	**0.0**	**0.0**	0.0	**0.0**	**0.0**	0.0	0.0	0.0	0.0

Table 6 shows the route inflow pattern for OD pair (1, 4), which is located outside the road pricing area. We can see from Table 6 that compared to Scheme S3, Schemes S1 and S2 shift the route inflows from Routes 4 and 5, which traverse the toll cordon, to Routes 3 and 6, which bypass the cordon toll area. This indicates that road pricing would not only make a direct and significant impact on the choice of road users traveling to destinations that are located within the road pricing area, but would also have an indirect impact on travel within OD pairs that fall outside the road pricing area.

Finally, we demonstrate how inequity occurs in the example network, and how it influences the OD demand level or network congestion level. Figure 7 shows the change in OD disutility and the corresponding realized OD demand against variation in the equity parameter β. It can be seen that when $\beta < 1.20$, the increase of β can lead to a dramatic increase in the disutility for OD pair (1, 3), and thus a dramatic decrease in the corresponding OD demand according to Equation (36). However, as β increases, the disutility for OD pair (1, 4) increases slightly, and therefore the corresponding demand for this OD pair decreases very little.

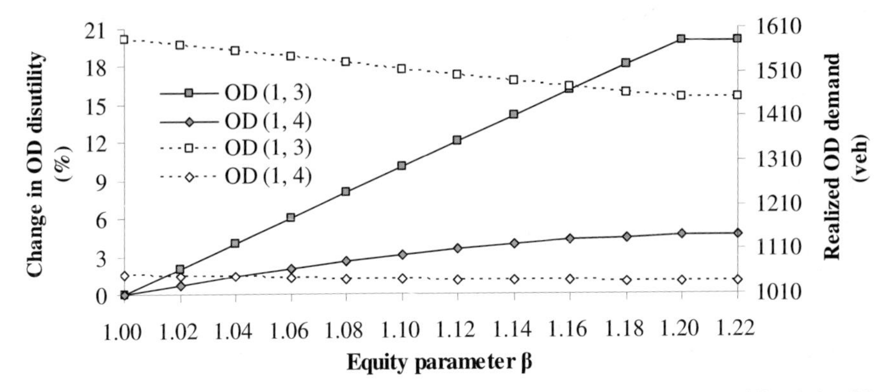

Figure 7 Change in OD disutility (solid lines) and realized OD demand (dotted lines) for OD pairs (1, 3) and (1, 4) versus the equity parameter

The substantial difference between the changes in the OD disutility for OD pairs (1, 3) and (1, 4) causes the spatial equity issue, as stated in previous sections. This is because the equity constraint (28) is binding for OD pair (1, 3) when $\beta < 1.20$, which is the critical value of the equity parameter that corresponds to the optimal pricing scheme without the equity constraint. As β increases, the equity constraint is gradually relaxed and higher tolls are allowed. This causes a dramatic increase in the disutility for OD pair (1, 3) because the destination falls within the toll cordon, whereas the disutility for OD pair (1, 4) increases slightly because road users traveling within this OD pair are able to complete their journey by choosing bypass routes that avoid the cordon toll area. It can also be seen in Figure 7 that when $\beta \geq 1.20$, the relaxation of the equity constraint does not lead to a further increase in the OD disutility for OD pair (1, 3) or decrease in the realized demand for this OD pair. This is because all of the equity constraints become inactive when β exceeds the critical value of 1.20.

CONCLUSIONS

In this paper, we investigated the equilibrium problem of road pricing and parking charging in the form of competition between the road authority and the car park operator, while taking into consideration the responses of road users and the equity issue in a time-dependent transportation network. We proposed a multiobjective bilevel programming model in which the lower level is a time-dependent network equilibrium problem with elastic demand that simultaneously considers road users' choices of departure time, route, and parking location, and the upper level is a multiobjective programming that maximizes the social welfare of the system with an equity constraint and the total revenue of the car park operator. To solve the bilevel problem, we developed a penalty function method that was embedded by a multiobjective simulated annealing algorithm. The proposed modeling approach provides a useful tool for finding an efficient pricing scheme and for evaluating the effect of various pricing schemes on travel behavior at a strategic level.

It was shown in the numerical example that the proposed solution algorithm could effectively solve the time-dependent network equilibrium problem, and construct the Pareto efficient solutions for the multiobjective programming problem with various equity levels. The time-differential pricing scheme was found to be generally superior to the uniform pricing scheme as the former could provide a more flexible temporal and spatial alternative to travelers. We also found that road pricing not only has a direct and significant impact on the choice behavior of road users traveling to destinations that are located within the road pricing area, but also has an indirect effect on road users traveling to destinations that lie outside of the road pricing area. Although the numerical results could give interesting insights on the topic concerned, we had to emphasize that they were drawn from a small hypothetical network. It would be interesting to apply the methodology to a real case study to ascertain the reliability of these findings. This is a promising future avenue of research that would have policy implications for the planning and implementation of efficient yet politically acceptable road pricing schemes.

Some other relevant issues that we are investigating include the consideration of multiple heterogeneous user classes and various types of parking facilities (Yang and Zhang, 2002; Lam et al., 2006), the explicit treatment of the supply and demand uncertainty (Li and Huang, 2005), the investigation of spillback queuing on roads and parking bottlenecks (Daganzo, 1998), and the development of efficient solution algorithms for large-scale networks.

ACKNOWLEDGEMENTS

The work that is described in this paper was supported by grants from the Research Grants Council of the Hong Kong Special Administrative Region (Project No. PolyU 5084/05E, PolyU 5143/03E and HKU 7126/04E), the National Natural Science Foundation of China (Project No. 50578006, 70521001) and the China Postdoctoral Science Foundation (20060400573). The first author would like to thank the Fudan University and the Hong Kong

Polytechnic University for their support so that he can carry out part of his postdoctoral research works under the supervision of the third author in Hong Kong.

REFERENCES

Arnott, R., A. De Palma and R. Lindsey (1990). Departure time and route choice for the morning commute. *Transp. Res.*, **24B**, 209-228.

Axhausen, K. W. and J. Polak (1991). Choice of parking: stated preference approach. *Transportation*, **18**, 59-81.

Button, K. J. and E. T. Verhoef (1998). *Road Pricing, Traffic Congestion, and the Environment: Issues of Efficiency and Social Feasibility*. Edward Elgar, Cheltenham.

Chen, H. K. (1998). *Dynamic Travel Choice Models: A Variational Inequality Approach*. Springer, Berlin.

Chen, H. K. and C. F. Hsueh (1998). A model and an algorithm for the dynamic user-optimal route choice problem. *Transp. Res.*, **32B**, 219-234.

Chen, H. K., C. W. Chang and M. S. Chang (1999). Comparison of link-based versus route-based algorithms in the dynamic user optimal route choice problem. *Transp. Res. Rec.*, **1667**, 114-120.

Daganzo, C. F. (1995). Properties of link travel time functions under dynamic loads. *Transp. Res.*, **29B**, 93-98.

Daganzo, C. F. (1998). Queue spillovers in transportation networks with a route choice. *Transp. Sci.*, **32**, 3-11.

Eglese, R. W. (1990). Simulated annealing: a tool for operational research. *Eur. J. Oper. Res.*, **46**, 271-281.

Friesz, T. L., H. J. Cho, N. J. Mehta, R. L. Tobin and G. Anandalingam (1992). A simulated annealing approach to the network design problem with variational inequality constraints. *Transp. Sci.*, **26**, 18-25.

Friesz, T. L., G. Anandalingam, N. J. Mehta, K. Nam, S. J. Shah and R. L. Tobin (1993). The multiobjective equilibrium network design problem revisited: a simulated annealing approach. *Eur. J. Oper. Res.*, **65**, 44-57.

Ho, H. W., S. C. Wong, H. Yang and B. P. Y. Loo (2005). Cordon-based congestion pricing in a continuum traffic equilibrium system. *Transp. Res.*, **39A**, 813-834.

Huang, H. J. and M. G. H. Bell (1998). Continuous equilibrium network design problem with elastic demand: derivative-free solution methods. In: *Transportation Networks: Recent Methodological Advances* (M.G.H. Bell, ed.), pp. 175-193. Elsevier, Oxford.

Huang, H. J. and W. H. K. Lam (2002). Modeling and solving the dynamic user equilibrium route and departure time choice problem in network with queues. *Transp. Res.*, **36B**, 253-273.

Lam, W. H. K., M. L. Tam, H. Yang and S. C. Wong (1999). Balance of demand and supply of parking spaces. In: *Proc. of the 14th Int. Symp. on Transportation and Traffic Theory* (A. Ceder, ed.), pp. 707-731. Elsevier, Oxford.

Lam, W. H. K., Z. C. Li, H. J. Huang and S. C. Wong (2006). Modeling time-dependent travel choice problems in road networks with multiple user classes and multiple parking facilities. *Transp. Res.*, **40B**, 368-395.

Li, Z. C., H. J. Huang (2005). Fixed-point model and schedule reliability of morning commuting in stochastic and time-dependent transport networks. *Lecture Notes in Computer Science*, **3828**, 777-787.

Lim, Y. and B.G. Heydecker (2005). Dynamic departure time and stochastic user equilibrium assignment. *Transp. Res.*, **39B**, 97-118.

Lo, H. K. (1999). A dynamic traffic assignment formulation that encapsulates the cell-transmission model. In: *Proc. of the 14th Int. Symp. on Transportation and Traffic Theory* (A. Ceder, ed.), pp. 327-350. Elsevier, Oxford.

Lo, H. K. and W. Y. Szeto (2005). Road pricing modeling for hyper-congestion. *Transp. Res.*, **39A**, 705-722.

Mahmassani, H. S. and S. Peeta (1993). Network performance under system optimal and user equilibrium dynamic assignments: implications for ATIS. *Transp. Res. Rec.*, **1408**, 83-93.

Mansouri, S. A. (2006). A simulated annealing approach to a bi-criteria sequencing problem in a two-stage supply chain. *Comput. Ind. Eng.*, **50**, 105-119.

Meng, Q. and H. Yang (2002). Benefit distribution and equity in road network design. *Transp. Res.*, **36B**, 19-35.

Miettinen, K. M. (1999). *Nonlinear Multiobjective Optimization*. Kluwer Academic Publishers, Dordrecht, MA.

Ran, B. and D. E. Boyce (1996). *Modeling Dynamic Transportation Networks: An Intelligent Transportation System Oriented Approach*. Second Revised Edition, Springer, Berlin.

Sheffi, Y. (1985). *Urban Transportation Networks: Equilibrium Analysis with Mathematical Programming Methods*. Prentice-Hall, Englewood Cliffs.

Suman, B. (2004). Study of simulated annealing based algorithms for multiobjective optimization of a constrained problem. *Comput. Chem. Eng.*, **28**, 1849-1871.

Szeto, W. Y. and H. K. Lo (2004). A cell-based simultaneous route and departure time choice model with elastic demand. *Transp. Res.*, **38B**, 593-612.

Szeto, W. Y. and H. K. Lo (2006). Transportation network improvement and tolling strategies: the issue of intergeneration equity. *Transp. Res.*, **40A**, 227-243.

Teghem, J., D. Tuyttens and E. L. Ulungu (2000). An interactive heuristic method for multiobjective combinatorial optimization. *Comput. Oper. Res.*, **27**, 621-634.

Tong C. O., S. C. Wong and W. W. T. Lau (2004). A demand-supply equilibrium model for parking services in Hong Kong. *Hong Kong Institute of Engineers Transactions*, **11**, 48-53.

Ulungu, E. L., J. Teghem, P. Fortemps and D. Tuyttens (1999). MOSA method: a tool for solving MOCO problems. *J. Multi-Criteria Decision Analysis*, **8**, 221-236.

Verhoef, E. T. (2002). Second-best congestion pricing in general static transportation networks with elastic demands. *Reg. Sci. Urban Econ.*, **32**, 281-310.

Wie, B. W., R. L. Tobin, T. L. Friesz and D. H. Bernstein (1995). A discrete time, nested cost operator approach to the dynamic network user equilibrium. *Transp. Sci.*, **29**, 79-92.

Wie, B. W., and R. L. Tobin (1998). Dynamic congestion pricing models for general traffic networks. *Transp. Res.*, **32B**, 313-327.

Wong, W. K. I., R. B. Noland and M. G. H. Bell (2005). The theory and practice of congestion charging. *Transp. Res.*, **39A**, 567-570.

Yang, H. and M. G. H. Bell (1997). Traffic restraint, road pricing and network equilibrium. *Transp. Res.*, **31B**, 303-314.

Yang H. and H. J. Huang (1997). Analysis of time-varying of a bottleneck with elastic demand using optimal control theory. *Transp. Res.*, **31B**, 425-440.

Yang, H. and Q. Meng (1998). Departure time route choice and congestion toll in a queuing network with elastic demand. *Transp. Res.*, **32B**, 247-260.

Yang, H. and W. H. K. Lam (1996). Optimal road tolls under conditions of queuing and congestion. *Transp. Res.*, **30A**, 319-332.

Yang, H. and X. Zhang (2002). Multiclass network toll design problem with social and spatial equity constraints. *J. Transp. Eng.*, **128**, 420-428.

Yang, H. and H. J. Huang (2005). *Mathematical and Economic Theory of Road Pricing*. Elsevier Ltd. Oxford, UK.

Young, W., R. G. Thompson and M. A. P. Taylor (1991). A review of urban parking models. *Transport Reviews*, **11**, 63-84.

Transportation and Traffic Theory 2007
Edited by R.E. Allsop, M.G.H. Bell and B.G. Heydecker

4

DYNAMIC RAMP CONTROL STRATEGIES FOR RISK AVERSE SYSTEM OPTIMAL ASSIGNMENT

Takashi Akamatsu, Graduate School of Information Sciences, Tohoku University, Sendai, Miyagi, Japan
Takeshi Nagae, Graduate School of Science and Technology, Kobe University, Kobe, Hyogo, Japan

INTRODUCTION

This paper explores dynamic ramp control strategies for a simple transportation network with stochastic travel time. The objective of the ramp metering considered here is not only to mitigate congestion in a freeway but also to achieve *Dynamic System Optimal* (DSO) assignment in a network with two parallel links; one of the links is a freeway with a single bottleneck, and the other is a local bypass link that can be regarded as an aggregation of a local street network. Travel time on the bypass link is assumed to follow a stochastic process due to many factors that cannot be controlled or predicted. A road network manager is expected to control the inflow rate to the freeway at each time point so as to attain the DSO assignment for a certain time horizon.

We approach this problem by formulating it as a *stochastic control problem* (SCP), and then provide *feedback* ('*state contingent*') *control rules* that exploit the real-time observation of the realization of the stochastic state variable (*i.e.* random travel times). We then find that the optimal control (ramp metering) strategies at each time period can be classified into seven patterns, depending on the realization of the queue length in the freeway and the observed travel time of the bypass link. In order to obtain more detailed properties of the optimal control strategies, we need to solve the problem numerically. For this purpose, we reveal that the optimality conditions of the problem (*i.e. Hamilton-Jacobi-Bellman* (HJB) *equations*) can be equivalently stated as a dynamical system of *Generalized Complementarity Problems* (GCP). Based on this reformulation, we provide an efficient and robust algorithm for obtaining quantitative results for the control problem. Furthermore, we present results from systematic numerical experiments, which reveal how the uncertainty in the travel time affects the optimal control policies.

Although there have been some studies on the DSO assignment in recent years, conventional approaches cannot be used to tackle the DSO assignment problem in this paper. Indeed, as we show below, little is known about the theoretical properties of the DSO assignment problem in deterministic models as well as stochastic environments in this paper. Friesz et al.(1989) study DSO assignment on a network with general topology. They present a deterministic optimal control formulation of the DSO assignment model, in which an 'exit function' is assumed for describing an outflow rate of a link. However, this modeling approach has the serious drawback of violating First-In-First-Out (FIFO) conditions according to which the traffic flow should be satisfied in each link. Ziliaskopoulous (2000) provides a linear programming (LP) formulation of the DSO assignment on a network with a one-to-many OD pattern, in which the cell transmission model (CTM) of Daganzo (1994) is employed to describe traffic flow propagation. However, in order for the LP formulation to be consistent with the CTM, we have to assume that the position of any vehicle can always be controlled (*eg*. stopped; the assumption of 'holding'). This is a problematic assumption to implement, and hence it is questionable to think that this model represents a natural DSO assignment. We should also note that analysis for general networks might face difficulties due to the non-convexity of the DSO assignment problem even if we could provide a sound model of the DSO assignment preserving FIFO conditions (for this point, see Lovel and Daganzo (2000) and Erera et al.(2002)). In view of these studies, it seems that analyzing the model for general networks in one leap is not a very fruitful way to understand the theoretical properties of the DSO assignment. More recently, Kuwahara et al.(2000) and Munoz and Laval (2006) study the properties of the optimal control (ramp metering) in simple parallel-link networks. They show a graphical solution method based on the concept of dynamic marginal cost, which provides useful insights into the DSO assignment. However, this method cannot be extended systematically to the DSO assignment problem with stochastic travel time. Thus, our novel approach based on stochastic control theory as well as the theoretical findings contribute to the studies on dynamic traffic assignment and control.

This paper is organized as follows: After presenting the stochastic control formulation of our DSO assignment model in the next section, we derive the HJB equations of the SCP in the third section. We then show some qualitative properties of the optimal control policies. In the fourth section, we reformulate the HJB equations for the DSO assignment as a system of GCPs, which enables us to develop an efficient solution method for the DSO assignment. In the fifth section, we provide an illustrative example of the proposed control method. The final section summarizes the paper.

DYNAMIC SYSTEM OPTIMAL RAMP METERING UNDER UNCERTAINTY

Networks and Link Models

We consider a dynamic system optimal (DSO) assignment problem defined on a simple network with two parallel links connecting a single origin-destination (OD) pair (see **Figure 1**). One of the links, *link 1*, is a freeway with a single bottleneck of capacity μ, and the other, *link 2*, is a local bypass link with a very large capacity; link 2 may be regarded as a virtual link that is an aggregation of a local street network. We suppose that the number of vehicles arriving at the origin of the network until time t, $Q(t)$, is known for all t in a fixed time period $[0,T]$.

In the DSO assignment, a road network manager is assumed to control the inflow rate to the freeway, $u(t)$, at each time point t; this implicitly determines the inflow rate to link 2 as $q(t)-u(t)$, where $q(t)$ is the OD flow rate at time t defined as $q(t) = dQ(t)/dt$. Thus, the DSO assignment considered in this paper can be viewed as a ramp-metering problem with a single freeway on-ramp.

We suppose that the travel time of link 1 at time t, $c(t)$, is determined simply by the queue length, $x(t)$, at the bottleneck:

$$c(t) = x(t)/\mu, \tag{1}$$

and the queue evolution is governed by the following state equation (*i.e.* the point-queue model is assumed):

$$\dot{x}(t) = \begin{cases} u(t)-\mu & \text{if} \quad x(t)>0 \\ \max[u(t)-\mu, 0] & \text{if} \quad x(t)=0 \end{cases}, \quad x(0)=0, \tag{2}$$

where $u(t)$ is the controlled inflow rate into link 1. The travel time of link 2, $m(t)$, is supposed to be just a function of time, which implies that $m(t)$ can not be controlled by a road manager's metering strategies. We also assume that $m(t)$ evolves *unpredictably* over time due to many factors (such as fluctuation of OD flows into the local street network or traffic accidents) that cannot be controlled or predicted. We model the stochastic dynamics of $m(t)$ as

$$dm/m(t) = \alpha(t)dt + \sigma dW, \quad m(0) = m_0, \tag{3}$$

where $W(t)$ is a standard Wiener process; $\alpha(t)$ and σ are an exogenously given (time-dependent) function and a volatility parameter, respectively. To illustrate the intuitive meanings of (3), in **Figure 2**, we provide an example of a sample path of the stochastic process $m(t)$.

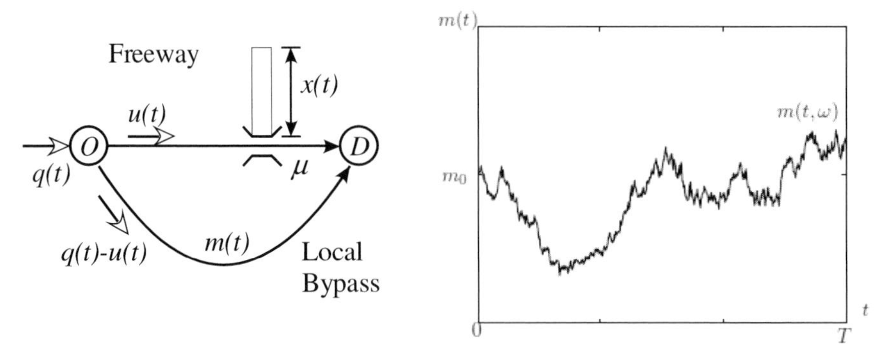

Figure 1: A parallel link network

Figure 2: A sample path of the stochastic process $m(t)$

Stochastic Feedback Control Formulation

The objective of the road manager is to minimize 'social cost' due to the total travel time spent in the network for a certain time horizon $[0,T]$. Before providing a formal definition of the social cost, we first introduce the total travel time spent in the network by vehicles arriving at the origin at time t:

$$C(t) \equiv c(t) \cdot u(t) + m(t) \cdot (q(t) - u(t)) . \tag{4}$$

Then the total travel time for the whole control period $[0,T]$ is given by

$$C \equiv \int_0^T C(t)\, dt .$$

Note here that both the total travel time C as well as $C(t)$ are random variables because the travel time of link 2, $m(t)$, follows a stochastic process given in (3). For this stochastic setting, the simplest definition of the social cost in this network is the expected total travel time:

$$E_0\left[\int_0^T C(t)\, dt \,\middle|\, m(0) = m_0, x(0) = x_0 \right],$$

where $E_t[]$ denotes the expectation operator conditional on the information available at time t. However, employing this definition of the social cost as the road manager's objective could cause a serious problem: it allows the use of '*risky*' control strategies in which the total travel time C in the worst case is very large (*i.e.* variance of C is large) while the expectation of C is small. Therefore, we need to consider the trade-off between 'risk' and 'return' in controlling the stochastic social cost $C(t)$.

This trade-off can be modelled in an expected utility maximization framework; we evaluate the total travel time *in terms of utility* $U(-C(t))$, rather than $C(t)$ itself, and then the expected utility is maximized. Thus, the road manager's problem is formulated as the following *stochastic control problem*:

$$\text{[SCP]} \qquad \max_{\{0 \le u(t) \le q(t)\}} E_0 \left[\int_0^T U(-C(\tau))d\tau - \frac{x^2(T)}{2\mu} \,\middle|\, m(0) = m_0, x(0) = x_0 \right] \tag{5}$$

subject to eqs.(2) and (3); and $x(T)$ is free.

A few remarks are in order here: first, the problem formulated above, [SCP], is *not* an *open-loop control* problem, in which $\{u(t)\}$ is determined in advance before controlling and observing the realization of state variables $\{x(t)\}$ and $\{m(t)\}$; rather the problem [SCP] gives a *feedback* (*'state contingent'*) *control*, in which the optimal inflow rate $u(t)$ is a function of state variables, $x(t)$ and $m(t)$, observed at time t. This implies that the optimal control for [SCP] exploits the real-time observation of the realization of the stochastic state variable (*i.e.* random travel time $m(t)$). Second, the optimal control depends on a degree of *risk aversion* towards potentially '*risky*' control strategies that exhibit a large variance of C; a risk averse manager should prefer a ramp control strategy that exhibits less risky control to a more risky one with the expectation of C being equal. The degree of risk aversion is reflected by a concave utility function $U(-C)$ in our model; the higher the curvature of $U(-C)$, the higher the risk aversion of the road manager.

OPTIMALITY CONDITIONS

Hamilton-Jacobi-Bellman Equations

We shall derive the optimality conditions for the SCP formulated in the previous section. We first define the value function $V(t, x, m)$ by

$$V(t,x,m) \equiv \max_{\{u(t)\}} E_t \left[\int_t^T U(-C(\tau))d\tau \,\middle|\, x(t) = x, m(t) = m \right]. \tag{6}$$

By applying the dynamic programming (DP) principle, we have

$$V(t,x,m) = \max_{u(t)} E_t \left[\int_t^{t+\Delta} U(-C(\tau))d\tau + V(t,x,m) + \Delta V(t,x,m) \,\middle|\, x(t) = x, m(t) = m \right],$$

where $\Delta V(t,x,m) \equiv \int_t^{t+\Delta} dV(\tau)d\tau$. Taking the limit of $\Delta \to +0$ and using stochastic calculus (Ito's lemma):

$$dV = \frac{\partial V}{\partial t}dt + \frac{\partial V}{\partial x}dx + \frac{\partial V}{\partial m}dm + \frac{1}{2}\frac{\partial^2 V}{\partial m^2}(dm)^2, \tag{7}$$

we obtain the *Hamilton-Jacobi-Bellman* (HJB) equation:

$$0 = \max_{0 \le u(t) \le q(t)} \left[U(-C(t)) + \dot{x}(u(t)) \frac{\partial V}{\partial x} \right] + \frac{\partial V}{\partial t} + \alpha(t) m(t) \frac{\partial V}{\partial m} + \frac{1}{2} \sigma^2 m^2(t) \frac{\partial^2 V}{\partial m^2}$$

$$\textit{for all} \quad t \in [0, T). \qquad (8)$$

In order to avoid notational complexity, we denote this in a more compact form as

$$0 = \max_{0 \le u(t) \le q(t)} Z(u(t)) + L_0 V \qquad \textit{for all} \quad t \in [0, T), \qquad (9)$$

with an infinitesimal generator L_0 defined as

$$L_0 \equiv \frac{\partial}{\partial t} + \alpha(t) m(t) \frac{\partial}{\partial m} + \frac{1}{2} \sigma^2 m^2(t) \frac{\partial^2}{\partial m^2}, \qquad (10)$$

and the term $Z(u(t))$ involving the control variable $u(t)$:

$$Z(u(t)) \equiv U(-C(u(t))) + \dot{x}(u(t)) \frac{\partial V}{\partial x}$$
$$= \begin{cases} U(-C(u(t))) + (u(t) - \mu) V_x & \textit{if } x(t) > 0 \\ U(-C(u(t))) + \max.[u(t) - \mu, 0] V_x & \textit{if } x(t) = 0 \end{cases}, \qquad (11)$$

where the subscript of V denotes the partial derivative (*i.e.* $V_x \equiv \partial V / \partial x$).

Optimal Control Strategies

Optimal control can be derived by solving the maximization problem (with respect to $u(t)$) in the HJB equation (9). Since the objective function Z takes two distinct forms depending on whether $x(t) = 0$ or $x(t) > 0$, we will divide the derivation into two cases.

(a) The case of $x(t) = 0$

When there is no queue in the freeway, the function $Z(u)$ should be further classified into two cases due to the indifferentiability of the max. function in the state equation (2). For the first case in which $\max.[u(t) - \mu, 0] = 0$ (*i.e.* $u(t) < \mu$), the derivative of Z is always positive:

$$\frac{\partial Z}{\partial u} = f_0(u, m(t)) \equiv U'(-C(u)) \cdot m(t) > 0, \qquad (12)$$

It follows from this that the optimal control in this case is to assign all the OD flow into the freeway:

$$\textit{control } \mathbf{A}\text{:} \quad u(t) = q(t) \quad \textit{if} \quad x(t) = 0 \textit{ and } q(t) < \mu. \qquad (13)$$

Substituting this into (9), we obtain the HJB equation for this case as $0 = C_0 + L_0 V$ where

$C_0 \equiv U(0)$.

For the second case in which $\max.[u(t)-\mu, 0] = u(t)-\mu$ (*i.e.* $u(t) \geq \mu$), the optimality condition for a regular interior maximum to (9) is given by

$$\frac{\partial Z}{\partial u} = 0 = f_0(u,m) + V_x(t,0,m). \tag{14}$$

Defining the inverse function of $U'(\cdot)$ by $I(\cdot)$, we can represent (14) as

$$I(V_x(t,0,m)/m(t)) = -C(u).$$

Since $C(u) = (q(t)-u)\cdot m(t)$ for $x(t) = 0$, we obtain the interior solution $u(t) = v_0$ as

$$v_0 = q(t) + \frac{1}{m(t)} I\left(\frac{V_x(t,0,m(t))}{m(t)}\right). \tag{15}$$

It can be proved that the solution v_0 given by (15) is a monotone function of $m(t)$, and there exist m^* and m^{**} such that

$$\begin{cases} v_0 < \mu & \textit{for } m(t) < m^* \\ \mu \leq v_0 < q(t) & \textit{for } m^* \leq m(t) < m^{**} \\ v_0 \geq q(t) & \textit{for } m^{**} \leq m(t) \end{cases} \tag{16}$$

holds. Since the inflow rate $u(t)$ is restricted by $\mu \leq u(t) \leq q(t)$, (16) implies that the optimal control strategies for the case of $x(t) = 0$ and $q(t) \geq \mu$ are given by

$$\begin{cases} \textit{control } \boldsymbol{B} : u(t) = \mu & \textit{if } m(t) < m^* \\ \textit{control } \boldsymbol{C} : u(t) = v_0 & \textit{if } m^* \leq m(t) < m^{**} \\ \textit{control } \boldsymbol{D} : u(t) = q(t) & \textit{if } m^{**} \leq m(t) \end{cases}. \tag{17}$$

For these controls, the HJB equations that the value function V should satisfy can be obtained by substituting (17) into (9):

$$\begin{aligned} &C_1 + L_0 V = 0 && \text{for control } \boldsymbol{B}, \\ &N_0 V = 0 && \text{for control } \boldsymbol{C}, \\ &C_0 + L_1 V = 0 && \text{for control } \boldsymbol{D}, \end{aligned} \tag{18}$$

where $C_1 \equiv U\big(m(t)\cdot(\mu - q(t))\big)$, $L_1 \equiv \big(q(t)-\mu\big)\frac{\partial}{\partial x} + L_0$, $N_0 V \equiv (v_0 - \mu - \frac{1}{m(t)})\frac{\partial V}{\partial x} + L_0 V$.

Note that quantitative determination of m^* and m^{**} requires the identification of the value function by solving the HJB equations.

(b) The case of $x(t) > 0$

When there is a queue in the freeway, the optimality condition for an interior maximum to (9) is given by

$$\frac{\partial Z}{\partial u} = 0 = f_1(u,x,m) + V_x(t,x,m) \tag{19}$$

where the function $f_1(u, x, m)$ is defined by

$$f_1(u, x(t), m(t)) \equiv U'(-C(u)) \cdot d(t) \text{ and } d(t) \equiv m(t) - (x(t)/\mu).$$

Representing (19) as $I(V_x(t,x,m)/d(t)) = -C(u)$ and using $C(u) = -u \cdot d(t) + m(t)q(t)$, we obtain the interior solution as

$$v_1 = \frac{1}{d(t)}\left[m(t)q(t) + I\left(\frac{V_x(t,x(t),m(t))}{d(t)}\right)\right]. \tag{20}$$

For any fixed value of $x(t) > 0$, the solution v_1 given by (20) is a monotone function of $m(t)$, and there exist $\underline{m}$ and $\overline{m}$ such that

$$\begin{cases} v_1 < 0 & for\ \ m(t) < \underline{m} \\ 0 \le v_1 < q(t) & for\ \ \underline{m} \le m(t) < \overline{m} \\ v_1 \ge q(t) & for\ \ \overline{m} \le m(t) \end{cases} \tag{21}$$

holds (for the proof, see Akamatsu and Yamazaki (2006)). This, together with the constraint $0 \le u(t) \le q(t)$, implies that the optimal control for the case of $x(t) > 0$ can be characterized by the following three strategies:

$$\begin{cases} \text{control } \boldsymbol{E} : u(t) = 0 & if\ \ m(t) < \underline{m} \\ \text{control } \boldsymbol{F} : u(t) = v_1 & if\ \ \underline{m} \le m(t) < \overline{m}. \\ \text{control } \boldsymbol{G} : u(t) = q(t) & if\ \ \overline{m} \le m(t) \end{cases} \tag{22}$$

The corresponding HJB equations can be obtained by substituting these controls into (9):

$$\begin{aligned} C_2 + L_2 V &= 0 && \text{for control } \boldsymbol{E}, \\ N_1 V &= 0 && \text{for control } \boldsymbol{F}, \\ C_3 + L_1 V &= 0 && \text{for control } \boldsymbol{G}, \end{aligned} \tag{23}$$

where $\quad C_2 = U(-m(t)\cdot q(t)), \quad C_3 = U(-(x(t)/\mu)\cdot q(t)),$

$$L_2 \equiv -\mu\frac{\partial}{\partial x} + L_0, \qquad N_1 V \equiv -\frac{1}{d(t)}\frac{\partial V}{\partial x} + (v_1 - \mu)\frac{\partial V}{\partial x} + L_0 V.$$

Note that these boundaries, $\underline{m}$ and $\overline{m}$, are functions of $(t, x(t))$; this implies that the optimal control at time t is selected from either one of the controls **E**, **F**, or **G**, depending on the realization of both state variables $x(t)$ and $m(t)$.

Table 1 and **Figure 3** summarize these results. **Table 1** shows all the optimal control patterns (from ***A*** to ***F***) and the corresponding HJB equations for possible queuing states of the freeway. **Figure 3** illustrates a typical pattern of the optimal control strategies at time t (when the OD flow rate $q(t)$ is greater than the bottleneck capacity μ). This figure shows that the optimal control strategies at each time period can be classified into 6 patterns (from control ***B*** to ***F***), depending on the values of two state variables realized at time t, $m(t)$ and $x(t)$. Suppose, for example, that some moderate length of queue x_e is observed at time t and the realized travel time of link 2 at that time is m_e (which is below the lower boundary curve). We then see from the figure that the optimal inflow rate into the freeway should be zero (*i.e.* all the OD flow should be assigned to the local bypass link). The two boundary curves dividing the $m(t)-x(t)$ plane into 6 regions in the figure are determined as a function of time and some parameters in the model (*eg.* $q(t)$, μ, $\alpha(t)$, σ and the risk aversion parameter of $U(C)$). The detailed properties of these curves are revealed by numerical experiments in the later section.

Table 1. Optimality Conditions (HJB equations) and Optimal Inflow Rate into the Freeway

Queuing State of Link 1		Optimality Conditions	$u(t)$	Control ID
$x(t)=0$	$\dot{x}(t)=0$ $(q(t)<\mu)$	$L_0\, V(t,x(t),m(t)) + C_0 = 0$	$q(t)$	**A**
	$\dot{x}(t)=u(t)-\mu$ $(q(t)\geq\mu)$	$L_0\, V(t,x(t),m(t)) + C_1 = 0$	μ	**B**
		$N_0\, V(t,x(t),m(t)) = 0$	v_0	**C**
		$L_1\, V(t,x(t),m(t)) + C_0 = 0$	$q(t)$	**D**
$x(t)>0$	$\dot{x}(t)=u(t)-\mu$	$L_2\, V(t,x(t),m(t)) + C_2 = 0$	0	**E**
		$N_1\, V(t,x(t),m(t)) = 0$	v_1	**F**
		$L_1\, V(t,x(t),m(t)) + C_3 = 0$	$q(t)$	**G**

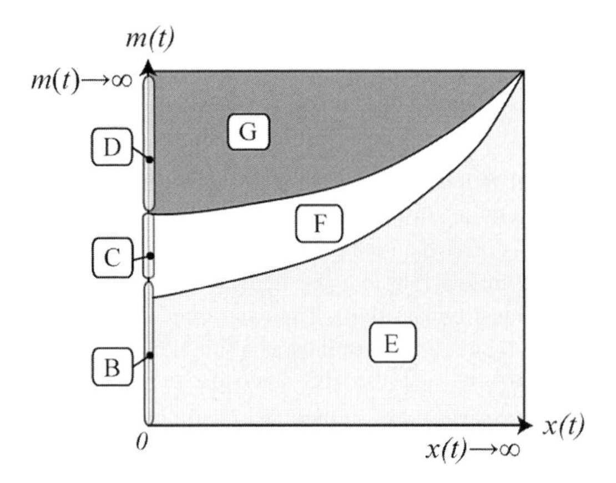

Figure 3: The optimal control strategies at time t in the state space ($m(t)$, $x(t)$)

Comparisons with the Risk-neutral Strategies

It is worthwhile to discuss the optimal controls for the '*risk neutral*' case in which the utility function is linear (*i.e.* the objective function of [SCP] reduces to only the expected total travel time). For the linear utility function, the objective function $Z(u(t))$ defined in (11) reduces to

$$Z(u(t)) = \begin{cases} -C(u(t)) + (u(t) - \mu)V_x & \text{if } x(t) > 0 \\ -C(u(t)) + \max.[u(t) - \mu, 0]V_x & \text{if } x(t) = 0 \end{cases}, \tag{11'}$$

In a similar vein to the discussion of the previous subsection, the optimal control for this case may be divided according to whether the freeway queue exists or not (*i.e.* $x = 0$ or $x \geq 0$).

When $x(t) = 0$, the derivative of $Z(u)$ is given by

$$\frac{\partial Z}{\partial u} = \begin{cases} -m & \text{if } u(t) < \mu \\ -m + V_x & \text{if } u(t) \geq \mu \end{cases}.$$

This implies that the function $Z(u)$ is piecewise linear with respect to u, and the maximum is attained at either $u(t) = q(t)$ or $u(t) = \mu$, depending on the sign of $\partial Z / \partial u$. Thus, the optimal metering strategy should be

$$\begin{cases} \text{control } \mathbf{B} : u(t) = \mu & \text{if } m(t) < V_x(t,x,m) \\ \text{control } \mathbf{D} : u(t) = q(t) & \text{if } m(t) \geq V_x(t,x,m) \end{cases} \tag{17'}$$

Comparing this with the optimal control given in (17), we see that the control ***C*** vanishes in the risk-neutral case, and the control rule results in 'bang-bang control'.

When $x(t) > 0$, the derivative of $Z(u)$ is given by

$$\frac{\partial Z}{\partial u} = \begin{cases} (x/\mu) - m & \text{if } u(t) < \mu \\ (x/\mu) - m + V_x & \text{if } u(t) \geq \mu \end{cases}.$$

This implies that the optimal control for this case is given by

$$\begin{cases} \text{control } \boldsymbol{E}: u(t) = 0 & \text{if } m(t) < (x/\mu) + V_x(t,x,m) \\ \text{control } \boldsymbol{G}: u(t) = q(t) & \text{if } m(t) \geq (x/\mu) + V_x(t,x,m) \end{cases} \qquad (22')$$

Similar to the case of $x(t) = 0$, this shows that the intermediate control $\boldsymbol{F}$ in (22) disappears. These results are summarized by **Figures 4** and **5**. The former (**Figure 4**) shows the relationship between u and m when $x = 0$, in which (**a**) gives the optimal control for a general risk-averse utility function and (**b**) the control for the risk-neutral case. Similarly, the latter (**Figure 5**) depicts the optimal control rule when $x \geq 0$. We can conclude from these figures that the introduction of risk aversion has the effect of 'smoothing' the optimal control.

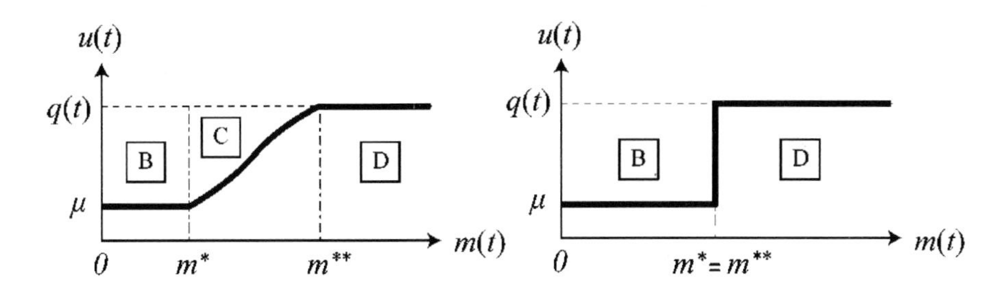

(**a**) risk averse utility (**b**) risk-neutral (linear) utility

Figure 4: The optimal control strategies for the case of $x(t) = 0$.

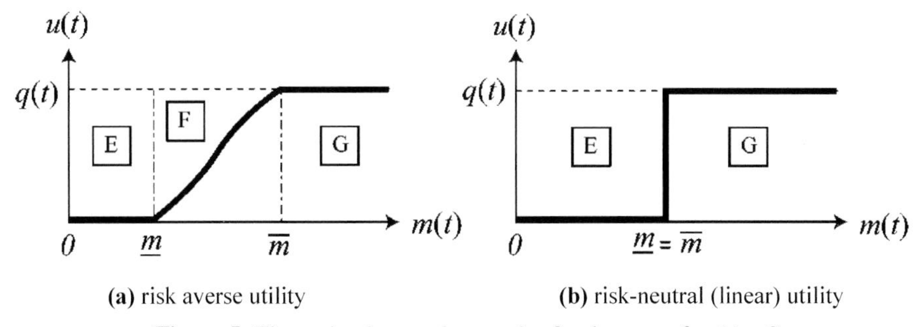

(**a**) risk averse utility (**b**) risk-neutral (linear) utility

Figure 5: The optimal control strategies for the case of $x(t) > 0$.

REFORMULATION AND NUMERICAL ALGORITHMS

In order to examine the detailed properties of the optimal control strategies (*i.e.* properties of v_0, v_1, and the two boundary curves in the state space), we need to solve the problem numerically. This section presents an efficient numerical algorithm for solving the SCP. For this purpose, we reveal that the optimality conditions of our problem (*i.e.* HJB equations) can be reformulated as a dynamical system of GCP. We then show that the system is decomposed with respect to time under an appropriate discretization framework. This enables us to reduce our problem to successively solving the sub-problems, each of which is formulated as a finite-dimensional GCP. We also provide an algorithm for solving the sub-problem. Due to space limitation, the technical details of the algorithm are relegated to Nagae and Akamatsu (2006a,b) and Akamatsu and Yamazaki (2006).

Reformulation as a Dynamical System of Nonlinear Complementarity Problems

(a) Optimality Conditions for the Inner Region

At any time $t \in [0, T)$ in which $x(t) > 0$ holds, the optimal control is given by either ***E***, ***F***, or ***G*** (see Table 1). The HJB equations in (23) for these controls are also mutually exclusive. More concretely, suppose one of the controls, assume control ***E***, is optimal at time *t*. Then, only one of the HJB equations in (23), $C_2 + L_2 V = 0$ for control ***E***, holds and the other HJB equations do not hold; it can be easily verified from the definition of the HJB equation (9) that $N_1 V \geq 0$ and $C_3 + L_1 V \geq 0$ when control ***E*** is optimal. This mutual exclusiveness of the HJB equations can be naturally expressed by the following dynamical system of GCP:

[GCP2]

$$\begin{cases} [C_2 + L_2 V] \cdot [N_1 V] \cdot [C_3 + L_1 V] = 0 \\ C_2 + L_2 V \geq 0, \; N_1 V \geq 0, \; C_3 + L_1 V \geq 0 \end{cases}$$

$$\forall (t, x, m) \in [0, T) \times R_+ \times R_+ \qquad (24a)$$

or equivalently,

$$\min.[C_2 + L_2 V(t, x, m), \; N_1 V(t, x, m), \; C_2 + L_1 V(t, x, m)] = 0,$$

$$\forall (t, x, m) \in [0, T) \times R_+ \times R_+ \qquad (24b)$$

(b) Boundary Conditions

For [GCP2] to be a well-posed problem, the value function $V(t)$ at any time t should satisfy some appropriate boundary conditions; we should consider the conditions on the following four boundaries of the state space ($x(t)$, $m(t)$) (see **Figure 6** for a conceptual illustration of these conditions).

i) Upper boundary of $m(t)$: $m(t) \to +\infty$

When $m(t)$ tends to infinity, the optimal control is to assign all the OD flow into link 1 (the freeway); hence, the value function should satisfy

$$V(t,x,m \mid m \to +\infty) = \frac{1}{\mu}\int_t^T q(\tau)x(\tau)d\tau, \qquad \forall (t,x) \in [0,T) \times R_+. \tag{25}$$

ii) Lower boundary of $m(t)$: $m(t) = 0$

Clearly, the optimal control for $m(t) = 0$ is to assign all the OD flow into link 2 whose travel time is zero; hence, the value function also should be zero:

$$V(t,x,m \mid m \to 0) = 0, \qquad \forall (t,x) \in [0,T] \times R_+. \tag{26}$$

iii) Upper boundary of $x(t)$: $x(t) \to +\infty$

When $x(t)$ tends to infinity, the optimal control is to assign all the OD flow into link 2; hence, the value function should satisfy

$$V(t,x,m \mid x \to +\infty) = E_t\left[\int_t^T q(\tau)m(\tau)d\tau \,\middle|\, m(t) = m\right],$$

$$\forall (t,m) \in [0,T) \times R_+. \tag{27}$$

From the Feynman-Kac formula, the value function satisfying (27) can be obtained as the solution of the following partial differential equation (PDE):

$$C_3(t) + L_0 V(t,x,m \mid x \to \infty) = 0, \qquad \forall (t,m) \in [0,T) \times R_+, \tag{28}$$

with a terminal condition: $V(T,x,m \mid x \to +\infty) = 0 \quad \forall m \in R_+$.

iv) Lower boundary of $x(t)$: $x(t) = 0$

The optimality conditions for this boundary has already been shown in the previous section (*ie*. (13), (17) and (18)), and they are classified into two cases according to whether or not the OD flow rate exceeds the link capacity μ. When the OD flow rate is less than the link capacity (*i.e.* $x(t) = 0$ *and* $q(t) < \mu$), the optimal control is given by ***control A*** in Table 1. The optimal control is $u(t) = q(t)$, and the HJB equation reduces to the following PDE:

[PDE1] $$C_0 + L_0 V(t,x,m \mid x = 0) = 0, \qquad \forall (t,m) \in [0,T) \times R_+. \tag{29}$$

When the OD flow rate is greater than the link capacity (*i.e.* $x(t) = 0$ *and* $q(t) \geq \mu$), the optimal control is given by any one of ***B***, ***C*** or ***D*** in Table 1, depending on the level of $m(t)$; the HJB equations in (18) for these controls are also mutually exclusive. In a similar vein to [GCP2], this mutual exclusiveness of the HJB equations can be expressed as

[GCP1] $$\begin{cases} [C_1 + L_0 V]\cdot[N_0 V]\cdot[C_0 + L_1 V] = 0 \\ C_1 + L_0 V \geq 0,\ N_0 V \geq 0,\ C_0 + L_1 V \geq 0 \end{cases} \qquad \forall(t,m)\in[0,T)\times R_+, \qquad (30a)$$

or equivalently,

$$\min.[C_1 + L_0 V(t,x,m),\ N_0 V(t,x,m),\ C_0 + L_1 V(t,x,m)] = 0,$$

$$\forall(t,m)\in[0,T)\times R_+. \qquad (30b)$$

Finally, we should impose a terminal condition that must be satisfied by the value function at the terminal time T:

$$V(T,x,m) = 0 \qquad \forall(x,m)\in R_+\times R_+. \qquad (31)$$

Thus, the SCP has been reformulated as a generalized complementarity problem [GCP], which consists of [GCP2] with four boundary conditions (*i.e.* eqs.(25), (26), (27), and [PDE1]/[GCP1]) and a terminal condition (31).

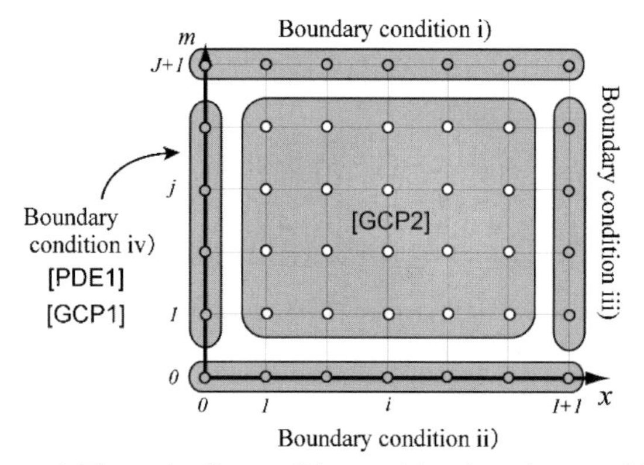

Figure 6: The optimality conditions and four boundary conditions that should hold at time t in the state space $(m(t), x(t))$

Discretization

In order to develop a numerical algorithm for solving [GCP], it is convenient to represent the problem in a discrete (time-state) framework. We consider a discrete grid in the time-state space $[0,X]\times[0,M]\times[0,T]$ with increments dx, dm, and dt. Let (x_i, m_j, t_k) be each point of the grid, where the indices i (= 0,1,..., I+1), j (= 0,1,..., J+1), and k (k = 0,1,…,K) characterize the locations of the point with respect to state variables x, m, and time t, respectively. We also denote the value of $V(t, x, m)$ at a grid point (x_i, m_j, t_k) by $V^{i,j}(k)$. Using this grid point representation, we can approximate the value function $V(t, x, m)$ at time t_k by a (column) vector $\mathbf{V}(k)$ whose elements are $[V^{i,j}(k) \mid i=0,1,...,I+1; j=0,1,...,J+1]$.

In this framework, infinitesimal operator L_n (n = 0,1,2) can be approximated as a matrix $\mathbf{L}_n$ by using an appropriate finite difference scheme (*eg*., that of Crank–Nicholson). Similarly, infinitesimal operator N_n (n = 0 or 1) can be approximated by a set of nonlinear functions $\mathbf{N}_n$ of $\mathbf{V}(k)$. Thus, the three operators that appear in the HJB equations (23) for controls ***E***, ***F***, and ***G*** (*i.e.* $C_2 + L_2V$, N_1V, and $C_3 + L_1V$) at time t_k can be represented as

$$\begin{cases} \mathbf{E}(k) \equiv \mathbf{L}_2(k+1)\mathbf{V}(k+1) - \mathbf{L}_2(k)\mathbf{V}(k) + \mathbf{C}_2(k) \\ \mathbf{F}(k) \equiv \mathbf{N}_1(\mathbf{V}(k+1)) - \mathbf{N}_1(\mathbf{V}(k)) \\ \mathbf{G}(k) \equiv \mathbf{L}_1(k+1)\mathbf{V}(k+1) - \mathbf{L}_1(k)\mathbf{V}(k) + \mathbf{C}_3(k) \end{cases}. \quad (32)$$

We are now in a position to express [GCP] obtained in the previous subsection as a *finite -dimensional* GCP. To begin with, we represent the main problem [GCP2] (for time t_k) by

[GCP2(k)] $$\min.[\mathbf{E}(k), \mathbf{F}(k), \mathbf{G}(k)] = \mathbf{0} \quad (33)$$

Next, the four boundary conditions from *i)* to *iv)* in **4.1** can be easily fitted into this discrete framework as follows. The first (original) boundary condition (for $m \to +\infty$) is governed by (25) and the state equation (2). By using the discrete counterpart of (25):

$$\mathrm{V}^{i,J+1}(k) = \frac{1}{\mu}\sum_{n=k}^{K} q(n)x(n \mid x(k) = x_i)\Delta t \quad (34)$$

we easily obtain the values of $\{\mathrm{V}^{i,J+1}(k);\ i = 0,1,\ldots,I\}$. The second condition (for $m = 0$) is given by (26), whose discrete correspondence is only to set

$$\{\mathrm{V}^{i,0}(k) = 0;\ i = 0,1,\ldots,I\}. \quad (35)$$

The third condition (for $x \to +\infty$) is given by (27), which reduces to solving the PDE in (28). The discrete counterpart of this PDE is

$$\mathbf{A}(k) + \mathbf{C}_3(k) = \mathbf{0}, \quad (36)$$

where $\mathbf{A}(k)$ is a discrete approximation of L_0V, which is defined as

$$\mathbf{A}(k) \equiv \mathbf{L}_0(k+1)\mathbf{V}(k+1) - \mathbf{L}_0(k)\mathbf{V}(k).$$

For given $\mathbf{V}(k+1)$, the system of linear equations (36) can be solved, which determines the value of $\{\mathrm{V}^{I+1,j}(k);\ j = 0,1,\ldots,J\}$. The final boundary condition (for $x = 0$) is governed by [PDE1] and [GCP1]. The former reduces to the following system of linear equations:

[PDE1(k)] $$\mathbf{A}(k) + \mathbf{C}_0(k) = \mathbf{0}, \quad (37)$$

and the latter is represented by

[GCP1(k)] $$\min.[\mathbf{B}(k), \mathbf{C}(k), \mathbf{D}(k)] = \mathbf{0}, \quad (38)$$

where $\mathbf{B}(k)$, $\mathbf{C}(k)$ and $\mathbf{D}(k)$ are defined by

$$\begin{cases} \mathbf{B}(k) \equiv \mathbf{L}_0(k+1)\mathbf{V}(k+1) - \mathbf{L}_0(k)\mathbf{V}(k) + \mathbf{C}_1(k) \\ \mathbf{C}(k) \equiv \mathbf{N}_0(\mathbf{V}(k+1)) - \mathbf{N}_0(\mathbf{V}(k)) \\ \mathbf{D}(k) \equiv \mathbf{L}_1(k+1)\mathbf{V}(k+1) - \mathbf{L}_1(k)\mathbf{V}(k) + \mathbf{C}_0(k) \end{cases}. \quad (39)$$

The solution of (37)/(38) gives the value of $\mathbf{V}(k)$ on this boundary, $\{V^{0,j}(k);\ j=1,2,\ldots,J\}$. In summary, [GCP] is thus expressed as a dynamical system of finite dimensional GCPs:

$$\text{[GCP-D]} \qquad \min.[\mathbf{E}(k), \mathbf{F}(k), \mathbf{G}(k)] = \mathbf{0} \qquad \text{for } k = 0,1,\ldots,K \tag{40}$$

with the four boundary conditions:

i) $\{V^{i,J+1}(k);\ i=0,1,\ldots,I\}$ is given by (34),
ii) $\{V^{i,0}(k) = 0;\ i=0,1,\ldots,I\}$,
iii) $\{V^{I+1,j}(k);\ j=0,1,\ldots,J\}$ is given by (36),

iv) $\begin{cases} \mathbf{A}(k) = \mathbf{0} & \text{if } q(t_k) < \mu \\ \min.[\mathbf{B}(k), \mathbf{C}(k), \mathbf{D}(k)] = \mathbf{0} & \text{if } q(t_k) \geq \mu \end{cases}$,

and a terminal condition: $\mathbf{V}(K) = \mathbf{0}$.

Algorithm

The problem [GCP-D] has a convenient property that the sub-problem [GCP(k)] is independent from other sub-problems [GCP(l)] ($k \neq l$) when $\mathbf{V}(k+1)$ is known. This implies that the series of sub-problems $\{[\text{GCP}(k)] \mid k = 0,1,\ldots K\}$ can be solved in a successive manner: using the terminal condition for $\mathbf{V}(K)$, we first solve the sub-problem [GCP(K–1)] and obtain the solution $\mathbf{V}(K-1)$; using $\mathbf{V}(K-1)$ as a given constant, we solve the sub-problem [GCP(K–2)] and obtain $\mathbf{V}(K-1)$; and by repeating the procedure recursively, we obtain the entire value function $\{\mathbf{V}(k) \mid k=0,1,2,\ldots,K\}$. Since each sub-problem [GCP(k)] consists of [GCP2(k)] and the four boundary conditions, the procedure for obtaining $\mathbf{V}(k)$ is naturally divided into computation of $\mathbf{V}(k)$ on the boundaries and solving [GCP2(k)]. Thus, the outline of the algorithm for solving [GCP] can be summarized as follows:

***Step* 0.** Set the terminal condition: $\mathbf{V}(K) := \mathbf{0}$; Set time counter $k := K-1$.

***Step* 1.** Compute $\mathbf{V}(k)$ for the state space boundaries *i)* $\{V^{i,J+1}(k);\ i=0,1,\ldots,I\}$,
ii) $\{V^{i,0}(k) = 0;\ i=0,1,\ldots,I\}$, and *iii)* $\{V^{I+1,j}(k);\ j=0,1,\ldots,J\}$.

***Step* 2.** Compute $\mathbf{V}(k)$ for the boundary *iv)* (*i.e.* $\{V^{0,j}(k);\ j=1,2,\ldots,J\}$):
Given $\mathbf{V}(k+1)$ and $(V^{0,0}(k), V^{0,J+1}(k))$,
solve [PDE1(k)] if $q(k) \leq \mu$, [GCP1(k)] otherwise.

***Step* 3.** Compute $\mathbf{V}(k)$ for the inner region (*i.e.* $\{V^{i,j}(k);\ i=1,2,\ldots,I;\ j=1,2,\ldots,J\}$):
Given $\mathbf{V}(k+1)$ and $\mathbf{V}(k)$ for all boundaries obtained in *Step 1* and *Step 2*,
solve [GCP2(k)].

***Step* 4.** If $k = 0$ terminate, otherwise $k := k-1$ and return to *Step 1*.

In the algorithm above, we need an efficient procedure for solving sub-problems [GCP1(k)] and [GCP2(k)], which are formulated as a finite-dimensional GCP. For this, we use a smoothing function approach developed by Peng (1998), Qi and Liao (1999), and Peng and Lin

(1999). This approach is not only the state-of-the-art technique but is also suitable for our problems with a special sparse Jacobian matrix from the view point of efficiency, as discussed in Akamatsu and Yamazaki (2006) and Nagae and Akamatsu (2006a,b).

In the smoothing function approach, one solves the following system of nonlinear equations:

$$\mathbf{H}(\mathbf{V}) \equiv \min.\{\mathbf{F}_1(\mathbf{V}), \mathbf{F}_2(\mathbf{V}), \mathbf{F}_3(\mathbf{V})\} = \mathbf{0}, \tag{41}$$

where $\min.\{\mathbf{F}_1(\mathbf{V}), \mathbf{F}_2(\mathbf{V}), \mathbf{F}_3(\mathbf{V})\}$ is a vector operator whose j th element is defined as $\min.\{F_1^j, F_2^j, F_3^j\}$. Note that the equations system, $\mathbf{H}(\mathbf{V}) = \mathbf{0}$, cannot be solved by naive methods, since $\mathbf{H}(\mathbf{V})$ is indifferentiable. The key idea of the smoothing approach, in order to overcome difficulties due to the indifferentiability of $\mathbf{H}$, is to transform the original problem into a system of smooth equations via a so-called smoothing function $\mathbf{S}(\mathbf{V},\eta)$ with j th component,

$$S^j(\mathbf{V},\eta) \equiv -\eta \ln \sum_j \exp[-F^j(\mathbf{V})/\eta] \tag{42}$$

where $\eta \geq 0$ is referred to as the smoothing parameter. This type of function in eq.(42) is also known as an expected minimum cost (or a LOG-sum function) for a LOGIT model in the random utility theory. It is well known that the smoothing function has two desirable properties for developing an efficient algorithm: First,

$$\mathbf{S}(\mathbf{V},+0) \equiv \lim_{\eta \to +0} \mathbf{S}(\mathbf{V},\eta) = \mathbf{H}(\mathbf{V}).$$

In other words, the solution of the smooth equations system $\mathbf{S}(\mathbf{V},\eta) = \mathbf{0}$ is equivalent to the solution of (41), $\mathbf{H}(\mathbf{V})$ = 0, at the limit of $\eta \to 0$; second, $\mathbf{S}(\mathbf{V},\eta)$ is a continuously differentiable function of $\mathbf{V}$ for all $\eta > 0$. The former property ensures that the present algorithm provides a good approximation to the solution of (41), whereas the latter property is exploited to guarantee the efficiency of the algorithm.

The smoothing approach-based algorithm generates a solution set of the smooth equations system, forming a path $\{(\mathbf{V},\eta \mid \mathbf{S}(\mathbf{V},\eta) = \mathbf{0})\}$ as the parameter η tends to zero. This path is usually referred to as the smoothing path. Let $\eta^{(n)}$ denote the smoothing parameter in the n th iteration, and $\mathbf{V}^{(n)}$ be a solution of the corresponding smooth equation $\mathbf{S}(\mathbf{V},\eta^{(n)}) = \mathbf{0}$. For this notation, we can summarize the procedure for generating the smoothing path as

> ***Step* 0.** Choose $\eta^{(1)} \in R_+$. Set iteration counter $n := 1$;
> ***Step* 1.** If $\mathbf{H}(\mathbf{V}^{(n)}) = \mathbf{0}$ terminate; $\mathbf{V}^{(n)}$ is the solution of the GLCP;
> ***Step* 2.** Solve the smooth equations system $\mathbf{S}(\mathbf{V}^{(n)},\eta^{(n)}) = \mathbf{0}$;
> ***Step* 3.** Choose the next smoothing parameter $\eta^{(n+1)} \in [0,\eta^{(n)})$;
> ***Step* 4.** Set $n := n+1$; return to ***Step* 1**.

It is easy to verify that any accumulation point of the smoothing path $\{\mathbf{V}^{(n)},\eta^{(n)}\}$ generated by the algorithm above is the solution of (41), since the first property of the smoothing function $\mathbf{S}(\mathbf{V},+0) = \mathbf{H}(\mathbf{V})$ and the condition applicable to the smoothing parameters, $\eta^{(n)} > \eta^{(n+1)} \geq 0$ is satisfied. The global convergence of the generic algorithm has been established (*e.g.* Peng and Lin (1999)): Any smoothing path $\{\mathbf{V}^{(n)}, \eta^{(n)}\}$ generated by the algorithm converges to $(\mathbf{V}^{*}, +0)$ globally, when i) $\nabla\mathbf{S}^{(n)} \equiv \nabla\mathbf{S}(\mathbf{V}^{(n)}, \eta^{(n)})$ is nonsingular, and ii) the norm of $[\nabla\mathbf{S}^{(n)}]^{-1}$ is finite for all n. Since both the conditions are naturally satisfied in our framework, the smoothing path globally converges to the solution of (41).

ILLUSTRATIVE EXAMPLES OF CONTROL OPERATIONS

An Example Illustrating the Proposed Method

For illustrating how the proposed method works, we show a numerical example obtained by applying the algorithm presented in the previous section. We first assume the OD flow rate $q(t)$ to be a step function:

$$q(t) = \begin{cases} q_0 & if \ \ 0 \leq t < \tau \\ q_1 & if \ \ \tau \leq t < T \end{cases},$$

where q_0 and q_1 are given as $q_0 < \mu < q_1$ and τ is given as $0 < \tau < T$. We then assume that the road manager has a constant absolute risk aversion (CARA) utility function:

$$U(-C(t)) \equiv -\exp[-\theta(-C(t))]/\theta,$$

where θ is a given parameter that represents risk averseness of the road manager; the larger it is, the more risk averse the road manager becomes.

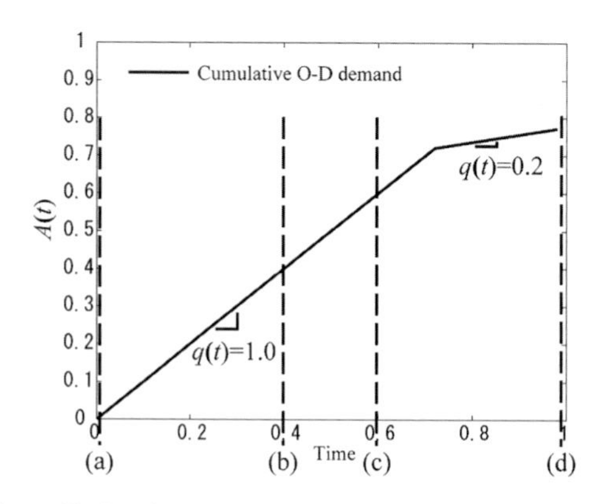

Figure 7: OD flow profile for the numerical experiment

The base case parameters are as follows: The length of the control horizon is $T = 1$. The OD flow rate is initially $q_0 = 1.0$ and switches to $q_1 = 0.2$ at time $\tau = 0.7$. **Figure 7** shows the cumulative OD flow. The capacity of the bottleneck is $\mu = 0.5$. For the sake of simplifying the discussion, we assume $\alpha(t) = 0 \quad \forall t \in [0, T]$, which implies that the *expected* travel time of the local bypass (link 2) is fixed at the initial value m_0. The volatility of the local bypass travel time is $\sigma = 0.3$. The road manager's risk aversion is assumed to be $\theta = 0.5$.

As we discussed in the third section, the optimal control strategy is characterized by two boundary (threshold) curves, $\overline{m}(t, x)$ and $\underline{m}(t, x)$, each of which is a function of $(t, x(t))$. At time t, these boundary curves can be described as functions of $x(t)$, as illustrated in **Figure 3**. **Figure 8** shows these boundary curves obtained by applying our algorithm in the setting described above. In this figure, the diagrams (a), (b), (c), and (d) are the boundary curves at t= 0.01, 0.4, 0.6 and 0.99, respectively. Observe that both boundary curves $\overline{m}(t, x)$ and $\underline{m}(t, x)$ decrease with respect to t. This can be interpreted as follows: the effect of increasing x at t on the value function (*i.e.* the expected total travel time in the remaining duration evaluated in terms of utility) decreases as t increases because the remaining duration (and thus the total travel time for the duration) decreases with respect to time.

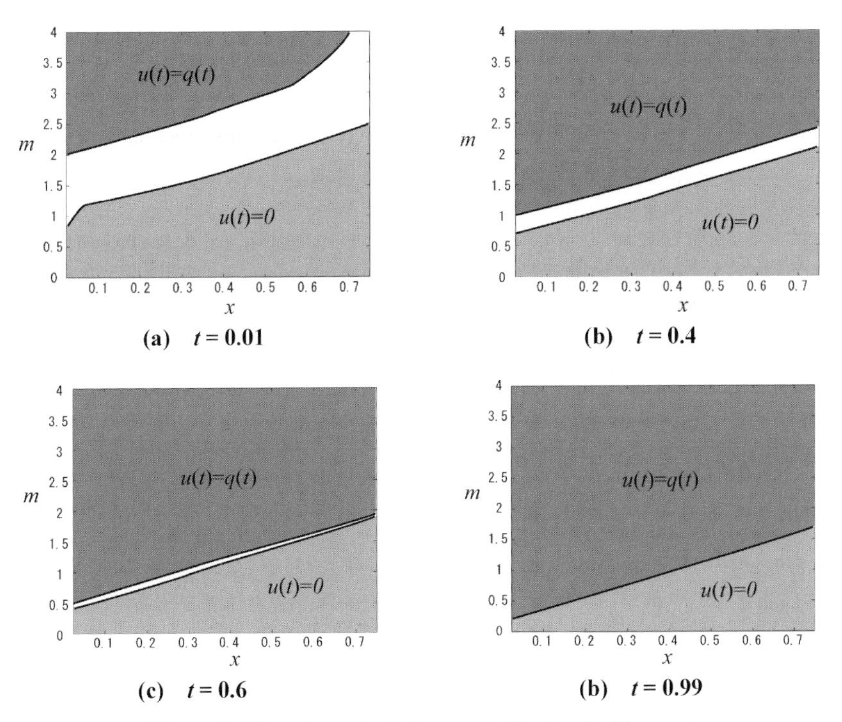

Figure 8: Time-dependent changes in the threshold curves

Figures 9,10, and 11 illustrate how the two boundary curves $\overline{m}(t,x)$ and $\underline{m}(t,x)$ can be used to decide the optimal ramp-metering for a particular sample path of the stochastic travel time $m(t)$. **Figure 9** shows a sample path of $m(t)$ and the corresponding two boundary curves $\overline{m}$ and $\underline{m}$ as functions of time. **Figures 10** and **11** represent the optimal freeway inflow rate $u(t)$ and the queue length under the optimal control, $x^*(t)$, for this sample path, respectively. In these figures, we use the notation $x^*(t)$ in order to emphasize that the queue length at t is obtained as the result of the optimal ramp-metering before t, $\{u(s) \mid s \in [0,t)\}$, corresponding to the sample path $m(t)$. In **Figure 9**, the solid line represents the sample path of $m(t)$, and two dotted lines are the boundary curves, $\overline{m}[t,x^*(t)]$ and $\underline{m}[t,x^*(t)]$. Note that both the boundary curves are plotted as functions of time. This is because the queue length $x^*(t)$ at t is *automatically* decided from both the sample path of the travel time $\{m(s) \mid s \in [0,t)\}$ and corresponding optimal inflow rate $\{u(s) \mid s \in [0,t)\}$. **Figure 9** also indicates that the travel time $m(t)$ hits the lower and upper boundary curves at t_1 and t_2, respectively. In **Figure 10**, we can see that the basic property of the optimal inflow rate $u(t)$ switches at these two hitting times as denoted by (22). That is, $u(t) = 0$ before the first hitting time t_1 (*i.e.* $m(t)$ is below the lower boundary $\underline{m}$), $u(t) = q(t)$ after the second hitting time t_2 (*i.e.* $m(t)$ exceeds the upper boundary $\overline{m}$), and $0 < u(t) < q(t)$ during $t_1 < t < t_2$ (*i.e.* $m(t)$ remains between the two boundary curves). Observe that the inflow rate fluctuates during $t \in (t_1, t_2)$ unlike other time windows. This reflects the fact that the optimal inflow rate for this period should be a function of $u(t) = v_1(t, x(t), m(t), q(t))$, as denoted by (20).

Properties of the Threshold Curves

Using the numerical algorithm developed above, we explore the effects of uncertainty on our control strategies. More specifically, from numerical experiments, we show the manner in which the boundary curves change through the controlled period (1) when the uncertainty in the travel time of the local bypass link (*i.e.* the volatility σ of the travel time) is increased, and (2) when the degree of risk aversion (*i.e.* the risk aversion measure θ) for the control strategy is increased.

Figure 12 shows the dependence of the boundary curves at time $t = 0$ on the volatility σ. In this figure, the two dotted lines represent the upper and lower boundary curves, $\overline{m}(0,x)$ and $\underline{m}(0,x)$, for $\sigma = 0.0$; further, the solid lines and chain lines are the boundary curves for $\sigma = 0.3$ and 0.5, respectively. Observe that both the upper and lower boundary curves shift downward when σ increases. The reason for this is that the greater volatility implies a higher chance of encountering small values of $m(t)$ in the future, in which the road manager can improve his/her utility by decreasing the queue length. Therefore, higher σ increases the optimal queue length $x^*(t)$ and shifts both boundary curves downward. In other words, the road manager has *an option to feedback (state-contingent) management of the queue length*, and the economic value of the option increases as the travel time volatility σ increases.

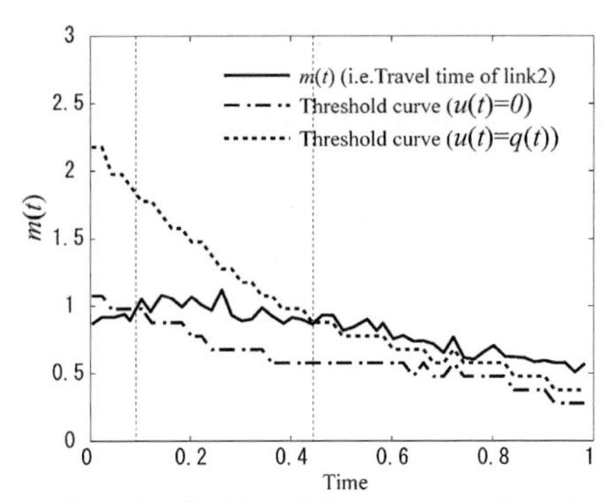

Figure 9: A sample path of $m(t)$ and the corresponding threshold curves for the control period [0,1]

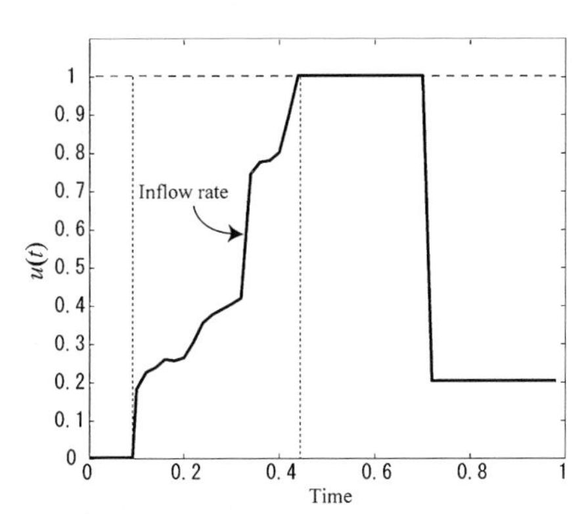

Figure 10: Optimal inflow rate for the sample path of $m(t)$ in **Figure 9**.

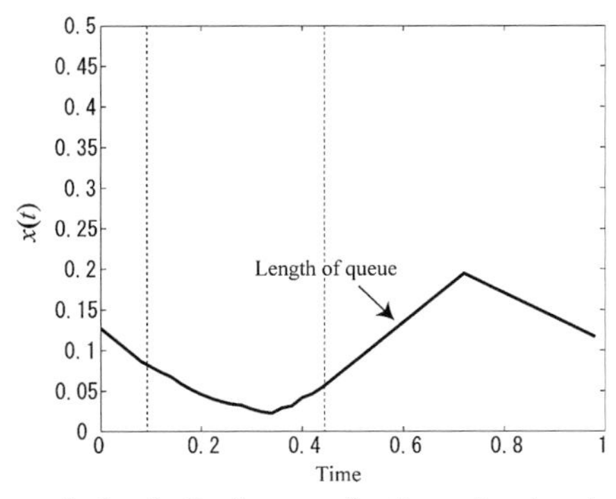

Figure 11: Queue evolution in the freeway for the optimal metering in **Figure 10**.

Figure 13 shows the way in which the boundary curves change with respect to the road manager's risk averseness, θ. In this figure, each pair of solid lines, dotted lines, and chain lines represents the boundary curves at $t = 0$ for $\theta = 0.1$, 0.3, and 0.9, respectively. We can see that the upper boundary curve increases with respect to θ, whereas the lower boundary curve decreases. This implies that when the manager is more risk averse, an extreme control, *i.e.* either $u(t) = 0$ or $u(t) = q(t)$, becomes not optimal.

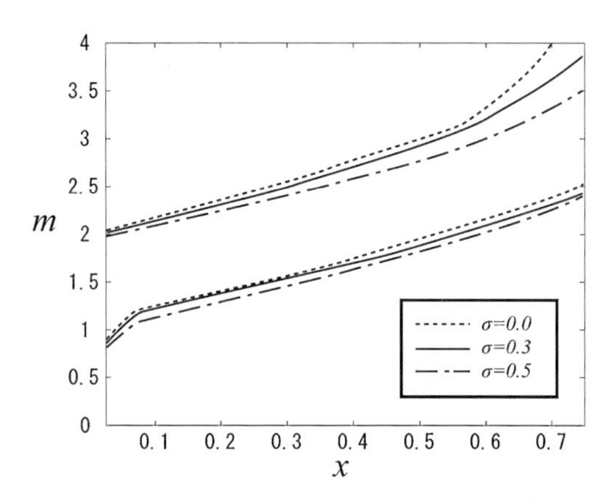

Figure 12: The threshold curves for various levels of volatility

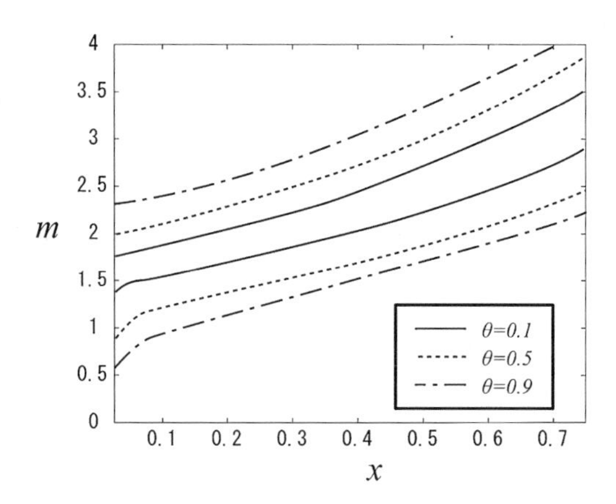

Figure 13: The threshold curves for various levels of risk-aversion

CONCLUDING REMARKS

This paper provides ramp control strategies that achieve *dynamic system optimal* assignment on a network with two parallel links; one of the links is a freeway with a single bottleneck, and the other is a local bypass link (or an aggregation of a local street network) whose travel time follows a stochastic process. Formulating the model as a continuous-time stochastic control problem, we provide *feedback* (*'state contingent'*) *control rules* that exploit the real-time observation of the realization of the stochastic travel time. Our theoretical analysis shows that the optimal ramp control strategies at each time period can be classified into seven patterns (as summarized in **Table 1** and **Figure 3**), depending on the realization of queue length in the freeway and the observed travel time of the bypass link. We further reveal that the optimality conditions of the problem can be reformulated as *a dynamical system of generalized complementarity problems*, which enables us to provide an efficient and robust algorithm for obtaining quantitative results for the control problem. Finally, we provide an illustrative example of the proposed control method, and present results from systematic numerical experiments, which reveal how the uncertainty in travel time affects the optimal control policies.

ACKNOWLEDGMENT

The authors are grateful to Shuichi Yamazaki for his significant assistance in the numerical experiments presented in this paper.

REFERENCES

Akamatsu, T. and S. Yamazaki (2006). An efficient algorithm for risk averse dynamic system optimal traffic assignment. *Infrastructure Planning Review*, **23**, 963-972.

Cottle, R. W., and G. B. Dantzig (1970). A generalization of the linear complementarity problem. *Journal of Combinatorial Theory*, **8**, 79–90.

Erera, A.L., C.F. Daganzo, and D.J. Lovell (2002). Access control problem on capacitated FIFO networks with unique origin-destination paths is hard. *Operations Research*, **50**, 736-743.

Ferris, M.C., and J.-S. Pang (1997). Engineering and economic applications of complemen -tarity problems. *SIAM Review*, **39**, 669–713.

Friesz, T.L., J. Luque, R. Tobin and B. Wie (1989). Dynamic network traffic assignment considered as a continuous time optimal control problem. *Operations Research*, **37**, 893-901.

Kuwahara, M., T. Yoshii and K. Kumagai (2001). An analysis on dynamic system optimal assignment and ramp control on a simple network. *JSCE Journal of Infrastructure Planning and Management*, **667/IV-50**, 59-71.

Lovell, D.J. and C.F. Daganzo (2000). Access control on networks with unique origin-destination paths. *Transportation Research*, **34B**, 185-202.

Munoz, J.C. and J.A. Laval (2006). System optimal dynamic traffic assignment graphical solution method for a congested freeway and one destination. *Transportation Research*, **40B**, 1-15.

Nagae, T. and T. Akamatsu (2006a). A generalized complementarity approach to solving real option problems. *Journal of Economic Dynamics and Control* (*in press*).

Nagae, T. and T. Akamatsu (2006b). Dynamic system optimal traffic assignment exploiting information on real-time traffic conditions. *JSCE Journal of Infrastructure Planning and Management* (*in press*).

Peng, J.-M. (1999). A smoothing function and its applications, in: M. Fukushima and Qi Liqun (*Eds*.), *Reformulation*: *Nonsmooth, Piecewise Smooth, Semismooth and Smoothing Methods*. Kluwer Academic Publishers, 293–316.

Peng, J.-M., and Z. Lin (1999). A non-interior continuation method for generalized linear complementarity problem. *Mathematical Programming*, **86**, 533-563.

Qi, H.-D. and L.-Z. Liao (1999). A smoothing Newton method for extended vertical linear complementarity problem. *SIAM Journal on Matrix Analysis and Applications*, **21**, 45-66.

Ziliaskopoulos, A.K. (2000). A linear programming model for the single destination system optimum dynamic traffic assignment problem. *Transportation Science*, **34**, 1-12.

Transportation and Traffic Theory 2007
Edited by R.E. Allsop, M.G.H. Bell and B.G. Heydecker

5

A ROBUST APPROACH TO CONTINUOUS NETWORK DESIGNS WITH DEMAND UNCERTAINTY

Yafeng Yin, Department of Civil and Coastal Engineering, University of Florida, Gainesville, FL, USA
Siriphong Lawphongpanich, Department of Industrial and Systems Engineering, University of Florida, Gainesville, FL, USA

SUMMARY

In this paper, we consider a robust optimization approach to solve a continuous network design problem with demand uncertainty. We assume that the travel demands belong to a convex and compact uncertainty set instead of having them follow some probability distributions and traffic flows on the underlying network are in user equilibrium. For a given demand realization, our problem reduces to a mathematical program with equilibrium constraints. The algorithm we propose for the problem converges under certain conditions. However, numerical results using two networks from the literature empirically demonstrate that the algorithm is effective and has the potential to solve realistic problems.

INTRODUCTION

The objective of a continuous network design (CND) problem is to determine a (continuous) capacity expansion plan for an existing transportation network within a given budgetary constraint. Because of the continuous nature of the problem, the number of feasible plans is infinite and it is typical to choose one with the best system performance measure. When travel demands are fixed, one such performance measure is the system delay or the total travel time and the expansion plan yielding the least system delay is desirable.

Many articles in transportation science assume that travel demands (either fixed or elastic) are deterministic and formulate CND problems as bi-level optimization problems or mathematical programs with equilibrium constraints or MPEC (see, e.g., Yang and Bell, 1998, and Chiou, 2005, for recent reviews). However, only a small number of articles deal with uncertain demands. Barnhart (2000) proposes a scenario-based stochastic model to solve a network design problem with uncertain demands to distribute crops in Mexico. Waller and Zilliaskopoulos (2001) use an inflated demand for a dynamic network design problem to address the uncertainty. They also suggest a two-stage stochastic optimization model with recourse for the same problem. Chen et al. (2003) develop a multi-objective bi-level mean-variance model to determine the optimal toll and capacity in a build-operate-transfer roadway. In an attempt to balance efficiency and robustness in a network design problem with demand uncertainty, Yin et al. (2005) offer two optimization models to achieve trade-offs that reflect decision makers' attitude towards risks. In optimization literature, some (see, e.g., Patriksson and Wynter, 1999) extend stochastic programming approaches to bi-level or MPEC problems and, in our setting, these approaches generally require knowing the probability distributions associated with the travel demand for each origin-destination (O-D) pair. Although our CND problem can be formulated as stochastic bi-level or MPEC problem, it is not clear that such approach is practical because realistic problems have a large number of O-D pairs and the relationships between them are often difficult to estimate due in part to the lack of data.

In this paper, we assume that travel demands are uncertain to account for the intrinsic day-to-day or within-day fluctuations and adopt the robust optimization approach in, e.g., Ben-Tal and Nemirovski (1999, 2002). Instead of using probability distributions to model demand uncertainties, the robust optimization approach assumes that demands belong to an uncertainty set denoted as Q. Although it may be possible to make Q contain all possible demand realizations, doing so may lead to decisions too conservative or models too computationally complex to solve. Generally, Q need not contain all possible demand realizations and, in some models (see, e.g., the portfolio example in Ben-Tal and Nemirovski, 1999), Q does not represent the support of the demand distributions. Instead, the choice of Q should make the resulting model computationally tractable and reflect the decision maker's attitude toward risk.

Two common forms of uncertainty sets in robust optimization are polyhedra and ellipsoids. To illustrate them in our setting, let W be the set of O-D pairs with cardinality m (or $m = |W|$) and d denote a vector of demands, d_w, where $w \in W$. Then, an example of a polyhedral uncertainty set is a box uncertainty set (see, e.g., Bertsimas and Sim, 2003). Mathematically, $Q = \{d \mid d_w^{mn} \leq d_w \leq d_w^{mx}, \forall w \in W\}$, where d_w^{mn} and d_w^{mx} represent the minimum and maximum travel demands for O-D pair w, respectively. In its most familiar form, an ellipsoidal uncertainty set can be represented as follows:

$$Q = \{d \mid \sum_w \frac{(d_w - d_w^0)^2}{s_w^2} \leq \theta^2\},$$

where d_w^0 and s_w are the nominal demand and possible variation for O-D pair w and θ is a subjective value representing the decision maker's attitude toward risk. Given d_w^{mn} and d_w^{mx}, one set of parameters for the above ellipsoid is:

$d_w^0 = \frac{1}{2}(d_w^{mn} + d_w^{mx})$ and $s_w = \frac{1}{2}(d_w^{mx} - d_w^{mn})$.

When $\theta = 0$, Q with the above parameters reduces to the singleton $\{d^0\}$, where d^0 is a vector of d_w^0. On the other hand, Q becomes the largest volume ellipsoid inside and the smallest volume ellipsoid inscribing the box when θ = 1 and $\sqrt{m}$ respectively. Thus, Q does not include all the demands inside the box defined above when $\theta \leq 1$. When $\theta \geq \sqrt{m}$, there are demands in Q not in the box. In general, larger θ means the decision maker is more risk averse.

Between the two forms, Ben-Tal and Nemirovski (1999) demonstrate via a simple portfolio problem that a box uncertainty set can lead to a decision too conservative to be of interest. The same article also discusses a linear program with uncertain coefficients motivated by Soyster (1973) in which a polyhedral uncertainty set leads to a solution that can be obtained by solving a (deterministic) linear program with the uncertain coefficients at their worst values, an extremely rare event. For our CND problem, setting the travel demands at their worst values, i.e., setting $d_w = d_w^{mx}$, may not produce the desired result. The counterexample in Fisk (1979) shows that the largest demand scenario may not lead to a solution with the worst or largest system delay. Moreover, a design against such a worst case scenario would also be too conservative.

In this paper, we adopt the ellipsoidal uncertainty set for our CND problem for its flexibility and others (see, e.g., Ben-Tal and Nemirovski, 2002, and Chen et al., 2006) have used it successfully in various applications. For the remainder, the following sections formulate a robust network design problem and propose an algorithm to solve the problem. This algorithm is convergent under a given set of conditions. Using ellipsoidal uncertainty sets, we then discuss how to implement the algorithm and summarize results from a numerical experiment involving two transportation networks from the literature. Finally, some concluding remarks are given.

ROBUST NETWORK DESIGN PROBLEM

To formulate our robust network design problem, let N and L denote the sets of nodes and links in the transportation network. For a given demand vector $d \in Q$, $V(d)$ denotes the associated set of all feasible flow distributions and can be mathematically expressed as follows:

$$V(d) = \{v \mid v = \sum_w x^w, Ax^w = E^w d_w, x^w \geq 0, \forall w \in W\},$$

where x^w is the vector of link flows associated with O-D pair w, v is the vector of total or aggregated link flows, and A is the node-link incidence matrix associated with the network. To specify the origins and destinations of link flows, E^w is a vector in $R^{|N|}$ with two nonzero components, one has a value 1 in the component corresponding the origin node of O-D pair w and the other has a value -1 in the component for the destination. For each link $a \in L$, $t_a(v_a, c_a)$ denotes the travel cost (time) as a function of link flow v_a and the expansion amount c_a. Below, $t(v, c)$ denotes a vector whose components consist of individual link travel cost functions, $t_a(v_a, c_a)$.

Using the above notation, the robust network design (RND) problem can be formulated as follows:

$$
\begin{array}{llll}
\text{RND:} & \min_{(c,y,v^d)} & y & \\
& \text{s.t.} & \sum_{a \in A} h_a(c_a) \le B & \\
& & v^d \in V(d), & \forall d \in Q \\
& & t(v^d, c)^T (u - v^d) \ge 0, & \forall u \in V(d),\ d \in Q \\
& & t(v^d, c)^T v^d \le y, & \forall d \in Q \\
& & 0 \le c_a \le c_a^{\max}, & \forall a \in A
\end{array}
$$

When combined with the last constraint, the variable y in the objective function represents the maximum total travel time (or system delay) over Q. Thus, the objective is to minimize the maximum possible total travel time associated with the capacity expansion vector c. In the first constraint, $h_a(c_a)$ is the expansion cost function for link a and B is the available budget. Thus, the constraint guarantees that the total cost of expansion is within the available budget. In combination, the second and third sets of constraints ensure that v^d is a user equilibrium flow distribution with respect to each demand vector $d \in Q$. In particular, the user equilibrium condition is stated as a variational inequality in the third set of constraints, where u denotes any feasible flow distribution with respect to a demand vector d. As stated, the numbers of constraints and decision variables (v^d in particular) in these two sets are infinite. The fourth set of constraints, in conjunction with the objective function, calculates the maximum total travel time over Q associated with a capacity expansion vector c. Finally, the last set requires the amount of expansion on each arc to be nonnegative and not to exceed its maximum.

As formulated above, the RND problem is a MPEC (see Luo et al., 1996), a class of nonconvex optimization problems difficult to solve. Compounding to this difficulty is the fact that the RND problem has an infinite number of constraints and decision variables. The next section proposes an algorithm for the RND problem and shows that it converges to an optimal solution under some assumptions.

AN ALGORITHM

The algorithm proposed below solves a sequence of relaxed RND problems, each one better approximating the original RND problem than its predecessors. In the literature, some, e.g., Lawphongpanich and Hearn (2004) refer to this approach as a cutting constraint or plane algorithm (see also Kelly, 1960, and Marcotte, 1983).

Assume that d^1, d^2, …, d^n are elements of Q. Then, a relaxed robust network design (RRND) problem can be written as:

$$
\begin{array}{llll}
\text{RRND:} & \min_{(c,y,v^i)} & y & \\
& \text{s.t.} & \sum_{a\in A} h_a(c_a) \le B & \\
& & v^i \in V(d^i), & \forall i = 1, \ldots, n \\
& & t(v^i, c)^T(u - v^i) \ge 0, & \forall u \in V(d^i), i = 1, \ldots, n \\
& & t(v^i, c)^T v^i \le y, & \forall\, i = 1, \ldots, n \\
& & 0 \le c_a \le c_a^{\max}, & \forall a \in A
\end{array}
$$

The RRND problem stated above is simply the original RND problem with Q approximated by the discrete demand set $\hat{Q} = \{d^1, \ldots, d^n\}$.

Let $(\hat{c}, \hat{y})$ be a global optimal solution to RRND. (To simplify our presentation, we ignore the v^i component of the solution vector.) Then, $(\hat{c}, \hat{y})$ solves the original RND problem only if the user equilibrium, $v(d)$, associated with every $d \in Q$ has a total travel time no larger than $\hat{y}$, i.e., $t(v(d),\hat{c})^T v(d) \le \hat{y}$. To verify this inequality computationally, consider the following worst-case demand (WCD) problem with $\hat{c}$:

$$
\begin{array}{lll}
\text{WCD}(\hat{c}): & \max_d & t(v(d), \hat{c})^T v(d) \\
& \text{s.t.} & v(d) \in \Omega(d, \hat{c}) \\
& & d \in Q
\end{array}
$$

where $\Omega(d, \hat{c})$ is the set of user equilibrium distributions associated with d and $\hat{c}$, i.e., $\Omega(d, \hat{c}) = \{v \mid t(v, \hat{c})^T(u - v) \ge 0, \forall u \in V(d)\}$. Among the two constraints in the problem, the first ensures that $v(d)$ is a user equilibrium distribution and the other forces d to be in Q. For a given capacity expansion vector $\hat{c}$, the objective of WCD($\hat{c}$) is to find a demand vector in Q whose user equilibrium flow distribution yields the maximum total travel time. If $\hat{d}$, a global optimal solution to WCD($\hat{c}$), is such that $t(v(\hat{d}), \hat{c})^T v(\hat{d}) \le \hat{y}$, then $\hat{c}$ is an optimal capacity expansion vector. On the other hand, if $t(v(\hat{d}), \hat{c})^T v(\hat{d}) > \hat{y}$, then an improved solution may be obtained by solving the RRND problem with an expanded discrete demand set $\tilde{Q} \cup \{\hat{d}\}$.

Similar to the RND problem, WCD($\hat{c}$) is a difficult MPEC problem. Later, we also let $\Psi(\hat{c})$ denote the set of globally optimal solutions to the WCD($\hat{c}$) problems, i.e., $\hat{d} \in \Psi(\hat{c})$. Below, we state the approach outlined above as the Demand Generation (DG) Algorithm.

Demand Generation Algorithm

Step 0: Choose an initial demand vector $d^1 \in Q$. Set $n = 1$ and $Q^1 = \{d^1\}$.

Step 1: Solve the RRND problem with the discrete demand set Q^n and let (c^n, y^n) denote the resulting optimal solution.

Step 2: Solve WCD(c^n) and let $d^{(n+1)}$ denote the resulting optimal solution.

Step 3: If $t(v(d^{(n+1)}), c^n)^T v(d^{(n+1)}) \le y^n$, stop and c^n is an optimal capacity expansion vector. Otherwise, set $Q^{(n+1)} = Q^n \cup \{d^{(n+1)}\}$ and $n = n + 1$. Go to Step 1.

Assume that both (c^n, y^n) and $d^{(n+1)}$ globally solve the RRND and WCD(c^n) problems in Steps 1 and 2, respectively. Additionally, let (c^*, d^*) be a global optimal solution to the RND problem. Because the RRND problem is a relaxation of RND, $y^n \le y^*$ for all n. If the above algorithm stops at some finite iteration n, then $t(v(d^{(n+1)}), c^n)^T v(d^{(n+1)}) \le y^n$ implying that (c^n, y^n) is feasible to the RND problem. When combined with $y^n \le y^*$, the optimality of (c^*, y^*) implies that $y^n = y^*$, i.e., (c^n, y^n) is optimal to RND.

When the demand generation algorithm generates an infinite sequence, the theorem below gives conditions under which any of its subsequential limits is optimal to the RND problem. In particular, the theorem views the solution sets $\Omega(d, c)$ and $\Psi(c)$ of the user equilibrium and the WCD(c) problems as mappings and assumes that both are *closed* (see, e.g., Chapter 7 of Bazaraa et al, 1993). When each consists of a single element for all d and c, both $\Omega(d, c)$ and $\Psi(c)$ becomes functions and being closed reduces to being continuous.

Theorem: Assume that (c^n, y^n) and $d^{(n+1)}$ globally solve the RRND and WCD(c^n) problems in Steps 1 and 2, respectively, and the mappings $\Omega(d, c)$ and $\Psi(c)$ are closed. If the demand generation algorithm generates an infinite sequence $\{(c^n, y^n)\}_n$, then any of its subsequential limits globally solves the RND problem.

Proof: First, note that the demand vectors generated in Step 2 are distinct. If $d^{(n+1)} = d^j$ for some $j \le n$, then $t(v(d^{(n+1)}), c^n)^T v(d^{(n+1)}) = t(v(d^j), c^n)^T v(d^j) \le y^n$. Therefore, the stopping criterion in Step 3 is satisfied and the algorithm must stop. This contradicts the hypothesis that the algorithm generates an infinite sequence.

Because the RRND problem in iteration $(n + 1)$ has one more constraint that the one in iteration n, $y^n \le y^{(n+1)}$. As mentioned previously, $y^n \le y^*$ for all n. So, the sequence $\{y^n\}_n$ is bounded and non-decreasing and must therefore converge. Let $y^\infty = \lim_{n\to\infty} y^n$. Then, $y^\infty \le y^*$.

Observe that $\{c^n\}_n$ is a sequence in the compact set $\{c_a: 0 \le c_a \le c_a^{\max}\}$. Thus, there must exist a convergent subsequence, i.e., $c^\infty = \lim_{n\in\Phi} c^n$, for some $\Phi \subseteq \{1, \ldots, \infty\}$. For each n, the following hold:

$$t(v(d^i), c^n)^T v(d^i) \le y^n < t(v(d^{(n+1)}), c^n)^T v(d^{(n+1)}), \quad \forall\, i = 1, \ldots, n.$$

(Previously, $v(d^i)$ is written as v^i in the DG algorithm for brevity. However, the notation $v(d^i)$ is more convenient here.) Because Q is compact, there exists $d^\infty = \lim_{n\in\hat{\Phi}} d^{(n+1)}$, where $\hat{\Phi} \subseteq \Phi$. Taking the limit with respect to $\hat{\Phi}$, the above yields

$$t(v(d^i), c^n)^T v(d^i) \le y^\infty \le t(v(d^\infty), c^\infty)^T v(d^\infty), \qquad \forall\, i = 1, \ldots, \infty.$$

The closeness property of $\Omega(d, c)$ and $\Psi(c)$ implies that d^∞ solves the WCD(c^∞) problem and $v(d^\infty)$ is a user equilibrium with respect to c^∞ and d^∞. Taking the limit of the first term with respect to set $\{(i+1, n): i \in \hat{\Phi}, n \in \hat{\Phi}\}$, the above inequality leads to

$$t(v(d^\infty), c^\infty)^T v(d^\infty) \le y^\infty \le t(v(d^\infty), c^\infty)^T v(d^\infty)$$

or $t(v(d^\infty), c^\infty)^T v(d^\infty) = y^\infty \le y^*$. As before, (c^∞, y^∞) is optimal to the RND problem because the worse-case demand d^∞ satisfies the stopping criterion in Step 3, thereby implying that (c^∞, y^∞) is feasible to the RND problem. The optimality of (c^*, y^*) further implies that $y^* = y^\infty$. □

The above convergence proof relies on a number of assumptions that are either difficult to verify or satisfy in practice. Our main purpose of the above discussion is to demonstrate that the DG algorithm is not totally heuristic. In particular, it is difficult both theoretically and computationally to ensure both (c^n, y^n) and $d^{(n+1)}$ in Steps 1 and 2 globally solve the RRND and WCD(c^n). (See, e.g., Vavasis, 1995 and references cited therein.) In addition, $\Psi(c)$ is a solution set to an MPEC, a difficult class of problems to solve and explore theoretically. In fact, theories concerning the optimality conditions of MPECs are still in early stages (see, e.g., Luo et al., 1996). On the other hand, the numerically results in the next section demonstrate empirically that the DG algorithm is effective at providing a good solution to the CND problem despite not being able to verify the assumptions stated above.

IMPLEMENTATION AND EXAMPLES

To investigate its effectiveness, we implemented the DG algorithm and solved two problems from the literature. We used a 300 MHz IBM SP2 computer with 512 MB of RAM, an algebraic modeling system called GAMS (Brooke, et al., 1992) and CONOPT Version 3.14 (Drud, 1992) to solve the RRND and WCD(c) problems. In Step 3, it is more practical to terminate the algorithm when $(y^n - t(v(d^{(n+1)}), c^n)^T v(d^{(n+1)})) \le \varepsilon$ and $\varepsilon = 5 \times 10^{-3}$ for the results reported below. Additional details concerning the solutions to problems in Steps 1 and

2 of the algorithm are presented in the following two subsections and then results from our experiments are summarized.

Solving the Relaxed Robust Network Design Problem

The third set of constraints in the RRND problem consists of n variational inequalities, one for each $i = 1, \ldots, n$, of the form:

$$t(v^i, c)^T(u - v^i) \geq 0, \ \forall u \in V(d^i). \tag{1}$$

For each i, this variational inequality is equivalent to requiring v^i to satisfy the following set of constraints:

$$t(v^i, c) \geq A^T \rho^{i,w}, \ \forall\, w \in W,$$
$$t(v^i, c)^T v^i \leq \sum\nolimits_{w \in W} d^i_w E^T_w \rho^{i,w} + \mu.$$

For each demand vector i and O-D pair w, $\rho^{i,w}$ is a Lagrangian multiplier associated with the flow balance constraints: $Ax^w = E^w d^i_w$. When μ is zero, the above constraints are KKT conditions associated with the variational inequality in (1) (see, e.g., Luo et al, 1996, and Lawphongpanich and Hearn, 2004). When μ is a small positive constant, the constraints allow v^i to satisfy the variational inequality and the KKT conditions approximately (see, Bai et al., 2006).

In our implementation, $\mu = (10^{-4}) \cdot t(v(d^n), c^{(n-1)})^T v(d^n)$ in iteration n. Because the CONOPT solver can only provide local optimal solutions, we solved the RRND problem ten times, each time with a different randomly generated initial solution. The best local optimal solution among these ten is taken to be (c^n, y^n) in Step 1.

Solving the Worst-Case Demand Problem

Instead of solving the WCD(c^n) problem in Step 2 optimally, our strategy is to find $d^{(n+1)}$ that yields $v(d^{(n+1)})$, a user equilibrium distribution associated with c^n and $d^{(n+1)}$, satisfying $t(v(d^{(n+1)}), c^n)^T v(d^{(n+1)}) > y^n$. Our approach is to randomly select ten demand vectors along the boundary of Q. For each random demand vector, say $\tilde{d}$, we solved the user equilibrium problem associated with c^n and $\tilde{d}$ using MINOS Version 5.0 (Murtagh and Saunder, 1983). Let $v(\tilde{d})$ be the user equilibrium distribution associated with c^n and $\tilde{d}$. If the associate system delay, $t(v(\tilde{d}), c^n)^T v(\tilde{d})$, is greater than y^n by at least 0.5%, set $d^{(n+1)} = \tilde{d}$ and go to Step 3. Otherwise, we used CONOPT to solve the WCD(c^n) problem with $\tilde{d}$ as an initial solution. If CONOPT gives a demand vector whose user equilibrium travel time exceeds y^n, ignore the remaining random demand vectors, set $d^{(n+1)}$ equal to the demand vector from CONOPT, and go to Step 3. If none of the ten random demand vectors produces a new demand vector with system delay larger y^n, then we assume that the stopping criterion in Step 3 is met and terminate the demand generation algorithm.

Numerical Results

In our experiment, we consider two networks from the literature, nine node (see Hearn and Ramana, 1998) and Sioux Falls (see, LeBlanc et al., 1975).

Nine-node problem: Figure 1 displays the underlying network. For each link, the travel cost function is of the form

$$t_a(v_a, c_a) = t_a^0 \cdot \left(1 + 0.15 \cdot \left(\frac{v_a}{c_a^0 + c_a}\right)^4\right) \tag{2}$$

where the values for (t_a^0, c_a^0) are as listed in Hearn and Ramana (1998). There are four O-D pairs for the nine-node problem and they include 1 – 3, 1 – 4, 2 – 3, and 2 – 4. We assume that all O-D pairs have uncertain demands and the ellipsoid uncertainty set is of the form:

$$Q = \{d \in R^{|W|} \mid \sum_{w=1}^{|W|}\left(\frac{d_w - d_w^0}{s_w}\right)^2 \leq 1\} \tag{3}$$

where $d_{1-3}^0 = 10$, $d_{1-4}^0 = 20$, $d_{2-3}^0 = 30$, $d_{2-4}^0 = 40$ and $s_w = 0.5 \times d_w^0$. Finally, the expansion cost function is $h_a(c_a) = c_a$ for all link a.

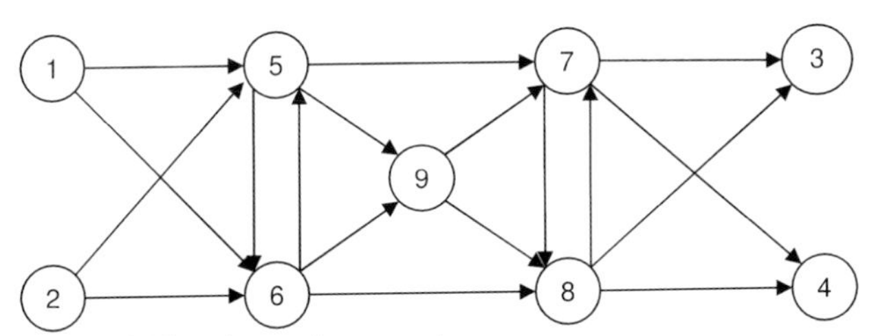

Figure 1: The nine-node network

Results for the nine-node problem are presented below. Figure 2 displays the amount of CPU time required to approximately solve the RND problem in the manner described in Subsections 4.1 and 4.2 with the total budget, *B*, varied from 10 to 100. The figure indicates that problems with moderate budgets require more CPU times. When the budget is small, possibilities are limited, thereby making the expansion decision simpler. With a budget sufficient large, the optimal expansion decision is also simple and involves expanding the capacity of each utilized link until the travel time reaches its free-flow level. The expansion decision becomes more difficult with a moderate budget because more competing alternatives must be considered in order to find the one that minimizes the maximum travel time over the uncertainty set.

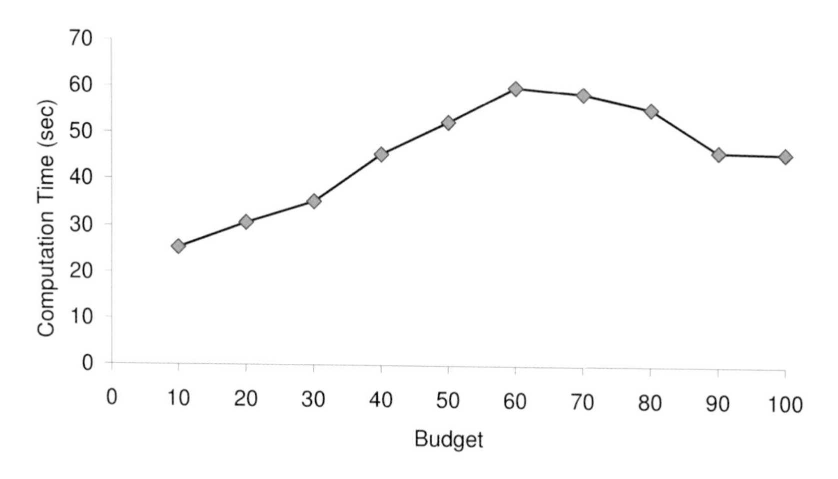

Figure 2: Computation times for the nine-node network

Table 1 and Figure 3 compare the (approximate) solutions to the RND problem (or the robust plans) against the solutions from solving continuous network design problem (or the nominal plans) assuming that the travel demand for each O-D pair is d_w^0. As an example, Table 1 shows the two expansion plans when the available budget, B, is 50. Figure 3 plots the differences between the two plans for B ranging from 10 to 100, where the difference between the robust, c^R, and nominal, c^N, is defined as $\left\|c^R - c^N\right\|_2 / \left\|c^R\right\|_2$. On average, the difference between robust and nominal expansion plans is approximately 31% of $\|c^R\|_2$. In addition, the difference tends to decrease as the available budget increases. In particular, the differences when B = 80, 90, and 100 are relatively small. When B is sufficiently large, $c^R = c^N$ because, as mentioned previously, it is optimal to expand the capacity of every utilized link until its travel time equals to its free flow level.

Table 1: Nominal and robust capacity expansion plans with B = 50

Link	1-6	2-5	5-7	6-8	7-3	Total
Nominal	0	7.7	36.2	0	6.1	50
Robust	2.1	12.6	15.2	8.7	11.5	50

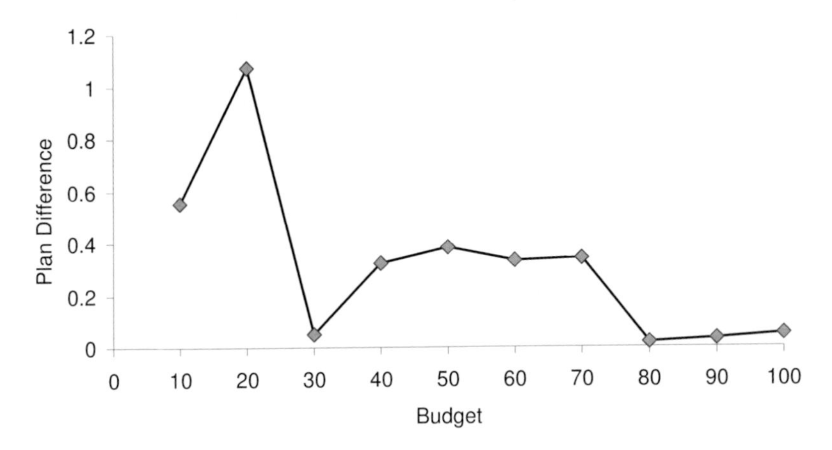

Figure 3: Differences between c^R and c^N for the nine-node problem

To compare the operating behaviors of the two capacity expansion plans, robust and nominal, we randomly generated 300 travel demand vectors in which the demand of each O-D pair is uniformly distributed between $0.5d_w^0$ and $1.5d_w^0$. For each random demand vector $\tilde{d}$, two user equilibrium distributions, one ($v^R(\tilde{d})$) associated with capacity expansion c^R and the other ($v^N(\tilde{d})$) with c^N, are computed. In total, there are 600 user equilibrium distributions, 300 are of the form $v^R(\tilde{d})$ and the remaining are of the form $v^N(\tilde{d})$. The average and standard deviation of the total travel times associated with the user equilibrium distributions in these two sets are computed and compared in Figure 4.

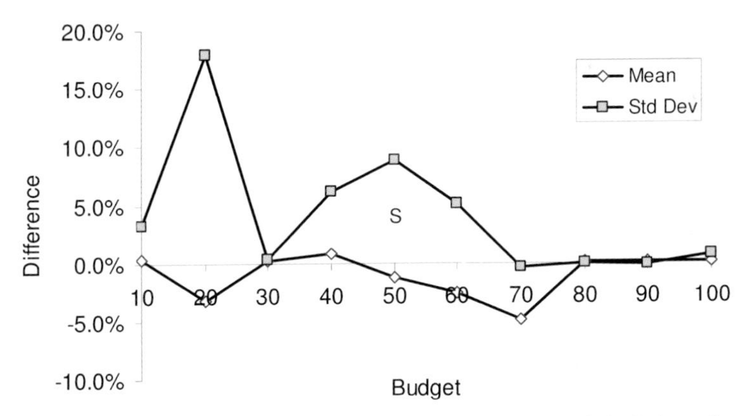

Figure 4: Percent differences between the mean and standard deviation of total travel times associated with nominal and robust expansion plans for the nine-node problem

There are two graphs in Figure 4, one for the relative percent difference in the mean system delays, or (mean[c^N] – mean[c^R])/mean[c^R], and the other for the difference in the standard deviations, or (std[c^N] – std[c^R])/std[c^R]. Being more conservative, c^R yields slightly higher mean system delay for the nine-node problem on average. However, the standard deviation associated with c^R is generally smaller than those associated with c^N. In other words, the total travel time associated with c^R is less volatile.

The optimal objective value of the RRND problem, denoted as $y^*(c^R)$, provides an estimate of the system delay over the uncertainty set Q. As stated previously, the capacity expansion plan c^N is computed based on the nominal demand vector d^0. Let $y^*(c^N)$ be the total travel time associated with the user equilibrium distribution with capacity expansion c^N and demand d^0. In one interpretation, $y^*(c^N)$ gives an estimate of the system delay associated with c^N over the uncertainty set Q. Among all 300 random travel demand vectors ($\tilde{d}^i$, $i = 1, \ldots, 300$), we compare total travel times associated $v^R(\tilde{d}^i)$ and $v^N(\tilde{d}^i)$ against the system delay estimates $y^*(c^R)$ and $y^*(c^N)$, respectively. Table 2 lists the percentages for which $t(v^R(\tilde{d}^i), c^n)^T v^R(\tilde{d}^i) \leq y^*(c^R)$ and $t(v^N(\tilde{d}^i), c^n)^T v^N(\tilde{d}^i) \leq y^*(c^N)$.

Table 2: Reliability of total travel time estimates associated with robust and nominal capacity expansion plans.

Budget	$y^*(c^N)$	$t(v^R(\tilde{d}^i),c^n)^T v^R(\tilde{d}^i) \leq y^*(c^N)$	$y^*(c^R)$	$t(v^N(\tilde{d}^i),c^n)^T v^N(\tilde{d}^i) \leq y^*(c^R)$
10	2273	43.7%	3450	92.3%
20	2119	45.3%	3227	92.0%
30	1971	43.0%	3058	92.7%
40	1898	43.3%	2801	91.7%
50	1729	42.3%	2651	91.7%
60	1612	42.3%	2543	91.7%
70	1529	42.3%	2465	91.3%
80	1469	42.7%	2270	91.0%
90	1426	42.3%	2138	91.3%
100	1394	43.0%	2089	92.7%

On average, approximately 43% or 129 random demands out of the 300 yields total travel times no larger than the nominal estimates or $y^*(c^N)$. In other words, the nominal system delay estimate, $y^*(c^N)$, is 43% reliable. On the other hand, the average reliability of the robust system delay estimate, $y^*(c^R)$, is nearly 92%. Because c^N is computed from a single demand vector, d^0, the above comparison may be unfair. However, the DG algorithm generated no more than seven demand vectors for the nine-node problem, i.e., c^R is computed from at most seven demand vectors when the uncertainty set is an ellipsoid. Thus, computing an expansion plan with a highly reliable system delay estimate does not require a large number of demand vectors from the uncertainty set.

Sioux Falls Problem: Our main objective for solving this problem is to demonstrate the potential of the demand generation algorithm in solving realistic RND problems. The Sioux Falls network (see, LeBlanc et al., 1975) contains 76 links, 24 nodes, and 528 O-D pairs. The original parameters in LeBlanc et al (1975) are transformed into (t_a^0, c_a^0) by associating the original parameters with those in equation (2). We also set d^0, the nominal travel demand vector, to be original demands divided 13 and assume that O-D pairs with nominal demands greater one (thousand trips) have uncertain travel demands. Based on this criterion, there are 53 O-D pairs with uncertain demands. As before, the uncertainty set of these O-D demands is an ellipsoid (see equation 3) with $s_w = d_w^0$ and $h_a(c_a) = c_a$ for all $a \in L$.

In general, the demand generation algorithm required two or three iterations or between 869 and 2917 CPU seconds (see Figure 5) to solve for the Sioux Fall problem with the available budget ranging from 50 to 200.

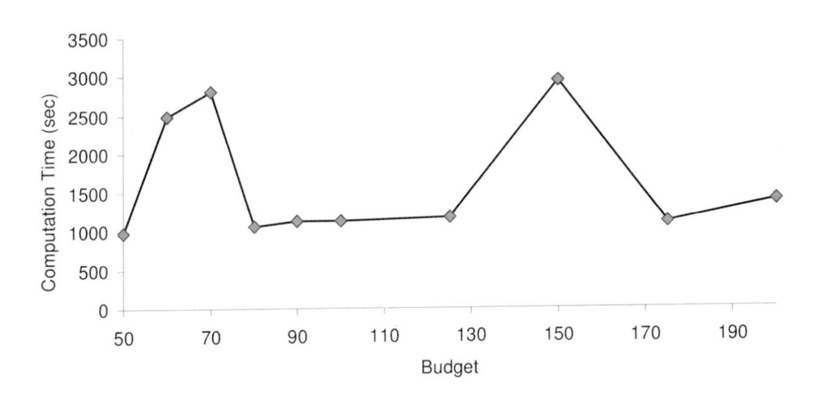

Figure 5: Computational times for the Sioux Fall network

Figures 6 and 7 illustrate that the differences between the robust and nominal plans. The first one, Figure 6, displays the differences between the two expansion vectors, c^R and c^N, i.e., $\|c^R - c^N\|_2 / \|c^R\|_2$. On the other hand, Figure 7 shows the difference in the operating behaviors based 100 random demand vectors generated using the Normal distribution with mean d_w^0 and standard deviation $0.5 \cdot d_w^0$. (Recall that only 53 O-D pairs have uncertain demands.) Similar to the nine-node problem, the total travel times associated with the robust expansion vector is slightly less volatile than the ones associated with the nominal counterpart.

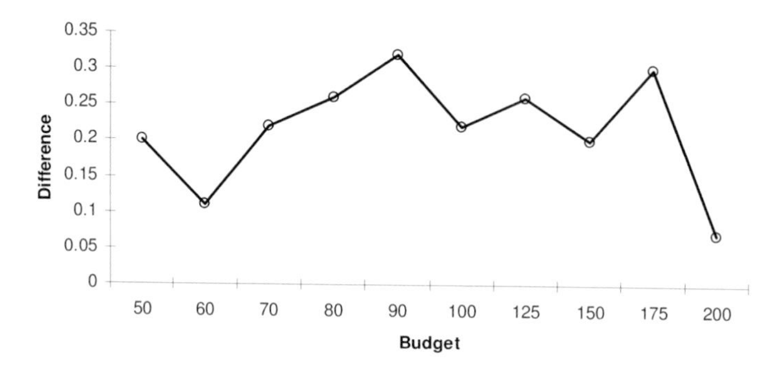

Figure 6: Difference between c^R and c^N for Sioux Falls

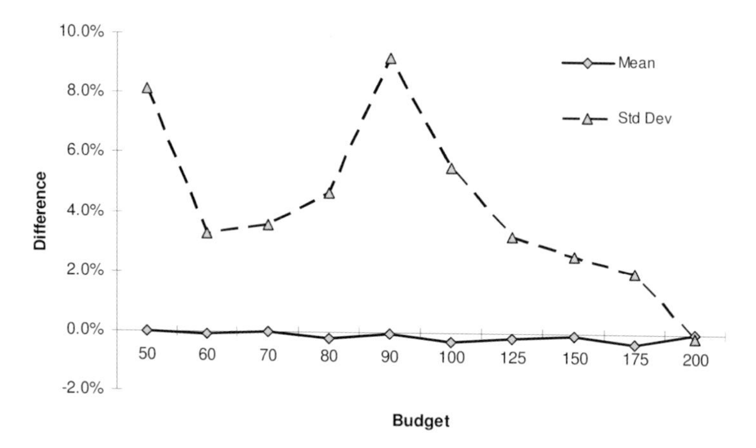

Figure 7: Percent differences between the mean and standard deviation of total travel times associated with nominal and robust expansion plans for Sioux Falls

CONCLUSION

In this paper, we consider a robust optimization approach and formulate a continuous network design problem with demand uncertainty as a mathematical program with equilibrium constraints. The problem assumes that flows on the underlying network are in user equilibrium and travel demands belong to a convex and compact uncertainty set. We propose an algorithm, called the demand generation algorithm, to solve the problem. During each iteration, the algorithm solves two problems. One is a relaxed continuous network design problem based on a discrete set of demand vectors and the other is a problem that generates a worst-case demand. In theory, the algorithm converges under a set of conditions. However, numerical results from the nine-node and Sioux Falls networks empirically demonstrate that the algorithm is effective in practice and has the potential to solve realistic network design problems. Moreover, the results from these two networks indicate that capacity expansion plans from our robust formulation are less volatile and produce reliable system delays estimates when compared to the plan based only on nominal demands. Furthermore, the framework for our model is general and can be extended to accommodate other forms of equilibria, elastic demands and other types of uncertainty such as incident-induced travel time uncertainty.

ACKNOWLEDGEMENTS

The authors thank the three anonymous referees for their constructive comments and numerous suggestions that greatly improved the presentation of this paper. The research of S. Lawphongpanich is partially supported by grants from the National Science Foundation (DMI-0300316) and Volvo Research and Educational Foundations (SP-2004-5),

REFERENCES

Bai, L., Hearn, D.W., and Lawphongpanich, S. (2006) Relaxed Toll Sets for Congestion Pricing Problems. In *Mathematical and Computational Models for Congestion Charging*, S. Lawphongpanich, D.W. Hearn, and M.J. Smith (Eds), Springer, New York, 23 – 44.

Barnhart, C. (2000) *Planning and Control of Transportation Systems: Stochastic Optimization for Robust Planning in Transportation.* A Final Report to the New England (Region One) UTC, Cambridge, Massachusetts.

Bazaraa, M.S., Sherali, H.D. and Shetty, C.M. (1993) *Nonlinear Programming: Theory and Algorithms.* Second Edition, John Wiley & Sons, New York, New York.

Ben-Tal, A. and Nemirovski, A. (1999) Robust Solutions to Uncertain Linear Programs. *Operations Research Letters*, 25, 1-13.

Ben-Tal, A. and Nemirovski, A. (2002) Robust Optimization – Methodology and Applications. *Mathematical Programming Series B*, 92, 453-380.

Bertsimas, D. and Sim, M. (2003) Robust Discrete Optimization and Network Flows. *Mathematical Programming Series B*, 98, 49 – 71.

Brooke, A., Kendirck, D. and Meeraus, A. (1992) *GAMS: A User's Guide*, The Scientific Press, South San Francisco, California.

Chen, A., Subprasom, K. and Ji, Z. (2003) Mean-Variance Model for the Build-Operate-Transfer Scheme under Demand Uncertainty. *Transportation Research Record*, 1857, National Research Council, Washington, D.C., 93-101.

Chen, Y., Ordonez, F. and Palmer, K. (2006) Confidence Intervals for OD Demand Estimation. USC-ISE Working paper #2006-01, University of Southern California.

Chiou, S-W (2005) Bilevel Programming for the Continuous Transport Network Design Problem. *Transportation Research*, Vol.39B, 361-383.

Drud, A. (1992) A Large-Scale GRG Code. *ORSA Journal on Computing*, 6, 207 – 216.

El Ghaoui, L. (2003) A Tutorial on Robust Optimization, Presentation at the Institute of Mathematics and Its Applications, March 11, 2003. Available at http://robotics.eecs.berkeley.edu/~elghaoui/

Fisk, C. (1979) More Paradoxes in the Equilibrium Assignment Problem. *Transportation Research*, Vol.13B, 305-309.

Hearn, D. W. and Ramana, M. V. (1998) Solving Congestion Toll Pricing Models. *Equilibrium and Advanced Transportation Modeling*, P. Marcotte and S. Nguyen (Eds.), Kluwer Academic Publishers, Boston, 109-124.

Kelly, J.E. (1960) The Cutting Plane Method for Solving Convex Programs. *SIAM Journal of Industrial and Applied Mathematics*, 8, 703 – 712.

Lawphongpanich, S. and Hearn, D. W. (2004) An MPEC Approach to Second-Best Toll Pricing. *Mathematical Programming Series B*, 7, 33-55.

LeBlanc, L. J., Morlok, E. K. and Pierskalla, W. P. (1975) An Efficient Approach to Solving the Road Network Equilibrium Traffic Assignment Problem. *Transportation Research*, 9, 309-318.

Luo, Z.-Q., Pang, J.-S., and Ralph, D. (1996) *Mathematical Programs with Equilibrium Constraints*, Cambridge University Press, Cambridge, England.

Marcotte, P. (1983) Network Optimization with Continuous Control Parameters. *Transportation Science*, 17 181 – 197.

Murtagh, B. A. and Saunder, M.A. (1983) *MINOS 5.0 User's Guide*. Report SOL 83-20R, Department of Operations Research, Stanford University, Palo Alto, California.

Patriksson, M. and Wynter, L. (1999) Stochastic Mathematical Programs with Equilibrium Constraints. *Operations Research Letters*, Vol.25, No.4, 159-167.

Soyster, A.L. (1973) Convex Programming with Set-Inclusive Constraints and Applications to Inexact Linear Programming. *Operations Research*, 1154 – 1157.

Vavasis, S.A. (1995) Complexity Issues in Global Optimization: A Survey. *Handbook of Global Optimization*, R. Horst and P.M. Pardalos (Eds.), Kluwer Academic Publishers, Boston, 27 – 41.

Waller, S.T. and Ziliaskopoulos, A.K. (2001) Stochastic Dynamic Network Design Problem. *Transportation Research Record*, 1771, 106-113.

Yang, H. and Bell, M.G.H. (1998) Models and Algorithms for Road Network Design: A Review and Some New Developments. *Transport Review*, 18, No.3, 257-278.

Yin, Y., Madanat, S.M. and Lu, X.Y. (2005) Robust Improvement Schemes for Road Networks under Demand Uncertainty. *The 84th Annual Meeting of the Transportation Research Board, Compendium of Papers CD-ROM, No. 05-1724, January 9 –13, 2005.*

Transportation and Traffic Theory 2007
Edited by R.E. Allsop, M.G.H. Bell and B.G. Heydecker

6

GENERALISATION OF THE RISK-AVERSE TRAFFIC ASSIGNMENT

W.Y. Szeto, Liam O'Brien, Margaret O'Mahony, Centre for Transport Research, Department of Civil, Structural and Environmental Engineering, Trinity College Dublin, Dublin 2, Ireland.

SUMMARY

Traditionally, the risk-averse traffic assignment is described by a game played between network users who seek minimum cost routes and demons that seek to impose maximum costs on the network users by damaging links in the network. This problem assumes the presence of one and only one active OD specific demon in each OD pair, and furthermore assumes the capacity reduction to be 50% if the link is selected for damage by one or more OD specific demons. In this paper, we relax these two assumptions and propose a multiple network demon formulation in which each demon is free to select any link to damage. Numerical studies are carried out to examine the effects of relaxing the two assumptions on expected network cost, give an insight into the network demon behaviour in selecting links to damage, demonstrate the existence of multiple solutions to the proposed game, and compare the link selection behaviour of the OD specific and network demons and their impacts on expected network cost. Overall the results indicate the importance of the assumptions used to expected network cost and reliability measures, and provide some further insights into the nature of the route choice game.

INTRODUCTION

Transport networks cannot often operate at their full capacities due to many traffic disruptions. These disruptions can range from irregular and random incidents, like adverse weather, traffic accidents, breakdowns, signal failures and road-works to disasters like landslides and earthquakes. These unpredictable disruptions not only reduce link capacities and increase travel times of transport network users, but also result in travel time uncertainty. Faced with uncertainty about the travel times in their route choice, the transport network users are required to make a trade-off between the travel cost (including travel time, early or late arrival penalty, and so on) and its uncertainty. This behaviour is known as risk-taking behaviour (Yin et al., 2004). Many studies (e.g. Mirchandani and Soroush, 1987; Uchida and Iida, 1993; Boyce et al., 1998; Chen and Recker, 2000; Yin and Ieda, 2001; Bell and Cassir, 2002; Yin et al., 2002; Lo and Tung, 2003; Chan and Lam, 2005; Sumalee et al., 2005) have been performed to model this risk-taking behaviour of route choices. One approach is the game theoretic approach (e.g. Bell, 2000; Bell and Cassir, 2002; Szeto et al., 2006).

The game theoretic approach was first proposed in Bell (2000) as an approach to assess the performance reliability of transport networks. This approach uses the network cost resulting from a two-player zero-sum non-cooperative game as a measure of the performance reliability. Bell and Cassir (2002) extended this game concept to a multiplayer non-cooperative game to model the risk-averse behaviour in route choice assignment with fixed demand. Szeto et al. (2006) extended the Bell and Cassir (2002) formulation to incorporate the elastic behaviour of demand.

The game theoretic approach enjoys at least three advantages over the more conventional approaches according to Bell (2000) and Bell and Cassir (2002). First, the game theoretic approach can determine links where the network users are most vulnerable to link failure under the assumption that the users are highly pessimistic or naturally very negative about the state of the road network. Second, the total expected network cost of the game gives a useful measure of network reliability. Different designs may be compared on the basis of this measure. Third, statistical distributions for link performance (such as delay, travel time or capacity) are not required, unlike the more conventional approaches. This information is very often either absent or not accurate enough to be used.

The main idea behind the game theoretic approach is based on the notion of a fictitious game played between on the one hand users who seek minimum cost routes, and on the other hand, an evil entity or demon that seeks to maximise the total expected network cost to the users by damaging links in the network. The traffic assignment problem when formulated based on this approach, known as risk-averse traffic assignment, consists of two sub-problems: the user problem and the demon problem. The user problem describes the non-cooperative behaviour of network users, whereas the demon problem describes the evil behaviour in the sense of trying to cause maximum damage to the users. However, the formulation developed based on this approach only allows exactly one demon per OD pair in which each OD-specific demon is not free to damage links outside of its own specific OD pair. This one demon per OD pair assumption is not general enough. For example, it does not allow two links to be damaged on one OD pair but no link to be damaged on other OD pairs. It also restricts the number of demons or links to be damaged to be equal to the number of OD pairs and does not allow the number of links to be damaged in a network to be greater or smaller than the number of OD pairs. Furthermore, the capacity reduction when link damage occurs is typically 50%. In other words only two operational states are accounted for – congested and un-congested. This clearly does not model the wide range of capacity levels.

In this paper we propose a more general game theoretic formulation by relaxing the above two assumptions. We extend the existing game theoretic formulation to allow demons (called network-specific demons or simply network demons) to be free to select any links to damage and to impose no restriction on the number of demons or number of links to be damaged. We assess the impact of this extension on the total expected network cost. Moreover we challenge the effect of the assumption of 50% capacity reduction, especially when there are multiple demons involved, and when more than one demon selects a particular link to damage. In addition, we examine the effect of using or relaxing this assumption with particular reference to the implications for network (or performance) reliability. We also conduct a brief study to provide a better insight into the demon behaviour in selecting links to damage, illustrate the existence of multiple solutions, and compare the results obtained from OD specific and network demons.

The remainder of this paper is organised as follows: section 2 proposes a multiple network demon formulation; section 3 details the numerical studies while section 4 provides conclusions and suggests some important future research directions.

FORMULATION

A general transportation network with multiple links, routes, and Origin-Destination (OD) pairs is considered. There are N^{rs} homogenous players between each OD pair rs, each whom it is assumed seeks minimum cost routes among all other alternatives (hereafter we refer to them as the 'users'). There are also M demons with the freedom to roam the network without restriction and with the capability to select any link to damage, including links already selected for damage by other demon(s). It is assumed that the network-specific demons are intent on causing maximum delays (and hence increased costs) to the users by damaging link(s) in the network. By so doing the demons maximise their expected pay-off which is the total network cost faced by the users. Moreover, we assume that no user knows which link(s) the demon(s) will choose to damage, and none of the demons know in advance which link a user may decide to use. In general we assume that each link is assumed to have two costs depending on whether it is damaged or not. However for the purposes of some of our studies we relax this assumption so that the link costs increase with an increasing number of demons that select a particular link to damage. This is due to the assumption that the link capacity reduces linearly or non-linearly with respect to the number of demons choosing that particular link. Furthermore, both N^{rs} and M are assumed to be fixed. With this setting, we can formulate the proposed risk-averse user equilibrium traffic assignment problem with multiple network demons by applying the results in Nash (1951), Bell and Cassir (2002), and Szeto et al. (2006).

Similar to Bell and Cassir (2002) and Szeto et al. (2006), the proposed formulation comprises two main sub-problems: the user problem and the demon problem. The user problem can be viewed as a non-cooperative game in which each homogenous player tries to select the route with minimum expected trip cost. The user problem can be approximated to deterministic traffic assignment when the number of homogenous users is large (Bell and Cassir, 2002). The first order condition of deterministic traffic assignment can be expressed as the following non-linear complementarity conditions:

$$h_j^{rs}\left[\sum_k g_{jk}^{rs}(\mathbf{h})q_k - \min_{\mathrm{d}}\left(\sum_k g_{dk}^{rs}(\mathbf{h})q_k\right)\right]=0, \tag{1}$$

$$\sum_k g_{jk}^{rs}(\mathbf{h})q_k - \min_{\mathrm{d}}\left(\sum_k g_{dk}^{rs}(\mathbf{h})q_k\right)\geq 0, \tag{2}$$

$$h_j^{rs}\geq 0, \tag{3}$$

where h_j^{rs} is the flow on route j between OD pair rs; $\mathbf{h}$ is the route flow vector; $g_{jk}^{rs}(\mathbf{h})$ represents the cost of route j between OD pair rs in scenario k based on the flow vector $\mathbf{h}$; d is the minimum expected travel cost route between OD pair rs; q_k is the probability of scenario k occurring.

According to (1), if route j connecting OD pair rs carries flow ($h_j^{rs}>0$), the corresponding expected route travel cost $\sum_k g_{jk}^{rs}(\mathbf{h})q_k$ must be equal to the minimum expected travel cost $\min_{\mathrm{d}}\left(\sum_k g_{dk}^{rs}(\mathbf{h})q_k\right)$ between OD pair rs, as the term in the square brackets in (1) must equal zero. If route j carries no flow ($h_j^{rs}=0$), the corresponding expected route travel cost must be greater than or equal to the minimum expected travel cost based on (2). Condition (3) is the non-negativity condition of route flows.

The route flow h_j^{rs} in (1)-(3) must satisfy flow conservation, expressed as:

$$N^{rs}=\sum_j \delta_j^{rs}h_j^{rs}, \tag{4}$$

where N^{rs} is the travel demand of OD pair rs; δ_j^{rs} is the route-OD incidence indicator. $\delta_j^{rs}=1$ if j connects OD pair rs, and $\delta_j^{rs}=0$ otherwise.

The cost of route j between OD pair rs in scenario k, $g_{jk}^{rs}(\mathbf{h})$, in (1) and (2) is the total cost on every link that is on this route:

$$g_{jk}^{rs}(\mathbf{h}) = \sum_{a} \delta_{ja} t_{ak}(v_a), \tag{5}$$

where δ_{ja} is 1 if link a is on route j, and 0 otherwise; $t_{ak}(v_a)$ denotes the flow-dependent cost on link a in scenario k and is defined by the Bureau of Public Roads (BPR) function in (6) below:

$$t_{ak}(v_a) = t_a^0 \left(1 + 0.15\left(\frac{v_a}{c_{ak}}\right)^4\right), \tag{6}$$

where t_a^0 is the free-flow travel time of link a; c_{ak} is the capacity of link a in scenario k; v_a is the link flow given in (7) below:

$$v_a = \sum_{rs} \sum_{j} \delta_{ja} h_j^{rs}. \tag{7}$$

The above equation states that the link flow is simply the sum of the route flows using that link.

The scenario probability in (1)-(2), q_k, depends on the link selection probability of all demons. Let $k = (l_1, \ldots, l_m, \ldots, l_M)$, l_m be the link selected for damage by demon m, $p_{l_m}^m$ be the probability of demon m selecting l_m to damage, and M be the number of network-specific demons. The scenario probability q_k is defined by:

$$q_k = \prod_{m=1}^{M} p_{l_m}^m. \tag{8}$$

The demon problem describes the objectives of the demons. All demons seek the mixed strategy to maximise their individual expected pay-offs or the total expected network cost to the users. The problem for multiple network demons can be written as the following nonlinear complementarity conditions:

$$p_{l_m}^m \left\{ \begin{aligned} &\max_{w_m} \sum_{l_1=1}^{n} \cdots \sum_{l_q=1, q\neq m}^{n} \cdots \sum_{l_M=1}^{n} \left(\prod_{i=1, i\neq m}^{M} p_{l_i}^i \right) \left[\sum_j g_{j,l_1,\ldots,w_m,\ldots,l_M}^{rs}(\mathbf{h}) h_j^{rs} \right] \\ &- \sum_{l_1=1}^{n} \cdots \sum_{l_q=1, q\neq m}^{n} \cdots \sum_{l_M=1}^{n} \left(\prod_{i=1, i\neq m}^{M} p_{l_i}^i \right) \left[\sum_j g_{j,l_1,\ldots,l_m,\ldots,l_M}^{rs}(\mathbf{h}) h_j^{rs} \right] \end{aligned} \right\} = 0, m = 1, \ldots, M, l_m = 1, \ldots, n, \tag{9}$$

$$\left\{\begin{aligned}&\max_{w_m}\sum_{l_1=1}^{n}\cdots\sum_{l_q=1,q\neq m}^{n}\cdots\sum_{l_M=1}^{n}\left(\prod_{i=1,i\neq m}^{M}p_{l_i}^{i}\right)\left[\sum_{j}g_{j,l_1,\ldots,w_m,\ldots,l_M}^{rs}(\mathbf{h})h_j^{rs}\right]\\&-\sum_{l_1=1}^{n}\cdots\sum_{l_q=1,q\neq m}^{n}\cdots\sum_{l_M=1}^{n}\left(\prod_{i=1,i\neq m}^{M}p_{l_i}^{i}\right)\left[\sum_{j}g_{j,l_1,\ldots,l_m,\ldots,l_M}^{rs}(\mathbf{h})h_j^{rs}\right]\end{aligned}\right\}\geq 0, m=1,\ldots,M, l_m=1,\ldots n, \tag{10}$$

$$p_{l_m}^{m}\geq 0, \tag{11}$$

where n is the number of links in the network; $g_{j,l_1,\ldots,l_m,\ldots,l_M}^{rs}(\mathbf{h})$ is the cost of route j between OD pair rs in scenario $k=(l_1,\ldots,l_m,\ldots,l_M)$; l_m and M follow earlier definitions; w_m is the link selected for damage by demon m that gives the demon maximum expected payoff; i and q are indices for demons. $p_{l_m}^{m}$ is the probability of demon m selecting l_m to damage, must satisfy:

$$\sum_{l_m}p_{l_m}^{m}-1=0, \tag{12}$$

and be non-negative by definition as shown in (11).

Conditions (9)-(10) represent the necessary and sufficient conditions of the Nash equilibrium of a non-cooperative, mixed strategy game for the demons (see Nash 1951). The second square bracket term in (9) is the total network cost in scenario $(l_1,\ldots,l_m,\ldots,l_M)$. The second term is the total expected network cost when demon m selects link l_m to damage (or the expected payoff to demon m when it selects link l_m to damage) and considers the link selection strategies of all other demons. The first term is the maximum expected payoff to demon m or the maximum total expected network cost. When demon m selects a particular link to damage with some probability ($p_{l_m}^{m}>0$) then the term inside the braces must be zero and the maximum expected pay-off to that demon (the first term inside the braces of (9)) must be equal to the total expected network cost (the second term inside the braces of (9)) when the demon selects that link to damage. If a particular demon does not choose a particular link l_m to damage ($p_{l_m}^{m}=0$), then the maximum expected pay-off to the demon must be greater than or equal to the total expected network cost when the demon selects link l_m to damage according to (10).

The risk-averse traffic assignment with multiple network-specific demons is to find h_j^{rs} and $p_{l_m}^{m}$ such that (1)-(12) are satisfied. This problem can be re-expressed as a Non-linear Complementarity Problem (NCP). To simplify our NCP, we introduce two notations. We let the minimum expected OD travel costs faced by users (or the second term in (2)) be ϕ^{rs}:

$$\phi^{rs} = \min_{d} \left(\sum_{k} g_{dk}^{rs}(\mathbf{h}) q_k \right), \tag{13}$$

and let the maximum expected pay-off to demon m (the first term inside the braces of (9) and (10)) be π^m:

$$\pi^m = \max_{w_m} \sum_{l_1=1}^{n} \cdots \sum_{l_q=1, q \neq m}^{n} \cdots \sum_{l_M=1}^{n} \left(\prod_{i=1, i\neq m}^{M} p_{l_i}^{i} \right) \left[\sum_{j} g_{j,l_1,\ldots,w_m,\ldots,l_M}^{rs}(\mathbf{h}) h_j \right]. \tag{14}$$

The risk-averse traffic assignment with multiple network-specific demons can then be expressed as a NCP: to find $\mathbf{x}^*$ such that:

$$\mathbf{x}^* \geq \mathbf{0},\ \mathbf{F}(\mathbf{x}^*) \geq \mathbf{0}, \text{ and} \tag{15}$$

$$\mathbf{x}^{*\mathrm{T}} \cdot \mathbf{F}(\mathbf{x}^*) = 0, \tag{16}$$

where the asterisk associated with the variable refers to the optimal solution;

$$\mathbf{x} = \begin{bmatrix} h_j^{rs}, \forall j, rs \\ \phi^{rs}, \forall rs \\ p_{l_m}^{m}, \forall l_m, m \\ \pi^m, \forall m \end{bmatrix} \text{ and} \tag{17}$$

$$\mathbf{F}(\mathbf{x}) = \begin{bmatrix} \sum_{k} g_{jk}^{rs}(\mathbf{h}) q_k - \phi^{rs} \\ \sum_{j} \delta_j^{rs} h_j^{rs} - N^{rs} \\ \pi^m - \sum_{l_1=1}^{n} \cdots \sum_{l_q=1, q\neq m}^{n} \cdots \sum_{l_M=1}^{n} \left(\prod_{i=1, i\neq m}^{M} p_{l_i}^{i} \right) \left[\sum_{j} g_{j,l_1,\ldots,l_m,\ldots,l_M}^{rs}(\mathbf{h}) h_j^{rs} \right] \\ \sum_{l_m} p_{l_m}^{m} - 1 \end{bmatrix}. \tag{18}$$

It should be clear that when $\mathbf{x}$ satisfies the NCP (15)-(18), the h_j^{rs} and $p_{l_m}^{m}$ in $\mathbf{x}$ must satisfy the risk-averse traffic assignment (1)-(12). Moreover, the NCP (15)-(18) must have at least one solution. The proof can be found in the Appendix. Furthermore, this NCP can have more than one solution. This will be shown in the numerical studies.

The NCP (15)-(18) can be transformed to a minimization problem via a gap function. Let Ω be the solution set to the NCP formulation and $\psi = \{\mathbf{x} \geq \mathbf{0}, \mathbf{F}(\mathbf{x}) \geq \mathbf{0}\}$. A function $G(\mathbf{x})$ is the Gap function for the NCP if the following three conditions are satisfied:

(i) $G(\mathbf{x}) \geq 0$;

(ii) $G(\mathbf{x}) = 0 \Leftrightarrow \mathbf{x} \in \Omega$, and;

(iii) $\min_{\mathbf{x} \in \psi} G(\mathbf{x}) = 0$ is a global minimum.

These properties mean that if we can find $G(\mathbf{x})$ to be zero, we have a solution to the NCP. On the other hand, if the NCP has no solution, then $G(\mathbf{x})$ must have global solutions with positive objective values (Theorem 3.1 in Kanzow and Fukushima, 1996).

In this paper, we adopt the one in Lo and Chen (2000):

$$G(\mathbf{x}) = \sum_i \varphi(x_i, F_i(\mathbf{x})), \tag{19}$$

where

$$\varphi(a,b) = \frac{1}{2}\phi^2(a,b), \text{ and} \tag{20}$$

$\phi(a,b)$ is the *Fischer Function* defined as follows:

$$\phi(a,b) = \sqrt{(a^2 + b^2)} - (a+b). \tag{21}$$

By adopting the gap function (19), the NCP (15)-(18) can be reformulated into a minimization program:

$$\min_{\mathbf{x}} G(\mathbf{x}), \tag{22}$$

where $G(\mathbf{x})$ is defined by (19)-(21). This minimization program can then be solved by a number of existing algorithms. In this paper, this program is solved by the Generalised Reduced Gradient (GRG) algorithm (Abadie and Carpentier, 1969).

NUMERICAL STUDIES

The effect of introducing multiple network-specific demons

In this study we consider the effect of introducing multiple network-specific demons into the game. The simple test network used to generate results is shown in figure 1 below and consists of one OD pair with three parallel links. Since this network has only one OD pair, the demon(s) here can be considered as OD-specific demon(s), and the results here can also be used to explain the effect of introducing more than one demon per OD pair. So, in this study, we use the term 'demons' instead of 'network demons'. The following network parameters also apply:

Free-flow travel times: $t_1^0 = t_2^0 = t_3^0 = 10$ minutes.

Link capacities: $c_{1k} = c_{2k} = c_{3k} = 4000$ Vehicles/hour, if a link is not selected for damage by any demon;
$c_{1k} = c_{2k} = c_{3k} = 2000$ Vehicles/hour, if a link is selected for damage by one or more demons.

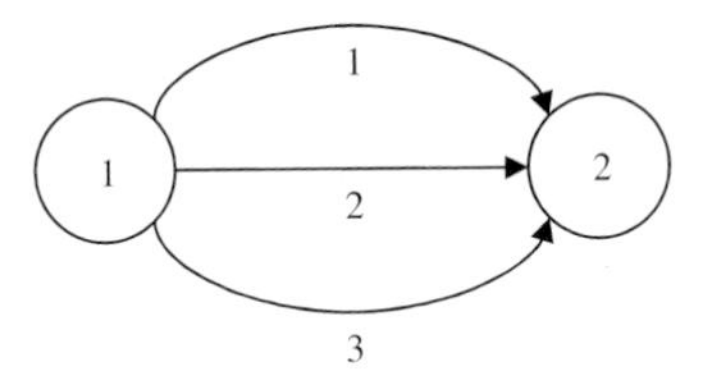

Figure 1: Test network 1

While the network chosen need not be symmetric, it is used here for illustrative purposes. In this example, we assume that the capacity is always reduced by 50% regardless of how many (network or OD) demons select the same link to damage simultaneously. Again the reason for this is to simplify the analysis. Later we will challenge this assumption by studying the effects of relaxing it, that is, by allowing for a greater capacity reduction when multiple demons select the same link to damage. We also assume as much freedom as possible for demons to damage links so that on the one hand demons are forced to damage at least one link per day, while on the other hand they are not restricted from selecting the same links to damage that other demons have selected. Therefore the number of possible scenarios of link damage increases with the number of demons in the network. In our study we calculate the total expected network cost for the cases when there are one, two and then three demons in the network giving us respectively, 3, 9 and 27 possible scenarios of link damage. For comparison purposes, we also consider the case without demons (i.e., there is no capacity uncertainty).

Figure 2 shows the total expected network cost for each case under various levels of network congestion. The level of network congestion is represented by the congestion index (C.I.):

$$\text{Congestion Index} = \frac{\text{Total Number of Network Users}}{\text{Total Network Capacity}},$$

where the total network capacity can be determined using the following rules:

a) For networks with links in series only, the total network capacity is equal to the minimum link capacity of all links.
b) For networks with links in parallel only, the total network capacity is simply equal to the sum of all the individual link capacities.
c) For other networks, the total network capacity can be obtained by first dividing the networks into many (strongly connected) subnetworks with only links in series or links in parallel and then by considering each subnetwork as an 'artificial' link and applying rules a) and b) repeatedly until the total network capacity is obtained.

Replacing some links in the original network by many 'artificial' links with equal capacity may be necessary before dividing the network into many subnetworks.

In general, the higher the value of C.I., the more congested the network is.

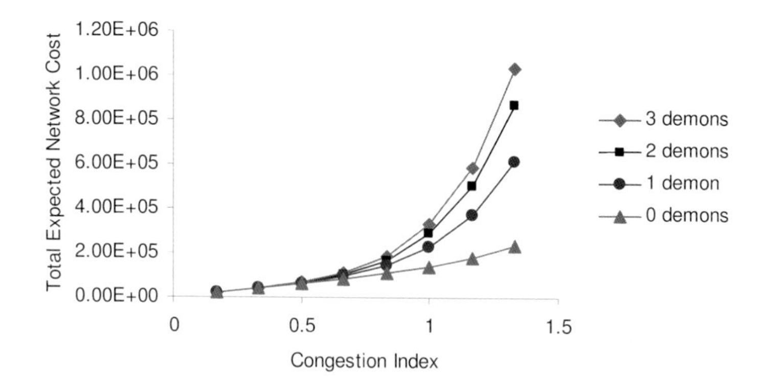

Figure 2: The total expected network cost for a range of congested conditions

According to figure 2, we can see that when the number of demons is fixed, the total expected network cost increases with rising levels of congestion in the network, which is expected. Moreover, the total expected network cost tends to increase with the number of demons present in the network. Again, this is expected. The reasons are as follows: First, the network cost under capacity degradation must be higher than the one without as a lower capacity leads to a higher travel cost. Therefore, the total expected network cost in the multiple-demon case must be higher than the one in the no-demon case. Second, more demons allows more links to be degraded at the same time as shown in table 1, and hence a lower network capacity and a higher total expected network cost. This is why two or more demons can certainly impose a higher network cost on the users than one demon.

Table 1a: Link capacities for the one demon case

	Scenario		
	1	2	3
Link 1	2000	4000	4000
Link 2	4000	2000	4000
Link 3	4000	4000	2000

Table 1b: Link capacities for the two demon case

	Scenario								
	1	2	3	4	5	6	7	8	9
Link 1	2000	4000	4000	2000	2000	2000	4000	2000	4000
Link 2	4000	2000	4000	2000	4000	2000	2000	4000	2000
Link 3	4000	4000	2000	4000	2000	4000	2000	2000	2000

It is also clear from figure 2 that the differences in the total expected network cost produced by different numbers of demons become very significant as the number of users (the travel demand) tends to the capacity of the network, i.e. as the network becomes more congested. At low levels of demand relative to the capacity (un-congested conditions), figure 2 shows that the number of demons we assume to be active has little or no impact on the total expected network cost. This result makes sense when we consider some damage to a link such as, for example, a lane closure due to road repairs. This would have the effect of halving the capacity of the link but would not cause that much delay to users when the travel demand is low. However if the travel demand were high then a lane closure would have a serious impact on the congestion levels. In fact, in this example at the highest demand level used (i.e., 16000 users, C.I. =1.33), we found that the differences between the total expected network costs is very significant ranging from about 41% when we go from assuming one demon to two demons in the network, up to about 68.5% when we go from assuming one demon to three demons. According to Bell (2000) the total expected network cost provides a useful measure of the network reliability since it reflects the expectations of the pessimistic users. Therefore we can conclude that the number of demons we assume will have a significant impact on the total expected network cost that we use to measure the network reliability.

The obvious question at this point is – how can we determine exactly how many demons we should choose to be active in the network (or in an OD pair)? This is a difficult question to answer since it is also linked to the other question of how much we should assume that each demon contributes to the link capacity reduction when the demon selects a link to damage. However one possible method may be to choose the number of demons that are suitable based on some available data on incidents that have led to past link failures in the network. Very often however this kind of data may be unavailable.

Actually, in this study we consider an even more uncertain network than the ones considered in Bell and Cassir (2002) or in Szeto et al. (2006), since we consider more demons, which results in greater uncertainty in link capacity. If only one demon is active then we have three scenarios of link damage (see table 1a) whereas if we have two demons there are nine scenarios of link damage (see table 1b). These two tables illustrate how greater uncertainty arises. It is clear from the two-demon case that there are many more scenarios to choose from in which the capacities of links 1, 2, or 3 (or all) are reduced by 50%. Take scenarios 3 to 9 for example - here two link capacities are reduced by 50% since the two demons select different links to damage.

To conclude, in this study we have ascertained the effect of ignoring multiple demons which really means ignoring greater uncertainty in link capacity (or ignoring greater network capacity uncertainty) and pointed out the implications to network reliability, however we leave the question of exactly how many demons to assume in the model of risk-averse traffic assignment to future studies.

The effect of introducing capacity reduction in some proportion to the number of demons

In previous work on risk-averse traffic assignment (Bell and Cassir, 2002; Szeto et al., 2006) the conventional approach to modelling the damaged capacity was to assume that if a demon selects a link to damage, then the capacity of that particular link is half the value of its initial capacity. The rationale for such an assumption is based on a link having two operational states – congested and un-congested. Even when a number of OD-specific demons select the same link to damage at the same time, the capacity is still reduced by half. Because of this, we used a similar assumption in our first numerical study above: when more than one network-specific demon selects the same link to damage at the same time, the capacity is still reduced by 50%. Indeed, it is reasonable to assume that if many (network or OD-specific) demons select the same link to damage at the same time, the resultant link capacity reduction is in some proportion to the number of active demons selecting this particular link to damage, since there are many more than just two operational states in real road networks. Therefore in this study, we relax the assumptions on link capacity degradation due to the presence of multiple demons and examine the effects of introducing the capacity reduction of a particular link on the maximum expected pay-off to the demon(s).

We use the same network as shown in figure 1 above and therefore the results here can be used to explain the effects of 'greater capacity degradation' due to either multiple network-specific demons or multiple OD-specific demons. The free-flow travel times and undamaged link capacities are the same as in our first study (above) but the damaged link capacities are set based on the relaxed assumptions. The values for the damaged link capacities are given in table 2. We consider two different cases of this: Case one (two) capacity reduction is linear (non-linear) with respect to the number of demons. We repeat the calculations that we carried out in the first numerical study using the same travel demand range, and obtain the maximum expected pay-offs to the demon(s) for different congestion levels in the network represented again by the congestion index that we defined earlier.

Table 2: The damaged link capacities

	No. of demons selecting a link to damage			
	0	1	2	3
Link Capacity (vph) for Case 1	4000	3000	2000	1000
Link Capacity (vph) for Case 2	4000	2000	1000	500

Case 1: linear capacity reduction

Figure 3 shows the total expected network cost in the case of linear capacity reduction. The trend is similar to the one in the first numerical study (i.e., base case). The total expected network cost tends to increase with the number of demons present in the network. Moreover, the total expected network cost increases as the congestion levels rise for any number of demons. Furthermore, like before, the differences in the total expected network costs produced by different numbers of demons are negligible when the congestion levels are low, regardless of the number of demons present. However, when the network is congested, the demons have a much greater impact on the total expected network cost and these differences are quite substantial as we can see from the graph in figure 3. For the highest level of demand shown here of 16,000 users (C.I=1.33), the total expected network cost varies greatly depending on how many demons we assume are active. In fact, going from assuming one demon to assuming three demons produces a massive 339.9% increase in the total expected network cost. Contrast this to the first numerical study in which we used a cruder assumption on capacity reduction, we find only a corresponding 68.5% difference in the total expected network cost when we go from having one to three active demons in the network. This has important implications to network reliability defined based on expected network cost. The total expected network cost resulting from multiple demons in the network will be quite different (higher) than the case with one demon and will therefore result in a different value that we use to measure the network reliability.

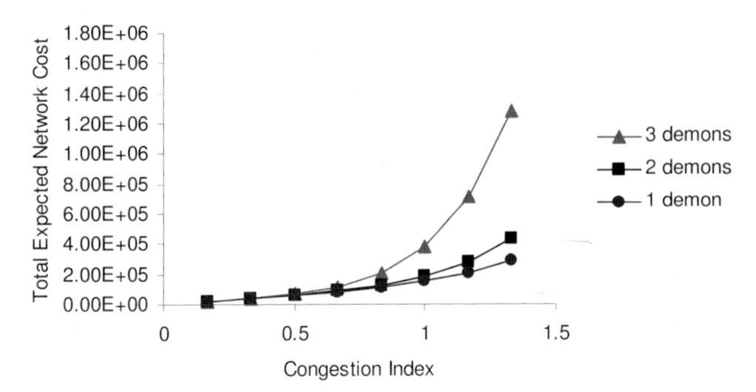

Figure 3: The total expected network cost for a range of demons from free-flow to congested conditions when we assume a linear capacity reduction

Case 2: non-linear capacity reduction

Figure 4 reveals the total expected network cost in the case of non-linear capacity reduction. The observation is similar but the manner in which the total expected network cost increases with increasing levels of congestion differs and the differences in total expected network cost produced by different number of demons are even more substantial when compared with those in the base and linear capacity reduction cases we have previously examined.

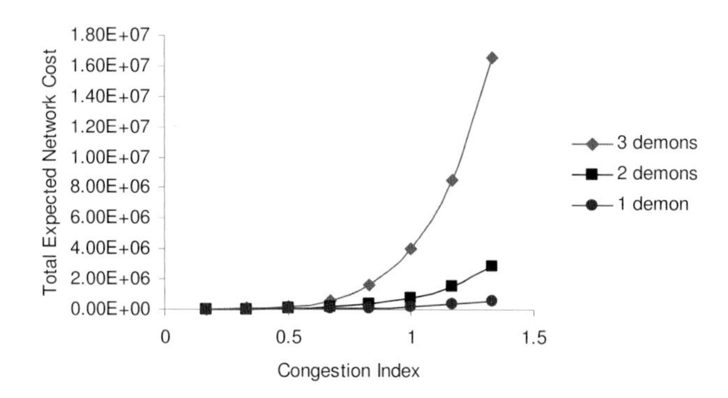

Figure 4: The total expected network cost for a range of demons from free-flow to congested conditions when we assume a non-linear capacity reduction

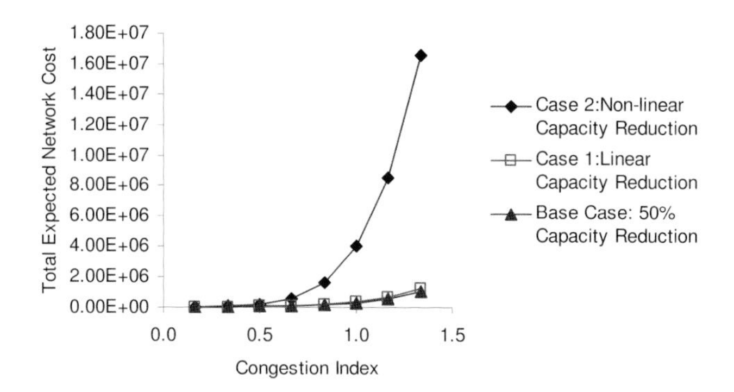

Figure 5: The total expected network cost vs. congestion levels for the three different assumptions on link capacity degradation for the three demon case

To illustrate this clearly, we plot figure 5, which clearly shows the total expected network cost against congestion levels for the three cases on link capacity degradation for the three demon case. From this figure, we can see that the total expected network cost in the non-linear capacity reduction case is much larger than those in the base and linear capacity reduction cases when C.I. is greater than or equal to 1. At the highest level of demand (i.e., C.I. = 1.33), the total expected network cost in the nonlinear capacity reduction case is 16 times larger than that in the base case. This large difference not only is significant but also poses more questions than it necessarily answers regarding how much uncertainty we should consider (i.e., how many demons) and further how much link degradation we should assume. From both this numerical study and the first one, it is quite clear that the assumptions regarding capacity degradation may have more significance than those regarding how many demons to select. However when both of these assumptions are relaxed, they produce significantly higher network cost for users to face and are worthy of further study due to the implications for the network reliability.

The Demon Probabilities – the demons' evil behaviour

It is worthwhile to have a closer look at the probabilities of demons selecting particular links to damage so as to gain a better understanding of the demons' evil behaviour (evil in the sense of how they maximise their expected pay-offs). This is of importance since it gives us a sense for which links get damaged with higher probabilities, what scenarios are more likely to occur and also provides us with some insight into the problem. For these purposes we choose a simple test network of just one OD pair with two parallel links. This sample network is shown in figure 6.

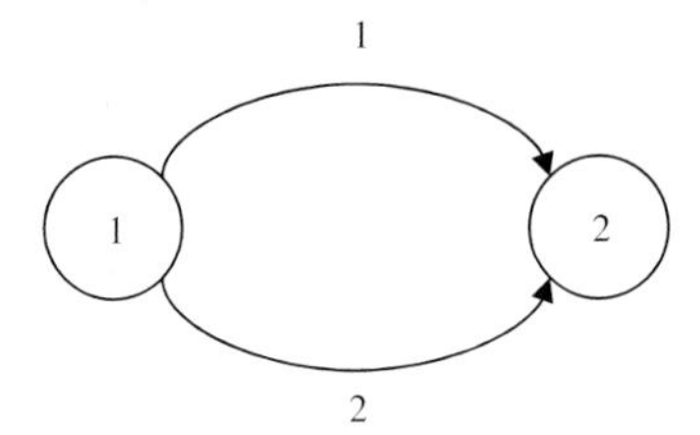

Figure 6: Test network 2

The following network parameters also apply:

Free-flow travel times: $t_1^0 = 5$ mins; $t_2^0 = 10$ mins.

Link capacities: $c_{1k} = c_{2k} = 4000$ vehicles/hour if there is no damage and 2000 vehicles per hour if demon(s) select these links to damage.

Case 1: the one demon case

Firstly we consider a simple case where there is only one demon active in the network, because this case will provide us with some insight into the 'evil behaviour' and increases our understanding of what happens when there are two or more demons involved. In this simple case, there are only two resulting scenarios. These two scenarios are respectively: demon 1 selects link 1 to damage; demon 1 selects link 2 to damage. The links that the demon selects to damage varies in their probabilities over a travel demand range (here we examine from 2,000-10,000 users). We expect that the sole demon will damage the link with the most flow with the highest probability. In this way the demon can impose the highest network cost on most users thereby maximising its maximum expected pay-off.

Figure 7 shows the risk-averse flows in the network. Interpreting figure 7 is relatively straightforward since all flow will choose link 1 initially because it is the minimum cost route. As link 1 becomes congested (volume/capacity =0.75, or congestion index = 0.375), i.e. when it tends towards its capacity of 4000 vph, the users begin to slowly choose link 2. From a demand of about 6,000 users onwards (or congestion index= 0.75), the users begin to choose both routes fairly evenly although there is of course still more flow on link 1 since it is the lowest cost route.

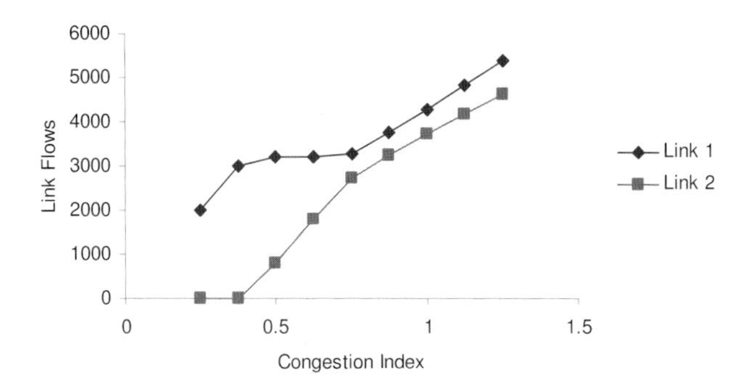

Figure 7: The link flow patterns for the case of one active demon

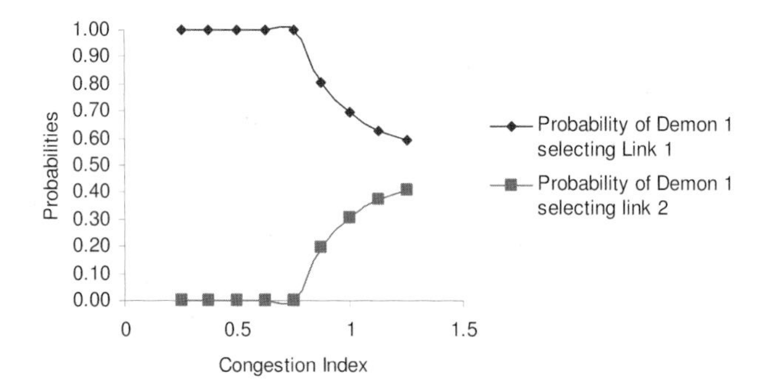

Figure 8: The link damage probabilities for the one demon case

The probabilities of demon 1 selecting links 1 and 2 to damage closely mirror the flow patterns. These probabilities are shown in a plot against the rising congestion levels in figure 8. The sole demon will choose link 1 to damage until the demand reaches a level approaching capacity of that facility. At this point the demand is high enough to force some users to choose link 2. However at this point there is still not that much flow on link 2 so the demon continues to damage link 1 almost exclusively. At about 6,000 users (or congestion index= 0.75), the flow on link 2 becomes significant enough to attract the demon to choose link 2 to damage with a somewhat lower probability and link 1 still with a higher probability. However the values of the probabilities of the demon selecting links 1 and 2 to damage tend towards approx 0.55 and 0.45 respectively once congestion begins to occur. These values seem reasonable as they reflect how the slightly higher probability of the demon selecting link 1 to damage is due to the fact that there are more users on this link. In this way the demon can maximise the network cost faced by the user and hence, in the process, maximise its own expected pay-off, as we expected.

Case 2: the two demon case

Secondly, we consider two demons, which is beneficial as no one has presented and discussed the results about the case of two or more (network-specific or OD-specific) demons. We adopt the same network and parameters as in the case just above where we considered one demon. As before, the demons are not restricted from damaging the same links at the same time. Table 3 sets out the possible scenarios of link damage.

Table 3: Scenarios for two demons in network 2

Scenario	Link 1	Link 2
1	Demons 1 & 2	0
2	0	Demons 1 & 2
3	Demon 1	Demon 2
4	Demon 2	Demon 1

In figure 9, the link flows for the users who face the maximum (total expected) network cost imposed by two demons are plotted for increasing levels of congestion in the network. These link flow patterns follow a broadly similar trend to flows in the network with one demon (see figure 7), and that is, when the network is un-congested the users tend to choose the minimum cost route until the demand rises to a level that approaches the capacity of that facility. After this point the users begin to choose route 2 but do so slightly more readily than in the previous case with just one demon. In general, however, the flow patterns are very similar between the two cases. In some ways like before the demon behaviour in selecting links to damage reflects the popularity of the routes chosen. When we assumed only one active demon in the network, this demon damaged link 1 with a probability of one at low demand levels, since link 1 was the minimum cost route and it attracted all of the users.

Similarly, in the case of two demons, according to table 4, one of the demons damages link 1 with a probability of one when the demand is low (but both routes carry flows) for the same reason as we have just mentioned in the one demon case, while the other demon will damage the other link since there is no further benefit or pay-off to be gained from damaging the same link as the first demon. The explanation for this is relatively simple. If demon 1 damages link 1 with a probability of one which in un-congested conditions imposes maximum (total expected) network cost on all users then demon 2 receives a higher pay-off in damaging link 2 with a probability of one so as to impose the network cost on the remaining users and in the process maximise its expected pay-off. There is no benefit to demon 2 to further damage link 1, since under our capacity assumptions the capacity of the link will not reduce any further by doing so. For rising congestion levels when users begin to choose route 2, the pay-off to the demons remain at a maximum since they each individually damage the two links and therefore impose the highest network cost possible on all users.

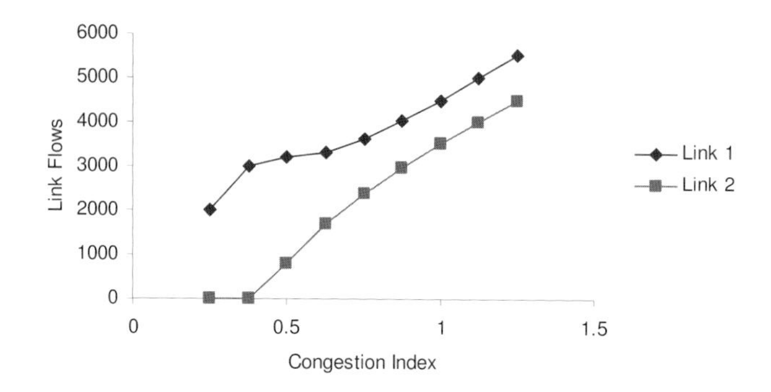

Figure 9: The link flow patterns for the case of two active demons

Table 4 also reveals that the demon selection strategies change with the congestion level. For example, demon 1 selects link 1 to damage with probability of one when the congestion indices are 0.25, 0.375, 0.5, 0.875, and 1.25 but selects link 1 to damage with zero probability when the congestion indices are 0.625, 0.75, 1.0, and 1.125. However, as long as one demon selects a different link to damage from the other demon, their payoffs must be maximized. This suggests that the 'flip-flop' solution as shown in table 4 is only one of the optimal solutions, which will be further discussed in the next subsection.

Table 4: The demons' link selection probabilities for the two demon case

Congestion Index	Link selection probabilities			
	Demon 1 Selects		Demon 2 Selects	
	Link 1	Link 2	Link 1	Link 2
0.25	1	0	0	1
0.375	1	0	0	1
0.5	1	0	0	1
0.625	0	1	1	0
0.75	0	1	1	0
0.875	1	0	0	1
1	0	1	1	0
1.125	0	1	1	0
1.25	1	0	0	1

One point is worthwhile to mention. In a sense the overall trend is similar to the case with one demon where the sole demon begins to damage the link with less flow with a lower probability so as to impose costs on all users. However, significantly, the sole demon only begins to damage link 2 when the number of users on that link is greater than about 3000 vph (congestion index = 0.375). Compare this to the evil behaviour of the two demons in this case, where the other demon always damages link 2 with a probability of one no matter how many users are on that link. These slightly different strategies may help to explain why the maximum expected pay-off is higher when there are two demons instead of one since they quite literally have the resources to impose maximum damage on the users by damaging more links and therefore in the process maximising their expected pay-offs. In fact the percentage increase in the maximum expected pay-off when we assume two demons instead of one is approximately 63%. These results follow a similar trend to our earlier studies involving multiple demons on network 1. Figure 10 below shows the maximum expected pay-offs to the demon(s) for rising congestion levels. As before the significant differences occur when the network is very congested. In general we can conclude that whenever both links are damaged this implies the optimal solution.

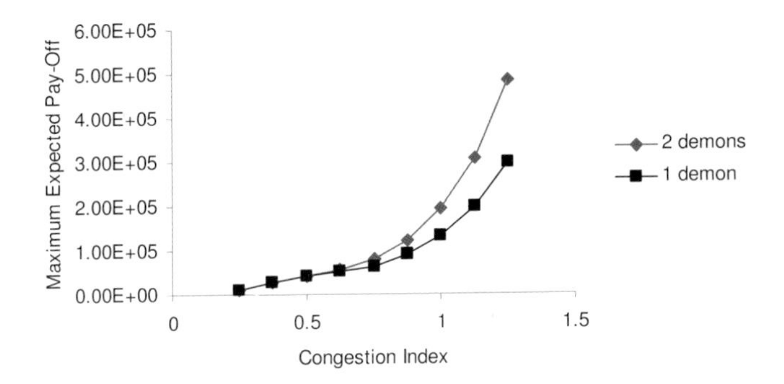

Figure 10: The maximum expected pay-off vs. level of congestion in the network

Similar effects and trends in the demon probabilities can be observed for a three links network with 3 network-specific demons but due to space constraints we cannot show them here. The simpler network used here with two network-specific demons makes it easier to visualise the results. In general, the trends that we have shown hold well but it must be mentioned however that due to the nature of the problem, there can be multiple solutions as shown in the following example.

The existence of multiple solutions

Table 5 shows another set of the optimal link selection probabilities for the demons when link 2 carries flow. The overall effect in terms of the total expected network cost is the same, since as the table shows that both demons still select different links to damage for each level of congestion shown. In other words we have demonstrated that the demons choose exactly the opposite links to damage between the solution shown here and those shown in the previous numerical study.

To summarise the results shown in tables 4 and 5, when link 2 carries flow then both demons have a probability of one in different links. However with one small exception which occurs when the network is very un-congested. In this scenario there are other possible solutions. These other possible probabilities occur when there is no flow on link 2 (i.e., when the demand is lower than the capacity of the minimum cost route). In this case it is possible for both demons to select link 1 (the minimum cost route) to damage with a probability of one. This does not affect the maximum expected pay-off since there is no real benefit in terms of getting a higher network cost for the other demon to damage link 2 when there is no flow on this link so the overall expected pay-offs to the demons remains the same. We can say therefore that when link 2 has no flow, at least one of the demons has a probability of 1 in selecting link 1 to damage.

Table 5: Another set of the optimal demons' link selection probabilities

Congestion Index	Link selection probabilities			
	Demon 1 Selects		Demon 2 Selects	
	Link 1	Link 2	Link 1	Link 2
0.25	0	1	1	0
0.375	0	1	1	0
0.5	0	1	1	0
0.625	1	0	0	1
0.75	1	0	0	1
0.875	0	1	1	0
1	1	0	0	1
1.125	1	0	0	1
1.25	0	1	1	0

OD Specific vs. Network Specific Demons in a Multi-Commodity Network

The purpose of this study is to highlight the impacts of the OD specific demons as formulated in Bell and Cassir (2002) and the network demons as described by this paper since in previous examples we considered only one OD pair which does not allow us to consider more than one OD specific demon and does not allow for other possibilities such as all network demons damaging one OD pair only. To have a fair comparison between the two types of demons, we assume that the link capacity is reduced by 50% when it is selected for damage by one or more demons. This assumption is necessary here since there is no possibility of more than one OD specific demon selecting the same link to damage in the same OD pair. Furthermore we consider only two network specific demons to be active in the network since there are only two OD specific demons present (one for each OD pair). The network used here is shown in figure 11 and consists of two OD pairs: OD pairs (1,2) and (2,3). OD pair (1,2) connects origin 1 to destination 2 via two parallel and symmetric links. OD pair (2,3) connects origin 2 to destination 3 via one link only – link 3.

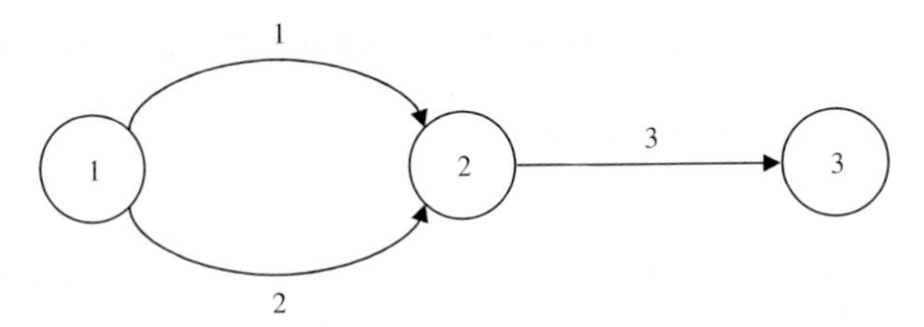

Figure 11: Test network 3

The following network parameters chosen for illustrative purposes also apply:

Free-flow travel times: $t_1^0 = t_2^0 = t_3^0 = 10$ mins.

Link capacities: $c_{1k} = c_{2k} = c_{3k} = 4000$ Vehicles/hour, if a link is not selected for damage by any demon;
$c_{1k} = c_{2k} = c_{3k} = 2000$ Vehicles/hour, if a link is selected for damage by one or more demons.

OD travel demand: $N^{12} = 6000$ users,

We consider two demand levels for OD pair (2,3): 2000 users (case 1) and 4000 users (case 2)

Case 1: $N^{23} = 2000$

The link selection strategies of the demons for this case are shown in table 6. Due to the two parallel and symmetric links, the OD specific demon for OD pair (1,2) will always have an equal probability of selecting links 1 and 2 to damage to maximise the total expected network cost to the users. The O-D specific demon for OD pair (2,3) will always damage link 3 with a probability of one since under the assumptions in Bell and Cassir (2002) that the O-D specific demon for OD pair (2,3) is forced to damage at least one link per day and only link 3 connects the OD pair (2,3). Regardless of the demand for travel and network parameters, the OD specific demons will always behave in this way. However, the same cannot be said for the network demons. The formulation proposed in this paper allows the network demons to have the freedom to 'roam' the network choosing whatever links they wish to damage so as to maximise the total expected network cost to the users. For the fixed demand level of 2000 users here, both network specific demons concentrate entirely on damaging links between OD pair (1,2) and they do so in such a way that both demons select different links to damage. Whether demon 1 selects link 1 to damage and demon 2 selects link 2 to damage or vice-versa does not matter as the total expected network cost to the users remains the same.

Table 6: Link selection probabilities of the demons for case 1

OD Pair	Link	OD-Specific Demons		Network Demons	
		Demon 1	Demon 2	Demon 1	Demon 2
1-3	1	0.5	0	1	0
	2	0.5	0	0	1
2-3	3	0	1	0	0

It is clear from this example that the network demons are more flexible and can choose to concentrate on one OD pair only, in this case OD pair (1,2), which seems appropriate since this OD pair has a higher demand and therefore more users can be affected by damaging these links. The OD specific demons do not have this freedom however, and are *forced* to damage different OD pairs. This example shows that this may not necessarily be the best strategy since there are not as many users affected by link 3 being damaged, and therefore the total expected network cost imposed on the users by the OD specific demons is consequently lower than those imposed by the network demons in this case. In this particular example, the total expected network cost imposed on the users by the network specific demons is 14.75% higher than that of the OD specific demons. This is a significant difference and underlines how the network demons give the worse case scenario of link damage in this case. This has important implications for the network reliability since we define the network reliability based on these values.

Case 2: $N^{23} = 4000$

In this case, we consider the same network and network parameters as before. However the demand parameters are changed somewhat with OD pair (2,3) having an increased travel demand of 4000 users. The resulting demons' link selection strategies are shown in table 7. As before the link selection strategies of the OD specific demons remains the same (as we would expect) but the network demons select different links to damage than before. In this case the two links in OD pair (1,2) are selected for damage with equal probability by one network demon and the sole link in OD pair (2,3) is selected for damage with a probability of one by the second network demon. This means that in this case the link damage probabilities under the two assumptions are the same and hence the OD-specific and network specific demons impose the exact same network cost on the users.

Table 7: Link selection probabilities of the demons for case 2

OD Pair	Link	OD-Specific Demons		Network Demons	
		Demon 1	Demon 2	Demon 1	Demon 2
1-3	1	0.5	0	0.5	0
	2	0.5	0	0.5	0
2-3	3	0	1	0	1

By comparing the two cases, we can see that the network demons can give the same link damage probabilities as the OD specific demons. This implies that the network demon approach is more general in the sense that this approach can capture the results obtained by the OD specific demon approach. This is reasonable as network demons can behave as if they are OD specific demons. Moreover, the network demons can at least yield the same total expected network cost as the OD specific demons. This means that the network demon approach can always give the worst case scenario in network reliability analysis. This is again reasonable as network demons have freedom to choose links anywhere in the network to damage, regardless of what OD pair those links connect.

CONCLUSIONS

This paper develops a formulation of the risk-averse traffic assignment that allows for the presence of multiple network-specific demons and therefore allows us to consider the effect of greater network uncertainty in terms of capacity degradation. Through numerical studies we demonstrate just how important the assumption on how many demons to use is. A numerical study is also performed to consider the effect of relaxing the assumption on capacity reduction when links are selected for damage by one or more demons. The results show just how important it is to accurately capture the link capacity degradations. Relaxing this assumption also allows for more operational states to be modelled. Apart from these two contributions, we also provide some further insights into the demon behaviour in selecting links to damage and what their optimal damage strategies are, demonstrate the existence of multiple solutions to this route choice game, and compare the network demon approach proposed here and the traditional OD specific demon approach.

This paper opens up many future research directions. First, the proposed multiple network demon formulation and the ones in Bell and Cassir (2002) and Szeto et al. (2006) could be formulated into one model of risk-averse traffic assignment. This would have the benefit of providing one clearly defined framework for the problem. Second, the formulation is defined as a Nash game where the demon cannot predict user response to link damage. An alternative to this model would be to formulate the problem as a Stackelberg game in which the demon is placed at the upper level. This extension may allow for a better approximation of the worst-case expected travel cost. Third, the very important question of how many demons are appropriate to assume remains unanswered. Extensive studies on real networks involving data on previous incidents are required to rectify this. Fourth, the studies here on the behaviour of two demons are carried out on small test networks. Different, larger networks are required to further investigate the demons behaviour and to determine if they always select certain links to damage over others in the network with a higher probability, in particular those critical links near the origin and destination nodes. Fifth, by assuming a greater number of demons and more capacity degradation states, the complexity of the problem increases since the number of scenarios increases exponentially according to the number of demons used and the number of capacity degradation states. The resultant formulation is difficult to solve for large networks. This should be addressed in the future. Sixth, the route set is assumed to be known in advance in this paper. One important future extension is to develop a link-based formulation so that the route set is not required to be known in advance and that the number of variables is reduced (since there are considerably less links than routes in large networks). Finally, using the risk-averse framework proposed in this study, one can assess the reliability of transport networks, develop a reliable network design, and analyse the effect of assumptions on traveller behaviour in each approach on network design and reliability. This may bring up some interesting and important future research directions.

ACKNOWLEDGEMENT

This research is funded under the Programme for Research in Third-Level Institutions (PRTLI), administered by the Irish Higher Education Authority. The authors are grateful for the constructive comments of the referees.

REFERENCES

Abadie, J., and J. Carpentier, (1969). Generalisation of the Wolfe reduced gradient method to the case of non-linear constraints, in R. Fletcher (ed.) *Optimisation*, pp. 37-47. New York: Academia Press.

Bell, M.G.H., (2000). A game theory approach to measuring the performance reliability of transport networks. *Transportation Research* **34B**, 533–546.

Bell, M.G.H. and C. Cassir, (2002). Risk-averse user equilibrium traffic assignment: an application of game theory. *Transportation Research* **36B**, 671-681.

Boyce, D.E., B. Ran, Y. Li, (1998). Considering travellers' risk-taking behavior in dynamic traffic assignment. In: Bell, M.G.H. (Ed.), *Transportation Networks: Recent Methodological Advances*. pp. 67–81. Oxford: Elsevier.

Chan, K.S. and W.H.K. Lam, (2005). Impact of road pricing on the network reliability. *Journal of the Eastern Asia Society for Transportation Studies* **6**, pp. 2060 – 2075.

Chen, A. and W. Recker, (2000). Considering risk-taking behaviour in travel time reliability. Paper Presented in *Matsuyama Workshop on Transport Network Analysis*, Matsuyama, Japan.

Kanzow C. and M. Fukushima, (1996). Equivalence of the generalized complementarity problem to differentiable unconstrained minimization. *Journal of Optimization Theory and Applications* **90** (2), 581-603.

Lo H. and A. Chen. (2000). Traffic equilibrium problem with route-specific costs: Formulation and algorithms. *Transportation Research* **34B**, 493-513.

Lo, H. and Y.K. Tung. (2003). Network with Degradable Links: Capacity Analysis and Design. *Transportation Research* **37B**, 345–363.

Mirchandani, P., H. Soroush, (1987). Generalized traffic equilibrium with probabilistic travel times and perceptions. *Transportation Science* **21**, 133–152.

Nagurney, A. (1993). *Network Economics: A Variational Inequality Approach*. Kluwer Academic Publishers. Norwell, Massachusetts, USA.

Nash, J., (1951). Non-cooperative games. *Annals of Mathematics* **54**, 286–295.

Sumalee A., R. Connors, and R. Batley (2005). The applicability of prospect theory to the analysis of transport networks. *Proc. of European Transport Conference*, Strasbourg, France.

Szeto, W.Y., L. O'Brien, M. O'Mahony (2006). Risk-Averse Traffic Assignment with Elastic Demands: NCP Formulation and Solution Method for Assessing Performance Reliability. *Network and Spatial Economics* **6**, 313-332.

Uchida, T., Y. Iida, (1993). Risk assignment: a new traffic assignment model considering risk of travel time variation. In: Daganzo, C.F. (Ed.), *Proceedings of the 12th International Symposium on Transportation and Traffic Theory*. pp. 89–105. Amsterdam: Elsevier.

Yin, Y., H. Ieda, (2001). Assessing performance reliability of road networks under non-recurrent congestion. *Transportation Research Record* **1771**, 148–155.

Yin, Y., W.H.K Lam., and H. Ieda, (2002). Modeling risk-taking behavior in queuing networks with advanced traveler information systems. *Transportation and Traffic Theory in the 21st Century*, Edited by Taylor, M.A.P., 2002, pp. 309-328. Oxford: Elsevier.

Yin, Y., W.H.K. Lam, and H. Ieda (2004). New technology and the modelling of risk taking behaviour in congested road networks. *Transportation Research*, **12C**, 171-192.

APPENDIX: THE PROOF OF THE EXISTENCE OF SOLUTIONS

We prove the existence of solutions of the NCP (15)-(18) by using two important results in Nagurney (1993):

1. A NCP can be reformulated into a Variational Inequality Problem (VIP) as the VIP contains the NCP as a special case (Proposition 1.4), and
2. A VIP must have at least one solution if the mapping function $\mathbf{F}(\mathbf{x})$ is continuous with respect to $\mathbf{x}$ and if the solution set is the compact convex set and bounded (Theorems 1.4 and 1.5).

Based on Proposition 1.4 in Nagurney (1993), we have the following proposition:

Proposition 1: The NCP (15)-(18) can be reformulated into the following VIP.

Find $\mathbf{x}^*$ *such that:*

$$(\mathbf{x}-\mathbf{x}^*)^{\mathrm{T}}\mathbf{F}(\mathbf{x}^*) \geq 0, \ \forall \mathbf{x} \in \Omega, \tag{23}$$

where $\mathbf{x}$ *and* $\mathbf{F}(\mathbf{x})$ *are defined as in (17) and (18), respectively;* $\Omega = R_+^u$ *is the solution set; u is the dimension of* $\mathbf{x}$, *which is equal to the sum of the number of paths in the network, the number of OD pairs, the product of the number of demons and the number of links, and the number of demons.*

This proposition implies that if the VIP (23) has a solution, the NCP (15)-(18) will also have the same solution. The following proves that the VIP (23) indeed has a solution using the result in Theorems 1.4 and 1.5 in Nagurney (1993).

Proposition 2: A solution to the VIP (23) exists.
Proof:
We first show that the solution set Ω can be reduced to a bounded compact convex set. By definition, the link selection probability $p_{l_m}^m$ must not exceed one and the route flow h_j^{rs} must be less than or equal to the corresponding demand:

$$p_{l_m}^m \leq 1 \text{ and} \tag{24}$$

$$h_j^{rs} \leq N^{rs}. \tag{25}$$

Using (24) and (25), we have:

$$\sum_k g_{jk}^{rs}(\mathbf{h}) q_k \leq \sum_k g_{jk}^{rs}(\mathbf{N}) q_k \leq \sum_k g_{jk}^{rs}(\mathbf{N}), \tag{26}$$

where $\mathbf{N} = \left[h_j^{rs}, \forall j, rs \mid h_j^{rs} = N^{rs} \right]$.

Since by definitions $N^{rs} \leq \max_{rs} N^{rs}$ and $\sum_k g_{jk}^{rs}(\mathbf{N}) \leq \max_j \sum_k g_{jk}^{rs}(\mathbf{N})$, (25) and (26) can respectively be rewritten as:

$$h_j^{rs} \leq \max_{rs} N^{rs} \text{ and} \tag{27}$$

$$\sum_k g_{jk}^{rs}(\mathbf{h}) q_k \leq \max_j \sum_k g_{jk}^{rs}(\mathbf{N}). \tag{28}$$

By (28) and the fact that the minimum expected OD travel cost must be less than or equal to the expected route travel cost between that OD pair (i.e., $\phi^{rs} = \min_{\mathrm{d}} \left(\sum_k g_{dk}^{rs}(\mathbf{h}) q_k \right) \leq \sum_k g_{jk}^{rs}(\mathbf{h}) q_k$), we know that the minimum expected OD travel cost is bounded:

$$\phi^{rs} \leq \max_j \sum_k g_{jk}^{rs}(\mathbf{N}). \tag{29}$$

Moreover, by (14), (24) and (25), we have:

$$\begin{aligned} \pi^m &= \max_{\mathrm{w}_m} \sum_{l_1=1}^{n} \cdots \sum_{l_q=1, q\neq m}^{n} \cdots \sum_{l_M=1}^{n} \left(\prod_{i=1, i\neq m}^{M} p_{l_i}^i \right) \left[\sum_j g_{j,l_1,\ldots,w_m,\ldots,l_M}^{rs}(\mathbf{h}) h_j \right] \\ &\leq \max_{\mathrm{w}_m} \sum_{l_1=1}^{n} \cdots \sum_{l_q=1, q\neq m}^{n} \cdots \sum_{l_M=1}^{n} \left[\sum_j g_{j,l_1,\ldots,w_m,\ldots,l_M}^{rs}(\mathbf{N}) N^{rs} \right] \\ &\leq \max_m \left\{ \max_{\mathrm{w}_m} \sum_{l_1=1}^{n} \cdots \sum_{l_q=1, q\neq m}^{n} \cdots \sum_{l_M=1}^{n} \left[\sum_j g_{j,l_1,\ldots,w_m,\ldots,l_M}^{rs}(\mathbf{N}) N^{rs} \right] \right\}. \end{aligned} \tag{30}$$

From (24), (27), (29), (30), we can define an upper bound $\breve{x}$ for each element x_i of vector $\mathbf{x}$ to be: $\breve{x} = \max \left\{ 1, \max_{rs} N^{rs}, \max_j \sum_k g_{jk}^{rs}(\mathbf{N}), \Phi \right\}$, where $\Phi = \max_{m}\left\{\max_{w_m}\sum_{l_1=1}^{n}\cdots\sum_{l_q=1,q\neq m}^{n}\cdots\sum_{l_M=1}^{n}\left[\sum_j g^{rs}_{j,l_1,\dots,w_m,\dots,l_M}(\mathbf{N})N^{rs}\right]\right\}$.

With the upper bound $\breve{x}$, we can define a closed ball $B_r(\mathbf{0})$ centred at $\mathbf{0}$ with the radius $r=\breve{x}$ so that the solution set $\Omega = R^u_+$ to be a bounded compact convex set.

We now show that $\mathbf{F}(\mathbf{x})$ is a continuous function of $\mathbf{x}$. From (5)-(8), it is not difficult for us to see that $g^{rs}_{jk}(\mathbf{h})$ is a continuous function of h^{rs}_j and q_k is a continuous function of $p^m_{l_m}$. Consequently, we have $\sum_k g^{rs}_{jk}(\mathbf{h})q_k - \phi^{rs}$ to be a continuous function of ϕ^{rs}, $p^m_{l_m}$, and h^{rs}_j. Moreover, it is not difficult for us to see the following:

1. $\pi^m - \sum_{l_1=1}^{n}\cdots\sum_{l_q=1,q\neq m}^{n}\cdots\sum_{l_M=1}^{n}\left(\prod_{i=1,i\neq m}^{M} p^i_{l_i}\right)\left[\sum_j g^{rs}_{j,l_1\dots l_m\dots l_M}(\mathbf{h})h^{rs}_j\right]$ is a continuous function of π^m, $p^m_{l_m}$, and h^{rs}_j;
2. $\sum_j \delta^{rs}_j h^{rs}_j - N^{rs}$ is a continuous function of h^{rs}_j, and;
3. $\sum_{l_m} p^m_{l_m} - 1$ is a continuous function of $p^m_{l_m}$.

We can therefore say that all elements of $\mathbf{F}(\mathbf{x})$ are continuous functions of $\mathbf{x}$ and hence $\mathbf{F}(\mathbf{x})$ is a continuous function of $\mathbf{x}$.

Based on propositions 1 and 2, we can conclude that the NCP (15)-(18) must have at least one solution.

Transportation and Traffic Theory 2007
Edited by R.E. Allsop, M.G.H. Bell and B.G. Heydecker

7

EMPIRICAL STUDIES ON ROAD TRAFFIC RESPONSE TO CAPACITY REDUCTION

Richard Clegg, Department of Electronics and Electrical Engineering, University College London, UK

SUMMARY

This paper uses empirical data collected on street to provide insight into how real traffic systems are behaving. The data are fitted to statistical models and these statistical models show how traffic systems respond when the capacity of the system is reduced. This research follows on from the general interest in highway capacity reduction following Cairns et al (1998).

A great deal of effort has been spent on models (mathematical and computational) that simulate the day-to-day behaviour of road traffic. Many of these models include parameters that adjust how quickly users will respond by changing their behaviour. This paper provides a starting point for the calibration of such models by careful statistical analysis of real-life data collected in the city of York (UK). The data were collected to track a real-life capacity-reducing event that occurred in the city (a partial road closure). The paper begins with a brief discussion of the context of the research, other studies providing evidence in this area and the modelling context.

In the second section of the paper the survey performed is described. The survey took place in the morning peak and collected licence plate data. It considers the effect of a road closure. Before and during data are available for several days at several sites to establish the ambient variability in the system and then to compare this with the response to the change in the system. The data collected are freely available to other researchers.

The third section of the paper considers day-to-day variability in the traffic pattern and models "recurrence rate" (the percentage of drivers seen on one day who reappear on another day). A statistical model is fitted to the data showing that the recurrence rate starts lower than many researchers might expect for a morning rush hour and falls off quickly with subsequent days. A marked day of the week effect is noted showing that recurrence is higher on the same day of the week. This can be thought of as the "See you next Wednesday" effect, that is, the effect on the traffic system of a pool of drivers who travel on the same day or the same set of days.

The fourth section of the paper considers the transient response of the drivers to a change in a road system. The data studied show a "healing" effect. The initial (day one) response to the change is strong but the effect becomes weaker in subsequent days. This response can be seen in both travel times and link flows. This can be thought of as analogous to the road traffic engineering rule of thumb that a change that seems disastrous on Monday will "be alright by Friday".

The final section of the paper considers the effects of rerouting as seen in the data and measures the route choice response to an incident. Again this can be separated into an initial response and a later "settling down" consistent with the "It'll be alright by Friday" rule.

While the results presented here are just on one road system, the analysis is intended as a starting point for people wishing to calibrate network models in which drivers change route in response to congestion and changes to the network. The conclusion section of this paper describes how these results could have implications for modelling and further research.

INTRODUCTION

The aim of this paper is to make a rigorous statistical examination of an urban road traffic network. The statistical analysis takes account of the behaviour down to the individual vehicle level. In particular the real-life effect of reduction in road capacity is analysed. To this end two surveys were undertaken in the city of York looking at interventions leading to road capacity reduction. Note that throughout this paper the term intervention is used to refer to a planned alteration to a road network such as a road or bridge closure.

It should be stressed that the motivation of the surveys undertaken and of this paper is not to provide insight into the detailed effects of specific interventions in a specific town. It is hoped that in the future, through technologies such as GPS and licence plate recognition, large data sets that can identify individual vehicles will become more readily available. Methods such as those used in this paper could be used to identify general characteristics of road traffic networks and, in turn, to provide insight for mathematical and computational modelling. The techniques given in this paper could be applied such data set and this statistical analysis could do much to inform efforts in creating accurate simulation models of the urban traffic environment.

The data set considered here is also examined in Clegg (2005), chapter five of Clegg (2005b) and Clegg (2006).

Background and Previous Research

Responses to congestion and to road capacity reduction are of great interest both to the road traffic engineer and to the modeller. Cairns et al (1998) and a summary update in Cairns et al (2002), reviewed on-street evidence from over ninety cases of road capacity reduction (in some cases this was deliberate reduction due to a planned scheme and in some cases due to accidents which closed roads for long periods). Two important conclusions of this report were, *"...overall, the two responses - changing route and changing journey time – seem to be the most universal"* (Cairns et al 1998 p28) and also that in the *"short term...it is the common experience that, after an adjustment period, traffic alters to take account of the new conditions. Reference to a 'settling down' period has been made"* (Cairns et al 1998 p36).

Note that a number of other driver responses to congestion are possible including a no-travel decision, destination switching and mode switching. These possibilities are not investigated here since the data could not produce firm conclusions on these travel choices.

The effect of roadworks on congestion has been investigated by a number of researchers, for example, Marvin and Slater (1997) and Goodwin (2005) although the emphasis is on calculating the cost to the economy of the delays imposed by roadworks. In the UK there is some governmental interest in this subject following the Traffic Management Act 2004 that regulates roadworks in the UK. More information can be found at
www.dft.gov.uk/stellent/groups/dft_roads/documents/divisionhomepage/612852.hcsp

Ambient Variability

Before considering the response to a change to a network it would be useful to know how much to expect networks to be in a state of change anyway. Do drivers habitually travel at the same time via the same route most days or are drivers constantly experimenting with new routes and travelling at different times of day? This is sometimes referred to in the literature as *churn*.

Bonsall *et al* (1984) report on the collection of a large number of licence plates from roadside surveys undertaken in Leeds. They report that over the hours they surveyed then considering only those drivers who travelled at the absolute height of the morning rush hour, only 45% (but see later note about a 15% increase) of those drivers were seen in a two and a quarter hour period around the rush hour on a subsequent day. Over the course of the full rush hour it seems that, for these survey results at least, more than half the drivers seen will not travel on the next day. Their results also showed that of those drivers who were seen on a subsequent day, many of them were not travelling within a quarter of an hour window of the journey on the original day. It is also important to note that they stress the recording process was unavoidably error prone when looking at the number of matches they *"assume a 15% increase in the number of matches"* (p387) due to missed matches from incorrectly recorded data.

However, Stephenson and Tepley (1984) conclude that when comparing two days from the before-period: *"60% of drivers travelled at the same time (+/- 5 minutes) every day during uncongested conditions"* but it is unclear from their paper if they intend this to mean 60% of those drivers seen on both of the two days or 60% of the drivers seen on the first day.

Jan et al (1999) report on the use of GPS data to understand route choice. Their data set was GPS data from 100 households (216 drivers) over a one-week period. They reported that *"the path chosen on a trip most often differs considerably from the shortest time path across the network"* (p1) and also that *"travelers habitually follow the same path for the same trip"*" (p12).

A review from the point of view of Global Positioning System (GPS) data is given by Pendyala (2003) and the author analyses several small data sets of between sixteen and thirty-two individuals, concluding that, *"The percentage of individuals in each sample who exhibit the same characteristic across all days...is extremely small... [often] zero."*

One important result comes from Huff and Hanson (1986) who studied data from individual drivers collected over a 35-day period. A major conclusion of their work is that *"observations taken for a single day in the travel history of an individual are not likely to be representative of the range of daily travel patterns exhibited by that person over a more extended time period, and we are led to reject the view that travel is highly routinized in the restricted sense that every weekday is assumed to look much like every other weekday"* (p108)."

Evidence about Road Capacity Reduction

As mentioned, Cairns et al, (1998) investigated over ninety reports of capacity reducing incidents, but few of these reports produced quantitative results about route choice. Stephenson and Tepley (1984) is a notable exception. They examined data obtained in Edmonton after the closure of the Kinnaird Bridge. They considered how drivers changed their route as a result of the closure and showed a "knock-on" effect of drivers not directly affected by the closure shifting their route to avoid the congestion. In addition the paper makes reference to a settling down period, that is an initial response of the closure that lessens as time goes on. However, the paper does not provide any quantification of this effect.

Daugherty et al (1999) reports on a number of bus priority schemes implemented in Great Britain. They conclude that drivers often change their route in response to these schemes and this can cause problems if the new routes are unable to absorb the traffic. Unfortunately, the reports tend to be qualitative not quantitative in nature.

Many authors report peak spreading as a response to increased congestion, but offer little in the way of qualitative evidence. Such peak spreading could be either a result of a departure time choice response on the part of the drivers or a result of changes in travel time in other parts of the network as a result of the intervention.

SUMMARY OF SURVEYS

While not of general interest, description of the surveys performed is included here in order to provide context for the following discussion. If desired, a fuller description of the surveys can be found in Clegg (2005), Clegg (2005b) and Clegg (2006). The modelling here supersedes the modelling performed in those references. The data can all be downloaded from www.richardclegg.org/route/

In fact, two surveys took place but the first of these, a survey of the closure of Lendal Bridge in York, was affected both by the UK "fuel crisis" (protests which blockaded refineries and caused a temporary fuel shortage throughout the UK) and also by flooding in York. No results from this survey are given here but they can be found in the above website.

The survey recorded licence plate data for the morning rush hour for several days over several sites. The data recorded is in the form of partial licence plates from all vehicles at sites. The survey monitored the temporary closure of one lane of the inner ring road in York on a road called "Fishergate".

General Survey Information

The survey took place over the course of several weeks and recorded the morning rush hour on weekdays. Several sites were simultaneously monitored and licence plates of all vehicles recorded over those sites. The plates were recorded manually (audio recording with a phonetic alphabet for commonly confused letters) and transcribed later.

The timing of the surveys was chosen to meet several aims. It was important to get a good estimate of the ambient variability of traffic over days and weeks by monitoring the traffic before the intervention. It was considered important to estimate the transient response of the traffic to an intervention in the weekdays immediately following. It was desirable to get some estimate of the longer-term impact over more than one week. Weekdays only were monitored. While it was recognised that day of the week effects do occur the difference in traffic patterns between weekdays are widely recognised to be less than the differences between weekdays and weekends. The morning rush hour was picked over the evening rush hour as, in York, this has more traffic and occurs at a more predictable time of day.

Other considerations were taken into account when picking the geographical location. It was of primary importance to survey those sites directly affected by the closure. Secondarily, those sites that would have knock-on effects by being directly before or after the closure should be monitored. It was important to try to monitor potential rerouting locations where possible. There was a recognised trade off between monitoring sites close to the site of the intervention and monitoring sites far from the site. Nearby sites would have more vehicles which were affected by the intervention but more distant sites would provide more information per vehicle matched with a vehicle seen at the intervention site. A compromise between these two cases was sought.

The time was recorded at intervals of approximately five minutes. Surveyors were supplied with synchronised watches at the beginning of the surveys. The times for data between each time stamp are interpolated so, for example, if there are ten plates between a time stamp at 8:10 and one at 8:19 they will be split so that one plate is seen in each minute. Because of this interpolation and possible rounding of the time, the times recorded can only be assumed to be accurate to within five minutes, however, it is thought that the timing is accurate to a much greater resolution than this.

The surveys were mainly undertaken by audiotape (that is, surveyors recording licence plates onto audio tape which were later transcribed). The possibility of recording or transcription errors should be taken seriously (this is discussed in a later section).

The Fishergate Survey

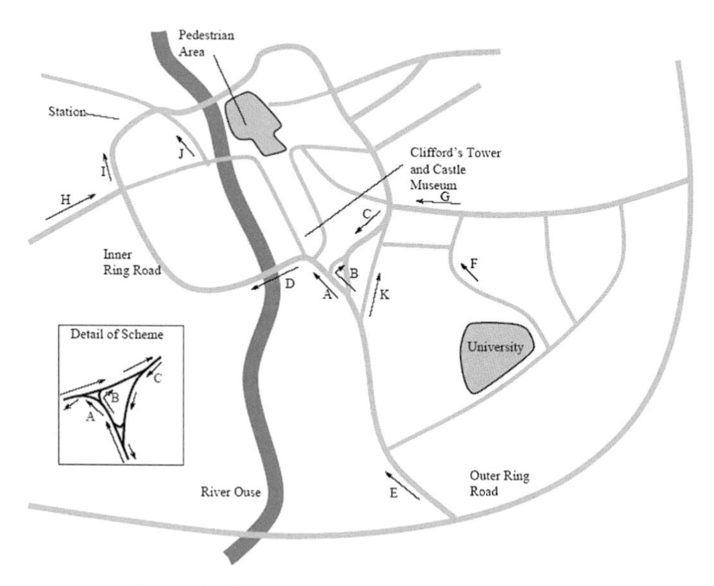

Figure 1: Fishergate Survey showing survey sites.

Figure 1 shows the survey sites used for the Fishergate Survey. All sites are monitored only in the direction of the arrows given. This survey was based around works to repair a collapsed sewer at site A. The repair work involved a partial closure of site A, essentially one lane being removed from the road.

The closure was originally scheduled to last only two weeks and therefore the plan was to survey for one week before, one week during and one week after the closure. However, the closure was extended to four weeks and therefore no true after survey data is available. A possible exception is the 13th of July when the closure was suspended for one day to allow for the increase in traffic due to a major horseracing event that weekend (the extra traffic due to the race-goers is thought not to have had a great effect on traffic during the morning peak). The days surveyed (all in the summer of 2001) were as followed:

1. 25th, 26th, 27th, 28th, 29th June and 2nd July – before surveys.
2. 3rd, 4th, 5th, 6th, 11th, 12th July – during surveys.
3. 13th July – temporary removal of roadworks (could be seen as after).
4. 16th July – during survey (roadworks back in place).

For the Fishergate survey, for most sites, the traffic was monitored at most sites from 7:45 to 9:15. This was in order to catch all of the rush hour traffic and a quarter of an hour window either side. However, at selected sites, this window was adjusted to monitor from 8:00 to 9:30. This happened at those sites that would be reached last on a journey (for example, in Figure 1 site J would always be reached after site A). This was decided since the travel time between some pairs of sites was of the order of half an hour. Without such an offset some of the survey time would otherwise be wasted since the earliest (or latest) parts of the data could not be expected to match with data at any other site. The sites that were surveyed from 8:00 to 9:30 were sites A, I and J.

Data and Analysis Techniques Used

In any project of this nature considerable pre-processing of data is necessary before fitting models. The details of this pre-processing are not of general interest but are briefly recorded here as a guide for other researchers who might undertake similar work. Since the aim of this work is to consider the behaviour of individuals it is necessary to infer matches between observations. These matches may, for example, be a car seen on two different days at the same site or a car seen on the same day at different sites.

Statistical methods used

The statistical methods used in this paper are the standard t-tests and linear modelling. Both of these are well-known techniques and are only summarised briefly here. Full descriptions can be found in most standard statistical text, for example. Comparisons between sample means are done using the two-sided Welch t-test (this allows small sample sizes and the possibility that the first mean may be smaller or larger than the second).

The linear models used are general linear models (not generalised linear models, although the results do not differ if generalised linear models are used instead). The models take the form

$$y = \beta_0 + \sum_{i=1}^{k} \beta_i x_i + \varepsilon,$$

where the observed variable y is dependent on the observed variables $x_1, x_2, \ldots, x_k$, and the β_i are the parameters of the model (to be fitted). The errors ε are assumed to be independent and identically distributed with a normal distribution, mean zero and with variance independent of the x_i.

In this paper significances are assigned at the 10%, 5%, 1% or 0.1% levels indicating that the significance of that parameter is at the level given or lower. If a result is stated as having "low" significance then that parameter is not significant at the 10% level. This indicates that the data analysed do not support the inclusion of this parameter in the model.

A note on errors in recording and transcription of data

In any data gathering experiment some sources of error are to be expected. Attempts have been made to compensate for some sources of error. In other cases, the errors have been ignored and no systematic attempt has been made to deal with them.

In some cases surveys began late, ended early, had missing data or no data due to failures of the surveyors or their equipment. On these occasions data for that site and day were completely removed from the results presented here. In addition analysis of modelling residuals showed that Fishergate survey site H had an unusually high flow of traffic (50% more than usual) on 16th July 2001. No particular reason is known for this anomalous high flow. While it did not appear to change modelling results a great deal, that day of data at that site was omitted from analysis.

Secondly, errors could be made either in the recording or transcription process either by vehicles being omitted or recorded or transcribed incorrectly. The surveyors were encouraged to use a phonetic alphabet to reduce such errors. In addition, any plates that appeared to have been only partly recorded and foreign plates (likely to be recorded inconsistently by different surveyors) were removed from the data. However, no systematic method has been used to estimate such errors. As previously mentioned, Bonsall et al (1984) estimate that in looking at matches on their licence plate data, one must *"assume a 15% increase in the number of matches"* (p387) due to missed matches from incorrectly recorded data. In the data reported here, it would be inappropriate to assume a particular percentage increase should be applied across the board. It is certain that these recording errors will have occurred and will affect the absolute levels of the figures measured. However, the models fitted are fitted in terms of trends and increases and decreases which would be less affected by such errors.

False matches

A final source of error even if the data is correctly recorded is that of false matches. Since the licence plates recorded are only partial number plates then a match may not necessarily indicate that the same vehicle has been seen but, instead, that two different vehicles with the same partial licence plate have been seen. This problem is something of a classic problem in road traffic surveys. For more details on possible solutions see Hauer (1979), Maher(1985), Watling and Maher (1988 and 1992) and Watling (1994). For this paper, two approaches are used. When it can be assumed that vehicles have journeyed between two sites on a single day then a maximum likelihood estimation method based upon assumptions about journey time following Watling (1994). When an estimate of a number of matches between two sites is required and journey time information is irrelevant (for example, when wanting the number of matches between one site on two different days) then a simple probabilistic correction is applied similar to the approach in Hauer (1979).

Adjustment of times considered

Sometimes the particular times considered can be of critical importance. Consider, asking the question "what percentage of traffic at site A is seen again at the downstream site B?" Assume that under normal conditions it takes 20 minutes to get between site A and site B. If the comparison is made from all data between 8:00 and 9:00 at both sites then the vehicles seen at site A between 8:40 and 9:00 are unlikely to be seen at site B not because they do not get there but because they do not get there *in the time considered.* Now, if this percentage as measured on two different days and is seen to go down on the second day it could indicate that a smaller percentage of the vehicles at site A travel on to site B on the second day. However, it could also indicate that the same percentage will eventually reach site B on the second day but they took longer to get there. This effect can be avoided by not considering the final half hour of data at site A (assuming half an hour is the maximum time that could be expected to be taken from site A to site B). This process of omitting data will be referred to in this paper as "trimming".

Normalisation by site

Finally, it is sometimes necessary to compare measurements between different sites. For example, it might be necessary to measure whether a particular effect increases or decreases traffic flow. However, the flows would naturally be expected to differ considerably between sites and this effect could overwhelm the effect being considered. Therefore, where appropriate, the data has been normalised, in this case by, for each site, taking the raw measurement and, for each site, subtracting the mean at that site and dividing by the standard deviation over that site. This means that the measurement for each site will be mean zero and variance one. When this has been done the data will be described in the text as having been "normalised by site".

Initial data analysis

Basic information about the data including the number of vehicles recorded on every day surveyed and histograms showing the distribution throughout the survey period can be found in Clegg (2005b).

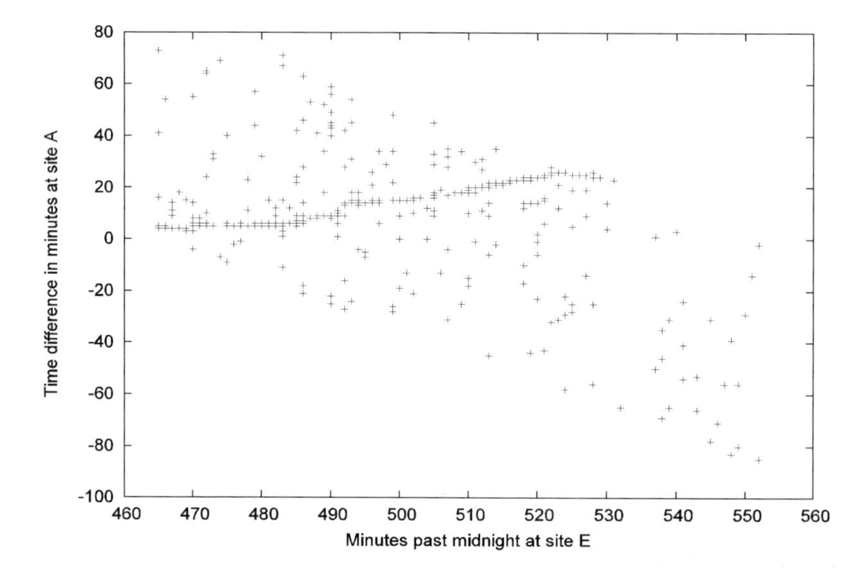

Figure 2: Matches between vehicles at Fishergate sites E and A on first during day.

Figure 2 gives in insight into the nature of the data being used. The plot is from the first day where the closure is in place (3rd July 2001). A cross represents an observation of a partial licence plate at both site E and A. The x-axis is the time (in minutes past midnight) when the plate was seen at site E (480 is 8:00 am and 540 is 9:00 am). The y-axis is the time difference between the sighting at E and the sighting at A. Assuming that the match is a genuine one, this will be the time taken to travel from E to A. The darker line of crosses in the centre of the diagram represents the journeys most likely to be genuine. As can be seen, the travel time from E to A increases steadily from 8:00am until 9:00am). The travel time seems to decrease slightly after this. This could be genuine (as a result of the rush hour finishing) or it could be a result of the fact that the data only goes up to 9:15 (535 minutes after midnight). The crosses off this main diagonal are likely to be false matches as discussed earlier. The blank areas in the bottom left and top right of the plot are areas that were not surveyed because they were before the start of the survey or after the end of the survey respectively.

ANALYSIS OF RECURRENCE RATE

Consider two sets of observations made at two disjoint time periods T_1 and T_2 (it does not matter, for the purposes of this definition, whether the observations are made at the same geographical location). The recurrence rate R is given by

$$R(T_1,T_2) = \frac{V(T_1,T_2)}{V(T_1)},$$

where $V(T_1,T_2)$ is the number of vehicles seen in both sets of observations and $V(T_2)$ is the number of vehicles seen in time period one. In other words R is the percentage of vehicles seen in the set of observations at time period T_1 also seen in time period T_2. In the case of the data observed here, due to the problem of false matches the quantity $V(T_1,T_2)$ can only be estimated. The simple probabilistic correction method is used. To increase the reported recurrence rate, "trimming" is performed as described earlier so the first site is only measured between 8:20 and 8:40 whereas the second site is analysed over the whole data period. This time interval is certainly sufficient for a driver to travel between the most distant site pairs surveyed. The general conclusions are not sensitive to the trimming process and similar conclusions follow if the exact time period used is changed.

	25/6/01	**26/6/01**	**27/6/01**	**28/6/01**
25/6/01	**123.7 (107.7)**	57.5 (42.2)	55.8 (40.3)	55.4 (39.1)
26/6/01	54.7 (38.8)	**116.0 (100.7)**	61.0 (45.5)	62.6 (46.3)
27/6/01	54.9 (38.9)	55.3 (40.0)	**116.6 (101.1)**	64.2 (47.9)
28/6/01	56.3 (40.3)	59.2 (43.8)	61.0 (45.5)	**130.5 (114.2)**
29/6/01	54.1 (38.1)	48.6 (33.2)	53.9 (38.4)	57.8 (41.5)
2/7/01	56.2 (40.2)	55.1 (39.8)	54.0 (38.5)	57.7 (41.4)
3/7/01	50.6 (34.7)	61.0 (45.7)	55.9 (40.5)	59.0 (42.7)
4/7/01	50.0 (34.0)	54.7 (39.4)	56.2 (40.7)	58.9 (42.6)
5/7/01	50.4 (34.4)	49.6 (34.3)	50.4 (34.9)	61.0 (44.7)
6/7/01	44.5 (28.5)	51.4 (36.1)	50.9 (35.4)	52.4 (36.1)
11/7/01	44.1 (28.2)	49.7 (34.4)	54.5 (39.0)	56.9 (40.6)
12/7/01	43.9 (27.9)	46.4 (31.0)	48.6 (33.1)	52.8 (36.5)
13/7/01	42.7 (26.8)	39.6 (24.3)	46.3 (30.8)	46.7 (30.4)
16/7/01	50.4 (34.4)	44.4 (29.1)	46.9 (31.4)	48.9 (32.6)

Table 1: Recurrence rate as percentage at Fishergate site A.

A partial table for Fishergate site A is shown in Table 1. The raw recurrence rate (before correction for false matches) is the main figure and the figure in brackets is adjusted by the probabilistic correction. The diagonal (in bold) is the match of a day with itself and should (after correction) be exactly 100% – note that the definition of recurrence rate above would not allow for a recurrence rate above 100% but the figures in brackets are only estimates of the correct recurrence rate. The reason it is over 100% in the uncorrected data is that a plate may match with more than one plate in the other data set. Note that the recurrence rate of a day with itself is not considered in the analysis in the next section.

A number of features are evident from this table. Even for days close together the adjusted recurrence rate is always lower than 50%. The majority of the rush hour in a relatively congested city does not seem to consist of the same drivers travelling day after day. This corresponds with the work of other authors as reported in the introduction.

Modelling recurrence rates

A linear model is fitted to the recurrence rate data,

$$E[R] = \beta_0 + \beta_1|d| + \beta_2 I_w + \beta_3 I_d,$$

where R is the recurrence rate, d is the number of days separating the two surveyed days (not including weekends), I_w is an indicator variable which is 1 if the two survey days are in different weeks and I_d is an indicator variable which is 1 if the two survey days are on the same day of the week. The parameters can, therefore be interpreted as follows:
β_0 – intercept (the theoretical day zero recurrence rate with no other effects),
β_1 – correction for number of days between two days considered,
β_2 – correction added if the two days are on different days of the week and
β_3 – correction added if the two days are the same day of the week.

The results of the model fitting are shown in Table 2. At the majority of sites, all modelled parameters were significant. The exceptions are site B which has extremely low traffic and sites F and H. Apart from these sites, the models all have quite high R^2 values indicating that the model, simple as it is, explains a good deal of the variability of recurrence rate. The exception is the final row. The model parameters all have the signs expected, that is, recurrence rates are higher if the two days are on the same day of the week, lower if they are significantly far apart in time and lower still if they are in different weeks.

The final row shows a combined model for all sites. This has all parameters significant but a lower R^2 value than the other successful models. This could indicate that some of the variability in recurrence rate is site dependant and the final model does not capture this. This hypothesis is backed up by the fact that the intercept parameter is very different across the different sites. It is possible that some particular geographical features make some sites more likely to have higher recurrence rates.

Discussion

The results presented here are broadly consistent with those of Bonsall et al (1984). With that study suggesting a 45% match from one day to the next and this study suggesting a 38% decrease (39% reduced by 0.66% according to parameter β_1) which varies greatly between sites. It should also be noted that Bonsall et al match against a longer survey time on the second day and hence the figures are not strictly comparable. However, these numbers certainly do back up those authors who claim that the make-up of rush hour drivers changes greatly from day-to-day.

Site	β_0	β_1	β_2	β_3	R^2	R_a^2	p-value
A	42.2 (0.1%)	-0.800 (0.1%)	-3.35 (1%)	5.41 (0.1%)	0.560	0.545	1.73e-15
B	24.6 (0.1%)	-0.476 (low)	-3.40 (low)	0.0966 (low)	0.060	0.022	0.198
C	37.6 (0.1%)	-0.510 (0.1%)	-3.00 (5%)	4.39 (0.1%)	0.407	0.386	6.79e-10
D	37.4 (0.1%)	-0.780 (0.1%)	-2.35 (10%)	5.18 (0.1%)	0.567	0.556	1.32e-11
E	53.5 (0.1%)	-0.731 (0.1%)	-5.36 (0.1%)	2.90 (1%)	0.656	0.644	<1e-15
F	34.2 (0.1%)	-0.597 (0.1%)	-2.46 (low)	2.01 (low)	0.267	0.242	5.36-6
G	40.1 (0.1%)	-1.00 (0.1%)	-4.32 (1%)	5.11 (0.1%)	0.630	0.614	6.18e-16
H	25.5 (0.1%)	-0.360 (1%)	-1.23 (low)	3.35 (1%)	0.234	0.203	0.000180
I	35.9 (0.1%)	-0.831 (0.1%)	-3.23 (1%)	3.70 (1%)	0.562	0.540	1.36e-15
J	39.2 (0.1%)	-0.412 (1%)	-5.41 (0.1%)	2.49 (10%)	0.413	0.389	1.27e-8
K	42.7 (0.1%)	-0.719 (0.1%)	-4.87 (0.1%)	3.49 (1%)	0.650	0.636	<1e-15
All	39.0 (0.1%)	-0.666 (0.1%)	-3.71 (0.1%)	3.72 (0.1%)	0.226	0.222	<1e-15

Table 2: Model fitting for Fishergate matches 8:20-8:40 am against full data.

The modelling in this section shows evidence for an effect that might be thought of as the "see you next Wednesday" effect. That is, a consistent increase in recurrence related to the day of the week. This could be a pool of drivers who only drive on one day every week or it could be a pool of drivers with a consistent working pattern that happens to have a daily component (for example, they drive to work on Mondays through to Thursdays but car share as a passenger on Fridays). There is certainly not enough information available to definitively settle this issue with this data set.

The recurrence rate of traffic falls off extremely quickly even in the short period of time measured. The model shows a reasonable fit to a linear fall off with weekday and an additional term if the days are in separate weeks. It cannot be ascertained whether this is due to route choice, departure time-choice that moves the driver outside the measured period or some other choice element such as destination choice, mode choice or even a choice not to travel.

The implications for modellers developing a simulation model of an urban system are twofold. Firstly, if the model is a learning model, it may be that the period a driver has to learn about a route is quite short. The low recurrence rate which falls off quickly may well mean that the typical driver does not have long to learn about a journey before making a change. Also it may be that there are a considerable number of drivers who are not habitual drivers on that network even in the rush hour. Of course, the source of the fall off in recurrence rate is unknown. Some of it may be connected with vehicle replacement and some of it may be associated with shifts in destination and overall vehicle behaviour.

TRANSIENT RESPONSES TO INTERVENTION

Road traffic engineers often assume that the initial "day one" results of an intervention will lessen in subsequent days. This is sometimes thought of as the "It'll be alright by Friday" effect, the idea that a poor performance on Monday may well be acceptable by Friday as the network somehow adjusts to the change. In this section the changes to flows and travel times as a result of the Fishergate intervention are considered. For the purposes of the analysis here the 13th July 2001 is considered to be a day without intervention (since the closure was removed on that day). Repeating the analysis with this day removed does not substantially change the results as presented here.

Analysis of Site Flows

Firstly the raw flows are normalised by site (so each site has mean zero and unit variance). The question is then does the flow increase or decrease as a result of the intervention. A t-test is used to compare the mean flow when the closure is in place and when it is not.

The mean normalised flow is -0.0811 with the closure in place and 0.0811 when it is not. However, a 95% confidence interval for the mean when the closure is not in place is given by (-0.153, 0.478) and the p-value is 0.312 indicating that there is no statistically significant change in flow over all sites as a result of the closure. This is unsurprising since some sites were chosen as rerouting sites (that is sites where the flow would be expected to increase with the closure in place because drivers switch routes to use that site).

One approach is to separate the sites into those that might be expected to have a decrease in flow, as they would be directly affected by the closure, and those that might be expected to have an increase in flow, as they are potential rerouting sites. Sites A, C and D were chosen as those most likely to have a decrease in flow and sites F, G and K were chosen as potential rerouting sites.

A t-test comparing the flow at sites A, C and D during the closure with the flow at other sites and times showed that the mean flow decreased as expected, from 0.121 before closure to -0.764 afterwards. The p-value for the test was 8.19e-6 indicated a strongly significant difference in the means indicating that it is likely that due to this closure the flow was reduced overall at these sites.

A t-test comparing the flow at sites F, G and K during the closure with the flow at other sites and times showed that the mean increased as expected from -0.058 to 0.39 with a p-value of 0.031. This indicates that, overall, the flow seemed to increase over these sites.

Finally, the effects at each site were separated using a linear model as follows:

$$E[f] = \beta_0 + \beta_A I_A + \beta_B I_B + ... + \beta_K I_K,$$

where *E[f]* is the expected value of the flow, the β variables are the various parameters of the model and I_A, I_B etc are indicator variables which are 1 at the site in question if the closure in place and 0 otherwise. The results of fitting a linear model are given in Table 3.

β_0	β_A	β_B	β_C	β_D	β_E	β_F	β_G
0.081 (low)	-0.995 1%	-0.607 10%	-0.719 5%	-0.818 5%	0.285 (low)	0.129 (low)	0.029 (low)
β_H	β_I	β_J	β_K	**R^2**	**R_a^2**	**df**	**p-value**
0.501 (low)	0.219 (low)	-0.428 (low)	0.785 5%	0.175	0.108	11, 134	0.005

Table 3: Flow changes by site at Fishergate

A model with so many parameters should be treated with caution. However, from this modelling it seems that sites A, B, C and D have a statistically significant reduction in flow and site K has a statistically significant increase. These directions coincide with what might be expected considering their physical location. Site K is the most obvious rerouting as a result of the intervention and sites A, B, C and D are those sites most directly affected. The R^2 statistic is relatively low indicating that this model explains only a small amount of the variation in flows observed. However, this would certainly be expected in such a simplistic statistical model. Note also that the model is over-specified (it has a parameter for each site as well as an intercept).

Analysis of Site Pairs

In this section the transient response to the intervention is considered with a simple linear model. This model is fitted to the travel times and flows between site pairs. These times and flows are calculated using the Maximum Likelihood Estimator method as described earlier. The site pairs were chosen as those pairs that seem to be likely to be most affected by the intervention. In making these comparisons the data has been "trimmed" by removing the last half hour of observations from the first site of the pair to avoid the incomplete journeys problem as described earlier.

Flow and travel time data for each site pair was produced using the Maximum Likelihood Estimator and these results then individually fitted to a linear model for each site pair considered. The model used was

$$E[y] = \beta_0 + \beta_1 I_c + \beta_2 D,$$

where y represents either flow or travel time I_c is an indicator variable which is 1 if the closure is in place and 0 otherwise and D is a variable indicating the number of days since the closure occurred (not including weekends) or 0 if the closure is not in place. For clarity, the variable D has the value 1 on the 3rd July, 2 on the 4th July and 10 on the last day surveyed, the 16th July. It is expected that the parameters β_1 and β_2 will have different signs since one represents the effect of the closure and the second represents the return to normal effect.

The parameters are to be interpreted as follows,
β_0 – the model intercept, that is the flow or travel time with no other effects occurring,
β_1 – the effect of the closure itself on flow/travel time,
β_2 – the effect of the number of days since the closure, if positive, the dependent variable will rise following the closure as time goes on and if negative the dependent variable will fall following the closure.

The results of the modelling for flow across the site pairs are shown in Table 4. The rows shaded are those models where parameters other than the intercept had statistical significance. The other site pairs cannot be said to show a statistically significant response to the intervention. In some ways this is not surprising since a three-parameter model is being fitted to at most fourteen data points. Site pair E-K shows an increase in flow as a result of the intervention which is as expected since E-K is a potential rerouting. Similarly, site pairs A-D, A-J and D-I show decreases in flow and again, this might be expected since those pairs are directly affected by the closure. D-I shows a return to normal effect with a 10% statistical significance. However, this result should be treated with caution, when fitting so many models, some results are likely to show as having statistical significance at this level.

Note that site pair A-D was the only site pair that had statistical significance for either β_1 or β_2 but had the two in the same direction. That is, it showed the effect progressing as time went on rather than a return to normal. However, the effect was not of statistical significance.

The results shown in Table 5 show the equivalent results to Table 4 but considering travel time instead of flow. Again the shaded rows show those models with statistically significant parameters other than the intercept. Site pairs E-A, C-A and F-A all showed an increase in travel time as a result of the intervention. These sites were leading up to the intervention and hence such an increase would be expected. At site pair C-A, one of the pairs likely to be most strongly affected by the intervention, a return to normal effect was observed with a strong significance (0.1% level).

Site pairs A-D, H-I and D-I showed a decrease in travel time and a "return to normal" effect though only A-D and H-I had statistically significant return to normal effects and at the lowest level of significance considered. These site pairs were "after" the intervention on the road network and it might be expected that travel times would decrease since the reduced flow would allow those drivers past the intervention to travel more quickly.

Site Pair	β_0	β_1	β_2	R^2	R_z^2	df	p-value
E-A	214 0.1%	-18.5 (low)	2.72 (low)	0.228	0.087	: 2, 11	0.241
E-B	13.5 0.1%	-2.56 (low)	0.257 (low)	0.122	-0.054	2, 10	0.523
E-F	48.0 0.1%	4.50 (low)	-0.519 (low)	0.014	-0.165	2,11	0.926
E-K	216 0.1%	55.4 1%	-3.47 (low)	0.658	0.589	2,10	0.005
A-D	736 0.1%	-61.4 5%	-1.38 (low)	0.622	0.538	2,9	0.013
F-G	26.8 0.1%	7.05 (low)	-0.836 (low)	0.158	-0.011	2,10	0.424
C-A	574 0.1%	-20.9 (low)	-1.55 (low)	0.292	0.164	2,11	0.149
A-J	138 0.1%	-27.2 1%	1.28 (low)	0.587	0.504	2,10	0.012
G-C	349 0.1%	-28.7 (low)	1.48 (low)	0.250	0.100	2,10	0.238
D-I	77.9 0.1%	-20.8 5%	2.29 10%	0.431	0.305	2,9	0.079
H-I	211 0.1%	19.2 (low)	-1.28 (low)	0.205	0.028	2,9	0.357
F-A	83.6 0.1%	6.24 (low)	-2.11 (low)	0.254	0.119	2,11	0.199

Table 4: Linear Model of flow response for Fishergate site-pairs

Discussion

Overall, the effects of the intervention on flows and travel times were subtle but detectable in statistical modelling. The statistical models showed that some sites experienced an increase in traffic as a result of the closure and some sites experienced a decrease, however, six of the eleven sites considered had no significant change of flow as a result of the intervention being studied.

Modelling was carried out to attempt to fit a model that contained both an initial response and a "return to normal" effect. Although the model was only successful on a few of the site pairs considered this would be expected given the small effect on flows previously established.

Site Pair	β_0	β_1	β_2	R^2	R_z^2	df	p-value
E-A	7.19 0.1%	3.72 5%	-0.311 (low)	0.453	0.353	2,11	0.036
E-B	7.92 0.1%	1.43 (low)	0.005 (low)	0.130	-0.044	2,10	0.497
E-F	7.36 0.1%	-0.285 (low)	-0.037 (low)	0.299	0.171	2,11	0.142
E-K	6.39 0.1%	-0.814 (low)	-0.114 (low)	0.268	0.121	2,10	0.211
A-D	0.443 0.1%	-0.126 (low)	0.051 10%	0.321	0.171	2,9	0.175
F-G	2.59 0.1%	-0.927 (low)	0.008 (low)	0.380	0.256	2,10	0.092
C-A	1.08 0.1%	2.06 0.1%	-0.162 0.1%	0.883	0.861	2,11	<0.001
A-J	4.27 0.1%	-0.393 (low)	-0.015 (low)	0.230	0.076	2,10	0.271
G-C	1.71 0.1%	0.650 (low)	0.062 (low)	0.591	0.509	2,10	0.012
D-I	5.16 0.1%	-1.57 10%	0.177 (low)	0.309	0.156	2,9	0.189
H-I	1.22 0.1%	-0.174 (low)	0.055 10%	0.358	0.215	2,9	0.136
F-A	3.77 0.1%	1.67 5%	-0.130 (low)	0.441	0.339	2,11	0.042

Table 5: Linear Model of travel time response for Fishergate site pairs.

Four models on site pairs were found with a statistically significant "return to normal" effect and one of these was significant at the 0.1% level. Further this was an effect on the travel time for a site pair that might reasonably be expected to be amongst the most strongly affected. This provides reasonable evidence for the "it'll be alright by Friday" effect. The latest observation here was only ten weekdays after the initial intervention. Further studies would certainly be necessary to establish the nature of a longer-term effect. The model here considered only a linear "return to normal" effect. While this is certainly not the case over a longer period of time, it was not considered that sufficient data were available to consider a more sophisticated model.

The statistical model described here could have considerable implication for simulation modelling of transport systems. Most of the urban traffic models available today are aimed at answering questions about the results of a change and some would be able to provide an estimate as to how long the results of a change would take to settle down. However, as far as the author is aware, no models have had the duration of these transient effects calibrated against real data. If the modelling of transient responses is to be improved then such calibration via studies of this type is vital.

DRIVER REROUTING RESPONSE

Finally, a model was created to attempt to assess the level of rerouting as a result of the intervention. The switch of drivers from site A to site K was modelled. These sites were chosen because A was the site of the intervention and K was the only site that had been shown to have a significant increase in traffic after the intervention.

The rerouting problem was considered by looking at the recurrence rate between site A and K on separate days. As with the previous recurrence rate experiment, the data is "trimmed" by reducing the data to just that from 8:20 to 8:40 on the first day in order to increase the recurrence rate seen. That is, the percentage of vehicles seen at site A on one day and site K on a different day. If this increased during the intervention then this could be a result of drivers from A rerouting to K as a result of the closure. Let $R(d_1,d_2)$ be the recurrence rate between site A on day d_1 and site K on day d_2. Because no (or extremely few) vehicles would be seen at A and K on the same day then measurements are not made where $d_1 = d_2$. The following model was then fitted

$$E[R] = \beta_0 + \beta_1 I_{c1} + \beta_2 D_1 + \beta_3 I_{c2} + \beta_4 D_2 \ ,$$

where R is the recurrence rate, I_{c1} is an indicator which is 1 if d_1 is a closure day D_1 is the number of days d_1 is since the closure (or 0 if it is an open day) as described in the previous section. I_{c2} and D_2 are the equivalent quantities for d_2. The parameters can be interpreted as follows,

β_0 – the intercept, or recurrence rate with no other effects in place,

β_1 – the effect of the closure at site A. If negative, this indicates that fewer drivers switch from site A on one day to site K on another when the closure is in place on the day they were at site A.

β_2 – the effect of the time since the closure at site A. If it has the opposite sign to β_1, this indicates that the effect measured by parameter β_1 declines with the time since closure.

β_3 – the effect of the closure at site K. If positive, this indicates that more drivers switch to site K from site A when the closure is in place on the day they were seen at site K,

β_4 – the effect of the time since the closure at site K. If it has the opposite sign to β_3, this indicates that the effect measured by parameter β_3 declines with the time since closure.

The results of the model fitting are seen in Table 6.

β_0	β_1	β_2	β_3	β_4	R^2	R_a^2	df	p-value
1.15e-4	-1.46e-5	3.1e-6	2.55e-5	-1.73e-6	0.223	0.204	4,164	1.87e-8
0.1%	1%	0.1%	0.1%	5%				

Table 6: Linear model of recurrence rates from site A to site K.

As can be seen, all the parameters of the model are statistically significant. The negative value for β_1 indicates that the recurrence rate decreases if the first day considered is in a closure period. The positive value for β_3 indicates that the recurrence rate increases if the second day considered is in the closure period. In other words, the percentage of drivers switching from A to K increases if we look at A before the closure and K during the closure. This is consistent with drivers switching route from site A to site K as a result of the closure. Meanwhile the parameters β_2 and β_4 have the opposite signs to β_1 and β_3 respectively. This indicates that the effects both decrease if the closure is in place for longer. This is consistent with the idea that initially a certain number of drivers swap their route from A to K when the closure occurs but fewer drivers do so as the time continues.

β_0	β_1	β_2	β_3	β_4	R^2	R_a^2	df	p-value
21.1	-6.33	0.421	9.90	-0.398	0.490	0.478	4,164	<1e-15
0.1%	0.1%	5%	0.1%	5%				

Table 7: Linear model of travellers from site A to site K.

If the same model is fitted to the estimated number of matches between the two sites instead of the recurrence rate then similar results are seen. The results are summarised in Table 7. This table gives some idea how small the effects being measured are. The estimated number of drivers that would take A before the closure and K after the first day of closure would only be approximately 10 according to this model. As can be seen, the effects being estimated in this model are very small and hard to pick out from the noise.

Discussion

The model fitted here shows evidence consistent with a small number of drivers changing route from site A to site K as a result of intervention. Perhaps even more interestingly, it shows evidence of a return to normal effect as time goes on. It provides further evidence for an "It'll be alright by Friday" effect. It should be stressed that all these recurrence rate effects may be subject to the decays in recurrence rate described in the previous section about recurrence rate.

The results in this section have an interesting implication for modellers. These results appear to show an initial "over-reaction" effect followed by a settling down effect. Some dynamic day-to-day models could well capture this behaviour but more real-life data analysis of this type would be needed to calibrate how many days it would take for this rerouting and settling down to occur.

CONCLUSION

The three main sections of this paper look at fitting different statistical models to data gathered on a traffic system. The aim is to better understand the behaviour of the traffic system. None of the results described are particularly unexpected but they are, nonetheless, important things to quantify if researchers are to produce calibrated simulation models of urban road networks. Three separate modelling results were obtained.

Firstly, recurrence rate at a site was shown to reduce sharply as the number of weekdays between the two observations increased. This recurrence rate reduced further if the days observed were in different weeks and increased if the days were the same day of the week (the "See you next Wednesday" effect. The recurrence rate was observed to differ between different sites.

Secondly, the effect on flows and travel times of a capacity reducing intervention was investigated. It was shown that in some cases this effect could be separated into an initial response and a dying down of this initial effect. This corresponds to the engineering rule of thumb that "It'll be alright by Friday", the idea that a change producing a large effect when initially implemented may well settle down within a short space of time.

Thirdly, the effect on rerouting of the intervention was investigated. Recurrence rates of traffic between two different sites likely to be the main two sites for rerouting was fitted to a statistical model. The model was shown to be consistent with an initial rerouting and a dying down of this response as the intervention continues. This is consistent with the idea that the initial rerouting was "too much" and this was subsequently corrected as time continued, as if the system over-reacts and then corrects.

Critical discussion

The models described here are certainly far from perfect. Due to the relative scarcity of data (and uncertainty in some measurements) the author deliberately kept the number of parameters modelled small and avoided using heavily parameterised models. For example, the linear dependence of recurrence rate on the difference in weekdays between the two days on which measurements take place is certainly only valid for a short time period (apart from anything else, this model would predict large negative recurrence rates after a few months). More data could build more sophisticated models of how the recurrence rate falls off in longer time periods.

All the work here depends on investigation of two interventions in a single city both within a year of each other. There is no way of knowing how similar such measurements would be if made while studying other cities and other interventions.
The data worked with is prone to a number of measurement errors (as previously described). This will certainly affect the results given. While it is accepted that, for example, the absolute level of the recurrence rate will be subject to some correction (perhaps large) due to these recording errors, it is hard to see how this could systematically affect the direction of changes. Therefore, while the absolute levels of the parameters in this paper may not be correct in all cases, the direction of the changes given seem more certain.

Finally, it should again be stressed that the author does not expect a general audience to have a particular interest in the exact behaviour of the traffic as a result of this particular intervention in this particular city. Indeed, these studies could be considered as a single data point with very different results would be obtained in different cities or at different times. A main aim of this paper is to stimulate further modelling and research along the lines described and, in this way, to understand how well these results generalise.

Implications for modelling and further research

Perhaps what is of most interest in this paper is that the techniques used could easily be used on subsequent data sets (as GPS and licence plate recognition cameras become more common it may well become easier to get good data sets for this purpose). This would establish the generality of these results. In turn this could then be used to inform the development of models of an urban road network.

In particular it may become very useful for models to be able to predict the transient response of a scheme on the earliest days after implementation and to know, not just what is likely to happen on day one but how long it will take for the situation to improve (assuming that an improvement is predicted).

This paper points to a number of directions for further research. If, in the future, data sets, from GPS or licence plate cameras become more readily available then this type of modelling can be used to investigate both the ambient variability of road traffic and also how driver behaviour is affected by interventions.

If network modellers truly wish to capture the on-street behaviour of drivers who learn and change their route as a result of changes to the network then those models must be calibrated against real data. The statistical models in this paper provide a starting point for modellers who want their learning drivers to behave in a realistic way, including the fact that drivers do not travel at the same time to the same destination every day. Instead driver behaviour will change as time goes on and that day of the week effects may have importance.

In addition, modellers may wish to be able to model how drivers on a network will respond to an intervention. If drivers do, as the evidence here seems to suggest, have an initial response and a settling down period then it may be important to know how long this period lasts and how severe the intermediate effects might be.

If these results could be backed up by further studies they would give a sound empirical basis for setting learning parameters and memory parameters in models that consider how individual drivers reroute as day follows day.

REFERENCES

Bonsall, P., Montgomery, F. And Jones, C. (1984) Deriving the constancy of traffic flow composition from vehicle registration data. *Traffic Engineering and Control*, 25 (7/8), pp386-391.

Cairns, S., Carmen, H-K. and Goodwin, P. (1998) *Traffic Impact of Highway Capacity Reductions: Assessment of the Evidence*. Landor Publishing.

Cairns, S., Atikins, S. and Goodwin, P. (2002) Disappearing Traffic? The Story so Far *Proceedings of the Institution of Civil Engineers, Municipal Engineer* 151(1) pp13-22

Clegg, R. G. (2005). An Empirical Study of Day-to-Day Variability in Driver Travel Behaviour. *Presented at Universities Transport Studies Group* 2005 and available online at: www.richardclegg.org/pubs/rgc_utsg2005.doc

Clegg, R. G. (2005b). *The Statistics of Dynamic Networks*. PhD Thesis, Department of Mathematics, University of York. Available online at: www.richardclegg.org/pubs/thesis.pdf

Clegg, R. G. (2006). "It'll be Alright by Friday": Traffic Response to Capacity Reduction. *Presented at Universities Transport Studies Group* 2006 and available online at: www.richardclegg.org/pubs/rgc_utsg2006.doc

Daugherty, G.G., Balcombe, R.J. and Astrop, A.J. (1999) A comparative assessment of major bus priority schemes in Great Britain. *TRL Report 409*.

Goodwin, P. (2005). *Final report of "Utilities' Street Works and the Cost of Traffic Congestion"* EPSRC project, available online at: www.transport.uwe.ac.uk/research/projects/njugcongestionreportfinal4goodwin.pdf

Hauer, E. (1979). Correction of licence plate surveys for spurious matches, *Transportation Research A* 13A pp 71—78.

Huff, J. and Hanson, S. (1986) Repetition and variability in urban travel. *Geographical Analysis*, 18 (2), pp97-114.

Jan, O., Horowitz, A.J. and Peng Z-R. (1999) Using GPS data to understand variations in path choice. *Paper presented at the 78th meeting of the Transportation Research Board, Washington DC*, January 1999.

Maher, M. J. (1985). The analysis of partial registration-plate data, *Traffic Engineering and Control* 26 (10) pp 495—497.

Marvin, S. and Slater, S. (1997) Urban Infrastructure: The Contemporary Conflict Between Roads and Utilities, *Progress in Planning*, Vol. 48, Part 4. pp. 247-318.

Mendenhall, W. and Sincich, J. (2000). *Statistics for Engineering and the Sciences*. Prentice-Hall (Englewood Cliffs, N.J.)

Pendyala, R. M. (2003). *Measuring day-to-day variability in travel behavior using GPS data.* Report for Federal Highway Administration. Available online at: www.fhwa.dot.gov/ohim/gps/index.html

Stephenson, B. and Tepley, S. (1984) The verification of CONTRAM in Edmonton, *Traffic Engineering and Control*, 25 (7/8), pp376-385.

Watling, D. P. and Maher, M. J. (1988). A graphical procedure for analysing partial registration-plate data, *Traffic Engineering and Control* 29 (10) pp515—519

Watling, D. P. and Maher, M. J. (1992). A statistical procedure for estimating a mean origin-destination matrix from a partial registration plate survey, *Transportation Research B* 26B(3) pp171—193

Watling, D. P. (1994): Maximum likelihood estimation of an origin-destination matrix from a partial registration plate survey, *Transportation Research B* 28B(3) pp289—314

Transportation and Traffic Theory 2007
Edited by R.E. Allsop, M.G.H. Bell and B.G. Heydecker

8

VARIATION OF VALUE OF TRAVEL TIME SAVINGS OVER TRAVEL TIME IN URBAN COMMUTING: THEORETICAL AND EMPIRICAL ANALYSIS

Hironori Kato, Department of Civil Engineering, The University of Tokyo, Japan

INTRODUCTION

In measuring the benefit stemming from a particular transport project, the starting point is generally the traveller's willingness to pay: the amount of money each individual would be willing to pay for the resulting change in her or his circumstances. Typically, the dominant component of the benefit derived from transport investment is a saving in travel time. There have been many empirical and theoretical studies of the value of travel time saving (VTTS) following the introduction of the economic theory of time allocation in the 1960s. It was Becker (1965) who first suggested that a consumer gains utility from the consumption of both time and goods, and not from consumed goods only. Following on from Becker's work, several researchers such as DeSerpa (1971), Evans (1972) and Small (1982) have developed time-allocation models in which utility to the consumer is maximized with respect to the consumption of time and goods under the constraints of available time and money. Simultaneously, several different definitions of VTTS have been proposed (Jara-Diaz, 2000). DeSerpa's definition of VTTS is particularly important as it includes two distinct values of time: the value of time as a resource and its value as a commodity.

In the traditional time-allocation framework, we focus on the individual's behaviour over the short term. This means that individuals never change jobs and they remain in the same houses. However, as Small (1999) points out, many factors such as constant job turnover, house-moving, family status and habits may have an effect on travel time savings in the long run. In general, whether a consumer model should be formulated from a short-term or a long-term viewpoint depends on the behavioural context that a modeller assumes in her or his analysis. For example, if we set out to analyze an individual's leisure travel on a daily basis, we probably do not need to consider choice of residential location. This is because we generally expect that individuals neglect the location of their residence in choosing leisure activities in which to participate. However, consider the act of commuting. Can we expect commuters to make decisions from a short-term viewpoint only? The answer is probably 'no'. We may need to take account of where the commuter chooses to live, because individuals often consider both choices simultaneously. In such cases, individuals choose travel time not only from the viewpoint of the trade-off between travel time and travel cost, but also take into account the trade-off between travel time and cost of housing. Thus residential location choice may lead to different VTTS results. In this paper, VTTS analysis is carried out with explicit consideration of choice of residential location.

Regarding empirical analysis of VTTS, models based on disaggregate discrete choice have so far been the most popular approach. Train and McFadden (1978), using the choice of mode for the home to workplace trip, show that the conditional indirect utility function formulated in discrete choice theory will give the value of travel time savings as the marginal substitution rate between travel time and travel cost. In a similar manner, Truong and Hensher (1985) and later discussions (Bates, 1987; Truong and Hensher, 1987) show how the Becker and DeSerpa models can be incorporated into VTTS estimation within the discrete choice model framework. These days, in practical transport planning, VTTS is usually estimated using a discrete choice model, such as the multinomial logit model. Constant values of VTTS estimated with a discrete choice model are often used in practical transport planning; this constancy derives simply from an assumption of a linear utility function. However, this linearity assumption may not be acceptable, as De Lapparent *et al.* (2004) point out.

By using a non-linear utility function, we can derive a non-constant VTTS with respect to travel time. There have been a number of investigations studying the relationship between VTTS and travel time (Hague Consulting Group and Accent Marketing & Research, 1996; Hensher, 1997; Wardman, 1998; 2001; 2004; Hultkrantz and Mortazavi, 2001; Axhausen *et al.*, 2004). This paper adds new evidence to the body of research related to non-constant VTTS.

In this investigation, the variation of VTTS over the travel time of urban commuters is examined both theoretically and empirically by incorporating choice of residential location and a non-linear utility function. The following section of the paper describes the formulation of a model for an individual's time allocation allowing for residential location choice. We then derive VTTS from the model and analyze it under market equilibrium. Next, we provide a theoretical analysis of the variation of VTTS over travel time. This is followed by an empirical analysis using the revealed preference data of urban rail users in Tokyo. We analyze the variation of VTTS over travel time empirically. Finally, we summarize the results and discuss further topics of research.

VTTS DERIVED FROM TIME ALLOCATION MODEL WITH RESIDENTIAL LOCATION CHOICE

Model formulation and derivation of VTTS

We formulate a consumer's behaviour with respect to time allocation and residential location choice with the theory of urban economics and a time allocation model (Alonso, 1964; Solow, 1973; Becker, 1965). Basically, we follow the simple residential land use model presented by Kanemoto (1980).

Suppose a simple city that has a unique central business district (CBD) at its core. All households are identical. They have the same preferences and the same number of workers. For simplicity, we assume that each household has one worker. An individual resides at only one location. This assumption eliminates, for example, households with an apartment in the city and a house in the suburbs; the actual number of such households is so small that they can safely be ignored. Housing capital can be instantaneously adjusted. Although housing is in reality a durable good, we assume that all the characteristics of houses such as lot size and building area can be changed instantaneously. Ours is, therefore, a city in an imaginary long-term equilibrium state in which the capital-land ratio is always perfectly adjusted. Travel cost stems from commuting cost only.

We assume an individual gains utility from leisure time, travel time, work time, consumption of a composite good and consumption of land for housing. The individual is constrained by available time and budget and there is a constraint consisting of a fixed minimum travel time. We also assume utility maximization as the behavioural principle. Let the utility of the individual be u. An individual's behaviour can then be described as:

$$\max_{z,h,T,T_w,t} \quad u = u(z,h,T,T_w,t) \tag{1}$$

$$\text{subject to } R\cdot h + P\cdot z + c = \omega\cdot T_w \tag{2a}$$

$$T + T_w + t = T^o, \quad t \geq \bar{t} \tag{2b,c}$$

where u is a utility function; z is the composite good; h is land area; T is time available for leisure; t is travel time; R is rent; P is the price of the composite good; c is travel cost; ω is wage rate; T_w is the work duration; T^o is available time and $\bar{t}$ is the minimum travel time. We assume that the marginal utility with respect to travel time is negative.

Let a Lagrange function for this optimization problem be,

$$L = u(z,h,T,T_w,t) + \lambda(\omega\cdot T_w - R\cdot h - P\cdot z - c) + \mu(T^o - T - T_w - t) + \kappa(t - \bar{t}) \tag{3}$$

where λ, μ, κ are the Lagrange multipliers. Then, from the first-order optimality conditions, we can derive

$$\frac{\partial u}{\partial z} = \lambda\cdot P, \quad \frac{\partial u}{\partial h} = \lambda\cdot R, \quad \frac{\partial u}{\partial T} = \mu \tag{4a,b,c}$$

$$\frac{\partial u}{\partial t} = \mu - k = 0, \quad \frac{\partial u}{\partial T_w} = \mu - \omega\cdot\lambda \tag{4d,e}$$

In general, we can expect that rent R and travel cost c are some functions of travel time. More exactly, they are functions of minimum travel time. In the same way, we assume that the wage rate also depends on minimum travel time. Then, we can describe the individual's indirect utility function, which is derived from utility maximization, as $v(R(\bar{t}), Y(\bar{t}) - c(\bar{t}))$, where $Y \equiv \omega(\bar{t}) \cdot T_w - P \cdot z$.

Then we can derive the following two equations from the Envelope Theorem (Varian, 1987):

$$\frac{\partial v}{\partial \bar{t}} = \frac{\partial u}{\partial \bar{t}} + \frac{\partial \lambda (Y - c - R \cdot h)}{\partial \bar{t}} + \frac{\partial \kappa (t - \bar{t})}{\partial \bar{t}} = \lambda \left(\frac{\partial \omega}{\partial \bar{t}} T_w - \left(\frac{\partial R}{\partial \bar{t}} \cdot h + \frac{\partial c}{\partial \bar{t}} \right) \right) - \kappa \tag{5}$$

$$\frac{\partial v}{\partial c} = \frac{\partial u}{\partial c} + \frac{\partial \lambda (Y - c - R \cdot h)}{\partial c} = -\lambda \tag{6}$$

If we define the value of travel time savings as "the willingness to pay to recover the utility level to the original situation when the minimum travel time is reduced" as DeSerpa shows, we can derive VTTS from our model as

$$VTTS = \frac{\partial v / \partial \bar{t}}{\partial v / \partial c} = \left\{ \left(\frac{\partial R}{\partial \bar{t}} \cdot h + \frac{\partial c}{\partial \bar{t}} \right) - \frac{\partial \omega}{\partial \bar{t}} \cdot T_w \right\} + \frac{\kappa}{\lambda} \tag{7}$$

This means that VTTS consists of four components. The first is the value of relaxing the constraint of minimum travel time, the second is the value of income change stemming from the change of wage rate caused by travel time saving, the third is the value of cost saving caused by travel time saving, and the fourth is the value of land-cost change caused by travel time saving.

VTTS under the market equilibrium

Since households are identical in our model, the utility level at equilibrium must be the same everywhere in the city. Otherwise, households at a place of lower utility level would have an incentive to relocate and the allocation would not meet the requirements of a market equilibrium. Therefore, under market equilibrium, the indirect utility function should satisfy

$$v(R(\bar{t}), Y(\bar{t}) - c(\bar{t})) = u(const.) \tag{8}$$

By solving this equation with respect to $R(\bar{t})$, we can derive the bid rent function as

$$R(\bar{t}) = R(Y(\bar{t}) - c(\bar{t}), u) \tag{9}$$

This describes the maximum rent that a household can pay at a particular distance from the centre if it is to receive the given utility level. Here, define $I = Y(\bar{t}) - c(\bar{t})$. By total differentiation, Equation (8) yields

$$\frac{\partial v}{\partial R} dR + \frac{\partial v}{\partial I} dI = 0 \tag{10}$$

From Equation (10), we obtain

$$\frac{dR}{dI} = \frac{\partial R}{\partial I} = -\frac{\partial v / \partial I}{\partial v / \partial R} \tag{11}$$

On the other hand, from Roy's Identity, the indirect utility function (8) should satisfy

$$-\frac{\partial v / \partial R}{\partial v / \partial I} = h \tag{12}$$

As $\partial I/\partial \bar{t} = (\partial \omega/\partial \bar{t})T_w - \partial c/\partial \bar{t}$ should be satisfied, finally we derive the following equation from Equations (11) and (12),

$$\frac{\partial R}{\partial \bar{t}} = \frac{\partial R}{\partial I}\cdot\frac{\partial I}{\partial \bar{t}} = \frac{1}{h}\cdot\left(\frac{\partial \omega}{\partial \bar{t}}\cdot T_w - \frac{\partial c}{\partial \bar{t}}\right) \tag{13}$$

This shows that the sum of marginal income and marginal travel cost with respect to minimum travel time should be equal to the marginal land-cost with respect to minimum travel time under market equilibrium.

By substituting Equation (13) into the VTTS formula (7), we find that VTTS is reduced to κ/λ. As shown later, this is the same as the value of VTTS derived from the simple time allocation model. Under market equilibrium, the formula for VTTS reduces to a simple equation including the value of relaxing the constraint of minimum travel time alone. This result may prove very useful in VTTS research, because it means we can neglect individual location choice in VTTS discussions. Moreover, the above result is quite general. It is not affected by the functional forms of the utility function, the rent function, the travel cost function or the wage rate function. However, we should bear in mind the basic assumptions of this model, especially the assumption of the simple form of city.

THEORETICAL ANALYSIS OF VTTS OVER TRAVEL TIME

Time allocation model and VTTS

From this section onward, we assume equilibrium in the land market. By making this assumption, we are able to neglect choice of residential location, as shown earlier. Land will be treated as one component of the composite good. Then, we consider that an individual gains utility from leisure time, work time, travel time and consumption of the composite good. Let the utility of the individual be u. The time-allocation model can then be formulated as

$$\max_{z,T,T_w,t} \; u = u(z,T,T_w,t) \tag{14}$$

$$\text{subject to} \quad z + c = \omega\cdot T_w, \quad T + T_w + t = T^o, \quad t \ge \bar{t} \tag{15a,b,c}$$

The Lagrange function corresponding to the above time-allocation model is shown as

$$L = u(z,T,T_w,t) + \lambda(\omega\cdot T_w - z - c) + \mu(T^o - T - T_w - t) + \kappa(t - \bar{t}) \tag{16}$$

where λ, μ and κ are the Lagrange multipliers. The first-order conditions of optimality are derived as

$$\frac{\partial u}{\partial z} = \lambda, \quad \frac{\partial u}{\partial T} = \mu, \quad \frac{\partial u}{\partial T_w} = \mu - \omega\lambda, \quad \frac{\partial u}{\partial t} = \mu - \kappa \tag{17a,b,c,d}$$

and equations (15a) to (15c).

Next, let the indirect utility function of the individual be $v(c, T^o, \bar{t})$. Then, by applying the Envelope Theorem to the above utility maximization problem, we obtain

$$\frac{\partial v}{\partial c} = \frac{\partial u}{\partial c} + \lambda \frac{\partial(\omega \cdot T_w - z - c)}{\partial c} = -\lambda \tag{18}$$

$$\frac{\partial v}{\partial \bar{t}} = \frac{\partial u}{\partial \bar{t}} + \kappa \frac{\partial(t - \bar{t})}{\partial \bar{t}} = -\kappa \tag{19}$$

As VTTS can be defined as the willingness to pay for travel time savings, VTTS can be derived from Equations (18) and (19) as

$$VTTS = \frac{\partial v / \partial \bar{t}}{\partial v / \partial c} = \frac{\kappa}{\lambda} \tag{20}$$

On the other hand, from the first-order optimality conditions, we can obtain VTTS as

$$VTTS = \frac{k}{\lambda} = \omega + \frac{\partial u / \partial T_w}{\lambda} - \frac{\partial u / \partial t}{\lambda} \tag{21}$$

Comparative static analysis

We next examine the variation of VTTS over travel time based on the time-allocation model. In order to see this variation, we must visualize the impact of travel time on the components of VTTS shown in Equation (20), which satisfy the first-order optimality conditions of Equations (15) and (17). We carry out the analysis using a method of comparative static analysis presented by Kono and Morisugi (2000) and Jiang and Morikawa (2004).

First, by total differential of Equations (15a) to (15c), we obtain

$$\begin{bmatrix} \frac{\partial^2 u}{\partial z^2} & \frac{\partial^2 u}{\partial z \partial T} & \frac{\partial^2 u}{\partial z \partial T_w} & \frac{\partial^2 u}{\partial z \partial t} \\ \frac{\partial^2 u}{\partial T \partial z} & \frac{\partial^2 u}{\partial T^2} & \frac{\partial^2 u}{\partial T \partial T_w} & \frac{\partial^2 u}{\partial T \partial t} \\ \frac{\partial^2 u}{\partial T_w \partial z} & \frac{\partial^2 u}{\partial T_w \partial T} & \frac{\partial^2 u}{\partial T_w^2} & \frac{\partial^2 u}{\partial T_w \partial t} \\ \frac{\partial^2 u}{\partial t \partial z} & \frac{\partial^2 u}{\partial t \partial T} & \frac{\partial^2 u}{\partial \partial t \partial T_w} & \frac{\partial^2 u}{\partial t^2} \end{bmatrix} \begin{bmatrix} \frac{\partial z}{\partial \lambda} & \frac{\partial z}{\partial \mu} & \frac{\partial z}{\partial \kappa} \\ \frac{\partial T}{\partial \lambda} & \frac{\partial T}{\partial \mu} & \frac{\partial T}{\partial \kappa} \\ \frac{\partial T_w}{\partial \lambda} & \frac{\partial T_w}{\partial \mu} & \frac{\partial T_w}{\partial \kappa} \\ \frac{\partial t}{\partial \lambda} & \frac{\partial t}{\partial \mu} & \frac{\partial t}{\partial \kappa} \end{bmatrix} = \begin{bmatrix} 1 & 0 & 0 \\ 0 & 1 & 0 \\ -\omega & 1 & 0 \\ 0 & 1 & -1 \end{bmatrix} \tag{22}$$

To simplify the analysis, we assume that the utility function is additive separable with respect to z, T, T_w, t. This means,

$$\frac{\partial^2 u}{\partial z \partial T} = \frac{\partial^2 u}{\partial z \partial T_w} = \frac{\partial^2 u}{\partial z \partial t} = \frac{\partial^2 u}{\partial T \partial T_w} = \frac{\partial^2 u}{\partial T \partial t} = \frac{\partial^2 u}{\partial T_w \partial t} = 0 \tag{23}$$

We also assume that the marginal utility, both with respect to consumption of the composite good z and with respect to leisure time T, is positive and decreasing, according to neoclassical microeconomics theory. Therefore, there are

$$\frac{\partial u}{\partial z} > 0, \ \frac{\partial^2 u}{\partial z^2} < 0, \ \frac{\partial u}{\partial T} > 0, \ \frac{\partial^2 u}{\partial T^2} < 0 \tag{24}$$

Then, Equations (22) and (24) lead to

$$\frac{\partial z}{\partial \lambda} < 0, \ \frac{\partial T}{\partial \mu} < 0 \tag{25}$$

We assume that the marginal utility with respect to work time is negative according to past research (Greenven *et al.*, 2005; Jara-Diaz *et al.*, 2004).

$$\frac{\partial u}{\partial T_w} < 0 \tag{26}$$

On the other hand, we cannot assume a priori whether the marginal utility with respect to work time and the marginal utility with respect to travel time increase or decrease. That means we have to consider the following two cases:

$$if \ \frac{\partial^2 u}{\partial T_w^2} > 0 \Rightarrow \frac{\partial T_w}{\partial \lambda} < 0, \ \frac{\partial T_w}{\partial \mu} > 0, \quad else\,if \ \frac{\partial^2 u}{\partial T_w^2} < 0 \Rightarrow \frac{\partial T_w}{\partial \lambda} > 0, \ \frac{\partial T_w}{\partial \mu} < 0 \tag{27a}$$

$$if \ \frac{\partial^2 u}{\partial t^2} > 0 \Rightarrow \frac{\partial t}{\partial \mu} > 0, \ \frac{\partial t}{\partial \kappa} < 0, \quad else\,if \ \frac{\partial^2 u}{\partial t^2} < 0 \Rightarrow \frac{\partial t}{\partial \mu} < 0, \ \frac{\partial t}{\partial \kappa} > 0 \tag{27b}$$

From above considerations, we can describe the equations (15a), (15b) and (15c) as

$$z(\lambda) + c = \omega \cdot T_w(\lambda, \mu) \tag{28a}$$

$$T(\mu) + T_w(\lambda, \mu) + t(\mu, \kappa) = T^o \tag{28b}$$

$$t(\mu, \kappa) = \bar{t} \tag{28c}$$

where we assume that the travel time is equal to the minimum travel time.

Let the wage rate and travel time follow the function of minimum travel time, as the earlier analysis assumes. We can then derive the following equations from the total differential of Equations (28a), (28b) and (28c) with respect to $\bar{t}$:

$$\frac{\partial z}{\partial \lambda} \cdot \frac{\partial \lambda}{\partial \bar{t}} + \frac{\partial c}{\partial \bar{t}} = \frac{\partial \omega}{\partial \bar{t}} \cdot T_w + \omega \cdot \left(\frac{\partial T_w}{\partial \lambda} \cdot \frac{\partial \lambda}{\partial \bar{t}} + \frac{\partial T_w}{\partial \mu} \cdot \frac{\partial \mu}{\partial \bar{t}} \right) \tag{29a}$$

$$\frac{\partial T}{\partial \mu} \cdot \frac{\partial \mu}{\partial \bar{t}} + \frac{\partial T_w}{\partial \lambda} \cdot \frac{\partial \lambda}{\partial \bar{t}} + \frac{\partial T_w}{\partial \mu} \cdot \frac{\partial \mu}{\partial \bar{t}} + \frac{\partial t}{\partial \mu} \cdot \frac{\partial \mu}{\partial \bar{t}} + \frac{\partial t}{\partial \kappa} \cdot \frac{\partial \kappa}{\partial \bar{t}} = 0 \tag{29b}$$

$$\frac{\partial t}{\partial \mu} \cdot \frac{\partial \mu}{\partial \bar{t}} + \frac{\partial t}{\partial \kappa} \cdot \frac{\partial \kappa}{\partial \bar{t}} = 1 \tag{29c}$$

By solving these equations, we obtain

$$\frac{\partial \mu}{\partial \bar{t}} = \frac{\omega \dfrac{\partial T_w}{\partial \lambda} \dfrac{\partial T}{\partial \mu} - \dfrac{\partial z}{\partial \lambda} \dfrac{\partial T}{\partial \mu} - \dfrac{\partial z}{\partial \lambda} \dfrac{\partial T_w}{\partial \mu}}{T_w \dfrac{\partial T_w}{\partial \lambda} \dfrac{\partial w}{\partial \bar{t}} - \dfrac{\partial T_w}{\partial \lambda} \dfrac{\partial c}{\partial \bar{t}} - \omega \dfrac{\partial T_w}{\partial \lambda} + \dfrac{\partial z}{\partial \lambda}} \tag{30a}$$

$$\frac{\partial \lambda}{\partial \bar{t}} = -\frac{\left(\dfrac{\partial T}{\partial \mu} + \dfrac{\partial T_w}{\partial \mu} \right) \dfrac{\partial \mu}{\partial \bar{t}} + 1}{\dfrac{\partial T_w}{\partial \lambda}} \tag{30b}$$

$$\frac{\partial \kappa}{\partial \bar{t}} = \frac{1 - \dfrac{\partial t}{\partial \mu} \dfrac{\partial \mu}{\partial \bar{t}}}{\dfrac{\partial t}{\partial \kappa}} \tag{30c}$$

In order to clarify the signs of the above three equations, we assume the following:

- The marginal travel cost with respect to minimum travel time is positive: $\partial c / \partial \bar{t} > 0$.
- The marginal wage rate with respect to minimum travel time is negative: $\partial \omega / \partial \bar{t} > 0$.

The first assumption is quite natural, because in general travel cost increases as travel time increases. The second assumption may be dependent on market characteristics. We follow the empirical results of Yoshida and Endo (1999), which show average wage rate decreasing as travel time increases in the Tokyo Metropolitan Area.

Then, from Equation (22), we can derive $\partial\mu/\partial\bar{t} > 0$ if the utility function satisfies $\partial^2 u/\partial T_w^2 < 0$. However, we cannot fix the sign of $\partial\mu/\partial\bar{t}$ if $\partial^2 u/\partial T_w^2 > 0$. Even if we were to assume $\partial\omega/\partial\bar{t} \geq 0$, we would not be able to fix the sign of $\partial\mu/\partial\bar{t}$. In the same way, we can derive $\partial\kappa/\partial\bar{t} > 0$ if the utility function satisfies $\partial^2 u/\partial T_w^2 < 0$ and $\partial^2 u/\partial t^2 < 0$. However, we cannot fix the sign of $\partial\kappa/\partial\bar{t}$ in other cases. We cannot fix the sign of $\partial\lambda/\partial\bar{t}$.

We summarize the signs of $\partial\lambda/\partial\bar{t}$, $\partial\mu/\partial\bar{t}$ and $\partial\kappa/\partial\bar{t}$ derived from our analysis in Table 1. As Table 1 shows, we cannot fix the signs of all elements of VTTS simultaneously. In other words, we have to assume a specific utility function in order to fix the sign of the marginal VTTS with respect to minimum travel time.

As a special case, we may point out a case which satisfies the following three conditions simultaneously:
- The marginal utility with respect to work time is decreasing: $\partial^2 u/\partial T_w^2 < 0$.
- The marginal utility with respect to (minimum) travel time is decreasing: $\partial^2 u/\partial\bar{t}^2 < 0$.
- The marginal utility with respect to income will decrease as the (minimum) travel time increases: $\partial\lambda/\partial\bar{t} < 0$.

This case leads to a simple result; that is, VTTS will increase monotonously as (minimum) travel time increases. We will examine the condition $\partial\lambda/\partial\bar{t} < 0$ more deeply for this special case. Since the utility function satisfies $\partial^2 u/\partial T_w^2 < 0$, the work time should satisfy $\partial T_w/\partial\lambda > 0$ and $\partial T_w/\partial\mu < 0$ as Equation (27a) shows. Then, from Equation (30b), the condition of $\partial\lambda/\partial\bar{t} < 0$ is derived as:

$$\left(\frac{\partial T}{\partial\mu} + \frac{\partial T_w}{\partial\mu}\right)\frac{\partial\mu}{\partial\bar{t}} > -1 \tag{31}$$

The left-hand side of Equation (31) is negative, because the three terms on the left-hand side satisfy $\partial T/\partial\mu < 0$, $\partial T_w/\partial\mu < 0$ and, $\partial\mu/\partial\bar{t} > 0$, respectively, where $\partial\mu/\partial\bar{t} > 0$ can be read from Table 1. However, we cannot judge whether the left-hand side of Equation (31) is greater than -1 or not. Consequently, we can summarize the conditions for the above special case as

Table 1 Signs of three elements of marginal VTTS with respect to minimum travel time

$\frac{\partial^2 u}{\partial T_w^2}$	$\frac{\partial^2 u}{\partial t^2}$	$\frac{\partial\lambda}{\partial\bar{t}}$	$\frac{\partial\mu}{\partial\bar{t}}$	$\frac{\partial\kappa}{\partial\bar{t}}$
+	+	?	?	?
	-	?	?	?
-	+	?	+	?
	-	?	+	+

follows: if the utility function satisfies $\partial^2 u/\partial T_w^2 < 0$, $\partial^2 u/\partial \bar{t}^2 < 0$ as well as Equation (31), VTTS will increase as (minimum) travel time increases. Note that the above conditions are not the only ones that give a monotonic relation between VTTS and travel time.

From this comparative static analysis, we conclude that it is difficult to obtain a simple result for the relationship between travel time and VTTS, even if we assume an additive separable utility function. Moreover, as Kono and Morisugi (2000) point out, there is no reason why we should use an additive separable utility function. Hence, in the next section, we will analyze the variation of VTTS over travel time not from a theoretical viewpoint, but using an empirical approach. In this empirical analysis, we will not assume additive separability of the utility function, but rather will approximate the utility function. Using this approximation, the analysis may not provide a strict solution for the relationship between VTTS and travel time, but we will be able to examine it with richer implications.

EMPIRICAL ANALYSIS OF VTTS OVER TRAVEL TIME

Derivation of VTTS based on the discrete-choice model with non-linear utility function

The discrete choice model system is used to clarify variations in VTTS over travel time. The discrete choice model assumes utility maximization by individuals under the condition that a specific service is chosen to the exclusion of all others. Suppose an individual chooses transport service i. Let u_i be the conditional utility function. Then, the conditional utility maximization is given as

$$\max_{z,T,T_w,t_i} \; u_i = u(z,T,T_w,t_i) \tag{32}$$

$$\text{subject to, } z + c_i = \omega \cdot T_w, T + T_w + t_i = T^o, t_i \geq \bar{t}_i \tag{33a,b,c}$$

We can obtain the following equations from the first optimality condition:

$$\frac{\partial u_i}{\partial z} = \lambda^*, \; \frac{\partial u_i}{\partial T} = \mu^*, \; \frac{\partial u_i}{\partial T_w} = \mu^* - \omega\lambda^*, \; \frac{\partial u_i}{\partial t_i} = \mu^* - \kappa^* \tag{34a,b,c,d}$$

Then, we derive an approximation of the direct utility function and the indirect utility function. There have been some investigations that formulate the non-linear utility function by approximation (for example, Axhausen *et al.*, 2004; De Lapparent *et al.*, 2002). Here, we follow a method shown by Blayac and Causse (2001). First, we obtain the first-order approximation from the Taylor expansion of the direct utility function as

$$u_i(z,T,T_w,t_i) = \frac{\partial u_i}{\partial z} z + \frac{\partial u_i}{\partial T} T + \frac{\partial u_i}{\partial T_w} T_w + \frac{\partial u_i}{\partial t_i} t_i + X_{1\_i} \tag{35}$$

By substituting the first-order optimality conditions into Equation (35), an approximated indirect utility function is derived as

$$\begin{aligned} v_i(c_i,\bar{t}_i,T^o) &= \lambda^*(\omega T_w - c_i) + \mu^*(T^o - T_w - \bar{t}_i) + (\mu^* - \omega\lambda^*)T_w + (\mu^* - \kappa_i^*)\bar{t}_i + X_{1\_i} \\ &= -\lambda^* c_i - \kappa_i^* \bar{t}_i + \mu^* T^o + Z_{1\_i} \end{aligned} \tag{36}$$

As Ben-Akiva and Lerman (1985) show, the generic variables among the conditional indirect utility functions cannot affect the individual's probability of making a specific choice option.

Thus, by omitting the generic variables from Equation (36), we can rewrite the conditional indirect utility function as

$$v_{1\_i} = \theta_{1\_c} c_i + \theta_{1\_t,i} \bar{t}_i + \theta_{1\_i} \tag{37}$$

This indicates that the parameters of the conditional indirect utility function should be dependent on the service chosen. However, in our empirical analysis we assume that the parameters are common among the various services. This is because one individual has a different choice set from another individual in our empirical analysis.

We can derive the quadratic indirect utility function in the same way as the first-order approximation. It is shown as

$$v_{2\_i} = \theta_{2\_c} c_i + \theta_{2\_t} \bar{t}_i + \theta_{2\_c2} c_i^2 + \theta_{2\_t2} \bar{t}_i^2 + \theta_{2\_ct} c_i \bar{t}_i + \theta_{2\_i} \tag{38}$$

When travel cost follows a function of minimum travel time, such as $c_i(\bar{t}_i)$, the approximated VTTSs are derived as

$$VTTS_{1\_i} = \frac{\theta_{1\_t} + \theta_{1\_c}\left(\partial c_i / \partial \bar{t}_i\right)}{\theta_{1\_c}} \tag{39a}$$

$$VTTS_{2\_i} = \frac{\theta_{2\_t} + 2\theta_{2\_t2}\bar{t}_i + \theta_{2\_c}\left(\partial c_i / \partial \bar{t}_i\right) + 2\theta_{2\_c2} c_i \left(\partial c_i / \partial \bar{t}_i\right) + \theta_{2\_ct}\left(c_i + \bar{t}_i\left(\partial c_i / \partial \bar{t}_i\right)\right)}{\theta_{2\_c} + 2\theta_{2\_c2} c_i + \theta_{2\_ct}\bar{t}_i} \tag{39b}$$

Empirical Analysis

The empirical analysis makes use of revealed preference data collected during the 8th Tokyo Metropolitan Transport Census. This data was originally collected through a paper-based questionnaire in the Tokyo Metropolitan Area in October and November 1995. The original data includes origin and destination information about urban rail users, purpose of journey, chosen route including first station, final station and transfer points and users' socio-demographic data. Sample data are selected using the following procedure. First, only commuter journeys by workers are selected. This is simply because our model deals with commuter journeys. Second, we select commuters who have two or more than two alternative routes. This is because we want to analyze route choice behaviour. Third, we select commuters whose destinations are located within the central business district (CBD). This is because our model basically assumes a city with a unique CBD in which all residents commute to the CBD. These selection criteria lead ultimately to data on 1,218 journeys.

For these selected journeys, we prepare data relating to level of service. We determine the travel cost and travel time for home-to-workplace journeys, including access from home to the origin stations, travel from the destination station to the workplace, and rail travel between the stations. As a part of the route choice, we consider an origin station choice as well as a rail route choice. That is, we prepare data for alternative routes between different pairs of stations where available. The available stations and routes are selected as follows: First, we set access links from centres of origin zones to adjacent stations. The maximum number of access links is five for each zone. Second, we calculate the home-to-workplace travel time for all available routes. Third, we identify three routes for each origin-destination pair by selecting the shortest, the second shortest and the third shortest home-to-workplace travel-time routes. Consequently, two or more than two origin stations may be included in the route choice set. We select the three shortest travel-time routes in the route choice set, because the empirical analysis in the Tokyo Metropolitan Area (Kosuda *et al.*, 2004) shows that 99 % of all rail-use travellers choose either of the three shortest travel-time routes.

Before considering parameter estimation, we will analyze the relationship between travel time and travel cost. Figure 1 shows the travel time and cost of routes chosen by commuters included in the data used for parameter estimation. We find that the marginal travel cost with respect to travel time seems to be increasing as the travel time increases. There are three reasons for this. The first is the influence of initial cost in rail-use travel. There are about twenty private rail operators in Tokyo alone. Each sets its own fare table independently. However, all generally have a fare structure consisting of a fixed initial fare that must be paid for any journey. That is, the fare between any two stations comprises this initial fare and a distance-based fare. Someone travelling only a short distance pays just the initial fare, while another rail user going further has to pay more. Moreover, a rail user travelling a long way may need to use more than one rail operator. In such a case of a multi- operator journey, the user has to pay the initial fare several times. This existence of an initial fare causes a non-linearity in the travel cost curve. The second reason for the rising marginal travel cost with respect to travel time is the influence of non-charged or low-cost transport modes such as walking and bicycle. The travel time shown in Figure 1 includes access time from home to station and egress time from station to workplace. When a rail user's journey takes little time, the marginal travel cost with respect to travel time is nearly constant, because the ratio of travel time by foot or bicycle to the total travel time is quite high. On the other hand, when a user has a longer journey, the influence of the walked or cycled portion decreases and the marginal cost may become positive. The third reason is the influence of commuter's choice of rail-service type. In Tokyo, many rail operators provide various types of express services in the urban rail network. Some types of express service do not require an additional charge as compared to the local service. When an express service is available for no extra charge, long-

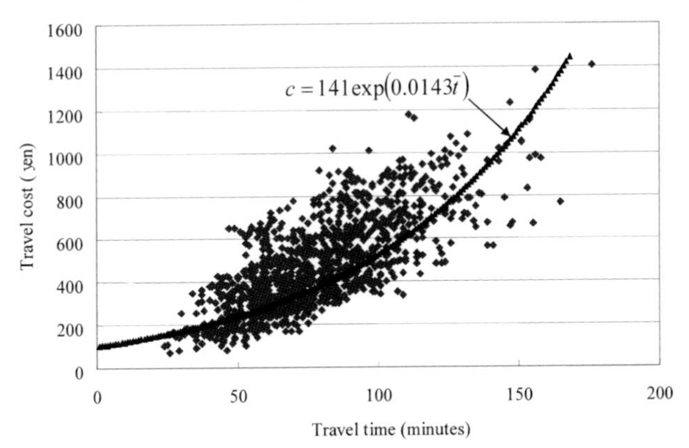

Figure 1 Travel time vs. travel cost of urban rail service in Tokyo.

distance rail users tend to choose the express service more than short-distance users. This may also result in non-linearity in the travel cost curve.

Based on the above considerations, we assume an exponential curve for the relationship between travel time and travel cost. We estimate the two parameters of the curve as

$$c = 141\exp(0.0143t) \quad R^2 = 0.530 \tag{34}$$

where c means travel cost (in yen) and t means travel time (in minutes). It is clear from this that the initial travel cost is estimated as 141 yen.

Results of Empirical Analysis

Table 2 shows the results of parameter estimation with the multinomial logit model. As Yai *et al.* (1997) and Morichi *et al.* (2001) point out, the similarity of the routes may be observed in the urban rail network in the Tokyo. However, our model does not consider the similarity of the routes, because Kosuda *et al.* (2004) show that the estimated results of the urban rail route choice model with the Multinomial logit model are not significantly different from the estimated results with the Mixed logit model that considers the similarity of the routes. We use only travel time and travel cost as explanatory variables in the conditional indirect utility function. As Kato *et al.* (2003) shows, the other factors including the number of transfers, the in-vehicle congestion and the number of companies used may also influence the rail route choice, we neglect them due to the analytical simplicity. Two cases of parameters are estimated: one using a linear utility function and the other using a quadratic utility function. We find that the estimation results are quite good in both cases. All variables comfortably pass statistical tests and the model fit is also quite good. The signs of all variables are also reasonable.

We examine the value of travel time savings based on these results. First, we evaluate VTTS using the linear and quadratic models using the observed data. The travel times and travel costs used are the same as those used in parameter estimation. Figure 2 shows the results. Using the linear utility function, VTTS increases as travel time increases. This is because we

Table 2 Parameter estimation results

			linear model		quadratic model	
variables	unit		parameter	t-value	parameter	t-value
travel cost	yen	c	-0.00229	(-5.62***)	-0.00234	(-1.79*)
travel time	min	t	-0.0963	(-18.8***)	-0.138	(-16.5***)
(travel cost)^2	yen^2	c^2			0.00000205	(-1.79*)
(travel time)*(travel cost)	yen*min	tc			-0.0000194	(-2.32**)
(travel time)^2	min^2	t^2			0.000136	(-18.5***)
initial log likelihood		L(0)	-1336.8		-1336.8	
maximum log likelihood		$L^*(x)$	-881.9		-782.1	
likelihood ratio			0.339		0.414	
number of samples			1218		1218	

*** means significance in 99% degree, ** means significance in 95% degree
and * means significance in 90% degree

use Equation (34) in the evaluation. With the linear approximation, VTTS seems stable at about 50 yen/minute, whereas when the quadratic utility function is used it seems to vary considerably in the range 50 to 100 yen/minute. In 1995, the year of the census data, the average wage rate in Tokyo was 51.7 yen/minute, so this evaluation shows that VTTS is almost the same or slightly more than the average wage rate.

Since there is no clear relationship between VTTS and travel time when using the quadratic approximation, we simulate VTTS by inputting travel times from 0 to 180 minutes into the quadratic VTTS formula (Equation 39b) directly. We again use Equation (34) for this

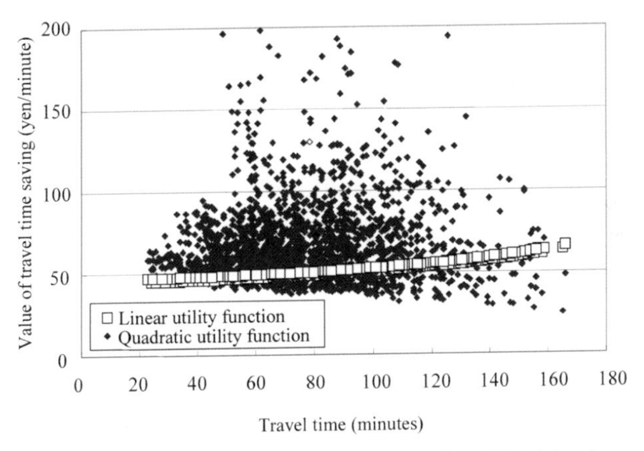

Figure 2 Evaluation results of VTTS with the estimated model and the observed data.

simulation. Figure 3 shows results. For travel times of less than about 85 minutes, VTTS decreases as travel time increases. On the other hand, it increases if travel time is over 85 minutes. Note that the average travel time of rail-user commuters in the Tokyo Metropolitan Area is about 65 minutes. Two questions are of interest here: "why does the marginal VTTS decrease when the travel time is short?" and "why does the marginal VTTS increase when the travel time is long?" The second question was covered by the special case described earlier in our theoretical analysis. Hence, the first question is addressed here. According to the theoretical analysis, there is no simple condition in which VTTS decreases monotonously as travel time increases. Therefore, only a hypothetical consideration is possible. We propose the hypothesis that the absolute value of marginal wage rate with respect to travel time is much larger than other factors if the travel time is short. If the wage rate function satisfies $\partial w/\partial \bar{t} << 0$, VTTS may fall as travel time increases, as Equation (21) illustrates. This explanation, as well as the analysis of the second question given earlier, remains purely hypothetical. In order to be more specific, we may need to carry out further examinations with additional data.

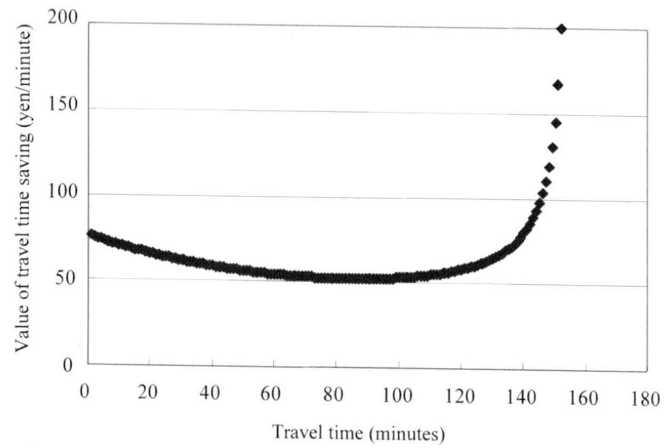

Figure 3 Simulation result of VTTS with the quadratic utility function.

CONCLUSIONS

We have examined the variation of VTTS over travel time for urban commuters both theoretically and empirically. Our model incorporates the non-linearity of the utility function as well as choice of residential location. The theoretical formulation shows that a joint model of time allocation and residential location choice can be reduced to a time-allocation model under market equilibrium. The comparative static analysis with the time-allocation model concludes that the properties of VTTS over travel time cannot be specified in a simple manner. The empirical analysis demonstrates that the sign of marginal VTTS with respect to travel time may be negative for specific travel times, but may become positive for other specific travel times. Although we have not yet found explicit reasons for these empirical results, we have discussed possible reasons. The results of our analysis indicate that the mechanism of the variation of VTTS over travel time is quite complicated. A more careful examination using additional empirical data is required.

This work points to some possible future research. The conclusions of our VTTS analysis by the theoretical approach may be improved with some specific modifications. First, our model is formulated using the assumption that the city has a single CBD. However, there are many cities around the world with multiple sub-centres. As a matter of fact, Tokyo has several urban sub-centres around the CBD. To more properly present the distribution of residential location, we could reformulate the city structure to give it sub-centres, as shown by Helsley and Sullivan (1991) and Zhang and Sasaki (1997). Second, the model presented in this paper assumes identical individuals, leading to a constant utility level for all residents in the urban area. However, individual heterogeneity may result in a more complicated market equilibrium. Classical urban economic theory (for example, Alonso, 1964) shows that variations in income level may lead to separation of residential areas by income level. We can improve the model by considering residential location choice at various income levels. Third, our model assumes that commuters always consume urban rail services. This reflects the fact of very high modal share of rail ride in Tokyo. However, in general, people choose from among several transport modes, including automobile, bus, tram, bicycle and walking. The choice of transport mode is highly influenced by wage rate and residential location (Anas and Moses, 1979; Sasaki, 1990; and DeSalvo and Hug, 1996).

Regarding the empirical analysis, our model is based on a simple multinomial logit model under the homogeneity of individual preference. To take into account the heterogeneity of individual preference, we could use the random coefficient model such as the mixed logit model (Train, 2003). There has been some VTTS research involving the mixed logit model such as Hess *et al.* (2005) and Sillano and Ortuzar (2005). In addition to the simple formulation, we use only two explanatory variables: travel time and travel cost. As Hess *et al.* (2005) point out, bias would be introduced if we chose the wrong variables for the model. Rail-use commuters may consider not only travel time and travel cost, but also transfers, congestion and train comfort in making a route choice as Kato *et al.* (2003) indicate. Incorporating these factors, even as approximations, would make the model more complicated but may be worth trying.

ACKNOWLEDGEMENT

This study was funded by the Japan Society for the Promotion of Science (Grant-in-Aide for Scientific Research). The author greatly appreciates the useful comments made by Professor Kay W. Axhausen (Swiss Federal Institute of Technology Zurich, ETHZ) regarding my March 2006 presentation about this research in Zurich.

REFERENCES

Alonso, W. (1964). *Location and land use*. Harvard University Press, Cambridge, MA.

Anas, A. and L. N. Moses (1979). Mode choice, transport structure and urban land use. *Journal of Urban Economics*, **6**, 228-246.

Axhausen, K.W., A. König, G. Abay, J.J. Bates and M. Bierlaire (2004). Swiss value of travel time savings. Presented at the 2004 European Transport Conference, Strasbourg, 2004.

Bates, J. J. (1987). Measuring travel time values with a discrete choice model: A note. *The Economic Journal*, **97**, 493-498.

Becker, G. (1965). A Theory of the allocation of time. *The Economic Journal*, **75**, 493-517.

Ben-Akiva, M. and S. Lerman (1985). *Discrete choice analysis: Theory and application to travel demand*, MIT Press.

Blayac, T. and A. Causse (2001). Value of travel time: A theoretical legitimization of some nonlinear representative utility in discrete choice models. *Transportation Research*, **35B**, 391-400.

De Lapparent, M., A. de Palma and C. Fontan (2002). Non-linearities in the valuation of time estimates. Presented at the 2002 European Transport Conference, Strasbourg, 2002.

DeSalvo, J. S. and M. Huq (1996). Income, residential location, and mode choice. *Journal of Urban Economics*, **40**, 84-99.

De Serpa, A.C. (1971). A theory of the economics of time. *The Economic Journal*, **81**, 828-846.

De Serpa, A.C. (1973). Microeconomic theory and the valuation of travel time: Some clarification. *Regional and Urban Economics*, **2**, 401-410.

Evans. A. (1972). On the theory of the valuation and allocation of time. *Scottish Journal of Political Economy*, **19**, 1-17.

Greeven, P., S. R. Jara-Díaz, M. Munizaga and K.W. Axhausen (2005). Calibration of a joint time assignment and mode choice model system. *Arbeitsbericht Verkehrs- und Raumplanung*, **299**, IVT, ETH Zürich, Zürich.

Hague Consulting Group and Accent Marketing & Research (1996). *The value of travel time on UK roads 1994.*

Helsley, R. and A. Sullivan (1991). Urban subcenter formation. *Regional Science and Urban Economics*, **21**, 255-275.

Hensher, D. A. (1997). Behavioral value of travel time savings in personal and commercial automobile travel. *The full costs and benefits of transportation.* Eds. Greene, D. L., Jones, D. W. and Delucchi, M. A., Springer.

Hultkrantz, L. and R. Mortazavi (2001). Anomalies in the value of travel-time changes. *Journal of Transport Economics and Policy*, **35**, 285-300.

Hess, S., M. Bierlaire and J. W. Polak (2005). Estimating value-of-time using mixed logit models. Presented at 84th Annual Meeting of the Transportation Research Board, Washington D.C., January, 2005.

Jara-Díaz, S. D. (2000). Allocation and valuation of travel-time savings. *Handbook of Transport Modelling*, Hensher, D. A. and Button, K. J. eds., Elsevier Science Ltd, 303-318.

Jara-Díaz, S.R., M. Munizaga, P. Greeven, and P. Romero (2004). The activities time assignment model system: The value of work and leisure for Germans and Chileans. Presented at the 2004 European Transport Conference, Strasbourg, 2004.

Jiang, M. and T. Morikawa (2004). Theoretical analysis on the variation of value of travel time savings. *Transportation Research*, **38A**, 551-571.

Kanemoto, Y. (1980). *Theories of urban externalities*, North Holland.

Kato, H., M. Itoh, S. Kato, and H. Ishida (2003). Cost-benefit analysis for improvement of transfer at urban railway stations. *World Transport Research: Selected Proceedings of the 9th World Conference on Transport Research.*

Kosuda, K., H. Asami, H. Suzawa and H. Kato (2004). Empirical analysis on user's choice set problem in urban rail route choice. Presented at 30th Conference of Infrastructure Planning, Ube, November 2004 (in Japanese).

Kono, T. and H. Morisugi (2000). Theoretical examination on value of time for private trips. *Journal of Infrastructure Planning and Management*, **639(46)**, 53-64 (in Japanese).

Morichi, S., S. Iwakura, S. Morishige, M. Itoh and S. Hayasaki (2001). Tokyo Metropolitan rail network long-range plan for the 21st century. Presented at 80th Annual Meeting of the Transportation Research Board, Washington D.C., January, 2001.

Sasaki, K. (1990). Income class, modal choice, and urban spatial structure. *Journal of Urban Economics*, **27**, 322-343.

Sillano, M. and J. de D. Ortuzar (2005). Willingness-to-pay estimation with mixed logit models: Some new evidence. *Environment and Planning A*, **37**, 525-550.

Small, K. A. (1999). Project evaluation. *Essays in transportation economics and policy: A handbook in honor of John R. Meyer*, eds. by Gomez-Ibanez, J., Tye, W. B. and Winston, C., The Brookings Institution, Washington D.C.

Small, K. A. (1982). Scheduling of consumer activities: work trips. *American Economic Review*, **72**, 467-479.

Solow, R.M. (1973). On equilibrium models of urban location. *Essays in modern economics*, ed. by M. Parkin, Longman, London.

Train, K. (2003). *Discrete choice methods with simulation.* Cambridge University Press, UK.

Train, K. and D. McFadden (1978). The goods/leisure trade-off and disaggregate work trip mode choice models. *Transportation Research*, **12**, 349-353.

Truong, T. P. and D. A. Hensher (1985). Measurement of travel time values and opportunity cost from a discrete-choice model. *The Economic Journal*, **95**, 438-451.

Wardman, M. (1998). The value of travel time — A review of British evidence. *Journal of Transport Economics and Policy*, **32**, 285-316.

Wardman, M. (2001). A review of British evidence on time and service quality valuation. *Transportation Research*, **37E**, 107-128.

Wardman, M. (2004). Public transport values of time. *Transport Policy*, **11**, 363-377.

Wheaton, W. (1977). Income and urban residence: An analysis of consumer demand for location. *American Economics Review*, **67**, 620-631.

Yai, T., S. Iwakura and S. Morichi (1997). Multinomial probit with structured covariance for route choice behaviour. *Transportation Research*, **31B**, 195-207.

Yoshida, A. and H. Endo (1999). Spatial distribution of per-capita income and occupations in the Tokyo Metropolitan Area. *Applied Regional Science*, **4**, 15-25 (in Japanese).

Varian, H. R. (1992). *Microeconomic analysis*, third edition, W. W. Norton & Company, Inc.

Zhang, Y. and K. Sasaki (1997). Effects of subcenter formation on urban spatial structure. *Regional Science and Urban Economics*, **27**, 297-324.

Transportation and Traffic Theory 2007
Edited by R.E. Allsop, M.G.H. Bell and B.G. Heydecker

9

A GAME THEORETICAL APPROACH FOR MODELLING MERGING AND YIELDING BEHAVIOUR AT FREEWAY ON-RAMP SECTIONS

Henry X. Liu & Wuping Xin, Civil Engineering Department, University of Minnesota, USA
Zain M. Adam, George Butler Associates, Inc., Lenexa, Kansas, USA
Jeff X. Ban, Institute of Transportation Studies, UC-Berkeley, USA

SUMMARY

Traffic conflicts between merging and through vehicles are typical phenomena near freeway on-ramp sections, yet few microscopic models describing the interaction of these vehicles in the merging process have been proposed. In this paper, vehicle interactions during merging process are modeled under an enhanced game-theoretic framework. Freeway on-coming through vehicle and on-ramp merging vehicle are considered as competing players that seek to maximize their respective rewards during the merging process. As the freeway vehicle aims to maintain their initial car-following state and minimize speed variations, the on-ramp merging vehicle strives to join mainline traffic in the minimal time possible subject to safety constraints. Considering non-cooperative nature of the game, drivers at the merging section would eventually adopt strategies that form Nash equilibrium. To assess the model parameters, we propose a bi-level estimation methodology with the upper level as a least square problem and the lower level a linear complementarity problem searching for the equilibria. Applicability of the proposed model is examined and validated using trajectory data collected from field. Testing results indicate that this framework can effectively capture vehicle interactions at freeway merging sections while achieving a relatively high accuracy of predicting vehicles' actions.

INTRODUCTION

Traffic conflicts between merging and through vehicles are typical phenomena near freeway on-ramp sections. Such conflicts often slow down the freeway vehicles and trigger shockwaves that propagate and dissipate over time and space, or result in localized congestion that could evolve into long-lasting bottlenecks throughout entire peak periods. A clear understanding of the merging process renders indispensable technical foundations for designing freeway on-ramps and developing sophisticated traffic management strategies such as ramp control. In previous studies, freeway merging process was often described in a "one-way" fashion, focusing on the influence of through traffic on merging vehicles in terms of the latter's gap acceptance behavior (Hidas, 2002; Kita, 1993; Kosonen, 1999; Owen and Zhang, 1998; Yang and Koutsopoulos, 1996). However, in real life it is frequently observed that a through vehicle performs courtesy yield or accelerates when seeing a merging vehicle on the ramp, while an on-ramp vehicle accelerates or decelerates looking for appropriate gaps in response to the movements of through traffic. Clearly, there exists mutual influence between the merging and through vehicles, i.e., they are not independent of but affecting each other's decision in an "interactive" manner. Such pattern has been noted as typical in freeway merging area and considered a dominant factor affecting traffic characteristics near on-ramp sections (Kita et al., 1999).

A few research efforts have been devoted to model freeway merging process taking into account vehicular interactions. Troutbeck (1999) analyzed give-way behavior of mainline vehicles before approaching on-ramps. Rysgaard and Nielsen (1998) performed a study of motorway merging-giveway behavior in Europe. In addition to these studies, freeway merging process with vehicle interactions has also been studied from macroscopic perspective using aggregate variables (Cassidy et al., 1990; Vermijs, 1991). Notably, Kita et al. (1999, 2002) were one of the first to model vehicles merging interactions as a "game", where each involving vehicle determines its final action by considering each other's alternatives. Specifically, Kita's model considers collision risk as an incentive factor to build players' payoff functions. That is, in order to minimize collision risks, freeway through vehicle can opt to give way to ramp merging vehicle, while ramp merging vehicle will choose to merge to freeway mainline or stay on the ramp merging section. However, in Kita's model vehicle speeds are assumed constant during the merging process, which is not true in reality. In addition, as pointed out by Troutbeck (1999), freeway vehicle usually perform giveaway behavior even before the appearance of merging vehicle, so the alleged interaction may not exist.

In this paper, merging and yielding behavior at freeway on-ramp sections are modeled under an improved game-theoretic framework. Vehicle speeds are no longer assumed constant as in Kita's study, while minimum safety gaps are explicitly considered in players' payoff functions. Comparing to previous modeling efforts, more realistic behavioral rules are proposed in this study to describe typical behavior at merging sections. To be sure, it is assumed that during the merging process freeway on-coming vehicles would try to maintain their initial car-following state and minimize speed variations, while on-ramp merging vehicles would strive to join mainline traffic in the minimum time possible subject to safety constraints. These behavioral rules are incorporated into respective payoff functions of conflicting vehicles, and the resultant pair of their actions is formulated as an equilibrium solution. Finally, an estimation methodology based on bi-level programming technique is proposed for assessing model parameters. Applicability of the proposed model is examined and validated using trajectory data collected from field. Testing results indicate that this framework can effectively capture vehicle interactions at freeway merging sections while achieving a relatively high accuracy of predicting vehicles' actions.

GAME THEORETICAL MERGING MODEL

Game Definition

Consider a typical merging situation shown in Figure 1, where vehicles involved include a merging vehicle, a lag vehicle (i.e., oncoming through vehicle) and possibly a lead vehicle. As illustrated in this figure, merging vehicle is that vehicle in acceleration lane trying to join the freeway; lag vehicle is the vehicle in the target lane just behind the merging vehicle, and lead vehicle is the one immediately in front of the lag vehicle in the target lane. The process is modeled as an independent game by taking merging and lag vehicles as players. This means, immediately upon seeing each other, both the merging vehicle and lag vehicle have to decide a set of moves to maximize their respective rewards in the game. The decisions are based on their instantaneous states including speed and acceleration rates as well as their predictions on the interaction situation. The action strategies for each player (vehicle) are assumed as follows:

1. The merging vehicle can select either to merge into the mainline traffic immediately or to wait until the next available gap;
2. The lag vehicle's options are whether to keep its current car-following state or decelerate to yield in order to facilitate a smooth merge.

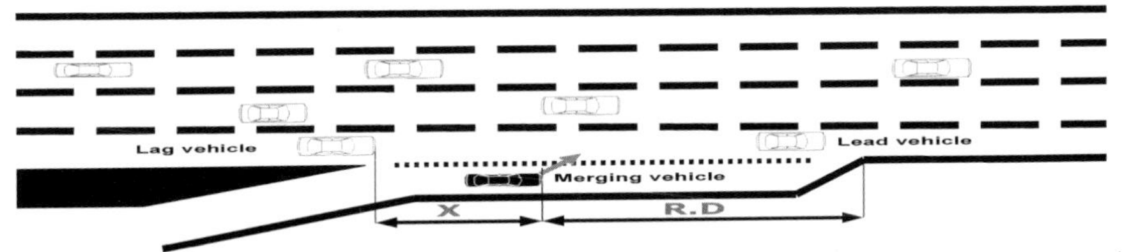

X represents the headway distance between merging and lag vehicle; *RD* represents the remaining distance on acceleration lane.

Figure 1. Vehicles in a typical merging situation

It is important to clarify that albeit the lead vehicle is not directly considered as a player, its influence is implicitly accounted for by incorporating it into the payoff functions of the lag vehicle. This should become clear when payoff functions of each player are detailed in forthcoming sections.

For the game defined above, equilibrium is achieved when no player can unilaterally increase his *expected* payoff by changing his probability of selecting a particular strategy. This essentially gives Nash equilibrium:

$$\begin{aligned} E_1(\boldsymbol{p}^*,\boldsymbol{q}^*) &\geq E_1(\boldsymbol{p},\boldsymbol{q}^*) \\ E_2(\boldsymbol{p}^*,\boldsymbol{q}^*) &\geq E_2(\boldsymbol{p}^*,\boldsymbol{q}) \end{aligned} \tag{1}$$

where E_1 and E_2 is the expected payoff at equilibrium, and $\boldsymbol{p}^*$ and $\boldsymbol{q}^*$ represent the equilibrium strategy sets for merging and lag vehicles respectively. In case of multiple equilibriums, a superior solution is considered to be the one that gives the highest payoffs for both players.

Payoff Functions Formulation

In the literature, payoff functions are often formulated assuming that minimizing collision risks is the behavioral goal for each individual player (Kita 1993; Kita and Fukuyama 1999). This assumption may result in trivial equilibrium solutions as collision risks are affecting both players thus each player would have similar magnitude of effects on game equilibrium (Kita 1993). In this study, it is assumed that freeway through vehicle's objective is to minimize speed variations, i.e., try to employ a lowest possible acceleration rate during the merging process. By contrast, merging vehicle's objective is to minimize the time spent in acceleration lane subject to safety constraints. These rules are mapped to mathematical functions describing each player's payoffs. Prior to detailing these functions, it should be noted that lag vehicle's payoffs are in the unit of acceleration (ft/s^2), and merging vehicle's, in the unit of time (second). Also the typical merging scenario is assumed as follows: (1) prior to approaching the merging section, the *lag* and *lead* vehicles are interacting with each other as in a normal car-following situation, and (2) lag and merging vehicle immediately constructs their respective payoff matrix once the merging vehicle appears on the acceleration lane *and* the distance between lag and merging vehicle is less than 200 feet. Vehicle beyond this distance are assumed out of the interaction range (Toledo, 2003). The time at which they construct payoff matrix and make decisions will be referred to as *decision time* henceforth.

Payoffs for the lag vehicle

At the decision time, lag vehicle needs to decide whether to perform a courtesy yield, or maintain its current car following state as dictated by its lead vehicle. First, consider the case if the lag vehicle chooses to maintain its current car-following state with instantaneous acceleration rate a_l. If its opponent on the acceleration lane chooses to wait for the next available gap, then the lag vehicle can indeed maintain its current state as it desires. In this case, the lag vehicle's payoff is a_l, i.e., it can keep the acceleration rate dictated by the lead vehicle. Note a_l is directly observable at the decision time. However, if the merging vehicle

decides to merge right away, the lag vehicle may have to apply an unexpected braking to avoid potential collision in response to the sudden cutting-in of the merging vehicle. This sharp and unfavorable deceleration rate is not directly observable at the decision time thus has to be projected based on the instantaneous states of both lag and merging vehicles. The *projected time* is the time at which the lag vehicle *anticipates* the merging vehicle enters the freeway. The initial states at the decision time are denoted by the following:
v_m: Instantaneous speed of the merging vehicle at decision time;
v_l : Instantaneous speed of the lag vehicle at decision time;
a_m: Instantaneous acceleration of the merging vehicle at decision time;
a_l: Instantaneous acceleration of the lag vehicle at decision time;
RD: Remaining distance on the acceleration lane for the merging vehicle at decision time;
X: Initial gap distance between lag and merging vehicles at decision time.

The *projected* states, from the lag vehicle's perspective are as follows:
v'_m: Instantaneous speed of the merging vehicle at projected time;
v'_l : Instantaneous speed of the lag vehicle at projected time;
t'_m: The time duration that the lag vehicle anticipates the merging vehicle would need to complete the remaining distance (RD) on the acceleration lane;
X': Gap distance between lag and merging vehicle when the latter joins the freeway.

Given the initial states at the decision time, projected states can be computed as:

$$v'_m = \sqrt{(v_m)^2 + 2a_m RD} \tag{2}$$

$$t'_m = \frac{v'_m - v_m}{a_m} \tag{3}$$

$$v'_l = v_l + a_l t'_m \tag{4}$$

$$X' = X - \frac{(v'_l)^2 - (v_l)^2}{2a_l} + RD \tag{5}$$

With these projected state variables, the lag vehicle is able to estimate the braking rate needed to avoid a potential collision when the merging vehicle suddenly cuts in. Using X' to approximate the braking distance the payoff of the lag vehicle a_s can be estimated as:

$$a_s = \begin{cases} \beta_1 \dfrac{2(X' - v'_l t_b)}{t_b^2} + \beta_2, if\ X' > 0 \\ a_l, if\ X' \le 0 \end{cases} \tag{6}$$

where t_b is the braking time anticipated by the lag vehicle; β_1 and β_2 are free coefficients to be estimated from data. $X' \le 0$ indicates the lag vehicle should have surpassed the merging vehicle when the latter joints the freeway. In this case, there is no need for the lag vehicle to brake and it just keeps its initial car-following state a_l. Also it should be stressed that a_s is essentially a quantity assumed to be "*perceived*" as necessary by the lag vehicle at the decision time; it reflects the lag vehicle's prediction about possible interactions if lag vehicle maintains its car-following state while merging vehicle selects to merge anyway. Moreover, even though various approaches are available for obtaining a_s analytically, the relationship assumed in Equation (6) has been found to yield best accuracy with X', v'_l and t_b, which are directly obtainable from the collected trajectory data.

The other option for the lag vehicle is to conduct an early courtesy yield, giving a clear sign of invitation for the on-ramp vehicle to merge. This yielding action produces a gentle decrease in speed by applying a comfortable deceleration rate, therefore the payoff a_y is determined using the following equation:

$$a_y = \beta_3 \max[\frac{v_m - v_l}{t'_m - 1.0}, -10] \ ft/s^2 \tag{7}$$

where β_3 is a parameter to be calibrated from observation data, and 1.0 is the assumed safety time margin. Equation (7) gives a braking rate that ensures the lag vehicle to achieve relatively low speed some time before the merging vehicle joins the freeway. If the merging vehicle's speed v_m is lower than the lag vehicle's speed v_l, the lag vehicle will brake at a rate bounded by -10 $ft/\sec^2$, which is the limit of comfortable deceleration rate suggested in Traffic Engineering Handbook. Note when the lag vehicle takes yielding action, its payoff is regardless of merging vehicle's action.

The payoff matrix of the lag vehicle is summarized in Table 1.

Table 1. Payoff matrix of freeway lag vehicle

Players	**Lag Vehicle**		
Merging Vehicle	**Actions**	**Yield**	**Not Yield**
	Merge	a_y	a_s
	Wait	a_y	a_l

Payoffs for the merging vehicle

The merging vehicle driver creates its payoff matrix as soon as he/she enters the acceleration ramp and recognizes freeway conditions. The payoff functions proposed here are the times required to join freeway traffic. These times are calculated based on initial conditions of both vehicles as well as anticipated actions to be carried by the freeway vehicle. As being pointed out earlier, these times are associated with the acceleration/deceleration rates the merging vehicle anticipates the lag vehicle to adopt, therefore reflecting strong interactions between both decision makers.

First consider the situation where the merging vehicle decides to merge instead of waiting for the next available gap. If the freeway lag vehicle selects to yield, then the merging vehicle can smoothly join the freeway traffic using a comfortable acceleration rate $a_{comfort}$ as there would be no conflict at the merging point. On the other hand, if the lag vehicle chooses not to yield, the merging vehicle would need to adopt a more aggressive acceleration rate $a_{\max}$ in order to arrive at the merging point earlier than the lag vehicle to avoid potential collision risks. The specific payoffs are expressed as follows:

$$t_{m-y} = \beta_4 \frac{-v_m + \sqrt{v_m^2 + 2a_{comfort}RD}}{a_{comfort}} + \beta_5 \tag{8}$$

$$t_{m-ny} = \beta_6 \frac{-v_m + \sqrt{v_m^2 + 2a_{\max}RD}}{a_{\max}} + \beta_7 \tag{9}$$

where

t_{m-y}: The payoff that the merging vehicle needs to join freeway traffic if lag vehicle selects to yield;

t_{m-ny}: The payoff that the merging vehicle needs to join freeway traffic if lag vehicle selects not to yield;

$\beta_4\ \beta_5\ \beta_6\ \beta_7$: Free coefficients to be calibrated from observation data;

v_m: Merging vehicle's initial speed when entering the acceleration lane;

RD: Remaining distance on the acceleration lane;

$a_{comfort}$: Comfortable acceleration rate merging vehicle adopts if lag vehicle selects to yield;

$a_{\max}$: Maximum acceleration rate merging vehicle adopts if lag vehicle selects not to yield;

Alternatively, merging vehicle can also select to wait for the next available gap rather than competing with the freeway lag vehicle. In this case, if the lag vehicle still performs a courtesy yield, signaling a clear invitation for the merging vehicle to take the move first, the latter doesn't really have to wait till next available gap, rather, it will wait for a while till recognizing the yielding gesture, then accelerate and merge immediately with a comfortable acceleration rate $a_{comfort}$. However, if the lag vehicle selects not to yield but keeps its initial acceleration dictated by the car-following situation, the merging vehicle needs to wait till the lag vehicle overpasses it, and takes the next immediate gap with a more aggressive acceleration rate $a_{\max}$. The payoffs for this situation are expressed as:

$$t_{w-y} = \beta_8.t_0 + \beta_9 \frac{-v_m + \sqrt{v_m^2 + 2a_{comfort}(RD - v_m t_0)}}{a_{comfort}} + \beta_{10} \tag{10}$$

$$t_{w-ny} = \beta_{11}t'_0 + \beta_{12} \frac{-v_m + \sqrt{v_m^2 + 2a_{\max}(RD - v_m t'_0)}}{a_{\max}} + \beta_{13} \tag{11}$$

$$t'_0 = \frac{(v_m - v_l) + \sqrt{(v_m - v_l)^2 + 2a_l X}}{a_l} + 1.0 \tag{12}$$

where

t_{w-y} : The payoff that the merging vehicle spends on acceleration lane before joining freeway traffic if the lag vehicle selects to yield;
t_{w-ny} : The payoff that the merging vehicle spends on acceleration lane before joining freeway traffic if the lag vehicle selects not to yield;
β_8 β_9 β_{10} β_{11} β_{12} : Free coefficients to be calibrated from observation data;
v_m : Merging vehicle's initial speed when entering the acceleration lane;
t_0 : Waiting time merging vehicle has to wait before recognizing lag vehicle's yielding gesture;
t'_0 : Waiting time merging vehicle has to wait till lag vehicle overpasses it;
is safety margin of time headway;
X : Initial lag distance;
a_l : Lag vehicle's initial acceleration rate;
RD : Remaining distance on the acceleration lane for the merging vehicle;
$a_{comfort}$: Comfortable acceleration rate merging vehicle adopts if lag vehicle selects to yield;
$a_{\max}$: Maximum acceleration rate merging vehicle adopts if lag vehicle selects not to yield;

The payoff matrix of the merging vehicle is summarized in Table 2.

Table 2. Payoff matrix of on-ramp merging vehicle

Players	Lag Vehicle		
Merging Vehicle	**Actions**	**Yield**	**Not Yield**
	Merge	t_{m-y}	t_{m-ny}
	Wait	t_{w-y}	t_{w-ny}

Summarizing from the above, a payoff bi-matrix can be constructed in Table 3:

Table 3. Merging-yielding game in normal form

Players	Lag Vehicle			
Merging vehicle	Actions		Yield	Not Yield
		Probability	q	$1-q$
	Merge	p	(a_y, t_{m-y})	(a_s, t_{m-ny})
	Wait	$1-p$	(a_y, t_{w-y})	(a_l, t_{w-ny})

Parameter Estimation

Parameters estimation of the proposed model is achieved by solving a bi-level programming problem. The upper level is a non-linear programming problem minimizing system total deviation from actual observed actions:

$$\min_{(p,q)} \sum_{i=1}^{n}[(Q_i - \hat{Q}_i(p,q))^2 + (P_i - \hat{P}_i(p,q))^2] \tag{13}$$

where i is the index of observations, Q_i is the observed choice of through vehicle (1 yield, 0 otherwise), P_i represents the observed choice of merging vehicle (1 merge, 0 otherwise), while $\hat{Q}_i$ is the model predicted choice of through vehicle (1 yield, 0 otherwise), $\hat{P}_i$ is the model predicted choice of merging vehicle (1 merge, 0 otherwise). Both $\hat{Q}_i$ and $\hat{P}_i$ are functions of yielding and merging probabilities p and q, which are the optimizers for the upper level programming problem. The optimal value of p and q should minimize the square difference between observed choices and model predicted choices.

The lower level program seeks solution for Nash equilibrium. The bi-matrix game may have several equilibrium solutions in pure strategies, as well in mixed strategies. The non-uniqueness of Nash equilibrium of bi-matrix games is a serious theoretical and practical problem. For our modeling purpose, any local solution should suffice. Since the lower level problem is actually a two-player game, Stengel (1999) has showed that there exists an equivalent linear complementarity formulation as follows:

$$0 \le (e - M \cdot S) \perp S \ge 0 \tag{14}$$

Here $S = [S_1 \quad S_2]' \in R^4$ is an auxiliary variable with $S_1, S_2 \in R^2$. Also, e is a vector of all 1's with a proper dimension, and $M = \begin{bmatrix} 0 & A \\ T^T & 0 \end{bmatrix}$ with A and T being the freeway vehicle and merging vehicle payoff matrix respectively.

Therefore, given the payoff matrices A and T, S can be obtained by solving (14). The probability for choosing each strategy can then be computed as:

$$\begin{cases} q = \dfrac{S_1}{e^T S_1} \\ p = \dfrac{S_2}{e^T S_2} \end{cases} \tag{15}$$

Provided equations (14) and (15), the bi-level program (13) can be formulated as a mathematical program with complementarity constraints (MPCC). MPCC has been recently extensively studied, and Ferris et al. (2002) implemented a solver of nonlinear program with equilibrium constraints (NLPEC) as sub-system of GAMS (General Algebraic Modeling System). NLPEC can automatically convert MPCC into an equivalent single level NLP using a number of reformulation techniques. We adopt NLPEC in this study for the parameter estimation. Figure 2 illustrates the schematic workflow for this bi-level programming process.

It should be noted that the parameter estimation method above is consistent with the method of "probability of equilibrium selection" originally developed by Kita and Fukuyama (2002), as there may exist multiple equlibria. Our method can jointly estimate the payoff and the probability of equilibrium selection so that multiple equilibria can be accommodated. Table 3 lists how the probabilities of different equilibrium strategies are defined and Equation (13) demonstrates the objective function of the estimation model that incorporates these probabilities. The method does not require any selection criteria form their resultant actions (i.e., no need for identifying the correspondence between realized equilibrium and the values of the explanatory variables).

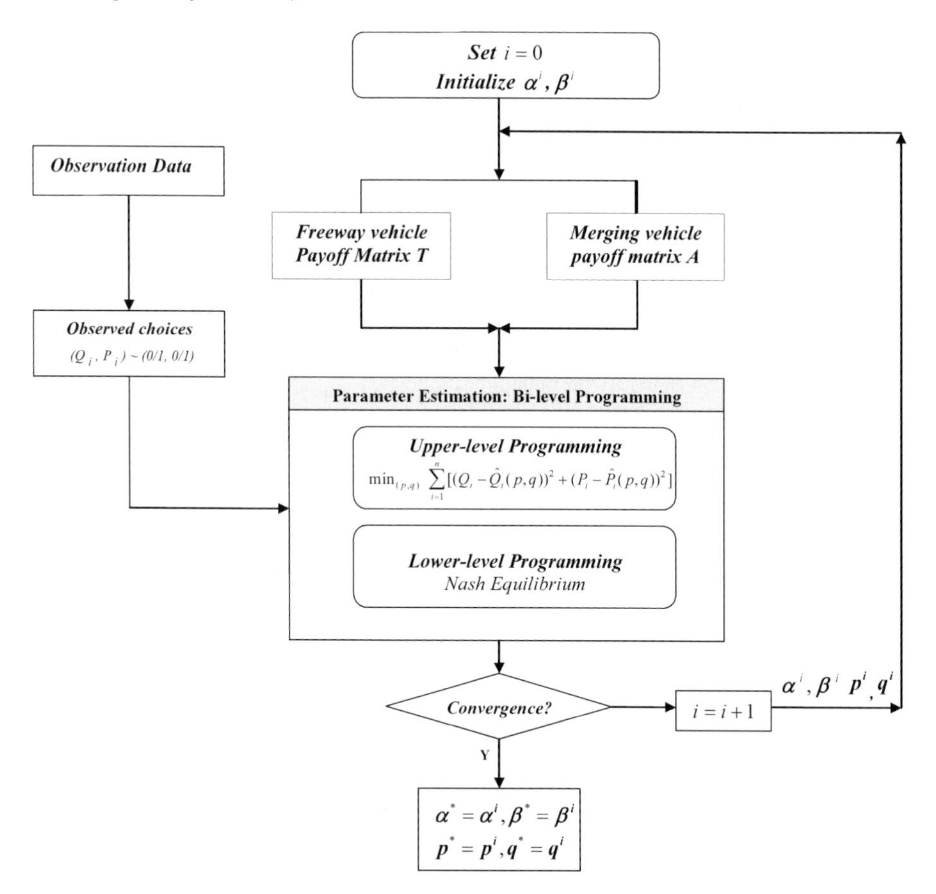

Figure 2. Schematic workflow for bi-level programming

Model Validation

Model validation is the process of quantifying predicting capability of the calibrated model using validation data set. The following metrics are employed in model validation:

Root Mean Square Error (RMSE)

$$RMSE = \sqrt{\frac{1}{n}\sum_{i=1}^{n}(x_i - y_i)^2} \tag{16}$$

Where x_i is the model predicted value indexed by i;
y_i is the actual observation indexed by i;
n is the number of total observations.

Correlation Coefficient

$$R = \frac{1}{n-1}\sum_{i=1}^{n}\frac{(x_i - \overline{x})(y_i - \overline{y})}{\sigma_x \sigma_y} \tag{17}$$

where $\overline{x}$ is the mean of model predicted values;
$\overline{y}$ is the mean of the actual observed values;
σ_x is the standard deviation of model predicted values;
σ_y is the standard deviation of the actual observed values.

Mean Absolute Error (MAE)

$$MAE = \frac{1}{n}\sum_{i=1}^{n}| y_i - x_i | \tag{18}$$

CASE STUDY

The field observation data were used to calibrate and validate the proposed model. These data were obtained from *Freeway Data Collection for Studying Vehicle Interactions* (DCSVI) project conducted by FHWA in 1983. The data collection site was an on-ramp section of I-405 at Roscoe Boulevard, Van Nuys, California. This site includes a 4-lane freeway section that is 1728 feet in length with a metered entrance ramp. The length of the ramp acceleration lane after meter signal light is about 400 feet. A full-frame 33 mm motion picture camera was mounted on a fixed-wing, short-take-off-and-landing aircraft. The site was then filmed at one frame per second with the aircraft flying clockwise at altitudes ranging between 2,500 and 4,500 feet. Individual vehicle trajectories were then extracted from the film at 1-second resolution. The extracted data set contains 200,000 data records (vehicle-seconds), including detailed information about vehicle speed, acceleration, front and lag distances and other variables. Figure 3 demonstrates the geometry of the test site.

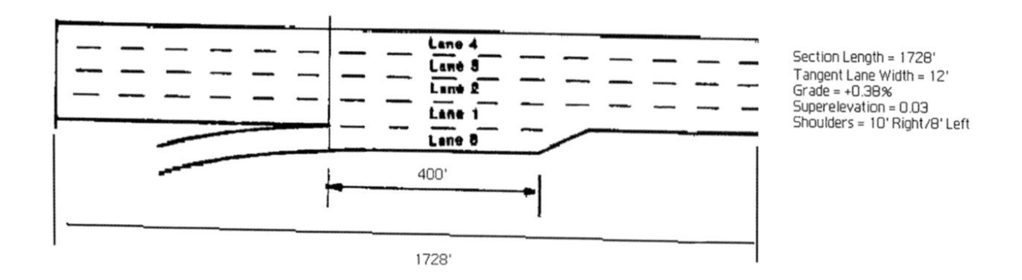

Figure 3. Geometry of the test site

The extracted data set was carefully examined to identify merging cases consistent with the assumptions prescribed earlier. With each identified case, trajectories of involved vehicles were traced back from their actual choices (i.e., yield/not yield, wait/merge) while examining their respective speed and acceleration profiles. In the end this screening process identified a total of 86 merging cases that meet model assumptions. For each merging case, the lead, lag, and merging vehicle's trajectories are meticulously investigated and analyzed. Figure 4 illustrates an example of speed profiles for merging and lag vehicles in four typical merging scenarios. In this figure, the time when the merging vehicle joins freeway mainline is marked as "merge point" by the vertical line, and the first point in merging vehicle's speed profile represents the decision point for all the players. This way, merging situation can be effectively deduced and reconstructed using speed profiles between the merge and decision point. For instance, in the scenario where lag vehicle choose to yield and merging vehicle chooses to wait (see Figure 4(d)), it can be clearly seen from the figure that lag vehicle slowly decelerates signalling his yielding intention while the merging vehicle has a relatively long accelerating time frame suggesting the latter selects not to merge immediately but rather to wait. Likewise, when the lag vehicle chooses not to yield and merging vehicle chooses to merge immediately (see Figure 4(b)), corresponding speed profiles depict a steep increase for merging vehicle, while a drop in lag vehicle's speed can be seen right before the merge point. This indicates that a sudden cutting in of the merging vehicle causes the lag vehicle to decelerate unexpectedly.

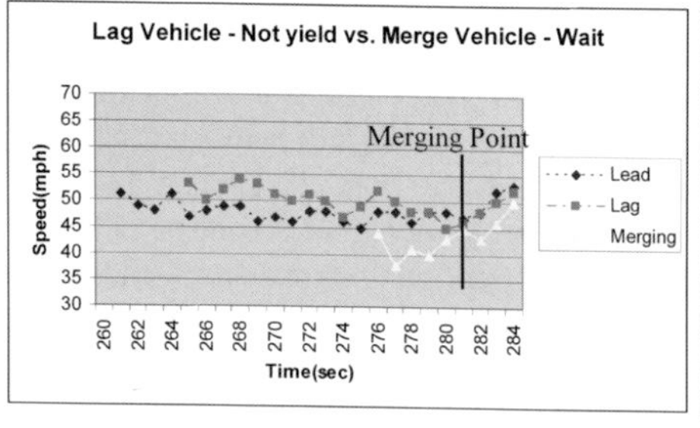

4(a)

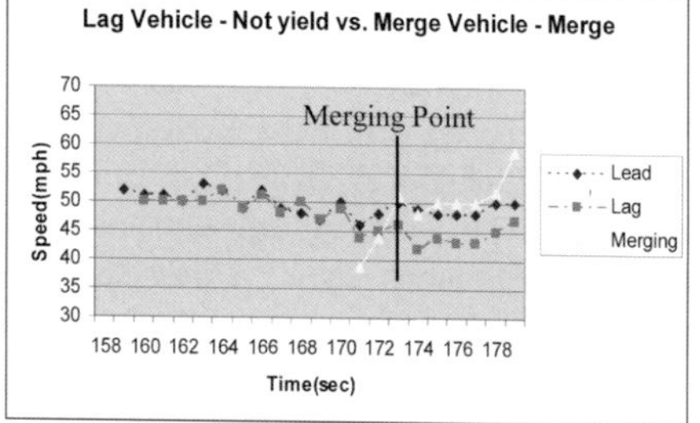

4(b)

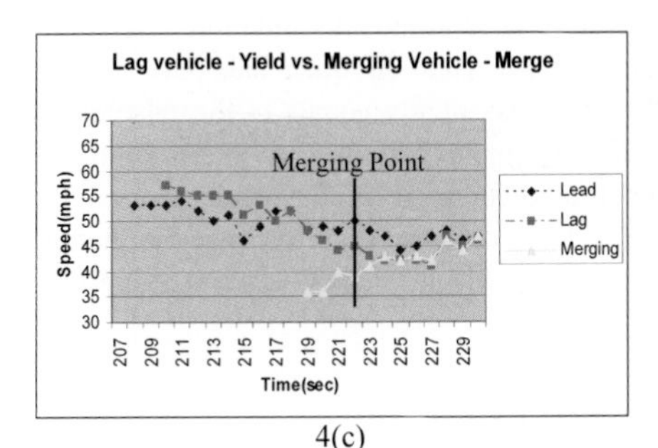

4(c)

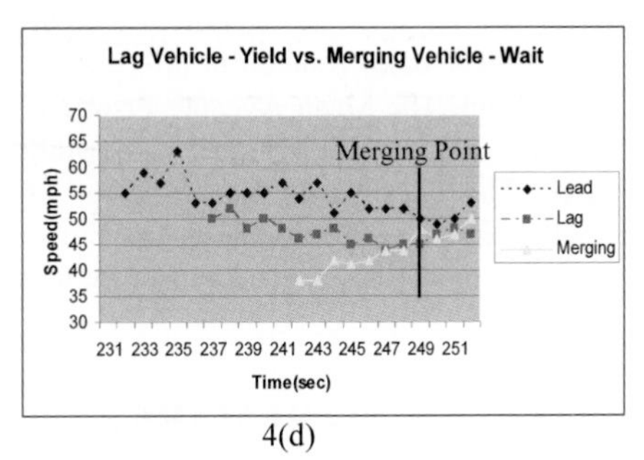

4(d)

Figure 4. Vehicle speed profiles in the merging process

From the total 86 cases, the extracted trajectories from 63 cases were used in calibrating model parameters via bi-level programming, while the rest of 23 cases were used for validation. Table 4 summarizes the calibrated parameters and Table 5 summarizes the validation results.

Table 4. Calibrated Model Parameters

Freeway Lag Vehicle	β_1	1.0658166
	β_2	-0.01398846
	β_3	6.06633624
On-Ramp Merging Vehicle	β_4	0.00432315
	β_5	1.10E+02
	β_6	-0.00589191
	β_7	2.06E+03
	β_8	-0.00000247
	β_9	0.00266124
	β_{10}	1.10E+02
	β_{11}	-0.0000699
	β_{12}	-0.00717059
	β_{13}	2.06E+03

Table 5. Validation results

No. of Validation Cases	23
No. of Pure Strategy Equilibrium	12
No. of Mixed Strategy Equilibrium	11
Mean Average Error	0.087
Root Mean Square Error	0.280
Correlation Coefficient	0.915

As shown in Table 5, a total of 23 merging cases were used in model validation. About one half of these cases resulted in mixed strategy equilibrium while the other half pure strategy equilibrium. The Mean Average Error is 0.087, which essentially equals to the false alarm rate, in other words, the model successfully predicted vehicles actions with 91.3% of all the cases. The Root Mean Square Error of model predictions is 0.289 while Correlation Coefficient between real choices and model predicted choices is 0.915. This indicates that the proposed model has a good capability to replicate and predicate vehicle actions at merging sections.

CONCLUDING REMARKS

This paper describes a game-theoretical framework that can model driver's behavior during the complex merging maneuver. In the game, freeway on-coming vehicles aims at maintaining their initial car-following state and minimize speed variations, while on-ramp merging vehicles strive to join mainline traffic in the minimum time possible subject to safety constraints. These behavioral rules are incorporated into respective payoff functions of conflicting vehicles, and the selected actions become the outcome of a game with each player trying to maximize his own rewards. An estimation methodology based on bi-level programming technique is proposed for assessing model parameters. Vehicle trajectory data from the field is used for the estimation and validation of the proposed model.

This study is an attempt to better understand merging behavior from game perspective. The proposed model can be implemented in simulation package to improve current modeling technique. Currently only two players, each with two alternatives are considered in the game, yet the framework could be further expanded "horizontally" to include more players, or "vertically" to consider multiple choices and sequential moves for each player. Albeit a game with more than 4 players may involve tremendous amount of computation thus practically not feasible, however a 3-player game with multiple options for each player merits further exploration. For example, the lag vehicle can have more options including accelerating to surpass the merging vehicle or conducting a lane change to avoid potential conflicts. This should also include collecting new high-resolution trajectory data to aid detailed analysis and model calibration/validation. Such work is the subject of a research project currently underway by the authors.

REFERENCE

Cassidy, M., Chan, P., Robinson, B., and May, A. D. (1990). "A proposed analytical technique for the design and analysis of major freeway weaving sections." UCB-ITS-RR-90-16, Institute of Transportation Studies, University of California, Berkeley.

Chang, G.-L., and Cao, Y.-M. (1991). "An empirical investigation of macroscopic lane changing characteristics on uncongested multilane freeways." Transportation Research Part A: Policy and Practice, 25(6), 375-389.

Ferris, M. C., Dirkse, S. P., and Meeraus, A. (2002). "Mathematical programs with equilibrium constraints: Automatic reformulation and solution via constrained optimization." Frontiers in Applied General Equilibrium Modeling, T.-J. Kehoe, T.-N. Srinivasan, and J.Whalley, eds., Cambridge University Press.

Hidas, P. (2002). "Modeling lane changing and merging in microscopic traffic simulation." Transportation Research Part C: Emerging Technology, 10, 351-371.

Kita, H. (1993). "Effect of merging lance length on merging behavior at expressway on-ramps." Proceedings of the 12th International Symposium on Transportation and Traffic Theory, C. Daganzo, ed., Elsevier, Amsterdam, 37-51.

Kita, H., and Fukuyama, K. (1999). "A merging-giveaway behavior model considering interactions at expressway on-ramps." Proceedings of the 14th International Symposium on Transportation and Traffic Theory, A. Cedar, ed., Pergamon, Amsterdam.

Kita, H., and Keishi, T. (2002). "A game theoretic analysis of merging-giveaway interaction: A joint estimation model." Transportation and Traffic Theory in the 21st Century: Proceedings of the 15th International Symposium on Transportation and Traffic Theory, M. A. P. Taylor, ed., Elsevier, Adelaide, Australia, 503-518.

Kita, H., Tanimoto, K., and Fukuyama, K. (1999). "An inverse analysis of interactive travel behavior." International Symposium on Structural Change in Transportation and Communications in Knowledge Society: Implications for Theory, Modeling and Data, Boston University.

Kosonen, I. (1999). "HUTSIM-Urban traffic simulation and control model: Principles and applications," PhD Dissertation, Helsinki University of Technology, Helsinki.

Owen, L. E., and Zhang, Y. "A multi-regime microscopic traffic simulation approach." the 5th International Conference on Applications of Advanced Technologies in Transportation Engineering, Newport Beach, California, 199-206.

Rysgaard, R., and Nielsen, M. A. "Merging contra give way when entering a motorway." Proceedings of the 3rd International Symposium on Highway Capacity, Copenhagen, 873-882.

Stengel, B.V. (1999) "Computing equilibria for two-person games". In Handbook of Game Theory, vol. 3, R.J.Aumann and S. Hart (eds.), North-Holland, Amsterdam.

Troutbeck, R. J. (1999). "Capacity of limited-priority merge." Transportation Research Record, 1678, 269-276.

Toledo, T. (2003). "Integrated Driving Behavior Modeling." PhD Dissertation, Department of Civil and Environmental Engineering, MIT, Cambridge, Massachusetts.

Vermijs, R. (1991). "The use of micro simulation for the design of weaving sections." Highway Capacity and Level of Service, U. Brannolte, ed., Balkema, Rotterdam, Karlsruhe, Germany, 419-427.

Yang, Q., and Koutsopoulos, H. N. (1996). "A microscopic traffic simulator for evaluation of dynamic traffic management systems." Transportation Research Part C: Emerging Technology, 4(3), 113-129.

Transportation and Traffic Theory 2007
Edited by R.E. Allsop, M.G.H. Bell and B.G. Heydecker

10

RANDOM SUPPLY AND STRATEGIC BEHAVIOUR IN STATIC TRAFFIC ASSIGNMENT

Giuseppe Bellei and Guido Gentile, Dipartimento di Idraulica, Trasporti e Strade Università degli Studi di Roma "La Sapienza", Italy

SUMMARY

In this paper the modelling framework of static traffic assignment is extended to the case where random arc costs are jointly distributed, although implicitly through a local correlation structure at nodes. The resulting network loading problem is then solved through a Monte Carlo simulation at each node processed in topological order. This way of reproducing the day-by-day randomness of transport supply allows highlighting the role of en-route information to drivers, who are induced to make adaptive choices. The two limit situations where maximum level of information is supplied by a route guidance system and minimum level is self-obtained are considered.

INTRODUCTION

Deterministic traffic assignment is based on the assumption that each user is perfectly informed about the topology of the network and the costs that he would incur by travelling on any road arc. In this context, as a rational decision maker he will choose to travel on a shortest path, i.e. one among the paths connecting on the network his origin to his destination with the least cost. Any congestion phenomena is reproduced by means of an arc cost function, through which the disutility of travelling throughout each arc depends in general on the flows of all the arcs of the network. This induces an equilibrium, where no user finds convenient to change unilaterally his path choice.

With stochastic traffic assignment some degree of randomness in the cost pattern is introduced. The users are assumed to choose among paths on the basis of their costs, represented by random variables, whose residual reproduces the perception error by the user and/or the modelling error by the analyst (e.g. data inaccuracy, unrepresented variables and drivers' heterogeneity within each user class). More specifically, path costs are associated with independent identically distributed Gumbel random residuals in the Logit model and with correlated normal residuals in the Probit model. Path cost correlation within the Probit model arises from path overlapping, assuming independent normal residuals for arc costs. When such randomness is taken into account together with arc cost dependence on flows, stochastic equilibrium is defined, where users travelling between the same origin and destination will split among paths consistently with the probability of being perceived (and/or evaluated) as minimum cost ones.

However, such a demand-oriented interpretation of supply randomness does not seem to constitute an appropriate modelling framework for reproducing many phenomena occurring in our everyday traffic experience, corresponding to the occurrence both of intrinsically random events and of events perceived by drivers as random, independently on their nature, such as:

- the downgrading of capacity, due to accidents, occasional road works, illegal parking, and so on, which is typically perceived as random, although may be known to take place with a certain frequency;
- the downgrading of speed, due to weather conditions, which is only partially appraised by drivers, although may be known to take place under specific circumstances;
- the day-by-day variability of trip generation, due to the individual mobility needs related to cyclic and non-systematic activities, of which drivers are not fully aware;
- the within-day variability of supply performances, which cannot be represented in a static model, but yet has to be accounted here, since the daily cyclic pattern of travel costs is known by most drivers only in rather aggregate terms and the rapid changes of travel times when queues are building up, or vanishing, may look as random to them;
- the current phase of the traffic light when the driver approaches an intersection is considered as random by the user and can induce to an adaptive behaviour – for instance, if equivalent left-turn opportunities are given at traffic lights on an arterial, many drivers adopt the strategy to take the first left-turn that is green when reached.

Among the randomness sources listed above there are some local ones, that take place at the arc/node level, and some global ones that take place at network level, and the proposed approach is not meant to analyse and represent them separately. Moreover, although these random arc costs may be independent in some cases, in general correlations are relevant and have to be considered. Global sources, like demand variation, affect in most cases the whole network, determining a non-trivial correlation pattern among arc costs, but also local independent sources, like capacity downgrading due to accidents, can give rise to cost correlations, because the queue may spillback on upstream arcs. In particular, the correlation among arc costs is relevant with respect to the role of information, since drivers utilize information about such costs through the knowledge of the correlation pattern coming from experience.

The representation and evaluation of drivers' information systems is one of the main issues faced in the last years by research on transportation, since they are seen as a tool to mitigate road congestion in urban areas at a lesser investment cost and impact on urban environment than new infrastructures. The proper modelling framework to design and operate route guidance systems is Dynamic Traffic Assignment which requires a huge amount of input data, often unavailable or roughly approximated. However, if the role of information is highlighted within static traffic assignment, the utilization of such a less demanding modelling tool can be envisaged, both in the evaluation stage and in some early design stages of many different information systems, with specific formulations to be tailored to the features of such different systems. A traffic assignment model which includes the representation of drivers' route choice behaviour with respect to information acquisition and processing when no information system is present is also useful, to get a consistent "do nothing" scenario for information systems evaluation, or if empirical evidence is found that this improves the fitting of the observed flows.

The aim of this work is thus to formulate and solve two novel traffic assignment models in the static framework, characterized respectively by a maximum and a minimum level of information, which we named the Full Information Model (FIM) and the Self Information Model (SIM). In both cases there are two key factors to be taken into account. The first one is the day-by-day randomness of transport supply in road networks, which allows drivers to take advantage of information concerning the actual travel costs they are going to face during a specific trip. The second one is the adaptive nature of the travel choices made by a rational decision maker when such information is obtained en-route, and the consequent adoption of a strategic behaviour.

STATE OF THE ART AND PROPOSED MODELS

The issue of supply randomness has been set up by Mirchandani and Soroush (1987), who extended the traditional "demand-side" Stochastic User Equilibrium (SUE) definition to include inherent travel time randomness by introducing, in addition to the usual error term of random utility discrete choice models, applied to route choice and identified as a perception error, a further random term, representative of actual travel time variability. They also introduce both linear and nonlinear disutility functions of perceived actual travel time to model risk-averse, or risk-prone, drivers' behaviour. In the case of linear functions model and solution algorithm development parallels traditional SUE and arc travel time independence is not required, although no scope for considering their correlation is given. Following their work, considerable effort has been spent on taking supply randomness into account, but this effort mainly focused on the impact of such randomness on network reliability, as surveyed by Clark and Watling (2005), when addressing the problem of identifying the distribution of total travel time in the light of day-to-day demand variation, as a mean to evaluate overall network reliability in case of capacity downgrading on some arc.

The proposed approach falls into the "behavioural reliability" class, among the ones identified by Clark and Watling as classes of modelling techniques, quantifying the impact on variable network performance of elements like the abovementioned randomness sources and the more severe performance degradation occurring in emergencies. The issue this class of modelling techniques deals with is, following almost literally Clark and Watling, how to represent the impact on the "typical" route choice pattern, or on other responses such as departure time choice, which is presumed to arise from the modified mean behaviour of drivers in their attitude to the unpredictable variation of travel times and/or the "risk" perceived. The aspect of drivers' behaviour which has been more frequently addressed is risk averseness, which, beside the nonlinear disutility approach of Mirchandani and Soroush, is addressed by Uchida an Iida (1993) utilising risk analysis and by Bell and Cassir (2002) utilising game theory. Another perspective on supply randomness came from Watling (2002), who discussed consistency of SUE paradigm, observing that random flows, consistent with a random utility route choice model, should be made consistent also with random costs. The answer to this consistency problem is the definition of a class of Generalized SUE models, relying on representing drivers' information acquisition process as a day by day sampling from random cost distributions.

All these approaches, however, consider only pre-trip drivers' choices, thus allowing for pre-trip information systems modelling (although none explicitly addresses this issue), while modelling of en-route information is possible only within a strategic, framework, where the transport system users take both pre-trip choices with regard to their travelling strategy and adaptive choices with regard to the implementation of the chosen strategy, where en-route information may play a role.

This framework is commonly adopted in transit assignment, where the strategy (Spiess and Florian, 1989) is represented by a subgraph of the transit network graph, grouping a set of origin destination paths with suitable properties, named hyperpath (Nguyen and Pallottino, 1988). The pre-trip choice is the hyperpath choice made prior to departure, including the choice of a set of lines to board at any stop and the stop where to alight once a line had been boarded. The adaptive behaviour is usually represented as boarding the first transit line arriving within the set. Such a behaviour is not necessarily defined as a choice based on information, as it can be seen simply as a rule describing the utilization of lines calling at a stop. An overview of methods for representing line choice/utilization at stops, together with a model for the case of vehicle overcrowding, can be found in Bouzaïene-Ayari et al. (2001). A choice model has to be developed if availability of real-time user information at stops is taken into account, as proposed by Gentile et al., (2005). Beside transit assignment, a model where adaptive choices are considered within general capacitated networks with queues has been proposed by Marcotte and Nguyen (1998), but no specific reference to supply randomness and information is made. By our knowledge, thus, an approach dealing with supply randomness in road networks by representing route choice within a strategic behaviour framework, hasn't been yet developed. Since adopting such a framework is a straightforward way to represent the role of information within drivers' choices we believe that such a development is of both theoretical and practical interest.

The proposed model is based on several simplifying assumptions, to help focusing on the issue of relations between random supply and strategic behaviour: pre-trip information on actual costs, however obtained, is ignored, as well as risk-adverse, or risk-prone, behaviour. A deterministic choice model is considered, and extension to Deterministic User Equilibrium (DUE) is only sketched, although an assignment procedure dealing with cost dependence on flow is defined and utilized, following the steps of consolidated approaches for passing from network loading to equilibrium. The model formally presented thus qualifies as a Deterministic Network Loading (DNL) on a stochastic network, with a single user class to further simplify the model. Extension to Stochastic User Equilibrium (SUE) will not be addressed at all, since it would require a more sophisticated approach, distinguishing between endogenous supply randomness, the one arising from demand randomness which both Mirchandani and Soroush (1987) and Watling (2002) considered, and exogenous supply randomness, considered here.

A sequential route choice model is adopted, like the one developed by Nguyen et al. (1998) for a logit-type Stochastic Network Loading (SNL) to transit networks, and that developed by Gentile and Papola (2006) for road networks. The drivers are assumed to choose, sequentially, at any node, starting from the origin, the exiting arc which leads to their destination at the minimum cost, instead of choosing jointly a minimum cost path. While seemingly rather different from usual behavioural assumptions, this sequential choice is trivially coincident with DNL minimum path choice, and a similar result holds also for SNL, since the well known Dial's method implements a logit path choice model, which can be shown to be equivalent to a sequential choice model.

CHOICE MODELS AND NETWORK LOADING

The road network is represented, as usual, by a directed graph $G = (N, A)$, where N is the set of the nodes, and $A \subseteq N \times N$ is the set of the arcs. A generic arc from node i to node j is denoted as ij. The forward and backward star of node $i \in N$ are defined as $F(i) = \{j \in N: \exists ij \in A\}$ and $B(i) = \{j \in N: \exists ji \in A\}$, respectively. An efficient subgraph $G^d(N, A^d)$ for any destination d is considered where only the efficient arcs $ij \in A^d = \{ij \in A: TO^d(i) > TO^d(i)\}$ which get closer to the destination are included, for ease of computation. Indeed, the topological order $TO^d(i)$ is some "distance" measure on the network from the generic node $i \in N$ to destination d. The efficient forward and backward star of node i for each destination d are then defined as the set of nodes $F^d(i) = \{j \in N: \exists ij \in A^d\}$ and $B^d(i) = \{j \in N: \exists ji \in A^d\}$, respectively. There are two special subsets of the nodes: the centroid nodes $C \subseteq N$, where trips have their origins and destinations, and the information nodes $I \subseteq N$, where adaptive route choices may take place. At nodes $i \notin I$ drivers' choices are based on average costs, in such a way that, if $I = \varnothing$ and costs are independent by flows the model reduces to a DNL, while at nodes $i \in I$ drivers get, during the trip, some information about actual costs and utilize such information to decide the next node j on their way to the destination, belonging to node i efficient forward star $F^d(i)$.

The random arc costs correlation structure is not explicitly defined. It is represented as drivers perceive it within a sequential model, where route choices are assumed to be made at the local arc/node level, and is implicitly defined by marginal joint distributions representing:

- serial correlation between the cost γ_{ij} of each arc $ij \in A$ and the cost ω_j^d to reach the destination d from final node j of the arc, mainly arising from queue spillover and depending on arc length;
- parallel correlation at each node $i \in I$ among the costs ω_k^d to reach the destination d from every node $k \in F^d(i)$, mainly arising from demand fluctuations and depending on path overlapping.

In particular, the marginal joint distribution, independent on d, of the arc cost γ_{ij} and the node cost ω_j^d is assumed to be a bivariate normal variable, while the marginal joint distribution of the node costs ω_k^d belonging to the efficient forward star of node i is assumed to be a multivariate normal variable. Denoting by:

- c_{ij}, σ_{ij} the mean and standard deviation of random variable γ_{ij};
- w_k^d, σ_k^d the mean and standard deviation of random variable ω_k^d;
- ρ_{ij} the correlation coefficient for the bivariate distribution of γ_{ij} and ω_j^d;
- $\boldsymbol{\omega}_i^d$, $\mathbf{w}_i^d$ the vectors of random and mean cost from nodes $k \in F^d(i)$ to the destination d;
- Σ_i^d the variance-covariance matrix of $\boldsymbol{\omega}_i^d$;

the assumptions on how cost distribution is perceived by drivers are formally expressed as:

$$\begin{bmatrix} \gamma_{ij} \\ \omega_j^d \end{bmatrix} \sim N\left(\begin{bmatrix} c_{ij} \\ w_j^d \end{bmatrix}, \begin{bmatrix} \sigma_{ij}^2 & \rho_{ij} \\ \rho_{ij} & \sigma_j^{d2} \end{bmatrix} \right) \quad i \in N, j \in F^d(i) \tag{1}$$

$$\boldsymbol{\omega}_i^d \sim N\left(\mathbf{w}_i^d, \boldsymbol{\Sigma}_i^d\right) \quad i \in I \tag{2}$$

The expression (2) for the random actual costs from nodes $k \in F^d(i)$ to the destination d is written in compact and general, but rather implicit, form. In most cases, however, the cardinality of $F^d(i)$ is three or less. When $F^d(i) = \{h, j, k\}$, for example, denoting by σ_h^d, σ_j^d, σ_k^d the standard deviations of variables in $\boldsymbol{\omega}_i^d$ and by ρ_{hj}^d, ρ_{hk}^d, ρ_{jk}^d the correlation coefficients among such variables, the (2) can be explicitly written as:

$$\begin{bmatrix} \omega_h^d \\ \omega_j^d \\ \omega_k^d \end{bmatrix} \sim N\left(\begin{bmatrix} w_h^d \\ w_j^d \\ w_k^d \end{bmatrix}, \begin{bmatrix} \sigma_h^{d2} & \rho_{hj}^d \sigma_h^d \sigma_j^d & \rho_{hk}^d \sigma_h^d \sigma_k^d \\ \rho_{hj}^d \sigma_h^d \sigma_j^d & \sigma_j^{d2} & \rho_{jk}^d \sigma_j^d \\ \rho_{hk}^d \sigma_h^d \sigma_k^d & \rho_{jk}^d \sigma_j^d \sigma_k^d & \sigma_k^{d2} \end{bmatrix} \right) \tag{3}$$

The cost γ_{ij} of arc $ij \in A$ is its actual travel cost, while the cost ω_j^d is the expected value of travel cost from node j down to destination d deriving by sequential choices made en-route, which is in general different from the cost that drivers evaluate at node i to make the choice.

The costs that drivers are assumed to evaluate at nodes $i \notin I$ are in fact simply average costs, and the next node, with respect to i, on the way to destination d, is node j such that:

$$j = \arg\min_{k \in F^d(i)} \left\{ w_k^d \right\} \quad i \in N, i \notin I \tag{4}$$

Even if only average cost is considered for sequential choice at nodes $i \notin I$, random cost to destination d from node i, as perceived by drivers, has to be calculated from its joint distribution (1) with the cost of arc j in (4). This ensures that cost to destination is evaluated taking properly into account serial correlation when making adaptive choices at nodes $i \in I$:

$$\omega_i^d = \omega_j^d + \gamma_{ij} \quad i \in N, i \notin I \tag{5}$$

It results then, applying to (5) the formula for the linear combination of bivariate normal random variables, that marginal distribution of ω_i^d is univariate normal with mean and variance derived from the parameters of the joint distribution of ω_j^d and γ_{ij}:

$$\omega_i^d \sim N\left(w_j^d + c_{ij},\ \sigma_{ij}^2 + \sigma_j^{d2} + 2\rho_{ij}\sigma_{ij}\sigma_j^d\right) \tag{6}$$

The drivers are thus assumed to be aware of the random nature of cost to destination, with reference to serial correlation, also at nodes $i \notin I$, where no information is available on the cost of alternative routes to destination and choices are made on the basis of average cost.
At nodes $i \in I$ the drivers are assumed to use some information on actual cost to evaluate the cost of continuing their trip from node i to destination d passing through any node $k \in F^d(i)$, here defined as continuation cost from node i to node k. Denoting by ξ_{ik}^d these continuation costs, which are random variables as well, the probability that node $k \in F^d(i)$ is chosen to reach destination d, P_{ik}^d, is defined by:

$$P_{ik}^d = \Pr\left\{\xi_{ik}^d < \xi_{ih}^d\right\} \quad k, h \in F^d(i), k \neq h, i \in I \tag{7}$$

Two alternative specifications of the continuation costs are given, corresponding to the different choice models. In the case of full information it is assumed that drivers are informed at nodes $i \in I$ about the actual travel cost from node i to destination d corresponding to any choice of continuation node k. In the case of self information it is assumed that drivers get informed at nodes $i \in I$ about the actual cost of the arc leading to node k, only if such cost is higher than a given threshold, and infer continuation costs to all nodes $k \in F^d(i)$ from their knowledge about average costs and correlation.

Full Information Model

In the FIM continuation cost at node $i \in I$ to each node $k \in F^d(i)$ is simply the sum of the actual cost to destination from node k and the actual cost of arc ik:

$$\xi_{ik}^d = \omega_k^d + \gamma_{ik} \quad \forall k \in F^d(i), i \in I \tag{8}$$

The distribution of continuation costs at node $i \in I$ is thus multivariate normal, being a linear combination of the costs of arcs ik and the costs to destination from nodes k, with $k \in F^d(i)$. Denoting by $\boldsymbol{\xi}_i^d$ the continuation cost vector at $i \in I$, by $\boldsymbol{c}_i^d$ the arc cost vector and by $\boldsymbol{\Sigma}^{\xi^d}{}_i$ the variance-covariance matrix of $\boldsymbol{\xi}_i^d$ we get, assuming that arc costs γ_{ik} are uncorrelated, the following expressions for continuation cost, correspondent to (2) and (3):

$$\boldsymbol{\xi}_i^d \sim N\left(\boldsymbol{w}_i^d + \boldsymbol{c}_i^d, \boldsymbol{\Sigma}_i^{\xi^d}\right) \quad i \in I \tag{9}$$

$$\begin{bmatrix} \xi_{ih}^{d} \\ \xi_{ij}^{d} \\ \xi_{ik}^{d} \end{bmatrix} \sim N\left(\begin{bmatrix} w_{h}^{d} \\ w_{j}^{d} \\ w_{k}^{d} \end{bmatrix} + \begin{bmatrix} c_{ih} \\ c_{ij} \\ c_{ik} \end{bmatrix}, \begin{bmatrix} \sigma^{\xi_{h}^{d}2} & \rho_{hj}^{d}\sigma_{h}^{d}\sigma_{j}^{d} & \rho_{hk}^{d}\sigma_{h}^{d}\sigma_{k}^{d} \\ \rho_{hj}^{d}\sigma_{h}^{d}\sigma_{j}^{d} & \sigma^{\xi_{j}^{d}2} & \rho_{jk}^{d}\sigma_{j}^{d}\sigma_{k}^{d} \\ \rho_{hk}^{d}\sigma_{h}^{d}\sigma_{k}^{d} & \rho_{jk}^{d}\sigma_{j}^{d}\sigma_{k}^{d} & \sigma^{\xi_{k}^{d}2} \end{bmatrix} \right) \tag{10}$$

where the terms on the diagonal of $\mathbf{\Sigma}\xi_i^d$ are the variances of the sum of the arc costs and the costs to destination from the final node of each arc $k \in F^d(i)$:

$$\sigma^{\xi_{k}^{d}2} = \sigma_{ik}^{2} + \sigma_{j}^{d2} + 2\rho_{ik}\sigma_{ik}\sigma_{k}^{d} \tag{11}$$

It is worth noting that in the FIM minimum actual cost paths would be chosen only if $I = N$, while information results to be, as a matter of fact, limited if $I \subset N$. If $I = N$ the model can represent a route guidance system, but user classes representative of equipped and non equipped drivers should be considered, which would complicate somewhat the model, since cost to destination should be class specific. Within this work, drivers are homogeneous, thus in the FIM cost to destination d from nodes $i \in I$ coincides with minimum continuation cost:

$$\omega_{i}^{d} = \min_{j \in F^{d}(i)} \left\{ \xi_{ij}^{d} \right\} \tag{12}$$

Assumptions (1) and (2) imply that marginal distribution of each ω_i^d, ω_j^d and γ_{ij} ,for $j \in F^d(i)$, is normal, hence there is admittedly a lack of consistency with respect to (8) and (12). It is, however, the same lack of consistency which is inherent to widely adopted Clark's approximation (Clark, 1961). Expressions (7) and (12) define continuation probabilities and the random variable cost from node $i \in I$ to destination d. Neither expression allows analytical calculation, so that continuation probabilities are obtained by numerical calculation, based on sampling from distribution (9) and approximating each continuation probability P_{ij}^d by the relative frequency of cases when the correspondent continuation cost ξ_{ij}^d results to be the minimum one. Denoting by $\boldsymbol{P}_i^d$ the continuation probability vector at node $i \in I$ such numerical calculation is formally expressed as:

$$\boldsymbol{P}_{i}^{d} = \mathrm{X}^{\mathrm{P}}\left(\boldsymbol{\xi}_{i}^{d}\right) \tag{13}$$

The same sampling allows approximating mean and variance of ω_i^d distribution by the mean and variance of minimum continuation cost in the sample, built with a Montecarlo simulation approach. These numerical calculations are formally expressed as:

$$w_{i}^{d} = \mathrm{X}^{\mathrm{W}}\left(\boldsymbol{\xi}_{i}^{d}\right); \quad \sigma_{i}^{d} = \mathrm{X}^{\mathrm{S}}\left(\boldsymbol{\xi}_{i}^{d}\right) \tag{14}$$

Self Information Model

In the SIM the continuation cost vector ξ_i^d at node $i \in I$ is given a more complex definition, since drivers are assumed to go through a two stage forecasting process, triggered by observed congestion on outgoing arcs and implying that: the cost to destination from final nodes of arcs where congestion is observed is estimated utilizing their cost and serial correlation; cost to destination from final nodes of other arcs is estimated utilizing estimates for cost to destination from final nodes of these arcs and parallel correlation. This process is formalized in the following.

First, if the actual cost γ_{ij} of some arc ij is higher than a predefined threshold value $(1+k_1)c_{ij}$, cost γ_{ij} is attributed to arc ij, cost to destination from node j is estimated as the mean of distribution (1), conditional on γ_{ij}, and continuation cost $\xi_{ij}{}^d$ is evaluated as their sum.

$$\omega_j^d = w_j^d + \rho_{ij}\frac{\sigma_j^d}{\sigma_{ij}}\left(\gamma_{ij} - c_{ij}\right); \quad \xi_{ij}^d = w_j^d + \rho_{ij}\frac{\sigma_j^d}{\sigma_{ij}}\left(\gamma_{ij} - c_{ij}\right) + \gamma_{ij} \tag{15}$$

Second, continuation costs $\xi_{ih}{}^d$ towards nodes $h \in F^d(i)$ such that the cost of the arc ih is lower than, or equal to, $(1+k_1)c_{ih}$, are evaluated by attributing to each arc its mean cost c_{ih} and estimating cost to destination from nodes h as the means of the corresponding components of distribution (2), conditional on estimated costs to destination from nodes $j \in F^d(i)$ such that cost of arc ij is higher than $(1+k_1)c_{ij}$.

Let's then partition the vector $\boldsymbol{\omega}_i{}^d$ and the matrix $\boldsymbol{\Sigma}_i{}^d$ with respect to the final nodes of arcs whose cost is above the threshold (conditioning arcs) and below the threshold, or equal to it (conditional arcs). If the vector components and the matrix rows, or columns, corresponding to conditional arcs are denoted by a minus sign superscript and the ones corresponding to conditioning arcs are denoted by a plus sign superscript, suitably rearranging the rows and the columns of the vector $\boldsymbol{\omega}_i{}^d$ and of the matrix $\boldsymbol{\Sigma}_i{}^d$ the partition is as follows:

$$\boldsymbol{\omega}_i^d = \begin{bmatrix} \boldsymbol{\omega}_i^{d-} \\ \boldsymbol{\omega}_i^{d+} \end{bmatrix}; \boldsymbol{\Sigma}_i^d = \begin{bmatrix} \boldsymbol{\Sigma}_i^{d--} & \boldsymbol{\Sigma}_i^{d-+} \\ \boldsymbol{\Sigma}_i^{d+-} & \boldsymbol{\Sigma}_i^{d++} \end{bmatrix} \tag{16}$$

The costs to destination $\boldsymbol{\omega}_i{}^{d+}$ and continuation costs $\boldsymbol{\xi}_i{}^{d+}$ corresponding to conditioning arcs are given by (15), while the vector of costs to destination $\boldsymbol{\omega}_i{}^{d-}$, conditional on $\boldsymbol{\omega}_i{}^{d+}$, is normally distributed as follows:

$$\boldsymbol{\omega}_i^{d-} \sim N\left(\mathbf{w}_i^{d-} + \boldsymbol{\Sigma}_i^{d-+} \cdot \boldsymbol{\Sigma}_i^{d++^{-1}} \cdot \left(\boldsymbol{\omega}_i^{d+} - \mathbf{w}_i^{d+}\right), \boldsymbol{\Sigma}_i^{d--} - \boldsymbol{\Sigma}_i^{d-+} \cdot \boldsymbol{\Sigma}_i^{d++^{-1}} \cdot \boldsymbol{\Sigma}_i^{d+-}\right) \tag{17}$$

Finally, denoting by $\mathbf{c}_i$ the mean arc cost vector, partitioned as $\boldsymbol{\omega}_i{}^d$, continuation costs corresponding to conditional arcs are given as:

$$\boldsymbol{\xi}_i^{d-} = \mathbf{w}_i^{d-} + \mathbf{c}_i^{-} + \boldsymbol{\Sigma}_i^{d-+} \cdot \boldsymbol{\Sigma}_i^{d++^{-1}} \cdot \left(\boldsymbol{\omega}_i^{d+} - \mathbf{w}_i^{d+}\right) \tag{18}$$

Expressions (16)÷(18) assume that both $\boldsymbol{\omega}_i{}^{d+}$ and $\boldsymbol{\omega}_i{}^{d-}$ exist, but also cases when no arc cost, or every arc cost, is above the threshold have to be taken into account. If we consider again that $F^d(i) = \{h, j, k\}$, the (18) can be made explicit, and the continuation costs can be derived in all the possible cases, which are four: 1 - no arc cost above the threshold; 2 - only the cost of one arc, let it be j, above the threshold; 3 - only the cost of one arc, let it be j, below, or equal to the threshold; 4 - all arc costs above the threshold. The elements of $\boldsymbol{\xi}_i{}^d$ are thus:

Case 1 $\gamma_{ih} \le (1+k_1)c_{ih}$, $\gamma_{ij} \le (1+k_1)c_{ij}$, $\gamma_{ik} \le (1+k_1)c_{ik}$

$$\xi_{ih}^d = w_h^d + c_{ih}; \quad \xi_{ij}^d = w_j^d + c_{ij}; \quad \xi_{ik}^d = w_k^d + c_{ik} \tag{19}$$

Case 2 $\gamma_{ih} \le (1+k_1)c_{ih}$, $\gamma_{ij} > (1+k_1)c_{ij}$, $\gamma_{ik} \le (1+k_1)c_{ik}$

$$\begin{aligned}
\xi_{ih}^d &= w_h^d + c_{ih} + \rho_{hj}^d \frac{\sigma_h^d}{\sigma_{ij}}\left(\gamma_{ij} - c_{ij}\right) \\
\xi_{ij}^d &= w_j^d + \rho_{ij} \frac{\sigma_j^d}{\sigma_{ij}}\left(\gamma_{ij} - c_{ij}\right) + \gamma_{ij} \\
\xi_{ik}^d &= w_k^d + c_{ik} + \rho_{kj}^d \frac{\sigma_k^d}{\sigma_{ij}}\left(\gamma_{ij} - c_{ij}\right)
\end{aligned} \tag{20}$$

Case 3 $\gamma_{ih} > (1+k_1)c_{ih}$, $\gamma_{ij} \le (1+k_1)c_{ij}$, $\gamma_{ik} > (1+k_1)c_{ik}$

$$\begin{aligned}
\xi_{ih}^d &= w_h^d + \rho_{ih} \frac{\sigma_h^d}{\sigma_{ih}}\left(\gamma_{ih} - c_{ih}\right) + \gamma_{ih} \\
\xi_{ij}^d &= w_j^d + c_{ij} + \frac{\left(\rho_{hj}^d - \rho_{hk}^d \rho_{jk}^d\right)\rho_{ih}\sigma_j^d}{\left(1 - \rho_{hk}^{d\,2}\right)\sigma_{ih}}\left(\gamma_{ih} - c_{ih}\right) + \frac{\left(\rho_{jk}^d - \rho_{hk}^d \rho_{hj}^d\right)\rho_{ik}\sigma_j^d}{\left(1 - \rho_{hk}^{d\,2}\right)\sigma_{ik}}\left(\gamma_{ik} - c_{ik}\right) \\
\xi_{ik}^d &= w_k^d + \rho_{ik} \frac{\sigma_k^d}{\sigma_{ik}}\left(\gamma_{ik} - c_{ik}\right) + \gamma_{ik}
\end{aligned} \tag{21}$$

Case 4 $\gamma_{ih} > (1+k_1)c_{ih}$, $\gamma_{ij} > (1+k_1)c_{ij}$, $\gamma_{ik} > (1+k_1)c_{ik}$

$$\begin{aligned}
\xi_{ih}^d &= w_h^d + \rho_{ih} \frac{\sigma_h^d}{\sigma_{ih}}\left(\gamma_{ih} - c_{ih}\right) + \gamma_{ih} \\
\xi_{ij}^d &= w_j^d + \rho_{ij} \frac{\sigma_j^d}{\sigma_{ij}}\left(\gamma_{ij} - c_{ij}\right) + \gamma_{ij} \\
\xi_{ik}^d &= w_k^d + \rho_{ik} \frac{\sigma_k^d}{\sigma_{ik}}\left(\gamma_{ik} - c_{ik}\right) + \gamma_{ik}
\end{aligned} \tag{22}$$

The calculation of continuation cost in the SIM aims at representing the behaviour of drivers when faced to unusual congestion, by assuming that some information can be visually obtained in such cases, for instance by observing that some queue on downstream arcs is close to spillback. Since this phenomenon is not the only one which can be observed by drivers, we have taken a more general approach, also allowing to use most common cost functions.

User classes could be introduced, in the SIM, to distinguish between commuters, whose experience helps adapting their choices to changing network performance, and occasional users, less prone to adaptive behaviour.

A conceptual difference arises between SIM and FIM when we note that, even if user classes are not considered, actual and perceived random cost have to be considered. In the FIM, in fact, the costs ξ_i^d, determining continuation choice probabilities are the same drivers will face on their route to destination, allowing to use them to calculate also the mean and variance of ω_i^d distribution, while in the SIM the same costs systematically differ from actually experienced ones, which are the actual costs of chosen alternative. The distribution of cost to destination ω_i^d, at each node $i \in I$, is thus a distribution conditional on the choices made, whose mean and variance are defined as follow:

$$w_i^d = \mathrm{E}\left[\omega_i^d\right] = \mathrm{E}\left[\omega_j^d + \gamma_{ij} \middle| j = \arg\min\left\{\xi_{ih}^d : h \in F^d(i)\right\}\right] \tag{23}$$

$$\sigma_i^d = \mathrm{V}\left[\omega_i^d\right] = \mathrm{V}\left[\omega_j^d + \gamma_{ij} \middle| j = \arg\min\left\{\xi_{ih}^d : h \in F^d(i)\right\}\right] \tag{24}$$

No equation like (12) for the FIM is defined for the SIM, since equations (23) and (24) define only the parameters of ω_i^d distribution, while equation (7) for continuation probabilities still holds, taking into account, of course, that continuation costs are given by (18). It is clear, however, that the assumption of a normal marginal distribution for ω_i^d is not supported, also in this case, by its derivation and it is adopted to facilitate calculating cost distributions upstream. As for FIM, numerical calculation of both continuation probability and mean and variance of cost to destination is needed.

The formal expressions (13) and (14) for numerical calculation of continuation probability vector at node $i \in I$, mean and variance of distribution conditional to choices made become:

$$\boldsymbol{P}_i^d = \Upsilon^{\mathrm{P}}\left(\xi_i^d\right) \tag{25}$$

$$w_i^d = \Upsilon^{\mathrm{W}}\left(\omega_i^d, \gamma_i^d\right); \quad \sigma_i^d = \Upsilon^{\mathrm{S}}\left(\omega_i^d, \gamma_i^d\right) \tag{26}$$

having denoted by γ_i^d the vector of random costs of arcs ij, $j \in F^d(i)$.

The dependence of continuation probabilities on continuation costs in (25) doesn't mean that sampling takes place on their distribution, which is, as a matter of fact, unknown. The explicit expressions (19)÷(22) show how continuation costs can be determined from the costs of conditioning arcs, but such dependence represents the two-stage drivers' cost forecasting process, while the sampling process, which has two-stages as well, goes in the opposite direction. Cost to destination vector is sampled first from distribution (2), then the cost of each arc ij is sampled, deriving from (1) the distribution of γ_{ij}, conditional on sampled cost to destination from node j, denoted as ϖ_j^d, that is it is sampled from:

$$\gamma_{ij} \middle| \varpi_j^d \sim N\left(c_{ij} + \rho_{ij}\frac{\sigma_{ij}}{\sigma_j^d}\left(\varpi_j^d - w_j^d\right), \left(1 - \rho_{ij}^2\right)\sigma_{ij}^2\right) \tag{27}$$

Deterministic Network Loading

Both choice models defined above are easily integrated into a recursive procedure for the calculation of cost to destination and continuation probability vector at each node, that is followed by a network loading based on the equations for demand consistency and conservation of flow by destination at nodes. Such equations, denoting by $f^d{}_{ij}$ the flow on arc ij having d as its destination and by D_{id} the demand flow from node i to destination d, are expressed as:

$$\sum_{j\in F(i)} f_{ij}^{d} = \sum_{h\in B(i)} f_{hi}^{d} + D_{id} \ \forall i \in N \tag{28}$$

and define, together with relation with arc flows and non-negativity constraints:

$$f_{ij} = \sum_{d\in C} f_{ij}^{d} \ \forall ij \in A \tag{29}$$

$$f_{ij}^{d} \geq 0 \ \forall d \in C; \forall ij \in A \tag{30}$$

the feasible arc flows which are loaded to the network consistently with previously defined choice models.

To make the definition of recursive procedures most compact the (4)÷(6) for nodes $i \notin I$ are written by setting $P_{ij}{}^{d} = 1$ for $j = s^{d}(i)$. Costs and continuation probabilities for the FIM are calculated in the topological order from destination d as follows:

$$\begin{aligned} & w_i^d = 0; \sigma_i^d = 0 \quad if\ i = d \\ & w_i^d = \min_{h\in F^d(i)}\left\{w_h^d\right\} + c_{ij};\ P_{ij}^d = 1;\ \sigma_i^{d\,2} = \sigma_j^{d\,2} + \sigma_{ij}{}^2 + 2\rho_{ij}\sigma_{ij}\sigma_j^d \quad i \notin I \\ & \boldsymbol{P}_i^d = \mathrm{X}^{\mathrm{P}}\left(\xi_i^d\right); w_i^d = \mathrm{X}^{\mathrm{W}}\left(\xi_i^d\right); \sigma_i^{d\,2} = \mathrm{X}^{\mathrm{S}}\left(\xi_i^d\right) \quad i \in I \end{aligned} \tag{31}$$

The analogous procedure for the SIM differs only with respect to formal expressions for numerical calculations, since (23)-(24) replace (13)-(14):

$$\begin{aligned} & w_i^d = 0; \sigma_i^d = 0 \quad if\ i = d \\ & w_i^d = \min_{h\in F^d(i)}\left\{w_h^d\right\} + c_{ij};\ P_{ij}^d = 1;\ \sigma_i^{d\,2} = \sigma_j^{d\,2} + \sigma_{ij}{}^2 + 2\rho_{ij}\sigma_{ij}\sigma_j^d \quad i \notin I \\ & \boldsymbol{P}_i^d = \mathrm{Y}^{\mathrm{P}}\left(\xi_i^d\right); w_i^d = \mathrm{Y}^{\mathrm{W}}\left(\omega_i^d, \gamma_i^d\right); \sigma_i^{d\,2} = \mathrm{Y}^{\mathrm{S}}\left(\omega_i^d, \gamma_i^d\right) \quad i \in I \end{aligned} \tag{32}$$

Once recursion defined by (31), or (32), has been executed for one destination d, arc flows bound to that destination can be loaded to the network by applying continuation probabilities to conservation equations in the inverse topological order:

$$f_{ij}^{d} = P_{ij}^{d}\left(\sum_{h\in B^d(i)} f_{hi}^{d} + d_{id}\right) \forall j \in F^d(i); \forall i \in N \tag{33}$$

and arc flows defined by (29) are simply obtained by executing recursion (31), or (32), and loading (33) for any destination, each time summing the flows bound to that destination to the flows bound to other destinations previously loaded.

Taken together, the random cost parameters/ continuation probability recursion **X**, defined by (31), or **Y**, defined by (32) and the flow loading **L**, defined by (33), are the network loading map (NLM) of full information, or self information, respectively. The main feature of random supply NLM is that, in addition to mean arc cost vector $\boldsymbol{c}$, also arc standard deviation $\boldsymbol{\sigma}^A$ and correlation coefficient $\boldsymbol{\rho}$ vectors are input to the NLM. The vectors of mean cost to destination $\boldsymbol{w}$ and of its standard deviation $\boldsymbol{\sigma}^N$, instead, are endogenous with respect to **X** and **Y**, being calculated and utilized within such recursions, while the continuation probabilities $\boldsymbol{P}$, output from **X** and **Y** and input to **L**, are endogenous to the NLM.

An algorithm implementing such NLM on general networks has been developed. An equilibrium model could be easily formulated by expressing consistency of the NLM and the performance function $\mathbf{c}(\boldsymbol{f})$, supplying mean arc costs $\boldsymbol{c}$, as a function of arc flows $\boldsymbol{f}$. Within this work, we limit ourselves to formalize NLM definition, since some problems are likely to arise if the properties of a random supply equilibrium are investigated, as usual, within a fixed point formulation framework.

It is easy to see, in fact, that NLM is increasing with respect to cost variance, which is exogenous and fixed here, while it should rather be assumed to increase with mean cost.
The assumption of a monotone decreasing NLM, underlying existence and uniqueness proofs of DUE and SUE models, is thus strongly complicated by contrasting effects of mean cost and variance, or it should be based on a fixed variance assumption, rather simplistic in a random supply framework. The random supply equilibrium is thus only graphically represented as follows and the solution algorithm proposed is to be regarded only as an heuristic seeking for consistency of NLM and performance function:

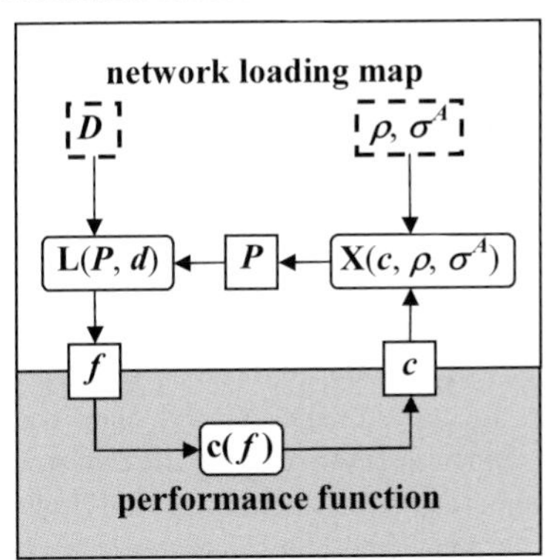

Figure 1 Equilibrium with random supply – Full-Info

The equivalent representation for the SIM is simply obtained by substituting **Y** to **X**, being however the arguments of the two recursions identical at this level of aggregation.

The algorithm utilized to achieve consistency is a standard Method of Successive Averages (MSA), which could be easily extended to cope with demand elasticity by considering that vector $\boldsymbol{w}$ calculated by recursions **X** and **Y** contains detailed information on road network performance, such that demand elasticity may be represented as a function $\mathbf{d}(\boldsymbol{w})$.

It may be interesting, when concluding the part devoted to model formulation, to note that the strategic behaviour has not been explicitly formalized, as usual, in terms of choice among hyperpaths. Indeed, we adopted a sequential approach to route choice, whose strategic aspect is implicit in considering expected values of the cost to destination at nodes where full information is not available. Only in the case of the FIM with $I = N$, expected costs are never taken into account since there is one obvious optimal hyperpath, including all efficient arcs to the destination.

NUMERICAL EXPERIMENTS

Some numerical experiments have been performed utilizing the MSA algorithm based on the NLM described by (31)-(32) and (33), to evaluate how adaptive choices influence the flow pattern on a simple network, by comparing them with flow pattern obtained by applying the same algorithm to DUE, SUE Logit and SUE Probit. To this purpose, the parameters of variance-covariance matrices (1) and (2) have to be specified. This issue is faced, for the purpose of carrying out numerical experiments, by deriving fixed variance and correlation parameters from some network features and from a reference cost pattern, determined by a preliminary assignment, which is used also for other purposes, from defining efficient arcs to destination, to setting variances for the SUE models to be compared with the proposed ones. The way the parameters have been derived is illustrated here below, while the network utilized and calculations made, as well as the results obtained, are reported in the following.

Determination of cost distribution parameters

The arc variances σ_{ij} are taken as proportional to the reference arc costs g_{ij}, determined by a preliminary DUE assignment, that is $\sigma_{ij} = k_2 g_{ij}$ for each $ij \in A$. The serial correlation coefficients ρ_{ij} could be taken as dependent on arc storage capacity, that is the product of arc length, number of lanes and maximum vehicle linear density, since queue spillback, causing serial correlation, represented by such coefficients, takes place as more frequently as smaller it is the arc storage capacity. Such a dependence can be specified by assuming perfect correlation in correspondence to null storage capacity, then decreasing and taking a given correlation value in correspondence to another representative storage capacity. Since the arcs of the network utilized for numerical experiments can't be distinguished by each other with this respect, ρ_{ij} is assumed to be dependent on g_{ij} in the same way, fixing the arc cost at which it is $\rho_{ij} = 0.5$.

The parallel correlation coefficients ρ_j^d are determined from the overlapping of efficient paths, weighted with path utilization, accordingly to the method presented in Gentile and Papola (2006), which allows to accomplish such task without explicit path enumeration. Here we just recall the founding equations of such method, denoting by:

- $P^d(h)$ the set of efficient paths from node h to destination d, that is the set of paths from h to d on graph $G^d(N, A^d)$, based on topological order $TO^d(h)$ for each node h;
- P_q the probability of using path q, which is determined by preliminary assignment;
- $P^\circ{}_{hj}{}^d$ the conditional probability of using arc ij to reach destination d from node i, as determined by preliminary assignment;
- $\Delta_{ij}{}^q$ the generic element of the incidence matrix, equal to 1 if link ij belongs to path q and equal to 0 otherwise.

The topological order of node h, with respect to destination d, is determined by taking as a measure of distance the cost of shortest paths with arc costs g_{ij}. The overlapping of efficient paths to destination d from nodes h and k, $r_{hk}{}^d$, can thus be defined equivalently as the sum of the reference arc costs which are common to each couple of paths $q \in P^d(h)$ and $r \in P^d(k)$ weighted by the joint probability to use path q leaving from node h and path r leaving from node k:

$$r_{hk}^d = \sum_{q\in P^d(h)} \sum_{r\in P^d(k)} P_q P_r \sum_{ij\in A^d} \Delta_{ij}^q \Delta_{ij}^r g_{ij} \tag{34}$$

or as the sum of each reference arc cost, weighted by the joint probability to use it leaving from node h and leaving from node k:

$$r_{hk}^d = \sum_{ij\in A^d} g_{ij} \sum_{q\in P^d(h)} \Delta_{ij}^q P_q \sum_{r\in P^d(k)} \Delta_{ij}^r P_r \tag{35}$$

The average reference cost from h to d, is by definition:

$$r_{hh}^d = \sum_{q\in P^d(h)} P_q \sum_{ij\in A^d} \Delta_{ij}^q g_{ij} \tag{36}$$

It is denoted by $r_{hh}{}^d$ since it coincides with the definition of the overlapping of $P^d(h)$ with itself according to (34), being in this case equal to P_q the joint probability of using q and r, if $q = r$, and null otherwise. The probability P_q of using the generic path $q \in P^d(h)$ for users leaving from node h and directed to destination d is given by the product of the conditional probabilities of its arcs:

$$P_q = \prod_{ij\in A^d} \Delta_{ij}^q P_{ij}^d \tag{37}$$

while $P_q = 0$ if $q \notin P^d(h)$. On these bases the following recursive formulas are proved to hold true, and can be easily solved by visiting each node h in topological order:

$$r_{hk}^d = \sum_{j\in F^d(h)} P^\circ{}_{hj}^d r_{jk}^d \quad \text{if } TO^d(h) > TO^d(k) \quad \text{otherwise } r_{hk}^d = r_{kh}^d \tag{38}$$

$$r_{hh}^d = \sum_{j\in F^d(h)} P^\circ{}_{hj}^d \left(g_{hj} + r_{jj}^d\right) \tag{39}$$

Once all the $r_{hk}{}^d$ and $r_{hh}{}^d$ have been determined by (38) and (39), we set:

$$\rho_{hk}^d = \frac{r_{hk}^d}{\sqrt{r_{hh}^d + r_{kk}^d}} \tag{40}$$

for any each node pair h, k such that $\exists i |\ h, k \in F^d(i)$ and for each destination d.

Also the variance parameters for stochastic models utilized for comparisons is derived from preliminary assignment costs, in the sense that the same σ_{ij} as for the proposed models is utilized within Probit SUE and the variance parameter of Logit SUE is determined in such a way to be the same as the Probit one in correspondence to the preliminary assignment equilibrium origin destination cost.

Test network, calculations and results

The developed algorithm has been applied to the network defined in Table 1, where both the network topology and the arc cost functions parameters are represented. The separable cost function associated to each arc is $c_{ij}(f_{ij}) = T0_{ij} + T1_{ij} \cdot (f_{ij} / Q_{ij})^{\beta_{ij}}$, of the usual BPR type.

The arcs have identical parameters, except for 4-5 and 5-6, whose *T0* and *T1* are half the ones of the other arcs, so that the network is characterized by the presence of a central North-South fast corridor from node 4 to node 6. There is only one origin-destination demand component from node 1 to node 9.

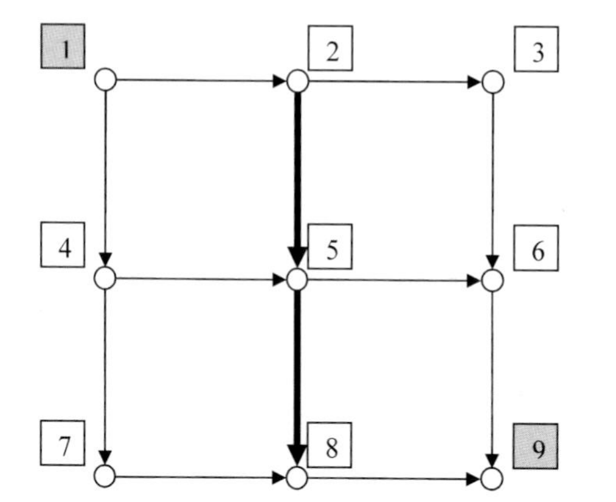

i	j	T0	T1	β	Q
1	2	1	1	4	1000
1	4	1	1	4	1000
2	3	1	1	4	1000
2	5	1	1	4	1000
3	6	1	1	4	1000
4	5	0.5	0.5	4	1000
4	7	1	1	4	1000
5	6	0.5	0.5	4	1000
5	8	1	1	4	1000
6	9	1	1	4	1000
7	8	1	1	4	1000
8	9	1	1	4	1000

Table 1. The network used in the numerical experiments.

A reference case for numerical experiments has been defined by setting the main model parameters as follows. The arc variance factor is $k_2 = 0.2$. The preliminary assignment cost g_{ij} at which serial correlation coefficient ρ_{ij} is one half is assumed to be 1, that is the cost of arcs 2-3, 3-6, 4-7, and 7-8; since cost g_{ij} is lower than 1 for the north-south corridor and higher for the other arcs, we have correlation coefficients respectively higher and lower than one half. The demand d_{19} is equal to 2000 veh./h.

A sensitivity analysis has then been carried out with respect to these parameters by varying them in such a way to cover a range of reasonable values. The arc variance factor has been varied from 0.1 to 0.3 with an increment of 0.05. The serial correlation has been varied fixing the arc cost at which ρ_{ij} is one half, in such a way that the serial correlation spanned, approximately, from 0 to 1 with an increment of 0.25. The demand D_{19} has been varied from 1000 to 3000 with an increment of 500.

In addition to arc flows, the expected origin destination cost w_1^9 is analyzed as a measure of network performance in correspondence to different assumptions about the information available to drivers, the strength of serial correlation and the awareness of parallel correlation. In Table 2 all the arc flows are reported for the reference case, considering five different network loading models: Deterministic, Logit, Probit, the FIM and the SIM with $I = N$.

The sensitivity analysis has been carried out only with respect to the proposed models and the results are presented in the form of diagrams, where the expected origin destination cost are depicted, together with the share of demand flow travelling on the North-South corridor evaluated through the mean flow of the two arcs 2-5 and 5-8.

The reference cost pattern for the stochastic models is obtained by performing 1000 MSA iterations to solve DUE, while 100 MSA network loading iterations are performed within the sensitivity analysis to approach equilibrium, each one involving 1000 local Montecarlo random drawings to find out at every node continuation probabilities, as well as mean and variance of cost to destination distribution.

i node	j node	Determ.	Logit	Probit	Full-Info	Self-Info
1	2	936	951	950	1002	935
1	4	1064	1049	1050	998	1065
2	3	224	294	302	387	180
2	5	712	657	648	615	755
3	6	224	294	302	387	180
4	5	840	755	745	612	821
4	7	224	294	306	386	244
5	6	840	755	747	611	838
5	8	712	657	646	616	739
6	9	1064	1049	1049	998	1017
7	8	224	294	306	386	244
8	9	936	951	951	1002	983

Table 2. The arc flows for different network loading models - Reference case.

The most evident feature of the FIM arc flow pattern is that the share of flow on the North-South corridor is the lowest, not only with respect to DUE, which was expected, but also with respect to SUE Logit and SUE Probit, and up to the point that the East-West corridor becomes slightly more loaded than the North-South one. Although at the other demand levels it is the North-South corridor that is more loaded, differences are always negligible.

Also arcs 2-3, 3-6, 4-7, and 7-8, belonging to outer paths, are significantly more loaded by the FIM than by DUE and SUE, although still less than inner paths.

The flow pattern of the SIM closely resembles DUE and only small changes towards FIM can be noted, much lower than those of SUE models.

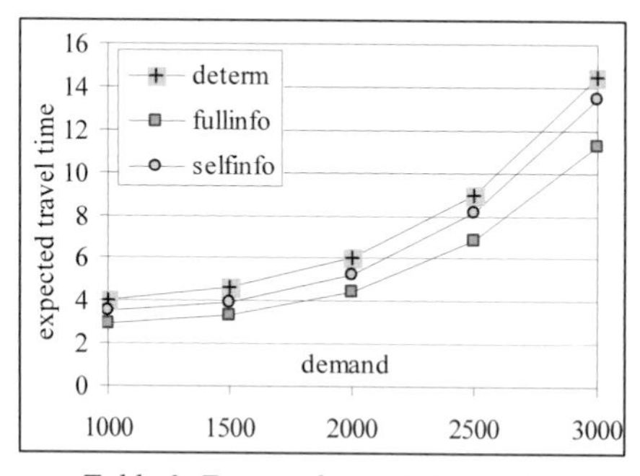

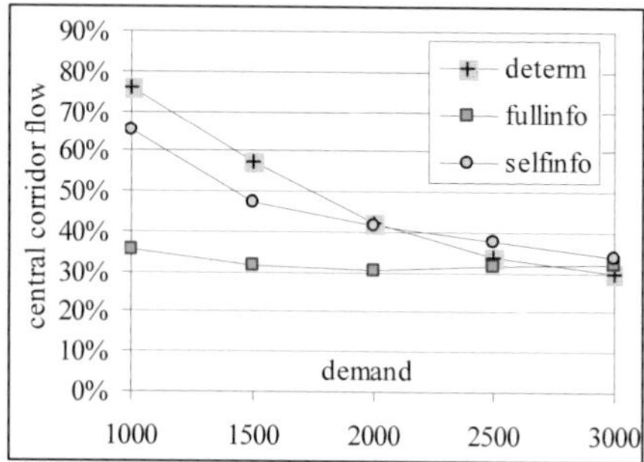

Table 3. Expected travel time and central corridor flow share with varying demand.

The variation of the level of service with the demand, on the left side diagram, is much more regular than the flow pattern variation, on the right side. The expected origin-destination travel time is lower for FIM than for DUE on the whole demand range considered, with a difference that is steadily increasing with demand. Travel time is also lower for SIM than for DUE, but the difference is much smaller and almost insensitive to demand. The central corridor flow share is steadily decreasing with the demand for the SIM, as well for DUE, although less rapidly; while it is much less sensitive to demand for the FIM and it seems to converge to the same value for highest demand levels, where we can suppose that congestion doesn't allow anymore taking advantage of adaptive behaviour.

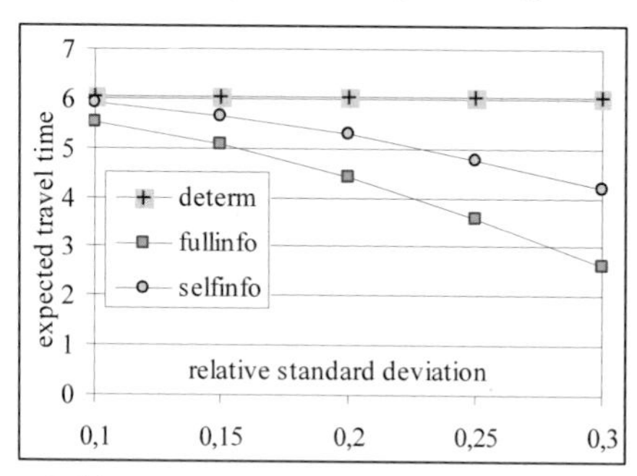

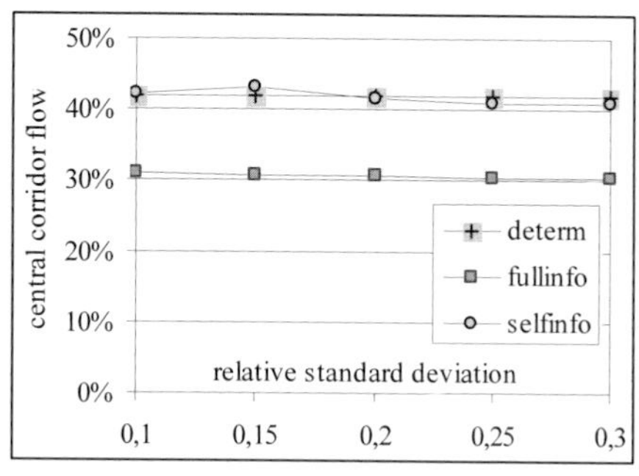

Table 4. Expected travel time and central corridor flow share with varying cost variance.

In this table and in the following, DUE doesn't depend on the parameters, which represent random supply, and is reported as a reference. The analysis of sensitivity to arc variance is quite interesting, because it shows that differences in flow pattern may be minimal, as it happens to be for FIM with respect to variance, or for SIM with respect to DUE, even if steadily variable and remarkable gains are obtained from adaptive behaviour in both cases.

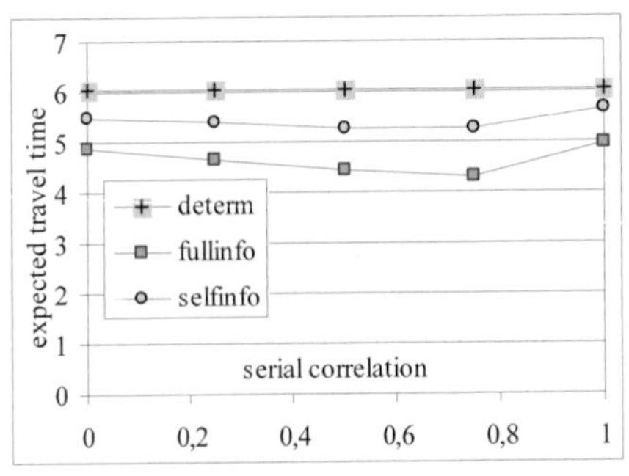

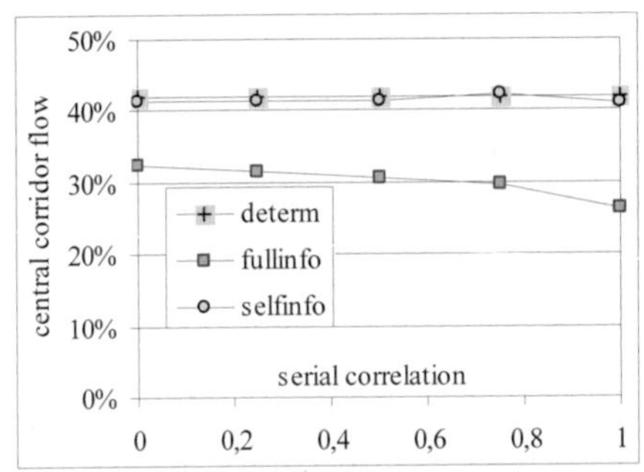

Table 4. Expected travel time and central corridor flow share with varying serial correlation.

The influence of serial correlation on numerical results seems to be rather limited, even if a decreasing expected travel time up to highest correlation levels is in line with expectations, since adaptive behaviour relies on correlation, and also the final increase may be explained by a decrease of overall network efficiency when the arcs of fast corridor become strongly correlated and, on the average, underutilized because their costs are often high together and drivers get informed about, as shown by left side diagram.

CONCLUSIONS

The basic concepts for the definition of a framework where random supply and drivers' adaptive choices based on information can be jointly represented has been set up, by introducing two alternative route choice models and developing the correspondent NLM. Although the properties of equilibrium have not yet been studied, an algorithm attaining consistency of NLM and performance function has been developed and applied to a test network.

Numerical results show that significantly different flow patterns are obtained from deterministic models, or stochastic models representing demand randomness, and from the random supply models proposed in this paper, at least when full information is available to drivers and congestion is not too high. The flow pattern resulting when only self obtained information are available is not significantly different from that of a deterministic model Utilizing such model as an alternative to existing assignment models is thus probably not worth the additional data requirements and computational burden. The network performances, instead, look always significantly different from those measured by a deterministic model, both in case of self obtained and full information, implying that the use of the two proposed models requiring additional data and computational burden may be justified to evaluate the benefits of a driver information system.

The role of arc cost variance resulted to be more important in determining model outputs, and in particular network performance, than arc cost correlation. This may be due to the small dimension of the test network, or by the emphasis on local representation of cost distribution of our approach, which may lead, despite our effort to "think globally and model locally", to hinder the role of arc cost correlation.

Indeed, several issues emerge as necessary to complete what we recognize to be an exploratory work, from the extension of numerical experiments to more complex networks, to developing a sound equilibrium formulation, not excluding further investigation on the overall modelling implications of the proposed approach. For example: what is the sense of an equilibrium where the link flows and the corresponding travel times express the expected values among potentially very different day-by-day traffic patterns? Are we allowed to evaluate the average flow pattern as the composition of the expected local route choice at nodes each one evaluated independently from the others?

Once these issues are dealt with, a number of extensions can be envisaged, from introducing users' classes to integrating within the same framework demand randomness and supply randomness with adaptive behaviour and to improving the representation of information systems that provide a partial information such as variable message signs. A perhaps more decisive step, both from a theoretical and from an application point of view, would be to transfer into a within day dynamic framework the representation of supply randomness and drivers' adaptive behaviour in presence of information.

REFERENCES

Bell M.G.H.. Cassir C. (2002) Risk-averse user equilibrium traffic assignment: an application of game theory, *Transpn Res B* **36** 671-682

Bouzaïene-Ayari B, Gendreau M, Nguyen S (2001) Modeling Bus Stops in Transit Networks: A Survey and New Formulations. *Transpn Sci* **35** 304-321

Clark C. E. (1961) The greatest of a finite set of random variables *Operations Research* 9 145-162

Clark S., Watling D. (2005) Modelling network travel time reliability under stochastic demand, *Transpn Res B* **39** 119-140

Gentile G, Nguyen S, Pallottino S (2005) Passenger's route choice on networks with on-line information at stops. *Transpn Sci* **39** 287-297

Gentile G., Papola A. (2006) An alternative approach to route choice simulation: the sequential models, in *Proceedings of the European Transport Conference 2006* – ETC 2006, Strasbourg, France;

Mirchandani P., Soroush H. (1987) Generalized traffic equilibrium with probabilistic travel times and perceptions *Transpn Sci* **21** 133-152

Marcotte, P., Nguyen S. (1998). Hyperpath Formulations of Traffic Assignment Problems, in *Equilibrium and Advanced Transportation Modelling*, (P. Marcotte and S. Nguyen eds.) Kluwer Academic Publishers, Dordrecht 175-199.

Nguyen S, Pallottino S, (1988) Equilibrium Traffic Assignment for Large Scale Transit Networks *European Journal of Operations Research* **37** **22** 176-186

Nguyen S, Pallottino S, Gendreau M (1998) Implicit enumeration of hyperpaths in logit models for transit networks. *Transpn Sci* **32** 54–64

Spiess H, Florian M (1989) Optimal strategies: A new assignment model for transit networks. *Transpn Res B* **23** 83-102

Uchida T., Iida Y. (1993) Risk Assignment: a new traffic assignment model considering risk of travel time variation in *Proceedings of the 12th International Symposium on Transportation and Traffic Theory* (C. F. Daganzo ed.) Elsevier, Amsterdam 89-105

Watling (2002) A second order stochastic network equilibrium model *Transpn Sci* **32** 54–64

Transportation and Traffic Theory 2007
Edited by R.E. Allsop, M.G.H. Bell and B.G. Heydecker

11

ASSESSING NETWORK VULNERABILITY OF DEGRADABLE TRANSPORTATION SYSTEMS: AN ACCESSIBILITY BASED APPROACH

Anthony Chen[a], Sirisak Kongsomsaksakul[a], Zhong Zhou[a], Ming Lee[b], and Will Recker[c]
[a] *Department of Civil & Environmental Engineering, Utah State University, Logan, UT 84322-4110, USA*
[b] *Department of Civil and Environmental Engineering, University of Alaska, Fairbanks AK 99775-5900,, USA*
[c] *Department of Civil Engineering, University of California, Irvine, CA 92697-3600, USA*

SUMMARY

Transportation networks are an indispensable component of everyday life in modern society. Disruption to the networks can make peoples' daily lives extremely difficult as well as seriously cripple economic productivity. In this paper, we develop network-based accessibility measures for assessing vulnerability of degradable transportation networks. The accessibility-based vulnerability measures explicitly consider the interaction between the disrupted network and the multi-dimensional travel responses of the network users. To model different dimensions of travel behavioral responses, a combined travel demand model formulated as a variational inequality problem is adopted to estimate the utility-based accessibility measure that is consistent with random utility theory. Numerical examples are conducted to demonstrate the feasibility of the proposed network-based accessibility measures for assessing vulnerability of degradable transportation networks. The results indicate that the accessibility measures derived from the combined travel demand model are capable of measuring the consequences of both demand and supply changes in the network and have the flexibility to reflect the effects of different travel choice dimensions on the network vulnerability.

INTRODUCTION

Transportation networks are an indispensable component of everyday life in the modern society. Platt (1995) refers to such physical or virtual networks (i.e., road networks, power lines, water distribution networks, communication networks, and the Internet) that are vital to people's health, safety, comfort, and economic activities as *lifelines*. Disruption to these lifelines can seriously damage the economic productivity of the society as well as making peoples' daily lives extremely difficult (Miller, 2003). Transportation networks as one of the lifelines demand meticulous security consideration especially in the aftermath of recent disastrous events around the world. The terrorist attacks of September 11th expose not only our vulnerability to transportation network disruptions—crippling air travel in the immediate aftermath and perhaps forever changing the convenience of air travel—but also our lack of knowledge about the extent and impact of these disruptions—in this case, impacting the economic viability of segments of the air travel industry. The terrorist bombings on London's subway and bus systems provide a solemn illustration of how disruption to the lifelines can impact people's lives in the city. Months after the attack, signs of a city of 7.2 million people straining to adjust to the fractured lifelines were everywhere. Since the bombings, subway passenger numbers had dropped by 30% on weekends and between 5% and 15% on weekdays on some periods (BBC News, Aug 4, 2005). Tourist activities and retail figures were also down. Most recently, Hurricanes Katrina and Rita highlight the importance of planning and preparedness for natural disasters.

Given that transportation networks are so critical to the functioning of modern society and yet are so fragile, it is important to understand their vulnerability and the consequences to disruption in order to manage risks associated with critical events. Despite the significance of the subject, the current knowledge based on the subject is limited due to the lack of empirical insights, models, data, and decision support tools. Current efforts in transportation research to characterize network vulnerability tend to be qualitative due to the absence of well-defined quantitative measures. While these qualitative indices are useful in communicating the risk of threats to the public, they do not possess the necessary basis for comparison of various threats and the trade-off among potential response measures. A few quantitative approaches have been suggested to measure network vulnerability (Chang, 2003; Chang and Nojima, 2001; D'Este and Taylor, 2001, 2003; Jenelius et al., 2006; Lleras-Echeverri and Sanchez-Silva, 2001, Nicholson and Dalziell, 2003; Sohn 2006; Taylor et al., 2006). However, these approaches overlook many factors that may influence actual changes in travel choices (e.g., congestion, alternatives available to travelers at different damage locations including changes to route choice, mode choice, destination choice, and travel choice). The development of a theoretical quantitative measure that considers different travel behavioral responses of network users when the network is disrupted is needed.

Transportation network vulnerability is not only related to terrorist attacks, but also to natural disasters and traffic accidents, which will degrade the service of a transportation network. All these are considered as abnormal events that will occur in a transportation network with different probabilities, and the consequences of these abnormal events are also different. Due to the complexity of the probability and uncertainty in the vulnerability related events, the evaluation of transportation network vulnerability can be extremely difficult. The aim of this paper is to develop network-based accessibility measures for assessing vulnerability of degradable transportation networks. Accessibility is a fundamental concept in transportation analysis and planning. Typically, accessibility refers to the 'ease' of reaching opportunities for activities and services and can be used to evaluate the performance of a transportation and urban system. These network-based accessibility measures consider the consequence of one or more links failures in terms of network travel time (or generalized travel cost) increase as well as the enforced changes in travel behavior. In this analytical context, transportation network vulnerability is considered as a problem of reduced accessibility due to disruptions. To model different dimensions of travel behavioral responses (e.g., change in route choice, mode choice, destination choice, and travel choice), a combined travel demand model formulated as a variational inequality (*VI*) problem is adopted to estimate the utility-based accessibility measure that is consistent with random utility theory. Numerical examples are conducted to demonstrate the feasibility of the proposed network-based accessibility measures for assessing vulnerability of degradable transportation networks.

BACKGROUND

Because network vulnerability is a relatively new research topic in the transportation field, there are no universal definitions and quantitative measures for assessing transportation network vulnerability. Recently, a few quantitative approaches have been suggested to measure network vulnerability. Berdica (2002) defined vulnerability as "a susceptibility to incidents that can result in considerable reductions in road network serviceability." Services of transportation networks are to provide means for moving passengers and goods to different places in different times. Incidents (or network disruptions) are events that can directly or indirectly result in considerable reductions or interruptions in the serviceability of a link/route/road network. Transportation network vulnerability can be regarded as a problem of reduced accessibility due to different disruptions.

D'Este and Taylor (2001, 2003) and Taylor et al. (2006) used the Australian national strategic transportation network to illustrate that the standard approaches to network reliability based on probabilities may not be adequate in dealing with the 'weak spots' in a network, where failure of some part of the transportation infrastructure can have adverse consequences on accessibility between specific locations and overall system performance. Because the network is sparse at the national level, failure of a link or a subset of links can significantly reduce accessibility, causing delays and detours with significant social, economic, and environmental consequences. Two definitions of vulnerability for the network analysis were used. One is the *connective vulnerability* between two nodes and the other is the *access vulnerability* of a node. *Connective vulnerability* considers the consequence of network degradation. It uses loss of utility (or increase of the generalized travel cost) as the measure of vulnerability (i.e., the connection between two nodes is vulnerable if it gains generalized travel cost after link failure). The concept of *access vulnerability* is built on the concept of accessibility, which quantifies an individual's freedom to participate in activities in the environment. A node is vulnerable if the accessibility of the node is reduced by link disruption. To determine the weaknesses in the network, D'Este and Taylor (2003) used link choice probabilities, which can be calculated in a stochastic traffic assignment model (Bell, 1995), to measure the relative performance of alternative paths and the consequences of network failure using links with a higher link choice probability as candidate links for studying the effects of network degradation through cutting those candidate links and assessing the impacts on network operation of those cuts.

Lleras-Echeverri and Sanchez-Silva (2001) proposed a method to analyze the consequences of the failure of each link. They defined a classification parameter (CP) for each link in the network. A critical link is defined as a link whose failure causes the highest increase in the generalized travel cost (i.e., a critical link is associated with a high CP value). A centroid-based accessibility index, which is defined as the sum of generalized costs of the shortest path from one centroid to all the other centroids, is employed as the measure for the consequences of link failures. Chang and Nojima (2001) applied the distance-based accessibility measure (i.e., without congestion effects) to assess the transportation system performance for an earthquake scenario. Chang (2003) also used the distance-based accessibility measure for evaluating restoration strategies after the Hansin earthquake. The travel time-based accessibility measure was adopted for assessing a hypothetical earthquake scenario on the potential bridge damage in Seattle (Chang, 2003). Jenelius et al. (2006) proposed the concepts of link importance and site exposure to assess the vulnerability of transport networks. Both link importance indices and site exposure indices were derived based on the generalized travel cost increase and travel demand increase to reflect equal opportunities and social efficiency. Application of the proposed measures to the road network of northern Sweden showed that the results could be used in the planning of transport networks. Sohn (2006) employed the weighted accessibility measure by distance and traffic volume to prioritize the retrofit plans for highway links under the event of a flood disaster. Nicholson and Dalziell (2003) applied standard risk assessment method to evaluate a range of natural disasters to the regional highway in New Zealand. However, the above approaches overlook many factors that may influence actual changes in travel choices – factors such as congestion, alternatives available to travelers at different damage locations (including changes to different modes, origin-destination flows, and destination opportunities), or even canceling of trips. Ignoring these factors could lead to biased estimates of network vulnerability. Hence, the purpose of this paper is to develop a vulnerability analysis framework that is capable of assessing the consequences of both demand and supply changes in the network and have the flexibility to reflect the effects of different travel choice dimensions (e.g., changing route, switching mode, changing destination, and canceling or postponing trip) on network vulnerability. In this paper, we extend our previous work (Chen et al., forthcoming) by formulating the combined travel demand model as a *VI* problem to consider the interactions of different modes. This extension is significant, because it offers a more complete assessment of the consequences of network disruptions.

MATHEMATICAL FORMULATION

The accessibility-based vulnerability measures proposed in this paper consider the consequences of one or more links failures in terms of the increase in network travel time (or generalized travel cost). Within this analytical context, the interaction between the enforced transportation network disruptions and the enforced responses of network users can also be considered. According to Nicholson and Dalziell (2003), when the intended travel routes are closed, the behavioral responses of the travelers include:

1. Canceling their trip (i.e., change in trip generation);
2. Postponing their trip (i.e., change in the temporal distribution of trips);
3. Choosing another destination (i.e., change in the spatial distribution of trips);
4. Choosing another mode (i.e., change in mode split);
5. Choosing a different route (i.e., change in trip assignment).

To model different dimensions of travel behavioral responses, a combined travel demand model (*CTDM*) is adopted to estimate the long-term equilibrium network condition due to network disruptions. Over the past three decades, several combined (or integrated) models that can consider the interrelated travelers' decisions (trip generation, trip distribution, mode split, and trip assignment) have been developed (e.g., Evans, 1976; Dafermos and Nagurney, 1984; Safwat and Magnanti, 1988; Lam and Huang, 1992; Oppenheim, 1995; Bar-Gera and Boyce, 2003; Boyce and Bar-Gera, 2004). In this paper, we develop a variational inequality (*VI*) formulation for a combined travel demand model that integrates the travel-destination-mode-route choice. The model is based on the well-established random utility theory in microeconomics, and provides an explicit, rigorous framework where each individual traveler is regarded as a consumer of urban trips such that their travel behavior can be interpreted as outcome of rational decision-making process. Thus, the proposed formulation not only allows a systematic and consistent treatment of travel choice over different dimensions but also behavioral richness. In the following, we present the description of the combined travel-destination-mode-route model.

Notation

i	Origin index
j	Destination index
m	Mode index
r	Route index
a_m	Link index of mode m
β	Parameters in the combined travel demand model: β_r, β_m, β_d, and β_t are positive parameters associated with the route, mode, destination, and travel choice, respectively.
β'	Rescaled parameters, where $\frac{1}{\beta'_m}=\frac{1}{\beta_m}-\frac{1}{\beta_r}$; $\frac{1}{\beta'_d}=\frac{1}{\beta_d}-\frac{1}{\beta_m}$; $\frac{1}{\beta'_t}=\frac{1}{\beta_t}-\frac{1}{\beta_d}$

τ A scalar attached to travel time in the utility function (value of time)
h_i is the traveling propensity of origin i;
h_{ij} is the traveling propensity of destination j from origin i
h_{ijm} is the traveling propensity of mode m between i and j
$\delta_{ijr}^{a_m}$ Link-route incidence indicator, 1 if link a_m on the route r from origin i to destination j on mode m, 0 otherwise
v_a^m Flow on link a of mode m: $v_a^m = \sum_i \sum_j \sum_r T_{ijmr} \delta_{ijr}^{a_m}$
v_a Flow on link a: $v_a = \sum_m v_a^m, \forall a$
c_a^m Fixed cost of travel on link a of mode m
c_{ijmr} Fixed cost of taking route r on mode m from origin i to destination j: $c_{ijmr} = \sum_{a_m} c_a^m \delta_{ijr}^{a_m}$
$t_a^m(\cdot)$ Travel time function for link a of mode m
$g_a^m(\cdot)$ Generalized link travel cost function for link a of mode m: $g_a^m(\cdot) = \tau t_a^m(\cdot) + c_a^m$
g_{ijmr} Generalized travel cost of taking route r on mode m from origin i to destination j: $g_{ijmr} = \sum_{a_m} g_a^m(\cdot)\delta_{ijr}^{a_m}$
P_{xy} (Unconditional) joint probability of x and y (e.g., P_{ijmr} is the probability that a traveler in origin i travels to destination j on mode m through route r)
$P_{y|x}$ Conditional probability of choosing y given x (e.g., $P_{r|ijm}$ is the probability of choosing route r given that a traveler in origin i has chosen to travel to destination j on mode m)
N_i Number of potential travelers from origin i in a given time period
T_i Number of travelers from origin i in a given time period
T_{i0} Number of non-travelers in origin i in a given time period
T_{ij} Number of travelers from origin i to destination j in a given time period
T_{ijm} Number of travelers using mode m from origin i to destination j in a given time period
T_{ijmr} Number of travelers taking route r on mode m from origin i to destination j in a given time period
$\tilde{W}_i$ Expected received utility of origin i
$\tilde{W}_{t/i}$ Expected received utility of travel given a traveler in origin i
$\tilde{W}_{j/i}$ Expected received utility of choosing destination j given that a traveler in origin i has chosen to travel
$\tilde{W}_{m/ij}$ Expected received utility of choosing mode m given that a traveler in origin i has chosen to travel to destination j

Combined Travel Demand Model

Following Oppenheim (1995), each traveler's decision process is assumed to have the following top-down structure:

1. Given an individual at location i, a given time period (hour, day, etc.), and an activity (e.g., shopping, work, recreation, etc.), a potential traveler first decides whether to travel or not. $P_{t|i}$ is the probability that a potential traveler makes a trip in the study time period. Since departure time choice is not considered in this combined model, this step is assumed to account for the first two behavioral responses (i.e., canceling and postponing their trips).
2. Given the choice made at the first level, the conditional probability that an individual will choose destination j to conduct the activity is $P_{j|i}$. This step accounts for the change in destination choice.
3. Given the outcomes from the first two decisions, the conditional probability that an individual will choose mode m (for traveling from i to j) to conduct the activity is $P_{m|ij}$. This step accounts for the change in mode choice.
4. Given the outcomes from the preceding decisions, the conditional probability that an individual will choose route r (for traveling from i to j on mode m) to conduct the activity is $P_{r|ijm}$. This step accounts for the change in route choice.

The above hierarchical structure of the traveler choice process can be represented as in Figure 1. This "nested" structure is the basis for constructing the conditional probabilities at each stage based on the multinomial logit choice function, which is also the foundation for developing a variational inequality formulation for the combined travel-destination-mode-route model.

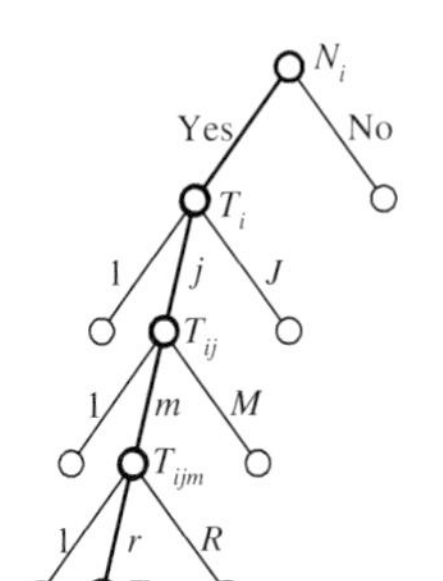

N_i is the potential travelers in origin i
$P_{t|i}$ is the probability of making a trip given N_i

$T_i = N_i\, P_{t|i}$ is the travel demand in origin i
$P_{j|i}$ is the probability of choosing destination j given T_i

$T_{ij} = T_i\, P_{j|i}$ is the travel demand from origin i to destination j
$P_{m|ij}$ is the probability of choosing mode m given T_{ij}

$T_{ijm} = T_{ij}\, P_{m|ij}$ is the travel demand from origin i to destination j on mode m
$P_{r|ijm}$ is the probability of choosing route r given T_{ijm}

$T_{ijmr} = T_{ijm}\, P_{r|ijm}$ is the travel demand taking route r from origin i to destination j on mode m

Figure 1 Hierarchical structure of a combined travel-destination-mode-route model

Given the number of potential travelers N_i at origin i in a given time period (hour, day, etc.), the number of travelers taking route r on mode m from origin i to destination j in the study period can be computed by multiplying the conditional probability at each stage in a nested structure from the route choice stage up to the travel decision stage are

$$T_{ijmr} = N_i \cdot P_{ijmr} = N_i \cdot P_{t/i} \cdot P_{j/i} \cdot P_{m/ij} \cdot P_{r/ijm}, \quad \forall i, j, m, r \tag{1}$$

At each stage, the traveler's choice follows a logit-based probability expression as follows.

$$\frac{T_i}{N_i} = P_{t|i} = \frac{e^{\beta_t (h_i + \tilde{W}_{t|i})}}{1 + e^{\beta_t (h_i + \tilde{W}_{t|i})}}, \ \forall i, \qquad (2)$$

$$\frac{T_{ij}}{T_i} = P_{j|i} = \frac{e^{\beta_d (h_{ij} + \tilde{W}_{j|i})}}{\sum_l e^{\beta_d (h_{il} + \tilde{W}_{l|i})}}, \ \forall i, j, \qquad (3)$$

$$\frac{T_{ijm}}{T_{ij}} = P_{m|ij} = \frac{e^{\beta_m (h_{ijm} + \tilde{W}_{m|ij})}}{\sum_n e^{\beta_m (h_{ijn} + \tilde{W}_{n|ij})}}, \ \forall i, j, m, \qquad (4)$$

$$\frac{T_{ijmr}}{T_{ijm}} = P_{r|ijm} = \frac{e^{-\beta_r g_{ijmr}}}{\sum_k e^{-\beta_r g_{ijmk}}}, \ \forall i, j, m, r, \qquad (5)$$

where $\tilde{W}_{t/i}$, $\tilde{W}_{j/i}$, $\tilde{W}_{m/ij}$ are the expected received utilities of choices at each stage respectively. These expected received utilities are calculated recursively starting from the last stage (i.e., route choice stage) and up to the first stage (i.e., travel decision stage).

$$\tilde{W}_{m/ij} = \frac{1}{\beta_r} \ln \sum_r e^{-\beta_r g_{ijmr}}, \forall i, j, m \qquad (6)$$

$$\tilde{W}_{j/i} = \frac{1}{\beta_m} \ln \sum_m e^{\beta_m (h_{ijm} + \tilde{W}_{m/ij})}, \forall i, j \qquad (7)$$

$$\tilde{W}_{t/i} = \frac{1}{\beta_d} \ln \sum_j e^{\beta_d (h_{ij} + \tilde{W}_{j/i})}, \forall i \qquad (8)$$

We also define the expected received utilities of each origin as

$$\tilde{W}_i = \frac{1}{\beta_t} \ln\left(1 + e^{\beta_t (h_i + \tilde{W}_{t/i})}\right), \forall i \qquad (9)$$

In addition to the conditional probabilities specified above, conservation and non-negativity constraints must hold at each stage:

$$T_{ijm} = \sum_r T_{ijmr}, \ \forall i, j, m, \qquad (10)$$

$$T_{ij} = \sum_m T_{ijm}, \ \forall i, j, \qquad (11)$$

$$T_i = \sum_j T_{ij}, \ \forall i, \qquad (12)$$

$$N_i = T_i + T_{i0},\ \forall i, \tag{13}$$

$$T_{i0} \geq 0,\ T_i \geq 0,\ T_{ij} \geq 0,\ T_{ijm} \geq 0,\ T_{ijmr} \geq 0,\ \forall i, j, m, r. \tag{14}$$

Thus, the equilibrium conditions of the proposed combined travel demand model are defined by Eqs. (1)-(14). In the following, we propose a variational inequality (*VI*) formulation and prove that the solution of the *VI* problem is exactly the equilibrium solution of the combined travel-destination-mode-route model.

Let Ω denote the feasible set satisfying constraints (10)-(14). Then, the variational inequality problem is to find $\left(T_{ijmr}^*, T_{ijm}^*, T_{ij}^*, T_i^*, T_{i0}^*\right) \in \Omega$, such that

$$\begin{aligned}
&\sum_{ijmr}\left[g_{ijmr}\left(\mathbf{T}_{ijmr}^*\right)+\frac{1}{\beta_r}\ln T_{ijmr}^* - \frac{1}{\beta_r}\ln T_{ijm}^*\right]\left(T_{ijmr}-T_{ijmr}^*\right)\\
&+\sum_{ijm}\left[\frac{1}{\beta_m}\ln T_{ijm}^* - h_{ijm} - \frac{1}{\beta_m}\ln T_{ij}^*\right]\left(T_{ijm}-T_{ijm}^*\right)\\
&+\sum_{ij}\left[\frac{1}{\beta_d}\ln T_{ij}^* - h_{ij} - \frac{1}{\beta_d}\ln T_i^*\right]\left(T_{ij}-T_{ij}^*\right)\\
&+\sum_{i}\left[\frac{1}{\beta_t}\ln T_i^* - h_i\right]\left(T_i-T_i^*\right)+\sum_i\left[\frac{1}{\beta_t}\ln T_{i0}^*\right]\left(T_{i0}-T_{i0}^*\right)\geq 0
\end{aligned} \tag{15}$$

for all $\left(T_i, T_{i0}, T_{ij}, T_{ijm}, T_{ijmr}\right) \in \Omega$. Let

$$\begin{aligned}
c_{ijmr} &= g_{ijmr}\left(T_{ijmr}\right)+\frac{1}{\beta_r}\ln T_{ijmr} - \frac{1}{\beta_r}\ln T_{ijm}, \forall i, j, m, r;\\
c_{ijm} &= \frac{1}{\beta_m}\ln T_{ijm} - h_{ijm} - \frac{1}{\beta_m}\ln T_{ij}, \forall i, j, m;\\
c_{ij} &= \frac{1}{\beta_d}\ln T_{ij} - h_{ij} - \frac{1}{\beta_d}\ln T_i, \forall i, j;\\
c_i &= \frac{1}{\beta_t}\ln T_i - h_i, \forall i; c_{i0} = \frac{1}{\beta_t}\ln T_{i0}, \forall i.
\end{aligned} \tag{16}$$

and the corresponding vectors $\mathbf{c}_{ijmr}, \mathbf{c}_{ijm}, \mathbf{c}_{ij}, \mathbf{c}_i, \mathbf{c}_{i0}$. The *VI* can be simplified to the standard form:

$$F(\mathbf{x}^*)^T(\mathbf{x}-\mathbf{x}^*) \geq 0, \forall \mathbf{x} \in \Omega, \tag{17}$$

where $F(\mathrm{x}) = \left[\mathbf{c}_{ijmr}, \mathbf{c}_{ijm}, \mathbf{c}_{ij}, \mathbf{c}_i, \mathbf{c}_{i0}\right]^T$ and $\mathbf{x} = \left[\mathbf{T}_{ijmr}, \mathbf{T}_{ijm}, \mathbf{T}_{ij}, \mathbf{T}_i, \mathbf{T}_{i0}\right]^T$.

Note that the proposed *VI* formulation is based on the generalized link cost function, which depends on the entire link flow pattern. That means the proposed *VI* formulation has the ability to deal with link interactions and includes the mathematical programming formulations (Evans, 1976; Dafermos and Nagurney, 1984; Safwat and Magnanti, 1988; Lam and Huang, 1992; Oppenheim, 1995; Bar-Gera and Boyce, 2003; Boyce and Bar-Gera, 2004) of the combined travel demand model as a special case.

In the following, we give some qualitative properties of the proposed *VI* formulation.

Definition 1. A mapping $H: K \subseteq R^n \rightarrow R^n$ is said to be strictly monotone on *K*, if

$$\left(H(x)-H(y)\right)^T(x-y)>0, \qquad \forall x, y \in K \ and \ x \neq y. \tag{18}$$

Theorem 1. Suppose $\mathbf{x}^* \in \Omega$ is a solution of the *VI* problem (15), then it is an equilibrium solution of the combined travel demand model.

Proof. See Appendix.

Theorem 2. (Existence of solution) Suppose the link cost function of mode *m* (i.e., $g_a^m(\mathbf{v})$) is continuous. Then the *VI* problem (15) has at least one solution.

Proof. See Appendix.

Theorem 3. (Uniqueness of solution) Suppose the link cost function of mode *m* (i.e., $g_a^m(\mathbf{v})$) is monotone and continuous, then the *VI* problem (15) has a unique solution.

Proof. See Appendix.

An Accessibility Based Approach for Network Vulnerability Analysis

Accessibility is a fundamental concept in transportation analysis and urban planning (Miller, 1999). Typically, accessibility refers to the 'ease' of reach and is frequently measured as a function of the available opportunities moderated by some measure of impedance. Despite the popularity of the concept, accessibility has historically been measured in different ways depending on the context of the applications (see Bhat et. al, 2000 and references therein for a comprehensive review). Accessibility measures considered in the literature are of five primary types: spatial separation, cumulative opportunities, gravity, utility, and time-space. Spatial separation measures use the distance between a location and every other location in the study area as the measurement of accessibility. Cumulative opportunities measures consider the opportunities within a specified travel time, or distance, and use the value of cumulative opportunities as the measurement of accessibility. A third type of accessibility measures is gravity measures. These are continuous measures that sum attractions in a study area but discount them with increasing time or distance from the origin. Utility measures are based on an individual's perceived utility for different travel choices. These measures take the form of the natural log of the sum of the travel choices. Time-space measures add a third dimension to the conceptual framework of accessibility. They take into account the time constraints of the individuals being considered.

A few vulnerability studies have employed accessibility as an indicator (Chang, 2003; Chang and Nojima, 2001; D'Este and Taylor, 2001, 2003; Jenelius et al., 2006; Lleras-Echeverri and Sanchez-Silva, 2001, Sohn 2006; Taylor et al., 2006). In this paper, the utility measure is adopted to quantify the accessibility of different travel choice dimensions that is consistent with random utility theory used in the combined travel demand model described above. Ben-Akiva and Lerman (1979) defined the accessibility measure derived from random utility theory as follows:

$$A = E\left(\underset{k\in K}{Max}\, U_k\right), \tag{19}$$

where E = expectation operator,
K = a set of feasible alternatives,
U_k = random utility of alternative k.

For an individual, accessibility was defined as the expected value of the maximum random utility of alternative k in choice set K. They also proved two important properties for the accessibility measure in Eq. (19):

1. *Monotonicity with respect to choice set size.* This property implies that any addition to a person's choice set leaves the individual no worse off than before the addition.

2. *Monotonicity with respect to the systematic utilities.* This property implies that the measure of accessibility does not decrease if the systematic utility of any of the alternatives in the choice set increases.

With the two important properties, accessibility measures can be determined to assess network vulnerability due to disruptions. For example, under a certain abnormal event, the capacities of some links in the network may be significantly degraded, making them unattractive to some road users. This will reduce the size of the choice set (i.e., smaller number of alternative routes). Under the first property, the accessibility measure for route choice should not increase due to the unattractiveness (or unavailability) of some degraded links. Furthermore, due to congestion caused by the reduced number of available links, travel time of the private car mode will increase and hence will decrease the utility of the private car mode. According to the second property, the accessibility measure of the private car mode should not increase.

Furthermore, if the random variable is assumed to follow a Gumbel distribution, the accessibility measure will have a closed-form expression. Let $U_k = V_k + \varepsilon_k$ and $V_k = E(U_k)$, where ε_k is a Gumbel random variate. The accessibility measure defined in Eq. (19) becomes

$$A = E\left(\underset{k \in K}{Max} U_k \right) = \frac{1}{\mu} \ln \sum_{k \in K} e^{\mu V_k}, \tag{20}$$

where μ is a positive scale parameter. In this paper, we adopt Eq. (20) as the accessibility measure for the combined travel-destination-mode-route model to evaluate the consequences of network disruptions. Using the expected received utilities ($\tilde{W}_{m/ij}$, $\tilde{W}_{j/i}$, $\tilde{W}_{t/i}$, and $\tilde{W}_i$) given in Eqs. (6) to (9), the accessibility measure for each stage is defined as follows:

Network accessibility: $$A = \sum_i N_i \tilde{W}_i \Big/ \sum_i N_i, \tag{21}$$

Zonal accessibility: $$A_i = \tilde{W}_{t|i}, \forall i, \tag{22}$$

O-D accessibility: $$A_{ij} = \tilde{W}_{j|i}, \forall i, j, \tag{23}$$

O-D accessibility by mode m: $$A_{ijm} = \tilde{W}_{m|ij}, \forall i, j, m. \tag{24}$$

The accessibility measures given in Eqs. (21) to (24) provide a quantitative measure to assess the effects of network disruptions at different travel choice dimensions, ranging from the level of O-D accessibility by mode to the aggregate level of network accessibility.

VULNERABILTIY ASSESSMENT PROCEDURE

In this section, a vulnerability assessment procedure is provided to compute the accessibility measures and to perform the quantitative network vulnerability analysis. The steps are briefly summarized as follows. The overall procedure is presented in Figure 2 and can be summarized as follows.

Step 1: Define a degraded scenario.

Step 2: Solve the *VI* problem (Eq. 15) to obtain the equilibrium solutions for each travel choice dimension (i.e., route flows, O-D flows by mode, O-D flows, and trips generated in each origin) for both normal and degraded scenarios.

Step 3: Using Eqs. (21) – (24), compute the accessibility measures for each travel choice dimension (i.e., O-D accessibility by mode, O-D accessibility, zonal accessibility, and network accessibility) for both normal and degraded scenarios.

Step 4: Perform the vulnerability assessment analysis based on the accessibility measures from Step 3.

Step 5: Report the results of the vulnerability assessment analysis.

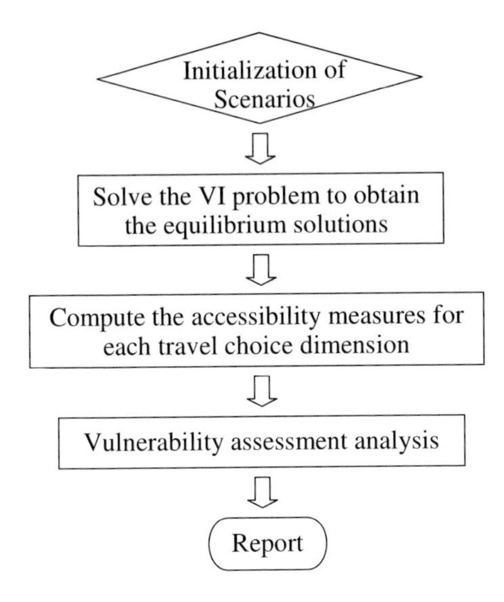

Figure 2 Vulnerability assessment procedure

The core of the above procedure is to solve the *VI* combined travel demand model (Eq. (15)). There are many iterative methods that have been developed for solving the *VI* problem, such as the projection methods (Bertsekas and Gafni, 1982; Cui and He, 1999), the nonlinear Jacobian methods (Pang and Chan, 1982), the successive overrelaxation methods (Pang, 1985), the proximal point methods (Auslender *et al.*, 1999), and the Newton-type methods (Bonnans, 1994; Harker and Pang, 1990). Some of the recent developments are also reviewed by Facchinei and Pang (2003). Among these iterative methods, the projection methods have received much attention due to its global convergence and simplicity of implementation. In this study, we adopt a new self-adaptive Goldstein-Levitin-Polyak (*GLP*) projection algorithm developed by Han (2002) and Zhou and Chen (2003) to solve the *VI* problem (15). A unique feature of this algorithm is that the stepsize is self-adaptive using the information derived from the previous iterations. This feature is designed to effectively minimize the expensive line searches and to guarantee global convergence. Readers may refer to Han (2002) and Zhou and Chen (2003) for the details of the self-adaptive *GLP* projection method for solving the combined travel demand model.

NUMERICAL EXPERIMENTS

In this section, we illustrate the network vulnerability assessment approach proposed in this paper using a simplified network as depicted in Figure 3. The network consists of 6 nodes and 7 links. There are two modes available: car and transit. The free-flow travel time and capacity of each link are given in Table 1 with subscripts '*c*' and '*t*' for car and transit, respectively.

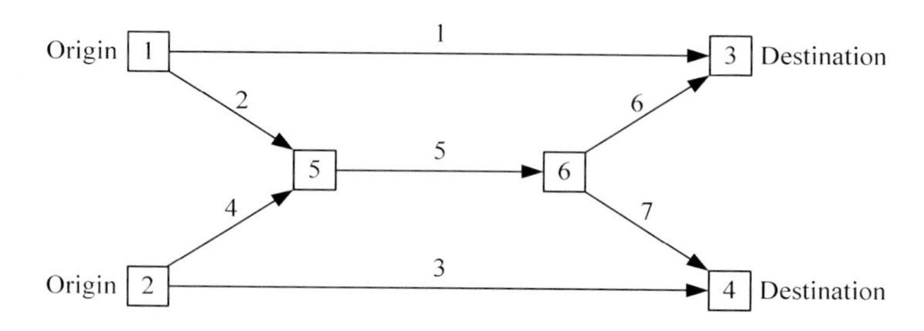

Figure 3 Example network

Table 1 Network characteristics

Link	$t^0_{a_c}$	C_{a_c}	$t^0_{a_t}$	C_{a_t}
1	10.00	100.00	12.00	100.00
2	4.00	80.00	4.80	80.00
3	12.00	80.00	14.40	80.00
4	4.00	50.00	4.80	50.00
5	5.00	120.00	6.00	120.00
6	5.00	50.00	6.00	50.00
7	4.00	50.00	4.80	50.00

The link cost function for both car network and transit network adopted for this study are from Chen and Bernstein (2004) and can be expressed as follows:

$$t_{a_c}(v_{a_c}) = t^0_{a_c}\left[1+\alpha_c\left(\frac{v_{a_c}+0.5E\cdot v_{a_t}}{C_{a_c}}\right)^{\gamma_c}\right], \tag{25}$$

$$t_{a_t}(v_{a_t}) = t^0_{a_t}\left[1+\alpha_t\left(\frac{0.5v_{a_c}+E\cdot v_{a_t}}{C_{a_t}}\right)^{\gamma_t}\right], \tag{26}$$

where v_{a_c} and v_{a_t} are link flows for car and transit; $t^0_{a_c}$ and $t^0_{a_t}$ are free-flow travel times for car and transit (e.g., $t^0_{a_t} = 1.2t^0_{a_c}$); C_{a_c} and C_{a_t} are link capacities for car and transit; E is the passenger car equivalence to public transit (e.g., bus); α_c and α_t are set to 0.15; and γ_c and γ_t are set to 4.0. E is also used to model link flow interaction between the two modes. In the experiments, E is assumed to be 4.0.

The zonal input data are provided in Tables 2 to 5. The route, mode, destination, and travel choice parameters, $\beta_r, \beta_m, \beta_d$, and β_t, are assumed as 2.0, 1.0, 0.5, 0.1, respectively. For illustration, we consider two scenarios: S1 and S2. S1 is the normal scenario without disruption; S2 is the degraded scenario with the capacity of link 5 degraded to half of its original capacity.

Table 2 Zonal data

Zone (i)	Number of Potential Travelers (N_i)	Traveling propensity (h_i)
1	150	12.9
2	150	13.5

Table 3 Traveling propensity of O-D pairs (h_{ij})

O\D	3	4
1	3.4	4.2
2	9.1	8.7

Table 4 Traveling propensity of O-D pairs by car (h_{ijc})

O\D	3	4
1	4.1	3.8
2	3.3	2.9

Table 5 Traveling propensity of O-D pairs by transit (h_{ijt})

O\D	3	4
1	9.5	8.8
2	9.9	10.1

Equilibrium Solutions of the Combined Travel Demand Model

The equilibrium solutions of each travel choice dimension of the two scenarios are shown in Table 6. The results show that the combined travel demand model is capable of quantifying the effects of degraded capacity of link 5 on the different travel choice dimensions. Under the degraded scenario S2, trip productions from both origins drop slightly (i.e., number of trips cancelled or postponed increases due to the reduced capacity of link 5). At the trip distribution step, OD pairs (1,4) and (2,3) are directly affected by the capacity disruption of link 5 since link 5 is part of the only route that connects these two OD pairs. Because of the reduction of OD flows between OD pairs (1,4) and (2,3), part of the total flows generated in the trip generation step is shifted to alternate destinations. Hence, the OD flows between OD pairs (1,3) and (2,4) are increased as a result of the degradation of link 5. In terms of mode choice, the results are similar to the trip distribution step (i.e., OD flows for both modes are decreased for OD pairs (1,4) and (2,3) and slightly increased for OD pairs (1,3) and (2,4)). At the route choice step, flows on routes that pass through link 5 are reduced because of lower level-of-service (higher travel times) due to the reduced capacity of link 5.

Equilibrium link flows of both scenarios are presented in Figure 4. Figure 4(a) depicts the results of the car network while Figure 4(b) depicts the results of the transit network. In each figure, there are two values for each link. The value in the parenthesis is for the degraded scenario. As expected, flows on link 5 for both modes are decreased due to reduced capacity of link 5 in the degraded scenario. These aggregate link-flow patterns are consistent with the equilibrium solutions shown in Table 6.

Table 6 Equilibrium solutions of combined travel demand model

Choice Dimensions	Variables	S1	S2
Trip Generation (T_i and T_{0i})	T_{01}	41.42	43.98
	T_{02}	35.63	37.56
	T_1	108.58	106.02
	T_2	114.37	112.44
Trip Distribution (T_{ij})	T_{13}	78.32	85.40
	T_{14}	30.27	20.63
	T_{23}	38.82	33.25
	T_{24}	75.55	79.19
Mode Choice (T_{ijm})	T_{13c}	50.90	58.09
	T_{14c}	19.04	15.55
	T_{23c}	26.90	24.61
	T_{24c}	50.01	52.40
	T_{13t}	27.41	27.30
	T_{14t}	11.22	5.07
	T_{23t}	11.93	8.64
	T_{24t}	25.53	26.79
Route Choice (T_{ijmr})	T_{13c1}	50.85	58.08
	T_{13c2}	0.05	0.01
	T_{14c1}	19.04	15.55
	T_{23c1}	26.90	24.61
	T_{24c1}	5.98	4.15
	T_{24c2}	44.03	48.25
	T_{13t1}	27.24	27.30
	T_{13t2}	0.18	0.00
	T_{14t1}	11.22	5.07
	T_{23t1}	11.93	8.64
	T_{24t1}	3.11	4.40
	T_{24t2}	22.43	22.39

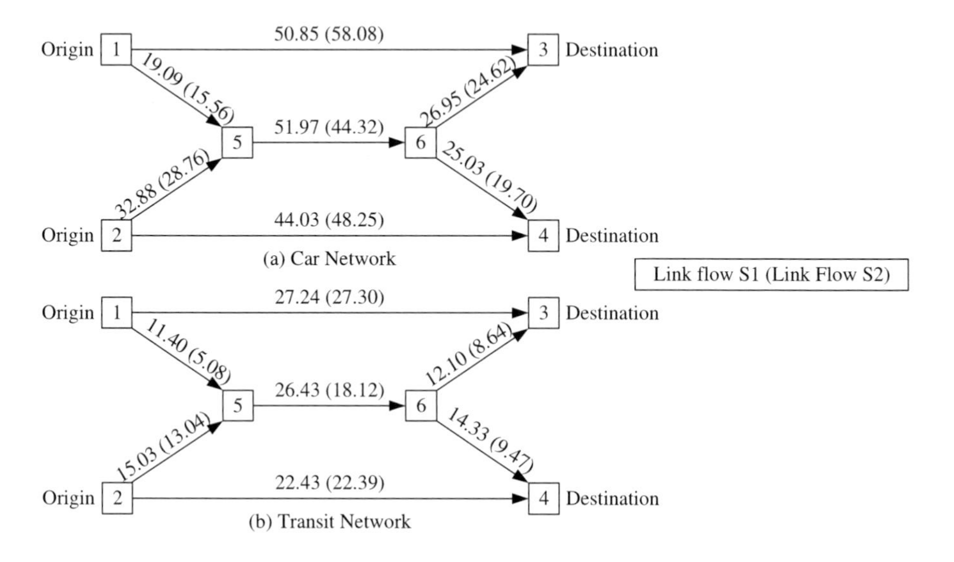

Figure 4 Equilibrium link flows of scenarios S1 and S2

Accessibility Measures and Vulnerability Analysis

Using the equilibrium solutions obtained from solving the combined travel demand model, accessibility measures can be computed using Eqs. (21) to (24). Table 7 shows the accessibility measures of both scenarios and the changes in accessibility measures (ΔAs) between S2 and S1 for different travel choice dimensions. In both scenarios, all four dimensions of users' responses are modeled. Accessibility measures at all levels decrease when link 5 is degraded as indicated by the last column in Table 7. To explain the results, let's start from the bottom of the hierarchical choice structure at the OD mode level. For OD pair (1,4), the deterioration of accessibility for transit is more than that of car (also see Figures 6(a) and 6(b)). This is due to the different link cost functions used in the experiment. The congestion externality of transit links is higher than that of car links (i.e., $t_{a_t}^0 = 1.2 t_{a_c}^0$). As a result, the transit network appears to be more vulnerable than the car network. Similar results can also be observed for other OD pairs. At the OD level, we also observe that OD pair (1,4) is more vulnerable (also see Figures 5). The upper-level accessibility is a logsum of the lower-level accessibilities (see Eqs. (6) to (9)). In this hierarchical structure, the vulnerability information from the lower level will be transferred to the upper level. This is one of the advantages of using the combined travel demand model and the corresponding utility-based accessibility measures. At the origin level, the change in accessibility is in between the changes in accessibility of its OD pairs, because users can choose among the available destinations. At the network level, the change in accessibility is less than that of both origins, because some potential travelers can choose not to travel. Overall, the vulnerability of the network is related to the travel choice dimensions available for the users. More travel choice dimensions available will result in a lesser vulnerable network under degraded conditions.

Table 7 Accessibility measures for different travel choice dimensions

Accessibility Level	Variables	S1	S2	ΔA (S2-S1)
Network (A)	A	13.62	13.06	-0.56
Zonal (A_i)	A_1	-3.26	-4.10	-0.84
	A_2	-1.84	-2.54	-0.70
O-D (A_{ij}) See Figure 4	A_{13}	-7.31	-7.93	-0.62
	A_{14}	-10.02	-11.57	-1.56
	A_{23}	-13.10	-14.07	-0.97
	A_{24}	-11.37	-11.94	-0.57
O-D by mode (A_{ijm}) See Figures 5(a) and 5(b)	A_{13c}	-11.85	-12.42	-0.57
	A_{14c}	-14.28	-15.66	-1.38
	A_{23c}	-16.77	-17.67	-0.91
	A_{24c}	-14.68	-15.25	-0.57
	A_{13t}	-17.86	-18.57	-0.71
	A_{14t}	-19.81	-21.78	-1.97
	A_{23t}	-24.18	-25.32	-1.14
	A_{24t}	-22.55	-23.12	-0.57

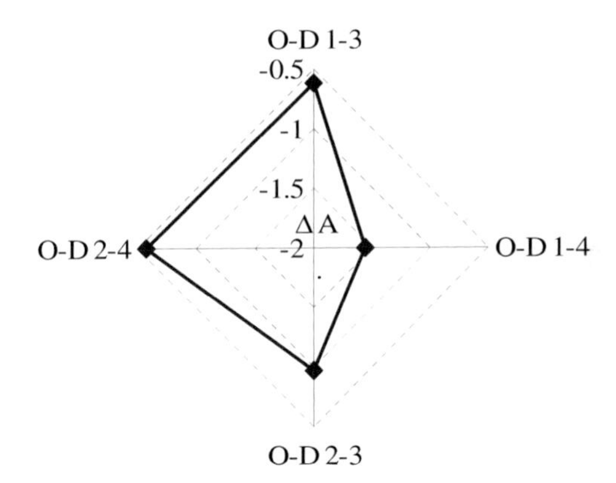

Note: ΔA refers to changes in accessibility measures

Figure 5 Changes in accessibility measures on O-D level

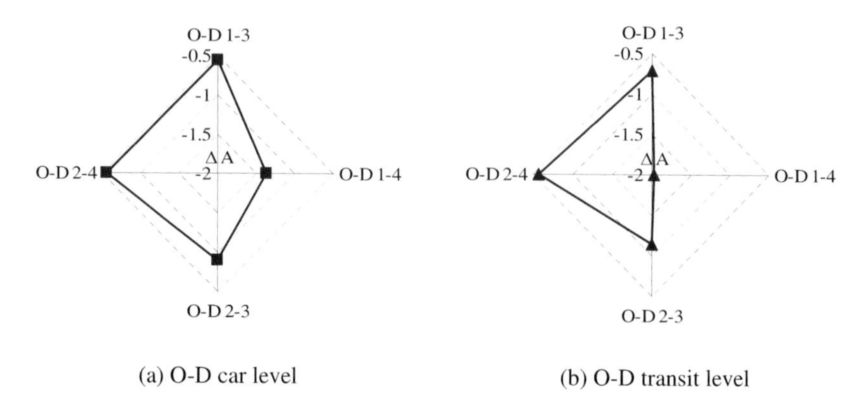

(a) O-D car level (b) O-D transit level

Figure 6 Changes in accessibility measures on different O-D mode levels

Sensitivity Analysis

In this section, three sensitivity analyses are conducted with respect to the degree of link flow interactions (E), the amount of capacity degraded on link 5, and the number of travel choice dimensions on the accessibility measures. Figure 7 shows that the effect of the degree of link flow interactions (E value) on the accessibility measures for the normal scenario. Figure 8 shows the effect of the amount of degraded capacity on link 5 on the reductions of accessibility measures. Figure 9 shows the effect of the number of travel choice dimensions on the changes in accessibility measures at the network level for both modes. From these figures, the following observations can be drawn:

- Ignoring link flow interactions (i.e., without interactions in Figure 7) overestimates the accessibility measures for all travel choice dimensions. The overestimation of accessibility measures increases as the value of E increases (though at a decreasing rate). Also, the amount of accessibility overestimated is different for different accessibility levels as well as within the same accessibility level.
- As the amount of degraded capacity of link 5 increases, the reduction of accessibility measures for all levels increases. At the network level, disruption to link 5 can significantly reduce the network accessibility as link 5 is a critical link to the network. At the origin level, origin 2 is slightly more vulnerable compared to origin 1 because of a slightly larger trip production from origin 2. At the OD level, OD pairs (1,4) and (2,3) are the most vulnerable as there is no alternate route without using link 5.
- Using less travel choice dimensions can overestimate the vulnerability measures as indicated by Figure 9. The changes in accessibility measures are more drastic if only the route choice dimension is used to assess network vulnerability. Regardless of the number of travel dimensions used in the network vulnerability assessment procedure, all travel choice models indicate OD pairs (1,4) and (2,3) are more vulnerable to the disruption of

link 5. Between the two modes, the transit mode appears to be more vulnerable than the car mode.

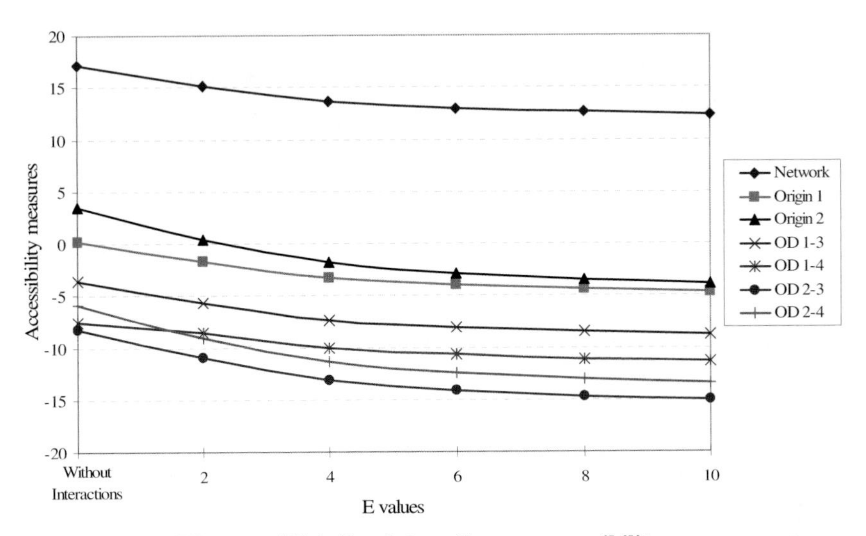

Figure 7 Degree of link flow interactions on accessibility measures

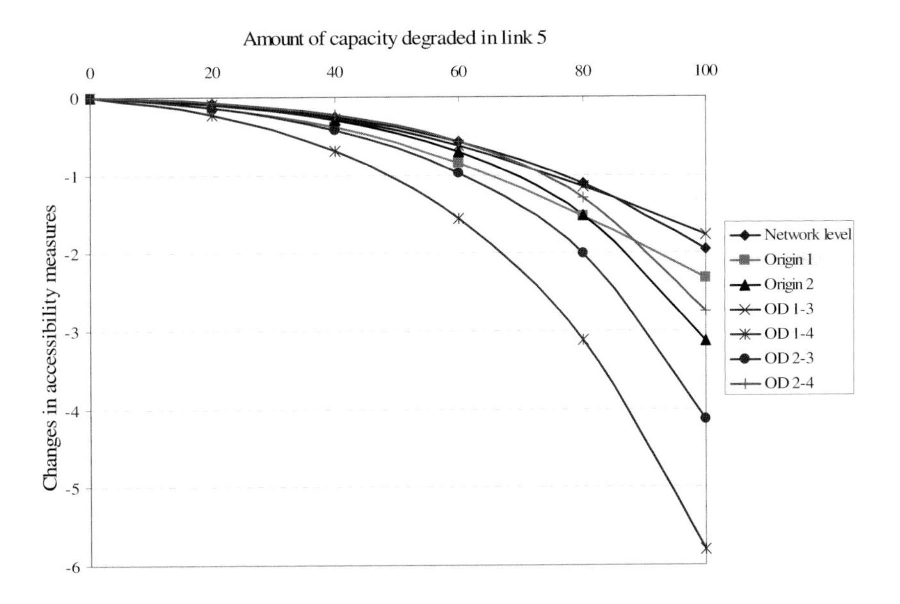

Figure 8 Changes in accessibility measures under different capacity degradation

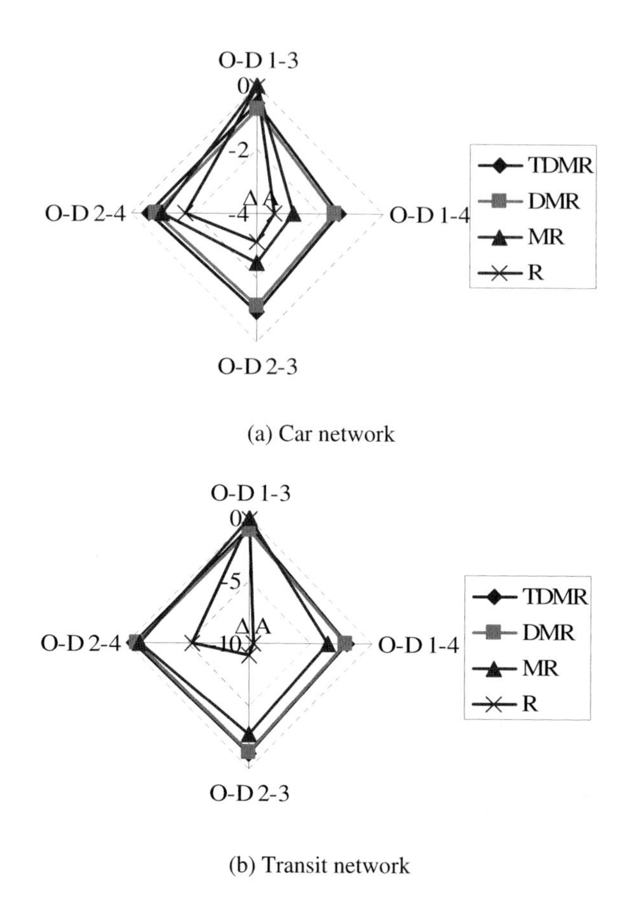

Figure 9 Changes in accessibility measures on the number of travel choice dimensions

CONCLUSIONS

This paper has presented a network-based accessibility measures for assessing vulnerability of degradable transportation networks. The network vulnerability assessment approach employed a combined travel demand model formulated as a variational inequality problem to explicitly consider the interactions between modes (i.e., car and transit) and between the disrupted network and the multi-dimensional travel responses of the network users (e.g., changing route, switching mode, changing destination, and canceling or postponing trip). Network-based accessibility measures, derived from the combined travel demand model that is consistent with random utility theory, were used to quantify the vulnerability of degradable transportation networks. Numerical results, albeit using a small network, indicated that the proposed vulnerability assessment approach is capable of measuring the consequences of both demand and supply changes in the network and has the flexibility to reflect the effects of different travel choice dimensions on network vulnerability.

ACKNOWLEDGEMENTS

The work described in this paper is supported by the Community/University Research Initiative (CURI) grant from the State of Utah and the CAREER grant from the National Science Foundation (CMS-0134161).

REFERENCES

Auslender, A., M. Teboulle, and S. Ben-Tiba (1999). Interior proximal and multiplier methods based on second order homogeneous kernels, *Math of Operations Res*, **24**, 645-668.

Bar-Gera, H. and D. Boyce (2003). Origin-based algorithms for combined travel forecasting models. *Transp Research*, **37B**, 403-422.

BBC News, UK Edition. (2005). Tube passenger numbers drop 30%. *http://news.bbc.co.uk/1/hi/uk/4744139.stm.*

Bell, M.G.H. (1995). Alternative to Dial's logit assignment algorithm. *Transp Research*, **29B**, 287-295

Ben-Akiva, M. and S. R. Lerman (1979). Disaggregate travel and mobility-choice models and measures of accessibility. *Proceedings of the 3rd International Conference on Behavioural Travel Modelling*, 654-679.

Berdica, K. (2002). An introduction to road vulnerability: what has been done, is done and should be done. *Transp Policy,* **9**, 117-127.

Bertsekas, D.P. and E. M. Gafni (1982). Projection methods for variational inequalities with applications to the traffic assignment problem. *Math Programming Study*, **17**, 139-159.

Bhat, C., S. Handy, K. Kockelman, H. Mahmassani, Q. Chen and L. Weston (2000). Urban accessibility index: literature review. *Technical Report TX-01/7-4938-1*, Center for Transportation Research, The University of Texas at Austin.

Bonnans, J.F. (1994). Local analysis of Newton-type methods for variational inequalities and nonlinear programming. *Applied Math and Optimization*, **29**, 161-186.

Boyce D. and H. Bar-Gera (2004). Multiclass combined models for urban travel forecasting. *Network and Spatial Economics*, **4**, 115-124.

Chang, S. E. (2003). Transportation planning for disasters: an accessibility approach. *Environment and Planning A*, **35**, 1051-1072.

Chang, S. E. and N. Nojima (2001). Measuring post-disaster transportation system performance: the 1995 Kobe earthquake in comparative perspective. *Transp Res*, **35A**, 475-494.

Chen, A., C. Yang, S. Kongsomsaksakul and M. Lee (2007). Network-based accessibility measures for vulnerability analysis of degradable transportation networks. *Network and Spatial Economics* (in press).

Chen, M. and D.H. Bernstein (2004). Solving the toll design problem with multiple user groups. *Transp Res*, **38B**, 61-79.

Cui, Y. and B. S. He (1999). A class of projection and contraction methods for asymmetric linear variational inequalities and their relations to Fukushima's descent method. *Computers and Math with Applications*, **38**, 151-164.

D'Este, G. M. and M. A. P. Taylor (2001). Modelling network vulnerability at the level of the national strategic transport network. *J. of the Eastern Asia Society for Transp Studies,* **4**, 1-14.

D'Este, G. M. and M. A. P. Taylor (2003). Network vulnerability: an approach to reliability analysis at the level of national strategic transport network. In: *The Network Reliability of Transport* (Bell, M.G.H. and Y. Iida, ed.), pp. 23-44. Pergamon, Oxford.

Dafermos, S. and A. Nagurney (1984). Stability and sensitivity analysis for the general network equilibrium-travel choice model. *Proceedings of the 9th International Symposium on Transportation and Traffic Theory*, pp. 217-231.

Daganzo, C., and M. Kusnic (1993). Two properties of the nested logit model. *Transportation Science*, **27**, 395-400.

Evans, S. (1976). Derivation and analysis of some models for combining trip distribution and assignment. *Transp Res*, **9**, 241-246.

Facchinei, F. and J. S. Pang (2003). *Finite-Dimensional Variational Inequalities and Complementarity Problems*. Springer-Verlag, New York.

Han, D. R. (2002). Some new numerical methods for variational inequality problems, *Ph.D. Dissertation*, Department of Mathematics, Nanjing University, Nanjing, China.

Harker, P. T. and J. S. Pang (1990). A damped-Newton method for the linear complementarity problem. *Lectures in Applied Math*, **26**, 265-284.

Jenelius, E., T. Petersen, and L. Mattsson (2006). Importance and exposure in road network vulnerability analysis. *Transp Res*, **40A**, 537-560.

Lam, W. H. K. and H. J. Huang (1992). A combined trip distribution and assignment model for multiple user classes. *Transp Res*, **26B,** 275-287.

Lleras-Echeverri, G. and M. Sanchez-Silva (2001). Vulnerability analysis of highway networks, methodology and case study. *Proceedings of the Institution of Civil Engineers*, *Transport*, **147**, 223-230.

Miller, H.J. (1999). Measuring space-time accessibility benefits within transportation networks: basic theory and computational procedures. *Geographical Analysis*, **31**, 187-212.

Miller, H. J. (2003). Transportation and communication lifeline disruption. In: *The Geographic Dimensions of Terrorism* (Cutter, S. L., D. B. Richardson, and T. Wilbanks, ed.), pp. 145-152. Routledge.

Nagurney, A. (1993). *Network Economics: A Variational Inequality Approach*. Kluwer Academic Publishers, Dordrecht, The Netherlands.

Nicholson, A. and E. Dalziell (2003). Risk evaluation and management: a road network reliability study. In: *The Network Reliability of Transport* (Bell, M.G.H. and Y. Iida, ed.), pp. 45-59. Pergamon, Oxford.

Oppenheim, N. (1995). *Urban Travel Demand Modeling*. John Wiley and Sons, Inc., New York.

Pang, J.S. (1985). Asymmetric variational inequality problems over product sets: Applications and iterative methods. *Math Programming*, **31**, 206-219.

Pang, J.S. and D. Chan (1982) Iterative methods for variational and complementarity problems. *Math Programming*, **24**, 284-313.

Platt, R.H. (1995). Lifelines: An emergency management priority for the United States in the 1990s. *Disasters*, **15**, 172-176.

Safwat, K.N. and T.L. Magnanti (1988). A combined trip generation, trip distribution, modal split, and trip assignment model. *Transp Sci*, **18**, 14-30.

Sohn, J. (2006). Evaluating the significance of highway network links under the flood damage: An accessibility approach. *Transp Res*, **40A**, 491-506.

Taylor, M.A.P., S.V.C. Sekhar and G.M. D'Este (2006). Application of accessibility based methods for vulnerability analysis of strategic road networks. *Network and Spatial Economic*, **6**, 267-291.

Zhou, Z. and A. Chen (2003). A self-adaptive scaling technique embedded in the projection traffic assignment algorithm. *J. of the Eastern Asia Society for Transp Studies*, **5**, 1647-1662.

APPENDIX

I. Proof of Theorem 1: Let π_{ijm}, μ_{ij}, λ_i, φ_i be the dual variables associated with equality constraints (10)-(14) respectively. The *KKT* conditions of the *VI* problem (15) are given as below (Facchinei and Pang, 2003):

$$\left[g_{ijmr}\left(T_{ijmr}\right)+\frac{1}{\beta_r}\ln T_{ijmr}-\frac{1}{\beta_r}\ln T_{ijm}-\pi_{ijm}\right]T_{ijmr}=0,\forall i,j,m,r \tag{A.1}$$

$$g_{ijmr}\left(T_{ijmr}\right)+\frac{1}{\beta_r}\ln T_{ijmr}-\frac{1}{\beta_r}\ln T_{ijm}-\pi_{ijm}\geq 0,\forall i,j,m,r \tag{A.2}$$

$$\left[\frac{1}{\beta_m}\ln T_{ijm}-h_{ijm}-\frac{1}{\beta_m}\ln T_{ij}+\pi_{ijm}-\mu_{ij}\right]T_{ijm}=0,\forall i,j,m \tag{A.3}$$

$$\frac{1}{\beta_m}\ln T_{ijm}-h_{ijm}-\frac{1}{\beta_m}\ln T_{ij}+\pi_{ijm}-\mu_{ij}\geq 0,\forall i,j,m \tag{A.4}$$

$$\left[\frac{1}{\beta_d}\ln T_{ij} - h_{ij} - \frac{1}{\beta_d}\ln T_i + \mu_{ij} - \lambda_i\right]T_{ij} = 0, \forall i, j \tag{A.5}$$

$$\frac{1}{\beta_d}\ln T_{ij} - h_{ij} - \frac{1}{\beta_d}\ln T_i + \mu_{ij} - \lambda_i \geq 0, \forall i, j \tag{A.6}$$

$$\left[\frac{1}{\beta_t}\ln T_i - h_i + \lambda_i - \varphi_i\right]T_i = 0, \frac{1}{\beta_t}\ln T_i - h_i + \lambda_i - \varphi_i \geq 0, \forall i \tag{A.7}$$

$$\left[\frac{1}{\beta_t}\ln T_{i0} - \varphi_i\right]T_{i0} = 0, \frac{1}{\beta_t}\ln T_{i0} - \varphi_i \geq 0, \forall i. \tag{A.8}$$

Note that at equilibrium, the solutions of the combined travel demand model are positive. Without loss of generality, we can assume $T_{i0} > 0,\ T_i > 0,\ T_{ij} > 0,\ T_{ijm} > 0,\ T_{ijmr} > 0$ for all *i*, *j*, *m*, *r*. From (A.1) and (A.2), we have

$$g_{ijmr}\left(T_{ijmr}\right) + \frac{1}{\beta_r}\ln T_{ijmr} - \frac{1}{\beta_r}\ln T_{ijm} - \pi_{ijm} = 0, \forall i, j, m, r \tag{A.9}$$

with constraint (10), we have $\pi_{ijm} = -\tilde{W}_{m/ij}$, so

$$\ln\left(\frac{T_{ijmr}}{T_{ijm}}\right) = -\beta_r g_{ijmr} - \beta_r \tilde{W}_{m/ij} \tag{A.10}$$

$$\frac{T_{ijmr}}{T_{ijm}} = P_{r/ijm} = \frac{e^{-\beta_r g_{ijmr}}}{\sum_k e^{-\beta_r g_{ijmk}}}, \forall i, j, m, r. \tag{A.11}$$

Similarly, from (A.3), (A.4), and (11), we have

$$\frac{T_{ijm}}{T_{ij}} = P_{m/ij} = \frac{e^{\beta_m (h_{ijm} + \tilde{W}_{m/ij})}}{\sum_k e^{\beta_m (h_{ijk} + \tilde{W}_{k/ij})}}, \forall i, j, m. \tag{A.12}$$

From (A.5), (A.6), and (12), we have

$$\frac{T_{ij}}{T_i} = P_{j|i} = \frac{e^{\beta_d(h_{ij}+\tilde{W}_{j|i})}}{\sum_l e^{\beta_d(h_{il}+\tilde{W}_{l|i})}}, \quad \forall i, j. \tag{A.13}$$

From (A.7), (A.8), and (13), we have

$$\frac{T_i}{N_i} = P_{t|i} = \frac{e^{\beta_t(h_i+\tilde{W}_{t|i})}}{1+e^{\beta_t(h_i+\tilde{W}_{t|i})}}, \quad \forall i. \tag{A.14}$$

The above derivations show that the solution of the *VI* problem (15) satisfies the equilibrium conditions of the combined travel demand model. The proof is completed. □

II. Proof of Theorem 2: According to the relationship,

$$g_{ijmr}\left(T_{ijmr}\right) = \sum_{a_m} g_a^m(T_{ijmr})\delta_{ijr}^{a_m}.$$

g_{ijmr} is also continuous. Then, $F(\mathbf{x})$ is continuous. Since the set Ω is convex and compact, existence of a solution is guaranteed (Theorem 1.4, Nagurney, 1993). □

III. Proof of Theorem 3: By rearranging the left-hand side of (15), we obtain an alternative *VI* formulation as shown below:

$$\begin{aligned}
&\sum_{ijmr}\left[g_{ijmr}\left(T_{ijmr}^*\right)+\frac{1}{\beta_r}\ln T_{ijmr}^*\right]\left(T_{ijmr}-T_{ijmr}^*\right) \\
&+\sum_{ijm}\left[\frac{1}{\beta_m'}\ln T_{ijm}^*-h_{ijm}\right]\left(T_{ijm}-T_{ijm}^*\right) \\
&+\sum_{ij}\left[\frac{1}{\beta_d'}\ln T_{ij}^*-h_{ij}\right]\left(T_{ij}-T_{ij}^*\right) \\
&+\sum_{i}\left[\frac{1}{\beta_t'}\ln T_i^*-h_i\right]\left(T_i-T_i^*\right)+\sum_{i}\left[\frac{1}{\beta_t}\ln T_{i0}^*\right]\left(T_{i0}-T_{i0}^*\right)\geq 0
\end{aligned} \tag{A.15}$$

for all $\left(T_i, T_{i0}, T_{ij}, T_{ijm}, T_{ijmr}\right) \in \Omega$.

Here, $\frac{1}{\beta'_m}=\frac{1}{\beta_m}-\frac{1}{\beta_r}$; $\frac{1}{\beta'_d}=\frac{1}{\beta_d}-\frac{1}{\beta_m}$; $\frac{1}{\beta'_t}=\frac{1}{\beta_t}-\frac{1}{\beta_d}$. Without loss of generality, we assume $\beta_t<\beta_d<\beta_m<\beta_r$ (Daganzo and Kusnic, 1993), which implies β'_m, β'_d and β'_t are positive.

Let

$$
\begin{aligned}
&\bar{c}_{ijmr}=g_{ijmr}\left(T_{ijmr}\right)+\frac{1}{\beta_r}\ln T_{ijmr},\forall i,j,m,r;\\
&\bar{c}_{ijm}=\frac{1}{\beta'_m}\ln T_{ijm}-h_{ijm},\forall i,j,m;\\
&\bar{c}_{ij}=\frac{1}{\beta'_d}\ln T_{ij}-h_{ij},\forall i,j;\\
&\bar{c}_{i}=\frac{1}{\beta'_t}\ln T_{i}-h_{i},\forall i;\bar{c}_{i0}=\frac{1}{\beta_t}\ln T_{i0},\forall i.
\end{aligned}
\tag{A.16}
$$

and the corresponding vectors $\bar{\mathbf{c}}_{ijmr},\bar{\mathbf{c}}_{ijm},\bar{\mathbf{c}}_{ij},\bar{\mathbf{c}}_{i},\bar{\mathbf{c}}_{i0}$. The *VI* (A.15) can be rewritten as

$$
\bar{F}(\mathbf{x}^*)^T(\mathbf{x}-\mathbf{x}^*)\geq 0,\forall\mathbf{x}\in\Omega, \tag{A.17}
$$

where $\bar{F}(\mathrm{x})=\left[\bar{\mathbf{c}}_{ijmr},\bar{\mathbf{c}}_{ijm},\bar{\mathbf{c}}_{ij},\bar{\mathbf{c}}_{i},\bar{\mathbf{c}}_{i0}\right]^T$ and $\mathbf{x}=\left[\mathbf{T}_{ijmr},\mathbf{T}_{ijm},\mathbf{T}_{ij},\mathbf{T}_{i},\mathbf{T}_{i0}\right]^T$.

According to Theorem 2, there exists at least one solution. Since $g_{ijmr}\left(T_{ijmr}\right)=\sum_{a_m}g_a^m(T_{ijmr})\delta_{ijr}^{a_m}$, according to the assumption, $g_{ijmr}\left(T_{ijmr}\right)$ is monotone. Furthermore, the logarithm function is strictly monotone on R_{++} (i.e., positive orthant of R). Thus, it is straightforward to verify that mapping $\bar{F}$ is strictly monotone. Thus, the uniqueness of solution can be obtained easily following a similar proof of Theorem 1.6 by Nagurney (1993). □

Transportation and Traffic Theory 2007
Edited by R.E. Allsop, M.G.H. Bell and B.G. Heydecker

12

THEORETICAL BOUNDS OF CONGESTION-PRICING EFFICIENCY FOR A CONTINUUM TRANSPORTATION SYSTEM

H.W. Ho and S.C. Wong, Department of Civil Engineering
Timothy D. Hau, School of Economics and Finance
The University of Hong Kong

SUMMARY

Consider a city of an arbitrary shape where users are distributed continuously over the city region. Within this region, the road network is dense and can be represented as a continuum: users patronize a two-dimensional continuum network to the central business district (CBD). A congestion-pricing technique is applied to this continuum transportation system to maximize its total social welfare. This paper presents the theoretical bounds for the efficiency of this congestion-pricing scheme for both the fixed and elastic demand cases. The ratio of total costs and total social welfare between the system-optimal and user-optimal travel patterns are taken as the measures of efficiency in the fixed and elastic demand cases, respectively. With a better understanding of the efficiency of congestion-pricing, traffic planners can better comprehend the likely benefit that is obtainable from a congestion-pricing scheme.

INTRODUCTION

In the literature, the modeling of traffic equilibrium problems is classified into two general approaches: the discrete modeling approach and the continuum modeling approach. The discrete modeling approach, in which each road link within the network is modeled separately and the demand is assumed to be concentrated at hypothetical zone centroids, is commonly adopted for detailed planning. The continuum modeling approach, in contrast, is used for the initial phase of planning and modeling in broad-scale regional studies, in which the focus is on the general trend and pattern of the distribution and travel choice of users at the macroscopic level, rather than at the detailed level. In the continuum approach, a dense network is approximated as a continuum in which users are free to choose their routes in a two-dimensional space. The fundamental assumption is that the differences in modeling characteristics, such as the travel cost and the demand pattern, between adjacent areas within a network are relatively small as compared to the variation over the entire network. Hence, the characteristics of a network, such as the flow intensity, demand, and travel cost, can be represented by smooth mathematical functions (Vaughan, 1987). A promising extension to modeling an arbitrary city shape has recently been made, and a solution algorithm that uses the finite element method (FEM) to solve the resultant continuum model has been published (Wong *et al.*, 1998; Ho *et al.*, 2006).

The continuum modeling approach has various advantages over the discrete modeling approach in macroscopic studies with very dense transportation systems (Blumenfeld, 1977; Taguchi and Iri, 1982; Sasaki *et al.*, 1990; Gwinner, 1998). First, it reduces the problem size for dense transportation networks. The problem size in the continuum modeling approach depends on the method that is adopted to approximate the modeling region, and thus an effective approximation method, such as the FEM (Zienkiewicz and Taylor, 1989), can help to reduce the size of the problem. Second, less data is required to set up a continuum model. As a continuum model can be characterized by a small number of spatial variables, it can be set up with a much smaller amount of data in comparison to a discrete modeling approach, which requires data for all of the included links. This makes the continuum model suitable for macroscopic studies in the initial design phase, as the resources for the collection of data in this phase are very limited. Finally, the continuum modeling approach provides a better understanding of the global characteristics of a road network. As the numerical results of a continuum model can be visualized in a two-dimensional sense, the influence of different model parameters and the spatial interaction between locations can easily be detected and analyzed.

Traffic equilibrium problems are usually formulated as a mathematical program. By considering different objectives in the program, different equilibrium travel patterns, namely the user-optimal and system-optimal travel patterns, can be obtained. In the user-optimal travel pattern, all users in the transportation system choose their routes by minimizing the total travel cost incurred. As these users are not cooperating with each other, extra delays and congestion will be experienced, which reduces the total system benefit. In contrast, a system-optimal travel pattern, which maximizes the total system benefit, can be obtained by introducing proper regulatory measures, such as congestion-pricing, to the transportation system. Congestion-pricing is a method that is used to reduce traffic congestion and to raise revenue for funding transportation improvements. The underlying principle of congestion-pricing is one of regulating the choice behavior of users to approach a system-optimal travel pattern by imposing tolls on the users. From the system point of view, a system-optimal travel pattern is superior to a user-optimal pattern, because the system-optimal pattern maximizes the total system benefit. However, the question becomes one of whether the cost of implementing congestion-pricing can be compensated for by a gain in the system benefit.

To answer this question, it is necessary to ascertain how efficient the system-optimal travel pattern is in maximizing the total system benefit, or how inefficient the user-optimal travel pattern is in this context. Roughgarden and Tardos (2002) attempted to ascertain the theoretical bounds of the price of anarchy, which is a term that was introduced by Papadimitriou (2001) to represent the ratio of total cost for a user-optimal travel pattern compared to total cost for a system-optimal pattern, in a discrete transportation network with a fixed demand. Roughgarden and Tardos proved that the price of anarchy for networks with a linear link cost function is at most 4/3. In addition to this linear link cost function, the theoretical bounds of the price of anarchy for discrete networks with a more general link cost function were derived by Roughgarden (2003), who pointed out that the bounds of the price of anarchy are solely dependent on the link cost function and is independent of the network topography. Chau and Sim (2003) investigated the bound of the price of anarchy of the non-atomic congestion game with elastic demand, and concluded that it is more difficult to obtain the bound of the price of anarchy of the elastic demand when compared to the fixed demand case. Similar analyses of the price of anarchy can be found in the work of Correa *et al.* (2004) for capacitated networks, and in that of Guo and Yang (2005) for networks in the context of stochastic user-equilibrium.

Based on the idea of the price of anarchy in the discrete modeling approach, we study the theoretical bounds of congestion-pricing efficiency in a continuum transportation system. We consider a single class of users in a study region of arbitrary shape. The region has a central business district (CBD), which is the destination of all users. For this single class of users, both fixed and elastic demand cases are considered and the existence of a theoretical bound for these two cases is explored. To evaluate the efficiency of congestion-pricing for the determination of the theoretical bound, two different models are considered: the user-optimal model, which is based on the model that was introduced by Wong *et al.* (1998), and the system-optimal model, which is based on the first-best formulation of Ho *et al.* (2005). As congestion-pricing is adopted in the system-optimal model to ensure the system is best performed, the comparison of these the user-optimal and system-optimal models helps to shed light on the efficiency of congestion-pricing for improving the system performance. Due to the differences in the formulation of the fixed and elastic demand in the system-optimal model, different measures of efficiency will be considered. In the fixed demand case, the measure of efficiency is taken as the ratio of total travel cost of a user-optimal pattern to that of a system-optimal pattern, whereas in the elastic demand case the measure of efficiency is taken as the ratio of social welfare of a system-optimal pattern to that of a user-optimal pattern. By considering the interrelation of different variables, such as the flow and total travel cost, at the optimal points of the aforementioned problems, the theoretical bounds of congestion-pricing efficiency will be obtained.

In the remainder of this paper, we first introduce the notation and the specific equations that are used to obtain the theoretical bounds for the fixed and elastic demand cases. Based on these definitions, the proofs for the theoretical bounds of congestion-pricing efficiency in a continuum transportation system are presented for both of the fixed and elastic demand cases.

DEFINITIONS AND NOTATION

General

In this study, we consider a two-dimensional city with an arbitrary shape and denote the city region as Ω. The outer boundary of the city is denoted as Γ with $\Gamma \subset \Omega$, and the CBD is located at O which is encircled by the inner boundary of the city Γ_c with $\Gamma_c \subset \Omega$. The transportation cost per unit distance of travel at location $(x, y) \in \Omega$ is denoted as $c(x, y, \mathbf{f})$, which has the following functional form in terms of the traffic flow vector, $\mathbf{f}(x, y)$, at that location:

$$c(x, y, \mathbf{f}) = a(x, y) + b(x, y)|\mathbf{f}(x, y)|^{\gamma}, \tag{1}$$

where $a(x, y)$ and $b(x, y)$, which are strictly positive scalar functions of the cost-flow relationship that reflects the local characteristics of the streets at location $(x, y) \in \Omega$, are the free-flow travel cost and congestion-related parameter of the cost-flow relationship. γ, which is a positively defined constant for the entire region, is the non-linearity parameter for this cost-flow relationship. $\mathbf{f}(x, y) = (f_x(x, y), f_y(x, y))$ is a vector that represents the flow state at location (x, y) in the study area, and $f_x(x, y)$ and $f_y(x, y)$ are the flow flux in the directions x and y, respectively, and $|\mathbf{f}(x, y)| = \sqrt{f_x(x, y)^2 + f_y(x, y)^2}$ is the norm of the flow vector that measures the traffic flow intensity at (x, y). Ho *et al.* (2005) proved that for a continuum transportation system with the unit transportation cost function that is defined in equation (1), the marginal cost function $c_{\mathrm{m}}(x, y, \mathbf{f})$ at location $(x, y) \in \Omega$ can be expressed as:

$$c_{\mathrm{m}}(x, y, \mathbf{f}) = a(x, y) + (\gamma + 1)b(x, y)|\mathbf{f}(x, y)|^{\gamma}. \tag{2}$$

Based on the unit transportation cost in equation (1) and the marginal cost function in equation (2), the path travel cost and path marginal cost from a home location $\mathrm{H} \in \Omega$ to the CBD (O) along the chosen path P can be expressed by the following equations:

$$u(x, y, \mathbf{f}) = \int_P c(x, y, \mathbf{f}) \mathrm{d}s, \tag{3}$$

$$u_{\mathrm{m}}(x, y, \mathbf{f}) = \int_P c_{\mathrm{m}}(x, y, \mathbf{f}) \mathrm{d}s. \tag{4}$$

For brevity, the location parameters, x and y, of the forgoing variables and functions will be omitted in the forthcoming discussions. As the efficiencies of congestion-pricing in both of the fixed and elastic demand cases in a continuum transportation system are to be studied in this paper, the formulation of the user-optimal and system-optimal models, which are introduced in Wong *et al.* (1998) and Ho *et al.* (2005), are introduced in the following subsections.

Fixed demand models

In fixed demand models, the demand at location (x, y) is denoted as $q(x, y)$, which is a strictly positive scalar function. The user-optimal model, $S_{uf}(\Omega, q, c)$, for the fixed demand case is formulated into a minimization problem with the following objective function:

$$Z_{uf}(\mathbf{f}) = \iint_{\Omega}\left[\int_0^{|\mathbf{f}|} c(\xi)\,d\xi\right]d\Omega, \tag{5}$$

where subscript 'u' represents a user-optimal model and the subscript 'f' indicates that it is a fixed demand case. This minimization problem is subject to the following two constraints:

$$\nabla\cdot\mathbf{f} - q = 0,\ \forall(x, y)\in\Omega, \tag{6}$$

$$\mathbf{f} = 0,\ \forall(x, y)\in\Gamma, \tag{7}$$

where equation (6) is the flow conservation relationship, and equation (7) is the boundary condition at the boundary Γ of the study region Ω. For this minimization problem, the following differential equations are satisfied at its optimal solution $\Psi_{uf} = (\mathbf{f}_{uf}, u_{uf})$:

$$\left(a + b|\mathbf{f}_{uf}|^{\gamma}\right)\frac{\mathbf{f}_{uf}}{|\mathbf{f}_{uf}|} + \nabla u_{uf} = 0,\ \ \forall(x, y)\in\Omega, \tag{8a}$$

$$\nabla\cdot\mathbf{f}_{uf} - q = 0,\ \ \forall(x, y)\in\Omega, \tag{8b}$$

$$\mathbf{f}_{uf} = 0,\ \ \forall(x, y)\in\Gamma, \tag{8c}$$

$$u_{uf} = 0,\ \ \forall(x, y)\in\Gamma_c. \tag{8d}$$

The derivation of these differential equations that are satisfied at the optimal point of the optimization problem can be found in Wong *et al.* (1998). At the optimal solution Ψ_{uf}, the path travel cost is denoted as $u(\mathbf{f}_{uf})$, which is dependent on the flow pattern $\mathbf{f}_{uf}$. However, for brevity, this path travel cost is denoted as u_{uf}, which also holds for the elastic demand case. From equation (8a) we can easily observe that $\mathbf{f}_{uf} // -\nabla u_{uf}$, and $c(\mathbf{f}_{uf}) = a + b|\mathbf{f}_{uf}|^{\gamma} = |\nabla u_{uf}|$. The system-optimal model, $S_{sf}(\Omega, q, c)$, for the fixed demand case is formulated into a minimization problem with the following objective function:

$$Z_{sf}(\mathbf{f}) = \iint_{\Omega}\left[|\mathbf{f}|c(\mathbf{f})\right]d\Omega, \tag{9}$$

where subscript 's' represents a system-optimal model. This minimization problem is also subject to the same constraints that are shown in equations (6) and (7). At the optimal solution $\Psi_{sf} = (\mathbf{f}_{sf}, u_{m,sf})$ for this minimization problem, the following differential equation,

$$\left(a + (\gamma+1)b|\mathbf{f}_{sf}|^{\gamma}\right)\frac{\mathbf{f}_{sf}}{|\mathbf{f}_{sf}|} + \nabla u_{m,sf} = 0,\ \ \forall(x, y)\in\Omega, \tag{10}$$

together with equations (8b), (8c) and (8d), are satisfied. The derivation of these differential equations that are satisfied at the optimal point of the optimization problem can be found in Ho *et al.* (2005). At the optimal solution Ψ_{sf}, the path marginal cost is denoted as $u_m(\mathbf{f}_{sf})$, which is dependent on the flow pattern $\mathbf{f}_{sf}$. However, for brevity, the path marginal cost is denoted as $u_{\mathrm{m,sf}}$, which also holds for the elastic demand case. From equation (10), we can easily that $\mathbf{f}_{\mathrm{sf}} // -\nabla u_{\mathrm{m,sf}}$, and $c_{\mathrm{m}}(\mathbf{f}_{\mathrm{sf}}) = a + (\gamma+1)b|\mathbf{f}_{\mathrm{sf}}|^{\gamma} = |\nabla u_{\mathrm{m,sf}}|$. The total social cost and total marginal user cost for the study area can be expressed as:

$$C_{\mathrm{u}}(\Psi) = \iint_{\Omega} uq\,\mathrm{d}\Omega, \tag{11}$$

$$C_{\mathrm{m}}(\Psi) = \iint_{\Omega} u_{\mathrm{m}}q\,\mathrm{d}\Omega. \tag{12}$$

As an indicator for measuring the efficiency of the congestion-pricing method in a continuum transportation system for the fixed demand case, the following ratio is defined:

$$\rho_{\mathrm{f}}(\Omega, q, c) = \frac{C_{\mathrm{u}}(\Psi_{\mathrm{uf}})}{C_{\mathrm{u}}(\Psi_{\mathrm{sf}})}. \tag{13}$$

This ratio is specific for the given study region Ω, fixed demand pattern q, and unit transportation cost function c.

Elastic demand models

In the user-optimal model with elastic demand, the demand at location (x, y) is denoted as $q(x, y, u)$, and its relation to the path travel cost $u(\mathbf{f})$ is specified as:

$$q(x, y, u) = D(x, y, u(\mathbf{f})). \tag{14}$$

Similarly, for the system-optimal model with elastic demand, the demand at location (x, y) is defined as:

$$q(x, y, u_m) = D(x, y, u_m(\mathbf{f})), \tag{15}$$

where in this case the demand is dependent on the path marginal cost $u_m(\mathbf{f})$ at location (x, y).

To ensure the existence of the inverse function, D^{-1}, the elastic demand function is assumed to be a monotone function. Similar to the fixed demand case, the user-optimal model, $\mathrm{S}_{\mathrm{ue}}(\Omega, D, c)$, for the elastic demand case is formulated into a minimization problem with the following objective function:

$$Z_{\mathrm{ue}}(\mathbf{f}, q) = \iint_{\Omega}\left[\int_0^{|\mathbf{f}|} c(\xi)\,\mathrm{d}\xi - \int_0^{q} D^{-1}(\xi)\,\mathrm{d}\xi\right]\mathrm{d}\Omega, \tag{16}$$

where subscript ‘e’ indicates the elastic demand case. This minimization problem is also subject to the same constraints in equations (6) and (7). At the optimal solution $\Psi_{\mathrm{ue}} = (\mathbf{f}_{\mathrm{ue}}, u_{\mathrm{ue}}, q_{ue})$ of this minimization problem, the following differential equation

$$u_{\mathrm{ue}} - D^{-1}(q_{\mathrm{ue}}) = 0, \quad \forall (x, y) \in \Omega, \tag{17}$$

together with equations (8a), (8b), (8c) and (8d) are satisfied (Wong *et al.*, 1998). The system-optimal model, $S_{se}(\Omega, D, c)$, for the elastic demand case is formulated into a maximization problem with the following objective function:

$$Z_{se}(\mathbf{f}, q) = \iint_{\Omega} \left[\int_0^q D^{-1}(\xi)\,\mathrm{d}\xi - |\mathbf{f}|c(\mathbf{f}) \right] \mathrm{d}\Omega \,. \tag{18}$$

This maximization problem is also subject to the same constraints in equations (6) and (7). At the optimal solution $\Psi_{se} = (\mathbf{f}_{se}, u_{m,se}, q_{se})$ of this maximization problem, equations (10), (17), (8b), (8c) and (8d) are satisfied (Ho *et al.*, 2005). The total user benefit of the system is defined as:

$$B(\Psi) = \iint_{\Omega} \int_0^q D^{-1}(\xi)\mathrm{d}\xi \mathrm{d}\Omega \,. \tag{19}$$

Based on the total social cost that is defined in equation (11) and the total user benefit in equation (19), the social welfare of the system is defined as:

$$W(\Psi) = B(\Psi) - C_{u}(\Psi). \tag{20}$$

As an indicator for measuring the efficiency of the congestion-pricing scheme in a continuum transportation system with elastic demand case, the following ratio is defined:

$$\rho_{e}(\Omega, D, c) = \frac{W(\Psi_{se})}{W(\Psi_{ue})}. \tag{21}$$

Similarly, this ratio is specific to the given study region Ω, elastic demand function *D*, and unit transportation cost function *c*.

In addition to the optimal solutions Ψ_{ue}, Ψ_{uf}, Ψ_{se}, and Ψ_{sf} that are obtained by solving the corresponding optimization problems, two other sets of solutions, $\Psi_{cf} = (\mathbf{f}_{cf}, u_{cf})$ and $\Psi_{ce} = (\mathbf{f}_{ce}, u_{ce}, q_{ce})$, are considered for the fixed and elastic demand cases. These sets of solutions, which are indicated by subscript 'c', are the arbitrary feasible solutions for the forgoing fixed and elastic demand models. These solutions satisfy constraints (6) and (7) only, and do not guarantee optimality with respect to any objective function.

THE THORETICAL BOUND FOR THE FIXED DEMAND CASE

In this section, the theoretical bound for the effectiveness of the system-optimal scheme compared to the user-optimal scheme in fixed demand case is presented. The proof follows the steps that were introduced by Roughgarden and Tardos (2002) for discrete networks. First, we consider the following lemmas that concern the total travel costs of used and unused path in the continuum transportation system:

LEMMA 1A. *A flow* **f** *is at user optimal (UO) if and only if for each OD pair the path travel cost for all used paths are equal, and*

$$\int_P c(\mathbf{f})\mathrm{d}s \le \int_{P'} c(\mathbf{f})\mathrm{d}s \quad or \quad \int_P a + b|\mathbf{f}|^{\gamma}\mathrm{d}s \le \int_{P'} a + b|\mathbf{f}|^{\gamma}\mathrm{d}s\,,$$

where P is the used path and P' is any unused path.

PROOF Necessity – By the definition of Wardrop's first principle of equilibrium.

Sufficiency – For used path P, using equations (8a) and (8d), and the fact that flow $\mathbf{f}_{\mathrm{uf}}$ is parallel to P, we have:

$$\int_P c(\mathbf{f}_{\mathrm{uf}})\mathrm{d}s = \int_P \left(a + b|\mathbf{f}_{\mathrm{uf}}|^{\gamma}\right)\frac{\mathbf{f}_{\mathrm{uf}}}{|\mathbf{f}_{\mathrm{uf}}|}\cdot\mathrm{d}s = -\int_P \nabla u_{\mathrm{uf}}\cdot\mathrm{d}s = -\left(u_{\mathrm{uf}}(\mathrm{O}) - u_{\mathrm{uf}}(\mathrm{H})\right) = u_{\mathrm{uf}}(\mathrm{H})$$

Thus, the path travel cost is independent of the path taken. For any unused path P', using equations (8a) and (8d), and the fact that flow $\mathbf{f}_{\mathrm{uf}}$ is not parallel to P', we have:

$$\int_{P'} c(\mathbf{f}_{\mathrm{uf}})\mathrm{d}s \ge \int_{P'} \left(a + b|\mathbf{f}_{\mathrm{uf}}|^{\gamma}\right)\frac{\mathbf{f}_{\mathrm{uf}}}{|\mathbf{f}_{\mathrm{uf}}|}\cdot\mathrm{d}s = -\int_{P'} \nabla u_{\mathrm{uf}}\cdot\mathrm{d}s = -\left(u_{\mathrm{uf}}(\mathrm{O}) - u_{\mathrm{uf}}(\mathrm{H})\right) = u_{\mathrm{uf}}(\mathrm{H})$$

Combining these two equations, we have:

$$\int_P c(\mathbf{f})\mathrm{d}s = u_{\mathrm{uf}}(\mathrm{H}) \le \int_{P'} c(\mathbf{f})\mathrm{d}s$$

This completes the proof.

LEMMA 1B. *A flow* **f** *is at system optimal (SO) if and only if for each OD pair the path marginal cost for all used paths are equal, and*

$$\int_P c_{\mathrm{m}}(\mathbf{f})\mathrm{d}s \le \int_{P'} c_{\mathrm{m}}(\mathbf{f})\mathrm{d}s \quad or \quad \int_P a + (\gamma+1)b|\mathbf{f}|^{\gamma}\mathrm{d}s \le \int_{P'} a + (\gamma+1)b|\mathbf{f}|^{\gamma}\mathrm{d}s\,,$$

where P is the used path and P' is any unused path.

PROOF Necessity – By Lemma 1A (with the assumption that the path marginal cost for all used paths between each OD pair are equal and $\int_P c_{\mathrm{m}}(\mathbf{f})\mathrm{d}s \le \int_{P'} c_{\mathrm{m}}(\mathbf{f})\mathrm{d}s$), flow **f** is a user optimal flow pattern with respect to the marginal cost function c_{m}. Thus, the objective function,

$$Z(\mathbf{f}) = \iint_{\Omega}\left[\int_0^{|\mathbf{f}|} c_{\mathrm{m}}(\xi)\,\mathrm{d}\xi\right]\mathrm{d}\Omega$$

of the user-optimal model for the fixed demand case is minimized subject to constraints (6) and (7) for flow **f**. Therefore, by integrating the marginal cost function c_{m} with respect to $|\mathbf{f}|$, this objective function is equivalent to equation (9). This implies that the system-optimal model for the fixed demand case is solved by this flow pattern **f**.

Sufficiency – The proof is similar to that in Lemma 1A, but we use the marginal cost function c_{m} and the path marginal cost u_{m} (instead of the unit transportation cost function c and total travel cost u that are used in the proof of Lemma 1A). This completes the proof.

Similar proofs can also be found in Wong *et al.* (1998) and Ho *et al.* (2005). In addition to the total travel cost, we consider the following lemmas for the definition of the total social cost for the UE travel pattern and the total marginal user cost for the SO travel pattern.

LEMMA 2A. *The total social cost,* $C_u(\Psi_{uf})$, *of a user-optimal (UO) model is defined as:*

$$C_u(\Psi_{uf}) = \iint_\Omega |\mathbf{f}_{uf}| c(\mathbf{f}_{uf}) d\Omega .$$

PROOF First, we consider the rightmost part of Lemma 2A, and by equation (8a) for the fixed demand UO flow pattern, we have:

$$\iint_\Omega |\mathbf{f}_{uf}| c(\mathbf{f}_{uf}) d\Omega = \iint_\Omega |\mathbf{f}_{uf}| |\nabla u_{uf}| d\Omega .$$

As $\mathbf{f}_{uf} // -\nabla u_{uf}$, and by Green's theorem, we have:

$$\iint_\Omega |\mathbf{f}_{uf}| c(\mathbf{f}_{uf}) d\Omega = \iint_\Omega u_{uf} \nabla \cdot \mathbf{f}_{uf} d\Omega - \int_\Gamma u_{uf} \mathbf{f}_{uf} \cdot \mathbf{n} d\Gamma - \int_{\Gamma_c} u_{uf} \mathbf{f}_{uf} \cdot \mathbf{n}_c d\Gamma .$$

where $\mathbf{n}$ and $\mathbf{n}_c$ are respectively the unit normal unit vector for boundaries Γ and Γ_c. From equations (8b), (8c), (8d) and (11):

$$\iint_\Omega |\mathbf{f}_{uf}| c(\mathbf{f}_{uf}) d\Omega = \iint_\Omega u_{uf} q d\Omega = C_u(\Psi_{uf}) .$$

This completes the proof.

LEMMA 2B. *The total marginal user cost,* $C_m(\Psi_{sf})$, *of a system-optimal (SO) model is defined as:*

$$C_m(\Psi_{sf}) = \iint_\Omega |\mathbf{f}_{sf}| c_m(\mathbf{f}_{sf}) d\Omega .$$

PROOF The proof is similar to that for Lemma 2A, but we use the set of differential equations for the system-optimal model with fixed demand.

The above lemmas define the total social cost and total marginal user cost for the UO and SO schemes, respectively. As this study focuses on the effectiveness of congestion-pricing, which is a comparison of the user-optimal and system-optimal schemes within the same study area, the following lemma establishes a linkage between the UO and SO schemes through the flow pattern.

LEMMA 3. *Suppose that flow pattern* $\mathbf{f}_{uf}$ *is the solution of the problem* $S_{uf}(\Omega, q, c)$, *then the flow pattern* $\mathbf{f}_{uf} / (\gamma+1)^{1/\gamma}$ *is the solution of the problem* $S_{sf}(\Omega, q/(\gamma+1)^{1/\gamma}, c)$.

PROOF Substituting the flow pattern $\mathbf{f}_{\text{uf}}/(\gamma+1)^{1/\gamma}$ into the left-hand side of the inequality in Lemma 1B, we have:

$$\int_P a+(\gamma+1)b\left|\frac{\mathbf{f}_{\text{uf}}}{(\gamma+1)^{1/\gamma}}\right|^{\gamma} \mathrm{d}s = \int_P a+(\gamma+1)b\frac{|\mathbf{f}_{\text{uf}}|^{\gamma}}{(\gamma+1)}\mathrm{d}s = \int_P a+b|\mathbf{f}_{\text{uf}}|^{\gamma}\,\mathrm{d}s .$$

Together with Lemma 1A, we have:

$$\int_P a+(\gamma+1)b\left|\frac{\mathbf{f}_{\text{uf}}}{(\gamma+1)^{1/\gamma}}\right|^{\gamma} \mathrm{d}s \le \int_{P'} a+b|\mathbf{f}_{\text{uf}}|^{\gamma}\,\mathrm{d}s = \int_{P'} a+(\gamma+1)b\left|\frac{\mathbf{f}_{\text{uf}}}{(\gamma+1)^{1/\gamma}}\right|^{\gamma} \mathrm{d}s .$$

Thus, it is shown that if $\mathbf{f}_{\text{uf}}$ is a UO flow pattern then $\mathbf{f}_{\text{uf}}/(\gamma+1)^{1/\gamma}$ is a SO flow pattern. As the flow is scaled down by a factor of $(\gamma+1)^{1/\gamma}$, the fixed demand of this SO problem is also scaled down by the same factor (that is with the fixed demand at the level of $q/(\gamma+1)^{1/\gamma}$) for the continuity equation to hold and this completes the proof.

Based on the linkage of the flow pattern between the UO and SO schemes within the same study area Ω and the unit transportation cost function c, the following corollary and lemma establish the relationships of the path travel cost, path marginal cost, total social cost and total marginal user cost between the UO and SO schemes.

COROLLARY 4. *The path marginal cost along a path P with respect to the flow* $\mathbf{f}_{\text{uf}}/(\gamma+1)^{1/\gamma}$*, which is the solution of the problem* $\mathrm{S}_{\text{sf}}\left(\Omega, q/(\gamma+1)^{1/\gamma}, c\right)$*, is equal to the path travel cost on the same path P with respect to the flow* $\mathbf{f}_{\text{uf}}$*, which is the solution of the problem* $\mathrm{S}_{\text{uf}}(\Omega, q, c)$.

PROOF Substitute the flow pattern $\mathbf{f}_{\text{uf}}/(\gamma+1)^{1/\gamma}$ into equation (4) and by equation (3), we have:

$$u_{\text{m}}\left(\frac{\mathbf{f}_{\text{uf}}}{(\gamma+1)^{1/\gamma}}\right) = \int_P a+(\gamma+1)b\left|\frac{\mathbf{f}_{\text{uf}}}{(\gamma+1)^{1/\gamma}}\right|^{\gamma} \mathrm{d}s = \int_P a+b|\mathbf{f}_{\text{uf}}|^{\gamma}\,\mathrm{d}s = u_{\text{uf}} .$$

This completes the proof.

LEMMA 5. *For any* $\delta > 0$, *an arbitrary feasible solution* Ψ_{cf} *for the problem* $\mathrm{S_{uf}}(\Omega,(1+\delta)q,c)$ *has a total social cost of at least* $C_{\mathrm{u}}(\Psi_{\mathrm{sf}})+\delta C_{\mathrm{m}}(\Psi_{\mathrm{sf}})$, *where* Ψ_{sf} *is the solution of the problem* $\mathrm{S_{sf}}(\Omega,q,c)$.

PROOF Consider the convexity of the function $|\mathbf{f}|c(\mathbf{f}) = a|\mathbf{f}| + b|\mathbf{f}|^{\gamma+1}$, where:

$$\begin{aligned} |\mathbf{f}_{\mathrm{cf}}|c(\mathbf{f}_{\mathrm{cf}}) &\geq |\mathbf{f}_{\mathrm{sf}}|c(\mathbf{f}_{\mathrm{sf}}) + (|\mathbf{f}_{\mathrm{cf}}| - |\mathbf{f}_{\mathrm{sf}}|)\left.\frac{\partial |\mathbf{f}|c(\mathbf{f})}{\partial |\mathbf{f}|}\right|_{\mathbf{f}=\mathbf{f}_{\mathrm{sf}}} \\ &= |\mathbf{f}_{\mathrm{sf}}|c(\mathbf{f}_{\mathrm{sf}}) + (|\mathbf{f}_{\mathrm{cf}}| - |\mathbf{f}_{\mathrm{sf}}|)c_{\mathrm{m}}(\mathbf{f}_{\mathrm{sf}}). \end{aligned} \quad (22)$$

Evaluating the total social cost that is associated with flow pattern $\mathbf{f}_{\mathrm{cf}}$ and using equation (22), we have:

$$\begin{aligned} C_{\mathrm{u}}(\Psi_{\mathrm{cf}}) &= \iint_{\Omega} |\mathbf{f}_{\mathrm{cf}}|c(\mathbf{f}_{\mathrm{cf}})\mathrm{d}\Omega \\ &\geq C_{\mathrm{u}}(\Psi_{\mathrm{sf}}) + \iint_{\Omega} |\mathbf{f}_{\mathrm{cf}}|c_{\mathrm{m}}(\mathbf{f}_{\mathrm{sf}})\mathrm{d}\Omega - \iint_{\Omega} |\mathbf{f}_{\mathrm{sf}}|c_{\mathrm{m}}(\mathbf{f}_{\mathrm{sf}})\mathrm{d}\Omega. \end{aligned} \quad (23)$$

Consider the second term $\iint_{\Omega} |\mathbf{f}_{\mathrm{cf}}|c_{\mathrm{m}}(\mathbf{f}_{\mathrm{sf}})\mathrm{d}\Omega$ on the right-hand side of the inequality in equation (23). By equation (10) and the fact that $\mathbf{f}_{\mathrm{cf}}$ is not necessarily parallel to ∇u_{sf}, we have:

$$\iint_{\Omega} |\mathbf{f}_{\mathrm{cf}}|c_{\mathrm{m}}(\mathbf{f}_{\mathrm{sf}})\mathrm{d}\Omega \geq -\iint_{\Omega} \nabla u_{\mathrm{m,sf}} \cdot \mathbf{f}_{\mathrm{cf}}\,\mathrm{d}\Omega.$$

By Green's theorem and with equations (8b), (8c), and (8d) for the system-optimal model with fixed demand, and after rearranging, we have:

$$\iint_{\Omega} |\mathbf{f}_{\mathrm{cf}}|c_{\mathrm{m}}(\mathbf{f}_{\mathrm{sf}})\mathrm{d}\Omega \geq \iint_{\Omega} (1+\delta)u_{\mathrm{m,sf}}\,q\,\mathrm{d}\Omega.$$

Thus, substituting the above inequality back to equation (23), and by Lemma 2B and equation (12), we have:

$$C_{\mathrm{u}}(\Psi_{\mathrm{cf}}) \geq C_{\mathrm{u}}(\Psi_{\mathrm{sf}}) + \delta C_{\mathrm{m}}(\Psi_{\mathrm{sf}}).$$

This completes the proof.

With the definition of the total social cost and total marginal user cost in Lemma 2A and 2B, and the relationship of different costs between UO and SO scheme in Corollary 4 and Lemma 5, the theoretical bound for the effectiveness of congestion-pricing in the fixed demand case is given in the following theorem.

THEOREM 6. *For any continuum transportation network with fixed demand and the unit transportation cost function taking the form of* $c(\mathbf{f}) = a + b|\mathbf{f}|^{\gamma}$, *the ratio* $\rho_{\mathrm{f}}(\Omega,q,c)$ *should be at most* $1+\gamma\left((1+\gamma)^{(\gamma+1)/\gamma} - \gamma\right)^{-1}$, *of which* γ *is the parameter for the non-linearity of flow intensity in the unit transportation cost function.*

PROOF Let Ψ_{sf} be the solution of the problem $\mathrm{S}_{\mathrm{sf}}\left(\Omega,(1+\delta)q/(\gamma+1)^{1/\gamma},c\right)$. As Ψ_{sf} is also an arbitrary feasible solution of $\mathrm{S}_{\mathrm{uf}}\left(\Omega,(1+\delta)q/(\gamma+1)^{1/\gamma},c\right)$, and from Lemma 5, we have:

$$C_{\mathrm{u}}(\Psi_{\mathrm{sf}})\geq C_{\mathrm{u}}(\Psi_{\mathrm{sf1}})+\delta C_{\mathrm{m}}(\Psi_{\mathrm{sf1}}), \tag{24}$$

where Ψ_{sf1} is the solution of the problem $\mathrm{S}_{\mathrm{sf}}\left(\Omega,q/(\gamma+1)^{1/\gamma},c\right)$. By taking $\delta=(\gamma+1)^{1/\gamma}-1$, Ψ_{sf} becomes the solution of the problem $\mathrm{S}_{\mathrm{sf}}(\Omega,q,c)$, and equation (24) can be written as $C_{\mathrm{u}}(\Psi_{\mathrm{sf}})\geq C_{\mathrm{u}}(\Psi_{\mathrm{sf1}})+\left((\gamma+1)^{1/\gamma}-1\right)C_{\mathrm{m}}(\Psi_{\mathrm{sf1}})$. By equation (12), Corollary 4 and then equation (11), the above inequality can be modified as:

$$C_{\mathrm{u}}(\Psi_{\mathrm{sf}})\geq C_{\mathrm{u}}(\Psi_{\mathrm{sf1}})+\left(1-\frac{1}{(\gamma+1)^{1/\gamma}}\right)C_{\mathrm{u}}(\Psi_{\mathrm{uf}}),$$

where $\mathbf{f}_{\mathrm{uf}}$ is the flow pattern of problem $\mathrm{S}_{\mathrm{uf}}(\Omega,q,c)$. Applying Lemma 3 and Lemma 2A to the first term on the right-hand side of the above inequality and applying the definition of the unit transportation cost c, we can show that:

$$\begin{aligned}
C_{\mathrm{u}}(\Psi_{\mathrm{sf}}) &\geq \iint_{\Omega}\frac{a|\mathbf{f}_{\mathrm{uf}}|}{(\gamma+1)^{1/\gamma}}+\frac{b|\mathbf{f}_{\mathrm{uf}}|^{\gamma+1}}{(\gamma+1)^{(\gamma+1)/\gamma}}\mathrm{d}\Omega+\left(1-\frac{1}{(\gamma+1)^{1/\gamma}}\right)C_{\mathrm{u}}(\Psi_{\mathrm{uf}})\\
&\geq \iint_{\Omega}\frac{a|\mathbf{f}_{\mathrm{uf}}|}{(\gamma+1)^{(\gamma+1)/\gamma}}+\frac{b|\mathbf{f}_{\mathrm{uf}}|^{\gamma+1}}{(\gamma+1)^{(\gamma+1)/\gamma}}\mathrm{d}\Omega+\left(1-\frac{1}{(\gamma+1)^{1/\gamma}}\right)C_{\mathrm{u}}(\Psi_{\mathrm{uf}})\\
&=\frac{1}{(\gamma+1)^{(\gamma+1)/\gamma}}\iint_{\Omega}\left(a+b|\mathbf{f}_{\mathrm{uf}}|^{\gamma}\right)|\mathbf{f}_{\mathrm{uf}}|\mathrm{d}\Omega+\left(1-\frac{1}{(\gamma+1)^{1/\gamma}}\right)C_{\mathrm{u}}(\Psi_{\mathrm{uf}})\\
&=\frac{1}{(\gamma+1)^{(\gamma+1)/\gamma}}C_{\mathrm{u}}(\Psi_{\mathrm{uf}})+\left(1-\frac{1}{(\gamma+1)^{1/\gamma}}\right)C_{\mathrm{u}}(\Psi_{\mathrm{uf}})\\
&=\left(1-\frac{\gamma}{(\gamma+1)^{(\gamma+1)/\gamma}}\right)C_{\mathrm{u}}(\Psi_{\mathrm{uf}}).
\end{aligned}$$

Thus, $\rho_{\mathrm{f}}(\Omega,q,c)=\dfrac{C_{\mathrm{u}}(\Psi_{\mathrm{uf}})}{C_{\mathrm{u}}(\Psi_{\mathrm{sf}})}\leq 1+\dfrac{\gamma}{(1+\gamma)^{(\gamma+1)/\gamma}-\gamma}$. This completes the proof.

In this section, we prove that the theoretical bound for the effectiveness of the system-optimal scheme over the user-optimal scheme (or the efficiency of congestion-pricing) for a continuum transportation network with fixed demand and specific type of unit transportation cost function is $1+\gamma\left((1+\gamma)^{(\gamma+1)/\gamma}-\gamma\right)^{-1}$, where γ is the parameter for the non-linearity of flow intensity in the unit transportation cost function. The following assumptions are made in this proof: only a single destination (CBD) is considered, no destination charge or toll for users to access the CBD, and the specific type of unit transportation cost function, $c(\mathbf{f})=a+b|\mathbf{f}|^{\gamma}$, is used. Also, the above theoretical bound for the efficiency of congestion-pricing in the continuum transportation system is analogous to that of the discrete networks with polynomial cost function (Roughgarden, 2003).

THE THORETICAL BOUND FOR THE ELASTIC DEMAND CASE

This section derives the theoretical bound of effectiveness for the elastic demand case in a continuum transportation system. In this section, the proof of the theoretical bound follows the steps that were introduced by Yang and Huang (2005) for discrete networks. Firstly, a general theoretical bound is found for a decreasing elastic demand function, and then based on this result a more specific bound is found for a particular type of elastic demand function.

General elastic demand function

In this section, the upper bound of congestion-pricing efficiency is considered for a general decreasing demand function. In order to find this upper bound, the relationship of the user-optimal solution, Ψ_{ue}, and any arbitrary feasible solution, Ψ_{ce}, is first considered in terms of demand and inverse of demand function as shown in the following lemma.

LEMMA 7 *For any decreasing elastic demand function, $D(u)$, the following inequality should be satisfied:*

$$\iint_{\Omega}\int_0^{q_{ce}} D^{-1}(\xi)\mathrm{d}\xi\,\mathrm{d}\Omega \le \iint_{\Omega}\int_0^{q_{ue}} D^{-1}(\xi)\mathrm{d}\xi + D^{-1}(q_{ue})(q_{ce}-q_{ue})\mathrm{d}\Omega .$$

PROOF Two different cases, $q_{ce} \ge q_{ue}$ and $q_{ce} < q_{ue}$, are considered. In the case of $q_{ce} \ge q_{ue}$, as the demand function is a decreasing function, its inverse is also a decreasing function, and thus we have:

$$(q_{ce}-q_{ue})u_{ue} \ge \int_{q_{ue}}^{q_{ce}} D^{-1}(\xi)\mathrm{d}\xi .$$

By equation (17) and integrating over the entire study area:

$$\iint_{\Omega}\int_0^{q_{ce}} D^{-1}(\xi)\mathrm{d}\xi\,\mathrm{d}\Omega \le \iint_{\Omega}\int_0^{q_{ue}} D^{-1}(\xi)\mathrm{d}\xi + D^{-1}(q_{ue})(q_{ce}-q_{ue})\mathrm{d}\Omega .$$

Thus, this lemma holds in the case of $q_{ce} \ge q_{ue}$. In the case of $q_{ce} < q_{ue}$ and as D^{-1} is a decreasing function:

$$(q_{ue}-q_{ce})u_{ue} < \int_{q_{ce}}^{q_{ue}} D^{-1}(\xi)\mathrm{d}\xi .$$

By equation (17) and integrating over the entire study area:

$$\iint_{\Omega}\int_0^{q_{ce}} D^{-1}(\xi)\mathrm{d}\xi\,\mathrm{d}\Omega < \iint_{\Omega}\int_0^{q_{ue}} D^{-1}(\xi)\mathrm{d}\xi + D^{-1}(q_{ue})(q_{ce}-q_{ue})\mathrm{d}\Omega .$$

Therefore, this lemma also holds for the case of $q_{ce} < q_{ue}$. By combining the two cases of $q_{ce} \ge q_{ue}$ and $q_{ce} < q_{uc}$, the proof is completed.

Apart from the linkage in the inverse demand function, the relationship between the user-optimal solution and any arbitrary feasible solution is also considered in terms of the unit transportation cost function c as follows.

LEMMA 8. *For the unit transportation cost function in the form of* $c(\mathbf{f}) = a + b|\mathbf{f}|^{\gamma}$, *the following inequality should be satisfied:*

$$R(\gamma)c(\mathbf{f}_{\text{ue}})|\mathbf{f}_{\text{ue}}| \geq (c(\mathbf{f}_{\text{ue}}) - c(\mathbf{f}_{\text{ce}}))|\mathbf{f}_{\text{ce}}|,$$

where $R(\gamma) = \dfrac{\gamma}{(\gamma+1)^{(\gamma+1)/\gamma}}$.

PROOF Consider a ratio $r(\mathbf{f})$ such that:

$$r(\mathbf{f}_{\text{ce}}) = \frac{(c(\mathbf{f}_{\text{ue}}) - c(\mathbf{f}_{\text{ce}}))|\mathbf{f}_{\text{ce}}|}{c(\mathbf{f}_{\text{ue}})|\mathbf{f}_{\text{ue}}|}. \tag{25}$$

Differentiating equation (25) with respect to $|\mathbf{f}_{\text{ce}}|$ and by the definition of the unit transportation cost c, we have:

$$\frac{\mathrm{d}r(\mathbf{f}_{\text{ce}})}{\mathrm{d}|\mathbf{f}_{\text{ce}}|} = \frac{b|\mathbf{f}_{\text{ue}}|^{\gamma} - (\gamma+1)b|\mathbf{f}_{\text{ce}}|^{\gamma}}{c(\mathbf{f}_{\text{ue}})|\mathbf{f}_{\text{ue}}|}. \tag{26}$$

Differentiating again with respect to $|\mathbf{f}_{\text{ce}}|$, we have:

$$\frac{\mathrm{d}^2 r(\mathbf{f}_{\text{ce}})}{\mathrm{d}|\mathbf{f}_{\text{ce}}|^2} = \frac{-\gamma(\gamma+1)b|\mathbf{f}_{\text{ce}}|^{\gamma-1}}{c(\mathbf{f}_{\text{ue}})|\mathbf{f}_{\text{ue}}|} \leq 0.$$

Thus, for the function $r(\mathbf{f}_{\text{ce}})$, maximum point(s) exist. To find the flow pattern of this maximum point, we let equation (26) vanish, which gives $|\mathbf{f}_{\text{ce}}| = (\gamma+1)^{-1/\gamma}|\mathbf{f}_{\text{ue}}|$. As this is the maximum point for the function $r(\mathbf{f})$, we have:

$$r\left(\frac{1}{(\gamma+1)^{1/\gamma}}\mathbf{f}_{\text{ue}}\right) \geq r(\mathbf{f}_{\text{ce}}).$$

Using equation (25) and rearranging, we have:

$$\frac{\gamma c(\mathbf{f}_{\text{ue}})|\mathbf{f}_{\text{ue}}|}{(\gamma+1)^{(\gamma+1)/\gamma}} \geq (c(\mathbf{f}_{\text{ue}}) - c(\mathbf{f}_{\text{ce}}))|\mathbf{f}_{\text{ce}}|.$$

This completes the proof.

LEMMA 9. *For a user-optimal (UO) flow pattern,* $\mathbf{f}_{ue}$ *, which solved the model* $S_{ue}(\Omega, D, c)$*, the following inequality should be satisfied:*

$$\iint_\Omega c(\mathbf{f}_{ue})(|\mathbf{f}_{ce}| - |\mathbf{f}_{ue}|) - D^{-1}(q_{ue})(q_{ce} - q_{ue})\mathrm{d}\Omega \geq 0 .$$

PROOF From equation (17) and rearranging, we have:

$$\iint_\Omega q_{ce}(u_{ue} - D^{-1}(q_{ue})) - q_{ue}(u_{ue} - D^{-1}(q_{ue}))\mathrm{d}\Omega = 0 .$$

By equations (6) and (8b), the above equation can be modified as:

$$\iint_\Omega u_{ue}\nabla \cdot \mathbf{f}_{ce} - u_{ue}\nabla \cdot \mathbf{f}_{ue} - D^{-1}(q_{ue})(q_{ce} - q_{ue})\mathrm{d}\Omega = 0 .$$

By Green's theorem and with equations (8c) and (8d), we have:

$$\iint_\Omega \mathbf{f}_{ue} \cdot \nabla u_{ue} - \mathbf{f}_{ce} \cdot \nabla u_{ue} - D^{-1}(q_{ue})(q_{ce} - q_{ue})\mathrm{d}\Omega = 0 .$$

Applying equation (8a) and using the fact that $-\mathbf{f}_{ce} \cdot \nabla u_{ue} \leq |\mathbf{f}_{ce}||\nabla u_{ue}| = |\mathbf{f}_{ce}|c(\mathbf{f}_{ue})$ gives:

$$\iint_\Omega c(\mathbf{f}_{ue})(|\mathbf{f}_{ce}| - |\mathbf{f}_{ue}|) - D^{-1}(q_{ue})(q_{ce} - q_{ue})\mathrm{d}\Omega \geq 0 .$$

This completes the proof.

Based on all of the three lemmas developed previously in this section, the theoretical bound of the effectiveness of congestion-pricing with elastic demand is given in the following theorem.

THEOREM 10 *For any continuum transportation network with elastic demand and where the unit transportation cost function takes the form of* $c(\mathbf{f}) = a + b|\mathbf{f}|^\gamma$*, the ratio* $\rho_e(\Omega, D, c)$ *should be at most* $1 + R(\gamma)(B(\Psi_{ue})/W(\Psi_{ue}) - 1)$.

PROOF From Lemma 9 and assuming that $\mathbf{f}_{ce} = \mathbf{f}_{se}$ of which $\mathbf{f}_{se}$ is a special case of $\mathbf{f}_{ce}$ that maximizes the social welfare of the system, we have:

$$\iint_\Omega c(\mathbf{f}_{ue})(|\mathbf{f}_{se}| - |\mathbf{f}_{ue}|) - D^{-1}(q_{ue})(q_{se} - q_{ue})\mathrm{d}\Omega \geq 0 .$$

By Lemma 7 and rearranging, the above inequality becomes:

$$\iint_\Omega (c(\mathbf{f}_{ue}) - c(\mathbf{f}_{se}))|\mathbf{f}_{se}| + \int_0^{q_{ue}} D^{-1}(\xi)\mathrm{d}\xi - c(\mathbf{f}_{ue})|\mathbf{f}_{ue}|$$
$$- \int_0^{q_{se}} D^{-1}(\xi)\mathrm{d}\xi + c(\mathbf{f}_{se})|\mathbf{f}_{se}|\mathrm{d}\Omega \geq 0.$$

By equations (19) and (20), Lemma 2A, and Lemma 8, and after rearranging, we have:

$$\rho_e(\Omega, D, c) \leq 1 + R(\gamma)\left(\frac{B(\Psi_{ue})}{W(\Psi_{ue})} - 1\right).$$

This completes the proof.

In this section, the theoretical bound for the effectiveness of the system-optimal scheme over the user-optimal scheme (or the efficiency of congestion-pricing) for a continuum transportation network with elastic demand is found to be $1 + R(\gamma)\left(\frac{B(\Psi_{ue})}{W(\Psi_{ue})} - 1\right)$, which is dependent on the non-linearity parameter γ of the flow intensity in the unit transportation cost function and the ratio of the total user benefit $B(\Psi_{ue})$ to the social welfare $W(\Psi_{ue})$ of the system at the user-optimal solution. In addition to the three assumptions made in the proof of the fixed demand case, a decreasing elastic demand function, D, is assumed in this proof.

Specific elastic demand function

The previous section considered the theoretical bound of congestion-pricing effectiveness for a decreasing demand function, in which the bound is dependent on the total user benefit and social welfare of the system, which means that this bound is based on the result of the user-optimal problem. In this section, we develop a more explicit theoretical bound for the following specific form of elastic demand function:

$$q = D(u) = K(u + \beta)^{-\alpha}, \tag{27}$$

where K and β are positive real numbers and $\alpha \geq 1$. We choose the functional form in equation (27) because this function decreases with respect to the path travel cost or path marginal cost and the function has a finite value of $K\beta^{-\alpha}$ when the cost vanishes, which are essential properties for typical demand functions that are commonly used. More importantly, a close form solution for the theoretical bound can be obtained for this demand function, which helps to shed light on the analytical properties of the elastic demand case. From equation (27), the following inverse function of demand, D^{-1}, can be derived:

$$u = D^{-1}(q) = K^{1/\alpha} q^{-1/\alpha} - \beta. \tag{28}$$

To obtain the user benefit at a particular point with this elastic demand function, the following integration of the inverse demand function is considered:

$$\int_0^q D^{-1}(\xi)\,\mathrm{d}\xi = \int_0^q K^{1/\alpha}\xi^{-1/\alpha} - \beta\,\mathrm{d}\xi = \frac{\alpha K^{1/\alpha} q^{(\alpha-1)/\alpha}}{\alpha - 1} - \beta q. \tag{29}$$

Consider the following ratio $\int_0^{q_{ue}} D^{-1}(\xi)\,\mathrm{d}\xi \Big/ \left(\int_0^{q_{ue}} D^{-1}(\xi)\,\mathrm{d}\xi - u_{ue} q_{ue} \right)$. By equation (29) and the fact that β and q_{ue} is non-negative, we have:

$$\frac{\int_0^{q_{ue}} D^{-1}(\xi)\,\mathrm{d}\xi}{\int_0^{q_{ue}} D^{-1}(\xi)\,\mathrm{d}\xi - u_{ue} q_{ue}} \leq \alpha . \tag{30}$$

Rearranging equation (30) and integrating over the entire study area gives:

$$\frac{\iint_\Omega \int_0^{q_{ue}} D^{-1}(\xi)\,\mathrm{d}\xi\,\mathrm{d}\Omega}{\iint_\Omega \int_0^{q_{ue}} D^{-1}(\xi)\,\mathrm{d}\xi - u_{ue} q_{ue}\,\mathrm{d}\Omega} = \frac{B(\Psi_{ue})}{W(\Psi_{ue})} \leq \alpha .$$

Applying this result to Theorem 10 gives the following:

$$\rho_e(\Omega, D, c) \leq 1 + R(\gamma)\left(\frac{B(\Psi_{ue})}{W(\Psi_{ue})} - 1 \right) \leq 1 + R(\gamma)(\alpha - 1) .$$

Therefore, for any continuum transportation network with an elastic demand function that takes the form $q = D(u) = K(u + \beta)^{-\alpha}$, and with a unit transportation cost function that takes the form $c(\mathbf{f}) = a + b|\mathbf{f}|^{\gamma}$, the ratio $\rho_e(\Omega, D, c)$ is at most $1 + R(\gamma)(\alpha - 1)$.

CONCLUSIONS

We have derived the theoretical bounds for the efficiency of congestion-pricing in a continuum transportation system. The proofs have been given for both the fixed and elastic demand cases. In the fixed demand case, it has been proven that if a particular type of unit transportation cost function is considered, the upper bound of the efficiency of congestion-pricing is a function of the non-linearity parameter of the unit transportation cost function. In the elastic demand case, it is proven that if a decreasing elastic demand function is considered, the theoretical bound is based on the user benefit and social welfare of the user-optimal solution. In addition to the general forms of decreasing elastic demand function, this paper has also proved the upper bound of the efficiency of congestion-pricing for a specific type of elastic demand function. It is found that the theoretical bound is solely dependent on the non-linearity parameters of the unit transportation cost and elastic demand functions.

ACKNOWLEDGEMENTS

The work described in this paper was supported by a grant from the Research Grants Council of the Hong Kong Special Administrative Region, China (Project No. HKU 7126/04E).

REFERENCES

Blumenfeld, D.E. (1977). Modeling the joint distribution of home and workplace location in a city. *Transportation Science,* **11**(4), 307-377.

Chau, C.K. and Sim, K.M. (2003). The price of anarchy for non-atomic congestion games with symmetric cost maps and elastic demands. *Operations Research Letters,* **31**, 327-334

Correa, J.R., A.S. Schulz and N.E. Stier-Moses (2004). Selfish routing in capacitated networks. *Mathematics of Operations Research*, **29**(4), 961-976.

Guo, X and H. Yang (2005). Analysis of the inefficiency of stochastic user equilibrium. In: *Proceedings of the 10th International Conference of Hong Kong Society for Transportation Studies*, Hong Kong, 10 December, (W.H.K. Lam and J. Yan, eds.), 63-72.

Gwinner, J. (1998). On continuum modeling of large dense networks in urban road traffic. **In**: *Mathematics in Transport Planning and Control: Proceedings of the Third IMA International Conference on Mathematics in Transport Planning and Control* (J.D. Griffiths ed.), 321-330, Pergamon Publishing Company, New York.

Ho, H.W., S.C. Wong, H. Yang and B.P.Y. Loo (2005). Cordon-based congestion pricing in a continuum traffic equilibrium system. *Transportation Research*, **39A**, 813-834.

Ho, H.W., S.C. Wong and B.P.Y. Loo (2006). Combined distribution and assignment model for a continuum traffic equilibrium problem with multiple user classes. *Transportation Research*, **40B**, 633-650.

Papadimitriou, C.H. (2001). Algorithms, games, and the internet. *Proceedings of the 33rd Annual ACM Symposium on the Theory of Computing*, 749-753.

Roughgarden, T. (2003). The price of anarchy is independent of the network topology. *Journal of Computer and System Sciences,* **67,** 341-364.

Roughgarden, T. and E. Tardos (2002). How bad is selfish routing? *Journal of ACM*, **49**(2), 236-259.

Sasaki, T., Y. Iida and H. Yang (1990). User-equilibrium traffic assignment by continuum approximation of network flow. *Proceedings of 11th International Symposium on Transportation and Traffic Theory*, Yokohama Japan, July, 233-252.

Taguchi, A and M. Iri (1982). Continuum approximation to dense networks and its application to the analysis of urban road networks. *Mathematical Programming Study,* **20**, 178-217.

Vaughan, R.J. (1987). *Urban Spatial Traffic Patterns.* Pion, London.

Wong, S.C., C.K. Lee and C.O. Tong (1998). Finite element solution for the continuum traffic equilibrium problems. *International Journal for Numerical Methods in Engineering*, **43**, 1253-1273.

Yang, H. and H.J. Huang (2005). *Mathematical and Economic Theory of Road Pricing.* Elsevier, Amsterdam.

Zienkiewicz, O.C. and R.L. Taylor (1989). *The Finite Element Method.* McGraw-Hill, International Editions.

Transportation and Traffic Theory 2007
Edited by R.E. Allsop, M.G.H. Bell and B.G. Heydecker

13

UNIQUENESS OF EQUILIBRIUM IN STEADY STATE AND DYNAMIC TRAFFIC NETWORKS

Richard Mounce and Mike Smith, Department of Mathematics, University of York, UK.

SUMMARY

This paper addresses the issue of uniqueness of equilibrium in traffic networks, which are considered to be directed graphs in which traffic flows along acyclic paths, which we call routes, connecting origin-destination (OD) pairs. The demand for travel between each OD pair is assumed to be rigid. It is shown that in the steady state model, provided that each link cost function is a non-decreasing function of link flow, costs at equilibrium are unique. The paper then goes on to consider dynamic user equilibrium in dynamic traffic models, with special attention given to the bottleneck model. In the dynamic bottleneck queueing model, the route cost vector is not a monotone function of the route flow vector; an example network is given to illustrate this. An alternative definition of what constitutes an increasing function (of a function) is then given; and the cost function is shown to satisfy this condition in the bottleneck model. A number of additional properties are then put forward that must be satisfied in order for the equilibrium flow pattern to be essentially unique in the single OD pair case; the bottleneck model is shown to satisfy these properties.

INTRODUCTION

The steady state model

In the steady state model, the set D of feasible vectors consists of all those non-negative vectors that meet the non-negative (rigid) demand between each OD pair. Link flows are found by summing route flows. Each link has a corresponding link cost function and route costs are found by simply summing the link costs along that particular route. Further details of the model can be found in Smith (1979).

Theorem 1. Costs at equilibrium are unique in the steady state model provided that each link cost $c_i(x)$ (or indeed $c_i(x_i)$ since the cost for link i depends only on the flow on link i) is a non-decreasing function of its corresponding link flow x_i.

Proof. Suppose that there are two equilibrium flow vectors X and Y, with different route cost vectors, i.e. $C(X) \neq C(Y)$. The link flow vectors must also be different, i.e.
$c(x) \neq c(y)$, since $c(x) = c(y)$ implies that $C(X) = C(Y)$. Since each link cost function is a non-decreasing function of link flow, $c_i(x_i) > c_i(y_i) \Rightarrow x_i > y_i$ and $c_i(x_i) < c_i(y_i) \Rightarrow x_i < y_i$. Therefore, if we let the dot represent the vector dot product,

$$c(x) - c(y) \cdot (x - y) = \sum_i (c_i(x_i) - c_i(y_i))(x_i - y_i) > 0 . \tag{1}$$

Since X is an equilibrium,

$$-C(X) \cdot (Z - X) \leq 0 \tag{2}$$

for all vectors Z in the set of feasible vectors D. Since Y is also an equilibrium,

$$-C(Y) \cdot (Z - Y) \leq 0 \tag{3}$$

for all $Z \in D$. However, if we choose $Z = Y$ in (2) and $Z = X$ in (3) we obtain
$(c(x) - c(y)) \cdot (x - y) = (C(X) - C(Y))(X - Y) \leq 0$,
which clearly contradicts (1). Consequently, link costs (and hence route costs) must be unique at equilibrium.

The dynamic traffic assignment model

Within-day time is represented by the interval [0,1]. Each route inflow, say X_r, is considered to be a real-valued, non-negative, essentially bounded and measurable function (this may or may not be continuous). The null sets are then quotiented out (i.e. $X_r = Y_r$ means that the two functions agree for almost all time $t \in [0,1]$) so that each route inflow is in $L^\infty[0,1]$. All of these route inflows are components in the route flow vector X. Demand for travel between a given OD-pair k is considered to be a fixed function $\rho_k \in L^\infty[0,1]$. Therefore the set of feasible route flow vectors is

$$D = \{X : X_r \geq 0, X_r \in L^\infty[0,1], \sum_{r \in R_k} X_r = \rho_k\}$$

where R_k is the set of routes connecting OD-pair k.

Given any link inflow function x_i we suppose that the cost to traverse link i if entered at time t is the sum of a constant (congestion-free) travel time c_i and a delay $d_i^x(t)$. Constant (monetary) prices could also be incorporated without making any difference to the results throughout the paper (these prices could be converted into cost in time units). Although the delay depends upon the whole link inflow function, it is reasonable to require that traffic entering the link be not affected by traffic entering later than it. It is also reasonable to require first-in first-out (FIFO) at each link, i.e. traffic cannot exit a link earlier than traffic entering the link earlier than it. The cost to traverse route r, C_r^X can then be found by summing all of the link costs at the times that each link is reached, i.e.

$$C_r(X)(t) = \sum_{i:i \in r} c_i^x(A_{ir}^X(t))$$

where $A_{ir}^X(t)$ is the arrival time at link i if route r is entered at time t and the route inflow vector is X.

We now denote by x_i the inflow to link i if the route flow vector is X. If we let x_{ir} be the inflow at link i of traffic on route r, then clearly

$$\sum_{r:i \in r} x_{ir} = x_i$$

where $i \in r$ means that link i is a link on route r. If we let Ox_{ir} represent the outflow from link i of flow on route r, then

$$\int_0^t x_{ir}(u)du = \int_0^{t+d_i^x(t)} Ox_{ir}(u + c_i)du \tag{1}$$

since traffic entering at time t exits at time $t + c_i + d_i^x(t)$. Notice that there is interdependency between the route flows and the link flows and delays. Given a particular route flow vector X, the associated link flow vector (consisting of all the link inflow functions) is defined to be the solution of the integral equations (1).

If the network is at dynamical user equilibrium, more costly routes are unused for all within-day time, i.e. for all routes r and s connecting the same OD-pair,

$C_r^X(t) > C_s^X(t) \Rightarrow X_r(t) = 0$.

A natural day-to-day swap vector is given by $\phi(X)$ defined by

$$\phi(X)(t) = \sum_{r,s:r \sim s} X_r(t)[C_r^X(t) - C_s^X(t)]_+ \delta_{rs}$$

for each within-day $t \in [0,1]$, where $r \sim s$ means that routes r and s connect the same OD pair, $[x]_+ = \max\{x,0\}$ and δ_{rs} is the swap from route r to route s vector (i.e. it has -1 in the rth place, 1 in the sth place but zeros everywhere else). If day-to-day time τ is considered continuous, we have the dynamical system

$$\frac{dX(\tau)}{d\tau} = \phi(X), \quad X(0) = X_0 \tag{2}$$

where $\tau \geq 0$ and X_0 is any initial route inflow vector in D. This dynamical system evolves continuously over day-to-day time with each element being a within-day inflow function giving all inflow rates to all routes at all within-day times.

The Bottleneck Model

In the bottleneck model, queueing occurs vertically at link exits when traffic flow exceeds capacity (in which case we say that the bottleneck is congested). The bottleneck delay d_i on link i is connected to the bottleneck capacity s_i and the bottleneck inflow x_i by the following integral equation:

$$\int_0^t x_i(u)du = \int_{t_0}^{t+d_i^x(t)} s_i(u)du$$

for all t in some congested period $[t_0, t_1]$.

Smith and Wisten (1995) used Schauder's fixed point theorem to prove existence of equilibrium of the dynamical system (2) provided that route cost is continuous (as a function of route inflow) and that the feasible set D is convex and compact. It is clear that D is convex. Mounce (2006) establishes that the feasible set D is compact and Mounce (2005) uses an implicit function theorem to show that the route cost vector is indeed a continuous function of the route flow vector.

Akamatsu (2000) was able to obtain an analytical solution to the DUE problem in the bottleneck model for a single origin network or a single destination network; but only assuming that for each time all links have both positive inflows and queues.

The dynamical system (2) is globally convergent if for any starting vector X_0 the dynamical system converges to the set of equilibria as $\tau \to \infty$. Mounce (2006) shows that in the bottleneck model this occurs when each route passes through at most one bottleneck.

MONOTONICITY OF THE COST OPERATOR (WITH RESPECT TO AN INNER PRODUCT)

Monotonicity in function space

In the steady state case, the route cost vector $C(X)$ is a non-decreasing function of the route flow vector X if and only if for all route flow vectors X and Y,

$$(C(X) - C(Y)) \cdot (X - Y) \geq 0 \tag{3}$$

where the dot represents the vector dot product. In the dynamic case, the route cost vector is an operator, i.e. the route cost vector function $C(X)$ depends upon the (whole) route flow vector function X. A natural generalisation of (3) in the dynamic case is to say that $C(X)$ is a monotone function of X if and only if

$$\langle C(X) - C(Y), X - Y \rangle = \sum_r \int_0^t (C_r^X(t) - C_r^Y(t))(X_r(t) - Y_r(t))dt \geq 0$$

for all route flow vectors X and Y.

Monotonicity in the bottleneck model

Smith and Ghali (1990a) showed that in the bottleneck model, route cost is monotone in the single bottleneck per route case when link capacities are constant; and Mounce (2006) shows that this is also the case when link capacities are non-decreasing functions of within-day time. However, route cost monotonicity does not necessarily hold in networks with routes passing through more than one bottleneck. This was noted by Smith and Ghali (1990b) and an example network given in Mounce (2001).

A counterexample to monotonicity in the bottleneck model

Figure 1: The network

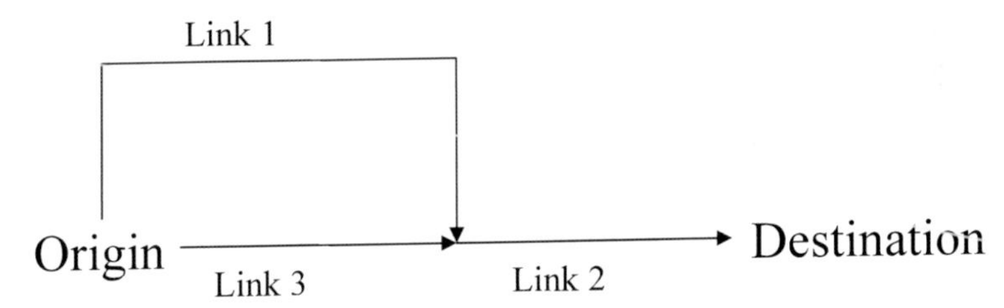

Figure 1 shows a network with two routes; route 1 traverses links 1 then link 2, whereas route 2 traverses link 3 followed by link 2. Link 1 has congestion-free cost c_1 and delay function d_1^x; and similarly for the other links. In this example, within-day time is represented by the interval $[0,T]$, where T is chosen so that all traffic will reach the destination (and note that T is greater than 1 here only to ease the calculations; a rescaling could easily be implemented). Let $c_1 = 1$ and $c_2 = c_3 = 0$. Then choose the route flow vectors X and Y as follows:

	[0,0.5]	[0.5,1]	[1,1.75]	[1.75,2]
$X_1(t)$	1.5	0.5	0	0
$X_2(t)$	1	2	1	3
$Y_1(t)$	1	1	0	0
$Y_2(t)$	1.5	1.5	1	3

Figure 2: Congestion for traffic entering link 1 during [0,1]

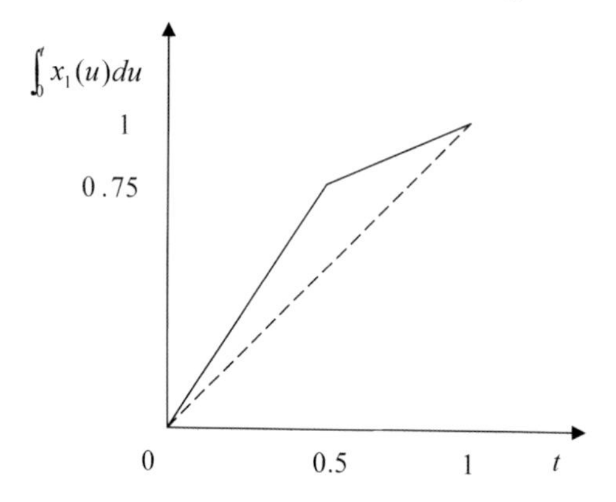

Figures 2 and 3 show levels of congestion at the two bottlenecks during the congested time periods. In Figure 2 the dashed line has gradient 1 and represents both the inflow for Y and the outflow for both X and Y. In Figure 3 the dashed line has gradient 2.

Figure 3: Congestion for traffic entering link 2 during [1,2]

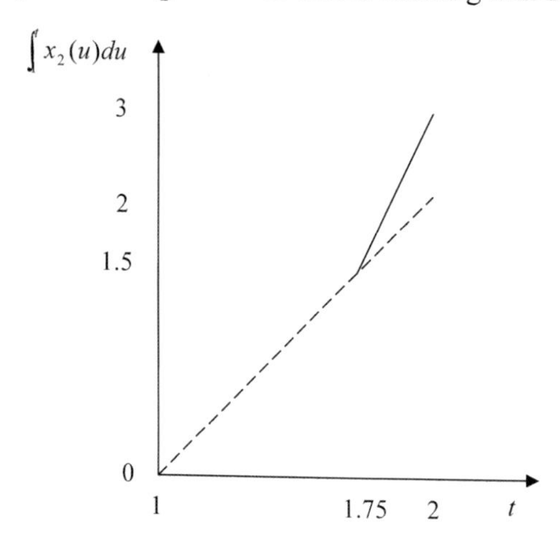

Since the congestion-free costs do not vary with time or traffic volume, clearly the contribution to $\langle C(X)-C(Y), X-Y\rangle$ will be zero and therefore

$$\langle C(X)-C(Y), X-Y\rangle = \sum_{i,r} \int_0^T d_i^x(A_{ir}^X(t)) - d_i^y(A_{ir}^Y(t))(X_r(t)-Y_r(t))dt\,.$$

Obviously only the time periods when the route flows X and Y differ need to be considered. Since link 1 is the first link on route 1, $A_{11}^X(t) = A_{11}^Y(t) = t+1$. Clearly $d_1^y(t)=0$ for all t, whereas $d_1^x(t) = t/2$ for $t \in [0,1/2]$ and $d_1^x(t) = (1-t)/2$ for $t \in [1/2,1]$. Therefore

$$\int_0^1 (d_1^x(A_{11}^X(t)) - d_1^y(A_{11}^Y(t)))(X_1(t)-Y_1(t))dt = \int_0^{1/2} \frac{t}{2}\cdot\frac{1}{2}dt + \int_{1/2}^1 \frac{1-t}{2}\cdot -\frac{1}{2}dt = 0\,.$$

Since $A_{22}^X(t) = t$ for all t for both X and Y, and $d_2^x(t) = d_2^y(t) = 0$ for $t \in [0,1]$,

$$\int_0^1 (d_2^x(A_{22}^X(t)) - d_2^y(A_{22}^Y(t)))(X_2(t)-Y_2(t))dt = 0$$

Now note that $d_2^x(1+t+d_1^x(t)) = 0$ for $t \in [0,1/2]$ and $d_2^y(1+t+d_1^y(t)) = 0$ for $t \in [0,3/4]$. Since the flows are linear, the average of $d_2^x(1+t+d_1^x(t))$ on $[1/2,1]$ and the average of $d_2^y(1+t+d_1^y(t))$ on $[3/4,1]$ is the same and is given by

$$\frac{1}{1-3/4}\int_{3/4}^1 d_2^y(1+t+d_1^x(t)) = 4\int_{3/4}^1 (t-3/4)dt = \frac{1}{8}\,.$$

Therefore

$$\int_0^1 (d_2^x(A_{21}^X(t)) - d_2^y(A_{21}^Y(t)))(X_1(t)-Y_1(t))dt$$

$$= -\frac{1}{2}\int_{1/2}^1 d_2^x(1+t+d_1^x(t))dt + \frac{1}{2}\int_{1/2}^1 d_2^y(1+t+d_1^y(t))dt = -\frac{1}{2}\cdot\frac{1}{2}\cdot\frac{1}{8} + \frac{1}{2}\cdot\frac{1}{4}\cdot\frac{1}{8} = -\frac{1}{64}.$$

When all of these terms are summed, we obtain

$$\langle C(X)-C(Y), X-Y\rangle = -\frac{1}{64} < 0$$

and therefore monotonicity does not hold in general for networks with routes passing through more than one bottleneck.

COST AS A NON-DECREASING OPERATOR WITH RESPECT TO A PARTIAL ORDER

Link cost as a non-decreasing operator

We shall start by showing that link delay (and hence link cost) is a one-to-one function of link flow when the link is congested. We define a congested period to be the largest interval $[t_0, t_1]$ for which $d_i^x(t) > 0$ for all $t \in (t_0, t_1)$. In the bottleneck model this is clear. In other models, the cost function can be restructured so that $c_i^x(t) = c_i + d_i^x(t)$. Here we prove the results for the bottleneck model.

Lemma 1. Link delays are a one-to-one function of both the congested periods and the inflows during the congested periods (together).

Proof. Suppose that $d_i^x(t) = d_i^y(t)$ for all $t \in [0,1]$. Clearly the congested periods for x and y must coincide, otherwise one delay would be zero and the other would be positive (and hence they would be unequal). On any congested period $[t_0, t_1]$,

$$\int_{t_0}^{t} x_i(u)du = \int_{t_0}^{t+d_i^x(t)} s_i(u)du = \int_{t_0}^{t+d_i^y(t)} s_i(u)du = \int_{t_0}^{t} y_i(u)du. \qquad \square$$

Definition 1. By using a partial order it is possible to define the link cost being a non-decreasing function of link flow if and only if

$$\int^{t'} x_i(u)du \geq \int^{t'} y_i(u)du \quad \forall t,t' \in [0,1] \text{ with } t \leq t' \Rightarrow c_i^x(t) \geq c_i^y(t) \text{ for all } t \in [0,1] \qquad (4)$$

where clearly $c_i^x(t) \geq c_i^y(t) \Leftrightarrow d_i^x(t) \geq d_i^y(t)$.

If causality is assumed in our model (i.e. traffic delays at time t are only affected by traffic entering the link up until time t), then from (4) it can be deduced that

$$x_i(t) \geq y_i(t) \text{ for almost all } t \in [0,t_0] \Rightarrow c_i^x(t) \geq c_i^y(t) \text{ for all } t \in [0,t_0]. \qquad (5)$$

Theorem 2. Link delay (and hence also link cost) is a non-decreasing function of link flow.

Proof. First define $q_i^x(t)$ to be the queue on link i at time t and let $b_i^x(t) = \sup\{u \in [0,t] : q_i^x(u) = 0\}$, i.e. the last time that the link became congested (if the link is uncongested at time t then $b_i^x(t) = t$). Now suppose that $\int^{t'} x_i(u)du \geq \int^{t'} y_i(u)du$ $\forall t,t' \in [0,1]$ with $t \leq t'$. Then

$$\int_{b_i^x(t)}^{t} x_i(u)du \geq \int_{b_i^x(t)}^{t} y_i(u)du,$$

which can be rewritten as

$$\int_{b_i^x(t)}^{t+d_i^x(t)} s_i(u)du = \int_{b_i^x(t)}^{t} x_i(u)du \geq \int_{b_i^x(t)}^{b_i^y(t)} s_i(u)du + \int_{b_i^y(t)}^{t} y_i(u)du = \int_{b_i^x(t)}^{t+d_i^y(t)} s_i(u)du \qquad (6)$$

It then immediately follows that $d_i^x(t) \geq d_i^y(t)$. $\square$

Of more practical use is the following result:

Lemma 2. If $d_i^x(t) = d_i^y(t)$ for $t \in [0,t_0]$ and $\int_0^t x_i(u)du \geq \int_0^t y_i(u)du$ for all $t \in (t_0, t_1]$, then $d_i^x(t) \geq d_i^y(t)$ for t in some interval $(t_0, t_2]$.

Proof. Firstly, suppose that $b_i^x(t_0) < t_0$ (and hence $b_i^y(t) = b_i^x(t_0) < t_0$). Then it must be possible to choose an interval $(t_0, t_2]$ such that $b_i^x(t) = b_i^x(t_0)$ for $t \in (t_0, t_2]$. Clearly $\int_{b_i^x(t_0)}^{t_0} x_i(u)du = \int_{b_i^x(t_0)}^{t_0} y_i(u)du$ by Lemma 1 and therefore for $t \in (t_0, t_2]$,

$$\int_{b_i^x(t)}^{t+d_i^x(t)} s_i(u)du = \int_{b_i^x(t_0)}^{t+d_i^x(t)} s_i(u)du = \int_{b_i^x(t_0)}^{t} x_i(u)du \geq \int_{b_i^y(t_0)}^{t} y_i(u)du = \int_{b_i^y(t)}^{t+d_i^y(t)} s_i(u)du\,,$$

from which it immediately follows that $d_i^x(t) \geq d_i^y(t)$ for $t \in (t_0, t_2]$. Alternatively if $b_i^x(t_0) = t_0$ then either $b_i^x(t) = t_0$ on some interval $(t_0, t_2]$ in which case (6) holds with $b_i^x(t) = t_0$, or otherwise $b_i^x(t) = t$ for t in some interval $(t_0, t_2]$ in which case $d_i^x(t) = d_i^y(t) = 0$ on the interval $(t_0, t_2]$. □

An alternative viewpoint is to say that, on a congested period, link inflows are an increasing function of link delays:

Lemma 3. If $d_i^x(t) = d_i^y(t)$ for $t \in [0, t_0]$ but $d_i^x(t) > d_i^y(t)$ on some interval $(t_0, t_1]$, then $\int_0^t x_i(u)du > \int_0^t y_i(u)du$ for all $t \in (t_0, t_1]$.

Proof. Suppose that $d_i^x(t) = d_i^y(t)$ for $t \in [0, t_0]$ but $d_i^x(t) > d_i^y(t)$ on some interval $(t_0, t_1]$. Then either $b_i^x(t_0) < t_0$ or $b_i^x(t_0) = t_o$. Firstly, suppose that $b_i^x(t_0) < t_0$, in which case $b_i^x(t_0) = b_i^y(t_0)$ (otherwise delays would be unequal before time t_0) and therefore $\int_{b_i^x(t_0)}^{t_0} x_i(u)du = \int_{b_i^x(t_0)}^{t_0} y_i(u)du$ by Lemma 1. Then for $t \in (t_0, t_1]$,

$$\int_{t_0}^{t} x_i(u)du = \int_{b_i^x(t_0)}^{t} x_i(u)du - \int_{b_i^x(t_0)}^{t_0} x_i(u)du = \int_{b_i^x(t)}^{t+d_i^x(t)} s_i(u)du - \int_{b_i^x(t)}^{t_0} y_i(u)du$$

$$> \int_{b_i^x(t_0)}^{t+d_i^y(t)} s_i(u)du - \int_{b_i^x(t_0)}^{t_0} y_i(u)du$$

$$= \int_{b_i^x(t_0)}^{t} y_i(u)du - \int_{b_i^x(t_0)}^{t_0} y_i(u)du = \int_{t_0}^{t} y_i(u)du\,.$$

Otherwise, if $b_i^x(t_0) = t_o$ then, since $d_i^x(t) > d_i^y(t)$ on $(t_0, t_1]$, it must be true that $b_i^x(t) = t_o$ for $t \in [t_0, t_1]$. Then for $t \in (t_0, t_1]$,

$$\int_{t_0}^{t} x_i(u)du = \int_{t_0}^{t+d_i^x(t)} s_i(u)du > \int_{t_0}^{t+d_i^y(t)} s_i(u)du = \int_{b_i^y(t)}^{t} y_i(u)du + \int_{t_0}^{b_i^y(t)} s_i(u)du > \int_{t_0}^{t} y_i(u)du\,.$$ □

Link output as a non-decreasing operator

Theorem 3. The cumulative outflow from a link is a non-decreasing function of the link inflow.

Proof. Suppose that $\int^{t'} x_i(u)du \geq \int^{t'} y_i(u)du \ \forall t,t' \in [0,1]$, in which case it is clear that $b_i^x(t) \leq b_i^y(t)$ for all $t \in [0,1]$. Then for all $t \in [0,1]$,

$$\int_{b_i^x(t)}^{b_i^y(t)} x_i(u)du \geq \int_{b_i^x(t)}^{b_i^y(t)} s_i(u)du \geq \int_{b_i^x(t)}^{b_i^y(t)} y_i(u)du \tag{7}$$

and hence

$$\begin{aligned}\int_0^t Ox_i(u+c_i)du = \int_0^{b_i^x(t)} x_i(u)du + \int_{b_i^x(t)}^t s_i(u)du &\geq \int_0^{b_i^x(t)} y_i(u)du + \int_{b_i^x(t)}^t s_i(u)du \\ &\geq \int_0^{b_i^x(t)} y_i(u)du + \int_{b_i^x(t)}^{b_i^y(t)} y_i(u)du + \int_{b_i^y(t)}^t s_i(u)du \\ &\geq \int_0^{b_i^y(t)} y_i(u)du + \int_{b_i^y(t)}^t s_i(u)du \\ &= \int_0^t Oy_i(u+c_i)du.\end{aligned}$$

□

Of more practical use is the following result:

Lemma 4. If $d_i^x(t) = d_i^y(t)$ for $t \in [0,t_0]$ and $\int_0^t x_i(u)du \geq \int_0^t y_i(u)du$ for t in some interval $[t_0,t_1]$, then for all $t \in [t_0 + d_i^x(t_0), t_1 + d_i^y(t_1)]$,

$$\int_{t_0+d_i^x(t_0)}^t Ox_i(u+c_i)du \geq \int_{t_0+d_i^x(t_0)}^t Oy_i(u+c_i)du.$$

Proof. Suppose that $d_i^x(t) = d_i^y(t)$ for $t \in [0,t_0]$ and $\int_0^t x_i(u)du \geq \int_0^t y_i(u)du$ for $t \in [t_0,t_1]$. Consider $t \in [t_0 + d_i^x(t_0), t_1 + d_i^y(t_1)]$. Firstly, suppose that $b_i^x(t) \leq t_0 + d_i^x(t_0)$ and therefore also $b_i^y(t) = b_i^x(t) \leq t_0 + d_i^x(t_0)$, in which case clearly

$$\int_{t_0+d_i^x(t_0)}^t Ox_i(u+c_i)du = \int_{t_0+d_i^x(t_0)}^t s_i(u)du = \int_{t_0+d_i^x(t_0)}^t Oy_i(u+c_i)du.$$

Then, again for $t \in [t_0 + d_i^x(t_0), t_1 + d_i^y(t_1)]$, suppose that $b_i^x(t) > t_0 + d_i^x(t_0)$ and therefore $b_i^y(t) \geq b_i^x(t) > t_0 + d_i^x(t_0)$.

Using (7),

$$\begin{aligned}\int_{t_0+d_i^x(t_0)}^{t} Ox_i(u+c_i)du &= \int_{t_0}^{b_i^x(t)} x_i(u)du + \int_{b_i^x(t)}^{t} s_i(u)du \\ &\geq \int_{t_0}^{b_i^x(t)} y_i(u)du + \int_{b_i^x(t)}^{t} s_i(u)du \\ &\geq \int_{t_0}^{b_i^y(t)} y_i(u)du + \int_{b_i^y(t)}^{t} s_i(u)du = \int_{t_0}^{t} Oy_i(u+c_i)du.\end{aligned}$$

□

In addition, link outflow during a congested period is a non-decreasing function of link delay:

Corollary 1. If $d_i^x(t) = d_i^y(t)$ for $t \in [0, t_0]$ and $d_i^x(t) \geq d_i^y(t)$ on some interval $(t_0, t_1]$, then, for all $t \in [t_0 + d_i^x(t_0), t_1 + d_i^y(t_1)]$,

$$\int_{t_0+d_i^x(t_0)}^{t} Ox_i(u+c_i)du \geq \int_{t_0+d_i^x(t_0)}^{t} Oy_i(u+c_i)du .$$

Proof. Suppose that $d_i^x(t) = d_i^y(t)$ for $t \in [0, t_0]$ but $d_i^x(t) > d_i^y(t)$ on some interval $(t_0, t_1]$. Then by Lemma 3, $\int_0^t x_i(u)du > \int_0^t y_i(u)du$ for all $t \in (t_0, t_1]$, and then by Lemma 4 the result follows immediately. □

UNIQUENESS OF COSTS AT EQUILIBRIUM

We already know that equilibrium exists (Smith and Wisten, 1995, Mounce, 2005) and in this section we show that costs at equilibrium are unique in the single OD pair case. Firstly, we introduce the notion of a branch as follows:

Definition 2. Given any link, say l, in the network, the branch associated with link l is defined to be the longest (acyclic) path, say p, traversing link l that is the unique (acyclic) path between the start node and end node of path p.

Lemma 5. Suppose that branch b is such that $C_b^X(t) = C_b^Y(t)$ for $t \in [0, t_0]$ and

$$\int_{t_0}^{t} X_b(u)du \geq \int_{t_0}^{t} Y_b(u)du$$

for $t \in [t_0, t_1]$ (where X_b is the flow into branch b) then it is possible to choose t_2, with $t_0 < t_2 \leq t_1$, such that $C_b^X(t) \geq C_b^Y(t)$ for $t \in [t_0, t_2]$.

Proof. If link 1 is the first link of branch b, then since $X_b = x_1$, clearly

$$\int_{t_0}^{t} x_1(u)du \geq \int_{t_0}^{t} y_1(u)du$$

and then by Lemma 2, $c_1^x(t) \geq c_1^y(t)$ for all $t \in [t_0, t_1]$, and also by Lemma 4,

$$\int_{t_0+d_i^x(t_0)}^{t} Ox_i(u+c_i)du \geq \int_{t_0+d_i^x(t_0)}^{t} Oy_i(u+c_i)du\,.$$

for all $t \in [t_0 + d_1^x(t_0), t_1 + d_1^y(t_1)]$. If link 2 is the next link on branch b, then this is equivalent to

$$\int_{t_0+c_1+d_1^x(t_0)}^{t} x_2(u)du \geq \int_{t_0+c_1+d_1^x(t_0)}^{t} y_2(u)du$$

for all $t \in [t_0 + d_1^x(t_0), t_1 + d_1^y(t_1)]$. It is possible to choose t_2, with $t_2 \leq t_1$, such that $t_2 + d_1^x(t_2) = t_1 + d_1^y(t_1)$. Then since link exit time is a non-decreasing function of link arrival time (see Lemma 1 of Mounce (2005)),
$t + c_1^x(t) + c_2^x(t + c_1^x(t)) \geq t + c_1^y(t) + c_2^x(t + c_1^y(t)) \geq t + c_1^y(t) + c_2^y(t + c_1^y(t))$.
The result then follows by induction. □

Lemma 6. Suppose that branch b is such that $C_b^X(t) = C_b^Y(t)$ for all $t \in [0, t_0]$ but $C_b^X(t) > C_b^Y(t)$ for all $t \in (t_0, t_1]$. Then

$$\int_{t_0}^{t} X_b(u)du > \int_{t_0}^{t} Y_b(u)du$$

for all $t \in [t_0, t_1]$, and therefore from Lemma 5, if $e_b^x(t_0)$ is the exit time from branch b,

$$\int_{e_b^x(t_0)}^{t} OX_b(u)du \geq \int_{e_b^x(t_0)}^{t} OY_b(u)du$$

for all t in some interval $[e_b^x(t_0), t_2]$ with $e_b^x(t_0) < t_2 \leq t_1$.

Proof. Suppose that branch b is such that $C_b^X(t) = C_b^Y(t)$ for all $t \in [0, t_0]$ but $C_b^X(t) > C_b^Y(t)$ for all $t \in (t_0, t_1]$. For each $t \in (t_0, t_1]$, there must be a first link i along branch b such that $d_i^x(t) > d_i^y(t)$ and then applying Lemma 3 on each interval gives

$$\int_{t_0}^{t} X_b(u)du > \int_{t_0}^{t} Y_b(u)du$$

for all $t \in (t_0, t_1]$. □

Now we come to the main result of the paper:

Theorem 4. In the bottleneck model, costs at dynamic user equilibrium are unique in the single OD pair case.

Proof. Suppose that there exist two equilibria X and Y with different route costs and hence different link costs (for some within-day time). As noted by Kuwahara (1993), at equilibrium the travel time between any two nodes is the same for all used paths connecting the two nodes, and therefore we can define $A_i^X(t)$ to be the arrival time at link i when the route flow vector is X. Without loss of generality, suppose that for some $t_0 \in [0,1])$, $c_i^x(A_i^X(t)) = c_i^y(A_i^Y(t))$ for all $t \in [0,t_0]$ and for all links i in the network (in which case clearly $A_i^X(t) = A_i^Y(t)$ for all $t \in [0,t_0]$), but that costs are unequal for at least one link in the network after time t_0.

Define a sequence of (sets of) branches as follows:
Step 1. Let S_1 be the set of branches from the origin (i.e. with the origin as the start node).
Step 2. Let S_2 consist of all the branches with start node equal to the tail node of some branch in S_1; and also all those branches in S_1 with the destination as the tail node.
Step 3. For any i, define T_i to be the set of branches in S_i that terminate at a node, say n, for which there is a branch that is not in $\bigcup_{j=1}^{i} S_j$ that terminates at node n. Then let S_3 consist of all the branches with start node equal to the tail node of some branch in $S_2 \setminus T_2$, all those branches with tail node equal to the destination and all the branches in T_2.

Step m. Let S_m consist of all the branches with start node equal to the tail node of some branch in $S_{m-1} \setminus T_{m-1}$, all those branches with their tail node equal to the destination and all the branches in T_{m-1}.

Figure 4: Example of the process that generates a sequence of cuts

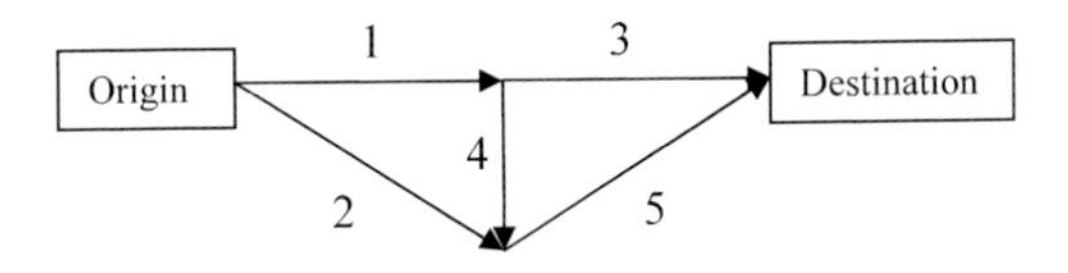

An example is shown in Figure 4 for which $S_1 = \{1,2\}$, $S_2 = \{2,3,4\}$ and $S_3 = \{3,5\}$.

Let $m = \min\{j : \sup\{t \in [0,1] : C_b^X(A_b^X(t)) = C_b^Y(A_b^Y(t)) \forall i \in S_j\} = t_0\}$ where $A_b^X(t)$ is the arrival time at the first link of branch b, i.e. m is the first stage at which the branch costs are different immediately after time t_0. Consequently, it is possible to choose $t_2 \in (t_0,1]$ such that at all stages $1,2,\ldots m-1$, $C_b^X(A_b^X(t)) = C_b^Y(A_b^Y(t))$ for all $t \in [0,t_2]$. Now we let S_m^+ be the set of branches in S_m with $C_b^X(A_b^X(t)) > C_b^Y(A_b^Y(t))$ on some interval to the right of t_0, and S_m^- be the set of branches in S_m with $C_b^X(A_b^X(t)) \le C_b^Y(A_b^Y(t))$ on some interval to the right of t_0 (and clearly $S_m = S_m^- \cup S_m^+$). It is possible to choose t_3, with $t_3 \le t_2$, such that for all $b \in S_m^+$, $C_b^X(A_b^X(t)) > C_b^Y(A_b^X(t))$ for all $t \in (t_0, t_3]$, and furthermore such that there is a path p, with the same terminal node as branch b, that is used for X during $t \in [t_0, t_3]$. Since branch costs are equal for X and Y for all branches upstream of branch b, $A_b^X(t) = A_b^Y(t)$ for all $t \in [0, t_2]$. Then by Lemma 1 of Mounce (2005),

$$C_p^X(t) = A_b^X(t) + C_b^X(A_b^X(t)) > A_b^Y(t) + C_b^Y(A_b^Y(t)) = C_p^Y(t)$$

for all $t \in [t_0, t_3]$.

Given branch $b \in S_m^+$, since m is the first stage where costs (and hence arrival times) are different, $A_b^X(t) = A_b^Y(t)$ for all $t \in [0, t_3]$. Therefore by Lemma 6,

$$\int_{A_b^X(t_0)}^{t} X_b(u)du > \int_{A_b^Y(t_0)}^{t} Y_b(u)du$$

for all $t \in (A_b^X(t_0), A_b^X(t_3)]$, or equivalently

$$\sum_{r:b\in r} \int_0^t X_r(u)du > \sum_{r:b\in r} \int_0^t Y_r(u)du$$

(where $b \in r$ means that route r traverses branch b) and summing these inequalities over all branches in S_m^+ gives, for all $t \in (t_0, t_3]$,

$$\sum_{b\in S_m^+} \int_{A_b^X(t_0)}^{t} X_b(u)du > \sum_{b\in S_m^+} \int_{A_b^X(t_0)}^{t} Y_b(u)du \tag{8}$$

and

$$\sum_{b\in S_m^+} \sum_{r:b\in r} \int_0^t X_r(u)du > \sum_{b\in S_m^+} \sum_{r:b\in r} \int_0^t Y_r(u)du . \tag{9}$$

Since the set S_m in the above process forms a cut of the network (and the total flow between the origin and destination is fixed), by (9) it is clear that, for all $t \in (t_0, t_3]$,

$$\sum_{b\in S_m^-} \sum_{r:b\in r} \int_0^t X_r(u)du < \sum_{b\in S_m^-} \sum_{r:b\in r} \int_0^t Y_r(u)du$$

and since $A_b^X(t) = A_b^Y(t)$ for all $t \in (t_0, t_3]$,

$$\sum_{b\in S_m^-} \int_{A_b^X(t_0)}^{t} X_b(u)du < \sum_{b\in S_m^-} \int_{A_b^X(t_0)}^{t} Y_b(u)du .$$

Hence it is possible to choose t_4, with $t_4 \le t_3$, such that for some branch $b \in S_m^-$,

$$\int_{A_b^X(t_0)}^{t} X_b(u)du < \int_{A_b^X(t_0)}^{t} Y_b(u)du$$

for all $t \in (A_b^X(t_0), A_b^X(t_4)]$. Then by Lemma 5, it is possible to choose t_5 with $t_5 \le t_1$ such that for $t \in [t_0, t_5]$, $C_b^X(A_b^X(t)) \le C_b^Y(A_b^Y(t))$ and such that there is a path p that traverses branch b that is used for Y during $(t_0, t_5]$. Since branch costs are equal for X and Y for all branches upstream of branch b, $A_b^X(t) = A_b^Y(t)$ for all $t \in [0, t_1]$.

Then by Lemma 1 of Mounce (2005),
$C_p^X(t) = A_b^X(t) + C_b^X(A_b^X(t)) \le A_b^Y(t) + C_b^Y(A_b^Y(t)) = C_p^Y(t)$
for all $t \in [t_0, t_1]$.

Therefore at stage m there is at least one path p_1 that is higher cost, and also used, for X during $[t_0, t_1]$ and also at least one path p_2 that has cost no greater for X during $[t_0, t_1]$ that is used for Y. Suppose that these two paths were to meet at a node. Since p_2 is used for Y during $[t_0, t_1]$ and Y is an equilibrium (and at equilibrium all used paths to any given node must be least costly for all within-day time) it must be true that $C_{p_2}^Y(t) \le C_{p_1}^Y(t)$. Then combining the cost inequalities yields,

$$C_{p_1}^X(t) > C_{p_1}^Y(t) \ge C_{p_2}^Y(t) \ge C_{p_2}^X(t) \tag{10}$$

for $t \in (t_0, t_1]$, which contradicts the fact that X is an equilibrium.

Since all paths must eventually meet at the destination, in order to avoid a contradiction, one of the following must occur at some stage subsequent to stage m:
i) All paths that are equal or less costly for X must become more costly for X during the relevant time period; or
ii) All paths that are more costly for X must become equal or less costly for X during the relevant time period.

When considering whether (i) can occur, first note that on any interval, if a branch is used for Y, it must be used for X (except where there are two paths that are uncongested for both X and Y that connect the same two nodes) otherwise there would be a less costly unused route for X. Assume that $C_r^X(t) > C_r^Y(t)$ for all routes that are used for X on some interval $(t_0, t_1]$ (with costs equal before time t_0). Each route r that is used for X must reach a branch b for which $C_b^X(t) > C_b^Y(t)$ on some interval $(A_b^X(t_0), A_b^X(t_1)]$, and therefore for all $t \in (A_b^x(t_0), A_b^x(t_1)]$,

$$\int_{A_b^X(t_0)}^{t} X_b(u)du > \int_{A_b^X(t_0)}^{t} Y_b(u)du$$

which implies that

$$\sum_{r:b\in r} \int_0^t X_r(u)du > \sum_{r:b\in r} \int_0^t Y_r(u)du$$

for all $t \in (t_0, t_1)$. We can follow a modified version of the process above that defines S_m, that stops when it reaches such branches, to give a cut S of the used routes with

$$\sum_{b\in S} \int_0^t X_b(u)du > \sum_{b\in S} \int_0^t Y_b(u)du,$$

for $t \in (t_0, t_1)$, from which we can deduce that

$$\sum_{r} \int_0^t X_r(u)du > \sum_{r} \int_0^t Y_r(u)du$$

for $t \in (t_0, t_1)$, which is a contradiction, since there is a fixed volume of traffic departing from the origin on any interval of time.

Now consider whether (ii) can occur. Firstly, note that by (10) if a path p_2 that is equal or less costly for X joins with a path p_1 that is higher cost for X, then path p_2 must be unused for Y, and then for all $t \in (t_0, t_1]$,

$$\int_0^t X_{p_2}(u)du \geq 0 = \int_0^t Y_{p_2}(u)du.$$

Hence paths that are equal or less costly for X cannot affect (by combining with) the cost along paths that are higher cost for X. Secondly, from (8) and Lemma 6, it can be deduced that the total outflow from S_m^+ must be no less for X than Y during an interval for traffic leaving the origin after t_0. Therefore the flow into at least one branch b out of S_m^+ must be higher on some interval, and therefore the cost of a resulting path p with the same terminal node as branch b must satisfy

$$C_p^X(t) = A_b^X(t) + C_b^X(A_b^X(t)) > A_b^Y(t) + C_b^Y(A_b^Y(t)) = C_p^Y(t)$$

on some half open interval to the right of t_0. One might think that this argument can be repeated at each subsequent stage. However, it is possible for the total inflow to a node to be greater for X than for Y but the total outflow from the node to be less for X on some appropriate intervals of time. This is only possible, say at stage k, if a branch $b_1 \in S_k^+$ from say node m_1 to node n_1 that satisfies $C_{b_1}^X(P_1^X(t)) > C_{b_1}^Y(P_1^Y(t))$ and

$\int_{P_1^X(t_0)}^{P_1^X(t)} X_{b_1}(u)du > \int_{P_1^X(t_0)}^{P_1^X(t)} Y_{b_1}(u)du$ for t in some interval $(t_0, t_1]$, intersects with a branch b_2 into node n_1 with $\int_{P_2^X(t_0)}^{P_2^X(t)} X_{b_2}(u)du < \int_{P_2^X(t_0)}^{P_2^X(t)} Y_{b_2}(u)du$ for $t \in (t_0, t_1]$ that is uncongested for X. Since the cost on all used paths connecting any two nodes must be the same, branch b_2 cannot originate from node m_1, but instead from say node m_2. Branch b_2 must originate from a branch that is higher cost and higher flow for X and therefore there must be another branch b_3 that is higher flow for X and such that all used paths to the terminal node of branch b_3 are higher cost for X. Then since branch b_1 is higher flow for X, there must be another branch b_4 such that for all $t \in (t_0, t_1]$, $\int_{P_4^X(t_0)}^{P_4^X(t)} X_{b_4}(u)du < \int_{P_4^X(t_0)}^{P_4^X(t)} Y_{b_4}(u)du$.

Now we show that it is not possible for branches b_3 and b_4 to intersect at node n_2 to give lower total outflow for X during some interval to the right of $A_{n_2}^X(t_0)$ (where $A_{n_2}^X(t_0)$ is the arrival time at node n_2 for traffic departing the origin at time t_0). Figure 5 shows the configuration of the network if they were to intersect.

Figure 5: The resulting network configuration

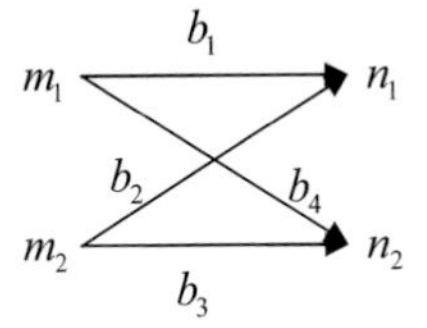

In order for the total outflow to be lower for X, branch b_4 would have to be uncongested during the relevant time interval. Using Lemma 5 and the fact that any used path to a node must be least costly, for t in some interval to the right of t_0, the following hold:

$$P_1^Y(t)+C_{b_1}^X(P_1^Y(t)) > P_1^Y(t)+C_{b_1}^Y(P_1^Y(t)) \geq P_2^Y(t)+C_{b_2}^Y(P_2^Y(t)) \geq P_2^Y(t)+C_{b_2}^X(P_2^Y(t)) \quad (11)$$

$$P_2^Y(t)+C_{b_3}^X(P_2^Y(t)) > P_2^Y(t)+C_{b_3}^Y(P_2^Y(t)) \geq P_1^Y(t)+C_{b_4}^Y(P_1^Y(t)) \geq P_1^Y(t)+C_{b_4}^X(P_1^Y(t)) \quad (12)$$

Summing the right hand sides and left hand sides of (11) and (12) respectively (and cancelling common terms) yields

$$C_{b_1}^X(P_1^Y(t))+C_{b_3}^X(P_2^Y(t)) > C_{b_2}^X(P_2^Y(t))+C_{b_4}^X(P_1^Y(t)) \quad (13)$$

whereas from the equilibrium conditions we can derive the following:

$$C_{b_1}^X(P_1^X(t))+C_{b_3}^X(P_2^X(t)) \leq C_{b_2}^X(P_2^X(t))+C_{b_4}^X(P_1^X(t)) \quad (14)$$

both holding on some interval to the right of t_0. For both (13) and (14) to hold on some interval to the right of t_0, equality has to hold in (14) at time t_0, and then since

$P_1^X(t), P_2^X(t), P_1^Y(t)$ and $P_2^Y(t)$ are all non-decreasing, on some interval to the right of $P_1^X(t_0)$, either:

a) $C_{b_1}^X(t)-C_{b_1}^X(P_1^X(t_0)) > C_{b_4}^X(t)-C_{b_4}^X(P_1^X(t_0))=0$ and

$C_{b_3}^X(t)-C_{b_3}^X(P_2^X(t_0)) < C_{b_2}^X(t)-C_{b_2}^X(P_1^X(t_0))=0$

with $P_1^Y(t) > P_1^X(t)$, contradicting $P_1^Y(t) \leq P_1^X(t)$; or

b) $C_{b_3}^X(t)-C_{b_3}^X(P_2^X(t_0)) > C_{b_2}^X(t)-C_{b_2}^X(P_1^X(t_0))=0$ and

$C_{b_1}^X(t)-C_{b_1}^X(P_1^X(t_0)) < C_{b_4}^X(t)-C_{b_4}^X(P_1^X(t_0))=0$

with $P_2^Y(t) > P_2^X(t)$, contradicting $P_2^Y(t) \leq P_2^X(t)$.

Therefore branch b_3 cannot join with branch b_4. A similar argument can be applied if there are more than two pairs of nodes.

Since (i) or (ii) above can never occur at any stage subsequent to stage m, the contradiction in (10) cannot be avoided since all routes must eventually meet at the destination. Therefore the initial assumption that the costs are different for X and Y must be rejected. Hence both route costs and link costs at dynamical user equilibrium are unique in the single OD pair case.

□

Since Lemma 1 showed that link costs are a one-to-one function of inflow when links are congested, in the single OD pair case, link flows are unique at equilibrium at least when there is sufficient traffic to cause congestion.

CONCLUSIONS AND FURTHER RESEARCH

In the steady state model, it is clear what is meant by an increasing function. Costs at user equilibrium were shown to be unique provided that each link cost is a non-decreasing function of link flow. In the dynamic model, the traffic assignment is taking place in function space and there are different interpretations of increasing. We first considered a natural generalisation of a monotone function in terms of a dot product. In the dynamic bottleneck queueing model, the route cost vector is not a monotone function of the route flow vector and an example network was given to illustrate this. Alternative definitions for increasing and non-decreasing operators in function space were then given. The bottleneck model was shown to satisfy these properties and then these properties were used to show that costs at dynamical user equilibrium are unique in the single OD pair case.

The issue of unique equilibrium is still open in the case where there are many origins and many destinations. However, this research should facilitate a proof of the uniqueness of costs at equilibrium in the single destination case and also in the single
destination case. In the multiple origins and multiple destinations case, equilibrium may well not be unique.

Finally, it should be stressed that although the results in the paper were constructed around the bottleneck model, one would expect that these techniques can be applied to other dynamic traffic assignment models.

REFERENCES

Akamatsu, T. (2000). A dynamic traffic equilibrium assignment paradox. *Transportation Research*, **34B**, 515-531.

Kuwahara, M., Akamatsu, T. (1993). Dynamic equilibrium assignment with queues for a one-to-many OD pattern. *Proceedings of the 12th International Symposium on Transportation and Traffic Theory*, pp 185-204.

Mounce, R. (2001). Non-monotonicity in dynamic traffic assignment networks. *Proceedings of the 33rd Universities Transport Study Group annual conference*, University of Oxford, UK.

Mounce, R. (2005). Dynamics and equilibrium in a continuous dynamic queueing model for traffic networks. *Proceedings of the 4th IMA International Conference on Mathematics in Transport*, University College London, UK.

Mounce, R. (2006). Convergence in a continuous dynamic queueing model for traffic networks. *Transportation Research*, **40B**, 779-791.

Smith, M.J. (1979). A continuous day-to-day traffic assignment model and the existence of a continuous dynamic user equilibrium. *Transportation Research,* **13B**, 295-304.

Smith, M.J. and Ghali, M.O. (1990). Dynamic traffic assignment and dynamic traffic control. *Proceedings of the 11th International Symposium on Transportation and Traffic Theory*, pp 273-290.

Smith, M.J. and Ghali, M.O. (1990). The dynamics of traffic assignment and control: a theoretical study. *Transportation Research*, **24B** 409-422.

Smith, M.J. and Wisten, M.B. (1995). A continuous day-to-day traffic assignment model and the existence of a continuous dynamic user equilibrium. *Annals of Operations Research,* 60, 59-79.

Transportation and Traffic Theory 2007
Edited by R.E. Allsop, M.G.H. Bell and B.G. Heydecker

14

ANALYSIS OF THE DYNAMIC SYSTEM OPTIMUM AND EXTERNALITIES WITH DEPARTURE TIME CHOICE

Andy H. F. Chow, Centre for Transport Studies, University College London, London, U.K.

INTRODUCTION

Dynamic traffic assignment models of route and departure time choice for travellers through congested networks provide important insight into the dynamics of peak periods and sensitivity of travellers' behaviour in response to a range of transport policy measures. In general, formulations of dynamic traffic assignment follow the extension of the two Wardrop's (1952) principles: user equilibrium and system optimum. The dynamic user equilibrium assignment has been the focus in the past two decades. As a result of previous research (see for example, Friesz et al., 1993; Friesz et al., 2001; Heydecker and Addison, 1996, 1998, 2005; Szeto and Lo, 2005), we have gained a substantial knowledge on the formulations, properties, and solution methods of dynamic equilibrium assignment.

This paper aims to analyse the dynamic system optimal assignment with departure time choice, which is an important, yet underdeveloped area. The dynamic system optimal assignment process suggests that there is a central "system manager" to distribute network traffic over time within a fixed horizon. Consequently, the total, rather than individual, travel cost of all travellers through the network is minimised. Although the system optimal assignment is not a realistic representation of network traffic, it provides a bound on how we can make the best use of the road system, and as such it is a useful benchmark for evaluating various transport policy measures. These measures include time-varying pricing (Yang and Huang, 1997), network access control (Smith and Ghali, 1990), and road capacity management (Ghali and Smith, 1993; Heydecker, 2002).

Proceeding after Heydecker and Addison (2005), the travel cost incurred by each traveller is considered to have three distinct components: a cost related to the travel time *en route*, and time-specific costs associated with the departure time of the traveller from the origin and the arrival time at the destination respectively. Given the assigned network flow, the associated travel times through the network are determined by a traffic model. The travel times then influence the arrival times of travellers and hence the travel costs incurred. Many previous analyses (see, for example, Friesz et al., 1989; Carey and Srinivasan, 1993; Yang and Huang, 1997) in the area of dynamic system optimal assignment adopted an optimal control theoretic formulation with Merchant and Nemhauser's (1978a,b) outflow traffic model. On the one hand, this formulation provides some attractive mathematical properties for analysis. On the other hand, however, it ignores the importance of ensuring proper flow propagation as first shown by Tobin (1993). In addition, the outflow models have also been widely criticized for their implausible traffic behaviour (see Astarita, 1996; Heydecker and Addison, 1998; Friesz and Bernstein, 2000; Mun, 2001). Following Daganzo (1995) and Mun (2001), to ensure the satisfaction of several necessary physical principles such as proper flow propagation (or consistency between flows and travel times), non-negativity of flow, first-in-first-out (FIFO) queue discipline, and causality, the traffic model adopted in this paper considers the travel time on each link to be a linear non-decreasing function of link traffic volume. Examples of the traffic models of this kind include deterministic queuing model, Friesz et al.'s (1993) linear whole-link traffic model, and Mun's (2002) divided linear travel time model. Detailed discussion of the traffic models and their properties can be referred to Astarita (1996), Mun (2001), and Carey (2004a,b).

In addition to the system optimizing flow, it is noted that each additional traveller, who enters the system at a certain time, imposes an additional travel cost on the others who enter the system at that time and thereafter. The additional travel cost imposed to the system by the additional traveller is called "externality". Understanding the nature of this externality in a dynamic setting is important in managing time-dependent networks. Nevertheless, much previous research on the externality was specific to certain kinds of traffic models. For example, Kuwahara (2001) investigated the dynamic externality, while the analysis is confined to deterministic queuing model. Some traffic models adopted in some previous studies were even now considered to be implausible for various reasons. For example, Carey (1987) and Carey and Srinvasan (1993) provided one of the first comprehensive analyses on system optimizing flow and dynamic externality using Kuhn-Tucker conditions. However, they adopted the Merchant and Nemhauser's outflow traffic model, which was later found to violate causality and unable to capture the flow propagation behaviour properly. This paper revisits the topic of dynamic externality in a more general and plausible way. We develop a novel sensitivity analysis of the traffic models, and apply it to derive the externality through an optimal control theoretic formulation.

This paper starts with deriving the formulation and necessary conditions of the dynamic system optimal assignment in the next section. The dynamic system optimal assignment problem is formulated as a state-dependent optimal control problem, which was first introduced by Friesz et al. (2001) to transport research area for analysing and solving dynamic user equilibrium assignment problem. As an extension to Friesz et al., (2001), the paper applied the state-dependent control theoretic formulation to analyse and solve dynamic system optimal assignment problem. To analyse the dynamic externalities, a novel sensitivity analysis of the traffic model with respect to the link inflow is adopted. The sensitivity analysis is developed through flow propagation mechanism and the analysis is not confined to a specific traffic model. Indeed, we apply the sensitivity analysis to deterministic queuing model and we are managed to restore previous analytical results achieved by Ghali and Smith (1993) and Kuwahara (2001). Then, solution algorithms are presented for implementing the sensitivity analysis and solving the dynamic traffic assignments. With the solution algorithms, we provide some numerical calculations and discuss the characteristics of the results. Finally, some concluding remarks including possible and necessary future research work are given.

DYNAMIC SYSTEM OPTIMAL ASSIGNMENT

In the present study, the formulation and analysis for system optimal assignment are restricted to networks in which origin-destination pairs are connected with single travel links. In addition, capacity limitations of different links are considered to be mutually distinct. The system optimal assignment with departure time choice for fixed travel demand is then formulated as the following optimal control problem, which looks for an optimal inflow profile $e_a(s)$, where s represents the time of entry of the traffic to the link. The optimal inflow profile minimizes the total system travel cost within the planning horizon, T, given a fixed amount of total throughput, J_{od}. The optimal control problem is formulated as:

$$\min_{e_a(s)} Z = \sum_{\forall a} \int_{s=0}^{T} C_a(s) e_a(s) ds \tag{1}$$

subject to:

$$\frac{dG_a[\tau_a(s)]}{ds} = e_a(s) \quad ,\forall s, \forall a \tag{2}$$

$$\frac{dx_a(s)}{ds} = e_a(s) - g_a(s) \quad ,\forall s, \forall a \tag{3}$$

$$\frac{dE_a(s)}{ds} = e_a(s) \quad ,\forall s, \forall a \tag{4}$$

$$\sum_{\forall a} E_a(T) = J_{od} \tag{5}$$

$$e_a(s) \geq 0 \quad ,\forall s, \forall a \tag{6}$$

The objective function was adopted by Merchant and Nemhauser (1978a,b), and by several other researchers since then. Proceeding after Heydecker and Addison (2005), this study considers the total travel cost $C_a(s)$ encountered by each traveller on the travel link has three distinct components. The first component is the time spent on travelling along the link, which is determined by the travel time model that is adopted. In addition to the travel time, we add a time-specific cost $f[\tau_a(s)]$ associated with arrival time $\tau_a(s)$ through link a at the destination for traffic which enters the link at time s. Finally, we add a time-specific cost $h(s)$ associated with departure from the origin at time s. Possible choices of these time-specific cost functions are investigated by Heydecker and Addison (2005). Consequently, the total travel cost $C_a(s)$ associated with entry time to link a at time s is determined as a linear combination of these costs as

$$C_a(s) = h(s) + [\tau_a(s) - s] + f[\tau_a(s)]. \tag{7}$$

Following Daganzo (1995) and Mun (2001), we consider the exit time $\tau_a(s)$ to be a linear non-decreasing of link traffic volume $x_a(s)$, hence FIFO queue discipline are structurally guaranteed. As a result, we do not need to add any explicit constraint for this and so by pass the associated computational problems as shown by Carey (1992). Following Friesz et al.'s (1993), we consider that $\tau_a(s)$ takes the functional form as

$$\tau_a(s) = s + \phi_a + \frac{x_a(s)}{Q_a}, \tag{8}$$

The notation ϕ_a and Q_a denote the free flow travel time and the capacity of the travel link respectively. This travel time model is chosen simply for ensuring the plausibility of the corresponding assignment results, it does not affect the generality of the analysis and the calculation in this paper except when the deterministic queuing model or the bottleneck model is adopted. For the bottleneck model (see Vickrey, 1969; Arnott, de Palma and Lindsey, 1998), the state variable, which is the amount of traffic in queue, is not differentiable at the point when the inflow equals to capacity. Arnott, de Palma and Lindsey (1998) derived the dynamic system optimal solution for the bottleneck model by intuitive reasoning that showed that the dynamic system optimal inflow profile is equal to the link capacity through the assignment period. A mathematical analysis and proof of conditions for dynamic system optimal assignment with the bottleneck model can be referred to Chow (2007) which adopted a *bang-bang* control theoretical formulation and analysis.

Equations (2) ensure the proper flow propagation along each link, in which $G_a(s)$ denotes the cumulative link outflow by the exit time $\tau_a(s)$. Equations (3) are the state equations that govern the evolution of link traffic, $x_a(s)$. The variables $e_a(s)$ and $g_a(s)$ represent the flow rates at time s of inflow and outflow respectively. Equations (4) define the cumulative link inflow $E_a(s)$. Equation (5) specifies the amount of total throughput J_{od} generated in the system within the time horizon T. Conditions (6) ensure the positivity of the control variable,

$e_a(s)$. Given a non-negative inflow $e_a(s)$, the corresponding outflow $g_a(s)$ and link traffic volume $x_a(s)$ is guaranteed to be non-negative (Mun, 2001). Hence, we do not add any additional constraints to ensure the non-negativity of $g_a(s)$ and $x_a(s)$.

One technical difficulty is that with the traffic models above, the time lag between changes to the control variable, $e_a(s)$, and the corresponding responses, $g_a(s)$, is state-dependent (Friesz et al, 2001). This state-dependent control theoretic formulation is unorthodox. Its properties and application to dynamic equilibrium were studied by Friesz et al. (2001). As an extension to Friesz et al. (2001), we derive the necessary conditions for the state-dependent system optimization problem and state them in the following the proposition.

Proposition 1: The necessary conditions for the optimization problem (1) – (6) can be derived as

$$e_a(s)\begin{cases} >0 \Rightarrow C_a(s)+\int_0^T\left[\left.\frac{\partial C_a}{\partial u_s}\right|_t e_a(t)\right]dt+\lambda_a(s)-\gamma_a(s)=\mu_a(s)=v_{od} \\ =0 \Rightarrow C_a(s)+\int_0^T\left[\left.\frac{\partial C_a}{\partial u_s}\right|_t e_a(t)\right]dt+\lambda_a(s)-\gamma_a(s)\geq \mu_a(s)=v_{od} \end{cases},\forall a,\forall s\in[0,T] \tag{9}$$

where $\mu_a(s)=v_{od}$ is constant with respect to time and its magnitude is determined by the predefined amount of throughput.

Proof:
See Appendix A. □

The first term on the left-hand-side of (9), $C_a(s)$, is the cost experienced by that additional traveller given the current traffic condition, and the integral in the second term on the left-hand-side of (9), $\Psi_a(s)=\int_0^T\left[\left.\frac{\partial C_a}{\partial u_s}\right|_t e_a(t)\right]dt$, is the additional travel cost, which is regarded as externality, imposed by an additional amount of traffic, u_s, at time s to existing travellers in the system. In this study, we consider parameters u_s of the form for which

$$\frac{de_a(t)}{du_s}=\begin{cases}1 & \text{if } t\in[s,s+ds) \\ 0 & \text{otherwise}\end{cases}, \tag{10}$$

in which ds represents the incremental time step*.

* The inflow $e_a(s)$ is a continuous quantity with respect to time. The value of $\partial e_a(s)$ is zero if we refer to only one particular instant, and hence it will not be effective on the cost $C_a(s)$. To validate the analysis, we adopt the notation ∂u_s to represent the change in inflow within a time interval rather than at a particular time instant.

The terms $\lambda_a(s)$ and $\gamma_a(s)$ denote the costate variables at times s. The costate variables $\lambda_a(s)$ and $\gamma_a(s)$ in the optimal control formulation represents the sensitivity of the value of the objective function with respect to the changes in the state variables $x_a(s)$ and $g_a(s)$ in the corresponding constraints at the associated time s (Dorfman, 1969). Thus, the value of the costate variables in the system optimal control formulation equals to the total change in the value of the total system travel cost Z with respect to slight changes in the state variables (i.e. link traffic volume $x_a(s)$ and outflow profile $g_a(s)$) at time s. The costate variable $\lambda_a(s)$ is given by

$$\lambda_a(s) = \frac{1}{Q_a} \int_{t=s}^{T} (1 + f'[\tau_a(t)]) e_a(t) dt . \tag{11}$$

and $\gamma_a(s) = \lambda_a[\tau_a(s)]$ [†] can then be determined accordingly. The difference between the costate variables $\lambda_a(s)$ and $\gamma_a(s)$ can also be calculated as

$$\lambda_a(s) - \gamma_a(s) = \lambda_a(s) - \lambda_a[\tau_a(s)] = \frac{1}{Q_a} \int_{t=s}^{\tau_a(t)} (1 + f'[\tau_a(t)]) e_a(t) dt \qquad , \forall a, \forall s . \tag{12}$$

The quantity in (12) is interpreted as the *external cost* to be imposed on a traveller who enters the link at time s and leaves at time $\tau_a(s)$ such that the system can be transformed from a user equilibrium state to a decentralized system optimal state. Following this, similar to their static counterparts (see Sheffi, 1985), proposition 1 shows that the dynamic system optimal assignment can be reduced to an equivalent dynamic user equilibrium assignment formulation in which additional components of the cost, $[\Psi_a(s) + \lambda_a(s) - \gamma_a(s)]$, are introduced. In the optimality conditions, the cost components $C_a(s)$ and $\Psi_a(s)$ are generated within the system, while last two cost components (i.e. the costate variables) are external to the system.

SENSITIVITY ANALYSIS OF TRAFFIC MODELS

As shown in the previous section, knowing the externality, $\Psi_a(s)$, is important in managing road networks in dynamic setting. It requires determining the sensitivity of the total travel cost $\left.\frac{\partial C_a}{\partial u_s}\right|_t$ for each departure time s with respect to a change of u_s in the link inflow a particular time s. We further note that if we differentiate both sides of (7) with respect to u_s, we have

---

[†] See Appendix A for derivation.

$$\left.\frac{\partial C_a}{\partial u_s}\right|_t = \left(1 + f'[\tau_a(t)]\right)\left.\frac{\partial \tau_a}{\partial u_s}\right|_t . \tag{13}$$

As a result, calculating the externality $\Psi_a(s)$ requires the sensitivity of traffic models with respect to perturbations in link traffic inflow. Consequently, the section derives a novel expression for the sensitivity of the time of exit with respect to such perturbations in inflow, which is given in the following proposition.

> ***Proposition 2:*** Suppose there is a change of u_s in the link inflow rate at a particular time s, the sensitivity of the time of exit at a time s with respect to this perturbation can be calculated as
>
> $$\left.\frac{\partial \tau_a}{\partial u_s}\right|_t = \frac{1}{Q_a}\left\{ \int_{\kappa=\sigma_a(t)}^{t} \frac{de_a(\kappa)}{du_s} d\kappa + g_a(t) \left.\frac{\partial \tau_a}{\partial u_s}\right|_{\sigma_a(t)} \right\}, \tag{14}$$
>
> in which $\sigma_a(t)$ is the time of entry to the link that leads to exit at time t. Indeed, $\sigma_a(\cdot)$ is defined as the inverse function of $\tau_a(\cdot)$.
>
> *Proof:*
> See Appendix B. ☐

Discussion

The derivative of exit time function at time t with respect to the change of parameter u_s in inflow at time s is then expressed in terms of the dependence of the inflow profile $e_a(\kappa)$ in which κ lies between t and $\sigma_a(t)$, the outflow $g_a(t)$ at time t, and the value of the derivative at time $\sigma_a(t)$.

When the deterministic queuing model is adopted (see for example, Vickrey, 1969; Arnott, de Palma, Lindsey, 1998), the externality will be zero when the travel link is uncongested. When the travel link is congested, the externality will be greater than zero, and traffic will be discharged with an outflow rate $g_a(t)$ that equals to the link capacity Q_a for all times t. Substituting $g_a(t) = Q_a$ for all times t into (14) reduces the equation to

$$\left.\frac{\partial \tau_a}{\partial u_s}\right|_t = \frac{1}{Q_a} \int_{\kappa=0}^{t} \frac{de_a(\kappa)}{du_s} d\kappa . \tag{15}$$

Equation (15) implies that in the particular case of the deterministic queuing model, the derivative $\left.\frac{\partial \tau_a}{\partial u_s}\right|_t$ takes the value of zero for all times t before the time of perturbation s, and $\left.\frac{\partial \tau_a}{\partial u_s}\right|_t$ equals to $\frac{1}{Q_a}$ for all times t after the time of perturbation s. This agrees with the previous analyses on the sensitivity of the deterministic queuing model (see for example, Ghali and Smith, 1993; Kuwahara, 2001). However, the sensitivity analysis developed in this paper allows for other mechanisms of delay and flow propagation, and hence is more general so that it can be applied to other traffic models.

SOLUTION ALGORITHMS

In this section, we first present Algorithm 1 for solving the dynamic user equilibrium assignment. Algorithm 2 is used to evaluate the derivatives given in proposition 2. Finally, Algorithm 3 is used to solve dynamic system optimal assignment.

Algorithm 1: Calculate dynamic user equilibrium assignment

Step 0: Initialisation

0.1 Choose an initial equilibrium cost C^*_{od};

0.2 Set the overall iteration counter $n := 1$;

0.3 Set $e_a(k) := 0$ for all links a, and all times k, $k \in [0, K]$. The notation $e_a(k)$ represents the assigned inflow to link a between times $k\Delta s$ and $(k+1)\Delta s$. The total number of simulated time steps is denoted as $K = T / \Delta s$;

0.4 Set the link index $a := 1$;

0.5 Set the time index $k := 0$;

0.6 Set the inner iteration counter $n^i := 1$.

Step 1: Network loading

Find $\tau_a(k+1)$ by loading the travel link using the inflow $e_a(k)$ at the current iteration. The network loading algorithm "Algorithm D2" described in Nie and Zhang (2005) was adopted for this purpose.

Step 2: Update the inflow

2.1 Calculate

$$C_a(k+1) = h(k+1) + [\tau_a(k+1) - (k+1)] + f[\tau_a(k+1)];$$

2.2 Calculate $\Omega = \dfrac{C_a(k+1) - C_a(k)}{\Delta s}$ and $\Omega' = \dfrac{\partial \Omega}{\partial e_a(k)} = (1 + f'[\tau_a(k+1)])\dfrac{1}{Q_a}$, in which $f'[\tau_a(k)] \approx \dfrac{f[\tau_a(k+1)] - f[\tau_a(k)]}{\tau_a(k+1) - \tau_a(k)}$ using a finite difference approximation.

We note the equilibrium is achieved if and only if $\Omega = 0$ for all positive inflow $e_a(k)$;

2.3 Update the inflow as $e_a(k) := \max[(e_p(k) + \pi d), 0]$ using Newton's method. The second-order searching direction is denoted by $d = -\Omega/\Omega'$ and the step size π, which is interpolated linearly as

$$\pi = \frac{C^*_{od} - C^0_a(k+1)}{C^1_a(k+1) - C^0_a(k+1)},$$

where $C^1_a(k+1)$ and $C^0_a(k+1)$ represent the corresponding values of $C_a(k+1)$ when $e^*_a(k)$ is being updated with π is taken as 1 and 0 respectively. To determine π, two network loadings are required to calculate the values of $C^1_a(k+1)$ and $C^0_a(k+1)$ respectively.

<u>*Step 3: Stopping criteria*</u>

3.1. Check if $\left|C_a(k+1) - C^*_{od}\right| \le \varepsilon$ or n^i is greater than the predefined maximum number of inner iterations, then go to step 3.2; otherwise, set $n^i := n^i + 1$ and go to step 1;

3.2. If $k = K$, then go to step 3.3; otherwise $k := k + 1$ and go to step 1;

3.3. If $a = A$, then go to step 3.4; otherwise $a := a + 1$ and go to step 0.5;

3.4 Define $\xi = \dfrac{\sum_{k\in K}\sum_{a\in A} e_a(k)\left|C_a(k+1) - C^*_{od}\right|}{\sum_{k\in K}\sum_{a\in A} e_a(k) C^*_{od}}$ as a measure of disequilibrium, which is equal to zero at equilibrium. If n is greater than the predefined maximum number of overall iterations or ξ is sufficiently small, i.e. $\xi \le \varepsilon$ where ε is a test value, then go to step 3.5; otherwise set $n := n+1$ and go to step 1.2;

3.5. Check if the total throughput $E_{od} = \sum_{\forall a}\sum_{\forall k} e_a(k)$ from the system is equal to the predefined total demand J_{od} for the *o-d* pair. If yes, then *terminate* the algorithm; otherwise update $C^*_{od} := C^*_{od} + \left[\dfrac{J_{od} - E_{od}}{\dfrac{dE_{od}}{dC^*}}\right]$, and go back to step 0.3. The derivative $\dfrac{dE_{od}}{dC^*_{od}}$ is given by Heydecker (2002) as

$$\frac{dE_{od}}{dC^*_{od}} = \sum_{a\in A}\left(\frac{\left[h'(s^0_a) + f'[\tau_a(s^0_a)]\right] - \left[h'(s^1_a) + f'[\tau_a(s^1_a)]\right]}{\left[h'(s^0_a) + f'[\tau_a(s^0_a)]\right]\left[h'(s^1_a) + f'[\tau_a(s^1_a)]\right]}\right) Q_a, \tag{16}$$

where s_a^0 and s_a^1 respectively denote the first and last times of entry to link a.

Algorithm 2: Calculate externality

Step 1: Initialisation for calculating the derivatives of link exit time

1.1 Set the link index $a := 1$;

1.2 Set the time index $k := 0$, to represent the time when the inflow is perturbed;

1.3 Set the time index $\omega := 0$ to represent time at which we consider the change in exit time due to the perturbation in inflow at time k.

1.4: Calculate the derivatives of link exit time:

$$\text{If } \omega < k \text{, then } \left.\frac{d\tau_a}{du_k}\right|_{\omega} := 0;$$

$$\text{else if } k \le \omega \le \lceil \tau_a(k) \rceil \text{, then } \left.\frac{d\tau_a}{du_k}\right|_{\omega} := \frac{1}{Q_a};$$

$$\text{else } \left.\frac{\partial \tau_a}{\partial u_k}\right|_{\omega} := \frac{g_a(\omega)}{Q_a} \left.\frac{\partial \tau_a}{\partial u_k}\right|_{\sigma_a(\omega)}.$$

1.5 If $\omega = K$, then go to step 1.6; otherwise $\omega := \omega + 1$ and go to step 1.4;

1.6 If $k = K$, then go to step 1.7; otherwise $k := k + 1$ and go to step 1.3;

1.7 If $a = A$, then go to step 2; otherwise $a := a + 1$ and go to step 1.2.

Step 2: Calculate the derivatives of total travel cost function

2.1 Set the link index $a := 1$;

2.2 Set the time index $k := 0$;

2.3 Set the time index $\omega := 0$;

2.4 Calculate $\left.\frac{dC_a}{du_k}\right|_{\omega} = \left(1 + f'[\tau_a(\omega)]\right) \left.\frac{d\tau_a}{du_k}\right|_{\omega}$;

2.5 If $\omega = K$, then go to step 2.6; otherwise $\omega := \omega + 1$ and go to step 2.4;

2.6 If $k = K$, then go to step 2.7; otherwise $k := k + 1$ and go to step 2.3;

2.7 If $a = A$, then go to step 3; otherwise $a := a + 1$ and go to step 2.2.

Step 3: Calculate the externality

3.1 Set the link index $a := 1$;

3.2 Set the time index $k := 0$;

3.3 Initialise $\Psi_a(k) := 0$;

3.4 Set the time index $\omega := 0$;

3.5 Calculate $\Psi_a(k) = \Psi_a(k) + e_a(\omega) \left.\frac{dC_a}{du_k}\right|_{\omega}$;

3.6 If $\omega = K$, then go to step 3.7; otherwise $\omega := \omega + 1$ and go to step 3.5;

3.7 If $k = K$, then go to step 3.8; otherwise $k := k + 1$ and go to step 3.3;
3.8 If $a = A$, then stop; otherwise $a := a + 1$ and go to step 3.2.

Algorithm 3: Calculate dynamic system optimal assignment

Step 0: Initialisation
0.1 Choose an initial equilibrium cost C_{od}^{*};
0.2 Set the overall iteration counter $n := 1$;
0.3 Set $e_a(k) := 0$ for all links a, and all times k, $k \in [0, K]$;
0.4 Set costates $\lambda_a(k) := 0$ for all times $k \in [0, K]$;
0.5 Set the link index $a := 1$;
0.6 Set the time index $k := 0$;
0.7 Set the inner iteration counter $n^i := 1$.

Step 1: Network loading
Find $\tau_a(k+1)$ by loading the travel link using the inflow $e_a(k)$ at the current iteration. The network loading algorithm "Algorithm D2" described in Nie and Zhang (2005) was adopted for this purpose.

Step 2: Calculate externality
Use **Algorithm 2** to calculate the externality $\Psi_a(k)$ associated with each $e_a(k)$.

Step 3: Determine the auxiliary inflow
3.1 Calculate

$$C_a(k+1) = h(k+1) + [\tau_a(k+1) - (k+1)] + f[\tau_a(k+1)] + \Psi_a(k+1) + \lambda_a(k) - \lambda_a[\tau_a(k)];$$

3.2 Calculate $\Omega(k) = \dfrac{C_a(k+1) - C_a(k)}{\Delta s}$ and $\Omega'(k) = \dfrac{\partial \Omega(k)}{\partial e_a(k)} = (1 + f'[\tau_a(k+1)])\dfrac{1}{Q_a}$;

3.3 Calculate the auxiliary inflow $d_a(k) = -\Omega_a(k) / \Omega_a'(k)$;

3.4. If $a = A$, then go to step 3.5; otherwise $a := a + 1$ and go to step 0.7;
3.5. If $k = K$, then go to step 4; otherwise $k := k + 1$ and go to step 0.6.

Step 4: Determine step size for inflow
Search for θ, for all a and k, by golden section method such that $e_a(k) := \max\{[e_a(k) + \theta[d_a(k) - e_a(k)]], 0\}$ gives the minimum total travel cost.

Step 5: Calculate the associated costate variables
5.1 Set the link index $a := 1$;
5.2 Set $\lambda_a(K) = 0$;
5.3 Set the time index $k := K - 1$;

5.4 Compute $\lambda_a(k) = \lambda_a(k+1) + (1 + f'[\tau_a(k)])\frac{e_a(k)}{Q_a}\Delta s$;

5.5 Calculate $\lambda_a[\tau_a(k)]$ from $\lambda_a(k)$ and $\tau_a(k)$ using linear interpolation as
$\lambda_a[\tau_a(k)] \approx \lambda_a|_{\lfloor \tau_a(k) \rfloor} + (\lambda_a|_{\lceil \tau_a(k) \rceil} - \lambda_a|_{\lfloor \tau_a(k) \rfloor})(\tau_a(k) - \lfloor \tau_a(k) \rfloor)$;

5.6. If $k = 0$, then go to step 6.7; otherwise k:= k - 1 and go to step 5.2;

5.7. If $a = A$, then go to step 7; otherwise a:= a + 1 and go to step 5.1.

Step 6: Overall stopping criteria

6.1 Define $\xi = \frac{\sum_{\forall a}\sum_{\forall k} e_a(k)|C_a(k+1) - C^*_{od}|}{\sum_{\forall a}\sum_{\forall k} e_a(k)C^*_{od}}$ as a measure of disequilibrium, which is equal to zero at system optimum. If n is greater than the predefined maximum number of overall iterations or ξ is sufficiently small, i.e. $\xi \le \varepsilon$ where ε is a test value, then go to Step 6.2; otherwise set n:=n+1 and go to step 0.5;

6.2. Check if the total throughput $E_{od} = \sum_{\forall a}\sum_{\forall k} e_a(k)$ from the system is equal to the predefined total demand J_{od} for the *o-d* pair. If yes, then *terminate* the algorithm; otherwise update $C^*_{od} := C^*_{od} + \left[\frac{J_{od} - E_{od}}{\frac{dE_{od}}{dC^*}}\right]$, and go back to step 0.2. The derivative $\frac{dE_{od}}{dC^*_{od}}$ is given by (16).

Discussion

In Algorithm 2, step 1.4, we note that the function $\sigma_a(\omega)$ does not necessarily give an integral value. To implement the sensitivity analysis into computer, a interpolation is needed to determine the value of $\left.\frac{d\tau_a}{du_k}\right|_{\sigma_a(\omega)}$. This study adopts a linear interpolation which approximates the value of $\left.\frac{d\tau_a}{du_k}\right|_{\sigma_a(\omega)}$ as

$$\left.\frac{d\tau_a}{du_k}\right|_{\sigma_a(\omega)} \approx \left.\frac{d\tau_a}{du_k}\right|_{\lfloor \sigma_a(\omega) \rfloor} + \left(\left.\frac{d\tau_a}{du_k}\right|_{\lceil \sigma_a(\omega) \rceil} - \left.\frac{d\tau_a}{du_k}\right|_{\lfloor \sigma_a(\omega) \rfloor}\right)(\sigma_a(\omega) - \lfloor \sigma_a(\omega) \rfloor), \tag{17}$$

where the notation $\lceil \sigma_a(\omega) \rceil$ represent the smallest integer not smaller than $\sigma_a(\omega)$, and $\lfloor \sigma_a(\omega) \rfloor$ is the greatest integer not larger than $\sigma_a(\omega)$.

In addition, in Algorithms 1 and 3, the inflow profile is solved by a forward dynamic programme in the order of departure time interval. It is possible due to the causal property of the travel time models that ensures the travel cost experienced by the traffic that departs from an origin at time s is independent of the traffic departing from that origin after time s. The study period in continuous time, T, is discretized into K intervals of length Δs. Following this, the instantaneous flow in continuous time formulation is represented by the amount of traffic over a discrete time interval k, which represents the interval $[k\Delta s,(k+1)\Delta s)$. Within each departure time interval k, the equilibrium inflow is calculated by using Newton method, which converges with an order of convergence at least two (Luenberger, 1989, p202). Nevertheless, when we calculate the costate variables in Algorithm 3, we have to calculate them backward in time due to the transversality condition of the costates is given at the end of the study horizon. The auxiliary flows are calculated based on the traffic conditions at the last iteration, while the costate variables are calculated based on the traffic conditions at the current iteration. As a result, they are not consistent, and we adopt a step size search (Step 4) in Algorithm 3 as a kind of heuristics to accommodate this.

Finally, a crucial point in developing numerical algorithms for solving dynamic traffic assignments is to discretize the continuous time formulation (Heydecker and Verlander. 1999). The main difference between the continue time formulation and the discrete time solution algorithm is that the continuous time formulation treats quantities of flow rate (e.g. inflow, $e_a(s)$) at a time instant s, whilst the discrete time solution algorithm has to typically consider the flow rate $e_a(k)$ over a time interval k, which represents the time interval $[k\Delta s,(k+1)\Delta s)$. The discretization process brings in difficulties in deciding the time instant at when the associated costs should be considered. Heydecker and Verlander (1999) suggested that a predictive manner should be adopted for plausible assignment results. Following Heydecker and Verlander (1999), in a discrete time algorithm the travel cost, which are calculated forward in time, associated with the flow should be considered at the end of the interval (i.e. at the time $(k+1)\Delta s$), instead of at the start of the interval (i.e. at time $k\Delta s$). However, when we work with the costate variables (Step 3.1 in Algorithm 3), we should look at the value of costate at the start of the time interval $k\Delta s$ instead of the end of the interval. This is because the costate variables are calculated backward in time. The consequence of considering the costs at an inappropriate time was illustrated by Heydecker and Verlander (1999).

NUMERICAL CALCULATIONS

We first consider a single link, which has a free flow time 3 mins and a capacity 20 vehs/min, connecting a single origin-destination pair. The size of discretized time interval Δs is taken as 1 min. We first show the numerical solutions of the whole link traffic model. A parabolic profile, which is specified as (17), of inflow is loaded into the travel link.

$$e_a(s) = \begin{cases} \frac{1}{8}(40 - s)s & \text{if } 0 \le s \le 40 \\ 0 & \text{otherwise} \end{cases} . \tag{18}$$

This profile has a peak inflow rate of 50 vehs/min, which equals to 2.5 times of the link capacity. The function of link traffic model is to determine the corresponding profile of link outflow and the link travel time, given the profile of link inflow. The traffic model adopted in this example is Friesz et al's (1993) linear whole-link traffic model. Figure 1 depicts the profile of link outflow. We note that the outflow will approach to, but not exceed, the link capacity for a high inflow rate. This is an important feature for a plausible traffic model. Figure 2 plots the associated link travel times.

Then, we investigate the accuracy of the sensitivity analysis in proposition 2. We consider the parabolic inflow profile is perturbed at time 1, and the associated variations in travel time are plotted in Figure 3. The "analytical" variations are calculated according to Equation (14). The "numerical" variations are determined by using direct numerical finite difference method, and they are plotted in the same figure for comparison. To calculate the finite difference, one extra unit of inflow is added at time 1, while the inflow profile remains unchanged at other times. The "numerical" variations in travel times are then calculated by subtracting the link travel time loaded by the original inflow profile from that loaded by the perturbed inflow profile. The result shows that the analytical variations given by Equation (14) can represent the true numerical variations in travel time reasonably well. Both numerical and analytical variations drop to zero at time 83 when all traffic is cleared from the link.

Next, we calculate the dynamic traffic assignments. We consider a network with a single origin-destination pair connected with two parallel travel routes consisting of one single link as shown in Figure 4. Link 1 has free flow time 3 mins and capacity 20 vehs/min, and link 2 has free flow time 4 mins and capacity 30 vehs/min. Furthermore, the origin-specific cost is specified to be a monotone linear function of time with a slope -0.4. The destination cost function is piecewise linear, with no penalty for arrivals before the preferred arrival time $t^* = 50$, and increases with a rate 2 afterwards. The length of the planning horizon $[0, T]$, where T=100, is set such that all traffic can be cleared by time T. The total amount of traffic J_{od} is taken as 800 vehs. Figure 5 shows the corresponding profiles of link inflows and the total travel cost at equilibrium. The traffic is assigned to the route 1 during times 18 and 49, and to route 2 during times 21 and 49. The link flow volumes using route 1 and route 2 are 380.25 (vehs) and 419.75 (vehs) respectively. The measure of disequilibrium ξ achieved is below 10^{-17}. At dynamic user equilibrium, the total system travel cost is 12,465.2 veh-hr.

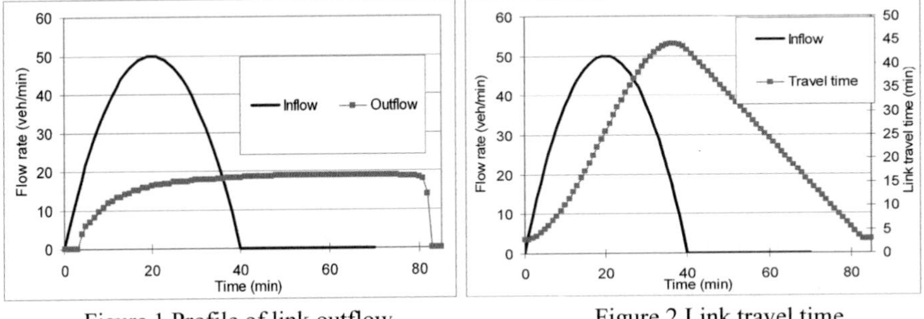

Figure 1 Profile of link outflow

Figure 2 Link travel time

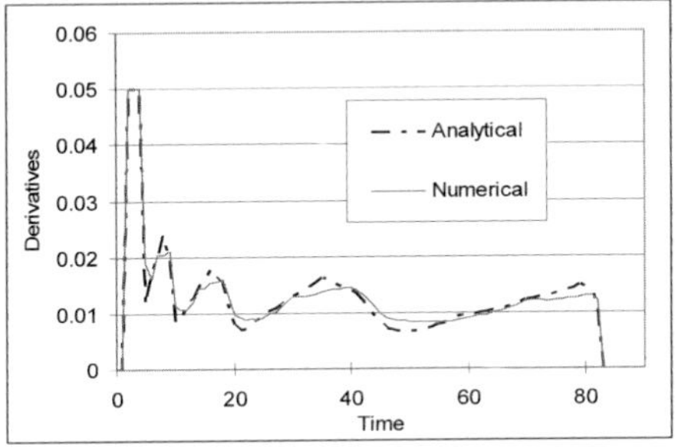

Figure 3 Sensitivity of travel time with respect to a perturbation in inflow

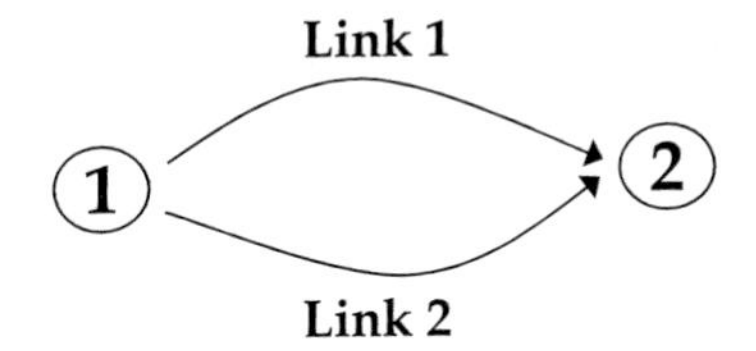

Figure 4 Example network

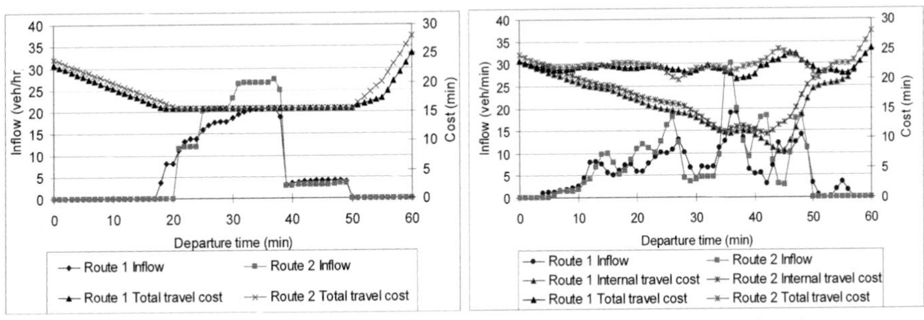

Figure 5 Equilibrium assignment

Figure 6 System optimal assignment

Figure 6 shows the assignment of the dynamic system optimum. With the same total throughput J_{od}, the period of assignment to link 1 expands from times [18, 49] to times [4, 56], while that to link 2 expands from times [21, 49] to times [6, 50]. In general, the profiles of link inflows are more spread at system optimum in order to reduce the intensity of congestion on the links, whilst maintaining the same volume of travel. The associated total system travel cost at system optimum is decreased from 12,465.2 veh-hr in user equilibrium to 11,447.3 veh-hr in system optimum. However, due to the addition of externality and the costate variables, the marginal social cost at which travel takes place increases from 15.58 min at user equilibrium to 21.78 min at system optimum, although the system optimizing flow causes the decrease in total system cost.

To illustrate the cause of the decrease in system travel cost, Figure 7 shows the profiles of the link traffic volumes at user equilibrium and system optimum respectively. Interestingly, yet importantly, the results show that, with Friesz et al's (1993) travel time model, the system optimal assignment has to allow queuing even at system optimum. The system optimal assignment can only manage and minimize congestion. This implies that the previous analyses on dynamic system optimum using the deterministic queuing model in which congestion can be eliminated (see for example, Vickrey, 1969; Arnott, de Palma, Lindsey, 1998) do not apply in general. Furthermore, the dynamic system optimizing tolls (i.e. the sum of externality and external costs: $[\Psi_a(s)+\lambda_a(s)-\lambda_a[\tau_a(s)]]$) which are to be imposed to the travellers to decentralize the system optimal flows are calculated and shown in Figure 8. The results show that the dynamic tolls generally increase with time for travellers who arrive at the destination before the preferred arrival time, and decrease afterwards.

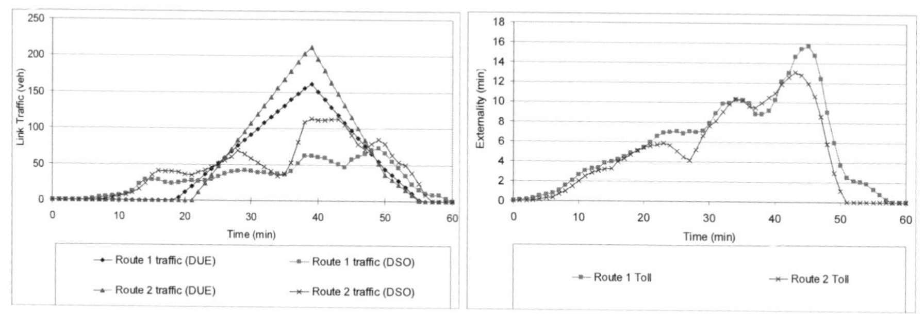

Figure 7 Link traffic volumes

Figure 8 System optimizing tolls

Finally, the progress of the system optimization algorithm is shown by Figure 9 and Figure 10, which illustrate the reduction of the total system cost and the measure of disequilibrium over iteration respectively. The results agree with the analysis that the total system cost reduces as the measure of disequilibrium drops. However, the measure of disequilibrium can only reach 0.04 in this calculation. It is because indeed solving the dynamic system optimal assignment is difficult, since the solution procedure involves solving two dynamic programmes simultaneously and consistently: solving the network loading forward in time for the state variables and solving the costate equations backward in time for the costate variables. We are exploring better strategies for better quality solution.

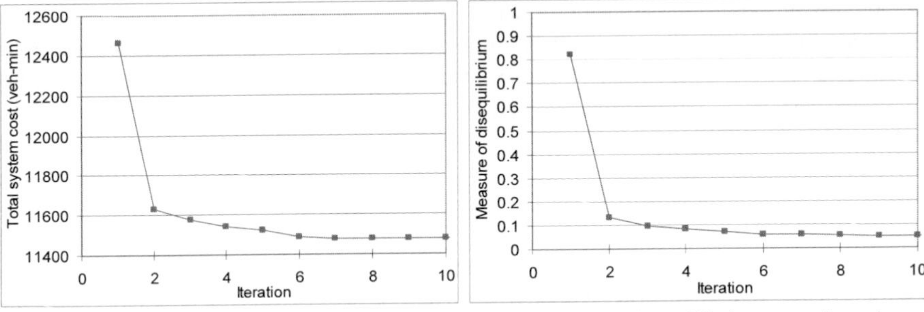

Figure 9 Total system cost over iteration

Figure 10 Disequilibrium over iteration

CONCLUDING REMARKS

The main contribution of this paper is the necessary conditions for dynamic system optimizing flow and the analysis of dynamic externalities. This study gives us a deeper understanding of the nature of dynamic system optimal assignment through a general and plausible framework. We developed a novel sensitivity analysis of traffic models with respect to perturbations in link inflow. The analytical results developed can be used to recover some earlier more specific ones of Vickrey (1969) where the bottleneck model was used. However, the present results are substantially more general. The knowledge of system optimizing flows and dynamic externalities provide important insight into the management of dynamics of peak periods and travellers' behaviour.

We also presented solution algorithms for implementing the sensitivity analysis and solving the dynamic traffic assignments. We also applied the algorithms to numerical calculations. The characteristics of the results were discussed. With Friesz et al's (1993) linear traffic model, the system optimal assignment has to allow queuing, and the externality that each traveller imposes on the others is not zero even at system optimum. We can only manage and minimize queuing and externality of each traveller imposes on the others. This implies that the previous analyses on dynamic system optimum using the deterministic queuing model do not apply generally. Further study is required to improve the performance of the solution algorithm for the system optimal assignment.

In the present study, the formulation and analysis presented are restricted to networks in which capacity limitations of different routes are mutually distinct. We are currently exploring ways in which this analysis can be extended to consider shared bottlenecks in general networks. In case of networks that have multiple origin-destination pairs with overlapping routes, traffic entering the network during the journey time of a traveller from other origins downstream can influence the travel time of travellers from its upstream. As a result, some special computational technique, for example Guass-Seidel relaxation (see Sheffi, 1985), seems likely to be required. The basic idea of such relaxation scheme is to decompose the assignment problem for networks with overlapping routes connecting multiple origin-destination pairs into several sub-problems. In each sub-problem, we calculate the assignments for one origin-destination pair, and temporarily neglect the influences from the flows between other origin-destination pairs. When equilibrium or system optimum is reached for the current origin-destination pair, we proceed with calculations for another pair. The procedure is repeated until equilibrium or system optimum is reached in the whole network. The relaxation scheme is not guaranteed to converge, but if it does, the solution will be the final assignment pattern (see Sheffi, 1985, p217). In case of routes with multiple links, difficulties are introduced when we have to calculate the derivatives of *route* exit time (see for example Balijepalli, 2005; Balijepalli and Watling, 2005). Following Proposition 2, changing the inflow to a link on the route during one time interval will induce perturbations in the link travel time, the link outflow, and hence the inflow to subsequent link(s) in several succeeding time intervals. Hence, the dimension of time intervals to be considered in calculating the derivatives will expand exponentially along the route. We are currently investigating the strategies to cope with this *curse of dimensionality*. Efficient computing methods for system optimal assignments in general networks are still under investigation, however, the work reported in the present paper provide a solid and necessary foundation for future research.

ACKNOWLEDGEMENTS

The helpful and constructive comments from all anonymous referees are gratefully acknowledged. I would like to thank Professor Benjamin Heydecker and Dr. JD Addison for their continuing encouragement and supervision of this study. An earlier version of this paper was presented at the 1st international symposium on dynamic traffic assignment, Leeds, 2006. I would like to thank Professor Terry Friesz for a stimulating exchange of ideas. Comments from other audiences are also gratefully appreciated.

REFERENCES

Arnott, R., de Palma, A. and Lindsey, R. (1998) Recent developments in the bottleneck model. K.J. Button and E.T. Verhoef, eds. *Road Pricing, Traffic Congestion and the Environment: Issues of Efficiency and Social Feasibility.* Edward Elgar, Aldershot, Hampshire, U.K., 79-110.

Astarita, A (1996) A continuous time link model for dynamic network loading based on travel time functions. J-B Lesort, ed. *Transportation and Traffic Theory*. Pergamon, Oxford, 79-102.

Balijepalli, NC (2005) Doubly Dynamic Equilibrium Distribution Approximation Model for Dynamic Traffic Assignment. *Proceedings of the 37th Annual Conference of Universities Transport Study Group*, January 5-7. Bristol, UK.

Balijepalli, NC, Watling, DP (2005) Doubly Dynamic Equilibrium Distribution Approximation Model for Dynamic Traffic Assignment. *Proceedings of the 16th International Symposium on Transportation and Traffic Theory*, Maryland, USA, 19-21 July, 741-760, Pergamon, Oxford.

Carey, M. (1987) Optimal time-varying flows on congested networks. *Operations research*, **35**, 58-69.

Carey, M (1992) Non-convexity of the dynamic traffic assignment problem. *Transportation Research,* **26B**(2), 127-133.

Carey M. and Srinivasan, A. (1993) Externalities, average and marginal costs, and tolls on congested networks with time-varying flows. *Operations Research* **41**, 217-231.

Carey, M. (2004a) Link travel times I: desirable properties. *Network and Spatial Economics*, **4**, 257-268.

Carey, M. (2004b) Link travel times II: properties derived from traffic-flow models. *Network and Spatial Economics*, **4**, 379-402.

Chow, AHF (2007) System optimizing flows in time-dependent road network. *PhD thesis*, University of London, *in progress.*

Daganzo, CF (1995) Properties of link travel time functions under dynamic loads. *Transportation Research*, **29B**(2), 95-98.

Dorfman, R (1969) An economic interpretation of optimal control theory, *The American Economics Review*, **59**, 817-831.

Friesz, TL, Luque, F.J., Tobin, RL and Wie, BW (1989) Dynamic network assignment considered as a continuous time optimal control problem. *Operations Research*, **37**, 893-901.

Friesz, TL, Bernstein, D, Smith, TE, Tobin, RL and Wie BW (1993) A variational inequality formulation of the dynamic network user equilibrium problem. *Operations Research* **41**(1), 179-191.

Friesz, T.L. and Bernstein, D. (2000) Analytical dynamic traffic assignment models. D.A. Hensher amd K.J. Button, eds. *Handbook of transport modelling*, 181-195.

Friesz, TL, Bernstein, D, Suo, Z and Tobin, RL (2001) Dynamic network user equilibrium with state-dependent time lags. *Network and Spatial Economics*, **1**(3/4), 319-347.

Ghali, MO and Smith, MJ (1993) Traffic assignment, traffic control and road pricing. CF Daganzo, ed. *Transportation and Traffic Theory*. Elsevier, Amsterdam, 147-170.

Heydecker, BG (2002) Dynamic equilibrium network design. MAP Taylor, ed. *Transportation and Traffic Theory*. Pergamon, Oxford, 349-370.

Heydecker, BG and Addison, JD (1996) An exact solution of dynamic equilibrium. JB Lesort, ed. *Transportation and Traffic Theory*. Pergamon, Oxford, 359-384.

Heydecker, BG and Addison, JD (1998) Analysis of traffic models for dynamic equilibrium traffic assignment. M.G.H. Bell, ed. *Transportation Networks: Recent methodological Advance*. Pergamon, Oxford, 35-49.

Heydecker, BG and Addison, JD (2005) Analysis of dynamic traffic equilibrium with departure time choice. *Transportation Science,* **39**(1), 39–57.

Heydecker, BG and Verlander, NQ (1999) Calculation of dynamic traffic equilibrium assignments. *Proceedings of the European Transport Conference, Seminar F, P343,* London, 79-91.

Kuwahara, M (2001) A theoretical analysis on dynamic marginal cost pricing. *Proceedings of the Sixth Conference of Hong Kong Society for Transportation Studies*, 28-39.

Luenberger, DG (1984) *Linear and nonlinear programming.* Addison-Wesley, Menlo Park, California.

Merchant, DK and Nemhauser, GL (1978a) A model and an algorithm for the dynamic traffic assignment problem. *Transportation Science,* **12**(3), 183-199.

Merchant, DK and Nemhauser, GL (1978b) Optimality conditions for a dynamic traffic assignment model. *Transportation Science,* **12**(3), 200-207.

Mun, JS (2001) A divided linear travel time model for dynamic traffic assignment. *9th World Conference on Transport Research*, Seoul, S.Korea.

Nie, X and Zhang HM (2005) Delay-function-based link models: their properties and computational issues. *Transportation Research,* **39B**(8), 729-752.

Sheffi, Y (1985) Urban transportation networks: equilibrium analysis with mathematical programming methods. Englewood Cliffs, NJ: Prentice-Hall.

Smith, MJ and Ghali, MO (1990) Dynamic traffic assignment and dynamic traffic control. M Koshi, ed. *Transportation and Traffic Theory*. Elsevier, New York, 223-263.

Szeto, WY and Lo, H. (2005) Non-equilibrium dynamic traffic assignment. *Proceedings of the 16th International Symposium on Transportation and Traffic Theory*, Maryland, USA, 19-21 July, 427-445, Pergamon, Oxford.

Tobin, RL (1993) Notes on flow propagation constraints. *Working paper*, **93-10**, Network Analysis Laboratory, George Mason University, USA.

Vickrey, WS (1969) Congestion theory and transport investment. *American Economics Review*, **72**(3), 467-479.

Wardrop, JG (1952) Some theoretical aspects of road traffic research. *Proceedings of the Institute of Civil Engineers Part II*, 325-78.

Yang, H and Huang, HJ (1997) Analysis of the time-varying pricing of a bottleneck with elastic demand using optimal control theory. *Transportation Research* **31B**, 425-440.

APPENDIX A: DERIVATION OF THE OPTIMALITY CONDITIONS OF DYNAMIC SYSTEM OPTIMAL ASSIGNMENT

This appendix derives the necessary conditions for the dynamic system optimization problem (1) – (6) by using a calculus of variations technique.

The objective function Z is first augmented with the constraints to form the following Lagrangian:

$$\min_{e_a^*(s)} Z^* = \sum_{\forall a} \int_0^T \left\{ \begin{aligned} & C_a(s)e_a(s) + \lambda_a(s)\left\{[e_a(s) - g_a(s)] - \frac{dx_a(s)}{ds}\right\} + \mu_a(s)\left\{\frac{dE_a(s)}{ds} - e_a(s)\right\} \\ & + \gamma_a(s)\left\{\frac{dG_a[\tau_a(s)]}{ds} - e_a(s)\right\} - \rho_a(s)e_a(s) \end{aligned} \right\} ds, \\ + \nu_{od}\left(\sum_{\forall a} E_a(T) - J_{od}\right) \tag{A-1}$$

where $\lambda_a(s)$ and $\gamma_a(s)$ are the respective costate variables for the flow conservation and flow propagation constraints; and $\mu_a(s)$ and $\rho_a(s)$ are the associated multipliers on the cumulative and the non-negativity constraints of the control variables respectively. Finally, ν_{od} is the multiplier associated with the total throughput. Using integration by parts, the terms involving $\frac{dx_a(s)}{ds}$ and $\frac{dE_a(s)}{ds}$ in the integrand over time can be rewritten as

$$\int_0^T \lambda_a(s)\frac{dx_a(s)}{ds}ds = \int_0^T \lambda_a(s)dx_a(s) = \lambda_a(T)x_a(T) - \lambda_a(0)x_a(0) - \int_0^T x_a(s)\frac{d\lambda_a(s)}{ds}ds, \tag{A-2}$$

and

$$\int_0^T \mu_a(s)\frac{dE_a(s)}{ds}ds = \int_0^T \mu_a(s)dE_a(s) = \mu_a(T)E_a(T) - \mu_a(0)E_a(0) - \int_0^T E_a(s)\frac{d\mu_a(s)}{ds}ds, \tag{A-3}$$

in which the initial values $x_a(0)$ and $E_a(0)$ are considered to be zero. Consequently, the Lagrangian Z^* becomes

$$Z^* = \sum_{\forall a}\left[-\lambda_a(T)x_a(T) + \mu_a(T)E_a(T)\right] + \nu_{od}\left(\sum_{\forall a} E_a(T) - J_{od}\right) \\ + \sum_{\forall a}\int_0^T \left[H_a(s) + \frac{d\lambda_a(s)}{ds}x_a(s) - \frac{d\mu_a(s)}{ds}E_a(s)\right]ds, \tag{A-4}$$

in which we define the Hamiltonian function:

$$H_a(s) = C_a(s)e_a(s) + \lambda_a(s)[e_a(s) - g_a(s)] - \mu_a(s)e_a(s) \\ + \gamma_a(s)\left\{g_a[\tau_a(s)]\frac{d\tau_a(s)}{ds} - e_a(s)\right\} - \rho_a(s)e_a(s). \tag{A-5}$$

Then, the variation δZ^* of Z^* with respect to its variables is derived as

$$\delta Z^{*}=\sum_{\forall a}\left[-\lambda_a(T)\delta x_a(T)+\mu_a(T)\delta E_a(T)\right]+\sum_{\forall a}\int_0^T\left[\frac{\partial H_a(s)}{\partial e_a(s)}\delta e_a(s)\right]ds$$
$$+\sum_{\forall a}\int_0^T\left[\frac{\partial H_a(s)}{\partial g_a(s)}\delta g_a(s)+\frac{\partial H_a(s)}{\partial g_a[\tau_a(s)]}\delta g_a[\tau_a(s)]+\left(\frac{\partial H_a(s)}{\partial x_a(s)}+\frac{d\lambda_a(s)}{ds}\right)\delta x_a(s)\right]ds, \quad \text{(A-6)}$$
$$-\sum_{\forall a}\int_0^T\frac{d\mu_a(s)}{ds}\delta E_a(s)ds+v_{od}\sum_{\forall a}\delta E_a(T)$$

in which $\frac{\partial H_a}{\partial e_a(s)}$, $\frac{\partial H_a}{\partial g_a(s)}$, $\frac{\partial H_a}{\partial g_a[\tau_a(s)]}$, and $\frac{\partial H_a}{\partial x_a(s)}$ represent the derivatives of Hamiltonian function with respect to its corresponding variables.

Applying the change of variables, $t=\tau_a(s)\Rightarrow dt=\dot{\tau}_a(s)ds$, the bounds of the integral are changed accordingly: $s=0\Rightarrow t=\tau_a(0)$ and $s=T\Rightarrow t=\tau_a(T)$.

The variation with respect to $g_a[\tau_a(s)]$ can now be transformed to

$$\int_0^T\left[\frac{\partial H_a}{\partial g_a[\tau_a(s)]}\delta g_a[\tau_a(s)]\right]ds=\int_{\tau_a(0)}^{\tau_a(T)}\left[\frac{\partial H_a}{\partial g_a[\tau_a(s)]}\delta g_a(t)\right]\frac{dt}{\dot{\tau}_a(s)}$$
$$=\int_{\tau_a(0)}^{\tau_a(T)}\frac{\partial H_a}{\partial g_a[\tau_a[\sigma_a(s)]]}\frac{1}{\dot{\tau}_a[\sigma_a(s)]}\delta g_a(s)ds, \quad \text{(A-7)}$$

in which $\sigma_a(s)$ is the time of entry to the link that leads to exit at time s.

The time horizon T is taken such that it is long enough for all traffic to be cleared by the end of it, the integral on the right hand side in Equation (A-7) only need to be calculated up to time T as

$$\int_0^T\left[\frac{\partial H_a}{\partial g_a[\tau_a(s)]}\delta g_a[\tau_a(s)]\right]ds=\int_{\tau_a(0)}^{T}\frac{\partial H_a[\sigma_a(s)]}{\partial g_a[\tau_a[\sigma_a(s)]]}\frac{1}{\dot{\tau}_a[\sigma_a(s)]}\delta g_a(s)ds. \quad \text{(A-8)}$$

Finally, δZ^{*} becomes

$$\delta Z^* = \sum_{\forall a}\left[-\lambda_a(T)\delta x_a(T) + \mu_a(T)\delta E_a(T)\right] + \sum_{\forall a}\int_0^T\left[\frac{\partial H_a(s)}{\partial e_a(s)}\delta e_a(s)\right]ds + \sum_{\forall a}\int_0^{\tau_a(0)}\left[\frac{\partial H_a}{\partial g_a(s)}\delta g_a(s)\right]ds$$
$$+\sum_{\forall a}\int_{\tau_a(0)}^{T}\left[\frac{\partial H_a(s)}{\partial g_a(s)} + \frac{\partial H_a[\sigma_a(s)]}{\partial g_a[\tau_a[\sigma_a(s)]]}\frac{1}{\dot{\tau}_a[\sigma_a(s)]}\right]\delta g_a(s)ds$$
$$+\sum_{\forall a}\int_0^T\left[\left(\frac{\partial H_a(s)}{\partial x_a(s)} + \frac{d\lambda_a(s)}{ds}\right)\delta x_a(s)\right]ds - \sum_{\forall a}\int_0^T\frac{d\mu_a(s)}{ds}\delta E_a(s)ds + v_{od}\sum_{\forall a}\delta E_a(T) \quad \text{(A-9)}$$

The optimality is achieved when Z^* is stationary (i.e. $\delta Z^* = 0$) with respect to all variations. This can only be guaranteed to happen when the following stationarity conditions are satisfied simultaneously:

$$\frac{\partial H_a}{\partial e_a(s)} = C_a(s) + \int_0^T \left.\frac{\partial C_a}{\partial u_s}\right|_t e_a(t)dt + \lambda_a(s) - \gamma_a(s) - \mu_a(s) - \rho_a(s) = 0 \quad ,\forall a, \forall s; \quad \text{(A-10)}$$

$$\frac{\partial H_a}{\partial g_a(s)} + \frac{\partial H_a}{\partial g_a[\tau_a[\sigma_a(s)]]}\frac{1}{\dot{\tau}_a[\sigma_a(s)]} = -\lambda_a(s) + \gamma_a[\sigma_a(s)] = 0 \quad ,\forall a, \forall s \in [\tau_a(0), T]; \quad \text{(A-11)}$$

$$\frac{\partial H_a}{\partial x_a(s)} + \frac{d\lambda_a(s)}{ds} = \left(1 + f'[\tau_a(s)]\right)\frac{e_a(s)}{Q_a} + \frac{d\lambda_a(s)}{ds} = 0 \quad ,\forall a, \forall s; \quad \text{(A-12)}$$

$$\lambda_a(T) = 0 \quad ,\forall a; \quad \text{(A-13)}$$

$$\frac{\partial H_a(s)}{\partial E_a(s)} = -\frac{d\mu_a(s)}{ds} = 0 \quad ,\forall a, \forall s; \quad \text{(A-14)}$$

$$\mu_a(T) - v_{od} = 0 \quad ,\forall a. \quad \text{(A-15)}$$

We also have the following Karush-Kuhn-Tucker (KKT) conditions hold for the non-negativity constraints on the control on all links and all times:

$$e_a(s) \geq 0; \quad \text{(A-16)}$$
$$e_a(s)\rho_a(s) = 0; \quad \text{(A-17)}$$
$$\rho_a(s) \geq 0. \quad \text{(A-18)}$$

With equations (A-14) and (A-15), we can deduce that $\mu_a(s)$ will remain constant at $\mu_a(T) = v_{od}$ for all s within T since $\frac{d\mu_a(s)}{ds} = 0$.

Furthermore, equation (A-11) can be written equivalently as

$$\lambda_a[\tau_a(s)] = \gamma_a(s) \quad ,\forall a, \forall s \in [\tau_a(0), T], \tag{A-19}$$

and the evolution of the costate variable $\lambda_a(s)$ is governed by (A-12) and (A-13) for all s. Finally, combining (A-10) and the KKT conditions (A-16), (A-17), and (A-18), we then get the following conditions for system optimum for a single travel link represented by the whole-link traffic model:

$$e_a(s)\begin{cases} \geq 0 \Rightarrow C_a(s) + \int_0^T \left.\frac{\partial C_a}{\partial u_s}\right|_t e_a(t)dt + \lambda_a(s) - \lambda_a[\tau_a(s)] = \mu_a(s) = \nu_{od} \\ = 0 \Rightarrow C_a(s) + \int_0^T \left.\frac{\partial C_a}{\partial u_s}\right|_t e_a(t)dt + \lambda_a(s) - \lambda_a[\tau_a(s)] \geq \mu_a(s) = \nu_{od} \end{cases} ,\forall a, \forall s\,. \tag{A-20}$$

Using the costate equation (A-12) and the transversality condition (A-13), the costate variable $\lambda_a(s)$ for any time s can be calculated backward in time as

$$\begin{aligned} \lambda_a(s) &= -\frac{1}{Q_a}\int_{t=T}^{s} (1 + f'[\tau_a(t)])e_a(t)dt \\ &= \frac{1}{Q_a}\int_{t=s}^{T} (1 + f'[\tau_a(t)])e_a(t)dt \qquad ,\forall a, \forall s \end{aligned} \tag{A-21}$$

APPENDIX B: DERIVATION OF THE SENSITIVITY OF TRAVEL TIME WITH RESPECT TO LINK INFLOW

This appendix derives equation (14).

The traffic volume on the travel link, $x_a(t)$, at time t can be expressed as

$$x_a(t) = E_a(t) - G_a(t) = E_a(t) - E_a[\sigma_a(t)] = \int_{\kappa=\sigma_a(t)}^{t} e_a(\kappa)d\kappa\,, \tag{B-1}$$

The expression for the time of exit for the entry time t then becomes

$$\tau_a(t) = t + \phi_a + \frac{1}{Q_a}\int_{\kappa=\sigma_a(t)}^{t} e_a(\kappa)d\kappa\,. \tag{B-2}$$

Considered that there is a small change u_s induced in the profile of inflow at a particular time s, the associated change in the value of the function of the time of exit at a time t can be deduced as

$$\begin{aligned}\left.\frac{d\tau_a}{du_s}\right|_t &= \frac{d}{du_s}\left(t+\phi_a+\frac{1}{Q_a}\int_{\kappa=\sigma_a(t)}^{t} e_a(\kappa)d\kappa\right)\\ &= \frac{1}{Q_a}\frac{d}{du_s}\left(\int_{\kappa=\sigma_a(t)}^{t} e_a(\kappa)d\kappa\right)\\ &= \frac{1}{Q_a}\left\{\int_{\kappa=\sigma_a(t)}^{t}\frac{de_a(\kappa)}{du_s}d\kappa-\frac{d\sigma_a(t)}{du_s}e_a[\sigma_a(t)]\right\}\end{aligned} \tag{B-3}$$

The first term is the bracket can be calculated directly.

To calculate the second term in (B-3), we first apply the definitional relationship,

$$\tau_a[\sigma_a(t)] = t. \tag{B-4}$$

Differentiating the left hand side with respect to u_s and by using chain rule, the left-hand-side of (B-4) can be written as

$$\left.\frac{d\tau_a}{du_s}\right|_{\sigma_a(t)} = \left.\frac{\partial\tau_a}{\partial u_s}\right|_{\sigma_a(t)} + \frac{\partial\tau_a[\sigma_a(t)]}{\partial\sigma_a(t)}\frac{\partial\sigma_a(t)}{\partial u_s}. \tag{B-5}$$

Likewise, differentiating the right hand side with respect to u_s, it gives

$$\left.\frac{d\tau_a}{du_s}\right|_{\sigma_a(t)} = \frac{dt}{du_s} = 0, \tag{B-6}$$

which is because the time s is not affected by the change of u_s in the inflow.

Hence, combining (B-5) and (B-6), it can be deduced that

$$\left.\frac{\partial\tau_a}{\partial u_s}\right|_{\sigma_a(t)} + \frac{\partial\tau_a[\sigma_a(t)]}{\partial\sigma_a(t)}\frac{\partial\sigma_a(t)}{\partial u_s} = 0. \tag{B-7}$$

Furthermore, since $\sigma_a(\cdot)$ is an inverse function of $\tau_a(\cdot)$, it follows that

$$\frac{\partial \tau_a[\sigma_a(t)]}{\partial \sigma_a(t)} = \frac{1}{\dfrac{\partial \sigma_a(t)}{\partial t}}. \tag{B-8}$$

Therefore, combining (B-7) and (B-8), and after rearranging terms, it gives

$$\frac{\partial \sigma_a(t)}{\partial u_s} = -\left(\frac{\partial \tau_a[\sigma_a(t)]}{\partial \sigma_a(t)}\right)^{-1} \left.\frac{\partial \tau_a}{\partial u_s}\right|_{\sigma_a(t)} = -\frac{d\sigma_a(t)}{dt}\left.\frac{\partial \tau_a}{\partial u_s}\right|_{\sigma_a(t)}. \tag{B-9}$$

Finally, substituting (B-9) into (B-3) gives

$$\begin{aligned}
\left.\frac{\partial \tau_a}{\partial u_s}\right|_s &= \frac{1}{Q_a}\left\{\int_{\kappa=\sigma_a(t)}^{t} \frac{de_a(\kappa)}{du_s}d\kappa - \frac{d\sigma_a(t)}{du_s}e_a[\sigma_a(t)]\right\} \\
&= \frac{1}{Q_a}\left\{\int_{\kappa=\sigma_a(t)}^{t} \frac{de_a(\kappa)}{du_s}d\kappa + e_a[\sigma_a(t)]\frac{d\sigma_a(t)}{dt}\left.\frac{\partial \tau_a}{\partial u_s}\right|_{\sigma_a(t)}\right\}. \\
&= \frac{1}{Q_a}\left\{\int_{\kappa=\sigma_a(t)}^{t} \frac{de_a(\kappa)}{du_s}d\kappa + g_a(t)\left.\frac{\partial \tau_a}{\partial u_s}\right|_{\sigma_a(t)}\right\}
\end{aligned} \tag{B-10}$$

Transportation and Traffic Theory 2007
Edited by R.E. Allsop, M.G.H. Bell and B.G. Heydecker

15

ON PATH MARGINAL COST ANALYSIS AND ITS RELATION TO DYNAMIC SYSTEM-OPTIMAL TRAFFIC ASSIGNMENT

Wei Shen, Yu Nie, and H. Michael Zhang, Department of Civil and Environmental Engineering, University of California, Davis, USA

SUMMARY

This paper studies the problem of evaluating path marginal cost in system-optimal dynamic traffic assignment (SO-DTA) models. Through a series of examples, we demonstrate that path marginal costs are not simple additions of the corresponding link marginal costs unless the flow perturbation travels with the vehicle unit that initiated the perturbation. We further show that one can compute efficiently path marginal costs for networks with a special structure (i.e. mono-centric), and proposed an evaluation method that decomposes path marginal costs to link marginal costs for such networks. Our preliminary numerical experiments indicate that this solution scheme can generate numerical solutions close to analytical solutions where they are known. At present, our method applies to networks without diverges and the embedded traffic flow models are restricted to those not considering queue spillback. The relaxation of either aspect brings in additional challenges in predicting path flow perturbation propagation, hence additional difficulties in the evaluation of path marginal cost, and is worthy of further research.

INTRODUCTION

The system-optimal dynamic traffic assignment (SO-DTA) problem, which determines the time-dependent traffic evolution pattern in a transportation network resulting in the minimal total system cost, is of great importance to traffic congestion management. The minimal total system cost calculated from the SO-DTA problem provides the best network performance that can be achieved, serving as a benchmark to evaluate the benefits of practical traffic management measures. The resulting optimal traffic flow pattern, on the other hand, provides valuable insights for designing congestion alleviation strategies such as congestion pricing schemes, advanced traveler information systems (ATIS), advanced transportation management systems (ATMS), or staging and routing plans for emergency evacuations.

Early SO-DTA studies focus on idealized networks, such as a single route network (Vickrey 1969, Hendrickson and Kocur 1981), a network with parallel routes (Arnott *et al.* 1990) or a freeway corridor network (Munoz and Laval 2005). Elegant solutions can be derived by exploiting the special features of such networks. A major limitation of these studies is that their results and conclusions can hardly be extended to other networks with different topology types. This limitation has served as the primary motivation for studies to extend the SO-DTA problem to general networks.

Following Merchant and Nemhauser's (1978a, 1978b) pioneering work, the last decade witnessed the birth of a wide variety of SO-DTA models on general networks. The majority of these models, referred to as link-based analytical SO-DTA models, has a similar model structure represented by link flow variables, with the objective function minimizing the total system cost and constraints describing traffic dynamics and flow conservation. The major difference among various studies in this category is the traffic dynamics models incorporated, such as the exit flow function and its variants (Merchant and Nemhauser 1978a, b, Ho 1980, Carey 1986, 1987, Friesz *et al.* 1989, Wie *et al.* 1994, 1995, Wie and Tobin 1998, Wie 1998), the link performance function model (Carey and Subrahmanian 2000), and the cell transmission model (Ziliaskopoulos 2000, Li 2001). No matter what type of traffic flow model is embedded, these studies usually run into two obstacles: 1) the variables and constraints of the problem are proportional to the network size and taking advantages of network structures to improve the solution procedure is difficult, thus incurring prohibitively expensive computational overhead when applied to large-scale networks; 2) analytically representing traffic dynamics as convex constraints is not easy because the embeded traffic flow models and the *first-in-first-out (FIFO) requirement (see Carey 1992) for multi-commodities at the link level* often call for non-linear equality constraints, which destroys the convexity of the constraint set.

To avoid the aforementioned two sources of non-convexity, constraints describing traffic flow propagations in a link-based SO-DTA model are often relaxed into inequality constraints (e.g., Carey 1987, Ziliaskopoulos 2000) and most SO-DTA studies in this category are restricted to the many-origin-to-single-destination type of networks. The mathematical elegance and tractability of these relaxed models, however, do not come without a price. It substantially limits the application of the SO-DTA models within the category of networks with only a single destination. In addition, the relaxed models may lead to a traffic evolution pattern involving "vehicle holding", which is not practical since to arbitrarily hold vehicles on any link in a network is extremely labour-consuming if not impossible to implement.

In view of the intrinsic non-convexity resulted from traffic propagation rules, Chang *et al.* (1988) and Yang and Meng (1998) proposed a time-space expansion network (STEN) scheme to solve the SO-DTA problem. In their method, a STEN is carefully constructed to endogenously represent the deterministic queuing processes, and hence the SO-DTA problem is transformed into a static system-optimal traffic assignment problem. However, introducing this type of STEN gives rise to some new problems and difficulties, such as the high computational overhead associated with network expansions. In addition, as addressed by Yang and Meng (1998), the FIFO requirements for multiple commodities at the link level still cannot be ensured.

In recent years, there is a renewed interest in formulating the DTA problem directly with path flows and cast the traffic flow dynamics into a path cost mapping (Lo 1999). For the SO-DTA problem, the motivations of this redirection of efforts can be summarized as follows: 1) constraints formed with path variables concern with only flow conservation, making the feasible set polyhedral; 2) since both traffic propagation and FIFO can be taken care of by the path cost mapping, the model may have the potential to deal with problems with multiple destinations and to eliminate "vehicle holding" in the solution; 3) Special network structures can be exploited to improve the solution procedure by taking advantage of shortest path searching algorithms, thus leading to solution algorithms not sensitive to network size.

Despite of the potential advantages of path-based SO-DTA models, research along this line is rather limited. The major reason is that solving path-based SO-DTA models usually requires gradients, here the change in the total system cost with respect to the unit change in the path flow, which we refer to as the path marginal cost (PMC) hereafter. Since the path cost mapping usually does not have an explicit functional form, the evaluation of PMC is not straightforward. Instead of numerically evaluating PMCs by perturbing the path flow pattern and performing a dynamic network loading (DNL), which is prohibitively expensive and may involve significant rounding errors, a detailed examination on how to exploit properties of the non-closed-form mapping (the path cost mapping in this case) to obtain PMCs efficiently may provide insights to the solution of not only the path-based SO-DTA problem but other time-dependent optimization problems in transportation as well. Such problems could include the design of signal timing plans or ramp metering schemes, since the need to obtain gradient information for mappings without a closed form also lies in the core of these optimization problems in transportation.

To date the only studies related to PMC evaluation are due to Ghali and Smith (1995) and Peeta and Mahmassani (1995). Ghali and Smith (1995) proposed a PMC evaluation method by summing up the link marginal cost along the path according to link traversal times, and the link marginal cost was evaluated based on link cumulative curves obtained from DNL results. Peeta and Mahmassani (1995) adopted a similar evaluation procedure, except that the link marginal cost is evaluated by constructing approximate link performance functions based on DNL results. Unfortunately, because of their additivity assumption on PMC, severe deficiencies seem to exist in both methods, causing them fail to provide accurate path marginal costs even for simple networks. As we shall show later, although the additivity assumption of PMCs seems to be a natural extension to that in the static case at the first glance, it turns out to be invalid in the dynamic case.

In view of the drawbacks in the existing path-based SO-DTA studies, especially their PMC evaluation methods, this paper makes a thorough investigation of the path-based SO-DTA problem, including its formulation and solution procedures, with an emphasis on the evaluation PMCs. We shall clarify misconceptions in current PMC evaluation methods, identify the associated difficulties and propose an improved PMC evaluation scheme for networks with a special structure.

According to the definition, PMCs can be evaluated by adding an additional unit flow (i.e., a small flow perturbation) to the specific path flow and evaluate its impact on the total system cost. To avoid conducting DNL repeatedly, a natural simplification is to trace the propagation of the path flow perturbation over all the links and evaluate the travel cost change on all the links. The traditionally adopted additivity assumption assumes that the perturbation travels simultaneously with the additional unit of flow along the path. However, due to the bottleneck effect, this additivity assumption is actually not true because 1) for two sequential links or two successive links of a merge, the flow perturbation on the upstream link can not reach the downstream link as long as a queue is present on the upstream link; 2) for two sequential links of a diverge, the change in the inflow rate on the upstream link may not just affect the inflow rates of the downstream link on the perturbed path. I can also affect the inflow rates of other diverging branches as well. In view of the nature of these violations of additivity, a new PMC evaluation method for networks in monocentric cities (i.e., networks without diverges) can be designed by tracing the propagation of a path flow perturbation along the path (Note that at the current stage, our method excludes networks with diverges. Methods for more general networks with diverges will be reported elsewhere). This new PMC evaluation method can be embedded into the time-dependent minimal cost path (TDMCP) searching algorithm to find the time-dependent least marginal cost path, by replacing the link-traversal time in the compact STEN (Space-Time Extended Network) with the path flow perturbation propagation time lag. Many algorithms for solving equilibrium problems, such as the heuristic method of successive averages (MSA), can then be applied to solve the path-based SO-DTA problem. Through several numerical examples, we show that solving path-based SO-DTA models based on our new PMC marginal cost evaluation method can converge to results very close to analytical SO solutions. But the heuristic MSA algorithm based on the PMC evaluation method fails to converge to the analytical SO solutions.

The remainder of this paper is structured as follows: Section 2 introduces the formulation of the path-based SO-DTA model. The optimal conditions that resemble Wardrop's second principle (Wardrop 1952) are provided and the importance of the PMC in solving path-based SO-DTA models is emphasized. The PMC evaluation problem is discussed thoroughly in Section 3, including the traditional additivity assumption, the deficiencies in the traditional evaluation method, as well as our new PMC evaluation method. Section 4 introduces the solution procedures for the path-based SO-DTA problem for mono-centric networks. The critical module of finding least marginal cost paths is also discussed. Computational results and discussions are reported in Section 5, and Section 6 presents conclusions and future research directions.

THE PATH-BASED SO-DTA MODEL

We consider a general transportation network with multiple origin-destination (OD) flows. The whole study period T_d is discretized into T intervals with length δ. We assume that T_d is long enough for all the traffic flows to clear the network. The goal of the model is to find the optimal departure time and path flow pattern such that the travel cost of the whole system including travel time cost and schedule delay cost is minimized.

The following notations are used throughout this paper:

a) Set notations

RS	set of OD pairs
T_d	the whole departure time horizon, $T_d = \{1, 2, ..., T\}$
P^{rs}	set of routes connecting OD pair rs
Ω	feasible set of path flows

b) Indices

rs	OD pair, $rs \in RS$
p	route between OD pair rs, $p \in P^{rs}$
t	index for departure time, $t \in T_d$

c) Variables to be determined

f_{pt}^{rs}	flow entering path $p \in P^{rs}$ at time t
$\mathbf{f}$	path flow vector, $\mathbf{f} = \{f_{pt}^{rs}\}$ with dimension $n = T\sum_{rs\in RS} \lvert P^{rs} \rvert$

d) Functions of path flow $\mathbf{f}$

$c_{pt}^{rs}(\mathbf{f})$	actual path travel time for flow entering path $p \in P^{rs}$ during time t, which is a unique mapping with respect to $\mathbf{f}$
$\phi_{pt}^{rs}(\mathbf{f})$	generalized cost incurred by flow entering path $p \in P^{rs}$ during time t, which is a unique mapping with respect to $\mathbf{f}$
q_t^{rs}	flow between OD pair rs departing during time t

e) Inputs and parameters

Q^{rs}	total demand for OD pair rs during the study horizon
$c^s(t)$	schedule delay cost for travelers arriving at destinations during time t
$\tilde{t}^s$	desired arrival time for travelers going to destination s, $\tilde{t}^s \in T_d$
Δ^s	arrival time flexibility for travelers going to destination s, $\Delta^s \geq 0$
α^s	unit cost of travel time for travelers going to destination s, $\alpha^s > 0$
β^s	unit cost of schedule delay caused by the early arrival of travelers at destination s, $\beta^s > 0$
γ^s	unit cost of schedule delay caused by the late arrival of travelers at destination s, $\gamma^s > 0$.

According to empirical data, $\gamma^s > \alpha^s > \beta^s$ (Small 1982), and we have the following relationships:

$$\phi_{pt}^{rs}(\mathbf{f}) = \alpha c_{pt}^{rs}(\mathbf{f}) + c^s[t + c_{pt}^{rs}(\mathbf{f})], \tag{3}$$

where the schedule delay cost function $c^s(t)$ is piecewise linear and can be represented by:

$$c^s(t) = \begin{cases} \beta^s[(\tilde{t}^s - \Delta^s) - t] & \text{if } t < \tilde{t}^s - \Delta^s \\ 0 & \text{if } \tilde{t}^s - \Delta^s \leq t \leq \tilde{t}^s + \Delta^s \\ \gamma^s[t - (\tilde{t}^s + \Delta^s)] & \text{if } t > \tilde{t}^s + \Delta^s . \end{cases} \tag{4}$$

Using the defined path variables and functions, the SO-DTA problem optimizing both departure time and route choice can be formulated as the following minimization problem:

[Model M1]

$$\min_{\mathbf{f}\in\Omega} TC(\mathbf{f}) = \sum_{t\in T_d} \sum_{rs\in RS} \sum_{p\in P^{rs}} f_{pt}^{rs} \cdot \phi_{pt}^{rs}(\mathbf{f}) \tag{5}$$

subject to

$$\sum_{p\in P^{rs}} f_{pt}^{rs} = q_t^{rs}, \forall rs \in RS, t \in T_d \tag{6}$$

$$\sum_{t\in T_d} q_t^{rs} = Q^{rs}(\text{given}), \forall rs \in RS \tag{7}$$

$$f_{pt}^{rs} \geq 0, \forall rs \in RS, k \in K_{rs}, t \in T_d. \tag{8}$$

Note that for cases in which we do not consider departure time choice, the corresponding path-based SO-DTA model is similar to M1[(5)-(8)], except that we replace $\phi_{pt}^{rs}(\mathbf{f})$ in the objective function by $c_{pt}^{rs}(\mathbf{f})$, remove the constraints $\sum_{t\in T_d} q_t^{rs} = Q^{rs}$ and treat q_t^{rs} as given. Namely, we have the following formulation:

[Model M2]

$$\min_{\mathbf{f}\in\Omega} TC(\mathbf{f}) = \sum_{t\in T_d}\sum_{rs\in RS}\sum_{p\in P^{rs}} f_{pt}^{rs}\cdot c_{pt}^{rs}(\mathbf{f}) \tag{9}$$

subject to

$$\sum_{p\in P^{rs}} f_{pt}^{rs} = q_t^{rs}\,(\text{given}), \forall rs\in RS, t\in T_d \tag{10}$$

$$f_{pt}^{rs}\geq 0, \forall rs\in RS, k\in K_{rs}, t\in T_d \tag{11}$$

Because of the similarity of these two models, our following discussions will only focus on the model optimizing both departure time and route choices, unless stated otherwise.

As shown in Model M1 and M2, the feasible set in terms of path flows $\{f_{pt}^{rs}\}$ is polyhedral, and the formulation itself is free of specific traffic dynamics models since the traffic dynamics is endogenously taken care of by the path cost mapping $\mathbf{c}(\mathbf{f}) = \{c_{pt}^{rs}(\mathbf{f})\}$.

If we attach multipliers μ^{rs} to constraints (7) and substitute constraints (6) into constraints (7), the Lagrangian of the model M1 is as follows:

$$L(\mathbf{f},\boldsymbol{\mu}) = TC(\mathbf{f}) + \sum_{rs\in RS}\mu^{rs}(Q^{rs} - \sum_{t\in T_d}\sum_{p\in P^{rs}} f_{pt}^{rs}). \tag{12}$$

According to Karush-Kuhn-Tucker (KKT) conditions, the first-order necessary conditions of optimality for M1 are constraints (6) - (8) plus $f_{pt}^{rs}\frac{\partial L(\mathbf{f},\mathbf{u})}{\partial f_{pt}^{rs}} = 0, \forall r,s,p,t$ and $\frac{\partial L(\mathbf{f},\mathbf{u})}{\partial f_{pt}^{rs}}\geq 0$. Namely,

$$f_{pt}^{rs}\left(\frac{\partial TC(\mathbf{f})}{\partial f_{pt}^{rs}} - \mu^{rs}\right) = 0, \forall rs\in RS, p\in P^{rs}, t\in T_d \tag{13}$$

$$\frac{\partial TC(\mathbf{f})}{\partial f_{pt}^{rs}} - \mu^{rs}\geq 0, \forall rs\in RS, p\in P^{rs}, t\in T_d \tag{14}$$

$$\sum_{t\in T_d}\sum_{p\in P^{rs}} f_{pt}^{rs} - Q^{rs} = 0, \forall rs\in RS \tag{15}$$

$$f_{pt}^{rs}\geq 0, \forall rs\in RS, k\in K_{rs}, t\in T_d \tag{16}$$

To facilitate further discussion, we provide the mathematical definition for PMC below.

Definition 1 (Path marginal cost) *Given a specific path flow pattern* $\mathbf{f} = \{f_{pt}^{rs}, \forall p,t,rs\}$, *the path marginal cost (PMC) for path* p *at time* t *represents the increase in the total system cost when the path inflow on* p *at time* t *is increased by one unit. Namely,*

$$PMC_{pt}^{rs}(\mathbf{f}) = \frac{\partial TC(\mathbf{f})}{\partial f_{pt}^{rs}} = \frac{\sum_{\tau\in T_d}\sum_{rs\in RS}\sum_{p\in RS} f_{p\tau}^{rs}\phi_{p\tau}^{rs}(\mathbf{f})}{\partial f_{pt}^{rs}}. \tag{17}$$

Evidently, eqn (13) and eqn (14) convey the Wardrop second principle in terms of time-dependent path marginal cost, i.e., at dynamic system optimum, the time-dependent marginal cost on all the paths actually used are equal and less than the marginal cost on any unused path. Consequently, if we can efficiently evaluate path marginal cost $PMC_{pt}^{rs}, \forall r,s,p,t$, algorithms for solving equilibrium problems may be applied to solve the SO-DTA problem, at least approximately[1].

PATH MARGINAL COST EVALUATION

As shown in Sheffi(1985), the PMC in the static case is given by,

$$PMC_p^{rs}(\mathbf{f}) = \sum_{a \in A} \left[c_a(x_a) + x_a \frac{dc_a(x_a)}{dx_a} \right] \delta_{ap}^{rs}$$

where x_a is the traffic flow on link a, $c_a(x_a)$ is the corresponding link cost, and δ_{ap}^{rs} is the path-link incidence indicator. Namely, the PMC is equal to the sum of the marginal cost of the links on that path.

In the dynamic case, the evaluation of the PMC is much more complicated since path flows are not assigned to links simultaneously. However, for traffic dynamics models which do not consider queue spillback, such as the point queue, the exit flow function and the link performance function models, a decomposition scheme from path marginal cost to link marginal cost is still possible. To illustrate this, we further introduce the following additional notations:

u_{ak} flow entering link a during time k
$\mathbf{u}_a$ link inflow vector $\mathbf{u}_a = \{u_{ak}, \forall k \in T_d\}$
c_{ak} link travel time for flow entering link a during time k
u_k^s flow arriving at destination s during time k

For traffic dynamics models not considering link interactions, c_{ak} is uniquely determined by the inflow pattern $\mathbf{u}_a$ on link a. Hence, we can treat c_{ak} as a function of $\mathbf{u}_a$, i.e., $c_{ak} = c_{ak}(\mathbf{u}_a)$. The total travel cost $TC(\mathbf{f})$ can then be written as follows:

$$TC(\mathbf{f}) = \sum_{a \in A} \sum_{k \in T_d} u_{ak} c_{ak}(\mathbf{u}_a) + \sum_{s \in S} \sum_{k \in T_d} u_k^s c^s(k) \tag{18}$$

---

[1]Since $\phi_{pt}^{rs}(\mathbf{f})$ does not have an explicit functional form, its convexity property is unknown. Hence, the KKT condition may not be sufficient to guarantee a global optimal solution.

Substituting eqn (18) into eqn (17), we get

$$PMC_{pt}^{rs}(\mathbf{f}) = \frac{\partial \sum_{a\in A}\sum_{k\in T_d} u_{ak}c_{ak}(\mathbf{u}_a)}{\partial f_{pt}^{rs}} + \frac{\partial \sum_{s\in S}\sum_{k\in T_d} u_k^s c^s(k)}{\partial f_{pt}^{rs}}$$

$$= \sum_{a\in A} \frac{\partial \sum_{k\in T_d} u_{ak}c_{ak}(\mathbf{u}_a)}{\partial f_{pt}^{rs}} + \sum_{s\in S} \frac{\partial \sum_{k\in T_d} u_k^s c^s(k)}{\partial f_{pt}^{rs}}$$

Using chain rule, we obtain the following relationship,

$$PMC_{pt}^{rs}(\mathbf{f}) = \sum_{a\in A} \frac{\partial \sum_{k\in T_d} u_{ak}c_{ak}(\mathbf{u}_a)}{\partial f_{pt}^{rs}} + \sum_{s\in S} \frac{\partial \sum_{k\in T_d} u_k^s c^s(k)}{\partial f_{pt}^{rs}}$$
$$= \sum_{a\in A}\left(\sum_{\tau\in T_d} \frac{\partial \sum_{k\in T_d} u_{ak}c_{ak}(\mathbf{u}_a)}{\partial u_{a\tau}} \frac{\partial u_{a\tau}}{\partial f_{pt}^{rs}} \right) + \sum_{s\in S}\sum_{\tau\in T_d} c^s(\tau)\frac{\partial u_\tau^s}{\partial f_{pt}^{rs}} \tag{19}$$

Before proceeding further, let us introduce two additional definitions to facilitate the discussion:

Definition 2 (Link marginal cost) *Given a specific link inflow pattern* $\mathbf{u}_a = \{u_{ak}, k \in T_d\}$ *for link* a *, the link marginal cost for link* a *at time* τ *represents the change in the total link cost when the link inflow at time* τ *is increased by one unit. Namely*

$$LMC_{a\tau}(\mathbf{u}_a) := \frac{\partial \sum_{k\in T_d} u_{ak}c_{ak}(\mathbf{u}_a)}{\partial u_{a\tau}}, \forall a \in A, \tau \in T_d \tag{20}$$

Definition 3 (Path flow perturbation propagation index) *Given a specific path flow pattern* $\mathbf{f}$ *and the corresponding link inflow pattern* $\mathbf{u}$ *, the path flow perturbation propagation index* $Ind_{aprs}^{t\tau}(\mathbf{f})$ *represents the change in the inflow of link* a *at time* τ *when the path flow at time* t *is increased by one unit. Namely,*

$$Ind_{aprs}^{t\tau}(\mathbf{f}) = \frac{\partial u_{a\tau}}{\partial f_{pt}^{rs}} \tag{21}$$

Using Definition 2 and 3, the PMC formula can be simplified as follows:

$$PMC_{pt}^{rs}(\mathbf{f}) = \sum_{a\in A}\sum_{\tau\in T_d} LMC_{a\tau}(\mathbf{u}_a) Ind_{ap}^{rst\tau}(\mathbf{f}) + \sum_{s\in S}\sum_{\tau\in T_d} c^s(\tau) Ind_{sp}^{rst\tau}(\mathbf{f}) \tag{22}$$

According to Equation (22), PMC for traffic dynamics models without queue spillback can still be regarded as *additive* as long as the path flow perturbation propagation is correctly captured.

The following two subsections discuss how to evaluate link marginal cost $LMC_{a\tau}(\mathbf{f})$ and path flow perturbation propagation index $Ind_{ap}^{rst\tau}(\mathbf{f})$.

Evaluation of link marginal cost ($LMC_{a\tau}(\mathbf{f})$)

As mentioned earlier, existing methods of evaluating $LMC_{at}(\mathbf{u}_a)$ include Ghali and Smith 1995 and Peeta and Mahmassani 1995. Ghali and Smith (1995) provides a sound analytical formulation for LMCs based on the link cumulative curves for the point queue model. Suppose the cumulative curves at the downstream bottleneck of link a are what depicted in Figure 2, where T^s is the time when the queue begins to form, T^e is the time at which the queue vanishes, and T^f is the link free flow travel time. If $\tau + T^f \notin [T^s, T^e]$, i.e., the additional flow unit will not encounter a queue on the link, the LMC is equal to the free flow travel time of one flow unit; when the bottleneck is active, the LMC (the thick portion of line in Figure 1[2]) includes the travel time of that additional flow unit ($T^f + c(\tau + T^f)$) and the incurred delay of later arrival flows ($T^e - [\tau + T^f + c(\tau + T^f)]$).

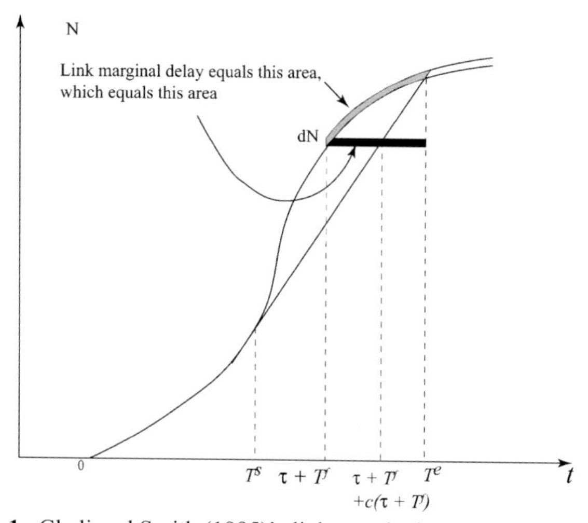

Figure 1. Ghali and Smith (1995)'s link marginal cost evaluation method

Namely,

$$LMC_{a\tau}(\mathbf{u}_a) = \begin{cases} T^f & \text{if } \tau + T^f \notin [T^s, T^e] \\ T^f + T^e - (\tau + T^f) = T^e - \tau & \text{if } \tau + T^f \in [T^s, T^e] \end{cases} \tag{23}$$

[2] Using cumulative curves, the total link delay is the area between the arrival curve and departure curve. Hence, the incurred delay is the extra area due to the change in the curves.

Peeta and Mahmassani (1995) approached this problem based on a different idea. They assume that the link travel time $c_{a\tau}$ for a vehicle entering link a during time τ is only determined by the link inflow rate at time τ, namely, $c_{ak} = c_{ak}(u_{ak}), \forall k \in T_d$. Then LMC can be simplified as follows:

$$LMC_{a\tau}(\mathbf{u}_a) = \frac{\partial \sum_{k\in T_d} u_{ak} c_{ak}(u_{ak})}{\partial u_{a\tau}} = \frac{\partial u_{a\tau} c_{a\tau}(u_{a\tau})}{\partial u_{a\tau}}$$
$$= c_{a\tau}(u_{a\tau}) + u_{a\tau} \frac{dc_{a\tau}(u_{a\tau})}{du_{a\tau}} \tag{24}$$

where $u_{a\tau}$ and $c_{a\tau}$ are obtained from the DNL results (by traffic simulator DYNASMART). The link performance function (LPF) $c_{a\tau}(u_{a\tau})$ is constructed approximately to numerically obtain $\frac{dc_{a\tau}(u_{a\tau})}{du_{a\tau}}$, based on the loading results of $(u_{a,\tau-1}, c_{a,\tau-1})$, $(u_{a,\tau}, c_{a,\tau})$, and $(u_{a,\tau+1}, c_{a,\tau+1})$. The major deficiency of this LPF model is that its link travel time for a vehicle entering link a at time τ is solely determined by the link inflow rate at time τ. This underestimate link travel time when the link is congested.

In the development of our PMC evaluation procedure, we adopt Ghali and Smith's method as applied to the point queue model, because we believe that the point queue model is a reasonable comprise between mathematical tractability and realism in capturing the main form of congestion: queuing at bottlenecks (see Nie and Zhang 2005a,b for a comparison of several link models used in dynamic traffic assignment and an in-depth analysis of the delay-function based link model).

Tracing the path flow perturbation ($Ind_{ap}^{rst\tau}(\mathbf{f})$)

The problem of tracing the path flow perturbation seems to be neglected in most existing path-based SO-DTA studies. Researchers (e.g., Ghali and Smith (1995), Peeta and Mahmassani (1995), etc.) simply assume that the path flow perturbation travels along the path at the same speed as that of the additional flow unit. In other words, they assume that

$$Ind_{ap}^{rst\tau} = \delta_{ap}^{rst\tau}, \forall a \in A, rs \in RS, p \in P_{rs}, t \in T_d, \tau \in T_d$$

where $\delta_{ap}^{rst\tau}$ is the dynamic path-link incidence indicator, i.e., $\delta_{ap}^{rst\tau} = 1$ if traffic departing origin r at time t heading for destination on path $p \in P_{rs}$ arrives link a during time τ, and 0 otherwise. Correspondingly,

$$PMC_{pt}^{rs}(\mathbf{f}) = \sum_{a\in A}\sum_{\tau\in T_d} LMC_{a\tau}(\mathbf{u}_a)\delta_{ap}^{rst\tau}(\mathbf{f}) + \sum_{s\in S}\sum_{\tau\in T_d} c^s(\tau)\delta_p^{rst\tau}(\mathbf{f}). \tag{25}$$

For narrative convenience, this PMC evaluation method is referred to as the link traversal time(LTT) method hereafter. Unfortunately, as we shall see in the following discussion, this assumption is actually NOT true in the dynamic case due to bottleneck restrictions. We demonstrate this claim by showing in three simple example networks how path flow perturbation propagates along two sequential links, through a merge, and through a diverge, respectively.

1) Propagation of a path flow perturbation in two sequential links

To examine the propagation of a path flow perturbation along two sequential links, let's look at the following small example network (Figure 2).

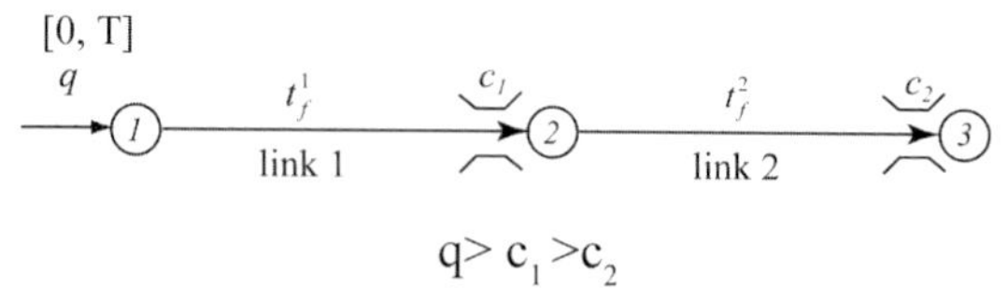

Figure 2. Illustration network I with two sequential links

The illustration network in Figure 2 contains two links, 1 and 2, and each link has a bottleneck at its downstream end. The capacities of the bottlenecks at links 1 and 2 are c_1 and c_2, respectively, and free flow travel times of links 1 and 2 are t_f^1 and t_f^2, respectively. During time $[0,T]$, vehicles enter the network from link 1 at a constant flow rate q. We assume that $q > c_1 > c_2$. Obviously, queues will develop at both bottlenecks. The cumulative curves for these two links are illustrated in Figure 3, where t_e^1 and t_e^2 represent the respective times that the queues on link 1 and link 2 vanish, and N is the total number of vehicles.

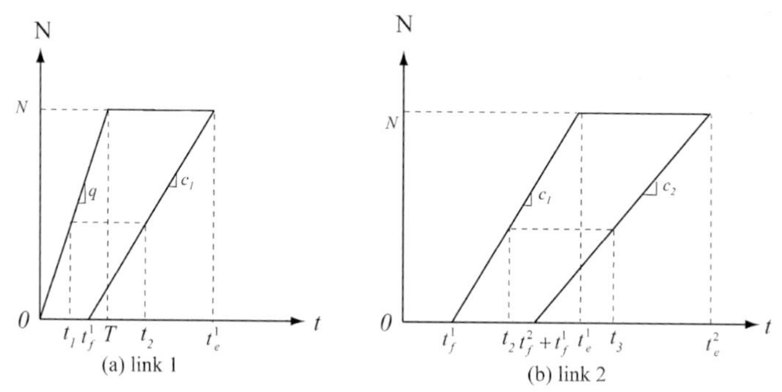

Figure 3. Cumulative curves for illustration network I

According to the cumulative curves in Figure 3, the vehicle entering link 1 at time $t_1 \in [0, T_d]$ will enter link 2 at time t_2 and leave at time t_3. Suppose we want to evaluate the propagation of the path flow perturbation at time t_1 and to calculate $PMC_{t_1}(\mathbf{f})$[3], $t_1 \in [0,T]$. To simplify the discussion, no schedule delay cost is considered. Based on the definitions, both the propagation of the flow perturbation and the PMC can be evaluated by constructing the new cumulative curves for links 1 and 2 with an additional flow unit entering link 1 at time t_1 (Figure 4).

According to Figure 4, the propagation of the flow perturbation cannot arrive at the same time when the additional unit of flow enters link 2. Instead, the perturbation reaches link 2 at the time when the queue on link 1 dissipates. Namely,

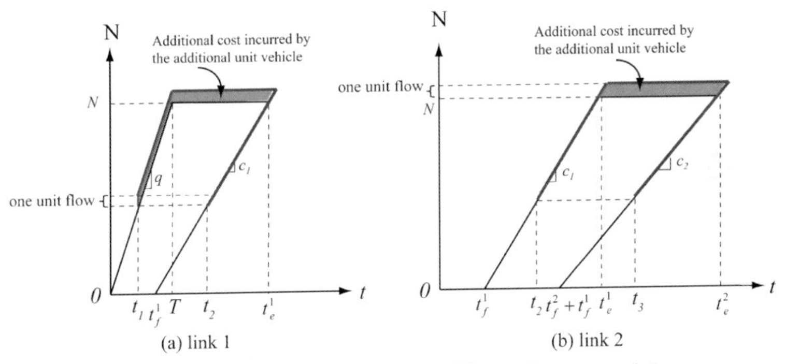

Figure 4. Path marginal cost for illustration network I

$$Ind_1^{\eta\tau} = \begin{cases} 1 & \text{if } \tau = t_1 \\ 0 & \text{otherwise} \end{cases}, Ind_2^{\eta\tau} = \begin{cases} 1 & \text{if } \tau = t_e^1 \\ 0 & \text{otherwise} \end{cases}$$

Figure 4 also shows that the additional cost incurred on link 1 and link 2 are $t_e^1 - t_1$ and $t_e^2 - t_e^1$, respectively. Hence,

$$\begin{aligned} PMC_{t_1}(\mathbf{f}) &= LMC_{1,t_1}(\mathbf{f}) + LMC_{2,t_e^1}(\mathbf{f}) \\ &= (t_e^1 - t_1) + (t_e^2 - t_e^1) \\ &= t_e^2 - t_1 \end{aligned} \tag{26}$$

where $t_3 - t_1$ is the cost incurred by the marginal user and $t_e^2 - t_3$ is the cost imposed on others. However,

$$\delta_1^{\eta\tau} = \begin{cases} 1 & \text{if } \tau = t_1 \\ 0 & \text{otherwise} \end{cases}, \quad \delta_2^{\eta\tau} = \begin{cases} 1 & \text{if } \tau = t_2 \\ 0 & \text{otherwise} \end{cases}$$

---

[3]Note that in this simple network, since there is only one OD pair $1-3$ and one path, the OD pair index *rs* and path index *p* is omitted for *PMC*; the same for *Ind*.

Hence, the LTT method yields

$$
\begin{aligned}
PMC'_{t_1}(\mathbf{f}) &= LMC_{1,t_1}(\mathbf{u}_1) + LMC_{2,t_2}(\mathbf{u}_2) \\
&= t_e^1 - t_1 + t_e^2 - t_2 \\
&> PMC_{t_1}(\mathbf{f}).
\end{aligned}
\tag{27}
$$

Equations (26) and (27) show that $PMC_{t_1}(\mathbf{f})$ is larger than $PMC'_{t_1}(\mathbf{f})$ by $(t_e^1 - t_2)$. In other words, the LTT method tends to overestimate the PMC for two sequential links in this simple network. The reason of the overestimation is that the path flow perturbation actually travels more slowly than the additional flow unit in two sequential links because of the capacity restriction at bottlenecks. More specifically, in the above simple network, the perturbation caused by an additional unit flow entering link 1 at time t_1 will not propagate onto link 2 so long there is a queue present on link 1.

2) Path flow perturbation propagation through a merge

How the path flow perturbation propagates through a merge is similar to the case of two sequential links. Again, we illustrate this by a simple network example (Figure 5).

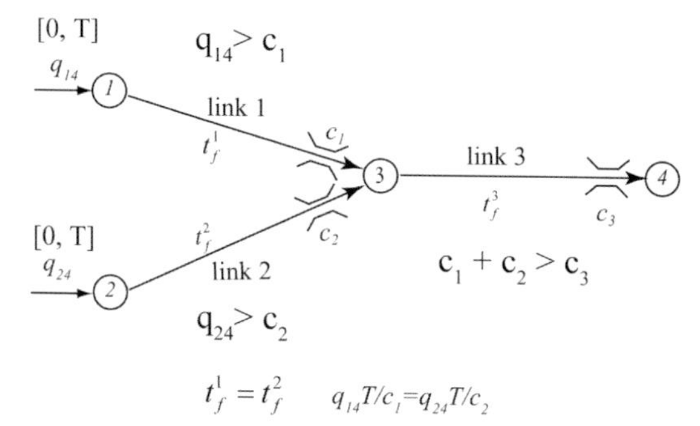

Figure 5. Illustration network II with a merge

The network consists of three links, 1, 2 and 3, and each has a bottleneck with capacity $c_i, i = 1,2,3$ at its downstream end. The link free flow travel times are $t_f^i, i = 1,2,3$, respectively. During time $[0, T]$, vehicles enter the network from links 1 and 2 at constant flow rates q_{14} and q_{24}, respectively. We further assume that $t_f^1 = t_f^2$, $q_{14} > c_1$, $q_{24} > c_2$, $q_{14}T / c_1 = q_{24}T / c_2$, and $c_1 + c_2 > c_3$.

Since $q_{14} > c_1$, $q_{24} > c_2$, there will be queues at both bottlenecks of links 1 and 2. Because $t_f^1 = t_f^2$ and $q_{14}T / c_1 = q_{24}T / c_2$, the inflow rate of link 3 will be $c_1 + c_2$ from time t_f^1 (or t_f^2) to time $t_e^1 = q_{14}T / c_1 + t_f^1$ (or $t_e^2 = q_{24}T / c_2 + t_f^2$). Since $c_1 + c_2 > c_3$, a queue will also develop at the bottleneck of link 3. Figure 6 shows the cumulative curves for the three links.

According to the cumulative curves, the vehicle entering link 1 at time $t_1 \in [0,T]$ will enter link 3 at time t_2. We now evaluate $PMC_{t_1}^{14}(\mathbf{f})$ by tracing the path flow perturbation in the network [4]. The new cumulative curves with an additional one unit of inflow at time t_1 as well as the additional costs incurred in all the links are shown in Figure 7.

Just like the case with two sequential links, the path flow perturbation will arrive at link 3 at time t_e^1 when the queue on link 1 dissipates. Namely,

$$Ind_{t_1 14}^{1\tau} = \begin{cases} 1 & \text{if } \tau = t_1 \\ 0 & \text{otherwise} \end{cases}, \quad Ind_{t_1 14}^{2\tau} = 0 \ \forall \tau, \quad Ind_{t_1 14}^{3\tau} = \begin{cases} 1 & \text{if } \tau = t_e^1 \\ 0 & \text{otherwise} \end{cases}.$$

Hence, the additional cost in link 1 and link 3 are $t_e^1 - t_1$ and $t_e^3 - t_e^1$, respectively. Namely,

$$PMC_{t_1}^{14}(\mathbf{f}) = t_e^1 - t_1 + (t_e^3 - t_e^1) = t_e^3 - t_1 .$$

Again, $PMC_{t_1}^{14}(\mathbf{f}) < PMC_{t_1}^{\prime 14}(\mathbf{f}) = LMC_{1,t_1}(\mathbf{u}_1) + LMC_{3,t_2}(\mathbf{u}_3) = t_e^1 - t_1 + t_e^3 - t_2$.

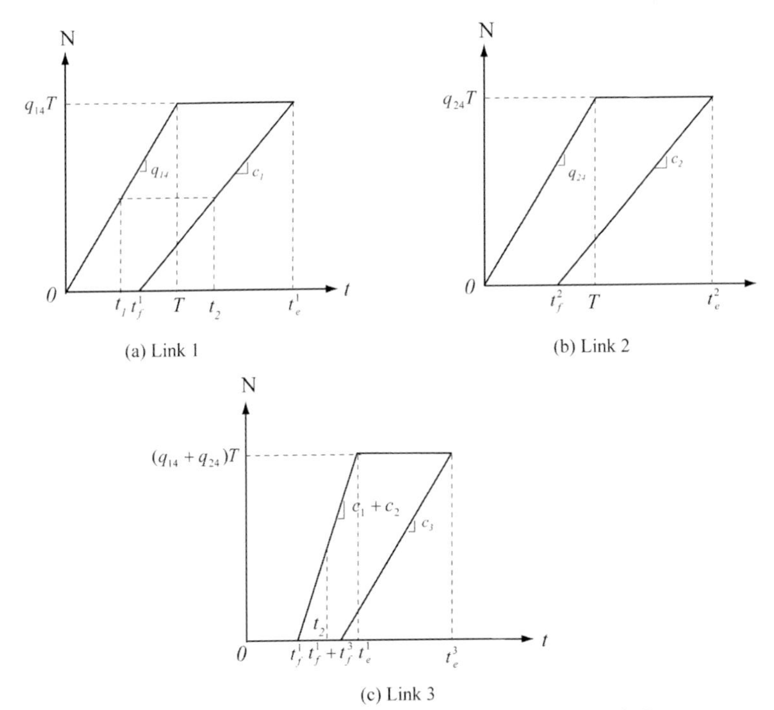

Figure 6. Cumulative curves for illustration network II

---

[4] Since there is only one path for each OD pair, the path index p is omitted for PMC ; the same for Ind .

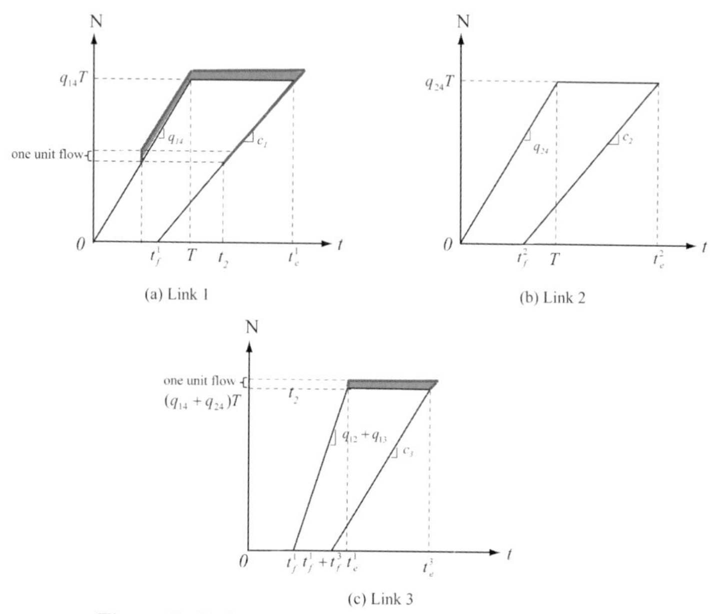

Figure 7. Path marginal cost for illustration network II

3) Path flow perturbation propagation through a diverge

How the path flow perturbation travels through a diverge is different from the cases of sequential links and of a merge. We use the following network (Figure 8) to illustrate this.

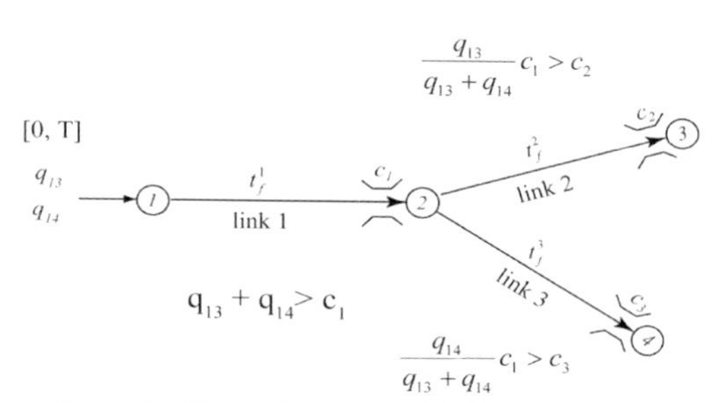

Figure 8. Illustration network III with a diverge

The network is made up of three links 1, 2 and 3 with bottleneck capacities $c_i, i=1,2,3$ and free flow travel time $t_f^i, i=1,2,3$. During time $[0,T]$, vehicles originated from node 1 and heading to node 3 enters link 1 at a constant rate q_{13} , and vehicles originated from node 1 and heading to node 4 enters link 1 at a constant rate q_{14}. We further assume that $q_{13}+q_{14}>c_1, \frac{q_{13}}{q_{13}+q_{14}}c_1>c_2, \frac{q_{14}}{q_{13}+q_{14}}c_1>c_3$, and thus queues will develop on all the three links. Figure 9 shows the cumulative curves of all the three links.

According to the cumulative curves, the vehicle entering link 1 at time $t_1 \in [0,T]$ and heading to node 3 will enter link 2 at time t_2. Again, we want to evaluate $PMC_{t_1}^{13}(\mathbf{f})$[5]. First we draw the cumulative curves when there is an additional unit flow heading to node 3 at time t_1 (Figure 10).

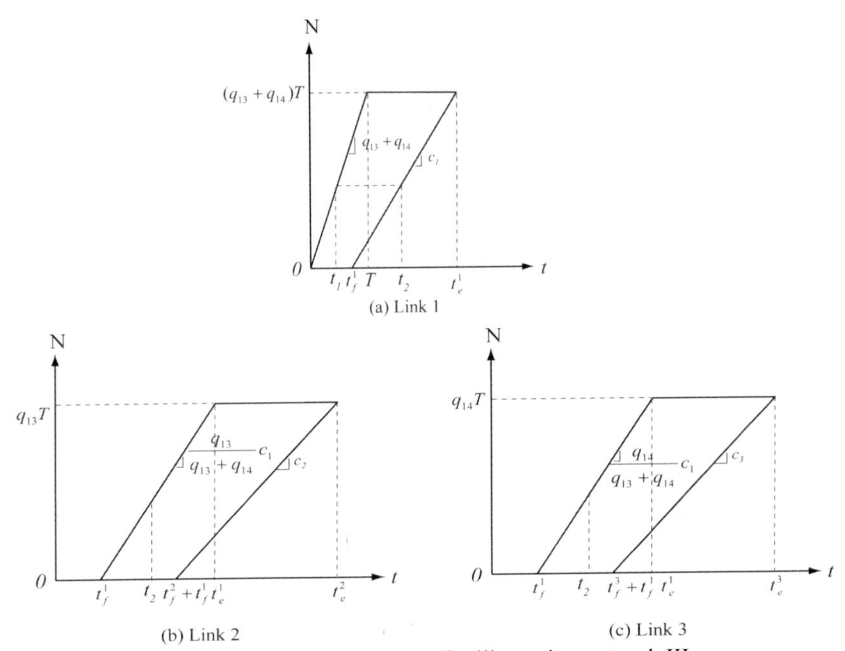

(b) Link 2 (c) Link 3

Figure 9. Cumulative curves for illustration network III

The new cumulative curves show that the perturbation of the flow on the path made up of links 1 and 2 will not only cause flow surge on links 1 and 2 , but also affect the inflow rates on link 3. A quick calculation shows that $Ind_{t_1,13}^{1\tau}, Ind_{t_1,13}^{2\tau}$ and $Ind_{t_1,13}^{3\tau}$ satisfies the following relationships:

$$Ind_{t_1,13}^{1\tau}=\begin{cases}1 & \text{if } \tau=t_1\\ 0 & \text{otherwise}\end{cases}, Ind_{t_1,13}^{2\tau}=\begin{cases}1 & \text{if } \tau=t_2\\ 0 & \text{otherwise}\end{cases}, Ind_{t_1,13}^{3\tau}=\begin{cases}-1 & \text{if } \tau=t_2\\ 1 & \text{if } \tau=t_e^1\\ 0 & \text{otherwise}\end{cases}$$

[5]Note that since there is only one path for each OD pair, the path index p is omitted for *PMC*; the same for *Ind*.

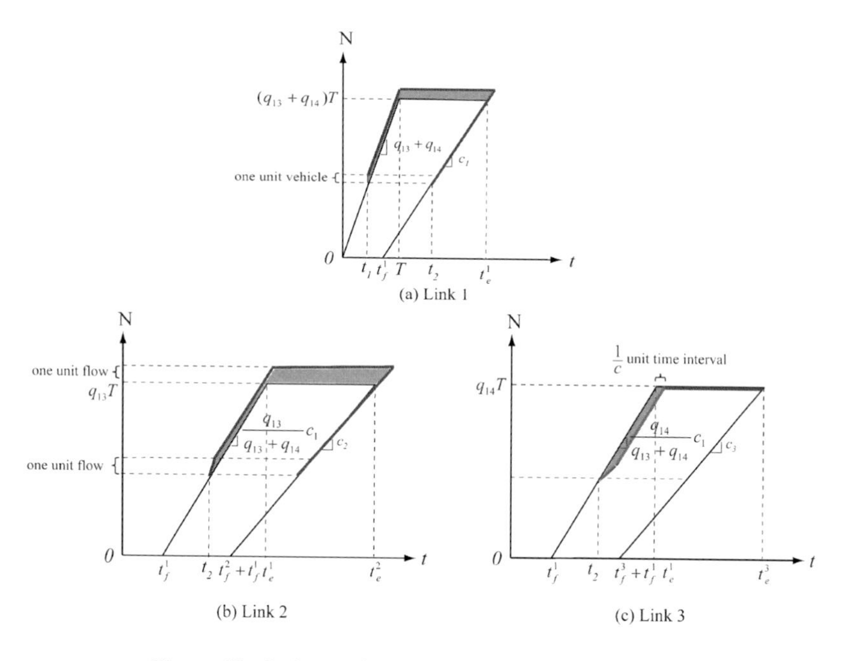

Figure 10. Path marginal cost for illustration network III

Consequently, the additional cost incurred links 1, 2 and 3 are $t_e^1 - t^1$, $t_e^2 - t_2$, and $-(t_e^1 - t_2)$, respectively, and the path marginal cost is as follows:

$$PMC_{12,t_1}(\mathbf{f}) = t_e^1 - t_1 + (t_e^2 - t_2) - (t_e^1 - t_2)$$
$$= t_e^2 - t_1$$

which is still larger than $PMC'_{12,t_1}(\mathbf{f}) = LMC_{1,t_1}(\mathbf{u}_1) + LMC_{2,t_2}(\mathbf{u}_2) = t_e^1 - t_1 + t_e^2 - t_2$ by $t_e^1 - t_2$.

Based on the analyses in three simple networks, we can see that the path flow perturbation may not travel along with the additional vehicle at the same speed, because on the upstream link the marginal vehicle already slightly delayed the vehicles joining the queue later. Summing the link marginal cost along the time-space path of the marginal vehicle omits this "knock-on" costs that persist after the marginal vehicle has exited from the queue. In fact, for networks without diverges, the perturbation will actually reach the corresponding downstream link at the time that the queue on the upstream link vanishes. For the network with a diverge, the conditions are far more complicated since the path flow perturbation also affects links that are not on the path.

A new path marginal cost evaluation method for networks without diverges

The previous discussions on the evaluation of PMC reveals a major deficiency in the existing evaluation methods, i.e., the problematic assumption that the path flow perturbation will travel at the same speed as the additional unit of flow. In this subsection, we present a new path marginal cost evaluation method for networks in mono-centric cities, i.e., networks without diverges. The method for more general networks with diverges is much more complicated and will be reported elsewhere.

As we have demonstrated earlier, to evaluate PMCs, we need to keep track of the path flow perturbation propagation on links. We now articulate this point in a mathematical manner. Suppose a path p is made up of m links $a_1,\ldots a_m$. Denote $d_{a_i}(t)$ as the actual time that the perturbation of the path flow departing at time t reaches link a_i, and $d_s(t)$ as the time that the perturbation reaches the destination s, we have the following relationship:

$$Ind_{a_i p}^{rst\tau}(\mathbf{f}) = \begin{cases} 1 & \text{if } \tau = d_{a_i}(t) \\ 0 & \text{otherwise} \end{cases} \quad \forall i \in [1,m], \quad Ind_p^{rs\tau t}(\mathbf{f}) = \begin{cases} 1 & \text{if } \tau = d_s(t) \\ 0 & \text{otherwise} \end{cases} \tag{28}$$

Substituting these relationships into the PMC formula (23), we get:

$$PMC_{pt}^{rs}(\mathbf{f}) = \sum_{i=1}^{m} LMC_{a_i, d_{a_i}(t)}(\mathbf{u}_{a_i}) + c^s[d_s(t)] \tag{29}$$

Namely, the PMC can still be regarded as "additive" according to the actual time that *the path flow perturbation* reaches link a_i . We refer to this new PMC evaluation method as the path propagation time(PPT) method.

Based on DNL results, $d_{a_i}(t)$ can be derived by the following recursive relationships:

$$\begin{aligned} &d_{a_1}(t) = t \\ &d_{a_i}(t) = w_{a_{i-1}}[d_{a_{i-1}}(t)], i = 2,\ldots,m \\ &d_s(t) = w_{a_m}[d_{a_m}(t)] \end{aligned} \tag{30}$$

where $w_{a_i}(t)$ is the earliest time after $t + c_{a_i t}$ when the queue on link a_i vanishes. Evidently, $w_{a_i}(t) = t + LMC_{a_i t}(\mathbf{u}_{a_i})$. Below is a numerical procedure for deriving $w_a(t)$ (and $LMC_{at}(\mathbf{u}_a)$ as well) from cumulative curves:

[Algorithm 1]

Step 0: Preprocessing: Perform a DNL and obtain the time dependent link traversal times $c_{at}, \forall a \in A, t \in T_d$;

Step 1: Deriving $w_a(t)$ and $LMC_{at}(\mathbf{u}_a)$

for $a \in A$

for $t = 0$ to T_d

let $\tau = t$

if $c_{at} > t_a^f$ (free flow travel time) , set $t = t+1$;endif;

$w_a(\tau) = t + c_{at}$, $LMC_{a\tau}(\mathbf{u}_a) = w_a(\tau) - \tau$;

endfor

endfor

In summary, to evaluate the path marginal cost for a specific path and a specific departure time, we check from the first link of the path. If there is no queue present when the additional unit flow reaches the link's bottleneck, the path flow perturbation will enter the second link simultaneously with the additional flow unit; otherwise, the path flow perturbation will enter the second link at the time when the queue on link 1 vanishes. The same procedure is repeated for successive links till we reach the destination.

A SOLUTION SCHEME FOR THE PATH-BASED SO-DTA MODEL

The heuristic MSA algorithm

Once PMCs are available, we can transform the path-based SO-DTA model into an equilibrium problem and solve it using solution methods for variational inequalities. It is well known that equilibrium conditions like the first order optimality conditions of the path-based SO-DTA model can be transformed into the following variational inequality (VI).

$$\sum_{t \in T_d} \sum_{rs \in RS} \sum_{p \in P^{rs}} PMC_{pt}^{rs}(\mathbf{f}^*)[f_{pt}^{rs} - f_{pt}^{rs*}] \geq 0, \forall \mathbf{f} \in \Omega \tag{31}$$

where Ω is a polyhedron defined by constraints (15) and (16).

Since Friesz et al. (1993) and Smith (1993) proposed the VI formulation for the predictive dynamic user equilibrium traffic assignment problem, the solution algorithms to this type of dynamic equilibrium problems in transportation have been studied extensively, and many are suggested, including heuristic algorithms (e.g., Tong and Wong 2000, Huang and Lam 2002, etc.), projection-type algorithms (e.g., Wu *et al.* 1998, Facchinei and Pang 2003, etc.), and algorithms based on ascend directions (e.g., Zhu and Marcotte 1993, etc.). Since the comparison of the performance of different algorithms is beyond the scope of this paper (interested readers may refer to Nie and Zhang 2005 for detailed comparisons), in this study, we adopted the heuristic MSA algorithm to solve the path-based SO-DTA problem.

We describe the complete steps of the MSA algorithm for solving the path-based SO-DTA problem in networks in mono-centric cities as follows:

[Algorithm 2]

Step 0: Select an initial path flow pattern $\mathbf{f_0}$ and set the iteration index $k=0$;

Step 1: Load $\mathbf{f^k}$ into the network;

Step 2: For all $rs \in RS$, search for the time-dependent path $[p^*,t^*]$ with the least marginal cost, i.e., $[p^*,t^*] = \arg\min_{p\in P^{rs}, t\in T_d} PMC_{pt}^{rs}(\mathbf{f})$;

Step 3: Obtain the auxiliary path flow pattern $\mathbf{g}(\mathbf{f^k})$ by assigning all the demands $Q^{rs}, \forall rs \in RS$ onto $[p^*,t^*]$;

Step 4: Set $\lambda = 1/k$ and update the solution by setting $\mathbf{f^{k+1}} = (1-\lambda)\mathbf{f^k} + \lambda g(\mathbf{f^k})$;

Step 5: Check if $\mathbf{f^{k+1}}$ satisfies the convergence criterion. If yes, stop; otherwise, set $k = k+1$ and return to step 1.

The searching algorithm for time-dependent least marginal cost path

One major step in the above heuristic MSA algorithm is to search for the time-dependent least marginal cost path. Algorithms searching for time-dependent shortest paths have been studied for a very long time. Many algorithms, such as algorithms based on extensions of static shortest path algorithms(Ziliaskopoulos and Mahmassani 1993), algorithms based on decreasing order of time (DOT) (Chabini 1998), algorithms based on space-time network expansion (STEN) (Pallottino and Scutell 1998), have been proposed, among which, the DOT algorithm by Chabini (1998), was proven to have the least computational complexity. Our time-dependent least marginal cost path searching algorithm is developed based on the DOT algorithm. Modifications are made to the original DOT algorithms to take care of the correct path perturbation propagation. We resort to the STEN to illustrate our modifications to DOT.

It is well recognized that searching for the time-dependent shortest path is equivalent to searching for the minimal cost path in the STEN (either fully expanded or compacted). For example, suppose we have a simple network as illustrated in Figure 11, the corresponding time-space expansion network is shown in Figure 12. The time-dependent nodes are connected to adjacent nodes according to the time-dependent link traversal times. In our case of searching for the time-dependent least marginal cost path, however, the transitions between links are not according to c_{at} but $w_a(t)-t$ instead. Therefore, each link $[(i,t),(j,t+c_{ijt})$ in the original STEN is replaced by $[(i,t),(j,w_{ij}(t))]$. For instance, for the simple network shown in Figure 13, the original links $[(1,t),(2,t+c_{12t})]$ and $[(2,t),(3,t+c_{23t})]$ (the solid lines) are replaced by links $[(1,t),(2,w_{12}(t))]$ and $[(2,t),(3,w_{23}(t))]$ (the dashed links), respectively. Once this change is made, the DOT algorithm can still be applied to solve for the least marginal cost path problem.

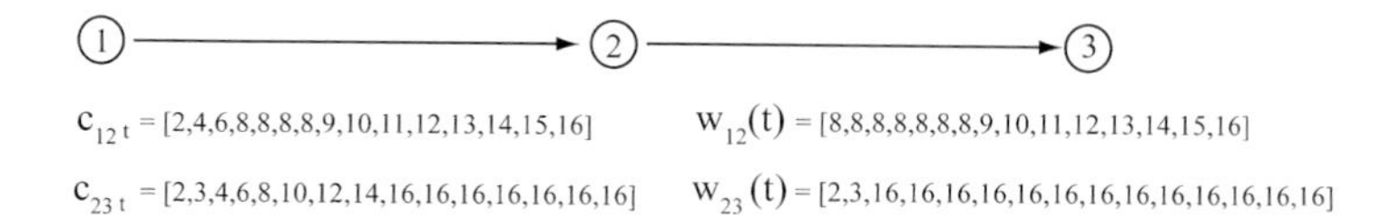

Figure 11. An illustration network for constructing a STEN

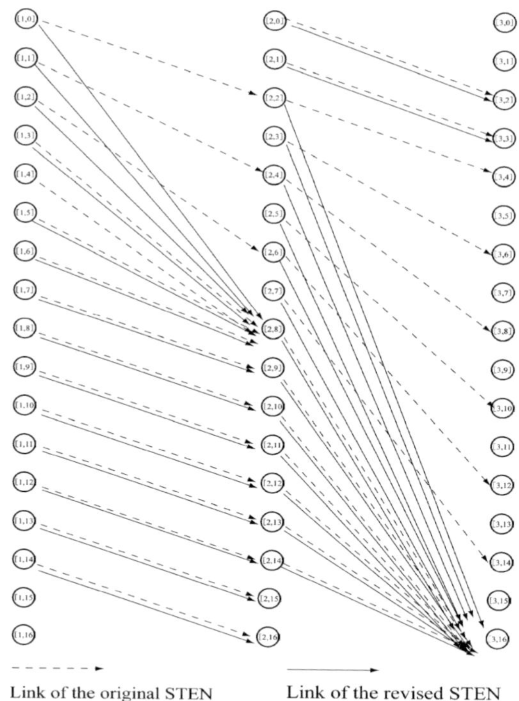

Figure 12. The space-time expansion network

For completeness, we present our least marginal cost path searching algorithm as follows:

[Algorithm 3]

Step 0: Preprocessing: derive $w_a(t)$ and $LMC_{at}(\mathbf{u}_a), \forall a \in A, t \in T_d$ using Algorithm 1;

Step 1: Initialization: set $D_i(t) = \infty, \forall i \neq s$ and $D_s(t) = 0, \forall t < T$, where s denote a destination node. Set $p_s(t) := 0, \forall t$;

Step 2: Set $D_i(T) :=$ the static shortest path tree rooted at s with all costs defined by $c_{ij}(T)$. Furthermore, note that $D_i(t) = D_i(T), \forall t \geq T$;

Step 3: for $t = T-1$ down to 0

```
        for (i, j) ∈ A
            if D_i(t) > LMC_ij(t) + D_j[w_ij(t)] ;
                D_i(t) := LMC_ij(t) + D_j[w_ij(t)];
                p_j[w_ij(t)] := [i,t] ;
            endif
        endfor
      endfor
```

NUMERICAL RESULTS

In this section, we first validate the PPT method by comparing the PMCs generated by the PPT method with those by performing DNL repeatedly. Next, we give two numerical examples to demonstrate how the proposed algorithm for path-based SO-DTA models perform. All the algorithms are coded in MS-VC++ and run on a Windows-XP PC (Intel Pentium M 1.60 GHz, 768 MB of RAM) .

Numerical validation of the PPT method

The test network, shown in Figure 13. , consists of two sequential links. The demand pattern shown in Figure 14 is loaded onto the network. The PMCs corresponding to each time interval by the PPT method is depicted in Figure 15. For comparison, the numerical PMC generated by performing a DNL with a perturbed path flow pattern is also depicted. As shown, the PPT method produces results very close to the numerical PMC evaluation results especially when the amount of perturbed flow in the DNL method gets smaller.

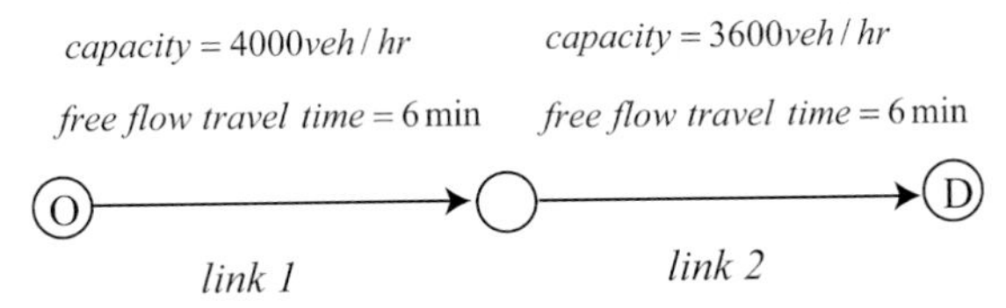

Figure 13. Test network for the PPT method validation

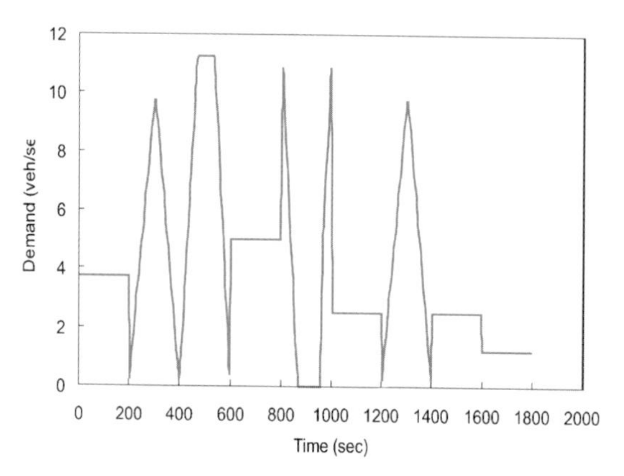

Figure 14. Demand pattern for the test network

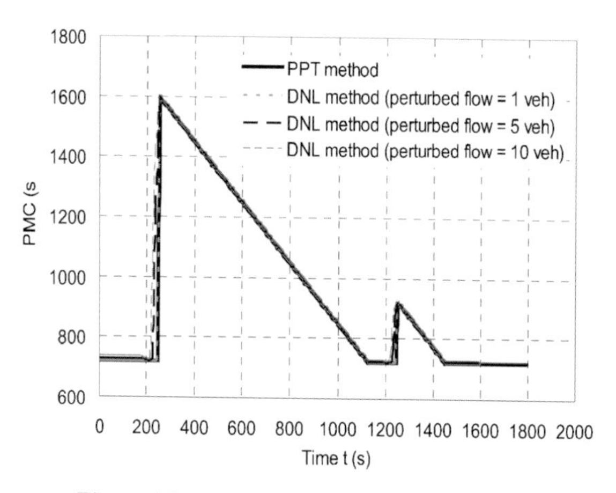

Figure 15. PMCs calculated by different methods

SO-DTA Numerical Example I

To demonstrate how the prediction of path flow perturbation propagation affects the accuracy of the final system optimum solution, an example network which contains two routes in parallel (Figure 16) is constructed. To simplify the discussion, in this example we only focus on the system optimal route choice, and the time-dependent departure rates are assumed to be given. The free flow travel times of route 1 and route 2 are 60 min and 12 min respectively. Vehicles depart from the origin at a constant departure rate $q = 3000$ veh/hr for one hour.

Route 1 does not have any bottleneck. Three scenarios which differ from each other in the number of bottlenecks on route 2 are designed. The capacity characteristics of the three scenarios are summarized in Table 1. We expect that the more bottlenecks on a route, the larger error incurred by the deficient PMC evaluation method.

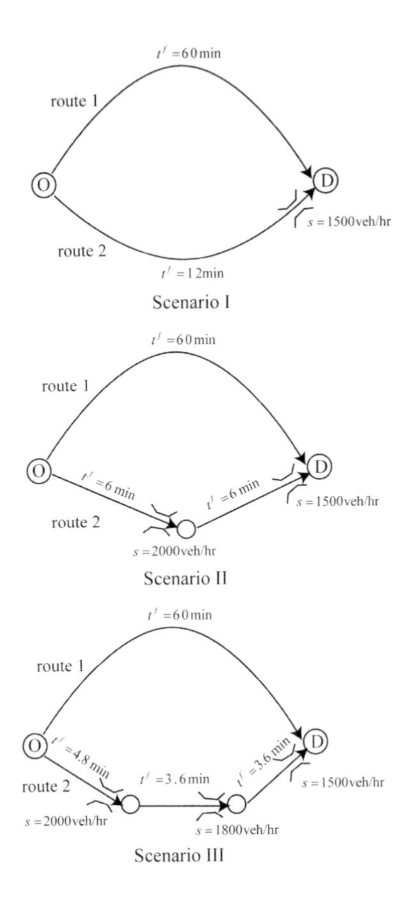

Figure 16. Example network I

Before presenting the numerical results, we first derive the analytical solution for this example network. In all the three scenarios, the smallest bottleneck is the one that locates at the downstream end of route 2, which controls the travel time on route 2. Hence, the analytical SO solution in terms of the route choice pattern for all the scenarios are actually the same. Hence, it suffices to derive the analytical solution for one scenario (scenario I, for instance).

Table 1. Network characteristics

Scenario I	route 1	no bottleneck, $t^f = 60\,\text{min}$
	route 2	bottleneck I: $t^f = 12\,\text{min}, s = 1500\text{veh/hr}$
Scenario II	route 1	no bottleneck, $t^f = 60\,\text{min}$
	route 2	bottleneck I: $t^f = 6\,\text{min}, s = 2000\text{veh/hr}$
		bottleneck II: $t^f = 12\,\text{min}, s = 1500\text{veh/hr}$
Scenario III	route 1	no bottleneck, $t^f = 60\,\text{min}$
	route 2	bottleneck I: $t^f = 4.8\,\text{min}, s = 2000\text{veh/hr}$
		bottleneck I: $t^f = 8.4\,\text{min}, s = 1800\text{veh/hr}$
		bottleneck II: $t^f = 12\,\text{min}, s = 1500\text{veh/hr}$

As we know, at system optimum the PMCs are the same on utilized paths. Since route 1 always operates at the free flow condition, $PMC_{1,t}(\mathbf{f}) \equiv 60$ min $\forall t$. According to the optimality conditions, we must have the following relationship (where t^* is the last time that vehicles enter route 1):

$$PMC_{1,t^*} = 60\text{min} = PMC_{2,t^*} = T^e - t^* \tag{32}$$

From time t^* to $T = 60$ min, the inflow rate on route 2 will always be 3000 veh/hr and the inflow rate on route 1 will always be 0. From time 0 to time t^*, it is obvious that route 2 will move as much vehicles as possible. However, once the inflow rate on route 1 exceeds its capacity $s = 1500$ veh/hr, its path marginal cost will be larger than the path marginal cost on route 1 since $PMC_{1,t}(\mathbf{f}) = T^e - t > T^e - t^* = 60$ min. Consequently, at system optimum, the inflow rate on route 2 during time $[0, t^*]$ should be equal to its capacity $s = 1500$ veh/hr, and the inflow rate on route 1 is equal to $3000 - 1500 = 1500$ veh/hr. Now we only need to determine T^e and t^*.

Since the queue on route 2 start to build up from time $t^* + 12$ min. we have the following relationship for t^* and T^e:

$$\frac{3000}{60}(60 - t^*) = \frac{1500}{60}[T^e - (t^* + 12)] \tag{33}$$

Based on eqn (32) and eqn (33), we get $t^* = 36$ min, and $T^e = 96$ min. In summary, the system optimal route choice for this example is as follows:

$$\text{For}\,[0, 36\text{min}]: \; d_1(t) = 1500\text{veh/hr}, d_2(t) = 1500\text{veh/hr}$$

$$\text{For}\,[36\text{min}, 60\text{min}]: \; d_1(t) = 0, d_2(t) = 3000\text{veh/hr}$$

Both the PPT and LTT methods are applied and combined with the heuristic MSA algorithm to solve the SO-DTA problem. The convergence curves for both methods in the three scenarios are depicted in Figures 17 and 18.

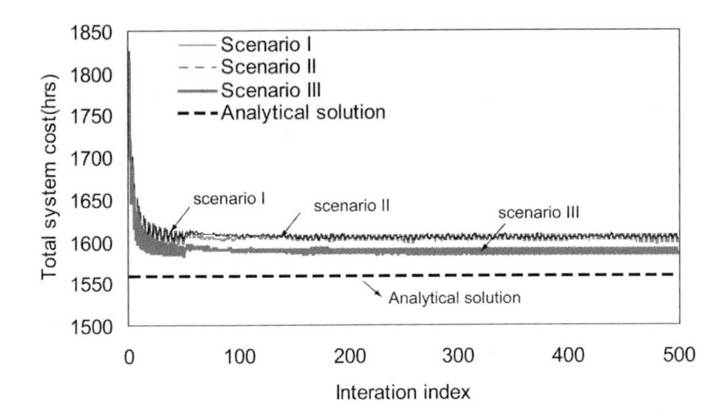

Figure 17. Convergence curves for the algorithm based on the PPT method

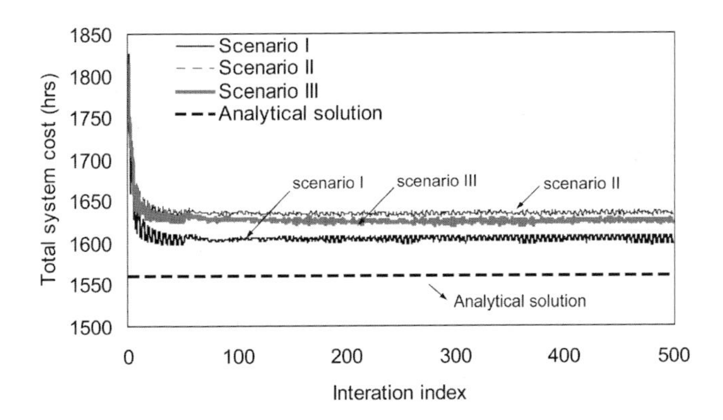

Figure 18. Convergence curves for the algorithm based on the LTT method

From Figures 17 and 18 we can see that the objective function, i.e., the total system cost, stop decreasing after about 100 iterations. Table 2 lists the total system cost for both methods in the three scenarios. Figures 19 and 20 visualize the resulting route choice pattern for both methods. All these solutions are obtained after performing 500 iterations of the corresponding algorithm.

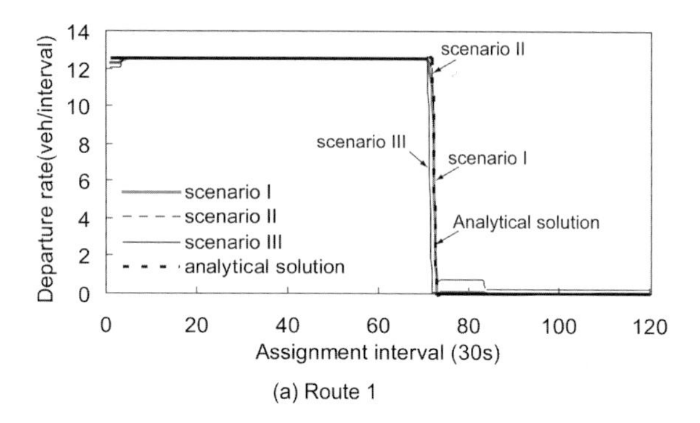

(a) Route 1

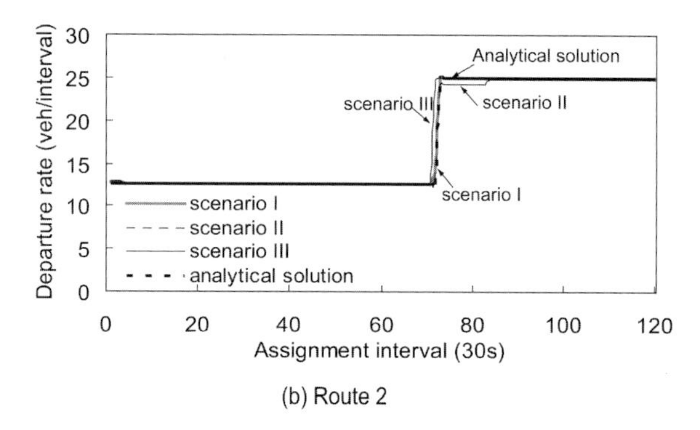

(b) Route 2

Figure 19. Numerical solutions based on the PPT method

As we can see from Table 2, and Figures 19 and 20, in scenario I the numerical solutions based on both the PPT and LTT methods are identical and very close to the analytical solution. This is not a surprise because when there is only one bottleneck on route 2, the PMCs and LMCs are identical. In scenarios II and III, the PPT method can still achieve very good accuracy compared to the analytical solution, while the numerical solutions based on the LTT method have distinct deviations from the analytical solution. Nevertheless, we should point out that the PMC becomes discontinuous at the point where a queue just starts to form, and this slightly changes the convergence pattern near the optimal solution. A more rigorous treatment of this problem should apply variational analysis theory to obtain the subgradients at these points. With the subgradients, a more accurate solution procedure can be developed to solve the path based SO-DTA problem. Due to space limitations, the variational treatment of this problem will be reported elsewhere.

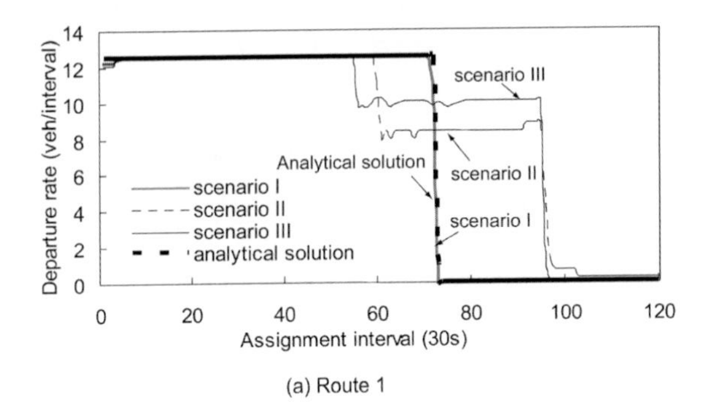

(a) Route 1

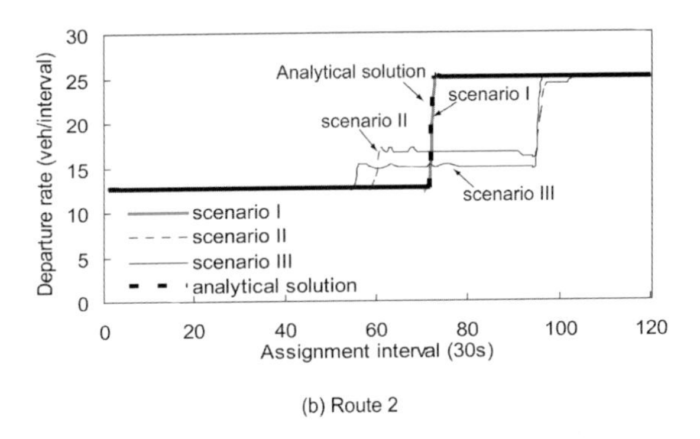

(b) Route 2

Figure 20. Numerical solutions based on the LTT method

Table 2. Total system cost (Hrs)

Scenario	PPT	LTT
Scenario I	1597.92 (102.43%)	1597.92 (102.43%)
Scenario II	1601.38 (102.65%)	1631.69 (104.60%)
Scenario III	1584.04 (101.54%)	1623.04 (104.04%)
Analytical solution	1560	1560

SO-DTA Numerical Example II

The second numerical example is designed to test whether the heuristic MSA method based on the PPT method can generate an accurate system optimal departure time pattern. The test network contains only one link (Figure 21) and our main objective is to derive the system optimal departure time pattern. The link free flow travel time is $t^f = 10$ min, and there is a bottleneck with capacity $s = 1800$ veh/hr at the downstream end of the link. The total demand is 1500 veh. The desired arrival time is $\tilde{t} = 7:00$ am. The schedule delay parameters are $\Delta = 0, \alpha = 1, \beta = 0.8, \gamma = 1.2$.

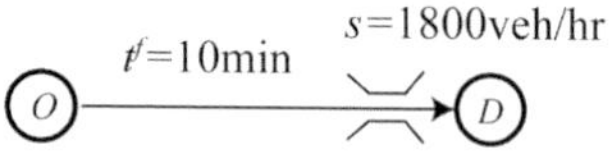

t^f=10min

s=1800veh/hr

O

D

Figure 21. Example network II

Similar as before, we first derive the analytical SO solution for this example.

In order to achieve minimal system cost, it can be easily shown that whenever there are commuters traversing the bottleneck, no queue is present and the bottleneck serves at its capacity. Namely, during the morning peak, the system optimal departure rate is always constant and equal to s . Consequently, it suffices to only determine the starting time t_s and ending time t_e of the departure process, which can be obtained via solving the following optimization problem:

$$\min_{\{t_s, t_e\}} TC(N) = \int_{t_s}^{\tilde{t}-t^f} \beta[\tilde{t} - (t + t^f)]sdt + \int_{\tilde{t}-t^f}^{t_e} \gamma[(t + t^f) - \tilde{t}]sdt$$

subject to

$$N = (t_e - t_s)s$$
$$t_s + t^f \le \tilde{t}, t_e + t^f \ge \tilde{t} .$$

The optimal solution is as follows:

$$t_s = \tilde{t} - t^f - \frac{\gamma}{\beta + \gamma} \frac{N}{s} = 6:20 \text{ am} \tag{34}$$

$$t_e = \tilde{t} - t^f + \frac{\beta}{\beta + \gamma} \frac{N}{s} = 7:10 \text{ am} \tag{35}$$

$$\text{Departure rate } a(t) = s = 1800 \text{veh/hr} . \tag{36}$$

The convergence patterns for both methods are very similar to those in numerical example I and hence are not shown here. We only show the optimal departure time choice patterns obtained based on both methods in Figure 22 (t = 0 corresponds to the time 6:00 am), in comparison with the analytical solution computed directly from eqn (34) - eqn (36).

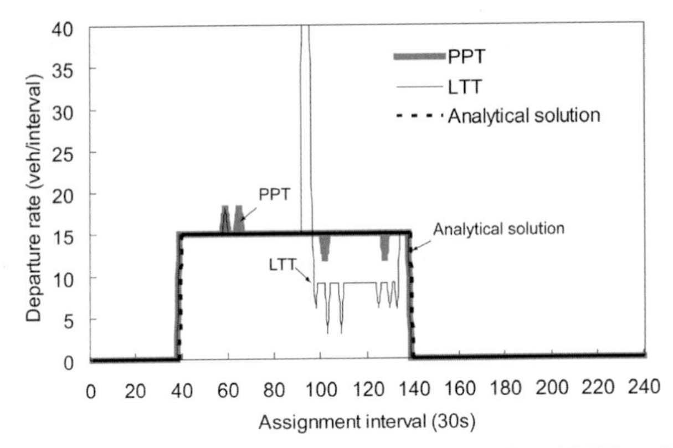

Figure 22. Numerical solutions based on both PPT and LTT methods

As we can see from the results, the heuristic MSA algorithm based on the PPT method still converges to the analytical solution while the same algorithm based on the LTT method cannot. This is understandable because in the LTT method, its deficiency in predicting the propagation of path flow perturbation yields inaccurate marginal schedule delay at the destination D.

CONCLUSIONS

In this paper we have studied the path-based SO-DTA problem with an emphasis on its first-order optimality conditions and the computation of its path marginal costs. Through a series of examples, we demonstrated that path marginal costs are not simple additions of the corresponding link marginal costs unless the flow perturbation travels with the vehicle unit that initiated the perturbation. We further show that one can compute efficiently path marginal costs for networks with a special structure (i.e., no diverges), and proposed an evaluation method that decomposes PMCs to LMCs for such networks. This decomposition scheme involves two major tasks: 1) the evaluation of LMC and 2) the trace of the path flow perturbation through the network. The former is accomplished with the assistance of cumulative flow diagrams, and the latter with DNL. Finally, we employed the method of successive averages with a modified time-dependent minimum cost algorithm to solve the SO-DTA problem for a mono-centric network.

Our preliminary numerical results indicate that

1) The proposed heuristic MSA algorithm based on our path marginal cost computational procedure can generate numerical solutions very close to analytical solutions for SO-DTA problems with and without departure time choice.

2) On the other hand, the heuristic MSA algorithm based on the path marginal cost computational procedure that directly sums up link marginal costs, fails to converge to the analytical solutions when there are more than one bottleneck on the path, or the problem involves departure time choice, because this path marginal cost evaluation procedure does not correctly track the path flow perturbation.

It should be pointed out that our proposed algorithm for solving path-based SO-DTA models has its own limitations. At present the method can be applied to networks without diverges and the embedded traffic flow models are restricted to those not considering queue spillback. The relaxation of either aspect may bring in additional challenges in predicting path flow perturbation propagation and is worth further investigation. Moreover, the path cost mapping under certain traffic flow models may not be convex or even continuous, hence the KKT optimality conditions may not be sufficient to guarantee a global optimum solution when such models are used to describe traffic evolution. Finally, the computational efficiency of this method when applied to medium or large size networks is also worthy of a careful study. Nevertheless, we hope that our analysis of the SO-DTA problem and the evaluation procedures for PMCs for mono-centric networks can provide insights to the solution of the SO-DTA problem for general networks, as well as other optimization problems in transportation that involve the computation of gradients of mappings without a closed form.

ACKNOWLEDGEMENTS

This research is supported in part by a grant from the National Science Foundation under the contract number CMS#9984239, and a grant from the University Transportation Center at UC Davis. We also wish to thank two anonymous referees for their suggestions to improve the paper. The views are those of the authors alone.

REFERENCES

Arnott, R, A de Palma and R Lindsey (1990). Departure time and route choice for the morning commute. *Transportation Research*, **23B**(3), 209-228.

Carey, M (1986). A constraint qualification for a dynamic traffic assignment model. *Transportation Science*, **20**(1). 55-58.

Carey, M (1987). Optimal time-varying flows on congested networks, *Operations Research*, **35**(1), 58-69.

Carey, M (1992). Nonconvexity of the dynamic traffic assignment problem, *Transportation Research,* **26B**, 127-133.

Carey, M and E Subrahmanian (2000). An approach to modelling time-varying flows on congested networks. *Transportation Research,* **34B**, 157-183.

Chabini, I (1998). Discrete dynamic shortest path problems in transportation applications. *Transportation Research Record* **1645**, 170-175.

Chang, G-L, H S Mahmassani and M L Engquist (1988). System-optimal trip scheduling and routing in commuting networks. *Transportation Research Record*, **1251**, 54-65.

Daganzo, C F (1995). The cell transmission model, part II: network traffic. *Transportation Research*, **29B**(2), 79-93.

Facchinei, F and J S Pang (2003). *Finite-dimensional variational inequalities and complementarity problems, Vol. I, II*. Springer, New York, U.S.A.

Friesz, T, D Bernstein, T Smith, R Tobin and B-W Wie (1993). A variational inequality formulation of the dynamic network user equilibrium problem. *Operations Research* **41**(1), 179-191.

Friesz, T L, J Luque, R L Tobin and B-W Wie (1989). Dynamic network traffic assignment considered as a continuous time optimal control problem. *Operations Research*, **37**(6), 893-901.

Ghali, M O and M J Smith (1995). A model for the dynamic system optimum traffic assignment problem. *Transportation Research*, **29B**(3), 155-170.

Hendrickson, C and G Kocur (1981). Schedule delay and departure time decisions in a deterministic model. *Transportation Science*, **15**(1), 62-77.

Ho, J K (1980). A successive linear optimization approach to the dynamic traffic assignment problem. *Transportation Science*, **14**(4), 295-305.

Huang, H-J and Lam, W H K (2002). Modelling and solving dynamic user equilibrium route and departure time choice problem in networks with queues. *Transportation Research*, **36B**, 253-273.

Lebacque, J P (1996). The Godunov scheme and what it means for first order traffic flow models. *Proceedings of the 13th International Symposium on Transportation and Traffic Theory*, 647-677.

Li, Y (2001). Development of dynamic traffic assignment models for planning applications, PhD thesis, Northwestern University.

Lo, H (1999). A dynamic traffic assignment formulation that encapsulates the cell-transmission model. *Proceedings of the 14th International Symposium on Transportation and Traffic Theory*, 327-350.

Merchant, D K and G L Nemhauser (1978*a*). A model and an algorithm for the dynamic traffic assignment problems. *Transportation Science*, **12**(3), 183-199.

Merchant, D K and Nemhauser, G L (1978*b*). Optimality conditions for a dynamic traffic assignment model. *Transportation Science*, **12**(3), 200-207.

Munoz, J C and J A Laval (2005). System optimum dynamic traffic assignment graphical solution method for a congested freeway and one destination. *Transportation Research*, **40B**(1), 1-15.

Nie, X and H M Zhang (2005a). A comparative study of some macroscopic link models used in dynamic traffic assignment. *Networks and Spatial Economics*, **5**(1), 89-115.

Nie, X and H M Zhang (2005b). Delay-function-based link models: their properties and computational issues. *Transportation* Research, **39B**, 729-751.

Nie, Y and H M Zhang (2005). Equilibrium analysis of the morning commute problem: numerical solution procedures *(working paper)*.

Pallottino, S and M Scutell (1998). Shortest path algorithms in transportation models: classical and innovative aspects. In: *Equilibrium and Advanced Transportation Modelling*, Kluwer.

Peeta, S and H S Mahmassani (1995). System optimal and user equilibrium time-dependent traffic assignment in congested networks. *Annals of Operations Research*, **60**, 81-113.

Sheffi, Y (1985). Urban transportation networks: equilibrium analysis with mathematical programming methods. Englewood Cliffs, NJ: Prentice-Hall.

Small, K (1982). The scheduling of consumer activities: work trips. *American Economic Review*, **72**, 467-479.

Smith, M J (1993). A new dynamic traffic model and the existence and calculation of dynamic user equilibria on congested capacity-constrained road networks. *Transportation Research*, **26B**, 49-63.

Tong, C and S Wong (2000). A predictive dynamic traffic assignment model in congested capacity-constrained road networks. *Transportation Research*, **34B**, 625-644.

Vickrey, W S (1969). Congestion theory and transport investment. *American Economic Association*, **59**, 251-261.

Wardrop, J G (1952). Some theoretical aspects of road traffic research. *Proceedings of the Institute of Civil Engineering, Part II* **1**, 325-378.

Wie, B-W (1998). A convex control model of dynamic system-optimal traffic assignment. *Control Engineering Practice*, **6**. 745-753.

Wie, B-W and R L Tobin (1998). Dynamic congestion pricing models for general traffic networks. *Transportation Research,* **32B**(5), 313-327.

Wie, B-W, R L Tobin, D Bernstein and T L Friesz (1995). A comparison of system optimum and user equilibrium dynamic traffic assignment with schedule delays. *Transportation Research*, **3C**(6), 389-411.

Wie, B-W, R L Tobin, and T L Friesz (1994). The augmented Lagrangian method for solving dynamic network traffic assignment models in discrete time. *Transportation Science*, **28**(3), 204-220.

Wu, J H, M Florian, Y W Xu and J M Rubio-Ardannaz (1998). A projection algorithm for the dynamic network equilibrium problem. *Proceedings of the International Conference on Traffic and Transportation Studies*, 379-390.

Yang, H and Q Meng (1998). Departure time, route choice and congestion toll in a queuing network with elastic demand. *Transportation Research*, **32**(4), 247-260.

Zhu, D L and P Marcotte (1993). Modified descent direction methods for solving the monotone variational inequality problem. *Operations Research Letters*, **14**, 111-120.

Ziliaskopoulos, A K (2000). A linear programming model for the single destination system optimum dynamic traffic assignment problem. *Transportation Science*, **34**(1), 37-49.

Ziliaskopoulos, A K and H S Mahmassani (1993). Time-dependent shortest path algorithms for real-time intelligent vehicle highway system applications. *Transportation Research Record*, **1408**, 94-100.

Transportation and Traffic Theory 2007
Edited by R.E. Allsop, M.G.H. Bell and B.G. Heydecker

16

COMMUTING EQUILIBRIA ON A MASS TRANSIT SYSTEM WITH CAPACITY CONSTRAINTS

Qiong Tian and Hai-Jun Huang, School of Management, Beijing University of Aeronautics and Astronautics, China; Hai Yang, Department of Civil Engineering, The Hong Kong University of Science and Technology, China

INTRODUCTION

Urban mass transit system plays a vital role in reducing road traffic congestion through offering alternative means of travel. It in fact directly influences the quality of urban life. The planning and operation of transit systems have become important issues of long standing interests to economists and transportation scientists. In the end of 1960s, scholars started to optimize the inter-station spacing of a rapid transit system (Vuchic and Newell, 1968; Vuchic, 1969). Mohring (1972, 1976) developed a microeconomic foundation for studying the public transportation services with fixed demand. He proposed the well-known 'square root rule' for the determination of optimal bus service frequency (i.e., the optimal bus frequency should be proportional to the square root of travel demand). This rule is in line with the work by Vickrey (1955) on the implications of marginal cost pricing for public utilities. De Cea and Fernandez (1993) proposed an equilibrium transit assignment model in the limited line capacity networks. In this model, for finishing a whole journey from origin to destination, a passenger must determine a sequence of transfer nodes, and the congestion cost at each transfer node is dependent upon both the boarding flow and the flow already on the line. The latest studies concerning the transit system modeling can be found in De Cea and Fernandez (2000), Guan et al. (2005) and Vuchic (2005).

There are few attentions paid to the problem of modeling commuters' departure time choice for urban transit services. Sumi et al. (1990) presented a stochastic model for optimizing commuters' departure time and route choices in a mass transit system. They assumed that the departure time is mainly dependent on the system's operational features and the travelers' preferred time of arrival at the destination. Alfa and Chen (1995) examined a public transportation system with multiple origins and destinations and proposed an algorithm for calculating the peak-hour departure time of commuters, where commuters ride on the first coming bus in a random order. Recently, Kraus and Yoshida (2002) and Kraus (2003) provided economic analyses about the time-of-use decisions of commuters in which the boarding behavior follows the first-in-first-out (FIFO) principle. Cascetta and Papola (2003) proposed a joint mode-transit model considering the choices of departure time and transit mode. Cominetti and Correa (2001) and Cepeda et al. (2006) investigated the congestion of transit system by means of a simplified bulk queue model.

To explicitly take into account the effects of in-vehicle crowding on commuters' departure time choice, Huang et al. (2004) introduced a in-vehicle crowding cost function for modeling urban mass transit services, i.e., $C(n,\tau)=g(n)\tau$, where τ is the in-vehicle riding time of a commuter, n is the number of commuters in the vehicle, and $g(n)$ is the average crowding cost per unit time. It is believed that the in-vehicle crowding cost is mainly contributed by privacy loss, body touch (uncomfortable physical proximity), air pollution and venture of being stolen, and can be regarded as an increasing function of n, with $g(n)\geq 0$ and $g(0)=0$. With this crowding cost function, commuters make a trade-off between increasing crowding cost from traveling closer to the peak of the rush-hour and reduced early/late arrival penalty cost at workplace in determining their optimal departure times. Using this function, Huang et al. (2004) developed an equilibrium departure time choice model for an urban mass transit system with a single origin and a single destination. In equilibrium, the individual's total generalized cost is identical for all actually chosen transit services during the peak-period. The crowding cost function adopted by Huang et al. (2004) is of the property that $\lim_{n\to N_0} g(n)=\infty$, where N_0 is the maximum physical capacity of a transit run. Huang et al. (2004, 2005) further investigated the optimal pricing and service frequency of mass transit services with elastic demand in alternative market settings, such as monopoly and competitive transit services.

Although the relative importance of radial travel toward the central business district (CBD) in urban areas has been decreasing in recent years, this movement remains the most concentrated one, in both space and time, and therefore represents one of the most critical urban transportation problems. With the use of in-vehicle crowding function, Tian et al. (2006) proposed a model to examine the equilibrium properties of the peak-period commuting in a mass transit system with multiple origins and a single destination. They obtained some important findings about the boarding behaviors of commuters deeply. However, this model doesn't consider the physical and seat capacities explicitly, thus the effects of overflow queues at stations and the different discomforts experienced by sitting and standing commuters cannot be revealed.

In this paper, we make one important step forward to the work of Tian et al. (2006) by explicitly formulating the vehicle capacity constraints into transit corridor commuting modeling with crowding effects. Two types of capacities, namely physical capacity (the maximum number of passengers that a vehicle can safely load) and seat capacity (the number of seats designed in a vehicle), are taken into account. By explicitly introducing the physical capacity constraint, we are able to model the effects of overflow queues at stations in a highly congested transit system. Commuters who want to board a capacitated transit run should arrive earlier and wait for more time to get the boarding priority and thus experience peak-period queuing at the stop. By introducing the in-vehicle seat capacity, we can investigate the asymmetric property of utilizing transit service, i.e., the different degree of discomfort experienced by sitting and standing commuters. Passengers who board a transit run at upstream stops can have seats and will not be affected by the crowding condition at downstream stops. This asymmetrical property is a challenging but important subject in transit modeling.

In the next section, a basic mathematical programming model which incorporates the physical capacity constraint only is formulated. The model consists of multiple interrelated minimization problems, each correspond to one transit stop. We then derive some equilibrium properties of commuting by analyzing the model's solution and give another mathematical programming model which considers the physical and seat capacity constraints simultaneously. A solution method for obtaining the equilibrium departure pattern of commuters is presented and tested in two examples. We finally compare the commuters' travel costs and behaviors caused by physical and seat capacity constraints.

THE MODEL

As shown in Figure 1, we consider a mass transit line with multiple origins and a single destination. Transit runs depart from the most distant residential location H_1 (original stop) and pass through H_2, ..., H_{K-1} and H_K home locations or stops in the order of decreasing distance to the workplace/destination W. For simplicity and without loss of generality, we suppose each transit run consists of only one vehicle in this paper. In each morning commuting peak period, there are N_1, N_2, ..., and N_K commuters, who use the transit line from stops H_1, H_2, ..., and H_K all the way to workplace W, respectively. Each transit run vehicle has the same limiting physical capacity denoted by N_0. Because of the physical capacity constraint, some transit runs may be fully occupied before they arrive at the destination stop W. We call these runs as "saturated runs" in this paper. With no fare variation across transit runs, the equilibrium entails such arrival patterns at stops that commuters who experience less schedule delay in "saturated runs" must wait for more time at the stop.

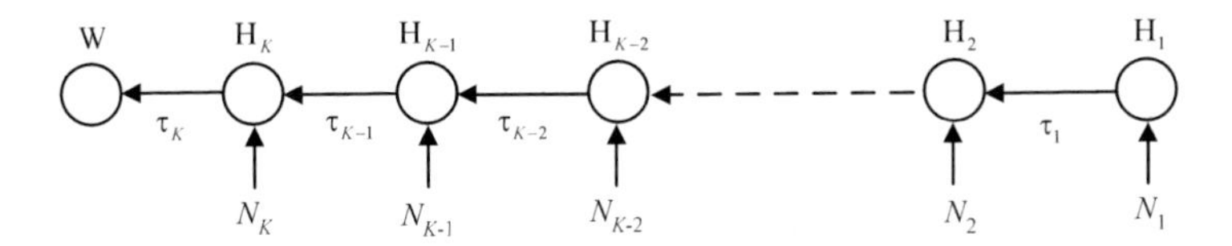

Figure 1. The transit line with multiple origins and a single destination.

If a transit run reaches this limiting capacity before arriving at a stop, the commuters originated from this stop would not board this run and have to wait until a subsequent unsaturated transit run arrives. Thus, not all commuters can choose their most desirable transit runs, even if they would like to wait at the stop for boarding. In this case, they will depart late or early and thus arrive at workplace late or early in order to avoid queuing at the stops.

The generalized commuting cost experienced by a commuter consists of waiting time cost at stops, in-vehicle travel time, in-vehicle crowding cost and penalty cost for early or late arrival. For having analytical tractability, we assume that all commuters are homogeneous in the sense that they have the same desirable arrival time, the same value of time and the valuation of in-vehicle crowding. Because a long-term commuting equilibrium is concerned, commuters are assumed to have full information about the transit system timetable. The generalized commuting cost of a commuter who ride on transit run j at H_i is given by

$$TC_j^i = p_i + \alpha T^i + \lambda_j^i + \delta(j) + \rho_j^i, \tag{1}$$

where p_i is the transit fare from H_i to W and is assumed to be a constant (time-varying fare is not considered), αT^i is the in-vehicle time cost, with α being the unit in-vehicle time cost and T^i being the total in-vehicle moving time from H_i to W. Assume that the vehicle moving speed is constant, then the moving time in each line segment from H_1 to W (including the stopping time at each station for boarding) is constant and denoted by $\tau_1, \tau_2, \cdots, \tau_K$, respectively. Clearly, $T^i = \sum_{m=i}^{K} \tau_m$ is constant too.

The third term λ_j^i in Eq. (1) is the total crowding cost of a commuter taking transit run j at stop H_i. The crowding cost is assumed to be a function of the degree of crowding effects and the in-vehicle travel time. We then have

$$\lambda_j^i = \sum_{s=i}^{K} g\left(\sum_{m=1}^{s} n_j^m\right)\tau_s, \tag{2}$$

where n_j^m is the number of commuters who take transit run j at H_m, τ_s is the in-vehicle moving time from H_s to H_{s+1}, and $g(n)$ is the crowding cost function (crowding cost per unit in-vehicle travel time) which is assumed to be a monotonically increasing function of the number of commuters in the transit vehicle, n. The crowding cost is zero when the vehicle is empty (no passenger), i.e., $g(0)=0$.

The fourth term $\delta(j)$ in Eq. (1) is the early/late arrival penalty when taking transit run j. Suppose in this study that the timetable of the transit system is predefined and the transit headways or departure time intervals from original stop are identical for any two successive runs. Because the moving time in each line segment from H_1 to W is fixed to be $\tau_1, \tau_2, \cdots, \tau_K$, thus all transit runs arrive at workplace W at a uniform interval t. There exists a so-called one-to-one mapping between the commuters' platform departure times and destination arrival times. Let $Z = \{\xi, \cdots, 2, 1, 0, -1, -2, \cdots, -\zeta\}$ be the set of transit runs arriving at the workplace, where ξ and ζ are sufficiently large to ensure that all commuters can arrive at the workplace during the rush hour considered. We can assume that only one transit run arrives at the workplace W on time (arrival at work-start time) and this run is denoted by 0. In this way, $j > 0$ denotes the runs arriving early and the early arrival time is $j \times t$, while $j < 0$ denotes the runs arriving late and the late arrival time is $|j| \times t$. In our study, $\delta(j)$ is given below

$$\delta(j) = \begin{cases} \beta j t, & j > 0 \\ 0, & j = 0 \\ \gamma |j| t, & j < 0 \end{cases} \quad j \in Z, \tag{3}$$

where β and γ are the schedule delay penalties of unit early and late arrival time at workplace W, respectively.

The fifth term ρ_j^i in Eq. (1) is the aforementioned queuing time cost for taking transit run *j* at station H_i. If run *j* is unsaturated after leaving station H_i or $\sum_{s=1}^{i} n_j^s < N_0$, then $\rho_j^i = 0$, i.e., commuters at H_i doesn't bear (overflow) queuing time cost for riding on this run; if, however, run *j* is saturated after leaving H_i or $\sum_{s=1}^{i} n_j^s = N_0$, then commuters at H_i may bear (overflow) extra queuing time cost or $\rho_j^i \geq 0$. We thus have

$$\begin{cases} \rho_j^i = 0, & \text{if } \sum_{s=1}^{i} n_j^s < N_0 \\ \rho_j^i \geq 0, & \text{if } \sum_{s=1}^{i} n_j^s = N_0 \end{cases} \quad j \in Z,\ i = 1, 2, \cdots, K. \tag{4}$$

Because all commuters have identical value of in-vehicle travel time α, for all commuters from the same station H_i, αT^i is constant. Like the transit fare p_i, αT^i doesn't affect the departure time choice of commuters departing from the same station. Without loss of generality, we can simply assume that $p_i + \alpha T^i = 0$. Suppose all commuters attempt to minimize their individual total generalized commuting costs in choosing their departure times (or transit services). At equilibrium, commuters departing from the same station should have identical total generalized travel cost, no one has incentive to alter his/her departure time. Mathematically, this requirement can be expressed as

$$\begin{cases} TC_j^i = TC^i, & \text{if } n_j^i > 0 \\ TC_j^i \geq TC^i, & \text{if } n_j^i = 0 \end{cases} \quad j \in Z,\ i = 1, 2, \cdots, K, \tag{5}$$

where TC^i is the equilibrium total generalized commuting cost from H_i to W. This equation states that, for a given transit station H_i, the travel cost by run *j* is equal to the equilibrium cost if run *j* is utilized by some commuters from that station; and the travel cost by run *j* is not less than the equilibrium cost if it is not used by any commuters from that station.

With a given transit timetable, the equilibrium departure distribution of the commuters from all stops, $\{n_j^i \mid i = 1, 2, \cdots, K, j \in Z\}$, can be obtained by solving the following *K* interrelated minimization problems

$$\min L(\mathbf{n}_i) = \sum_{s=1}^{K} \left[\sum_{j \in Z} G\left(\sum_{m=1}^{s} n_j^m \right) \right] \tau_s + \sum_{j \in Z} \left(\sum_{i=1}^{K} n_j^i \right) \delta(j), \quad i = 1, 2, \cdots, K, \tag{6}$$

subject to

$$\sum_{j \in Z} n_j^i = N_i, \tag{7}$$

$$n_j^i \geq 0, \quad j \in Z, \tag{8}$$

$$\sum_{s=1}^{i} n_j^s \leq N_0, \quad j \in Z, \tag{9}$$

where $G(x) = \int_0^x g(\omega)\,d\omega$ and $\mathbf{n}_i = \{n_j^i \mid j \in Z\}$. The objective function (6) represents the sum of the integral of all commuters' in-vehicle crowding costs over number of commuters and their total late/early arrival penalty costs. Similar to the conventional user equilibrium model, the integral $G(x)$ doesn't have any economic interpretation. Constraint (7) states that the sum of the numbers of commuters using all transit services is equal to the total number of commuters during the whole rush hour. Constraint (8) is simply a nonnegative constraint for all decision variables and Constraint (9) is the physical capacity constraint.

The above K minimization problems describe the optimal departure time (or transit service) choice decisions made by commuters at each stop when the departure patterns at other stops are given. Clearly, each minimization problem has a convex objective function and a convex set of constraints in its own decision variables, and hence has a unique solution of transit service choices. Note that the simultaneous optimization of K problems may have not unique solution.

Now we show that the optimality conditions of the K minimization problems (6)-(9) constitute the equilibrium requirement (5). The first-order optimality conditions of the ith minimization problem are given as follows

$$n_j^i\left[\lambda_j^i+\delta(j)+u_j^i-v^i\right]=0, \quad j\in Z, \tag{10}$$

$$\lambda_j^i+\delta(j)+u_j^i-v^i\geq 0, \quad j\in Z, \tag{11}$$

$$u_j^i\left(N_0-\sum_{s=1}^{i}n_j^s\right)=0, \quad j\in Z, \tag{12}$$

$$u_j^i\geq 0, \quad j\in Z, \tag{13}$$

$$\sum_{s=1}^{i}n_j^s\leq N_0, \quad j\in Z, \tag{14}$$

$$\sum_{j\in Z}n_j^i=N_i, \tag{15}$$

$$n_j^i\geq 0, \quad j\in Z, \tag{16}$$

where v^i and u_j^i are the Lagrange multipliers associated with Constraints (7) and (9), respectively. Eqs. (10) and (11) indicate that if some commuters takes run j from stop H_i, i.e., $n_j^i>0$, then their total generalized travel costs are identical and equal to v^i, and otherwise, not less than v^i. Thus, v^i can be regarded as the equilibrium commuting cost TC^i defined in (5). Furthermore, Eqs. (12) and (13) describe the relationship between the station queuing time and the transit vehicle's physical capacity. If $\sum_{s=1}^{i}n_j^s<N_0$, i.e., the transit vehicle is unsaturated after leaving stop H_i, then the queuing time cost is zero (i.e., $u_j^i=0$); otherwise, $u_j^i\geq 0$ if $\sum_{s=1}^{i}n_j^s=N_0$. Therefore, the Lagrange u_j^i can be regarded as the equilibrium queuing time cost at stop H_i ρ_j^i defined in (4).

PROPERTIES OF THE EQUILIBRIUM

Some theorems

We now examine the basic properties of the equilibrium solution by looking into the relation of departure times of commuters from the stops at various distances from the workplace. Let c_j^i denote the crowding and schedule delay cost for taking transit run j at stop H$_i$, i.e.,

$$c_j^i = \lambda_j^i + \delta(j). \tag{17}$$

Combining Eq. (17) with Eqs. (10)-(13) generates

$$\begin{cases} c_j^i = v^i,\ u_j^i = 0, & \text{if } n_j^i > 0,\ \sum_{s=1}^{i} n_j^s < N_0 \\ c_j^i + u_j^i = v^i,\ u_j^i \geq 0, & \text{if } n_j^i > 0,\ \sum_{s=1}^{i} n_j^s = N_0 \\ c_j^i \geq v^i,\ u_j^i = 0, & \text{if } n_j^i = 0,\ \sum_{s=1}^{i} n_j^s < N_0 \\ c_j^i + u_j^i \geq v^i,\ u_j^i \geq 0, & \text{if } n_j^i = 0,\ \sum_{s=1}^{i} n_j^s = N_0 \end{cases} \quad j \in Z,\ i = 1, 2, \cdots, K, \tag{18}$$

where n_j^i is the number of commuters who board transit run j at stop H$_i$. When $\sum_{s=1}^{i} n_j^s = N_0$, we have $u_j^i = v^i - c_j^i$. Let n_j^{i-1} denote the number of commuters who board transit run j at stop H$_{i-1}$, $1 < i \leq K$. According to the definition (17), we have

$$c_j^i + g\left(\sum_{m=1}^{i-1} n_j^m\right)\tau_{i-1} = c_j^{i-1}. \tag{19}$$

We define some sets as follows. Let $J^i = \{j \mid n_j^i > 0, \forall j \in Z\}$ be the set of transit runs having positive boarding passenger flows at H$_i$, $A = \left\{j \,\middle|\, \sum_{s=1}^{K} n_j^s = N_0, \forall j \in Z\right\}$ the set of transit runs which are saturated when arriving at W in the peak period, $B = \left\{j \,\middle|\, \sum_{s=1}^{K} n_j^s < N_0, \forall j \in Z\right\}$ the set of transit runs which are unsaturated when arriving in W; $A_i = \left\{j \,\middle|\, \sum_{s=1}^{i} n_j^s = N_0, \forall j \in Z\right\}$ the set of transit runs which are saturated when leaving the stop H$_i$, and $B_i = \left\{j \,\middle|\, \sum_{s=1}^{i} n_j^s < N_0, \forall j \in Z\right\}$ the set of transit runs which are unsaturated when leaving the stop H$_i$.

Theorem 1. *At equilibrium,* $\forall a \in A_i$, $\forall b \in B_i$, $n_a^i > 0$, $n_b^i > 0$, *we have* $\delta(a) < \delta(b)$.

Proof. According to Eq.(16) and the definitions of A_i and B_i, we have $u_a^i \geq 0$ and $u_b^i = 0$. Combining with Eq. (18) leads to

$$c_a^i + u_a^i = v^i, \tag{20}$$

$$c_b^i = v^i. \tag{21}$$

Using Eqs. (2) and (17), we rewrite the relation given by Eqs. (20) and (21) as follows,

$$\sum_{s=i}^{K}\left[g\left(\sum_{m=1}^{s} n_b^m\right)\tau_s\right]+\delta(b)=\sum_{s=i}^{K}\left[g\left(\sum_{m=1}^{s} n_a^m\right)\tau_s\right]+\delta(a)+u_a^i. \quad (22)$$

The definitions of A_i and B_i give

$$\sum_{s=i}^{K}\left[g\left(\sum_{m=1}^{s} n_a^m\right)\tau_s\right]=g(N_0)\sum_{s=i}^{K}\tau_s, \quad (23)$$

$$\sum_{s=i}^{K}\left[g\left(\sum_{m=1}^{s} n_b^m\right)\tau_s\right]<g(N_0)\sum_{s=i}^{K}\tau_s. \quad (24)$$

Substituting Eqs.(23) and (24) into Eq.(22), we have $\delta(b)>\delta(a)$. ■

Theorem 1 states that at equilibrium, if there are some transit runs leaving the stop H_i in saturated state, the commuters who can ride on these runs at this stop experience lower schedule delay cost. Clearly, this theorem continuously holds when *i*=*K*, which leads to the following result.

Corollary 1. *At equilibrium,* $\forall a \in A, b \in B$, *we have* $\delta(a)<\delta(b)$.

Corollary 1 states that if there are some transit runs arriving at destination W in saturated state, these saturated runs would concentrate in a period around the work-start time so as to reduce the schedule delay cost. We define this period during which arrival transit runs are fully occupied as "saturated period".

Theorem 2. *At equilibrium, if there exists a stop* $H_i (i>1)$ *such that* $J^i \cap A \neq \varnothing$, *then* $J^{i-1} \cap A \neq \varnothing$.

Proof. Since $J^i \cap A \neq \varnothing$, for any $j \in J^i \cap A$, the following equations hold

$$c_j^i + u_j^i = v^i, \quad (25)$$

$$n_j^i > 0. \quad (26)$$

We provide a proof of the theorem by contradiction. Suppose $J^{i-1} \cap A = \varnothing$, then $J^{i-1} \cap B \neq \varnothing$. $\forall a \in J^i \cap A$, $\forall b \in J^{i-1}$, we have $n_a^i > 0$, $n_b^{i-1} > 0$, $n_a^{i-1} = 0$ and then $\sum_{s=1}^{i-1} n_a^s < N_0$, $\sum_{s=1}^{i-1} n_b^s < N_0$. Combining these with Eq. (18) generates

$$n_a^i > 0,\ u_a^i \geq 0, \quad (27)$$

$$n_b^i \geq 0,\ u_b^i = 0, \quad (28)$$

$$n_a^{i-1} = 0,\ u_a^{i-1} = 0, \quad (29)$$

$$n_b^{i-1} > 0,\ u_b^{i-1} = 0. \quad (30)$$

According to Eq. (18), we have

$$c_a^i = v^i, \quad (31)$$

$$c_b^i \geq v^i, \tag{32}$$
$$c_a^{i-1} \geq v^{i-1}, \tag{33}$$
$$c_b^{i-1} = v^{i-1}. \tag{34}$$

Substituting Eq. (19) into Eqs. (33) and (34) yields

$$c_a^i + g\left(\sum_{m=1}^{i-1} n_a^m\right)\tau_{i-1} \geq v^{i-1}, \tag{35}$$

$$c_b^i + g\left(\sum_{m=1}^{i-1} n_b^m\right)\tau_{i-1} = v^{i-1}. \tag{36}$$

Comparing Eqs. (31) and (32), we have $c_a^i \leq c_b^i$. Substituting this into Eqs. (35) and (36) yields

$$g\left(\sum_{m=1}^{i-1} n_a^m\right)\tau_{i-1} \geq g\left(\sum_{m=1}^{i-1} n_b^m\right)\tau_{i-1}. \tag{37}$$

According to the property of the $g(x)$, Eq. (37) becomes

$$\sum_{m=1}^{i-1} n_a^m \geq \sum_{m=1}^{i-1} n_b^m. \tag{38}$$

Substituting $n_a^{i-1} = 0$ into Eq. (38) and noting $n_b^{i-1} > 0$, we have

$$\sum_{m=1}^{i-2} n_a^m > \sum_{m=1}^{i-2} n_b^m. \tag{39}$$

Eqs. (33) and (34) state $c_a^{i-1} \geq c_b^{i-1}$. Using the definition (19), we conclude

$$c_a^{i-2} > c_b^{i-2}. \tag{40}$$

This means $n_a^{i-2} = 0$ and $n_b^{i-2} \geq 0$. Substituting this into Eq. (39) yields

$$\sum_{m=1}^{i-3} n_a^m > \sum_{m=1}^{i-3} n_b^m. \tag{41}$$

Repeating the above process, we conclude

$$n_a^s = 0, \qquad \forall s \leq i-1, \tag{42}$$
$$n_b^s \geq 0, \qquad \forall s \leq i-1, \tag{43}$$

and

$$\sum_{m=1}^{i-1} n_b^m > \sum_{m=1}^{i-1} n_a^m = 0. \tag{44}$$

This contradicts with Eq. (38). Therefore, $J^{i-1} \cap A = \varnothing$ is impossible. ■

Using Theorem 2, we can easily derive the following corollaries:

Corollary 2. *At equilibrium, if there exists a stop* $H_i\,(i < K)$ *such that* $J^i \cap A = \varnothing$, *then* $J^{i+1} \cap A = \varnothing$.

Corollary 3. *At equilibrium, if there exists a stop* $H_i\,(i < K)$ *such that* $J^i \cap A = \varnothing$, *then* $J^{i+1} \cap B \neq \varnothing$.

Corollary 4. *At equilibrium, if there exists a stop* $H_i\left(i<K\right)$ *such that* $J^i \cap A = \varnothing$, *then* $J^s \cap A = \varnothing$, $\forall s > i$

Corollary 4 states that if there exits a stop where no commuter boards the saturated vehicles, then all downstream commuters of that stop would avoid commuting in the saturated period. This result is different from that given in the case without capacity constraint (Tian et al., 2005). We here show that each upstream commuter has a priority of boarding the capacitated vehicles in the peak period.

Theorem 3. *At equilibrium, if there exists a stop* $H_i\left(i<K\right)$ *such that* $J^i \cap A \neq \varnothing$, *and a transit run* $a(\in A)$ *such that* $\sum_{s=1}^{i} n_a^s < N_0$, *then* $J^{i+1} \cap A \neq \varnothing$.

Proof. We provide a proof of the theorem by contradiction. Suppose $J^{i+1} \cap A = \varnothing$, according to Corollary 4, we then have $\forall s > i+1$, $J^s \cap A = \varnothing$. Thus $\sum_{s=i+1}^{K} n_a^s = 0$. We have

$$\sum_{s=1}^{K} n_a^s = \sum_{s=1}^{i} n_a^s + \sum_{s=i+1}^{K} n_a^s < N_0. \tag{45}$$

This contradicts with the definition of set A, which needs $\sum_{s=1}^{K} n_a^s = N_0$. Therefore, $J^{i+1} \cap A = \varnothing$ is impossible. ■

Theorem 3 states that the upstream commuters would firstly board the transit vehicles which leave the original stop during the saturated period.

Theorem 4. *At equilibrium, if there exists a stop* $H_i\left(i>1\right)$ *such that* $J^i \cap B \neq \varnothing$, $J^{i-1} \cap B \neq \varnothing$, *and a transit run* $b(\in B)$ with $n_b^i > 0$, *then* $n_b^{i-1} > 0$.

Proof. We provide a proof of the theorem by contradiction. Suppose $n_b^{i-1} = 0$. Let c be such a transit run ($c \in J^{i-1} \cap B$) with $n_c^{i-1} > 0$. Combining $\sum_{s=1}^{K} n_b^s < N_0$ and $\sum_{s=1}^{K} n_c^s < N_0$ with Eq. (18), we have

$$c_b^i = v^i, \tag{46}$$

$$c_c^i \geq v^i, \tag{47}$$

$$c_b^{i-1} \geq v^{i-1}, \tag{48}$$

$$c_c^{i-1} = v^{i-1}. \tag{49}$$

Substituting Eq. (19) into Eqs. (48) and (49), yields

$$c_b^i + g\left(\sum_{m=1}^{i-1} n_b^m\right)\tau_{i-1} \geq c_c^i + g\left(\sum_{m=1}^{i-1} n_c^m\right)\tau_{i-1}. \tag{50}$$

From Eqs. (46) and (47), we have $c_b^i \le c_c^i$. Substituting this into Eqs.(48) and (49), yields

$$g\left(\sum_{m=1}^{i-1} n_b^m\right)\tau_{i-1} \ge g\left(\sum_{m=1}^{i-1} n_c^m\right)\tau_{i-1}. \tag{51}$$

According to the property of the function $g(x)$, we have

$$\sum_{m=1}^{i-1} n_b^m \ge \sum_{m=1}^{i-1} n_c^m. \tag{52}$$

Because $n_c^{i-1} > n_b^{i-1} = 0$, we can derive

$$\sum_{m=1}^{i-2} n_b^m > \sum_{m=1}^{i-2} n_c^m. \tag{53}$$

Thus, we conclude that

$$c_b^{i-2} > c_c^{i-2}, \tag{54}$$

and then,

$$n_b^{i-2} = 0 \text{ and } n_c^{i-2} \ge 0. \tag{55}$$

Thus, no commuter will take the transit run b at stop H_{i-2}. Repeating this process, it can be proved that no commuter will take run b at stop H_s ($s \le i-2$), i.e., $\sum_{m=1}^{i-1} n_b^m = 0$. So, $\sum_{m=1}^{i-1} n_c^m > \sum_{m=1}^{i-1} n_b^m = 0$. This contradicts with Eq. (52). Therefore, $n_b^{i-1} = 0$ is impossible. ■

Theorem 4 states that for the stops where commuters can not entirely board the saturated vehicles, the departure time duration of commuters is increasing monotonically with the distance of the boarding stops from the workplace.

Theorem 5. *At equilibrium, if there exists a stop $H_i\,(i>1)$ such that $J^i \cap B \ne \varnothing$ and a transit run $a(\in A)$ with $\sum_{s=1}^{i} n_a^s < N_0$, then $n_a^i > 0$.*

Proof. We provide a proof of the theorem by contradiction. Suppose $n_a^i = 0$, $\forall a \in A$, and $\sum_{s=1}^{i} n_a^s < N_0$. Let b be a transit run, $n_b^i > 0$, $\forall b \in J^i \cap B$. According to Eqs. (16-19), we have

$$c_a^i \ge v_i, \tag{56}$$

$$c_b^i = v_i. \tag{57}$$

As $a \in A$, then $\sum_{i=1}^{K} n_a^i = N_0$. Due to $\sum_{m=1}^{i} n_a^m < N_0$, there must exist an intermediate stop l ($l > i$) with $n_a^l > 0$. Denote this stop as $l = \min\left\{s \middle| n_a^s > 0, s > i\right\}$. Then we have

$$c_a^l = v_l, \tag{58}$$

$$c_b^l \ge v_l. \tag{59}$$

Substituting Eq. (19) into Eq. (58) yields

$$c_a^i = v_l + g\left(\sum_{s=1}^{i} n_a^s\right)\sum_{m=i}^{l-1}\tau_m . \tag{60}$$

Combining Eq. (60) with Eq. (56) yields

$$v_i \le v_l + g\left(\sum_{s=1}^{i} n_a^s\right)\sum_{m=i}^{l-1}\tau_m . \tag{61}$$

Substituting Eq. (19) into Eq. (59) yields

$$c_b^i \ge v_l + g\left(\sum_{s=1}^{i} n_b^s\right)\sum_{m=i}^{l-1}\tau_m . \tag{62}$$

Combining Eq. (62) with Eq. (57) yields

$$v_i \ge v_l + g\left(\sum_{s=1}^{i} n_b^s\right)\sum_{m=i}^{l-1}\tau_m . \tag{63}$$

From Eqs. (61) and (63), we have

$$g\left(\sum_{s=1}^{i} n_b^s\right)\sum_{m=i}^{l-1}\tau_m \le g\left(\sum_{s=1}^{i} n_a^s\right)\sum_{m=i}^{l-1}\tau_m , \tag{64}$$

and then

$$\sum_{s=1}^{i} n_b^s \le \sum_{s=1}^{i} n_a^s . \tag{65}$$

Because $n_b^i > 0$ and $n_a^i = 0$, Eq. (65) leads to

$$\sum_{s=1}^{i-1} n_b^s < \sum_{s=1}^{i-1} n_a^s . \tag{66}$$

Rewrite Eqs. (56) and (57) as follows

$$c_b^i + g\left(\sum_{s=1}^{i-1} n_b^s\right)\tau_{i-1} \le c_a^i + g\left(\sum_{s=1}^{i-1} n_a^s\right)\tau_{i-1} . \tag{67}$$

Substituting Eq. (19) into Eq. (67) yields

$$c_b^{i-1} \le c_a^{i-1} . \tag{68}$$

Hence, $n_a^{i-1} = 0$ and $n_b^{i-1} \ge 0$. This means that no commuter takes the transit run a at stop H_{i-1}. Repeating this process, it can be concluded that no commuter will take transit run a at stop H_s ($s < i-1$), i.e., $\sum_{s=1}^{i-1} n_a^s = 0$. Thus, $\sum_{s=1}^{i-1} n_b^s > \sum_{s=1}^{i-1} n_a^s = 0$. This contradicts with Eq. (66). Therefore, $n_a^i = 0$ is impossible. ■

Theorem 5 states that for the stops where commuters can not entirely board the saturated vehicles, the commuters of that stop will firstly manage to board the saturated vehicle if it is available. As a result, these commuters, boarding saturated vehicle, should arrive earlier and wait in queue for their desired vehicle at equilibrium.

Extension to the case with seat capacity

In reality, transit operator usually provides some seats in the vehicle. Just as Tian et al. (2006) have mentioned, the seats capacity can be regarded as a constraint in model. We now formulate another mathematical programming model which can consider the physical capacity and in-vehicle seat capacity simultaneously. Suppose that the sitting passengers do not experience the in-vehicle crowding effect. With assumption that $p_i + \alpha T^i = 0$, the total generalized travel costs of standing and sitting passengers who take transit run j at stop H_i are

$$TC_j^i = \sum_{s=i}^{K} g\left(\sum_{m=1}^{s} n_j^m\right)\tau_s + \delta(j) + \rho_j^i, \tag{69}$$

$$\overline{TC}_j^i = q_j^i + \delta(j), \tag{70}$$

respectively. In Eq. (69), n_j^i is the number of passengers who take transit run j at stop H_i but is standing in the vehicle and ρ_j^i is the extra queuing time cost mentioned in previous section. As sitting is generally more comfortable than standing, here we assume that the sitting commuters do not suffer the in-vehicle crowding cost. In Eq. (70), q_j^i is the additional cost paid by a passenger in getting priority for a seat in the vehicle (Kraus and Yoshida, 2002). q_j^i is similar to ρ_j^i but larger than it under the FIFO queuing system. This reflects such a phenomenon in reality that some commuters in the front of the queue prefer to wait for boarding the next transit run when the current run has no seats available while some commuters standing behind them would like to board the run even they have to be standing during their journey. Thus, q_j^i exists only when all seats in the run j are occupied after leaving stop H_i and some commuters have to stand in it.

Let $\overline{n}_j^i$ denote the number of passengers who take transit run j at stop H_i and get seats in the vehicle, and N_s the number of seats designed in the vehicle. An equilibrium state is reached when no commuter can reduce his/her total generalized travel cost by unilaterally changing his/her choice of the transit services, which can be mathematically stated below

$$\begin{cases} TC_j^i \geq TC^i, \text{ if } n_j^i = 0 \\ TC_j^i = TC^i, \text{ if } n_j^i > 0 \\ \overline{TC}_j^i \geq TC^i, \text{ if } \overline{n}_j^i = 0 \\ \overline{TC}_j^i = TC^i, \text{ if } \overline{n}_j^i > 0 \end{cases} \quad j \in Z,\ i = 1, 2, \cdots, K, \tag{71}$$

where TC^i is a constant representing the identical equilibrium cost. The first two conditions in Eq. (71) are the same with that in Eq. (5). Note that the third condition in Eq. (71) covers two different situations: on the one hand, $\sum_{s=1}^{i-1} \overline{n}_j^s = N_s$ and then $\overline{n}_j^i = 0$, this means there are no seats available for commuters boarding at stop H_i, then q_j^i would be very large such that $\overline{TC}_j^i \geq TC^i$; on the other hand, $\sum_{s=1}^{i-1} \overline{n}_j^s < N_s$ but $\overline{n}_j^i = 0$, this means commuters do not board this run although there are seats available for them because the schedule delay cost of this run is very high such that $\overline{TC}_j^i \geq TC^i$ (although $q_j^i = 0$).

With the above considerations, we can formulate a model which results in a commuting equilibrium of the transit services with both physical capacity and in-vehicle seat capacity constraints. For a given transit vehicle departure interval t, the equilibrium departure distribution of the commuters from all stops, $\left\{n_j^i, \bar{n}_j^i \mid i=1,2,\cdots,K, j \in Z\right\}$, can be obtained by solving the following K interrelated minimization problems

$$\min L(\mathbf{n}_i) = \sum_{s=1}^{K}\left[\sum_{j\in Z} G\left(\sum_{m=1}^{s} n_j^m\right)\right]\tau_s + \sum_{j\in Z}\left(\sum_{i=1}^{K}\left(n_j^i + \bar{n}_j^i\right)\right)\delta(j), \quad i=1,2,\cdots,K, \tag{72}$$

subject to

$$\sum_{j\in Z}\left(n_j^i + \bar{n}_j^i\right) = N_i, \tag{73}$$

$$n_j^i \geq 0, \bar{n}_j^i \geq 0, \quad j \in Z, \tag{74}$$

$$\sum_{s=1}^{i} n_j^s \leq N_u, \ j \in Z, \tag{75}$$

$$\sum_{s=1}^{i} \bar{n}_j^s \leq N_s, \ j \in Z, \tag{76}$$

where $G(x) = \int_0^x g(\omega)\,\mathrm{d}\omega$, $\mathbf{n}_i = \left\{n_j^i, \bar{n}_j^i \mid i=1,2,\cdots,K, j \in Z\right\}$ and N_u is the physical capacity of standing in vehicle. Let κ denote the transfer parameter of a seat space to standing space. The total vehicle size N_0 can be regarded as $\kappa N_s + N_u$ standing position or $N_s + N_u/\kappa$ sitting position. As sitting usually takes more space than standing, we hence set $\kappa \geq 1$ in this paper. It is easy to show that the first-order conditions of each stop's minimization problem are equivalent to the equilibrium conditions (71) which govern the transit service choice at each stop.

The model consists of multiple minimization problems, each having a unique solution. However, the simultaneous optimization of K problems, i.e., the whole model, may have not unique solution. It is not easy to analytically derive the equilibrium properties of the model with two types of capacity constraints as done in the case with physical capacity constraint only. In this paper, we will give a numerical example to explore some properties.

SOLUTION ALGORITHM

In this section, a multi-stage algorithm is developed to solve the model proposed in this paper. At every stage, a convex optimization problem associated with one stop of the transit line is solved. The mathematical programs of the examples presented in this paper are directly solved by a standard FMINCON function in the Matlab 6.5 Optimization Toolbox. The step-by-step procedure of the multi-stage algorithm is given below.

Step 1 Initialization. Set the iteration index m=1. For each transit stop and transit run, choose an initial boarding flow $\mathbf{n}^{(m)}$.

Step 2 Optimization and assignment.

Step 2.1 Fix the boarding flows at all stops except the original stop H_1, denoted by $\mathbf{n}_{(-1)}^{(m)} = \left\{n_j^{i(m)} \middle| j \in Z, 1 < i \leq K\right\}$, solve the minimization problem (6)-(9) and obtain the equilibrium boarding flow at stop H_1, denoted by $\mathbf{n}_1^{(m+1)}$.

•
•
•

Step 2.*k* Fixed boarding flows all stops except the stop H_k, denoted by $\mathbf{n}_{(-k)}^{(m)} = \left\{ n_j^{1(m+1)}, \ldots, n_j^{k-1(m+1)}, n_j^{k+1(m)}, \ldots n_j^{K(m)} \mid j \in Z \right\}$, solve the minimization problem (6)-(9) and get the equilibrium boarding flow at H_k, denoted by $\mathbf{n}_k^{(m+1)}$.

Step 2.*K* With $\mathbf{n}_{(-K)}^{(m)} = \left\{ n_j^{i(m+1)} \mid j \in Z, 1 \le i < K \right\}$, solve the minimization problem (6)-(9) and denote the equilibrium boarding flow at stop H_k as $\mathbf{n}_K^{(m+1)}$.

Step 3 Update. Set $\mathbf{n}^{(m+1)} = \left\{ n_j^{i(m+1)} \mid j \in Z, 1 \le i \le K \right\}$.

Step 4 The iteration terminates if $\sum_{1 \le i \le K} \sum_{j \in Z} \left| n_j^{i(m+1)} - n_j^{i(m)} \right| < \varepsilon$ (ε=0.001 in this paper), otherwise, set m=m+1 and return to Step 2.

NUMERICAL EXAMPLES

In this section, we present numerical results on two examples for exploring the commuting characteristics. Example 1 is the case with physical capacity only. We will investigate how the capacity affects the departure behaviors of commuters from different locations. In Example 2, we fix the vehicle size and turn to investigate the impacts of in-vehicle seat capacity on travel behavior. Note that the total passenger capacity of a vehicle may change when seat capacity is changed.

Example 1—with physical capacity constraint only

This example sets the following input parameters: the number of stops in the transit line K=3 (stops), the time interval of dispatching transit runs t=0.05 (h), the work-start time t^*=8:00AM, the penalty cost of unit time arrival early $\beta=10$ (RMB/h), the penalty cost of unit time arrival late γ =30 (RMB/h), τ_1 =0.25 (h), τ_2 =0.2 (h), τ_3 =0.3 (h), $N_1 = N_2 = N_3 = 600$ (persons). A linear in-vehicle crowding cost function $g(n) = 10n / N_0$ (RMB/h) is adopted, where N_0 is the physical capacity. Here, the symbol RMB (pronunciation 'RenMinBi') stands for the monetary unit of the Chinese currency (8 RMB $\approx$ 1 US dollar).

Figure 2 shows that the physical capacity can greatly influence the commuting costs of passengers. Note that the transit fare and the constant travel time cost are not included in these commuting costs. It can be seen that all commuters from different stops can benefit from increasing the vehicle capacity. However, the marginal contribution of capacity expansion is decreasing and the downstream commuters enjoy more cost reduction than the upstream commuters. This is because the upstream commuters have priorities for boarding, but these priorities are gradually weakened by the capacity expansions. In an extreme situation where the physical capacity is large enough so that all commuters along the corridor can be accommodated in one run to get the workplace on time, then, the boarding priority is identical for every commuter.

Figure 3 depicts the boarding flows of all transit runs at different stops when the physical capacity is 700. It can be seen that the physical capacity is not reached even for the on-time transit run. The transit run arriving at 8:00 serves 102, 160 and 400 passengers at stop 1, stop 2 and stop 3, respectively, all amounting to 662 less than the capacity 700. This distribution pattern is identical to that shown in Tian et al. (2006) where no capacity constraint is considered in the model. Although the physical capacity of the on-time transit run is not completely utilized, some commuters turn to select other transit runs due to the increasing in-vehicle crowding cost.

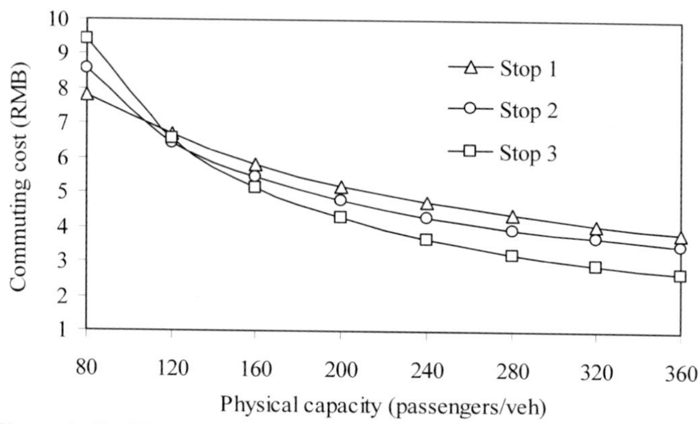

Figure 2. Equilibrium commuting costs with various physical capacities.

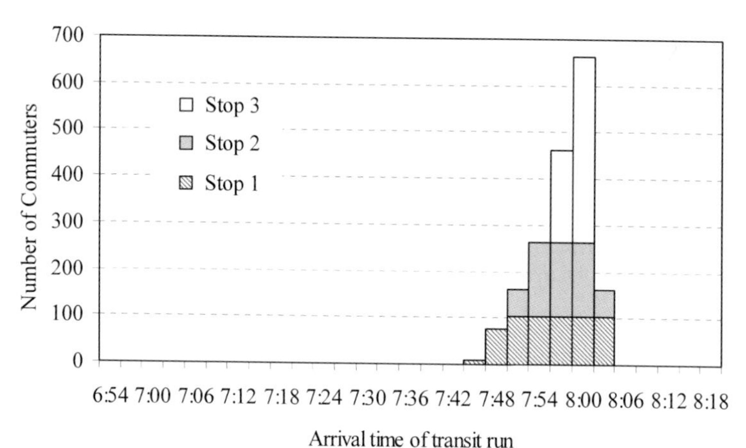

Figure 3. Boarding flows when the physical capacity is 700.

Figure 4 give the result when the physical capacity is reduced to 400. Two transit runs, arriving at 7:57 and 8:00, are fully utilized. These two runs serve the same number of commuters at stop 1, but do not so at stops 2 and 3. It is verified again that the capacity change greatly affects the boarding behavior of downstream commuters; the more the commuters live close to the workplace, the more their behaviors are affected by the capacity. In Figure 4, commuters from stop 2 can select transit runs according to their will since all transit runs have not fully utilized yet. But, some commuters from stop 3 are forced to use the runs excluding the two fully utilized ones, which makes the commuters who can board the two fully utilized runs have to spend queuing time at the platform for equalizing the individual costs among all commuter required by equilibrium condition.

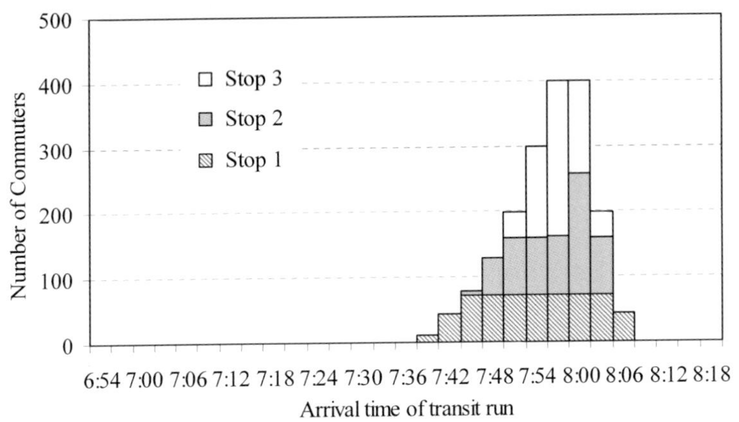

Figure 4. Boarding flows when the physical capacity is 400.

Figure 5 depicts the boarding flows of all chosen transit runs at the three stops when the physical capacity is further reduced to 200. It can be seen that the number of fully utilized transit runs increases to six. There are no commuters from stop 3 who can arrive at W on time. Note that the distribution of commuters at stop 1 is significantly different from that observed in Figure 4. The boarding priority for commuters from stop 1 is visible while some commuters from stop 2 cannot freely get on the two runs arriving at W at 7:57 and 8:00 (they have to spend waiting time). The boarding duration at stop 1 is the longest, lasting from 7:30 to 8:09.

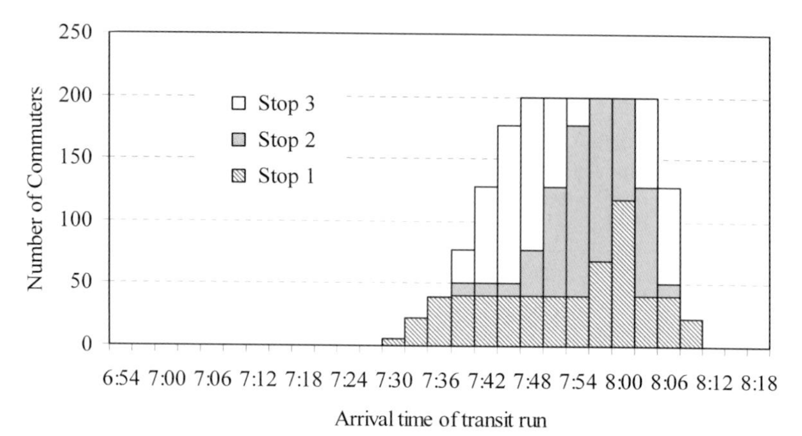

Figure 5. Boarding flows when the physical capacity is 200.

More congested distribution can be observed if the physical capacity decreases further, as shown in Figure 6 where the capacity is 100. It can be seen that more transit runs are fully utilized and the boarding duration has to be extended (in Figures 5, only 14 runs are actually chosen by commuter, but 21 runs in Figure 6). Commuters from stop 1 board the runs which arrive at W at the times tightly close to the work-start time. These commuters certainly have such priorities. Commuters from Stop 3 board the runs which arrive at W at the times far to the work-start time. This is because small physical capacity strengths the boarding priority for upstream commuters. This explains the phenomenon shown in Figure 2, i.e., the downstream commuters like more the capacity expansion than the upstream commuters.

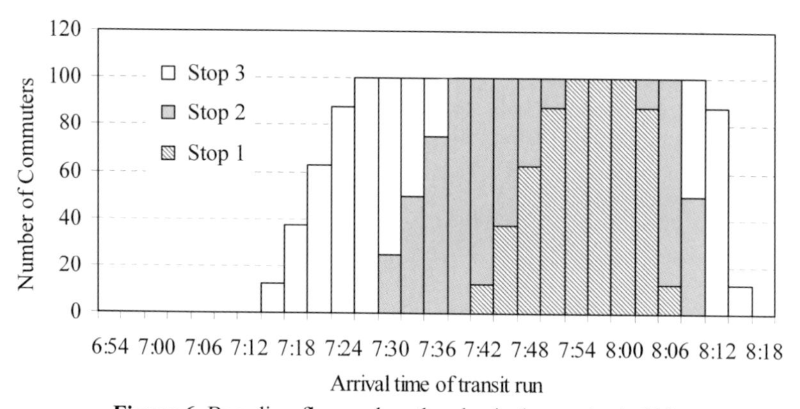

Figure 6. Boarding flows when the physical capacity is 100.

Example 2—with both physical capacity and seat capacity constraints

In this example, we set the total transit vehicle size N_0=400 (seats) and vary the in-vehicle seat capacity N_s to investigate the model solutions. The in-vehicle standing capacity is $N_u = \kappa(N_0 - N_s)$ where $\kappa = 1.5$. Other model parameters are: K=3 (stops), t=0.05 (h), t^* =8:00AM, β =10 (RMB/h), γ =30 (RMB/h), τ_1 =0.2 (h), τ_2 =0.2 (h), τ_3 =0.2 (h), $N_1 = N_2 = N_3 = 1{,}500$ (persons). The linear in-vehicle crowding cost function, $g(n) = 20n / N_u$ (RMB/h), is adopted in this example.

Figure 7 gives the equilibrium commuting costs of passengers boarding at different stops, against various seat capacities. Note that the transit fare and the constant travel time cost are not included in this commuting cost. It can be seen that if more seats are provided in the vehicle, commuters from stops 1 and 2 benefit while those from stop 3 suffer a light loose. This is explained as follows. With given total transit vehicle size, increasing the number of seats will lead the capacity for standing to come down. The upstream commuters have the priority to use the seats and the downstream commuters have to stand in the vehicle meanwhile experience more in-vehicle crowding costs with smaller space for standing (see Figures 9 and 10), or board the earliest and latest arrival transit runs (see Figures 10 and 11).

Figure 8 depicts the boarding flows when there are 100 seats in each vehicle. The standing capacity becomes 450 and the total number of commuters that a vehicle can serve is 550. Stop 1 has the longest commuting peak period [7:36, 8:06], all seats of the used runs are occupied by commuters from this stop and some ones have to be standing in their journeys. The commuters from other stops behave like that without any seats available.

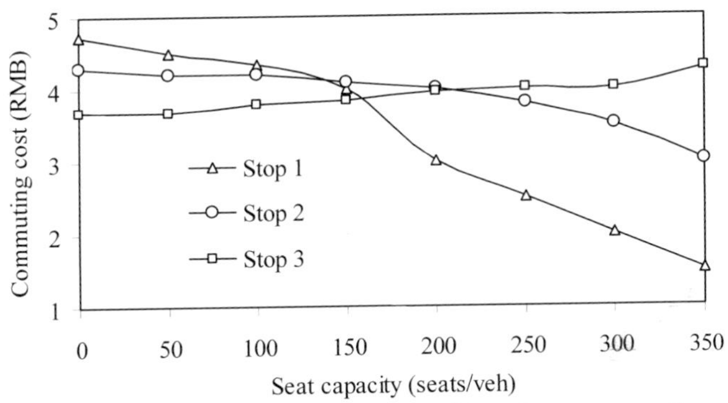

Figure 7. Equilibrium commuting cost with various seat supplies.

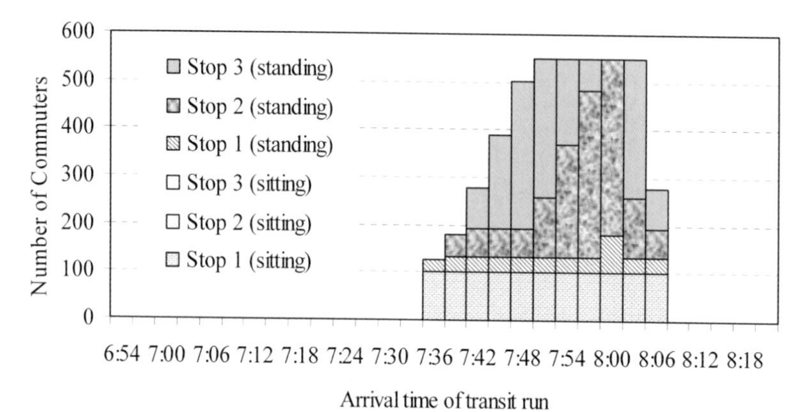

Figure 8. Boarding flows when there are 100 seats in the vehicle.

Figure 9 shows the result when there are 200 seats in each vehicle. The standing capacity becomes 300 and the total number of commuters that a vehicle can serve is 500. All commuters from stop 1 have seats and board the runs during a narrow period tightly close to the work-start time. Some commuters from stop 2 have seats too but board the runs arriving earlier or later than those runs chosen by commuters from stop 1. Even so, some commuters from stop 2 have to stand up. All commuters from stop 3 have to be standing in vehicle and endure the crowding costs. Moreover, the arrival time duration of commuters from stop 3 extends from [7:42, 8:06] in Figure 8 to [7:39, 8:06] in Figure 9, which leads these commuters to endure more schedule delay cost.

As shown in Figure 10, when there are 300 seats in each vehicle, the number of commuters that a vehicle can serve decreases further. More commuters from stop 2 can find seats. The earliest arrival run is utilized by commuters from stop 3. Commuters from stop 3 board all runs but few of them board the on-time arrival run which is occupied by commuters from stops 1 and 2.

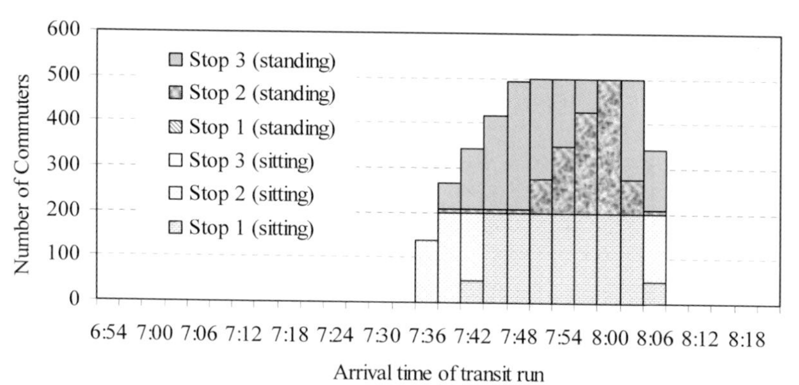

Figure 9. Boarding flows when there are 200 seats in each vehicle.

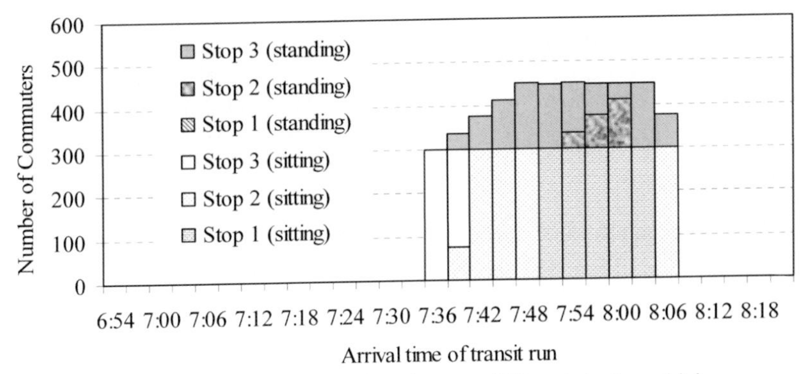

Figure10. Boarding flows when there are 300 seats in the vehicle.

If standing in vehicle is not allowed, the boarding flows are as shown in Figure 11 (there are 400 seats in each vehicle). Commuters from upstream stops concentrate into the transit runs arriving at W around the work-start time in an order from inner to out. This distribution is the same with that given by Alfa and Chen (1995), not considering the in-vehicle crowding cost.

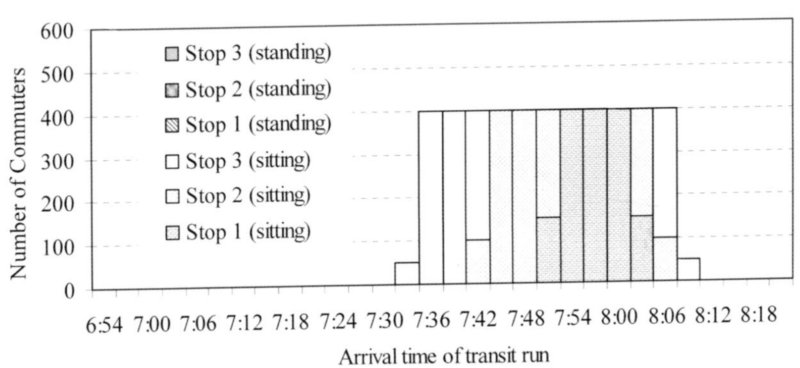

Figure 11. Boarding flows when there are 400 seats in each vehicle.

CONCLUSIONS

In this paper, we have investigated the morning peak-period commuting problem in a mono-centric city on a transit system with capacity constraints. Commuters are assumed to choose their optimal time-of-use decision from various stops/home-locations to a single destination (workplace) by trading off the travel time and in-vehicle crowding cost against the schedule delay cost. Two types of capacities, namely physical capacity (the maximum number of passengers that a transit vehicle can safely load) and seat capacity (the number of seats designed in a transit vehicle), are taken into account. Mathematical programming models are proposed to characterize the equilibrium properties of commuting, in which no commuter can reduce his/her total commuting cost by unilaterally changing his/her departure time or transit service. We compare the commuters' travel costs and behaviors caused by different physical and seat capacity constraints. It is found that increasing physical capacity can benefit all commuters from different stops along the transit line and weaken the boarding priority of upstream commuters. However, providing more seats in the vehicle would strength the boarding priority of upstream commuters and may increase the commuting cost of downstream commuters.

The work presented in this paper can be extended in several aspects. For example, the demand elasticity, variable transit service frequency and fare pricing can be incorporated into the equilibrium analyses. It is of interest to study the commuting equilibrium in a corridor network when transit vehicles and private cars share the road commonly (Huang et al., 2006). As our work is limited in the many-to-one transit system, extending it to general networks is valuable and challengeable.

ACKNOWLEDGEMENTS

The research described in this paper was substantially supported by grants from the National Natural Science Foundation of China (70429001, 50578006, 70521001), the National Basic Research Program of China (2006CB705503) and the Research Grants Council of the Hong Kong Special Administrative Region (HKUST6033/02E). The authors would like to thank anonymous referees for their helpful comments and valuable suggestions which improved the content and composition substantially.

REFERENCES

Alfa, A.S. and Chen, M. (1995). Temporal distribution of public transport demand during the peak period. *European J of Operational Research*, **83**, 137-153.

Arnott, R. (2004). The corridor problem. Presented at the WCTR 2004 Istanbul Conference (available from: http://fmwww.bc.edu/ec/Arnott.php).

Beckmann M., McGuire, C.B. and Winsten, C.B. (1956). *Studies in the Economics of Transportation*. Yale University Press.

Cascetta, E. and Papola, A. (2003). A joint mode - transit service choice model incorporating the effect of regional transport service timetables. *Transportation Research Part B*, **37**, 595 -614.

Cepeda, M., Cominetti, R. and Florian, M. (2006). A frequency based assignment model for congested transit networks with strict capacity constraints: Characterization and computation of equilibria. *Transportation Research Part B*, **42**, 437-59.

Cominetti, R. and Correa, J. (2001). Common lines and passenger assignment in congested transit networks. *Transportation Science*, **35**: 250-67.

De Cea, J. and Fernandez, J.E. (1993). Transit assignment for congested public transport systems: An equilibrium model. *Transportation Science*, **27**, 133-147.

De Cea, J. and Fernandez, J.E. (2000). Transit assignment models. In: *Handbook of Transport Modeling* (D.A. Hensher and K.J. Button, eds.), 497-508. Elsevier, Amsterdam.

Huang, H.J., Tian, Q. Yang, H. and Gao, Z.Y. (2004). Modeling the equilibrium bus riding behavior in morning rush hour. In: *Proceedings of The 9th Annual Conference of the Hong Kong Society of Transportation Studies*, pp. 434-442. HKSTS, Hong Kong.

Huang, H.J., Tian, Q. and Gao, Z.Y. (2005). An equilibrium model in urban transit riding and fare polices. *Lecture Notes in Computer Science*, **3521**, 112-121. Springer Verlag.

Huang, H.J., Tian, Q. Yang, H. and Gao, Z.Y. (2006). Modal split and commuting pattern on a bottleneck-constrained highway. *Transportation Research Part E,* doi:10.1016/j.tre. 2005.12.003.

Jehiel, P. (1993). Equilibrium on a traffic corridor with several congested modes. *Transportation Science*, **27**, 16-24.

Kraus, M. (2003). A new look at the two-mode problem. *Journal of Urban Economics*, **54**, 511–530.

Kraus, M. and Yoshida, Y. (2002). The commuter's time-of-use decision and optimal pricing and service in urban mass transit. *Journal of Urban Economics*, **51**, 170-195.

Guan, J.F., Yang, H. and Wirasinghe, S.C. (2005). Simultaneous optimization of transit line configuration and passenger line assignment on a minimum spanning tree network. *Transportation Research Part B*, **41**, 885-902.

Mohring, H. (1972). Optimization and scale economies in urban bus transportation. *American Economic Review*, **62**, 591-604.

Mohring, H. (1976). *Transportation Economics*. Cambridge, MA: Ballinger.

Ross, S. and Yinger, J. (2000). Timing equilibria in an urban model with congestion. *Journal of Urban Economics*, **47**, 390–413.

Sumi, T., Matsumoto, Y. and Miyaki, Y. (1990). Departure time and route choice of commuters on mass transit systems. *Transportation Research Part B*, **24**, 247-262.

Tian, Q. and Huang, H.J. (2004). An equilibrium ride model for subway passengers with arrival early penalty. *Journal of Transportation Systems Engineering and Information Technology*, **4**, 108-112 (in Chinese).

Tian, Q., Huang, H.J. and H. Yang (2006). Equilibrium properties of the morning peak-period commuting in a many-to-one mass transit system. *Transportation Research Part B*, doi:10.1010/j.trb.2006.10.003.

Vickrey, W.S. (1955). Some implications of marginal cost pricing for public utilities. *American Economic Review*, **45**, 605-620.

Vickrey, W.S. (1969). Congestion theory and transport investment. *American Economic Review*, **34**, 414-431.

Vuchic, V.R. (1969). Rapid transit interstation spacings for maximum number of passengers. *Transportation Science*, **3**, 214-232.

Vuchic, V.R. (2005). *Urban Transit: Operations, Planning and Economics*. Wiley, New York.

Vuchic, V.R. and Newell, G.F. (1968). Rapid transit interstation spacings for minimum travel time. *Transportation Science*, **2**, 303-339.

Wang, J.Y.T., Yang, H. and Lindsey, R. (2004). Locating and pricing park-and-ride facilities in a linear monocentric city with deterministic mode choice. *Transportation Research Part B*, **38**, 709-731.

Wang, J.Y.T., Yang, H. and Lindsey, R. (2003). Modeling park-and-ride service in a linear monocentric city. *Journal of the Eastern Asia Society for Transportation Studies*, **5**, 1377-1392.

Transportation and Traffic Theory 2007
Edited by R.E. Allsop, M.G.H. Bell and B.G. Heydecker

17

OPTIMAL SINGLE-ROUTE TRANSIT-VEHICLE SCHEDULING

Avishai (Avi) Ceder, Transportation Research Institute, Civil and Environmental Engineering Faculty, Technion-Israel Institute of Technology

INTRODUCTION

The transit-operation planning process commonly includes four basic activities, usually performed in sequence: (1) network route design, (2) timetable development, (3) vehicle scheduling, and (4) crew scheduling. The output of each activity positioned higher in the sequence becomes an important input for lower-level decisions. However it is desirable for all four activities to be planned simultaneously in order to exploit the system's capability to the greatest extent and to maximize the system's productivity and efficiency. However, since this planning process, especially for medium to large transit fleets, is extremely cumbersome and complex, separate treatment is required for each activity, with the outcome of one fed as an input to the next.

This work has two major aims: (1) to describe the task of vehicle scheduling and possible math-programming solutions for a single transit route; (2) to proffer a graphical technique that is easy to interact with and responds to practical concerns. Single transit routes deal with both fixed and variable schedules. In fixed schedules, departure times cannot be changed. In practical transit scheduling, however, schedulers should attempt to allocate vehicles in the most efficient manner possible, including the employment of small shifts in departure times.

This work contains three main parts following an introductory section, and a literature review section. First, a formula is derived to find the minimum fleet size required for a single route without deadheading (DH) trips and for a fixed schedule. Second, a graphical person-computer interactive approach, based on a step function called deficit function, is proffered for minimizing single-route fleet size and creating vehicle schedules with DH trip insertions. Third, the formula and procedure to attain single-route minimum fleet size are extended to include possible shifts in trip-departure times for given shifting tolerances.

LITERATURE REVIEW

Vehicle scheduling refers to the problem of determining the optimal allocation of vehicles to carry out all the trips in a given transit timetable. A chain of trips is assigned to each vehicle although some of them may be deadheading (DH) or empty trips in order to reach optimality. Vehicle scheduling on fixed routes is also related to finding the best dispatching policy for transit vehicles. Thus, this section reviews two groups of research: (1) studies on dispatching-policy problem, and (2) studies on vehicle scheduling models.

The first group was investigated by, for example, Newell (1971), Osana and Newell (1972), Hurdle (1973), and Wirasinghe (1990, 2003). Newell (1971) assumed a given passenger-arrival rate as a smooth function of time, with the objective of minimizing total passenger waiting time. He showed analytically that the frequency of transit vehicles with large capacities (in order to serve all waiting passengers) and the number of passengers served per vehicle each varied with time approximately as the square root of the arrival rate of passengers. Osana and Newell (1972) developed control strategies for either holding back a transit vehicle or dispatching it immediately, based on a given number of vehicles, random round-trip travel times with known distribution functions, and uniform passenger-arrival rates with a minimum waiting-time objective. Using dynamic programming, they found that the optimal strategy for two vehicles and a small coefficient of variation of trip time retained nearly equally spaced dispatch times. Hurdle (1973), investigating a similar problem, used a continuum fluid-flow model to derive an optimal dispatching policy while attempting to minimize the total cost of passenger waiting time and vehicle operation.

Wirasinghe (1990, 2003) examined and extended Newell's dispatching policy while considering the cost components initially used by Newell (1973). Wirasinghe considered the average value of a unit waiting time per passenger (C_1) and the cost of dispatching a vehicle (C_2) to show that the passenger-arrival rate in Newell's square root formula is multiplied by ($C_1/2C_2$). Wirasinghe also showed how to derive the equations of total mean cost per unit of time by using both uniform headway policy and Newell's variable-dispatching policy.

The second group of studies that are related directly to vehicle scheduling, was researched by, for example, Dell Amico *et al.* (1993), Löbel (1998, 1999), Mesquita and Paixao (1999), Banihashemi and Haghani (2000), Freling *et al.* (2001), Haghani and Banihashemi (2002), Haghani *et al.* (2003), and Huisman *et al.* (2004).

Dell Amico *et al.* (1993) developed several heuristic formulations, based on a shortest-path problem, that seek to minimize the number of required vehicles in a multiple-depot schedule. The algorithm presented is performed in stages, in each of which the duty of a new vehicle is determined. In each such stage, a set of forbidden arcs is defined, and then a feasible circuit through the network is sought that does not use any of the forbidden arcs. Computational efficiency is obtained by searching for the shortest path across a subset of all arcs in the network, rather than searching the entire network. Several modifications to the basic algorithms are offered that save computer time by substituting parts of the full problem with problems of a reduced size. These modifications include, for instance, solving the re-assignment of trips as a single-depot problem; an attempt to swap parts of duty segments; and an internal re-assignment of trips within each pair of vehicles associated with different depots.

Löbel (1998, 1999) discussed the multiple-depot vehicle scheduling problem and its relaxation into a linear programming formulation that can be tackled using the branch-and-cut method. A special multi-commodity flow formulation is presented, which, unlike most other such formulations, is not arc-oriented. A column-generation solution technique is developed, called Lagrangean pricing; it is based on two different Lagrangean relaxations. Heuristics are used within the procedure to determine the upper and lower bounds of the solution, but the final solution is proved to be the real optimum.

Mesquita and Paixao (1999) used a tree-search procedure, based on a multi-commodity network flow formulation, to obtain an exact solution for the multi-depot vehicle scheduling problem. The methodology employs two different types of decision variables. The first type describes connections between trips in order to obtain the vehicle blocks, and the other relates to the assignment of trips to depots. The procedure includes creating a more compact, multi-commodity network flow formulation that contains just one type of variables and a smaller amount of constraints, which are then solved using a branch-and-bound algorithm.

Banihashemi and Haghani (2000) and Haghani and Banihashemi (2002) focused on the solvability of real-world, large-scale, multiple-depot vehicle scheduling problems. The case presented includes additional constraints on route time in order to account for realistic operational restrictions such as fuel consumption. The authors proposed a formulation of the problem and the constraints, as well as an exact solution algorithm. In addition, they described several heuristic solution procedures. Among the differences between the exact approach and the heuristics is the replacement of each incorrect block of trips with a legal block in each iteration of the heuristics. Applications of the procedures in large cities are shown to require a reduction in the number of variables and constraints. Techniques for reducing the size of the problem are introduced, using such modifications as converting the problem into a series of single-depot problems.

Freling *et al.* (2001) discussed the case of single-depot with identical vehicles, concentrating on quasi-assignment formulations and auction algorithms. A quasi-assignment is a reduced-size, linear problem in which some of the nodes and their corresponding arcs are not considered. An auction algorithm is an iterative procedure in which neither the primal nor the dual costs are obliged to show an improvement after each iteration. The authors proposed four different algorithms, and compared their performance: an existing auction algorithm for the asymmetric assignment problem; a new auction algorithm for the quasi-assignment problem; an alternative, two-phase, asymmetric assignment formulation (valid in a special case), in which vehicle blocks are determined first and combined afterwards; and a core-oriented approach for reducing the problem size.

Haghani *et al.* (2003) compared three vehicle scheduling models: one multiple-depot (presented by Banihashemi and Haghani, 2002) and two single-depot formulations which are special cases of the multiple-depot problem. The analysis showed that a single-depot vehicle scheduling model performed better under certain conditions. A sensitivity analysis with respect to some important parameters is also performed; the results indicated that the travel speed in the DH trip was a very influential parameter.

Huisman *et al.* (2004) proposed a dynamic formulation of the multi-depot vehicle scheduling problem. The traditional, static vehicle scheduling problem assumes that travel times are a fixed input that enters the solution procedure only once; the dynamic formulation relaxes this assumption by solving a sequence of optimization problems for shorter periods. The dynamic approach enables an analysis based on other objectives except for the traditional minimization of the number of vehicles; that is, by minimizing the number of trips starting late and minimizing the overall cost of delays. The authors showed that a solution that required only a slight increase in the number of vehicles could also satisfy the minimum late starts and minimum delay-cost objectives. To solve the dynamic problem, a "cluster re-schedule" heuristic was used; it started with a static problem in which trips were assigned to depots, and then it solved many dynamic single-depot problems. The optimization itself was formulated through standard mathematical programming in a way that could use standard software.

FLEET SIZE REQUIRED FOR A SINGLE ROUTE

This section considers a case in which interlinings and deadheading (DH) trips are not allowed and each route operates separately. Given the average round-trip time and chosen layover time, the minimum fleet size for a radial route can be found according to the formula derived by Salzborn (1972, 1974). Specifically, let T_r be the average round-trip time, including layover and turn-around times, of a radial route r (departure and arrival points are same). The minimum fleet size is equal to the largest number of vehicles that departs within T_r.

Although Salzborn's model provides the base for fleet-size calculation, it relies on three assumptions that do not hold up in practice: (i) vehicle-departure rate is a continuous function of time, (ii) T_r is the same throughout the period under consideration, and (iii) route r is a radial route starting at a major point (e.g., CBD). In practice, departure times are discrete, average trip time is usually dependent on time-of-day, and a single transit route usually has different timetables for each direction of travel. For that reason, this section broadens Salzborn's model to account for practical operations planning.

Let route r have two end points: *a* and *b*. Let T_{ria} and T_{rjb} be the average trip time on route r for vehicles departing at t_{ia} and t_{jb} from *a* and *b*, respectively, including layover time at their respective arrival points. Let n_{ia} be the number of departures from *a* between t_{ia}, in which departure i*a* is included, and $t_{i'a}$ in which departure i'*a* is excluded. Thus, i*a* arrives to terminal *b*, then continues with trip j*b*, the latter being the *first* feasible departure from *b* to *a* at a time greater than or equal to the time $t_{ia}+ T_{ria}$; $t_{i'a}$ is the *first* feasible departure from *a* to *b* at a time greater than or equal to $t_{jb} + T_{rjb}$. Similarly n_{jb} may be defined for a trip j from *b*.

Lemma 1: In the case of no deadheading (DH) trips, n_{ia} departures must be performed by different vehicles at *a*, and n_{jb} must be performed by different vehicles at *b*, for all i*a* and j*b* in the timetables of r.

Proof: The proof is actually based on a contradiction. Let us assume that the same vehicle can perform two departures included in n_{ia} at *a*. However, in order to complete a full round trip, including layover times, this vehicle can only pick up the i'*a* departure at *a,* which is not included (by definition) in n_{ia}; hence, it is impossible for same vehicle to execute two departures within n_{ia}. Q.E.D.

Theorem 1: In the case with no interlining (between routes) and no DH trips, the minimum fleet size required for route r is

$$N^{r}_{min} = \max \{\max_i n_{ia} , \max_j n_{jb}\} \qquad (1)$$

Proof: Based on *Lemma* 1, $\max_i n_{ia}$ and $\max_j n_{jb}$ represent the maximum number of vehicles required to execute the timetables at *a* and *b*, respectively. Consequently the minimum fleet size for r is the greater of the two. Q.E.D.

An example of deriving the required fleet size for a single transit route r is shown in Figure 1. In this figure, a single average travel time $T_{ria} = T_{rjb} = 15$ minutes is used throughout the timetable for both directions of r. The timetables contain 12 departures at *b* and 10 at *a*. The calculations for n_{ia} and n_{jb} are shown by arrows; starting with the departures at *a* for n_{ia} (using $T_{ria} = 15$), and starting at *b* for n_{jb}. The solid line in Figure 1 represents the direction from the starting time to the first feasible connection (after 15 minutes), and the dashed line in the opposite direction links to the first feasible connection (also after 15 minutes) from the starting point. This leads to a determination of both n_{ia} and n_{jb}, and eventually the minimum fleet size, $N^{r}_{min} = 5$, according to Equation (1). It should be mentioned that the same T_{ria} and T_{rjb} are used throughout the example only for the sake of simplicity. Varied T_{ria} and T_{rjb} can be utilized in the same manner for each departure.

Vehicle chains (blocks) can be constructed by using the FIFO (first-in, first-out) rule. That is, a block will start at a depot for the first assigned scheduled trip, and then will make the first feasible connection with a departure (based on the route's timetable) at the other end point of the route, and so on. The block usually ends with a trip back to the depot. The trips to and from the depot are often deadheading trips. In the example illustrated in Figure 1, the five blocks can be constructed starting with the first departure (5:00) at *b* and using the FIFO (first-feasible connection) rule, then deleting the departures selected and continuing with another block until all departures are used. At the start of each step (at *b*), a check is made to see whether the next (in time) departure can be connected to an earlier unused departure at *a* and, if so, whether this connection can be allowed. The five blocks, therefore, are as follows: [5:00(at *b*) – 6:00(*a*) – 6:30(*b*) – 6:45(*a*) – 7:05(*b*) – 7:20(*a*) – 7:40(*b*) – 8:00(*a*)] ; [5:30(*b*) – 6:15(*a*) – 6:50(*b*) – 7:10(*a*) – 7:30(*b*)] ; [6:00(*b*) – 6:30(*a*) – 7:10(*b*) – 7:25(*a*) – 8:00(*b*)] ; [7:00(*a*) – 7:15(*b*) – 7:40(*a*)] ; [7:20(*b*)]. An earlier connection, linking the 7:15 departure at *b* to the 7:00 departure at *a,* is possible only in the fourth block. The above FIFO process can certainly start at *a,* as well. Note that the last block has only one trip; the five blocks can undergo changes, including swapping trips, between blocks. Each block can start and end at a depot or can be used as a segment in a larger block.

Finally when deadheading (DH) trips between the ends of two routes and/or slightly shifting departure times are allowed, it is more complex to use the formulation developed above. Instead, the solution can be found using the graphical method presented in the next section.

Calculating n_{ia} T_{rja} = 15 minutes			Calculating n_{jb} T_{rjb} = 15 minutes		
n_{ia}	Timetable at *a*	Timetable at *b*	Timetable at *a*	Timetable at *b*	n_{jb}
3	6:00	5:00	6:00	5:00	3
2	6:15	5:30	6:15	5:30	2
3	6:30	6:00	6:30	6:00	1
3	6:45	6:30	6:45	6:30	2
4	7:00	6:50	7:00	6:50	5
4	7:10	7:05	7:10	7:05	5
3	7:20	7:10	7:20	7:10	4
2	7:25	7:15	7:25	7:15	4
2	7:40	7:20	7:40	7:20	3
—	8:00	7:30	8:00	7:30	3
		7:40		7:40	2
		8:00		8:00	—
max n_{ia}	**4**				
max n_{jb}			**5**		
N^r_{min}	max (4,5) = **5**				

Figure 1 Example of derivation of single-route fleet size with no deadheading trips

DEFICIT-FUNCTION MODEL WITH DEAD-HEADING TRIP INSERTION

The minimum-fleet-size problem may be approached with and without DH trips. A DH trip is an empty trip between two terminuses and is usually inserted into the schedule (i) to ensure that the schedule is balanced at the start and end of the day and (ii) to transfer a vehicle from one terminal where it is not needed to another where it is needed to service a required trip. When DH is allowed, the counter-intuitive result of decreasing the required resources (fleet size) by introducing more work into the system (adding DH trips) is attained. This assumes that the capital cost of saving a vehicle far outweighs the cost of any increased operational cost (driver and vehicle travel cost) imposed by the introduction of DH trips. This section presents a graphical person-computer interactive approach, based on a step function called deficit function.

Definitions and Minimum Fleet Size

Let $I = \{i: i = 1, ..., n\}$ denote a set of required trips. The trips are conducted between a set of terminals $K = \{k: k = 1, ..., q\}$, each trip to be serviced by a single vehicle, and each vehicle able to service any trip. Each trip i can be represented as a 4-tuple (p^i, t_s^i, q^i, t_e^i), in which the ordered elements denote departure terminal, departure (start) time, arrival terminal, and arrival (end) time. It is assumed that each trip i lies within a schedule horizon $[T_1, T_2]$ i.e., $T_1 \le t_s^i \le t_e^i \le T_2$. The set of all trips $S = \{(p^i, t_s^i, q^i, t_e^i): p^i, q^i \in K, i \in I\}$ constitutes the timetable. Two trips i, j may be serviced sequentially (feasibly joined) by the same vehicle if and only if (a) $t_e^i \le t_s^j$ and (b) $q^i = p^j$.

If i is feasibly joined to j, then i is said to be the predecessor of j, and j the successor of i. A sequence of trips $i_1, i_2, \ldots, i_w$ ordered in such a way that each adjacent pair of trips satisfies (a) and (b) is called a chain or block. It follows that a chain is a set of trips that can be serviced by a single vehicle. A set of chains in which each trip i is included in I exactly once is said to constitute a vehicle schedule. The problem of finding the minimum number of chains for a fixed schedule S is defined as the minimum fleet- size problem.

Let us define a DH trip as an empty trip from some terminal p to some terminal q in time $\tau(p,q)$. If it is permissible to introduce DH trips into the schedule, then conditions (a) and (b) for the feasible joining of two trips, i, j, may be replaced by the following:

$$t_e^i + \tau(q^i, p^i) \le t_s^j \qquad (2)$$

Now let us introduce a deficit-function-based model.

A deficit function (DF) is a step function defined across the schedule horizon that increases by one at the time of each trip departure and decreases by one at the time of each trip arrival. This step function is called a deficit function (DF) because it represents the deficit number of vehicles required at a particular terminal in a multi-terminal transit system. To construct a set of DFs, the only information needed is a timetable of required trips. The main advantage of the DF is its visual nature. Let d(*k*,t,S) denote the DF for terminal *k* at time t for schedule S. The value of d(*k*,t,S) represents the total number of departures minus the total number of trip arrivals at terminal *k*, up to and including time t. The maximum value of d(*k*,t,S) over the schedule horizon [T_1, T_2], designated D(*k*,S), depicts the deficit number of vehicles required at *k*.

The DF notations are presented in Figure 2, in which [T_1, T_2] = [5:00, 8:30]. It is possible to partition the schedule horizon of d(*k*,t) into a sequence of alternating hollow and maximum intervals $\left(H_0^k, M_1^k, H_1^k, .., H_j^k, M_{j+1}^k, ..., M_{n(k)}^k, H_{n(k)}^k\right)$. Note that S will be deleted when it is clear which underlying schedule is being considered. Maximum intervals $M_j^k = [s_j^k, e_j^k]$, j=1,2,…,n(*k*) define the intervals of time over which d(*k*,t) takes on its maximum value. Index j represents the j-th maximum intervals from the left; n(*k*) represents the total number of maximal intervals in d(*k*,t), where s_j^k is the departure time for a trip leaving terminal *k* and e_j^k is the time of arrival at terminal *k* for this trip. The one exception occurs when the DF reaches its maximum value at $M_{n(k)}^k$ and is not followed by an arrival, in which case $e_{n(k)}^k = T_2$.

A hollow interval H_j^k, j = 0,1,2,…,n(*k*) is defined as the interval between two maximum intervals: this includes the first hollow, from T_1 to the first maximum interval, $H_0^k = [T_1, s_1^k]$; and the last hollow, which is from the last interval to T_2, $H_{n(k)}^k = [e_{n(k)}^k, T_2]$. Hollows may contain only one point; if this case is not on the schedule horizon boundaries (T_1 or T_2), the graphical representation of d(*k*,t) is emphasized by a clear dot.

The sum of all DFs over *k* is defined as the overall DF, $g(t) = \sum_{k \in K} d(k,t)$. This function g(t) represents the number of trips that are simultaneously in operation; i.e., a count, from a bird's-eye view at time t, of the number of transit vehicles in actual service over the entire transit network of routes. The maximum value of g(t), G(S), is exploited for a determination of the lower bound on the fleet size. An example of a two-terminal operation, a fixed schedule of trips, and the corresponding set of DFs and notations is illustrated in Figure 2.

Determining the minimum fleet size, D(S), from the set of DFs is simple enough - one merely adds up the deficits of all the terminals. In the example in Figure 2 without DH trips, D(S) = D(*a*) + D(*b*) = 4. This result was apparently derived independently by Bartlett (1957), Salzborn (1972, 1974), and Gertsbach & Gurevich (1977). It is formally stated as Theorem 2.

Theorem 2: If, for a set of terminals K and a fixed set of required trips I, all trips start and end within the schedule horizon [T1,T2] and no DH insertions are allowed, then the minimum number of vehicles required to service all trips in I is equal to the sum of all the deficits.

$$\text{Min N} = \sum_{k \in K} D(k) = \sum_{k \in K} \max_{t \in [T_1, T_2]} d(k, t) \tag{3}$$

Proof: Let F_k = the number of vehicles present in terminal k at the start of the schedule horizon T_1; let s(k, t) and e(k, t) be the cumulative number of trips starting and ending at k from T_1 up to and including time t. The number of vehicles remaining at k at time $t \geq T_1$ is F_k - s(k, t) + e(k, t).

In order to service all trips leaving k, the above expression must be non-negative; i.e., $F_k \geq s(k, t) - e(k, t)$, $T_1 \leq t \leq T_2$. The minimum number of vehicles required at k is then equal to the maximum deficit at k. Min F_k = $\text{Max}_t[s(k, t) - e(k, t)] = \text{Max}_t d(k, t)$. Hence, the minimum number of vehicles required for all terminals in the system is equal to the total deficit $\text{Min N} = \sum_{k \in K} \text{Min} F_k = \sum_{k \in K} d(k) = D(S)$. Q.E.D.

DH Trip Insertion

A DH trip is an empty trip between the ends of a single route. We start by asking, Where and when is such trip needed? Usually, a trip schedule received from operating personnel includes such deadheading trips, and it is easy to apply the fleet-size formula to determine the minimum fleet size, followed by the first in-first out rule to construct each vehicle's schedule. The assumption is that the trip schedule S has been purged of all DH trips, leaving only required trips. From this point, the question of how to insert deadheading trips into the schedule in order to further reduce the fleet size will be examined. At first, it seems counter-intuitive that this can be achieved, since it implies that increased work (adding trips to the schedule) can be carried out with decreased resources (fewer vehicles). This section will show through an examination of the effect of such deadheading trip insertions on deficit functions that this is indeed possible.

Consider the example in Figure 3. In its present configuration, according to the fleet-size formula, five vehicles are required at terminal a, and six at terminal b for a fleet size of eleven. The dashed arrows in the figure represent the insertion of DH trip from a to b. After the introduction of this DH trip into the schedule, the DFs at both terminals are shown updated by the dotted lines. The net effect is a reduction in fleet size by one unit at terminal b. In fact Ceder (2002, 2003) shows that a chain of DH trips may be required for the reduction of the fleet size by one.

All successful DH trip chains follow a common pattern. The initial DH trip is introduced to arrive in the first hollow of a terminal in which a reduction is desired. This DH trip must depart from some hollow of another terminal. Moving to the end of this hollow, another DH trip is inserted, such that its arrival epoch will compensate for the departure epoch added by the first DH trip. This is followed by additional compensating trips; however, in order to

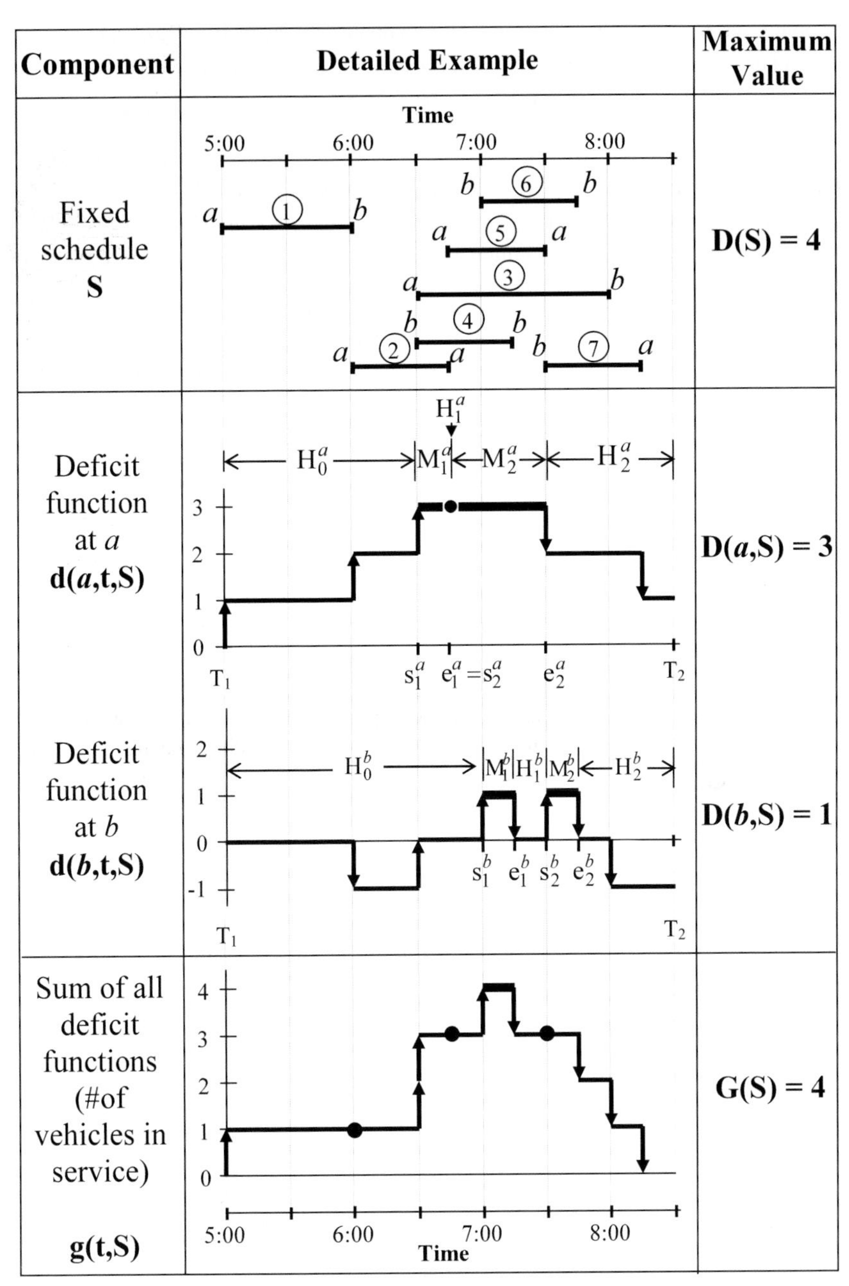

Figure 2 Illustration of two-terminal fixed schedule with associated deficit functions and their sum, including notations and definitions

avoid looping, no more than one DH trip will be allowed to depart from the same hollow. Each time a DH trip is inserted (from p to q) to arrive at the end of a hollow H_i^q from the start of a hollow H_j^q, it must pass a feasibility test; i.e., $e_j^q + \tau(p,q) \leq s_{i+1}^q$. If the inequality is true with <, then there will be some slack time, during which the DH trip can be shifted. Let this slack time be defined as $\delta_{pq} = s_{i+k}^q - [e_j^p + \tau(p,q)]$. In practice, if the DH time plus the slack time are greater than or equal to the average service travel time, then a service trip may replace the DH trip. In this way, an additional service trip is introduced, thereby resulting in higher frequency (i.e., an improved level of service) at usually the same operational cost.

The process ends when a final hollow of some terminal q is reached (i.e., $H_i^q = H_{n(q)}^q$), after which no compensation is necessary. It is possible to arrive at a point where no feasible compensating DH trips can be inserted, in which case the procedure terminates or one may back track to the arrival point of the last DH trip added and try to replace it with another. This procedure results in a sequence of DH trips known as a unit reduction dead-heading chain (URDHC) if it ends successfully (i.e., if it reduces the fleet size by a unit amount). Clearly, the continued reward for such a search must stop, and the Lower Bound Theorem (Ceder, 2002) provides a condition when it is futile to continue this search; this lower bound is based on the sum of DFs, g(t), and its maximum value G(S).

DEFICIT-FUNCTION MODEL WITH SHIFTS IN DEPARTURE TIMES

A small amount of shifting in scheduled departure times becomes almost common in practice when attempting to minimize fleet size or the number of vehicles required. This section presents two methods by which single routes may realize a variable trip schedule in an efficient manner. The first method ascertains the minimum fleet size required for a given single route, taking into account possible shifts in departure times for given backward and forward shifting tolerances (in minutes) for each trip. The second method, based on the deficit-function model, develops a formal algorithm to handle the complexities of shifting departure times. The algorithm is intended for both automatic and man-computer conversational modes. A secondary objective considered (for both methods) is to minimize the length of the shifting within their given tolerances.

Single-Route Minimum Fleet Size

In practice, departure times are shifted without any systematic method. Shifting tolerances are usually determined by rule of thumb although it makes sense to correlate them with the headways between departures. A proposed method appears in Table 1, in which the length of the shifting tolerance is headway dependent. That is to say, the longer the headway, the shorter is the tolerance. If the shifting is backward, the preceding headway is considered as H; if it is forward, the next headway is considered.

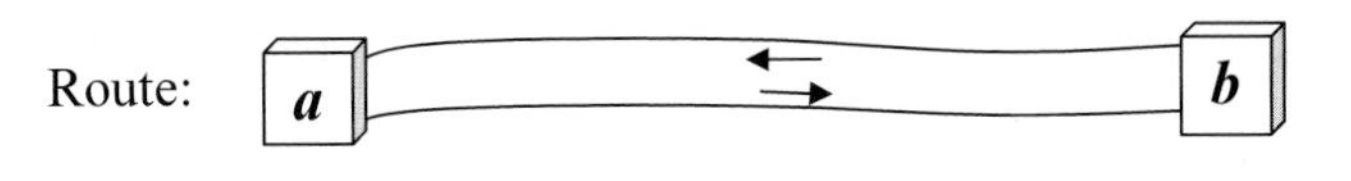

a → *b*		*b* → *a*	
a	***b***	***b***	***a***
7:00	7:40	7:00	7:35
7:10	7:50	7:15	7:50
7:25	8:05	7:20	7:55
7:35	8:15	7:25	8:00
7:40	8:20	7:30	8:05
7:45	8:30	7:40	8:15
7:50	8:35	7:50	8:25
8:00	8:45	8:05	8:45
8:15	9:00	8:15	8:55
8:25	9:10	8:20	9:00
8:40	9:25	8:25	9:05

Travel time (both directions) = 40 minutes

DH time (both directions) = 25 minutes

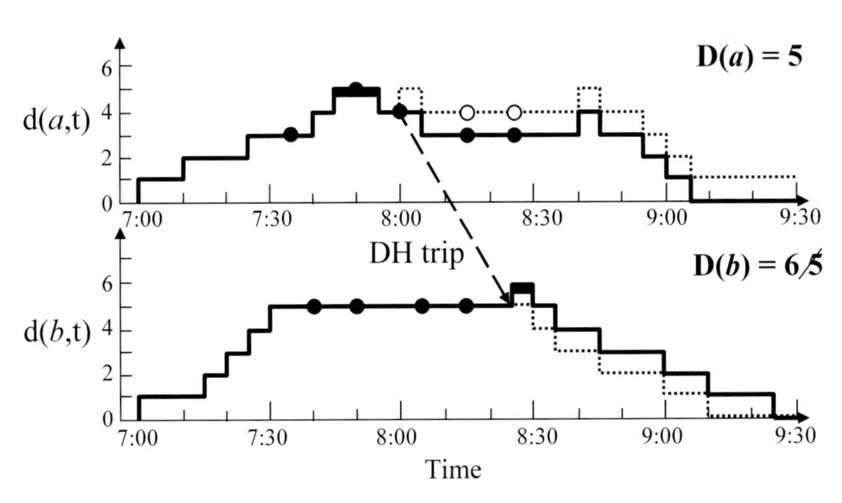

Figure 3 Example of a timetable of two-way route and its deficit functions; insertion of one DH trip reduces the required fleet size from 11 to 10 vehicles

Table 1 Shifting tolerances as headway dependent

Headway (H, in minutes)	**Percentage of H for tolerance determination (%)**	**Tolerance length as H-dependent (minutes)**
< 10	50	0.5H
10-20	40	0.4H
21-40	30	0.3H
>40	20	0.2H

A new process needs to be designed for applying the shifting capability of departure times on single routes. This process simply attempts, through possible shifting of the relevant departure times, to reduce the minimum fleet size required. We will use the same notation for route r as in the third section of this work, but the symbol r is deleted, because it is clear which underlying route is being referred to. Thus, *a* and *b* are the end points; T_{ia} and T_{jb} are the average trip times on the route for vehicles departing at t_{ia} and t_{jb} from *a* and *b*, respectively, including layover time at their respective arrival points; n_{ia} is the number of departures from *a* between t_{ia}, in which departure i*a* is included, and $t_{i'a}$, in which departure i'*a* is excluded. Trip i*a* arrives at terminal *b*, then continues with trip j*b*, the latter being the *first* feasible departure from *b* to *a* at a time greater than or equal to the time $t_{ia}+T_{ia}$; and $t_{i'a}$ is the *first* feasible departure from *a* to *b* at a time greater than or equal to $t_{jb}+T_{jb}$. Similar notations are defined for a trip starting from *b*.

Let $[t_{ia}-\Delta^{ia(-)}, t_{ia}+\Delta^{ia(+)}]$ be the tolerance time interval of the departure time of trip i*a*, in which: $\Delta^{ia(-)}$ = maximum advance of the trip's scheduled departure time (the case of an early departure), and $\Delta^{ia(+)}$ = maximum delay from the scheduled departure time (the case of a late departure). Note that $t_{ik}+\Delta^{ik(+)} < t_{(i+1)k}$ and $t_{ik}-\Delta^{ik(-)} > t_{(i-1)k}$, for all $k \in K$. The minimum fleet size, N_{min}, is then attained by construction, using the procedure illustrated in a flow diagram in Figures 4(a) and 4(b). The procedure described fits the case of Equation (1), in which $N_{min} = \max_i n_{ia}$. For the case in which $N_{min} = \max_j n_{jb}$ (determined by a trip starting from *b*), the same procedure is applied, but with *b* replacing *a* and j replacing i. The procedure first identifies the departure i*a* (or one of a few) referring to $N_{min} = n_{ia}$; then it attempts through shifting t_{ia} to arrive at *b* before or at $t_{(j-1)b}$, and most important - to arrive before or at $t_{(i'-1)a}$. If the process manages to reduce n_{ia} by one or more units, it looks for the next $n_{ia} = N_{min}$ or $n_{jb} = N_{min}$ to continue. A successful process is that in which N_{min} is reduced. In addition, the procedure depicted in Figures 4(a) and 4(b) minimizes the length of shifting departure times, except for the shifting of the first departure, t_{ia}.

The interpretation of the shifting procedure may be assisted by the example in Figure 1. Here, $N_{min} = 5$, resulting from the fifth and sixth departures from *b*. When *b* replaces *a* and j replaces i in Figures 4(a) and 4(b), we can then use the procedure described and start with the 6:50*b* departure. Given $\Delta^{6:50b(-)} = 5$ minutes, then $\Delta^{7:00a(+)} = \Delta^{7:20b(+)} = \Delta^{7:15b(+)} = \Delta^{7:10b(+)} =$ 3 minutes. The first check in the Figure 7.3 example, results in shifting 6:50 to 6:45 from *b*

having $\Delta^{6:45b(-)} = 0$ minutes. Then from *a*, the first feasible connection is at 7:03 (including a forward tolerance). The second check, 7:00 ≥ 6:45+15, results in setting the departure time from *a* at 7:00 with $\Delta^{7:00a(-)} = 0$. The third check, 7:00+15 ≤ 7:20+3, leads to finding the first feasible connection to 7:00 to be from *a*. That is, min [7:20+3, 7:15+3, 7:10+3] ≥ 7:15 is 7:18. In the fourth check, $t_{(j'-1)b}$=7:15. Hence, $n_{6:45b} = 3$, instead of the previously $n_{6:50b} = 5$.
We now move to the 7:05 departure from b, in which $n_{7:05b} = 5$. Given are $\Delta^{7:05b(-)} = 0$ minutes and $\Delta^{7:10a(+)} = \Delta^{7:20b(-)} = \Delta^{7:30b(+)} = 5$ minutes. In the first check, an early departure from *a* is impossible. The second check, 7:20−0+15 ≤ 7:30+5, results in setting $t_{(j'-1)b}$=7:35, and $n_{7:05b} = 4$. The result is a multi-case of $N_{min} = n_{ia} = n_{jb} = 4$, but without any further possibility of improvement, because $n_{7:05b}$ cannot be further reduced; this case is based on the procedure constructed in Figure 4(a) and 4(b).

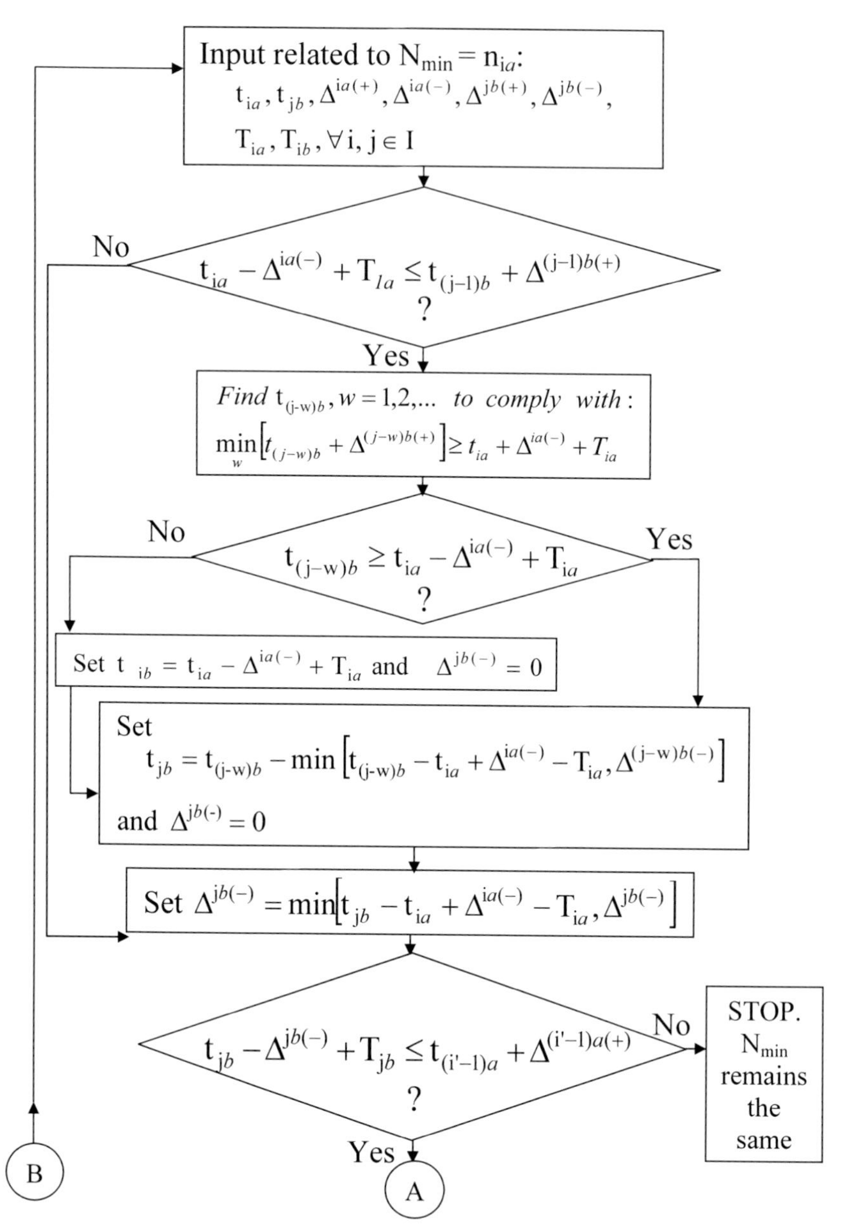

Figure 4 (a) Flow diagram of the shifting departure-times process for reducing the minimum fleet size N_{min} determined at terminal *a* (for terminal *b*, the same process is used with a change of symbols)

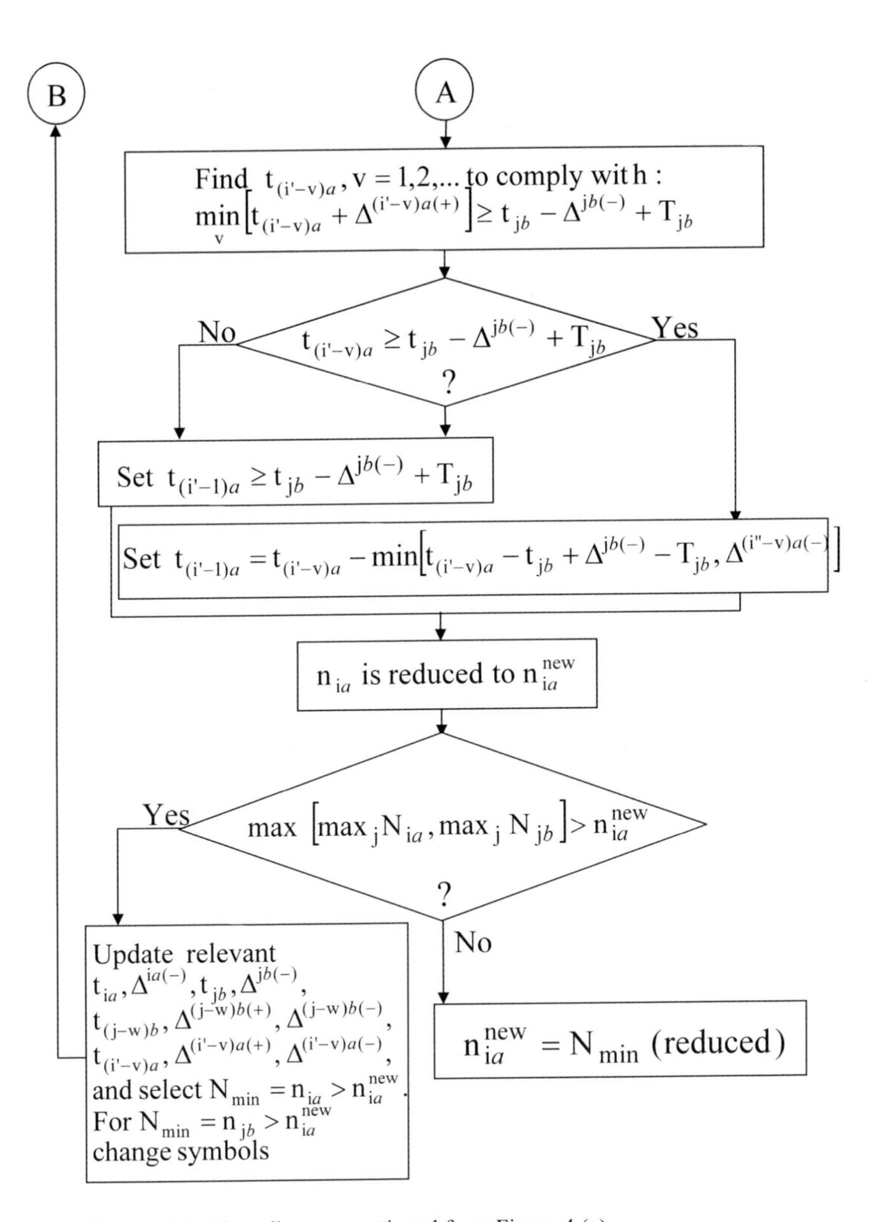

Figure 4 (b) Flow diagram continued from Figure 4 (a)

Variable Scheduling Using Deficit Functions

The second approach to handle possible shifting in departure time employs the DF model. A developed formal algorithm to deal with the complexities of shifting departure times appears in Ceder (2002). This section will discuss the characteristics of variable single-route schedule using simple examples.

Let us look closely at the 22-trip example in Figure 3 while assuming that the tolerances of this example are $\Delta^{i(+)} = \Delta^{i(-)} = 5$ minutes for all trips in the schedule. The question arises as to whether D(*a*)=5 and D(*b*)=6 can be reduced by shifting departure times within the given tolerances. In terminal *b* the two trips that construct the maximal interval are [7:45 (from *a*) – 8:30 (arrives to *b*)], and [8:25 (from *b*) – 9:05 (arrives to *a*)]. Because of the connection of the first trip to the maximal interval of terminal *a*, we shift the 8:25 departure forward by 5 minutes (to arrive to *a* at 9:10). At the other terminal *a* four trips are related to the maximal interval, two of which arrive and depart at 7:50. The solution will be to shift backward and forward by 5 minutes the two trips which arrive and depart at 7:50. That is, the trip of [7:15 (from *b*) - 7:50 (arrives to *a*)] becomes [7:10 – 7:45], and the trip [7:50 (from *a*) – 8:35 (arrives to *b*)] becomes [7:55 – 8:40]. The three shifts created manage to reduce the fleet size from 11 to 9, without any DH trip insertion.

Another seven-trip example of a single route is depicted in Figure 5. It demonstrates a possible chaining effect of shifting departure times. In this example, $\Delta^{i(+)} = \Delta^{i(-)} = 10$ minutes. Hence, in the beginning, it is possible to reduce D(*b*) by one unit through the shifts of t_s^3 to the right and t_s^4 to the left. However, these shifts increase D(*a*) and the net saving is zero. Consequently, another iteration is needed in which t_s^7 is shifted to the right. Only then, we obtain a total saving of one vehicle. Given the desire to reduce a maximal interval M_j^k by shifting maximum two trips, the following three cases exist (Ceder, 2002): (1) shift only trip i to the right, (2) shift only trip i' to the left, and (3) shift both trips i and i' in opposite directions (see Figure 1 for definitions used).

CONCLUDING REMARK

A common practice in vehicle scheduling is to use time-space diagrams. Each line in the diagram represents a trip moving over time (x-axis) at the same average commercial speed represented by the slope of the line. Although many schedulers became accustomed to this description, it is cumbersome, if not impossible, to use these diagrams to make changes and improvements in the scheduling. It is also difficult to use different average speeds for different route segments, in which the lines in the time-space diagram can cross one another; this is not to mention the inconvenience of using these diagrams manually for inserting deadheading trips and/or shifting departure times. These limitations of the time-space diagram caused us to look more closely into more appealing approaches – those that are presented in this work.

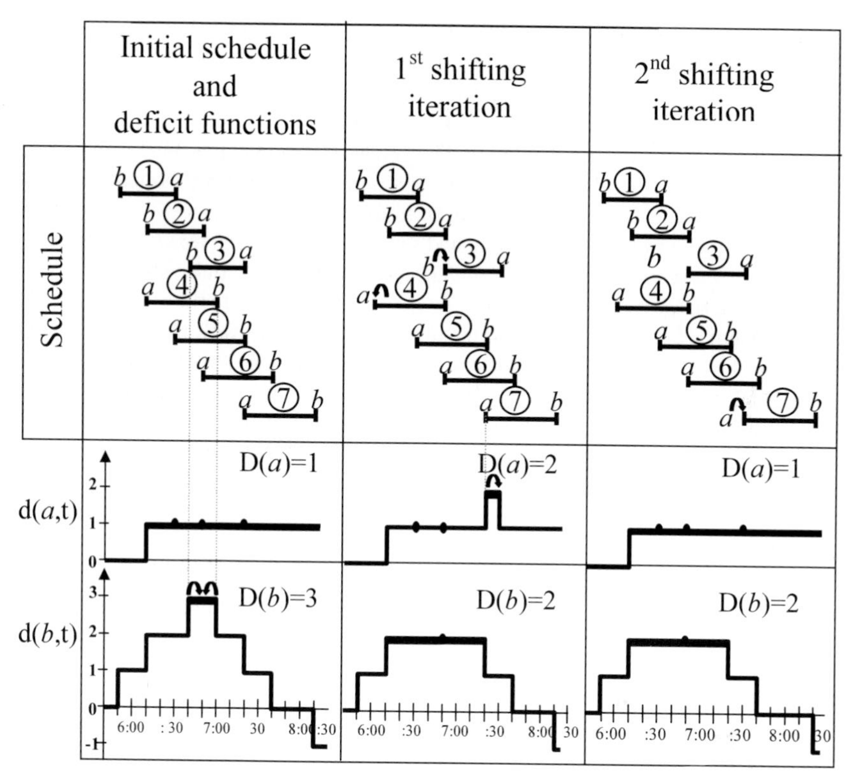

Figure 5 Example of two shifting iterations to reduce the required fleet size from 4 to 3

In practical single-route transit-vehicle scheduling, schedulers should attempt to allocate vehicles in the most efficient manner possible, including the insertion of deadheading (DH) trips and the employment of small shifts in departure times. For example, in Egged, Israel's national bus carrier (with 4000 buses), the schedulers consider a variable instead of a fixed schedule in addition to deadheading (DH) trip insertions. Moreover, some DH trip insertions are combined with small shifts in the departure times to allow these insertions; thus, reducing the fleet size and vehicle cost required. This work accentuates both DH insertions and shifts in departure times for single-route vehicle scheduling by two developed approaches: (1) math-programming solutions; and (2) a graphical (deficit function) technique that is easy to interact with and responds to practical concerns.

REFERENCES

Banihashemi, M. and Haghani, A. (2000). Optimization model for large-scale bus transit scheduling problems. *Transportation Research Record*, **1733**, 23-30.

Bartlett, T. E. (1957). An Algorithm for the minimum number of transport units to maintain a fixed schedule. *Naval Res. Logist. Quart.* **4**, 139-149.

Ceder, A. (2002). A Step Function for Improving Transit Operations Planning Using Fixed and Variable Scheduling. In *Transportation & Traffic Theory* (M.A.P. Taylor, ed.), pp. 1-21, Pergamon Imprint, Elsevier Science, Oxford, UK.

Ceder, A. (2003). Public transport timetabling and vehicle scheduling. In *Advanced Modeling for Transit Operations and Service Planning* (W. Lam and M. Bell, eds.), pp. 31-57, Pergamon Imprint, Elsevier Science.

Ceder, A. and Stern, H. I. (1981). Deficit Function Bus Scheduling with Deadheading Trip Insertion for Fleet Size Reduction. *Transportation Science* **15 (4)**, 338-363.

Dell Amico, M., Fischett, M., and Toth, P. (1993). heuristic algorithms for the multiple depot vehicle scheduling problem, *Management Science,* **39** (1), 115-123.

Freling, R., Wagelmans, A. P. M., and Paixao, J. M. P. (2001). Models and algorithms for single-depot vehicle scheduling, *Transportation Science*, **35 (2)**, 165-180.

Gertsbach, I. and Gurevich, Y. (1977). Constructing an optimal fleet for transportation schedule. *Transportation Science,* **11**, 20-36.

Haghani, A. and Banihashemo, M. (2002). Heuristic approaches for solving large-scale bus transit vehicle scheduling problem with route time constraints. *Transportation Research*, **36A**, 309-333.

Haghani, A., Banihashemi, M., and Chiang, K. H. (2003). A comparative analysis of bus transit vehicle scheduling models, *Transportation Research,* **37B**, 301-322.

Huisman, D., Freling, R., and Wagelmans, A.O.M. (2004). A robust solution approach to the dynamic vehicle scheduling problem. *Transportation Science,* **38 (4)**, 447-458.

Hurdle, V.F. (1973). Minimum cost schedules for a public transportation route. *Transportation Science,* **7**(2), 109-157.

Löbel, A. (1998), Vehicle scheduling in public transit and lagrangean pricing. *Management Science*, **44 (12)**, 1637-1649.

Löbel, A. (1999). Solving large-scale multiple-depot vehicle scheduling problems. In *Computer-Aided Transit Scheduling. Lecture Notes in Economics and Mathematical Systems,* **471** (N. H. M. Wilson, ed.), pp. 193-220, Springer-Verlag.

Mesquita, M. and Paixao, J.M.P. (1999). Exact algorithms for the multi-depot vehicle scheduling problem based on multicommodity network flow type formulations. In *Computer-Aided Transit Scheduling. Lecture Notes in Economics and Mathematical Systems,* **471** (N. H. M. Wilson, ed.), pp. 221-243, Springer-Verlag.

Newell, G. F. (1971). Dispatching policies for a transportation route. *Transportation Science*, **5**, 91-105.

Osana, E.E. and Newell, G. F. (1972). Control strategies for an idealized public transportation system. *Transportation Science* **6**(1), 52-72.

Salzborn. F. J. M. (1972). Optimum bus scheduling. *Transportation Science,* **6**, 137-148.

Salzborn. F. J. M. (1974). Minimum fleet size models for transportation systems. In *Transportation and Traffic Theory* (D. J. Buckley, ed.), pp. 607-624, Reed, Sydney.

Wirasinghe, S. C. (1990). Re-examination of Newell's dispatching policy and extension to a public transportation route with many to many time varying demand. In *Transportation and Traffic Theory* (M. Koshi, ed.), pp. 363-378 Elsevier Ltd.

Wirasinghe, S. C. (2003). Initial planning for an urban transit system. In *Advanced Modeling for Transit Operations and Service Planning* (W. Lam and M. Bell, eds.), pp. 1-29, Pergamon Imprint, Elsevier Science Ltd.

Transportation and Traffic Theory 2007
Edited by R.E. Allsop, M.G.H. Bell and B.G. Heydecker

18

A FREQUENCY BASED TRANSIT MODEL FOR DYNAMIC TRAFFIC ASSIGNMENT TO MULTIMODAL NETWORKS

Lorenzo Meschini, Guido Gentile, Natale Papola
Dipartimento di Idraulica Trasporti e Strade, Università "La Sapienza", Rome, Italy

INTRODUCTION

The pricing and rationing measures applied to discourage the use of private cars, in order to alleviate the increasing road congestion and the consequent worsening pollution, are not always coupled with a consistent policy aimed at improving the performance, or at least the capacity, of the transit system. As a result, in many modern cities the problem of full transit carriers is becoming increasingly prelevant. Although this situation should be avoided through a correct design of the transit network by suitably increasing the line capacities, it is still important to properly simulate the current scenario in order to justify the resources needed to carry out appropriate interventions.

Thus, we are interested here in modelling the dynamic behaviours of heavy congested, urban, multimodal (transit and road) networks, where the service is so irregular or so frequent that there is no point for passengers to synchronize their arrival at the stop with the scheduled time of carriers, if any is published. Within this context, we aim at properly reproducing the important dynamic congestion phenomenon of the temporary *over saturation* of roads and transit lines; that is, both the formation and dispersion of car and transit carrier queues on road arcs, and the formation and dispersion of passenger queues at transit stops, where passengers wait for the first run of the chosen line actually available to them.

The *frequency-based* static assignment models commonly used to simulate and plan urban transit networks are suited to represent systems (metro, tramways, buses) where it is generally assumed that a passenger, after reaching a stop, waits for the first attractive carrier among some common lines. This leads to the concept of *optimal strategy* (Spiess and Florian, 1989) which can be formally expressed in terms of a shortest *hyperpath* (Nguyen and Pallottino, 1988). These models allow to suitably represent the effect of congestion on travel choices and passengers flows, but cannot represent properly the dynamic congestion phenomenon introduced above. In fact, the traditional approach to reproduce this congestion phenomenon in a static framework is based on the concept of *effective frequency* (DeCea and Fernandez, 1993), stating that the line frequency perceived by the passengers waiting at a stop decreases as the probability of not boarding its first arriving carrier increases. Since the residual capacity of a run available to passengers waiting at the stop depends on the number of people that are already onboard, who do not experience the cost of queuing, then, in order to apply properly the effective frequency approach, an asymmetric arc cost function is to be introduced, as in Bellei *et al.* (2003).

An alternative and well established approach to represent transit systems, which are intrinsically discrete in time, adopts a *diachronic graph* (Nuzzolo *et al.*, 2003), where each run is modelled through a specific sub-graph whose nodes have space and time coordinates according to the timetable. The main advantage of transit models based on diachronic graphs lays in the fact that, even with an explicit representation of the time dimension, they can be reduced to static assignments on space-time networks. Then, a similar approach to that of effective frequency can be adopted in order to represent congestion due to limited capacity on transit carriers (Crisalli, 1999; Nguyen *et al.*, 2001); even though this makes it possible, as in the static case, to simulate the priority of passengers onboard, a distortion on the cost pattern is introduced: at the equilibrium, the cost for the passengers who board the arriving run is equal to that suffered by those who must wait at the stop for a successive run. Moreover, when using this approach a compromise is to be made between numerical convergence and accuracy in constraint satisfaction, because, if the waiting cost increases too strongly when the onboard flow approaches the residual capacity, then the assignment algorithm becomes unstable. Finally, when applied to congested multimodal urban network, these models present some significant drawbacks:

- on the supply side, diachronic graphs are not suited to represent congestion effects on travel times, since the graph structure itself must vary with the flow pattern;
- on the demand side, since in urban transit networks with high frequency passengers perceive lines as unitary supply facilities, it is not necessary to represent the single runs explicitly; this circumstance is widely exploited in the existing static models for transit assignment (De Cea and Fernandez, 1993; Wu *et al.*, 1994; Nguyen *et al.*, 1998);
- on the algorithm side, the complexity of the assignment problem increases more than linearly with transit line frequencies, due to the grow of graph dimension.

These drawbacks are overcome in this paper, were we present a new model and algorithm, aimed at solving the multimodal *Dynamic Traffic assignment* (DTA). The proposed approach extends an existing DTA model for road networks, presented in Bellei *et al.* (2005) and in Gentile *et al.* (2006), based on a macroscopic representation of time-continuous flows. Here, the multimodal DTA is regarded as a Dynamic User Equilibrium[1] and is formalized as a system between a dynamic arc performance function and a dynamic Network Loading Map[1] (NLM), thus avoiding the introduction of the Dynamic Network Loading[1] (DNL) model and the need of explicit path enumeration. The equilibrium model results in a fixed-point problem in terms of arc flow and transit frequency temporal profiles, similar to the static multimodal equilibrium model presented in Bellei *et al.* (2003), and is schematically depicted in Figure 1.

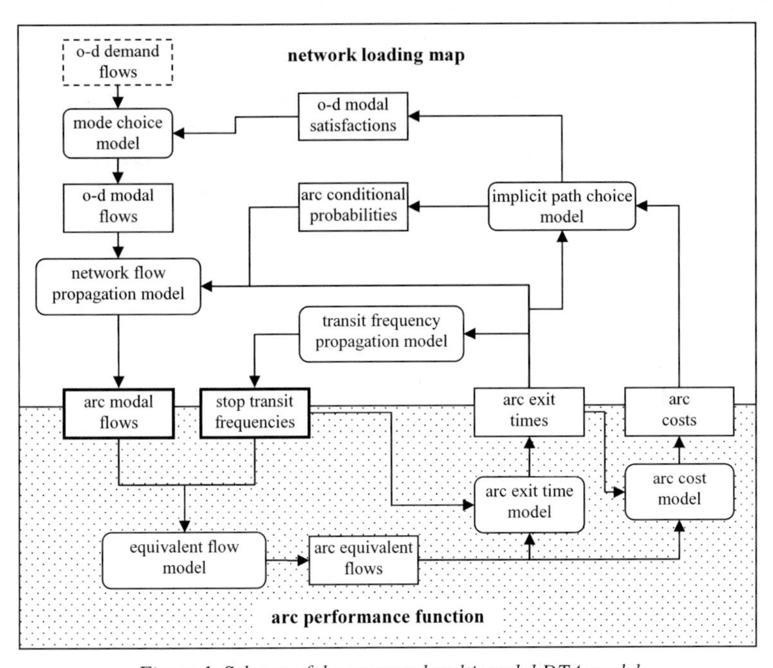

Figure 1. Scheme of the proposed multi modal DTA model

The main innovation in the multimodal assignment model proposed here is to represent the dynamic behavior of transit supply using a *frequency* approach, instead of a *run* approach, thus not requiring a diachronic graph. By so doing, intra modal congestion effects can be efficiently and effectively simulated, particularly those produced by the capacity constraints

[1] For reader's convenience we recall that a) the *Dynamic User Equilibrium* is a state of the network where, at each instant, no user can reduce his perceived travel cost by unilaterally changing route, which implies that users associate to each path its actual cost (the cost that would be actually experienced travelling along the path), and then choose a minimum actual cost path between their origin and destination; b) the *Network Loading Map* is the problem of loading all user trips on the chosen paths yielding the arc inflows corresponding to the given arc travel time and cost pattern; c) the *Dynamic Network Loading* is the problem of loading the network with given path flows in such a way that the resulting arc flow temporal profiles are consistent, through an arc performance model, with the corresponding travel time temporal profiles. The latter problem, raised by the presence of the temporal dimension affecting the DTA, is so important that in the literature much attention has been devoted to its specific analysis (see, for instance, Xu *et al.*, 1999).

of transit carriers, both in terms of waiting delays and passenger queues at stops. The proposed approach is simply to represent these phenomena within a suitable arc performance function, which yields waiting time temporal profiles, consistent with a FIFO representation of passenger queues, for any given passenger flow and transit frequency pattern. The detail of the waiting time of single runs at stops, which cannot be represented any more, is replaced here by the *average line access time* temporal profile, which indeed corresponds to the expected waiting time at a stop in a frequency-based transit service; this way, calculations are simplified, while no notable error is introduced when calculating waiting time temporal profiles, as will be clarified in the algorithm section.

The road congestion will be simulated with a suitable macroscopic dynamic arc performance model, such as the one presented in Gentile *et al.* (2005). Inter-modal congestion phenomena occurring whenever cars and transit carriers share the same facility are simulated, on the road side, through the concept of *equivalent flows*, representing the contribution of the transit system to road congestion; on the transit side, the dependence of line carriers travel times on road traffic conditions, which may affect transit frequencies and waiting times, is reproduced introducing a new *transit frequency propagation* model.

The demand model, based on random utility theory, has two main choice levels: mode choice (Road and Transit) and route choice. With reference to the latter, we present both a deterministic and a stochastic Logit model, based on the results presented in Gentile and Meschini (2006) and in Bellei *et al.* (2005), respectively; moreover, we assume a completely preventive user's behavior.
While the above model has a time continuous formulation, its numerical solution requires, as usual, a time discretization. However, a key feature of the above approach is that it does not exploit the acyclic graph characterizing the corresponding discrete time version of the problem (Gentile *et al.*, 2006), so that no upper bound is set on the interval length by the solution method itself; in fact, this approach is intended to work with time intervals of several minutes, and allows the modeller to choose the time discretization based on the best trade-off between results accuracy and calculation times.

In summary, this model inherits from the existing dynamic traffic assignment model presented in Bellei *et al.* (2005) and Gentile *et al.* (2006) several key features, consisting of:
- formalizing the problem as a system of two functions (namely, the NLM and the arc performance function), instead of as a system of a function and of the parametric solution to a problem (namely, the demand function and the DNL);
- achieving, jointly with the equilibrium, both the temporal consistency of the supply model and the demand-supply consistency, since it is no longer necessary to achieve the first through a DNL;
- formulating DTA through an implicit path approach;
- the possibility of defining "long time intervals" (5-10 min), which allows overcoming the difficulty of solving DTA instances on large networks and long period of analysis;
- devising, on these bases, an efficient dynamic assignment algorithm, whose complexity is equal to the one resulting in the static case multiplied by the number of long time intervals introduced.

Moreover, with respects to the representation of heavy congested multimodal urban transportation systems, it presents the following advantages and innovations:

- the dynamic behavior of transit supply is represented using a *frequency* approach, instead of a *run* approach, thus not requiring a diachronic graph representation of transit supply;
- intra and inter modal congestion effects can be efficiently and effectively simulated, particularly those produced by the capacity constraints of transit carriers and affecting waiting and travel times;
- on the algorithm side, the complexity of the assignment problem is independent of the transit line frequencies.

MULTIMODAL NETWORK FORMALIZATION

The aim of this section is twofold: firstly, we want to achieve a representation of the multimodal network such that the relations between road and transit elements, which are necessary to formalize non-separable cost functions modeling intra and inter modal congestion, can be correctly and univocally identified; secondly, we want to define the minimum amount of information needed to apply the proposed multimodal assignment model, highlighting also the operation of converting the input data, usually organized in a GIS database, into the *assignment graph* handled by the model. Without loss of generality, in the following we will represent two travelling modes: the *road mode R* and *the transit mode T*, and we will refer to the generic mode $m \in M = \{R, T\}$.

To this end, a *base network* is introduced, represented by a directed graph $H = (V, E)$, where $V \subset \aleph$ is the set of *vertices* ($\aleph$ is the set of positive integer numbers), and $E \subseteq V \times V$ is the set of *edges*. The generic edge ε is univocally identified by its initial vertex $IV(\varepsilon)$ and its final vertex $FV(\varepsilon)$, that is $\varepsilon = (IV(\varepsilon), FV(\varepsilon))$. The set of origins and destinations of passenger and car trips, referred to as *centroids*, is a subset $Z \subseteq V$ of the vertices.

The generic vertex $v \in V$ is associated with a location in space that can be accessed by passengers or cars, which is characterized by geographic coordinates $(\lambda_v, \theta_v) \in \Re^2$ ($\Re$ is the set of real numbers), while the generic edge $\varepsilon \in E$ is characterized by a length $L_\varepsilon \in \Re_+$ ($\Re_+$ is the set of non-negative real numbers).

Not each edge is allowed for all modes belonging to the road and transit systems (that is: pedestrians, transit carriers, private cars); therefore, three Boolean *car, pedestrian, transit line allowed-edge* variables $CAE(\varepsilon)$, $PAE(\varepsilon)$, $LAE(\varepsilon) \in \{0,1\}$ are introduced, where each one is equal to 1, if the corresponding mode is allowed on edge $\varepsilon \in E$, and to 0, otherwise.

The *road network* is represented associating to each edge ε: $CAE(\varepsilon) = 1$ an *exit capacity* $Q_\varepsilon \in \Re_{++}$ ($\Re_{++}$ is the set of positive real numbers), which is the maximum vehicular flow that can exit it, an *under saturation speed* $S_\varepsilon \in \Re_{++}$, which is the average speed on the edge when no queue is present on it, that is its outflow is below the exit capacity, and a *road fare*

$RF_\varepsilon \in \Re_+$, which can be time varying. The road network is also characterized by a *car occupancy coefficient* $\gamma \in \Re_{++}$, allowing us to express car flows as a function of user flows.

The *line network* is represented by a set $\Im \subseteq \aleph$ of *lines*. The generic line $\ell \in \Im$ is characterized, from a topological point of view, by an ordered sequence of $\sigma_\ell \in \aleph$ *progressive points*, referred to as its *route*, each one corresponding to a different vertex:
$R(\ell) = \{r_n(\ell) \in \aleph\colon (IV(r_n(\ell)), FV(r_n(\ell))) \in E,\ LAE((IV(r_n(\ell)), FV(r_n(\ell))) = 1,\ n \in [1, \sigma_\ell] \subseteq \aleph\} \subseteq V$,
where we assume that consecutive progressive points are always connected by an edge permitted for transit carriers.

For any given vertex $v \in V$ and line $\ell \in \Im$, the function $n(v, \ell)$ yields, if it exists, the index n such that $r_n(\ell) = v$, and 0 otherwise. Without loss of generality, we assume that line *stops* correspond to progressive points; since not every progressive point is a stop, a Boolean *is-a-stop* variable $IS_n^{\ \ell} \in \{0,1\}$ is introduced, that is equal to 1, if the n-th progressive point of line ℓ is a stop, and 0 otherwise. We assume that the first progressive point of a route is always a stop, that is $IS_1^{\ \ell} = 1$.

Line carriers are characterized by:

- a *carrier capacity* $Q_\ell \in \Re_{++}$, which is the nominal capacity, usually expressed by the number of available seats, if standing in the carrier is not allowed; otherwise, it expresses the maximum number of passengers that can physically fit in the carrier;
- a *boarding and alighting capacity* $BQ_\ell \in \Re_{++}$ and $AQ_\ell \in \Re_{++}$, expressing the maximum flow of passengers that can get on/off the carrier;
- a time needed to open and close the doors $\delta_\ell \in \Re_+$;
- a *car equivalent coefficient* $\lambda_\ell \in \Re_{++}$, expressing the carrier contribution to road congestion in terms of an equivalent number of cars;
- an operative free-flow speed $S_\ell \in \Re_{++}$;
- an operative acceleration $AC_\ell \in \Re_{++}$; and
- an operative deceleration $DE_\ell \in \Re_{++}$.

Each line $\ell \in \Im$ is operated with a *base frequency* $\psi_\ell(\tau) \in \Re_{++}$, expressing the instantaneous carriers departure frequency from the terminal (that is from vertex $r_1(\ell)$) at time $\tau \in \Re_+$.

Regarding the fare schema, we attach to the n-th *section* of line $\ell \in \Im$, from the vertex $r_n(\ell)$ to the vertex $r_{n+1}(\ell)$, with $n \in [1, \sigma_\ell\text{-}1]$, a specific *section fare* $SF_n^{\ \ell} \in \Re_+$, which can be time varying, so as to obtain purely additive path costs, which allows implicit path enumeration in route choice computations.

To each mode $m \in M$ is associated a vector of attributes $\mathbf{X}_m \in \Re$ and a corresponding vector of coefficient $\boldsymbol{\beta}_m \in \Re$, characterizing it with respect to the modal choice performed by users. On this basis, the formal *multimodal network* handled by the assignment model can be represented by a directed graph $G = (N, A)$, where N is the set of the *nodes*, and A is the set of the *arcs*.

The generic node $x \in N$ is identified by an ordered couple, whose first element is the *node line*, denoted $NL(x) \subseteq \{0\} \cup \Im$, and the second element is the *node vertex*, denoted $NV(x) \subseteq -\Im \cup \Im$, that is $x = (NL(x), NV(x))$. This lets us to distinguish 4 different types of nodes, as depicted in Figure 2:

$RN = \{(0, v): v \in V\}$ road nodes;
$PN = \{(0, -v): v \in V\}$ pedestrian nodes;
$LN = \{(\ell, r_n(\ell): \ell \in \Im, n \in [2, \dots, \sigma_\ell] \subseteq \aleph\}$ line nodes;
$QN = \{(\ell, -r_n(\ell)): \ell \in \Im, n \in [1, \dots, \sigma_\ell\text{-}1] \subseteq \aleph, IS_n^\ell = 1\}$ queuing nodes;
so that we have: $N = RN \cup PN \cup LN \cup QN$.

As usual, the generic arc $a \in A$ is identified by an ordered pair of nodes, referred to respectively as the *tail*, denoted $TL(a) \subseteq N$, and the *head*, denoted $HD(a) \subseteq N$; that is $a = (TL(a), HD(a))$. As depicted in Figure 2, we distinguish 6 different types of arcs:

$RA = \{(\,(0, u)\,, (0, v)\,): (u, v) \in E, CAE((u, v)) = 1\}$ road arcs;
$PA = \{(\,(0, -u)\,, (0, -v)\,): (u, v) \in E, CAE((u, v)) = 1\}$ pedestrian arcs;
$LA = \{(\,(\ell, r_n(\ell))\,, (\ell, r_{n+1}(\ell))\,): \ell \in \Im, n \in [1, \sigma_\ell\text{-}1] \subseteq \aleph\}$ line arcs;
$QA = \{(\,(0, r_n(\ell))\,, (\ell, -r_n(\ell))\,): \ell \in \Im, n \in [1, \sigma_\ell\text{-}1] \subseteq \aleph, IS_n^\ell = 1\}$ queueing arcs;
$BA = \{(\,(\ell, -r_n(\ell))\,, (\ell, r_n(\ell))\,): \ell \in \Im, n \in [1, \sigma_\ell\text{-}1] \subseteq \aleph, IS_n^\ell = 1\}$ boarding arcs;
$AA = \{(\,(\ell, r_n(\ell)), (0, r_n(\ell))\,): \ell \in \Im, n \in [2, \sigma_\ell] \subseteq \aleph, IS_n^\ell = 1\}$ alighting arcs,
so that we have: $A = RA \cup PA \cup LA \cup QA \cup BA \cup AA$.

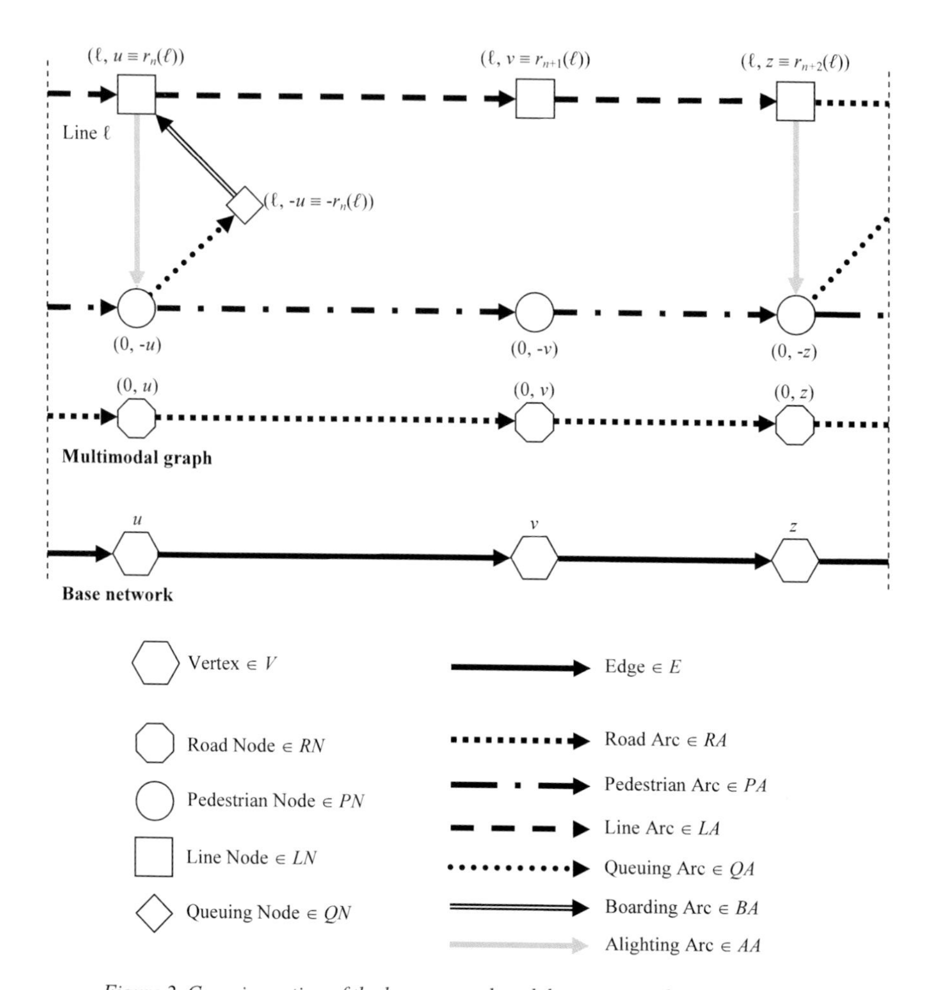

Figure 2. Generic portion of the base network and the corresponding modal graphs.

Note that more than one line may stop at the same pedestrian node, and that the structure of the pedestrian network can be very simple or very complex, depending on the modeling choices; Figure 2 does not illustrate these facts.

The proposed network formalization allows us to define the following relation among its elements:

- Each arc $a \in LA \cup QA \cup BA \cup AA$ is univocally associated with a line $\ell(a) \in \Im$; specifically, if $a \in AA$, then $\ell(a) = NL(TL(a))$, otherwise $\ell(a) = NL(HD(a))$. Moreover, we can denote as $n(a) = n(NV(TL(a)), \ell(a))$ the index of the associated route vertex;

- each arc $a \in LA$ is associated with the unique road arc, if it exists, sharing the same edge, that is:
  $LR(a) = \{b \in RA: ((0, VN(TL(a))) , (0, VN(HD(a)))) = b\} \subset RA \cup \varnothing$;
- each road arc $a \in RA$ is associated with the set of line arcs sharing the same edge, that is:
  $RL(a) = \{b \in LA: ((0, VN(TL(b))) , (0, VN(HD(b)))) = a \} \subset LA \cup \varnothing$;
- each arc $a \in A$ is associated with an edge, that is $\varepsilon(a) = (VN(TL(a)), VN(HD(a)))$.

Finally, we will denote with $N_m \subseteq N$ and $A_m \subseteq A$ the subset of nodes and arcs allowed for mode m, with $N_T = PN \cup LN \cup WN$, $A_T = PA \cup LA \cup QA \cup BA \cup AA$, and $N_R = RN$, $A_R = RA$.

THE ARC PERFORMANCE FUNCTION

We first introduce the notation for arc usage and arc performance variables:

$f_a(\tau)$	inflow of users on arc $a \in A$ at time τ;
$e_a(\tau)$	outflow of users from arc $a \in A$ at time τ;
$t_a(\tau)$, $c_a(\tau)$	exit time and cost for users entering arc $a \in A$ at time τ;
$\phi_\ell^n(\tau)$	transit frequency at the n-th progressive point of line ℓ at time τ;

In order to represent the inter-modal congestion phenomena occurring whenever cars and transit carriers share the same facility, we introduce the *equivalent inflow* $u_a(\tau)$ and the *equivalent outflow* $v_a(\tau)$. With reference to the generic road arc $a \in A_R$, we assume that its equivalent flows are a linear combination of car flows and transit frequencies of those lines using it, that is:

$$u_a(\tau) = f_a(\tau) / \gamma + \Sigma_{b \in RA(a)} \lambda_{\ell(b)} \cdot \phi_{\ell(b)}{}^{n(b)}(\tau) , \tag{1.a}$$

$$v_a(\tau) = e_a(\tau) / \gamma + \Sigma_{b \in RA(a)} \lambda_{\ell(b)} \cdot \phi_{\ell(b)}{}^{n(b)+1}(\tau) , \tag{1.b}$$

With reference to the generic transit arc $a \in A_T$, the equivalent flows coincide with the user flows.

Relations (1) can be expressed in a compact form by the following functional, expressing the *equivalent flow model*:

$$[\mathbf{u}, \mathbf{v}] = \Upsilon(\mathbf{f}, \mathbf{e}, \boldsymbol{\phi}) \tag{2}$$

where the arc components of $\mathbf{u}$, $\mathbf{v}$, $\mathbf{f}$, $\mathbf{e}$ and $\boldsymbol{\phi}$ are temporal profiles. On this basis, the *arc performance function* aims at determining the travel time temporal profile and the generalized cost temporal profile on each arc of the multimodal network as a function of the equivalent inflow and outflow temporal profiles and of the transit frequency temporal profiles of the adjacent arcs.

Introducing the function Γ, the arc performance function can be thus expressed in compact form, as:

$$[\mathbf{t}, \mathbf{c}] = \Gamma(\mathbf{u}, \mathbf{v}, \boldsymbol{\phi}) \tag{3}$$

where the arc components of $\mathbf{t}$ and $\mathbf{c}$ are temporal profiles.

Function (3) will be specified in the following sections with reference to the different arcs introduced above, so as to obtain a macroscopic arc performance function representing, with respect to the road network, the effect of limited road capacities, and with respect to the transit network, the effects of transit service discontinuity and transit carrier capacities in the context of transit line representation as continuous vehicle flows.

Road arcs

With reference to road arcs, function (3) can be specified utilizing a suitable macroscopic dynamic arc performance model, such the spatially separable model presented in Gentile *et al.* (2005), where arc performances are evaluated through an approximate solution of the Simplified Theory of Kinematic Waves and a parabolic-triangular fundamental diagram, or such the spatially non-separable network performance model presented in Gentile *et al.* (2006), which extends the previous one in order to represent spillback congestion.

Here, in order to focus on paper's topics, we will specify function (3) with reference to a simple link model, where the generic road arc $a \in RA$ represents a road link of length $L_{\varepsilon(a)}$; at the final section of the link we have a *bottleneck* with capacity $Q_{\varepsilon(a)}$ constant in time, strictly positive and bounded. When there is no queue, the vehicles travel along the arc with a constant *under saturation* speed $S_{\varepsilon(a)}$, so that their travel time is $L_{\varepsilon(a)}/S_{\varepsilon(a)}$. When the inflow $u_a(\tau)$ exceeds the exit capacity $Q_{\varepsilon(a)}$, an *over saturation* queue occurs; if the queue arises at time $\sigma + L_{\varepsilon(a)}/S_{\varepsilon(a)}$, the exit time of a vehicle entering the arc at time $\tau \geq \sigma$ is equal to that time, plus the time that takes for all the vehicles that entered the arc between σ and τ to pass through the bottleneck at the maximum rate $Q_{\varepsilon(a)}$. Based on the results achieved in Gentile *et al.* (2005), as depicted in Figure 3, the worst case dominates the others. Then, we have:

$$t_a(\tau) = \max\left\{\sigma + \frac{L_{\varepsilon(a)}}{S_{\varepsilon(a)}} + \frac{1}{Q_{\varepsilon(a)}} \cdot \int_\sigma^\tau u_a(x)dx : \sigma \leq \tau\right\}. \tag{4}$$

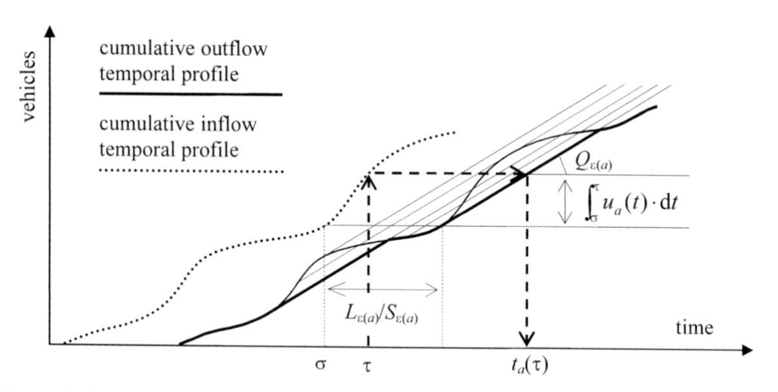

Figure 3. Arc exit time for a link with fixed under saturation speed and fixed exit capacity.

This model is consistent with the simplified kinematics waves theory (Daganzo, 1997) assuming a triangular shaped fundamental diagram, like the one depicted in Figure 4; in fact, in Gentile *et al.* (2005) it is proved that the density and the queue speed corresponding to the bottleneck capacity (respectively, point B' and $V(Q_{\varepsilon(a)})$ in Figure 4) are meaningless with respect to the arc travel times. Moreover, the model respects the non-strict version of the FIFO rule, that is:

$$t_{xy}(\tau') \geq t_{xy}(\tau) \text{ , for any } \tau' > \tau . \tag{5}$$

Indeed, if no vehicle enters the arc between any τ and $\tau' > \tau$, that is $\int_{\tau}^{\tau'} f_{xy}(t) \cdot \mathrm{d}t = 0$, while the queue is vanishing, that is $t_{xy}(\tau') > \tau' + L_{xy} / V_{xy}$, it results that $t_{xy}(\tau') = t_{xy}(\tau)$.

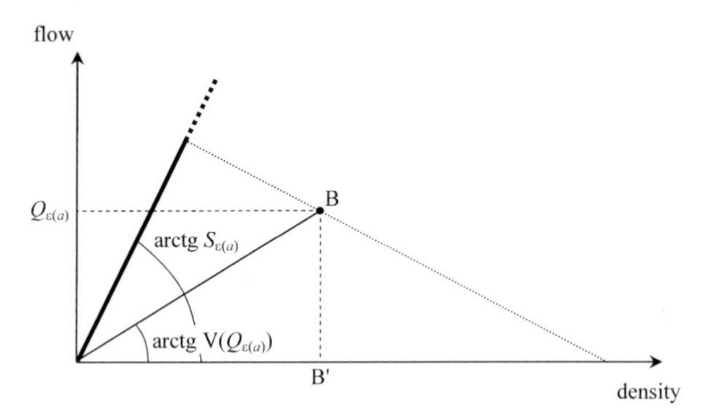

Figure 4. The proposed link model for road arcs is coherent with a triangular fundamental diagram.

The travel cost of the generic road arc $a \in RA$ is given by the sum of its travel time multiplied by the value of road time $\chi \in \Re_{+}$, and of the proper road fare:

$$c_a(\tau) = \chi \cdot (t_a(\tau) - \tau) + RF_{\varepsilon(a)}(\tau). \tag{6}$$

Queuing and boarding arcs

The queuing and boarding arcs of a given stop are aimed at representing the total waiting time suffered by a user in order to board the chosen line. In fact, when the line is congested, that is the flow willing to board is higher than the line available capacity at that stop, the waiting time results by the sum of two components, modelled respectively by the queuing arc and the boarding arc:

- the *over saturation* waiting time, which represents the time spent by users queuing at the stop and waiting that the service become actually available to them; it can be thought as the time spent by each user waiting until the next carrier arriving at the stop will be the one he can actually board;
- the *under saturation* delay, which represents the average delay due to the fact that the transit service is not continuously available over time; it can be thought as the additional delay suffered by each user waiting until the carrier that he will board arrives at the stop.

Obviously, if the line is not congested, the over saturation waiting time is null and the total waiting time coincides with the under saturation delay.

With reference to the generic boarding arc $a \in BA$, the under saturation delay is evaluated assuming it proportional to the inverse of the corresponding line frequency, evaluated at the boarding instant; thus we have:

$$t_a^{-1}(\tau) = \tau - \alpha_{\ell(a)} / \phi_a(\tau) \tag{7}$$

where $t_a^{-1}(\tau)$ is the entry time for the user exiting the arc $a \in BA$ at time τ and $\alpha_{\ell(a)}$ takes into account headway irregularity (in particular, $\alpha_{\ell(a)} = 0.5$ with uniform headway, $\alpha_{\ell(a)} = 1$ with Poissonian headway).

Then, for a given entry time τ, the corresponding exit time is given by:

$$t_a(\tau) = \tau' : t_a^{-1}(\tau') = \tau \tag{8}$$

and the exit time temporal profile is simply the inverse of the entry time temporal profile:

$$\mathrm{t}_{[a]} = [\mathrm{t}_{[a]}^{-1}]^{-1} \tag{9}$$

With reference to the generic queuing arc $a \in QA$, the over saturation waiting time is evaluated by means of a *bottleneck with time variable exit capacity*, whose exit capacity is related to the *available capacity* of line $\ell(a)$ at progressive point $n(a)$. At a given time τ, the available capacity is given by the line capacity at that point, depending on the carrier capacity and on the line frequency at τ, minus the flow already onboard at τ, that is:

$$AK_{n(a)}(\tau) = Q_{\ell(a)} \cdot \phi_{\ell(a)}{}^{n(a)}(\tau) - [v_b(\tau) - u_c(\tau)]$$

$$b \in LA: \ell(b) = \ell(a), n(b) = n(a) - 1 \ , \ c \in AA: \ell(b) = \ell(a), n(b) = n(a), \tag{10}$$

where the term in square brackets is the flow already onboard, which coincides with the onboard flow arriving at the stop minus the flow alighting at it (see Figure 2).
At a given time τ, the capacity actually available at the end of the queuing arc is not the available capacity at the same time, because of the presence of the under saturation delay on the boarding arc. Then, the bottleneck exit capacity temporal profile can be obtained propagating backward in time the available capacity temporal profile accordingly with the under saturation delay, the FIFO and the capacity conservation (Cascetta, 2001), that is:

$$\xi_a(t_b^{-1}(\tau)) = AK_{n(a)}(\tau) / \partial(t_b^{-1}(\tau))/\partial\tau \ , \quad b \in BA: TL(b) = HD(a) \tag{11}$$

Then, the problem of determining the *exit time* $t_a(\tau)$ for a user that enters the queuing arc $a \in QA$ at the generic time τ, in presence of a time-varying *exit capacity* $\xi_a(\sigma)$ for each time σ, shall be addressed identifying firstly the *cumulative exit flow* temporal profile, whose value $E_a(\tau)$ at time τ is given by:

$$E_a(\tau) = \min\{F_a(\sigma) + \Xi_a(\tau) - \Xi_a(\sigma) : \sigma \le \tau\} \ , \tag{12}$$

where $F_a(\tau)$ denotes the *cumulative inflow* at the generic time τ, that is the number of users that entered the arc until time τ:

$$F_a(\tau) = \int_{-\infty}^{\tau} u_a(\sigma) \cdot d\sigma \ , \tag{13}$$

and $\Xi_a(\tau)$ denotes the *cumulative exit capacity* at the generic time τ:

$$\Xi_a(\tau) = \int_{-\infty}^{\tau} \xi_a(\sigma) \cdot d\sigma \ . \tag{14}$$

The above expression (12) can be explained as follows. If there is no queue at a given time τ, the travel time is null and the cumulative exit flow is equal to the cumulative inflow. If a queue arises at time $\sigma < \tau$, from that instant until the queue vanishes the exit flow equals the exit capacity, and then, based on the FIFO rule, the cumulative exit flow $E_a(\tau)$ results from adding to the cumulative inflow at time σ the integral of the exit capacity between σ and τ, that is $\Xi_a(\tau) - \Xi_a(\sigma)$. Moreover, if there is no queue at time τ, the cumulative exit flow is the same as the case when the queue arises exactly at $\sigma = \tau$. The actual cumulative exit flow at time τ is the minimum among each cumulative exit flow that would occur if the queue began at any previous instant $\sigma \leq \tau$.

Based on the FIFO rule, the cumulative exit flow at the exit time $t_a(\tau)$ of a user that enters the arc at τ is equal to the cumulative inflow at time τ, that is:

$$E_a(t_a(\tau)) = F_a(\tau) \ . \tag{15}$$

However, in presence of intervals with null flow, the above implicit expression does not allow to obtain a univocal value of the travel time. To take this circumstances into account, once the cumulative exit flow temporal profile is known, the exit time temporal profile is calculated conventionally as:

$$t_a(\tau) = \max\{0, \min\{\sigma: E_a(\sigma) = F_a(\tau)\}\} \ . \tag{16}$$

Figure 5 depicts a graphical interpretation of equation (12), where the cumulative exit flow temporal profile $E_a(\tau)$ is the lower envelop of the following curves: a) the cumulative inflow $F_a(\tau)$; b) the family of functions $F_a(\sigma) + \Xi_a(\tau) - \Xi_a(\sigma)$ with $\tau > \sigma$, for every time σ, each one obtained from the vertical translation of the cumulative exit capacity temporal profile that goes through the point $(\sigma, F_a(\sigma))$. No queue is present when curve a) prevails; therefore, the queue arises at time σ' and vanishes at time σ''. In the same framework, the calculation of the exit time based on the cumulative inflow and exit flow temporal profiles is shown using thick arrows.

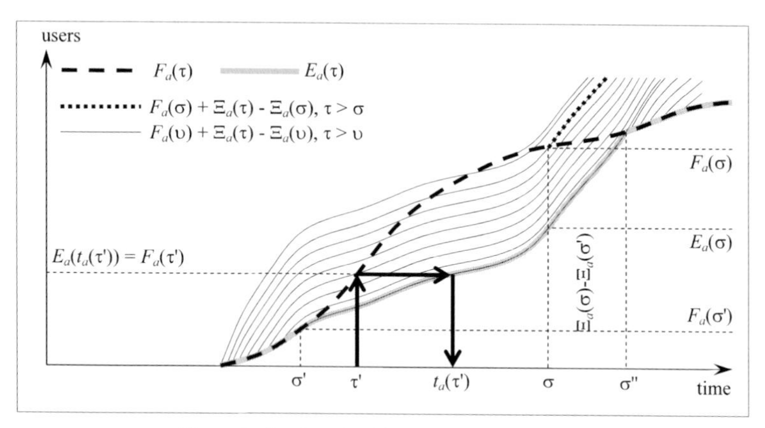

Figure 5. Bottleneck with time-varying capacity.

Finally, the cost for the generic arc $a \in QA \cup BA$ is given by its travel time multiplied by the value of waiting time $\mu \in \Re_+$:

$$c_a(\tau) = \mu \cdot (t_a(\tau) - \tau) . \tag{17}$$

Line arcs

The exit time from the generic arc $a \in LA$ at a given entry time τ is determined by the sum of three terms: the dwelling time at the stop DT_a, the uncongested travel time URT_a, and the delay CD_a due to road congestion, evaluated at time $\tau + DT_a(\tau)$ when the carrier leaves the stop, that is:

$$t_a(\tau) = \tau + DT_a(\tau) + URT_a + CD_a(\tau + DT_a(\tau)) \tag{18}$$

The dwelling time is determined by the time needed for passengers to alight and to board the carrier, plus time needed to open and close the doors:

$$t_a(\tau) = \tau + (u_b(\tau) / \phi_{\ell(a)}{}^{n(a)}(\tau)) / AQ_{\ell(a)} + (v_c(\tau) / \phi_{\ell(a)}{}^{n(a)}(\tau)) / BQ_{\ell(a)} + 2 \cdot \delta_{\ell(a)} ,$$

$$b \in AA\text{: } \ell(b) = \ell(a), n(b) = n(a), c \in BA\text{: } \ell(c) = \ell(a), n(c) = n(a) \tag{19}$$

where: $b \in AA$ and $c \in BA$ are respectively the alighting and boarding arcs corresponding to the same line and progressive point (see Figure 2), while $u_b(\tau) / \phi_{\ell(a)}{}^{n(a)}(\tau)$ and $v_c(\tau) / \phi_{\ell(a)}{}^{n(a)}(\tau)$ are the corresponding numbers of alighting and boarding passengers at time τ. The above specification of the dwelling time assumes that the doors are used first by alighting passengers, and then by boarding passengers. Alternatively, if the door usage is specified, the following expression can be adopted in place of (19):

$$t_a(\tau) = \tau + \max\{ (u_b(\tau) / \phi_{\ell(a)}{}^{n(a)}(\tau)) / AQ_{\ell(a)} , (v_c(\tau) / \phi_{\ell(a)}{}^{n(a)}(\tau)) / BQ_{\ell(a)} \} + 2 \cdot \delta_{\ell(a)} , \tag{20}$$

where $b \in AA$ and $c \in BA$ are the same as above.

To be noticed that the alighting and boarding capacities can be dependent from the line carrier congestion. This can be represented multiplying $AQ_{\ell(a)}$ and $BQ_{\ell(a)}$ by a function decreasing with the line occupancy rate, such: $1 - \alpha \cdot [f_a(\tau) / (Q_{\ell(a)} \cdot \phi_{\ell(a)}{}^{n(a)}(\tau))]^\beta$, $\alpha > 0$.

The uncongested travel time depends on the operative speed, the acceleration and the deceleration of the line carrier:

$$URT_a = L_a / S_{\ell(a)} + 0.5 \cdot S_{\ell(a)} \cdot (1 / AC_{\ell(a)} + 1 / DE_{\ell(a)}) \tag{21}$$

The delay due to road congestion, which is present only if the line is operated on a road arc allowed for cars, is equal to the grow of the road travel time with respect to the road free-flow travel time:

$$CD_a(\tau) = \begin{cases} 0 & \text{if } LR(a) = \varnothing \\ t_b(\tau) - \tau - su_b, \quad b \in LR(a) & \text{otherwise} \end{cases}, \tag{22}$$

where $t_b(\tau)$ and $su_b = L_{\varepsilon(b)} / V_{\varepsilon(b)}$ represent, respectively, the congested exit time and the under saturation travel time of the road arc associated with arc a.

The cost for the generic arc $a \in LA$ is given multiplying its travel time by the value of onboard time $\eta \in \Re_+$, and adding to it the proper section fare:

$$c_a(\tau) = \eta \cdot (t_a(\tau) - \tau) + SF_{n(a)}{}^{\ell(a)}(\tau). \tag{23}$$

Alighting arcs

The exit time from the generic arc $a \in AA$ at a given entry time τ is assumed to be determined by the time needed for passengers to alight the carrier, plus time lost to open the doors:

$$t_a(\tau) = (u_a(\tau) / \phi_{\ell(a)}{}^{n(a)}(\tau)) / AQ_{\ell(a)} + \delta_{\ell(a)}, \tag{24}$$

where $u_a(\tau) / \phi_{\ell(a)}{}^{n(a)}(\tau)$ is the number of alighting passengers at time τ.
As for the dwelling arc, the alighting capacity can be made dependent from the line carrier congestion.

The cost for the generic arc $a \in AA$ is given by multiplying its travel time by the value of alighting time $\pi \in \Re_+$:

$$c_a(\tau) = \pi \cdot (t_a(\tau) - \tau). \tag{25}$$

Pedestrian arcs

The uncongested exit time from the generic arc $a \in PA$ at a given entry time τ is simply:

$$t_a(\tau) = \tau + L_{\varepsilon(a)} / S_{\varepsilon(a)}, \tag{26}$$

while its cost is given by multiplying its travel time by the value of walking time $\zeta \in \Re_+$:

$$c_a(\tau) = \zeta \cdot (t_a(\tau) - \tau) . \tag{27}$$

THE TRANSIT FREQUENCY PROPAGATION MODEL

The *transit frequency propagation model* aims at finding line frequency temporal profiles as a function of the line frequency at terminal and of the line route travel time temporal profiles. In fact, contrary to the static case, the temporal profiles of the transit frequencies in general are not constant along the line. This is due, on the one hand, to the translation in space and time of the frequency at terminal due to the time needed by carriers to reach each line progressive point; on the other hand, to the variation in time of road arc travel times, which induces variations in the carrier headways (see the example in Figure 6). The variation of frequency temporal profiles along the line may be calculated based on road arc travel times, accordingly with the FIFO and vehicle conservation rules (Cascetta, 2001) applied to transit carriers, as follows.

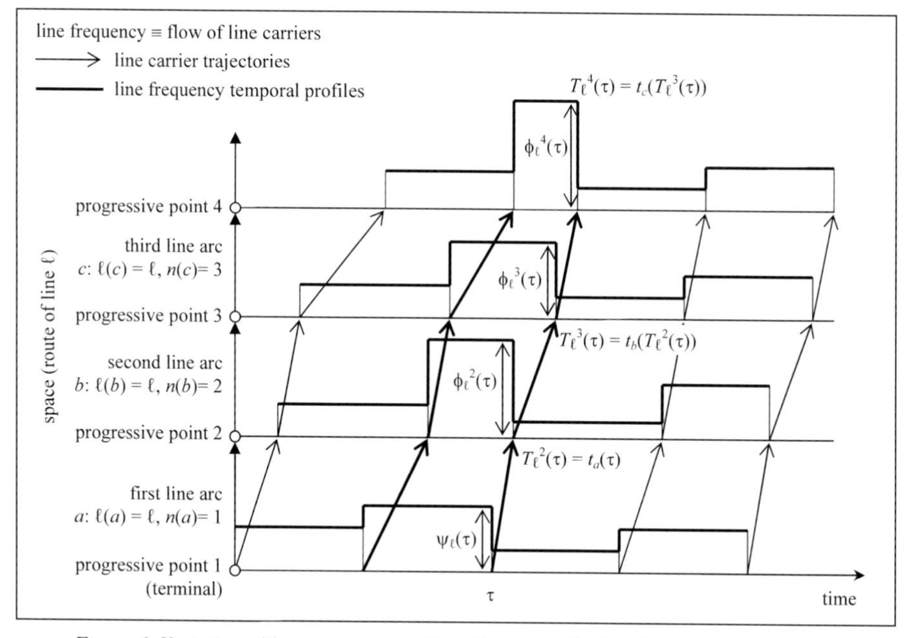

Figure 6. Variation of frequency temporal profiles along the line due to line travel times.

First, the instant $T_\ell^n(\tau)$ when the carrier operating line ℓ and departed from the first progressive point $r_1(\ell)$ at time τ reaches the n-th progressive point $r_n(\ell)$ can be determined recursively on the basis of the line arc exit times:

$$T_\ell^1(\tau) = \tau, \quad T_\ell^n(\tau) = t_a(T_\ell^{n-1}(\tau)), \; n \in [2, \sigma_\ell] \subseteq \aleph, \; a \in LA: \ell(a) = \ell, \; n(a) = n - 1 \tag{28}$$

Then, since the line frequency can be regarded as the flow of carriers operating the line, its propagation along the line route can be determined with the following expression, which is derived from the FIFO and vehicle conservation rules (Cascetta, 2001):

$$\phi_\ell^n(T_\ell^n(\tau)) = \psi_\ell(\tau) / (\partial T_\ell^n(\tau)/\partial\tau), \quad n\in[1, \sigma_\ell]\subseteq\aleph \,, \tag{29}$$

expressing the line frequency observed at the n-th progressive point as a function of the line frequency at terminal, and of the travel time from the terminal to that progressive point.
Relations (28) and (29) can be expressed in a compact form by the following functional:

$$\boldsymbol{\phi} = \phi(\mathbf{t}) \,, \tag{30}$$

where the arc components of $\boldsymbol{\phi}$ and $\mathbf{t}$ are temporal profiles.

THE NETWORK LOADING MAP

The *network loading map* (NLM) is complementary to the arc performance function in the sense that it aims at determining the inflow temporal profiles as a function of the travel time temporal profiles and of the generalized cost temporal profiles. We will outline two NLM, both allowing for implicit path enumeration: the one proposed in Bellei *et al.* (2005) for DTA on road networks, considering a Logit route choice model solved with a dynamic extension of the Dial's algorithm, and the formulation proposed in Gentile and Meschini (2006), considering a route choice model based on dynamic shortest path computations. In both cases, for seek of brevity we will simply present the resulting formulations, addressing the reader to the quoted papers for any insight on the models.

The route choice model

Dealing with implicit path enumeration, we have to introduce:

$w_x^{md}(\tau)$ *node satisfaction*, which is the opposite of the expected value of the minimum perceived cost to reach the destination $d\in Z$ with mode $m\in M$ being on node $x\in N\cup Z$ at time τ;

$p_a^{md}(\tau)$ *arc conditional probability*, which is the probability of choosing arc $a\in A$ to continue the trip towards the destination $d\in Z$ with mode $m\in M$ being on node $TL(a)\in N\cup Z$ at time τ.

In order to specify node satisfactions and arc conditional probabilities accordingly with a Logit route choice model, as in any Dial-like model we assume that users travel only on *efficient* arcs, that is they always near the destination with respect to a given node topological order TO_x^{md}, with $x\in N$, $d\in Z$, $m\in M$. Let then $FSE(x, d, m) = \{a\in A_m\colon TL(a) = x,\ TO_x^{md} > TO_{HD(a)}^{md}\}$ and $BSE(x, d, m) = \{a\in A_m\colon HD(a) = x,\ TO_{TL(a)}^{md} > TO_x^{md}\}$ be the efficient forward and backward star of node x with respect to destination d and mode m, respectively.

Based on the results achieved in Bellei *et al.* (2005), it is possible to express the node satisfactions and the arc conditional probabilities through the following recursive equations:

$$w_x^{md}(\tau) = \theta_m \ln(\sum_{a\in FSE(x,d,m)} \exp((-c_a(\tau) + w_{HD(a)}{}^d(t_a(\tau))) / \theta_m)) , \quad x\in N_m \cup Z \tag{31}$$

$$p_a^{md}(\tau) = \exp((-c_a(\tau) + w_{HD(a)}{}^{md}(t_a(\tau)) - w_{TL(a)}{}^d(\tau)) / \theta_m) , \quad a\in A_m , \tag{32}$$

where θ_m is the Logit parameter. Since users choose only efficient paths, the above system of equations can be solved by processing the nodes in topological order, while time instants may be processed in any order for each node.

The deterministic specification of the route choice model can be achieved as in Gentile and Meschini (2006). Again, let $FS(x, m) = \{a\in A_m: TL(a) = x\}$ and $BS(x, m) = \{a\in A_m: HD(a) = x\}$ be the forward and backward star of node x with respect to mode m, respectively; then, the node satisfactions and the arc conditional probabilities are expressed through the following recursive equations:

$$w_x^{md}(\tau) = \min\{c_a(\tau) + w_{HD(a)}{}^d(t_a(\tau))\}: a\in FS(x, m)\} , \quad x\in N_m \cup Z \tag{33}$$

$$p_a^{md}(\tau) \cdot [c_a(\tau) + w_{HD(a)}{}^d(t_a(\tau)) - w_{TL(a)}{}^d(\tau)] = 0 , \quad a\in A_m \tag{34}$$

$$\sum_{(a)\in FS(x,\, m)} p_a^{md}(\tau) = 1 , \tag{35}$$

$$p_a^{md}(\tau) \geq 0 , \tag{36}$$

The above system of equations can be solved processing time instants in reverse chronological order, while nodes may be processed in any order within each time instant. We can express the solution of the Logit route choice model (31)-(32) in compact form through the following functions:

$$\mathbf{w} = \mathrm{w}^{\mathrm{L}}(\mathbf{c}, \mathbf{t}) \tag{37}$$

$$\mathbf{p} = \mathrm{p}^{\mathrm{L}}(\mathbf{w}, \mathbf{c}, \mathbf{t}) , \tag{38}$$

With reference to the deterministic case, since when there is more than one arc exiting from a given node that yields the minimum cost to reach a destination, the arc conditional probability pattern solving the system (33)÷(36) is not unique, the deterministic route choice model is formally expressed through the following functional and point-to-set map:

$$\mathbf{w} = \mathrm{w}^{\mathrm{D}}(\mathbf{c}, \mathbf{t}) \tag{39}$$

$$\mathbf{p} \in \mathrm{p}^{\mathrm{D}}(\mathbf{w}, \mathbf{c}, \mathbf{t}) . \tag{40}$$

In both cases, the node components of **w** and the arc components of **p** are temporal profiles.

The mode choice model

With reference to users travelling from orgin $o\in Z$ toward the destination $d\in Z$, we define the following:

$V_m^{od}(\tau)$ systematic utility of mode m for users departing at time τ
$P_m^{od}(\tau)$ choice probability of mode m for users departing at time τ
$D^{od}(\tau)$ demand flow departing at time τ

$d_m^{od}(\tau)$ demand flow departing at time τ using mode m

Since we have only two alternatives, it is classical to reproduce the mode choice through a multinomial Logit model with parameter θ_M, that is:

$$P_m^{od}(\tau) = \exp(V_m^{od}(\tau) / \theta_M) / \Sigma_{m \in M} \exp(V_m^{od}(\tau) / \theta_M) \tag{41}$$

As usual, we assume:

$$V_m^{od}(\tau) = \boldsymbol{\beta}_m^{\mathrm{T}} \cdot \mathbf{X}_m + w_o^{md}(\tau) , \tag{42}$$

where the node satisfaction $w_o^{md}(\tau)$ plays the role of the inclusive utility associated to the path choice. The demand flow departing at time τ using mode m is:

$$d_m^{od}(\tau) = D^{od}(\tau) \cdot P_m^{od}(\tau) \tag{43}$$

Based on equations (41)÷(43), the mode choice model is expressed in a compact form by the following functional:

$$\mathbf{d} = \mathrm{d}(\mathbf{w}, \mathbf{D}) , \tag{44}$$

where the origin-destination components of **d** and **D** are temporal profiles.

The network flow propagation model

To formulate the network flow propagation model, it is useful to introduce inflow and outflow variables referred to passengers travelling toward a specific destination $d \in Z$:

$f_a^d(\tau)$ inflow of arc $a \in A$ at time τ directed to d ;

$e_a^d(\tau)$ outflow of arc $a \in A$ at time τ directed to d ;

With reference to the Logit formulation, the inflow $f_a^d(\tau)$ on arc $a \in A_m$ at time τ directed to destination $d \in Z$ is given by the arc conditional probability $p_a^{md}(\tau)$ multiplied by the flow exiting from node $TL(a)$ at time τ. The latter is given, in turn, by the sum of the outflow $e_b^d(\tau)$ from each arc $b \in BSE(TL(a), d, m)$ entering $TL(a)$, and of the demand flow $d_m^{TL(a)d}(\tau)$ from $TL(a)$ to d on mode m, which is null if $TL(a) \notin Z$. Then, we have:

$$f_a^d(\tau) = p_a^{md}(\tau) \cdot [d_m^{TL(a)d}(\tau) + \Sigma_{b \in BSE(TL(a), d, m)} e_b^d(\tau)] , \quad a \in A_m \tag{45}$$

Similarly, with reference to the deterministic formulation, the inflow $f_a^d(\tau)$ on arc $a \in A_m$ at time τ directed to destination $d \in Z$ is given by the arc conditional probability $p_a^{md}(\tau)$ multiplied by the flow exiting from node $TL(a)$ at time τ. The latter is given, in turn, by the sum of the outflow $e_b^d(\tau)$ from each arc $b \in BS(TL(a), m)$ entering $TL(a)$, and of the demand flow $d_m^{od}(\tau)$ from o to d on mode m, which is null if $TL(a) \notin Z$. Then, we have:

$$f_a^d(\tau) = p_a^{md}(\tau) \cdot [d_m^{TL(a)d}(\tau) + \Sigma_{b \in BS(TL(a), m)} e_b^d(\tau)] , \quad a \in A_m \tag{46}$$

In both cases, applying the FIFO and flow conservation rules (Cascetta, 2001) the outflow at time τ can be expressed in terms of the inflow at the entry time $t_b^{-1}(\tau)$:

$$e_b^d(\tau) = f_b^d(t_b^{-1}(\tau)) \, / \, [\mathrm{d}t_b(\tau)/\mathrm{d}\tau] \tag{47}$$

while the total flows entering and exiting arc $a \in A$ at time τ are:

$$f_a(\tau) = \sum_{d \in Z} f_a^{\,d}(\tau)\ ; \quad e_a(\tau) = \sum_{d \in Z} e_a^{\,d}(\tau) \tag{48}$$

Based on (45), (47) and (48), the Logit network flow propagation model is expressed by the following functional:

$$[\mathbf{f}, \mathbf{e}] = \omega^{\mathrm{L}}(\mathbf{p}, \mathbf{t}, \mathbf{d})\,, \tag{49}$$

while (46), (47) and (48) yield the deterministic network flow propagation functional:

$$[\mathbf{f}, \mathbf{e}] = \omega^{\mathrm{D}}(\mathbf{p}, \mathbf{t}, \mathbf{d})\,, \tag{50}$$

THE DYNAMIC EQUILIBRIUM MODEL

Extending to the dynamic case Wardrop's first principle, the DTA problem is here regarded as a Dynamic User Equilibrium (DUE), where no user can reduce his perceived travel cost by unilaterally changing path, under the assumption that the path cost is that actually experienced by the passenger while travelling on the network consistently with time-varying travel times and generalized costs. The formulation based on implicit path enumeration of the DUE model is synthetically depicted in Figure 7, which immediately highlights the possibility of formulating the model as a fixed point problem in terms of the arc inflow and outflow and line frequency temporal profiles **f**, **e** and $\boldsymbol{\phi}$.

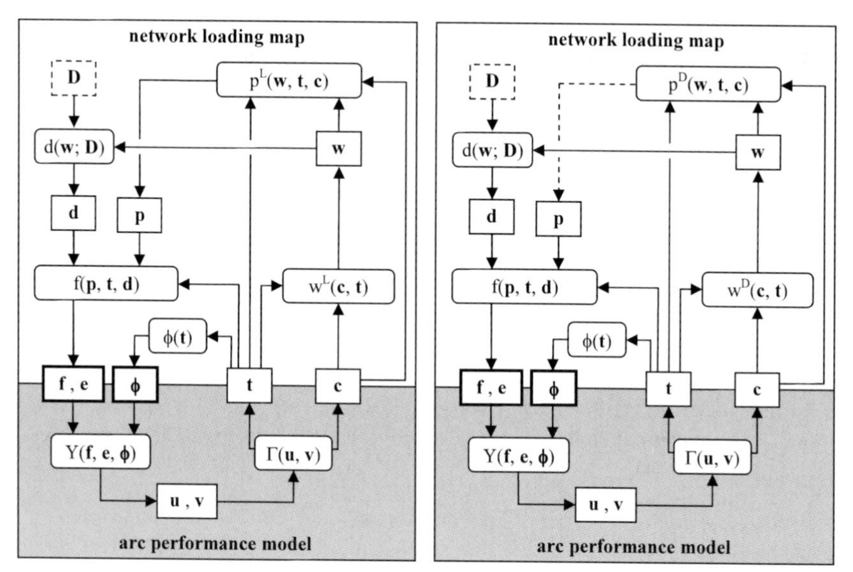

Figure 7. Formulation of the Logit (left) and deterministic (right) Dynamic User Equilibrium with implicit path enumeration. For the deterministic case, the dashed arrow indicate any solution of the choice map.

Specifically, combining the route choice model (37)-(38) and the mode choice model (44) with the flow propagation model (49) and the transit frequency propagation model (30) yields the formulation of the Logit NLM based on implicit path enumeration:

$$[\mathbf{f}, \mathbf{e}, \boldsymbol{\phi}] = [\omega^L(p^L(w^L(\mathbf{c}, \mathbf{t}), \mathbf{c}, \mathbf{t}), \mathbf{t}, d(w^L(\mathbf{c}, \mathbf{t}) ; \mathbf{D})) , \phi(\mathbf{t})] = f^L(\mathbf{c}, \mathbf{t} ; \mathbf{D}) , \tag{51}$$

while combining the route choice model (39)-(40) and the mode choice model (44) with the flow propagation model (50) and the transit frequency propagation model (30) yields the formulation of the deterministic NLM based on implicit path enumeration:

$$[\mathbf{f}, \mathbf{e}, \boldsymbol{\phi}] \in [\omega^D(p^D(w^D(\mathbf{c}, \mathbf{t}), \mathbf{c}, \mathbf{t}), \mathbf{t}, d(w^D(\mathbf{c}, \mathbf{t}) ; \mathbf{D})) , \phi(\mathbf{t})] = f^D(\mathbf{c}, \mathbf{t} ; \mathbf{D}) . \tag{52}$$

Combining the equivalent flow model (2) with the arc performance function (3) yields the formulation of the arc performance model:

$$[\mathbf{t}, \mathbf{c}] = \Gamma(Y(\mathbf{f}, \mathbf{e}, \boldsymbol{\phi}), \boldsymbol{\phi}) . \tag{53}$$

Finally, combining the Logit or deterministic NLM (51) or (52) with the arc performance model (53), we have respectively:

$$[\mathbf{f}, \mathbf{e}, \boldsymbol{\phi}] = f^L(\Gamma(Y(\mathbf{f}, \mathbf{e}, \boldsymbol{\phi}), \boldsymbol{\phi}) ; \mathbf{D}) = \Phi^L(\mathbf{f}, \mathbf{e}, \boldsymbol{\phi}) , \tag{54}$$

$$[\mathbf{f}, \mathbf{e}, \boldsymbol{\phi}] \in f^D(\Gamma(Y(\mathbf{f}, \mathbf{e}, \boldsymbol{\phi}), \boldsymbol{\phi}) ; \mathbf{D}) = \Phi^D(\mathbf{f}, \mathbf{e}, \boldsymbol{\phi}) . \tag{55}$$

SOLUTION ALGORITHM

To implement the proposed model, the simulation period is divided into I time intervals identified by the sequence of instants $\boldsymbol{\tau} = \{\tau^i \in \Re: i \in [0, I] \subseteq \aleph\}$, with $\tau^i < \tau^j$ for any $0 \le i < j \le I$. For computational convenience, we introduce also an additional instant $\tau^{I+1} = \infty$. In the following we approximate the generic temporal profile $g(\tau)$ of the performance and flow variables introduced in the previous sections, respectively, through a piecewise linear and a piecewise constant function, defined by the values $g^i = g(\tau^i)$ taken at each instant $\tau^i \in \boldsymbol{\tau}$. Under this assumption, for $\tau \in [\tau^i, \tau^{i+1})$, with $0 \le i \le I$, in the two cases we have, respectively:

$$g(\tau) = g^i + (\tau - \tau^i) \cdot (g^{i+1} - g^i) / (\tau^{i+1} - \tau^i) , \tag{56.a}$$

$$g(\tau) = g^i . \tag{56.b}$$

This way, the generic temporal profile $g(\tau)$ can be represented numerically through the $(1 \times I+1)$ row vector $\mathbf{g} = (g^0, \dots , g^i, \dots , g^I)$.

The state of the network at time τ^0 is assumed to be known; here, without loss of generality, an initially unloaded network is considered. Note that, since we assumed that time intervals are in the order of minutes, and thus comparable with urban transit headways, the error introduced assuming within each time interval a constant average line access time is negligible.

In this paper, we will present only the numerical methods solving the arc performance function, the transit frequency propagation model, and the mode choice model; in fact, procedures solving the route choice model and the network flow propagation model coincide with those presented in Bellei *et al.* (2005) and in Gentile and Meschini (2006) for the Logit and the deterministic NLM, respectively.

Arc performances

Given the flows and the frequencies **f**, **e** and $\boldsymbol{\phi}$, the computation of the equivalent flows and of the arc exit times and costs is straightforward, except for road, boarding and queuing arcs.

function $\Upsilon(\mathbf{f}, \mathbf{e}, \boldsymbol{\phi})$
  for each arc $a \in A$
    for each instant $\tau^i \in \boldsymbol{\tau}$
      compute u_a^i and v_a^i based on (1)

function $\Gamma(\mathbf{f})$
  for each arc $a \in RA$
    $t_a^0 = \tau^0 + L_{\varepsilon(a)} / V_{\varepsilon(a)}$; $c_a^0 = \chi \cdot (t_a^0 - \tau^0)$
    for each instant $\tau^i \in \boldsymbol{\tau} \setminus \tau^0$ in chronological order
      $t_a^i = \max\{\tau^i + L_{\varepsilon(a)} / V_{\varepsilon(a)}, t_a^{i-1} + (u_a^i - u_a^{i-1}) / Q_{\varepsilon(a)}\}$ (57)
      compute c_a^i based on (6)
  for each arc $a \in LA$
    for each instant $\tau^i \in \boldsymbol{\tau}$
      compute t_a^i and c_a^i based on (18) and (23)
  for each arc $a \in AA$
    for each instant $\tau^i \in \boldsymbol{\tau}$
      compute t_a^i and c_a^i based on (24) and (25)
  for each arc $a \in PA$
    for each instant $\tau^i \in \boldsymbol{\tau}$
      compute t_a^i and c_a^i based on (26) and (27)
  for each arc $a \in BA$
    for each instant $\tau^i \in \boldsymbol{\tau}$ in reverse chronological order
      $t_a^{-1\,i} = \min\{\tau^i - \alpha_{\ell(a)} / \phi_a(\tau^i), t_a^{-1\,i+1}\}$ (58)
    $j = 0$
    for each instant $\tau^i \in \boldsymbol{\tau}$ in chronological order
      until $t_a^{-1\,j} \geq \tau^i$ do $j = j+1$ (59)
      $t_a^i = \tau^{j-1} + (\tau^j - \tau^{j-1}) \cdot (\tau^i - t_a^{-1\,j-1}) / (t_a^{-1\,j} - t_a^{-1\,j-1})$ (60)
      compute c_a^i based on (17)
  for each arc $a \in QA$
    for each instant $\tau^i \in \boldsymbol{\tau}$
      compute $AK_{n(a)}^i$ accordingly with (10)
      $\xi'^{\,i}_a = AK_{n(a)}^i \cdot (\tau^i - \tau^{i-1}) / (t_a^{-1\,j} - t_a^{-1\,j-1})$ (61)
    $\xi_a = spread(\xi'_a, \mathbf{t}_a^{-1})$
    $F_a^0 = 0, E_a^0 = 0$
    for each instant $\tau^i \in \boldsymbol{\tau} \setminus \tau^0$ in chronological order
      $F_a^i = F_a^{i-1} + u_a^i \cdot (\tau^i - \tau^{i-1})$

$$E_a^{\,i} = \min\{F_a^{\,i}\,,\, E_a^{\,i\text{-}1} + \xi_a^{\,i}\cdot(\tau^{\,i} - \tau^{\,i\text{-}1})\} \tag{62}$$

$t_a^{\,0} = 0, j = 0$

for each instant $\tau^i \in \boldsymbol{\tau} \setminus \tau^0$ in chronological order

until $E_a^{\,j} \geq F_a^{\,i}$ do $j = j+1$ (63)

$$t_a^{\,i} = \max\ \{\tau^{\,i}\,,\, \tau^{j\text{-}1} + (F_a^{\,i} - E_a^{\,j\text{-}1})\cdot(\tau^j - \tau^{j\text{-}1}) \,/\, (E_a^{\,j} - E_a^{\,j\text{-}1})\} \tag{64}$$

compute $c_a^{\,i}$ based on (17)

With reference to road arcs, the recursive equation (57) determining the exit times is a specification of (4) complying with piecewise constants inflows, as for hypothesis (56.b).

With reference to boarding arcs, firstly the entry time temporal profile is computed by means of equation (58), which is a slight modification of (7) ensuring respect of the FIFO rule. Then, the exit time temporal profile is computed as the inverse of the entry time temporal profile; this is done with the line search (60) over the entry time profile, once the appropriate index j: $\tau^i \in (t_a^{\,-1\,j\text{-}1}\,,\, t_a^{\,-1\,j}]$ is identified by (59). As depicted in Figure 8, the resulting exit time profile (dashed line), complying with hypothesis (56.a), is not coincident with the entry time profile, yet it ensures the FIFO rule.

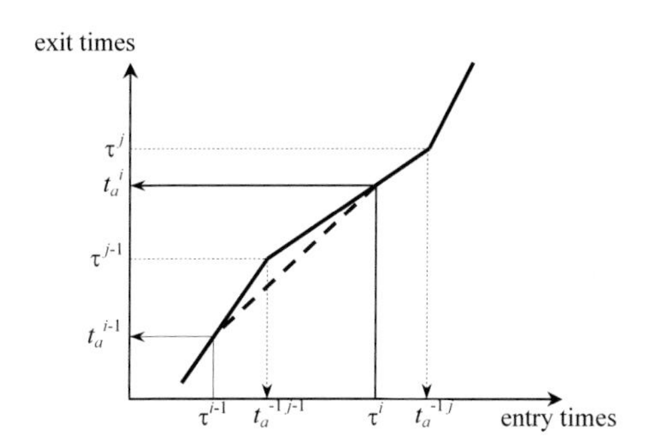

Figure 8. Piece-wise linear entry and exit time temporal profiles of the generic boarding arc.

With reference to queue arcs, first the temporal profile of the available capacity at the stop point is propagated backward to the head of the queuing arc by means of equation (61), which is a specification of (11) exploiting hypothesis (56) on exit capacity and entry time profiles. (61) yields a profile ξ'_a which is piecewise constant over instants $\mathbf{t}_a^{-1}$, thus not complying with (56.b); then, ξ'_a is transformed into an equivalent profile ξ_a piece-wise constant over instants $\boldsymbol{\tau}$, preserving vehicle conservation, by means of the function *spread* explained below. Then, the cumulative exit flow profile is determined by means of recursive equations (62), which is a specification of (12) exploiting hypothesis (56.b) on inflow and exit capacity profiles. Finally, as depicted in Figure 9, the exit time profile is determined by means of the line search (64), which is a specification (16) exploiting hypothesis (56.a), once the appropriate j: $E_a^{\,j\text{-}1} < F_a^{\,i} \leq E_a^{\,j}$ is identified by (63).

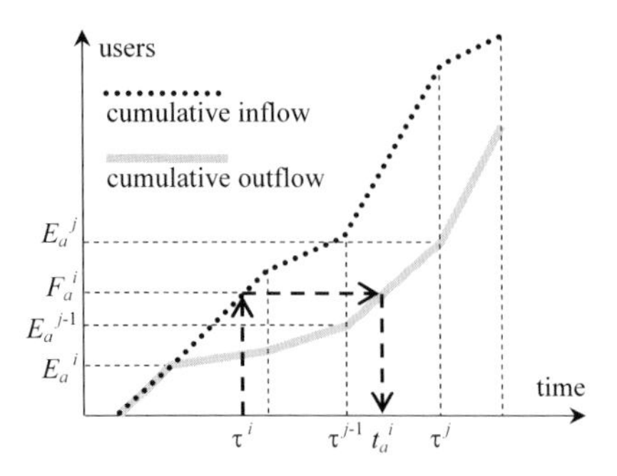

Figure 9. Exit flow and exit time for given piece-wise linear cumulative inflow and outflow.

The function *spread* evaluates the contribute that the i-th element x^i of a generic profile $\mathbf{x}$, piece-wise constant over a set of instants $\mathbf{t}$, gives to the generic j-th element y^j of a generic profile $\mathbf{y}$, piece-wise constant over predefined instants $\boldsymbol{\tau}$, proportionally to $(t^{i-1}, t^i] \cap (\tau^{j-1}, \tau^j]$, $i = 1, \ldots, I, j = 1, \ldots, I$. A graphical representation of this function is given in Figure 10.

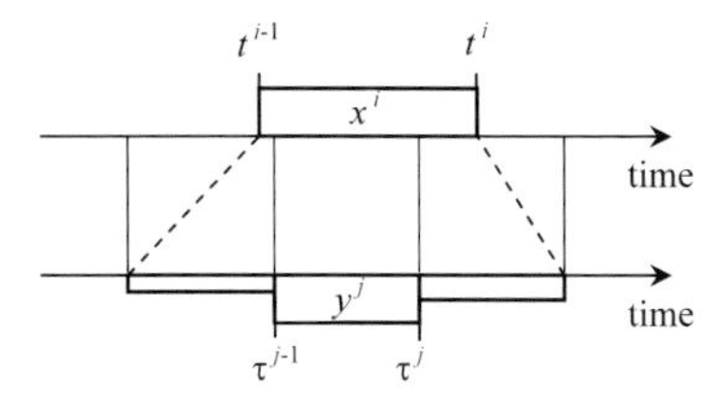

Figure 10. Graphical representation of the function spread

function *spread*($\mathbf{x}$, $\mathbf{t}$)

$j = 0$

until $\tau^j \geq t^0$ do $j = j+1$

for each instant $\tau^i \in \boldsymbol{\tau} \setminus \tau^0$

if $\tau^j \geq t^i$ then

$y^j = y^j + x^i \cdot (t^i - t^{i-1}) / (\tau^j - \tau^{j-1})$

else

$y^j = y^j + x^i \cdot (\tau^j - t^{i-1}) / (\tau^j - \tau^{j-1})$

$j = j + 1$

until $\tau^j \geq t^i$ do

$y^j = y^j + x^i$

$j = j + 1$

$y^j = y^j + x^i \cdot (t^i - \tau^{j-1}) / (\tau^j - \tau^{j-1})$

Transit frequency propagation

Given the exit times and the headway frequencies $\mathbf{t}$ and ψ, the computation of the transit frequencies at stops is as follows:

function $\phi(\mathbf{t})$

  for each $\ell \in \Im$

    for each instant $\tau^i \in \boldsymbol{\tau}$

      $T_\ell^{1i} = \tau^i$

    for each progressive point $n \in [2, \sigma_\ell]$ in the natural order

      $j = 0$

      for each instant $\tau^i \in \boldsymbol{\tau}$

        until $\tau^j \geq T_\ell^{n\text{-}1\, i}$ do $j = j+1$

        $$T_\ell^{n\,i} = t_a^{j\text{-}1} + (t_a^j - t_a^{j\text{-}1}) \cdot (T_\ell^{n\text{-}1\, i} - \tau^{j\text{-}1}) / (\tau^j - \tau^{j\text{-}1}) \tag{65}$$

        $$\phi'^{\,n\,i}_\ell = \psi_\ell^{\,i} \cdot (\tau^i - \tau^{i\text{-}1}) / (T_\ell^{n\,i} - T_\ell^{n\,i\text{-}1}) \tag{66}$$

      $\boldsymbol{\phi}_\ell^n = spread(\boldsymbol{\phi}'^{\,n}_\ell, \mathbf{T}_\ell^n)$

The arrival times at progressive points and the corresponding frequency temporal profiles are determined through equations (65) and (66), which are a specification of (28) and (29), respectively, exploiting hypothesis (56.a), once the appropriate j: $\tau^{j\text{-}1} < T_\ell^{n\text{-}1\ i} \leq \tau^j$ is identified; then, since $\boldsymbol{\phi}'^{\,n}_\ell$ yielded by (66) is piece-wise constant over the set of instants $\mathbf{T}_\ell^n = \{T_\ell^{n\ i} \in \Re: i \in [0, I] \subseteq \aleph\}$, an equivalent transit frequency profile $\boldsymbol{\phi}_\ell^n$ complying with hypothesis (56.b) is evaluated by means of the function *spread* already introduced.

Mode choice

Given the node satisfaction and the demand **w** and **D**, the computation of the mode flows is straightforward

function d(**w**, **D**)

  for each node $d \in Z$

    for each node $o \in Z$

      for each instant $\tau^i \in \boldsymbol{\tau}$

        for each mode $m \in M$

          compute mode systematic utility accordingly with (42)

        for each mode $m \in M$

          compute mode choice probability accordingly with (41)

          compute mode flows accordingly with (43)

Equilibrium

The dynamic equilibrium, expressed as a fixed point problem, can be solved through the following MSA algorithm, where ε and m_{max} are, respectively, the maximum relative error and the maximum number of iterations. The relative error is defined as $\mathbf{c} \cdot (\mathbf{f} - \mathbf{y}) / \mathbf{c} \cdot \mathbf{f}$ for the deterministic model, and as $\|\mathbf{f} - \mathbf{y}\|_\infty / \mathbf{f}$ for the stochastic Logit model

function DUE

  $m = 0$, $\mathbf{f} = \mathbf{0}$, $\mathbf{e} = \mathbf{0}$, $\boldsymbol{\phi} = \mathbf{0}$ *initialization*

until $m > m_{\max}$ do	*stop criterion*
$\quad m = m + 1$	*new iteration*
$\quad [\mathbf{u}, \mathbf{v}] = \Upsilon(\mathbf{f}, \mathbf{e}, \boldsymbol{\phi})$	*equivalent flows*
$\quad [\mathbf{t}, \mathbf{c}] = \Gamma(\mathbf{u}, \mathbf{v}, \boldsymbol{\phi})$	*arc performance function*
$\quad \mathbf{w} = w^{L}(\mathbf{c}, \mathbf{t}) \mid w^{D}(\mathbf{c}, \mathbf{t})$	*Logit|Deterministic node satisfactions*
$\quad \mathbf{p} = p^{L}(\mathbf{w}, \mathbf{c}, \mathbf{t}) \mid p^{D}(\mathbf{w}, \mathbf{c}, \mathbf{t})$	*Logit|Deterministic arc conditional probabilities*
$\quad \mathbf{d} = d(\mathbf{w}; \mathbf{D})$	*mode choice*
$\quad [\mathbf{x}, \mathbf{y}] = \omega^{L}(\mathbf{p}, \mathbf{t}, \mathbf{d}) \mid \omega^{D}(\mathbf{p}, \mathbf{t}, \mathbf{d})$	*Logit|Deterministic network flow propagation*
$\quad \boldsymbol{\varphi} = \phi(\mathbf{t})$	*transit frequency propagation*
$\quad \mathbf{e} = \mathbf{e} + 1/m \cdot (\mathbf{x} - \mathbf{e})$	*update arc outflows with MSA*
$\quad \mathbf{f} = \mathbf{f} + 1/m \cdot (\mathbf{y} - \mathbf{f})$	*update arc inflows with MSA*
$\quad \boldsymbol{\phi} = \boldsymbol{\phi} + 1/m \cdot (\boldsymbol{\varphi} - \boldsymbol{\phi})$	*update line frequencies with MSA*
if Logit_path_choice then	
if $\|\mathbf{f}-\mathbf{y}\|_{\infty} / \mathbf{f} < \varepsilon$ then end	*Logit stop criterion*
else	
if $\mathbf{c}\cdot(\mathbf{f}-\mathbf{y}) / \mathbf{c}\cdot\mathbf{f} < \varepsilon$ then end	*Deterministic stop criterion*

NUMERICAL RESULTS

The proposed algorithm was applied to a multimodal network (Sioux Falls), schematically depicted in Figure 11, consisting of 24 centroids, 76 road arcs and 5 transit lines operated by carriers having capacity of 2000 users and departure frequency from terminals of 8 veh/h. An available static demand matrix was multiplied by a suitable temporal profile simulating a morning peak hour, yielding a total demand of 901251 users.

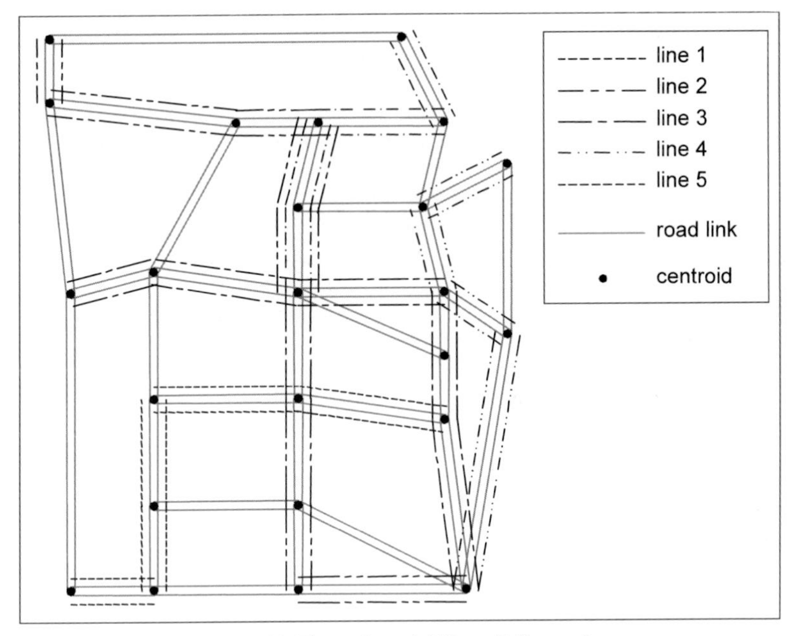

Figure 11. The multimodal Sioux Falls graph.

A deterministic DTA with mode choice between road and transit modes was performed over a period of analysis of 5 hours, which was divided into 30 time intervals of 10 min. A stop criterion $\varepsilon \leq 0.01$ was achieved with 186 iterations, while the calculation time was 22 sec on a PC with a 3.0 Ghz CPU.

With reference to the most loaded line, Figure 12 presents the frequency temporal profiles on some of its critical stops (left hand side), and the corresponding congestion delays between successive stops (right hand side). It can be noticed that the increment of travel time between stops 4 and 5 (due both to road congestion and to boarding and alighting congestion) induces a perturbation on carrier frequencies propagating in time forward along the line.

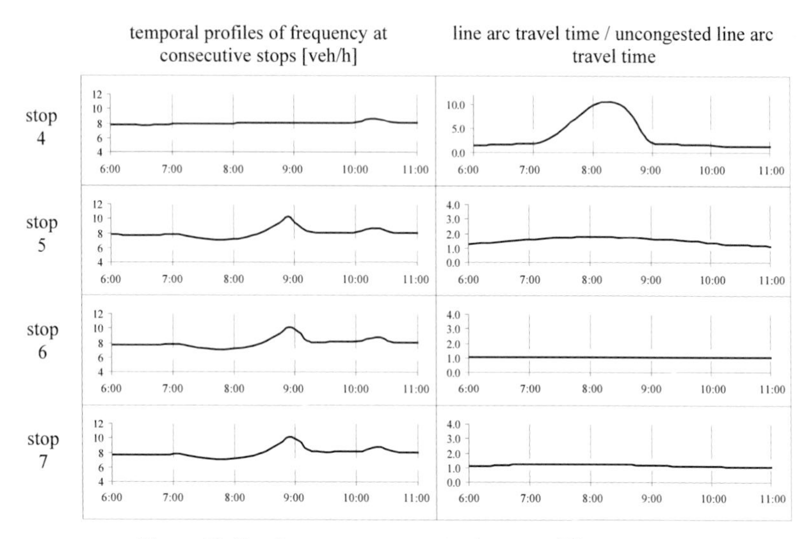

Figure 12. line frequency propagation between different stops.

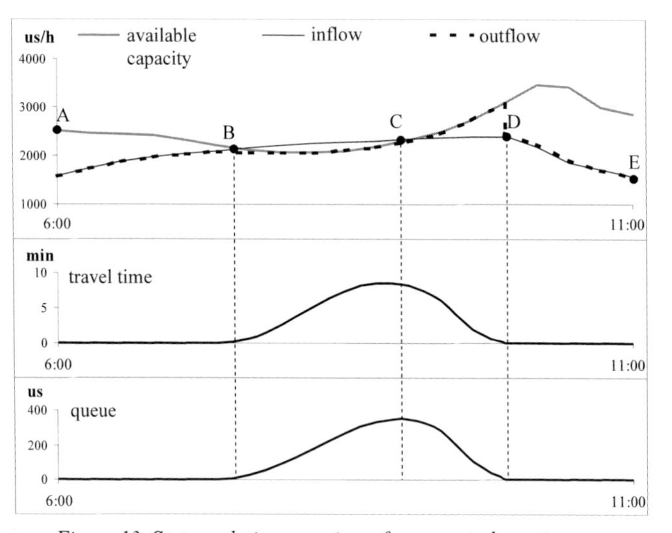

Figure 13. State evolution over time of a congested queuing arc.

Figure 13 represents the state evolution over time of a queuing arc where over saturation occurs. As long as the passenger's inflow remains below the available capacity (interval A-B), no queue is present; thus, the arc travel time is null and the outflow is equal to the inflow. When the inflow equals the available capacity a queue grows, along with the travel time, while the inflow is greater than the available capacity (interval B-C), and vice-versa decreases (interval C-D); as long as the queue is present, the outflow is equal to the available capacity. When the queue vanishes, the arc travel time returns to be null and the outflow is again equal to the inflow (interval D-E).

The above numerical results confirm that the approach proposed in this work in order to solve multi mode dynamic traffic assignment is valid: in fact, all congestion effects are properly taken into account, and calculation times are reasonable also on realistic networks.
However, more work is to be done in the future in order to improve the convergence of the proposed algorithm.

REFERENCES

Bellei G., Gentile G., Papola N. (2003). Assegnazione alle reti multimodali in presenza di congestione. In: *Metodi e Tecnologie dell'Ingegneria dei Trasporti. Seminario 2001* (G. Cantarella and F. Russo ed.s), pp. 117-134. Franco Angeli, Milano, Italy.

Bellei G., Gentile G., Papola N. (2005). A within-day dynamic traffic assignment model for urban road networks. *Transportation Research B* **39**, 1-29.

Cascetta E. (2001). *Transportation systems engineering: theory and methods*, pp. 384-386. Kluwer Academic Publisher, UK.

Crisalli U. (1999). Dynamic transit assignment algorithms for urban congested networks. In *Urban Transport and the Environment for the 21st century V* (L.J. Sucharov ed.), pp. 373-382. Computational Mechanics Publications.

Daganzo C. F. (1997). *Fundamentals of Transportation and Traffic Operations*, pp. 97-112. Pergamon, Oxford, UK.

DeCea J., Fernandez E. (1993). Transit assignment for congested public transport systems: an equilibrium model. *Transportation Science* **27**, 133-147.

Gentile G., Meschini L., Papola N. (2005). Macroscopic arc performance models with capacity constraints for within-day dynamic traffic assignment. *Transportation Research B* **39**, 319-338.

Gentile G., Meschini L. (2006). Fast heuristics for the continuous dynamic shortest path problem in traffic assignment. Submitted to the *AIRO Winter 2005 special issue of the European Journal of Operational Research on Network Flows*.

Gentile G., Meschini L., Papola N. (2006). Spillback congestion in dynamic traffic assignment: a macroscopic flow model with time-varying bottlenecks. Accepted for publication in *Transportation Research B*.

Nguyen S., Pallottino S. (1988). Equilibrium traffic assignment for large scale transit networks. *European Journal of Operational Research* **37**, 176-186.

Nguyen S., Pallottino S., Gendreau M. (1998). Implicit Enumeration of Hyperpaths in a Logit Model for Transit Networks. *Transportation Science* **32**, 54-64.

Nguyen S., Pallottino S., Malucelli F. (2001). A modelling framework for the passenger assignment on a transport network with time-tables. *Transportation Science* **35**, 238-249.

Nuzzolo A., Russo F. and Crisalli U. (2003). *Transit network modelling. The schedule-based dynamic approach*. Franco Angeli, Milano, Italy.

Spiess H., Florian M. (1989). Optimal strategies: a new assignment model for transit networks. *Transportation Research B* **23**, 83-102.

Wu J.H., Florian M., Marcotte P. (1994). Transit equilibrium assignment: a model and solution algorithms. *Transportation Science* **28**, 193-203.

Xu, Y.W., Wu, J.H., Florian, M., Marcotte, P., Zhu, L.H. (1999). Advances in the continuous dynamic network loading problem. *Transportation Science* **33**, 341-353.

Transportation and Traffic Theory 2007
Edited by R.E. Allsop, M.G.H. Bell and B.G. Heydecker

19

A THEORY ON MODELLING COLLABORATION IN LOGISTICS NETWORKS COMBINING NETWORK DESIGN THEORY AND TRANSACTION COSTS ECONOMICS

Bas Groothedde, Kirkman Company, Baarn, the Netherlands; Delft University of Technology, Delft, the Netherlands.
Piet Bovy, Delft University of Technology, Delft, the Netherlands

INTRODUCTION

The evolution of logistics networks during the last decades can be characterized by a strong rationalization of business processes. It has led to a constant search for economies of scale in the supply chain, which has been an important parallel development in line with the changes in globalisation and manufacturing. Companies can search for these economies of scale (and scope) in their own organization, they can benefit from the scale a third party can bring together, they can obtain scale through mergers and acquisitions or, finally, scale can be achieved through collaboration. The latter form is the subject of this paper. The choice between in-house organization, outsourcing, and collaboration is a fundamental decision in the logistics network and firms are naturally reluctant to transfer responsibility for an operational area as vital as logistics to a third party. Therefore an answer to the question: "*under what circumstances will organizations decide to collaborate with third parties?*" is crucial in order to be able to design a joint logistics network in which the different actors collaborate.

In the paper we focus on collaboration in logistics and transportation networks, hub network design and transaction costs economics. We extend on earlier work on collaboration in logistics by Lambert *et al.* (1997), on hub network design by of O'Kelly and Miller (1994) and Abdinnour-Helm (1998), and on transaction costs by Williamson (1975; 1981; 1985). We present a new modelling framework in which we combine *Network Design Theory* and Transaction Cost Economics (TCE) into a comprehensive network design approach suitable for dealing with collaboration among shippers (Groothedde, 2005). This design approach is based on economic objectives such as minimizing total logistics costs, minimizing the user-costs, and the preferred levels of service set by participants. In addition however, we incorporate the transaction costs of the involved actors (shippers, carriers, etc.) that influence their decision to participate in a collaborative network, seek an in-house solution, or outsource the activity. The logistics costs are of great importance in this choice. But when a company decides to switch from the current situation to the new network solution, costs are made to make this switch possible. Factors influencing this switch are for example the investments necessary for the alternative that is under consideration (referred to as asset specificity).

Another important factor is the frequency with which the decision needs to be reviewed and of course the uncertainty of the considered alternative. A methodology that can be used in this context is TCE, which is concerned with the minimization of the sum of production and governance costs. The set-up of the paper is as follows. We first describe a number of options for collaboration in logistics. Then, the factors determining the transaction costs of a firm due to switching from the current situation to a possible collaborative form are discussed. These costs play a crucial role in the firm's decision to participate in a collaborative undertaking. Using these basic notions a network design model is established for optimising the structure and capacities of a logistics service network given one or more options for collaboration among shippers. This design model is then adopted in two real-life cases to demonstrate the impacts of different forms of collaboration on the structure of the logistics networks and related business costs.

SEARCHING ECONOMIES OF SCALE

Companies have become more aware of the impact that their logistics organization can have on the costs of doing business and on the degree of satisfaction of their customers. This ongoing rationalization has led to a constant search for economies of scale in the supply chain, which has been an important parallel development in line with the changes in globalisation and manufacturing. The survey *2005 third-party logistics* (Cap Gemini *et al.* 2005) shows that there is an increasing pressure to reduce costs and enhance the customer service. For example, in 1996 87% of logistics executive's perceived pressure to reduce logistics costs, in 2005 96% indicated to perceive this pressure. In today's logistics market in Europe it looks as if it is to take-over or be taken-over. For example, Kuhne+Nagel acquired ACR Logistics, Exel was taken over by Deutsche Post, and the ambitions of UPS with the take over from TPG. A strategy that has become more and more apparent is seeking collaboration with partners in order to achieve the necessary scale and scope. When analysing its own performance a company should constantly look at the scale and scope they can attain and what a third party can bring about. When weighing these options there are several key issues to consider: what are the specific investments needed and with what frequency do we need to review this decision. What is, for example, the contract term? What are the uncertain factors in the market and in the relationship itself? For a company to weigh these options ex-ante is very important as theses decisions have far-reaching impact on the way of doing business, market share, and network structure. In Figure 1 we distinguish different options open to a company seeking economies of scale and scope. The first option is for a company to outsource its activities and hereby making use of the scale of this third party. When seeking a partner there are three options available. The first one is a Type I collaboration (*operational partnership*) that consists of organizations that recognize each other as partners and, on a limited basis, coordinate their activities and planning. The partnership usually has a short-term focus and involves only one division or functional area within each organization.

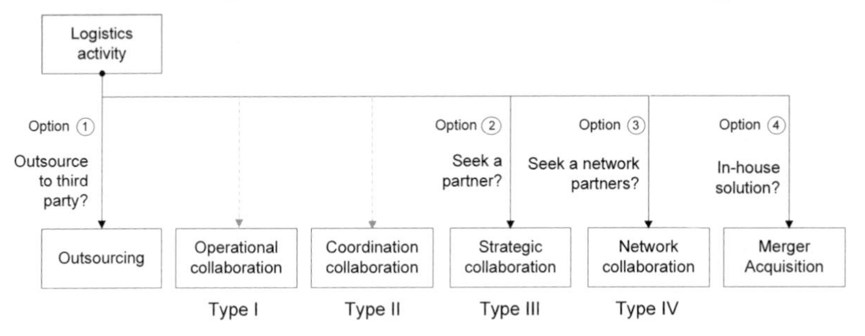

Figure 1: A selection of fundamental options open for a company to achieve economies of scale.

The second one is a Type II partnership (*coordination partnership*) in which the organizations progress beyond the coordination of activities to the integration of activities. Although not expected to last forever the partnership has a long-term horizon. Multiple divisions and functions within the firm are involved in the partnership. The third form we distinguish is a Type III partnership (*strategic partnership*), in which the organizations share a significant level of operational integration. Each party views the other as an extension of their own firm. Typically no end date for the partnership exists.

Finally, a fourth type of collaboration is introduced (*network collaboration*) that goes one step beyond a Type III collaboration in the sense that this type of collaboration is not only structural in nature but extends throughout the logistics network and involves multiple actors.

From Collaboration to a Hub Network Design

The question arises how to translate these different options into feasible logistics networks and what the impact of these different types of collaboration is on the performance of these networks (hub networks as we focus on these types of networks). We present a design methodology that is capable of translating these different options presented in Figure 1 into their consequences for the individual companies involved, providing an answer to the individual companies about the type of collaboration that suits their purposes.

Before the actual hub network design can start several key characteristics of the collaboration need to be listed. We distinguish the following characteristics: (1) *The Objective* of the relation (either asset or cost efficiencies; customer service, marketing advantage, or profit stability or growth); (2) the *Type* of collaboration (see Figure 1 for the four types we distinguish); (3) *Asset specificity*, in any collaboration the asset specificity for the participants is a key determinant and is the degree to which investments are specific for a certain relation. Asset specificity is usually defined as the extent to which the investments made to support a particular transaction have a higher value to that transaction than they would have if they were redeployed for any other purpose. Williamson (1975, 1985) argued that transaction-specific assets are non-redeployable physical and human investments that are specialized and unique to a task. There is a trade-off between the cost savings and the risk due to the assets' non-salvageable character; (4) *Uncertainty*, another important determinant is uncertainty and is considered to be a key determinant of the height of the transaction costs; (5) *Frequency*, refers to the frequency of occurrence and is considered an important attribute, from the point of view that the costs of specialized, expensive, governance structures will be easier to recover for large transactions of a recurring kind; (6) *Dominance* reflects its potential for influencing the actions and decisions of individuals and firms in the network and is a key issue in business logistics; and (7) *Transparency*, refers to the trust between the participating actors and the measurability of the costs, benefits, performance in the logistics network.

A preferred strategy, for example, to achieve economies of scale is to operate in a hub network in which the flows are consolidated and scale can be achieved. In these networks serving many origins and many destinations, one important function of the hubs or transportation terminals is to consolidate small shipments into vehicle loads. As O'Kelly and Bryan (1998) so aptly put it: "*economies of scale, due to the amalgamation of flows, provide a raison d'etre for hub systems*". Consolidation in these types of networks allows for more efficient and more frequent shipping by concentrating large flows onto relatively few links between hubs. Although use of indirect (that is via a hub) shipment may increase the distances travelled, the economies of scale due to the larger volume can reduce the total cost. These configurations can reduce and simplify network construction costs, centralize commodity handling and sorting and allow carriers to take advantage of economies of scale making the hub network well suited for collaboration.

To make the discussed translation between the type of collaboration and a hub network design possible we developed a network design methodology yielding feasible hub network solutions that has been applied in several network design projects two of which will be discussed in this paper. The first application is the Ricoh Family Group case, in which the European logistics network was designed and implemented. This is a typical example of an outsourcing relation. The second case study is the design and implementation of a collaborative logistics network in the Netherlands.

THE INTEGRAL LOGISTICS COSTS+ APPROACH

First we will elaborate on the logistics costs and the costs incurred of switching between different network alternatives. For example, the choice between the current logistics network solution and a collaborative network should be based on what we refer to as the integral logistics costs+. Looking at these different options the impact an alternative solution has on the logistics activities should be analyzed. When considering the different options open to a company (Figure 1), the logistics costs of these options should be considered but in addition the transaction costs should be incorporated (Figure 2). Transaction costs are the costs made in trying to attain a specific service not the costs of the service itself. Transaction costs economics (TCE) is concerned with the minimization of the sum of production and governance costs. Production costs are the costs for producing the product or service (wages, materials costs, equipment, etc.), while the governance costs represent both the bureaucratic costs of internal governance and the corresponding governance costs of markets, i.e. transaction costs.

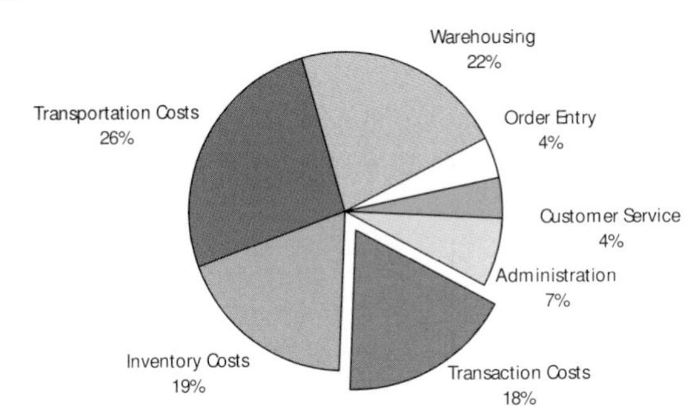

Figure 2: Typical breakdown of the total logistics costs of a company including the transaction costs when switching between alternatives. (Groothedde, 2005).

In the logistics context this would mean a minimization of the logistics costs (transportation, handling, and inventory) and the costs concerned with the governance of the network. Williamson (1975; 1981; 1985) presented a framework to analyse the costs mentioned above, which is based on economics, organization theory, and contractual law literature. The basic unit of analysis of TCE is the transaction laid down in a contract between two parties. According to the transactional view, a transaction can take place in the institutional framework of the market (using the price mechanism), or of a hierarchy (which requires a co-ordination of efforts), whatever allows it to be executed most efficiently (Kleas, 2000). The parcel delivery services of carriers like UPS, FedEx or DHL form a typical example of a hub network structure where economies of scale can be achieved. If a company decides to ship a single parcel, the economies of scale that can be achieved by UPS, FedEx or UPS usually outweigh the governance costs and transaction costs when outsourcing the delivery of the parcel to one of these companies.

In our definition the transaction costs consist of two parts: $\delta_i = \delta^I_{im} + \delta^{II}_{im}$ where index *i* refers to company *i* and index *m* refers to a considered business option by company *i*. The first (δ^I_{im}) is caused by the first round of information problems, while the second part (δ^{II}_{im}) indicates the consequential costs arising after the transaction has taken place. In Figure 3 *Finding a partner*, *Negotiation*, *Governance*, *Lawyer*, *External Advice* make up for the first round transaction costs (ex-ante) while the costs incurred for the *Increased Bid* and *Law-Suits* constitute the second round transaction costs (ex-post). The essence of the foregoing argument can be illustrated using simple example in which we consider only two organizational alternatives are considered: either a firm ships a parcel itself or it pays another company to do so. As goods and service become very close to unique (i.e. the asset specificity is high), economies of scale will no longer be realized in the market. The cost penalty for using internal organization can be severe however for standardized transactions for which market aggregation economies are great and where is low, as is the case in the example of the parcel delivery services. Next to outsourcing participation in a collaboration could be an option. If a partner is sought and the investments can be shared the asset specificity can be reduced.

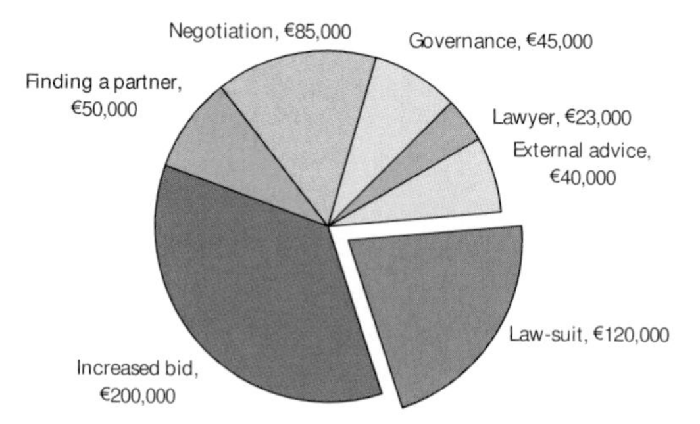

Figure 3: Typical breakdown of the transaction costs of a company when switching between two alternatives (Groothedde, 2005).

It is assumed that operating in a collaboration will lead to higher governance costs due to the increased number of partners (higher transaction costs). Then the type of collaboration highly influences the way in which the asset specificity and governance costs are divided amongst the individual participants. This in turn influences the potential cost reduction that can be achieved by a company when joining a collaboration and the structure of the logistics network (as we shall see later on in this paper).

THE DECISION TO PARTICIPATE

The key issue in modelling the choice of a candidate participant n to participate in collaboration m is the trade-off between the potential cost reduction (Δc_{im}) that can be obtained by joining a specific network solution and the costs incurred (δ^I_{im}) for making the switch from the current network to the newly designed network solution m. This trade-off is captured in the threshold Δc_{im} depicted in Figure 4, where the curves denote the minimum level of Δc_{im} that needs to be achieved before joining a collaboration. The different levels of uncertainty are denoted by Prob^1 and Prob^2. This figure depicts the cost advantage or disadvantage and revenue advantage or disadvantage a firm can have by deciding to join a collaborative network. If the effect of participating is a cost advantage the network solution can be positioned in either quadrant I or II. If there is a revenue advantage, the network solution can be positioned in either quadrant II or IV. If positioned in quadrant II there is a cost advantage as well as a revenue advantage. In this example the area indicates the cost-revenue ratio with which the firm decides to participate. It can be argued that if no cost and revenue advantages are to be gained, the firm is not willing to participate; a certain minimum level of benefits (Δc_{im}) needs to be present given the ratio between revenue and costs.

Next, due to the mentioned transaction costs (δ_{im}), based on the asset-specificity, frequency, etc. a company makes a decision to participate or not. We assume however, that there is a certain additional resistance to switch that firms have (not incorporated in the transaction costs) due to uncertainty, dominance and transparency issues, denoted $(\in_{im})$. If the cost advantage exceeds the revenue disadvantage or the revenue advantage exceeds the cost disadvantage a firm is still willing to participate. However, as the cost advantage and revenue disadvantage increase or the revenue advantage and cost disadvantage increase, yielding the same ratio, a firm becomes more hesitant to participate (under the same condition of uncertainty, Prob^1), resulting in $\Delta c_{i2} > \Delta c_{i1}$. We illustrated two different levels of uncertainty. These different levels of uncertainty could be caused by for example, different market characteristics, types of collaboration, or other external circumstances. If uncertainty increases (Prob^2) it follows that the reluctance to participate increases and the threshold rises accordingly ($\Delta c_{i3} > \Delta c_{i1}$). In our example the uncertainty has an increasing effect on the Δc_{im} but in addition we assume that Δc_{im} increases as the dominance of a trading partner increases, and Δc_{im} increases as well if transparency decreases (e.g. it becomes more difficult to measure the transaction and there is little trust between partners that information exchanged is accurate). The decision to participate in our framework is based on Δc_{im} and functions as a threshold. The ratio between the costs and revenues (the total net benefits) therefore determines the decision to participate.

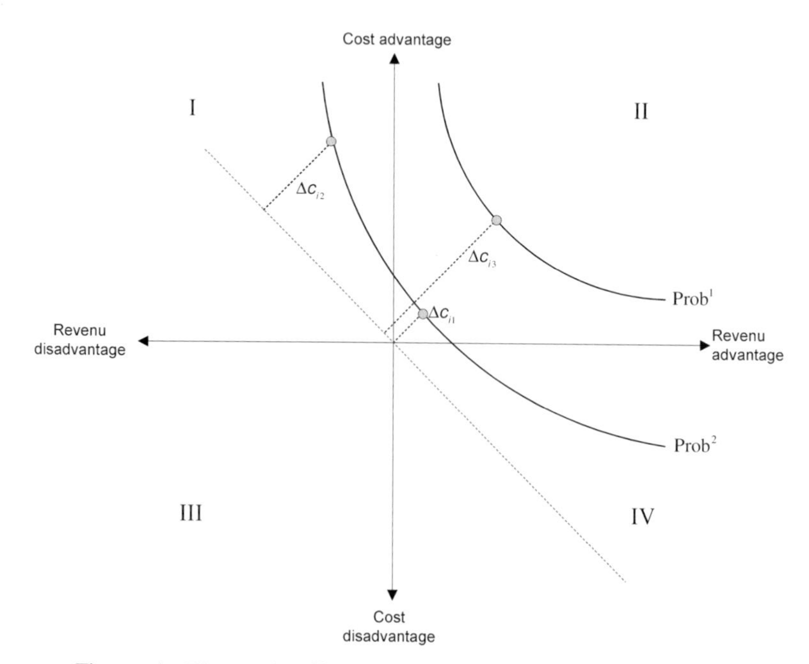

Figure 4: The trade-off between costs and revenues from the perspective of collaboration. Source: Ruijgrok and Groothedde (2005).

The level of Δc_{im} is determined by the costs of the current network solution (denoted c_{i0}), and in the example of a collaborative hub network the shipment costs for using the hub network (denoted c^{hub}), and the transaction costs associated with switching between alternative network solutions (δ_{im}). In addition we distinguish additional resistance ($\in_{im}$) that is associated with uncertainty, dominance and transparency. This final parameter can differ between participants, depending on their role in the logistics network (retailer, manufacturer, carrier, or logistics service provider) and the cost and or revenue advantage that is to be gained by participating in the collaboration. In Figure 5 the trade-off between the disadvantages and advantages is illustrated for the four actors in the network. In our example the service provider has the highest resistance to participate (Δc_L) due to the high asset specificity and uncertainty. The fact that the retailer has a dominant position in this constellation also increases the resistance of the other participants. It is therefore quite logical that the service provider will want safeguards.

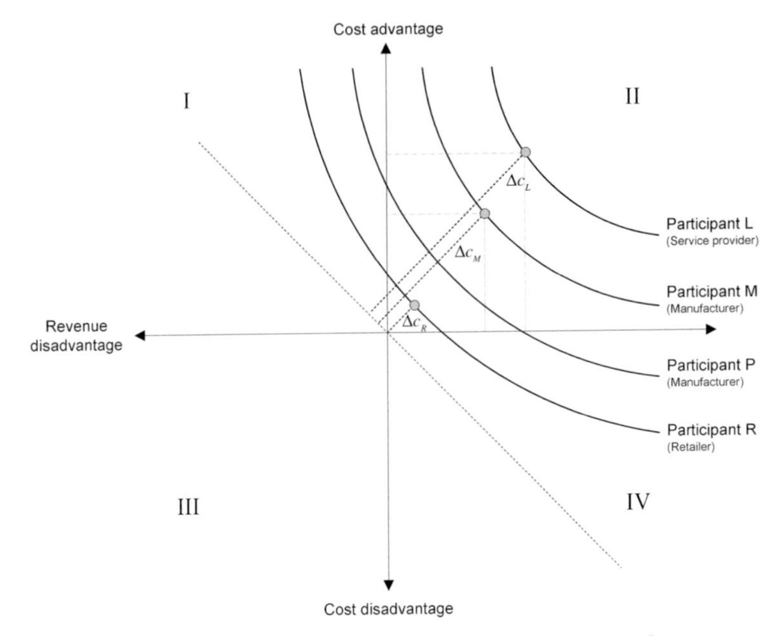

Figure 5: trade off between cost advantage and revenue advantage based on the role of the actor.

The two curves of both manufacturers are not similar due to the different situations they are in (in terms of volume, frequency, products) and therefore the costs and benefits will differ. To be able to understand whether participation at any point in the logistics network or a collaborative network is desirable or practical requires that the actors (retailers, manufacturers, service providers, etc.) understand whether the ownership and control of particular network resources will generate a sufficient financial return to make participation worthwhile. Thus, when optimising the logistics network of a focal company it is necessary to analyse its relations and interactions with other actors in the network, but also its position in the network and the dominance of other actors, both up and downstream. In our modelling approach these different roles of the participants (i) and accompanying thresholds are used to model the individual choice to participate in a hub network collaboration (m) and thus leading up to different entry levels for the different participants. In our case-studies we also used different roles and their influence on the resulting network.

THE LINK WITH NETWORK DESIGN THEORY

Our primary objective is to develop and demonstrate a design and evaluation methodology for logistics and transportation networks in which the participants collaborate, in particular, a collaborative hub-network. Generally speaking, the objectives we focus on in our network design methodology are cost minimization and service maximization, and in order to reach this objective we need to incorporate the following primary decision (design) variables: the number and location of the hubs, the services between these hubs, capacity of the hubs and services, inventory positions, and routing of the shipments.

Our network design procedure generates an 'optimal' network configuration given a particular type of intended collaboration. Collaboration thus is not an endogenous variable. Based on the type collaboration, for example a long term partnership, the implications on the transaction costs ($\delta_i = \delta^I_{im} + \delta^{II}_{im}$) are derived for all individual potential participants. In addition, during the network design procedure these are constantly updated; based on the δ_{im}, $\in_{im}$, and the Δc_{im}, the difference between the current situation (not participating) and the costs associated with participating in the generated network solution by the network design procedure.

There is considerable literature on a variety of problems closely related to discrete hub location problems. This includes research on continuous space hub location problems, where the hub locations are allowed to be located anywhere in a continuous region (Aykin, 1995; Aykin and Brown, 1992; O'Kelly, 1986; O'Kelly, 1992; O'Kelly and Miller, 1991, Suzuki and Drezner, 1997). One large area that is related to hub-network design is the research on designing hub network, but without the hub location component. The relevant models generally are those with two or model levels where the different levels form a hierarchy. There is considerable literature on network design problems in which the location of the hub (backbone) nodes is specified (Crainic *et al.* 2000; 1999; Klincewicz, 1998; Groothedde, 2005). We extend the existing hub-network design models by introducing capacity restrictions on the hub and on the inter-hub transfer and incorporate service network design techniques. In particular collaboration costs and benefits are introduced. In the following section we will present the overall network design approach, followed by a formal model formulation and a discussion on the extensions we made on existing hub network design approaches.

The model structure

In Figure 6 the structure of the overall modelling approach is illustrated. We distinguish seven steps, starting with the input data to acquire an accurate picture of the current network performance, ending with the implementation of the new network structure. The design starts with the description of the current situation of the candidate participants in terms of locations, flows, product characteristics and mode information of the potential participants, resulting in a description of the current logistics networks of potential participants. This description is used as a reference for the network design phase. Based on this input, the costs of the current situation are then calculated in step 2 (e.g. inventory, transportation, handling, warehousing, administration costs) for each individual participant. Given the mode of transport used on the origin-destination relations, the appropriate cost function to calculate the costs is used. Next to the transportation costs the facility, inventory, and handling costs are calculated. Resulting in an estimate of the total logistics costs which are then validated using information provided by the potential participants n (denoted A in Figure 6). The next step is to describe the type of intended collaboration m (B) and to translate the characteristics of non-tangible model parameters for all potential participants (e.g. translated into the transaction costs (δ_{im}), and the additional resistance, ($\in_{im}$). Note that these are indexed because the impact for all potential participants n is estimated for every business option m considered by company i.

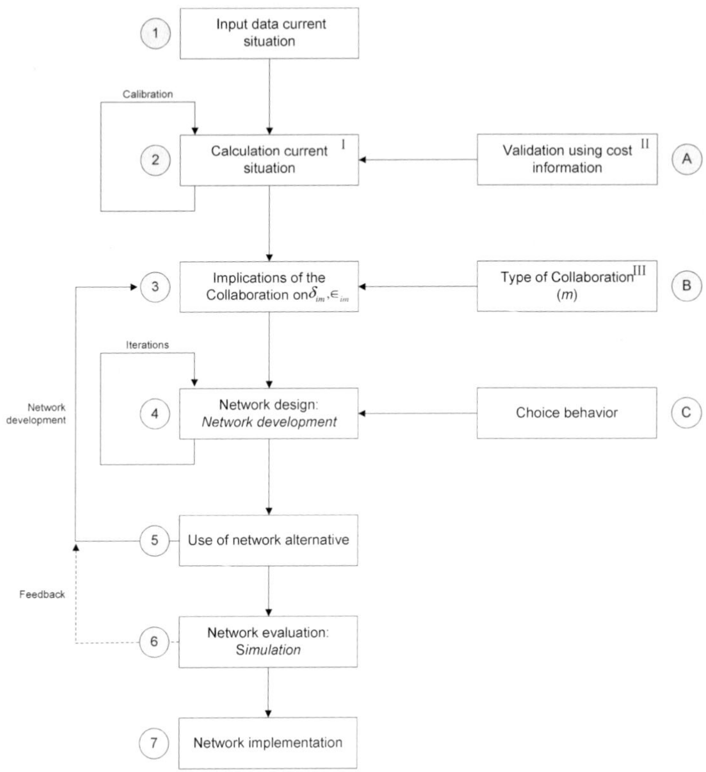

Figure 6: Comprehensive network design procedure for a collaborative logistics networks.

I) The methodology to calculate these costs is based on activity based costing and available benchmarks (NEA 2005; AC-Nielsen 2003).
II) These costs are expected to be provided by the participating firms and are used to validate and calibrate the cost functions used. If necessary the model is calibrated.
III) Based on the type of coordination, asset specificity, uncertainty, frequency, dominance, and type of contract. The type of collaboration (See Figure 1) is denoted m. Note that the index m is no longer used in the problem model formulation.

Having estimated the impact of the collaboration on the transaction costs and resistance it is then possible to estimate the costs and other implications of a new network solution for all potential participants facilitating to model their choice behaviour between the current situation and the new network solution. This choice behaviour can either be modelled deterministic or probabilistic (C). The fourth step consists of the procedure for the actual design of the network using a search procedure to find a feasible hub network. Every time an alteration is made to the network the implications of this new network solution on the use of the network is calculated followed by: (1) the transaction costs (δ_{im}), (2) the resistance ($\in_{im}$), (3) the costs of participation (c^{hub}), and (4) the resulting use (based on the costs) of the new network solution again needs to be calculated (step 5 in Figure 6). This is an iterative process resulting in the use of the network solution, the transaction costs (δ_{im}), and the resistance ($\in_{im}$) per individual participant *i*. For this search procedure the simulated annealing methodology may be used. For information on the solution procedure and performance on this large scale problem we refer to Groothedde (2005).

Once the criteria are met, or a stable solution is found the network is evaluated using the simulation module (6), for example the maximum number of iterations is reached or the improvement solution *w* and *w+1* is smaller than the predefined criteria. In this module the solution is tested and validated to assess the reliability, robustness and cost-effectiveness of the network solution based on actual order-information.

Finally in step 7 the network is implemented. This step is outside the scope of this paper though it is crucial to mention this phase since our objective is to find a feasible network solution but we also want to find a feasible network development, which is used during the implementation.

Model formulation

In addressing the hub network design question we make use of the standard formulation of a hub network problem, Our approach differs from most hub network design models in that we not only focus on the feasibility of the final network solution but are also interested in the development path towards this final configuration . We also incorporate restrictions on the path towards this final network solution in terms of the initial cost reduction, the capacity available, and the admittance of a new participants by the consortium. The underlying design problem can be formulated as a standard hub network problem in which the total logistics costs and transaction costs need to be minimized. In this section we present our formulation of the capacitated multiple allocation hub location problem. In this formulation there is also cost associated with the establishment of a hub. Consequently, the number of hubs for any given problem needs to be derived by the mathematical decision model. In Ebery *et al.* (2000) a capacity restriction is placed on the volume of traffic entering the hub via collection. In our version of the model we introduce capacity restrictions on the flows entering and leaving the hubs and on the inter-hub flows. Below the different types of variables are listed. The key design variables are the opening of the hubs, the sequences, the capacity of the hubs and sequences, and direct/hub shipments. In addition, the endogenous and exogenous variables are presented.

Design variables	
opening hubs	locations of the hubs (and consequently their number)
sequences	the inter-hub services connecting the hubs
capacity of the hubs	transshipment capacity in loading units/time unit
capacity of the sequences	the capacity deployed on the inter-hub services
direct/hub shipment	the choice between direct and the hub-network
participants	the actors joining the consortium

Endogenous variables	
integral costs/item	total cost (transport, handling, transaction costs, etc.)
lead-time	total shipment time between origin and destination
utilization of assets	the utilization of vehicles, equipment and assets
direct/hub shipment	fraction of the flow sent directly or via hub-network
choice to participate	the decision candidate participants to join the collaboration

Exogenous variables	
potential participants	the individual actors that can join the consortium
type of collaboration	type I, II, and III
potential hub locations	the nodes where a hub can be located
transport rates road transport	labour costs/hour, fuel costs, etc.
hub-network investments	investment in equipment, information technology,etc.
transaction cost	frequency, distribution of the asset specificity
additional resistance	additional costs incurred based on the collaboration
network characteristics	the network description, nodes, links, type, etc.
facility costs	investment costs in cost/m2
transshipment characteristics	technique and capacity on the hub
product characteristics	value, volume, weight, loading unit

We define the graph $\mathcal{G} = (\mathcal{V}, \mathcal{E})$ where $\mathcal{V} = \{1,...,v,...,|\mathcal{V}|\}$ is the node set of all origin and destination nodes. $\mathcal{E} = \mathcal{V}^2$ is the set of directed arcs while the set of potential hub nodes is $\mathcal{K} \subseteq \mathcal{E}$. The set of origin-destination pairs (OD) is $\mathcal{W} \subseteq \mathcal{V}^2$ and we consider a situation in which there are $\mathcal{I} = (1,...,i,...,|\mathcal{I}|)$ origins and $\mathcal{J}$ destinations indexed $\mathcal{J} = (1,..., j,....|\mathcal{J}|)$ Each potential participant is associated with an origin i. The volume on origin-destination pair (ij) is denoted d_{ij}. The cost of shipping d_{ij} units directly form origin i to destination j is denoted c_{ij} and is usually proportional to the distance between origin i and destination j. So as an alternative to direct *origin node-to destination-node* shipment there are $\mathcal{K} = (1,...,k,...,|\mathcal{K}|)$ possible hubs of which $\mathcal{S} = (1,...,s,...,|\mathcal{S}|)$ sequences can be made from the access-hub k near the origin (i) to the egress-hub l, $\mathcal{L} = (1,...,l,...,|\mathcal{L}|)$ near the destination (j), $\mathcal{L} = \mathcal{K}$. Let $s \subset \mathcal{K}$ denote a sequence of hubs $\mathcal{K}$. Sequence s consists of one or more segments *a,* and each sequence s consists of one or more hubs *k*.. The cost per unit flow for collection is denoted c_{ik}, for the transfer c_s^{hub} and distribution c_{lj}. In the hub network optimisation model (1-12) the decision variable X_{ijkl} denotes the fraction of flow that travels from node i to j via hubs located at k and l, the decision-variable X_{ij} denotes the fraction of flow from node i to j that is shipped directly

($X_{ijkl} + X_{ij} = 1$), the variable X_i, $\forall i \in \mathcal{I}$ is defined by $X_i = 1$ if a fraction of the flow of participant i is shipped via the hub network and $X_i = 0$, if otherwise.
This design variable indicates if a participants is joining the network. The decision-variable Y_k, $\forall k \in \mathcal{V}$ is defined by $Y_k = 1$ if node k is a hub and $Y_k = 0$ otherwise. F_k is the fixed cost associated with the establishment of a hub at node k and χ_k is its capacity. The decision variable Z_s, $\forall s \in \mathcal{S}$ is defined by $Z_s = 1$ if sequence s is present in the network and $Z_s = 0$ otherwise. ϕ_s is the capacity on sequence s.

$$\min_{x,z} F(x,z) = \sum_{i\in\mathcal{I}}\sum_{j\in\mathcal{J}}\sum_{k\in\mathcal{K}}\sum_{l\in\mathcal{L}}\Big(\big(c_{ij}(X)d_{ij}(1-X_{ijkl})\big)+(c_{ik}(X)+c_s^{hub}(X)+c_{lj}(X))d_{ij}X_{ijkl}\Big)+\sum_{k\in\mathcal{K}}(F_k)Y_k \quad (1)$$
$$+\sum_{i\in\mathcal{I}}(\delta_i(X)+\epsilon_i(X))X_i$$

Subject to:

$$X_{ij} + \sum_{k\in\mathcal{K}}\sum_{l\in\mathcal{L}} X_{ijkl} = 1 \qquad \forall i,j \in \mathcal{V} \quad (2)$$

$$\sum_{l\in\mathcal{L}} X_{ijkl} \le Y_k \qquad \forall i,j,k \in \mathcal{V} \quad (3)$$

$$\sum_{k\in\mathcal{K}} X_{ijkl} \le Y_l \qquad \forall i,j,k \in \mathcal{V} \quad (4)$$

$$\sum_{i\in\mathcal{I}}\sum_{j\in\mathcal{J}} d_{ij}\sum_{l\in\mathcal{L}} X_{ijkl} \le \chi_k Y_k \qquad \forall k \in \mathcal{V} \quad (5)$$

$$\sum_{i\in\mathcal{I}}\sum_{j\in\mathcal{J}} d_{ij}\sum_{k\in\mathcal{K}}\sum_{l\in\mathcal{L}} X_{ijkl} \le \phi_s Z_s \qquad \forall s \in \mathcal{S} \quad (6)$$

$$Y_k \in \{0,1\} \quad (7)$$

$$X_i \in \{0,1\} \quad (8)$$

$$Z_S \in \{0,1\} \quad (9)$$

$$0 \le X_{ijkl} \le 1 \quad (10)$$

$$\sum_{k\in\mathcal{K}}\sum_{l\in\mathcal{L}} X_{ijkl} \le 1 \quad (11)$$

$$0 \le X_{ij} \le 1 \quad (12)$$

In our problem formulation these costs are dependent of the flows and thus dependent of the decision variables (X_{ij}, X_{ijkl}, and Y_k). In turn, the decision to participate (captured by the decision variable) is determined by the costs. In our problem formulation we use the notation $c_{ik}(X)$, $c_s^{hub}(X)$, and $c_{lj}(X)$ to indicate that the costs (c_{ik}, c_s^{hub} and c_{lj}) are a function of the decision variables, denoted by X without subscript. The number of hubs open on a sequence

s is given by the cardinality of the set K. Paths in the graph are identified as a sequence of the nodes traversed.

In objective function (1) the first term represents the costs of shipping directly between origin and destination without using the hubs. These shippers do not participate in the collaboration.

The second terms represents the costs of collection when shipping through the hub network. The third term $c_s^{hub}(X)$ represents the total costs of the inter-hub-transfer. Note that these costs depend on the decision variables. The fourth term in our formulation represents the distribution costs (from the hub to the final destination). The fifth term in this formulation represents the costs of opening a hub in the network and incurs the associated fixed costs. In the previous sections the importance of the transaction costs (δ_i) and additional resistance (ϵ_i) to participate was discussed. In order to implement the effects of these factors on the network design and development path of the network, we introduce the hub design formulation including δ_i and ϵ_i presented in formula (1). In this formulation the costs of participating are incurred if $X_i = 1$, if potential participant i does not join the collaborative hub network $X_i = 0$.

Conservation equations (2) ensure that the total flow (d_{ij}) from i to j is transferred either through the hub network or using direct shipment from i to j, while equations (3) and (4) guarantee that transfers only occur via valid hubs. These equations (2)-(4) ensure that the total demand is sent through the network. Equation (5) ensures that the capacities on the hubs are being adhered to. Equation (6) ensures that the capacity on the sequences connecting the hubs is adhered to. Restrictions (7)-(12) define the decision variables.

With this model, the hubs are capacitated and there is a fixed cost associated with establishing a given node as a hub. Next to the capacity restriction on the hubs we introduce capacity restrictions on the inter-hub connections. In order to take into account the cost reductions that are obtained by consolidation at hub nodes, the technique used in the standard linear hub-location model has been to apply a so-called discount factor on the inter-hub links of the network, so that the per-unit price on inter-hub links is lower than that on external links of the network. It is clear, however, that the use of a linear cost function does not model scale economies, which requires that the marginal price decreases with increasing flow, in which case the cost function must be strictly concave increasing, rather than linear. Clearly, this simplification is costly in terms of accuracy of the solution since large and small flow values all receive the same discount. We explicitly incorporate economies of scale in our cost-functions (c_{ij}, c_{ik}, c_s^{hub}, and c_{lj})[1].

---

[1] The costs for shipping via the hub network, denoted c_s^{hub}, comprise the fixed costs (c_s^f) per time unit multiplied by the sum of the transport time and handling (T_s), the variable costs per time unit (c_s^v) multiplied by the transport time (t_s), and the utilization of the capacity available on sequence s (μ_s). We can formulate these costs as follows: $c_s^{hub} = \left(\left(c_s^f T_S + c_s^s t_s\right)\mu_s\right)$. An important determinant of the costs for shipment via the hub network is the

THE DIFFERENT BUSINESS MODELS

As we have seen relationships between organizations can range from arm's length relationships to complete vertical integration of both organizations. To illustrate the impact the type of relationship between firms has on the decision-making process and network design outcome we present two examples: (1) optimisation of the network with no transaction costs incurred, referred to as the lower bound (LB), and (2) a Type II collaboration with transaction costs. Let us first introduce the network design problem that was used to illustrate the impact of the transaction costs on the network design.

We start with a complex network of 700 locations (production facilities, warehouse locations, customers, and transport hub locations). The orders, volumes and shipments of 6 companies were included. These flows consist of apparel, consumer electronics, printers, faxes, parts, and components. The total flow between these origins and destinations is 1,578,000 shipments/year and a calculated average of €78.40/shipment. The shipment sizes range from 0.10 m^3 to 1.5 m^3 and the total calculated costs are €123,000,000 annually.

Using the model we first calculated the current situation (step 1 and step 2 in Figure 6) resulting in the routes, distances and logistics costs associated with the current network. These costs were then validated using the actual costs (inbound, outbound transportation costs, inventory costs, handling costs, warehousing costs).

Figure 7: Logistics network design with no transaction costs incurred.

---

utilization factor of the deployed equipment, denoted μ_s, a variable dependent of the capacity that is available on sequence s and the combined use of all participants. Using these costs c_s^{hub} and c_{ij} combined with the mentioned transaction costs δ_{im} and $\in_{im}$, the Δc_{im} is calculated to model the decision to participate. See for additional information on the incorporation of the economies of scale Groothedde (2005).

The objective of the algorithm is to minimize the total logistics costs and the decision-variables in this example are the number and location of the hubs, the inter-hub services, and direct or hub-shipment (in total 50 potential hub locations are used). The network design problem was solved using our *Simulated Annealing Heuristic* described in detail in Groothedde, 2005. The outcome of the hub network design phase without the transaction costs incurred is a hub network with 5 hub locations and total logistics costs of €98,000,000/year, yielding a cost reduction of €25,000,000/year, (-20%). The fraction of the shipments that is accommodated by the hub-network is 78% of the total 1,578,000. In total 5 hubs are included (Madrid, Paris, London, Düsseldorf, and Bologna), with 16 inter-hub services, of which 14 services directly connect 2 hubs, and 2 services make an intermediate stop (Madrid Paris Düsseldorf and Düsseldorf Paris Madrid).

In Figure 7 the logistics network solution is presented. The next step was to introduce transaction costs for the individual participants under the assumption that a Type II collaboration is started. The potential participant needs to invest in the network, based on volume, negotiate, search for partners, gather information and commit themselves for a period of 3 years (transaction costs vary between the €18,000 and €34,000). Figure 8 depicts the results of the design procedure when transaction costs for the individual participants are incurred. For certain (potential) participants the hub network is no longer a feasible network solution. The lower-bound calculation yielded a network solution in which the final network solution accommodated 78% of all flows. The network with the transaction costs yielded a result in which 54% of the flows are accommodated. The total reduction in this network solution is € 19,100,000 (-15.6%).

Figure 8: Network design with transaction costs (Type II collaboration).

In this case from the 468 participants only 270 remain (a reduction of 43%), illustrating the impact the transaction costs have on the structure and performance of the network. The number of inter-hub services is reduced from 18 services to 12 services. In addition to the

Type II collaboration a Type I (coordination collaboration) and a Type III (strategic collaboration) were calculated and again the implications of the type of collaboration, in terms of asset specificity, contract term, and transaction costs were calculated for the individual participants. In Figure 9 we see the total reduction of the costs that can be achieved in these scenario's.

The lower bound (LB) in which no transaction costs are incurred yields a reduction of the total logistics costs of €25,746,000. A Type III collaboration €12,378,000 (with 252 participants). If we look at the average cost reduction for the individual participants we see that the highest average cost reduction can be obtained in a Type II collaboration. So although the Type I yields the highest total costs reduction a Type II is more feasible in terms of cost reduction per participant. The Type III collaboration (strategic collaboration) results in the lowest total cost reduction and lowest average cost reduction. It is likely to be the most difficult collaboration to be implemented.

Illustrating the value of an instrument capable of assessing these options ex-ante and not only providing an indication of the cost reductions but also provides the network structure needed to achieve these cost reductions. In the next section we will illustrate the design methodology by presenting the results of two real life projects in which the methodology was used, leading up to the implementation of two logistics networks. The first case is the European logistic network design and implementation of the Ricoh Family Group, one of the largest manufacturers and supplier of office equipment. Te second example is a collaborative network in the Netherlands specifically designed to handle Fast Moving Consumer Goods (FMCG).

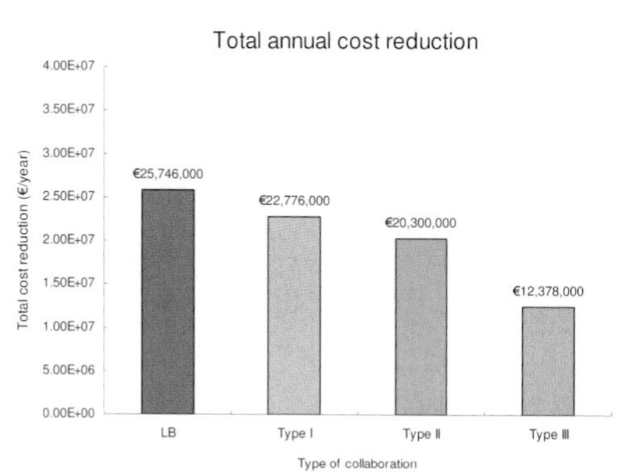

Figure 9: Total annual cost reduction depending on the type of collaboration.

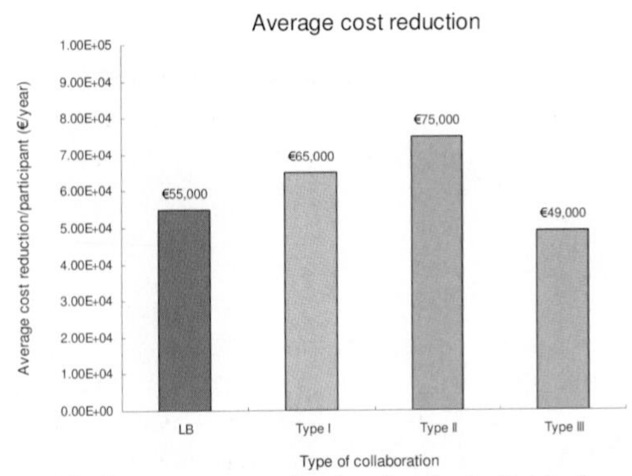

Figure 10: Average cost reduction for the individual participants in the network depending on the type collaboration.

APPLICATION OF THE METHODOLOGY

The first application of the model we discuss here is a project in which the European Logistics Network of the Ricoh Family Group was designed using the developed hub network design methodology. The Ricoh Family Group (RFG) is one of the world's largest manufacturers of office equipment with well known brands like Ricoh, Lanier and Nashuatec, RexRotary and Gestettner. With a total of €13.6 billion net sales (y/e march 2004), 73,000 employees, and 50 manufacturing and R&D facilities worldwide (Abeelen, 2005). RFG decided to reduce the complexity in their European distribution network structure through centralization of inventory in one new central European distribution center, and improving European manufacturing capabilities. The objective was to minimize the logistics costs of the European logistics network that consisted of 3 EDC operations, over 16 country warehouses and in addition improve the service levels: based on (1) the integral logistics costs of the European network, and (2) the service requirements set by Ricoh's customers the new hub network was designed. Using detailed order information, actual shipments to all RFG customers in Europe, the inbound from Asia and European manufacturing sites the current network structure was determined. These logistics costs and performance indicator were used as benchmark to assess the newly design network structures.

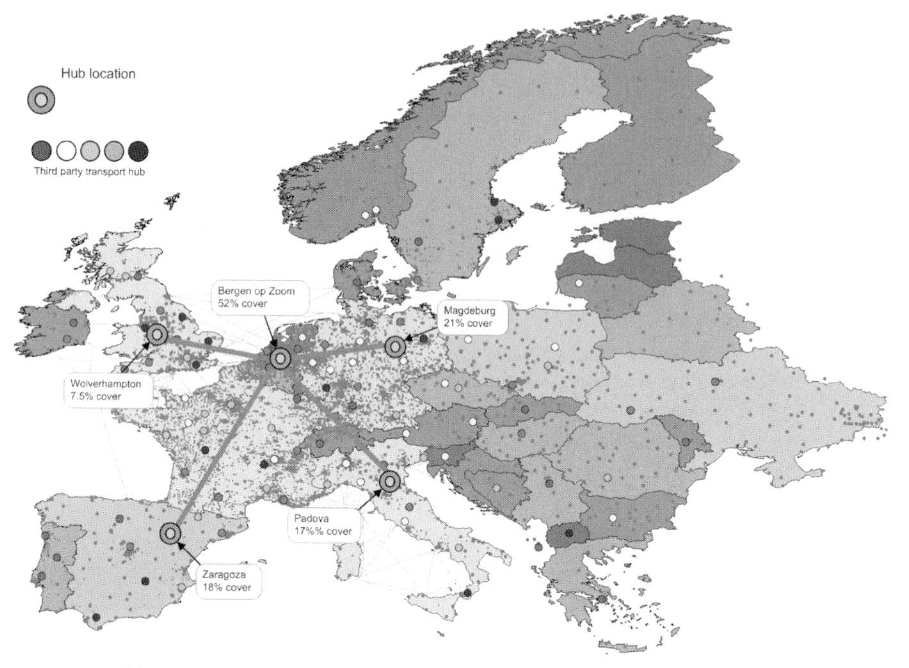

Figure 11: The Ricoh Logistics Network Design (Abeelen and Guis, 2005)

Based on the service requirements set by the customers of RFG the new network structure was designed. In Figure 11 an overview of the resulting network design is presented (Abeelen, 2005). This structure consists of a central hub location in Bergen op Zoom (covering 52% of the customers within a 24 hour radius) and 4 satellite locations (in Zaragoza, Padova, Magdeburg, and Wolverhampton). Different types of networks were evaluated but given the nature of the activities (transport, warehousing, and handling) outsourcing was found the most cost efficient option and make use of the economies of scale of several logistics service providers. In January 2003 the implementation was started, in October 2003 the construction of the central hub commenced, and in October 2004 the central hub location in Bergen op Zoom was fully operational. Since this milestone the remaining satellites are being implemented (Abeelen and Guis, 2005).

The second application of the model we discuss here consists of a collaborative network in which the necessary economies of scale could only be achieved through collaboration. This consisted of a network in the fast moving consumer goods' market. In the this project the aim was to develop an intermodal hub network with relatively small barges, capable of handling pallets, the loading unit most frequently used in the distribution of products in the FMCG sector (Groothedde *et al.*, 2005).

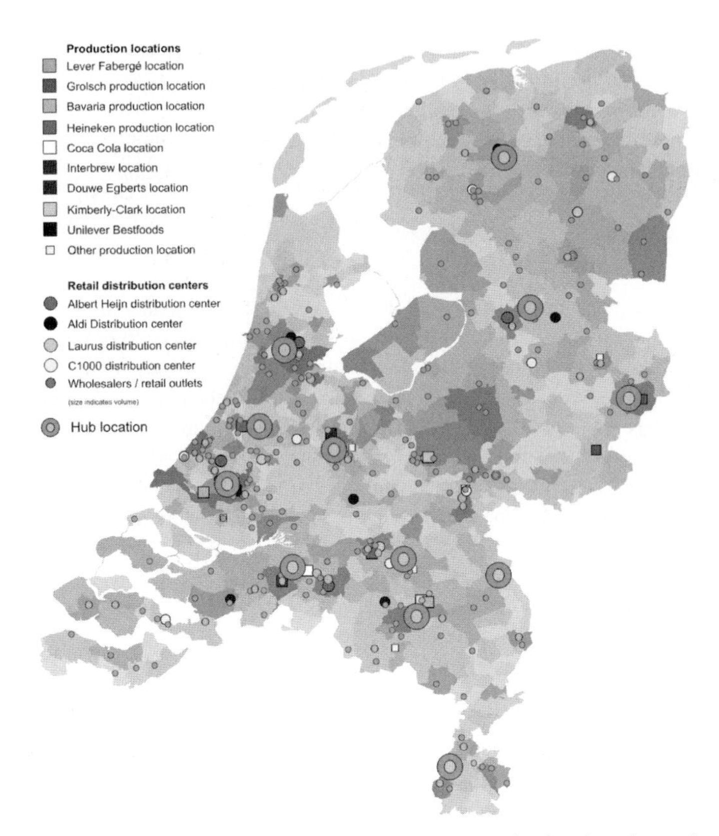

Figure 12: The retail distribution centers and production locations of the focus group. Source: Groothedde *et al.* (2005).

The group of companies that provided the information, used to conduct a detailed analysis of the hub network design methodology, consisted of nine key manufacturers in the Netherlands, two logistics service providers and the inland shipping carrier. In Figure 12 the locations of the manufacturing sites, production warehouses, retailers and additional customers are depicted. They provided detailed information on the orders and product flows between the manufacturing locations (20) and retail warehouses of four of the largest retailers in the food. Albert Heijn, Schuitema, Aldi, and Laurus, wholesalers, and additional customers, add up to in total 100 customer locations.

In total this detailed data set included 6,310,000 million pallets per year. The average sales (in pallets per week) of the focus group is on average 119,000, with a peak in week 32 of 25% above average and -25% below average in week 8.

Based on the flows and orders of these companies new hub networks were designed and different types of collaboration were evaluated. For a complete overview of the different types of collaboration that were evaluated we refer to Groothedde, (2005). It was found that the type II collaboration was the most cost efficient and provided the necessary lead-times. In this network, the average drop was 14.4 pallets with an average value of €1600/pallet. In total 12 hubs and 6 inter-hub services were calculated and 8% of the volume is sent through the collaborative hub network. The transaction costs are on average €63,000.

CONCLUSIONS

In this paper we presented a methodology in which the combination is made between transaction costs economics and network design theory that allows us to provide an answer to the question whether a company should outsource, seek a partner, or keep an in-house solution. Based on a literature survey, it was concluded that the existing network design and evaluation models were not sufficient. Instead of solely minimizing the total costs in the logistics network, or minimizing the costs of individual companies when designing a network, we incorporate the scope of the relationship and the type of relation, and then include the implications of the type and scope of the collaboration in the network design.

We developed a comprehensive framework that is based on economic objectives such as minimizing the total costs, minimizing the user costs and the preferred levels of service set by these users. We analyzed collaboration in logistics networks and the impact the different types of collaboration have on the network design and performance in terms of costs, lead-time and reliability. The elements of a specific collaboration (for example the scope, objective, frequency, uncertainty, transaction costs, etc.) are incorporated in the network design procedure. From a number of applications it can be concluded that these aspects appear to have a great impact on the optimal network configuration.

The hub network design problem we focused on is a multiple assignment problem (customers can use more than 1 hub), combined with direct transportation. We introduce capacity restrictions on the hubs and on the inter-hub transfers. In addition, we incorporate transaction costs that influence the decision to participate in the hub network of the actors, based on the performance of the network solutions (e.g. costs, lead-time, and reliability). This design methodology, yielding feasible network solutions and development paths, is new and is a considerable extension on the current network design approaches. When we compare the results of the network design when no transaction costs are incurred (lower-bound variant), the number of participants is relatively high. Introducing transaction costs changes the network structure and some participants then decide not to join this new network because of the higher logistics costs (excluding the transaction costs).

In our analysis we estimate these costs based on the interviews and information provided by the participants and were able to derive founded conclusions using a bandwidth for these costs. It is however, recommended that a stated preference survey or similar instrument is used to systematically list these costs.

Second, the impact of the distribution of the benefits within the consortium is of great importance. If we combine our network design framework with a gain-sharing instrument, optimizing the distribution of the benefits of the consortium, the participants or service providers know exactly what the optimal offer should be to maximize the cost reduction for the total consortium or of course, in a commercial setting, maximize their own profit.

A third recommendation is to apply this modeling framework in other network sectors. For example in the telecommunication sector, were customer switch between different networks and make this decision based on costs, service, but, in addition, on the transaction costs.

REFERENCES

Abdinnour-Helm, S., (1998). A hybrid heuristic for the uncapacitated hub location problem. *European Journal of Operational Research*, **106**(2/3), 489-499.

Abeelen, M. van, (2005), Mobility&Accessibility, National Distribution Day, HIDC, January 2005

Abeelen, M. van, E. Guis, (2005). Flexibility and Responsiveness in SCM, TIAS Business School.

AC-Nielsen, (2003). Direct Product Profitability, Version model 3.1, Diemen:AC- Nielsen.

Aykin, T. and G. Brown, (1992). Interacting new facilities and location-allocation problems. *Transportation Science*, **26**, 212-222.

Aykin, T. (1995). The hub location and routing problem. *European Journal of Operational Research*, **83**, 200 - 219.

CapGemini, Georgia Institute of Technology, SAP, DHL, (2005). 3PL 2005, 10th Annual Third-Party Logistics Study 2005. Available at: http://3plstudy.com/ [accessed October 2005].

Crainic, T.G. (2000). Service Network Design in freight transportation. *European Journal of Operational Research*, **122**, 272-288.

Ebery, J., Krishnamoorthy, M., Ernst, A. and Boland, N (2000). The capacitated multiple allocation hub location problem: Formulations and algorithms, *European Journal Of Operational Research*. **120** (3), 614-631.

Ellram, L. M., La Londe, B.J. and M.M. Weber (1989). 'Retail Logistics', *International Journal of Physical Distribution&Materials Management*., **19**(12), 29-39.

Groothedde, B., (2005), Collaborative Logistics and Transportation Networks, a modelling approach to network design. PhD Dissertation Delft University of Technology. Delft

Groothedde, B., Ruijgrok, C.J. and L.A. Tavasszy (2005). Towards Collaborative Hub Networks, a case study in consumer goods market, *Transportation Research Part E*. 2005, **41**(6), 567-583.

Kleas, M., (2000). The Birth of the Concept of Transaction Costs: Issues and Controversies. *Industrial and Corporate Change*, **9**(4), 567-593.

Klincewicz, J.G., (1998). Hub location in backbone/tributary network design: A review. *Location Science*, **6**, 307-335.

Lambert, D., M.Cooper, J. Pagh, (1997). Supply Chain Management: Implementation Issues and Research Opportunities. *The International Journal Of Logistics Management*, **9**(2), 1-16.

NEA, (2005). Development of the cost levels in domestic transportation in 2004 and 2005. Rijswijk: NEA. In Dutch

O'Kelly, M.E. and D.L. Bryan, (1998). Hub location with flow economies of scale. *Transportation Research Part B: Methodology*, **32**(8), 605-616.

O'Kelly, M.E., (1986), Activity Levels at Hub Facilities in Interacting Networks. *Geographical Analysis*, **18**(4), 343-356.

O'Kelly, M.E., (1992), Hub facility location with fixed costs. *Papers in Regional Science*, **71**(3), 293-306.

O'Kelly, M.E. and H. Miller, (1991). Solution strategies for the single facility minimax hub location problem. *Papers Regional Science Association*, **70**(4), 367-80.

O'Kelly, M.E. and H. Miller, (1994). The hub network design problem: a review and synthesis. *Journal of Transport Geography*, **2**(1), 31-40.

Pfohl, H, and R. Large, (1992). Research Activities in The Field of Logistics in Europe. The Hague: European Logistics Association.

Rackham, N., (2001). The Pitfalls of Partnering. *Sales and Marketing Management*, **153**(4), 32.

Ruijgrok, C.J. and B. Groothedde, (2005). Trading off advantages and disadvantages of collaboration in logistics (Het afwegen van voor- en nadelen van samenwerking in de logistiek), Proceedings Vervoerslogistieke werkdagen 2005. In Dutch)

Suzuki, S., Z. Drezner, (1996). the p-center location problem in an area. *Location Science*, **4**(1/2), 69-82.

Williamson, O.E., (1975). Markets and Hierarchies, analysis and antitrust implications. New York: Free Press.

Williamson, O.E., (1981). The Economics of Organization: The Transaction Cost Approach. *The American Journal of Sociology*, , **87**(3), 548-577.

Williamson, O.E., (1985). The Economic Institutions of Capitalism: Firms, Markets and Relational Contracting. New York: Free Press.

Transportation and Traffic Theory 2007
Edited by R.E. Allsop, M.G.H. Bell and B.G. Heydecker

20

ON-LINE AMBULANCE DISPATCHING HEURISTICS WITH THE CONSIDERATION OF TRIAGE

K I Wong, Department of Transportation Technology & Management, National Chiao Tung University, Taiwan;
Fumitaka Kurauchi, Department of Urban Management, Kyoto University, Japan; and
Michael G H Bell, Centre for Transport Studies, Imperial College London , UK

INTRODUCTION AND BACKGROUND

In emergency situations, a high priority must be placed on saving life. When there is an incident, emergency services, comprising police, fire fighters and ambulance services, must be provided at the scene as quickly as possible. This paper concerns ambulance dispatching. In many cities, there is a statutory obligation for the ambulance service to arrive on scene within a certain time limit. For example, the United States Emergency Medical Services Act requires that in urban areas 95% of requests should be reached within 10 minutes, while those in rural areas should be served within 30 minutes (see Ball and Lin, 1993). In United Kingdom, the target set by the London Ambulance Services for immediate life-threatening calls is 8 minutes (Thakore *et al.*, 2002). In Japan, average time of ambulance arrival on site is about 6 minutes. To help meet these targets, ambulance location, allocation and relocation models are necessary. In general, location models determine the server or facility locations where the ambulances are dispatched from; allocation models decide which vehicle to be dispatched for a specific call; and relocation models reposition idle vehicles to cover areas which are unprotected.

Ambulance location and relocation models were studied since 1970's, and can be broadly classified into deterministic models and probabilistic models (Brotcorne *et al.*, 2003; Galvao *et al.*, 2005). Deterministic models are usually used in the planning stage. An early model of this type is the location set covering model due to Toregas *et al.* (1971), with the objective of minimizing the number of ambulances needed to cover all demand points. Their model considers a set of demand points and a set of potential vehicle locations. Each demand point represents a geographic area to which service must be provided, and it is to decide the minimum number of locations that can cover all demand points within a specified distance and hence response time. A shortcoming of the model is that it ignores the unavailability of the ambulances when one is dispatched for a request. To rectify this limitation, Church and ReVelle (1974) suggested an alternative approach. In their maximum covering location model, the objective is to maximize the sum of demand covered. Since the number of ambulances is limited, the model may allow some of the demand points not to be covered. These were later extended for the case of multiple-coverage by Schilling *et al.* (1979), so vehicles of several types may be dispatched to the scene of an incident. To guarantee a better service with limited resources, Gendreau *et al.* (1997) suggested a double standard model. While all demand in the area concerned should be reachable within an acceptable time or distance, a certain proportion of the area must be reachable within a higher standard.

In contrast, probabilistic models are used at the operational level. Parameters, for example travel times, the locations of patients, the demand for and the availability of ambulances, are treated as random variables. One of the first probabilistic models for ambulance location is due to Daskin (1983), who maximized the expected coverage of the ambulances, each of which has a probability of being unavailable to answer a call. ReVelle and Hogan (1989) maximized the demand covered with a given probability.

With the designed positions and dimension of the vehicle fleets, such ambulance or emergency management system is usually operated with a decision support tool, having two sub-problems: an allocation problem and a redeployment problem (Gendreau *et al.*, 2001). The allocation problem, which is sometimes referred to as the dispatch problem, considers which ambulance to send to a patient or request. In the literature, heuristic rules are normally used for the dispatching decision. Assuming all incoming requests are urgent and of the same priority, an intuitive decision would be to send the nearest or quickest ambulance to the requests without any delay. Less urgent calls may be held manually or serviced subject to longer maximum waiting. To help meet the targets of adequate coverage and minimum service, a redeployment problem is employed to relocate the ambulances when idle.

There are many applications that motivate research in the field of real-time vehicle routing and dispatching. Such applications include dynamic fleet management, couriers, dial-a-ride, emergency services (police, fire fighting and ambulance services), and taxi cab services (Ghiani *et al.*, 2003). There are important similarities but also differences between these operations. Emergency and ambulance services are classified as strongly dynamic where response time should be minimized (Larson, 2000). With the latest advancements in Information and Communication Technologies (ICT) and Geographic Information Systems (GIS), it is now possible to develop efficient dispatching systems by using real-time, high quality and reliable positional information. The management centre can trace the coordinates and deploy the vehicles to strategic locations, through reposition or diversion. In recent years, ambulance relocation models have been extended to the dynamic case (Gendreau *et al.*, 2001). To insure a good coverage at all times, the vehicles are periodically relocated. In practice, ambulances are often repositioned between locations, like hospitals and medical service stations, when idle. Redeployment scenarios, which allow immediate decision-making when calls are received, are pre-computed. Tabu search, which was originally developed for static cases, has been extended for dynamic updating with re-optimization. Other models may consider diverting an ambulance to a more urgent call (Ichoua *et al.*, 2000).

MEDICAL PRIORITY DISPATCH SYSTEM

Ambulance services are set up to provide an immediate response to patients with life-threatening injuries or illnesses. To rectify the imbalance that exists between demand for and available resources of medical services, there is a need for prioritization of calls (Thakore *et al.*, 2002). Dispatching is often on a First-Come-First-Served basis. As a result, patients with critical illness who call later may have to wait whilst less serious cases are taken to hospital first. It is known that for patients with no heart beat, there is a 10% decrease in the chance of survival for every minute delay in treatment. Actually, by local and international standards, less than 50% calls require an immediate ambulance response, and on arrival at hospital less than 20% patients require to see a doctor immediately. Over 60% of ambulance patients can wait at least 30 to 60 minutes before seeing a doctor, and 30% can wait hours. Thus there is a need for the triage of patients, by prioritizing their need for an ambulance service.

In a Medical Priority Dispatch System (MPDS), all calls for ambulance services are categorized according to the seriousness of the patient's illness or injury. A priority-based dispatch system could reduce response times to those who are seriously ill. Such systems have been tested and tried in countries such as USA, UK and Australia, and have been found to be safe and effective. A key to the success is a centralized dispatch protocol system, hinging on the appropriateness and correctness of the priority assessment assigned by dispatchers (Palumbo *et al.*, 1996). While the process of determining the severity of the incident and its priority must be done by well-trained and experienced telephone operators, a quick and reliable computer-based dispatching system is necessary along with ambulance tracking to provide efficient dispatching and acquire historical information.

The existing literature on ambulance dispatching and repositioning has not discussed the above-mentioned issue of the prioritization of patients, or simply taken the subject into account in a heuristic way. On ambulance dispatching, Andersson and Varbrand (2006) considered the priorities of patients explicitly in their dispatch decision, solving iteratively with a relocation model. In their relocation model, preparedness was defined as a qualitative measure for each potential demand zone evaluating the ability to serve potential demand in the future. Without a call rejection mechanism, as acknowledged in their study, their model may fail to provide a solution when the call volumes are higher than the system can accept. Our approach, instead, is to develop a model with user urgency levels, which not only can decide which vehicle to dispatch but also if the service requested needs to be prolonged or rejected.

DYNAMIC DISPATCHING

In this paper we are interested to develop a dynamic dispatching model which solves the allocation and relocation subproblems simultaneously. Most of the previous research focused on the relocation of ambulances, while the decision on which ambulance to assign to a request is based on heuristic rules. When a call arrives, the natural dispatching rule is to send the ambulance which can arrive first, known as a Nearest Neighbourhood (NN) heuristic, when the objective is to minimize the response times. The First-Come-First-Served process is clearly not optimal (Carter *et al.*, 1972), especially in a priority-based dispatching system. By the same token, the NN heuristic is myopic and does not guarantee optimality under limited resources. In practice, the number of ambulances available is varying over time, and therefore to keep re-optimizing the relocation or redeployment model may not yield the best solution.

Ambulance dispatching is similar to taxi dispatching, in the sense that the locations of future requests are predictable probabilistically, and the requests normally ask for immediate service. Efficient dispatching strategies for taxi services have been developed in Bell *et al.* (2005) and Wong and Bell (2005). They utilize a rolling horizon approach to dispatching, whereby the next n dispatches are calculated to minimise expected passenger waiting time. Since the solution of an intractable dynamic programming problem was involved, a number of simple heuristics have been devised, and simulation experiments have shown that the number of jobs a given fleet can handle may be increased by about 10 percent through use of lookahead heuristics.

The objective of this paper is to develop a tool which helps dispatching vehicles efficiently for emergency medical services. A dispatching model is considered in a dynamic programming context, with the consideration of vehicle location and relocation, patient prioritization, and historical information about the locations of calls.

MODEL FORMULATION AND SPECIFICATION

On the supply side of the emergency medical services, we assume that in the planning stage the set of vehicle locations or medical centres have already been identified, and there are a limited number of ambulances associated with each of these medical centres. When a call is received, the EMS dispatching centre decides which ambulance or ambulances from which locations to assign to the call (we do not differentiate between the ambulances from a station or skills of medical crew). Once an ambulance is dispatched for a task it is not available until it finishes the task and returns to the centre. There is also a schedule of working hours for the medical crew. Therefore, the number of available ambulances varies over time. The vehicles may also be redeployed to another centre for improving the backup coverage.

The requests for medical services arrive over a day. Typically, a call arrives at the EMS dispatching centre, and the status of the patient is evaluated and prioritized for their level of severity. This triage process is different between countries or authorities. A typical procedure is to categorize the calls into three levels (Gendreau *et al.*, 2001; Andersson and Varbrand, 2006); urgent and life-threatening calls, less urgent calls which are not life-threatening, and non-urgent calls. A priority-based dispatch system responds urgent calls immediately, meeting the coverage target as set out by law. Less urgent calls and non-urgent calls can be treated with a looser response-time restriction, being revised periodically in practice. We assume that each of the requests is completely independent, but there is a tendency for the locations of future requests to be predictable probabilistically from historically data about past requests. The rate of calls is not constant, as there could be peak and non-peak periods in a day. Demands in future time periods can be forecast spatially and temporally in a stochastic manner.

Variable definitions

The dynamic problem is considered in a discrete time fashion over the discrete time instants $t = \{0, 1, ..., T\}$, where T is the length of planning horizon and 0 specifies the current stage. Assume a set of locations $i \in I$ where ambulances are dispatched from and a set of potential demand zones $j \in J$. A task (task and call for service will be used interchangeably) $a \in A$, located in $j_a \in J$, calls to the EMS dispatching centre requesting a service at a time t. We can further partition the set of tasks by their time of calling, as $A_t \in A$. In the notation of Dynamic Programming, t is the stage and i is the state of the system. Let

T = the number of periods in the planning horizon
t = discrete time instant, with $t = \{0,1,\ldots,T\}$
I = the set of locations where ambulances are dispatched from, indexed by i
A = the set of tasks over the planning horizon, indexed by a
J = the set of potential demand zones, indexed by j_a
A_t = the set of tasks which arrive at time t, with $\bigcup_{t \in T} A_t = A$

The decision variables of the model are the flows of vehicles to tasks and vehicle repositioning to another location or idling in the same location. Movement of vehicles involves a cost, and rewards are received if a task is served by a vehicle. New resources of vehicles may become available over time with the work schedule of the day. For each $i, k \in I$, $a \in A_t$ and $t \in T$, we define

x_{iat} = the number of ambulances from location i assigned to task a at time t
y_{ikt} = the number of empty ambulances from location i repositioned to location k at time t
τ_{iat} = the service time for an ambulance from i serving task a at time t
τ_{ikt} = the travel time for an ambulance travelling from location i to location k at time t
c_{iat} = the reward for assigning a vehicle from i to servicing a task a at time t
d_{ikt} = the costs of relocating a vehicle from i to k at time t
R_{it} = the number of new ambulances becoming available at location i at time t (i.e. schedule of workforce)

The value of x_{iat} takes a binary form of $\{0,1\}$ since we assume that only one vehicle is needed for each task. This can be relaxed by allowing a task to call for more than one vehicle, and in that case all vehicles must be originating from a same location. Otherwise we can model the task as a number of sub-tasks of one vehicle.

A deterministic model

Firstly we will present a deterministic dynamic model which incorporates both current and future demands. All demands in the first (i.e. the current) period are known, and that all demands in future time periods are forecast. While the locations of ambulances are associated with a set of medical centres, it is difficult to model the exact location of each forecast demand. For modelling purpose, the potential locations of forecast demand are aggregated into zones. The assignment of ambulances to requests and the reposition of ambulances to other locations can be captured by an assignment model as follows

$$\max \sum_{t=0}^{T} \sum_{i \in I} \left[\sum_{a \in A_t} c_{iat} x_{iat} - \sum_{k \in I} d_{ikt} y_{ikt} \right] \tag{1}$$

subject to

$$\left(\sum_{a \in A_t} x_{iat} + \sum_{k \in I} y_{ikt} \right) - \left(\sum_{a \in A_t} x_{ia,t-\tau_{iat}} + \sum_{k \in I} y_{ki,t-\tau_{kit}} \right) = R_{it} \text{ , } i \in I \text{ ,} t \in T \tag{2}$$

$$\sum_{i \in I} x_{iat} \leq 1 \text{, } a \in A_t \text{ , } t \in T \tag{3}$$

$$x_{iat}, y_{ikt} \geq 0 \quad i,k \in I \text{ , } a \in A_t \text{ , } t \in T \tag{4}$$

The above model is specified in a simultaneous form. The objective function Eq. (1) is to maximize the overall benefit or rewards of the job assignment minus costs of vehicle repositioning with a policy (x_{iat}, y_{ikt}) over the planning time horizon. Eq. (2) defines the conservation of vehicles during each time instant, where the number of assigned and relocated vehicles in each location should be equal to total available resources, including newly available vehicles and also those which completed their tasks or relocated from the previous time periods. It is assumed that a vehicle, once it has finished its task, returns to the centre from which it is dispatched. Eq. (3) guarantees each task will not be serviced more than once, and Eq. (4) is the non-negativity constraint of the control variables. The model also allows a call to be rejected if it is in a lower priority and the number of available ambulances is low. Since our aim is to determine the assignment policy at $t = 0$, as the demand for $t > 0$ is forecast, a large T would produce a good enough approximation on a rolling horizon basis. An example with a horizon of three is illustrated in Fig. 1.

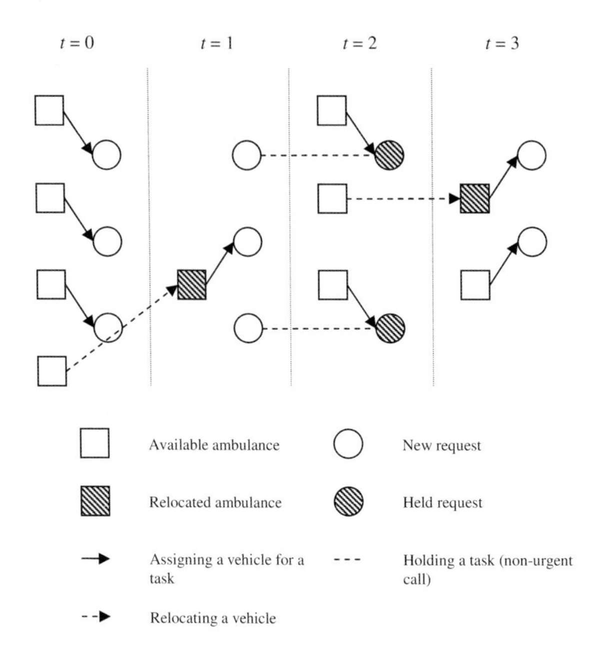

Figure 1. A dynamic ambulance allocation and relocation model

Each of the calls is associated with a priority or level of urgency. From the point of view of the dispatcher, the prioritization of calls can be weighted with the parameter c_{iat}, which defines the reward for assigning a vehicle from i for servicing a task a at time t. A typical aim of a dispatching centre is to minimize the overall delay to all calls, and therefore an intuitive definition of the reward is a reward for handling a task related to its priority minus a function of the delay from i to j_a. Let

q_a = the priority of task a, $q_a = \{1,2,3\}$

r_{q_a} = the reward for handling a task of priority q_a

h_{iat} = the travel time from location i to location j_a to serve task a at time t, converted into the unit of reward. We have

$$c_{iat} = r_{q_a} - h_{iat}, \; i \in I, \; a \in A_t, \; t \in T \tag{5}$$

The setting of c_{iat} should be positive for major elements. There must be some values of c_{iat} for specified a and t to be positive (i.e. r_{q_a} must be greater than some of h_{iat} for all $i \in I$), which implies that all possible j_a are adequately covered. The elements of c_{iat} could be zero or negative in the case that the priority of a task a is low and the distance from sending a vehicle from i to j_a is too far, such that the vehicle is better to be used for repositioning.

Predicted dynamic travel time information can be handled in our model as Eq. (5) is defined with a time index t. For the case of static travel times, the time index can be ignored. The dynamic model presented is equivalent to minimizing the response time to calls with prioritization.

Alternatively the model can be formulated as a recursive form, which will facilitate the analysis later on. Let S_{it} be the total number of vehicles to be available in location i at time t from all sources, and we have

$$S_{it} = R_{it} + \left(\sum_{a \in A} x_{ia,t-\tau_{iat}} + \sum_{k \in I} y_{ki,t-\tau_{kit}} \right) \tag{6}$$

Define J to be the expected cost of a dispatching policy. The model can be specified recursively as follows

$$J_t(S_t) = \max_{x_{iat}, y_{ikt}} \sum_{i \in I} \left[\sum_{a \in A} c_{iat} x_{iat} - \sum_{k \in I} d_{ikt} y_{ikt} \right] + J_{t+1}(S_{t+1}) \tag{7}$$

subject to

$$\left(\sum_{a \in A_t} x_{iat} + \sum_{k \in I} y_{ikt} \right) = S_{it}, \; i \in I \tag{8}$$

$$\sum_{i \in I} x_{iat} \le 1, \; a \in A_t \tag{9}$$

$$x_{iat}, y_{ikt} \ge 0 \;\; i, k \in I, \; a \in A_t \tag{10}$$

where $S_t = \{S_{it}, i \in I\}$.

So far we have presented a dynamic model which captures simultaneous vehicle allocation and relocation, together with patient prioritization, capturing both actual demands (in current period) and forecast demands (in future planning horizon). The decisions between allocating and relocating vehicles are weighted by the reward points that would be received. It is always a dilemma to ask if we should service a current non-urgent call or reserve the capacity for possible future urgent needs, and this is open to the operators. As will be discussed later in the solution algorithm, the model can dispatch vehicles for tasks if there will be more resources than requests in the future, or reject requests with less urgency if many of the vehicles have already been dispatched and are expected not to return in the short term. In contrast to previous models, we have a possibility of rejecting or holding a call, which will be practically kept track of by the operator in revising their priorities.

A stochastic model

In this section a stochastic model is introduced to handle uncertainties in forecast demand. Randomness in the demand is introduced as probabilistic locations of a call arriving in the future. The framework takes the recursive form so that the recourse function can be approximated in various ways in the solution algorithm. We will define the following additional variables:

x_{ia0} = the number of ambulances from location i assigned to task a in current period

y_{ik0} = the number of empty ambulances relocated from i to j in current period

S_{it} = the total number of vehicles to be available in location i at time t. This is the state variable of the system at the beginning of t, capturing the history of vehicle flows up to t; and $S_t = \{S_{it}, i \in I\}$

p_{jt} = the probability of a call arriving in period t being located in demand zone j; it represents a realization of the task A_t, i.e., $A_t(p_t)$, where $p_t = \{p_{jt}, j \in J\}$, and it is assumed that $\sum_{j \in J} p_{jt} = 1$

The stochastic model for the dynamic vehicle allocation and relocation can be written as follows

$$J_0(S_0) = \max_{x_{ia0}, y_{ik0}} \sum_{i \in I} \left[\sum_{a \in A_0} c_{iat} x_{iat} - \sum_{k \in I} d_{ikt} y_{ikt} \right] + E\left[J_1(S_1, p_1)\right] \tag{11}$$

subject to

$$\left(\sum_{a \in A_0} x_{ia0} + \sum_{k \in I} y_{ik0} \right) = R_{i0}, \; i \in I \tag{12}$$

$$\sum_{i \in I} x_{ia0} \le 1, a \in A_0 \tag{13}$$

$$x_{ia0}, y_{ik0} \geq 0 \ \ i,k \in I \ , \ a \in A_0 \tag{14}$$

where S_{it} is specified in Eq. (6) and $E[\cdot]$ is the expectation of the function. S_t defines the state of the system, which depends on the history of the process and decisions up to time *t*, and p_t represents a predicted distribution in generating the locations of tasks A_t. $J_t\left(S_t, p_t\right)$ is the recourse function, taking the state of the system and the estimated probability distributions in future tasks as the input, and is defined as follows.

$$J_t\left(S_t\right) = \max_{x_{iat}, y_{ikt}} \sum_{i \in I} \left[\sum_{a \in A_t} c_{iat} x_{iat} - \sum_{k \in I} d_{ikt} y_{ikt} \right] + E\left[J_{t+1}\left(S_{t+1}, p_{t+1}\right)\right] \tag{15}$$

subject to

$$\left(\sum_{a \in A_t} x_{iat} + \sum_{k \in I} y_{ikt} \right) = S_{it} \ , \ i \in I \tag{16}$$

$$\sum_{i \in I} x_{iat} \leq 1, \ a \in A_t \tag{17}$$

$$x_{iat}, y_{ikt} \geq 0 \ \ i,k \in I \ , \ a \in A_t \tag{18}$$

In the above program, $E\left[J_t\left(S_t, p_t\right)\right]$ is an expected recourse function. Given recursive structure of the model, Bellman's decomposition may be applied. However, finding the solution that maximizes the expected rewards (or minimizing the total delay time) over the rolling horizon remains computationally very demanding, and for large attribute spaces is impractical in real time due to Bellman's "curse of dimensionality". Exact solution is difficult to obtain, and a solution methodology is presented next for approximating this expected recourse function.

SOLUTION METHODOLOGY

A possible approach in solving dynamic programming model is to calculate the recourse function explicitly for each state. However, computational complexity of the presented model is $O\left((I \cdot S)^T J^T\right)$, and solving the model in practice is computational intractable. Algorithms for an approximate solution are usually used, and an efficient solution algorithm would be built upon how to approximate the expected recourse function in a neat way. A widely used method for approximating Eq. (11) is to replace the recourse function with a linear and separable approximation (see Bertsekas, 1995; Powell, 1996; Spivey and Powell, 2004). Spivey and Powell (2004) suggested a solution strategy for problems with small attribute spaces for the dynamic assignment problem, in which the recursive expression $J_{t+1}(S_{t+1}, p_{t+1})$ can be replaced with an approximation of the form

$$J_{t+1}(S_{t+1}, p_{t+1}) = \hat{J}^S \cdot S_{t+1} + \hat{J}^D(p_{t+1}) \tag{19}$$

where $\hat{J}^S$ and $\hat{J}^D$ are, respectively, the vectors of $\hat{J}^S_{i,t+1}$ and $\hat{J}^D_{j,t+1}$, consisting of the vehicle and task value approximations. The dynamic of state variables depend on Eq. (6) in which the available vehicles at time t depend on the allocation $x_{is,t-\tau_{ijt}}$ and relocation $y_{ki,t-\tau_{kit}}$ before t. Since S_{t+1} is the action variables of our problem, we are interested to derive the gradient of our vehicles as $\partial J_{t+1} / \partial S_{t+1} = \hat{J}^S$. For the stochastic gradient of the task value $\hat{J}^D$, the arrival of future demands is not related to our action variables, i.e., decisions of dispatching and repositioning, and it is very difficult to obtain. For the case of the deterministic problem, this gradient vanishes immediately. Several ways have been proposed for approximating this stochastic gradient in the literature. Powell (1996) mentioned that the problem can be formulated as a pure network if the random terms are discrete (which is true in our case of probabilistic locations). The expected recourse function can be represented by a cluster of "recourse links" which capture the expected marginal reward of each unit of task calling for a location at a time period. This method is analogue to the "one-step look ahead" policy in the general dynamic programming solution strategy, when we count only the next time step in the recourse links. It does not look deeper into the future, and we think this is not particularly useful in our problem. A solution strategy proposed by Godfrey and Powell (2002) looks more attractive to our stochastic problem. The idea is that the task gradient is not approximated directly, but the model is trained stochastically with a different Poisson sample drawn from the given probability at each iteration. The benefit is that it explores the whole solution set to a broader extent, but it may require more iterations to produce a good solution.

For notational simplicity and ease of presentation, we assume the travel time for each relocation takes one time interval. This is a reasonable assumption since the dispatcher may not like to relocation a vehicle to a distant location so that the vehicle is able to respond on the way. We further assume that the service time for a particular service is long and unknown before the request is actually received. In other words, it is difficult to determine when a vehicle will be next actionable at the time it is sent out. This is because it is normally not practical to assign a task to an ambulance which is still on duty. Nevertheless, this is not limited to the model and we do not restrict the duration between t and $t + 1$. Assuming τ_{iat} to be a large value and τ_{ikt} to be one, we can reformulate Eq. (6) as

$$S_{i,t+1} = R_{i,t+1} + \left(\sum_{k \in I} y_{kit} \right) \tag{20}$$

When maximizing the expectation of the recourse function, some terms will be dropped. This leads to an approximation of Eq. (15) as

$$J_t(S_t) = \max_{x_{iat}, y_{ikt}} \sum_{i \in I} \left[\sum_{a \in A_t} c_{iat} x_{iat} - \sum_{k \in I} d_{ikt} y_{ikt} + \sum_{k \in I} \hat{J}_{kt}^S y_{ikt} \right] \tag{21}$$

where the expectation has vanished. This simplifies the expectation which caused computational intractability. Now we can see that a quantity $\hat{J}_{it}^S$ is added to the objective function for each ambulance held or relocated, which represents the marginal value of an additional vehicle in a medical centre in the future. We can treat this as the contributions for not assigning the vehicles. The rest of the problem is to approximate the values of $\hat{J}^S$.

An adaptive dynamic programming algorithm

An algorithm using the gradient approximation developed above is presented as follows.

Step 0. Set a maximum number of iterations N, Set $\hat{z}_{it}^{S,0} = 0$ and $\bar{z}_{it}^{S,0} = 0$. Set n = 1 and t = 0.

Step 1. *Forward pass*. For the current n and t, solve the allocation and relocation problem:

$$\tilde{J}_t^k(S_t) = \max_{x_{iat}, y_{ikt}} \sum_{i \in I} \left[\sum_{a \in A_t} c_{iat} x_{iat} - \sum_{k \in I} \left(d_{ikt} - \bar{z}_{k,t+1}^{S,n-1} \right) y_{ikt} \right] \tag{22}$$

subject to

$$\left(\sum_{a \in A_t} x_{iat} + \sum_{k \in I} y_{ikt} \right) = S_{it}, \; i \in I \tag{23}$$

$$\sum_{i \in I} x_{iat} \leq 1, \; a \in A_t \tag{24}$$

$$x_{iat}, y_{ikt} \geq 0 \;\; i, k \in I \;\; , a \in A_t \tag{25}$$

Step 2. Once the x_{iat} and y_{ikt} in Step 1 is determined, update S_{t+1}; if $t < T$, then $t = t + 1$ and go back to Step 1. If $t = T$, go to Step 3.

Step 3. *Backward calculation of gradients*. For the current n and t, we update $\hat{z}_{it}^{S,n}$ with $\hat{z}_{it}^{S,n} = C(S_t, c_1) - C(S_t, c_2)$, where $C(S_t, c)$ is the maximum total rewards of vehicle assignment, for given vehicle repositioning from the previous iteration and S_t for the current iteration. We can obtain left or right gradient at the point of solution with c_1 and c_2 settings, which are vectors in the form of

$c = \{c_i, i \in I\}$. If there are no vehicles repositioned to location i at time $t\text{-}1$, i.e. $\sum_{k \in I} y_{ki,t-1} = 0$, $\hat{z}_{it}^{S,n}$ stands for the additional benefit of having an extra vehicle in location i at time t, and we have $c_1 = (0,\ldots,1,\ldots,0)$ with 1 for the ith element and 0 otherwise, and $c_2 = (0,\ldots,0)$. If $\sum_{k \in I} y_{ki,t-1} > 0$, $\hat{z}_{it}^{S,n}$ is computed as the negative consequences of removing a repositioned vehicle, by letting $c_1 = (0,\ldots,0)$ and $c_2 = (0,\ldots,-1,\ldots,0)$ with -1 for the ith element and 0 otherwise. $C(S_t, c)$ is obtained by solving the maximization problem:

$$C(S_t, c) = \max_{x_{iat}} \sum_{i \in I} \sum_{a \in A_t} c_{iat} x_{iat} \tag{26}$$

subject to

$$\sum_{a \in A_t} x_{iat} \le S_{it} + c_i,\ i \in I \tag{27}$$

$$\sum_{i \in I} x_{iat} \le 1,\ a \in A_t \tag{28}$$

$$x_{iat} \ge 0 \ \ i \in I\ ,\ a \in A_t \tag{29}$$

Step 4. *Smoothing*. Set $\bar{z}_{it}^{S,n} = \alpha^n \hat{z}_{it}^{S,n} + (1 - \alpha^n)\bar{z}_{it}^{S,n-1}$.

Step 5. If $t > 0$, then $t = t - 1$ and go back to Step 3.

Step 6. Terminate if $n = N$; otherwise set $n = n + 1$ and go to Step 1.

In Step 3 above, we can see that the value of $\hat{z}_{it}^{S,n}$ will be large when the certain region or whole of the system is busy in the next time period. However, $\hat{z}_{it}^{S,n}$ is not larger than the maximum setting of c_{iat} in any case, because an actual call is more important than a predicted one of the same priority in the future. On the other hand, if the system is not busy and there are plenty of vehicles idle, the value of $\hat{z}_{it}^{S,n}$ is diminishing. Step 4 adopts a stepsize smoothing function α^n in smoothing the approximation. It is a quantity between 0 and 1, and is referred to as the smoothing constant. A typical choice of the stepsize is $1/n$, a declining function with the iteration number.

The approximation of gradient using a linear marginal contribution function, instead of a nonlinear one, is important. It allows us to optimize the dispatching decisions without considering jointly the decision being made in other centres. The simplification will allow the model to be optimized in real time for problems of realistic size. The maximization problem of Forward Pass and Backward calculation can be solved easily with any off-the-shelf simplex code for linear programming. With the approximated gradient of the expected recourse function, a feature of the proposed iteration based algorithm is that the algorithm (and all LP subproblems) returns integer solutions naturally. This can be shown by the fact that both the objective and the constraints are linear with coefficients of one, while the right hand side of the constraints are integers. Therefore there must be an integer solution, or at least one integer solution (for the case of multiple solutions), for the subprograms. The Pseudo code of the proposed algorithm is shown in Table 1.

Table 1. Pseudo code of the adaptive dynamic programming algorithm

```
Initialize ẑ_it^{S,0} and z̄_it^{S,0} to be zero.
FOR n = 1 to Maximum number of iterations
        FOR t = 0 to Number of time periods
                Solve Forward Pass to obtain x_iat and y_ikt, with z̄_{i,t+1}^{S,n-1}.
                Update S_{t+1}.
        END FOR
        FOR t = Number of time periods down to 1
                Solve Backward calculation of gradients to obtain ẑ_it^{S,n}.
                Calculate z̄_it^{S,n} by smoothing ẑ_it^{S,n} with z̄_it^{S,n-1}
        END FOR
END FOR
```

EXPERIMENTAL RESULTS

To demonstrate the relative performance of the proposed methodology, a simulation model is setup to test against different scenarios. We will test the model against deterministic as well as stochastic settings of the problem. To capture the spatial manner of the ambulance dispatching, a network is created, in which nine potential demand zones and four medical centres are located as shown in Fig. 2. Once there is a call to the dispatching centre, the dispatcher will assign ambulances from one of the medical centres for the call (or reject it). Travel time on each of the links is assumed to be one unit (5 minutes here), and therefore all potential locations are covered by a quickest response of 5 minutes. The corners are subject to less (but acceptable) coverage than those on the boundary or in the centre of the network.

The experiment is simulated for a total period of 4 hours, with each time interval set at 5 minutes. Calls are generated randomly with assumed distributions for corresponding zones, with different levels of urgency. The probabilities of a call being Urgent, Less Urgent and Non-Urgent are assumed to be 0.2, 0.3 and 0.5 respectively. There are 10 ambulances located in each of the medical centres in the beginning of the simulation. Travelling between adjacent centres for repositioning takes one time interval (5 minutes), and an ambulance handling a request is assumed to take 6 intervals (30 minutes) of service time before returning to the centre. To discourage unnecessary repositioning, there is a cost of one fifth the travelling time for vehicles moving idle. We are interested to know the performance of the proposed model in relative to the myopic strategy, which is commonly used in practice, for different levels of demand level. The problem is tested against different numbers of calls, from 50 to 700, during a simulation period. The objective of the dispatcher is to minimize the delay to calls, but since calls are prioritized, we can transform the problem into a reward maximization problem. Each acceptance of a call receives a reward, which is an increasing function with the level of urgency and decreasing function with the delay to the call. We assume the rewards for handling a call of priority Urgent, Less Urgent and Non-Urgent to be 50, 20 and 10 units respectively, followed by one unit of decrease for each minute of delay reaching the patient.

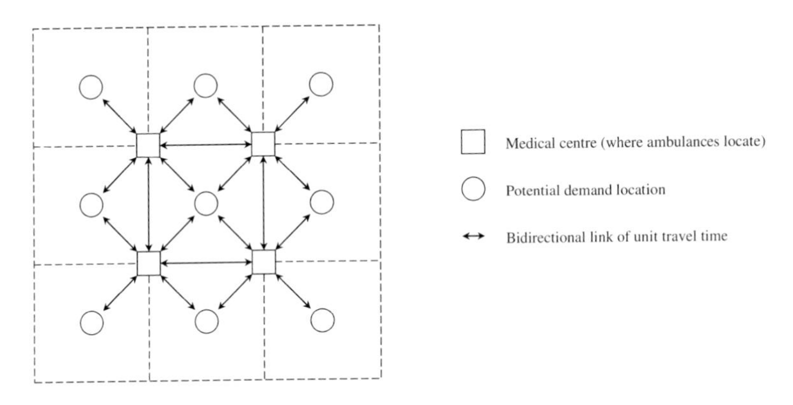

Figure 2. An example network

Deterministic runs

We first perform experiments on deterministic datasets. The dispatcher solves the whole period problem using forecast demand which is assumed to be deterministic. The gradients used in the adaptive algorithm are estimated from 50 training iterations. To eliminate the variation due to randomness, the evaluation of the solution in this section is computed from an average of 20 runs. Fig. 3 shows the total reward received in a simulation period against different demand intensity for myopic and adaptive strategies. Since the optimal solution is unknown to us, the reward is expressed as a percentage against the maximum possible reward received if no requests are rejected. The maximum possible reward is about linear to the number of requests. Both myopic and adaptive strategies reach 100%, i.e., the optimal solution, at a number of calls of 50. As the number of requests increases, the system is overloaded and not able to take up all the calls, and therefore the percentage of gain decreases, in which the rate of drop for the myopic strategy is steeper than that for the adaptive strategy.

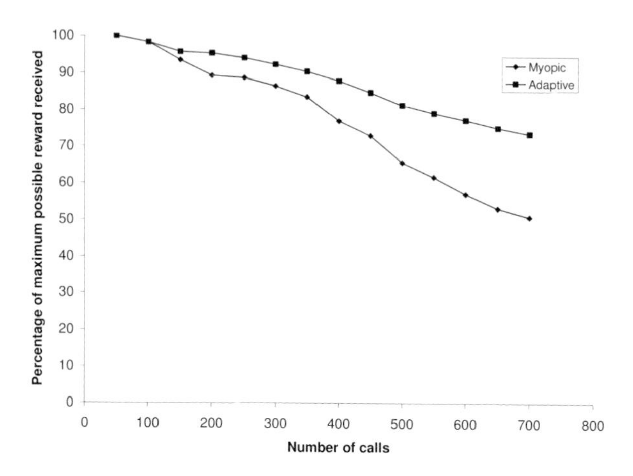

Figure 3. Percentage of maximum possible reward received against the demand intensity: Deterministic cases

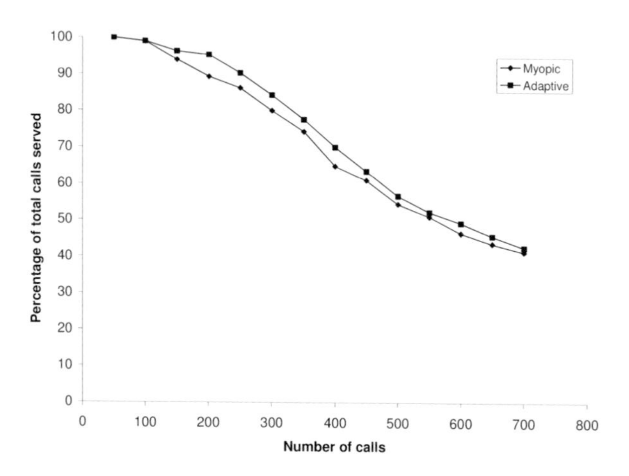

Figure 4. Percentage of total calls served against the demand intensity: Deterministic cases

The adaptive strategy outperforms the myopic one by improving the overall efficiency. It is shown in Fig. 4 that the adaptive algorithm can consistently service more calls compared to myopic. This is probably due to the relocation of ambulances, which can save the travel times in the future operations. Table 2 displays the quality of services for Urgent calls, represented by a target of waiting period for the both strategies tested. We can roughly estimate the capacity of the medical system as 320 calls, which is an overestimation of 40 vehicles each taking maximum 8 tasks in 4 hours without idling. At the demand level of 300, 96.5% and 98.8% of Urgent calls can be reached within 5 minutes and 10 minutes respectively with the adaptive strategy. These percentages are 80.9% and 96.1% with the myopic algorithm. When the demand is increased to 600, with the adaptive strategy the percentage for urgent calls dropped slightly to 92.9% and 97.9%, in contrast to the heavy decline to 56.1% and 66.7% for the myopic one. This confirms the purpose of our dynamic model that suggests holding resources for future critical instances, without sacrificing too much the benefit of the current events.

Table 2. Target of waiting period of 5 minutes and 10 minutes for Urgent calls: Deterministic cases

Number of calls	Myopic, 5 mins	Myopic, 10 mins	Adaptive, 5 mins	Adaptive, 10 mins
50	100	100	100	100
100	97.5	97.5	97.5	97.5
150	92.2	94.0	95.3	97.1
200	83.7	93.3	95.5	98.7
250	83.1	94.9	95.8	98.9
300	80.9	96.1	96.5	98.8
350	78.1	93.6	94.4	99.4
400	72.7	89.8	93.7	99.6
450	68.2	85.5	93.3	99.0
500	63.2	75.5	92.3	98.5
550	58.3	71.7	93.1	98.4
600	56.1	66.7	92.9	97.9
650	52.4	61.6	92.2	97.8
700	49.4	58.3	92.4	98.0

Stochastic Runs

We are more interested in the capability of the model in stochastic environment. In this section the runs were performed under uncertainty in forecast demand. The sets of future tasks are now given in probability and therefore unknown to the dispatcher at the time of making decisions. For the stochastic training, the problem is solved using resampling with a different Poisson sample at each iteration, where the Poisson samples are randomly generated with the given arrival probability. For each simulation, we performed 100 training iterations, a larger number than that in the deterministic case, before executing the simulation to solve the problem dataset which is drawn from the probability set. In fact we have tried different number of training iterations for the deterministic and stochastic tests, and found that the benefit of introducing more iteration for the deterministic problem diminishes fairly quickly, while the stochastic model needs more training since the time stage and space state are large. The results presented below are computed from an average of 20 runs.

Similar figures are computed as in the Deterministic runs. Fig. 5 displays the total reward received in a simulation period against different demand intensity. In general the adaptive strategy outperforms the myopic one, and compared to the deterministic cases as in Fig. 3, the improvements by adaptive strategy are very obvious when the number of calls is below 200.

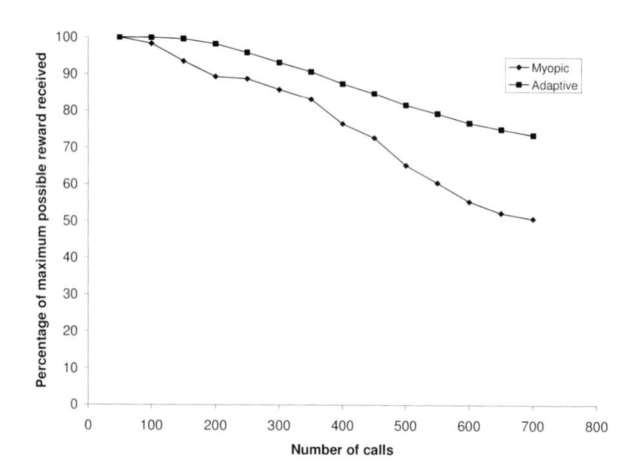

Figure 5. Percentage of maximum possible reward received against the demand intensity: Stochastic cases

Table 3. Target of waiting period of 5 minutes and 10 minutes for Urgent calls: Stochastic cases

Number of calls	Myopic, 5 mins	Myopic, 10 mins	Adaptive, 5 mins	Adaptive, 10 mins
50	100	100	100	100
100	97.5	97.5	100	100
150	92.2	94.0	99.8	100
200	83.7	93.3	97.6	99.8
250	83.1	94.9	96.2	99.6
300	80.2	95.8	93.2	99.4
350	77.6	93.3	91.2	99.5
400	72.1	89.4	90.3	99.4
450	68.1	85.6	89.4	99.2
500	63.0	75.5	89.7	98.7
550	57.0	71.4	89.0	98.5
600	54.9	64.8	86.6	97.6
650	51.8	60.9	88.4	97.1
700	50.0	58.7	88.2	98.2

Table 3 summarizes the quality of services for Urgent calls, with the target of 5 minutes and 10 minutes. The adaptive algorithm performs near-optimally for low demand cases when the number of calls is below 200, and the percentage of calls with delay under 5 minutes drops with the number of calls down to 93.2% at demand level of 300 and to 86.6% at demand level of 600. Coverage of 10 minutes is maintained for more than 97% for all cases.

Comparing the results between the deterministic cases and stochastic cases, it is interesting to see that, for the adaptive algorithm, more calls can be maintained at the 5 minutes target for the deterministic cases, while it provides a better coverage at the 10 minutes level in the stochastic experiments. A possible explanation for this is that the deterministic training estimates the dispatching action maximizing the overall reward targeting on the calls with 'highest value', i.e. nearest ambulances for most urgent calls. Benefiting from the random sampling, the stochastic training, on the other hand, distributes the available vehicles so strengthening the average coverage. Therefore, even though the total rewards received for the two experiments are very close, we can see that there exist differences in their inherent distributions of delay to patients. The stochastic model spreads out the opportunity cost of making a decision if it is incorrect, including the waste of vehicle times in repositioning and the extended delay of demand in the future.

On the computational efficiency, the number of computing steps of the adaptive algorithm is proportional to the number of training iterations and length of time horizons. In our numerical experiment, each run took 25 seconds for the stochastic case and 18 seconds for the deterministic case, executing on a Pentium 4 3.0GHz personal computer. The myopic scheme is similar to the adaptive scheme with only one iteration.

SUMMARY AND CONCLUSIONS

A dynamic allocation and relocation model is presented for ambulance dispatching with the consideration of patient triage, which is important when the need for medical services is more than the resources available. Prioritization of requests is becoming a standard procedure in practice, and this paper is one of the first in trying to incorporate this component into the ambulance dispatch problem. In the near-capacity situation, it is suggested that keeping or relocating vehicles to strategic locations could benefit the demand in the future. We formulate the problem with the objective of maximizing the reward gained for serving a request, showed as equivalent to minimizing the total delay incurred. The problem is formulated in a dynamic programming context, which suffers from the "curse of dimensionality" in applications of practical size. We propose a dynamic adaptive algorithm for solving the problem, approximating the gradient of the recursive subproblem in a linear form. Deterministic and stochastic settings of the problem are experimented with.

Previous models for locating emergency services focused on the case with the average number of calls, and may fail to provide a solution when the call volumes are high. This issue was acknowledged and can be unrealistic in practice. Our model provides a guideline for accepting or not a current call with the computation of values of resources currently and in the future. Of course, rejected calls which are lower in priority would be kept track of by the operators for possible upgrading if needed. This model is particular useful in near-capacity conditions of dispatching, in which the tactical locations of medical centres and fleet of ambulances are already fixed.

ACKNOWLEDGEMENTS

The first author gratefully acknowledges the final support from the National Science Council of Taiwan (Project No. NSC 95-2221-E-009-348).

REFERENCES

Andersson, T. and P. Varbrand (2006). Decision support tools for ambulance dispatch and relocation. *Journal of the Operational Research Society*, 1-7.

Ball, M.O. and L.F. Lin (1993). A reliability model applied to emergency service vehicle location. *Operations Research*, **41**, 18-36.

Bell, M.G.H., K.I. Wong and A.J. Nicholson (2005). A rolling horizon approach to the optimal dispatching of taxis. In H.S. Mahmassani (ed.) *Transportation and Traffic Theory: Flow, Dynamics and Human Interaction*, Elsevier Science Ltd., pp 629-648. (ISBN: 0-08-044680-9)

Bertsekas, D.P. (1995). *Dynamic programming and optimal control*, Volume 1. Athena Scientific.

Brotcorne, L., G. Laporte and F. Semet (2003). Ambulance location and relocation models. *European Journal of Operational Research*, **147**, 451-463.

Carter, G., J. Chaiken and E. Ignall (1972). Response areas for two emergency units. *Operations Research*, **20**(3), 571-594.

Church, R. and C. ReVelle (1974). The maximal covering location problem. *Papers of the Regional Science Association*, **32**, 101-118.

Daskin, M. (1983). The maximal expected covering location model: formulation, properties, and heuristic solution. *Transportation Science*, **17**, 48-70.

Galvao, R.D., F.Y. Chiyoshi and R. Morabito (2005). Towards unified formulations and extensions of two classical probabilistic location models. *Computers & Operations Research*, **32**, 15-33.

Gendreau, M., G. Laporte and F. Semet (1997). Solving an ambulance location model by tabu search. *Location Science*, **5**(2), 75-88.

Gendreau, M., G. Laporte and F. Semet (2001). A dynamic model and parallel tabu search heuristic for real-time ambulance relocation, *Parallel Computing*, **27**, 1641-1653.

Ghiani, G., F. Guerriero, G. Laporte and R. Musmanno (2003). Real-time vehicle routing: Solution concepts, algorithms and parallel computing strategies. *European Journal of Operational Research*, **151**, 1-11.

Godfrey, G.A. and W.B. Powell (2002). An adaptive dynamic programming algorithm for dynamic fleet management, I: single period travel times, II: multiperiod travel times. *Transportation Science*, **36**(1), 21-45.

Ichoua, S., M. Gendreau and J-Y. Potvin (2000). Diversion issues in real-time vehicle dispatching. *Transportation Science*, **34**, 426-438.

Larsen, A. (2000). *The dynamic vehicle routing problem*. Ph.D. thesis, Department of Mathematical Modelling (IMM) at the Technical University of Denmark (DTU). Available on the Internet.

Powell, W. (1996). A stochastic formulation of the dynamic assignment problem, with an application to truckload motor carriers. *Transportation Science*, **30**(3), 195-219.

Palumbo, L., J. Kubincanek, C. Emerman, N. Jouriles, R. Cydulka and B. Shade (1996). Performance of a system to determine EMS dispatch priorities. *American Journal of Emergency Medicine*, **14**(4), 388-390.

ReVelle, C. and K. Hogan (1989). The maximum availability location problem. *Transportation Science*, **23**, 192-200.

Schilling, D.A., D.J. Elzinga, J. Cohon, R.L. Church, and C. ReVelle (1979). The TEAM/FLEET models for simultaneous facility and equipment sitting. *Transportation Science*, **13**, 163–175.

Spivey, M.Z. and W. Powell (2004). The dynamic assignment problem. *Transportation Science*, **38**(4), 399-419.

Thakore, S., E.A. McGugan and W. Morrison (2002). Emergency ambulance dispatch: is there a case for triage? *Journal of the Royal Society of Medicine*, **95**(3): 126-129.

Toregas, C., R. Swain, C. ReVelle and L. Bergman (1971). The location of emergency service facilities. *Operations Research*, **19**, 1363-1373.

Wong, K.I. and M.G.H. Bell (2005). The optimal dispatching of taxis under congestion: a rolling horizon approach. *Journal of Advanced Transportation*, **40**(2), 203-220.

Transportation and Traffic Theory 2007
Edited by R.E. Allsop, M.G.H. Bell and B.G. Heydecker

21

A MARKOV PROCESS MODEL FOR CAPACITY CONSTRAINED TRANSIT ASSIGNMENT

Fitsum Teklu, David Watling and Richard Connors, Institute for Transport Studies, University of Leeds, UK

SUMMARY

Representing the finite capacity constraints of vehicles in a transit assignment model, for networks where the vehicle capacities are small and the buses do not operate to timetables, requires an accurate representation of the impacts on passenger costs and flows as well as their day-to-day variability. This paper presents a composite frequency-based and schedule-based *Markov process* model for transit assignment that considers the day-to-day dynamics of the transit network, whereby line frequencies are used to parameterise distributions of vehicle arrivals and the passenger flows are constrained to the *individual* vehicle capacities. A proof of the model's *Markov property* and *regularity*, its sensitivity to some parameters, and comparisons with the Cepeda et al. (2006) model are presented using a test network.

INTRODUCTION

The use of small (~12-20 passenger capacity) transit vehicles is common in some cities of the world. In such environments, it is often the case that passengers are unable to board a bus of their choice because of capacity constraints. It is also common, in some cities, for these services to not run to timetables. This study is concerned with such transit networks.

Frequency-based (FB) transit assignment models use a transit supply system represented in terms of average line frequencies and forecast passengers' cost and route choices (e.g. Spiess & Florian, 1989). On the other hand, schedule-based (SB) transit assignment methods consider each run of a transit line using timetables for forecasting route flows and costs (e.g. Nuzzolo et al., 2001). An assignment model for networks of the type described above needs to represent the impact of the strict vehicle capacities on waiting times and route flows, and the absence of timetables for passengers to base their route choices on.

In recognition of parallel services that might be available to passengers in transit networks, Spiess & Florian (1989) introduce the notion of optimal strategies that assumes passengers choose a *strategy*: i.e. they have a fixed subset of *attractive lines* for every stop they might encounter on their trip, and (at each stop) board the first arriving bus from that set. The attractive lines are identified by assuming passengers only consider lines that minimize their expected travel time. Nguyen & Pallottino (1988) provide a graph theoretic framework for Spiess and Florian's work that represents a strategy as a *hyperpath*. Albeit using a different network representation based on timetables, passengers' strategies have been included in SB models for high-frequency services as well (e.g. Nuzzolo et al., 2001). Slightly differently, De Cea & Fernandez (1989) define a route as a (fixed) sequence of transfer stops. The set of attractive lines is chosen between each pair of transfer stops to minimize expected travel time, making up what are called *route-sections*. This method only considers transit lines that 'visit' both ends of the route-section in the attractive lines set.

Passenger route flows and costs vary from day-to-day. On some days passengers will find it easier to get the bus of their choice, with spare capacity, than on some other days. Static models, although accounting for variations in waiting times through assumed headway distributions for passengers and buses, only use a deterministic representation of user costs and passenger flows. Information on the variability of these quantities helps planners make more informed decisions. This is especially true when the vehicle capacities are small and the lines are not operating to timetables.

Most FB models aiming to account for capacity constraints do so by using analytic congestion functions to constrain assigned flows to the aggregate capacity over some modelling period, without considering whether or not individual bus' capacities were exceeded. SB models that consider each run of a transit vehicle, on the other hand, provide a logical base for that but require timetable information to represent transit supply which is not available in networks of the type considered here.

Stochastic process (SP) models represent the day-to-day evolving interaction between transit system costs and passengers' information acquisition and choice processes based on assumptions of passengers' choice updating behaviour as well as learning and forecasting mechanisms. They have been applied in road traffic assignment (e.g. Cantarella & Cascetta, 1995) and are shown to subsume static and stochastic user equilibrium solutions as particular cases (Davis & Nihan, 1993). However, they have not had much application in transit assignment modelling. Nuzzolo et al. (2001) discuss the application of a similar approach for SB transit assignment. *Markov process models* are special classes of such models where the conditional probability distribution of the future states (describing the transit supply-demand interaction), given the present and all the past states, depends only upon the current state – the so called *Markov property*. Processes that are regular are guaranteed to have a stationary distribution of the state variable, and to converge to that distribution regardless of the initial conditions.

This paper presents a Markov process model for transit assignment that has strict capacity constraints and accounts for stochastic bus headways. Using a composite FB and SB approach, the proposed model uses aggregate line frequencies to parameterise bus headway distributions and a micro-simulator to enforce capacity constraints on individual vehicles and to model the variability in passengers' experience. The next chapter reviews models for capacity constrained transit assignment. In the following chapter, detailed discussion of the proposed model, the Monte Carlo simulation based approach developed to evaluate it, and a proof of its Markov property are given. Numerical experiments, sensitivity tests to model parameters, and a model comparison are then presented in the penultimate chapter after which conclusions are made.

HANDLING CAPACITY CONSTRAINTS

Most models of capacity constrained FB transit assignment employ passenger flow dependent *congestion functions* to represent effects on passenger waiting costs. BPR-type increasing functions, relating waiting costs with passenger flow to capacity ratios are proposed in De Cea and Fernandez (1993) and Wu et al. (1994). Considering the asymmetric nature of transit assignment problem, due to the dependence of passengers' waiting cost on the number of people in the buses from upstream stops, they include all passengers in the lines when computing the flow to capacity ratio. Wu et al. (1994) also use BPR-type in-vehicle travel cost functions to model crowding impacts. However, these models do not guarantee the capacity of the lines is not exceeded. They also use nominal frequencies to distribute passengers onto the attractive lines. Besides forecasting line flows inaccurately, this leads to erroneous calculation of passenger costs especially when the attractive lines offer different in-vehicle costs; this is carried through to the equilibrium calculations.

Hamdouch et al. (2004) account for the priority of passengers already in the buses and assume passengers' strategies specify an ordered set of attractive lines, at each stop, where the probability of boarding a line is proportional to the residual-line-capacity to demand ratio. Kurauchi et al. (2003) present a Markov chain absorbing method to incorporate line capacity constraints through failure-to-board probabilities for a FB network. To meet capacity constraints when demand exceeds capacity, passengers are sent on a "failure arc" to the next time period. However, both these models do not explain increased passenger waiting times due to congestion.

Cominetti & Correa (2001) make use of differentiable flow-dependent frequency functions, called effective-frequency functions, that fall with increasing passenger flow to model capacity constraints. Impacts on waiting times and passengers distribution onto attractive lines are included by using the effective frequencies instead of nominal frequencies. A queue theoretic support for the model for when there is only one strategy carrying equilibrium flow from each stop. Because of the complex dependency of waiting times on the whole strategy flows vector (due to the asymmetric nature of the problem) they comment the model is mathematically incorrect if that is not satisfied; however, they add the resulting errors are relatively small. Cominetti & Correa did not obtain a closed form solution for the effective frequency function. Cepeda et al. (2006) provided a Method of Successive Averages based solution algorithm for using the method on large networks. They also propose an effective frequency function in the paper – without any proof.

All the models above constrain passenger flows to aggregate capacity. For a (with-in day) dynamic, SB network, Poon et al. (2003) consider constraining passenger flows to individual vehicle capacities based on a simulation model. They propose an iterative procedure whereby the waiting times are calculated from passenger arrival and departure profiles at the different stops and passengers are assigned on minimum cost paths. The model's application on large scale networks is also presented. They, however, do not consider the stochasticity in bus arrival times and passenger strategies at transit stops.

In conclusion, analytic approaches used in FB methods suffer from the difficulty in calibrating congestion functions to consistently represent impacts on waiting times and passenger distributions throughout the network. Most models concentrate on capturing a particular impact of congestion, compromising some others. SB models that enforce strict capacity constraints on individual vehicles have been applied on large scale networks but they do not account for passenger strategies and bus headway stochasticity. The impacts on the day-to-day variability of passengers' cost experiences and flows have not had much consideration in the literature.

MODEL DESCRIPTION

As mentioned earlier, SP models have been shown to converge to unique stationary probability distributions without imposing any equilibrium conditions. They are particularly attractive for modelling asymmetric problems with multiple equilibria (Watling, 1996). In this chapter, SIMTRANSIT, a Markov process model for capacity constrained transit assignment, that uses Monte Carlo simulation to provide pseudorandom observations of the demand-supply interaction, is presented. The proposed model, shown in Figure 1, uses a micro-simulation based network model, explicitly considering demand and supply stochasticity. Each day, the cost experiences on the alternative routes, obtained from the network model, are used to update passengers route cost expectations through the learning process model. Based on the passengers' route cost expectations a random utility model is used to forecast passenger route choices assuming they perceive route costs differently. A detailed discussion of each of these components and the theoretical proof of the Markov property and regularity of this model are given in this chapter.

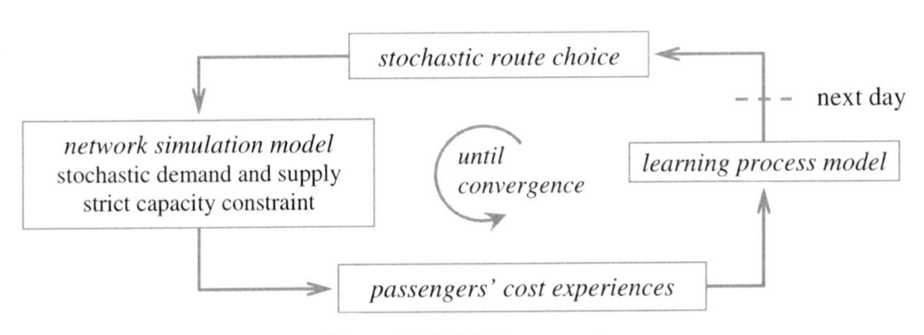

Figure 1 Model Framework

Network simulation model

A micro-simulation approach is used to model the buses' and the passengers' movements. This allows separate simulation of the different lines' runs (i.e. without aggregating all the runs over the modelling period) enabling better representation of the passengers' experiences. Using a simulation time step-length of ω, the model processes bus arrivals, passenger arrivals, and bus departures sequentially; each initiating the different procedures briefly described as follows. Firstly, at each time step, τ, due passengers and buses are generated based on assumed headway distributions. Then, the program checks if there are any buses arriving at the different stops. For each bus arriving at a stop, alighting passengers are allowed to get off, after which bus departure times would be assigned. Transferring passengers join the queue for their next service. Next, newly generated passengers are made to choose their route and join the back of the queue. Finally, a check for buses departing from the each stop at τ is made. For each departing bus, passengers queuing at the stop are made to board following the first-in-first-out rule, if the bus is in their attractive lines set and has spare capacity. Bus dwell times are linearly dependent on the number of passengers alighting and boarding through, the *passenger service time parameter*, π – a constant multiplier, that could be obtained from roadside surveys. The buses are then assigned arrival times to the next stop in their itinerary by adding the dwell times and stop-to-stop travel times. Although stochastic travel times could easily be incorporated, a constant stop-to-stop travel time is assumed in this paper. It should be noted that the bus headway distributions are used only at the first stops in the buses' itinerary, and the arrival distributions are not identical across all the stops. This cyclic process continues by incrementing τ by ω until the total simulation period, Γ, is reached.

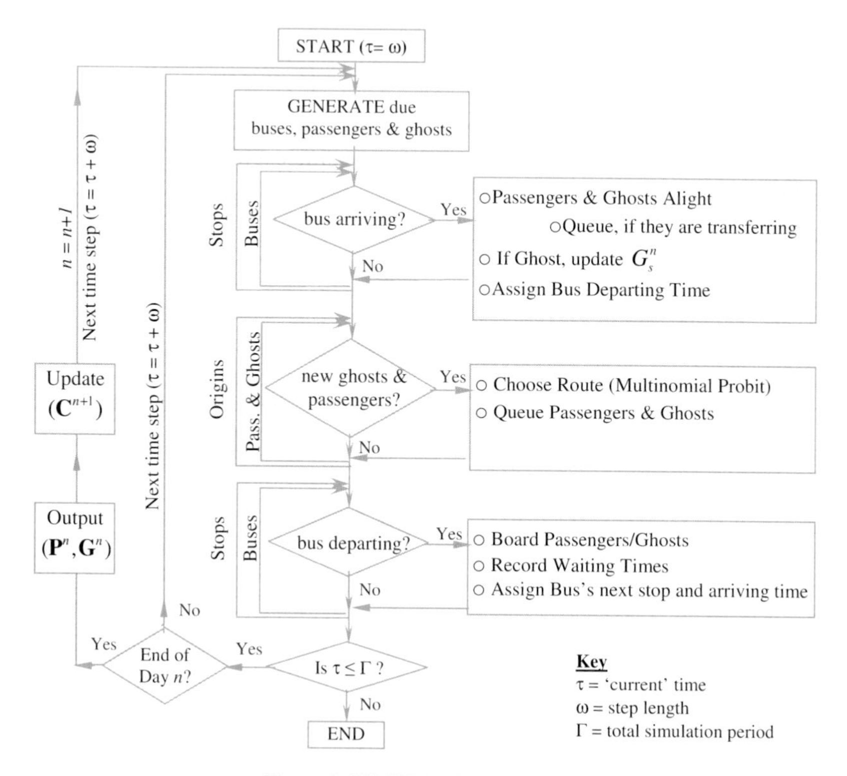

Figure 2 SIMTRANSIT flow chart

Route choice model

In this study, the network supply representation presented in De Cea & Fernandez (1993) is used. As noted earlier, this approach assumes passengers' *routes* are defined by a sequence of transfer stops between which alternative *route-sections* representing the attractive lines set for that leg of a journey are identified. Passengers are not allowed to change their choice of transfer stops once they set off on their journey. Consider the small test network shown in Figure 4a. Figure 4b shows all the route-sections for all the transfer stop pairs. For example, passengers on single line route-sections B and D only use the RED and GRN lines, respectively, to go from S1 to S2; those on C use whichever one comes first. The *sensible* routes, for the different OD pairs, are enumerated in Figure 4c. For example, passengers on route 2 wait for, and board the RED line at S1, alight at S2, and board the GRN line to complete their journey. Routes such as B & F and C & E, for travelling from S1 to S4, are not included in that set assuming passengers would rather choose A and L, respectively, which do not require 'alighting' from a route-section only to 'board' another one, with an identical attractive lines set, to finish their journey.

A stochastic route choice model is proposed assuming passengers do not have an exact knowledge of the line travel times and perceive waiting times differently. The set of routes considered is thus based on an *all route-sections choice menu*. Especially in multimodal, multi user class networks, passengers may perceive the different lines to be different choices due to different in-vehicle costs (e.g. travel time and fares). For example, including B, C, & D for the test network (Figure 4) allows passengers that perceive travel times between S1 and S2 differently, to choose their route accordingly. The active route-sections are determined through the Markov process model.

The stochastic transit assignment model is specified as follows. Let $\mathbf{D}$= vector of OD pairs and $\mathbf{R}_h$ = set of routes for OD pair $h \in \mathbf{D}$. For each route $r \in \mathbf{R}_h$, passengers' perceived cost (PC_r) is the sum of the deterministic generalised cost (GC_r) and an error term (ξ_r) as:

$$PC_r = GC_r + \xi_r , \quad \forall r \in \mathbf{R}_h \tag{1}$$

, for which the route choice probability could be written as:

$$\Pr(r) = \Pr(GC_r + \xi_r < GC_j + \xi_j \quad \forall j \neq r), \tag{2}$$

, where: $r, j \in \mathbf{R}_h$.

The deterministic generalized cost is composed of in-vehicle travel cost (T), waiting cost (W), and fares (F), see equation (2).

$$GC_r = \gamma_{(t)} \cdot T_r + \gamma_{(w)} \cdot W_r + F_r \tag{3}$$

where, $\gamma_{(t)}$ and $\gamma_{(w)}$ are the values of travel time and waiting time, respectively.

To obtain democratic cost sampling for GC_r across the alternative routes, *ghosts* (probes) are generated at a constant rate and made to travel along them – experiencing the costs without contributing to congestion. In SIMTRANSIT's bus and passenger simulation model, the ghosts are treated in the same way as the passengers, except in the buses where they are assumed not to take up space. As long as they do not have a "real" passenger ahead of them in the queue, they are allowed to board a full service. It is noted that although it would, alternatively, be possible to use real passenger experiences (rather than the ghosts) to build up the day-by-day travel costs, such an approach produces extremely slow convergence. This is due to the fact that a route, perceived in an initial period as unattractive, is not used for a long period by travellers sequentially, giving no data on which to update its costs, until some low probability random disturbance causes a traveller to experiment with it.

To maintain realism, the route choice model should consider the cost dependencies (in mean and variance) of the passengers' perceptions, starting from that between the different route-sections for each pair of transfer stops to that between the different routes for each OD pair. For example, a passenger's perceptions of the waiting and travel costs of route-sections that contain a particular line (for example, the RED line in route-sections B and C in Figure 4) should intuitively be correlated since the perceived quality of the RED line service will be of a significant contribution to the way both route-sections will be viewed by a passenger. Likewise, as a route-section could itself be part of different routes, the costs of the routes that contain it should be correlated as well; consider routes 3 and 4 in Figure 4. These requirements negate the use of discrete choice models such as the multinomial Logit model which suffer from the IIA property.

The ghosts automatically extract the cost dependencies in the mean costs of route-sections and routes. The variance in passengers' perceived costs is modelled using error terms starting from the lines, through the routes-sections, up to the routes as follows. Let $\mathbf{K}_r$ = the vector of route-sections for route $r \in \mathbf{R}_h$ and $\mathbf{A}_k$ = the set of attractive lines that constitute the route-section $k \in \mathbf{K}_r$. Then, the variance in the perceived cost on route-section $k \in \mathbf{K}_r$, σ_k^2, is given by equation (3). Assuming uniformly and exponentially distributed passenger and line headways[1], respectively, the average uncongested waiting costs and frequency-weighted average line travel costs are used as a basis for calculating σ_k^2:

[1] Note that this assumption is only made in order to get some kind of statistical model of cost mis-perceptions; such an assumption is not an integral part of the network loading module described later.

$$\sigma_k^{\ 2} = \eta \cdot \left(\frac{\gamma_{(w)} + \sum_{\forall l \in A_k} f_l \cdot \gamma_{(t)} \cdot t_l}{\sum_{\forall l \in A_k} f_l} \right), \quad \forall k \in \mathbf{K}_r, \forall r \in \mathbf{R}_h, \forall h \in \mathbf{D} \qquad (4)$$

where, $\eta > 0$, f_l= frequency of line l, and t_l = in-vehicle travel time on line l. Here η is a model parameter which can be interpreted as the variance of the perceived costs over a route-section of unit cost; it is exogenously defined. Admittedly, equation (4) neglects a third form of correlation, assuming that perceived waiting times for lines that are included in multiple route-sections are independent, which might not be realistic; this is an outstanding research issue which is not addressed in this study.

For simplicity of calibration, assuming the perceived costs of the *non-overlapping* route-sections of the alternative routes are independent, the elements of the covariance matrix for the routes are given by:

$$\mathrm{cov}(\xi_r, \xi_s) = \sum_{\forall k \in \mathbf{K}_r \cap \mathbf{K}_s} \sigma_k^{\ 2} \qquad , \forall r, s \in \mathbf{R}_h \, , \forall h \in \mathbf{D} \qquad (5)$$

and the variance in a route's perceived cost is given by:

$$\rho_r^{\ 2} = \mathrm{cov}(\xi_r, \xi_r) = \sum_{\forall k \in \mathbf{K}_r} \sigma_k^{\ 2} \quad , \forall r \in \mathbf{R}_h \, . \qquad (6)$$

Assuming additionally that the perception errors for the route-sections are independent and normally distributed, it follows that the joint distribution of the perception errors across all routes is multivariate normal, i.e. $MVN(0, \Omega)$ - where, Ω is the covariance matrix with elements given in (5) and (6). In this study, it is assumed that the covariance matrix is constant through out the evolution of the SP.

Learning process model

The basic assumption behind SP models is that the state of the system on a particular day n is random. In SIMTRANSIT, the random vector of the expected cost on the alternative routes, $\mathbf{C}^n$, which is based on the ghosts' experiences, is chosen to describe the state of the system on day n. The route flow probabilities vector $\mathbf{P}^n$ is based on $\mathbf{C}^{n-1}$, according to a random utility model. A Monte Carlo simulation model approach is followed to simulate the $\mathbf{P}^n$'s and $\mathbf{C}^n$'s over a long period, from which the stationary route flow and cost distributions are calculated.

The learning process model employs simple weighted averages to update $\mathbf{C}^{n+1}$ using the ghosts experiences from the previous day, $\mathbf{G}^n$. This captures the effect of passengers choices on day n on the route costs based on which passengers' update their expected costs. For a particular route $r \in \mathbf{R}_h$, G_r^n is calculated as a simple average of the experiences of all ghosts that used the route, as shown in equation (8). Using a *"learning parameter"* (ϕ) as weight, $\mathbf{G}^n$ is used to update $\mathbf{C}^{n+1}$, for the next day, as shown in equation (7). Even though, past experiences will have increasingly lesser impacts as the process evolves, all previous experiences influence future decisions through a nested functional dependence.

$$\mathbf{C}^{n+1} = \phi . \mathbf{G}^n + (1-\phi) \cdot \mathbf{C}^n \qquad (7)$$

where, $0 < \phi < 1$. The value of learning parameter, ϕ, represents the frequency with which passengers revise, on average, their route cost perceptions.

$$G_r^n = \sum_1^{g_r^n} GC_r^n \Big/ g_r^n \qquad (8)$$

where, g_r^n = number of ghosts finishing their journey on route *r* on day *n*. Fig. 2 shows how the models described in this chapter are integrated to provide estimates of route flow proportions and costs considering the day-to-day variation in system costs. The learning process model updates $\mathbf{C}^{n+1}$ and outputs $\mathbf{G}^n$ and $\mathbf{P}^n$ at the end of each day. In this model, a day is represented by one simulation hour to correspond with the passenger demand and line frequency data.

Frequency-based and schedule-based analyses

The approach adopted in the present paper is a composite between the FB and SB approaches. On the one hand, it takes as input aggregate line frequencies, which are used to parameterise the distributions of vehicle arrivals. When demand is sufficiently low that the capacity constraints are not active, then this approach follows De Cea & Fernandez (1989) in terms of its computation of travel costs and assignment of passengers to routes, the differences being an algorithmic one in that a simulation method is used here, rather than an analytic formula. Even in the under-capacity case, the proposed method differs in the way that it assigns passengers to attractive lines. In this model, the assignment onto routes is integrated with the assignment onto lines based on a consistent definition of travel costs, whereas in the typical FB models the strategy/route-section flows are divided onto the attractive lines based on frequencies. This distinction becomes especially important as demand is increased, when capacity constraints become important. In this case, the proposed method deviates from the FB philosophy in its handling of the capacity constraints. In particular, the constraints are applied at the disaggregate level of particular (simulated) vehicles, and not at an aggregate level, with no modification to the approach required to handle the capacity constraints; in contrast, adaptations are needed in the traditional FB approaches to deal with capacity constraints, introducing assumptions (e.g. about 'effective frequencies') that may not be consistent with the underlying model of strategy costs and strategy/line choices; this is a point we shall return to in the numerical experiments chapter. In this respect, the proposed approach has more similarities with the SB approaches, even though the method fundamentally takes as input, aggregate frequencies to parameterise a distribution. In fact, if a detailed timetable were available, this approach could be readily adapted to incorporate such information, and in that case be a pure SB approach. However, we have applications in mind where such a timetable may not be available (e.g. longer term forecasting or services not operating to a timetable), and have favoured the basic premise of using frequencies as the input. This discussion means that the approach is somewhat difficult to classify as either FB or SB, in the conventional understanding of these terms.

The Markov property

Adapting the approach used in Cantarella & Cascetta (1995) for road traffic assignment, this chapter gives a theoretical discussion of the Markov property of the model described in the preceding section.

Let $V()$ be the experienced route cost filter, proxied by the ghosts as shown in equation (8) and $L()$ the pooled passengers' learning filter (equation 7). Also let $S()$ be the passengers' route choice filter given by equation (2). The filters are assumed to be independent of time and stable over the SP – i.e. the flows and costs may change, but not the way in which they relate.

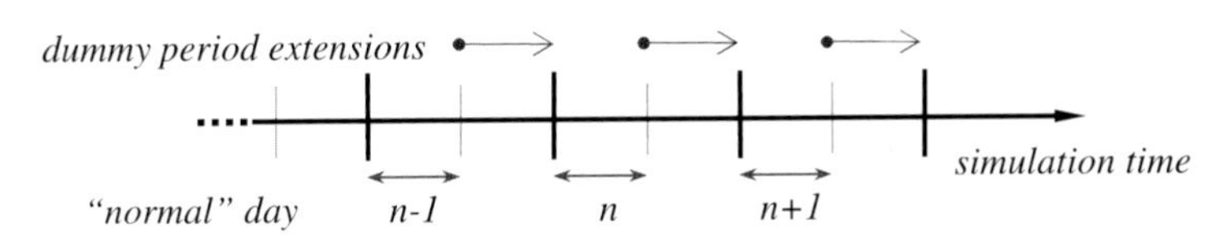

Figure 3 Dummy period extensions to allow ghosts finish their journeys

The state of the system at any day n is completely described by the average expected (i.e. forecasted) route costs vector, $\mathbf{C}^n$ – it drives passengers' route choice decisions and contains information on their past cost experiences. At the end of each day, it is calculated from the ghosts after the dummy period extensions are completed through the learning filter (equation 7). Dummy period extensions, Figure 3, are added at the end of each day to allow ghosts that started their journeys in the one-hour demand period, to reach their destination. This allows a full information on the days costs to be available in $\mathbf{G}^n$. Ghosts are not generated within the dummy period but passengers continue to be generated and travel the routes so that the ghosts en route experience the appropriate costs. Thus $\mathbf{C}^n$ has full cost information for the day. The distribution of the expected costs around the mean values is taken into account through random residuals in the route choice model, as shown in equation (6). The control parameters are the parameters in the simulation model – i.e. η and ϕ. The SP model is described by a recursive function (or set of functions) relating the state of the system on day n to that on day $n-1$ called the transition function, say $\mathbf{C}^n = \psi(\mathbf{C}^{n-1})$.

For each $h \in \mathbf{D}$, the number of passengers on the different routes on day n, $\mathbf{X}_h^n$, is modelled as a discrete random vector. It is assumed that each day n user, independently chooses a route $r \in R_h$ regardless of the routes chosen by other individuals on the previous day. This implies, conditional on the forecasted travel costs, the route flow vector can be seen as the sum of $|R_h|$ multinomial random vectors, one for each OD pair, $h \in \mathbf{D}$. The whole stochastic process can be specified as:

$$\mathbf{G}^{n-1} = V(\mathbf{X}^{n-1}) \qquad \text{[network loading]} \qquad (9)$$

$$\mathbf{C}^n = L(\mathbf{C}^{n-1}, \mathbf{G}^{n-1}) \qquad \text{[learning filter]} \qquad (10)$$

$$\mathbf{X}_h^n \mid \mathbf{C}_h^n \sim \mathit{Multinomial}(q_h, S(\mathbf{C}^n)) \qquad \text{[route choice model]} \qquad (11)$$

Once an initial state ($\mathbf{C}^0$) is given, this stochastic process can be easily solved by using the Monte Carlo simulation model described in the preceding chapters.

Let $Z \subset \mathbb{R}^N$ be the set of all feasible route costs, where $N = \sum_{h\in\mathbf{D}} |R_h|$. The learning filter is defined over the set $Z \times Z$. $\mathbf{C}^n \in Z$ describes the system state at day n.

The SP is a discrete-time continuous-space Markov process since the costs are updated every day (i.e. not continuously) and the expected route costs are continuous variables. Based on the probability distribution of the system state, it allows, at least theoretically, an investigation into the type of the convergence of the Markov process to be made – see for example, Cantarella & Cascetta (1995) and Stokey & Lucas (1989).

A stochastic process is called *stationary* if at least one stationary probability distribution exists. It is *ergodic* if stationary and exactly one stationary probability distribution exists, and it is *regular* if ergodic and its probability distribution converges to the one stationary probability distribution whatever the starting distribution (Meyn & Tweedie, 1993). As noted by Cantarella & Cascetta (1995), the regularity of the process ensures that one probability distribution of the system states can be associated to each system, independently of the starting configurations, and allows all statistical descriptions of the system state to be obtained from one pseudo realization of the process.

A necessary and sufficient condition for a Markov process to be regular is that the probability of a transition in a finite number of days, from any feasible state to any subset of the state space (or its complement) is greater than zero. For a given Markov process, let $\Pr^{\tau}(\mathbf{C}, E)$ be the probability of a transition in τ days from state $\mathbf{C} \in Z$ to the subset $E \subseteq Z$.

The Markov process is regular if and only if $\exists \varepsilon > 0, \tau \geq 1 : \forall E \subseteq Z$, $\Pr^{\tau}(\mathbf{C}, E) \geq \varepsilon, \forall \mathbf{C} \in Z$ or $\Pr^{\tau}(\mathbf{C}, Z - E) \geq \varepsilon, \forall \mathbf{C} \in Z$, where $Z - E$ is the complement of Z with respect to E.

Theorem: assuming each OD pair is connected by at least one route; the learning filter, L, is continuous (over Z x Z); and the cost filter, V, is continuous over the compact set of feasible route flows, then the resulting stochastic process is regular.

Proof: Each route flow is bounded below by zero, and above by the route capacity; the set of feasible route flows is therefore compact. The Probit-based stochastic route choice model ensures the transition probability to any nonempty set of path flow vectors from a given previous day state is positive. If B is a neighbourhood of $\mathbf{X}^1$ then

$$\forall \mathbf{C}^0 \in Z \Rightarrow \Pr(\mathbf{X}^1 \mid \mathbf{C}^0, \mathbf{X}^1 \in B) > 0 \quad \forall \mathbf{X}^1 \tag{12}$$

This is to say, since the Probit model assigns non-zero probability to any (set of) route flows regardless of the forecasted costs, it satisfies the state transition requirement above over just one day.

Given the previous day state, the corresponding cost (ghosts' experience) vector belongs to the set of feasible cost vectors:

$$\forall \mathbf{C}^0 \in Z \Rightarrow \mathbf{G}^1 = V(X(\mathbf{C}^0)) \in Z \tag{13}$$

By continuity of *V*, mapping route flows to route costs, and the continuity of the learning process mapping, *L*; the composite map $L\left(\mathbf{C}^0, V\left(\mathbf{X}^1\right)\right)$ from the compact space of feasible route flows to *Z* is continuous. Therefore, the transition probability from any initial cost vector $\mathbf{C}^0$ to *E* (or *Z-E*) is positively lower bounded by:

$$\varepsilon = \max\left\{\Pr\left[\mathbf{X}^1 \mid \mathbf{C}^0, \mathbf{X}^1 : L\left(\mathbf{C}^0, V\left(\mathbf{X}^1\right)\right) \in E\right], \Pr\left[\mathbf{X}^1 \mid \mathbf{C}^0, \mathbf{X}^1 : L\left(\mathbf{C}^0, V\left(\mathbf{X}^1\right)\right) \in Z - E\right]\right\} \tag{14}$$

Each term in the max operator corresponds to an integral of the Probit model; the first term is the total Probit probability of generating any route flows that map into E under the learning process, when starting from initial route costs $\mathbf{C}^0$.

In this chapter a composite FB and SB Markov process model for capacity constrained transit assignment is presented. The stochastic route choice model is based on an all-route-sections menu and accounts for cost correlations between alternative routes and route-sections. Although expensive in terms of the time it requires, the Monte Carlo simulation based approach provides a realistic estimate of user costs and flows considering the dynamic nature of demand-supply interaction. It is not a simple micro-simulation model either; it has some important mathematical properties as shown above, that provide the basis for developing analytical approximation to the model.

NUMERICAL EXPERIMENTS

In this chapter numerical experiments using the proposed model are presented for the test network shown in Figure 4a. The network has four stops that are served by two lines: RED and GRN with average frequencies of 8/hr and 10/hr, respectively. The 5 line sections, which are parts of the lines between adjacent stops, and their travel times are shown in Figure 4a. The buses have a maximum capacity of 20. Figure 4b shows all available route-sections that passengers could consider using between transfer stops. For the two OD pairs, S1-S4 and S2-S4, considered in this study, the average hourly demand is 100 and 200 passengers, respectively. It is worth noting that total demand does not exceed total capacity; the congestion impacts observed are due to capacity constraints on individual vehicle capacities and stochasticity in bus headways. Figure 4c lists all the sensible routes, defined as route-section sequences.

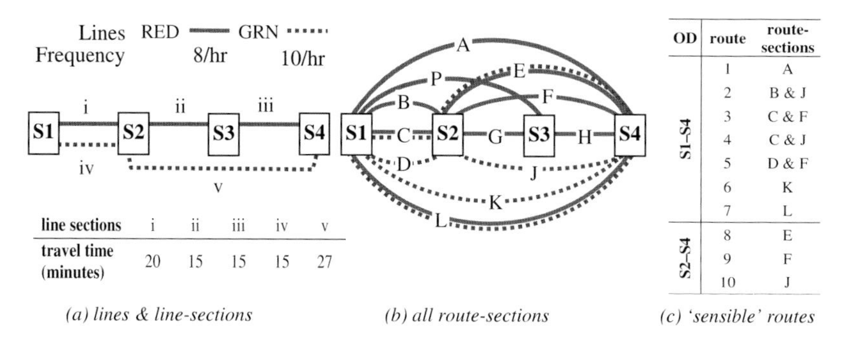

line sections	i	ii	iii	iv	v
travel time (minutes)	20	15	15	15	27

(a) lines & line-sections

(b) all route-sections

OD	route	route-sections
S1–S4	1	A
	2	B & J
	3	C & F
	4	C & J
	5	D & F
	6	K
	7	L
S2–S4	8	E
	9	F
	10	J

(c) 'sensible' routes

Figure 4 Test network

Exponential inter-arrival headway distributions are used for the buses and passengers. To maintain realism, the bus headways are truncated to a maximum of three times the average; for a line with an average frequency of 10/hr, the corresponding maximum headway is 18 minutes. Unless stated otherwise, the following parameter values are used for the different tests discussed hereafter: $\gamma_w = \gamma_t = 1$, $\phi = 0.1$, $\eta = 0.2$, and $\pi = 0$. The ghosts are generated every 3 minutes.

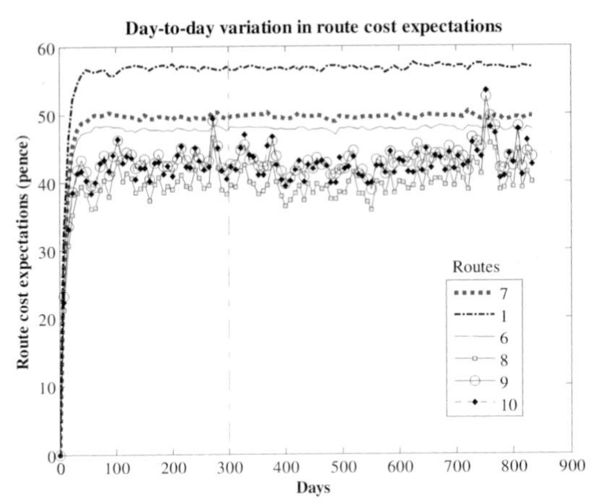

Figure 5 Evolution of the Markov process model

SIMTRANSIT is run for a total of 840 days; the day-to-day evolution of the system state variable, $\mathbf{C}^n$, is shown in Figure 5. Assuming the model has converged after a burn-in period of 300 days, the route cost expectations, costs experiences, and flow proportions afterwards are used to calculate the probability distributions of the same considering the day-to-day dynamics in the network. It is observed that passengers boarding at S1 only experience lower variability in their route cost expectations than those boarding at S2; this is because passengers from S1 only experience the stochasticity in the line headways while those boarding at the downstream stop, S2, experience the stochasticity in spare capacity, in addition.

Table 1 shows the forecasted route flow, expected cost and experienced cost (put simply as 'costs' in the tables) distributions and their sensitivity to initial conditions. It shows the mean values of these distributions and their standard deviations (in brackets) – this approach is adapted to represent such distributions in this paper. The difference in the three initial conditions is in the initial seed used and level of network congestion over the first 80 days. About 99% of S1 to S4 passengers use routes 6 and 7 – the latter (GRN line) is chosen by slightly more passengers. The remaining few use the more expensive route 1, perhaps not knowing of the cheaper routes that exist. About 68% of S2 to S4 passengers choose to board the first one of either the RED or GRN lines while the rest choose routes 9 and 10, and wait for only one of the lines. It is also observed that the Markov process model has converged to a stationary distribution regardless of the different initial conditions; the route cost expectations, the system state variable, have nearly identical distributions.

Table 1 Route costs and flow proportion variability* and sensitivity to initial conditions

Initial conditions	OD	S1-S4			S2-S4		
	Route	1	6	7	8	9	10
I	Costs - pence	56.77 (1.37)	47.81 (1.11)	49.57 (1.25)	39.85 (8.28)	43.23 (8.23)	42.80 (9.98)
	Flow proportions -%	1.0 (0.7)	64.6 (3.7)	34.4 (3.7)	68.8 (4.5)	14.0 (3.3)	17.2 (5.4)
	Cost expectations - pence	56.8 (0.3)	47.8 (0.2)	49.6 (0.3)	39.8 (2.3)	43.2 (2.3)	42.8 (2.4)
II	Costs - pence	56.85 (1.35)	47.78 (1.19)	49.49 (1.32)	39.86 (8.68)	43.20 (8.61)	42.71 (10.95)
	Flow proportions -%	0.8 (0.6)	64.5 (3.7)	34.6 (3.7)	67.9 (4.6)	14.3 (3.3)	17.9 (5.3)
	Cost expectations - pence	56.9 (0.3)	47.8 (0.3)	49.5 (0.4)	39.8 (3.1)	43.2 (3.1)	42.7 (3.4)
III	Costs - pence	56.88 (1.32)	47.88 (1.16)	49.62 (1.35)	39.74 (8.04)	43.06 (7.95)	42.57 (9.55)
	Flow proportions -%	0.9 (0.7)	64.8 (3.6)	34.3 (3.5)	68.0 (4.4)	14.4 (3.0)	17.6 (4.6)
	Cost expectations - pence	56.9 (0.3)	47.9 (0.2)	49.6 (0.3)	39.7 (2.0)	43.0 (2.0)	42.5 (2.2)

* The standard deviations are given in brackets, under the mean flow proportions and costs.

Sensitivity to the Probit variability parameter

To test the sensitivity of the route flows and costs to η, the model is run with $\eta = 0.05$ and $\eta = 0.5$ (Table 2); $\eta = 0.2$ for the ones in Table 1. It is observed that the passengers are distributed on more routes with increasing values of η: albeit small, the more expensive routes 3, 4, and 5 carry some flow for $\eta = 0.5$.

Table 2 Sensitivity of route flows and costs to the Probit variability parameter, η

η	OD	S1-S4						S2-S4		
	Route	1	3	4	5	6	7	8	9	10
0.05	Cost - pence	56.83 (1.33)	65.01 (8.30)	63.52 (10.06)	64.15 (8.66)	48.01 (1.32)	49.75 (1.45)	39.78 (8.35)	43.07 (8.39)	42.60 (9.72)
	Flow proportion - %	0.0 (0.0)	0.0 (0.0)	0.0 (0.0)	0.0 (0.0)	78.4 (3.6)	21.6 (3.7)	89.7 (4.2)	3.4 (1.6)	6.9 (4.5)
0.5	Cost – pence	56.83 (1.33)	64.79 (8.26)	63.44 (10.16)	63.95 (8.61)	47.82 (1.11)	49.59 (1.36)	39.55 (8.26)	43.07 (8.36)	42.47 (9.76)
	Flow proportion - %	5.0 (1.5)	0.2 (0.3)	0.4 (0.6)	0.4 (0.5)	56.4 (3.4)	37.5 (3.4)	56.5 (3.8)	19.9 (3.1)	23.6 (4.7)

Sensitivity to the learning parameter

To test the sensitivity of the model to the learning parameter, ϕ, the model was run with parameter values of 0.02, 0.1 and 0.2. Figure 6 shows the evolution of cost expectations for routes 6 and 8 – the two routes are typical of the uncongested and congested routes in the network, respectively. It is observed that the process reaches the "true" cost of the routes relatively early when using high values of ϕ; relatively longer burn-in periods might be required when using low values of ϕ. It is also observed that higher values of ϕ are associated with higher variability in the evolution of the system and the route cost expectations. This variability is pronounced in congested cases (for e.g. route 8 with $\phi = 0.2$) and may lead to unstable evolution of the system in some cases. There is not much effect on the distribution of flows and costs due to the different values of ϕ.

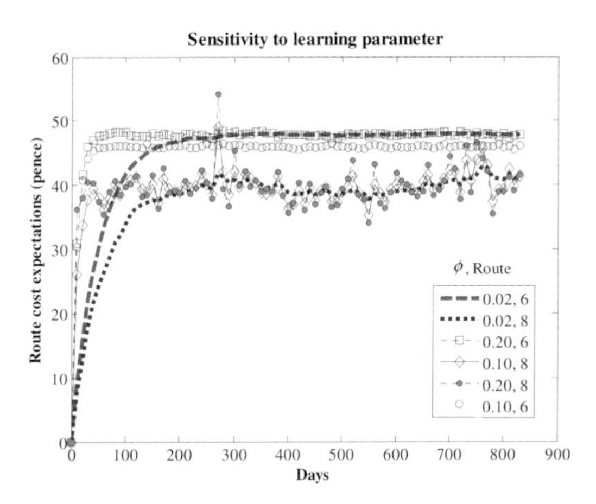

Figure 6 Sensitivity of the Markov process evolution to the learning parameter

Sensitivity to bus headway distributions

To test the effect of the stochasticity in bus headway distributions on the evolution of the system, the model was run assuming the buses travel at constant headways. It is observed that the evolution of the system (Figure 7) is less variable than when the bus headways are exponentially distributed (Figure 5). The route flow and cost distributions are given in Table 3. Especially when considering the congested routes (8, 9, & 10) the route costs are significantly smaller and less variable than when the bus headways are exponentially distributed.

Table 3 Route flow and cost distributions with constant bus headways

OD	S1-S4					S2-S4		
Routes	1	4	5	6	7	8	9	10
Cost – pence	54.67 (0.42)	53.63 (1.27)	53.68 (0.97)	45.88 (0.85)	48.34 (1.59)	32.83 (0.66)	34.68 (0.36)	32.85 (1.13)
Flow proportion - %	1.0 (0.7)	1.8 (0.9)	2.0 (1.0)	68.0 (3.4)	26.9 (3.3)	41.8 (2.6)	17.1 (1.9)	41.2 (2.6)

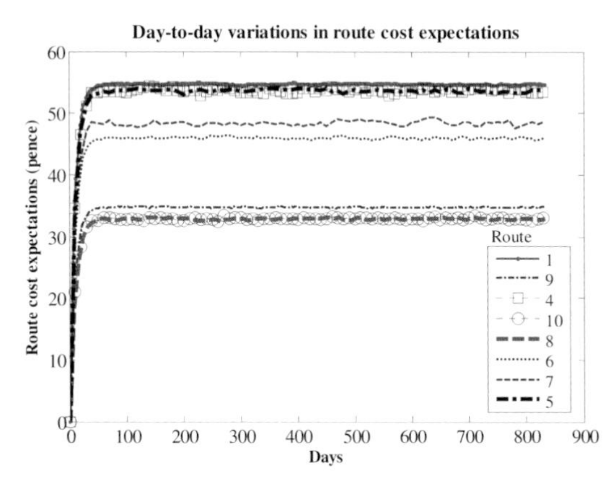

Figure 7 Evolution of the system with constant bus headways

Model Comparison

SIMTRANSIT is also compared with a Probit SUE version of the model presented in Cepeda et al. (2006) that uses the effective frequency approach – referred to as SUE_EF. Although they use the more general hyperpath based network representation (Nguyen & Pallottino, 1988) in that paper, for the test network considered (Figure 4) all sensible hyperpaths that passengers could be on are represented. This is because the attractive lines passengers might consider from the different stops, share the exit stops as well.

$$ef_a = f_a \cdot \left(1 - b_a^{\beta}\right) \tag{15}$$

$$w_k = \frac{60}{\sum_{a \in A_k} ef_a} = \frac{60}{\sum_{a \in A_k} f_a \cdot \left(1 - b_a^{\beta}\right)} \tag{16}$$

where, b_a = *number boarding* / (*capacity – line section volume + number boarding*) for line-section a; ef_a & f_a are effective and nominal line section frequencies; and w_k is the route-section waiting time.

Before making the model comparisons with SUE_EF, it is necessary to calibrate the effective frequency functions and obtain the value of the parameter β for the line sections, see equation (15). That is done by using the waiting times for single line route-sections such as A and J (see Figure 4b) which are defined as inverses of the effective frequencies as shown in equation (16). Running SIMTRANSIT with different demand levels, the number of passengers that boarded each line section, the total line section flow (including passengers that stayed on from the preceding section of the same line), and the waiting times at the end of each day are recorded. As could be observed in Figure 8a, a value of $\beta=0.2$ used in Cepeda et al. (2006) overestimates the waiting times across the different values of b. A regression curve is fitted using the sequential quadratic programming algorithm in the SPSS software to better represent the relationship between w, the waiting time, and b. Table 4 gives the values of β along with their associated standard error. It is observed that the value β decreases downstream indicating the higher waiting times experienced by passengers due to capacity constraints. The waiting times for multi-line route-sections (e.g. E) are defined as the inverse of the sum of the constituent lines' effective frequencies, equation (16). To check whether the values of β, calibrated using single line route-sections, also represent waiting times of multi-line route sections a 3D waiting time surface for route-section E is plotted against b_{ii} and b_v, corresponding to line-sections *ii* and *v* (Figure 8b). As could be seen in the figure, the surface compares well with the waiting times obtained from SIMTRANSIT for the different values b_{ii} and b_v.

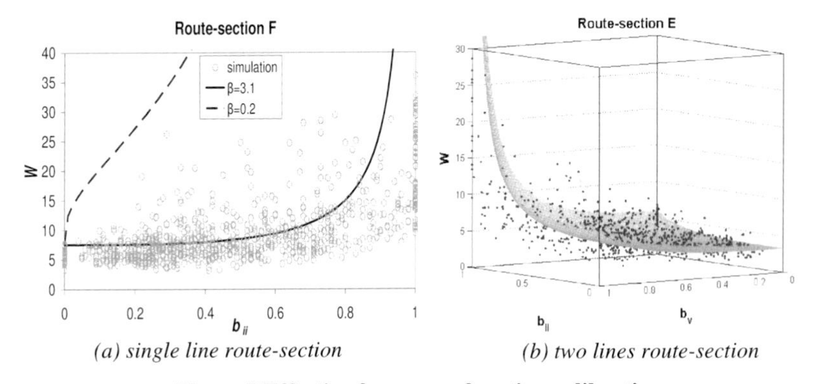

(a) single line route-section *(b) two lines route-section*

Figure 8 Effective frequency function calibration

Table 4 Calibrated values of β

Line section	i	ii	iii	iv	v
β (standard error)	5.669 (0.211)	3.095 (0.078)	1.739 (0.100)	4.261 (0.104)	2.538 (0.081)

After obtaining the values of β for the different route-sections, a Probit SUE version of the deterministic route choice model presented in Cepeda et al. (2006), presented below, is used.

i. *Initialization*: line-section flows vector $\boldsymbol{\mu}^0$, set counter h=0
ii. compute line-section effective frequencies, ef_a
iii. sample the perceived line-section travel times vector **t** assuming $t_a \sim Normal(t'_a, \eta \cdot t'_a)$, where t'_a is the mean travel time for line section a
iv. compute shortest hyperpaths to each destination
v. assign total demand onto minimum cost hyperpaths
vi. determine induced line-section flows ($\boldsymbol{\mu}'$) using the effective line-section frequencies as weights
vii. update $\boldsymbol{\mu}^h$ using method of successive averages
viii. if $h <$ maximum number of iterations, $h=h+1$, go to (ii)

After initialization, the method computes transit network equilibrium for the network at each iteration, obtained by fixing the effective frequencies at the values determined by current flows. In step (iv) total travel costs are calculated from the destinations to upstream stops, identifying the minimum perceived cost hyperpaths from each upstream stop. Waiting times are calculated using equation (16). At each stop, newly arriving passengers as well as those continuing their journeys from upstream line sections are aggregated together assigned to the identified minimum cost strategies from the origins to the destinations (step v). Following Sheffi and Powell (1981), a single sample of perceived line-section costs is used in step (iii), and step (v) is an all-or-nothing assignment. The model was run for 25000 iterations to ensure the convergence of the Monte Carlo simulation based Probit model. Check for convergence is done using different initial random number seeds.

Table 5 Route costs from SUE_EF

	S1-S4			S2-S4		
Route	1	6	7	8	9	10
Cost	57.5	48.0	48.9	34.8	44.3	38.8

Table 6 Line-section flow comparisons (passengers/hr)

Line-section	i	ii	iii	iv	v
SIMTRANSIT	17.09	131.61	131.61	86.47	178.69
SUE_EF	44.53	135.38	135.38	55.47	164.62

Table 5 shows the costs obtained from SUE_EF; these are different from the ones from SIMTRANSIT (Table 1) – the magnitude of these differences is larger in the congested parts of the network (routes 8, 9, and 10). This is despite the effective frequency functions being calibrated using SIMTRANSIT to account for the capacity constraints of individual vehicles.

The main reasons for the difference are the inability of the calibrated curves, which are applied on aggregate capacity, to represent the waiting times accurately at the different values of *b*, and the structural differences (e.g. static vs. dynamic model and link based costing vs. route based costing used in SUE_EF and SIMTRANSIT, respectively) in the two approaches. The line flows obtained from SIMTRANSIT and SUE_EF – the former of which is calculated from the average loading on each bus – are also different (Table 6) due the errors in the costs and the abovementioned issue with the calibrated curves.

CONCLUDING REMARKS

A composite frequency based and schedule-based, Markov process model for capacity constrained transit assignment is presented in this paper. It accounts for the day-to-day dynamics of a transit network and outputs a stationary probability distribution of route flows and costs. Line frequencies are used to model transit supply. Passenger flows are constrained to individual vehicle capacities using a micro-simulation model. Based on an all route-sections choice menu, a Probit based stochastic route choice model that considers the cost correlations between alternative routes is used to model passengers route choice. A proof of the Markov property of the stochastic process model and its regularity are given; these guarantee the model's convergence to a stationary equilibrium distribution regardless of the initial conditions. This was confirmed when the model was tested using three different initial conditions using a small test network.

Using the small test network, it was observed that passengers in the congested section experience highly variable costs, especially when the bus headways are random. This is because of the stochasticity associated in finding spare seats. The evolution of the Markov process is sensitive to the value of the learning parameter used, especially in congested networks. higher values of the Probit dispersion parameter seem to distribute passengers on more routes. In congested sections of the network, passengers were observed to incur higher costs and higher variability in their experiences due to higher stochasticity in the bus headways.

The model was also used to calibrate the effective frequency function proposed by Cepeda et al. (2006). The calibration showed the difficulty in consistently representing the impacts of capacity constraints across for different levels of congestion and across the network. The two models forecasted different route costs and line flows as well. Besides systematic differences between the two approaches, the main reason for the difference is the inability of the calibrated functions to represent, consistently, capacity constraints at the level of individual vehicles. More tests, with different types and sizes of network, are needed to confirm this.

Further research needs to look at convergence properties and stability of the resulting equilibrium distribution of the Markov process, particularly in congested conditions – perhaps for different route choice and learning process models. To enable the application of the model on networks of realistic size, a path selecting algorithm is also needed. The proposed, simulation-based, model is time consuming in nature; studies on its analytic approximation would help solutions to be obtained faster.

ACKNOWLEDGEMENTS

We would like to thank Universities UK and the Institute for Transport Studies, University of Leeds, for financially supporting Fitsum Teklu's PhD study. We would also like to thank three anonymous referees for their comments and suggestions.

REFERENCE

Cantarella, G.E. and E. Cascetta (1995). Dynamic processes and equilibrium in transportation networks: Towards a unifying theory. *Transportation Science* **29**, 305-29.

Cepeda, M., R. Cominetti and M. Florian (2006). A frequency based assignment model for congested transit networks with strict capacity constraints: Characterization and computation of equilibria. *Transportation Research Part B*, **40**, 437-59.

Cominetti, R. and J. Correa (2001). Common lines and passenger assignment in congested transit networks. *Transportation Science*, **35**, 250-67.

Davis, G.A. and N.L. Nihan (1993). Large population approximations of a general stochastic traffic assignment model. *Operations Research*, **41**, 169-78.

De Cea, J. and E. Fernandez (1989). Transit assignment to minimal routes: An efficient new algorithm. *Traffic Engineering & Control* **30**, 491-94.

De Cea, J. and E. Fernandez (1993). Transit assignment for congestion public transport networks: An equilibrium model. *Transportation Science*, **27**, 133-47.

Hamdouch, Y., P. Marcotte and S. Nguyen (2004). Capacitated transit assignment with loading priorities. *Mathematical Programming B*, **101** 205-30.

Kurauchi, F., M.G.H. Bell and J.-D. Schmöcker (2003). Capacity constrained transit assignment with common lines. *Journal of Mathematical Modelling and Algorithms* **2**, 309-27.

Meyn, S.P. and R.L. Tweedie (1993) *Markov chains and stochastic stability*. Springer-Verlang, London.

Nguyen, S. and S. Pallottino (1988). Equilibrium traffic assignment for large scale transit networks. *European Journal of Operational Research* **37**, 176-86.

Nuzzolo, A., F. Russo and U. Crisalli (2001). A doubly dynamic schedule-based assignment model for transit networks. *Transportation Science*, **35**, 268-85.

Poon, M.H., C.O. Tong and K.I. Wong (2003). Validation of a schedule based capcity restraint transit assignment model for a large scale network. *Journal of Advanced Transportation*, **38**, 5-26.

Sheffi, Y. and W.B. Powell (1981). A comparison of stochastic and deterministic traffic assignment over congested networks. *Transportation Research Part B*, **15**, 53-64.
Spiess, H. and M. Florian (1989). Optimal strategies: A new assignment model for transit networks. *Transportation Research Part B*, **23**, 83-102.
Stokey, N.L. and R.E. Lucas (1989) *Recursive methods in economic dynamics.* Harvard University Press, Cambridge, Mass.
Watling, D. (1996). Asymmetric problems and stochastic process models of traffic assignment. *Transportation Research Part B*, **30**, 339-57.
Wu, J.H., M. Florian and P. Marcotte (1994). Transit equilibrium assignment: A model and solution algorithms. *Transportation Science*, **28**, 193-203.

Transportation and Traffic Theory 2007
Edited by R.E. Allsop, M.G.H. Bell and B.G. Heydecker

22

A NOVEL FUZZY LOGIC CONTROLLER FOR TRANSIT SIGNAL PREEMPTION

Yu-Chiun Chiou, Institute of Traffic and Transportation, National Chiao Tung University
Ming-Te Wang, Institute of Transportation, Ministry of Transportation and Communications
Lawrence W Lan, Institute of Traffic and Transportation, National Chiao Tung University
TAIWAN

SUMMARY

This paper develops a novel iterative evolving fuzzy logic controller, called ant-genetic fuzzy logic controller (AGFLC), for transit signal preemption (TSP) wherein variation of traffic conditions and ambiguity of expert judgment are accounted for. The core logics of this iterative evolving AGFLC algorithm include learning the combination of rules by ant colony optimization and tuning the shapes of membership functions by genetic algorithm. We propose an AGFLC-based TSP model that provides conditional signal priority to the actuated transit vehicles to minimize the total person delays of the intersections studied. To realize the control performance, both exemplified and field cases are tested at an isolated intersection and consecutive intersections along an arterial. Compared with other models, including genetic fuzzy logic controller (GFLC)-based TSP model, net-benefit conditional TSP model, unconditional TSP model, and pre-timed signal without TSP, the results show that the proposed AGFLC-based TSP model has outperformed under different circumstances.

INTRODUCTION

Transit signal preemption (TSP) gives preferential treatment to transit vehicles, such as trams and buses, passing through signalized intersections on the surface roads. It has become prevalent in many metropolitan cities around the world since its first field implementation in 1968. The major benefits of TSP include reduction of transit operating costs and emissions and increase of transit operating efficiency and schedule reliability. However, TSP can also cause remarkable delays to the competing traffic. Without detailed timing design and careful evaluation, the benefits of TSP may be largely offset by its negative impacts. Although numerous TSP analytical models have been proposed and/or implemented (see, for example, Jacobson and Sheffi, 1981; Khasnabis *et al.*, 1993; Sunkari *et al.*, 1995; Cisco and Khasnabis, 1995; Chang *et al.*, 1996; Wu and Hounsell, 1998; Dion *et al.*, 2004), very few have been devoted to the optimal design of its signal timing due to the difficulty of developing an exact mathematical model. Whether or not to give TSP is a complex decision problem that may be viewed as a knowledge representation of uncertainty and imprecision. As the traffic compositions and volumes from different approaches are changing over time and the occupancies of different vehicle types are also varying, it is difficult to develop an exact mathematical model to determine the optimal timing for TSP. Fuzzy logic, first introduced by Zadeh (1973), has been effectively used in various fields in dealing with the knowledge representation problems, especially in a context of uncertainty and imprecision. One promising application of fuzzy logic in TSP timing design is perhaps the development of an adaptive fuzzy logic controller (FLC) that can make better decision in determining the signal changes in response to variation of traffic conditions. The underlying logic for an FLC-based TSP model is to use fuzzy logic rules to form a preferential signal control mechanism to approximate better judgment under given conditions.

A typical FLC system contains four major components: rule base, data base, inference engine, and defuzzification. The rule base is composed of finite IF-THEN rules from which an inference mechanism is formed. The data base is formed by the specific membership functions of linguistic variables that transform crisp inputs into fuzzy ones. The inference engine is formed by the operators within the logic rules. Defuzzification is for making decisions, the synthesis of inference results of all activated logic rules into crisp outputs. The methods used in inference engine and defuzzification are rather consistent in previous literature; however, the methods for formulating the rule base and data base are still too subjective in previous works. It was argued that without appropriately setting the rule and data bases, the performance of an FLC system can be greatly reduced, and its applicability can be limited. Therefore, the task of automatically defining the fuzzy rules and membership functions for a concrete application is considered a challenging issue, and a large number of new methods have been proposed, such as ad hoc data-driven methods (Bárdossy and Duckstein, 1995), neural networks (Gupta and Gorsalcany, 1992; Esobgue and Murrell, 1993; Nauck and Kruse, 1993; Du and Wolfe, 1995; Nauck *et al.*, 1997), fuzzy clustering (Babuška, 1998), genetic algorithms (Wang and Mendel, 1992; Linkens and Nie, 1993; Bonissone *et al.*, 1996; Hwang, 1998; Cordón *et al.*, 2001; Chiou *et al.*, 2003, 2005), and ant colony optimization (Casillas *et al.*, 2000, 2005; Parpinelli *et al.*, 2002).

Chiou *et al.* (2003, 2005) proposed genetic fuzzy logic controller (GFLC)-based TSP models to enhance the control performance and self-learning capability of an FLC. The core logic of a GFLC-based TSP model is to select the logic rules and tune the membership function by genetic algorithms (GAs) sequentially and iteratively so as to minimize the total person-delays at an isolated intersection and consecutive intersections along an arterial. In view of the potential superiority of ant colony optimization (ACO) technique in solving a combinatorial optimization problem, this paper attempts to further develop an ant-genetic fuzzy logic controller (AGFLC)-based TSP model, which employs ACO to select the combination of logic rules and then uses GAs to tune the membership functions. To test the performance of our proposed AGFLC-based TSP model, exemplified examples and field cases are conducted at an isolated intersection and consecutive intersections along an arterial. The control performances for five strategies -- pre-timed signal without TSP, unconditional TSP, net-benefit (NB) conditional TSP, GFLC-based conditional TSP, and AGFLC-based conditional TSP, are compared.

The rest of this paper is organized as follows. Section 2 describes the rationales for TSP and FLC. Section 3 formulates the AGFLC-based TSP model. Section 4 validates the effectiveness of the proposed model at an isolated intersection. Section 5 further evaluates the performance of the proposed model at consecutive intersections along an arterial. Finally, the concluding remarks and suggestions for future studies are addressed.

RATIONALE FOR TRANSIT SIGNAL PREEMPTION

A variety of signal priority strategies, including passive priority, green extension, red truncation (early green), actuated transit phase, phase insertion, phase rotation, and adaptive/real-time control, have been proposed or implemented while designing a TSP system (ITS America, 2004). This paper focuses only on the most commonly used ones -- green extension and red truncation strategies. The rationales for implementing green extension and red truncation strategies under unconditional and conditional TSP respectively are briefed below.

Figure 1 depicts the control logic for unconditional TSP. In the green phase, **IF** $GR < H$, **THEN** implement green extension strategy and let $G_{ext} = H - GR$, where GR represents the remaining green time at the moment when a transit vehicle actuates the detector. H represents the time needed for a transit vehicle travelling from the detector to the far-side stop line of the intersection. G_{ext} represents the green extension time. In the red phase, **IF** $(RR + AR) > L$, **THEN** implement red truncation strategy and let $R_{tru} = RR + AR - L$, where RR represents the remaining red time when a transit actuates the detector; and AR represents the all-red time. L represents the time needed for a transit vehicle travelling from the detector to the near-side stop line of the intersection. R_{tru} represents the red truncation time.

To avoid a serious distortion of original signal timing plans, these two strategies are implemented under the following conditions: (1) If the phase comes to a transition period, such as all-red, it will not activate any strategy. (2) The total green extension time should not exceed the maximal green time (G_{max}). (3) The red time after truncation should not be less than the minimal green time (G_{min}). (4) No compensation mechanism is provided. All parameters including H, L, AR, G_{max}, and G_{min} are given.

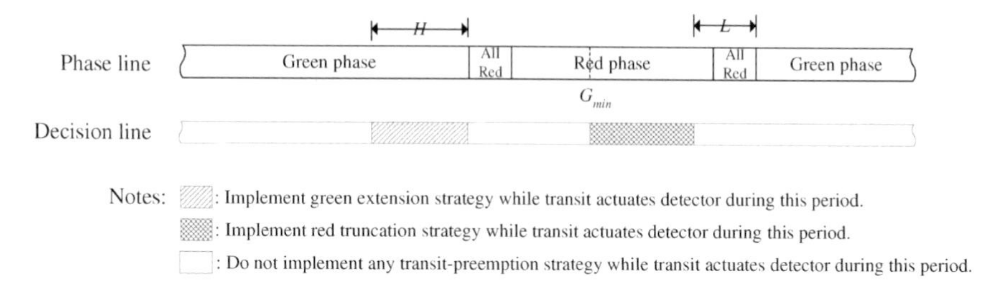

Figure 1. Unconditional TSP control logics.

In contrast to the above unconditional TSP control logics, our proposed AGFLC-based TSP model will provide conditional priority to the actuated transit vehicles. It concludes a decision by considering traffic situations in all approaches to minimize the total person delays of the intersection(s). Our proposed AGFLC-based conditional TSP control logics are depicted in Figure 2. In the green phase, **IF** $GR < H$ **AND** $NE \geq N_t$, **THEN** implement the green extension strategy and let $G_{ext} = H - GR$, where NE represents the degree of necessity to implement TSP, which is concluded by the AGFLC with a value ranging from 0 to 1. N_t represents the threshold value preset to determine whether the priority is provided or not. In the red phase, **IF** $(RR + AR) > L$ **AND** $NE \geq N_t$, **THEN** implement the red truncation strategy and let $R_{tru} = RR + AR - L$.

Note that the AGFLC inference in Figure 2 forms a control mechanism to approximate expert judgment under given information. Its rule base is composed of finite IF-THEN rules with state and control variables. In the present paper, we use total traffic flows (*TF*) at all approaches in the green phase and total queue length (*QL*) at all approaches in the red phase as the state variables and the degree of necessity for implementing TSP (*NE*) as the control variable to form the AGFLC. All of these variables are assumed with five linguistic degrees (*NL*: negative large, *NS*: negative small, *ZE*: zero, *PS*: positive small, *PL*: positive large) and are represented by triangular membership functions. This makes a total of 25 combinations in the antecedent part of the logic rule base. Moreover, the logic rules use AND as the connecting operator between the state variables. As such, the rule base is illustrated below:

Rule 1: IF $TF = NL$ AND $QL = NL$ THEN $NE = B_1$
Rule 2: IF $TF = NL$ AND $QL = NS$ THEN $NE = B_2$
Rule 3: IF $TF = NL$ AND $QL = ZE$ THEN $NE = B_3$

.

.

.

Rule 25: IF $TF = PL$ AND $QL = PL$ THEN $NE = B_{25}$
Where, $B_i \in \{NL, NS, ZE, PS, PL\}$, $i = 1 \sim 25$.

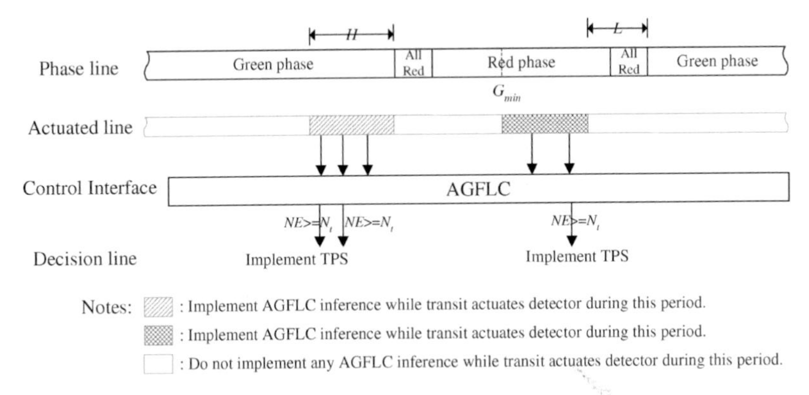

Figure 2. AGFLC-based conditional TSP control logics.

The implementation of the proposed AGFLC requires some real-time traffic information: arrival of transit, traffic flow in the green-phase direction, and queue length in the red-phase direction. Thus, the transit vehicles may require equipped with positioning devices, such as global positioning system (GPS), to provide the information on arrival. On the other hand, in order to collect the information on traffic flow and queue length of other vehicles, two sets of detectors, acting as check-in and check-out points, are also required on all lanes in all approaches. The former detector set can be located near the stop line of the intersection to count the number of departing vehicles; whereas the later detector set can be at a certain distant point from the stop line to count the number of arriving traffic. The queue length is thus determined by the difference between these two counting results. In practice, the distance between these two sets of detectors requires a proper design to accommodate the possible maximum queue length. With the emerging innovative detection and communication technologies, numerous advanced detectors are introduced that could facilitate the implementation of more sophisticated TSP systems.

THE ITERATIVE EVOLVING ALGORITHM

Our proposed iterative evolving AGFLC algorithm employs ant colony optimization (ACO) to select the logic rules and then utilizes genetic algorithm (GA) to tune the membership functions in an iterative manner. The steps are detailed below.

Step 0: Initialization. Set the values of all parameters and let evolution epoch $V = 1$.

Step 1: Rules selection. Select logic rules by ACO.

Step 1-1: Network formation. In order to adapt ACO to a rules selection problem, we reformulate the problem into a clustering problem, which divides the potential antecedents of logic rules into clusters that represent the consequents of logic rules. The object to be clustered is denoted as AR_i, i=1,..., I. I represents the total number of antecedents. Table 1 shows an example of two state variables, x_1 and x_2, each assumed with five linguistic degrees; thus, there are in total 25 potential antecedents.

Table 1 Antecedents of two state variables with five linguistic degrees

x_2	x_1				
	NL	*NS*	*ZE*	*PS*	*PL*
NL	AR_1	AR_6	AR_{11}	AR_{16}	AR_{21}
NS	AR_2	AR_7	AR_{12}	AR_{17}	AR_{22}
ZE	AR_3	AR_8	AR_{13}	AR_{18}	AR_{23}
PS	AR_4	AR_9	AR_{14}	AR_{19}	AR_{24}
PL	AR_5	AR_{10}	AR_{15}	AR_{20}	AR_{25}

Note: *NL*: negative large, *NS*: negative small, *ZE*: zero, *PS*: positive small, *PL*: positive large.

The potential antecedent AR_i will be linked to any one of the possible consequents, denoted C_j, j=1, 2,..., J. J is the number of linguistic degree of control variable. To exclude badly defined or conflicted rules, the antecedent AR_i could be possibly assigned to a cluster set, called exclusion set (C_{J+1}). Taking two state variables and one control variable as an example, if each variable has five linguistic degrees, a total of 25 objects (potential antecedents) can be grouped into six clusters, where C_j, j=1, 2,..., 5, stand for the consequents of y=*NL*, y=*NS*, y=*ZE*, y=*PS*, y=*PL*, respectively, and C_6 represents the exclusion set. All objects are fully connected to these six clusters as depicted in Figure 3.

Step 1-2: Pheromone initialization. For a minimization problem, the initial pheromone value (τ^0) can be set as the reciprocal of an objective function (E) of any initial solution (namely the predetermined rule base). For a rules learning problem with an input-output training dataset, the objective function could be to minimize the error between the observed outputs and the outputs concluded by the AGFLC. For a rules learning problem without the training dataset, the objective function could be defined as the performance index of the control system.

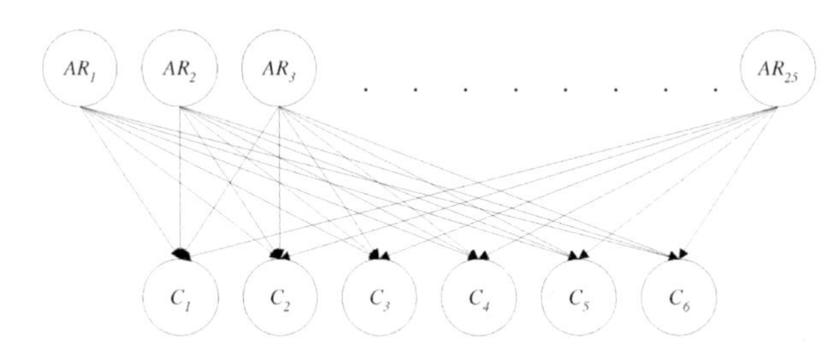

Figure 3. Clustering network of two state variables and one control variable (each with five linguistic degrees).

Step 1-3: Tour construction. To construct a complete solution, each ant successively visits each potential antecedent and chooses a consequent with transition probability depending on the heuristic information and pheromone level. Initially, each ant is put on one antecedent. That is, the number of ants is equal to the number of antecedents. After assigning an antecedent to a consequent for each ant, the ants move to the next antecedent in parallel. Thus, the choice of each ant would be affected by the previous tour construction step because of the local pheromone update rule.

In this paper, the reasonability of linking one antecedent to a consequent is taken as the heuristic information for tour construction. The reasonability information (θ_{ij}) is defined as the degree of similarity of assigning the result of a specific antecedent to a consequent with a predetermined assigning result. Thus, to obtain the reasonability information, a predetermined rule table must be established in advance. A higher information value shows that the selection result is more similar to the predetermined rules. For instance, assuming that AR_{21} (*i.e.*, IF x_1=*PL* and x_2=*NL*) is connected to cluster C_5 (y=*PL*) in the predetermined rule table, the reasonability information on this arc would have the highest value, followed by the arc connecting to C_4 (y=*PS*), C_3 (y=*ZE*), C_2 (y=*NS*), while the arc connecting to cluster C_1 (y=*NL*) would have the least value. In this paper, the value of reasonability information on the arc connecting to cluster C_6 is preset, and this value will serve as the threshold to exclude the rules. Without loss of generality, we assume that the maximum value of reasonability information equals to 1 and that the value is decreased by $1/J$ (J stands for the number of linguistic degrees of control variable) for each additional linguistic degree gap. If an antecedent AR_i is assigned to consequent C_j, but AR_i is connected to consequent C_p in a predetermined rule table, then the reasonability information can be expressed as

$$\theta_{ij} = 1 - \frac{|j-p|}{J} \tag{1}$$

where θ_{ij} represents reasonability information value on the arc connecting AR_i and C_j. Take J=5, C_j= C_2 (y=*NS*) and C_p= C_4 (y=*PS*); for instance, $\theta_{ij} = 1 - \frac{|2-4|}{5} = 0.6$.

After defining the reasonability information, ant k in antecedent r choosing consequent s can be determined by the following equations:

$$s = \arg\max_{j=1}^{J+1}\{[\theta_{rj}]^{\alpha}[\tau_{rj}]^{\beta}\},\ \text{if } q \le q_0 \text{ (exploitation)}, \tag{2}$$

or visit s with P_{rs}^k, if $q > q_0$ (exploration), where

$$P_{rs}^k = \frac{[\theta_{rs}]^{\alpha}[\tau_{rs}]^{\beta}}{\sum_{j=1}^{J+1}[\theta_{rj}]^{\alpha}[\tau_{rj}]^{\beta}} \tag{3}$$

where τ_{rj} is the amount of pheromone on the arc connecting AR_r and C_j. The symbols α and β are parameters that determine the relative importance of reasonability and pheromone, and P_{rs}^k is the probability of ant k assigning the antecedent r to consequent s.

Step 1-4: Pheromone updates. In this proposed AGFLC model, the pheromone levels on arcs are updated both locally and globally. The local pheromone update rule is applied immediately after one ant has crossed an arc (i, j) during the tour construction. It can be represented by

$$\tau_{ij} \leftarrow (1-\xi)\tau_{ij} + \xi\tau^0 \tag{4}$$

where $\xi \in (0,1)$ is a parameter making the pheromone not going too far beyond τ^0. After all ants have completed their tours, the global updating rule is to deposit a certain amount of pheromone ($\Delta\tau_{ij}$) on the arcs belonging to the best-so-far tour ($T^*(t)$) constructed by the best-so-far performed ant. The pheromone level of the tth iteration is updated by

$$\tau_{ij}(t+1) \leftarrow (1-\rho)\tau_{ij}(t) + \Delta\tau_{ij}(t) \quad if\ arc(i,j) \in T^*(t) \tag{5}$$

where $\tau_{ij}(t)$ and $\tau_{ij}(t+1)$ are the pheromone levels of the incumbent iteration and next iteration on the arc (i, j), respectively. $\Delta\tau_{ij}(t) = 1/E^*(t)$, where $E^*(t)$ is the objective function of $T^*(t)$. Finally, $\rho \in (0,1]$ is a parameter governing the evaporation of the pheromone trail.

Step 1-5: Incumbent tour updating. After an iteration (global updating) has been completed, the incumbent optimal solution is tested and updated as follows: If $\min_k\{E_k(t)\} = E^+(t) < E^*$, then let $E^* = E^+(t)$ and $T^* = T^+(t)$; otherwise, E^* and T^* remain unchanged, where $E_k(t)$ is the value of the objective function of ant k of iteration t; $E^+(t)$ is the value of objective function of the best tour $T^+(t)$ of iteration t; T^* is the best-so far tour, and E^* is its objective function.

Step 1-6: Testing of the stop condition. If the maximal iterations t_{max} has been reached, proceed to Step 2. Otherwise, go back to Step 1-3.

Step 2: Tuning membership function by GAs. This paper tunes the membership functions with the method proposed by Chiou and Lan (2005). It is briefly narrated as follows.

Step 2-1: Encoding membership function. Assume that the first and last degrees of fuzzy numbers are left- and right-skewed triangles, respectively, and that the others are isosceles triangles as shown in Figure 4. Therefore, a variable with five linguistic degrees has eight parameters to be calibrated, and their orders are:

$$c_{max} = c_5^c = c_5^r \geq c_4^r \geq \begin{matrix} c_5^l \\ c_3^r \end{matrix} \geq \begin{matrix} c_4^l \\ c_2^r \end{matrix} \geq \begin{matrix} c_3^l \\ c_1^r \end{matrix} \geq c_2^l \geq c_1^c = c_1^l = c_{min} \tag{6}$$

$$c_k^c = \frac{(c_k^r + c_k^l)}{2},\ k=2, 3, 4 \tag{7}$$

where c_{max} and c_{min} are the maximum and minimum values of the variable, respectively. The orders between c_5^l and c_3^r, c_4^l and c_2^r, as well as c_3^l and c_1^r are indeterminate. In order to tune these eight parameters, nine position variables $r_1 \sim r_9$ are designed as follows:

$$c_2^l = c_{min} + r_1 \times \omega \tag{8}$$

$$c_1^r = c_2^l + r_2 \times \omega \tag{9}$$

$$c_3^l = c_2^l + r_3 \times \omega \tag{10}$$

$$c_2^r = \max\{c_1^r, c_3^l\} + r_4 \times \omega \tag{11}$$

$$c_4^l = \max\{c_1^r, c_3^l\} + r_5 \times \omega \tag{12}$$

$$c_3^r = \max\{c_2^r, c_4^l\} + r_6 \times \omega \tag{13}$$

$$c_5^l = \max\{c_2^r, c_4^l\} + r_7 \times \omega \tag{14}$$

$$c_4^r = \max\{c_3^r, c_5^l\} + r_8 \times \omega \tag{15}$$

where $\omega = \dfrac{(c_{\max} - c_{\min})}{\sum_{i=1}^{9} r_i}$.

Each position variable is represented by four real-coding genes which are also depicted in Figure 4.

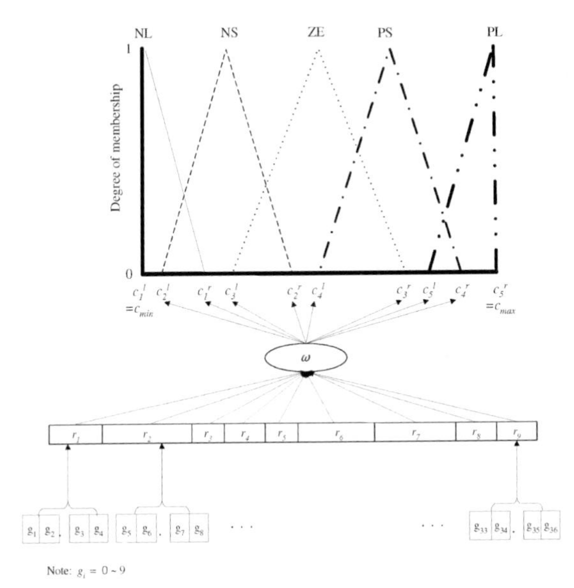

Figure 4. Encoding method for membership functions.

Step 2-2: Generating initial population. Randomly generate an initial population with p chromosomes. For an FLC with N state variables and one control variable, each chromosome has $36(N+1)$ genes, and each gene randomly takes one integer from [0, 9].

Step 2-3: Calculating fitness values. The fitness value set as the reciprocal of the objective function (E) of the problem to be minimized. The fitness value of each chromosome is calculated for the evaluation of the next step.

Step 2-4: Selection. Select the chromosomes for crossover and mutation by evaluating their fitness values with the Monte Carlo wheel method.

Step 2-5: Crossover. The max-min-arithmetical crossover proposed by Herrera *et al.* (1995) is employed. Let $G_w^t = \{ g_{w1}^t ,..., g_{wk}^t ,..., g_{wK}^t \}$ and $G_v^t = \{ g_{v1}^t ,..., g_{vk}^t ,..., g_{vK}^t \}$ be two chromosomes selected for crossover, and the following four descendants will be generated:

$$G_1^{t+1} = aG_w^t + (1-a)G_v^t \tag{16}$$
$$G_2^{t+1} = aG_v^t + (1-a)G_w^t \tag{17}$$
$$G_3^{t+1} \text{ with } g_{3k}^{t+1} = \min\{g_{wk}^t, g_{vk}^t\} \tag{18}$$
$$G_4^{t+1} \text{ with } g_{4k}^{t+1} = \max\{g_{wk}^t, g_{vk}^t\} \tag{19}$$

where a is a parameter ($0 < a < 1$), and t is the number of generations.

Step 2-6: Mutation. The non-uniform mutation proposed by Michalewicz (1992) is employed. Let $G_t = \{ g_1^t ,..., g_k^t ,..., g_K^t \}$ be a chromosome and the gene g_k^t be selected for mutation (the domain of g_k^t is $[g_k^l, g_k^u]$); the value of g_k^{t+1} after mutation can be computed as follows:

$$g_k^{t+1} = \begin{cases} g_k^t + \Delta(t, g_k^u - g_k^t) & \text{if } b = 0 \\ g_k^t - \Delta(t, g_k^t - g_k^l) & \text{if } b = 1 \end{cases} \tag{20}$$

where b randomly takes a binary value of 0 or 1. The function $\Delta(t,z)$ returns a value in the range of $[0, z]$ such that the probability of $\Delta(t,z)$ approaches to 0 as t increases:

$$\Delta(t,z) = z(1 - r^{(1-t/T)^h}) \tag{21}$$

where r is a random number in the interval [0,1], T is the maximum number of generations, and h is a given constant. In Equation (20), the value returned by $\Delta(t,z)$ will gradually decrease as the evolution progresses.

Step 2-7: Testing the stop condition. The stop condition is set based on whether the mature rate has reached a given constant δ. If so, proceed to Step 3; otherwise, go back to Step 2-4.

Step 3: Testing the stop condition. If $(f_V - f_{V-1}) \le \varepsilon$, where f_V and f_{V-1} are the best values of energy function for the Vth and V-1th evolution epoch, respectively, and ε is an arbitrary small number, then stop. The incumbent logic rules and membership functions are the optimal learning results. Otherwise, let $V=V+1$, and go to Step 1.

EXPERIMENTS AT AN ISOLATED INTERSECTION

An Example

Data and parameters

To investigate the effectiveness of the proposed AGFLC-based TSP model, an exemplified case is first conducted at an intersection with configuration shown in Figure 5. Assume that the intersection has two lanes in each approach with a saturation flow of 1,800 pcu/hr/lane. Ten-hour five-minute flow rates in the TSP control directions and the non-TSP directions are randomly generated at the range of 0.4 to 0.6 and 0.2 to 0.3 degree of saturation separately. Transit vehicles (i.e. buses) are assumed to be arriving in *Poission* distribution with $\lambda = 0.17$ veh/sec. The loading factors for a passenger car and a bus are assumed to be 2 and 40 persons, respectively. The cycle length and green time of the pre-timed signal are determined by Webster's minimum delay model as 156 and 100 seconds. The other parameters are set as G_{max}=130 seconds, G_{min}=20 seconds, AR=3 seconds, H=13 seconds, and L = 10 seconds.

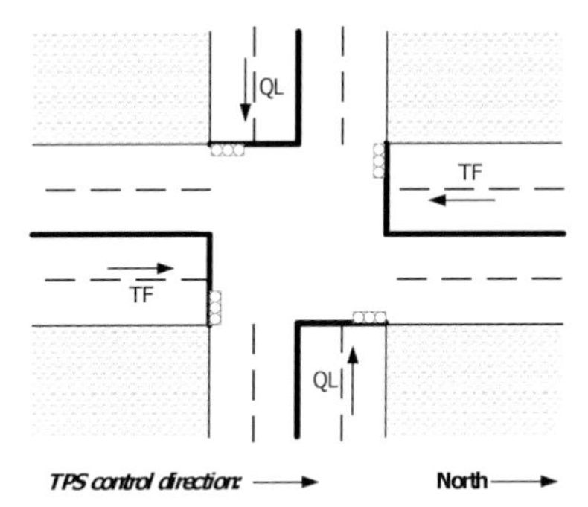

Figure 5. Configuration of an exemplified isolated intersection.

Under equally distributed membership functions, the learning results of ACO with various settings of parameters are investigated. The most appropriate values of these parameters are suggested as α=2, β=5, ξ=0.1, ρ=0.1, and q_0=0.3. The number of ants (K) is 25 (equal to the number of antecedents), and the maximal iteration (t_{max}) is 100. The predetermined rule table for computing the heuristic information, represented by the reasonability index, can be defined as Table 2 by referring to the MacVicar-Whelan rule base (MacVicar-Whelan, 1976). In this table, *TF* is regarded as having a positive proportion to *NE* and *QL* a negative proportion to *NE*. Furthermore, the initial pheromone (τ^0) is set as a reciprocal of the *TPD* performed by the predetermined rule table with equally distributed membership functions, and the heuristic information on the arcs connecting exclusion set with all antecedents (θ_{i6}) is assumed as 0.5. The parameters of the tuning membership function by GAs are set the same as those of Chiou and Lan (2005) which are as follows: population size=100, crossover rate=0.9, mutation rate=0.1, a=0.35, h=0.5, δ=80%, ε=0.001, NE_t=0.5, and the center of gravity method is employed for defuzzification.

This paper employs an analytical fluid approximation to estimate vehicle delays for each cycle under different TSP strategies for the exemplified isolated intersection. The estimation is depicted in Figure 6. Bus delays are evaluated one-by-one depending on whether they are stopped at the intersection or not. Then person delays can be calculated by multiplying the loading factors.

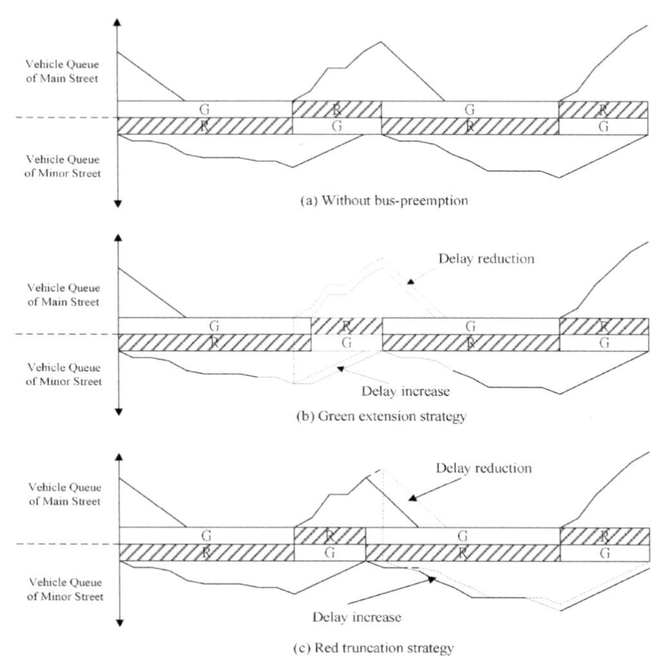

Figure 6. Estimation of vehicle delays for the exemplified isolated intersection.

Table 2. Predetermined rule table for heuristic information

y (*NE*)		$x_1(TF)$				
		NL	***NS***	***ZE***	***PS***	***PL***
$x_2(QL)$	***NL***	*ZE*	*PS*	*PL*	*PL*	*PL*
	NS	*NS*	*ZE*	*PS*	*PS*	*PL*
	ZE	*NL*	*NS*	*ZE*	*PS*	*PL*
	PS	*NL*	*NS*	*NS*	*ZE*	*PS*
	PL	*NL*	*NL*	*NL*	*NS*	*ZE*

Note: *NL*: negative large, *NS*: negative small, *ZE*: zero, *PS*: positive small, *PL*: positive large.

Learning results

To begin with equally distributed membership functions, which are intuitively designed, Figure 7 depicts the evolving processes of applying the AGFLC model to the green extension and red truncation TSP strategies. In green extension, the AGFLC converges after three iterative evolutions with a total of 346 ant iterations and genetic generations. The value of *TPD* slightly decreases from 1371.2 to 1350.0 person-hours; while in red truncation, the AGFLC converges after three iterative evolutions with a total of 376 ant iterations and genetic generations where the value of *TPD* considerably decreases from 2131.1 to 1276.0 person-hours. The evolving processes indicate that the converging variation of *TPD* in the learning processes is larger in the red truncation than in the green extension strategy.

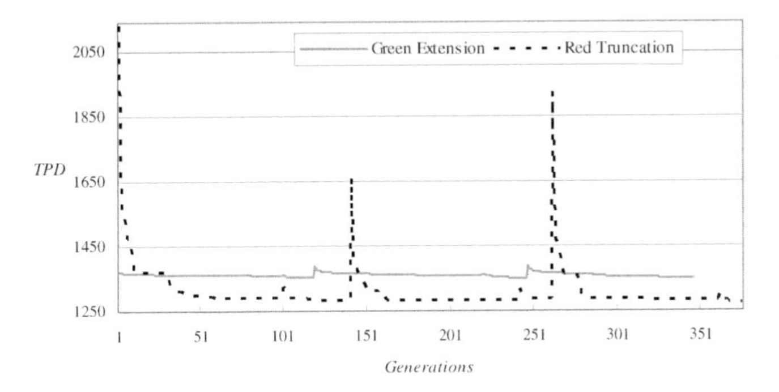

Figure 7. Evolving process of the exemplified isolated intersection.

Comparisons

To investigate the performance of the proposed AGFLC models, pre-timed signal without TSP, unconditional TSP, NB conditional TSP and GFLC-based conditional TSP are also simulated under the same contexts. The control logic of the NB conditional TSP is non-fuzzy wherein the transit priority is provided when the net benefit of extending the green phase or truncating the red phase is positive. The net benefit equals to the difference of person delays between the beneficial vehicles (including the approaching trams or buses) and the impaired vehicles due to the provision of transit priority. The GFLC-based conditional TSP is the same as the one in Chiou *et al.* (2003). Table 3 shows the simulation results of the green and red truncation. Comparing to a pre-timed signal, the AGFLC can curtail *TPD* by 8.17% and 13.20% in the green extension and red truncation, respectively, while the GFLC can reduce *TPD* by 8.11% and 13.01% and the NB can reduce *TPD* by 7.17% and 4.16%. Unconditional TSP can curtail *TPD* by 3.97% in green extension; however, it increases to more than six times of *TPD* when implementing red truncation due to incredible increase of person delay to other vehicles. The results indicate that the three conditional TSP approaches (AGFLC, GFLC, and NB) perform better than the unconditional TSP, of which the AGFLC performs even better than the GFLC and the NB. As anticipated, the fuzzy control methods perform better than non-fuzzy method, because they can synthesize all of the firing fuzzy rules to produce a continuous control output in a more diligent way, where non-fuzzy method can only deliver a discrete control output from one corresponding firing rule. In terms of bus person delay, unconditional TSP curtails person delay by 20.69% in green extension and 52.75% in red truncation. The red truncation strategy could curtail a larger transit person delay due to a great deduction in red time of arriving buses.

Table 3. Comparison of different TSP models (the exemplified isolated intersection)

TSP Strategy	Types of Vehicles	Without TSP	With TSP			
			Unconditional	Conditional		
				NB	GFLC	AGFLC
Green Extension	Buses	563.2	446.7 (20.69%)	427.3 (24.13%)	415.2 (26.28%)	410.4 (27.13%)
	Other vehicles	906.8	965.0 (-6.42%)	937.3 (-3.36%)	935.6 (-3.18%)	939.5 (-3.61%)
	All vehicles	1470.0	1411.7 (3.97%)	1364.6 (7.17%)	1350.8 (8.11%)	1349.9 (8.17%)
Red Truncation	Buses	563.2	266.1 (52.75%)	521.0 (7.49%)	403.2 (28.41%)	414.7 (26.37%)
	Other vehicles	906.8	10997.3 (-1112%)	879.1 (3.05%)	875.5 (3.45%)	861.3 (5.02%)
	All vehicles	1470.0	11263.4 (-666%)	1400.1 (4.76%)	1278.7 (13.01%)	1276.0 (13.20%)

Note: The unit of person delay is person-hour. Figures in parenthesis represent the percentages of person delay reduced in comparing to that of without TSP model.

Sensitivity analyses

To examine the robustness of the proposed model, sensitivity analyses on various traffic scenarios and bus loading factors are conducted. Traffic flow rates increase by 20% (called high traffic scenario) and decrease by 20% (called low traffic scenario), and the average bus loading factors are varied as 20, 30, 50, and 60 persons per bus. Table 4 shows the simulation results of green extension and red truncation for various traffic scenarios. When implementing green extension, as comparing to the pre-timed signal, the AGFLC model can curtail *TPD* by 6.16% to 12.34%, while the GFLC can reduce *TPD* by 5.98% to 12.27%. The NB can reduce *TPD* by 4.45% to 10.40% under medium and low traffic and slightly increase *TPD* by 0.48% to 3.21% under high traffic. The unconditional TSP can only reduce *TPD* by 3.97% to 10.11% and even increase *TPD* by 63.98% under high traffic. Note that both unconditional TSP and conditional TSP (including AGFLC, GFLC, and NB) perform better in low traffic than in high traffic when implementing green extension. These results reveal that the AGFLC outperforms in all scenarios, followed by the GFLC and the NB. When implementing red truncation, similar results are obtained. Moreover, focusing on the difference between green extension and red truncation, with the increase of traffic, green extension would perform better than red truncation due to the fact that the latter would cause a larger impact on the competing approaches as traffic increases. This indicates the advantages of implementing green extension under high traffic and red truncation under low traffic. Furthermore, Table 5 shows the person delay of green extension and red truncation for different bus loading factors. Comparing to the pre-timed signal without TSP, the AGFLC can curtail the largest percentage of *TPD*, followed by the GFLC and the NB under both green extension and red truncation. Unconditional TSP even worsens *TPD* when implementing red truncation. As expected, when the bus loading factor gets higher, the effectiveness in reducing *TPD* would be further enhanced for all unconditional and conditional TSP examined. It supports the advantage of implementing TSP in a high bus loading factor situation.

Table 4. Comparison of different TSP models under various scenarios (the exemplified isolated intersection)

TSP Strategy	Scenarios	Types of Vehicles	Without TSP	With TSP			
				Unconditional	Conditional		
					NB	GFLC	AGFLC
Green Extension	High Traffic	Buses	685.9	488.2 (28.82%)	541.8 (21.01%)	560.1 (18.34%)	556.7 (18.84%)
		Other vehicles	1824.8	3628.8 (-	1981.0 (-8.56%)	1800.4 (1.34%)	1799.3 (1.40%)
		All vehicles	2510.7	4117.0 (-	2522.8 (-0.48%)	2360.5 (5.98%)	2356.0 (6.16%)
	Medium Traffic	Buses	563.2	446.7 (20.69%)	427.3 (24.13%)	415.2 (26.28%)	410.4 (27.13%)
		Other vehicles	906.8	965.0 (-6.42%)	937.3 (-3.36%)	935.6 (-3.18%)	939.5 (-3.61%)
		All vehicles	1470.0	1411.7 (3.97%)	1364.6 (7.17%)	1350.8 (8.11%)	1349.9 (8.17%)
	Low Traffic	Buses	465.6	334.0 (28.26%)	336.8 (27.66%)	321.0 (31.06%)	320.3 (31.21%)
		Other vehicles	522.3	554.0 (-6.07%)	548.4 (-5.00%)	545.7 (-4.48%)	545.7 (-4.48%)
		All vehicles	987.9	888.0 (10.11%)	885.2 (10.40%)	866.7 (12.27%)	866.0 (12.34%)
Red Truncation	High Traffic	Buses	685.9	280.9 (59.05%)	633.8 (7.60%)	625.9 (8.75%)	621.9 (9.33%)
		Other vehicles	1824.8	68697.0 (-3664%)	1957.5 (-7.27%)	1791.4 (1.83%)	1794.5 (1.66%)
		All vehicles	2510.7	68977.9 (-2647%)	2591.3 (-3.21%)	2417.3 (3.72%)	2416.4 (3.76%)
	Medium Traffic	Buses	563.2	266.1 (52.75%)	521.0 (7.49%)	403.2 (28.41%)	414.7 (26.37%)
		Other vehicles	906.8	10997.3 (-1112%)	879.1 (3.05%)	875.5 (3.45%)	861.3 (5.02%)
		All vehicles	1470.0	11263.4 (-666%)	1400.1 (4.76%)	1278.7 (13.01%)	1276.0 (13.20%)
	Low Traffic	Buses	465.6	377.1 (19.01%)	440.0 (5.50%)	346.0 (25.69%)	343.6 (26.20%)
		Other vehicles	522.3	486.5 (6.85%)	503.9 (3.52%)	485.0 (7.14%)	485.9 (6.97%)
		All vehicles	987.9	863.6 (12.58%)	943.9 (4.45%)	831.0 (15.88%)	829.5 (16.03%)

Note: The unit of person delay is person-hour. Figures in parenthesis represent the percentages of person delay reduced in comparing to that of without TSP model.

A Field Case

For investigating the applicability of the AGFLC, a field case is conducted at the intersection of Hsin-I road and Ta-An road in Taipei City. Figure 8 demonstrates the basic configuration of this intersection. There are a total of six lanes containing two bus-exclusive lanes on Hsin-I road and two lanes on Ta-An road. The traffic data on bus, car, and motorcycle are collected by video cameras from 4:00 pm to 5:00 pm and then transformed into five-minute traffic flow data. The current signal phase of this intersection is pre-timed with green 140 seconds and red 60 seconds in the Hsin-I approaches. The parameters of ACO and GA, the predetermined rule table, the initial pheromone, and the loading factors of passenger cars and buses are assumed the same as those in the exemplified case. Referring to the current timing plan at the field intersection, the maximal and minimal green time is set as 170 seconds and 30 seconds, respectively.

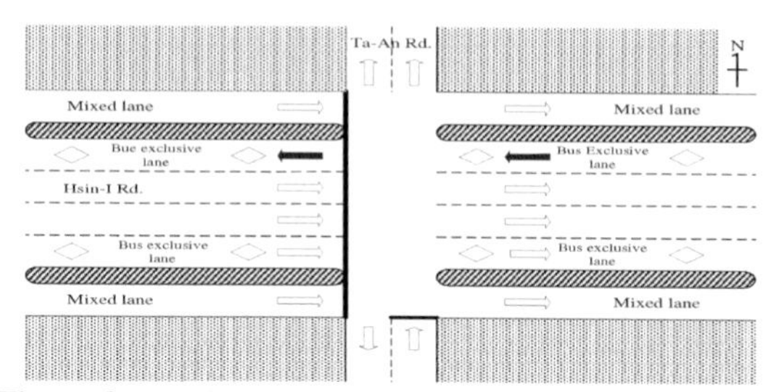

Figure 8. Configuration of the field isolated intersection.

Table 5. Comparison of different TSP models under various bus loading factors (the exemplified isolated intersection)

TSP Strategy	Loading Factor	Types of Vehicles	Without TSP	With TSP: Unconditional	With TSP: Conditional NB	With TSP: Conditional GFLC	With TSP: Conditional AGFLC
Green Extension	20	Buses	281.6	223.4 (20.69%)	228.4 (18.89%)	205.8 (26.92%)	205.7 (26.95%)
		Other vehicles	906.8	965.0 (-6.42%)	933.0 (-2.89%)	935.5 (-3.16%)	935.3 (-3.14%)
		All vehicles	1188.4	1188.4 (0.00%)	1161.4 (2.27%)	1141.3 (3.96%)	1141.0 (3.99%)
	30	Buses	422.4	335.0 (20.69%)	331.6 (21.50%)	310.7 (26.44%)	308.2 (27.04%)
		Other vehicles	906.8	965.0 (-6.42%)	937.6 (-3.40%)	937.2 (-3.35%)	938.7 (-3.52%)
		All vehicles	1329.2	1300.0 (2.19%)	1269.2 (4.51%)	1247.9 (6.12%)	1246.9 (6.19%)
	40	Buses	563.2	446.7 (20.69%)	427.3 (24.13%)	415.2 (26.28%)	410.4 (27.13%)
		Other vehicles	906.8	965.0 (-6.42%)	937.3 (-3.36%)	935.6 (-3.18%)	939.5 (-3.61%)
		All vehicles	1470.0	1411.7 (3.97%)	1364.6 (7.17%)	1350.8 (8.11%)	1349.9 (8.17%)
	50	Buses	704.0	558.4 (20.69%)	536.7 (23.76%)	513.3 (27.09%)	503.8 (28.44%)
		Other vehicles	906.8	965.0 (-6.42%)	941.0 (-3.77%)	936.5 (-3.28%)	937.4 (-3.37%)
		All vehicles	1610.8	1523.4 (5.43%)	1477.7 (8.26%)	1449.8 (10.00%)	1441.2 (10.53%)
	60	Buses	844.8	670.1 (20.69%)	631.7 (25.22%)	620.0 (26.61%)	610.3 (27.76%)
		Other vehicles	906.8	965.0 (-6.42%)	941.2 (-3.79%)	937.5 (-3.39%)	941.3 (-3.80%)
		All vehicles	1751.6	1635.1 (6.65%)	1572.9 (10.20%)	1557.5 (11.08%)	1551.6 (11.42%)
Red Truncation	20	Buses	281.6	133.1 (52.75%)	275.7 (2.10%)	216.4 (23.15%)	213.1 (24.33%)
		Other vehicles	906.8	10997.3 (-1112%)	883.2 (2.60%)	855.5 (5.66%)	858.6 (5.32%)
		All vehicles	1188.4	11130.4 (-836%)	1158.9 (2.48%)	1071.9 (9.80%)	1071.7 (9.82%)
	30	Buses	422.4	199.6 (52.75%)	406.1 (3.86%)	318.5 (24.60%)	301.6 (28.60%)
		Other vehicles	906.8	10997.3 (-1112%)	881.7 (2.77%)	858.9 (5.28%)	875.5 (3.45%)
		All vehicles	1329.2	11196.9 (-742%)	1287.8 (3.11%)	1177.4 (11.42%)	1177.1 (11.44%)
	40	Buses	563.2	266.1 (52.75%)	521.0 (7.49%)	403.2 (28.41%)	414.7 (26.37%)
		Other vehicles	906.8	10997.3 (-1112%)	879.1 (3.05%)	875.5 (3.45%)	861.3 (5.02%)
		All vehicles	1470.0	11263.4 (-666%)	1400.1 (4.76%)	1278.7 (13.01%)	1276.0 (13.20%)
	50	Buses	704.0	332.6 (52.75%)	651.9 (7.40%)	504.9 (28.28%)	505.8 (28.15%)
		Other vehicles	906.8	10997.3 (-1112%)	875.8 (3.42%)	881.1 (2.83%)	877.9 (3.19%)
		All vehicles	1610.8	11329.9 (-603%)	1527.7 (5.16%)	1386.0 (13.96%)	1383.7 (14.10%)
	60	Buses	844.8	399.2 (52.75%)	780.0 (7.67%)	608.9 (27.92%)	602.3 (28.71%)
		Other vehicles	906.8	10997.3 (-1112%)	875.1 (3.50%)	878.9 (3.08%)	878.5 (3.12%)
		All vehicles	1751.6	11396.5 (-550%)	1655.1 (5.51%)	1487.8 (15.06%)	1480.8 (15.46%)

Note: The unit of person delay is person-hour. Figures in parenthesis represent the percentages of person delay reduced in comparing to that of without TSP model.

Table 6 shows the simulation results of this field case. Comparing to the current pre-timed signal timing plan without TSP, the AGFLC can curtail *TPD* by 52.01% and 54.80% in green extension and red truncation, respectively; while the GFLC can reduce *TPD* by 51.50% and

53.26%, the NB can reduce *TPD* by 48.50% and 45.94%, and unconditional TSP can curtail *TPD* by 47.91% and 34.07%. The results indicate that the implementation of three different TSP could have a significant reduction on the *TPD*. However, the AGFLC still outperforms, followed by the GFLC and the NB. When implementing unconditional TSP, the performance of the green extension strategy is superior to red truncation strategy. When implementing conditional TSP approaches, however, the performances of the green extension and red truncation strategies do not significantly differ. As expected, unconditional TSP curtails more person delay for buses than conditional TSP because the latter does not provide absolute priority treatment to buses.

Table 6. Comparison of different TSP models (the field isolated intersection)

TSP Strategy	Types of Vehicles	Current Timing (Without TSP)	With TSP			
			Unconditional	Conditional		
				NB	GFLC	AGFLC
Green Extension	Buses	14.2	0.9 (93.66%)	4.5 (68.31%)	3.8 (73.24%)	2.7 (80.99%)
	Other vehicles	122.3	70.2 (42.60%)	65.8 (46.20%)	62.4 (48.98%)	62.8 (48.65%)
	All vehicles	136.5	71.1 (47.91%)	70.3 (48.50%)	66.2 (51.50%)	65.5 (52.01%)
Red Truncation	Buses	14.2	0.7 (95.07%)	8.9 (37.32%)	3.5 (75.35%)	3.1 (78.17%)
	Other vehicles	122.3	89.3 (26.98%)	64.9 (46.93%)	60.3 (50.70%)	58.6 (52.09%)
	All vehicles	136.5	90.0 (34.07%)	73.8 (45.93%)	63.8 (53.26%)	61.7 (54.80%)

Note: The unit of person delay is person-hour. Figures in parenthesis represent the percentages of person delay reduced in comparing to that of without TSP model.

EXPERIMENTS ALONG AN ARTERIAL

An Example

Data and parameters

To further investigate the performance of the proposed model to implement the TSP along an arterial, an exemplified case with two assumed consecutive intersections is conducted, which geometry configuration is shown in Figure 9. To synchronize the signal control of the consecutive intersections, three coordinated signal systems, simultaneous, alternate, and progressive, are considered. For ease of comparison, the cycle times of these three systems are assumed identical. The TSP strategy is implemented to determine the timing plan at the first intersection, while the timing plan at the succeeding intersection is determined by the coordinated system. The traffic flow conditions of downstream intersection are determined by the upstream traffic flows and upstream signal control results. The analytical fluid approximation is employed to estimate the total vehicle delays, which is the entire area between cumulative arrival and departure curves, as illustrated in Figure 10 with the case of simultaneous signal system. To simplify the analysis, this paper simply neglects the turning traffic and assumes the arrival traffic patterns at downstream intersections the same as the departure traffic patterns at upstream intersection. Then person delays can be acquired by multiplying the preset loading factors. Offset for progressive coordinated signal is assumed to be 20 seconds which is equal to the travelling time between the two intersections. The other parameters are set the same as those in the isolated intersection case.

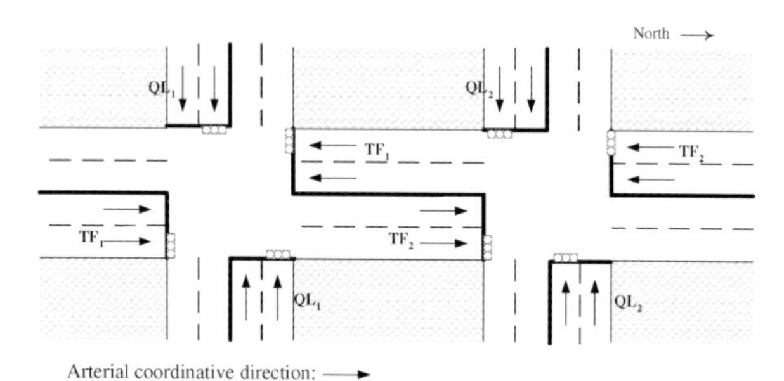

Figure 9. Configuration of the exemplified consecutive intersections.

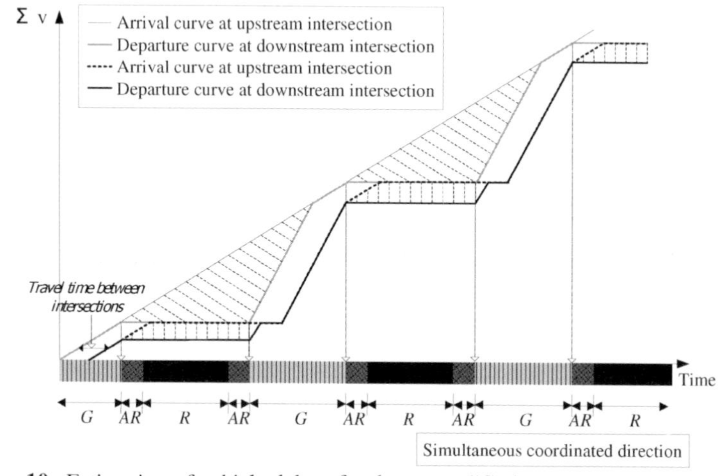

Figure 10. Estimation of vehicle delays for the exemplified consecutive intersections.

Learning results

The AGFLC conditional TSP is implemented under three coordinated signal systems and two TSP strategies (green extension and red truncation). Thus, six sets of AGFLC are required for optimization. Taking progressive system for instance, the learning processes of green extension and red truncation are depicted in Figure 11. In green extension, the AGFLC converges after two iterative evolutions with 243 ant iterations and genetic generations progressed. The value of *TPD* decreases from 2016.7 to 1992.1 person-hour. In red truncation, the AGFLC converges after five iterative evolutions with a total of 562 ant iterations and genetic generations where the value of *TPD* decreases from 2078.7 to 1993.4

person-hour. The evolving processes also indicate that the converging variation of *TPD* in the learning processes is larger in the red truncation than in the green extension strategy.

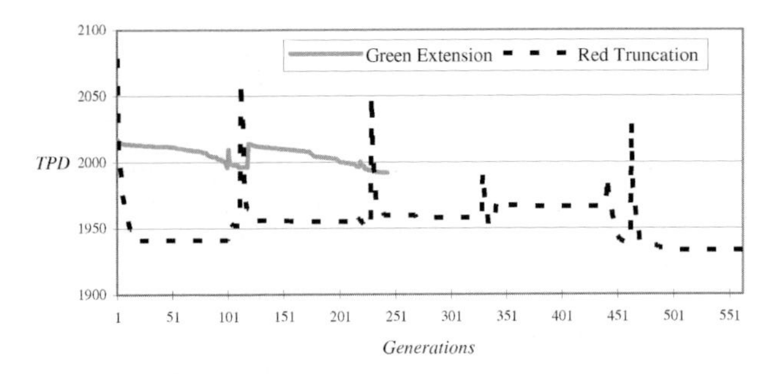

Figure 11. Evolving processes of the exemplified consecutive intersections.

Comparisons

To compare the control performances of the proposed AGFLC models, pre-timed signal without TSP, unconditional TSP, NB conditional TSP and GFLC-based conditional TSP are also simulated with the same traffic data under simultaneous, progressive, and alternate coordinated signal systems. Table 7 shows the simulation results of green extension and red truncation, respectively. In terms of total person delays, the progressive system outperforms, followed by the simultaneous system, and then by the alternate system. As comparing to the pre-timed signal without TSP, the AGFLC can curtail *TPD* at the largest amount, followed by the GFLC and the NB under simultaneous and progressive systems. While under the alternate system, the optimal control results determined by the AGFLC and GFLC are the same as the result of the pre-timed signal because providing priority to buses would not improve any system performance. These results reveal that the proposed AGFLC model could achieve an optimal control under three coordinated signal systems without deteriorating the system performance of the pre-timed signal timing plan. The results also show that the NB conditional TSP under alternate signal system and the unconditional TSP under those three coordinated signal systems would deteriorate the system performance.

Sensitivity analyses

Sensitivity analyses on various traffic scenarios and bus loading factors are also conducted under the progressive coordinated system. Tables 8 and 9 show the results for various traffic scenarios and for various bus loading factors, respectively. It is found that the results in the case of two consecutive intersections are consistent to those in the isolated intersection.

Table 7. Comparison of different TSP models under various coordinated systems (the exemplified consecutive intersections)

TSP Strategy	Coordinated systems	Types of Vehicles	Without TSP	With TSP			
				Unconditional	Conditional		
					NB	GFLC	AGFLC
Green Extension	Simultaneous	Buses	915.1	886.3 (3.15%)	842.3 (7.96%)	840.6 (8.14%)	837.1 (8.52%)
		Other vehicles	1509.3	1635.8 (-8.38%)	1582.4 (-4.84%)	1544.4 (-2.33%)	1547.0 (-2.50%)
		All vehicles	2424.4	2522.1 (-4.03%)	2424.7 (-0.01%)	2385.0 (1.63%)	2384.1 (1.66%)
	Progressive	Buses	563.2	446.7 (20.69%)	427.3 (24.13%)	450.2 (20.06%)	415.3 (26.26%)
		Other vehicles	1516.0	1634.0 (-7.78%)	1582.0 (-4.35%)	1546.7 (-2.03%)	1576.8 (-4.01%)
		All vehicles	2079.2	2080.7 (-0.07%)	2009.3 (3.36%)	1996.9 (3.96%)	1992.1 (4.19%)
	Alternate	Buses	3588.9	3705.3 (-3.24%)	3527.2 (1.72%)	3588.9 (0.00%)	3588.9 (0.00%)
		Other vehicles	134807.1	150616.5 (-11.73%)	146384.1 (-8.59%)	134807.1 (0.00%)	134807.1 (0.00%)
		All vehicles	138396.0	154321.8 (-11.51%)	149911.3 (-8.32%)	138396.0 (0.00%)	138396.0 (0.00%)
Red Truncation	Simultaneous	Buses	915.1	530.5 (42.03%)	847.8 (7.35%)	717.7 (21.57%)	727.1 (20.54%)
		Other vehicles	1509.3	16754.8 (-1010%)	1490.2 (1.27%)	1520.3 (-0.73%)	1509.1 (0.01%)
		All vehicles	2424.4	17285.3 (-612%)	2338.0 (3.56%)	2238.0 (7.69%)	2236.2 (7.76%)
	Progressive	Buses	563.2	314.4 (44.18%)	521.0 (7.49%)	448.1 (20.44%)	418.9 (25.62%)
		Other vehicles	1516.0	16742.0 (-1004%)	1488.4 (1.82%)	1485.5 (2.01%)	1514.5 (0.10%)
		All vehicles	2079.2	17056.4 (-720%)	2009.4 (3.36%)	1933.6 (7.00%)	1933.4 (7.01%)
	Alternate	Buses	3588.9	3277.5 (8.68%)	3541.3 (1.33%)	3588.9 (0.00%)	3588.9 (0.00%)
		Other vehicles	134807.1	212674.5 (-57.76%)	144838.1 (-7.44%)	134807.1 (0.00%)	134807.1 (0.00%)
		All vehicles	138396.0	215952.0 (-56.04%)	148379.4 (-7.21%)	138396.0 (0.00%)	138396.0 (0.00%)

Note: The unit of person delay is person-hour. Figures in parenthesis represent the percentages of person delay reduced in comparing to that of without TSP model.

Table 8. Comparison of different TSP models under various scenarios (the exemplified consecutive intersections under progressive coordinated system)

TSP Strategy	Scenarios	Types of Vehicles	Without TSP	With TSP			
				Unconditional	Conditional		
					NB	GFLC	AGFLC
Green Extension	High Traffic	Buses	685.9	488.2 (28.82%)	541.8 (21.01%)	561.5 (18.14%)	558.9 (18.52%)
		Other vehicles	2812.7	4906.4 (-74.44%)	3091.7 (-9.92%)	2813.3 (-0.02%)	2815.4 (-0.10%)
		All vehicles	3498.6	5394.6 (-54.19%)	3633.5 (-3.86%)	3374.8 (3.54%)	3374.3 (3.55%)
	Medium Traffic	Buses	563.2	446.7 (20.69%)	427.3 (24.13%)	450.2 (20.06%)	415.3 (26.26%)
		Other vehicles	1516.0	1634.0 (-7.78%)	1582.0 (-4.35%)	1546.7 (-2.03%)	1576.8 (-4.01%)
		All vehicles	2079.2	2080.7 (-0.07%)	2009.3 (3.36%)	1996.9 (3.96%)	1992.1 (4.19%)
	Low Traffic	Buses	465.6	334.0 (28.26%)	336.8 (27.66%)	330.6 (28.99%)	330.6 (28.99%)
		Other vehicles	903.4	977.1 (-8.16%)	964.7 (-6.79%)	959.0 (-6.15%)	959.0 (-6.15%)
		All vehicles	1369.0	1311.1 (4.23%)	1301.5 (4.93%)	1289.6 (5.80%)	1289.6 (5.80%)
Red Truncation	High Traffic	Buses	685.9	280.9 (59.05%)	633.8 (7.60%)	639.0 (6.84%)	637.2 (7.10%)
		Other vehicles	2812.7	126408.0 (-4394%)	3104.4 (-10.37%)	2810.2 (0.09%)	2807.1 (0.20%)
		All vehicles	3498.6	126688.9 (-3521%)	3738.2 (-6.85%)	3449.2 (1.41%)	3444.3 (1.55%)
	Medium Traffic	Buses	563.2	314.4 (44.18%)	521.0 (7.49%)	448.1 (20.44%)	418.9 (25.62%)
		Other vehicles	1516.0	16742.0 (-1004%)	1488.4 (1.82%)	1485.5 (2.01%)	1514.5 (0.10%)
		All vehicles	2079.2	17056.4 (-720%)	2009.4 (3.36%)	1933.6 (7.00%)	1933.4 (7.01%)
	Low Traffic	Buses	465.6	377.1 (19.01%)	432.9 (7.02%)	352.8 (24.23%)	352.1 (24.38%)
		Other vehicles	903.4	869.6 (3.74%)	885.9 (1.94%)	869.4 (3.76%)	869.8 (3.72%)
		All vehicles	1369.0	1246.7 (8.93%)	1318.8 (3.67%)	1222.2 (10.72%)	1221.9 (10.75%)

Note: The unit of person delay is person-hour. Figures in parenthesis represent the percentages of person delay reduced in comparing to that of without TSP model.

Table 9. Comparison of difference TSP models under various bus loading factors (the field consecutive intersections under progressive coordinated system)

TSP Strategy	Loading Factor	Types of Vehicles	Without TSP	With TSP			
				Unconditional	Conditional		
					NB	GFLC	AGFLC
Green Extension	20	Buses	281.6	223.4 (20.69%)	228.4 (18.89%)	230.9 (18.00%)	227.1 (19.35%)
		Other vehicles	1516.0	1634.0 (-7.78%)	1573.3 (-3.78%)	1541.6 (-1.69%)	1544.7 (-1.89%)
		All vehicles	1797.6	1857.4 (-3.32%)	1801.7 (-0.23%)	1772.5 (1.40%)	1771.8 (1.44%)
	30	Buses	422.4	335.0 (20.69%)	331.6 (21.50%)	347.8 (17.66%)	310.1 (26.59%)
		Other vehicles	1516.0	1634.0 (-7.78%)	1581.3 (-4.31%)	1544.0 (-1.85%)	1577.9 (-4.08%)
		All vehicles	1938.4	1969.0 (-1.58%)	1912.9 (1.32%)	1891.8 (2.40%)	1888.0 (2.60%)
	40	Buses	563.2	446.7 (20.69%)	427.3 (24.13%)	450.2 (20.06%)	415.3 (26.26%)
		Other vehicles	1516.0	1634.0 (-7.78%)	1582.0 (-4.35%)	1546.7 (-2.03%)	1576.8 (-4.01%)
		All vehicles	2079.2	2080.7 (-0.07%)	2009.3 (3.36%)	1996.9 (3.96%)	1992.1 (4.19%)
	50	Buses	704.0	558.4 (20.69%)	536.7 (23.76%)	519.8 (26.16%)	515.0 (26.85%)
		Other vehicles	1516.0	1634.0 (-7.78%)	1588.6 (-4.79%)	1579.3 (-4.18%)	1583.8 (-4.47%)
		All vehicles	2220.0	2192.4 (1.24%)	2125.3 (4.27%)	2099.1 (5.45%)	2098.8 (5.46%)
	60	Buses	844.8	670.1 (20.69%)	631.7 (25.22%)	624.3 (26.10%)	622.3 (26.34%)
		Other vehicles	1516.0	1634.0 (-7.78%)	1590.7 (-4.93%)	1577.1 (-4.03%)	1577.7 (-4.07%)
		All vehicles	2360.8	2304.1 (2.40%)	2222.4 (5.86%)	2201.4 (6.75%)	2200.0 (6.81%)
Red Truncation	20	Buses	281.6	157.2 (44.18%)	275.7 (2.10%)	231.6 (17.76%)	227.3 (19.28%)
		Other vehicles	1516.0	16742.0 (-1004%)	1494.5 (1.42%)	1476.2 (2.63%)	1476.2 (2.63%)
		All vehicles	1797.6	16899.2 (-840%)	1770.2 (1.52%)	1707.8 (5.00%)	1703.5 (5.23%)
	30	Buses	422.4	235.8 (44.18%)	406.1 (3.86%)	338.7 (19.82%)	332.5 (21.28%)
		Other vehicles	1516.0	16742.0 (-1004%)	1492.9 (1.52%)	1486.3 (1.96%)	1486.3 (1.96%)
		All vehicles	1938.4	16977.8 (-775%)	1899.0 (2.03%)	1825.0 (5.85%)	1818.8 (6.17%)
	40	Buses	563.2	314.4 (44.18%)	521.0 (7.49%)	448.1 (20.44%)	418.9 (25.62%)
		Other vehicles	1516.0	16742.0 (-1004%)	1488.4 (1.82%)	1485.5 (2.01%)	1514.5 (0.10%)
		All vehicles	2079.2	17056.4 (-720%)	2009.4 (3.36%)	1933.6 (7.00%)	1933.4 (7.01%)
	50	Buses	704.0	393.0 (44.18%)	651.9 (7.40%)	528.6 (24.91%)	529.9 (24.73%)
		Other vehicles	1516.0	16742.0 (-1004%)	1488.8 (1.79%)	1518.0 (-0.13%)	1512.5 (0.23%)
		All vehicles	2220.0	17135.0 (-671%)	2140.7 (3.57%)	2046.6 (7.81%)	2042.4 (8.00%)
	60	Buses	844.8	471.6 (44.18%)	780.0 (7.67%)	634.7 (24.87%)	629.4 (25.50%)
		Other vehicles	1516.0	16742.0 (-1004%)	1487.9 (1.85%)	1507.7 (0.55%)	1500.5 (1.02%)
		All vehicles	2360.8	17213.6 (-629%)	2267.9 (3.94%)	2142.4 (9.25%)	2129.9 (9.78%)

Note: The unit of person delay is person-hour. Figures in parenthesis represent the percentages of person delay reduced in comparing to that of without TSP model.

A Field Case

To examine the applicability of the proposed AGFLC along an arterial, a field case study is conducted in Ren-ai arterial intersected with Jinshan South Road and with Hangzhou South Road of Taipei City. Figure 12 depicts the configuration of the successive intersections, in which Ren-ai Road is a westbound one-way arterial, currently with eight lanes including two bus-exclusive lanes (one of which is in contra-flow direction); Jinshan S. Rd. has three northbound lanes and four southbound lanes; while Hangzhou S. Rd. is a northbound one-way street with three lanes. Five-minute flow rates during the morning peak hours from 7:00 a.m. to 9:00 a.m. are surveyed. The current timing plan in Ren-ai arterial is a progressive coordinated system with 60 seconds green, 50 seconds red, 3 seconds all-red, and 20 seconds offset. Referring to the current timing plan at the field intersections, the maximal and minimal green times are set 90 seconds and 20 seconds, respectively.

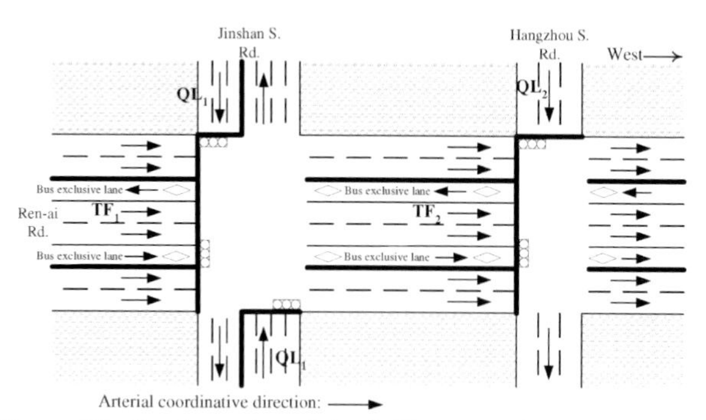

Figure 12. Configuration of the field consecutive intersections.

The simulated results are presented in Table 10. Comparing to the current progressive coordinated timing plan of the field case, the unconditional TSP model can curtail a considerable amount of person delays for buses; however, the overall performance deteriorates by 13.70% (green extension) and by 94.09% (red truncation) if all vehicles are taken into account. In contrast, the AGFLC can curtail *TPD* by 7.60% and 11.30% in green extension and red truncation, respectively, while the GFLC can reduce *TPD* by 7.10% and 10.61% and the NB can reduce *TPD* by 2.98% and 4.03%. The results indicate that the AGFLC and GFLC could perform better than the NB and the AGFLC is slightly superior to the GFLC.

Table 10. Comparisons of different TSP models (the field consecutive intersections under progressive coordinated system)

TSP Strategy	Types of Vehicles	Current Timing (Without TSP)	With TSP			
			Unconditional	Conditional		
				NB	GFLC	AGFLC
Green Extension	Buses	101.3	71.2 (29.71%)	70.7 (30.21%)	71.6 (29.32%)	72.1 (28.83%)
	Other vehicles	260.7	340.3 (-30.53%)	280.5 (-7.59%)	264.7 (-1.53%)	262.4 (-0.65%)
	All vehicles	362.0	411.5 (-13.67%)	351.2 (2.98%)	336.3 (7.10%)	334.5 (7.60%)
Red Truncation	Buses	101.3	58.9 (41.86%)	72.5 (28.43%)	75.3 (25.67%)	73.6 (27.34%)
	Other vehicles	260.7	643.7 (-146%)	274.9 (-5.45%)	248.2 (4.79%)	247.5 (5.06%)
	All vehicles	362.0	702.6 (-94.09%)	347.4 (4.03%)	323.5 (10.64%)	321.1 (11.30%)

Note: The unit of person delay is person-hour. Figures in parenthesis represent the percentages of person delay reduced in comparing to that of without TSP model.

CONCLUDING REMARKS

This paper has developed a novel AGFLC-based TSP model by considering the traffic conditions on all approaches at the signalized intersections to minimize the total person delays. Two TSP strategies including green extension and red truncation are analyzed. Exemplified and field cases are both tested at an isolated intersection and at consecutive intersections along an arterial. The control performances of five control strategies, without TSP, unconditional TSP, NB conditional TSP, GFLC-based conditional TSP, and the proposed AGFLC-based conditional TSP, are compared. The results have consistently shown that the AGFLC model outperforms in both tested contexts. The sensitivity analyses show that unconditional and conditional TSP would perform better in light traffic than in heavy traffic and better in high bus loading than in low loading. Moreover, green extension strategy performs better than red truncation strategy in heavy traffic; in contrast, red truncation has better performance than green extension in light traffic for both unconditional and conditional TSP. In any circumstances, our AGFLC model can perform slightly better than the GFLC model recently proposed by Chiou and Lan (2005), suggesting that it is more appropriate to optimally select logic rules by ACO than by GA.

Some directions for future studies can be identified. Firstly, the improvement of AGFLC over GFLC can vary depending on such factors as traffic composition and volume from different approaches and occupancy of different vehicle types; thus, testing for other intersections and for other times of day deserve further study. Secondly, it deserves to investigate the performance by further employing ACO to tune the membership functions, instead of GA. However, due to the network-based searching behaviours of ACO, remarkable modifications in algorithms of ACO might be needed. Thirdly, the proposed AGFLC model defines reasonability as the heuristic information in the tour construction of an ant. Other definitions of heuristic information might also be worthy of further exploration. Fourthly, to avoid time-consuming traffic simulation, the present paper employs analytical fluid approximation to estimating total person delays and simulating traffic behaviours. Other macroscopic traffic simulation models, such as cell transmission model proposed by Daganzo (1994), can be attempted to more accurately simulate the traffic flow behaviours. Fifthly, if TSP is solely from the transit agency's standpoint without considering the negative impacts to other private vehicles, minimization of bus delay or maximization of bus schedule reliability can serve as the objection function, which is easily incorporated into the proposed models to generate a more appealing control scheme. Last but not least, the interpretations of learning results of the proposed models, including the selected logic rules and tuned membership functions, are worthy of further investigation.

ACKNOWLEDGEMENTS

This paper is part of the research projects granted by the ROC National Science Council (NSC-91-2211-E-035-031, NSC-94-2211-E-009-028). Constructive comments given by three referees are highly appreciated.

REFERENCES

Babuška, R. (1998). *Fuzzy Modeling for Control.* Norwell, MA: Kluwer Academic.

Bárdossy, A. and L. Duckstein (1995). *Fuzzy Rule-Based Modeling with Application to Geophysical, Biological and Engineering Systems.* Boca Raton, FL: CRC Press.

Bonissone, P. P., P. S. Khedkar and Y. Chen (1996). Genetic algorithms for automated tuning of fuzzy controllers: a transportation application. In *Proceedings of Fifth International Conference On Fuzzy Systems* (FUZZ-IEEE'96), 674-680.

Casillas, J., O. Cordón and F. Herrera (2000). Learning fuzzy rules using ant colony optimization algorithms. **In**: *Proceedings 2nd International Workshop on Ant Algorithms*, Brussels, Belgium, 13-21.

Casillas, J., O. Cordón, I. F. Viana and F. Herrera (2005). Learning cooperative linguistic fuzzy rules using the best-worst ant system algorithm. *International Journal of Intelligent Systems*, **20**, 433-452.

Chang, G. L., M. Vasudevan and C. C. Su (1996). Modeling and evaluation adaptive bus-preemption control with and without automatic vehicle location systems. *Transportation Research,* **30A**(4), 251-268.

Chiou, Y. C. and L. W. Lan (2005). Genetic fuzzy logic controller: an iterative evolution algorithm with new encoding method. *Fuzzy Sets and Systems*, **152**, 617-635.

Chiou, Y. C., M. T. Wang and L. W. Lan (2003). Adaptive bus preemption signals with genetic fuzzy logic controller (GFLC). *Journal of the Eastern Asia Society for Transportation Studies*, **5**, 1745-1759.

Chiou, Y. C., M. T. Wang and L. W. Lan (2005). Coordinated transit preemption signal controllers along an arterial: iterative genetic fuzzy logic controller (GFLC) method. *Journal of the Eastern Asia Society for Transportation Studies*, **6**, 2321-2336.

Cisco, B. A. and S. Khasnabis (1995). Technique to assess delay and queue length consequence of bus pre-emption. *Transportation Research Record*, **1494**, 167-175.

Cordón, O., F. Herrera and L. Magdalena (2001). *Genetic Fuzzy Systems: Evolutionary Tuning and Learning of Fuzzy Knowledge Bases.* Singapore: World Scientific.

Daganzo, C. F. (1994). The cell transmission model: a dynamic representation of highway traffic consistent with the hydrodynamic theory. *Transportation Research*, **28B**(4), 269-287.

Dion, F., H. Rakha and Y. Zhang (2004). Evaluation of potential transit signal priority benefits along a fixed-time signalized arterial. *Journal of Transportation Engineering*, **130**(3), 294-303.

Du, T. C. H. and P. M. Wolfe (1995). The amalgamation of neural networks and fuzzy logic systems-a survey. *Computers and Industrial Engineering*, **29**(1), 193-197.

Esobgue, A. and J. Murrell (1993). A fuzzy adaptive controller using reinforcement learning neural networks. *IEEE International Conference on Fuzzy Systems*, 178-183.

Gupta, M. M. and M. B. Gorsalcany (1992). Fuzzy neural-computational technique and its application to modeling and control. *IEEE International Conference on Fuzzy Systems*, 1271-1274.

Herrera, F., M. Lozano and J. L. Verdegay (1995). Tuning fuzzy logic controllers by genetic algorithms. *International Journal of Approximate Reasoning*, **12**, 299-315.

Hwang, H. S. (1998). Control strategy for optimal compromise between trip time and energy consumption in a high-speed railway. *IEEE Transactions on Systems, Man and Cybernetics*, **28**(6) 791-802.

ITS America (2004), *An Overview of Transit Signal Priority*, Advanced Traffic Management Systems Committee and Advanced Public Transportation System Committee.

Jacobson, J. and Y. Sheffi (1981). Analytical model of traffic delays under bus signal preemption: theory and application. *Transportation Research*, **15B**(2), 127-138.

Khasnabis, S., G. V. Reddy and B. B. Chaudry (1993). Signal preemption as a priority treatment tool for transit demand management. *Vehicle Navigation and Information System Conference Proceeding*, Paper No. 912865, Dearbirn, MI.

Linkens, D. A. and J. Nie (1993). Fuzzifier RBF network-based learning control: structure and self-construction. *IEEE International Conference on Neural Networks*, 1016-1021.

Macvicar-Whelan, P. J. (1976). Fuzzy sets for man-machine interaction. *International Journal of Man-Machine Studies*, **8**, 687-697.

Michalewicz, Z. (1992). *Genetic Algorithms + Data Structures = Evolution Programs.* Springer, Berlin.

Nauck, D. and R. Kruse (1993). A fuzzy neural network learning fuzzy control rules and membership functions by fuzzy error backpropagation. *IEEE International Conference on Neural Networks*, 1022-1027.

Nauck, D., F. Klawonn and R. Kruse (1997). *Foundations of Neuro-Fuzzy Systems.* New York: Wiley.

Parpinelli, R. S., H. S. Lopes and A. A. Freitas (2002). Data mining with ant colony optimization algorithm. *IEEE Transactions on Evolutionary Computation*, **6**(4), 321-332.

Sunkari, S. R., P. S. Beasley, J. T. Urbanik and D. B. Fambro (1995). Model to evaluate the impacts of bus priority on signalized intersections. *Transportation Research Record*, **1494**, 117-123.

Wang, L.X. and J. Mendel (1992). Generating fuzzy rules by learning from examples. *IEEE Transactions on Systems, Man and Cybernetics*, **22**(6), 1414-1427.

Wu, J. and N. Hounsell (1998). Bus priority using pre-signal. *Transportation Research*, **32A**(8), 563-583.

Zadeh, L. (1973). Outline of a new approach to the analysis of complex systems and decision processes. *IEEE Transactions on Systems, Man and Cybernetics*, **3**, 28-44.

Transportation and Traffic Theory 2007
Edited by R.E. Allsop, M.G.H. Bell and B.G. Heydecker

23

RESERVE CAPACITY OF A SIGNAL-CONTROLLED NETWORK CONSIDERING THE EFFECT OF PHYSICAL QUEUING

C.K. Wong, Department of Building and Construction, City University of Hong Kong;
S.C. Wong, Department of Civil Engineering, The University of Hong Kong;
Hong K. Lo, Department of Civil Engineering, The Hong Kong University of Science and Technology

SUMMARY

Reserve capacity is a commonly used performance indicator for a signal-controlled system that can be obtained by maximizing a common flow multiplier. Transient overloading is very common in urban signal-controlled systems in which the peak demand may last a very short period of time. The reserve capacity could be overestimated in conventional point queue modeling framework assuming infinite holding capacities along road links. More realistically, vehicles are able to be held up in the form of spatial queues and fully dissipated if sufficient green times are provided. To model the effect of physical queuing, a signalized cell transmission model (CTM) is employed and a multi-phase signal optimization algorithm is integrated for determining the reserve capacity of a linked signal system. Starts of red and green times and their durations are key decision variables. The problem is formulated as a Binary-Mix-Integer-Linear-Program (BMILP) that can be solved by standard branch-and-bound routines. Optimization heuristics are also developed to speedup the solution process. A staggered junction with short link connections is given as a numerical example for illustrations.

INTRODUCTION

Reserve capacity has been established as an important engineering measure of the overall performances of signal-controlled junctions. It dedicates the maximum demand level that can enter a signal-controlled system in which all signal phases operate within certain given acceptable degree of saturation. The reserve capacity in the planning stage is always maximized in order to allow the largest flow increment in the system during practical operation with respect to the existing levels of design flows. The implementation of the resultant signal settings ensures that the random flow fluctuations can be well controlled without impairing the system. Only with positive reserve capacity, total delay incurred in the entire signal-controlled system can be maintained at a satisfactory level. Webster and Cobbe (1966) were first to calculate the reserve capacity by explicit formulae. Allsop (1972) then introduced a common flow multiplier, which is to be maximized, into a mathematical program for isolated signal-controlled junctions within the stage-based design framework. The largest common flow multiplier that can be found in the linear programming optimization, so that the degrees of saturation of all traffic streams do not exceed a prescribed level, represents the reserve capacity of the signal junction. Gallivan and Heydecker (1988) developed reserve capacity optimization in a more flexible group-based design context for isolated signal-controlled junctions. The concept was extended to roundabouts and priority junctions (Wong, 1996), and signal-controlled networks (Wong and Yang, 1997). All of these studies took the time-stationary traffic demands as inputs for analysis, and only fixed-time signal plans could be produced for operation. The reserve capacity so optimized may not be able to reflect the realistic junction performance under time-varying traffic conditions. Transient overloading are common phenomena during peak periods but the highest demand may temporarily occur during a very short period of time. In the static design framework, a very low or even negative reserve capacity could be obtained for the system which may underestimate its actual performance. More realistically, the excess demand flow could be held up in the form of spatial queues and could then be fully released in the next signal cycle when sufficient green duration is provided. In this case, the overall performance of the signal-controlled system is acceptable without inducing severe delays. A dynamic reserve capacity concept will be introduced in the present study to give a more accurate assessment for signal-controlled systems under the time-varying demand conditions.

The point queue modeling paradigm has been used to analyze and design traffic signal control. It is simple and convenient, and can draw upon results from classical queuing theory for analysis. However, the point queue approach assumes an infinite holding capacity for each road link in which vehicles in a queue are stacked vertically rather than spreading horizontally along the link, and the spillback effect that is caused by a long queue is not explicitly considered. This may overestimate the amount of traffic that can be held between two adjacent junctions, especially for short roadways. In such a case, the queue length and holding capacity become critical parameters to be considered in the signal optimization framework, especially for congested networks with relatively short roadways wherein spillback and blockage are not uncommon. To achieve this, one must encapsulate traffic flow models in the signal optimization that explicitly capture these physical queuing effects. One good traffic flow model for this purpose is the hydrodynamic theory of traffic flow.

Based on the hydrodynamic theory of traffic flow, the macroscopic Lighthill-Whitham-Richards (LWR) model was formulated by incorporating the continuity (or conservation) equation in the form of a partial differential equation and the fundamental relationship between flow and density (Lighthill and Whitham, 1955; Richards, 1956). Daganzo (1994, 1995a) developed a cell transmission model (CTM) for the LWR continuum model with a triangular or trapezoid fundamental diagram. In the CTM, a highway is divided into a number of homogeneous sections or cells, and the time horizon is partitioned into intervals or steps. The traffic flow that is transmitted from the upstream cell to the downstream cell is prescribed by relevant receiving and sending functions in terms of the free-flow speed, inflow capacity, jam density, and speed of backward shock wave. The concepts of merge and diverge cells in the CTM were introduced to replicate traffic movements across a junction, based on which a road network was constructed and represented by a series of cells, and a network traffic with multiple origin-destination pairs was modeled (Daganzo, 1995b). Lo (1999, 2001) further developed a mathematical programming approach to the CTM by introducing basic constraints for signal controls, in which the exit flow capacity of a signal cell was defined and controlled by the signal settings. Effective green durations were the key decision variables for network delay optimization. A dynamic intersection signal control optimization (DISCO) model that is based on the CTM was also developed and applied to practical designs (Lo et al., 2001). In contrast, Lin and Wang (2004) used a set of binary variables to represent the right-of-way between two conflicting traffic movements. However, all of these methods only considered two signal phases, and the intergreen structure for separating conflicting traffic movements was not explicitly specified. The difficult problem of signal sequencing for complicated junctions was also not effectively dealt with in their optimization framework. Gallivan and Heydecker (1988) and Heydecker (1992) developed a group-based optimization method for complicated signal-controlled junctions, in which a sophisticated mathematical programming approach was used to optimize the signal sequencing and other signal aspects. Wong and Wong (2003) and Wong et al. (2006) further extended the group-based optimization method to the lane-based method, in which both road markings and signal settings were simultaneously optimized in a mixed-integer mathematical program. The mixed-integer programming approach was also applied to model and solve the signal coordination problem on a two-lane highway with two closely spaced work zones, with each work zone having one lane to serve both directions of traffic that were controlled by traffic signals (Wong et al., 2005, 2006).

However, little research has been devoted to the definition of a reserve capacity in a CTM system. Maher (2005) made a first attempt to develop a trial-and-error procedure to estimate the reserve capacity for a traffic stream with respect to a fixed signal plan, in which the largest common flow multiplier was searched for until the degree of saturation of the traffic stream converged to a given limit. The maximum common flow multiplier depended largely on the chosen offset between two closely spaced signalized junctions.

In this study, we optimize the signal settings for maximizing the common flow multiplier and in turn the reserve capacity of a signal-controlled network using the CTM as the modeling platform. Time-varying traffic demands are taken into consideration. Due to the stochastic nature of traffic demand, a maximum acceptable degree of saturation is usually set in the design framework to provide a buffer to accommodate the short-term flow increase, without which substantial overflow delay may be incurred in successive signal cycles. Conventionally, maximum thresholds are specified for the flow-to-capacity (v/c) ratios for traffic streams. In the present CTM, maximum thresholds for the spatial occupancy of a link are adopted, which will provide spatial buffers for all the links especially the short links with a maximum holding capacity to reduce the chance of blocking back due to random flow fluctuations.

The formulation for traffic signal controls also, will be enhanced, whereby the right-of-way of each signal group in each time step will be defined as a binary variable (green = 1; red = 0), based on which the constraints for the start of green, minimum green, minimum red, clearance time, and maximum acceptable occupancy will be formulated in the form of linear inequalities. Apart from preserving the group-based control features, the binary-integer approach to the specification of a signal plan will provide a high degree of flexibility for the optimization of signal sequencing and other signal aspects, including the possibility of double-green in a signal cycle. Only starts and durations of green are required and defined as control variables in the present formulation.

With the holding capability in the CTM, vehicles in the form of a spatial queue can be packed and stored in cells in front of a traffic signal during the red period, and then discharged when the signal display changes to green. A longer green period implies the greater discharging capability of a signal group, and thus more vehicles can be stored during the red period. As a result, a larger common flow multiplier can be achieved. Cycle length is no longer a fixed parameter or is not needed as a design parameter in the present formulation because green durations, signal sequences, and offsets between the upstream and downstream signals, which will be optimized endogenously, all become critical parameters in the determination of the maximum reserve capacity of the signal-controlled system.

LIST OF SYMBOLS

l	Road link.
l'	Downstream road link (relative to road link l).
l^*	Upstream road link (relative to road link l).
O	Set of all demand input link(s).
D	Set of all exit link(s).
T	Total number of time steps in the study period.
T'	Total number of time steps to be considered in the optimization framework.
t	Time step (interval) where $t \in [0,T]$.
i, j	Traffic signal phase (i and j are two signal phases).

ψ	Set of conflicting traffic signal pairs $(i, j) \in \psi$ if i and j are mutually incompatible phases.
k	Cell.
K_l	Total number of cells on a road link l.
$\delta_{i,t}$	Signal display of phase i at time step t in which 1 represents green and 0 represents red.
$\Delta_{i,t}$	Given signal setting of phase i at time step t in which 1 represents green and 0 represents red (a fixed parameter used in the optimization heuristics).
$q_{l,t}$	Traffic demand in time step t on a road link l.
s_i	Saturation flow at signal phase i (the maximum number of vehicles that can move across a traffic signal in a green interval $\delta_{i,t}$), i.e. the exit flow capacity of a signal cell in the CTM, expressed as the number of vehicles.
χ_l	Saturation flow of a road link l (the maximum number of vehicles that can pass a point along the road link in a time interval t), i.e. the exit flow capacity of an ordinary cell in the CTM, expressed in number of vehicles.
ρ_l	Maximum acceptable spatial occupancy on a road link l.
$\sigma_{l,t}$	Sent flow at the end of a road link l at time step t.
g_i	Minimum green at signal phase i expressed as an integral number of time intervals.
r_i	Minimum red for signal phase i expressed as an integral number of time intervals.
$c_{i,j}$	Clearance time (or intergreen) between signal phases i and j expressed as an integral number of time intervals.
$N_{l,k,t}$	Cell holding capacity in cell k on link l at time step t.
$n_{l,k,t}$	Number of vehicles in cell k on link l at time step t.
$Q_{l,k,t}$	Exit flow capacity in cell k on link l at time step t.
V	Free flow speed.
W	Backward wave speed.
$f_{l,k,t}$	Inflow to cell k on link l at time step t.
μ	Common flow multiplier.
$\Gamma(l)$	Function of l to identify all downstream link(s) l' connecting to link l.
$\Lambda(l)$	Function of l to identify all upstream link(s) l^* connecting to link l.
$Z(l)$	Function of l to identify the associated signal phase i installed at the end of link l.
$p_{l,l'}$	Proportion of the sent flow from the end of a single upstream link l to enter all connecting downstream link(s) l' given by $\Gamma(l)$.
λ_{l,l^*}	Proportion of the sent flow from the end of all upstream link(s) l^* given by $\Lambda(l)$ to enter a single downstream link l.
L	Arbitrary large integer number used in linear constraints.
ω	Given numerical parameter.

CONSTRAINT SETS OF THE SIGNALIZED CELL TRANSMISSION MODEL

Minimum Red Duration in Signal Displays

To ensure safety, a minimum red time is usually required for traffic losing the right-of-way and to avoid frequent changes from one state to another (green to red, or red to green) in signal control. The constraint sets for the minimum red are given as follows.

$$-L(\delta_{i,t}-\delta_{i,t+1}-1) \geq \delta_{i,t+2}+\ldots+\delta_{i,t+r_i} \geq L(\delta_{i,t}-\delta_{i,t+1}-1), \qquad t=1,\ldots,T-r_i;\forall i \quad (1)$$

Minimum Green Duration in Signal Displays

The duration of green display in signal phase *i* also, is subject to minimum value g_i. The constraint can be set as

$$-L(\delta_{i,t+1}-\delta_{i,t}-1) \geq g_i-1-\delta_{i,t+2}-\ldots-\delta_{i,t+g_i} \geq L(\delta_{i,t+1}-\delta_{i,t}-1),\ t=1,\ldots,T-g_i;\forall i\,. \quad (2)$$

Clearance Time

To allow sufficient separation time for any pair of conflicting traffic movements controlled by signal phases *i* and *j* to clear the common area in the junction, the following clearance time constraint sets are required.

$$-L(\delta_{i,t}-\delta_{i,t+1}-1) \geq \delta_{j,t+1}+\delta_{j,t+2}+\ldots+\delta_{j,t+c_{i,j}} \geq L(\delta_{i,t}-\delta_{i,t+1}-1),$$
$$\forall t=1,\ldots T-c_{i,j};\forall(i,j)\in\psi \quad (3)$$

$$-L(\delta_{j,t}-\delta_{j,t+1}-1) \geq \delta_{i,t+1}+\delta_{i,t+2}+\ldots+\delta_{i,t+c_{j,i}} \geq L(\delta_{j,t}-\delta_{j,t+1}-1),$$
$$\forall t=1,\ldots T-c_{j,i};\forall(j,i)\in\psi \quad (4)$$

Once the signal display is changed from green to red in two consecutive time steps, constraint sets (3) and (4) become effective in providing sufficient red times to separate the conflicting signal phases by forcing all of its related δs to be 0. This ensures that the conflicting movement is prohibited in the following $c_{i,j}$ intervals, and thus a safety gap between the end of green in one signal phase and the start of green in another is created. ψ defines the set of conflicting traffic signal pairs.

Single Allowable Right-of-Way

The green times of any pair of conflicting signal phases must be displayed exclusively. Otherwise, the two involved traffic movements will receive the green signals and move into the common area of a junction simultaneously. The following constraint sets have to be given to ensure a single right-of-way among the conflicting signal pairs.

$$\delta_{i,t} + \delta_{j,t} \leq 1 , \qquad \forall t = 1,...T; (i, j) \in \psi \tag{5}$$

Total Travel Demand

The first cell of a demand input (source) link acts as a large hypothetical parking lot, which means that it stores the total demand that is scheduled to enter the signalized CTM system. Mathematically, the following constraint set is formulated,

$$n_{l,1,1} = \mu \sum_{t=1}^{T} q_{l,t} , \qquad \forall l \in O , \tag{6}$$

where $n_{l,1,1}$ denotes the number of vehicles in cell 1 of link l at time step t = 1. A common flow multiplier is applied to scale the total demand flows so that the reserve capacity of the signalized CTM system can be evaluated.

Scheduling the Traffic Demand Input

In the CTM, the traffic demand pattern that enters the system can be controlled by the exit flow capacity of the first cell, which already stores all of the traffic demands. A time-varying traffic demand pattern can simply be modeled to require time-varying exit flow capacities. With the following constraint sets, vehicles from different sources can enter the signalized CTM system according to their observed demand patterns,

$$Q_{l,1,t} = \mu q_{l,t} , \qquad t = 1,...,T; \forall l \in O , \tag{7}$$

where $Q_{l,1,t}$ is the exit flow capacity of cell 1 on link l at time step t. As a common flow multiplier has been applied to scale the total traffic demands, it also applies here for consistency.

Infinite Cell Holding Capacity at Exits

The last cell K_l on every exit link l ($\in D$) serves as a large reservoir to store all of the vehicles leaving the system. Its holding capacity is thus set to infinity to maintain a clear exit path on which vehicles can leave the signalized CTM system. The effect of limited space at the exit of a system can also be introduced by assigning a finite capacity. That is,

$$N_{l,K_l,t} = \infty, \qquad t = 1,\ldots,T; \forall l \in D. \tag{8}$$

Holding Capacity of Ordinary Cells

Except the last cells on various exit links with infinite holding capacities, a specific cell holding capacity ω should be assigned in all other cells in the signalized CTM system to model the physical spatial limit of a road link. The following three constraint sets are required.

$$N_{l,k,t} = \omega, \qquad t = 1,\ldots,T; k = 2,\ldots, K_l; \forall l \in O \tag{9}$$

$$N_{l,k,t} = \omega, \qquad t = 1,\ldots,T; k = 1,\ldots, K_l - 1; \forall l \in D \tag{10}$$

$$N_{l,k,t} = \omega, \qquad t = 1,\ldots,T; k = 1,\ldots, K_l; \forall l \notin \{O, D\} \tag{11}$$

Sent Flow at the End of Link

In the signalized CTM system, traffic is able to turn in different directions onto different downstream links through the junctions. Except the exit links that traffic is considered to leave the system while reaching the last cells, sent flows σ are defined in the last cells of other road links. To simplify the network representation in the present formulation, only one link is used to model a roadway that may consist of more than one traffic lane. When two or more traffic directions are permitted at the end of a link, a proportion $p_{l,l'}$ is applied to split the sent flow accordingly. l specifies the upstream link and function $\Gamma(l)$ identifies all connecting downstream link(s) l'. W and V are the backward wave speed and free flow speed, respectively.

$$\sigma_{l,t} \le n_{l,K_l,t}, \qquad \forall t = 1,\ldots,T; \forall l \notin D \tag{12}$$

$$\sigma_{l,t} \le Q_{l,K_l,t}, \qquad \forall t = 1,\ldots,T; \forall l \notin D \tag{13}$$

$$p_{l,l'}\sigma_{l,t} \le \frac{W}{V}\left(N_{l',1,t} - n_{l',1,t}\right), \qquad \forall t = 1,\ldots,T; \forall l \notin D \text{ and } \forall l' = \Gamma(l) \tag{14}$$

Flow Transmission in Consecutive Cells along a Link

In a system without turning movements, the CTM originally uses two nonlinear equations with an embedded minimization. The direct inclusion of constraints makes the program difficult to solve, but a commonly accepted way to avoid nonlinear constraints is to replace them with appropriate linear constraints (Lo, 1999). The following sets of linear constraints are standardized to ensure smooth flow transmission across consecutive cells, either along a road link or between two connecting links.

$$n_{l,k,t+1} = n_{l,k,t} + f_{l,k,t} - f_{l,k+1,t}, \qquad t = 1,\ldots,T-1; k = 1,\ldots,K_l - 1; \forall l \in O \tag{15}$$

$$n_{l,K_l,t+1} = n_{l,K_l,t} + f_{l,K_l,t} - \sigma_{l,t}, \qquad t = 1,\ldots,T-1; \forall l \notin D \tag{16}$$

$$n_{l,K_l,t+1} = n_{l,K_l,t} + f_{l,K_l,t}, \qquad t = 1,\ldots,T-1; \forall l \in D \tag{17}$$

$$f_{l,k,t} \le n_{l,k-1,t}, \qquad t = 1,\ldots,T; k = 2,\ldots,K_l; \forall l \tag{18}$$

$$f_{l,k,t} \le Q_{l,k-1,t}, \qquad t = 1,\ldots,T; k = 2,\ldots,K_l; \forall l \tag{19}$$

$$f_{l,k,t} \le (W/V)(N_{l,k,t} - n_{l,k,t}), \qquad t = 1,\ldots,T; k = 2,\ldots,K_l; \forall l \tag{20}$$

$$f_{l,1,t} = \sum_{l^*} \lambda_{l,l^*} \sigma_{l^*,t}, \qquad t = 1,\ldots,T; \forall l \notin O \text{ and } l^* = \Lambda(l) \tag{21}$$

Again, the present CTM formulation permits different turning directions, and therefore, the inflow f of the first cell on a downstream link may come from more than one upstream link through the corresponding sent flow σ. λ_{l,l^*} denotes the specific sent flow proportion from the end of all upstream link(s) l^*, identified by the function $\Lambda(l)$, to enter a single downstream link l.

Exit Flow Capacity

The exit flow capacity, that denotes the maximum possible number of vehicle passage, of all signal and ordinary cells can be set in the following constraint sets. Except for the exit links, the exit flow capacity at the end of a link is controlled by the signal display. If the signal display is green, then full exit capacity of the signal cell will be given. Conversely, the exit flow capacity becomes zero and vehicles must be stopped if the signal is red. The function $Z(l)$ is developed to identify the signal set at the end of link l.

$$Q_{l,K_l,t} = \delta_{i,t} s_i, \qquad t = 1,\ldots,T; \forall l \notin D \text{ and } i = Z(l) \tag{22}$$

$$Q_{l,k,t} = \chi_l, \qquad t = 1,\ldots,T; k = 2,\ldots,K_l - 1; \forall l \tag{23}$$

$$Q_{l,1,t} = \chi_l, \qquad t = 1,\ldots,T; \forall l \notin O \tag{24}$$

Maximum acceptable spatial occupancy on links

To prevent overflowing due to the stochastic nature of traffic demand, a maximum acceptable degree of saturation, the flow-to-capacity (v/c) ratio of traffic streams, is usually set in conventional design frameworks to provide a buffer that accommodates the short-term flow increase. In this study, we focus on analyzing the physical queue dynamics. It is more meaningful to introduce a maximum threshold for the link spatial occupancy, which will provide a spatial buffer for all the road links with a maximum holding capacity during the whole study period. The following constraint sets are developed to ensure that the maximum spatial utilization of a link, given by $\frac{\sum n}{\sum N}$, is always under a given limit ρ_l.

$$\frac{\sum_k n_{l,k,t}}{\sum_k N_{l,k,t}} \le \rho_l, \quad t = 1,\ldots,T; k = 2,\ldots,K_l; \forall l \in O \tag{25}$$

$$\frac{\sum_k n_{l,k,t}}{\sum_k N_{l,k,t}} \le \rho_l, \quad t = 1,\ldots,T; k = 1,\ldots,K_l - 1; \forall l \in D \tag{26}$$

$$\frac{\sum_k n_{l,k,t}}{\sum_k N_{l,k,t}} \le \rho_l, \quad t = 1,\ldots,T; k = 1,\ldots,K_l; \forall l \notin \{O, D\} \tag{27}$$

OBJECTIVE FOR OPTIMIZATION

The main objective of this study is to maximize the reserve capacity of an entire signal-controlled system, taking the spatial queue limitation into account. In the formulation, the total traffic demands have been given and stored in the first cells of the input links. The time-varying demands entering the CTM system, through corresponding input links, are governed by the relevant exit flow capacity constraints that are given in (7). A common flow multiplier has been explicitly introduced into the mathematical formulation as a flow scaling factor by which the existing flow levels are multiplied. Consequently, the maximum common flow multiplier can be achieved automatically by maximizing the total flow leaving the CTM system, which is the sum of flows in the last cells on all exit links at the end of study period T as given in (28) below, subject to all of the required constraint sets (1-27).

$$\text{Maximize} \sum_{l \in D} n_{l,K_l,T} \tag{28}$$

The problem is formulated as a Binary-Mix-Integer-Linear-Program (BMILP) that can be solved by a standard branch-and-bound technique. To simplify the presentation of the optimization heuristics outlined in the next section, the term "BMILP" is used to represent this linear programming problem. If the optimized common flow multiplier μ^*, by solving the BMILP, is greater than 1, then the system is considered to have $(\mu^* - 1)$*100% reserve capacity. Conversely, if μ^* is less than 1, then the system is overloaded by $(1 - \mu^*)$*100%.

OPTIMIZATION HEURISTICS

In previous sections, the objective function and relevant governing constraints have been presented to design the signalized CTM system, taking the spatial queue limit and time-varying traffic demand into considerations. A deterministic solution can be achieved by solving the complete BMILP directly using standard routines. One imminent challenge is to apply the present formulation to design a dynamic signal plan under a time-varying traffic condition. In general, traffic demand pattern for analysis should consist of pre-peak, peak, and post-peak periods. Sufficient modeling time must be given to put forward the vehicle queue formation and dissipation across these periods. In the present formulation, no mathematical constraint is developed to restrict the length of the whole study period T. Practically, if a finer time step (say 2 seconds/time step) is adopted to preserve a high precision in modeling, huge number of binary variables is required to define the signal settings. The resultant BMILP problem may require substantial computation efforts to be solved. To reduce this computational difficulty, the original optimization problem is divided into a series of smaller sub-problems and each of them contains only a manageable subset of the signal variables with a sub-period length of $T^{'}$ for optimization. The signal variables within the sub-period $T^{'}$ are to be optimized and those outside the sub-period (but still within the study period T) are held fixed. After one batch of signal variables is optimized, the resultant signal settings become fixed parameters and the following batch of signal parameters is relaxed as variables for optimization. The procedure repeats until the whole study period T is optimized. Detailed steps of the solution algorithm for the reserve capacity optimization are summarized below. Since the algorithm continuously modifies a subset of signal variables, a feasible signal plan has to be given as initial settings. Epoch is a counter to activate different batches of signal parameters as variables for optimization.

Step 0: Prepare a feasible signal plan such as the minimum green setting to be the fixed signal setting $\Delta_{i,t}$ in the whole study period, set Epoch = 0, and evaluate the common flow multiplier as an initial reference

Step 1: Set Epoch = 1

Step 2: If Epoch $< \frac{T}{T^{'}}$, then

Solve the BMILP with

$$\delta_{i,t} = \Delta_{i,t}, \forall i, t \notin \{(\text{Epoch}-1)T^{'}+1, \ldots, (\text{Epoch})T^{'}\}$$

Collect the common flow multiplier μ (Epoch) and the optimization results; update and set

$$\Delta_{i,t} = \delta_{i,t}, \forall i, t \in \{(\text{Epoch}-1)T^{'}+1, \ldots, (\text{Epoch})T^{'}\}$$

Step 3: If Epoch $> \frac{T}{T^{'}}$, then

Solve the BMILP with

$$\delta_{i,t} = \Delta_{i,t}, \forall i, t \notin \left\{(\text{Epoch}-\frac{T}{T'}-1)T^{'}+\frac{T^{'}}{2}+1, \ldots, (\text{Epoch}-\frac{T}{T^{'}})T^{'}+\frac{T^{'}}{2}\right\}$$

Collect the common flow multiplier μ (Epoch) and the optimization results; update and set

$$\Delta_{i,t} = \delta_{i,t}, \forall i, t \in \left\{(\text{Epoch}-\frac{T}{T'}-1)T^{'}+\frac{T^{'}}{2}+1, \ldots, (\text{Epoch}-\frac{T}{T^{'}})T^{'}+\frac{T^{'}}{2}\right\}$$

Step 4: Set Epoch = Epoch +1

Check if Epoch < $\frac{2T}{T'}$, then return to Step 2.

Step 5: Stop

Further implementation details are referred to in the numerical example section.

NUMERICAL EXAMPLE

Geometric layout of the signal-controlled system

To demonstrate how the reserve capacity of a signal-controlled system can be optimized, taking into account the effect of physical queuing, the cell transmission model (CTM) is applied to model a staggered signal-controlled junction system as shown in Figure 1. The staggered junction system contains 8 road links, in which links L3 and L8 are short links. Traffic signals are installed at the end of links L1, L2, L3, L6, L7, and L8 (the two exit links L4 and L5 are excluded) to control the conflicting movements involved. The arrows represent the lane markings that show the directions which are permitted on different road links. L1, L2, L6, and L7 are input (source) links on which traffic demands are generated and entered into the signalized CTM system.

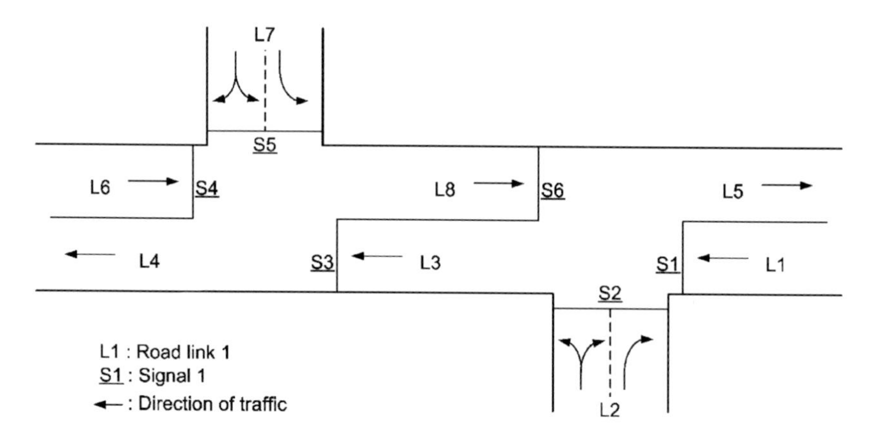

Figure 1 Example staggered junction.

In the CTM, each road link is discretized and represented by a series of homogenous cells, as given in Figure 2. Vehicles that are scheduling to enter a road link are stored in the first cell (Cell 1) of that link. After one time interval, those vehicles that are already in the first cell will move to the next downstream cell (Cell 2). Vehicles in Cell 2 will move to Cell 3 and so on up to Cell K_{l-1} to Cell K_l. The amount of traffic that can proceed forward depends on the downstream spatial availability and the link saturation flow. Instead of moving forward to next downstream cells, vehicles must hold up and stay in the current cell if there are not sufficient spaces available. Upstream traffic will even be blocked and held up simultaneously. A physical vehicle queue may then develop to realize the congestion effects.

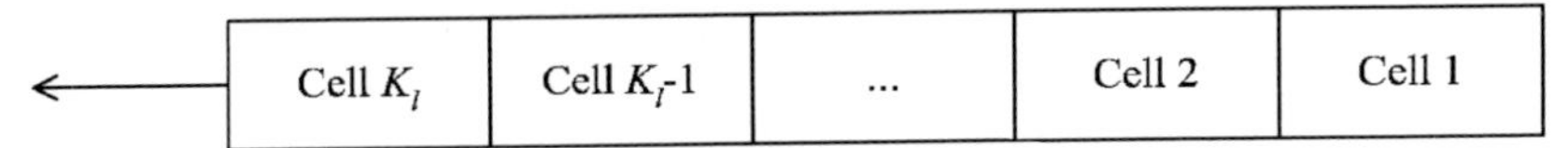

Figure 2 Cell representation of road link l.

The modeling details of the road links that are used in the example staggered junction are given in Table 1. Five cells are used to model all input links (L1, L2, L6, and L7). Three cells are given for all exit links (L4 and L5). Road links L3 and L8 are defined as short links that contain only a single cell. For all of the road links (except the two exit links) in the system, there are traffic signals, given by a function Z, installed at the ends of the associated links to control the conflicting junction traffic. Two other functions, Γ and Λ, are employed to identify, for different road links, all respective downstream and upstream connecting road links.

Table 1 Modeling details of the road links in the example junction system.

Link, l	Total no. of cells, K_l	Signal in cell no.	$i = Z(l)$	$l' = \Gamma(l)$	$l^* = \Lambda(l)$	Remarks
1	5	5	1	3		Input link, $\in O$
2	5	5	2	3,5		Input link, $\in O$
3	1	1	3	4	1,2	Short link
4	3	N/A			3,7	Exit link, $\in D$
5	3	N/A			2,8	Exit link, $\in D$
6	5	5	4	8		Input link, $\in O$
7	5	5	5	4,8		Input link, $\in O$
8	1	1	6	6	6,7	Short link

Input assumptions

Referring to the left-hand T-junction in Figure 1, it can be observed that traffic movements on link L7 contain both left and right turns that are conflicting with those on links L3 and L6. Correspondingly, a minimum 6-second (= 3 time steps) clearance time is assigned for the signal pairs $(i, j) = \{(3,5),(5,3),(4,5),(5,4)\}$. With these signal settings, only one signal sequence can be chosen for practical implementation. To demonstrate the design capability of the present formulation to deal with complicated signal sequences and settings, traffic signals S1, S2, and S6 are not allowed to display green signals concurrently in the right-hand T-junction. This implies that the traffic on links L1, L2, and L8 can only move across the junction exclusively. In this way, the feasible combinations and possible choices of signal sequences are increased. It is thus expected that a more complex signal sequence can be produced in the optimization framework. Details of the minimum clearance time that is adopted in the present design problem are tabulated in Table 2.

Table 2 Minimum clearance time matrix $c_{i,j}$ (in time intervals)

$c_{i,j}$		Signal phase j					
		1	2	3	4	5	6
Signal phase i	1		3				3
	2	3					3
	3					3	
	4					3	
	5			3	3		
	6	3	3				

Note: Signal pair $(i, j) \in \Psi$ if $c_{i,j} > 0$.

Another key input for the present signal optimization problem is the traffic demand pattern. The demand flows that enter the signalized CTM system through different input links are time-varying, covering pre-peak, peak, and post-peak periods, as used in the computer package OSCADY 5 developed by TRL (Binning and Meikle, 2003). The demand flow profile is synthesized and supposed to be survey data, whereby the pre-peak period and post-peak periods occur in the first and last 15 minutes respectively, at constant demand level, taking up only 75% of the hourly flow rate. The peak demand, occurring exactly at the middle of the study period, is 1.5 times higher than the pre- and post-peak demand levels (or 1.125 times higher than the hourly flow rate). The flow inputs are given in Table 3 below. Having identified the pre-peak, peak, and post-peak demand levels and their corresponding times of occurrence, a normal curve is fitted to generate a time-varying flow profile for every minute during the peak period. The demand flow starts to rise after the first 15 minutes (the pre-peak period) and reaches the peak at exactly the middle of the study period. After the peak, the demand flow drops immediately and remains constant for the last 15 minutes as the post-peak period. A minute-by-minute demand flow profile is constructed as an input into the signalized CTM system. A fixed proportion of 75% and 25% of the demand flow from link L2 will turn into downstream links L5 and L3, respectively. Similarly, 75% and 25% of the demand flow from link L7 will turn into links L8 and L4, respectively. It is also assumed that links L2 and L7 contain a single shared lane for both left- and right- turn traffic and thus all turning flows will be blocked if either one of the associated downstream cells are fully occupied.

Table 3 Input demand pattern.

Input link	L1	L2	L6	L7
Existing hourly flow rates (veh/hr)	225	250	300	250
Pre-peak and Post peak demands (veh/min)	2.81	3.13	3.75	3.13
Peak demands (veh/min)	4.22	4.70	5.63	4.70

In this example, we model and optimize the traffic flow pattern and signal settings for 1-hour comprising 1,800 time steps (= 3,600 seconds). The problem has been formulated as a

BMILP. To model the 6 signal phases for the present numerical example, 10,800 (= 1800 x 6) binary variables are required. This problem size is difficult to be solved computationally. From trial experiences, the optimization problem is still manageable if the problem size is trimmed to cover 180 time steps (= 6 minutes) at a time. Hence, the optimization heuristics are implemented with T = 1,800, $T^{'}$ = 180, and minimum green settings as the initial signal plan. Each of these reduced sub-problems, 389,393 linear constraints, 201,598 continuous variables, and 1,080 binary variables are defined. The required computation time is around 20 minutes running on a Pentium D processor of 3.6G Hz.

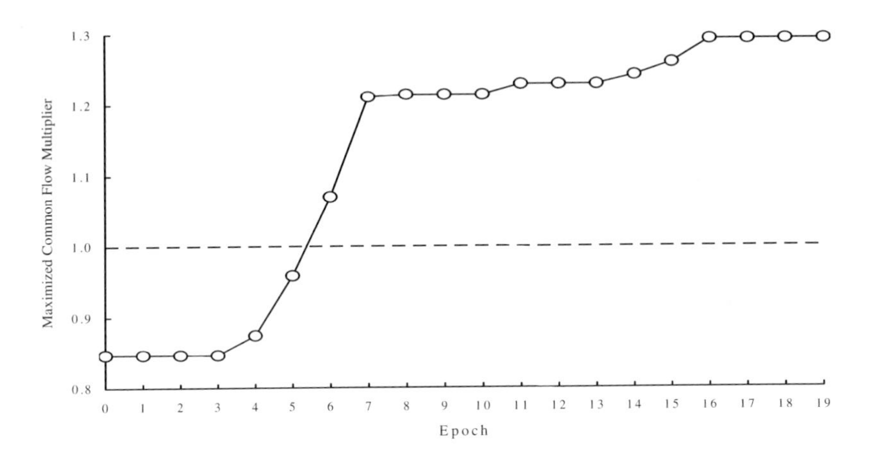

Figure 3 Optimization trend of the common flow multiplier.

Optimization results

Figure 3 plots the optimization trend of the maximized common flow multiplier against each epoch. At the initial epoch, the minimum green setting is applied to control the example junction system throughout the entire study period. The evaluated common flow multiplier is 0.85, thus indicating that the whole system is overloaded by 15.0%. This simply verifies that the minimum green setting is not good enough to control the given traffic demands in the example junction system. The proposed optimization heuristics then start to modify this initial signal setting. It can be observed that the maximized common flow multiplier rises after the third epoch and increases continuously until the seventh epoch. Within these time steps from 541 to 1,260, the traffic demands, scheduled to enter the signalized CTM system, are greatest according to the normal distributed demand profiles. During this critical period, the total number of vehicles inside the entire signalized CTM system is also the highest. This trend can be roughly traced from Figure 4, which provides the variation of the total number of vehicles against the time step. Once the signal settings are effectively optimized during this peak demand period to let the highest incoming traffic leave the system smoothly without inducing serious congestion, then the overall system capacity in terms of the common flow multiplier can be maximized. Still, a very slight increase in the common flow multiplier is found from the seventh to tenth epochs as the peak demands are prepared to leave the system and the lower post-peak traffic is again not critical enough to affect the maximized common flow multiplier. From the eleventh to the final epochs, the optimization heuristics are specifically designed to refine the signal settings across the transition periods (the break points between two successive epochs over the first ten epochs). Hence, only steady and marginal improvements on the overall system performance are perceived in terms of the maximized common flow multiplier. After completing the proposed solution algorithm, the common flow multiplier is found to be increased from 0.85 to 1.29, thus implying that the example staggered junction system possesses 29% reserve capacity with the implementation of the optimized signal setting.

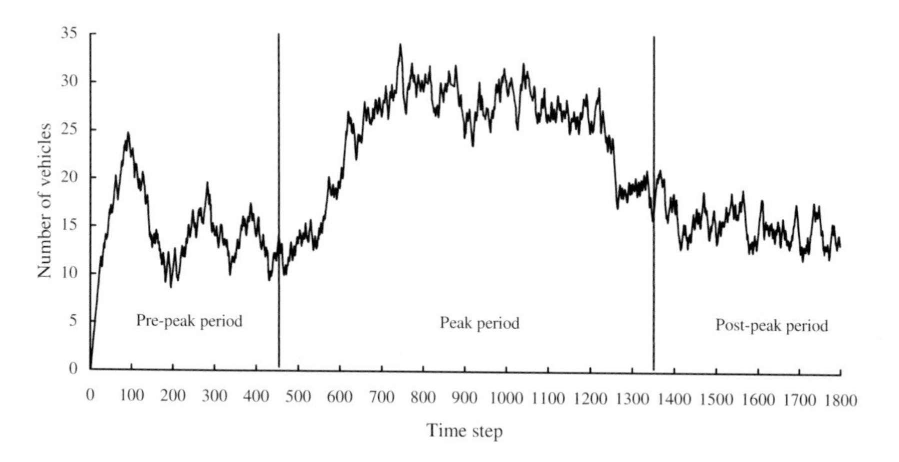

Figure 4 Total number of vehicles in the signalized CTM system.

Table 4 gives the details of the optimized cell contents from Links L1 to L4 during the most congested minute (the network contains the greatest number of vehicles within 30 time steps over the whole study period). During this most congested period, 4.44 (140.56-136.12) and 4.96 (157.43-152.47) vehicles enter the system through input links L1 and L2 respectively, and 5.75 (125.33-119.58) vehicles leave the system via exit link L4.

To further demonstrate the modeled queue dynamics in the most congested period (minute), variations of the queues on the two short links L3 and L8 are also plotted in Figure 5. Because the maximum acceptable spatial occupancy ρ takes on 1.0 and the cell holding capacity N equals 4.0 in the present example, the highest number of vehicles that appear on the short link L8 is 4.0 during time intervals from 741 to 742. As also given in Table 4, the maximum vehicle queue in the most congested period on link L3 is only 3.75 during time intervals from 739 to 746. For the whole study period (1,800 time intervals), there are 248 time intervals in which a maximum queue (= 4.0 vehicles) is formed on link L8. However, there are only 13 time intervals with a maximum queue on link L3. The average queues observed on links L3 and L8 are 0.86 and 1.76 vehicles respectively within the whole study period. With the current input demand profiles, it is consistent that short link L8 will be more congested and critical than short link L3 under the present control system.

Table 4 Partial flow pattern during the most congested minute (30 time steps).

t	Link L1, Cell					Link L2, Cell					Link L3, Cell	Link L4, Cell		
	1	2	3	4	5	1	2	3	4	5	1	1	2	3
723	140.56	0.15	0.15	3.29	4.00	157.43	0.34	0.00	2.71	4.00	1.00	0.75	0.00	119.58
724	140.41	0.15	0.15	3.44	4.00	157.26	0.51	0.00	2.71	4.00	1.00	0.00	0.75	119.58
725	140.25	0.15	0.15	3.59	4.00	157.09	0.68	0.00	2.71	4.00	0.00	1.00	0.00	120.33
726	140.10	0.31	0.00	3.75	4.00	156.92	0.17	0.68	2.71	4.00	0.00	0.00	1.00	120.33
727	139.95	0.15	0.31	3.75	4.00	156.75	0.17	0.86	2.71	4.00	0.00	0.00	0.00	121.33
728	139.79	0.15	0.20	4.00	4.00	156.58	0.17	1.03	2.71	4.00	0.00	0.00	0.00	121.33
729	139.64	0.15	0.36	4.00	4.00	156.41	0.17	0.20	3.71	3.00	0.25	0.00	0.00	121.33
730	139.49	0.15	0.51	4.00	4.00	156.24	0.17	0.17	2.91	3.00	0.50	0.00	0.00	121.33
731	139.33	0.15	0.66	4.00	4.00	156.06	0.17	0.17	2.08	3.00	0.25	0.50	0.00	121.33
732	139.18	0.15	0.82	4.00	4.00	155.89	0.17	0.17	1.25	3.00	0.50	0.50	0.00	121.33
733	139.03	0.15	0.97	4.00	4.00	155.72	0.17	0.17	0.42	3.00	0.75	0.50	0.00	121.33
734	138.88	0.15	1.12	4.00	4.00	155.55	0.34	0.00	0.17	3.42	0.75	0.50	0.00	121.33
735	138.72	0.15	1.27	4.00	4.00	155.38	0.17	0.34	0.00	3.60	0.75	0.50	0.00	121.33
736	138.57	0.15	1.43	4.00	4.00	155.21	0.17	0.17	0.34	3.60	0.75	0.50	0.00	121.33
737	138.42	0.15	1.58	4.00	3.00	155.04	0.17	0.17	0.17	3.94	1.75	0.50	0.00	121.33
738	138.26	0.15	1.73	3.00	3.00	154.87	0.17	0.17	0.28	4.00	2.75	0.00	0.50	121.33
739	138.11	0.15	0.89	3.00	3.00	154.69	0.17	0.17	0.45	4.00	3.75	0.25	0.00	121.83
740	137.96	0.15	0.15	2.89	4.00	154.52	0.17	0.17	0.62	4.00	3.75	0.25	0.25	121.83
741	137.81	0.15	0.15	3.04	4.00	154.35	0.17	0.17	0.79	4.00	3.75	0.25	0.25	122.08
742	137.65	0.15	0.15	3.19	4.00	154.18	0.17	0.17	0.97	4.00	3.75	0.25	0.00	122.33
743	137.50	0.15	0.15	3.34	4.00	154.01	0.17	0.17	1.14	4.00	3.75	0.25	0.00	122.33
744	137.35	0.15	0.15	3.50	4.00	153.84	0.17	0.17	1.31	4.00	3.75	0.00	0.25	122.33
745	137.19	0.15	0.15	3.65	4.00	153.67	0.17	0.17	1.48	4.00	3.75	0.00	0.00	122.58
746	137.04	0.15	0.15	3.80	4.00	153.50	0.17	0.17	1.65	4.00	3.75	0.00	0.00	122.58
747	136.89	0.15	0.15	3.96	4.00	153.33	0.34	0.17	1.65	4.00	3.00	0.75	0.00	122.58
748	136.74	0.15	0.26	4.00	4.00	153.15	0.17	0.34	1.82	4.00	2.00	1.00	0.75	122.58
749	136.58	0.15	0.41	4.00	4.00	152.98	0.17	0.17	2.16	4.00	2.00	0.00	1.00	123.33
750	136.43	0.15	0.57	4.00	4.00	152.81	0.17	0.17	2.33	4.00	1.00	1.00	0.00	124.33
751	136.28	0.15	0.72	4.00	4.00	152.64	0.17	0.17	2.51	4.00	1.00	0.00	1.00	124.33
752	136.12	0.16	0.87	4.00	4.00	152.47	0.17	0.17	2.68	4.00	1.00	0.00	0.00	125.33

As for the optimized signal settings, details of a 2-minute signal setting in three different time periods are extracted from the optimization results for discussion. Table 5 shows the timing details of the six signal phases during time steps 451-510, covering 120 seconds just after the pre-peak period. Table 6 gives the signal settings during time steps 871-930, which are another 120 seconds that exactly covers the highest peak in demand inputs. Table 7 illustrates the signal timings during time steps 1351-1410 at the beginning of the post-peak period. Very different signal settings are observed without fixed cycles and signal sequences. For instance, referring to the minimum clearance times given in Table 2, exclusive rights-of-way in signal phases 1, 2, and 6 are all separated by at least 3 time steps (= 6 seconds). The optimized signal sequence is [6-1-6-2] in the first two minutes when the peak period commences. The signal sequence becomes [6-2-6-1-2-1] right at the peak, and a more complicated signal sequence [6-1-2-6-2-6-1-6] occurs just after the peak period. Green duration patterns also vary greatly in all six signal phases. A minimum of 3 time steps (= 6 seconds) and a maximum of 18 time steps (= 36 seconds) of green durations are assigned in Signal 6 in different time periods.

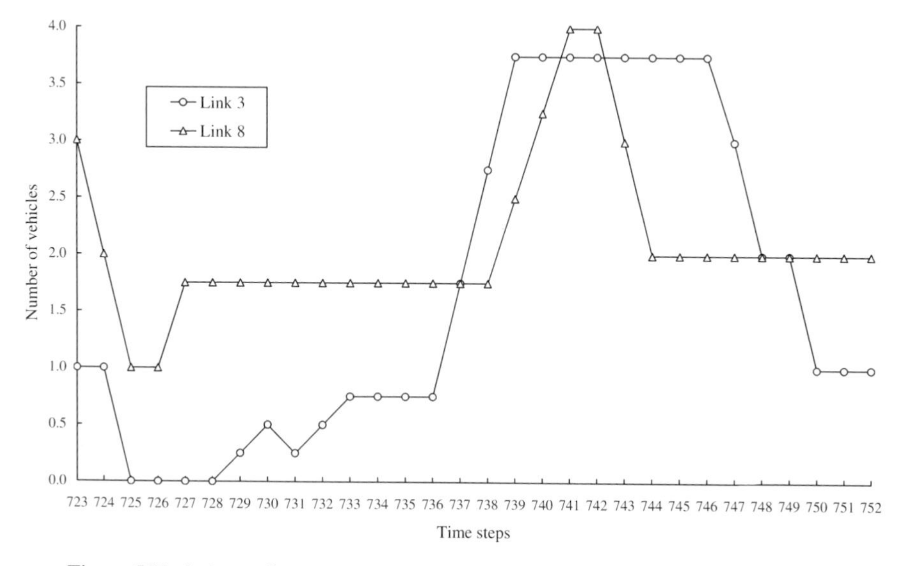

Figure 5 Variations of queues on the two short links in the most congested minute.

Table 5 Optimization results of signal settings when the peak demand period commences.

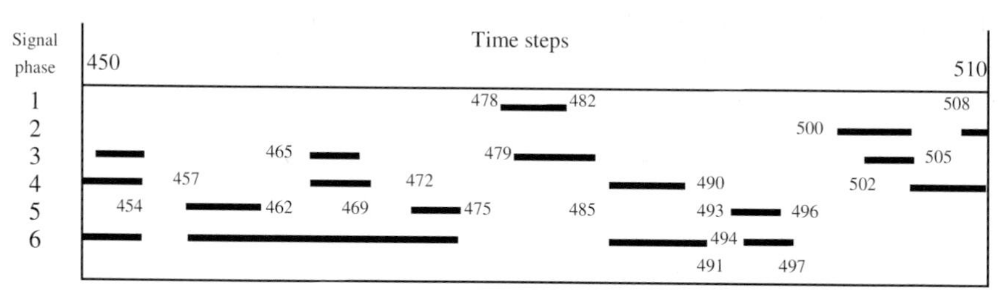

Table 6 Optimization results of signal settings right at the peak demand period.

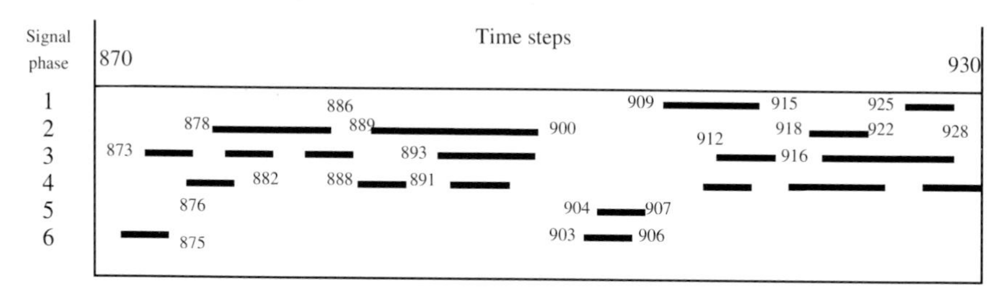

Table 7 Optimization results of signal settings upon the end of the peak demand period.

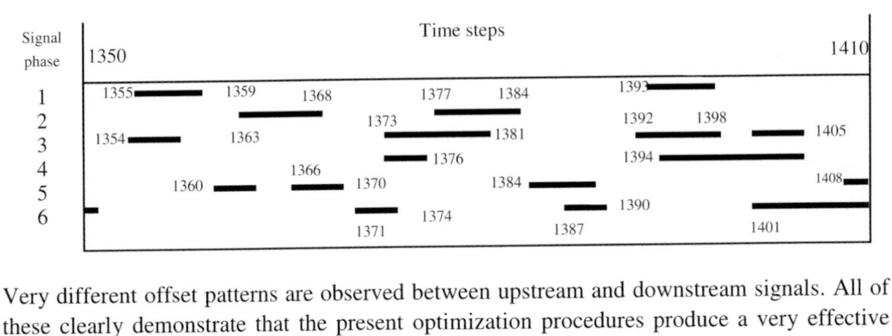

Very different offset patterns are observed between upstream and downstream signals. All of these clearly demonstrate that the present optimization procedures produce a very effective dynamic signal plan to control the time-varying traffic demands.

CONCLUSIONS

A reserve capacity optimization concept is applied in this study to design a signal-controlled network. The signalized cell transmission model is used as the modeling platform, taking the time-varying traffic demands and spatial queue dynamics into considerations. A common flow multiplier is defined as a flow scaling factor and introduced into the mathematical formulation. The maximum flow multiplier that represents the reserve capacity of a signalized system can be obtained by maximizing the total number of vehicles leaving the signalized CTM system at the end of the study period, which has been adopted as an objective function for optimization. Starts and durations of green are defined as control variables. The model constraints, including the control of the flexible group-based design parameters and relevant flow transmission in consecutive cells within the CTM, are developed in a linear framework. To realize the properties of the spatial queue limitation on short roadways, a set of constraints for the maximum acceptable spatial occupancy are included. The design problem is formulated as a Binary-Mix-Integer-Linear-Program (BMILP) problem that can be solved by standard branch-and-bound routines. Another challenge in the present study is to consider a large set of binary variables for the signal settings, which may create certain computational difficulties. Hence, optimization heuristics by considering a manageable subset of binary variables at a time until the whole study period is optimized are proposed which enhance the practical applicability of the formulation. A staggered junction with 8 traffic links and 6 signal phases is used as a numerical example for demonstrations. Promising solution results are obtained. Completing the present study, a reserve capacity of a signal-controlled system can be evaluated with the considerations of the spatial queue limitation along roadways. Vehicles can be held up and waiting for suitable green signals for dissipations. Traffic demands generated during peak periods can leave the system at the end utilizing some post-peak spare capacities. The reserve capacity can therefore be traded with the length of a given study period. A single reserve capacity may only represent one particular case. Should we consider in the future expanding the reserve capacity into a time series representation in order to give a more comprehensive assessment for a signal-controlled system?

ACKNOWLEDGEMENTS

Authors would like to thank the anonymous referees for their valuable comments and suggestions made to the earlier version of the paper.

The work described in this paper was jointly supported by grants from the City University of Hong Kong (Project Number: 7001967) and the Research Grants Council of the Hong Kong Special Administrative Region, China (Project Nos. HKU7031/02E, HKU7187/05E, and HKUST6283/04E).

REFERENCES

Allsop, R.E. (1972) Estimating the traffic capacity of a signalized road junction, *Transportation Research*, **6** (3) 245-255.

Binning, J. C. and Meikle, L. (2003) OSCADY 5.0 User Guide, *TRL Software Bureau.*

Daganzo, C.F. (1994) The cell transmission model: a dynamic representation of highway traffic consistent with the hydrodynamic theory, *Transportation Research*, **28B** (4) 269-287.

Daganzo, C.F. (1995a) A finite difference approximation of the kinematic wave model of traffic flow, *Transportation Research*, **29B** (4) 261-276.

Daganzo, C.F. (1995b) The cell transmission model, part II: network traffic, *Transportation Research*, **29B** (2) 79-93.

Gallivan, S. and Heydecker, B.G. (1988) Optimising the control performance of traffic signals at a signal junction, *Transportation Research*, **22B** (5) 357-370.

Heydecker, B.G. (1992) Sequencing of traffic signals, in J.D. Griffiths (Ed.) *Mathematics in Transport and Planning and Control*, 57-67, Clarendon Press, Oxford.

Lighthill, M.J. and Whitham, J.B. (1955) On kinematic waves. I. Flow movement in long rivers. II. A theory of traffic flow on long crowded road. *Proceedings of the Royal Society A229*, 281-345.

Lin, W.H. and Wang, C. (2004) An enhanced 0-1 mixed-integer LP formulation for traffic signal control, *IEEE Transactions on Intelligent Transportation Systems*, **5**(4) 238-245.

Lo, H.K. (1999) A novel traffic signal control formulation, *Transportation Research*, **33A** 433-448.

Lo, H.K. (2001) A cell-based traffic control formulation: strategies and benefits of dynamic timing plans, *Transportation Science*, **35** (2) 148-164.

Lo, H.K., Chang, E., and Chan, Y.C. (2001) Dynamic network traffic control, *Transportation Research*, **35A** 721-744.

Maher, M. (2005) Reserve capacity for a set of closely-spaced intersections. Paper presented at the *Fourth IMA International Conference on Mathematics in Transport*, 7-9 July, London, U.K..

Richards, P.I. (1956) Shockwaves on the highway. *Operation Research*, **4**, 42-51.

Webster, F.V. and Cobbe, B.M. (1966) Traffic signals. *Road Research Technical Paper No. 56*. HMSO, London.

Wong, C.K. and Wong, S.C. (2003) Lane-based optimization of signal timings for isolated junctions, *Transportation Research*, **37B** (1) 63-84.

Wong, C.K. and Wong, S.C. (2006) Lane-based optimization method for multi-period analysis of isolated signal control junctions, *Transportmetrica*, **2**, 53-85.

Wong, C.K., Wong, S.C., and Lo, H.K. (2005) A cell transmission model for signal timing optimization in work zones. Paper presented at the *Fourth IMA International Conference on Mathematics in Transport*, 7-9 July, London, U.K..

Wong, C.K., Wong, S.C., and Lo, H.K. (2006) A spatial queuing approach to optimize coordinated signal settings to obviate gridlock in adjacent work zones. *Journal of Advanced Transportation*, submitted.

Wong, S.C. (1996) On the reserve capacities of priority junctions and roundabouts, *Transportation Research*, **30B** (6) 441-453.

Wong, S.C. and Yang, H. (1997) Reserve capacity of a signal-controlled road network, *Transportation Research*, **31B** (5) 397-402.

Transportation and Traffic Theory 2007
Edited by R.E. Allsop, M.G.H. Bell and B.G. Heydecker

24

TIME DEPENDENT DELAY AT UNSIGNALIZED INTERSECTIONS

Werner Brilon, Ruhr-University Bochum, Germany

SUMMARY

Calculation of intersection delay is usually based on methods obtained from queuing theory. Due to the variability of traffic demand over time the estimation of delays for time-dependent flow and capacity, where also a temporary overload is allowed, is of primary interest. For solving this problem a variety of methods is used in current practice. All of these solutions are only approximations. One first step of approximation is the assumption that the priority system can be modeled by an M/M/1-queue. The second step is the so-called coordinate transformation technique. For this method three sub-groups can be defined. The paper investigates the background of the possible solutions and the quality of approximation. As a basis, a classification of potential delay formulas is defined. This classification accounts for the kind of sophistication of the approximation and for the kind of delay definition, which is treated as the average. Over all, nine useful classes of formulas can be defined. For each of these classes delay formulas are derived. Some of them correspond to well-known results. However, in addition to that the complete set of results offers new solutions - also for more realistic cases. Thus, also initial queues at the beginning of the observed peak period as well as different conditions in the post-peak period can be described. As methods for validating these formulas a Markov-chain formulation has been developed to produce numerically exact results. Also stochastic simulations and empirical data are used for comparison to check the approximate solutions against reality. As a result, a set of equations is available which can be applied to estimate average delays at unsignalized intersections for well-defined traffic conditions. The paper makes clear that instead of an uncritical use of delay formulas a well sophisticated selection of the adequate equation is also required in practice.

INTRODUCTION

For unsignalized intersections the average delay has been chosen as the representative measure of effectiveness for characterizing the performance of traffic flow by most of the guidelines used in practice (USA: HCM 2000, chapter 17; HBS 2001, chapter 7). Therefore a reliable and realistic estimation of average delay is of significant importance in traffic engineering. Guidelines use several different methods for calculating delays; e.g. the HCM uses the Akcelik, Troutbeck (1991) method whereas the German HBS recommends the rather complex formula proposed by Kimber, Hollis (1979; eq. 22). Both lead to different results. There is no clear insight, which of both is more correct or if alternative solutions might be even better.

The important aspect of the average delay estimation at intersections is the treatment of a temporary oversaturation. Here delay estimation turns out to be one specific application of a more general problem of mathematical queuing theory. The unfortunate fact is that useful analytical solutions are not offered from mathematical theory. Therefore engineering scientists have developed approximate formulas (Kimber, Hollis 1979; Akcelik, Troutbeck 1991 and other publications by Akcelik; Brilon, 1995).

Most of these approaches use the derivation of time dependent patterns for queue length as a starting point. A rather precise but still approximate method to estimate time-dependent queue length distributions are the differential equations given by Newell (1982; see also Troutbeck, Blogg 1998) derived from his so-called diffusion theory. Such time-dependent queue length functions can be integrated to derive delay parameters like average delay. Such an approach seems, however, to be too complicated for practical application. Moreover, if all solutions are still of approximate nature then for the sake of easy application it seems to be desirable to find solutions directly on the scale of delays.

This paper shows that the well-known solutions for average delays at unsignalized intersections are specific members of a larger family of approximate solutions, where other elements of the whole set of delay equations provide more advantages. A classification of different possibilities for average delay definition is also offered. Among all possible cases the traditional formulas only constitute a solution for rather specific and unrealistic cases. These considerations also make clear that a sound understanding of the formulas' background is crucial for a correct application. The key to the assessment of a useful delay estimation technique is, however, the comparison with exact results, which for this paper have been elaborated using Markov chain techniques, simulations, and some empirical data.

THE BASIC MODEL

Our considerations are concentrated on the simple case of one major stream and one minor stream where the vehicles from the minor stream have to give priority to major street vehicles (= "priority system"; cf. Figure 1). This priority system can be modeled by a queuing system where the first space for a vehicle next to the stop line is treated as the service counter whereas the further waiting positions for minor street vehicles form the queue. For the terminology within the paper we use:

s = service time = time spent by vehicles in the first position (= $1/c$, if we represent the priority system by a M/M/1-queue)

d = delay = time spent by vehicles in the queue (without the first position)

w = waiting time = time spent by minor street vehicles in the priority system $= d + s$ (definition following Heidemann's (2002) proposal)

We denote the averages by small letters, whereas capitals stand for the sum of all delays (D) and the sum of all waiting times (W) respectively.

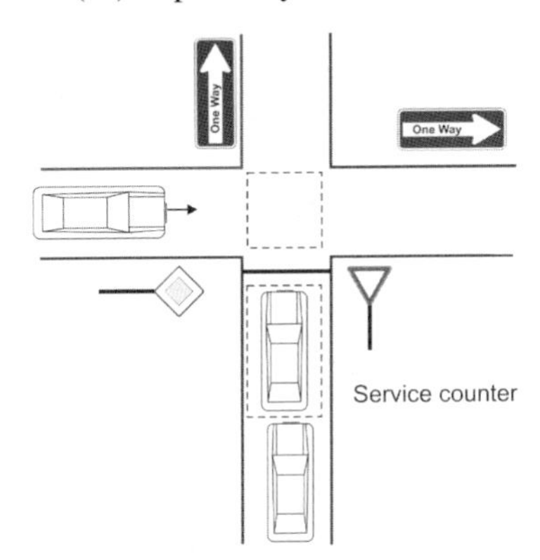

Figure 1: The "priority system"

There are different basic methods for estimating the capacity c, i.e. the maximum possible throughput for the minor stream, like the empirical regression theory (Kimber, Coombe, 1980), the critical gap theory, or - more recently - the conflict method (Brilon, Wu, 2002; Brilon, Miltner, 2003). These methods become even more complex if lower rank movements (e.g. left turning minor street movement) are treated. A good overview is given by Troutbeck, Brilon (2000) or Luttinen (2004). These methods should not be discussed here in detail. It is, however, desirable that the estimation of delay should become independent from the way of capacity calculation. Otherwise the solutions will become too complicated.

For undersaturated conditions, i.e. demand q is less than capacity c, the approximation of the priority system by an M/M/1-queue is rather popular among researchers (cf: Kimber, Hollis, 1979, Kimber et al (1986), or Heidemann, 2002). Figure 2 shows that the M/M/1-delay is not necessarily equal to the priority system delay. For this example illustration the average waiting time has been calculated based on gap acceptance theory using a set of equations given by Kremser (1962, 1964) and arranged as an equation for average waiting time by Brilon (1988). This arrangement is based on Yeo's results (Yeo, 1962; see also comments in Brilon, 1995). Kremser's equations have later been improved by Daganzo (1977). This improvement has not been used for this example due to the rather complicated form of the equations. For this comparison the following parameters have been chosen: t_c = critical gap = 6 s, t_f = follow-up time = 3 s, q_p = major street volume = 350 veh/h (Figure 2, left side) and q_p = 600 veh/h (right side).

For the M/M/1-queue the average time of customers in the system (i.e. the waiting time) is

$$w = \frac{1}{R} = \frac{1}{c-q} = \frac{1}{c \cdot (1-x)} \tag{1}$$

The average delay for the M/M/1-queue is

$$d = w - \frac{1}{c} = \frac{x}{c \cdot (1-x)} \tag{2}$$

where

w	=	average waiting time	[s]
d	=	average delay	[s]
R	=	reserve capacity = c - q	[veh/s]
c	=	capacity	[veh/s]
q	=	demand volume	[veh/s]
x	=	degree of saturation = q/c	[-]

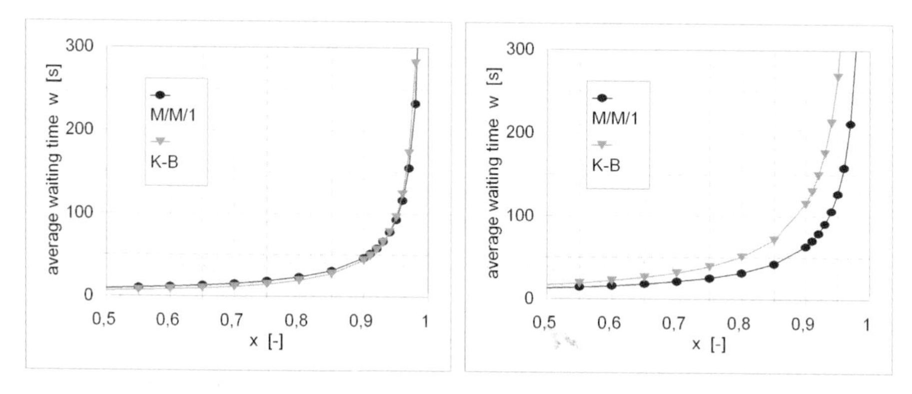

Figure 2 Comparison of the M/M/1-waiting time with the waiting time calculated by the Kremser/Brilon-method (K/B); left side: q_p = 350 veh/h, right side: q_p = 600 veh/h

Figure 2 shows that there are cases of complete compliance of average waiting time between the M/M/1-solution and the priority system. However, also significant differences might be possible where the difference could be even larger than Figure 2 suggests, depending on the driver's behavior characteristics (expressed by the critical gap t_c and the follow-up time t_f)

and on major street traffic volumes. Nevertheless, due to the similarities there is no useful alternative than to represent the priority system by a M/M/1-queue, if parameters of traffic performance like delays or queue lengths are to be described. Thus, the use of rather complicated equations for steady state priority delay is avoided. Moreover, the delay estimation becomes independent from the method of capacity calculation. It should, however, be noted that the representation of the priority system by the M/M/1-queue can cause a bias, especially for large major traffic volumes, i.e. for small capacities.

THE COORDINATE TRANSFORMATION METHOD

With this approximation by the M/M/1-queue only undersaturated conditions (i.e. $x = q/c < 1$) can be described. For practical purposes it is, however, also necessary to estimate delays for situations where the demand volume (q) is variable over time and where it could even exceed the capacity c during a specific peak period. To describe the average delay suffered by minor street drivers during such an oversaturated period no solutions which are based on an exact statistical theory are in use since they – if a solution would be available – would become too complicated. Instead, rather pragmatic approximations are in use.

These approximations go back to an idea by a researcher named Whiting who contributed much to traffic research but did not publish in his own name due to personal reasons (Kimber, Hollis, 1979, p. 6; Allsop, 1992). The idea was developed fully by Kimber, Hollis (1979). The method is also characterized as coordinate transformation by several authors. It is illustrated in Figure 3. There we see the M/M/1-waiting time (eq. 1) as a function of the degree of saturation ($= x = q/c$) and the deterministic delay d_d. The deterministic delay is valid for a D/D/1-queueing system. The idea for the approximate assumption is:

- For very low saturation of the system the time-dependent effects do not play a role since the relaxation time, during which the system adapts to a changing demand is very small compared to the duration T of the peak period. Thus, the time-dependent solution will be very close to the stationary solution which is represented by the M/M/1-queue.
- For extreme oversaturation (i.e. large x and long period T of oversaturation) the randomness of the system becomes less important. Effects of randomness then constitute only a very small part of the total delay. Thus, the average delay approaches the deterministic delay.

Therefore, the solution for the average delay in the time-dependent system should be a transition between the steady state delay (M/M/1) and the deterministic delay. This transition curve, since a solution determined by stochastic theory seems to be too complicated, is then based on an approximation.

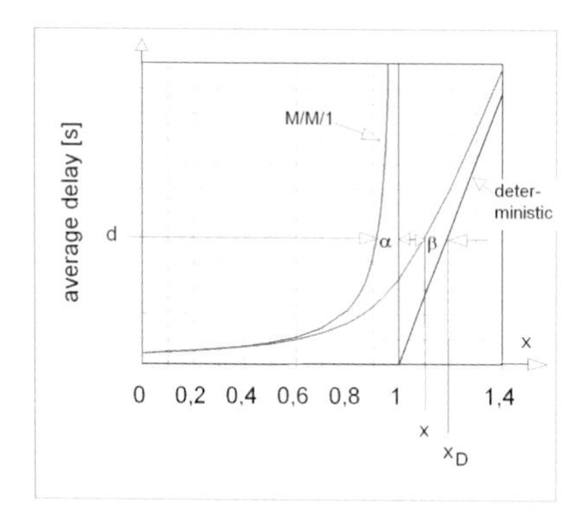

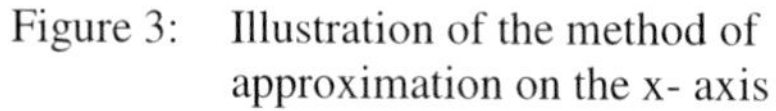
Figure 3: Illustration of the method of approximation on the x- axis

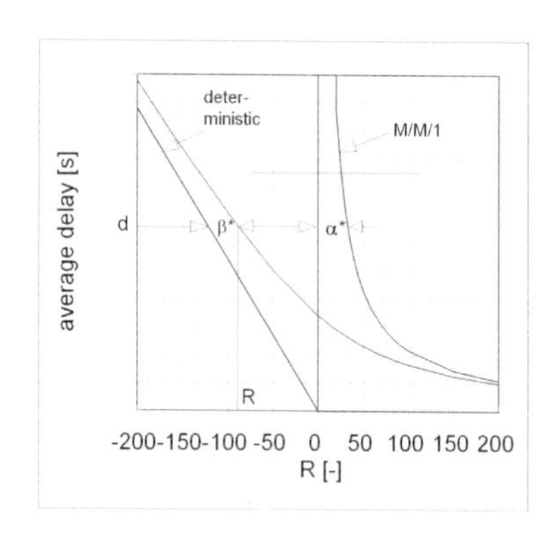

Figure 4: Illustration of the method of approximation on the R-axis

Three kinds of approximation may be used. The solution for each approach is derived from one of the following cases.

A1. additive; x-axis (see Figure 3) $\alpha = \beta$

A2. multiplicative; x-axis (see Figure 3) $\frac{\alpha}{1} = \frac{\beta}{x_D}$

A3. additive; R-axis (see Figure 4) $\alpha^* = \beta^*$

where

x = degree of saturation $= q / c$ [-]

R = reserve capacity $= c - q$ [veh/h]

α, β, α^*, β^* : parameters, see Figure 3 and 4

A potential fourth case, a multiplicative R-based approach gives no sense.

Each of these assumptions, constituting the fundament for the approximation, reveals a specific solution. There is no reasoning to prefer one of these approximations with regard to the basic sophistication.

DETERMINISTIC DELAY

To use the method of approximation we need to describe the deterministic delay in Figure 3 and Figure 4. As the D/D/1 system we understand a queuing system where all customers arrive with a headway of $1/q$ (q = demand volume) and where they are served with a constant service time of $s = 1/c$ (c = capacity). For $x = q/c < 1$ (i.e. $R > 0$) there are no delays for customers in such a system. The only time which they spend in such a system is the service time. The reason why the D/D/1-system is treated, is the fact that for a D/D/1-system also delays for a temporary oversaturation can be determined.

In each queuing system the sum of all delays is the area between the cumulative arrivals and cumulative departures each represented by their function over time. Figure 5 gives an illustration.

Figure 5a shows the demand volume q over time versus the capacity c, which is assumed to remain constant here. We see that during a peak period of duration T the demand exceeds the capacity, whereas the demand is assumed to be zero before and after the peak period. Then the sum of all arrived and departed vehicles, each as a pattern over time, is given in Figure 5b with a maximum difference N_T at the end of the peak period. The delay d for a vehicle arriving at time t can be obtained as the horizontal difference between the two curves. That means: the area included between both curves is the sum of all delays. The vertical difference between both curves, i.e. the queue length, is given in Figure 5c. Since the area between the two curves in Figure 5b is equal to the area under the queue length curve in Figure 5c, the sum D of all delays is the area below the curve for the length of the queue (Figure 5c).

Simple geometric considerations within Figure 5 reveal:

$$N_T = Max\begin{cases} (q-c)\cdot T = -R\cdot T = c\cdot(x-1)\cdot T \\ 0 \end{cases} \tag{3}$$

$$D = \frac{1}{2}\cdot c\cdot x\cdot(x-1)\cdot T^2 \tag{4}$$

Then the delay per vehicle averaged over all arriving vehicles for the deterministic case is

$$d = \frac{D}{q\cdot T} = \frac{(x-1)\cdot T}{2} \tag{5}$$

With approximation A1 we get a time-dependent solution for delay as:

$$\omega = \frac{T}{4}\cdot\left\{(x-1)+\sqrt{(x-1)^2+\frac{8\cdot z}{T\cdot c}}\right\} \tag{6}$$

where

ω	=	average delay d or average waiting time w	[s]
D	=	sum of total delay	[s]
x	=	degree of saturation = q/c	[-]
T	=	duration of the peak period	[s]
q	=	demand traffic volume	[veh/s]
c	=	capacity	[veh/s]
z	=	parameter	[-]

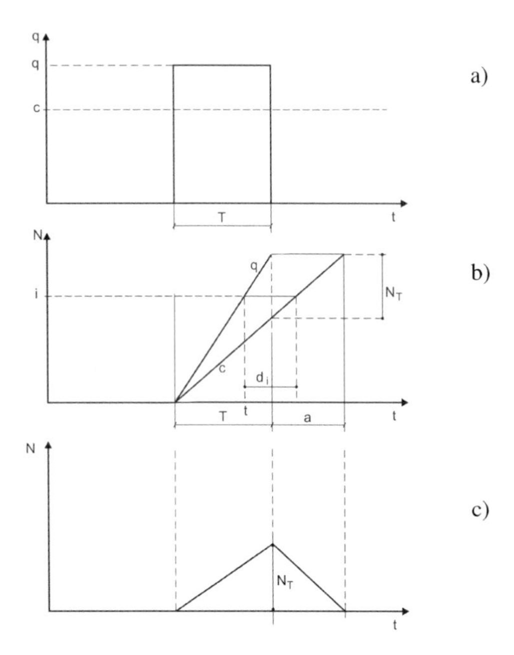

Figure 5: Derivation of the deterministic delay for the simple case

With $z = 1$ we get a function for $\omega(x)$ which constitutes a transition from waiting time w for undersaturated conditions to delay d for large x. On the other side, for $z = x$ eq. 6 converts into the well-known delay equation by Akcelik-Troutbeck (1991). This can be derived as a transition from random delay d for small x to deterministic waiting time w for over-saturation. Since the approximation converges on a term of different nature for both sides of the function (different for $x \to 0$ and for $x \to \infty$) this kind of solution shows a significant degree of inconsistency. This problem could be solved with the set of assumptions :

target function :	$x < 1$	$x > 1$
ω = delay d	$z = x$	$z = 1$
ω = waiting time w	$z = 1$	$z = x$

In addition, it is worth to notice that eq. 6 for any parameter z is only valid for the unrealistic case of no arriving traffic before and after the peak period as well as constant capacity.

The more general case for the traffic demand pattern over time is illustrated in Figure 6. For these more general circumstances also the capacity c is varied by a stepwise function. In addition an initial queue of length N_0 is assumed. Another difference to the previous case is that also after the observed peak period of duration T a continued traffic demand q_1 is assumed. Since we concentrate on peak intervals with a potential temporary oversaturation the conditions

$$\begin{aligned} 0 \le q_1 << c_1 \\ q_1 < q \end{aligned} \tag{7}$$

should be valid.

Two different clearance times for the queue can be defined. Period a is the time after which the last vehicle arriving during the peak departs. a_1 is the time after which the expected length of the deterministic queue becomes zero.

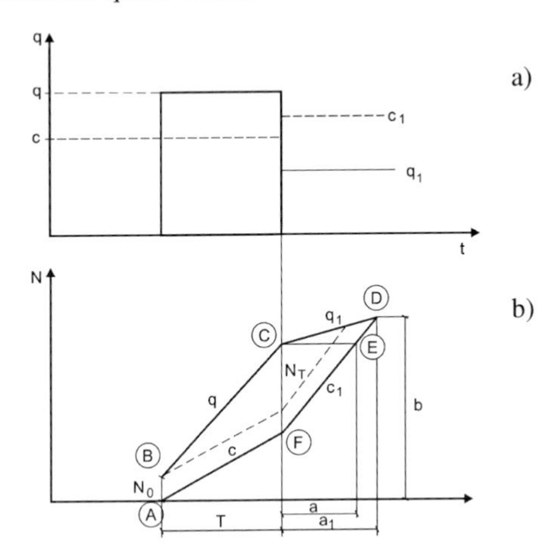

Figure 6: Deterministic delay for the general case.

Again, also in Figure 6 (like in Figure 5) the total delay is the area between the two cumulative curves. There are, however, several possibilities which part of the area should be regarded as the relevant sum of delays. Here, different cases for the deterministic delay D_D can be distinguished (table 1). As a general formula for the sum D_D of all delays we can use:

$$D_D = \begin{cases} \frac{1}{2} \cdot [N_0 \cdot T + N_T \cdot (T + a)] & \text{for } N_0 > T \cdot c \cdot (1 - x) \\ \frac{N_0^{\ 2}}{2 \cdot c \cdot (1 - x)} & \text{elsewhere} \end{cases} \tag{8}$$

$$N_T = Max \begin{cases} N_0 + c \cdot (x - 1) \cdot T \\ 0 \end{cases} \tag{9}$$

where

N_T	=	maximum queue length (deterministic)	[veh]
N_0	=	initial queue length	[veh]

a = time to dissolve the queue due to oversaturation (according to table 1) [s]

Table 1: Cases for the classification of total delay

case D	area in Figure 6b	equation for the term a (for the calculation of D_D)
D1	A B C F	$a = 0$
D2	A B C E F	$a = \frac{N_T}{c_1}$
D3	A B C D E F	$a = a_1 = \frac{N_T}{(c_1 - q_1)}$

Each case has advantages for specific purposes:

D1 This definition restricts the consideration on delays which do occur exactly during the relevant peak interval. This definition is the only one to be applied when delays from successive intervals are added, e.g. over all hours of a whole day.

D2 This is the delay, which engineers usually define, when they estimate delays by empirical methods. It avoids to integrate delays experienced by vehicles arriving after the peak. But the D2-definition contains delays experienced after the end of the peak period. For the special case of $N_0 = 0$ this definition is closely related to the solution of eq. 6.

D3 This is the total delay which is induced into the system by the temporary overload. But it contains delays experienced after the end of the peak. Even delays for vehicles arriving after the considered peak period are involved into the total delay. For an economic assessment of delays, caused by specific peak periods, this definition is the preferential one, since it represents the total consequences of the overload happening during the peak period.

It should be noted again that D_D covers delays – not waiting times.

From the sum D_D of total deterministic delay the average delay d has to be derived relating D_D to those vehicles which are exposed to become involved into the queue of waiting vehicles. This means

$$d_D = \frac{D_D}{N} \tag{10}$$

where

d_D = average deterministic delay [s]

D_D = sum of all deterministic delays [s]

N = number of vehicles exposed to contribute to D_D [-]

For undersaturated conditions (i.e. $x < 1$) N contains all vehicles which arrive during the relevant time period T with the consequence that $N = q \cdot T$.

For temporary oversaturation the calculation of N is not self-evident. Then N could comprise all vehicles which could be affected by the queue which is formed due to the oversaturation. Therefore, three cases can be formulated for the derivation of N:

N1: N contains those vehicles arriving during the time interval (t, t + T); i.e. $N = q \cdot T$

N2: N contains vehicles which arrive during the time of an existing queue; i.e. $N = q \cdot (T + a)$

N3: idem; $N = q \cdot (T + a_1)$

Of course, case N2 can only be combined with D2 and case N3 gives only sense with D3, whereas N1 can be combined with case D1, D2 and D3. Here it is preferred to relate all delays to the vehicles arriving during the peak period of duration T; i.e. we restrict ourselves to case N1.

One example for the deterministic delay depending on the degree of saturation x is shown in Figure 7. In Figure 8 the deterministic delay is shown how it depends on the reserve capacity R. In both figures the limiting case

$$\left(x = x_g = 1 - \frac{N_0}{T \cdot c} \longrightarrow R = R_g = \frac{N_0}{T} \quad with \quad d = \frac{T \cdot N_0}{2 \cdot (T \cdot c - N_0)} \right) \tag{11}$$

is marked. This is the point which - as a maximum - enables a dissipation of the initial queue (length N_0) within the peak period of duration T. Beyond this point (i.e. for $x > x_g$ or for $R < R_g$) the queue at the end of the peak period will be > 0 for the deterministic system. We see that N_0 has an influence on the shape of the curves for small x (i.e. $x < 1$). For $N_0 > 0$ the value of d increases to infinity for $x \rightarrow 0$, since due to N_0 there is always some delay experienced in interval (t , t+T). With eq. 10, for small x the number of N is also small. This will need some special treatment later in this paper. Only for $N_0 = 0$ the deterministic delay starts from the point ($x = 1$, $d_D = 0$). For case D2 the relation $D_D = Function(x)$ is always nearly linear. D3 is identical with D2 for $q_l = 0$. With increasing q_l the D3-curve becomes increasingly concave. In the limiting case of $q_l = c_l$ the curve for case D3 grows to infinity at $x = 1$. The D1-curve is always convex.

It would now be desirable to conduct each of the approximations A1 - A3 for every definition D1 to D3 of deterministic delay. This may, in principle, be a possible option. In practice this is, however, nearly impossible and it gives no real sense. The reason is: Equations 8 and 9 combined with Table 1 reveals rather complicated equations. For the approximations these equations have to be solved for x (and R in case A3). This will give extremely complicated solutions. Then, from the equations in Table 1 we get another set of equations which have to

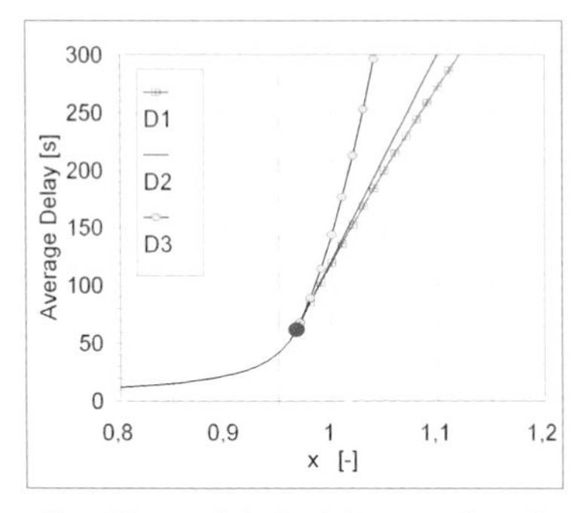

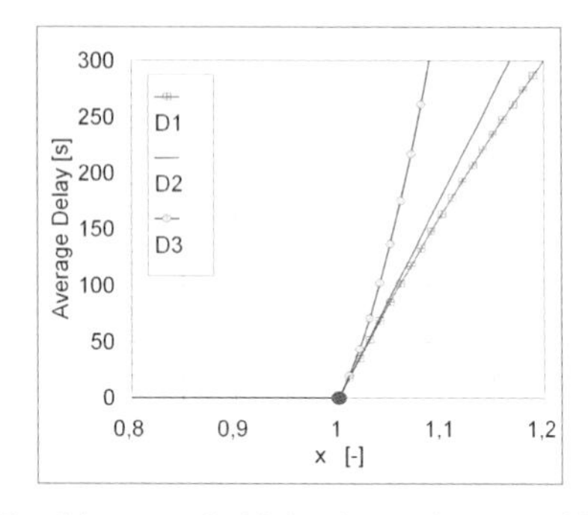

Figure 7: Deterministic delay as a function of x. For this example N_0 has been chosen as 20 (left side) and 0 (right side; both figures: c = 600 veh/h; c_1 = 600 veh/h; q_1 = 550 veh /h; T = 1 h).

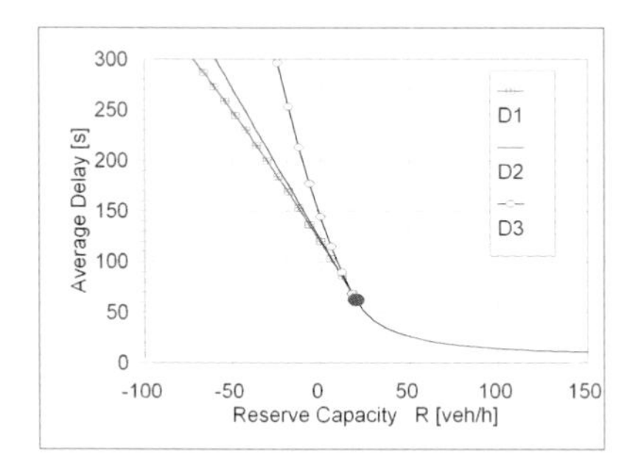

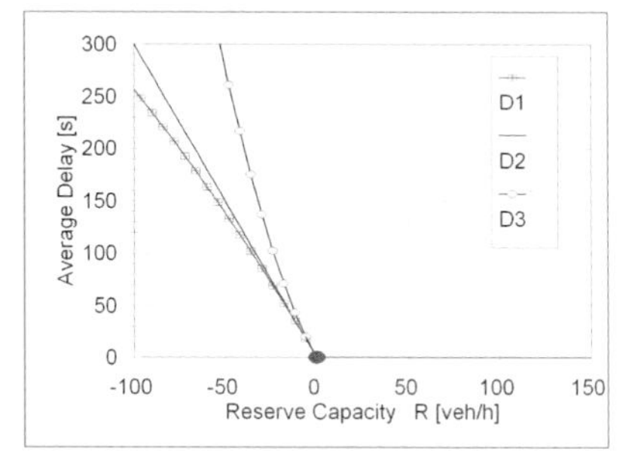

Figure 8: Deterministic delay as a function of the reserve capacity R. For this example N_0 has been chosen as 20 (left side) and 0 (right side; both figures: c = 600 veh/h; c_1 = 600 veh/h; q_1 = 550 veh /h; T = 1 h).

be solved for the delay d. The resulting equations will become even more complex, if all variables are used. It is, however, not very reasonable to develop extremely complex solutions if the result is still only an approximation.

Therefore, as a first approach we concentrate on the simple case where $N_0 = 0$ (i.e. no initial queue) and $c_1 = c$ (i.e. constant capacity). This simplified case is not relevant for case D1 and D3. Thus, first of all we can derive equations representing the average delay for case D2, which then will be used as a reference.

Case D2 + A1:

$$d = \frac{1}{4 \cdot c} \cdot \left[2 + c \cdot T \cdot (x - 1) + \sqrt{(2 + c \cdot T \cdot (x - 1))^2 + 8 \cdot x \cdot c \cdot T} \right] \tag{12}$$

This equation should be identical with the Akcelik-Troutbeck equation, which is not the case. The reason is: this equation is derived from the delay d , both within the M/M/1-queue and within the deterministic system. The original Akcelik-Troutbeck equation is derived as a transition from waiting time w in the M/M/1-case to the delay d in the deterministic case (see text in connection with eq. 6).

Case D2 + A2:

$$d = \frac{T}{4} \cdot \left[x - 1 + \sqrt{(x-1)^2 + \frac{8 \cdot x}{c \cdot T}} \right] \tag{13}$$

The surprising result is that this equation - resulting from the multiplicative approximation A2 - is exactly the Akcelik-Troutbeck equation.

Case D2 + A3:

$$d = -\frac{1}{4 \cdot c} \cdot \left[R \cdot T + 2 - \sqrt{(R \cdot T - 2)^2 + 8 \cdot c \cdot T} \right] \tag{14}$$

This equation is different from the corresponding solution by the author (Brilon, 1995, eq. 2.20). The reason is again that the older solution has been obtained by a transition from the waiting time w in the M/M/1-solution to the delay d in the deterministic case, which was not a consistent solution.

Even if these formulas look quite different from each other, the numerical results are rather similar. This is pointed out in Figure 9 for one example. But also with these small differences it is of interest which approach is the more realistic one. This is tested in the next section of this paper.

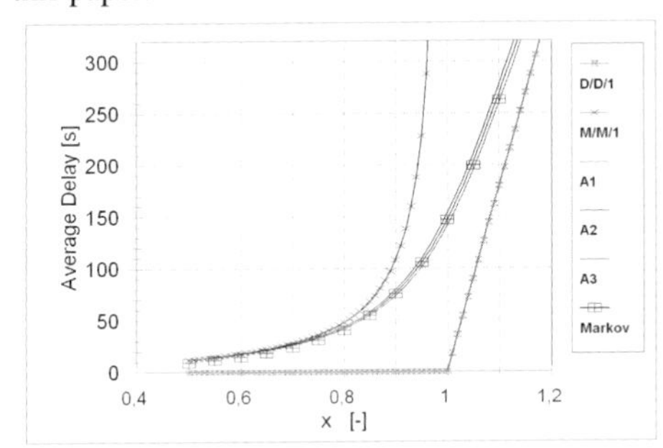

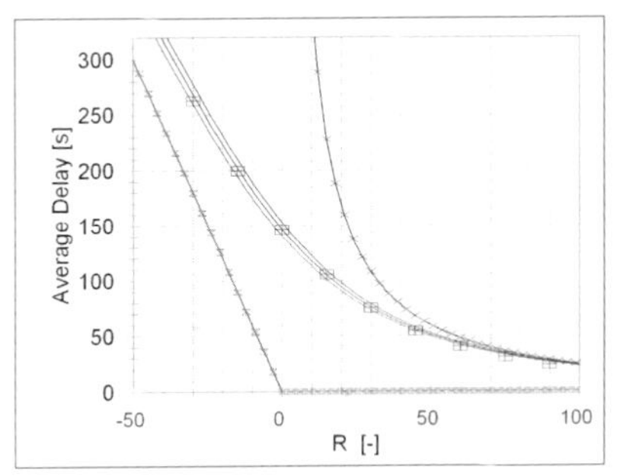

Figure 9: Average delay d as a function of the degree of saturation x (left side) and in relation to the reserve capacity R (right side). The figures compare the three kinds of approximation A1 – A3. Also the Markov chain results are indicated. Parameters for this example: $N_0 = 0$, $c = c_1 = 600$ veh/h.

ANALYTICAL SOLUTIONS

To check the desired approximate solutions for their correctness, methods for the determination of exact solutions for the time-dependent M/M/1-queue are desirable. Mathematical literature provides several solutions for the state probabilities or for the distribution function of the number of customers in the system also for dynamic (i.e. time dependent) queuing systems. One approach has been given by Heidemann (2002) by using LaPlace transformations for queue length distributions and delay distributions. With this approach Heidemann confirmed the Akcelik-Troutbeck solution. He did not, however, envision the other potential options for approximate equations. This analytical result contains the problem that no explicit formulas are used. Instead the LaPlace transforms have to be solved numerically to get applicable results.

Other candidates for equations to describe time dependent state probabilities within time dependent queuing systems can be found in Takacs (1960), Morse (1958/1976), or Tarabia (2000). They offer rather complicated equations containing trigonometric functions (sin, cos, ...). The critical point for all of these solutions is, however, that they do not allow input volumes q exceeding the capacity at any time. Thus, for temporary oversaturation the conditions for the analytical derivations are not fulfilled. As a consequence these analytical solutions are only of limited usefulness for application in traffic engineering, since here a temporary oversaturation of the system is the crucial case for application.

Newell (1982) proposed the so-called diffusion theory to estimate the average queue length as a pattern over time for a given capacity and demand pattern. Also the standard deviation of queue lengths can be determined. Troutbeck, Blogg (1998) have tested this approach. They confirm the quality of Newell's formula based on comparisons to stochastic simulations. However, they also underline the approximate nature of Newell's solution which produces biased results for small queue lengths. Troutbeck, Blogg do also identify the limits of the solution for the time dependent queue length estimation according to Kimber et al (1986). In any case all analytical solutions aim at an estimation of queue length from which average delays must be determined. Non of these approaches claims for an analytically exact solution neither for queue length nor for delays. It is, however, not desirable to check one approximate solution against another approximation as a reference.

MARKOV-CHAINS

Even if an analytical solution is not visible, there are possibilities to get exact results for the delay within each queuing system by using numerical methods. These are based on the Markov-properties of the priority process which are especially valid for the M/M/1-approximation.

The average number of vehicles in the system can be estimated by the following procedure. We observe the queuing system in intervals of $\Delta t = 1$ minute duration. Since arrivals and possible departures are Poisson, we get

$$\begin{aligned} a_i(t) &= e^{-q(t)\cdot\Delta t} \cdot \frac{\left(q(t)\cdot\Delta t\right)^i}{i!} \\ b_i(t) &= e^{-c(t)\cdot\Delta t} \frac{\left(c(t)\cdot\Delta t\right)^i}{i!} \end{aligned} \qquad (15)$$

where

$a_i(t)$	=	probability of i arrivals during the interval $(t, t+\Delta t)$	[-]
$b_i(t)$	=	probability that i departures are possible during the interval $(t, t+\Delta t)$	[-]
$q(t)$	=	traffic volume	[veh/min]
$c(t)$	=	capacity (q(t) and c(t) are assumed to remain constant with sufficient degree of approximation during the interval $(t, t+\Delta t)$).	[veh/min]
Δt	=	duration of the time interval (here: 1 minute). We assume that $T=k*\Delta t$ where k is any integer number.	

From the a_i and b_i we form two quadratic matrices $\overline{A}$ and $\overline{B}$ with the dimension n.

$$\overline{A}(t)=\begin{bmatrix} a_0(t) & a_1(t) & a_2(t) & a_3(t) & \dots \\ 0 & a_0(t) & a_1(t) & a_2(t) & \dots \\ 0 & 0 & a_0(t) & a_1(t) & \dots \\ 0 & 0 & 0 & a_0(t) & \dots \\ \vdots & \vdots & \vdots & \vdots & \dots \end{bmatrix} \tag{16}$$

$$\overline{B}(t)=\begin{bmatrix} 1 & 0 & 0 & 0 & \dots \\ 1-b_0(t) & b_0(t) & 0 & 0 & \dots \\ 1-b_0(t)-b_1(t) & b_1(t) & b_0(t) & 0 & \dots \\ 1-b_0(t)-b_1(t)-b_2(t) & b_2(t) & b_1(t) & b_0(t) & \dots \\ \vdots & \vdots & \vdots & \vdots & \dots \end{bmatrix} \tag{17}$$

Then the transition probabilities are given by the matrix $\overline{P}=\overline{A}\cdot\overline{B}$. This means: each term p_{ij} of the matrix $\overline{P}$ is calculated as

$$p_{ij}(t)=\sum_{k=0}^{n} a_{ik}(t)\cdot b_{kj}(t) \tag{18}$$

where

$p_{ij}(t)$	=	probability that the number of vehicle in the queue changes from i to j within the interval $(t, t+\Delta t)$
$a_{ik}(t)$	=	term of the matrix $\overline{A}$ at time t in row i and column k
$b_{kj}(t)$	=	term of the matrix $\overline{B}$ at time t in row k and column j
n	=	number for the range of numerical calculations. n should be significantly larger than the maximum possible queue length.

Then the state probabilities of the system will change over time according to

$$p_i(t+1) = \sum_{k=0}^{n} p_k(t) \cdot p_{ki}(t) \tag{19}$$

The equation makes it possible to calculate each of the state probabilities p_i at any time t + 1 out of the previous state probabilities at time t.

Then the average number N(t) of vehicles in the system at time t is

$$N(t) = \sum_{i=0}^{n} i \cdot p_i(t) \tag{20}$$

This is a numerically exact calculation for the dynamics of the expected number of vehicles in the system. Again – like in the case of the deterministic system – the area below the curve for N(t) is the sum of all delays.

$$D = \int_0^{\infty} N(t)dt \tag{21}$$

Or with stepwise constant demand q(t) and capacity c(t):

$$D = \sum_{t=0}^{t_{max}} N(t) \cdot \Delta t \tag{22}$$

This sum of delays then can be related to the number of arrivals during the peak period of duration T to estimate the average delay d.

$$d = \frac{\sum_{t=0}^{t_{max}} N(t) \cdot \Delta t}{\sum_{t=0}^{t_{max}} q(t) \cdot \Delta t} = \frac{\sum_{t=0}^{t_{max}} N(t)}{\sum_{t=0}^{t_{max}} q(t)} \tag{23}$$

Here t_{max} is the number of the final interval (duration Δt) where N(t) becomes 0 if case D3 (Table 1) is applied. For case D1 $t_{max} = T / \Delta t$. Case D2 can be constructed by setting $q(t) = q_1(t) = 0$ after time T.

Markov-chain calculations have been performed for several examples. Results for average delay according to case D2 were evaluated for several combinations of c and T with emphasis on longer peak periods in the range of T ~ 1 h. It turned out that on the scale of Figure 9 the relation for d = F(x) or d = F(R) estimated by eq. 12 – 14 matched quite well with Markov-chain results. The differences are so small that on the scale of Figure 9 they are not visible. To notice the differences between eq. 12 – 14 and Markov-chain results Figure 10 has been plotted. Here we see that the differences are quite small in absolute terms (upper part of the figure). The differences may become quite significant if we treat them in relative terms (bottom part of the figure). But the extremely large relative differences for small x-values are in the area of nearly zero delays, so that here the absolute differences are very small. Even if a remarkable deviation of the estimated average delays from the exact values could occur, the

Table 2: Comparison of results for average delay d: Residual standard deviation for Case D2, Cases A1 - A3 and eq. 6

	capacity	A1	A2	A3	eq. 6
compared to Markov-chain calculation [1)]	c = 100	14.63	13.24	24.38	15.75
	c = 300	9.36	4.31	3.07	2.02
	c = 600	7.84	5.38	3.58	4.02
compared to simulation [2)]	c = 100	81.7	76.8	70.8	78.0
	c = 300	42.9	40.8	39.3	41.1
	c = 600	28.0	27.4	27.1	27.8
	veh/h	s			

1) values = standard deviations for the difference between the result from eq. 12 - 14 and the Markov-chain result; compared over each $x \in \{0.5\ (0.05)\ 1.2\}$

2) values = standard deviations for the difference between results from eq. 12 - 14 and the simulated average delay; compared for 1000 1-hour simulation runs with various combinations of q and c within the interval (x= 0.5, 1.2)

degree of approximation seems to be quite acceptable for practical application. Similar figures can be plotted for other combinations of parameters. The tendency of the 3 curves remains similar, the absolute value of the differences may, however, vary. They become rather significant if the capacity has very low values.

In any case the A1- approximation (i.e. the additive approach over x) turns out to be of lowest quality. It is not very clear, which of the A2 or A3 is better. They seem to be equivalent with a small advantage for A3, the additive reserve capacity based solution. Also the solution for delays d according to eq. 6 shows a rather good performance.

SIMULATION

Another method for comparing the approximate results with "true" values is to use stochastic simulation. A computer program for simulating the priority system with constant critical gaps $t_c = 6$ s and $t_f = 3$ s has been written where drivers behaved in a consistent and homogeneous way. The program evaluates the average delay in peak periods of duration T (here T = 1 h) according to approach D2. A large number of repetitions is possible. Figure 11 shows that the delays do - on average - follow the calculated curves. The results for the 1-hour average of delay do, however, vary over a quite remarkably wide area. The standard deviation of delays (standard deviation between average delays over 1-hour intervals) is always in the range of 0.7 of the mean of the average delays.

The three curves represent the approximations for case A1 (upper curve, eq. 12), A2 (eq. 13), and A3 (lower curve, eq. 14). The residual standard deviations for the simulated points (relative to eq. 12 – 14) are also given in Table 2. If we try to interpret the small differences the results support the A3-solution. The difference to the A2-solution is, however, only quite

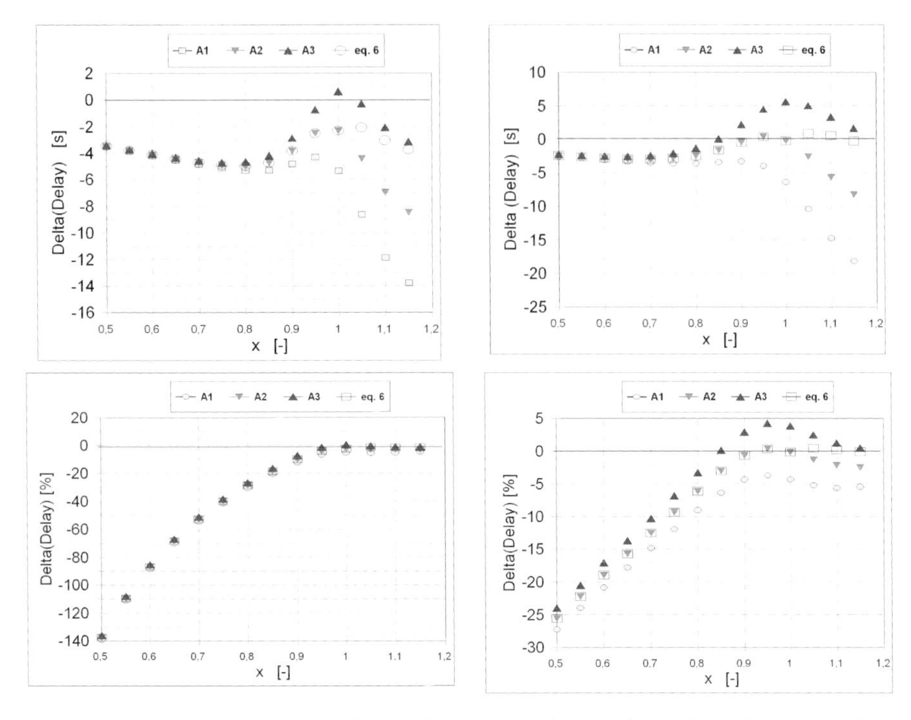

Figure 10: Difference in average delay d between Markov-chain results and the approximate solutions A1 - A3 as a function of the degree of saturation x. Parameters for this example: T = 1 h; $N_0 = 0$, $c = c_l = 600$ veh/h (left side) and $c = c_l = 300$ veh/h (right side).

small. With respect to the large variance of the average delay such small differences become meaningless in practical terms. Also eq. 6 provides an adequate degree of approximation.

All equations compared to simulation results show a tendency to slightly overestimate average delays in the range of x = 1. The validity of all equations might be improved if for the basic equations of approximation (cases A1 to A3) a factor would be used relating the values of a and b (or a* and b*) against each other. This may be subject of further research.

At this point it can be stated, also on the background of more example calculations, that the approximation method A3 gives the best correlation to simulation and Markov-chain results. The method A2 is, however, very close up. Based on this experience and knowing that the same result must not necessarily be obtained also for cases D1 and D3, method A1 is not further treated here.

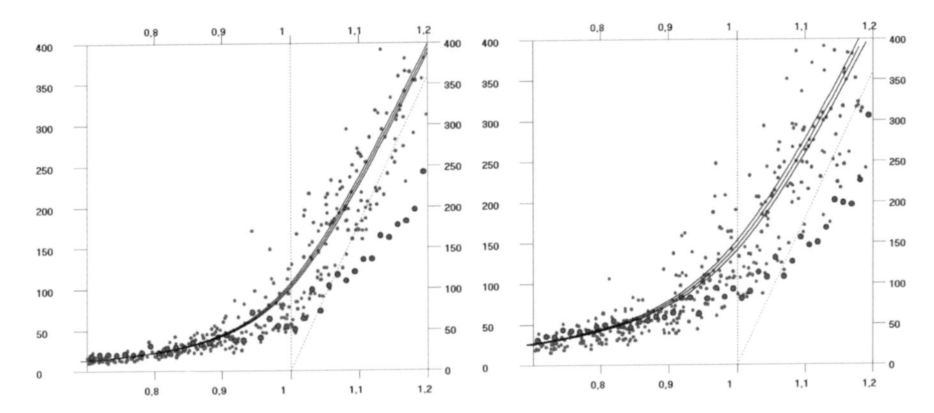

Figure 11: Simulation results for the priority system. Each point represents the average delay over a 1-hour-peak period according to case D2 in relation to the degree of saturation x. The larger points represent the standard deviation between the results for average delay in the corresponding range of x. Parameters for this example: T = 1 h; $N_0 = 0$, $c = c_l = 600$ veh/h (left side) and $c = c_l = 300$ veh/h (right side).

EFFECT OF N_0

Up to now we have studied the unrealistic simplified case of $N_0 = 0$, $c_l = c$, and $x_l = q_l/c_l = 0$. To adjust the solution to more realistic circumstances N_0, the initial queue length, should be allowed to have any positive integer value. Then the sum of deterministic delays assumes a function over x or over R like it is illustrated in Figure 12. The direct application of the principle of approximations A1 - A3 does not lead to useful results. Besides the fact that the equations assume unreasonable complicated functions, there is also the problem that the average delay does not only increase for large x (or small R) but also in the vicinity of $x \to 0$ (i.e. $R \to c$) due to eq. 10. If $x = 0$ the term $N = q \bullet T = x \bullet c \bullet T$ in the numerator of eq. 10 becomes 0 , such that with

$$D_0 = \frac{N_0^{\ 2}}{2 \cdot c} \tag{24}$$

where

D_0 = sum of delays for the case of no traffic arriving during the interval (t , t+T)

there will be a minimum sum of delay in the denominator. Such a function is not accessible to approximation A1 - A3.

One solution might be to perform the same type of approximation on the scale of D, i.e. with the sum D of delays instead of the average individual delay d. Trying this, the result is quite discouraging because the equations become unacceptably complicated including irrational functions. Nevertheless, a useful solution was found via the treatment of the sum of delays calculated from eq. 13 and 14 by multiplication with $N = q \bullet T = x \bullet c \bullet T$, which is the number of vehicles arriving during the peak interval (t , t+T). In addition to this sum of delays, the minimum amount of delays D_0 which goes back to the initial queue, has to be added. It can be obtained from Figure 7, that also the asymptote for the deterministic delay has to be transformed from $d_D = T/2 \cdot (x-1)$ (in case D2; $N_0 = 0$) to $d_D = T/2 \cdot (x - \hat{x})$ where

$$\hat{x} = 1 - \left(\frac{1}{T \cdot c} + \frac{1}{T \cdot c - N_0} \right) \cdot N_0 \tag{25}$$

Thus, the sum of delays can be described with a rather good approximation by the equation

$$D = D_0 + \frac{T^2}{4} \cdot c \cdot x \cdot \left(\Delta x + \sqrt{\Delta x^2 + \frac{8}{c \cdot T} \cdot x} \right) \tag{26}$$

where

$$\Delta x = x - \hat{x}$$

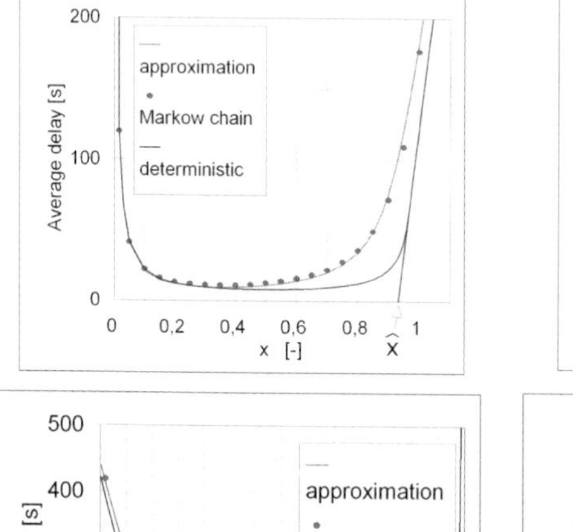

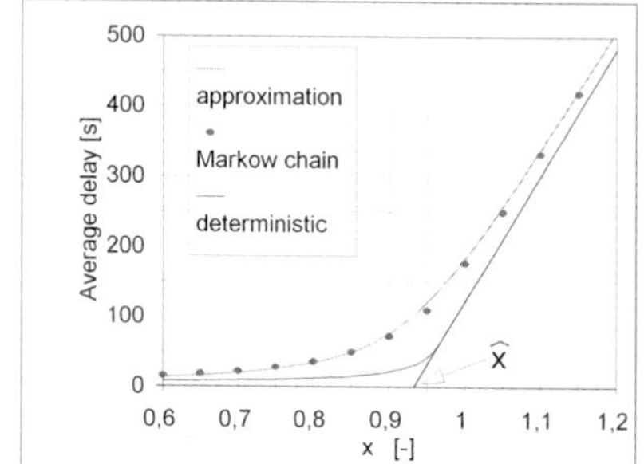

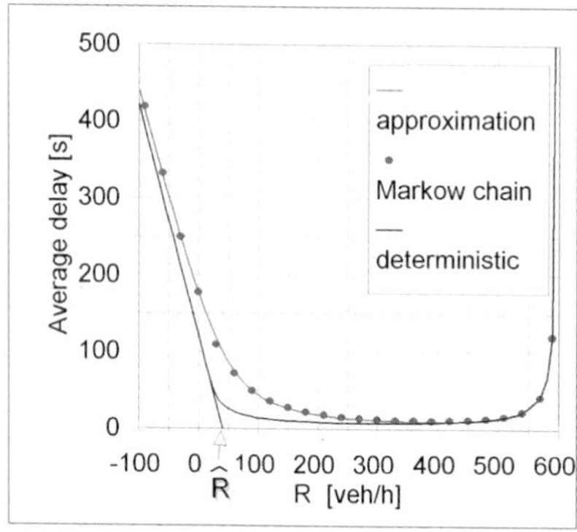

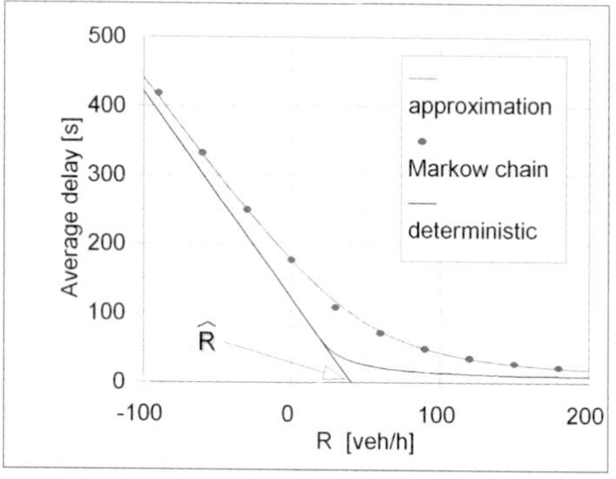

Figure 12: Average delay according to eq. 26 for an initial queue of N0 = 20 depending on x and according to eq. 27 depending on R. Parameters for this example: T = 1 h; N0 = 20, c = c1 = 600 veh/h. The figures on the left show results for the full scale of x and R whereas the right side is focusing on the more interesting area close to x = 1 (R = 0).

The similar result can be obtained by approximation A3 using the reserve capacity R.

$$D = D_0 - \frac{(c-R)\cdot T}{4\cdot c}\cdot\left(\Delta R\cdot T + 2 - \sqrt{(\Delta R\cdot T - 2)^2 + 8\cdot c\cdot T}\right) \tag{27}$$

where

$$\Delta R = R - \hat{R}$$

$$\hat{R} = \frac{N_0\cdot(2\cdot T\cdot c - N_0)}{T\cdot(T\cdot c - N_0)}$$

Then the average delay d is calculated from eq. 10 with D from eq. 26 or 27 and $N = q \bullet T = x\bullet c\bullet T$. Figure 12 shows the resulting average delay d for one example. This picture and other examples show: There is a rather good correspondence between the approximate formulas and the exact values represented by Markov-chain results. Only in the range of $x = 0.3$ to 0.7 there are smaller differences which increase if N_0 grows into unrealistic large values (e.g. above 30 vehicles). In any case the preciseness is quite sufficient for practice. For larger x the degree of approximation is quite good for any N_0-value.

EFFECTS OF THE POST-PEAK PERIOD

There is still one case which has not yet been solved. This is the influence of the capacity and demand, which is prevailing after the peak period and which has an influence on the delay for the vehicles arriving during the peak interval (t , t + T) in cases D2 and D3.

At first, for case D2 the effect of the post-peak capacity c_l (which may be different from the peak period capacity c) can be taken into account by adding the difference in deterministic delay to the sum of all delays. As a consequence, the sum of delays (eq. 26 and 27) has to be modified as

$$D_{cl} = D - \frac{N_T^{\,2}}{2}\cdot\left(\frac{1}{c} - \frac{1}{c_l}\right) \tag{28}$$

where

D_{cl}	=	sum of delays for $c_l \neq c$	[s]
D	=	sum of delays for $c_l = c$ (eq. 26 or 27)	[s]
q_l	=	post peak traffic demand	[veh/s]
c_l	=	post peak capacity	[veh/s]
N_T	=	max. deterministic queue length at the end of the peak period (cf. eq. 10)	[-]

In the similar way the additional sum of delay which is due to definition D3 is given by

$$D_{D3} = D_{cl} + \frac{N_T^{\,2}}{2\cdot c_l}\cdot\frac{q_l}{c_l - q_l} = D_{cl} + \frac{N_T^{\,2}}{2}\cdot\frac{x_l}{c_l\cdot(1-x_l)} = D_{cl} + \frac{N_T^{\,2}}{2}\cdot\frac{c_l - R_l}{R_l\cdot c_l} \tag{29}$$

where

x_I	=	degree of saturation after the peak period = q_I/c_I	[-]
R_I	=	reserve capacity after the peak period = c_I-q_I	[veh/s]

Finally, the whole set of equations, which can be equally recommended based on approximations A2 and A3, is given in Table 3 as an overview. The formulas in both columns of the table are alternative to each other. Here, the average waiting time w is calculated from the average delay d by adding the weighted average of the service times (s = 1/c = peak period service time, weighted by c•T, and s_I = 1/c_I = post-peak service time, weighted by c_I•a).

Table 3: Formulas for application

	A2 (using $x=\frac{q}{c}$)	A3 (using $R=q-c$)
D1	$D_1 = D_0 + \frac{c\cdot T^2}{4}\cdot\left(x-1-\frac{2\cdot x}{c\cdot T}+\sqrt{(x-1)^2+\frac{4\cdot x}{c\cdot T}\cdot\left(1+x+\frac{x}{c\cdot T}\right)}\right)$	-
D2	$D_{26} = D_0 + \frac{T^2}{4}\cdot c\cdot x\cdot\left(\Delta x+\sqrt{\Delta x^2+\frac{8\cdot x}{c\cdot T}}\right)$	$D_{27} = D_0 - \frac{(c-R)\cdot T}{4\cdot c}\cdot\left(\Delta R\cdot T+2-\sqrt{(\Delta R\cdot T-2)^2+8\cdot c\cdot T}\right)$
	$\Delta x = x-1+\left(\frac{1}{c\cdot T}+\frac{1}{c\cdot T-N_0}\right)\cdot N_0$	$\Delta R = R-\frac{N_0\cdot(2\cdot c\cdot T-N_0)}{T\cdot(c\cdot T-N_0)}$
	$a=\frac{N_T}{c_1}$	
D3	$D_3 = D_{26} + \frac{N_T^{\,2}}{2}\cdot\frac{x_I}{c_I\cdot(1-x_I)}$	$D_3 = D_{27} + \frac{N_T^{\,2}}{2}\cdot\frac{c_I-R_I}{R_I\cdot c_I}$
	$a=\frac{N_T}{c_I\cdot(1-x_I)}$	$a=\frac{N_T}{R_I}$
	$N_T = N_0 + c\cdot(x-1)\cdot T$	$D_0 = \frac{N_0^{\,2}}{2\cdot c}$
d	$d=\frac{D}{x\cdot c\cdot T}$	$d=\frac{D}{(c-R)\cdot T}$
w	$w = d + \frac{T+a}{c\cdot T+c_1\cdot a}$	

CASE D1

To describe case D1 by the similar degree of precision it turned out to be insufficient just to correct the D2-results by some specific terms. Thus, just the similar derivations like for case D2 had to be performed with special attention to the D1-conditions. As result the following equations for the average delay can be given for case D1 – A2.

$$D_1 = D_0 + \frac{c \cdot T^2}{4} \cdot \left(x - 1 - \frac{2 \cdot x}{c \cdot T} + \sqrt{(x-1)^2 + \frac{4 \cdot x}{c \cdot T} \cdot \left(1 + x + \frac{x}{c \cdot T} \right)} \right) \tag{30}$$

The solution for the case D1 – A3 is not possible. For the generalized case it leads to an undefined area in the vicinity of $R = \hat{R}$.

EMPIRICAL EVIDENCE

The measurement of average delays at unsignalized intersections is not a trivial task. For comparisons, due to the wide variance of delays, a large sample size is required. Thus, to get useful measurement data, long periods with constant traffic volumes would be needed which does hardly occur in reality. Moreover, it is not easy to find oversaturated priority intersections since under high traffic demand junctions usually are signalized. The observation of delays requires also a rather good overview since the end of the queue always has to be under control. Such observations have been performed and described by Brilon, Weinert (2002). In their sample there were 4 intersections (all T-junctions) at rural two-lane highways with a temporary overload. Here the left turner from the minor road (LTMR) was observed regarding delays. All the other movements were counted simultaneously. The comparisons here are made on the basis of 5-minute intervals (T = 5 minutes). To estimate x in each time interval the capacity c for the LTMR was estimated on the basis of the method in the German HCM (HBS, 2001). The calculation of average delays is according to definition D2 – A2.

We see an agreement of empirical and calculated delays as it could be expected on the background of the wide variance of waiting times during relatively short time slices. Apparently, at point 28 (left upper part of Figure 13) the calculation fails, since for higher degree of saturation the measured delays remain much below the calculated values. The videos from the measurements, however, made clear that a remarkable amount of gap forcing takes place at this junction during overloaded periods which, of course, reduces delays significantly below the modeled results. On all the other points a sufficient correspondence between measured and calculated delays is observed. A closer coincidence can not be expected, because of the large variation of delays and due to the fact that the estimated capacities must not necessarily represent the true maximum potential throughput during the observations correctly.

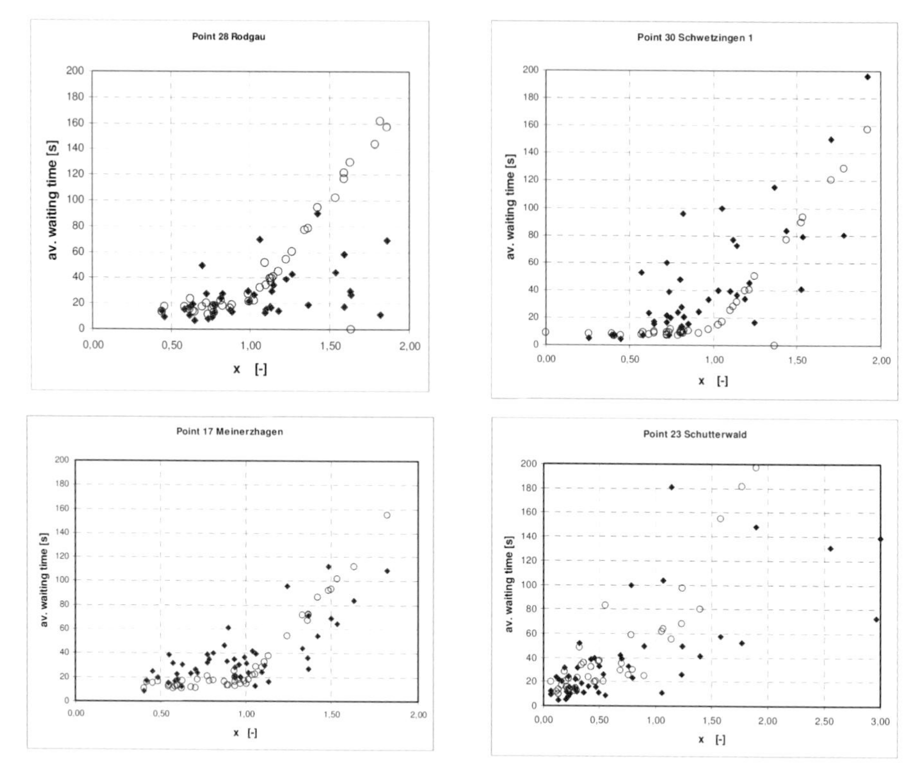

Figure 13: Comparison of measured waiting times with calculated values. The dark dots represent the average waiting time measured according to definition D2. The small circles represent the calculated waiting time for case D2-A2 for the same traffic data as during the measurement interval (T = 300 s).

OTHER CONSIDERATIONS

Of course, methods for the derivation of queue lengths are closely related to the subject of this paper. By Little's well-known formula ($\lambda = q \cdot d$) there is a relation between the average delay d and the average queue length λ. This formula is, however, only valid for stationary queues, i.e. for constant capacities c and demand flows q plus $x < 1$. As shown above (cf. Figure 5 and Figure 6) the queue length is not a steady state variable in case of time dependent demand and/or capacity and especially not during oversaturation, since then it is continuously growing. Thus, it gives only sense to describe the function $\lambda(t)$. To estimate this function several theoretical approaches have been published, e.g. Newell's diffusion equations (Newell, 1982) or Kimber's method (Kimber et al, 1986) (for comparisons see Troutbeck, Blogg, 1998) where Newell' method allows also to estimate standard deviations of the queue lengths based on approximate assumptions. For numerical calculations it is, however, more convenient to evaluate the exact results based on Markov-chain calculations. For one example this function λ (t) is illustrated in Figure 14. In addition to the expected queue length (in veh; without the vehicle in service) also percentiles of the queue length are given as a function of time.

It can be seen that the linear shape of the queue length over time, like it is assumed in Figure 5 and Figure 6, is not realistic. Instead the function λ (t) has a curved shape (cf. Troutbeck, Blogg, 1998). It is also questionable which parameter of these curves should be indicated by formulas for application in practice. Should it be the maximum of the percentile curves or some average of these curves? Estimations of queue length percentiles in practice are mainly based on Wu (1994) (cf. HCM, 2000 or HBS, 2001). It is imaginable that also queue length estimations might be accessible to another new consideration of systematic classification for time-dependent conditions.

Moreover, it will be desirable to extend the considerations in this paper to a fourth case D4, in which the pattern of traffic demand over time has also a time-dependency during the peak period.

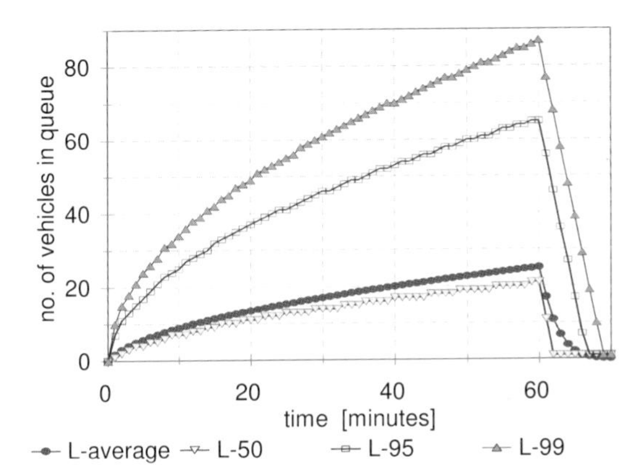

Figure 14: Profile of the average queue length and the 50-, the 95-, plus the 99-percentile queue length for T = 1 h, $c = c_1 = 600$ veh/h, q = 600 veh/h, $q_1 = 0$, $N_0 = 0$.

CONCLUSION

Average delay at unsignalized intersections for temporarily oversaturated conditions is usually calculated by approximate equations. On the one hand these approximations, in principle, do very well match with more precise estimates for average delay. The set of possible solutions is, however, wider than usually assumed. The most commonly applied formulas are just one special case out of a set of equally valid solutions. It must also be noticed that the most popular solutions, like the Akcelik-Troutbeck equation, are only valid for rather significantly simplified conditions which are not too representative for real world conditions.

The systematic classification of definitions for the average delay (case D1 - D3) and of the sophistication for the approximation (case A1 - A3) leads to some differentiation among the complete set of possible solutions. As a consequence, each application of such a delay formula needs some understanding of the background in definitions.

Each numerical calculation of performance measures at unsignalized intersections is based on a representation of the "priority system" by an M/M/1-queue. This first step of approximation does not produce a perfect fit but it is justified as a solution sufficient for practical application, since any alternative would provide a lack of practicability.

For the time-dependent M/M/1-queue the average delay can be calculated exactly by a numerical evaluation of the Markov-chain concept. Comparisons with the approximate equations show that the conventional additive concept A1 for the approximation sophistication is not the best. The best fit is achieved for concept A3; i.e. an additive concept using reserve capacity R. This is closely followed by concept A2 (multiplicative concept using x = degree of saturation).

These results are also confirmed by stochastic simulation. Here, we also get information about the high values of standard deviations of delays with a coefficient of variation for the 1-hour average delay in a range of 0.7 . Even if the precise measurement of average delays needs quite an effort, it can be shown that empirical observations are in coincidence with the theoretical results.

The extension of the simplified case D2 (no traffic before and after the peak period) is able to take into account the traffic conditions before (by the initial queue N_0) and after the peak period. Here formulas can be found which are only slightly more complicated than the traditional equations for the simple case. These approximate equations reveal a precision which is comparable to the traditional Akcelik-Troutbeck equation. Thus, they offer an extension of the traditional formulas into areas of more realistic conditions for practical application. Equations which are recommended for practical application are arranged in table 3. The Akcelik-Troutbeck formula receives an improvement by the use of eq. 6. This version is proposed instead of the addition of the service time (like in Akcelik, Troutbeck, 1991, eq. 5.4 - 5.6 and the HCM 2000, eq. 17-38).

As a result of the investigations a set of delay equations is available, which can be applied to estimate average delays at unsignalized intersections for well-defined traffic conditions. The paper makes clear that instead of an uncritical use of delay formulas a well sophisticated selection of the adequate equation is required also in practice.

REFERENCES

Akçelik, R., Troutbeck, R. (1991). Implementation of the Australian roundabout analysis method in SIDRA. in: *Highway Capacity and Level of Service*. Proceedings of the International Symposium on Highway Capacity Karlsruhe, Balkema

Allsop, R. (1992). personal communication

Brilon, W. (1988). Recent Developments in Calculation Methods for Unsignalized Intersections in West-Germany. In: *Intersections without Traffic Signals*. Proceedings of an International Workshop 1988, Springer, 1988

Brilon, W. (1995). Delays at Oversaturated Intersections Based on Reserve Capacities . *Transportation Research Record*, TRB, Washington DC, No. 1484, pp. 1 – 8,

Brilon, W. and Wu, N. (2002).Unsignalized Intersections - A Third Method for Analysis. In Taylor, A.P. (ed.): *Transportation and Traffic Theory in the 21st Century*, Proceedings of the 15th International Symposium on Transportation and Traffic. Elsevier

Brilon, W., Miltner, T. (2003). Verkehrsqualitaet unterschiedlicher Verkehrsteilnehmerarten an Knotenpunkten ohne Lichtsignalanlage (Traffic performance for different road users at unsignalized intersections). Bundesanstalt fuer Straßenwesen (BASt), No. V 100

HBS (2001). Handbuch zur Bemessung von Strassenverkehrsanlagen (German Highway Capacity Manual), chapter 7: Unsignalized Intersections, FGSV, Cologne

HCM (2000). Highway Capacity Manual. *Transportation Research Board*, Special Report 209, Washington D.C.

Kimber, R. M.; Coombe, R. D. (1980). The Traffic Capacity of Major / Minor Priority Junctions. *Transport and Road Research Laboratory*, Supplementary Report 582, Crowthorne

Kimber, R. M., Hollis, E. M. (1979). Traffic queues and delays at road junctions. *Transport and Road Research Laboratory*, Report No. LR 909, Crowthorne

Kimber, R. M., Summersgill, I., Burrow, I.J. (1986). Delay Processes at Unsignalized Junctions: The Interrelation between Geometric and Queuing Delay. *Transportation Research B*, Vol. 20B, No. 6, pp. 457 - 476

Kremser, H. (1962). Ein zusammengesetztes Wartezeitproblem bei poissonschen Verkehrsstroemen (A combined delay problem for poisson traffic streams) *Oesterreichisches Ingenieur-Archiv* (Austria), no. 16, pp. 231-252

Kremser, H. (1964). Wartezeiten und Warteschlangen bei Einfaedelung eines Poissonprozesses in einen anderen solchen Prozess (Delays and queues with one poisson process merging into another one). *Oesterreichisches Ingenieur-Archiv* (Austria), no. 18

Luttinen, T. (2004). Movement Capacity at Two-Way-Stop-Controlled Intersections. *Transportation Research Record 1883*. Washington D.C.

Morse, P.M. (1976). Queues - Inventories and Maintenance. 3. Edition, John Wiley & Sons, , revised version (first edition 1958)

Newell, G.F. (1982). Application of Queueing Theory. 2nd edition, Chapman Hall, London (first edition 1971)

Takacs, L. (1960). Introduction to the theory of queues. *Oxford University Press*

Tarabia, A.M.K. (2000). Transient Analysis of M/M/1/N Queue – An Alternative Approach. *Tamkang Journal of Science and Engineering*, Vol. 3, No. 4, pp. 263-264

Troutbeck, R. J., Blogg, M. (1998). Queueing at Congested Intersections. *Transportation Research Record 1646*, pp. 124 - 131

Troutbeck, R. J., Brilon, W. (1999). Unsignalized Intersection Theory. Chapter 8 in: *Monograph on Traffic Flow Theory*. Federal Highway Administration

Wu, N. (1994). An Approximation for the Distribution of Queue Lengths at Unsignalized Intersections. *Proceedings of the Second International Symposium on Highway Capacity*, Sydney, Vol. 2,

Yeo, G.F. (1962). Single Server Queues with Modified Service Mechanisms. *Journal of the Australian Mathematical Society*, vol. 2, pp. 499 - 502

JOLLY HOTEL ST ERMIN'S

2, Caxton Street LONDON SW1H OQW

Phone: +44 (0) 2072227888 Fax: +44 (0) 2072226914

E-mail: stermins.uk@jollyhotels.com

Transportation and Traffic Theory 2007
Edited by R.E. Allsop, M.G.H. Bell and B.G. Heydecker

25

PROPERTIES OF A MICROSCOPIC HETEROGENEOUS MULTI-ANTICIPATIVE TRAFFIC FLOW MODEL

Serge P. Hoogendoorn, Transport & Planning Department, Delft University of Technology
Saskia Ossen, Transport & Planning Department, Delft University of Technology
Marco Schreuder, Traffic Research Centre, Ministry of Transport, Public Works and Water Management, the Netherlands

INTRODUCTION

Realistic models describing the car-following task are important for many applications. In the last few decades, significant improvements have been achieved to better describe driver behavior and the resulting traffic flow dynamics. Good examples of such improvements are Zhang (1998), Lenz (1999), Tampère et al (2005), Kerner (2005), and Treiber et al (2006). The microscopic validity of these improvements is difficult to assess due to the absence of sufficient empirical microscopic data, which are furthermore accurate in terms of time and space. However, with advancing data collection technology, such data have become more widely available.

In a recent work, we have considered one specific mechanism the inclusion of which will make modeling more realistic, namely *multi-anticipative car-following behavior* (Hoogendoorn and Ossen, 2005), (Hoogendoorn et al, 2006). This behavior entails the fact that drivers anticipate traffic conditions further downstream by considering vehicles further downstream (besides the direct leader). The notion of multi-leader anticipation reaches back to the late sixties, when the well known car-following model of (Gazis et al, 1961) was extended by (Herman and Rothery,1965) and later by (Bexelius, 1968) to include multi-leader stimuli in the equations describing the response behavior of a driver. More recently, Lenz *et al.* (1999) extended the model of (Bando, 1995) to include multiple vehicle interactions.

Treiber *et al.* (2005) take a similar view and extend the Intelligent Driver Model (IDM) with (among other things) multi-vehicle interaction behavior.

In (Hoogendoorn et al, 2006), we have taken the models of (Chandler et al, 1961) and (Helly, 1959) as a starting point to investigate multi-anticipative car-following behavior from empirical trajectory data. In doing so, statistical evidence of multi-anticipative car-following behavior was provided, as well as insight into the kind of multi-anticipative stimuli to which drivers react. Furthermore, analyses of microscopic data led to the quantification of the *inter-driver differences* in the car-following parameters, which turned out to be very important to correctly describe multi-anticipative car-following behavior. The results presented in (Hoogendoorn et al, 2006) show that incorporating multi-anticipative behavior substantially improves the extent in which the models can explain driver behavior microscopically, while the best performing models include up to three leaders.

This contribution builds further upon these results, by showing that multi-anticipative behavior cannot be described by one, general car-following modeling paradigm: different classes of multi-anticipative car-following models are needed to correctly describe inter-driver driver differences in car-following, or that drivers may drive in different driving regimes (intra-driver differences). Based on variability in the parameter estimates for each considered model class, we show that differences between drivers whose behavior can be described by the same model class are also large. From this we can conclude that the extent in which drivers react to the second and the third leader can vary substantially between drivers. Also the differences in the reaction times are substantial.

Based on the estimation results, we propose a new *heterogeneous (or mixed-model) modeling approach* that includes the features identified from analyzing the trajectory data. The proposed model includes multi-anticipative behavior, as well as the inter-driver differences in sensitivity to the different stimuli. The model also includes the different modeling types required from a statistical point of view to correctly describe individual driving behavior, in particular including different versions of the Generalized Helly model (Hoogendoorn et al, 2006) and the modified model of Lenz (1999). The main properties of the resulting microscopic model will be analyzed. In particular, the equilibrium behavior will be considered (speed - density and flow - density curves). Besides the static model properties, we will furthermore briefly show the dynamic characteristics of the model. This will entail performing a stability analysis where the amplitude of a disturbance propagating through a platoon is analyzed. Finally, we will identify the different congestion patterns that occur for different main road demands – on-ramp demand scenarios.

Besides furthering empirical proof of multi-anticipative behavior and driver heterogeneity, and the car-following modeling it entails, the main contributions of the work presented here is that insight is gained the macroscopic characteristics of the stochastic multi-anticipative car-following model. It is shown that the estimated model is asymptotically stable. We also show that part of the scatter in the fundamental diagram can be explained by the modeled inter-driver differences.

MULTI-ANTICIPATIVE CAR-FOLLOWING MODELING

In (Hoogendoorn et al, 2006), we considered multi-anticipative car-following models of the generalized Helly type. Parameters were determined by fitting the models on empirical trajectory data collected from a helicopter. In this contribution, we will further study this model, as well as the so-called modified Lenz model. In this section, we briefly consider both.

Generalized Helly models

The different multi-anticipative models proposed in (Hoogendoorn et al, 2006) can be described by a single, generic model. This model describes car-following behavior, based on the difference between desired distance $S_j(v)$ of a driver having speed v with respect to the j-th vehicle ahead, and relative speeds w as follows:

$$a_{c.f.}(t) = \sum_{j=1}^{m_1} \alpha_j w_j(t-T_r) + \sum_{j=1}^{m_2} \beta_j \left(s_j(t-T_r) - S_j(v)\right) \quad (1)$$

where $a_{c.f.}(t)$ denotes the retarded car-following acceleration, with delay T_r. In Eq. (1), w_j denotes the relative speed of a vehicle with respect to the j-th vehicle ahead, for j = 1, …, m_1; S_j denotes the desired distance between the considered vehicle and the j-th vehicle ahead. For the desired distance, we generalize the model of Forbes (1958) to include multi-anticipative behavior as follows:

$$S_j(v) = s_0 + j \cdot Tv \quad (2)$$

where T denotes the minimum time headway, and s_0 denotes the gross stopping distances.

Based on the estimation results shown in (Hoogendoorn et al, 2006), we assume that the parameters of the generalized Helly model are in fact *driver-specific parameters* describing the heterogeneity in the driver population. Although intra-driver variations also have been observed (Tampère et al, 2005), (Hoogendoorn et al, 2005), these will not be considered explicitly in the remainder of this contribution.

Modified Lenz model

Lenz et al. (1999) consider a multi-leader generalization of the (Bando, 1995) model:

$$a(t) = \sum_{j=1}^{m} \kappa_j \left\{ V\left(\frac{s_j(t)}{j}\right) - v(t) \right\} \quad (3)$$

where V(s) is an equilibrium speed function describing the speed of the follower in relation to the distances to the vehicles ahead; the parameters κ_j denote the sensitivity to the jth leader. Please note the direct relation with the fundamental diagram describing the macroscopic properties of traffic flow.

Hoogendoorn and Ossen (2005) showed poor average model performance compared to the other car-following models. To improve performance, we propose here to include a true reaction time T_r as follows (Modified Lenz model):

$$a_{ML}(t) = \sum_{j=1}^{m} \kappa_j \left\{ V\left(\frac{s_j(t-T_r)}{j} \right) - v(t-T_r) \right\} \quad (4)$$

The following specification for the equilibrium speed V is used (Lenz et al, 1999) :

$$V(s) = v_0 \left\{ \left\{ 1+\exp\left(\frac{1000}{\gamma \cdot s} - \frac{10}{2.1} \right) \right\}^{-1} - 5.34 \cdot 10^{-9} \right\} \quad (5)$$

Where v_0 (free speed) and γ are parameters to be estimated from the data

Heterogeneous multi-anticipative theory and model

Inter-driver differences in car-following behavior are determined by many factors, such as experience and driving ability, trip purpose, vehicle characteristics, age, gender, attention level, etc. Note that the attention levels can vary between the drivers, but that the attention level of a single driver can also very during the trip (e.g. due to changing traffic conditions, increasing fatigue, etc.; see (Hoogendoorn and Ossen, 2005)). The latter will however not be considered further in this contribution.

As will be shown in the following section, car-following behavior of all drivers in a population cannot be captured by either the generalized Helly model or by the modified Lenz model alone. In the model proposed here, inter-driver differences are reflected by:

- Inter-driver differences in car-following models
- Inter-driver differences in parameters between drivers whose behavior can be described by the same model

To gain indication of the inter-driver differences in parameters, Figure 1 shows an example of estimating the parameters of the GH-1-1 model resulting from application of the estimation approach described in Hoogendoorn et al (2006) on the data described in the ensuing. Clearly, the inter-driver differences in the estimated parameter values are considerable.

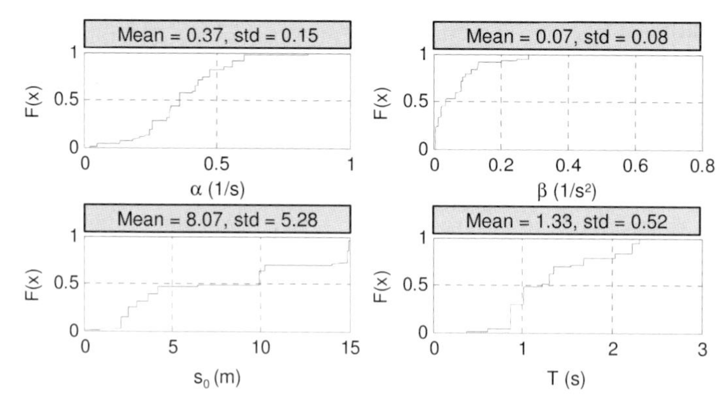

Figure 1 Example of parameter distributions for the GH-1-1 model, showing the individual-level estimates of the considered drivers. Note that especially the parameter β is skewed.

We propose the following HEterogeneous Multi-Anticipative car-following model HEMA, describing the behavior of driver i:

$$a^{(i)}(t) = (1-\chi^{(i)})\cdot a_{GH}(t\,|\,\vec{\theta}_{GH}^{(i)}) + \chi^{(i)}\cdot a_{ML}(t\,|\,\vec{\theta}_{ML}^{(i)}) \tag{6}$$

where $\chi^{(i)}$ follows a Bernouilli distribution with $\Pr(\chi^{(i)}=1) = \theta$ and $\Pr(\chi^{(i)}=0) = 1-\theta$ describing the binary choice for the one model or the other, and where the parameters $\vec{\theta}_{GH}^{(i)}$ and $\vec{\theta}_{GH}^{(i)}$ of the generalized Helly model and the modified Lenz model are randomly distributed. It is emphasized that the parameters are fixed for one driver, but vary amongst the drivers.

In the ensuing we will deal consecutively with the identification of the parameters of the HEMA model Eq. (6), and the resulting model properties. Note that the result will yield models that fit empirical trajectory data. Unlike for instance the model of Zhang et al. (1998), the model does not explain why the one submodel describes the behavior of the one driver, and the other submodel the behavior of the other. Furthermore, given the limited dataset that will be used for model identification, not all features of traffic flow dynamics can be reproduced.

PARAMETER IDENTIFICATION

Hoogendoorn et al. (2006) propose a new approach to identify the parameters of car-following models based on data from individually traced vehicles. The parameter identification approach entails estimating the parameters best describing the car-following behavior of individual drivers. The goodness-of-fit used for optimization is the likelihood of the individual samples.

In this contribution, we have further generalized the procedure to improve the suitability of the estimation results for simulation purposes. This is explained in the ensuing of this section.

Maximum Likelihood approach to parameter identification of car-following models

In (Hoogendoorn et al., 2006), the unknown parameters $\vec{\theta}$ of the considered car-following model are estimated by minimization of the likelihood of the observed speeds $v_{obs}(t_k)$ of a driver at instants $t_k = k\Delta t$, where Δt is the observation time step. The approach is based on the assumption that the observed speeds and the predicted speeds are related as follows:

$$v_{obs}(t_{k+1}) = v_{pred}(t_{k+1}\,|\,t_k,\vec{\theta}) + \varepsilon(t_k) \tag{7}$$

The error term $\varepsilon(t)$ is introduced to reflect errors in the modelling, similar to the error term used in multivariate linear regression. We assume that the error term is normally distributed with mean zero and (unknown) standard deviation σ. Based analyses of the residuals (after estimation), this choice is justifiable.

Hoogendoorn et al. (2006) propose that the maximum likelihood estimates can be determined by (constrained numerical) optimization:

$$\{\vec{\theta}^*,\sigma^*\} = \arg\max L(\vec{\theta},\sigma) \tag{8}$$

with the log-likelihood defined by:

$$L(\vec{\theta},\sigma) = -\frac{n}{2}\ln\left(2\pi\sigma^2\right) - \frac{1}{2\sigma^2}\sum_{k=1}^{n}\left(v_i^{obs}(t_{k+1}) - v_{pred}(t_{k+1} \mid t_k,\vec{\theta})\right)^2 \tag{9}$$

where n denotes the number of observations of the vehicle speed and position, and where $v_{pred}(t_{k+1} \mid t_k,\vec{\theta})$ denotes the one-step ahead prediction of the speed by the car following model, using the set of parameters $\vec{\theta}$ based on the *observed vehicle positions and speeds* at time instants t_k, t_{k-1}, ..., t_0. Note that the standard deviation can be determined analytically easily:

$$\frac{\partial L(\vec{\theta},\sigma)}{\partial\sigma^2} = 0 \quad\Rightarrow\quad \sigma^2 = \frac{1}{n}\sum_{k=1}^{n}\left(v_i^{obs}(t_{k+1}) - v_{pred}(t_{k+1} \mid t_k,\vec{\theta})\right)^2 \tag{10}$$

Hoogendoorn et al. (2006) show how the approach can be adapted to correct for the serial correlation in the data. It is also shown how the covariance matrix of the estimated parameters can be estimated using the so-called Cramér-Rao lower bound. Different models (of *different model complexity*) are cross-compared by application of the likelihood ratio test. By doing so, we can assess whether the one model is best compared to the others in a statistically significant way (see results in Table 1).

From local to global optimization

The log-likelihood (9) is determined based on one-step-ahead predictions. If the model is to be used for simulation purposes, the car-following model should not only reproduce the short-term dynamics, but rather yield the correct global, longer-term dynamics. This is why we propose a global, rolling horizon approach, where the *K-step-ahead prediction* is used to determine model performance. The proposed approach entails maximization of the modified log-likelihood function:

$$\tilde{L}(\vec{\theta}) = -\frac{n}{2}\ln\left(\frac{2\pi}{nK}\sum_{k=1}^{\lfloor n/K\rfloor}\sum_{p=1}^{K}\left(v_{obs}(t_{kK+p}) - v_{pred}(t_{kK+p} \mid t_{kK},\vec{\theta})\right)^2\right) - \frac{n}{2} \tag{11}$$

where $v_{pred}(t_{k+p} \mid t_k,\vec{\theta})$ denotes the p-step ahead prediction, for p = 1, ..., K.
The integer K defines the number of time steps between updates of the prediction. In this case, an update means that the predicted speed is replaced by an observed speed. In illustration, K = 1 implies that only one-step ahead predictions are made. On the other hand, K = n implies that only the initial speed of the modeled driver is used, and the predictions are not updated anymore.

Note that in line with (Hoogendoorn et al., 2006) we have used the differences between observed and predicted *speeds* to assess the performance of the model, to enable cross-comparison between the estimation results. Other criteria are however possible as well (e.g. differences in measured and predicted distances, accelerations, or combinations thereof).

Data used for model identification

The vehicle trajectory data used here was collected using a new data collection approach (Hoogendoorn and Van Zuylen, 2004) using an air-borne observation platform (a helicopter), mounted with a high-frequency digital camera and frame grabber. Using image processing software, the vehicles are detected and tracked as they move along the roadway. This yields trajectory data covering approximately 500 m of motorway stretch; the spatial resolution is smaller than 40 cm, while the temporal resolution is 0.1 s. The raw data is smoother using a Kalman filtering approach described in (Hoogendoorn and Van Zuylen, 2004), enabling derivation of smooth speed profiles from the data.

The dataset considered here was collected at the A2 motorway near the Dutch city of Utrecht and is characterized by stop-and-go flow conditions (see Figure 3). Figure 1 shows a sample from the dataset. The total dataset consist of 315 vehicle trajectories. Not all of these could however be used for estimation purposes, because many contain too little information to enable model identification (e.g. speeds which are nearly constant during the entire observation period, or observation period in which composition of the four vehicle platoon needed to estimate the parameters of a three leader multi-anticipative model was deemed too short). In the end, 144 trajectories were selected for further analysis.

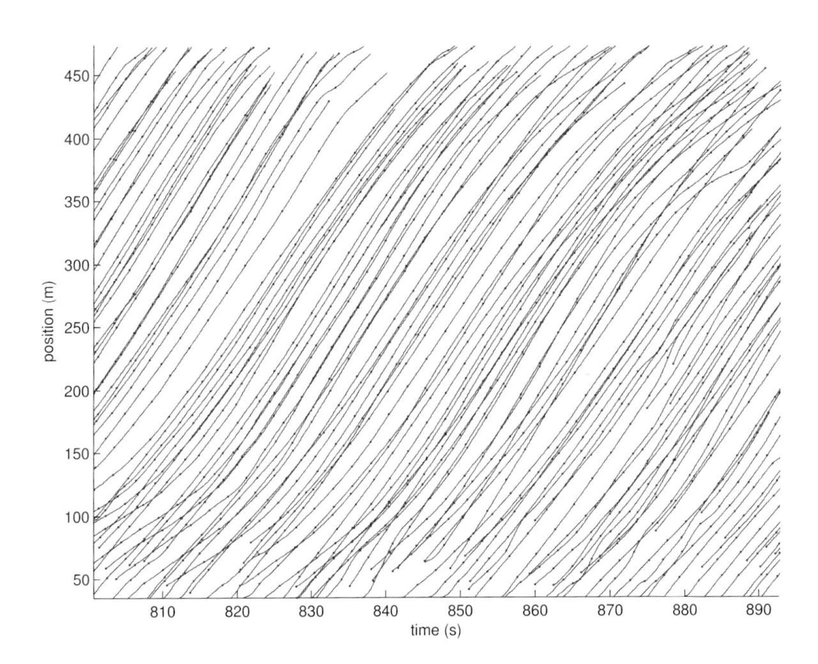

Figure 2 Sample of vehicle trajectories for data collected at A2 site. The small dots represent time instants which are 2.5 second apart; data is collected at a temporal resolution of 0.1 s.

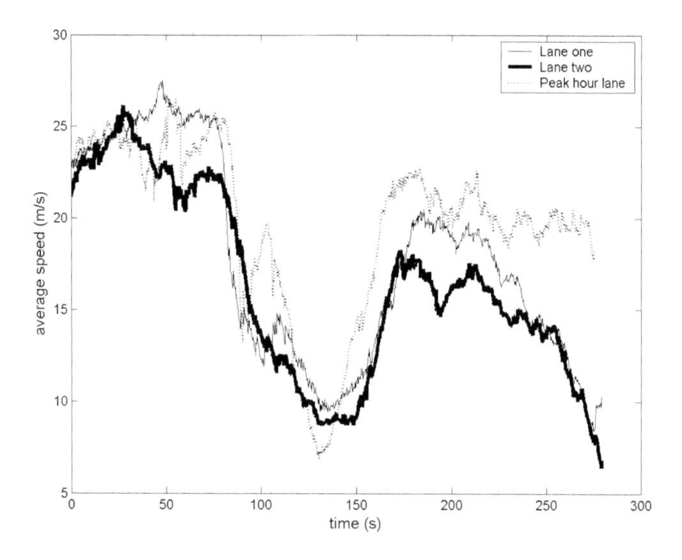

Figure 3 Space-mean speeds during the observation period (per lane), where 'lane one' is the median lane, 'lane two' is the right lane, and the 'peak hour lane' is a shoulder lane, used as a regular lane during peak hours.

Estimation results: model performance

The trajectory data were used to estimate different models of the generalized Helly type Eq. **Error! Reference source not found.** for different values of m_1 and m_2 (see Eq. (1)) and of the Lenz type (4). The different models are cross-compared using the likelihood-ratio test (Hoogendoorn et al., 2006). Note that this test favors models with fewer parameters, so that there is a trade-off between the self-evident performance increase of the more complex models and the number of parameters.

Table 1 provides an overview of the main estimation results for the cases for K = n (see Eq. (11)). Column (a) shows the percentage of all drivers for which the model performed best (based on the likelihood ratio test); column (b) shows the frequency with which the model was second best. Column (c) shows the average log-likelihood. The remaining columns pertain only to the situations in which the considered model performed best: column (d) and (e) respectively denote the average improvement compared to the default reference model (GH-1-0) and to the second best model; column (f) indicates the 2nd best model that occurred most frequently.

For instance, for the 19% of the situations in which the GH-3-1 model performed best. This means that the likelihood ratio test comparing the best and the second-best model was passed (in other words, the model is 'best' from a statistically significant viewpoint). The average improvement over the reference model was 214.3; the average improvement over the 2nd best model for these 19% was 29.1. The latter means that on average, the difference in performances of the best and the second best model is considerable. Furthermore, when the best model is the GH-3-1 model, the *second best model* is the GH-2-1 model most frequently. This is to be expected given that these models are alike, except for the inclusion of the driver reaction to the third leader.

Table 1 Estimation results for multi-anticipativemodels for K = n.

	(a)	(b)	(c)	(d)	(e)	(f)
Model	% best	% 2nd best	LL	$LL\text{-}LL_{ref}$	$LL\text{-}LL_{2nd}$	Most freq. 2nd best
GH-1-0 (reference)	0%	0%	-561.6	n/a	n/a	n/a
GH-2-0 (Bexelius)	1%	3%	-506.2	218.8	-1.1	GH-1-1
GH-1-1 (Helly)	10%	7%	-394.0	283.7	0.2	GH-2-1
GH-2-1	10%	24%	-358.4	285.9	-0.3	GH-3-1
GH-3-1	19%	11%	-348.4	214.3	29.1	GH-2-1
GH-1-2	13%	14%	-366.4	259.9	0.1	GH-1-3
GH-1-3	0%	13%	-368.0	n/a	n/a	n/a
GH-2-2	15%	13%	-391.3	302.7	24.2	GH-2-1
Bando	3%	10%	-495.9	260.2	53.8	Lenz
Lenz	29%	4%	-391.3	232.1	25.7	Bando
	100%	100%		198.4	27.3	

Without going into detail, Table 1 yields the following, generic conclusions:

- Including multi-anticipative behavior strongly improves the descriptive performance of the car-following models. This holds for both Helly and Lenz type models.
- The behavior of the individual drivers can only be captured by considering both the generalized Helly model family *and* the modified Lenz model family.
- Driver behavior of a specific driver is often best described by either of the model families (Helly or Lenz). E.g.: when the Lenz model best describes driver behavior, in many cases the Bando model is the second best model.
- For the Helly type models, including the relative speed of both the second and third leader (GH-3-1) on average yields the largest improvement.

Estimation results: parameter estimates

Let us now take a closer look at the parameter estimates. We will present the estimation results in two ways. First, we will show the average parameter value for a specific model averaged for all drivers (Table 1). Secondly, we will consider the average parameter value for all drivers for which the considered performed best (Table 2).

Note that the estimates shown in the two tables are quite distinct. On the one hand, this is caused by the fact that the averages in Table 2 are based on a smaller sample than those in Table 1. On the other hand, the average parameter values in Table 1 represent the values for a model that is 'forced' to describe behavior which another model describes better. In illustration: the average estimates for the GH-3-1 model for all drivers (Table 1) show that the sensitivity to the relative speeds of the vehicles ahead *decreases* when looking further downstream (sensitivities of $0.29s^{-1}$, $0.08s^{-1}$ and $0.06s^{-1}$ for speed difference with the first, second and third leader). Considering only the drivers for which the GH-3-1 model outperforms the other models (Table 2) yields a different result: in this case, the sensitivity to the third leader is just as large as the sensitivity to the second leader (both $0.10s^{-1}$).

The tables also provide insight into the magnitude of the reaction times for the different car-following models. Note that the reaction times are relatively small for the modified Lenz models (around 0.8s), compared to the generalized Helly models (around 1.0s).

Table 2 Average parameter values for the different estimated models determined by considering the individual estimations of all drivers in sample, irrespective of the predictive performance of that model.

Model type	T_r	α_1	α_2	α_3	β_1	β_2	β_3	s_0	T	n
GH-1-0	1.01	0.79						8.69	2.16	134
GH-2-0	1.16	0.47	0.16					8.70	2.44	134
GH-1-1	0.94	0.58			0.09			8.78	2.43	134
GH-2-1	1.06	0.30	0.14		0.06			8.82	2.26	134
GH-3-1	1.15	0.29	0.08	0.05	0.06			8.80	2.35	134
GH-1-2	0.96	0.41			0.07	0.03		8.89	2.42	134
GH-1-3	0.97	0.41			0.06	0.03	0.01	8.87	2.40	134
GH-2-2	1.09	0.29	0.12		0.06	0.02		8.86	2.37	134
		v_0	κ_1	κ_2	γ					
Bando	0.76	26.23	0.66	0.00	7.95					134
Lenz	0.79	28.13	0.36	0.30	5.19					134

Table 3 Average parameter values for the different estimated models determined by considering only the drivers for which the specific model outperforms the other models.

Model type	T_r	α_1	α_2	α_3	β_1	β_2	β_3	s_0	T	n
GH-1-0	-	-	-	-	-	-	-	-	-	0
GH-2-0	1.00	0.44	0.11					9.37	1.81	1
GH-1-1	0.83	0.35			0.09			8.56	2.61	13
GH-2-1	1.10	0.37	0.16		0.07			8.41	1.60	13
GH-3-1	1.09	0.25	0.10	0.10	0.07			8.87	2.03	26
GH-1-2	0.88	0.28			0.03	0.02		8.75	2.71	18
GH-1-3	-	-	-	-	-	-	-	-	-	0
GH-2-2	1.25	0.27	0.17		0.03	0.04		8.85	2.42	20
		v_0	κ_1	κ_2	γ					
Bando	1.55	31.61	0.17		17.31					4
Lenz	0.83	26.74	0.22	0.28	5.83					39

It is also noticed that the inter-driver co-variances of the model parameters are considerable. In illustration, Table 4 shows the estimation results for the GH-3-1 model for the 26 drivers for which the GH-3-1 model outperformed the other models. From the table, the large variability in the model parameters between the different drivers becomes clear. The correlation between the parameters is relatively small.

Table 4 Estimation results for GH-3-1 model. Table shows the average parameter estimates, the standard deviation and the inter-driver parameter correlations.

	T_r	α_1	α_2	α_3	β_1
average	1.09	0.25	0.10	0.10	0.07
std. dev.	0.28	0.31	0.11	0.13	0.08
	T_r	α_1	α_2	α_3	β_1
T_r	1.000	-0.300	-0.138	-0.027	-0.255
α_1		1.000	-0.161	-0.027	-0.056
α_2			1.000	-0.159	0.179
α_3				1.000	-0.100
β_1					1.000

In the remainder of this contribution, we will consider the characteristics of the traffic flow dynamics that the models discussed here will result in. This means that for each driver i = 1, …, n, first the model - say model j - is determined (the probability that a model is chosen is determined directly from the relative number of times it was appointed being 'best'), after which the model parameters are determined from the estimated parameter distributions for model j.

The analyses in the ensuing are performed is largely by means of simulation using the parameters estimated from the trajectory data. When necessary, hypothetical parameter settings will be considered to investigate the effect of specific model parameters.

FUNDAMENTAL DIAGRAM AND PHASE-SPACE PLOTS

We will start by considering the fundamental diagram of the heterogeneous multi-anticipative model. This can be done both analytically (by assuming a(t) = 0, in particular for the case of homogeneous platoons) and by means of simulation.

Equilibrium relations for the generalized Helly model

For the generalized Helly model, the fundamental diagram is determined directly by the relation between speed and distance headway under the assumption of stationary conditions:

$$a(t) = \min\left(a_{c.f.}, a_{free}\right) = 0 \tag{12}$$

This implies that:

$$\sum_{j=1}^{m_1} \alpha_j w_j(t-T_r) + \sum_{j=1}^{m_2} \beta_j \left(s_j(t-T_r) - S_j(v)\right) = 0 \quad \text{for} \quad v(t) < v^0 \tag{13}$$

Let us consider the deterministic case, i.e. car-following parameters are the same for all drivers i = 1, ..., n. If we consider the situation that all vehicles are driving at the same speed u (implying $w_j = 0$ by necessity), we have:

$$\sum_{j=1}^{m_2} \beta_j \left(s_j^{(i)} - s_0 - j \cdot Tu \right) = 0 \quad \forall i \tag{14}$$

Eq. (14) will hold for all drivers i, so we have:

$$\sum_{i=1}^{n} \sum_{j=1}^{m_2} \beta_j \left(s_j^{(i)} - s_0 - j \cdot Tu \right) = \sum_{j=1}^{m_2} \beta_j \sum_{i=1}^{n} \left(s_j^{(i)} - s_0 - j \cdot Tu \right) = 0 \tag{15}$$

Note that on average, we have:

$$s_j = \frac{j}{k} \tag{16}$$

where k denotes the density in veh/m. In combining Eq. (14) and Eq. (16), we find the following expression for the equilibrium relation between the density and the speed:

$$k(u) = \frac{\sum_{j=1}^{m_2} j\beta_j}{\sum_{j=1}^{m_2} \beta_j (s_0 + j \cdot Tu)} = \frac{B_1}{s_0 B_0 + TuB_1} \quad \text{for } u < v^0 \tag{17}$$

where the parameters B_0 and B_1 are defined by:

$$B_0 = \sum_{j=1}^{m_2} \beta_j \text{ and } B_1 = \sum_{j=1}^{m_2} j\beta_j \tag{18}$$

From Eq. (17), we see that for low speed values, the density will predominantly be determined by the stopping distances s_0, while for larger speeds, the minimum headway T and the speed risk factor will be the determining parameters.

We can easily invert Eq. (17) yielding the following relation between the mean speed u and the density k:

$$u(k) = \begin{cases} u_0 & k < k_c \\ \frac{1}{T}\left(\frac{1}{k} - s_0 \frac{B_0}{B_1} \right) & k \geq k_c \end{cases} \quad \text{where} \quad k_c = \frac{B_1}{u_0 TB_1 + s_0 B_0} \tag{19}$$

In turn, we find the following expression for the equilibrium volume q(k):

$$q(k) = \begin{cases} u_0 k & k < k_c \\ \frac{k}{T}\left(\frac{1}{k} - s_0 \frac{B_0}{B_1} \right) & k \geq k_c \end{cases} \tag{20}$$

The capacity and the jam density respectively equal:

$$C = \frac{u_0 B_1}{u_0 TB_1 + s_0 B_0} \approx \frac{1}{T} \text{ and } k_{jam} = \frac{1}{s_0} \tag{21}$$

Let us now consider a platoon of n drivers, whose car-following parameters are instances of random variables. The platoon leader is driving at speed u, as are the other vehicles in the platoon (stationary conditions). The total platoon length – determining the average density k(u) in the platoon – is now determined directly by the car-following parameters of the individual drivers i, in particular $u_0^{(i)}$, $s_0^{(i)}$, $T^{(i)}$ and $\beta_j^{(i)}$. As a result, the parameters of the fundamental diagram q(k) (Eq. (20)) are in fact also random variables, the standard deviation of which is determined by the number of cars n in the platoon, and the distributions of the parameters.

Equilibrium relations for the modified Lenz model

In the deterministic case, the modified Lenz model acts the same as the original model of Lenz et al (1999) under stationary conditions; the equilibrium speed is thus given by:

$$u(k) = V(1/k) \tag{22}$$

For the stochastic case, we will again see that the parameters describing u(k) (and consequently of q(k)) are in fact random variables stemming from the car-following parameters of the individual drivers.

Fundamental diagrams for mixed models without inter-driver parameter differences

In the remainder, the fundamental diagram is determined using simulation: the speed of the platoon leader u_{leader} is chosen equal to a specific value (e.g. 2, 4, ..., 32 m/s) and the behavior of the followers is simulated. After a while, the speed of all vehicles is equal to u_{leader} (in case of convergence, which appeared to occur in all cases described here); the distances between the vehicles are then used to determine the platoon density.

Figure 4 shows the fundamental diagram for the mixed generalized Helly model based on the average parameter estimates shown in Table 3. The probability that a specific vehicle is driving according to a specific car-following model is determined from the probability that the model outperforms the other models (see Table 1) (compared to the other Helly type models, that is). From the figure, we see that especially the capacity predicted by the model is substantially less than the capacity of one lane of a two-lane motorway (which is around 2200 veh/h in the Netherlands) due to the fact that the high estimated value for T (around 2.4s). This may be caused by the fact that all observations used for the estimation were collected during congestion, while drivers have adopted to a less efficient driving style (Tampère, 2004). Furthermore, the dataset consisted only of 134 drivers. The predicted jam density (of approximately 140 veh/km) is reasonable.

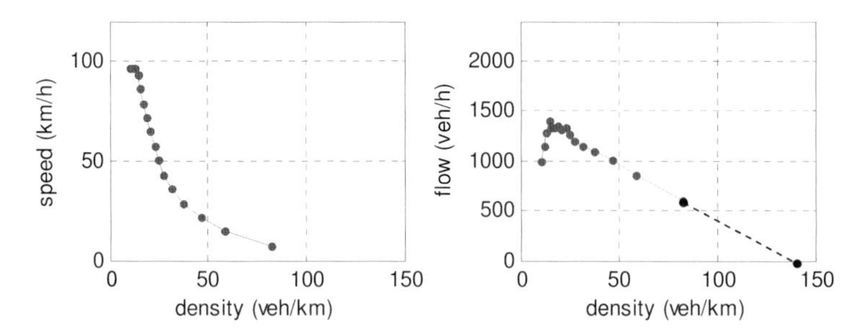

Figure 4a) speed – density and b) flow – density relation for the mixed generalized Helly model. Each dot indicates the average speed, density, and flow of the simulated platoon. Note that in all cases, the simulation converged to stationary conditions.

The same analysis has been performed for the modified Lenz model. Figure 5 shows the resulting fundamental diagram. We can again observe that the predicted capacity of around 1500 veh/h is low, but is consistent with the mixed Generalized Helly model. Note the large differences in the shapes of the resulting fundamental relations, in particular at high densities. This can partly be explained by the data that has been used for estimation of the parameters, featuring a speed range approximately 20 km/h and 80 km/h. The estimation procedure aims to reproduce the driver behavior in particular in this speed range. As we can see by comparing Figure 4 and Figure 5, the fundamental diagrams of both models are very alike in this speed range. Apparently, when extrapolating the models to other traffic conditions, they become quite dissimilar.

Figure 6 shows the fundamental diagram that results when both model types are combined, yielding the heterogeneous model presented in this contribution. Note that due to the random nature of the model, scatter is present especially in the congested branch of the fundamental diagram. Note that only the *type of model* is drawn randomly. For the model parameters, the average values in Table 2 are used. Note that the model will predict 'horizontal scatter' in the speed-density plane: for a fixed speed of the leading vehicle, different densities are predicted (due to the differences in car-following distances predicted by the two models); also see Figure 7 and Figure 8.

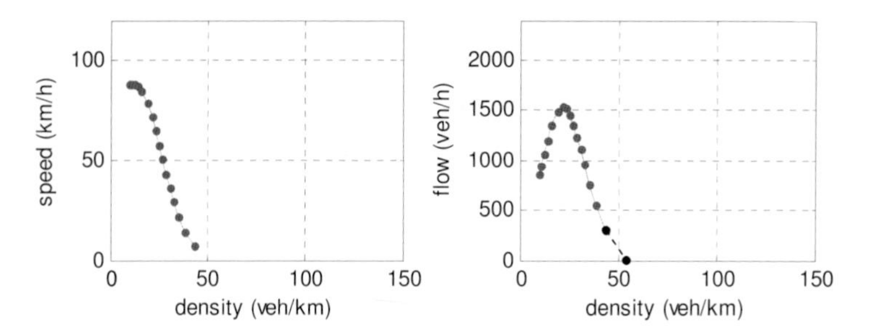

Figure 5 a) speed – density and b) flow – density relation for the modified Lenz model. Each dot indicates the average speed, density, and flow of the simulated platoon. Note that in all cases, the simulation converged to stationary conditions.

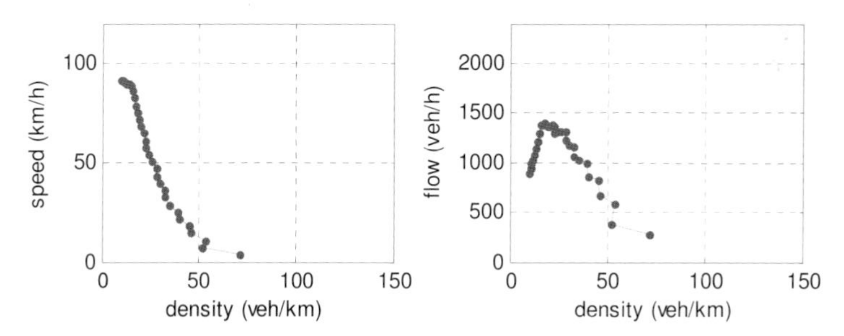

Figure 6 a) speed – density and b) flow – density relation for the HEMA model (with deterministic parameters per submodel). Each dot indicates the average speed, density, and flow of the simulated platoon. In all cases, the simulation converged to stationary conditions.

Fundamental diagrams for mixed models with inter-driver parameter differences

We will now consider the case where the car-following parameters of the drivers are stochastic rather than deterministic (i.e. different for each driver, but constant per driver). We will in particular focus on the Helly type models, in particular on the impact of stochastic values of the stopping distances s_0 and the minimum headway T. The results are shown in Figure 7 and Figure 8. Note that the impact of randomness in T is much more profound than the impact of randomness in s_0.

For the figures we can conclude that the observed heterogeneity of the model parameters as identified from the microscopic estimation to a certain extent explains the scatter in the empirically established phase-space plots. In this line of thought, the fact that the congested branch is a region rather than a line (see e.g. (Kerner, 2005)) could *partly* be explained by driver heterogeneity. Note that in contrast to the phases described by (Kerner, 2005), transition between phases in completely random in our case (since the phase is only determined by the coincidental composition of the platoon).

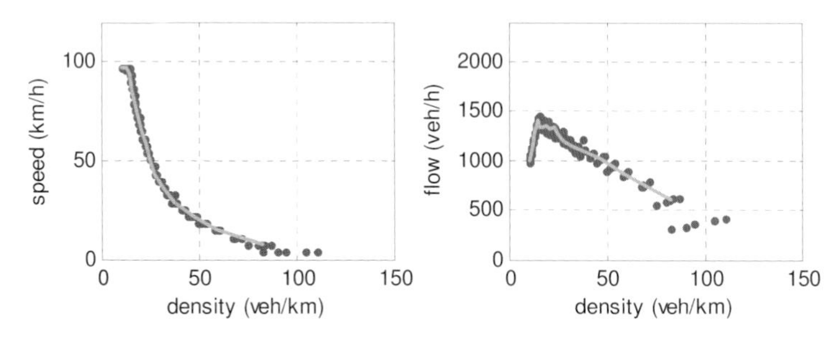

Figure 7 Effect of random stopping distances s_0. Note that the scatter becomes more profound at low speeds and high densities. The yellow line shows the equilibrium relation in case of deterministic parameters (from Figure 1).

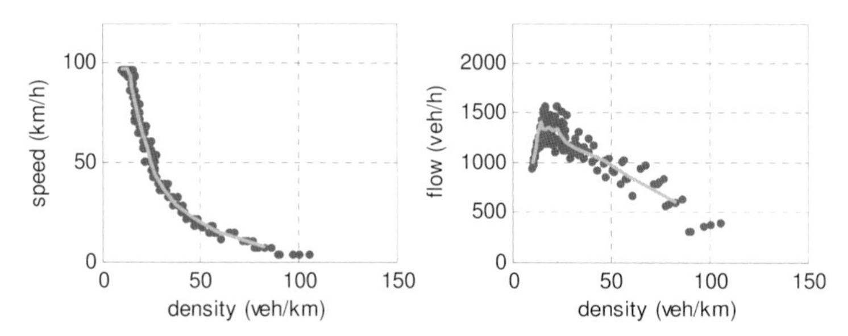

Figure 8 Effect of random minimum headways T. The yellow line shows the equilibrium relation in case of deterministic parameters (from Figure 1).

STABILITY ANALYSIS

In this section, we will consider the issue of stability. Recall that two types of stability can be distinguished. For the present study, we consider *platoon (or string, or asymptotic) stability*. This type of stability describes the way a disturbance propagates through the subsequent vehicles of a platoon. When the distribution damps out, we have platoon stability. Vice versa, when the perturbation grows, the platoon is unstable. For platoons of finite size, platoon instability does not necessarily imply that the disturbance will not damp out in the end.

A car-following model can also be *locally unstable*, depending on the dynamic response of a follower to the behavior of the leaders. The response can be damped, damped oscillatory, and undamped oscillatory (unstable). It is a well known fact that local stability is a necessary condition for platoon stability. Since local instability only has theoretical relevance - in practice all car-following will be locally stable - we will consider platoon stability only.

Platoon stability will be analyzed by means of simulation: a disturbance is applied (e.g. the speed of the lead vehicle is reduced temporarily) and the dynamic responses of the subsequent vehicles are monitored. The so-called *amplitude* A_i of the resulting response signal is determined for each vehicle i. Note that the platoon stability is guaranteed when A_i decreases over the platoon. Let us first mathematically define these indicators. The approach was thoroughly verified to ensure that influences of discretisation were kept sufficiently small.

Indicators for platoon instability

We consider a platoon of n vehicles. The speed dynamics of the leader are given. We assume that the leader will start at a certain speed u_0, and will instantaneously reduce his speed temporarily to u_1 between t_0 and t_1 with a fixed value (5 m/s). The car-following behavior of the following vehicles is determined by the car-following laws described in the preceding sections, considering up to three leaders. Note that the second vehicle and third vehicle in the platoon are treated differently, since these only have one or two leaders. The behavior of these vehicles will be the same for all considered experiments (irrespective of the considered models). These drivers are not considered in the analysis.

The *amplitude* of driver i is determined by the changes in the speed. More specifically, starting for equilibrium conditions ($v_i(t) = u_0$) – which are assumed to occur for t = t' onwards – the amplitude A_i is defined by:

$$A_i = \max_{t>t'} v_i(t) - \min_{t>t'} v_i(t) \tag{23}$$

For the platoon, the *mean amplitude growth factor* is given by:

$$\bar{\rho} = \frac{1}{N-3} \sum_{i=1}^{N-3} \frac{A_{i+1}}{A_i} \tag{24}$$

Note that we have skipped the first three vehicles in the platoon, since their behaviors are not described by the multi-anticipative car-following models that we aim to analyze. Furthermore, we compute the *maximum amplitude growth factor* over the platoon by:

$$\rho_{max} = \max_{i=1,\dots,N-3} \left(A_i / A_{N-2} \right) \tag{25}$$

We will also check if collisions occur (note that emergency braking was not included to ensure that the characteristics of the car-following models are studied, and not the characteristics of the emergency braking law). In case of random car-following parameters, the probability of a collision is computed. Note that although an unstable model will generally yield collisions (if the platoon is sufficiently long), a collision does not imply that the model is unstable (in a mathematical sense). On the contrary, in some cases a collision occurs because the car-following model is insensitive to the behavior of the leaders.

Stability analysis results

The model estimations will form the basis of our stability analysis. In particular, we will consider the stability of the individual models, and of the mixed model, with the parameters estimated from the empirical trajectory data. Additional analyses have been performed with hypothetical parameter values, with the aim to gain more insight into which of the factors influence platoon stability. In particular, the impact of the sensitivity parameters α, β, κ and the reaction time T_r have been considered.

Table 5 below shows an overview of the results of the stability analysis for the different models considered in this contribution, both in case of deterministic parameters. Note that for the specific models, the parameters used for the stability analysis are the average model parameters shown in Table 2, i.e. for all drivers in the sample. Also note that only the mixed HEMA model is in fact stochastic, since the model which is used to simulate driver behavior is drawn randomly.

Note that four of the calibrated car-following models have platoon instability (as can be seen from the value of the mean amplitude growth factor), namely the GH-1-0, the GH-2-0, the Bando model and the Lenz model. The mixed model has platoon stability.

Table 5 Overview stability analyses results for 50 vehicle platoon, with initial speed of 15 m/s. The speed drop used in the simulation is equal to 5 m/s.

Model	$\bar{\rho}$	ρ_{max}	P_{coll}	i_{coll}
GH-1-0	1.019	5.468	0	0
GH-2-0	1.018	5.488	1	85
GH-1-1	0.982	0.967	0	n/a
GH-2-1	0.979	0.945	0	n/a
GH-3-1	0.977	0.904	0	n/a
GH-1-2	0.984	0.971	0	n/a
GH-1-3	0.984	0.973	0	n/a
GH-2-2	0.981	0.939	0	n/a
Bando	1.031	5.447	1	96
Lenz	1.017	4.854	0	n/a
HEMA model	0.985	1.142	0	n/a

Let us now further consider the impacts of the different aspects that were found during model identification (in particular: multi-leader behavior, inter-driver differences, and model heterogeneity).

Single-leader vs. multi-leader

In general, inclusion of multi-anticipative behavior *increases* the stability of the car-following model (see (Bexelius, 1968), (Lenz et al, 1999), (Treiber et al, 2005)). Based on the analyses presented in this contribution, we can conclude the same: both the mean amplitude growth factor and the maximum amplitude growth factor decrease when multiple leaders are included. For instance, Table 5 shows that the GH-3-1 model features increased stability compared to the GH-2-1 and the GH-1-1 models (traditional Helly model). Additional simulations with different parameter settings support this finding.

Impact of stochastic parameters

We have also considered the effect of randomizing the car-following parameters on model stability. Overall, the impact of random parameters on platoon stability is present. When noise is applied to the sensitivities (i.e. α_j, β_j and κ_j), a decrease of the platoon stability is observed. In illustration: we have considered the case where the parameter α_1 of the GH-1-1 model was drawn from a uniform [0.2, 0.6] distribution and compared it to the case where $\alpha_1 = 0.4s^{-1}$ (with $\beta_1 = 0.1$). It turned out that randomizing α_1 increased the mean amplitude growth factor from 1.015 to 1.022. To illustrate the impact of this increase, note that by the time the disturbance reached the end of the platoon, its amplitude increased by a factor of $1.022^{100} = 8.81$, while in the deterministic case, it increased by $1.015^{100}=4.42$.

Apart from the increase in the average amplitude, note that the stability itself becomes a random variable. In other words, when multiple simulations are performed, some may yield platoon stability while others will not. This is caused by the coincidental order of the vehicles and their respective parameters. In the same line of thought, it was also found that random sensitivities increase the risk of a collision, due to possibility of an unsafe combination of subsequent drivers who are sensitive and insensitive to disturbances.

The findings are in line with the results reported in Ossen and Hoogendoorn (2006). Focussing on the GH-1-0 model (and comparing the results with analytical stability analysis resutls), clear differences between homogeneous platoons (deterministic values of α_1) and heterogeneous platoons (stochastic values of α_1) were found.

Sensitivity of stability to parameter values

To gain more insight into the stability characteristics of the mixed model, we have performed a sensitivity analysis by changing the magnitude of the different parameter types. Focusing on the heterogeneous generalized Helly model, we have applied different factors to the parameters α_i, β_i and T_r.

Figure 9 shows some of the results of this stability (for the heterogeneous Helly model). Note that since the model is in fact stochastic, multiple runs are required to get the average result for each case. The three pictures show the stable and unstable regions for fixed factors for T_r (in this case, T_r is multiplied by 0.5, 1.0 and 1.375 respectively). The thick black line indicates the approximate boundary between the stable and unstable regions.

From the figure, we can conclude that the heterogeneous generalized Helly model is *unstable* for small values of $\{\alpha_i\}$. This is to be expected, since for $\alpha_i = 0$ and $\beta_i > 0$, it can be mathematically proven that the model is (locally) unstable for all reaction times large than zero. In fact, the sensitivities to the relative speed have (at first) a damping effect. However, when the parameters $\{\alpha_i\}$ become too large, the model again becomes unstable. This becomes in particular apparent for larger reaction times (see Figure 9c).

For the parameters $\{\beta_i\}$, we see that in combination with small values of $\{\alpha_i\}$, the impact on platoon stability is moderate (the stability boundary on the 'left' of Figure 9a-c). For larger values of $\{\alpha_i\}$, the impact of changing $\{\beta_i\}$ is more profound. Figure 9c shows clearly that when the values of $\{\beta_i\}$ are increased, larger values of $\{\alpha_i\}$ will still ensure stability.

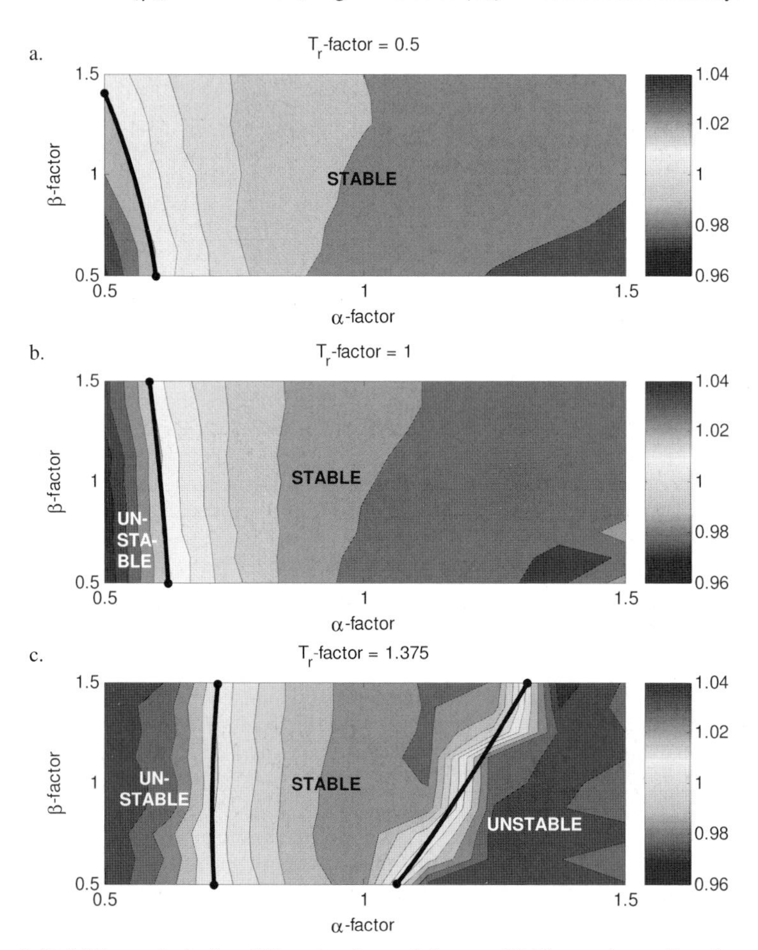

Figure 9 Stability analysis for different values of the sensitivities and reaction times for the heterogeneous generalized Helly model.

For the Lenz model, similar results holds: too large sensitivity values and / or reaction times cause platoon instability. This is not elaborated further here.

Impact of model heterogeneity

Figure 10 shows the stability analysis results for HEMA model (that is, including the Lenz model). For the sake of simplicity, the parameters of the Lenz model were kept constant. The figures clearly show how the stability of the model changes when incorporating the Lenz model. In illustration, Figure 10b (default value of the reaction time) shows a clear change of the stable region compared to Figure 9b. This is caused by the fact that the Lenz model is in fact unstable (recall results depicted in Table 5). Note especially that for larger reaction times (factor of 1.25), the stable region becomes very small indeed.

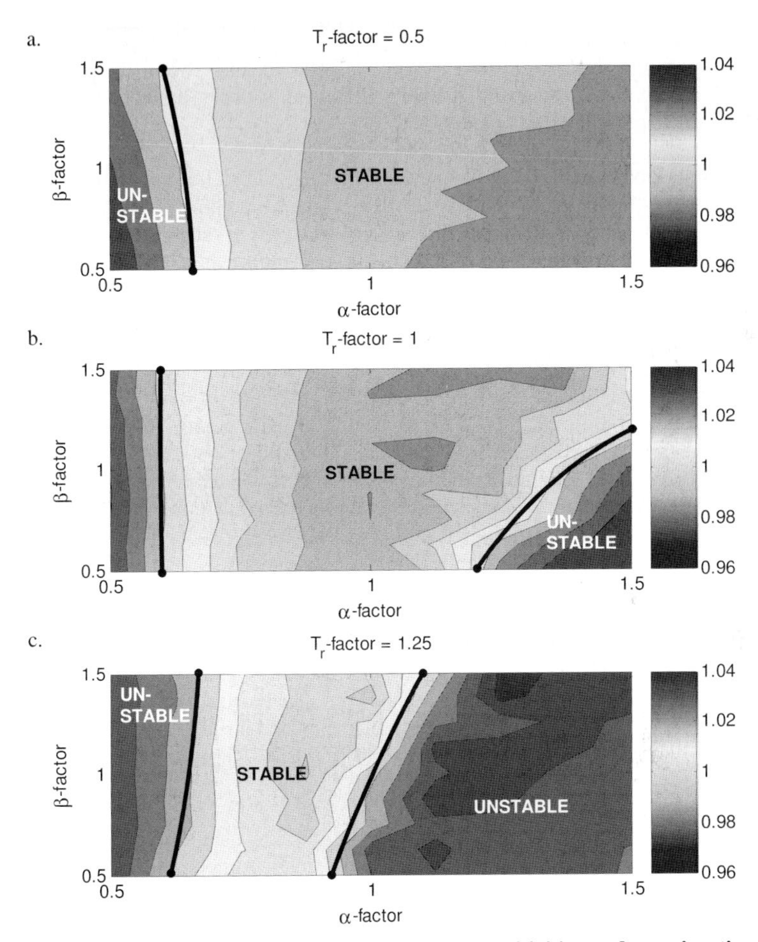

Figure 10 Stability analysis for different values of the sensitivities and reaction times for the heterogeneous mixed model (Helly and Lenz).

CONCLUSIONS

This contribution considered heterogeneous multi-anticipative car-following modeling established by estimating car-following models using empirical trajectory data. Based on the estimation results, we conclude that including multi-anticipative behavior strongly improves the descriptive performance of car-following models *at the microscopic level*. Furthermore, we have shown that the behavior of the individual drivers can only be captured by considering different car-following modeling paradigms, in this case the generalized Helly model family and the modified Lenz model family. Finally, it was concluded that the inter-driver differences were not only expressed by the suitability of different model types, but also by the large variability in the driver-specific car-following parameters. Based on this result, we proposed a HEterogeneous Multi-Anticipative (HEMA) model that includes the generalized Helly model and the modified Lenz model.

Given these microscopic results, we have established the traffic flow operations that result from application of the car-following models identified from the microscopic data. More specifically, we have shown to which extent specific aspects such as multi-anticipative behavior and driver heterogeneity (which turned out very important on the microscopic level) affect traffic operations on the macroscopic level. More specifically, we have established the impact of random parameters as well as multi-leader anticipation on the fundamental diagram, and asymptotic model stability.

With respect to the fundamental diagram, we know from empirical studies that especially in the congested branch of the fundamental diagram, flow-density points are very scattered. We have shown that this scatter can in part be explained by differences in driver behavior, in particular differences in the minimum time-headways maintained by drivers and the minimum stopping distances. Note that we have *not considered the causes* for these differences or whether these are intra- or inter-driver differences, but merely looked at the results from statistical fitting the car-following models on the microscopic data.

The stability analysis clearly shows the impact of including spatial anticipation on platoon stability: whereas the single-leader models estimated from the data are unstable, the multi-anticipative generalization of these models all feature enhanced platoon stability. Inclusion of randomness on the model parameters (in particular the sensitivities) has a considerable effect on platoon stability. The modified Lenz model alone appeared unstable. However, the HEMA model – mixing the generalized Helly model and the modified Lenz model – turned out to be stable again.

Overall, the macroscopic properties of the microscopically identified HEMA model are plausible[1]. The estimated capacity values are however small. Although this can be caused by the traffic regime under which the microscopic data was collected and the relatively few drivers considered in the analysis, the unrealistic capacity estimates point out that some care needs to be taken when scaling up the microscopic model to a macroscopic model.

---

[1] Note that additional experiments have been performed focusing on discontinuities. From these experiments, it turns out that the HEMA model features different congestion patterns, depending on the ratio between main-road and on-ramp flow, such as temporary disturbances, local non-moving cluster, moving localized cluster (start-stop waves in free-flowing traffic), oscillating congested traffic (start-stop waves in congested traffic) and homogeneous congested traffic.

ACKNOWLEDGEMENTS

The research presented in this paper is part of the research programme "Tracing Congestion Dynamics – with Innovative Traffic Data to a better Theory", sponsored by the Dutch Foundation of Scientific Research MaGW-NWO. The data was used at the courtesy of the Traffic Research Centre (AVV) of the Dutch Ministry of Transportation.

REFERENCES

Bando, M., et al., Dynamical Model of Traffic Congestion and Numerical Simulation. Physical Review, 1995. E(51): p. 1035-1042.

Bexelius, S., An extended model for car-following. Transportation Research, 1968. 2(1): p. 13-21.

Forbes, T., H. Zagorski, E. Holshouser, and W. Deterline (1958). Measurement of driver reactions to tunnel conditions. Highway Research Board Proceedings 37, 345-357.

Gazis, D.C., R. Herman, and R.W. Rothery, Nonlinear follow-the-leader models of traffic flow. Operation Research, 1961. 4(9): p. 545-567.

Helly, W. (1959) Simulation of Bottlenecks in Single Lane Traffic Flow. in International Symposium on the Theory of Traffic Flow. New York.

Herman, R. and R. W. Rothery (1965). *Car Following and Steady-State Flow.* Proceedings of the 2nd International Symposium on the Theory of Traffic Flow. Ed J. Almond, O.E.C.D., Paris.

Hoogendoorn, S. P. and H. J. van Zuylen (2004). Tracing Congestion Dynamics with Remote Sensing. Management and Information Systems 2004, Malaga, Spain, WIS

Hoogendoorn, S.P., and S. Ossen (2005). Parameter Estimation and Analysis of Car-Following Models. In: *Transportation and Traffic Flow Theory: Flow, Dynamics and Human Interactions*, H. Mahmassani (ed.), Elsevier, Oxford.

Hoogendoorn, S.P., Ossen, S., and Schreuder, M. (2006) Empirics of Multi-Anticipative Car-Following Behavior, Transportation Research Records (accepted for publication).

Kerner, B. (2005). Microscopic Three-Phase Traffic Theory and its applications for Freeway Traffic Control. Proceedings of the 16th International Symposium on Transportation and Traffic Theory, pp. 181-204.

Lenz, H., C.K. Wagner, and R. Sollacher, Multi-anticipative car-following model. The European Physical Journal, 1999. B(7): p. 331-335.

Ossen, S., and S.P. Hoogendoorn (2006). Multi-anticipation and heterogeneity in car-following: empirics and a first exploration of their implications. ITSC 2006, IEEE Toronto, Canada.

Schultz, G. and L. Rilett (2004). *An analysis of the distribution and calibration of car-following sensitivity parameters in microscopic traffic simulation models.* Transportation Research Board Annual Meeting, Washington D.C., Transportation Research Board.

Tampère, C.M.J. (2004), Human-kinetic traffic flow modelling for the analysis of Advanced Driver Assistance systems in congestion, Dissertation thesis, Delft University of Technology. Available through: http://www.kuleuven.ac.be/traffic/dwn/P2004C.pdf

Tampère, C., S. Hoogendoorn, and B. Van Arem (2005). A behavioural approach to instability, stop and go waves, wide jams and capacity drop. Proceedings of the 16th International Symposium on Transportation and Traffic Theory, pp. 205-229..

Treiber, M., A. Hennecke, and D. Helbing (2000). Congested Traffic States in Empirical Observations and Microscopic Simulation. Physical Review E 62, 1805-1824.

Treiber, M., A. Kesting, and D. Helbing (2006), *Delays, inaccuracies and anticipation in microscopic traffic models*. Physica A (in press).

Zhang, M., L.E. Owen, J.E. Clark (1998). Multiregime approach for microscopic traffic simulation. Transportation Research Record 1644, pp. 103-115.

Transportation and Traffic Theory 2007
Edited by R.E. Allsop, M.G.H. Bell and B.G. Heydecker

26

ANTICIPATIVE VEHICLE CONTROL ALGORITHM MITIGATING TEMPORAL INFORMATION DELAY

Yu Liu and François Dion, Department of Civil and Environmental Engineering, Michigan State University, USA

SUMMARY

This paper evaluates the impacts of information delay on automated vehicle control and explores the possibility of mitigating such impacts by considering information from multiple downstream vehicles in the vehicle control decision process. This evaluation is done through the use of a microscopic traffic simulation model that has been specifically developed to enable the simulation of information delays that may occur with Intelligent Transportation Systems (ITS) applications using onboard vehicle sensors or wireless communications. The impacts of communication delays are assessed through vehicle responses to simulated traffic events in three car-following scenarios: single-file platoon accelerating from standstill, vehicles responding to a slowdown by the platoon's lead vehicle, and vehicle responding to the sudden immobilization of the platoon's lead vehicle. These scenarios are first applied to situations in which only information about the vehicle immediately ahead is available, and then, to situations in which information is obtained from a specific number of lead vehicles. The simulation results reported in the paper clearly demonstrate that information delay has a negative impact on vehicle control, particularly when information from only one lead vehicle is considered. The results further show that improved vehicle control and reduced sensitivity to delays can be achieved by developing control systems considering information from at least two lead vehicles.

NOTATION

$a^{des}_{n(t)}$	Desired acceleration of vehicle n at time t
a^{min}_{n}	Maximum deceleration (braking capability) of vehicle n
a^{max}_{n}	Maximum acceleration (full throttle) of vehicle n
$a_{n-i\,(t)}$	Acceleration of the ith vehicle ahead of vehicle n at time t
f_a	Coefficient to incorporate vehicle acceleration in constant time headway model
$f_{n,i(t)}$	Adjustment factor for the desired time headway of vehicle n, based on its ith preceding vehicle at time t
h^{des}_{n}	Desired time headway for vehicle n, constant time headway model
$h^{des,adj}_{n}$	Adjusted desired time headway for vehicle n, constant time headway model
l_{n-i}	Vehicle length of the ith leader in front of vehicle n
T	Simulation time step
$v_{n\,(t)}$	Speed of vehicle n at time t
$v_{n-i\,(t)}$	Speed of ith vehicle ahead of vehicle n at time t
v_n^{des}	Desired speed of vehicle n
$x_{n\,(t)}$	Position of front bumper of vehicle n at time t
$x_{n-i\,(t)}$	Position of the ith vehicle ahead of vehicle n at time t, relative to front bumpers
$\Delta x_{(n-i)(t)}$	Distance between vehicle n and i^{th} vehicle ahead at time t
$\Delta x^{des}_{(n-i)(t)}$	Desired distance between vehicle n and i^{th} vehicle ahead at time t
$\Delta x^{min}_{(n-1)}$	Minimum safe distance between vehicle n and its preceding vehicle
$\varepsilon_{(n-i)(t)}$	Error between the positions of vehicle n and the ith vehicle ahead at time t
λ	Sliding coefficient in sliding mode control model
τ	Vehicle delay coefficient in first-order actuator delay model

INTRODUCTION

In an effort to provide enhanced driving experience and improve safety, the automobile industry is increasingly promoting the use of Intelligent Transportation Systems (ITS) applications designed to assist motorists with their driving tasks. Among the most prominent emerging applications in this area are adaptive cruise control systems. These systems are enhancements to existing fixed-speed cruise control systems in that they allow a vehicle to automatically adjust its speed to that of the vehicle ahead. In most cases, speed adjustments are based on a need to maintain a preset minimum distance with the vehicle ahead.

While range sensors are typically used to measure distance to vehicles ahead, vehicle control applications based solely on this technology may negatively impact traffic capacity and performance. This is due to the fact that drivers commonly maintain time gaps with vehicles ahead that are shorter than what is required for safe driving. For instance, drivers have often been observed to maintain headways around 1.3 s (Reichart et al., 1997), with values as low as 1 s (Ayres *et al.*, 2001). A safe following rule imposing a minimum headway ranging between 1.4 s and 2 s (Shladover *et al.*, 2001), as is commonly adopted by vehicle control systems, could then significantly reduce roadway capacities and lead to increased congestion when compared to current traffic behaviour. Another problem associated with large headways is to invite vehicle cut-ins, particularly on busy roadways with aggressive driving. In addition to increasing risks of collisions, frequent cut-ins may negatively be perceived and lead to reduced acceptance of the technology by the public.

Another limitation is the inability of range sensors to see past the vehicle immediately ahead. Many drivers do not typically concentrate their attention on only one vehicle but rather consider what the two, three or four vehicles ahead are doing, depending on how far they can see. As an example, drivers often ignore small speed fluctuations by the vehicle immediately ahead when it is apparent that those fluctuations are not reflective of general traffic behaviour. However, a vehicle control system may interpret these fluctuations differently if it has no information about the general traffic behaviour. In such a context, designing control logics relying on the monitoring of a single vehicle may create overly reactive systems that may in turn lead to less stable traffic flow patterns.

One potential solution is to develop cooperative control systems in which wireless communication technologies are used to obtain information from surrounding vehicles. In the simplest systems, data exchanges would only occur between vehicles present within a direct communication range. In more complex systems, data could be propagated along or across traffic streams by using vehicles as intermediary transmission nodes. In both cases, the ability of communicating with surrounding vehicles would allow vision beyond the vehicle immediately ahead and enable more complex anticipative decision-making processes. This extended vision range could in turn translate into an opportunity to reduce required headways between vehicles and an ability to increase roadway throughput and capacity.

Although wireless technologies allow traffic data communication among nearby vehicles, vehicle control algorithms that could utilize the transmitted data to improve traffic flow are still under development. Most identified studies in the field of multiple vehicle coupling have been focused on the development of car-following model attempting to capture the aforementioned multi-anticipative driver behaviour. For example, Herman *et al.* (1959) discussed the stability of car-following models with all vehicles in platoon identical and each reacting to the two nearest vehicles ahead according to relative velocity law. Similarly, Bexelius (1968) studied the expansion of the Gazis-Herman-Rothery model (Gazis *et al.*, 1961) to preview two preceding vehicles. Both studies established stability criteria expressed in terms of sensitivity parameters and reaction delay, which would allow perturbations propagated along vehicle platoon to diminish. More recent studies on multi-anticipative car-following model include an expansion of the Optimal Velocity model (Bando *et al.*, 1995) by Lenz *et al.* (1999) and an expansion of the IDM model (Treiber *et al.*, 2000) by Treiber *et al.* (2004), both to consider multiple preceding vehicles. Compared to single leader models, these recent multi-anticipative models have better stability characteristics and could replicate more closely the synchronized traffic flows (Kerner and Rehborn, 1997) that are observed in reality.

Although wireless technologies allow direct communication links between vehicles, these links typically never provide instantaneous data exchanges. While communication delays are generally not a concern when collecting informational data, such as travel maps or business directories, such delays can critically impact vehicle safety when the collected information is used to control the speed and position of the vehicle. Previous studies on communication delays have already revealed that information delay between a pair of vehicles could range from a few milliseconds to a few seconds (Biswas *et al.*, 2006). While a delay of a few milliseconds may not significantly affect vehicle control, delays close to one second may introduce significant errors between the reported and actual position of a vehicle. Increased collision risks could for instance arise from a simple delay in the application of brakes based on an erroneous assessment of the true location of the vehicle ahead. This may be particularly significant at high-speeds, when delayed breaking may lead to the need to impose harsh deceleration rates beyond the capability of the host vehicle.

While previous research has lead to the development of car-following models considering multiple preceding vehicles, these models primarily focused on characterizing human driver behaviour. None of these models are suitable for application in automated control situations in which information about surrounding vehicles may be characterized by some reception delay. A first objective of this paper is thus to evaluate how information delay may affect the operation of automated vehicle control systems. This is accomplished by using simulation to compare how vehicles equipped with an automated controller considering a single lead vehicle respond to specific traffic events in the absence and presence of information delay. A second objective is to evaluate whether information delay impacts can be mitigated by increasing the number of lead vehicles considered by the vehicle controller.

All the investigations are conducted using a specialized microscopic traffic simulation model that was designed to enable the simulation of wireless-based ITS applications. At the center of this model is the ability to simulate various communication setups. Vehicles may be allowed to communicate only with the vehicle immediately ahead, thus effectively simulating the use of onboard range sensors, or with any vehicle present within a specified communication range or number of communication hops. Another unique feature is its ability to simulate collisions that may result from the application of improper acceleration and speed control decisions. Vehicle control can finally be simulated using traditional driver behaviour models (for instance, Gipps (1981)) or custom-built automated control algorithms.

The remaining sections of this paper successively describe the automated vehicle controller used in this study, the simulation model used to conduct the evaluations, the simulation scenarios and evaluation parameters that were selected to perform the analysis, the results of the simulations, and the main conclusions of the study.

AUTOMATED VEHICLE CONTROLLER

The vehicle control algorithm used in this study is based on the principles of sliding mode control. Sliding mode control is a type of variable control structure with feedback in which high-speed switching control commands are used to adjust the dynamics of a nonlinear system to approach a desired status and then stay close to this status. The advantage of sliding mode control over traditional closed-loop control is the independence of system performance from system modelling accuracy. This allows mitigating the effects of potential mismatches between a mathematical model used to describe a particular system and the system's reality. In this case, the development of a vehicle control logic based on sliding mode control avoids more specifically the difficulty introduced by the potential need to consider variable wireless communication delays. As such, this approach is recognized as a promising one to realize stable vehicle control in the presence of information delay.

Single-Lead Model

To illustrate how sliding mode control principles were used to develop the study's vehicle control logic, first consider a situation in which a vehicle is following another one on a given roadway. For this situation, the spacing between the rear bumper of the lead vehicle and the front bumper of the following vehicle at time t is defined by:

$$\Delta x_{(n-i)(t)} = x_{n-i(t)} - x_{n(t)} - l_{n-i} \, , \tag{1}$$

where n is the index of the vehicle for which vehicle control is being determined, and i denotes the ith vehicle ahead of vehicle n. When i is equal to 1, Equation 1 simply calculates the distance between a pair of succeeding vehicles.

When constant time headway is adopted as a following rule, each vehicle tries to maintain a separation distance with the preceding vehicle that generally increases with speed. In its simplest form, the desired separation distance under such a rule is calculated as:

$$\Delta x_{(n-1)(t)}^{des} = h_n^{des} \, v_{n(t)} \; . \tag{2}$$

If information about the current acceleration level of the following vehicle is known, Equation 2 can be expanded into Equation 3 to account for the tendency of this vehicle to either increase or decrease its following distance from one time interval to the next.

$$\Delta x_{(n-1)(t)}^{des} = h_n^{des} \, v_{n(t)} + f_a a_{n(t)} \; . \tag{3}$$

In Equation 3, the second term on the right increases the desired following distance when the following vehicle is accelerating and decreases it when the vehicle is braking. The factor f_a is used to convert the observed acceleration at time t into a following distance adjustment. This factor can take various forms, depending on the importance put on the effect of vehicle acceleration and on the model being considered.

At any given time, the error in the following distance between two vehicles can be defined as the difference between the actual and desired distances between the two vehicles. Using the definition of desired following distance provided by Equation 3, the error in following distance at time t can then be mathematically expressed as:

$$\varepsilon_{(n-1)(t)} = \Delta x_{(n-1)(t)} - \Delta x_{(n-1)(t)}^{des} = \Delta x_{(n-1)(t)} - h_n^{des} \, v_n - f_a \, a_{n(t)} \; . \tag{4}$$

The ultimate goal of well-designed vehicle control logic is to reduce the value of ε and maintain it as close as possible to zero. In the context of sliding mode control, the minimization of vehicle spacing error is realized by choosing a sliding surface along which ε would tend to zero or some negligible value. The sliding surface is defined here as:

$$\dot{\varepsilon}_{(n-1)(t)} = -\lambda \, \varepsilon_{(n-1)(t)} \; , \tag{5}$$

where λ is the sliding coefficient and its value is always positive to allow the gradient of ε to always points to zero.

While the above equations represent the fundamental principles of sliding mode vehicle control, vehicle actuator lag should also be considered. A commonly adopted form is the first-order delay model shown in Equation 6, where τ is the vehicle delay coefficient.

$$\tau \, \dot{a}_{n(t)} + a_{n(t)} = a_{n(t)}^{des} \, . \tag{6}$$

Differentiating Equation 4 with respect to time yields:

$$\dot{\varepsilon}_{(n-1)(t)} = \Delta \dot{x}_{(n-1)(t)} - h_n^{des} \, \dot{v}_n - f_a \, \dot{a}_{n(t)} \; . \tag{7}$$

Rearranging Equation 6 and substituting into Equation 7 further yields:

$$\dot{\varepsilon}_{(n-1)(t)} = \Delta \dot{x}_{(n-1)(t)} - h_n^{des} \, \dot{v}_n - f_a \, \frac{a_{n(t)}^{des} - a_{n(t)}}{\tau} \; . \tag{8}$$

Substituting Equation 5 into Equation 8 then yields:

$$\Delta \dot{x}_{(n-1)(t)} - h_n^{des}\, \dot{v}_n - f_a \frac{a_{n(t)}^{des} - a_{n(t)}}{\tau} = -\lambda\, \varepsilon_{(n-1)(t)} \,. \tag{9}$$

Finally, rearranging Equation 9 for $a^{des}{}_n$ produces the following relationship:

$$a_{n(t)}^{des} = \left(1 - \frac{\tau\, h_n^{des}}{f_a}\right) a_{n(t)} + \frac{\tau}{f_a} \Delta \dot{x}_{(n-1)(t)} + \frac{\tau \lambda}{f_a} \varepsilon_{(n-1)(t)} \,. \tag{10}$$

This is the vehicle control logic used by Zhou and Peng (2004) in a previous study on adaptive cruise control logic. It represents a basic vehicle control logic considering information from a single lead vehicle. The relationship of desired acceleration and actual acceleration achieved on vehicles is specified by Equation 6. The actual acceleration can be expressed by solving Equation 6, which leads to the following relationship:

$$a_{n(t+T)} = a_{n(t)} + \left(\frac{a_{n(t)}^{des} - a_{n(t)}}{\tau} + a_{n(t)} \right) e^{-\left(\frac{T}{\tau}\right)} \,. \tag{11}$$

Using the definition of Equation 10, the final vehicle control algorithm used to update vehicle acceleration is given by Equation 12.

$$a_{n(t+T)} = a_{n(t)} + \left\{ \begin{matrix} \left(f_a - h_{n-1}^{des}\right) a_{n(t)} + \left(v_{n-1(t)} - v_{n(t)}\right) \\ + \lambda \left[\left(x_{n-1(t)} - x_{n(t)} - l_{n-i}\right) - h_{n-1}^{des}\, v_{n(t)} - f_a\, a_{n(t)} \right] \end{matrix} \right\} \frac{e^{\left(-\frac{T}{\tau}\right)}}{f_a} \tag{12}$$

In a last step, calculated vehicle accelerations are checked with vehicle acceleration and braking limits. If a calculated acceleration or deceleration exceeds the physical limit of a vehicle, the actual vehicle acceleration is replaced with either its maximum acceleration $a^{max}{}_n$ or maximum deceleration $a^{min}{}_n$, depending on whether the vehicle is accelerating or braking. In addition to maximum acceleration and deceleration rates, the rate of change in acceleration is also restricted by a deceleration jerk limit. This parameter represents the harshest braking action possible by a vehicle attempting to avoid a collision.

Multiple-Lead Model

When data from multiple downstream vehicles become available, there is hypothetically better chance to improve vehicle control performance if the additional information is appropriately used. However, few studies have been found on multiple vehicle following. In this study, the approach selected to incorporate traffic information from multiple downstream vehicles is to use an adjustment factor applied to the desired headway, h^{des}. The headway adjustment factor $f_{n,i(t)}$ used for this purpose is defined as:

$$f_{n,i(t)} = (-1) \frac{a_{n-i(t)}}{a_{n-i}^{\min}} \,, \tag{13}$$

where vehicle i is again the ith lead vehicle ahead of vehicle n. The sign of factor f is positive when a specific preceding vehicle is accelerating and negative when braking. The absolute value of factor f indicates the magnitude of the acceleration or braking of the corresponding preceding vehicle. Its value reverts zero when the preceding vehicle in question is stopped or moving at constant speed. For a given subject vehicle, there may exist one or more adjustment factors, depending on the number of lead vehicles whose traffic data are accessible. In such a case, the factor to apply is obtained by taking the average of all the individual factors. The resulting factor is then used to adjust the desired headway of the subject vehicle according to Equation 14.

$$h_n^{des,adj} = h_n^{des}\left(1 - mean\left(f_{n,i(t)}\right)\right) \tag{14}$$

The average of headway adjustment factors allows a host vehicle to recognize the general acceleration or deceleration pattern exhibited by preceding vehicles within data accessible range. Other type of evaluation functions selecting either the maximum or minimum of headway adjustment factors can also be used. However, these functions would typically result in excessively aggressive or conservative vehicle control.

When compared to its nominal value, the desired time headway is generally shortened when the lead vehicles are exhibiting accelerating patterns, to counteract the fact that the accelerations tend to increase following distances, and elongated when the lead vehicles are braking, to counteract the tendency to push vehicles closer. Within this context, it is anticipated that using lead vehicle information may allow a control logic to be more proactive than if it were to consider only one lead vehicle.

Introducing the adjusted desired headway in Equation 12, Equation 15 is finally obtained to delineate vehicle acceleration with the use of information from multiple lead vehicles.

$$a_{n(t+T)} = a_{n(t)} + \left\{ \begin{aligned} &\left(f_a - h_{n-1}^{des,adj}\right) a_{n(t)} + \left(v_{n-1(t)} - v_{n(t)}\right) \\ &+ \lambda \left[\left(x_{n-1(t)} - x_{n(t)} - l_{n-i}\right) - h_{n-1}^{des,adj}\, v_{n(t)} - f_a\, a_{n(t)}\right] \end{aligned} \right\} \frac{e^{\left(-\frac{T}{\tau}\right)}}{f_a} \tag{15}$$

At equilibrium, the values of all acceleration terms are zero, all vehicles share the same speed, and all headway adjustment factors are one. In such a context, it can easily be seen that Equation 15 then reduces to the basic expression of constant time headway logic.

TRAFFIC SIMULATION MODEL

Well established models for the evaluation of emerging transportation applications include AIMSUN, VISSIM and PARAMICS. These models all feature an application programming interface (API) allowing the addition of functionalities enabling data retrieval from the simulation and interactions with various driver behavior processes. Despite these features, the modeling of vehicle-to-vehicle wireless communications remains a difficult task within each model, as exemplified by the following elements:

- Since access to simulation code is generally restricted, wireless communication functionalities must typically be developed outside the traffic simulation environment. This may create problems if the API functions available to do not allow all the desired data to be retrieved from the simulation or restrict which parameters can be passed to the simulated vehicles. Another difficulty is the need to develop external databases tracking the information received and stored by each simulated vehicle.

- Vehicle status is typically updated using a sequential process. The process starts with the most downstream vehicle in the network and ends with the most upstream one. For each vehicle, control decisions are based on the status of the vehicle in the previous time step and the status of the vehicles ahead in the current time step. In this case, the use of information from the current time step is equivalent to assuming that instantaneous data exchanges occur between vehicles, which may not be correct.

- The shortest time step is typically 0.1 s, or 100 ms (Gettman and Head, 2003). Since delays of less than 100 ms can be observed, a minimum interval of 100 ms may already be too long for adequately evaluating the impacts of information delays associated with ITS applications. In addition, existing simulation models also often do not typically allow data retrievals at intervals of less than 1 s, which further question the adequately of the models in simulating communication delays.

- All established models are finally intrinsically collision-free. Accidents can typically only be staged by putting invisible static vehicles on the roadway at specific times. While this is appropriate for evaluating the impacts of incidents on traffic operations, it does not allow use of the models for the development of automated vehicle control algorithms, where the ability to simulate accidents may be valuable in identifying inadequate vehicle control decisions.

The need to use a simulation model in which all simulation processes would be accessible and addressing the above problems resulted in the development of the Communication and Traffic Simulation (CATSIM) microscopic traffic simulation model. This a Java-based model that is built upon the open-source Intelligent Driver Model (IDM) microscopic traffic simulation model developed by Treiber *et al.* (2000) and Treiber and Helbing (2002). Using the available code as a starting point, various modifications were made to the original IDM model to develop the functionalities described below. Only the most important model features relevant to this study are described below.

Communication delay within the CATSIM model is modelled by a process allowing data retrieval from preceding simulation time slices. Figure 1 illustrates the three-dimensional data array used for this process. This array stores vehicle parameters along its depth, information about different vehicles along its width, and data from successive time steps along its height. The figure illustrates more specifically how communication delay would affect control decisions for the nth vehicle in the current time step. As indicated, the vehicle's desired speed and acceleration would first be based on its status S_o in the previous time step. If it is hypothetically assumed that information about the two vehicles immediately ahead can be received after a delay equal to one simulation time step, the statuses S_1 and S_2 from the previous time step will then be considered for vehicles *n-1* and *n-2*. However, if additional delay is assumed to exist, as is the case for vehicles *n-3* and *n-4*, the corresponding data will instead be retrieved from time steps that are further back in time. In all cases, the time step from which data will be retrieved will correspond to the level of communication and/or processing delay assumed to exist.

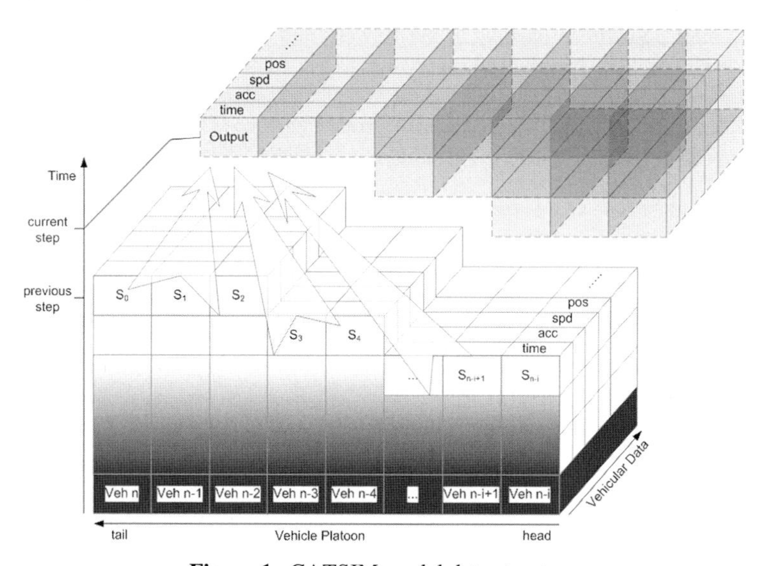

Figure 1. CATSIM model data structure

To remove constraints imposed by minimum time step durations, the CATSIM model is further allowed to simulate traffic with time steps as short as 0.01 s, or 10 ms, and definable in increments of 0.01 s. A routine has also been embedded into the model to determine whether following vehicles collide with each other. The routine assumes that collisions occur when the predicted spacing between a vehicle and its predecessor would fall in the next time step below a threshold corresponding to the physical length of the lead vehicle. If a collision is deemed to occur, the two vehicles involved are immediately assumed to stall and become a static obstacle for the approaching traffic for an indefinite or user-defined period.

The CATSIM model was validated by examining the speed-flow diagrams produced by the model when using various car-following models. Attention was more particularly put on the diagrams produced when using traditional models reflecting human driver behaviour. An example of a speed-flow diagram resulting from the use of the Gipps model (Gipps, 1981) is shown in Figure 2. This diagram was produced for a scenario considering traffic travelling along a one-mile circular highway. Multiple runs were executed with different flow levels to achieve various traffic densities. In each case, traffic was simulated until traffic equilibrium was achieved, i.e., until all vehicles had speed falling within a narrow range. Reported results do not include data from the first 80 s of each run to ensure that traffic instability in the initial portion of each simulation did not affect the validation. As can be observed, the resulting speed-flow diagram replicates the well-known shape of the relationship, consisting here of a parabolic bottom and an almost flat top portion.

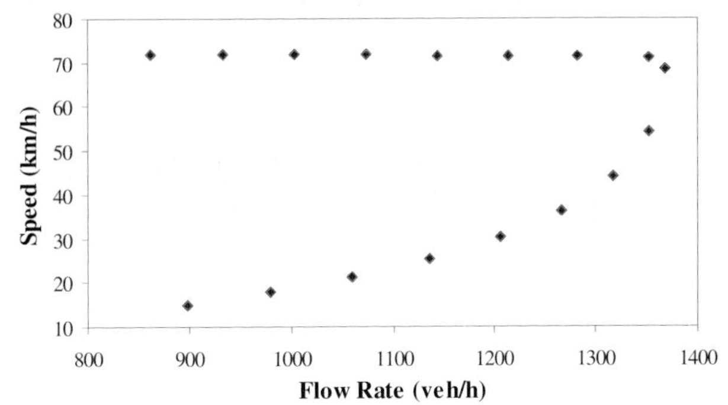

Figure 2. Simulated speed-flow diagram based on Gipps' model

EVALUATION SCENARIOS

Three car-following scenarios were developed to investigate the impacts of information delay on vehicle and traffic flow behavior. All scenarios consider a platoon of 40 vehicles travelling in a single file. Each scenario simulates various traffic events to test specific types of vehicle responses. These three scenarios are illustrated by the speed profiles of Figure 3, which each illustrates the speed trace of the lead and first 15 following vehicles in the platoon. These scenarios are explained in greater details in the following paragraphs, with Table 1 providing numerical values associated with the main scenario parameters.

The first scenario considers vehicles travelling in a single file on a one-lane circular road. Its purpose is to evaluate how quickly equilibrium can be achieved by a given control logic under various communication delay and information sharing assumptions. All vehicles are assumed to be initially immobilized. At a given time, the first vehicle in the platoon starts to accelerate and continue to do so with a parabolically diminishing rate until it reaches its desired speed. In turn, all the vehicles behind the lead vehicle start to accelerate, considering both their desire to attain a certain speed and the constraints imposed by the vehicles ahead.

The use of a circular road implies that at some point the lead vehicle in the platoon will start to interact with the last vehicle in the platoon. In most cases, this results in a need for the lead vehicle to decelerate as it generally travels across the circular path before the last platoon vehicle has fully accelerated. This creates a complex situation in which shockwaves created by accelerations and decelerations along the road keep propagating around the circle, thus creating the oscillation speed pattern observed in Figure 3(a). To achieve equilibrium in such a situation, a control algorithm must have the ability to dampen shockwaves. Equilibrium is assumed to be reached when the speeds of all vehicles in the platoon are maintained within ±0.5 m/s of the space mean speed of all vehicles in the platoon.

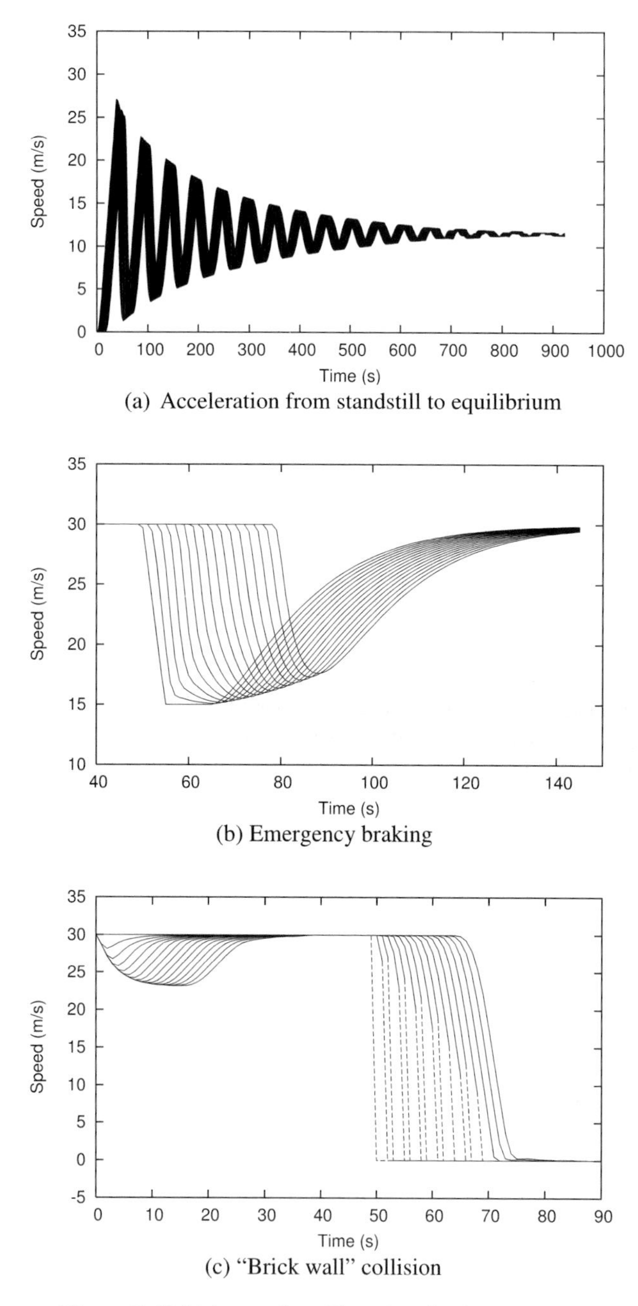

(a) Acceleration from standstill to equilibrium

(b) Emergency braking

(c) "Brick wall" collision

Figure 3. Vehicle speed profiles of evaluation scenarios

Table 1. Vehicle control and simulation parameters.

Variable	Standstill to Equilibrium	Emergency Braking	"Brick Wall" Collision
Initial vehicle speed	0 m/s	30 m/s	30 m/s
Initial headway	--	1.0 s	1.0 s
Road length	1000 m	--	--
Platoon size		40 veh	
Desired speed		30 m/s	
Desired headway		1.0 s	
Spacing between stopped vehicles		4.0 m	
Vehicle length		6.0 m	
Maximum Acceleration		0.8 m/s^2	
Maximum Deceleration		-5 m/s^2	
Maximum Braking Jerk		-30 m/s^3	
Information delay (wireless communication delay + on-board device processing time)		0.3 s	
Sliding mode model parameter (T_a)		0.1	
Sliding mode model parameter (τ)		0.8	
Sliding mode model parameter (λ)		0.3	
Critical time-to-collision (TTC*)		5 s	

The second scenario simulates emergency braking along a straight, open highway section. Its purpose is to evaluate the ability of vehicles to respond to typical collision threats and return traffic to equilibrium following a sudden disturbance. As illustrated in Figure 3(b), all vehicles are initially assumed to travel at a constant speed. At a given time, the lead platoon vehicle initiates a harsh deceleration. The vehicle decelerates at a constant rate until it reaches a speed of half its initial speed. Following a few seconds of travel at a constant speed, the vehicle starts accelerating again until it reaches back its desired speed. All following vehicles then successively respond to the deceleration and acceleration of the lead vehicles according to their imbedded control logic.

The third scenario pushes the threat of collision to the highest level by staging a so-called "brick wall" collision scenario. This scenario is similar to the second one, except for the fact that the lead platoon vehicle is assumed to come to an instantaneous stop following a hypothetical collision with an object on the road. After this event, all following vehicles will then try to decelerate to avoid colliding with the lead vehicle. However, given the suddenness of the stop, it is expected that a certain number of vehicles will be unable to avoid colliding. In Figure 3(c), colliding vehicles are those for which the corresponding speed trace line suddenly ends and is completed by a dotted line. By comparing the number of vehicles colliding and the speed at which they collide, the effectiveness of various vehicle control logic in anticipating deceleration needs can then be assessed.

The communication delay associated with a given situation generally depends on a number of parameters. Among the most prominent are: the number of simultaneous broadcasts being attempted in a given area, the robustness of routing protocols, the rate of communication errors at transmission nodes requiring rebroadcasts, and the level of background data traffic from other applications using the same communication channel. In all scenarios, it is assumed for simplicity that communication delays can be modeled as an arbitrary fixed 0.3 s delay per communication link. Variable delays are not considered due to the high level of complexity associated with such a situation, although future studies on the topic are being planned.

EVALUATION PARAMETERS

The impacts of communication delays on vehicle control are evaluated by considering both performance and safety parameters. For the evaluation of performance, the following two parameters are considered:

- Time to equilibrium, and
- Equilibrium speed.

For the safety assessment, a number of evaluation parameters based on the concept of time-to-collision (TTC) are used (Liu *et al.*, 2006):

- Number of vehicles collided and collision velocity,
- Time-to-collision (*TTC*) and critical time-to-collision (TTC^*),
- Time exposed time-to-collision (*TET*) and time integrated time-to-collision (*TIT*),
- Cumulative threatening frequency (*CTF*),
- Cumulative jerk frequency (*CJF*), and
- Acceleration noise.

Each of the above parameters is described in more details in the paragraphs that follow.

Time to Equilibrium

The time to equilibrium measures the ability to quickly adjust to changing traffic conditions or dampen unstable conditions. Vehicle control logics exhibiting shorter time to reach equilibrium, particularly in the presence of information delay, will then be more desirable than those exhibiting longer time to equilibrium.

Equilibrium Speed

For a given situation, different control algorithms may lead to the establishment of traffic equilibriums with different average traffic speeds. Within this context, the most efficient algorithms will tend to create equilibrium at speeds that are the closest to the desired speed.

Number of Vehicles Collided, Collision Velocity

Soundly designed vehicle control systems should be intrinsically collision-free. However, since it is typically not possible to control all factors that can cause vehicles to collide, another important feature should be an ability to reduce the severity of unavoidable accidents. Any reduction in the speed at which collisions occur could then be interpreted as a positive sign even if there is no reduction in the number of collisions.

Time-to-Collision and Critical Time-to-Collision

To help assess the risks of potential collisions, Hydén (1996) defines the time-to-collision (TTC) as the time that remains until a potential collision between two vehicles if the vehicles' current courses and speed differences are maintained, as shown in Equation 16. A small TTC is representative of an imminent collision due to small vehicle spacing or a large speed difference. A desirable situation would thus be one in which the TTC is as high as possible.

$$TTC_{n(t)} = \frac{x_{n-1(t)} - x_{n(t)} - \Delta x_{(n-1)}^{\min}}{v_{n(t)} - v_{n-1(t)}}, \qquad \forall v_{n(t)} - v_{n-1(t)} > 0 \tag{16}$$

To differentiate between risk levels, the critical time-to-collision (TTC*) is defined as a threshold above which a vehicle may not be considered facing an imminent collision threat. Typical choices for TTC* vary between 2 s and 4 s (Hirst and Graham, 1997; Minderhoud and Bovy, 2001). Such values are selected to reduce the number of false warnings while maintaining a certain level of efficiency. In this study, since false warnings are of no concern, the TTC* is set at 5 s to ensure adequate consideration of all safety-critical events.

Time Exposed Time-to-Collision, Time Integrated Time-to-Collision

To measure exposure to collision risk, the time exposed time-to-collision (TET) is defined as the summation of all time periods for which TTC is below TTC* (Minderhoud and Bovy, 2001). However, while TTC measures the total period for which a collision threat exists, it does not fully reflect the severity of potential hazards. For instance, it does not indicate whether a vehicle faced several mild risks that could be avoided with relative ease or a significant threat that required significant effort, such as a harsh deceleration, to avoid.

To incorporate threat severity, the time integrated time-to-collision (TIT) is defined as the summation of differences between TTC and TTC* multiplied by the duration of intervals during which TTC is below TTC*, as shown in Equations 17 and 18. Since potential hazards become more severe with lower TTC values, a large TIT would thus be indicative of collision threats for which significant efforts are required to avoid.

$$TIT_n = \sum_{0}^{T_{end}} \delta_{n(t)} \left[TTC^* - TTC_{i(t)} \right] T \tag{17}$$

$$\delta_{n(t)} = \begin{cases} 0 & else \\ 1 & \forall 0 \le TTC_{n(t)} \le TTC^* . \end{cases} \tag{18}$$

Cumulative Threatening Frequency

While TET and TIT consider the number of intervals in which TTC is below a critical value, they do not assess how frequently collision threats occur. This assessment is provided by the cumulative threatening frequency (CTF), which counts the number of times the TTC drops from a value above TTC* to a value below TTC*. Well-designed control systems should produce CTF values as low as possible. Since this study considers a single hazard, vehicles should experience at most one potential threat and a CTF of 1. Greater values would indicate multiple threatening situations due to inappropriate responses from the vehicle control system.

Cumulative Jerk Frequency

The cumulative jerk frequency (CJF) counts the number of times a vehicle switches between acceleration and deceleration commands. Well-designed vehicle control systems should produce CJF values close to zero or reflective only of necessary speed changes.

Acceleration Noise

While CJF measures the number of times a vehicle switches between acceleration and deceleration, the vehicle's overall stability is reflected by the acceleration noise, which is a measure of the magnitude of speed changes over time (Jones and Potts, 1962). The mathematical expression of this parameter is given by Equations 19 and 20.

$$a_{n(avg)} = \frac{1}{(T_{end} - T_{start})} \int_{T_{start}}^{T_{end}} a_{n(t)} \, dt = \frac{1}{(T_{end} - T_{start})} \left[v_{n(T_{end})} - v_{n(T_{start})} \right] \quad (19)$$

$$A_n = \sqrt{\frac{1}{(T_{end} - T_{start})} \int_{T_{start}}^{T_{end}} \left[a_{n(t)} - a_{n(avg)} \right]^2 dt} \quad (20)$$

Acceleration noise is always positive, with a lowest value of zero representing constant speed, constant acceleration, or deceleration, over the entire evaluation period. Systems with low acceleration noise will then tend to generate smoother riding than systems with high values.

IMPACTS OF INFORMATION DELAY ON SINGLE-LEAD VEHICLE CONTROL

This section presents the results of the evaluations focusing on the impacts of information delay on vehicle control when decisions are based uniquely on information about the position, speed and acceleration level of the vehicle immediately ahead. Results are presented for the three scenarios described earlier.

Scenario 1: Standstill to Equilibrium

For the standstill to equilibrium scenario, Table 2 shows that the time for the simulated platoon to reach equilibrium is more than doubled when an information delay of 0.3 s is introduced into the control loop. In the absence of delay, equilibrium, as defined by a speed differential of less than 0.5 m/s, is reached in just over 5.5 minutes. With a delay of 0.3 s, it takes more than 11 minutes to reach the equilibrium from an identical initial condition. An analysis of simulation results clearly indicate that this extended time is primary the result of the inability of individual vehicles to obtain precise information regarding the position, speed and acceleration of its lead vehicle in the presence of information delay. This results in situations in which control decisions more frequently overshoot or undershoot their true target, thus creating a need for extra time to reach equilibrium.

Table 2. Time to reach equilibrium from standstill, following single leader scenario.

Number of Lead Vehicles	Information Delay (ms)	Duration to Reach Equilibrium (s)	Equilibrium Speed (m/s)
1	0	331	15.0
1	100	438	13.6
1	200	555	12.5
1	300	684	11.5
1	400	880	10.7
1	500	1040	10.0

Another interesting result is a reduction of the average vehicle speed at equilibrium. In the absence of information delay, the equilibrium speed is 15.0 m/s (54.0 km/h). With a delay of 0.3 s, the speed drops to 11.5 m/s (41.4 km/h). This reduction is again explained by the use of erroneous information in the control loop. With a 0.3 s communication delay, vehicle control decisions are typically based on the position that the each preceding vehicle occupied 0.3 s ago. At a speed of 11.5 m/s, this translates into assuming that lead vehicles are 3.5 m closer than they are in reality. To maintain a desired headway, a speed reduction is then imposed to increase spacing with the lead vehicle, thus leading to a lower equilibrium speed.

An interesting coincidence in the data of Table 2 is that the difference in distances travelled in 1 s at equilibrium with and without information delay equals the distance that a vehicle experiencing information delay travels during the assumed information delay. For example in the scenario of 0.3 s delay, the difference in distance, $(15\,m/s - 11.5\,m/s)\times 1\,s = 3.5\,m$, corresponds to the distance that a vehicle experiencing information delay travels during the assumed information delay, $11.5\,m/s \times 0.3\,s = 3.5\,m$. Further tests considering varying levels of information delay also support this apparent relationship.

Scenario 2: Emergency Braking

The focus of the emergency braking scenario is on the safety impacts of information delay. Table 3 presents for this scenario the results of the evaluation of the surrogate safety measures described earlier. Only TTC and acceleration noise values are shown as the TTC values produced for each vehicle were all above 5 s. Since the critical time-to-collision (TTC*) is set at 5 s, this results in all TTC-based indicators having a value of zero.

For each parameter, Table 3 shows the values associated with the vehicles returning the lowest TTC or highest acceleration noise, as well as cumulative sums reflecting the values assessed for all vehicles in the platoon. Vehicles are identified according to their position in the platoon relative to the lead vehicle. From these results, it can be observed that the introduction of communication delay unexpectedly results in higher TTC values and in an assessment that information delay may potentially improve traffic safety. Although such an assessment is contradictive to the common assumption of negative impacts, it should be noted that safety-critical situations did not occur in both simulated alternatives. The true cause of the increases in TTC is related to the weights assigned to the errors regarding vehicle spacing error and speed in the control algorithm. In the absence of information delay, vehicles have few difficulties maintaining a given spacing to satisfy the constant 1-s headway separation rule. However, as was explained before, vehicles tend to perceive shorter vehicle spacing with information delay, thus pushing them to maintain longer separation distances. This increase in separation distance then results in lower collision threats and increased assessed TTC values.

For this scenario, a better performance indicator seems to be the acceleration noise, as this parameter focuses on risks created by speed variations and is not dependent on the presence of particular safety threats. In Table 3, it can indeed be observed that the acceleration noise expectedly increases with the introduction of information delay, indicating an increase in speed variations and higher collision risks that contradict the TTC evaluations.

Table 3. Selected surrogate safety measures, following single leader and emergency braking scenario.

Information Delay	**TTC**			**Acceleration Noise**		
	Lowest		**Sum**	**Highest**		**Sum**
(s)	Veh	(s)	(s)	Veh	(m/s^2)	(m/s^2)
0	2	6.1	330	18-39	0.6	21.9
0.3	3	6.8	420	39	0.8	25.5

Figure 4 presents another look at the impact of information delay on acceleration noise. This figure shows the impact on each vehicle in the platoon. The figure clearly indicates that acceleration noise is almost the same with or without information delay for the first few vehicles. However, as we move further back into the platoon, vehicles are observed to experience marked increases in acceleration/deceleration activity. For the most part, these results can be explained by differences in shockwave propagation speed that are introduced by communication delay.

To illustrate the above effect, Figure 5 compares the trajectories of all simulated vehicles in the alternatives with and without information delay. As can be observed in Figure 5(a), the shockwaves created by the deceleration/acceleration cycle of the lead platoon vehicle meet in the absence of delay before reaching the end of the platoon. Consequently, no vehicle beyond the 20th platoon vehicle is affected by the deceleration/acceleration event of the lead vehicle. In Figure 5(b), which presents the alternative with information delay, the shockwaves are clearly seen extending throughout the entire platoon as a result of slower propagating speeds. Similar to the first scenario, this reduction in shockwave propagation can be explained by the increased difficulty to achieve equilibrium or quickly respond to traffic events caused the by the use of slightly erroneous data in the vehicle control process.

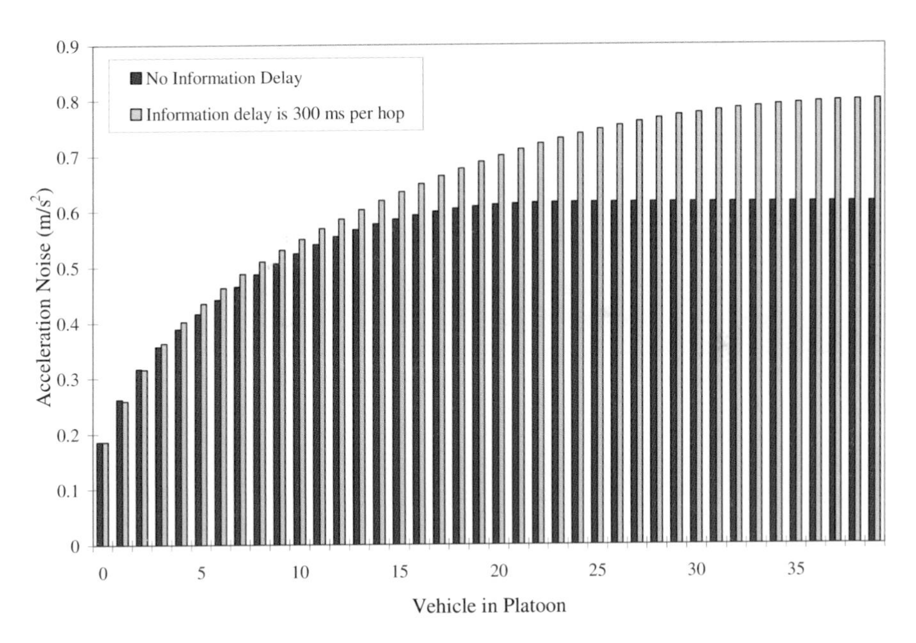

Figure 4. Acceleration noise on vehicles following one lead vehicle with and without information delay

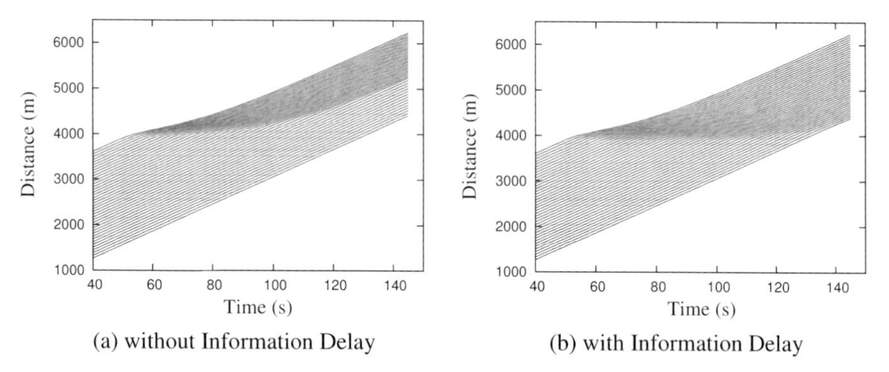

(a) without Information Delay (b) with Information Delay

Figure 5. Vehicle trajectories in the emergency braking scenario

Scenario 3: "Brick Wall" Collision

The third scenario focuses on the safety impacts of information delay in extreme circumstances. For this scenario, the simulation results indicate that 12 vehicles collide with each other following the sudden immobilization of the lead platoon vehicle in the presence and absence of information delay. As indicated by Table 4, collision speeds do not significantly change with the presence of information delay. This leads to an apparent lack of impacts on safety and this result is again explained by errors in vehicle information introduced by information delay. In the two previous scenarios, information errors caused vehicle controllers to believe that the lead vehicles were closer than they were in reality. This perception pushed for an increased in following distance. While the same perception exists here, information errors also lead control systems to believe that the vehicles ahead are travelling faster than in reality and that higher vehicle speeds may be allowed. While the perceived closer spacing pushes on one hand for a reduction in vehicle speed, the perceived higher speed of vehicles ahead pushes on the other hand for an increase in vehicle speed, thus muting the impacts of increased or reduced information delays.

Table 5 summarizes in a format similar to Table 3 the values estimated for the surrogate safety measures described earlier. In this case, the introduction of vehicle control delay clearly results in reduced safety. For instance, the introduction of information delay lowers the TTC value for the vehicle having the highest collision risk (13th platoon vehicle) from 2.1 to 1.9 s. Higher individual TET and TIT values are also produced, while no apparent changes are observed in individual CTF and CJF values. While the platoon TTC and CJF show improvement with information delay, all other parameters indicate an increase in safety risks consistent with the individual values. The improvements in cumulative TTC and CJF values with information delay are again explained by effects of erroneous information on vehicle control decisions.

Table 4. Colliding speeds, following single leader and "brick wall" collision scenario

Vehicle Position in Platoon	**Colliding Speed** (m/s)	
	No Delay	300 ms Delay
1	25.0	25.0
2	23.0	23.0
3	21.5	21.5
4	20.1	20.1
5	18.7	18.8
6	17.4	17.5
7	16.2	16.2
8	14.4	14.5
9	12.7	12.8
10	11.0	11.0
11	8.8	8.9
12	5.6	5.7

Table 5. Surrogate safety measures, following single leader and "brick wall" collision scenario.

Setup	TTC			TET			TIT			CTF			CJF		
	Lowest		**Sum**	**Highest**		**Sum**	**Highest**		**Sum**	**Highest**		**Sum**	**Highest**		**Sum**
n	Veh	(s)	(s)	Veh	(s)	(s)	Veh	(s^2)	(s^2)	Veh	(#)	(#)	Veh	(#)	(#)
1 [(a)]	13	2.1	152	13-19	2	16	13	3.7	14.1	13-21	1	9	14-16	2	6
1 [(b)]	13	1.9	184	13-21	3	33	13	6.5	30.9	13-25	1	13	14,15	2	5

(a) without delay; (b) 300 ms information delay

A comparison of the TTC evaluation results of Table 5 to those of Table 3 clearly indicates that the brick wall scenario presented higher safety risks than the emergency breaking scenario. This is an expected result that validates the use of selected performance measures. When going from the emergency to the brick wall scenarios, the lowest individual TTC drops from roughly 6 s to roughly 2 s. The cumulative platoon value also shows a similar drop, with a reduction from an initial range of 300-400 s to a range of 150 – 200 s.

IMPACTS OF MULTIPLE LEAD VEHICLES

This section presents the results of the evaluations focusing on the impacts of information delay on vehicle control when control decisions are based on information about a number of lead vehicles.

Scenario 1: Standstill to Equilibrium

For the standstill to equilibrium scenario, Table 6 shows that traffic equilibrium can be reached much faster in the presence of information delay by incorporating in the control logic information from a number of lead vehicles. These gains are primarily obtained by extending system vision from the vehicle immediately ahead to the first two lead vehicles. Extending vision beyond the first two vehicles did not provide additional benefits. Each increase in the number of lead vehicles considered resulted here in an identical equilibrium speed and a slight increase in time to reach equilibrium. The longer time to reach equilibrium is attributed to the increased complexity of achieving stability when having to deal with an increasing number of variables each marked with a certain degree of error.

Table 6. Time to reach equilibrium from standstill, following multiple leaders scenario.

Number of Lead Vehicles	Information Delay (ms)	Duration to Reach Equilibrium (s)	Equilibrium Speed (m/s)
1	300	684	11.5
2	300	136	11.5
3	300	142	11.5
4	300	146	11.5
5	300	149	11.5

Scenario 2: Emergency Braking

Similar to the emergency braking scenario with a single-lead vehicle, all individual vehicle TTC evaluations produced values above 5 s, thus resulting in zero values for all other TTC-based surrogate measures. In this case, the results of Table 7 clearly show that including information about additional lead vehicles can lead to safety improvements. This is demonstrated by the higher TTC and lower acceleration noise values produced for the alternatives with two or more lead vehicles. However, similar to the results of Table 6, the maximum benefits appear to be when system vision is extended from one to two lead vehicles only, as TTC values start to decrease while acceleration noise start to increase again when considering three or more lead vehicles.

Table 7. Selected surrogate safety measures, following multiple leaders and emergency braking scenario.

Setup	**TTC**			**Acceleration Noise**		
	Lowest		**Sum**	**Highest**		**Sum**
N	Veh	(s)	(s)	Veh	(m/s^2)	(m/s^2)
1	3	6.84	420	39	0.80	25.5
2	1	7.65	757	39	0.55	18.7
3	1	7.66	709	39	0.56	18.9
4	1	7.66	679	39	0.57	19.1
5	1	7.66	619	39	0.59	19.6

The estimated acceleration noise values for each vehicle in alternatives considering one and two lead vehicle are compared in Figure 6. This figure clearly shows that all vehicles experience a reduction in acceleration/deceleration activity, and thus less instability, when two lead vehicles are considered instead of one. Similar to Figure 4, the impact appears cumulative as moving back into the platoon, with vehicles further behind experiencing a more pronounced reduction in acceleration noise. When compared to Figure 4, it can further be observed that the acceleration noise with a control system considering two lead vehicles in the presence of 0.3 s delay is even lower than for a system considering a single lead vehicle with no delay. This is yet another strong indication of the potential safety benefits of designing vehicle control system considering information from more than one vehicle ahead.

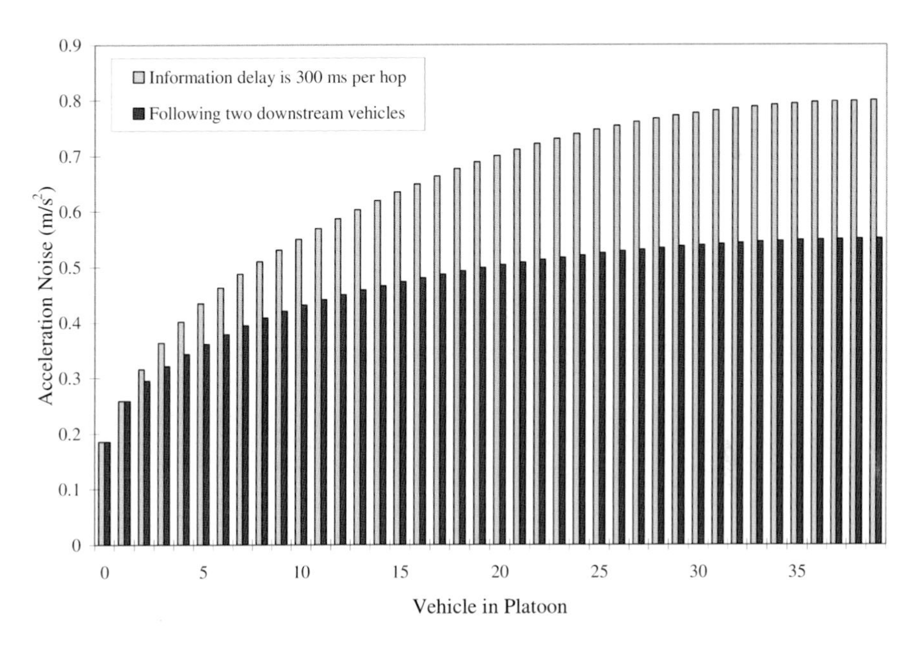

Figure 6. Acceleration noise on vehicles following one lead vehicle and two lead vehicles with information delay

Scenario 3: "Brick Wall" Collision

For the "brick wall" scenario, considering more than one lead vehicle results in a reduction of the number of vehicles colliding from 12 to 2. The change primarily occurs when expanding the control logic to consider two lead vehicles. This indicates a marked improvement in safety. The surrogate safety measures of Table 8 emphasize once more that the consideration of multiple lead vehicles has positive impacts on safety. For TTC, the best results in terms of individual vehicle performance are obtained when considering four or five lead vehicles, while the best platoon performance is with only two lead vehicles. For TET, the best individual performance is for three or four lead vehicles while the best platoon performance is for two lead vehicles. For TIT, a four lead vehicle setup is best, while a two-vehicle setup is best for CTF. The deterioration of CJF value with multiple lead vehicles is largely attributed to the chattering effect of sliding mode control around its sliding surface. While these differing results may be indicative of random and simulation effects, they again clearly demonstrate the potential safety benefits of including more than one lead vehicles in the vehicle control loop in scenarios considering information delay.

Table 8. Surrogate safety measures, following multiple leaders and "brick wall" collision scenario.

Setup	TTC			TET			TIT			CTF			CJF		
	Lowest		Sum	Highest		Sum	Highest		Sum	Highest		Sum	Highest		Sum
n	Veh	(s)	(s)	Veh	(s)	(s)	Veh	(s^2)	(s^2)	Veh	(#)	(#)	Veh	(#)	(#)
1	13	1.9	184	13-21	3	33	13	6.5	30.5	13-25	1	13	14,15	2	5
2	3	2.4	415	3	3	6	3	6.8	9.0	3,5,6	1	3	*(a)*	36	144
3	3	2.9	365	6-9	2	9	6	3.2	8.0	*(b)*	1	5	35,39	27	496
4	7	3.3	357	7-9	2	7	7	2.8	6.4	(c)	1	5	(d)	19	361
5	8	3.3	305	9	3	12	8	3.3	10.6	(e)	1	6	38	23	326

(a) 26,34,35,37; (b) 3,6,7,8,9; (c) 3,7,9, 9,10; (d) 33,34,36,37,38; (e) 3,8,9,10,11,12.

CONCLUSIONS AND RECOMMENDATIONS

The objective of this paper was to demonstrate in a first stage the impacts that information delay may have on vehicle behavior and safety, and in a second stage, the benefits that can be achieved by including information from more than one lead vehicle in the control loop. These impacts were evaluated using a simulation model specifically designed to consider wireless communication effects and simulation scenarios testing how quickly vehicles with automated control can reach equilibrium or reduce the safety impacts of a sudden deceleration or immobilization of the lead platoon vehicle.

As anticipated, the simulation results clearly showed that wireless communication delay has a significant impact on the performance of vehicle control system considering only information from the vehicle immediately ahead. These impacts typically results in an increased time to reach equilibrium, a lower equilibrium speed, and a reduction in the ability to avoid severe collisions or reduce traffic instability. However, in unavoidable collision incidents, the severity of crash is irrelevant to communication delay as colliding speed does not change with varying communication delay. Benefits were then obtained by extending the control logic to consider information from more than one lead vehicle, i.e., by increasing the vision space and anticipative nature of the system. The greatest benefits were obtained by extending the control logic to consider the nearest two lead vehicles. The inclusion of more than two lead vehicles generally resulted in significantly lower incremental gains, while some parameters even show decreased performance due to the complexity of reaching equilibrium or adequately responding to traffic events when considering information from multiple vehicles each marked with a certain degree of inaccuracy.

While the results of the study clearly show the benefits of designing vehicle control systems considering information from at least two lead vehicles and developing algorithms considering the potential impacts of information delay, it is suggested that further experiment be conducted to paint a clearer picture of the problem. In particular, the study did not look at the impacts of varying wireless delays. As indicated, various parameters may affect the delays associated with wirelessly data transmission. Depending on the situation, longer or shorter information delays may thus prevail from one moment to the next, leading to increased or decreased impacts on vehicle control systems. An important question to answer is then how much delay a control system can tolerate before its reduction in efficiency or safety reaches a critical point.

Another important factor not considered in the study is lane changing. The addition of a lane changing module in the simulation model would allow the testing of more complex vehicle control algorithms, including lane-changing functions. While it is anticipated that automated vehicle controls with both longitudinal and latitude functions are likely to be more promising than longitudinal control alone in promoting safety improvement on multi-lane roadway segments, uncertainty remains regarding the impact of information delay on lane changing safety.

A final element that should be explored is whether variations in the behaviour of preceding vehicles would affect the main results of the study. The current study essentially assumed that all vehicles would behave similarly. In reality, drivers exhibit slightly different behaviour that may translate into random accelerations and decelerations. Any robust vehicle control algorithm should then be able to identify these variations in behaviour and ignore them.

REFERENCES

Ayres, T. J., L. Li, D. Schleuning and D. Young (2001). preferred time-headway of highway drivers. *Proceedings, Intelligent Transportation Systems.*

Bando, M., K. Hasebe, A. Nakayama, A. Shibata and Y. Sugiyama (1995). Dynamic model of traffic congestion and numerical simulation. *Physical Review E*, **51**, 1035-1042.

Bexelius, S. (1968). An extended model for car-following. *Transportation Research*, **2**(1), 13-21.

Biswas, S., R. Tatchikou and F. Dion (2006). Vehicle-to-vehicle wireless communication protocols for enhancing highway traffic safety. *IEEE Communications Magazine*, **44**(1), 74-82.

Gazis, D.C., R. Herman and R.W. Rothery (1961). Nonlinear follow-the-leader models of traffic flow. *Operations Research*, **4**(9), 545-567.

Gettman, D. and L. Head (2003). *Surrogate Safety Measures from Traffic Simulation Models.* Report FHWA-RD-03-050, FHWA, U.S. Department of Transportation.

Gipps P.G. (1981). A behavioural car-following model for computer simulation. *Transportation Research*, **15B**, 105-111.

Herman, R., W. Montroll, R. Potts and R. Rothery (1959). Traffic dynamics: analysis of stability in car following. *Operations Research*, **7**, 86-106.

Hirst, S. and R. Graham (1997). The format and presentation of collision warning. *Ergonomics and Safety of Intelligent Driver Interfaces*, 203-219.

Hydén, C. (1996). Traffic conflicts technique: state of the art. Transportation Department, University Kaiserslautern.

Jones, T.R. and R.B. Potts (1962). The measurement of acceleration noise – a traffic parameter. *Operations Research*, **10**(6), 745-763.

Kerner, B.S. and H. Rehborn (1997). Experimental properties of phase transitions in traffic flow. *Physical Review Letters*, **79**, 4030-4033.

Lenz, H., C.K. Wanger and R. Sollacher (1999). Multi-anticipative car-following model. *The European Physical Journal B*, **7**, 331-335.

Liu, Y., F. Dion and S. Biswas (2005). Dedicated short-range wireless communications for intelligent transportation system applications: state of the art. *Transportation Research Record*, **1910**, 29-37.

Liu, Y., F. Dion and S. Biswas (2006). Safety assessment of information delay on intelligent vehicle control system performance. *Transportation Research Record*, **1944**, 16-25.

Minderhoud, M. and P. Bovy (2001). Extended time-to-collision measures for road traffic safety assessment. *Accident Analysis and Prevention*, **33**, 89-97.

Reichart, G., R. Haller and K. Naab (1997). Driver assistance: BMW solutions for the future of individual mobility. *Proceedings, Fourth World Congress on ITS*, Berlin, Germany.

Shladover, S., VanderWerf, J., Miller, M.A., Kourjanskaia, N. and Krishnan H. (2001). *Development and Performance Evaluation of AVCSS Deployment Sequences to Advance from Today's Driving Environment to Full Automation.* Report UCB-ITS-PRR-2001-18, California PATH Program, University of Califormia, Berkeley, CA.

Treiber M., D. Hennecke and D. Helbing (2000). Congested traffic states in empirical observation and numerical simulations. *Physical Review E*, **62**, 1805-1824.

Treiber, M. and D. Helbing (2002). Microsimulations of freeway traffic including control measures. eprint arXiv:cond-mat/0210096.

Treiber, M., A. Kesting and D. Helbing (2004). Multi-anticipative driving in microscopic traffic model. eprint arXiv:cond-mat/0404736.

Transportation and Traffic Theory 2007
Edited by R.E. Allsop, M.G.H. Bell and B.G. Heydecker

27

A NEW CONCEPT AND GENERAL ALGORITHM ARCHITECTURE TO IMPROVE AUTOMATED INCIDENT DETECTION

Kun Zhang and Michael A P Taylor, Transport Systems Centre, University of South Australia, Adelaide, Australia

INTRODUCTION

Road incidents and incident induced traffic congestion are a threat to the mobility and safety of our daily travel. Incidents are defined as random and nonrecurring events such as accidents, disabled vehicles, spilled loads, temporary maintenance and construction activities, and other unusual events that disrupt the normal flow of traffic. Timely and accurate incident detection using automated incident detection (AID) systems is essential to effectively tackle incident induced congestion problems and to improve traffic management. The core of an AID system is the incident detection algorithm which interprets real time traffic data and makes decision on incidents. The AID algorithms discussed in this paper aim to detect lane-blocking incidents when their effects are manifested by certain types of deterioration in traffic conditions. An incident that blocks the entire roadway (we call it link-blocking) is an extreme case of lane-blocking incidents.

The performance of AID algorithms is normally evaluated against three measures: detection rate (DR), false alarm rate (FAR) and mean time to detect (MTTD). The DR is defined as the ratio of the number of detected incidents to the recorded number of incidents in the test data set. The FAR is the ratio of the number of false alarms to the total number of intervals to which the algorithm is applied. These two measures are used to evaluate the effectiveness of an AID algorithm. The MTTD is the average time difference between the time an incident is detected by the algorithm and the actual time the incident occurs. The MTTD measures the efficiency of the algorithm.

The early California algorithms (Payne and Tignor 1978), the McMaster algorithm (Gall and Hall 1990; Hall et al. 1993) and the detection logic with smoothing (DELOS) algorithms (Chassiakos and Stephanedes 1993) were built on theoretical understanding about incidents on freeways. Incident detection based on these methods was performed in a serial fashion using detection rules. Following rapid development of computer technology, artificial neural networks were intensively applied to freeway incident detection to improve AID algorithm performance (Abdulhai and Ritchie 1999; Cheu and Ritchie 1995; Dia and Rose 1997; Jin et al. 2001; Ritchie and Cheu 1993; Srinivasan et al. 2004). More recently, the support vector machine (Yuan and Cheu 2003) and wavelet technique (Teng and Qi 2003) have been proposed for freeway incident detection. Despite the above improvements, high performance and strong transferability remain common issues concerning freeway AID algorithm development.

Urban arterial roads feature interrupted traffic flow, turning movements and a variety of traffic signal controls, and therefore provide a more challenging environment for incident detection. Early arterial road AID algorithm development focused on simple comparison methods using raw traffic data (Bell and Thancanamootoo 1988; Han and May 1989; Stephanedes and Vassilakis 1994). To enhance algorithm performance and to achieve real-time incident detection, advanced methods were suggested, which included image processing (Hoose et al. 1992), vehicle positioning (Sermons and Koppelman 1996), artificial neural networks such (Khan and Ritchie 1998; Thomas et al. 2001), support vector machines (Yuan and Cheu 2003) and data fusion (Ivan 1997; Thomas 1998). Although these published new methods represent significant improvements, algorithm performance stability remains a big issue concerning existing arterial road AID algorithms.

This paper introduces a new concept and general AID algorithm architecture to enhance the performance of incident detection on both freeways and urban arterial roads, and to improve algorithm transferability. We treat incident detection as a decision making problem rather than as pattern recognition. Hence, the focus of the AID algorithm design is shifted from precise incident pattern description and reduced traffic pattern misclassification to effective traffic knowledge representation and strong evidential reasoning capability of the algorithm. We use the Bayesian network technique to manage existing traffic knowledge and to perform coherent evidential reasoning for incident detection. Two new AID algorithms are presented in this paper, the TSC_fr algorithm (Zhang and Taylor 2004a; Zhang and Taylor 2006a) for freeways and the TSC_ar algorithm (Zhang and Taylor 2004b; Zhang and Taylor 2006b) for arterial roads. 'TSC' stands for Transport Systems Centre where the algorithms were developed. These two algorithms are used in this paper to demonstrate the feasibility and effectiveness of applying the new concept to AID algorithm development.

The paper is organized in six sections. The first section discusses the importance of "decision making", the new concept used to improve AID algorithm performance. The reasoning tool used to support this concept is Bayesian networks, which is detailed in section two. In the third section, the general AID algorithm architecture and the basic function specification of its building blocks are outlined. This algorithm architecture leads to two AID algorithms: the TSC_fr for freeways and the TSC_ar for urban arterial roads, whose main features and performance are discussed in the following two sections. In section six, conclusions are drawn on testing results of the above two algorithms, and our future research directions are indicated.

METHODOLOGY

Pattern Recognition

Incident detection is traditionally treated as a pattern recognition / classification problem. The well established freeway AID algorithms, such as the California algorithms, the DELOS algorithms and the McMaster algorithm, use detection rules to identify incident patterns from traffic data. A rule-based system consists of a library of rules (e.g. If A then B). These rules reflect essential relations within the domain under investigation, or ways to reason about the domain. When specific information about the domain becomes available, the rules are used to draw conclusions and to point out appropriate actions, which take place as a chain reaction. In incident detection application, detection rules and their associated thresholds attempt to form one *general* incident pattern for a *specific* road. If real time traffic data match this pattern after a serial rule testing, then an incident alarm is declared. Traffic systems are dynamic and stochastic, which implies that the causal relations reflected by detection rules are not absolutely certain. Meanwhile, gathered traffic information for inference is often subject to uncertainty as well. The combination of uncertainty and temporal variation is no longer a local phenomenon for which each detection rule is tailored; and it is challenging for rule-base AID algorithms to maintain performance stability by operating in a serial fashion. In addition, the strong site specific features of this type of algorithm tend to hamper algorithm transferability.

Neural networks perform pattern recognition / classification. Through proper training, neural networks can be used in domains that require uncertainty handling capability. The parallel processing mechanism and fast learning capability of neural networks made them popular for incident detection during the last decade. Apart from the relatively fixed architecture of a neural network (the number of layers and the number of nodes in each layer), the weights and thresholds actually determine the behaviour of the network. Incident pattern description in a neural network is vague as each perceptron[1] in the hidden (pattern) layers only has a meaning in the context of the functionality of the network. We will not know what assumptions about the traffic system have been made by the neural network and why a certain pattern fit value has been suggested from the observed traffic data. These assumptions (prior traffic

[1] A perceptron is a node along with its in-going edges.

knowledge) are crucial for traffic operators to make decision on an incident and to form appropriate responses to the incident. The key weakness of the neural network approach is that we are unable to utilize the expert traffic knowledge we might have in advance. Although probabilistic neural network (PNN), support vector machine (SVM, equivalent to PNN) and constructive probabilistic neural network (CPNN) do improve the adaptability of basic multi-layer feed-forward neural network (MLF) in terms of network retraining, the neural network still require substantial site specific training to establish appropriate weights and thresholds.

Decision Making

A freeway lane-blocking incident tends to block the upstream traffic and free the downstream one, if the reduced roadway capacity due to the incident could not handle the traffic demands. This fairly general incident pattern may vary substantially on urban arterial roads because of interrupted traffic flow, low speed manoeuvres, turning movements plus different traffic signal control strategies. Therefore, the basic question which an AID algorithm has to answer would be '*based on our prior knowledge, how likely is it that an incident might happen given the observed traffic data*' instead of '*how well do the observed traffic data fit in with certain predefined incident patterns*'. The fundamental difference between the above two questions is that the former places an emphasis on evidence based reasoning, while the latter focuses on pattern matching which heavily relies on precisely defined incident patterns. For incident detection, evidential reasoning is more stable because both the prior traffic knowledge and the observed traffic data interact with each other to produce a certainty measure (likelihood) associated with each decision making on incidents. Meanwhile, evidential reasoning is dynamic as a clearly described and fully accessible knowledge base can be easily adapted to different traffic sites.

We treat incident detection as a decision making process. The focus of AID algorithm design is then shifted from precise incident pattern description and reduced traffic pattern misclassification to effective traffic knowledge representation and strong evidential reasoning capability. We are seeking an effective way to represent existing traffic knowledge about incidents and subjectively build them into the algorithm rather than learn them from incident data. Meanwhile, we are trying to describe traffic condition in a more general and concise manner using real time traffic data, based on which evidential reasoning can be performed coherently using general traffic knowledge.

Experienced traffic operators can accurately detect incidents from data. In the human reasoning process, operators' general traffic knowledge are used to build a causal structure in which relations between traffic parameters (e.g. volume, occupancy, etc.) and traffic events (e.g. incident) are quantitatively described. Real time traffic measurements are used to determine the state of each traffic parameter, which is mainly based on their experience (site specific knowledge). Using traffic states as evidence, the likelihood of an incident can be sought from the causal structure. Clearly, the estimated incident probability given observed traffic data forms the base of each decision making. To mimic such human reasoning, we use a Bayesian network (Jensen 1996; Pearl 1988) to store prior traffic knowledge and to perform evidence based inference in our proposed AID algorithm architecture. Bayesian networks are causal probabilistic networks which are also called belief networks. The ability of the Bayesian network to coordinate bi-directional inferences filled a void in expert systems technology in early 1980s, and this method subsequently emerged as a general representation scheme for uncertain knowledge (Pearl 1988).

BAYESIAN NETWORK

In this section, we start with Bayesian inference and its key theorem (Bayes' theorem) which forms the mathematical basis for probability updating in the Bayesian networks.

Standard Statistical Inference vs Bayesian Inference

The standard statistical inference is widely used in the scientific community including transportation research. This method was developed largely by Ronald Fisher, Jerzy Neyman and Karl Pearson during the 1920s and 1930s, and aims to provide objective mathematical tests capable of falsifying theories (Matthews 1998). Statistical inference tries to work out *P(data | null hypothesis)*, which is the probability of getting at least as impressive data from an experiment given the null hypothesis.

Before the 1920s, another approach to statistical inference (Bayesian inference) was in general use. Bayesian approach is based on a result that flows directly from the axioms of probability. The key theorem behind it is Bayes' theorem:

$$P(B \mid A) = \frac{P(A \mid B)P(B)}{P(A)} \tag{1}$$

In Bayes' theorem, $P(B \mid A)$ is the posterior probability distribution of event B given that information about event A is available, $P(A \mid B)$ is the conditional probability distribution of A given B, which represents expert knowledge about the domain under investigation, and $P(B)$ and $P(A)$ are the prior probabilities of B and A respectively. Bayes' theorem becomes the basis of Bayesian inference when B is the event of a specific hypothesis being true, and A as the event of observing specific data. The power and importance of the Bayesian inference is immediately apparent in its solution to one of the central problems of standard statistical inference, which is that standard statistical inference does not tell us *P(hypothesis | data)*, the probability that the hypothesis really is correct given the data they observed, which is what incident detection is all about.

In earlier AID research (Thomas 1998), the Bayesian classifier was used to improve accuracy of traffic pattern classification by minimizing the conditional misclassification cost. The precision of traffic pattern description using traffic parameters for each class is crucial for this application. In contrast, the Bayesian network technique uses the Bayesian rule (theorem) to perform evidence base reasoning which relies on the clearly defined and quantified knowledge base (conditional probability tables of the network). This technique enable us to subjectively build prior traffic knowledge into the Bayesian network and modify them at any stage of AID algorithm implementation to cope with traffic environment changes. Hence, the Bayesian network approach for incident detection is a more advanced use of the Bayesian rule to enhance evidential reasoning capability of the AID algorithm

Bayesian Network

Bayesian networks are causal probabilistic networks. As shown in Figure 1, a Bayesian network consists of a set of nodes (the variables of interest) and a set of directed links between these nodes. Each variable has a finite set of mutually exclusive states. Links reflect cause-effect relations within the domain which the network models. Since these effects are normally not completely deterministic, the strength of an effect is modelled as a probability. The nodes together with the directed links form a directed acyclic graph (Jensen 1996).

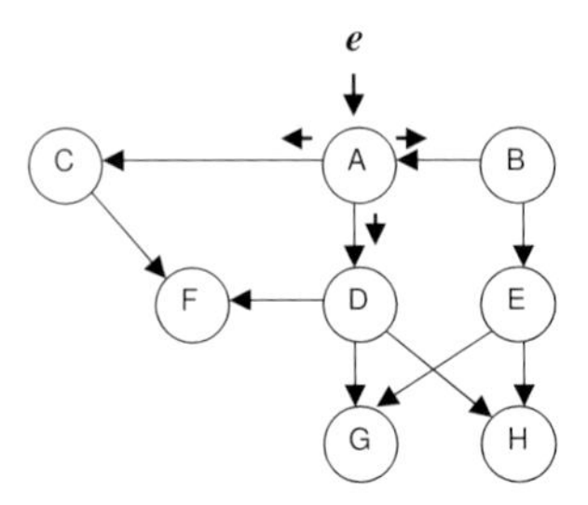

Figure 1 Bayesian network

All nodes of a Bayesian network represent concepts that are well defined with respect to the domain under investigation. If a node in a Bayesian network does not have any parents (i.e. no links pointing towards it), the node will contain a *marginal probability table*, a probability distribution over the states of the variable which the node represents. If a node does have parents (i.e. one or more links pointing towards it), the node contains a *conditional probability table* (CPT). Each cell in the CPT contains a conditional probability for the variable, which the node represents, being in a specific state given a specific configuration of the states of its parents. The CPTs of a Bayesian network are used to quantify the causal relations described in the network.

Fundamentally, a Bayesian network is used to update probabilities of certain variables whenever information on the other variables of the network becomes available. The inference (probability updating) in a Bayesian network can be thought of a message passing process. A message (e.g. ***e*** in Figure 1, which is the available state of variable *A*) can be passed along links in both directions. The tool for inferring in the opposite direction (e.g. from *A* to *B*) is

Bayes' theorem (1). The message passing process may be described mathematically as follows:

Let U be the universe of variables in the Bayesian network, $U = \{A_1, \ldots, A_n\}$. Here, A_1 to A_n could be thought as variables A to H in Figure 1. Let information e be the statement

> "the joint configuration of $A_1, \ldots, A_n$ is $(a_1, \ldots, a_n)$, which consists of information of several variables, and the information about each variable can be entered separately".

The posterior probability distribution $P(X \mid e)$ for all variables X in U (X represent the variables of our primary interest, such as incident), which is updated beliefs on X given information e, is calculated in the following way:

1. Use the chain rule (2) (Jensen 1996) to calculate $P(U)$, the joint probability table that provides the probabilities of all possible configurations of the universe U.

$$P(U) = \prod_i P(A_i \mid p(A_i)) \tag{2}$$

where $p(A_i)$ is the parent set of the node A_i, and $P(A_i \mid p(A_i))$ is the conditional probability table of A_i.

2. Enter information e, $e = (a_1, \ldots, a_n)$ into the Bayesian network to form $P(U,e)$, the part of $P(U)$ corresponding to the configuration $(a_1, \ldots, a_n)$.

$$P(U,e) = P(U) \cdot \underline{a_1} \cdots \underline{a_n} \tag{3}$$

where $\underline{a_i}$ is a m-dimensional table of zeros and ones corresponding to the information a_i on variable A_i which may have m possible states.

3. Marginalize $P(U, e)$ down to $P(X,e)$ for the variables X in U. For each state x of X, sum up all entries in $P(U,e)$ with X in state x

$$P(X,e) = \sum_{U \setminus \{X\}} P(U,e) \tag{4}$$

4. $P(X \mid e)$ is the result of normalizing $P(X,e)$, which is dividing $P(X,e)$ by the sum of all its entries. The Bayesian theorem (1) supports this calculation.

$$P(X \mid e) = \frac{P(X,e)}{P(e)} = \frac{P(X,e)}{\sum_X P(X,e)} \tag{5}$$

Hugin Propagation

The first algorithm proposed for probabilistic calculations in Bayesian networks used such message-passing architecture and was limited to trees (Kim and Pearl 1983). If the Bayesian network is not a tree, which is the case in Figure 1, the independence properties in the network need to be analysed to establish a set of clusters of variables and to construct a tree over the clusters. The resulting tree will have the junction tree property: for each pair of nodes *V*, *W* in the tree, all nodes on the path between *V* and *W* contains their intersection $V \cap W$. Figure 2 shows the junction tree corresponding to the Bayesian network presented in Figure 1. The node *BAE* represents a node cluster that contains the variables *A*, *B* and *E* in the Bayesian network. Inside the cluster, the causal dependencies among the three variables are organized in a tree structure. For each pair of nodes in the junction tree, such as the node pair *BAE* and *DEH*, the node *ADE* (in the path between the node pair) contains their intersection. The Hugin method (Jensen et al. 1990) for probability updating in the Bayesian networks, which is used in this research, is similar to the early tree propagation (Kim and Pearl 1983) but for junction trees. This method is a modification of a general method presented in Lauritsen and Spiegelhalter (1988).

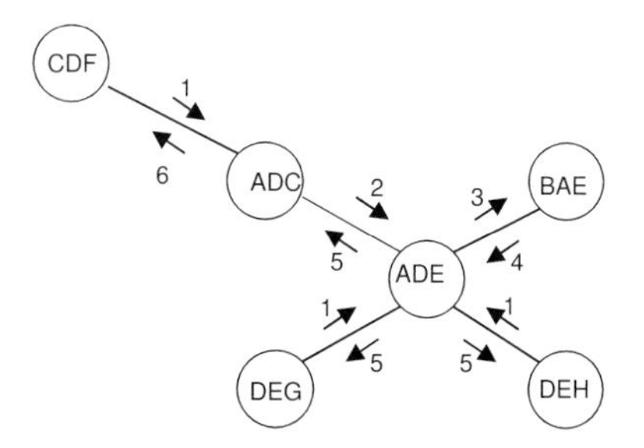

Figure 2 Junction tree and Hugin propagation

In Figure 2, each arrow and its associated number indicate the direction and the order of one specific message passing procedure that takes place in the corresponding link. The Hugin propagation starts with the root node *Rt* (i.e. the node *BAE* in this example) asking all its neighbours to send it a message. Note that a node *X* can send a message to a neighbour *Y* if *X* has received a message for all its other neighbours. If they are not allowed to do so, they recursively pass the request to all neighbours except the one from which the request came. This process is called *CollectEvidence*, which consists of message passing procedure 1 to 3. Then, *Rt* sends messages to all of its neighbours who recursively send messages to all neighbours except the one from which the message came. This process is called *DistributeEvidence*, which consists of message passing procedure 4 to 6. Whenever the Hugin propagation takes place, *CollectEvidence (Rt)* is called followed by a call of *DistributeEvidentce (Rt)*. When the calls are finished, the tables are normalized so that they sum to one. The Hugin method for probability updating in Bayesian networks uses a global perspective. Any node in the Bayesian network can receive information as the method does not distinguish between inference in or opposite to the direction of the links. Also, simultaneous input of information into several nodes will not affect the probability updating performance.

GENERAL AID ALGORITHM ARCHITECTURE

In the previous section, the discussions on the Bayesian networks concentrate on domain knowledge representation (causal structure and its CPTs) and evidential reasoning mechanism. The AID algorithm architecture design focuses more on efficient traffic knowledge management, which aims to improve performance stability and transferability of the algorithm. As shown in Figure 3, our proposed AID algorithm architecture consists of two modules: data processing module (DP) and incident detection module (ID).

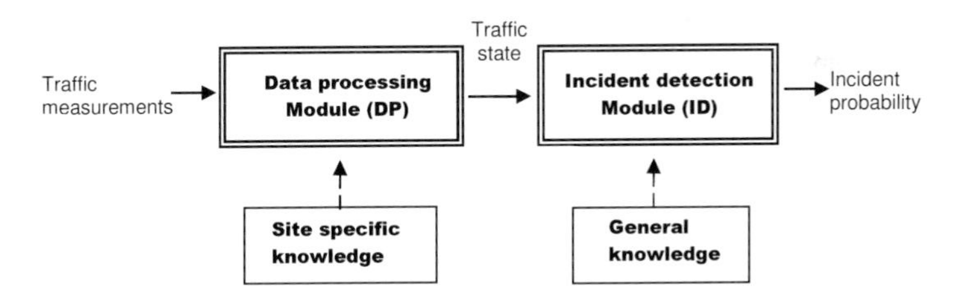

Figure 3 General AID algorithm architecture

The DP module could be treated as a traffic state generator. Real time traffic measurements on traffic parameters (e.g. volume, occupancy, etc.) are processed in the module and then are converted into their corresponding states (e.g. volume is High, Medium or Low). Site specific traffic knowledge (operators' experience about the specific road) is used in the module to set up the thresholds for traffic measurements conversion at each detection interval.

The ID module works as an inference engine. The Bayesian networks form the core of this module, which quantitatively model the causal dependencies among traffic parameters and traffic events (e.g. incident). Using the traffic state as evidence, the Bayesian networks continuously update the incident probability. If the estimated incident probability exceeds the predefined incident decision threshold, an incident alarm will be issued. General traffic knowledge about incidents is used to construct CPTs of the Bayesian networks, which could be shared between different sites.

Two direct benefits gained from this design are (1) both the general knowledge base and the evidential reasoning process are independent of site specific data processing, which makes the ID module universal, and (2) the conversion of traffic measurements into traffic states only requires local traffic knowledge, which could substantially reduce the algorithm implementation needs. This general algorithm architecture was used to develop two AID algorithms, the TSC_fr algorithm for freeway and the TSC_ar algorithm for urban arterial roads. The discussions in the following two sections will focus on the key Bayesian networks used in these two algorithms and the performance of these algorithms.

FREEWAY AID ALGORITHM TSC_fr

Detector Configuration and Data Processing

The TSC_fr algorithm was originally developed on Southern Expressway (SX) in Adelaide, Australia. The typical traffic detector configuration for incident detection on the SX is shown in Figure 4. Detector stations are evenly located (i.e. 500 metres apart) along the freeway. Each detection zone covers the road segment between each pair of detector stations (e.g. DS_2 at upstream and DS_3 at downstream). Lane volume, occupancy and speed at both upstream and downstream detector station are collected over each 20 seconds using loop detectors. The same data aggregation and transmission interval was also found on Tullamarine Freeway and South Eastern Freeway in Melbourne (Dia and Rose 1997). The detection interval for the TSC_fr algorithm was set to 20 seconds.

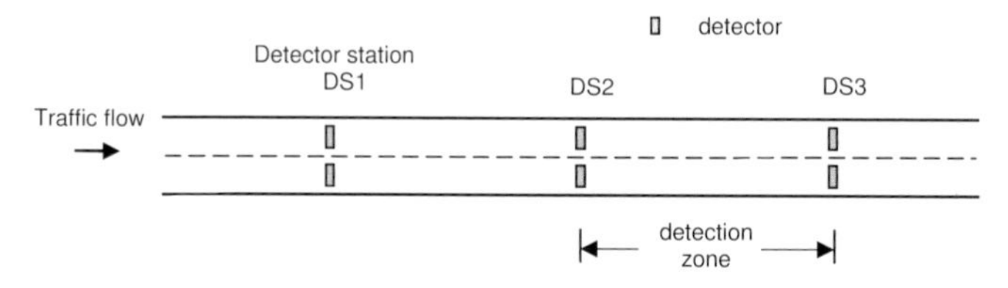

Figure 4 Traffic detector configuration for freeway incident detection

As shown in the algorithm architecture (Figure 3), raw traffic measurements (lane volume, occupancy and speed) are processed in the DP module. The link average of each traffic parameter is calculated at each detection interval. It is then compared with the predefined thresholds to determine the state of that parameter. The selected traffic parameters (not every one) with their states form an input for the DP module at each detection interval. The traffic parameter selection process will be discussed in the later section.

Typical Bayesian Network – Spatial Traffic Information

In the ID module of the TSC_fr algorithm, a typical Bayesian network is constructed to deal with the classic freeway incident detection. As shown in Figure 5, the variables of the Bayesian network include two traffic events (incident: *Inc1_1*, congestion: *Con1_1*) and seven traffic parameters (volumes: *Vol1_1* and *Vol2_1*, occupancies: *Occ1_1* and *Occ2_1*, speeds: *Spd1_1* and *Spd2_1*, and the occupancy difference between upstream and downstream: *Docc_1*). Here, '_1' stands for the 1st detection interval. The two traffic events are parent nodes. The impact of any traffic event can be observed from both upstream and downstream traffic parameters. The special traffic parameter *Docc_1* is also included in the Bayesian network, because both traffic events have great impact on its values but in different ways.

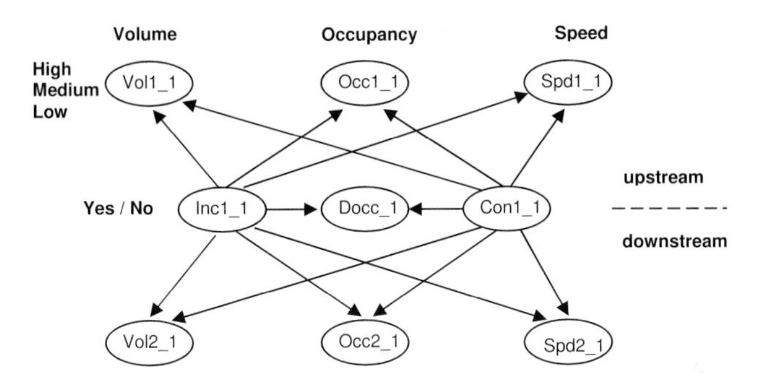

Figure 5 Typical Bayesian network for freeway incident detection

Prior traffic knowledge about incident detection is stored in each CPT of the Bayesian network. Table 1 shows the CPT of the variable *Spd1_1*. The first entry of the table (0.0, upper left hand corner) means that given an incident happened in the detection zone which was followed by incident induced congestion, it is almost impossible that the upstream speed would be high without any other reasons. This example shows how we convert existing traffic knowledge into each CPT.

Table 1 Conditional probability table of variable *Spd1_1*

Inc1_1	Yes		No	
Con1_1	Yes	No	Yes	No
High	0.0	0.1	0.0	0.6
Medium	0.2	0.6	0.2	0.3
Low	0.8	0.3	0.8	0.1

From above description of the Bayesian network and its CPTs one can see that the Bayesian network is a transparent causal structure. It constitutes a model of general environment and simulates the mechanism that nine variables act in the environment. That is the reason why the knowledge base of the Bayesian network can be easily adapted to new traffic environment through simply modification of the existing CPTs using site specific knowledge (if needed). Currently, we assign two states (*Yes* or *No*) to each traffic event, and three states (*High, Medium,* and *Low*) to each traffic parameter. This arrangement aims to simplify CPT construction and make the ID module of the algorithm more general.

The basic function of the Bayesian network is to update incident probability at each detection interval. Spatial traffic information, the state of volume, occupancy and speed at both upstream and downstream detector stations, is used by the Bayesian network as evidence to perform evidential reasoning. As shown in Figure 5, both congestion and incident probabilities are calculated at the same time. The updated congestion probability is not used as the precondition of incidents but the supporting information for decision making on incidents. In addition, the parallel two-way inference takes place in the Bayesian network, which makes use of all available states of traffic parameters simultaneously.

Dynamic Bayesian Network Structure – Temporal Traffic Information

The general way to reduce the FAR of an AID algorithm is the persistence test, which raise an incident alarm after multiple incidents have been detected by the algorithm at several consecutive detection intervals. To improve the efficiency of the TSC_fr algorithm, we use the dynamic Bayesian network structure as an alternative. As shown in Figure 6, the dynamic Bayesian network consists of two time slices. Each time slice represents one detection interval (*t-1* or *t*). The basic Bayesian network used in each time slice for incident probability updating is identical (see Figure 5). To make the following discussion simpler, the details of the basic Bayesian network topology for each time slice is hidden.

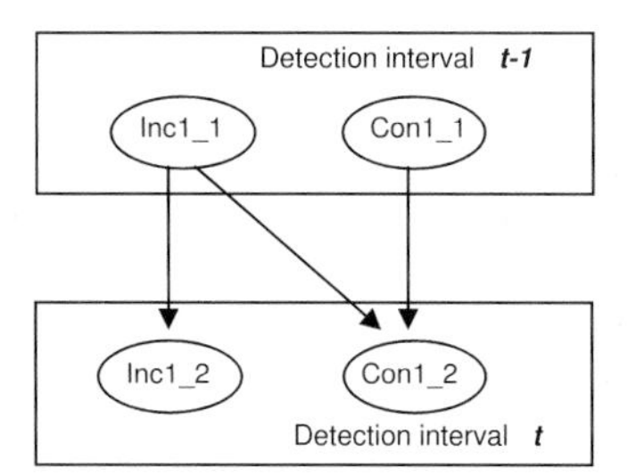

Figure 6 Dynamic Bayesian network structure

The causal links between the nodes in two different time slices try to model the evolving patterns of traffic events (incident or congestion) at two consecutive detection intervals. For time slice *t*, the incident probability updating is performed using both the states of traffic parameters at current interval *t* and the estimated incident and congestion probabilities at *t-1*(which are again based on the traffic states at *t-1*). Taking the advantages of bi-directional reasoning capability of the Bayesian network, similar reasoning can be performed for time slice *t-1*, which means the future states of both incident and congestion at next detection interval can be used to adjust the current incident probability estimate and make it more reliable. This is how temporal information is used in the TSC_fr algorithm to perform joint reasoning for incident detection.

In addition to the dynamic Bayesian network structure, we use estimated incident probability at the previous detection interval *t-1* as the traffic condition indicator for interval *t*. In the DP module, we select the most appropriate traffic parameters to perform incident detection at the current detection interval *t* based on the state of the traffic indicator (Zhang and Taylor 2006a). For example, if the estimated incident probability at *t-1* is very high, we may use both speed and occupancy data to detect incident at *t* and ignore the volume data for that interval. Lin and Daganzo (1997) proposed another effective method to improve incident detection reliability on homogeneous freeway segments. The cumulative occupancy data were used in the method to detect delay-inducing incidents whose disturbance generates congestion. This method is particular good for incident verification even though longer incident detection time might incur.

TSC_fr Algorithm Testing Results

The TSC_fr algorithm was first tested on Southern Expressway (SX) in Adelaide using simulated incident data (Zhang and Taylor 2004a). The SX is a novel one-way reversible direction expressway on which the TSC_fr algorithm was developed. Under the prevailing tidal flow nature, the SX is designed to operate northbound in the morning and southbound in the evening to relieve peak flow traffic on an alternative arterial route (Main South Road). A full description of the SX and its traffic control system is given in Taylor (2005). To compare the congestion parameters and emissions of the SX and Main South Road, and to investigate the impact of advanced traffic management implementation on the SX, a micro-simulation traffic model of the SX was constructed using the Paramics microsimulation package (Woolley et al. 2001b). The SX model has been validated using field operational data. A total of 36 different types of lane-blocking incidents were simulated using the SX model under different traffic conditions for algorithm testing. One incident was generated during each 45-minute simulation run. Incident duration varied from 15 to 25 minutes.

Following the simulation studies, a large number of field incident data sets obtained on Tullamarine Freeway and South Eastern Freeway (SEF) in Melbourne were used to evaluate the TSC_fr algorithm (Zhang and Taylor 2006a). This high quality incident database of real incidents had been used in early research to develop the neural network based MLF (multiple-layer feed-forward) algorithm (Dia and Rose 1997). The Tullamarine-SEF incident database contains 100 incidents which were collected on the Tullamarine (85 incidents) and the SEF (15 incidents) between February 1992 and March 1995. To meet the neural network training needs, the starting time of each incident in the database was estimated, which was based on both the manual inspection of each incident data set and on its corresponding descriptions recorded in operators' logs (Dia and Rose 1997).

TSC_fr Algorithm Performance

The performance of the TSC_fr algorithm on SX data and Tullamarine-SEF data is shown in Table 2. The simulation studies on the SX indicated that the TSC_fr algorithm had a perfect

DR (100 per cent) and a low FAR (0.07 per cent) when the decision threshold was set to 70 per cent. The FAR was calculated using the total number of incident free intervals of the 36 incident data sets. Meanwhile, the MTTD produced by the algorithm was 117 seconds, which was reasonably fast for field application.

Table 2 Performance of the TSC_fr Algorithm on SX Data (simulated) and Tullamarine-SEF data (real)

Test site	Incident decision threshold (%)	Incident detection performance		
		DR (%)	FAR (%)	MTTD (second)
SX (36 incidents)	60	100	0.35	117
	70	100	0.07	117
	85	100	0.07	128
Tullamarine & SEF (100 incidents)	55	92	0.143	158
	60	92	0.103	165
	70	92	0.087	175

When the Tullamarine-SEF data were used to test the TSC_fr algorithm, a very consistent algorithm performance was obtained: the DR was 92 per cent and the FAR was 0.087 per cent when the decision threshold was set to 70 per cent. The MTTD of the algorithm increased by about one minute (from 117 to 175 seconds) compared with the simulation results. Note that the estimated incident start time was used to calculate the MTTD instead of using the time that appeared in operator's log. As noted in Dia and Rose (1997), inspection of the log times corresponding to the evaluation data set (40 incidents) revealed that two incidents were detected by the operators before their impact on traffic was confirmed from the detector data. Only seven of the remaining 38 incidents were detected by the operators within 3 minutes of their occurrence. The average time taken by the operators to detect the 38 incidents was 6.9 minutes after their estimated occurrence times. This suggests that the TSC_fr algorithm has the potential to provide more than 50 per cent improvement in efficiency compared to the average time taken by the operators to detect incidents.

The eight undetected incidents belong to Tullamarine Freeway, which occurred at six different freeway sections. Five of them happened during peak periods. Visual inspection of the five peak-period incident data sets revealed a few common traffic features: (1) the average lane volume was still very high (around 1800 veh/h) during incident periods, (2) the average lane occupancy difference between the upstream and downstream detector was small (e.g. this value exceeded 0.05 for one incident only), and (3) the speed drops were experienced by both the upstream and downstream detector, no significant speed difference between the two detectors was observed (only one incident created a 15 km/h speed difference). For the three undetected off-peak incidents, one happened at early morning (6:03 am), another one occurred at evening (7:42 pm), and these two incidents did not generate apparent traffic disturbance. The third off-peak incident created a huge speed drop (from 90 km/h to 30 km/h) at both the upstream and downstream detectors, however, the volumes and occupancies experienced by the detectors were not changed.

The most distinct feature of the TSC_fr algorithm performance on Tullamarine-SEF data was that the DR was not sensitive to the decision threshold settings. When the decision threshold was set between 55 and 70 per cent, a stable DR of 92 per cent was obtained. Meanwhile, the FAR was slightly affected by the decision threshold, decreasing with the increasing values of the decision threshold (see Table 2). This feature was also found from the algorithm performance on SX data, which confirmed our belief that dynamic and concise traffic pattern description, general knowledgebase and strong evidential reasoning capability of the TSC_fr algorithm are the key to achieve stable incident detection.

As discussed in the first section, strong transferability is essential for any successful AID algorithm. Table 2 indicates a very consistent performance of the algorithm on two different incident databases. This result demonstrates a strong transferability of the TSC_fr algorithm. Note that when we adapted the original TSC_fr algorithm from the SX to the Tullamarine, no retraining was performed on the ID module, which means the original Bayesian network and its CPTs used by the original ID module were unchanged. Since we had little prior knowledge about the normal traffic on the two Melbourne freeways, we had to use field incident data to extract site specific knowledge for the DP module adaptation. We selected ten out of 85 Tullamarine Freeway incidents for manual inspection. Based on that, the original thresholds used for traffic data processing were adjusted. Alternatively, experienced traffic operators could perform such tasks without looking at those incident data sets. No further threshold modification was performed when the algorithm was tested on the SEF data. As such, the TSC_fr algorithm adaptation / implementation requirements had been reduced substantially.

Comparison Studies

To assess the competitiveness of the TSC_fr algorithm, we compare its performance against the most advanced freeway AID algorithms including the MLF, the PNN (probabilistic neural network), the CPNN (constructive probabilistic neural network), and the SVM_P (support vector machines with polynomial kernel function) algorithm. The direct performance comparison between the TSC_fr algorithm and the MLF algorithm is performed using the Tullamarine-SEF database (the evaluation set). This original evaluation data set contained 40 incidents, 25 from the Tullamarine and 15 from the SEF. None of these incidents had been used to train the MLF algorithm or to adapt the TSC_fr algorithm to the Tullamarine.

In general, it would be inappropriate to compare the performance of algorithms obtained on different data sets, especially when incident data based training and adaptation plays a significant role in algorithm performance. To assess the TSC_fr algorithm in a broad base, we cite a recent work in which three neural network models (MLF, PNN and CPNN) for freeway AID were compared directly on the same AYE databases from Singapore (Srinivasan et al. 2004). The Ayer Rajar Expressway (AYE) database contained 300 simulated incidents. This data set was used to evaluate the performance of the three neural network based algorithms. All three algorithms were thoroughly trained using half of the 300 incidents before they were tested on the other half of the incidents from AYE database. Using the MLF algorithm as a bridge, we may perform indirect comparison between the TSC_fr algorithm and the PNN and CPNN algorithms. In addition, we also include the performance of the SVM_P algorithm on the I-880 Freeway database from California, USA (Yuan and Cheu 2003) into our comparison studies. The I-880 Freeway database consisted of 45 field incidents. A total of 22 incidents from this database were used for the SVM_P algorithm training. The remaining 23 incidents were used to test algorithm performance. The performance of the above mentioned algorithms is shown in Table 3.

The figures in Table 3 indicate that, on the basis of the available data sets, the TSC_fr algorithm performed better than the MLF algorithm, no matter which decision threshold value (between 55 to 70 per cent) was selected. Given the similar low FAR (e.g. FAR < 0.07 per cent), both the DR and MTTD of the TSC_fr algorithm were superior than those of the MLF algorithm. Meanwhile, the TSC_fr algorithm was more stable in terms of the reduced positive correlation between the DR and FAR. As can be seen in Table 3, the DR of the TSC_fr algorithm was not sensitive to the decision threshold compared with the MLF algorithm, and the FAR of the TSC_fr algorithm improved slightly with increasing values of the decision threshold. The SVM_P algorithm also exhibited a superb performance which was comparable with that of the TSC_fr algorithm. However Yuan and Cheu (2003) noted that the speed of SVM_P training was found to increase exponentially with the number of training vectors in the data set. This fact might limit the SVM_P algorithm training and optimization process when a large number of field incident data are available. The performance of the former three algorithms is indicated in Figure 7.

Table 3 Performance of the TSC_fr, MLF, SVM_P, PNN and CPNN algorithm

Algorithm	Source	Data set	Number of incidents	Incident decision threshold / Number of persistence test	Performance		
					DR (%)	FAR (%)	MTTD (sec)
TSC_fr	Zhang and Taylor 2006a	Tullamarine & SEF	40 (field)	55 %	92.5	0.072	150
				60 %	92.5	0.057	163
				70 %	92.5	0.05	170
MLF	Dia and Rose 1997	Tullamarine & SEF	40 (field)	0.4	90	0.442	170
				0.5	87	0.273	181
				0.64	82.5	0.065	203
SVM_P	Yuan and Cheu 2003	I-880	23 (field)	0	91.3	0.17	138
				1	91.3	0.13	135
MLF	Srinivasan et al. 2004	AYE	150 (simulated)	1	90.2	0.18	139
				2	86	0.05	163
PNN	Srinivasan et al. 2004	AYE	150 (simulated)	1	87.3	0.46	147
				2	85.3	0.21	172
CPNN	Srinivasan et al. 2004	AYE	150 (simulated)	2	92	0.81	170
				3	86	0.34	188

Table 3 also shows that the MLF algorithm performed the best among the three neural network based algorithms on AYE data in terms of DR and FAR. Meanwhile, it produced a shorter MTTD when compared with the other two algorithms. Given the fact that the MLF model used in this work shared the same network structure and network training scheme with that in Dia's work (Dia and Rose 1997), the above results imply that the TSC_fr algorithm is also very competitive against both PNN and CPNN based algorithms.

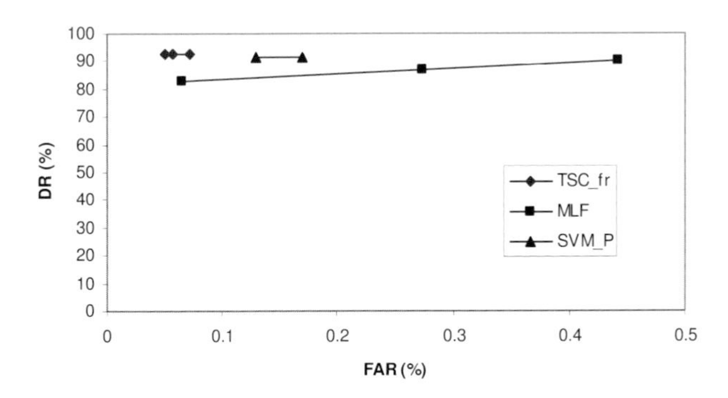

Figure 7 Performance of TSC_fr, MLF and SVM_P on filed data

URBAN ARTERIAL ROAD AID ALGORITHM TSC_ar

Detector Configuration and Data Processing

The TSC_ar algorithm was developed on Cross Road in Adelaide. As shown in Figure 8, the incident detection zone used in this research covers the upstream intersection and the roadway between the two adjacent intersections (between detector station a_1 and a_2). Under the SCATS (Hicks and Carter 2000; Lowrie 1982) signal control systems which are dominant in major Australian cities, traffic detectors are located on the approach side of the signalized intersection just next to the stop line (i.e. b_1 and b_2 in Figure 8). We propose the incident detectors to be located 50 metres away from the stop line further upstream for incident detection. The rationale for such detector configuration is that the detector a_1 can monitor the queue evolution during each signal cycle and it can indicate traffic demand better than stop-line detectors. In practice, one set of the currently available video detector (Nelson 2002) can provide such flexibility to monitor multiple loop detector zones of one approach (e.g. cover both a_1 and b_1) through on-screen detection zone configuration.

In the TSC_ar algorithm, we use the major traffic stream at the upstream intersection as an indicator to detect incidents downstream for each signal cycle. Traffic signal settings of both upstream and downstream intersection are incorporated into the DP module to perform traffic data extraction. The upstream traffic volume that corresponds to the major traffic stream of each signal cycle is extracted from b_1 data. Meanwhile, the upstream occupancy during the same major phase is extracted from a_1 data to provide the algorithm with concurrent queuing conditions. We only use a_2 data to extract downstream volume and occupancy which corresponds to the same traffic stream from upstream intersection. Incident detection interval for the TSC_ar algorithm depends on upstream traffic signal cycle time.

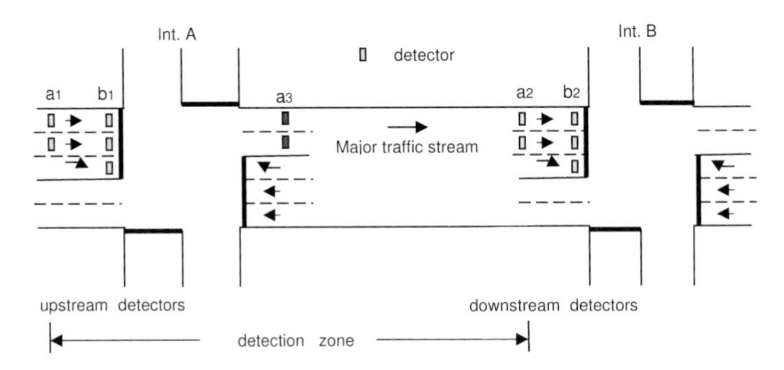

Figure 8 Detector configuration for arterial road incident detection

Our proposed detector configuration is different from traditional one (Khan and Ritchie 1998; Thomas et al. 2001; Yuan and Cheu 2003), in which the incident detection zone was defined as the road section between a_3 and a_2 (see Figure 8). Under the traditional detector configuration, traffic data processing is simpler, and incident detection can be performed at fixed time step (like freeway AID) without considering the actual traffic signal plans for the intersections. However, it is well known that the traffic signal plays an important role in traffic pattern formation (for both incident and incident-free patterns). It is very difficult to precisely describe incident patterns using both a_3 and a_2 data without considering the traffic signal plans and its impact on incident evolution, especially when adaptive traffic signal control is implemented at the intersections.

Basic Bayesian Network

The basic Bayesian network used in the ID module for arterial road incident detection is show in Figure 9. The network consists of three traffic events (incident: *Inc1_1*, congestion at both upstream and downstream intersection: *Con1_1* and *Con2_1*) and five traffic parameters (turning count at the upstream intersection: *Turn1_1*, volumes of both intersections: *Vol1_1* and *Vol2_1* (representing the major traffic stream), and occupancies of both intersections: *Occ1_1* and *Occ2_1*). Similar to the typical freeway Bayesian network, each traffic parameter has three states (High / Medium / Low) and each traffic event has two states (Yes / No).

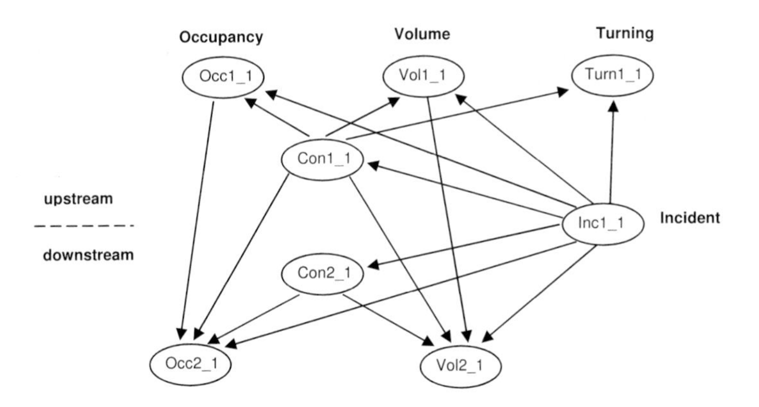

Figure 9 Basic Bayesian network for arterial road incident detection

Note that an arterial road traffic congestion is often characterized by queues surrounding certain intersections, the key points of an arterial road network (Taylor 1992). To better represent this feature, the concept of *node congestion* instead of *link congestion* is used to construct the Bayesian network. As shown in Figure 9, the node *Con1_1* and *Con2_1* are used to represent congestion at the upstream and downstream intersection respectively. Through this treatment, the impact from incident to two individual intersection congestion situations can be modelled.

An arterial road incident does not always block upstream traffic and free downstream one as the typical freeway incident does. In case of a mid-block lane-blocking incident, a certain proportion of the platoon, which was released from the upstream intersection and is interrupted and delayed by the incident, may encounter red signal at the downstream intersection (whereas normally it does not). Hence the incident induced long queue may appear downstream first. In addition to incidents themselves, the preceding traffic condition and the traffic signal control at both upstream and downstream intersections also influence incident pattern formation. To model these more complicated arterial road incident scenarios, the causal link between each pair of upstream-downstream traffic parameters (e.g. *Occ1_1* and *Occ2_1*) and the links from *Con1_1* to both *Occ2_1* and *Vol2_1* are built up. In addition, abnormal turning movements upstream are an important indicator of possible incidents, the node *Turn1_1* is included into the basic Bayesian network, whose value is extracted from upstream stop-line detector (b_1) data where possible.

Scenario-specific Bayesian Network

A link-blocking incident on arterial roads usually generates a distinct traffic pattern which is similar to the severe freeway incident pattern. However, this pattern differs from a normal capacity-reducing lane-blocking incident pattern. To pick up both lane-blocking and link-blocking incidents and to reduce the false alarm rate of the TSC_ar algorithm at the same time, we create two incident scenario specific Bayesian networks which work in parallel in the ID module. The two scenario specific Bayesian networks share the same network topology (see Figure 9), but each network has its respective CPTs.

Inside the ID module, the Bayesian network tailored for link-blocking incident scenario works as the main inference engine. The updated incident and congestion probability in this network will automatically be used to produce the final estimate of incident probability for each detection interval. Meanwhile, the updated incident probability in the other network (lane-blocking incident specific network) is used as a switch. If its value is higher than the predefined threshold and the incident probability produced by the link-blocking specific Bayesian network is not high enough, then the two Bayesian networks will swap at the next detection interval. The incident report will then be based on the reasoning results from the lane-blocking specific Bayesian network. The reason for this design is that the link-blocking incident scenario can be more clearly described. In addition, we can apply a narrowed tolerance region to some entries of the CPTs for the link-blocking specific Bayesian network with little ambiguous interpretation, which could reduce the MTTD without a large increase in false alarms.

TSC_ar Algorithm Testing Results

TSC_fr Algorithm Performance

In our early research (Zhang and Taylor 2004a; Zhang and Taylor 2005), microscopic traffic simulation with Paramics was the major tool used to develop the freeway AID algorithm TSC_fr. When the algorithm was evaluated using a large number of field incident data sets, its performance was very consistent with the one obtained from simulation study. Given the scarcity of high quality arterial road incident data and encouraged by the above results, simulated arterial road incident data were used to test the TSC_ar algorithm at the current stage of algorithm development (Zhang and Taylor 2006b). The test site was Cross Road, an urban arterial road in Adelaide, Australia. Cross Road is located in the Unley municipality which sits next to Adelaide CBD. As shown in Figure 10, our targeted road section is between Unley Road and Fullarton Road, which includes of three signalized intersections.

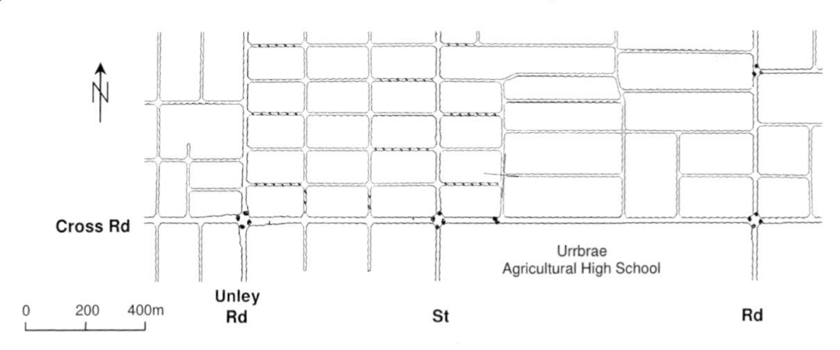

Figure 10 Targeted arterial road section of Cross Road

The Cross Road micro-simulation traffic model used in this research was part of the validated Paramics micro-simulation model that covered the entire Unley road network (Woolley et al. 2001a). This model was constructed for evaluating the 40 km/h urban speed limit scheme applied to residential streets in that area. Total 40 different types of arterial road incidents were simulated using the Cross Road model. One incident was generated during each one-hour simulation run, and incident duration varied from 10 to 35 minutes. We intended to limit the scale (not the complexity) of the algorithm test, as our primary interest was to know how effective our new approach for arterial road AID would be at this initial stage of algorithm development. The algorithm performance is shown in Table 4.

Table 4 TSC_ar algorithm performance on Cross Road data (simulated)

Incident decision threshold (%)	Incident detection performance		
	DR (%)	FAR (%)	MTTD (second)
65	88	0.78	175
70	88	0.62	178
80	83	0.57	203

The DR produced by the TSC_ar algorithm was 88 per cent and the FAR of the algorithm was 0.62 per cent when the incident threshold was set to 70 per cent, which were very encouraging. The MTTD of the algorithm (178 seconds) was reasonable for field application. Most excitingly, the DR of the TSC_ar algorithm reached a stable region (DR > 80 per cent) when the incident decision threshold was set between 65 and 80 per cent; meanwhile, the FAR of the algorithm was no longer sensitive to the decision threshold settings and improved slightly with the increases of the decision threshold. The above findings are very consistent with the performance of the TSC_fr algorithm, which demonstrates the capability of the Bayesian network approach in achieving the enhanced incident detection on arterial roads.

Comparison Studies

To assess the competitiveness of the TSC_ar algorithm, the Bayesian network method is compared with several advanced incident detection methods which include the vehicle positioning method (Sermons and Koppelman 1996), the neural networks method (Khan and Ritchie 1998; Thomas et al. 2001), support vector machine (Yuan and Cheu 2003), and the data fusion method (Ivan 1997). The review of the above literature suggested that detector configuration used for incident detection varied from site to site. Meanwhile, different traffic signal controls made arterial road incident detection even more site specific when compared with freeway incident detection. It is very difficult to perform a stringent performance comparison between two algorithms on the same data set, especially when they have different theoretical foundations. Hence, the algorithm performance shown in Table 5 was obtained from research literature which represented the best results of those methods. These figures only provide a fairly general indication on the effectiveness of these mentioned arterial road AID algorithms. To test the vehicle positing based algorithm, the short-term lane closures were substituted for spontaneous incidents. Except for this algorithm, the performances of the other algorithms shown in the table were produced using simulated incident data.

Table 5 Performance of the TSC_ar, MLF (basic and modular), PNN, SVM_P, vehicle positioning and data fusion algorithm

Algorithm	Source	Data set	Number of incidents	Incident decision threshold / Persistence test	Algorithm performance		
					DR (%)	FAR (%)	MTTD (sec)
TSC_ar	Zhang and Taylor 2006b	Cross Rd.	40	70 %	88	0.62	178
MLF	Yuan and Cheu 2003	Ave west-Clementi	324	PT=1	60.2	0.24	156
PNN				PT=1	77.2	0.89	155
SVM_P				PT=1	88.9	0.22	149
MLF (modular)	Thomas et al. 2001	Coronation Dr	13	PT=2	85	0.64	114
MLF (Basic)	Khan and Ritchie 1998	Anaheim	108	PT=0	76	1.16	205 (1.63 cycle)
				PT=1	60	0.23	(2.63 cycle)
Vehicle positioning	Sermons and Koppelman 1996	Chicago	56	Incident prior < 0.3	68	0	-
Data fusion	Ivan 1997	Chicago	90 (training)		93	0	-

The neural network based data fusion algorithm produced the best DR of 93 per cent with a zero FAR (Ivan 1997). Both fixed detector data and probe vehicle data were used for incident detection. The result implied a great potential of the data fusion method in tackling arterial road incident detection problems. Since this work focused on the algorithm output fusion network topology comparison, the algorithm performance shown in Table 5 was the best network training results rather than algorithm evaluation results. In addition to data fusion, the vehicle positioning based algorithm also produced the lowest FAR of zero. However, the relatively low DR of 68 per cent was generated by the algorithm at the same time when the incident prior was set to 0.15 to minimize the number of false alarms (Sermons and Koppelman 1996). It was also reported in the same literature that 'it would not be appropriate to increase the incident priors to enhance the DR as this would generate a large number of false alarms in an application to new data'.

The basic MLF algorithm had also produced the low FAR of 0.23 per cent by performing one-step persistence test (Khan and Ritchie 1998). On the other hand, the resultant DR of the algorithm became low (60 per cent) as well. This result was consistent with Yuan's later work (Yuan and Cheu 2003). To improve the DR of the algorithm and to maintain the low FAR, the modular neural networks was proposed by Khan and it was reported that the DR was improved to about 70 per cent (based on Figure 6 in that paper). This modular neural network architecture was used later in Thomas et al. (2001), in which both loop detector data and probe vehicle data were used to detect incidents, and the resultant DR was improved to 85 per cent.

The TSC_ar algorithm produced a good DR of 88 per cent, which was comparable with the SVM_P algorithm on fixed detector data and the modular MLF algorithm on multiple data sources. Meanwhile, the FAR was maintained at 0.62 per cent. The distinct feature of the TSC_fr algorithm was the DR of the TSC_ar algorithm reached a stable region (DR > 80 per cent) when the incident decision threshold was set between 65 and 80 per cent. We think this feature stems largely from the general knowledge base of the algorithm (CPTs of the Bayesian networks) and its strong reasoning capability. The transparent causal structure and fully accessible CPTs of the Bayesian networks could also facilitate the algorithm transfer from site to site, which is the other strength of our new approach to arterial incident detection and will be tested in our future research.

CONCLUSIONS

In this research, we treat incident detection problem as a decision making problem rather than pattern recognition. The focus of AID algorithm design is shifted from precise incident pattern description and reducing pattern misclassification to effective traffic knowledge management and strong evidential reasoning capability of the algorithm. We design a general incident detection module in which Bayesian networks are used to store general traffic knowledge and work as an inference engine for decision making on incidents. Incident reports are based on the updated incident probability estimated by the Bayesian networks at each detection interval. Meanwhile, we develop another site specific data processing module to convert real time traffic data to traffic states. This module provides the Bayesian networks with dynamic and concise traffic state information for evidential reasoning.

Two new AID algorithms, the TSC_fr algorithm for freeways and the TSC_ar algorithm for arterial roads, were developed from the above AID algorithm architecture. The TSC_fr algorithm was evaluated using high quality field incident data sets, and the TSC_ar algorithm was tested using simulation data. The TSC_fr algorithm demonstrated how the dynamic Bayesian network structure could make use of both spatial and temporal traffic information to perform fast and stable incident detection. Meanwhile, the TSC_ar algorithm showed the capability of multiple scenario-specific Bayesian networks approach in dealing with complicated arterial road incident detection problems. Most importantly, the performance of the two algorithms was very consistent, which demonstrated the effectiveness of our new concept for incident detection and the feasibility of the general AID algorithm architecture.

We are currently seeking opportunities to test the TSC_fr algorithm online. As mentioned in the previous section, the TSC_ar algorithm testing was restricted to one section of Cross Road at current stage of algorithm development. We are looking at more simulation studies at different road sections with varying traffic signal settings to test the stability and transferability for which the TSC_ar algorithm was designed. Incident detection is a decision making process under uncertainty. "More information less uncertainty" is also true for incident detection. The data fusion potential of the Bayesian network approach will be exploited further in our future research, not only in incident detection and management field, but also in our proposed proactive traffic signal control systems.

ACKNOWLEDGEMENTS

The authors would like to thank Hussein Dia and VicRoads for kindly providing the Tullamarine Freeway and South Eastern Freeway incident database, Jeremy Woolley and Branko Stazic for kind support with Southern Expressway and Cross Road micro-simulation traffic models, and MineLab Electronics Pty Ltd for their financial support for this research project. The comments of three anonymous reviewers are also gratefully acknowledged.

REFERENCES

Abdulhai, B. and Ritchie, S. G. (1999). Enhancing the universality and transferability of freeway incident detection using a Bayesian-based neural network. *Transportation Research C*, **7**, 261 -280.

Bell, M. G. H. and Thancanamootoo, B. (1988). Automatic incident detection within urban traffic control systems. *Roads and Traffic 2000*, Berlin, Germany, 35-38.

Chassiakos, A. P. and Stephanedes, Y. J. (1993). Smoothing algorithms for incident detection. *Transportation Research Record*, **1394**, 9 -16.

Cheu, R. L. and Ritchie, S. G. (1995). Automated detection of lane blocking freeway incidents using artificial neural networks. *Transportation Research C*, **3**, 371-388.

Dia, H. and Rose, G. (1997). Development and Evaluation of Neural Network Freeway Incident Detection Models Using Field Data. *Transportation Research C*, **5**, 313-331.

Gall, A. I. and Hall, F. L. (1990). Distinguishing between incident congestion and recurrent congestion: a proposed logic. *Transportation Research Record*, **1232**, 1 - 8.

Hall, F. L., Shi, Y. and Atala, G. (1993). On-line testing of the McMaster incident detection algorithm under recurrent congestion. *Transportation Research Record*, **1394**, 1 - 7.

Han, D. L. and May, A. D. (1989). Automatic detection of traffic operational problems on urban arterials. Institute of transportation studies, University of California, Berkeley.

Hicks, B. and Carter, M. (2000). What Have We Learned About Intelligent Transportation Systems? -- Arterial Management. US Department of Transportation / Federal Highway Administration.

Hoose, N., Vicencio, M. A. and Zhang, X. (1992). Incident detection in urban roads using computer image processing. *Traffic Engineering and Control*, **33**, 236-244.

Ivan, J. N. (1997). Neural network representations for arterial street incident detection data fusion. *Transportation Research C*, **5**, 245-254.

Jensen, F. V. (1996). *An Introduction to Bayesian Networks*, Springer-Verlag, London.

Jensen, F. V., Olesen, K. G. and Andersen, S. K. (1990). An algebra of Bayesian belief universes for Knowledge-based systems. *Networks*, **20**, 637 - 659.

Jin, X., Srinivasan, D. and Cheu, R. L. (2001). Classification of freeway traffic patterns for incident detection using constructive probabilistic neural networks. *IEEE transactions on neural networks*, **12 (5)**, 1173 - 1187.

Khan, S. I. and Ritchie, S. G. (1998). Statistical and neural classifiers to detect traffic operational problems on urban arterials. *Transportation Research C*, **6**, 291-314.

Kim, J. H. and Pearl, J. (1983). A computational model for causal and diagnostic reasoning in inference systems. *8th International joint conference on artificial intelligence*, Karlsruhe, Germany, 190 - 193.

Lauritsen, S. L. and Spiegelhalter, D. J. (1988). Local computations with probabilities on graphical structures and their application to expert systems. *Royal Statistical Society, Series B*, **50**, 157 - 224.

Lin, W.-H. and Daganzo, C. F. (1997). A simple detection scheme for delay-inducing freeway incidents. *Transportation Research A*, **31**, 141-155.

Lowrie, P. R. (1982). SCATS: The Sydney Co-ordinated Adaptive Traffic System. *IEE International conference on road traffic signalling*, London, UK.

Matthews, A. J. (1998). Facts versus Factions: the use and abuse of subjectivity in scientific research. *The European Science and Environment Forum*, Cambridge.

Nelson, L. J. (2002). Sensors Working Overtime. *Traffic Technology International*, March.

Payne, H. J. and Tignor, S. C. (1978). Freeway incident-detection algorithms based on decision tree with states. *Transportation Research Record*, **682**, 30 - 37.

Pearl, J. (1988). *Probabilistic Reasoning in Intelligent Systems*, Morgan Kaufmann, San Mateo, CA.

Ritchie, S. G. and Cheu, R. L. (1993). Simulation of freeway incident detection using artificial neural networks. *Transportation Research C*, **1**, 203-217.

Sermons, M. W. and Koppelman, F. S. (1996). Use of vehicle positioning data for arterial incident detection. *Transportation Research C*, **4**, 87 -96.

Srinivasan, D., Jin, X. and Cheu, R. L. (2004). Evaluation of adaptive neural network models for freeway incident detection. *IEEE transactions on intelligent transportation systems*, **5**, 1 - 11.

Stephanedes, Y. J. and Vassilakis, G. (1994). Intersection incident detection for IVHS. Transportation Research Board, 74th Annual Meeting, Washington DC, USA.

Taylor, M. A. P. (1992). Exploring the nature of urban traffic congestion: concepts, parameters, theories and models. *16th ARRB Conference*, Melbourne, Australia, 83 - 106.

Taylor, M. A. P. (2005). Adelaide: innovations in transport systems, infrastructure and services. In: *Advances in City Transport: Case Studies*, (S. Basbar, ed.), WIT Press: Southampton.

Teng, H. and Qi, Y. (2003). Application of wavelet technique to freeway incident detection. *Transportation Research C*, **11**, 289 - 308.

Thomas, K., Dia, H. and Cottman, N. (2001). Simulation of Arterial Incident Detection Using Neural Networks. *8th World Congress on ITS*, Sydney, Australia.

Thomas, N. E. (1998). Multi-state and multi-sensor incident detection systems for arterial streets. *Transportation Research C*, **6**, 337 - 357.

Woolley, J., Dyson, C. and Taylor, M. A. P. (2001a). Evaluation of a South Australian 40 km/h urban speed limit. *Transport Engineering in Australia*, **7** (1-2).

Woolley, J. E., Taylor, M. A. P. and Zito, R. (2001b). Modelling of the Southern Expressway using Paramics microsimulation software. *Journal of the Eastern Asia Society for Transportation Studies*, **4** (4), 279 - 295.

Yuan, F. and Cheu, R. L. (2003). Incident detection using support vector machines. *Transportation Research C*, **11**, 309 - 328.

Zhang, K. and Taylor, M. A. P. (2004a). Incident detection on freeway: a Bayesian network approach. *27th Australasian Transport Research Forum*, Adelaide, Australia.

Zhang, K. and Taylor, M. A. P. (2004b). A New Method for Incident Detection on Urban Arterial Roads. *11th World Congress on ITS*, Nagoya, Japan.

Zhang, K. and Taylor, M. A. P. (2005). Simulation of Freeway Incident Detection Using Bayesian Networks. *Intelligent Vehicles & Road Infrastructure Conference*, Melbourne, Australia.

Zhang, K. and Taylor, M. A. P. (2006a). Towards universal freeway incident detection algorithms. *Transportation Research C*, **14,** 68 - 80.

Zhang, K. and Taylor, M. A. P. (2006b). Effective arterial road incident detection: a Bayesian network based algorithm. *Transportation Research C*, (in press).

Transportation and Traffic Theory 2007
Edited by R.E. Allsop, M.G.H. Bell and B.G. Heydecker

28

STOCHASTIC MODELING AND SIMULATION OF MULTI-LANE TRAFFIC

Niamph Dundon and Alexandros Sopasakis, Department of Mathematics, University of Massachusetts, Amherst, USA

SUMMARY

A multi-lane traffic flow model based on stochastic noise driven dynamics is introduced and analyzed. The model employs conservative anisotropic Arrhenius spin-exchange (surface diffusion) dynamics. We generate an asymmetric simple exclusion process to model vehicle interactions. Vehicles react and advance based on the energy profile of their surrounding traffic through a novel look-ahead asymmetric interaction potential. The resulting vehicular traffic model is numerically implemented via kinetic Monte Carlo simulations and scrutinized under basic traffic flow situations.

INTRODUCTION

Modeling traffic during congested conditions is a long standing open problem which has intrigued and puzzled researchers and government officials alike. Although the first attempts in obtaining a traffic flow model date as far back as 1934, Greenshields (1934), the problem of describing the flow and overall behavior of traffic, for more than just a light traffic stream, still remains largely unsolved. The challenge is to accurately resolve the motion of a large number of interacting vehicles.

A traffic flow model which is capable of describing the state of traffic flow at all concentration regimes in real time would therefore be invaluable in analyzing problems, before they arise, and subsequently allowing for the appropriate course of action. A model of this type can also be useful in the planning or decision stages of building highways by allowing engineers to test the capacity of a given highway before it is even build. Although no such model currently exists there have been several efforts towards that direction in recent years in CA modeling (Nagel and Schreckenberg 1992; Knospe *et al.* 2000, 2002; Nagel and Paczuski 1995; Barlovic *et al.* 1998) in macroscopic PDE models (Whitham 1974; Newell 1961; Muramatsu and Nagatani 1999; Komatsu and Sasa 1995; Nagatani 2002; Kerner and Konhauser 1994; Jin and Liu 1994) or other types of models (Klar and Wegener 2000; Illner et al. 2003; Helbing and Treiber 1998; Treiber *et al.* 1999; Sopasakis 2003, 2002; Muramatsu and Nagatani 1999; Bando *et al.* 1995; Treiber *et al.* 1999) each of which can produce partially satisfying results depending on a specific range of prevailing traffic conditions for which the model is valid.

In this work we attempt a new approach to traffic modeling which results in a stochastic model for multi-lane vehicular traffic. The multi-lane model which we propose here is based on the principles first introduced in Sopasakis (2004). We construct a novel energy driven stochastic noise process in conjunction with an anisotropic type interaction potential in an attempt to describe the different phases of traffic and the vehicular behavior which arise for high concentration regimes. The resulting multi-lane model displays important similarities when compared to observed behavior from actual traffic data and as we will show is able to predict a number of key characteristics of vehicular traffic.

Although traffic states are quite complex we can, based on observations (Helbing 2001; Helbing *et al.* 2002; Kerner and Klenov 2002; Schadschneider 2002), categorize them into a small number of main phases: free flow, synchronized traffic, wide moving jams and congested traffic. Most everyone is familiar with free flow and congested traffic in road networks. Figure 1 depicts these well-known traffic states in a simple diagram. Congested traffic however has several different forms and includes the so called wide moving jams and synchronized traffic phenomena. In short, a wide moving jam, is a localized structure, such as traffic a waves, with width which is larger than its front and propagates unchanged against the direction of traffic. Synchronized traffic, on the other hand, is characterized by high vehicle flows which surprisingly can sustain also increasing vehicle velocities. Naturally this is a recipe for disaster since sooner or later the traffic must break down with a number of other interesting traffic states yet to be discovered. A meta-stable region, as also marked on our diagram in Figure 1, is a result of a type of synchronized traffic phenomenon where drivers try to attain their desired speed even though vehicle densities are increasing. Once again the observed capacity drop is inevitable.

It seems daunting for any *deterministic type* model of vehicular traffic (Nelson 2000; Schadschneider 2002; Helbing 1995; Phillips 1979) to manage and capture these multivalued and transitional effects observed for concentrations above c_{crit} . This is probably the reason that a number of such models have been criticized (Daganzo 1995; Nelson 2000; Rathi *et al.* 1987; Ross 1988; Newell 1989) thus leading to subsequent improvements more recently (Aw and Rascle 2000; Sopasakis 2003; Illner et al 2003). Nevertheless due to their nature deterministic models lack the descriptive effects which are possible through a stochastic model, and which we hope to entertain here.

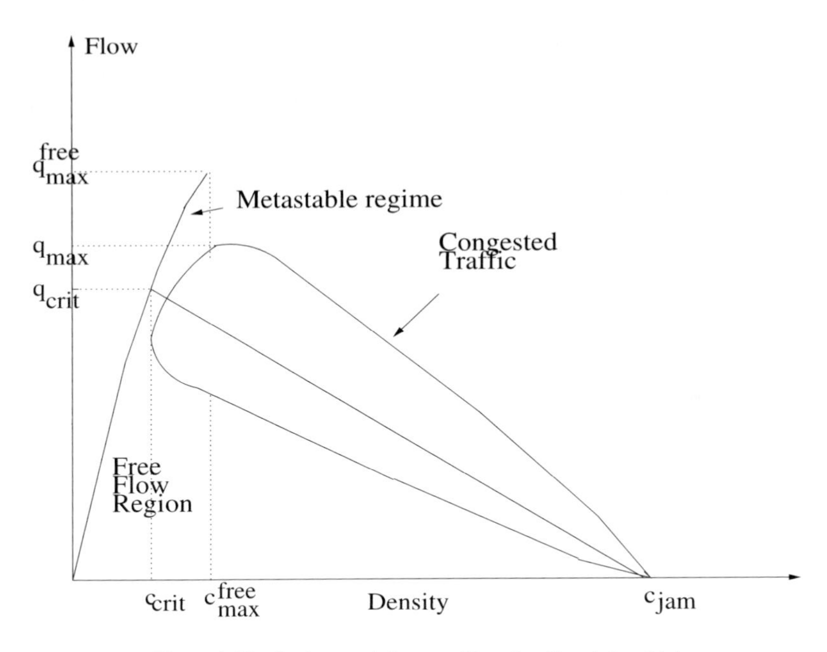

Figure 1. The fundamental diagram (flow-density relationship) depicting some of the major traffic states.

In contrast cellular automaton (CA) models, due to their microscopic structure, exhibit a number of these traffic phases with promise (Schreckenberg and Wolf 1998; Nagel and Schreckenberg 1992; Jiang and Wu 1998). CA models are discretized in space, time and (usually) velocity and rely heavily on computational resources while simple updating mechanisms trace the traffic conditions which prevail in time. To further enhance interactions and complex behavior however recent CA models include a small amount of *ad hoc* stochastic noise thus leading to more realistic and interesting traffic behavior (p159 of Schadschneider 2002).

The work is outlined as follows: we start by building all of the necessary mathematical tools for vehicle interactions in the next two sections. Overall we introduce an asymmetric simple exclusion process (ASEP) (Evals et al. 1998; Krug and Ferrari 1996; Landim and Kipnis 1999), guaranteeing that vehicles do not occupy the same site. Further vehicles are forced to move toward one direction since the dynamics, depending on spatial forward Arrhenius interactions, implement one-sided potentials and a look-ahead feature. A variety of ASEP models without look-ahead Arrhenius interactions have been studied by Gray and Griffeath (2001). The effects of passing and lane changing are relatively less near jam density than in near critical density. Clearly this is not the case in free flow traffic with few vehicles. A major advantage which further differentiates this model is that it relies on a small number of parameters which are directly linked to observed physical traffic characteristics. Therefore due to its simplicity this model can also lend itself to analysis.

Based on scaling and limit arguments as well as asymptotic expansions we show in the fourth section that up to leading order term our model is equivalent in form to Lighthill-Whitham-Richards, integrodifferential Burgers or higher order dispersive type PDE depending on the order of expansion carried out.

Last in order for any model of traffic flow to even have a chance of producing results resembling real traffic it must take into consideration the actual method used for data collection. There are several different methods for collecting traffic data depending on which traffic observables the researchers are interested in obtaining during their experiment. For comparison purposes therefore the simulated data must be collected in similar manner as actual data in order for differences or similarities to be detectable. The link between actual data collection and simulated traffic sampling is therefore important and is examined in the fifth section. The developing theory therefore also examines and account for measurement capability (Hall 1996).

Further this model can be implemented numerically in order to produce realistic traffic flow data via use of a kinetic Monte Carlo simulation. The resulting algorithm can handle heavy traffic and produces predictions in real time. Numerical simulations and test cases are presented in the fifth section. We test the resulting model under a variety of common traffic situations and observe or compare the developing flow against similar other models or equivalent data from Helbing *et al.* (2002); and Wiedemann (1974). Our findings are summarized in the conclusions section.

MICROSCOPIC MECHANISM

In this section we present the basic stochastic spin exchange dynamics which represents the vehicles on the highway and how they interact with each other and the environment. We start by providing the relevant notation. We define our physical space as a two-dimensional, periodic lattice, which is partitioned into n x m cells via,

$$L = \frac{1}{nm} Z^2 \cap [0,1]^2$$

where Z denotes the positive integers. The identification from particle dynamics to vehicles is understood in the simplest possible setting: on each of the lattice points $x \in L$ we define an order parameter $\sigma(x)$ via,

$$\sigma(x) = \begin{cases} 1 & \text{if a vehicle occupies site } x \\ 0 & \text{if site at } x \text{ is empty (no vehicles)} \end{cases} \tag{1}$$

A spin configuration σ is an element of the configuration space $\Sigma = \{0,1\}^L$ and we write $\sigma = \{\sigma(x) : x \in L\}$ denoted by $\sigma(x)$ the spin at x.

We extend the usual Ising type particle interactions to vehicles and derive a corresponding microscopic stochastic model. This procedure should be successful assuming we understand and can correctly capture and describe mathematically the interaction potential between vehicles.

We start by obtaining suitable interpretations of many of the usual parameters of Ising systems and in many instances absorb as many of those parameters as possible in order to simplify our model and adapt it better to vehicular traffic.

Since there are $|L|$ sites on the lattice then the system can be in any of $2^{|L|}$ possible states. The local situation at each of those states is appraised by the interaction potential J. We let J denote an asymmetric short range inter-vehicle interaction potential,

$$J(x,y) = V(\gamma(x-y)), \quad \text{for} \quad x, y \in L \tag{2}$$

where $\gamma = 1/L$ is a parameter prescribing the range of microscopic interactions and therefore L denotes the potential radius. Here we let $V : R \to R$ and we set,

$V(r) = 0, \quad r \in R^{-} \qquad \text{and} \qquad V(r) = 0 \quad \text{for} \quad r \geq 1$

In the simulations we choose a simple constant potential of the form,

$$V(r) = \begin{cases} J_0, & \text{if } 0 \prec r \prec 1 \\ 0, & \text{otherwise} \end{cases} \tag{3}$$

where J_0 is a parameter which based on its sign describes attractive, repulsive or no-interactions (note that we can adapt a non-constant value of J_0 which would be more realistic, such as a Maxwellian, or otherwise). Since naturally vehicles try to move toward the empty space in the highway we must implement repulsive interactions. In this work we implement, piecewise constant, $J_0 = 1$, short range local interactions. At the same time (3) indirectly enforces simple but necessary rules for vehicles, such as no negative speeds (since V is zero for moves backward). Similarly it is seen that vehicles only react to local, and based on the above discussion forward, traffic conditions and stimuli, as would be expected in real traffic situations, since we set $0 \prec r \prec 1$ (i.e. no long range interactions). It is possible and in fact easy however to change the potential so that interactions with vehicles behind can also be accounted for. This would be particularly interesting in the case of lane changing for instance. In the simulations which will follow we let L=4. This potential radius in effect corresponds to vehicles (or drivers) being able to 'look ahead' for up to 4 cells in order to evaluate the status of traffic in front of them and make a more sensible decision as to which cell to move to next based on the potential strength, (2).

Conservative Arrhenius dynamics

In this model we implement spin-exchange Arrhenius dynamics. We refer to Sopasakis (2004) and Liggett (1985) for other possibilities. Under this engine the simulation is driven based on the *energy barrier* a particle has to overcome in changing from one system state to another. During such a spin-exchange between nearest neighbor sites x and y the system will actually allow the order parameter $\sigma(x)$ at location x to exchange sign with the one at y. This is understood as advancing a vehicle from x to $y=N(x)$ where by $N(x)$ we denote the nearest neighbors of $x \in L$. The rate at which a process will do this for spin-exchange Arrhenius dynamics is defined Liggett (1985) as,

$$c(x,y,\sigma(x)) = \begin{cases} c_0 \exp[-U(x)] & \text{if } \sigma(x)=1 \text{ and } \sigma(y)=0 \\ 0, & \text{otherwise} \end{cases} \tag{4}$$

The parameters and variables comprising this exchange rate are as follows,

$$c_0 = 1/\tau \tag{5}$$

with τ the characteristic or relaxation time for the stochastic interaction process. As will become clear in the simulations section the two free parameters c_0 (or τ) and J_0 are directly related to known traffic observables and are therefore easy to calibrate. In (4) $U(x)$ denotes the inter-particle vehicle interaction potential and is comprised of contributions from short range exchange interactions U_e, and an external potential $h(x)$,

$$U(x) = U_e(x) + h(x) \tag{6}$$

Note that the external potential $h(x)$ could vary in both space and also time if so desired. In that case h can account for temporal and spatial traffic situations which in effect may simulate phenomena such as rush hour traffic, local weather anomalies etc. Although feasible we do not present any simulations with the influence of the external potential in this work. We set,

$$U_e(x) = \sum_{\substack{z \in L \\ z \neq x}} J(x,z)\sigma(z) \tag{7}$$

with J as in (2).

Clearly for one-lane traffic y in (4) corresponds to $x+1$. Note that the exchange, due to the specific construction of the interaction potential J in (2), can take effect if and only if the location at x is occupied while the adjacent location at y is not. In that respect vehicles are restricted from performing an exchange move backward – an unrealistic move for vehicular traffic. Similarly vehicles are only restricted to move to their nearest cells and no further.

Overall given (4) and the dynamics just described the probability of spin-exchange between x and y during time $[t, t+\Delta t]$ is, (Liggett 1985)

$$c(x, y, \sigma) = \Delta t + O(\Delta t^2) \tag{8}$$

In terms of traffic this is understood as the probability for a vehicle at location x to move to location at y. Note that if indeed there is already a vehicle at location y the corresponding probability will be zero, since we have designed an exclusion process, and therefore there will be no chance for this move to occur.

The stochastic process $\{\sigma_t\}_{t \geq 0}$ is a continuous time jump Markov process on $L^\infty(\Sigma; R)$ with generator (Liggett 1985),

$$(Mf)(\sigma) = \sum_{x,y \in L} c(x, y, \sigma)[f(\sigma^{x,y}) - f(\sigma)] \tag{9}$$

for any bounded test function $f \in L^\infty(\Sigma; R)$ with $c(x, y, \sigma)$ defined in (4). Here $\sigma^{x,y}$ denotes the configuration after an exchange of the spin between x and y,

$$c^{x,y}(z) = \begin{cases} \sigma(y) & \text{if } \ z = x \\ \sigma(x) & \text{if } \ z = y \\ \sigma(z) & \text{otherwise} \end{cases}$$

Therefore observables f (test functions) evolve with the rule (Liggett 1985)

$$\frac{d}{dt} Ef(\sigma_t) = E(Mf)(\sigma) \tag{10}$$

which is equivalent to the well known Dynkin's formula.

Considering a multi-lane highway requires different dynamics than those used in the one-lane case by Sopasakis (2004). Special attention to application of the proper vehicle interaction potentials is needed. We undertake this task next.

MULTI-LANE EXTENSIONS

In this section we develop the mechanism for lane changing. The lane changing dynamics which will be presented here can be adjusted for highways with American, European or other standards. The main difference being whether a vehicle is allowed to pass only on the left lane or on either lane. We follow ideas from Nagel et al. (1998) and Sparmann (1978) in terms of developing schemes which agree with observations of lane changing behavior.

Lane changing rules

There are two possibilities for passing. Either the drivers must use only the left lane at all times or, for example, an American type system, where the driver may also use the right lane for passing. In the first case we have a so-called asymmetric lane changing rule while in the second case we have a symmetric type of rule. Both such rules can be adapted in our model via a simple modular mechanism.

Lane changing can be incorporated in our model by essentially adjusting the vehicle interaction potential U to contain an extra anisotropic term.

Overall the interaction potential (6) is adjusted to differentiate between two possible directional mechanisms: forward or sideways streaming. This overall lane changing mechanism is included in (6) by augmenting the interaction potential with anisotropy interactions U_α

$$U(x) = U_e(x) + U_\alpha(x) + h(x) \tag{11}$$

where $h(x)$, as before, can account for both spatial and temporal effects in traffic and $U_e(x)$ as in (7). We define the anisotropy potential to be

$$U_\alpha(x) = \sum_{y=nn} \psi(x,y) \tag{12}$$

where nn signifies nearest neighbor and ψ denotes the preferred direction of the vehicle. Since vehicles may choose to move in three possible directions (forward, left and right) we set

$$\psi(x,y) = \begin{cases} k_l & \text{if } y = x+1 \\ k_r & \text{if } y = x-1 \\ k_f & \text{if } y = x+n \end{cases} \tag{13}$$

Here n is the number of lanes and k_l, k_r and k_f are anisotropy constants whose values influence how often drivers change lanes. If $y = x+1$, this corresponds to a vehicle switching into the lane to its left, if $y = x-1$, this corresponds to a right move, and if $y = x+n$ this signifies a move forward (without switching lanes). For realistic traffic conditions we always choose $k_f >> k_l$ and $k_f >> k_r$ which enforces the desire of motorists to move forward rather than switching lanes. Therefore, when cars have all possible choices of moving (left, right or forward), they would almost always elect to move forward.

Similarly we take here $k_l >> k_r$ which simulates the driver's preference to switch into the left rather than the right lane, ensuring the asymmetric lane-changing behavior which we wish to implement. Drivers will remain in the right lane until they meet a slow car, at which time they will most probably, but not always, move into the left lane based on (11). This also results in the natural division of the highway into slow and fast lanes. Note that we can modify (13) to implement symmetric lane-changing rules for American highways by setting $k_l = k_r$. In this way, a driver has equal probability of changing to either lane, given equivalent conditions in each lane.

In our simulations, three possible spin exchanges were permitted (left, right and forward) and the way vehicles move can therefore be compared to the simulations in Nagel et al. (1998) or Wiedemann (1974). Nevertheless our simulations can be very adapted to allow other exchanges, such as diagonal moves. This can be understood as a vehicle advancing forward and switching lanes simultaneously.

MACROSCOPIC MODEL EXTENSIONS

We now develop macroscopic PDE models of traffic based on our microscopic traffic model. The techniques and analysis from Sopasakis and Katsoulakis (2006) can also be applied here as follows: from our definition of a generator (9) we pick $f(\sigma) = \sigma(z)$ for z fixed in M and expand while simplifying as in Sopasakis and Katsoulakis (2006) to obtain the relation,

$$\frac{d}{dt} E\sigma_t(z) = -Ec_0\sigma(z)(1-\sigma(z+1))e^{-U(z,\sigma)} + Ec_0\sigma(z-1)(1-\sigma(z))e^{-U(z-1,\sigma)} \qquad (14)$$

Note that relation (14) is exact and can be used to evaluate the closures discussed below. However it is not yet a closed equation for $E\sigma_t(x) = \text{Prob}(\sigma_t(x) = 1)$.

Finite Difference Scheme

Suppose now that J (for J_0 fixed) in (6) has fairly long and weak interactions. We may assume that the stochastic process in (9) is a 'perturbation' of the simple exclusion process considered in Landim and Kipnis (1999). This process has a Bernoulli product invariant measure, thus at local equilibrium the probability measure is expected to be approximately a product measure. As in Penrose (1991) we assume 'propagation of chaos' for the microscopic system, in which case the fluctuations of the spins $\{\sigma(x); x \in L\}$ about their mean values are independent and the law of large numbers formally applies. Thus the fluctuations of $\sum_{y \neq x} J(y-x)\sigma(y)$ about their mean will be small such that in the long range interaction limit we have (Sopasakis and Katsoulakis 2006) the following *semi-discrete finite difference scheme*

$$\frac{du(z,t)}{dt} + F(z+1,t) - F(z,t) = 0, \textit{where} F(z,t) = c_0 u(z-1)(1-u(z,t))\exp(-J * u(z-1,t)) \qquad (15)$$

where $u(z,t) = E\sigma_t(z)$ denotes the probability that site z is occupied at time t.

Partial Differential Equation Limit

The resulting PDE can now be obtained by simply expanding the spatial variables in Taylor series. We also set $h = \Delta x$ and rescale time $t \to t/h$. Omitting the $O(h^2)$ terms we have $du/dt + c_0[u(1-u)\exp(-J*u)]_z = 0$, for $z \in R$ and $J*u(z) = \int_z^\infty J(y-z)u(y)dy$. The transport equation obtained is,

$$u_t + F(u)_z = 0 \quad \text{where} \quad F(u) = c_0 u(1-u)\exp(-J*u) \tag{16}$$

It is also interesting to point out here that the flux F above under the simplest case of no interactions ($J_0 = 0$) corresponds to a commonly used Lighthill-Whitham (Whitham 1974; Lighthill and Whitham 1955) type convex flux $\rho u(\rho)$ where ρ denotes density and $u(\rho)$ denotes equilibrium velocity. Note that this type of flux F in (16) produces a traffic stream formulation equivalent in form to Burgers equation $u_t + (u^2/2)_z = 0$.

Note that in fact (15), when fully discretized, provides a natural *finite difference scheme* for the PDE (16) above,

$$\begin{aligned} u(z,t^{n+1}) = u(z,t^n) + \frac{\Delta t}{h}[\ & u(z-1,t^n)(1-u(z,t^n))c_0\exp(-J*u(z-1,t^n)) \\ & -u(z,t^n)(1-u(z+1,t^n))c_0\exp(-J*u(z,t^n))] \end{aligned} \tag{17}$$

L	240	100	50	10	4	1
l_1 Relative Error	.0013	.0029	.0051	.0066	.0126	.02

Table 1. Relative error of final solutions. We compare the semi-discrete FD scheme (15) against the stochastic model (14) for different sizes of the potential radius L. The stochastic solution has been averaged over 500 realizations for this calculation. Other parameters: $\tau = .23$ and $J_0 = 6$.

In Table 1 we display the l_1 relative error estimates of the solutions for the semi-discrete (15) and stochastic model (10) based on different sizes of the interaction potential L at a given final time. To calculate the corresponding stochastic density u_{stoch} at that time we averaged over 500 realizations. We observe the smallest relative errors in Table 1 for the case of L=240. This is expected based on our assumption of long ranged potentials. We compare further the resulting stochastic microscopic (10) and finite difference models presented here against each other and provide possible connections with other well known traffic flow models in the following section.

Connections and comparisons between models and parameters

We now present connections between our microscopic, PDE and finite difference models (10), (16) and (17) respectively with other well known traffic flow models. We refer to Sopasakis and Katsoulakis (2006) for details in the one lane case.

We first present hierarchical connections with other well-known traffic flow models based on expansions of our underlying macroscopic equation (16). Expanding the convolution term $J*u$ in (16) gives the following higher order traffic flow model,

$$u_t + c_0[u(1-u)\exp(-J*u)]_z = c_0[J_1 u(1-u)\exp(-J_0 u)u_z]_z + c_0[J_2 u(1-u)\exp(-J_0 u)u_{zz}]_z \quad (18)$$

with J_0, J_1 and J_2 from the Taylor expansion of our potential. Note that (18) is a third order dispersive PDE with diffusion which is similar in form to the PDEs derived from optimal velocity models and usually referred to as 'modified' Korteweg-de-Vries (KdV) in Muramatsu and Nagatani (1999); Nagatani (2002); Newell (1961). We make general remarks below about the behavior of (18) as well as the equivalent but more general (16) under different scales and/or parameters:

a) Assuming first that there are no interactions $J = 0$ in the potential F of (16) we obtain $F(u) = c_0 u(1-u)$ which gives a commonly used (for traffic flow) version of the diffusive Lighthill-Whitham or Burgers equation flux.

b) In the opposite case however of long range interactions between vehicles, $L=N$, we obtain the following non-local flux from,

$$F(u) = c_0 u(1-u)\exp(-J_0 \bar{u}). \quad (19)$$

Based on Figure 2 under this long range interaction case the flux of the stochastic model and that of the PDE (16) agree.

c) Note further that the hyperbolic equation obtained by including terms up to J_0 in the convolution at (18), (disregarding J_1 etc...),

$$u_t + c_0[u(1-u)\exp(-J_0 u)]_z = 0 \quad (20)$$

has a non-convex flux. Indeed the right hand side in Figure 2 shows that if $J_0 \geq 3$ the flux is neither convex nor concave.

d) If on the other hand we include terms up to order J_1 in (18) then (18) takes the form of a nonlinear Lighthill-Whitham type equation with diffusion (Whitham 1974; Newell 1961; Nagatani 2002).

e) Returning to the higher order dispersive PDE (18) we note the similarities with other usual higher order traffic flow models found in Komatsu and Sasa (1995); Muramatsu and Nagatani (1999); Nagatani (2002); Kerner and Konhauser (1994); Kurtze and Hong (1995) although the coefficients obtained here include nonlinearities. Coherent structures can emerge as solutions of

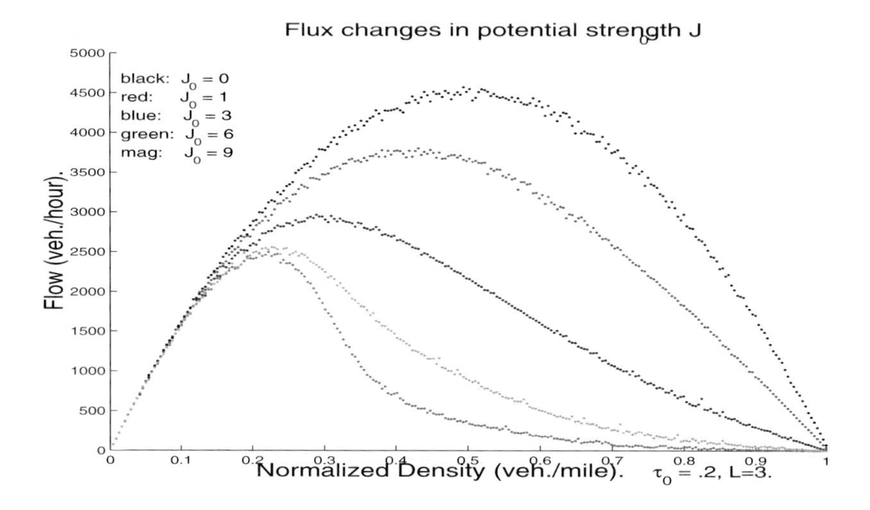

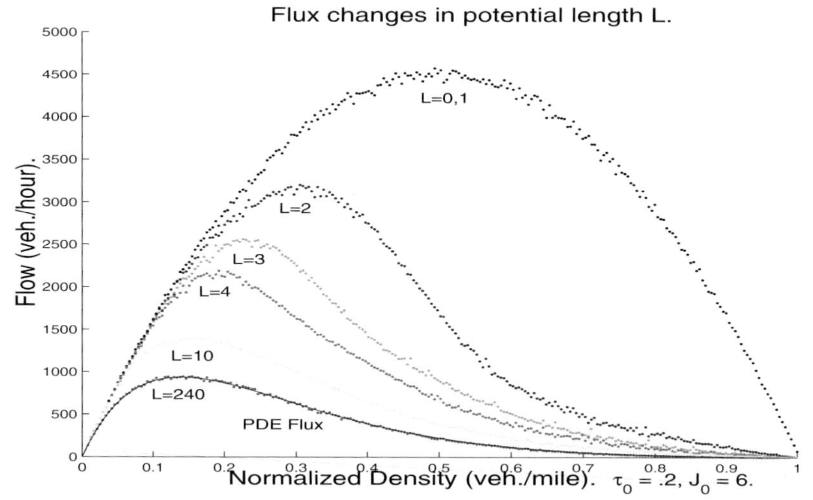

Figure 2. Long time averages. On the top, comparing flux (21) changes with respect to increasing L. On the bottom, comparing the influence of potential strength J_0 in the stochastic flux (21). We set $J_0 = 6, \tau = .23$ and run all microscopic simulations for the same total time and under the same initial conditions before plotting the flow per concentration.
Note that for long range interactions we observe that the PDE flux (16) coincides with the long range interaction (L=240) microscopic model flux (19) which fluctuates around it.

(18) which are similar to traffic waves or wide moving jams. It is known that traveling wave solutions of the Payne-Whitham model Kuhne (1984); Kerner and Konhauser (1994) with non-concave fundamental diagrams, which resemble the form of our higher order PDEs, are asymptotically stable under small perturbations for a sub-characteristic type of condition (Li and Liu 2005). It would be interesting to further examine traveling wave solutions of (16) as well as the higher order approximation (18) and compare them with observed soliton, kink-antikink or mixtures of other density solutions as have been noted in Komatsu and Sasa (1995); Muramatsu and Nagatani (1999); Kerner and Konhauser (1994).

f) Further, we remark that equations similar in form to (18) have also been studied in Jin and Liu (1994).

MONTE CARLO SIMULATIONS

We implement a kinetic Monte Carlo (KMC) algorithm (Bortz *et al.* 1975) and refer to Sopasakis (2004) for all the implementation details of this type of algorithm. Overall the spin-exchange algorithm with Arrhenius dynamics has the following form: we start by calculating all the transition rates for each vehicle based on (4) for moves to nearest neighbor locations. Note that automatically based on our design (2) the moves to occupied locations will always have rate 0. We calculate the total exchange rate for all vehicles (so as to create a measure) and we use a random number in order to choose one from among them. We perform the chosen move between locations x and y and update the simulation time by Δt which is equal to the inverse of the (already calculated) total rate for all vehicles. We repeat this process from the beginning until we have captured the dynamics of interest or simply have reached the end of time for our simulations (equilibration).

One of the biggest benefits of a KMC algorithm (Bortz et al. 1975) over a usual MC algorithm is that it produces no 'null' steps and therefore every iteration is a success. This is quite useful for the cases of high densities of vehicles or while reaching equilibration since in either of these situations a Monte Carlo algorithm would progressively slow down and move less and less often staying idle for long periods of time. In contrast the KMC algorithm continues to choose and move vehicles at every step by skipping the idle waiting and simply adjusting the simulation time by the appropriate amount (Sopasakis 2004), as if it had waited for that long.

The method used for recording and analyzing traffic observables in order to compare them to theoretical simulations is of paramount importance if we are to obtain conclusive statistics and must be clearly explained. In that respect we record quantities of interest such as flow, density and velocity from our simulations as shown in Sopasakis (2004). We formally follow studies which have been carried out by Athol (1972) and more recently Hall (1996); McShane and Roess

(1990); Nelson (2004) regarding proper data collecting procedures and underlying assumptions put forth in theory and in practice.

Regardless of detection method, the flow is measured as the number of vehicles $n(\tau)$ passing a detector at a given time interval τ via,

$$q = \frac{1}{\tau} n(\tau) \tag{21}$$

Based on this formulation flow cannot be found from a single snapshot of vehicles over an interval. Also note that another common point of contention is that usually flow is reported in units of number of vehicles per hour even though the actual time length of recorded observation is much smaller (1/2 to 2 minutes). As a result some concerns have been raised regarding sustainability of such high volumes when data corresponds to measurements over time intervals which are less than 15 minutes long (HCM 1985).

One of the most important quantities which we must account for in any data collection scheme is density. Density is a quantity which is quite hard to measure empirically (McShane and Roess 1990) and can only be measured along a length (Hall 1996). Based on the collection methods described above for flow and space mean speeds it is not uncommon to estimate the density from the well known *macroscopic* formula ('fundamental identity') as originally developed by Wardrop (1952),

$$q = c\bar{v} \tag{22}$$

The Monte Carlo simulations which we undertake here further allow for virtual detectors in an effort to reproduce real traffic data collection procedures during the simulation. Therefore it is actually possible to collect data from our simulations in the same manner as done in actual highways. We place our virtual detector across the highway so that data is collected for all lanes at that location. As an extra advantage which allows for more detailed information we could also record data per lane in our simulations which is usually not the case for usual traffic data gathering. In this way, flow can be calculated for each individual lane and data can be compiled to compute the overall flow for the highway. It is an advantage of the use of simulations that allows us to report both observables for all lanes in the highway as well as for individual lanes as in Figure 6. Specifically we compute flow from (21) while density is found through calculation of occupancy as explained above and in the same fashion as done in Nagel and Schreckenberg (1992). The data is averaged over small intervals of time as in real traffic data collection practices. We give all the details of the parameters used in each of the simulations which are examined in this section.

In this work we simulate a closed round road without entrances or exits which we initialize with a specified total density of randomly distributed vehicles. The fundamental diagrams (density-flow graphs) are constructed by collecting flow data for a given density. These diagrams therefore require several simulation runs in order to complete the complete density spectrum possible.

Calibration of parameters

Physically we let the actual length of each cell to be 22 feet. This allows for the average vehicle length plus safe distance. Therefore for a vehicle which has an average speed of 60 miles per hour we obtain a natural estimate of time to cross a cell,

$$\Delta t(cell) = \frac{22\,feet}{60 miles/hour} = \frac{1}{4}\,second$$

In all the one-lane examples considered below we allow L=4 nearest neighbors in the calculation of the interaction potential, U(x). Given the structure of this potential (one-sided) this implies that drivers are able to perceive traffic densities which are up to 4 vehicle lengths (plus safe distance) ahead of their vehicle (or 4x22 feet = 88 feet 'look ahead'). In the multi-lane examples, we multiply L by the number of lanes, so that drivers can 'look ahead' up to four vehicle lengths in each lane. However we have also run simulations (not presented here) with a 'look ahead' of 3 and 2 vehicle lengths and in the majority similar results were observed.

There are two free parameters J_0, and τ which we must calibrate before we start our simulations. The calibration itself is performed by simulating a free flow regime where we expect all vehicles to drive at their desired speed. We set such a speed to be 65 miles per hour. This is accomplished by the characteristic time τ which allows us to calibrate the maximum velocity vehicles would like to drive at assuming no other vehicles ahead. Naturally due to the stochasticity inherent in our simulation some vehicles will drive faster while some will drive slower than the set limit of 65 miles per hour. As pointed out earlier the free parameter J_0 indirectly influences how drivers react to conditions in front of them and subsequently allows us to set the velocity of an upstream front (which for some highways is found to be approximately -15 ± 5 km/hour (Helbing *et al.* 2002; Schadschneider 2002; Kerner and Rehborn 1997).

Note that for the chosen parameters $\tau = .23$ and $J_0 = 6$ we obtain the desired velocity of 65 miles per hour and velocity out of a jam of approximately –10 miles per hour for one lane traffic. Also note that other pairs of τ and J_0 are possible which adjust the traffic model for different standards set in other countries or regions.

We will now adjust our dynamics for a two-lane highway. We start by calibrating our code for the free parameters J_0, and τ by simulating a free flow regime where we expect all vehicles to drive at their desired speed. We pick $\tau = 1.8$ so that the desired vehicle speeds are approximately 71 miles/hour. Similarly we pick a value for $J_0 = 1$ so that the upstream shock velocity is approximately –11 miles per hour in agreement with Wiedemann (1974); Helbing (2001). The resulting simulation for this calibration is shown in Figure 3.

Note that the necessary safety distance, or gap, between moving vehicles is also automatically accounted for in Figure 3 by the potential (11). In fact, fast vehicles approaching a slow group will slow down thus keeping the proper safety distance up front. Mathematically this is implemented through (11) since an increased value of the potential will make it less likely for the vehicle to move to the free cell in front of it. This is interpreted as a slow down.

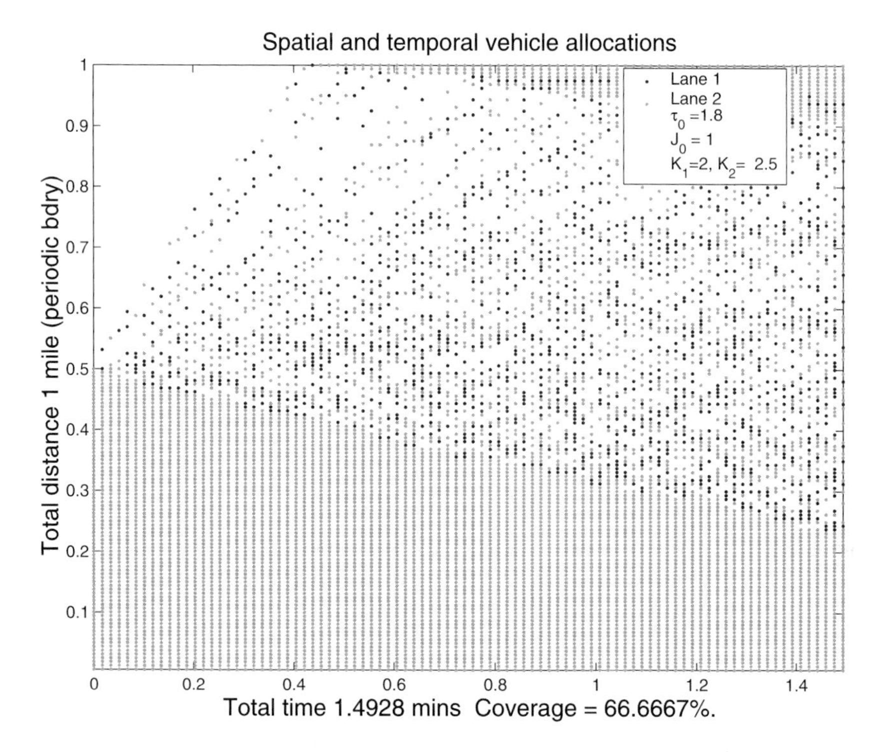

Figure 3. Calibration of parameters for a two-lane highway.
We choose τ and J_0 so that the desired vehicle speeds are set at approximately 71 miles/hour for free flow while the upstream shock velocity is set approximately at –11 miles/hour in order to agree with observations in Wiedemann (1974). Initially we assume a traffic release type condition.

We can calibrate for the parameters τ and J_0 in advance for any number of lanes as can be seen in Table 2. Note that these parameters once chosen do not need to be further adjusted and can be implemented in any traffic simulation which is described by these basic underlying characteristics. In the next few sections we will use the appropriate parameters in order to give examples of traffic behavior under a variety of traffic lanes and desired speeds. Also note that we have chosen the following values for $k_l = 2.5$, $k_r = 2$ and $k_f = 7.5$ although statistical testing must be undertaken here in order to identify the correct range for these parameters in actual highways. These values of k_r, k_l and k_f are used throughout our simulations in this work.

Number of Lanes	1	2	3	4
τ	.23	1.1	1.85	1.9
J_0	6	.7	.7	.5
Desired Velocity (mph)	65	62	68	72
Upstream Velocity (mph)	-10	-11	-12	-9.6

Table 2. Calibrated parameters τ and J_0 for multi-lane traffic with the resulting desired vehicle velocities for free flow and upstream front velocities. The look ahead parameter is chosen, based on physical considerations, to be $L = 4$, while the anisotropy constants for lane changing are $k_l = 2.5$, $k_r = 2$ and $k_f = 7.5$.

One-lane example

Using the corresponding one-lane calibrated parameters for J_0 and τ from Table 2 we now obtain the fundamental diagram (see the beginning of this section for details on collecting and reporting data from observations and simulations), the density-flow relationship, in Figure 4. In general we can implement a number of different types of initial conditions which we make specific for each example considered. For these figures we use a random initial vehicle distribution and observe the behavior of the traffic stream as density increases incrementally.

There is a number of very interesting observations that can be made from Figure 4. We compare our results with those of Nagel and Schreckenberg (1992) for one-lane traffic but also with observations from Wiedemann (1974) and observe qualitative agreement. Specifically, the region of free flow is clearly displayed up to approximately 50 vehicles/mile. Note here that the value of $c_{crit} = 50$ vehicles/mile is not forced on our simulation but instead is naturally created by the process dynamics through the calibration of the two parameters J_0 and τ. Similarly we observe a maximum vehicle flow of approximately 1900 vehicles/hour which also agrees with observations in Wiedemann (1974); Nagel *et al.* (1998) and Figure 1(b) from Helbing *et al.* (2002). The aggregation time of 1.65 minutes was selected so that we can compare with observed data in Nagel *et al.* (1998). The fluctuations in vehicle flows shown in Figures 4 are sizable for densities above c_{crit} and display a widely meta-stable 'stop and go' region.

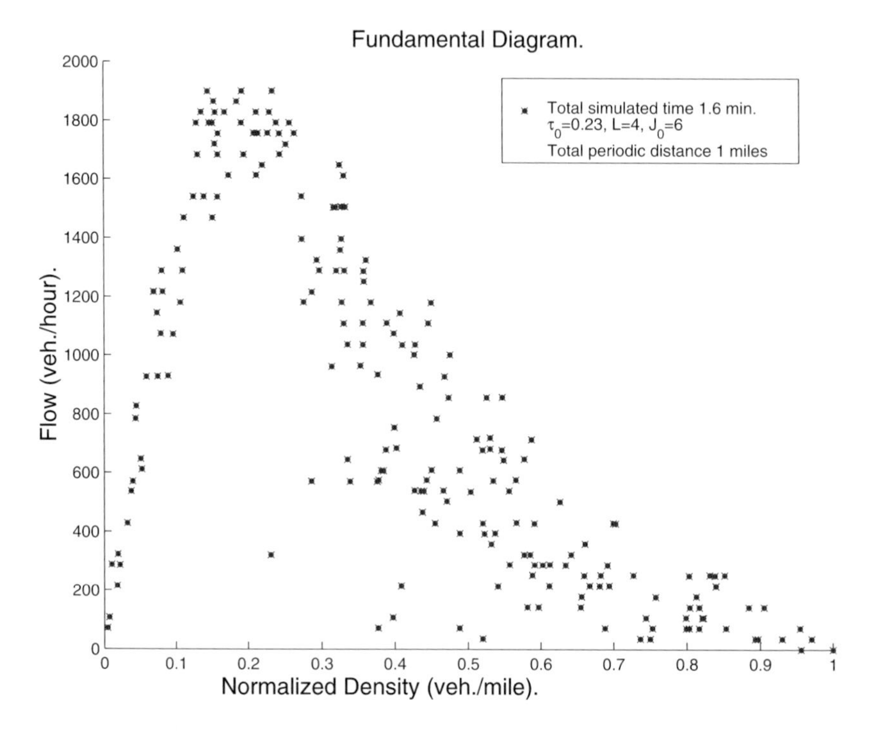

Figure 4. The flow versus density (fundamental diagram) relationship for 1-lane highway. Spatial periodic length of 1 mile, relaxation time $\tau = .23$, interaction strength $J_0 = 6$ and 'look-ahead' $L = 4$ cells. The aggregation time of 1.6 minutes was selected so that we can compare with observed data in Nagel *et al.* (1998).

Multi-lane examples

Using the parameters from the calibrated simulations presented in Table 2 we can now examine multi-lane highways. In the case of a 2-lane highway we present the fundamental diagram (the density-flow relationship) in Figure 5(a) and the corresponding velocity-flow relationship in Figure 5(b). There is a number of very interesting observations that can be made from this diagram. Note that once again we do obtain realistic values for c_{crit} and q_{crit} which are in agreement with equivalent 2-lane highway data.

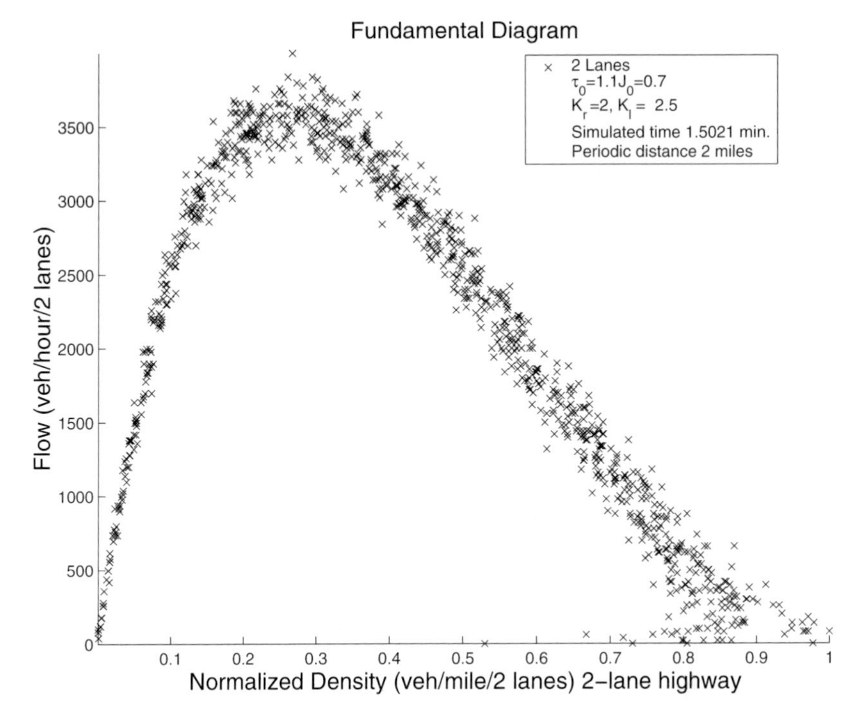

Figure 5(a). Flow versus concentration (fundamental diagram) for a 2-lane highway. Spatial periodic length of 2 miles, relaxation time $\tau = 1.1$, interaction strength of $J_0 = .7$ while the look ahead is maintained at $L = 4$.

Since the spatial distance is chosen to be 2 miles here there are 480 cells per lane. The form of these figures compares favorably with observations in Wiedemann (1974), Nagel *et al.* (1998) and Figures 2.10 and 2.14 in Hall (1996).

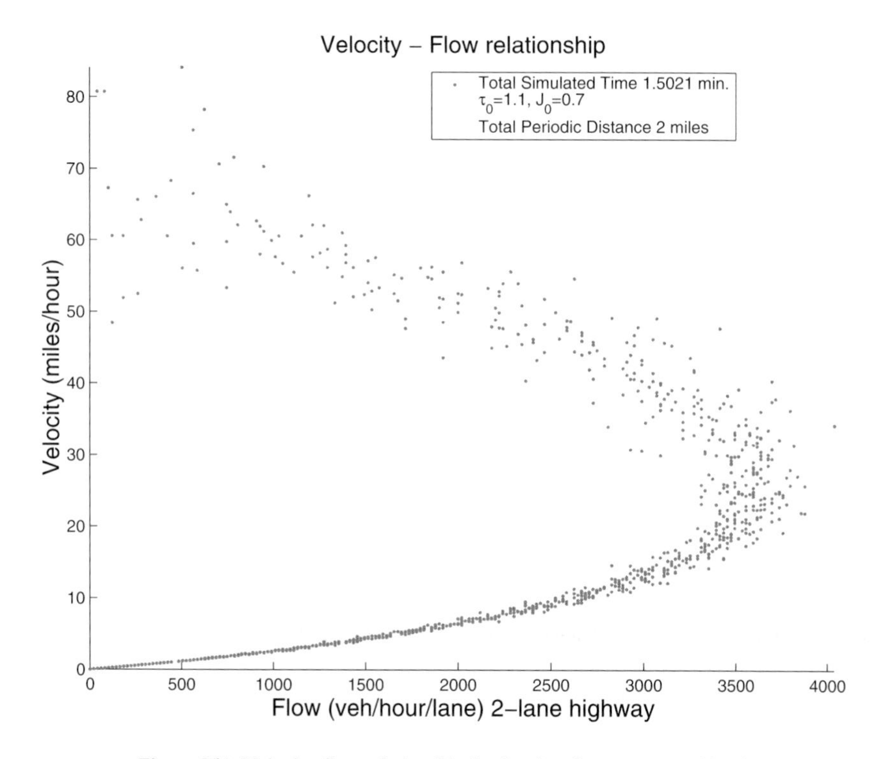

Figure 5(b). Velocity-flow relationship for the situation represented in Figure 5(a)

Similarly in the case of four lanes we obtain the results of Figures 6(a) and 6(b). As usual we have adjusted accordingly our main parameters for this four lane highway with a desired vehicle velocity for free flow at 72 miles per hour. Quantities of interest such as flow, velocity and density seem to agree both quantitatively and qualitatively with data.

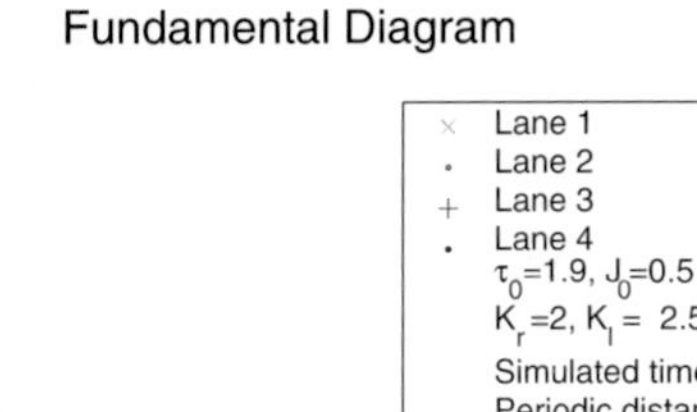
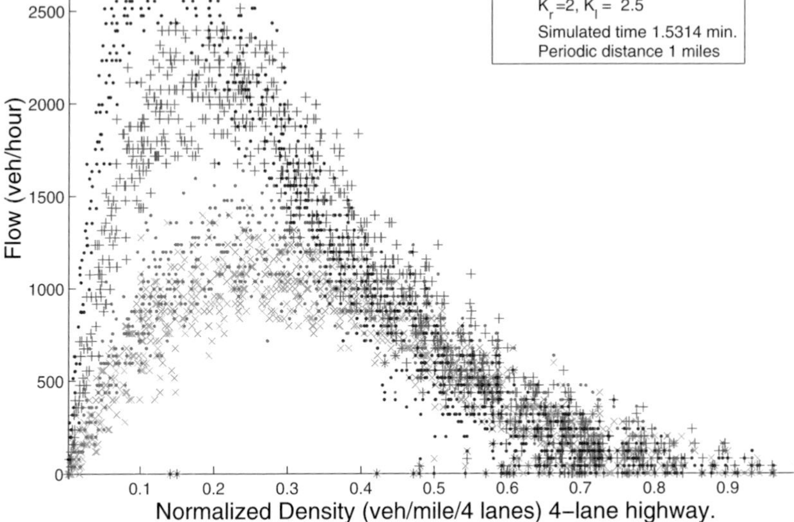

Figure 6(a). Flow versus concentration relationships for the individual lanes of a 4-lane highway. Spatial periodic length of 1 mile and 'look-ahead' of $L = 4$ cells. We pick the relaxation time τ and interaction strength of J_0 accordingly so that the desired vehicles speeds and upwind shock velocity are the same as before.

An incident test case

In this subsection we examine a case of an incident occurring on the highway in order to examine the effects of congestion. It is known that traffic congestion depends on road inhomogeneities as well as the history of previous perturbations due to hysteresis effects (Helbing *et al.* 2002). In this particular incident we will examine how congestion develops by simulating the distribution of traffic in the case of an accident for a three lane highway. A car in lane one crashing into a car in lane two was simulated by blocking one cell in each lane.

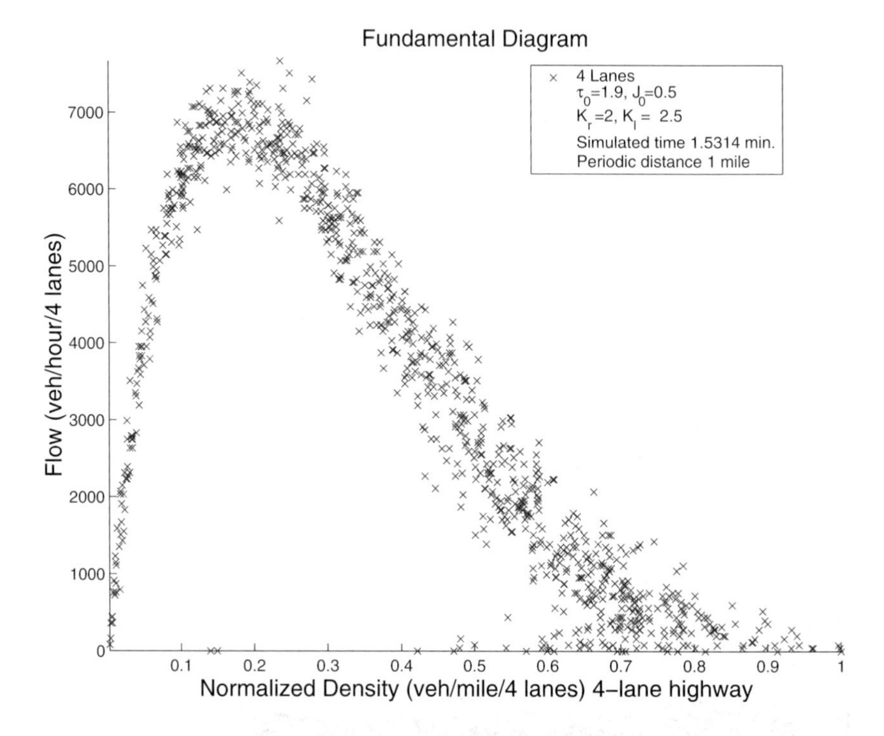

Figure 6(b). Flow versus concentration relationship for the whole traffic stream of the 4-lane highway in the situation represented in Figure 6(a).

It is very simple for the Monte Carlo simulation to impose this incident of an accident and closure of two lanes. To do this we first calculate the transition rate for each vehicle based on (11) and (4). Note that the rate for empty cells is zero by the construction of the interaction potential. For this purpose we select a specific cell in lane one and the adjacent cell in lane two and set their rates to zero for the duration of the simulation. Then we compute the total exchange rate for all vehicles and the simulation proceeds as outlined at the beginning of this section. Note that the vehicles which are initially placed into those two cells will be therefore automatically trapped and will never be able to move out since their rates are set to zero. Also, due to the built-in exclusion principle, no other vehicles will move into those two cells. We have therefore effectively simulated a car crash between these two vehicles.

As can be seen in Figures 7, 8 and 9, a number of shocks and rarefactions are generated as a result of this bottleneck on the highway. As expected the flow has dropped dramatically since vehicles try to change lanes and escape through the third lane. In general the overall traffic behavior of the simulated vehicles, once again, agrees with expectations.

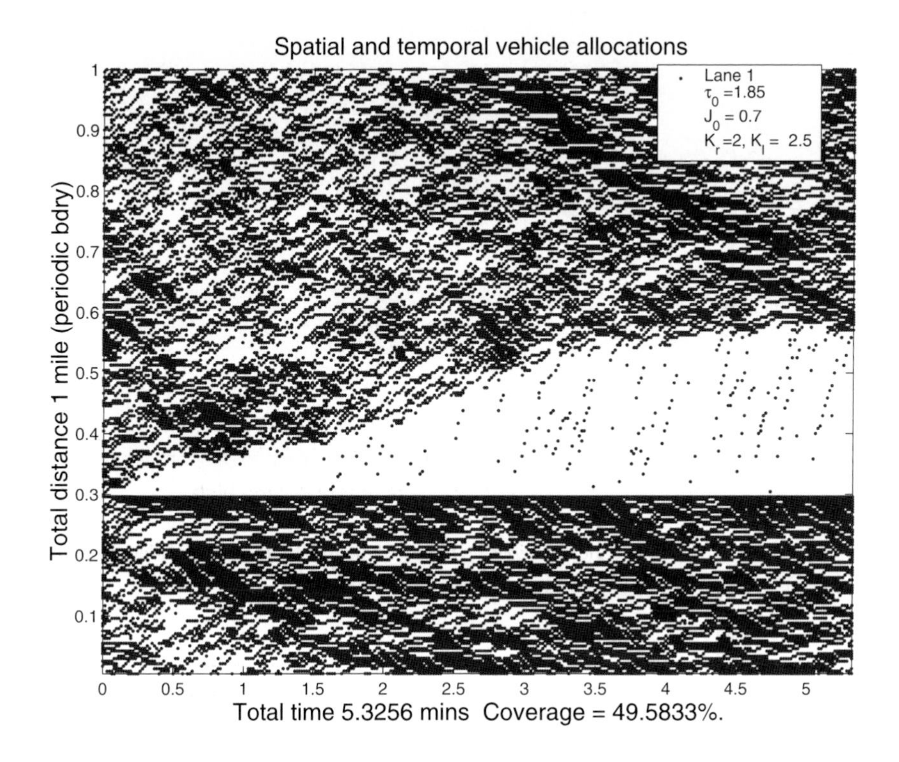

Figure 7. Lane 1 for the incident.
Note that the accident occurred at spatial location 0.3 and blocked lanes 1 and 2 of our 3 lane highway.
Parameters: $\tau = 1.85$, $J_0 = .7$ so that free flow speed is approximately 68 mph.
In this example vehicles were randomly initialized.

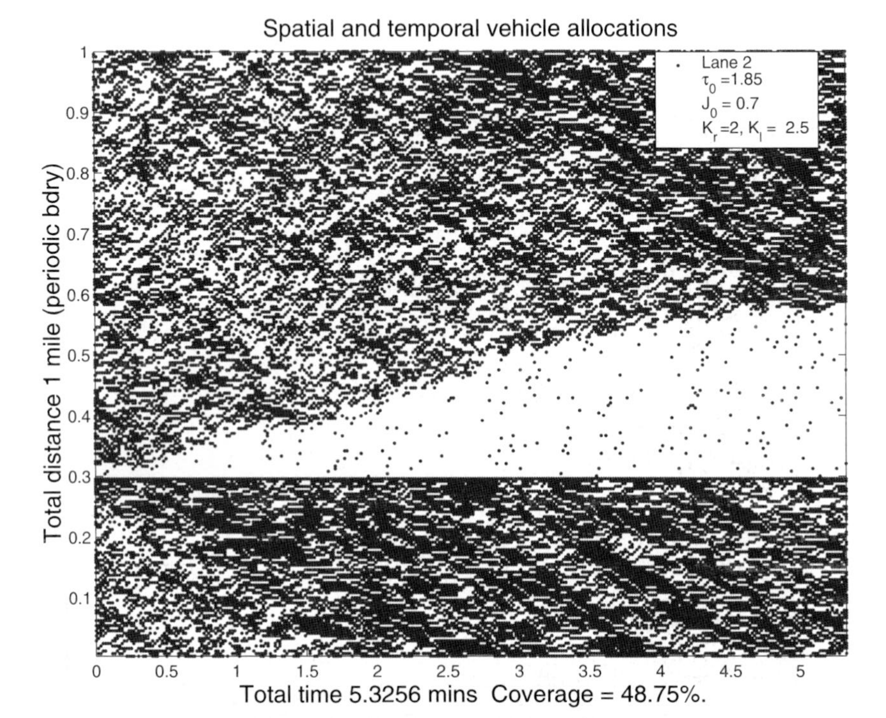

Figure 8. Lane 2 for the incident

Note that the accident occurred at spatial location 0.3 and blocked lanes 1 and 2 of our 3 lane highway.

Parameters: $\tau = 1.85$, $J_0 = .7$ so that free flow speed is approximately 68 mph.

In this example vehicles were randomly initialized.

CONCLUSIONS

In this work we developed a multi-lane stochastic traffic flow model which relies on microscopic conservative Arrhenius dynamics in order to realistically reproduce the behavior of actual traffic. Driver behavior and their reaction times toward their surrounding traffic are indirectly represented through the interaction potential. The model relies on only two calibrated traffic parameters J_0 and τ which directly link to known values from traffic observations. We further provide Monte Carlo simulations and produce solutions which compare favorably with similar real traffic data.

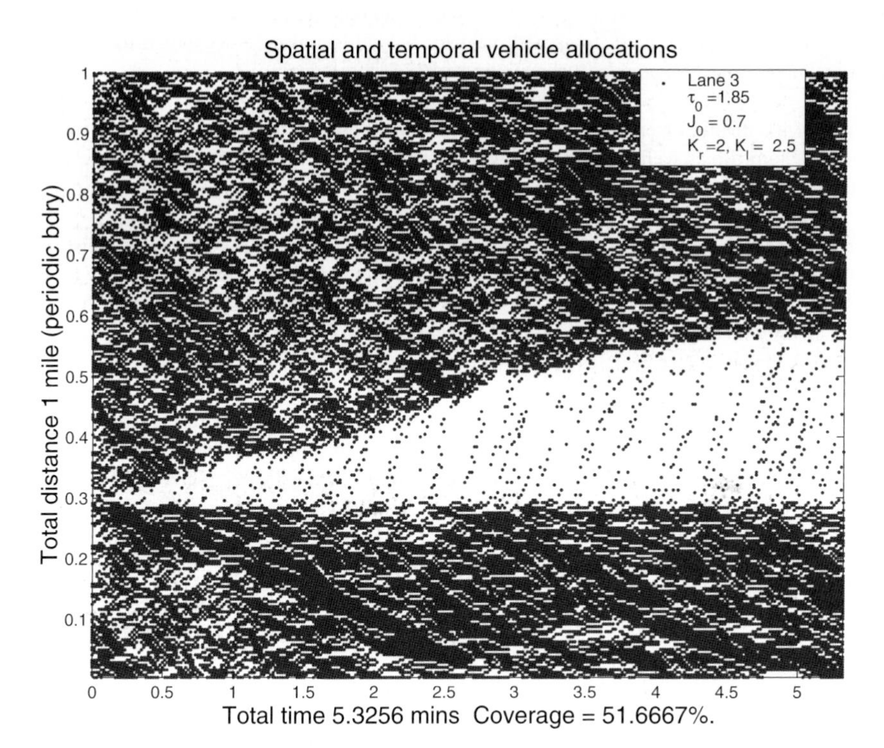

Figure 9. Lane 3 for the incident.
Note that although the accident occurred at spatial location 0.3 it has not blocked but nevertheless affected lane 3 of our 3 lane highway.
Parameters: $\tau = 1.85$, $J_0 = .7$ so that maximum free flow speed is set to be approximately 68 mph.
In this example vehicles were randomly initialized.

The multi-lane stochastic model presented here has been shown to have several important properties and advantages such as: a) lends itself to mathematical analysis in order to produce macroscopic traffic flow models (Sopasakis and Katsoulakis 2006) b) no *ad-hoc* noise is introduced c) timely braking d) retarded acceleration.

The stochastic model presented here can further be adapted in future work to account for even more realistic behavior. As an example we can calculate the interaction potential (11) through a bell-shaped (Gaussian) curve instead of the currently used uniform shape. In that respect the local interactions present in traffic flow would be stronger and therefore produce a much more realistic effect. At the same time interactions with vehicles behind can also be accounted for and in fact enhance the realism of the model especially in terms of lane changing.

As another interesting extension of the current model we propose to include entrances and exits. To do this, spin flip as well as spin exchange Arrhenius (adsorption/desorption) dynamics would be utilized to allow for different numbers of vehicles.

Furthermore another possible modification which can make this model even more realistic is to allow the look ahead parameter, L, to be a variable instead of a constant. It would be more realistic to have L vary with respect to the density of vehicles on the road. For example, if there is a low density of vehicles, then drivers can conceivably see further ahead than in a high density section of highway where vehicles are bumper-to-bumper.

Another consideration is how our model could handle non-homogeneous traffic. This distinction of different types of vehicles could potentially allow us to reproduce the empirical data from Wiedemann (1974) (which includes 10% trucks) even more realistically. There are two possible methods to incorporate this into our dynamics: we could count each truck as multiple passenger cars as is often done, or we could model the effect of trucks by giving 10% of vehicles a lower maximum velocity as is done in Nagel *et al.* (1998). Our model could be adapted for the latter approach through the parameter τ .

ACKNOWLEDGMENT

The research of A.S. is partially supported by NSF DMS 0606807.

REFERENCES

Athol, P. (1972). Interdependence of certain operational characteristics within a moving traffic stream. *Highway Research Record*, 58-97.

Aw, A., Rascle, M. (2000). Resurrection of "second order" models of traffic flow. *SIAM J. Appl. Math.*, **60**, 916.

Bando, M., Hasebe, K., Hakayama, A., Shibata, A., Sugiyama, Y. (1995). *Phys. Rev. E*, **51**, 1035.

Barlovic, R., Santen, L., Schadschneider, A., Schreckenberg, M. (1998). *Eur. J. Phys. B*, **5**, 793.

Bortz, A. B., Kalos, M.~H., Lebowitz, J. L. (1975). A new algorithm for Monte Carlo simulations of ising spin systems. *J. Comput. Phys.*, **17**, 10.

Daganzo, C. F., (1995). Requiem for second-order fluid approximations of traffic flow. *Transportation Research B*, **29**, 277.

Evans, M. R., Rajewsky, N., Speer, E. R. (1998). Exact solution of a cellular automaton for traffic. *arXiv:cond-mat/9810306*, **1**, 22.
Gray, L., Griffeath, D. (2001). The ergodic theory of traffic jams. *J. Statist. Phys.*, **105**, 413.
Greenshields, B. N. (1934). A study of traffic capacity. *Proceedings of the 14th Annual Meeting of the Highway Research Board*, 448-474.
Hall, F. L. (1996). *Traffic Flow Theory*. US Federal Highway Administration, Washington, D.C., Ch. 2, 2-34.
HCM, (1985). Highway capacity manual. *Transportation Research Board, special report 209.*
Helbing, D. (1995). Gas-kinetic derivation of Navier-Stokes-like traffic equations. *Phys. Rev. E*, **53**, 2366.
Helbing, D. (2001). Traffic and related self-driven many-particle systems. *Rev. Mod. Phys.*, **73**, 1067-1141.
Helbing, D., Hennecke, A., Shvetsov, V., Treiber, M. (2002). Micro and macro simulation of freeway traffic. Math. *Comp. Modelling*, **35**, 517.
Helbing, D., Treiber, M. (1998). Gas-kinetic-based traffic model explaining observed hysteretic phase transition. *Phys. Rev. Let.*, 3042-3045.
Illner, R., Klar, A., Materne, T. (2003). Vlasov-Fokker-Planck models for multilane traffic flow. *Commun. Math. Sci.*, **1**, 1.
Jiang, R., Wu, Q. S. (2003). Cellular automata models for synchronized traffic flow. *J. Phys. A*, **36**, 281.
Jin, S., Liu, J. G. (1994). Relaxation and diffusion enhanced dispersive waves. *Proc. Roy. Soc. London A*, **446**, 555-563.
Kerner, B. S., Klenov, S. L. (2002). A microscopic model for phase transitions in traffic flow. *J. Phys. A*, **35**, 31.
Kerner, B. S., Konhauser, P. (1994). Structure and parameters of clusters in traffic flow. *Phys. Rev. E*, **50**, 54-83.
Kerner, B. S., Rehborn, H. (1997). Experimental properties of phase transitions in traffic flow. *Phys. Rev. Let.*, **49**, 4030.
Klar, A., Wegener, R. (2000). Kinetic derivation of macroscopic anticipation models for vehicular traffic. *SIAM J. Appl. Math.*, **60**, 1749-1766.
Knospe, W., Santen, L., Schadschneider, A., Schreckenberg, M. (2000). *J. Phys. A: Math. Gen.*, **33**, L477.
Knospe, W., Santen, L., Schadschneider, A., Schreckenberg, M. (2002). *Phys. Rev. E*, **65**, 015101(R).
Komatsu, T., Sasa, S. (1995). Kink soliton characterizing traffic congestion. *Phys. Rev. E*, **52**, 5574-5582.
Krug, J., Ferrari, P. A. (1996). Phase transitions in driven diffusive systems with random rates. *J. Phys. A*, **29**, L465.

Kühne, R. D. (1984). Macroscopic freeway model for dense traffic – stop-start waves and incident detection. In: *Proceedings of the Ninth International Symposium on Transportation and Traffic Theory* J. Volmuller and R. Hamerslag, eds), 21-42. VNU Science Press, Utrecht.

Kurtze, D. A., Hong, D. S. (1995). Traffic jams, granular flow, and soliton selection. *Phys. Rev. E*, **52**, 218-221.

Landim, C., Kipnis, C. (1999). *Scaling limits of interacting particle systems*. Springer, Berlin.

Li, T., Liu, H. (2005). Stability of a traffic flow model with non-convex relaxation. *Commun. Math. Sci.*, **3**, 101-118.

Liggett, T. M. (1985). *Interacting Particle Systems*. Springer.

Lighthill, M. J., Whitham, G. B. (1955). On kinematic waves II - a theory of traffic flow on long crowded roads. *Proc. Roy. Soc. London A*, **229**, 317.

McShane, W. R., Roess, R. P. (1990). *Traffic Engineering*. Prentice-Hall, Englewood Cliff, New Jersey.

Muramatsu, M., Nagatani, T. (1999). *Phys. Rev. E*, **60**, 180.

Nagatani, T. (2002). The physics of traffic jams. *Rep. Prog. Phys.*, **65**, 1331.

Nagel, K., Paczuski, M. (1995). *Phys. Rev. E*, **51**, 2909.

Nagel, K., Schreckenberg, M. (1992). A cellular automaton model for freeway traffic. *J. Phys. I*, **2**, 2221.

Nagel, K., Wolf, D.~E., Wagner, P., Simon, P. (1998). Two-lane traffic rules for cellular automata: A systematic approach. *Phys. Rev. E*, **58**, 1425.

Nelson, P. (2000). *Phys. Rev. E*, **61**, 383.

Nelson, P. (2004). On two-regime flow, fundamental diagrams and kinematic-wave theory. (In Progress).

Newell, G. F. (1961). Nonlinear effects in theory of car following. *Operations Research*, **9**, 209-229.

Newell, G. F. (1989). Comments on traffic dynamics. *Transportation Research B*, **23**, 386.

Penrose, O. (1991). A mean-field equation of motion for the dynamic ising model. *J. Stat. Phys.*, **63**, 975-986.

Phillips, W. F. (1979). *Transp. Planning Technol.*, **5**, 131.

Rathi, A. K., Lieberman, E. B., Yedlin, M. (1987). Transp. Res. Rec. 61, 1112.

Ross, P. (1988). Traffic dynamics. *Transportation Research B*, **22**, 421.

Schadschneider, A. (2002). Traffic flow: a statistical physics point of view. *Physica A*, **312**, 153.

Schreckenberg, M., Wolf, D. E. (1998). *Traffic and Granular Flow*. Springer, Singapore.

Sopasakis, A. (2002). Unstable flow theory and modeling. *Math. Comp. Modelling*, **35**, 623.

Sopasakis, A. (2003). Formal asymptotic models of vehicular traffic. Model closures. *SIAM J. Appl. Math.*, **63**, 1561.

Sopasakis, A. (2004). Stochastic noise approach to traffic flow modeling. *Physica A*, **342**, 741-754.

Sopasakis, A., Katsoulakis, M. A. (2006). Stochastic modeling and simulation of traffic flow: Asep with arrhenius look-ahead dynamics. *SIAM J. Appl. Math.*, **66**, 921-944.

Sparmann, U. (1978). Spurwechselvorgänge auf zweispürigen BAB-Richtungsfahrbahnen. *Forschung Straßenbau und Straßenverkehrstechnik*. Bundesminister fur Verkehr, Bonn-Bad Godesberg.

Treiber, M., Hennecke, A., Helbing, D. (1999). Derivation, properties and simulation of a gas-kinetic based non-local traffic model. *Phys. Rev. E*, **59**, 239-253.

Wardrop, J. G. (1952). Some theoretical aspects of road traffic research. Proceeding of the Institution of Civil Engineers, Part II, **1**, 325.

Whitham, G. B. (1974). *Linear and Nonlinear Waves*. Wiley.

Wiedemann, R. (1974). Simulation des Straßenverkehrsflusses. *Schriftenreihe des Instituts für Verkehrswesen der Universität Karlsruhe*, **8**.

Transportation and Traffic Theory 2007
Edited by R.E. Allsop, M.G.H. Bell and B.G. Heydecker

29

FREEWAY TRAFFIC OSCILLATIONS AND VEHICLE LANE-CHANGE MANEUVERS

Soyoung Ahn, Arizona State University[1]*, USA*
Michael J. Cassidy, University of California, Berkeley, USA

SUMMARY

This work unveils the influence of vehicular lane-change maneuvers on oscillations in real freeway traffic. Measurements made upstream of bottlenecks reveal that oscillations formed in individual lanes when drivers squeezed their way in from neighboring lanes. Once oscillations had formed, moreover, lane changing caused the oscillations to at times grow in amplitude as they propagated upstream through queues.

The findings show that on (multi-lane) freeways where lane changing abounds, these maneuvers seemingly exert greater influence on the formation and growth of oscillations than do driver interactions that spontaneously arise in single-lane traffic. This is notable in light of the many attempts to explain oscillations as strictly a car-following phenomenon; and the findings motivate the need for theories of multi-lane traffic that describe lane changing in conjunction with car following.

---

[1] Work by the first author was performed at the University of California, Berkeley and at Portland State University.

INTRODUCTION

The present work solves some long-standing puzzles on the nature of oscillatory or "stop-and-go" driving conditions in real freeway traffic. Oscillations were observed to form in freeway queues due to vehicular lane-change maneuvers. Lane changes made into small vehicle spacings were especially prone to be the triggering events. Most formations occurred short distances in advance of bottlenecks. Once oscillations had formed, moreover, lane changing similarly caused the oscillations to grow in amplitude as they propagated upstream through queued traffic.

No evidence was found that oscillations formed or grew due to driver interactions that arose spontaneously in single-lane traffic, independent of vehicles in adjacent traffic streams.[2] The finding is incompatible with previous attempts to explain oscillations as strictly a car-following phenomenon. Theories formulated and used in some of these past attempts are reviewed in the following section of the paper. Observations from some additional studies are used here as well to tease-out clues that support our present findings.

Data for the present work were collected from two extended portions of queued freeway. Measurements came both from inductive loop detectors and from video images, as described in the third section.

Macro-level analyses of the loop data are provided in the fourth section. These analyses not only confirm some previously observed features of oscillatory traffic, they further imply that oscillations formed and grew due to events in individual lanes.

The nature of these events is unveiled in the fifth section by means of more detailed, micro-level analyses of the data taken from videos. We present the systematic method used to mine these data so as to pinpoint when and where oscillations formed or grew. We then furnish vehicle trajectories (measured from videos) to show that lane-change maneuvers were the triggering events.

Final remarks are offered in the sixth section. These include discussion on certain details of oscillatory traffic in need of further study. Implications of the present findings on traffic theory are discussed as well.

---

[2] Much like lane changing, vehicle merging and diverging maneuvers near ramps were observed to affect oscillation growth. Description of these merging and diverging effects is saved for a future paper.

BACKGROUND

Traffic theorists have long sought to describe oscillations using models of how a driver responds to the motion of the vehicle immediately in front. Some of the earliest of these car-following models have each driver responding to her spacing (Kometani and Sasaki, 1958) or to changes in her leader's speed (Chandler et al., 1958). Responses occur following a reaction time and the magnitude of a response depends upon the driver's "sensitivity," a parameter calibrated to data. These early models have undergone various modifications: New parameters have been introduced and model forms have been altered in attempts to match model predictions with real observations (e.g. Gazis et al., 1959; 1961; Edie, 1960).

The above-cited models exhibit instabilities: For certain values of the sensitivity parameter and the reaction time, the magnitude of driver responses successively amplify as each driver passes through a disturbance (Herman and Montroll, 1959). Other classes of car-following models display instabilities as well. Models that assume drivers continuously choose their speeds so as to eliminate the possibility of collision (e.g. Kometani and Sasaki, 1959) generate instabilities when drivers over-estimate the decelerations of their leaders (Gipps, 1981). Instabilities also arise in yet another model class whereby each driver presumably seeks to maintain both a speed equal to that of her leader and her desired spacing for that speed (Michaels, 1963). Within this latter theoretical framework, a driver's inability to promptly perceive speeds and spacings can cause her to enter into a perpetual cycle of accelerating and decelerating without ever reaching a steady state (Wiedeman, 1974).

These instabilities are commonly taken as descriptions of oscillatory traffic. However, model predictions of the former do not always match real measurements of the latter. For example, previous observations of real freeway traffic indicate that oscillations exhibit acceleration and deceleration periods that are several minutes in duration (Kerner and Rehborn, 1996; Mauch and Cassidy, 2002), and this is confirmed in the present work. Car-following models, on the other hand, reportedly produce instabilities with periods on the order of a driver reaction time (only several seconds; see again Herman and Montroll, 1959).

The earlier freeway studies just cited further report that oscillation amplitudes grew in the vicinity of busy ramps. This finding suggests that vehicle merging and diverging maneuvers play a role here (and we have unveiled additional details on this matter; see footnote 2).

Further clues concerning the nature of oscillations are evident in data presented in Treiterer and Myers (1974). In this latter work, the motions of platooned vehicles in a single freeway lane were traced from aerial photographs. Drivers reportedly exhibited little change in their spacings (densities within the platoons remained high) as they underwent accelerations. These measurements have been used to support various car-following models that assume drivers behave differently while accelerating than while decelerating (e.g. Aron, 1988; Ozaki, 1993). As it turns out, however, lane changing may have been the greatest influence here; Daganzo (2002) offers the following alternative interpretation of the Treiterer and Myers data.

During acceleration cycles, densities in the single-lane platoons stayed high because (i) drivers from the neighboring lane inserted themselves into the platoons (an observable detail in the data); and (ii) drivers in the platoons may have chosen to follow vehicles at tight spacings in attempts to ward-off these insertions. It further seems that a large disturbance in one of the platoons – formerly regarded as a puzzle – can be traced back to vehicle lane changing into and out of the platoon; this becomes evident by scrutinizing Figure 1 of Treiterer and Myers.

It is true, on the other hand, that oscillations have been observed on single-lane roads and test tracks (Smilowitz et al., 1999; Sugiyama et al., 2003) and in tunnels where lane changing was prohibited (Edie and Baverez, 1958). On these and perhaps other facilities, oscillations might be explained by the driver interactions described by car-following models (or something at least akin to these descriptions). What we provide in the present paper, however, is evidence of an important role played by lane-changing maneuvers on (multi-lane) freeways where these maneuvers abound. To our knowledge, it is the most compelling evidence of its kind offered to date.

DATA

Traffic data were collected in both travel directions of the freeway site shown in Fig. 1, a 6-km-long stretch of Interstate 80 in California's San Francisco Bay Area. During afternoon rush periods, vehicles in both directions encountered downstream bottlenecks, as labeled in the figure. The resulting queues filled the regular-use freeway lanes for much of the rush.[3] Freeway flows within these queues varied with location (from about 7,000 to 8,000 vph) due to inflows and outflows from the ramps.

Measurements came from two sources, the first being inductive loop detectors. These are located in every travel lane at (slightly irregular) intervals of about 0.5 km. Detector stations are numbered 1 – 8 in the figure. Vehicle counts, occupancies and time-mean speeds were collected over 30-sec sampling intervals and were used for the macro-level assessments presented in the following section.[4]

---

[3] The high-occupancy vehicle lanes (labeled with diamond-shaped icons in Fig. 1) remained freely flowing during much of the rush and were therefore excluded from analyses.

[4] Due to detector malfunctions, these data were not available for eastbound and westbound traffic in the shoulder lane at station 3 and for eastbound traffic in one of the center lanes at station 1.

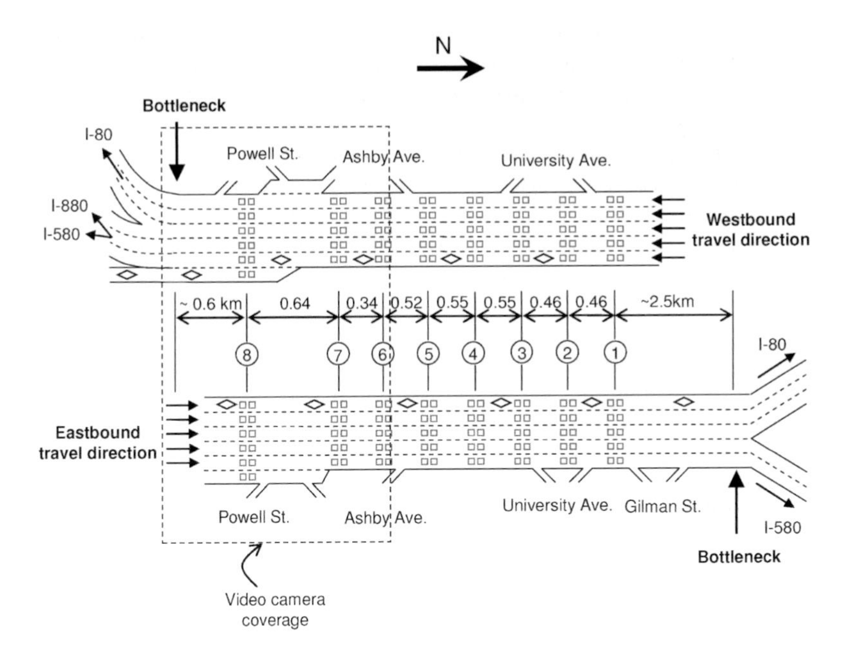

Figure 1: Interstate 80 near San Francisco, California

Additionally, a video surveillance system provided unobstructed views of traffic over the western-most freeway portions, as demarcated with dashed lines in Fig. 1. Traffic data extracted from these videos were used for the micro-level assessments in a later section.

MACRO-LEVEL ANALYSES

Next presented are features of oscillatory traffic observed in the loop detector data. We found that oscillations tended to form just upstream of bottlenecks. As in earlier studies, the oscillations exhibited periods of rather extended duration; they propagated upstream through queued traffic at a (nearly) constant wave speed; and they often grew in amplitude while doing so. Beyond confirming the above, we found that oscillations displayed certain patterns that were distinct across lanes.

These distinctions are clues to the lane-specific events (lane changing) that triggered formations and growths. The following macro-level analyses confirm the general features of oscillatory traffic described above and unveil the clues concerning the triggering mechanism.

Fig. 2(a) presents time series curves of vehicle speeds in the westbound travel direction. These were measured in each lane and at each detector station during a 50-minute period (on Aug. 19, 2002) when queues had filled the regular-use lanes. The freeway geometry for westbound travel is schematically shown left of the figure as a convenience for the reader. The numbering scheme used for the detectors is re-presented there as well.

The speed data in Fig. 2(a) were filtered to eliminate noise caused by driver differences and to retain longer-run trends. This was achieved in a simple way by plotting the vertical deviations between cumulative values of (time-mean) vehicle speeds at time, t, s(t), and the quantity $\bar{s}(t) \times (t - t_0)$ at all t, where $\bar{s}(t)$ is a longer-run average of the 30-sec time-mean speeds and t_0 is the start time of this 50-minute observation period.[5]

The resulting wiggles in each "speed deviation" curve are oscillations; they mark periods when vehicle speeds were higher (positive slopes) and lower (negative slopes) than the longer-run average. These wiggles confirm that oscillatory periods can persist for several minutes, and not just for short durations comparable to a driver reaction time.

Dotted arrows in the figure trace some kinematic waves. These appear straight and parallel, consistent with past reports that wave speed is independent of flow in queued traffic (e.g., Windover and Cassidy, 2001; Mauch and Cassidy, 2002).

Marked differences in oscillation amplitudes are evident when comparing wiggles at downstream-most detector 8 with those, for example, at detector 1. This indicates that oscillations generally grew as they propagated upstream. The trend is confirmed in Fig. 2(b). It shows for each detector station the Root Mean Squared Error (RMSE) of speed deviations measured for the 50-minute period.[6] Two curves display RMSEs in each of the center lanes; a third curve, shown in bold, is the average of all four regular-use lanes.

Inspection of the latter (bold) curve shows that the upward trend in (average) growth was interrupted only at detector 3. We attribute this interruption to high vehicular merging activity at

---

[5] The $\bar{s}(t)$ was computed as a moving average over a 5-minute period spanning each t; (t – 2.5 mins, t+2.5 mins). By using a moving average, we present shorter-run deviations from averages that gradually, but systematically, changed over time.

[6] Each RMSE was computed using 30-second samples as $\left[\frac{1}{T}\sum_{t=0}^{T}\left(s(t)-\bar{s}(t)\right)^2\right]^{1/2}$, where here T = 100 since the 50-min period shown in Fig. 2(a) is composed of 100 thirty-second intervals and $s(t)$ and $\bar{s}(t)$ are as previously defined.

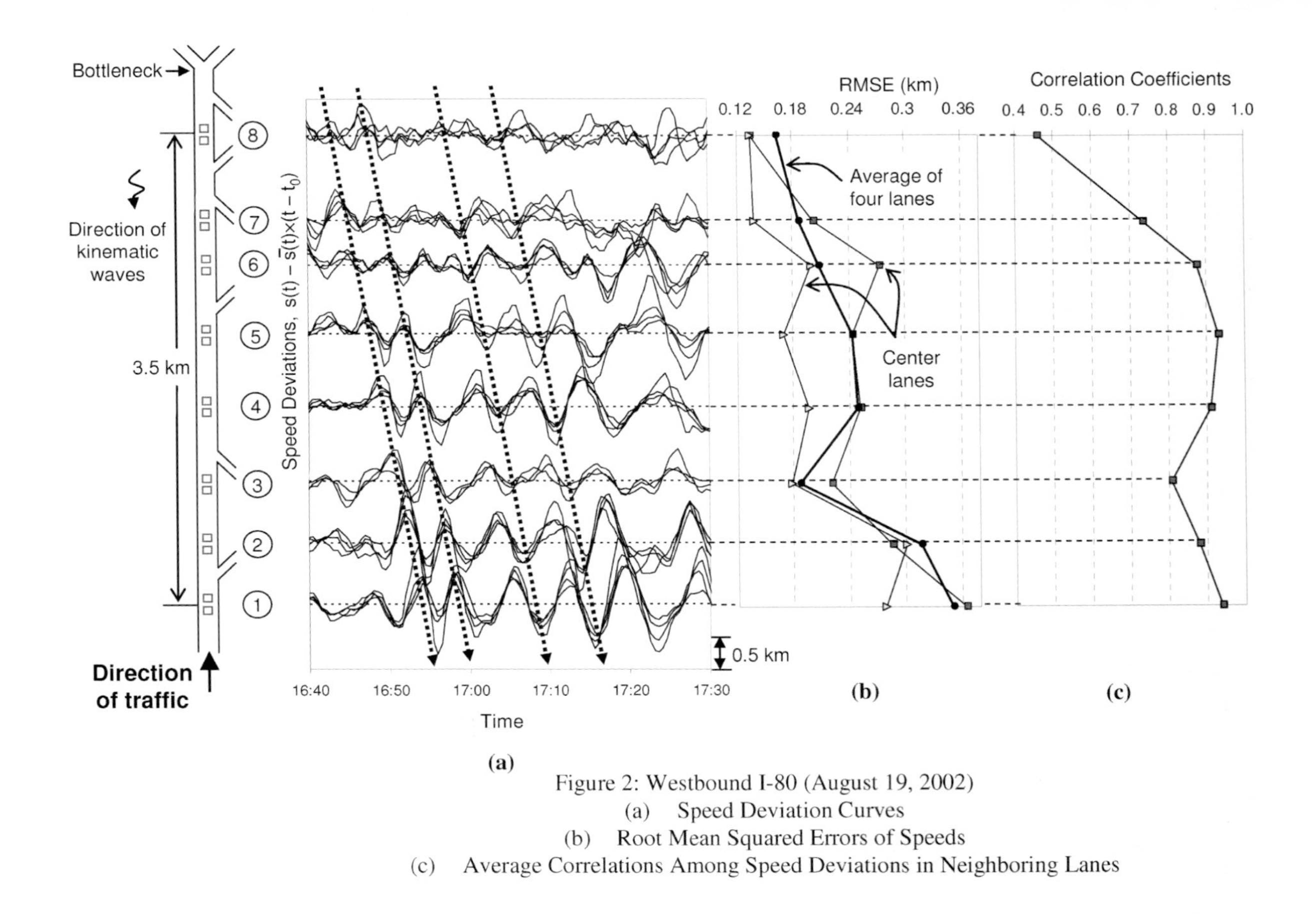

Figure 2: Westbound I-80 (August 19, 2002)
(a) Speed Deviation Curves
(b) Root Mean Squared Errors of Speeds
(c) Average Correlations Among Speed Deviations in Neighboring Lanes

the nearby on-ramp (see footnote 2). All three curves in Fig. 2(b) show relatively small RMSEs at downstream detector 8, suggesting that this location is about where most oscillations formed.

The reader will further note the distinctions in each of these RMSE curves. These distinctions are clues that oscillation growth was triggered by events in individual lanes.

Clues that lane-specific events also triggered oscillation formations are evident in the data as well. Referring again to Fig. 2(a), we see that speed deviation curves at downstream detector 8 reveal asynchronous oscillatory patterns across lanes; i.e., the emerging wiggles at this downstream location are not aligned across curves. Oscillations became more synchronized only after propagating to upstream detectors.

This synchronization pattern is also visible in Fig. 2(c). The figure displays correlation coefficients of speed deviations for all pairs of neighboring (regular-use) lanes; the values shown are averages of the pair-wise correlations over the 50-minute observation period. The relatively low correlation at detector 8 (where oscillations formed) implies that oscillations separately emerged in each lane, such that emergence was a result of the conditions in individual lanes. (A slight reduction in correlation near station 3 can again be attributed to large inflows from the on-ramp.)

Oscillations in the opposing (eastbound) travel direction display features that are qualitatively like those just described. Visual inspection of Figs. 3(a) – (c) attests to these similarities; each of these figures was constructed from detector data taken over a 1-hour period of queued traffic (on June 25, 2003).

These figures indicate that wiggles in the eastbound travel direction were more developed and more synchronized across lanes, even at downstream detectors (stations 1 and 2 for eastbound travel), than were their counterparts in westbound traffic. This difference was to be expected. The bottleneck for eastbound traffic resides relatively far downstream of the detectors (see Fig. 1). Thus upon their arrivals to these detectors, the oscillations in eastbound traffic had already become more fully formed and better synchronized.

The distance between detectors and bottleneck notwithstanding, the RMSEs (Fig. 3(b)) increase just upstream of detector 1 and display differences across lanes. And although the (average) pair-wise correlations (Fig. 3(c)) were already high at downstream detectors 1 and 2, these correlations increase at upstream locations. The features suggest that, once again, formations and growths were triggered by lane-specific events. The nature of these events is presented next.

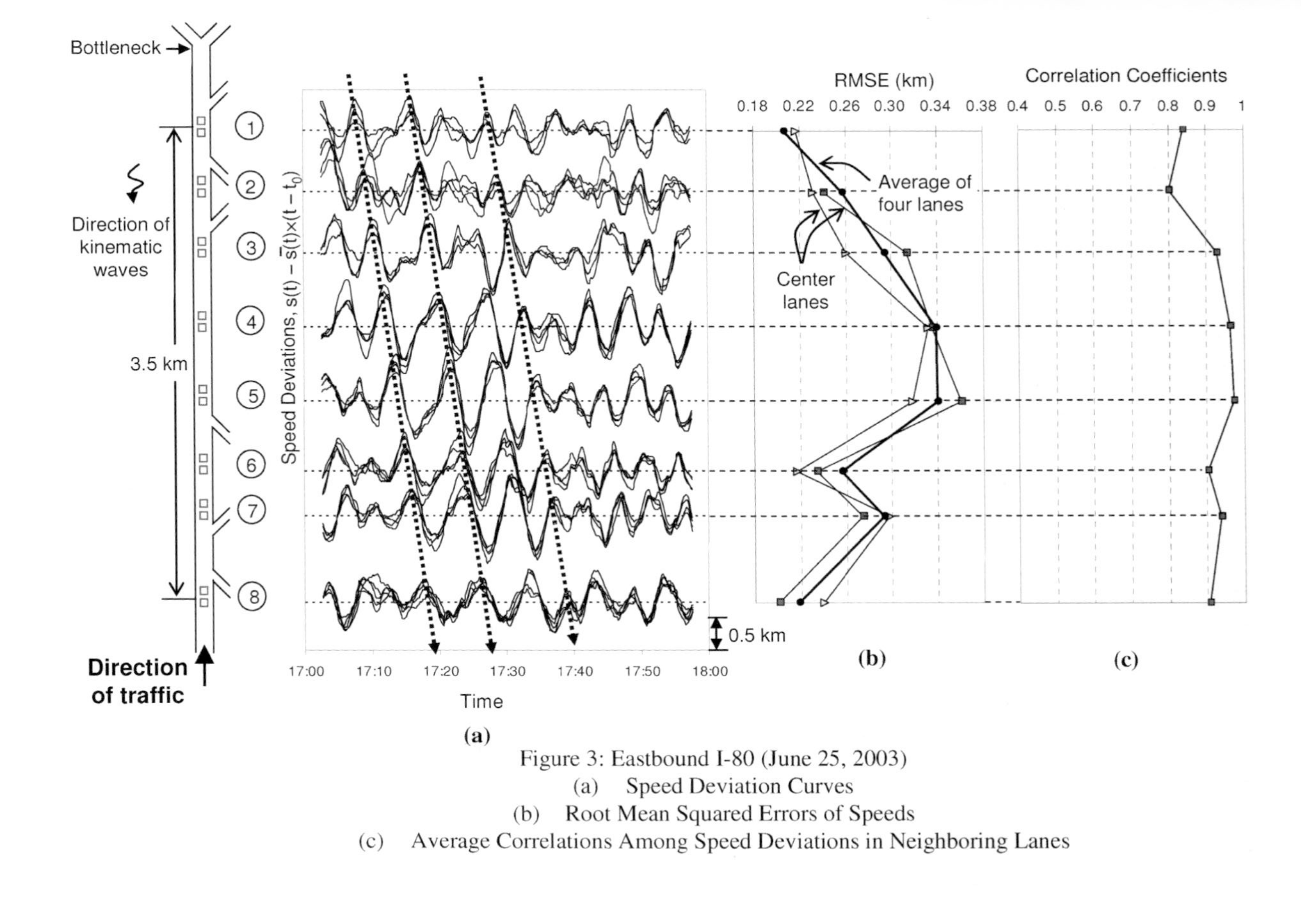

Figure 3: Eastbound I-80 (June 25, 2003)
(a) Speed Deviation Curves
(b) Root Mean Squared Errors of Speeds
(c) Average Correlations Among Speed Deviations in Neighboring Lanes

MICRO-LEVEL ANALYSES

An oscillation's formation is revealed when a vehicle's speed begins to vary while its leader's speed does not. Oscillation growth is revealed when a kinematic wave carries speed variations that increase at upstream locations.

To pinpoint when and where the above indicators of interest occurred, vehicle speeds were individually measured (from video) over short, contiguous freeway segments and were then compared across segments in ways that would detect all but perhaps the most subtle systematic variations. Lastly, vehicle trajectories were constructed for each time-space region that contained the indicators of formation or growth. These trajectories showed that lane changes were always the triggering events. Illustrations are provided below.

Formation

Fig. 4(a) illustrates a portion of westbound freeway near the downstream bottleneck; this is a location where oscillations often formed. Vehicles involved in formations were identified by measuring their speeds (trip times divided by distance) over contiguous 100-meter-long segments. Two such segments (labeled "upstream" and "downstream") are shown in the figure.

As an illustration, Fig. 4(b) displays speeds for 46 vehicles that were separately measured on the upstream segment (shown with circles) and on the downstream one (squares). These were measured in lane 2 (see Fig. 4(a)) and the vehicles represented in Fig. 4(b) are numbered 0 – 45 in the order of their entries into the upstream segment. (Only vehicles that traversed the upstream and downstream segments without changing lanes are represented in this figure so as to simplify the numbering scheme.) The dark line in the figure displays moving averages of speeds on the upstream segment; the lighter line shows moving averages on the downstream segment; and the moving average for each vehicle n was computed from the speeds in the vehicle set numbered (n–2, n+2).

The figure reveals the formation of an oscillation. Speeds on the upstream segment fell and then rose back to their initial values, as is characteristic of an oscillation. Visual inspection of the figure shows that this cycle began with a marked reduction in the speed of vehicle 11. Notably, no such cycle is evident on the downstream segment, indicating that the oscillation emerged on the upstream one.

Vehicle trajectories not only confirm this formation, but unveil its cause. Fig. 4(c) displays the trajectories for the vehicles numbered 2 – 26 and for an additional vehicle that triggered the

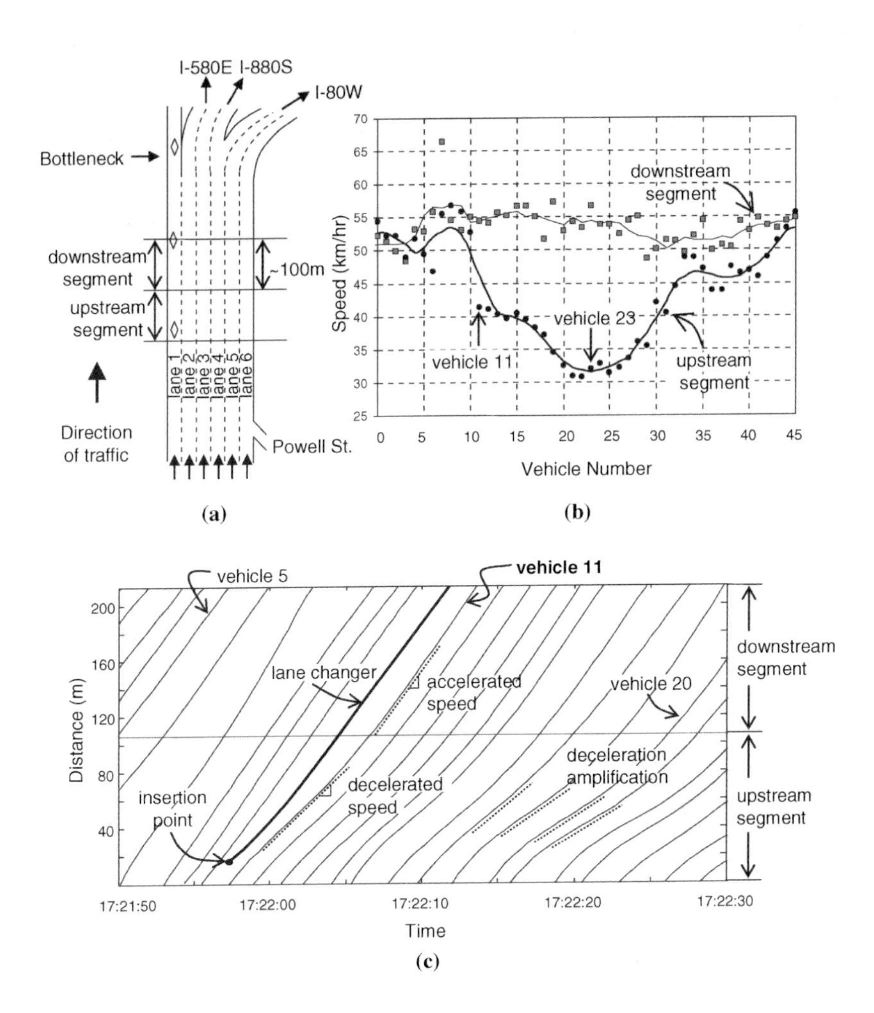

Figure 4: Formation of an Oscillation
(a) Westbound I-80
(b) Vehicle Speeds
(c) Vehicle Trajectories

formation by changing lanes.[7] The latter, drawn in bold, inserted itself directly in front of vehicle 11. From the trajectory of vehicle 11, we see that its driver decelerated soon after the insertion and then accelerated once she had recovered (approx) her earlier spacing.

One further sees in Fig. 4(c) that decelerations were amplified among higher numbered vehicles. These amplifications indicate that driver car-following behavior can contribute to oscillations. The point we make is that *the formation was triggered by a lane change* (the insertion in front of vehicle 11) and did not arise spontaneously.

The amplifications noted above caused speeds on the upstream segment to diminish from one vehicle to the next. This state gradually propagated backward and was eventually no longer felt on the upstream segment. The trajectories also show that the oscillation imparted little or no speed variations to vehicles on the downstream segment.

The reader will note that these effects so clearly evident in the trajectories are conveyed in Fig. 4(b) as well. Moving average speeds, like those in the latter-cited figure, were therefore used to search the data for instances of formations. In all, more than 1470 vehicle speeds were measured. These measurements were made in lanes 2 and 3 (see again Fig. 4(a)) and were taken over three contiguous 100-m-long segments. We judged that a formation occurred when the following two criteria were satisfied.

(i). The greatest difference in the moving average vehicle speeds on some segment (measured from the zenith to the nadir of a cycle like the one shown with the dark line in Fig. 4(b)) had to exceed 7 km/hr. Cycles marked by smaller differences showed no signs of propagating to upstream segments, leading us to conclude that these were merely statistical fluctuations.

A second criterion was established to ensure that instances of (i) actually emanated within the segment from which the measurements came, and had not instead formed downstream and propagated back to the subject section.

(ii). Where (i) was satisfied and the greatest speed variation was displayed by the moving average for vehicle n (e.g. n = 23 in Fig. 4(b)), we verified that (i) was not measured on the downstream segment among vehicles in the set numbered (n–20, n).

Limiting our check for (ii) to 20 vehicles seemed appropriate. Given the kinematic wave velocity estimated from our data, one would expect that a wave would, on average, propagate

[7] Each trajectory was constructed by measuring the vehicle's arrival times at a series of fixed reference points spaced at 15 m increments.

through less than 15 vehicles per 100 m (see Newell, 1993). The data further indicated that extending the check for (ii) beyond 20 vehicles ran a risk of inadvertently measuring the effects of a kinematic wave that carried some other oscillation.

Ten instances of formation were detected in the above fashion.[8] The trajectories then constructed for each time-space region containing a formation showed that lane changes were *always* the triggering events. There were no exceptions. (Trajectory plots for many of these formations are provided in Ahn, 2005.)

Finally, the lane changes that triggered formations tended to be those made into vehicle spacings that were small. As evidence, Fig. 5 shows the distributions of spacings filled by lane-change vehicles during a 10-minute period when video images were surveyed for lanes 2 and 3 of the freeway portion previously shown in Fig. 4(a). The darkened bars in Fig. 5 correspond to spacings that, when filled by a lane-change vehicle, triggered formations (9 observations in this period). The unshaded bars display all (other) spacings that were filled during the period without triggering formations or growths (18 observations). The median of the former is 27 m, while the median of the latter is 40 m. (The means are 32 m and 43 m.) The difference in these medians is

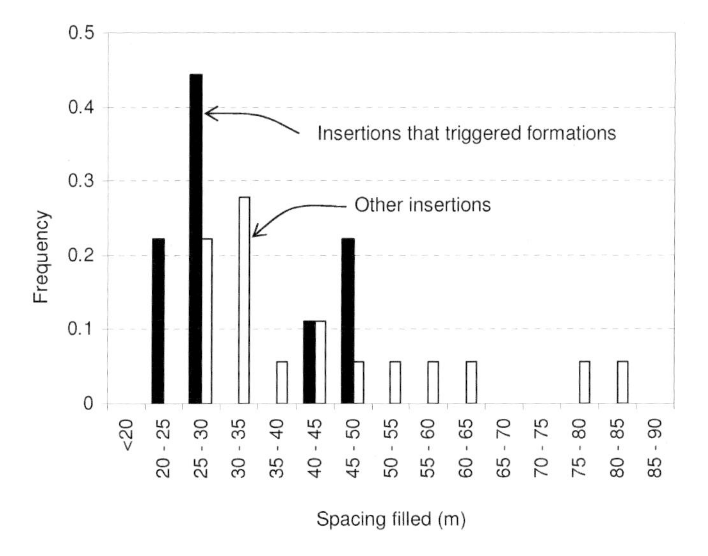

Figure 5: Distributions of Spacings Filled (Westbound I-80)

[8] Our search was limited by the time and cost of extracting data from videos.

statistically significant, as determined by the Wilcoxon two sample test at the 95% confidence level; see Rice, 1994[9]. This difference indicates that oscillations tended to emerge when drivers "squeezed their way" into neighboring lanes.

Growth

Details of oscillation growth were unveiled by studying eastbound traffic on the freeway portion shown in Fig. 6(a). Queued traffic at this location was commonly marked by well-formed oscillations that sometimes grew in amplitude as they propagated upstream. Vehicle speeds were once again measured over short, contiguous segments, including the two labeled "upstream" and "downstream" in the figure.

Fig. 6(b) displays the speeds of 51 vehicles in lane 2 on these two segments. The solid lines are, once again, moving averages taken over 5 vehicles.

Speeds on the downstream segment (squares) chart the fall-and-rise cycle that characterizes an oscillation. The speeds on the upstream segment (circles) do so as well. The reader can verify how speed changes downstream tend to be passed upstream to vehicles of higher arrival number, indicating that the oscillation propagated backward through traffic.

Notably, speeds on the upstream segment drop to lower values than do their downstream counterparts. This pattern indicates that the oscillation's amplitude increased (i.e., the oscillation "grew") as it propagated from one segment to the next. Further visual inspection of Fig. 6(b) shows that vehicle 29 was the first to display a speed on the upstream segment that was lower than any observed on the downstream segment.

The trajectories in Fig. 6(c) unveil the cause of this growth.[10] Two consecutive vehicles, shown with bold trajectories and labeled A and B, were inserted in front of vehicle 29. The (lightly drawn) trajectories of lower arrival number confirm that vehicles were already undergoing oscillatory motions prior to these insertions; e.g. the trajectory of vehicle 28 clearly displays a deceleration-acceleration pattern, though it was not affected by the insertions of A and B. What these insertions did was to amplify temporarily the (pre-exiting) decelerated state; the driver of vehicle 29 temporarily adopted a lower speed in response to the insertions and the resulting state propagated upstream through vehicles of higher arrival number.

[9] The Wilcoxon test was used in light of the small sample sizes.
[10] These trajectories were constructed from video images, as described in Hranac et al. (2004).

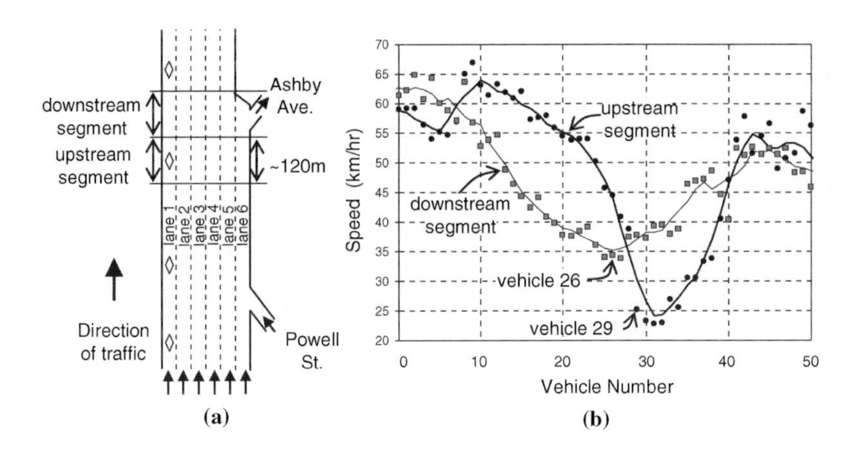

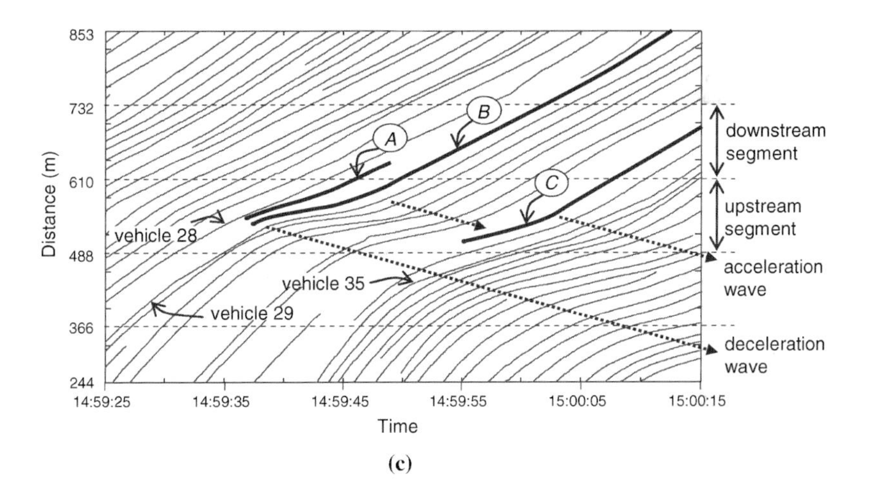

Figure 6: Growth of an Oscillation
(a) Eastbound I-80
(b) Vehicle Speeds
(c) Vehicle Trajectories

The eventual insertion of a third vehicle (labeled C in the figure) took place within a relatively large spacing and did not induce the drivers of upstream vehicles, such as vehicle 35, to adopt an even lower speed. The insertion did, however, prolong the period over which vehicles traveled at a lower speed; i.e., it displaced the acceleration wave in time, as shown with dotted lines in Fig. 6(c). It seems that lane changing may explain the relatively long oscillatory periods observed in real traffic, though the evidence of this is limited at present.

Our search for instances of growth consisted of measuring more than 1500 speeds. Measurements were taken from four contiguous segments and in four lanes (lanes 2 through 5). Oscillation growth was judged to have occurred when the moving average speeds satisfied the following.

(i). The greatest differences in the vehicle speeds on each segment (again measured from the zenith to the nadir of a cycle) exceeded 7 km/hr.

(ii). To ensure that instances of (i) grew systematically in amplitude as the oscillation propagated, we compared the lowest speed measured on a given segment with the lowest speed on the upstream neighboring one. Where the former was displayed by the moving average of vehicle n (n = 26 in Fig. 6(b)), we surveyed speeds on the upstream segment for vehicles in the set numbered (n, n+20). The difference between the two minimum speeds across the two segments had to be at least 1 km/hr.

With this threshold of 1 km/hr adopted in (ii), the lowest speed in an oscillatory cycle would diminish by less than 8.5 km/hr with every kilometer traveled by the kinematic wave. This constitutes very subtle growth; e.g. when emerging in lightly queued traffic where vehicle speeds approach free flow rates, an oscillation growing in this fashion would propagate at least 10 kilometers before devolving to a jammed state. Hence our threshold would fail to detect only the smallest and most subtle oscillatory growth.

Eleven separate instances of growth were thus detected. Trajectory plots showed that *every* instance was triggered by lane changing; Ahn (2005) presents plots for most of these instances.

The data indicate that growth was more likely to be spurred by lane changes made into spacings that were small. Fig. 7 presents distributions of spacings filled by lane-changing vehicles during a 10-minute period. These were measured in lane 2 of the freeway portion previously shown in Fig. 6(a). The darkened bars in Fig. 7 correspond to spacings that, when filled by a lane changer, triggered growth (8 observations). The unshaded bars are all the other spacings filled by lane changes in the period (41 observations). The median of the former is 30 m, while the median of the latter is 45 m. (The means are 34 m and 54 m.) As in the case of formation, these two medians were statistically different at the 95% confidence level (see footnote 9).

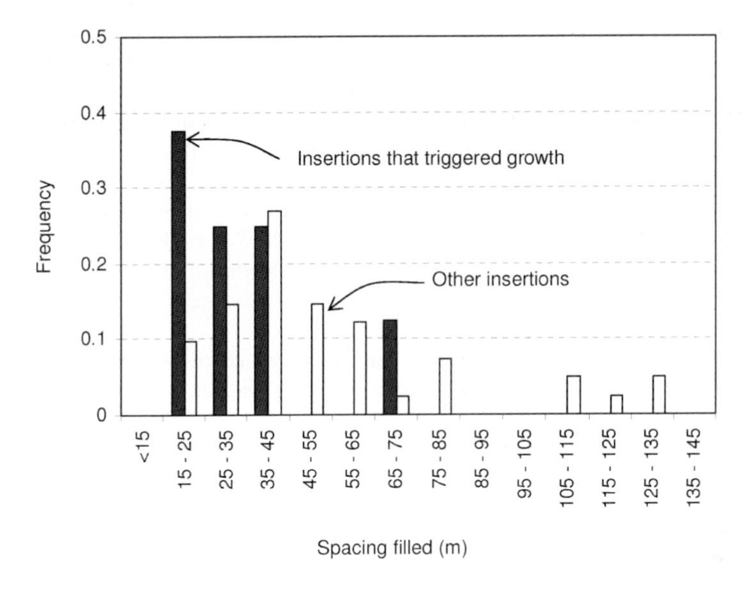

Figure 7: Distributions of Spacings Filled (Eastbound I-80)

CONCLUSIONS

The findings unveil a causal relation between vehicle lane-changing maneuvers and oscillatory traffic. Rational criteria were established to detect oscillation formations and even subtle instances of growth; and lane changing triggered every formation and growth thus observed (10 and 11 observations, respectively). Data illustrate how car following behavior also contributes to oscillation formation and growth. The point is that lane changing maneuvers always initiated these phenomena. We observed no instances of spontaneous formation or growth.

In retrospect, the findings may seem unsurprising; i.e., that stop-and-go conditions can be triggered by exogenous events like lane changing may even be what much of the driving public intuitively suspects. We argue, however, that the findings are notable in light of the large (scientific) literature that ignores any role of lane changing and instead views oscillations solely in terms of car-following. By having shown the need for theory that considers lane changing in conjunction with car following, we hope the present work will re-direct this line of thinking

reflected in the literature. We believe that this, in itself, would constitute a meaningful advancement in traffic flow theory.
Already a few theories of driver lane-changing behavior have been developed for multi-lane traffic (e.g. Gipps, 1986; Kerner, 2005). One such theory with notably few parameters (Laval and Daganzo, 2006) shows that lane changes in dense traffic can create voids between vehicles and that these voids can propagate forward through bottlenecks. The theory can thus explain the reductions in discharge flows that are commonly measured when certain types of freeway bottlenecks become active. In fact, the theory has been shown to match discharge rates and other traffic details at an active merge bottleneck (Laval *et al.*, 2005). Faithful descriptions of the oscillatory conditions that arise upstream of bottlenecks might come from a model of this kind, once it has been suitably refined for this purpose.

Identifying the needed model refinements might, however, require further observation and experiment. After all, certain details of oscillatory traffic remain puzzling. We have, for example, only very limited evidence that lane changing causes the relatively long oscillatory periods observed in this and other studies. Further, we do not yet know the mechanism by which propagating oscillations gradually synchronize across lanes, though here again lane changing could play an important role. Studies on these matters are ongoing.

REFERENCES

Ahn, S. (2005). Formation and Spatial Evolution of Traffic Oscillations. Doctoral Dissertation, Dept. of Civil and Environmental Engineering, University of California, Berkeley, U.S.A.

Aron, M. (1988). Car Following in an Urban Network: Simulation and Experiments. *In Proceedings. of Seminar D, 16th PTRC Meeting*, pp. 27-39.

Chandler, R. E., Herman, R. and Montroll, E. W. (1958). Traffic Dynamics: Studies in Car Following. *Operations Research*, **6**, pp. 165-184.

Daganzo, C. F. (1997). *Fundamentals of Transportation and Traffic Operations*, Pergamon-Elsevier, Oxford, U.K.

Daganzo, C. F. (2002). A Behavioral Theory of Multi-Lane Traffic Flow Part I: Long Homogeneous Freeway Sections, *Transportation Research B*, **36(2)**, pp. 131-158.

Edie, L. C. and Baverez, E. (1958). Generation and Propagation of Stop-Start Traffic Waves. *Proceedings of the 3rd International Symposium on Theory of Traffic Flow* (L.C. Edie, Ed.), American Elsevier Publishing Co., New York, pp. 26-37

Edie, L. C. (1960). Car Following and Steady State Theory for Non-Congested Traffic. *Operations Research*, **9**, pp. 66-76.

Gazis, D. C., Herman, R., and Potts, R. B. (1959). Car Following Theory of Steady State Traffic Flow. *Operations Research*, **7**, pp. 499-505.

Gazis, D. C., Herman, R., and Rothery, R. W. (1961). Nonlinear Follow the Leader Models of Traffic Flow. *Operations Research*, ***9***, pp. 545-567.

Gipps, P. G. (1981). A Behavioral Car-following Model for Computer Simulation. *Transportation Research B*, **15**, pp. 105-111.

Gipps, P.G. (1986). A Model for the Structure of Lane-changing Decisions. *Transportation Research B*, **39**, pp. 403-414.

Herman, R., Montroll, E. W., Potts, R. B., and Rothery R. W. (1959). Traffic Dynamics: Analysis of Stability in Car Following. *Operations Research*, **7**, pp. 86-106.

Hranac, R., Margiotta, R., and Alexiadis, V. (2004). Next Generation Simulation (NGSIM) High-Level Data Plan, Report to Federal Highway Administration, FHWA-HOP-06-011.

Kerner, B. (2005). Microscopic Three-phase Traffic Theory and Its Application for Freeway Traffic Control. *Proceedings of the 16th International Symposium of Transportation and Traffic Theory* (H. Mahmassani, Ed.), Maryland, pp. 181-203.

Kerner, B. S. and Rehborn, H. (1996). Experimental Features and Characteristics of Traffic Jams, *Physical Review E, Statistical Physics, Plasmas, Fluids, and Related Interdisciplinary Topics*, **53(2)**. The American Physical Society, R1297.

Kometani, E. and Sasaki, T. (1958). On the Stability of Traffic Flow. *J. Operations Research. Japan*, **2**, pp. 11-26.

Kometani, E. and Sasaki, T. (1959). Dynamic Behavior of Traffic with a Nonlinear Spacing-Speed Relationship. *Proceedings of the Symposium on Theory of Traffic Flow*. Research Laboratories, General Motors, Elsevier, pp.105-119.

Laval, J. A., Cassidy, M. J., and Daganzo, C. F. (2005). Impacts of lane changes at on-ramp bottlenecks: A theory and strategies to maximize capacity. In: Kuhne, R., Poschel, T., Schadschneider, A., Schreckenberg, M., Wolf, D. (Eds.), Forthcoming in *Traffic and Granular Flow '05*. Springer.

Laval J. A. and Daganzo C. F. (2006). Lane Changing in Traffic Streams. *Transportation Research B*, **40(3)**, pp. 251-264.

Lighthill, M. J. and Whitham, G. B. (1955). On Kinematic Waves I: Flow Movement in Long Rivers. II: A Theory of Traffic Flow on Long Crowded Roads, *Proceedings of Royal Society*. London, **A229(1178)**, pp. 317-345.

Michaels, R. M. (1963). Perceptual Factors in Car Following. *Proceedings. of the 2nd International Symposium on the Theory of Road Traffic Flow*, pp. 44-59. Paris: OECD

Mauch, M. and Cassidy, M. J. (2002). Freeway Traffic Oscillations: Observations and Predictions. *Proceedings of the 15th International.Symposium on Traffic and Transportation Theory*, M. P. Taylor (ed.)

Newell, G. F. (1993). A simplified theory of kinematic waves in highway traffic, I. General theory, II. Queueing at freeway bottlenecks, III. Multi-destination flows. *Transportation Research B*, **27**, pp. 281-313.

Ozaki, H. (1993). Reaction and Anticipation in the Car Following Behaviour. *Proceedings of the 13th International Symposium on Highway Capacity 2*, pp. 493-502.

Rice, J. (1994). *Mathematical Statistics and Data Analysis*. Duxbury Press, Belmont, California

Richards, P. I. (1956). Shock Waves on the Highway, *Operations Research*, **4**, pp. 42-51.

Smilowitz K., Daganzo, C., Cassidy M., and Bertini R. (1999). Some observations of highway traffic in long queues, *Transportation Research Record*, **1678**, pp. 225-233.

Sugiyama, Y., Nakayama, A., Fukui, M., Hasebe, K., Kikuchi, M., Nishinari, K., Tadaki, S. and Yukawa, S. (2003). Observation, Theory and Experiment for Freeway Traffic as Physics of Many-body System, In M. Schreckenberg P. Bovy, S. Hoogendoorn and D.E. Wolf, editors, *Traffic and granular flow '03*, pp. 45-59

Treiterer, J. and Myers, J. A. (1974). The hysteresis phenomenon in traffic flow. *Proceedings of the 6th International Symposium on Transportation and Traffic Theory*, D.J. Buckley (ed.)

Wiedemann, R. (1974). Simulation des Straßenverkehrsflußes. Technical Report. Institut für Verkehrswesen, Universität Karlsruhe, Karlsruhe, Germany.

Windover, J.R. and Cassidy, M.J. (2001). Some observed details of freeway traffic evolution. *Transportation Research A*, **35**, pp. 881-894.

Transportation and Traffic Theory 2007
Edited by R.E. Allsop, M.G.H. Bell and B.G. Heydecker

30

STATE DEPENDENCE IN LANE CHANGING MODELS

Charisma Choudhury[1], Moshe E. Ben-Akiva[1], Tomer Toledo[2], Anita Rao[1] and Gunwoo Lee[1]
[1]Department of Civil and Environmental Engineering, Massachusetts Institute of Technology, USA
[2]Faculty of Civil and Environmental Engineering, Technion - Israel Institute of Technology

INTRODUCTION

Driving behaviour models are used within microscopic traffic simulations to predict driving manoeuvres. With the increasing popularity of such tools, there has been extensive research in improving the key driving behaviour models: acceleration, lane changing and route choice. Existing models usually assume that drivers react to current traffic conditions and make instantaneous decisions. However, in reality, drivers may plan a set of actions based on previous, current and anticipated future conditions and make a sequence of choices to execute the chosen plan. For example, a driver who has decided to change lanes but cannot do that immediately may continue to attempt to change lanes by selecting a target gap and adapting his acceleration to facilitate lane changing into that gap. The actions of the driver are thus implementations of the prior decision to change lanes and the decision tree is state dependent. However, in most cases the decision state of the driver (e.g. the decision to attempt to change lanes) is unobserved and only lane action and acceleration manoeuvres of the driver are observed.

In most of the existing driving behaviour models, the drivers are assumed to be myopic (Gipps 1986, Benekohal and Treiterer 1988, Yang and Koutsopoulos 1996, Zhang et al. 1998, Ahmed 1999, Choudhury 2005). A 'partial short-term plan' based decision framework for lane changing and acceleration was proposed by Toledo (2003), where the effects of a driver's short term plan to execute a lane change through a chosen gap is reflected on his acceleration decisions. However, state dependency has been ignored in this model and it is assumed that the driver revaluates his short term plan at every instant regardless of his current or previous state. Wang et al. (2005) tested the sensitivity of model parameters for a similar partial short-term based gap selection and acceleration model for freeway merging situation within a simulation framework.

The above mentioned state dependency is thus not captured in existing lane changing models. As a consequence, application of these models in micro-simulation tools often result in unrealistic traffic flow characteristics in congested and incident situations where the decisions of the driver involve significant planning, cooperation and risk taking.

This paper presents a framework for modelling state dependency in lane changing behaviour of drivers and demonstrates it through an on-ramp merging model for congested freeway situations. The proposed model explicitly considers the anticipation of future conditions as a basis of decision-making and incorporates state dependence to capture the effects of past decisions the driver has made on his current decision-making process. The paper is structured as follows: the structure of the state dependent merging model is described first. The estimation data and the estimation methodology are presented next followed by the estimation results. The improvements in the proposed model are demonstrated by statistical comparisons of the model against an instantaneous model that is estimated with the same dataset ignoring state dependency.

MODELING STATE DEPENDENCE IN FREEWAY MERGES

Model Framework

In congested situations, acceptable gaps are often not available and more complex merging phenomena are observed. For example, drivers may merge through courtesy of the lag driver in the target lane or become impatient and decide to force in, compelling the lag driver to slow down. The execution of all types of merges involve acceptance of available gaps. The definition of acceptable gaps may depend on the merging mechanism.

Normal merge occurs when the available adjacent gaps are immediately acceptable and is therefore an instantaneous process. However, in case of courtesy lane change and forced merge, even after the driver has initiated the merge, the actual lane change may not be possible immediately. A driver who has initiated a forced (or courtesy) merge remains in the initiated forced (or courtesy) merge state and continues to evaluate the adjacent gaps for the chosen merging mechanism until they are acceptable. Thus the gap acceptance decisions the driver makes at any instant depend on his state.

The deicision to select the merging mechanism is a sequential process. The decision framework of the driver is summarized in Figure 1. The model hypothesizes four levels of decision-making: normal gap acceptance, gap anticipation and aniticipated gap acceptance (or decision to initiate a courtesy merging), decision whether to initiate a forced merging or not, and gap acceptance for courtesy/forced merging. The decision process is latent and only the end action of the driver (lane change to the target lane) is observed. Latent choices are shown in ovals and observed actions are shown in rectangles.

The merging driver first compares the available lead and lag gaps in the mainline to the corresponding minimum acceptable gaps (critical gaps) for normal gap acceptance. Critical gaps are modeled as random variables, their means being functions of explanatory variables. If both the lead and the lag gaps are greater than the critical gaps, a lane change can be executed using the existing gaps.

If the gaps are not acceptable, the merging vehicle evaluates the speed, acceleration and relative position of the through vehicles and tries to evaluate whether or not the lag driver is providing courtesy to him. The courtesy or discourtesy of the lag driver is reflected on the anticipated gap. If the lag driver has decided to provide courtesy to a merging vehicle and has started to decelerate, the anticipated gap increases. The anticipated gap of a particular driver also depends on the length of the time horizon over which the gap is estimated. Differences in perception and planning abilities among drivers are captured by the distribution of the length of the time horizon. If the anticipated gap is acceptable, the merging driver perceives that he is receiving courtesy from the lag driver and initiates a courtesy merge.

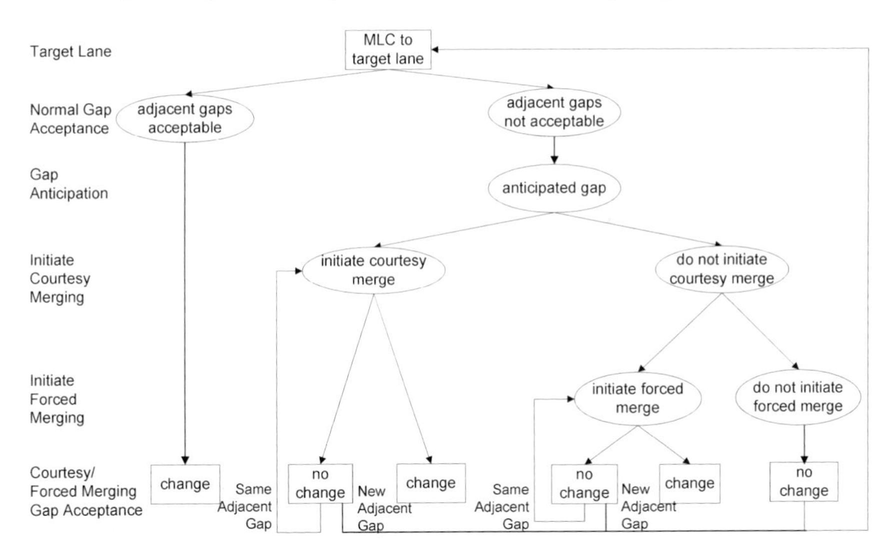

Figure 1: Framework of the merging model

If the anticipated gap is unacceptable, the driver decides whether to force the lag driver to slow down or not by nosing in. This decision can depend on the urgency of the merge, driver characteristics (e.g. risk averseness) and traffic conditions. If the driver does not initiate a courtesy or forced merge, he remains in the normal merging state and the entire decision process is repeated in the next time step.

A driver who has initiated a courtesy lane changing, compares the adjacent gaps against the courtesy merging critical gaps and makes the lane-change once these gaps are acceptable. The driver remains in the initiated courtesy merge state until the lane change is executed or the driver is adjacent to a new gap. Similarly, a driver who has initiated a forced merge remains in initiated forced merge state until the adjacent gaps are acceptable to execute the forced merge or the adjacent gap changes.

Therefore, the observed lane change (or no change) action at any instant is state dependent and can be the outcome of many possible decision sequences. Both the state of the driver and the decision sequence that led to the state are however unobserved/latent.

This paper focuses on formulation of the decision framework of the merging driver. The decisions of other drivers (e.g. decisions made by the lag driver whether or not to provide courtesy) are treated as external/observed variables in the model.

Model Components

Normal gap acceptance

Normal gap acceptance model indicates whether a lane change is possible or not using the existing gaps. The definition of related variables is presented in Figure 2. An available gap is acceptable if it is greater than the critical gap. Critical gaps are assumed to follow lognormal distributions, the mean gap being a function of explanatory variables. This can be expressed as follows:

$$\ln\left(G_{nt}^{Mg}\right) = \beta^{Mg^T} X_{nt} + \alpha^{Mg} \upsilon_n + \varepsilon_{nt}^{Mg} \quad g \in \{lead, lag\} \tag{1}$$

where G_{nt}^{Mg} is the critical gap g of individual n at time t for normal gap acceptance (M), $g \in \{lead, lag\}$, X_{nt} are explanatory variables, β^{Mg} is the corresponding vector of parameters for normal gap acceptance, υ_n is the individual specific random effect: $\upsilon_n \sim N(0,1)$ and α^{Mg} is the coefficient of the individual specific random term for normal gap acceptance, ε_{nt}^{Mg} is the random term for normal gap acceptance of individual n at time t: $\varepsilon_{nt}^{Mg} \sim N\left(0, \sigma_{Mg}^2\right)$.

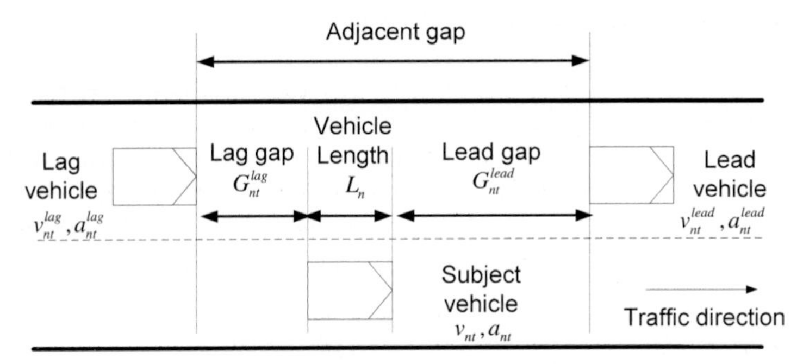

Figure 2: Vehicle relationships in a merging situation

The gap acceptance model assumes that the driver must accept both the lead and the lag gap to change lanes. If a merging vehicle is in normal state ($s_{t-1} = M$), i.e., he has not initiated a courtesy or forced merge, the probability of a lane change through normal gap acceptance, conditional on the individual specific term υ_n can be expressed as follows:

$$\begin{aligned} &P_n\left(l_t = 1 \middle| s_{t-1} = M, \upsilon_n\right) = \\ &P_n\left(accept\ lead\ gap \middle| s_{t-1} = M, \upsilon_n\right) P_n\left(accept\ lag\ gap | s_{t-1} = M, \upsilon_n\right) \\ &= P_n\left(G_{nt}^{lead} > G_{nt}^{M\,lead} \middle| s_{t-1} = M, \upsilon_n\right) P_n\left(G_{nt}^{lag} > G_{nt}^{M\,lag} \mid s_{t-1} = M, \upsilon_n\right) \end{aligned} \tag{2}$$

where, for driver *n* at time *t*, l_t is the lane changing indicator, 1 if a lane-change is performed, 0 otherwise. s_t denotes state of the driver (M=normal, C=courtesy, F=forced), G_{nt}^{lead} and G_{nt}^{lag} are the available lead and lag gaps respectively.

Assuming that critical gaps follow lognormal distributions, the conditional probabilities that gap $g \in \{lead, lag\}$ is acceptable can be expressed as follows:

$$\begin{aligned} &P_n\left(l_t = 1 \middle| s_{t-1} = M, \upsilon_n\right) \\ &= P\left(\ln\left(G_{nt}^{g}\right) > \ln\left(G_{nt}^{Mg}\right) \mid s_{t-1} = M, \upsilon_n\right) \\ &= \Phi\left[\frac{\ln\left(G_{nt}^{g}\right) - \left(\beta^{Mg^T} X_{nt} + \alpha^{Mg}\upsilon_n\right)}{\sigma_{Mg}}\right] \end{aligned} \tag{3}$$

$\Phi[\cdot]$ denotes the cumulative standard normal distribution.

If a driver has already initiated a courtesy or forced merge in a previous time step, he cannot decide to merge to the same adjacent gap under normal gap acceptance. Therefore, if a merging vehicle is in initiated courtesy/forced merging state at time *(t-1)*, the probability of a lane change through normal gap acceptance at *t* is zero, unless there is a new adjacent gap.

Anticipated gaps and decision to initiate courtesy yielding

If the adjacent gaps are not acceptable to make a normal merge, the merging vehicle evaluates the speed, acceleration and relative position of the through vehicles and approximates an expected/anticipated gap that is going to open up after time τ_n. Because of the difference in perception among individuals, the anticipation time τ_n may vary among individuals.

The anticipated/expected gap for individual n at time t can be expressed as follows:

$$\overline{G}_{nt}(\tau_n) = G_{nt}^{lead} + G_{nt}^{lag} + L_n + \tau_n(v_{nt}^{lead} - v_{nt}^{lag}) + \frac{1}{2}\tau_n^2(a_{nt}^{lead} - a_{nt}^{lag}) \tag{4}$$

where, for individual n at time t, $\overline{G}_{nt}$ is the anticipated gap, L_n is the length of the vehicle, v_{nt}^{lead} and v_{nt}^{lag} are the speeds of the lead and lag vehicles, a_{nt}^{lead} and a_{nt}^{lag} are the acceleration of the lead and lag vehicles respectively (Figure 2).

If this anticipated gap is acceptable, the driver decides to initiate a courtesy merge. The critical gap of the driver for the anticipated gap acceptance is assumed to follow a lognormal distribution and can be expressed as follows:

$$\ln\left(G_{nt}^{A}\right) = \beta^{A^T} X_{nt} + \alpha^A \upsilon_n + \varepsilon_{nt}^{A} \tag{5}$$

where, individual n at time t, G_{nt}^{A} is the critical gap for anticipated gap acceptance, β^A is the corresponding vector of parameters, ε_{nt}^{A} is the random term for anticipated gap acceptance: $\varepsilon_{nt}^{A} \sim N\left(0, \sigma_A^2\right)$.

If the driver has already initiated a courtesy merge in a previous time step and the adjacent gap has not changed, the probability of being in initiated courtesy merge state is 1. If the driver has already initiated a forced merge to the same gap in a previous time step, the probability of being in initiated courtesy merge state at current time step is 0. However, if the driver cannot complete the inititated courtesy merging within the time he is adjacent to the same gap and is adjacent to a new gap, the state of the driver is reset to the normal (not initiated courtesy or forced merging) state. This can be expressed as follows:

$$\begin{aligned}
&P_n\left(s_t = C \mid s_{t-1} = C, \upsilon_n, \tau_n\right) = \delta_{nt} + P_n\left(s_t = C \mid s_{t-1} = M, \upsilon_n, \tau_n\right)(1 - \delta_{nt}) \\
&P_n\left(s_t = C \mid s_{t-1} = M, \upsilon_n, \tau_n\right) = P_n\left(\overline{G}_{nt} > G_{nt}^{A} \mid s_{t-1} = M, \upsilon_n, \tau_n\right)\left[1 - P_n\left(l_t = 1 \middle| s_{t-1} = M, \upsilon_n\right)\right] \\
&P_n\left(s_t = C \mid s_{t-1} = F, \upsilon_n, \tau_n\right) = 0
\end{aligned} \tag{6}$$

where δ_{nt} =1 if driver n is adjacent to the same gap at time *(t-1)* and t, 0 otherwise.

The anticipation time is assumed to be truncated normally distributed with truncation on both sides. The distribution is given by:

$$f(\tau_n) = \begin{cases} \dfrac{\dfrac{1}{\sigma_\tau}\phi\left(\dfrac{\tau_n - \mu_\tau}{\sigma_\tau}\right)}{\Phi\left(\dfrac{\tau_{\max} - \mu_\tau}{\sigma_\tau}\right) - \Phi\left(\dfrac{\tau_{\min} - \mu_\tau}{\sigma_\tau}\right)} & \text{if } \tau_{\min} \leq \tau_n \leq \tau_{\max} \\ 0 & \text{otherwise} \end{cases} \tag{7}$$

where, μ_τ, σ_τ are the constant mean and standard deviations of the untruncated distribution, $\tau_{\min}$ and $\tau_{\max}$ are the minimum and maximum values of τ_n respectively. $\phi(\,)$ is the probability density function of a standard normal random variable and $\Phi(\,)$ is the cumulative distribution function of a standard normal random variable.

The advantage of using a truncated normal distribution is that it is not restricted to be skewed to a particular direction. This ensures that no a priori assumption is made on the probability of a driver being myopic or not.

Decision to initiate a forced merge

If the normal gaps are not acceptable and the driver perceives that he cannot merge through courtesy yielding (anticipated gap is not acceptable), he considers the decision whether to initiate forced merge $(s_t = F)$ or not $(s_t = M)$.

By initiating a forced merge, the merging driver takes a risk and imposes a deceleration on the lag vehicle in the mainline. The utility of initiating a forced merge can be expressed as follows:

$$U_{nt}^F = \beta^{F^T} X_{nt} + \alpha^F \upsilon_n + \varepsilon_{nt}^F \tag{8}$$

where, for individual n at time t, U_{nt}^F is the utility of initiating a forced merge, β^F is the corresponding vector of parameters, ε_{nt}^F is the random term for initiating forced merging, α^F is the coefficient of the driver specific random term for forced merging.

By assuming that the random error terms ε_{nt}^F are i.i.d. Gumbel distributed, this decision can be modelled as a logit model.

Similar to the initiation of the courtesy merge, the probability of the driver being in initiated forced merge state is conditional on his previous state: the probability being 1 if the driver had already initiated a forced merge to the same gap in a previous time step and 0 if the driver had already initiated a courtesy merge to the same gap in a previous time step. However, if the driver cannot finish the initiated forced merging within the time he is adjacent to the same gap and is adjacent to a new gap, the state of the driver is reset to the normal (not initiated courtesy or forced merging) state.

$$
\begin{aligned}
&P_n\left(s_t = F \mid s_{t-1} = F, \upsilon_n, \tau_n\right) = \delta_{nt} + P_n\left(s_t = F \mid s_{t-1} = M, \upsilon_n, \tau_n\right)\left(1-\delta_{nt}\right) \\
&P_n\left(s_t = F \mid s_{t-1} = M, \upsilon_n, \tau_n\right) \\
&= \left[\frac{1}{1+\exp\left(\left(-\beta^{F^T} X_{nt} - \alpha^F \upsilon_n\right) \mid \upsilon_n, \tau_n\right)}\right]\left[1 - P_n\left(s_t = C \middle| s_{t-1} = M, \upsilon_n\right)\right]\left[1 - P_n\left(l_t = 1 \middle| s_{t-1} = M, \upsilon_n\right)\right] \\
&P_n\left(s_t = F \mid s_{t-1} = C, \upsilon_n, \tau_n\right) = 0
\end{aligned}
\tag{9}
$$

where δ_{nt} =1 if the driver is adjacent to the same gap at time (t-1) and t, 0 otherwise.

Decision to make a courtesy/forced lane change

Even though a driver decides to initiate a courtesy/forced merge, the completion of the merge may take some time. That is, the actual merge is executed only when the available gaps are acceptable in comparison with the critical gaps for the respective merge. From the moment a driver initiates a forced merge up to T_n (the last time step the vehicle is observed as a merging vehicle), he is considered to be in initiated courtesy/forced merging state.

The functional form and variables influencing the critical gaps for courtesy and forced merging are assumed to be the same as in merging under normal gap acceptance, but the parameters are likely to be different.

State Transitions

At time *t* given an adjacent gap, driver *n*, can be in any one of the following states:

- Initiated courtesy merging ($s_t = C$)
- Initiated forced merging ($s_t = F$)
- Have not initiated courtesy/forced merging: normal ($s_t = M$)

Once a driver has initiated forced merging to an adjacent gap, he does not consider courtesy merging or normal gap acceptance in the subsequent time steps unless the gap changes. The decision in the subsequent time steps is only to decide whether or not to complete the forced merge in that time step. Thus once a transition is made from normal to forced merging state, the state cannot go back to normal and it cannot change to the initiated courtesy merging state unless the gap changes. Similarly, for a particular adjacent gap, once a transition is made from normal to initiated courtesy merging state, the state cannot change to initiated forced merging or normal. When the driver moves to a new adjacent gap, the state is reset to normal.

The possible decision state sequences are illustrated in Table 1 with two examples.

Table 1: Possible Decision State Sequences

Case 1: Same Adjacent Gap

Time Period	Observed Lane	Lane Action	State Sequences												
			1	2	3	⋯	T_n-1	T_n	T_n+1	T_n+2	T_n+3		$2T_n-1$	$2T_n$	$2T_n+1$
1	**CL**	**0**	C	M	M	⋯	M	M	F	M	M	⋯	M	M	M
2	**CL**	**0**	C	C	M	⋯	M	M	F	F	M	⋯	M	M	M
3	**CL**	**0**	C	C	C	⋯	M	M	F	F	F	⋯	M	M	M
⋮			⋮	⋮	⋮		⋮	⋮	⋮	⋮	⋮		⋮	⋮	⋮
T_n-1	**CL**	**0**	C	C	C	⋯	C	M	F	F	F	⋯	F	M	M
T_n	**CL**	**1**	C	C	C	⋯	C	C	F	F	F	⋯	F	F	M
T_n+1	**TL**														

Case 2: Two Adjacent Gaps

Gap	Time Period	Observed Lane	Lane ACTION	State Sequences											
				1	2	3	⋯	T_n-1	T_n	T_n+1	T_n+2	T_n+3		$2T_n-1$	$2T_n$
	1	**CL**	0	C	M	M	⋯	M	M	F	M	M	⋯	M	M
	2	**CL**	0	C	C	M	⋯	M	M	F	F	M	⋯	M	M
1	**3**	**CL**	0	C	C	C	⋯	M	M	F	F	F	⋯	M	M
	⋮			⋮	⋮	⋮		⋮	⋮	⋮	⋮	⋮		⋮	⋮
	T_n^1-1	**CL**	0	C	C	C	⋯	C	M	F	F	F	⋯	F	M
	T_n^1	**CL**	0	C	C	C	⋯	C	C	F	F	F	⋯	F	F
	1	**CL**	0	C	M	M	⋯	M	M	F	M	M	⋯	M	M
	2	**CL**	0	C	C	M	⋯	M	M	F	F	M	⋯	M	M
2	⋮			⋮	⋮	⋮		⋮	⋮	⋮	⋮	⋮		⋮	⋮
	T_n^2	**CL**	0	C	C	C	⋯	C	C	F	F	F	⋯	F	F
	T_n+1	**TL**													

C = Initiated courtesy merge $s_{nt} = C$, F = Initiated forced merge $s_{nt} = F$,

M = Normal (Had not initiated a courtesy or forced merge) $s_{nt} = M$,

CL = Current Lane, TL = Target Lane, 0 = No change, 1 = Change,

P_n = Total number of adjacent gaps of individual *n* (2 in this case),

T_n = Time individual *n* is observed as a merging vehicle,

T_n^p = Time individual *n* is adjacent to gap *p*, $T_n = T_n^1 + T_n^2$ in this case.

As observed in the table, when the driver is adjacent to the same gap in two subsequent time instants, the following state transitions are possible:

- Normal to Normal ($s_t = M \mid s_{t-1} = M$)
- Normal to Courtesy ($s_t = C \mid s_{t-1} = M$)
- Normal to Forced ($s_t = F \mid s_{t-1} = M$)
- Courtesy to Courtesy ($s_t = C \mid s_{t-1} = C$)
- Forced to Forced ($s_t = F \mid s_{t-1} = F$)

When the driver is adjacent to a new gap, the following transitions are possible.

- Normal to Normal ($s_t = M \mid s_{t-1} = M$)
- Courtesy to Normal ($s_t = M \mid s_{t-1} = C$)
- Forced to Normal ($s_t = M \mid s_{t-1} = F$)

The probabilities of each of these transitions can be calculated using equations (6) and (9).

MODEL ESTIMATION

Data

The disaggregate data used for estimating the merging model was collected from the northbound direction of Interstate-80 (I-80) in Emeryville, California (Figure 3). The data was collected and processed as part of the Federal Highway Administration's Next Generation Simulation (NGSIM) project. Vehicles were tracked over a length of 503 meters (merging needs to be completed by 200 meters). The vehicle trajectory data containing the coordinates of the various vehicles in the section were used to derive the required variables for estimation. The merging drivers entering from the on-ramp to the rightmost lane of the mainline were used for estimation. The resulting dataset included 17352 observations at a 1 second time resolution of 540 vehicles.

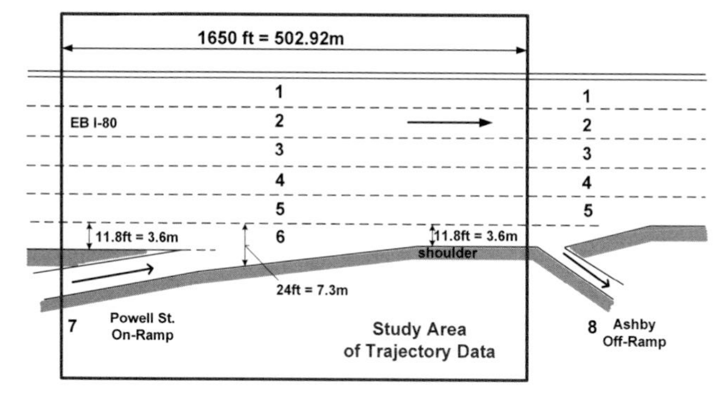

Figure 3: Data collection site

It may be noted that it was not possible to uniquely identify the state of the driver from the estimation data. For example, if there is an observation involving gap creation through deceleration of the lag vehicle, it is not difficult to determine whether it is the result of courtesy by the lag or the response to the merging vehicle forcing its way in. This motivated the latent choice formulation that has the flexibility to account for the various merge mechanisms without explicit knowledge of the mechanism that the driver has used.

Detailed analyses of the data and data processing methodology are presented in Cambridge Systematics (2005) and Choudhury et al. (2006a).

Likelihood of the Trajectory

All model parameters were estimated jointly using a maximum likelihood technique. The likelihood function that was maximized is presented in this section.

At any time *t*, an individual can be in courtesy merging ($s_t = C$), forced merging ($s_t = F$) or normal ($s_t = M$) state. The lane changing decision of the driver depends on his state. The state of the driver at any instant depends on his previous state and the lane changing decision at that state.

According to the first-order Markov assumption: the probability of individual *n* being in a particular decision state *j* at time *t* only depends on his decision state at time *(t-1)*.

Therefore, the fact that a person is in state *j* at time *t*, where $t<T_n$, indicates the following:

- He has made a transition to state *j* from state *i* at t^{th} time step , where $i, j \in M, C, F$
- He was at state *i* at time *t-1*
- He has not made any lane change when he was at state *i* at time *t-1* (since the observation for an individual ends when he makes a lane change)

The probability of being in state $s_t = j$ is therefore the product of probability of a transition from state *i* to state *j* at time *t*, the probability of being in state *i* at time *(t-1)* and is conditional on the lane actions at previous time periods. This can be expressed as follows:

$$P_n(s_t = j \mid l_{t-1}, \upsilon_n, \tau_n) = \sum_i [P_n(s_t = j \mid s_{t-1} = i, l_{t-1}, \upsilon_n, \tau_n) P_n(s_{t-1} = i \mid l_{t-2}, \upsilon_n, \tau_n)] \quad (10)$$

$i, j \in M, C, F$

It may be noted that $P_n(s_t = j)$ is thus the sum of probabilities of all possible paths to $s_t = j$ (Figure 4).

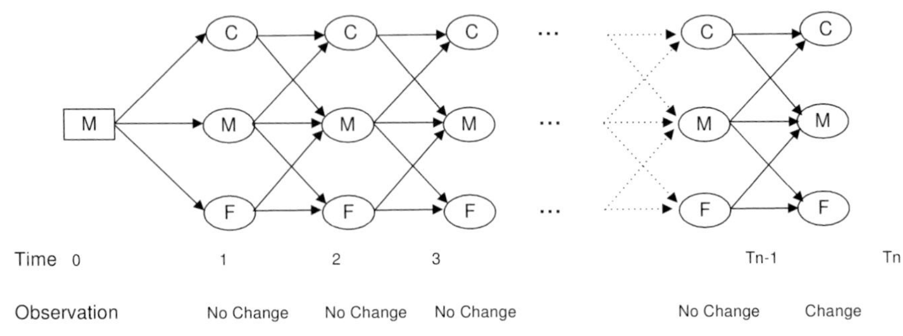

Figure 4: Decision state sequences

The state of the driver can thus be calculated recursively, the state of the driver at time t = 0 (when the driver first approaches the merging section) being normal.

Probability that at time t driver n executes lane changing decision l_t at state s_t is given by:

$$P_n(l_t, s_t \mid l_{t-1}, \upsilon_n, \tau_n) = P_n(l_t \mid s_t, \upsilon_n, \tau_n) P_n(s_t \mid l_{t-1}, \upsilon_n, \tau_n) \tag{11}$$

The state of the driver is not observed and only the lane changing actions are observed. Therefore, probability that driver n executes lane changing decision l_t at time t is given by:

$$P_n(l_t \mid l_{t-1}, \upsilon_n, \tau_n) = \sum_j P_n(l_t, s_t = j \mid l_{t-1}, \upsilon_n, \tau_n) \tag{12}$$

If driver n is observed over a sequence of T_n consecutive time intervals, the probability of observing his entire trajectory is the product of the probabilities given in equation (12) and can be expressed as:

$$P_n(\mathbf{l} \mid \upsilon_n, \tau_n) = \prod_{t=1}^{T_n} P_n(l_t \mid l_{t-1}, \upsilon_n, \tau_n) \tag{13}$$

The unconditional individual likelihood is given by:

$$L_n = \int_{\upsilon}\int_{\tau} P_n(\mathbf{l} \mid \upsilon_n, \tau_n)\, f(\upsilon) f(\tau)\, d\upsilon\, d\tau \tag{14}$$

where,
$f(\upsilon)$ is the standard normal probability density function, $f(\tau)$ is the probability density function of a doubly truncated normal distribution with mean μ_τ and variance σ_τ^2.

Maximum likelihood estimators of the model parameters can be found by maximizing this function.

Estimation Results

The estimation results estimated using the statistical estimation software Gauss7.0 are summarized in Table 2. The final log-likelihood is -1609.65 and the adjusted rho-bar square is 0.88.

Table 2. Estimation Results of State Dependence Model

Variable		Parameter Value	t-statistic
Normal Lead Gap			
Normal lead constant		-0.230	-0.33
*Relative average speed (positive) (m/sec)		0.521	0.81
*Relative lead speed (m/sec)		-0.505	-3.13
*Remaining distance function	Relative lead speed (negative) (m/sec)	1.32	3.64
	Remaining distance to MLC point (10 m)	0.420	0.89
	Remaining distance constant	0.355	1.68
σ^{MLead}		3.42	9.67
α^{MLead}		-0.819	-3.12
Normal Lag Gap			
Normal lag constant		0.198	2.87
*Relative lag speed (positive) (m/sec)		0.208	1.78
*Relative lag speed (negative) (m/sec)		0.184	1.63
*Remaining distance function	Remaining distance to MLC point (10 m)	0.439	5.09
	Remaining distance constant	0.0242	0.03
	$\alpha^{RemDisLag}$	0.000180	0.03
*Lag acceleration (positive) (m/sec^2)		0.0545	0.61
σ^{MLag}		0.840	3.03
α^{MLag}		-0.0000776	-0.01
Initiate Courtesy Merge			
Anticipated gap constant		1.82	1.00
Relative average speed (positive) (m/sec)		1.82	2.13
Relative lead speed (m/sec)		-0.153	-0.97
Remaining distance function	Distance to MLC point (10 m)	0.244	1.50
	Constant	0.449	0.49
	$\alpha^{RemDisA}$	0.360	0.18
σ_A		0.0106	0.07

Variable		Parameter Value	t-statistic
α^A		-0.231	-1.90
μ_τ		1.87	9.51
σ_τ		1.44	17.71
Courtesy Lead Gap			
Courtesy lead constant		-0.582	-0.20
*Relative average speed (positive) (m/sec)		0.521	0.81
*Relative lead speed (negative) (m/sec)		-0.505	-3.13
*Remaining distance function	Distance to MLC point (10 m)	1.32	3.64
	Constant	0.420	0.89
	$\alpha^{RemDisLead}$	0.355	1.68
σ^{CLead}		0.0109	0.08
α^{CLead}		-0.0540	-0.03
Courtesy Lag Gap			
Courtesy lag constant		-1.23	-0.07
*Relative lag speed (positive) (m/sec)		0.208	1.78
*Relative lag speed (negative) (m/sec)		0.184	1.63
*Remaining distance function	Distance to MLC point (10 m)	0.439	5.09
	Constant	0.0242	0.03
	$\alpha^{RemDisLag}$	0.000180	0.03
*Lag acceleration (positive) (m/sec^2)		0.0545	0.61
σ_{CLag}		0.554	0.05
α^{CLag}		-0.0226	-0.04
Initiate Forced Merge			
Initiate force constant		-6.41	-4.63
Heavy lag vehicle dummy		-1.25	-0.63
α^F		5.43	3.26
Forced Lead Gap			
Forced lead constant		3.11	2.11
*Relative average speed (positive) (m/sec)		0.521	0.81
*Relative lead speed (m/sec)		-0.505	-3.13
*Remaining distance function	Distance to MLC point (10 m)	1.32	3.64
	Constant	0.420	0.89

Variable		Parameter Value	t-statistic
	$\alpha^{RemDisLead}$	0.355	1.68
σ^{FLead}		7.95	5.82
α^{FLead}		-0.0401	-0.07
Forced Lag Gap			
Forced lag constant		-2.53	-3.42
*Relative lag speed (positive) (m/sec)		0.208	1.78
*Relative lag speed (negative) (m/sec)		0.184	1.63
*Remaining distance function	Distance to MLC point (10 m)	0.439	5.09
	Constant	0.0242	0.03
	$\alpha^{RemDisLag}$	0.000180	0.03
*Lag acceleration (positive) (m/sec²)		0.0545	0.61
σ^{FLag}		0.465	2.49
α^{FLag}		-0.0239	-0.19

* same coefficients in normal, courtesy and forced gap acceptance levels

The lead critical gap is a function of the average speed in the mainline relative to the subject vehicle's speed, the relative speed of the lead with respect to the subject and the remaining distance to the mandatory lane changing point. The lag critical gap is a function of the subject relative speed with respect to the lag vehicle, the remaining distance to the mandatory lane changing point and the acceleration of the lag vehicle.

The estimated lead and lag critical gaps for the normal gap acceptance are given by:

$$\begin{aligned} G_{nt}^{MLead} &= \exp\left(-0.230+0.521V_{nt}^{'}-0.505Min\left(0,\Delta V_{nt}^{lead}\right)+\frac{1.32}{1+\exp(0.420+0.355\upsilon_n)}d_{nt}-0.819\upsilon_n+\varepsilon_{nt}^{Mlead}\right) \\ G_{nt}^{MLag} &= \exp\left(\begin{array}{l} 0.198+0.208Max\left(0,\Delta V_{nt}^{lag}\right)+0.184Min\left(0,\Delta V_{nt}^{lag}\right)+\dfrac{0.439}{1+\exp(0.0242+0.00018\upsilon_n)}d_{nt} \\ +0.0545Max\left(0,a_{nt}^{lag}\right)-0.840\upsilon_n+\varepsilon_{nt}^{Mlag} \end{array}\right) \end{aligned} \quad (15)$$

where, G_{nt}^{Mlead} is the lead critical gap for the normal gap acceptance level (m), G_{nt}^{Mlag} lag critical gap for the normal gap acceptance level (m), $V_{nt}^{'}$ is the relative average speed factor (m/sec), ΔV_{nt}^{lead} relative speed of the lead vehicle with respect to the subject (m/sec), d_{nt} is the remaining distance to the mandatory lane changing point (10m), ΔV_{nt}^{lag} relative speed of the lag vehicle with respect to the subject (m/sec), a_{nt}^{lag} acceleration of the lag vehicle, ε_{nt}^{Mlead} *and* ε_{nt}^{Mlag} are random error terms with $\varepsilon_{nt}^{Mlead} \sim N\left(0,3.83^2\right)$ and $\varepsilon_{nt}^{Mlag} \sim N\left(0,0.532^2\right)$.

The lead critical gap increases with the increase in average speed of the mainline. As the mainline average speed increases, the driver needs larger critical gaps to adjust his speed to the speed of the mainstream. However, critical gap does not increase linearly with increasing average speeds in the mainline (Figure 5a), it rather increases as a diminishing function $\beta^{Mavg} V'_{nt}$, where, $V'_{nt} = \left(1 + \frac{1}{1+\exp\left(-Max\left(0, \Delta V_{nt}^{avg}\right)\right)}\right)$, ΔV_{nt}^{avg} being the relative speed of the average mainline speed with respect to the subject (m/sec).

The lead critical gap is larger when the lead vehicle is moving slower than the subject since the driver perceives an increased risk when the lead is slowing down and he gets closer to the lead vehicle (Figure 5b).

The lag critical gap increases with the relative lag speed: the faster the lag vehicle is relative to the subject, the larger the critical gap is (Figure 5c).The lag critical gap increases as the acceleration of the lag vehicle increases (Figure 5d), due to the higher perceived risk into merging onto the mainstream when the lag vehicle is accelerating.

Both the lead and lag critical gaps decrease as the remaining distance to the mandatory lane changing point decreases. This is because as the driver approaches the point where the ramp ends, his urgency to make the merge increases and he is willing to accept lower gaps to merge in to. To capture drivers' heterogeneity, an individual specific random term has been introduced in the coefficient of the remaining distance. Aggressive and timid drivers can thus have different critical gaps, the remaining distance being equal. The aggressiveness/timidness of the driver basically captures the heterogeneity among the driver population and is assumed to have a continuous distribution (truncated normal in this case) rather than discrete having a discrete class membership. For example, all other variables having no effect, the lead and lag critical gaps as a function of remaining distance for the aggressive drivers are much smaller than the gaps of timid drivers. Thus, aggressive drivers can find lead and lag gaps to be acceptable even when they are far from the MLC point. On the other hand, timid drivers have large critical gaps till they reach the end of the ramp, implying that they do not consider lane changes in the beginning of the on-ramp. The sensitivity of the lead and lag critical gaps as a function of the remaining distance according to the individual characteristics of the driver is shown in Figure 5e and Figure 5f respectively. The t-statistics for the linear part of the coefficient of remaining distance is found to be very significant both for lead and lag gaps.

Estimated coefficients of the unobserved driver characteristics (υ_n) are negative for both the lead and lag critical gaps. This implies that an aggressive driver requires smaller gaps for lane changing as compared to a timid driver.

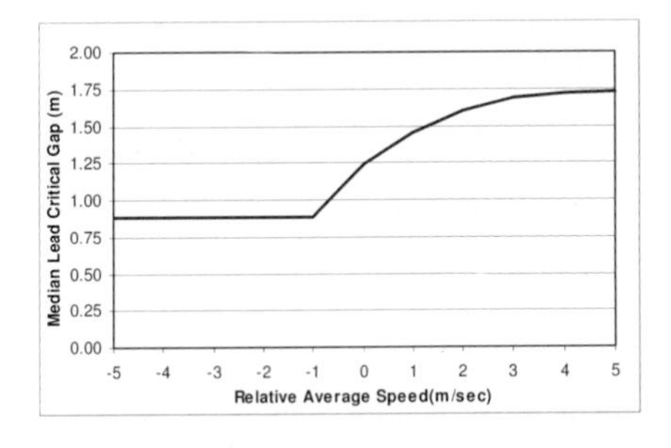

a. Lead critical gap as a function of relative average speed

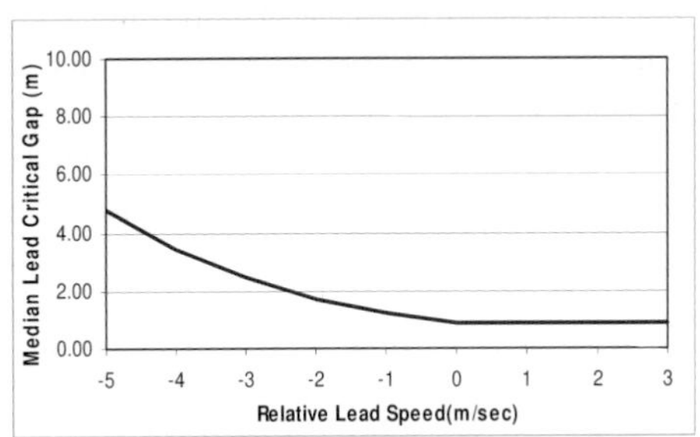

b. Lead critical gap as a function of relative lead speed

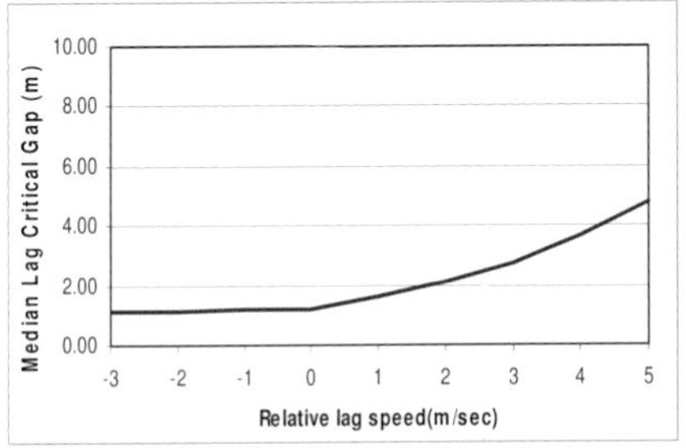

c. Lag critical gap as a function of relative lag speed

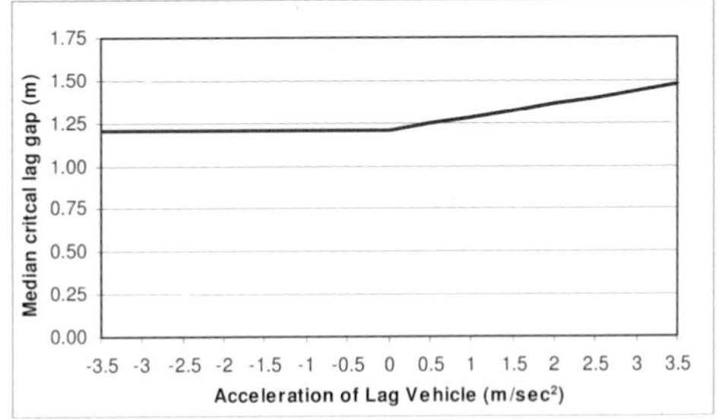

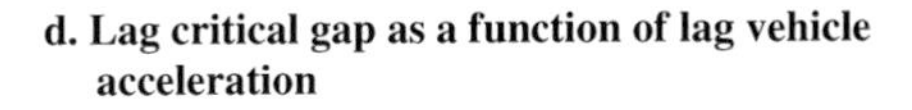

d. Lag critical gap as a function of lag vehicle acceleration

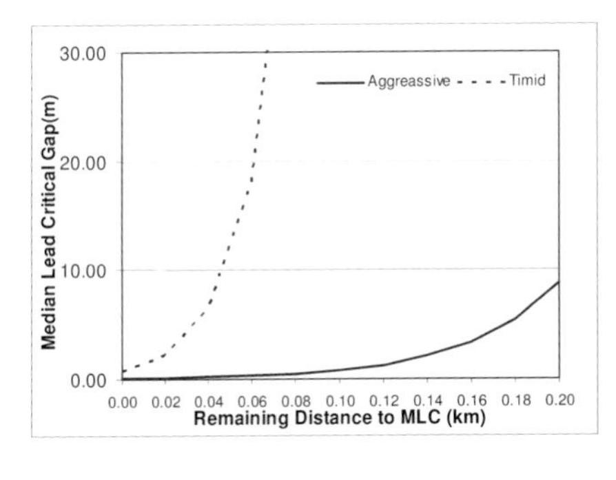

e. Lead critical gap as a function of remaining distance to MLC point

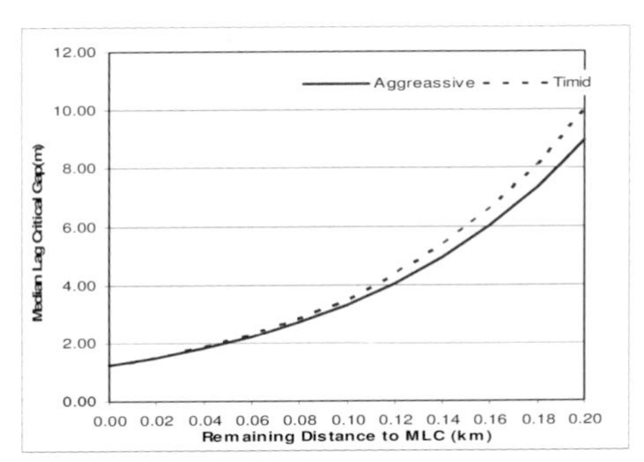

f. Lag critical gap as a function of remaining distance to MLC point

Figure 5: Median Lead and Lag gap variations

The anticipated gap acceptance (initiating courtesy) depends on lag speed, remaining distance and density of the traffic stream. The estimated critical anticipated gap is given by:

$$G_{nt}^{A} = \exp\left(1.82 + 1.81 Max\left(0, \Delta V_{nt}^{lag}\right) - 0.153\rho_{nt} + \frac{0.244}{1+\exp(0.449 + 0.360\upsilon_{n})} d_{nt} - 0.213\upsilon_{n} + \varepsilon_{nt}^{A}\right) \quad (16)$$

where, G_{nt}^{A} is the critical anticipated gap for initiating courtesy merge (m), ρ_{nt} is the density in the rightmost lane of the mainline (veh/10m), $\varepsilon_{nt}^{A} \sim N\left(0, 0.0106^{2}\right)$

Similar to normal critical gaps, the critical anticipated gap is higher at higher lag speeds. It decreases as the remaining distance decreases and it is smaller for aggressive drivers as compared to timid drivers. Courtesy yielding/merging more commonly occurs in dense traffic conditions and hence the probability to merge through courtesy increases as the density of traffic in the mainline increases. The critical anticipated gap therefore reduces with density of traffic in the rightmost lane of the mainline. Median critical anticipated gap as a function of density is presented in Figure 6.

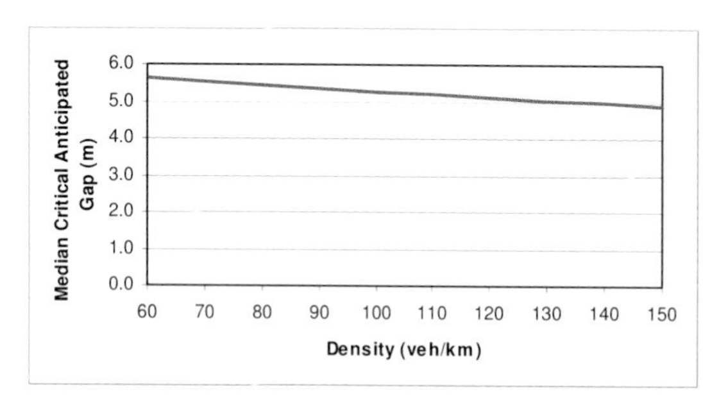

Figure 6: Median critical anticipated gap as a function of density in target lane

The decision to initiate a forced merge is dependent on whether the lag vehicle is a heavy vehicle or not. If the lag is a heavy vehicle, the probability of initiating a forced merge decreases, as the driver perceives a higher risk in undertaking such a manoeuvre.

The probability of initiating a forced merge is given by the following equation:

$$P_{nt}^{F} = \frac{1}{1+\exp\left(6.41 + 1.25\delta_{nt}^{hv} - 5.43\upsilon_{n}\right)} \quad (17)$$

where, δ_{nt}^{hv} is the heavy lag vehicle dummy, 1 if the lag vehicle is a heavy vehicle, 0 otherwise. It may be noted that the coefficient of aggressiveness has a significant impact on the decision to initiate a forced merge.

On initiating a courtesy/forced merge, the driver decides whether to complete the merge by accepting the available gap or not based on his respective lead and lag critical gaps. For identification purposes, except for the constant and the unobserved driver characteristics, the coefficients of variables in these levels are restricted to be the same as for the normal gap acceptance level.

Thus, the estimated lead and lag critical gaps can be given by the following equation:

$$G_{nt}^{CLead} = \exp\left(-0.582 + 0.521V_{nt}^{'} - 0.505Min\left(0, \Delta V_{nt}^{lead}\right) + \frac{1.32}{1+\exp(0.420+0.355\upsilon_n)} d_{nt} - 0.054\upsilon_n + \varepsilon_{nt}^{Clead}\right)$$

$$G_{nt}^{CLag} = \exp\left(\begin{array}{l} -1.23 + 0.208Max\left(0, \Delta V_{nt}^{lag}\right) + 0.184Min\left(0, \Delta V_{nt}^{lag}\right) \\ + \frac{0.439}{1+\exp(0.0242+0.00018\upsilon_n)} d_{nt} + 0.0545Max\left(0, a_{nt}^{lag}\right) - 0.554\upsilon_n + \varepsilon_{nt}^{Clag} \end{array}\right) \tag{18}$$

$$G_{nt}^{FLead} = \exp\left(3.11 + 0.521V_{nt}^{'} - 0.505Min\left(0, \Delta V_{nt}^{lead}\right) + \frac{1.32}{1+\exp(0.420+0.355\upsilon_n)} d_{nt} - 0.0401\upsilon_n + \varepsilon_{nt}^{Flead}\right)$$

$$G_{nt}^{FLag} = \exp\left(\begin{array}{l} -2.53 + 0.208Max\left(0, \Delta V_{nt}^{lag}\right) + 0.184Min\left(0, \Delta V_{nt}^{lag}\right) \\ + \frac{0.439}{1+\exp(0.0242+0.00018\upsilon_n)} d_{nt} + 0.0545Max\left(0, a_{nt}^{lag}\right) - 0.0239\upsilon_n + \varepsilon_{nt}^{Flag} \end{array}\right) \tag{19}$$

where, G_{nt}^{Clead} and G_{nt}^{Flead} are lead critical gaps for the courtesy and forced gap acceptance levels (m) respectively, G_{nt}^{Clag} and G_{nt}^{Flag} are lag critical gaps for the courtesy and forced gap acceptance levels (m) respectively, $\varepsilon_{nt}^{Clead}, \varepsilon_{nt}^{Clag}, \varepsilon_{nt}^{Flead}$ *and* ε_{nt}^{Flag} are random error terms: $\varepsilon_{nt}^{Clead} \sim N\left(0, 0.0109^2\right)$ and $\varepsilon_{nt}^{Clag} \sim N\left(0, 0.554^2\right)$, $\varepsilon_{nt}^{Flead} \sim N\left(0, 7.95^2\right)$ and $\varepsilon_{nt}^{Flag} \sim N\left(0, 0.465^2\right)$.

The estimation results showed that all other things held constant, a driver is more willing to accept smaller lead and lag gaps when he is in the courtesy merging state than in normal or forced merging state. This is intuitive since in case of courtesy merging, the lag vehicle is slowing down and therefore, a smaller buffer space is sufficient.

The constant term for the lag critical gap for forced merging is the smallest. However, the lead critical gap for the forced merging case is relatively large reflecting the fact that once the driver has initiated a forced merge (pushed his front bumper establishing his right of way), the merge is completed only when the lead gap is sufficiently large since the manoeuvre involves significantly higher risk as compared to the normal gap acceptance.

The anticipation time is normally distributed within 0 to 4 sec.[1] The estimated distribution of anticipation time is

---

[1] Different values between 0 to 6 sec were tested as the upper limit of anticipation time and the selected value (4 sec) provided the best goodness-of-fit.

$$f(\tau_n) = \begin{cases} \frac{1}{0.833}\phi\left(\frac{\tau_n - 1.87}{1.44}\right) & \text{if } 0 \le \tau_n \le 4 \\ 0 & \text{otherwise} \end{cases} \tag{20}$$

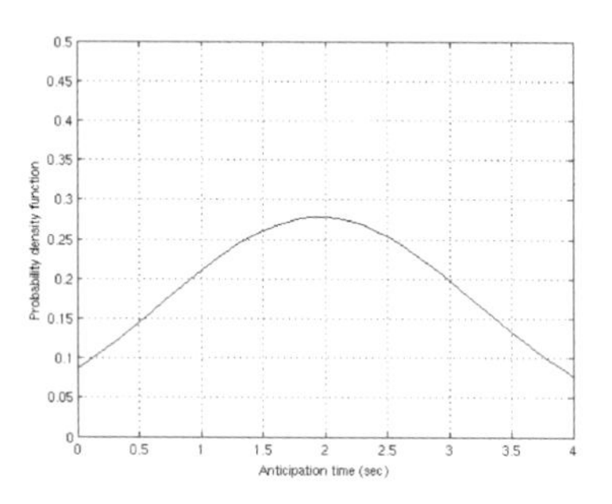

Figure 7: The distribution of anticipation time

MODEL COMPARISON

The state dependent merging model is compared against a simpler instantaneous model (Lee 2006) that does not capture the persistent behaviour of drivers and ignores state dependency. The instantaneous model aims at capturing the normal, forced and courtesy behaviour of drivers through a single gap acceptance level by including variables relevant to all three types of merges in a single critical gap function. The model structure is shown in Figure 8. The model is estimated with the same trajectory data.

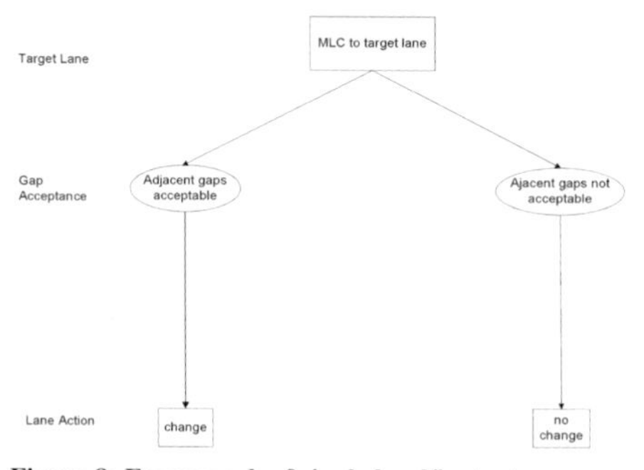

Figure 8: Framework of single level/instantaneous merging model (Lee 2006)

The state dependent model is an extension of the instantaneous model. The summary statistics of the estimation results for the two models, presented in Table 3, show a significant improvement in the fit of the model, even when accounting for the larger number of parameters in the state dependent model.

Table 3 – Model Comparison

Model	Likelihood Function Value	Number of Parameters
Instantaneous	-1639.69	17
State dependent model	-1609.65	42
$LR = 2\left[L(U) - L(R)\right] \sim \chi^2_{1-\alpha, k_U - k_R}$ $LR = 60.08 > \chi^2_{(0.95,25)} = 37.65$		

A likelihood ratio test was performed to select between the two alternative models. The likelihood ratio test results, also presented in Table 3, indicate that the unrestricted (U) state dependent model is significantly better than the restricted (R) instantaneous model. Therefore, the instantaneous model can be rejected as incorrect at 95% confidence interval.

The simulation capability of the state dependent model was compared with the performance of the instantaneous model within the microscopic traffic simulator MITSIMLab (Yang and Koutsopoulos 1996). Both models were implemented in MITSIMLab and the same merging section used for the model estimation (Interstate 80, California) was simulated. The comparison of the distribution of the actual travel time in the section and MITSIMLab simulations using each of the models are presented in Figure 9.

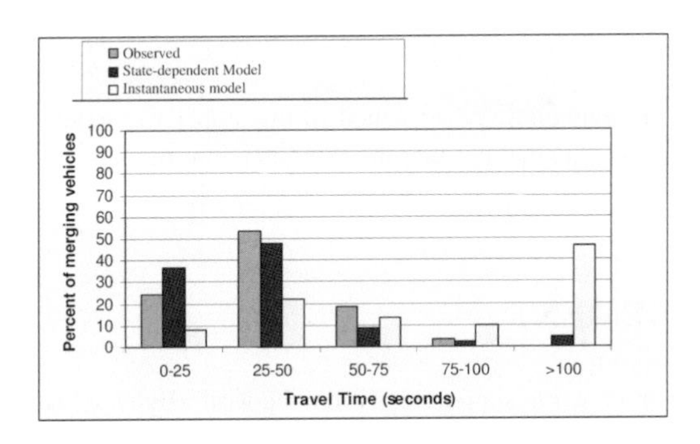

Figure 9: Observed and simulated travel times in the Interstate-80

As observed in Figure 9, the instantaneous model over predicts congestion in the merging section while the state dependent model has a much better replication of the reality. An extensive validation study to compare the simulation capability of the state dependent model using aggregate trajectory data collected from another site with a different ramp configuration is presented in Choudhury et al. (2006b).

CONCLUSIONS

In this paper, a methodology to model state dependency in lane changing behaviour has been demonstrated by applying it to model the merging behaviour of drivers in a congested freeway. The model has explicit normal, courtesy and forced merging components sequenced in a single decision framework. The decision to initiate a merge and the acceptance of gaps to complete the merge are affected by the decision state of the driver as well as neighbourhood variables and driver characteristics (agent effect). The model parameters for state-transition are estimated simultaneously with the parameters of the gap acceptance models with detailed vehicle trajectory data using maximum likelihood estimation technique.

The statistical model selection criteria using the estimation results showed that the proposed state dependent merging model is superior to a single level instantaneous model estimated with the same data ignoring state dependency. This result was further strengthened by a validation case study, which compared the results obtained from simulation runs from each of the model implementations in the microscopic traffic simulator MITSIMLab.

In the current model, only lateral decisions involved with the merging decision was modelled. The extent of the improvements obtained by incorporation of state dependency in the structure indicates the possibility of further enhancements through extension of the model to explicitly capture the state dependency between lane changing, target gap choice and acceleration decisions of the driver.

It may be noted that the methodology presented in this paper to model state dependency in merging behaviour can be extended to other driving behaviour models as well and this will be explored in future research.

ACKNOWLEDGEMENT

This paper is based upon work supported by the Federal Highway Administration under contract number DTFH61-02-C-00036. Any opinions, findings and conclusions or recommendations expressed in this publication are those of the authors and do not necessarily reflect the views of the Federal Highway Administration.

REFERENCES

Ahmed K.I. (1999), Modeling drivers' acceleration and lane changing behaviors, PhD thesis, Department of Civil and Environmental Engineering, MIT.

Benekohal R. and Treiterer J. (1988), Carsim: Car following model for simulation of traffic in normal and stop-and-go conditions. Transportation Research Record 1194, pp. 99-111

Cambridge Systematics Inc. (2005), NGSIM I-80 Data Analysis Report, prepared for Federal Highway Administration.

Choudhury C.F. (2005), Modeling Lane-changing Behavior in Presence of Exclusive Lanes, M.S thesis, Department of Civil and Environmental Engineering, MIT.

Choudhury C.F., Ben-Akiva M., Toledo T., Rao A., Lee G. (2006a), Cooperative Lane changing and Forced Merging Model: NGSIM Final Report, Federal Highway Administration.

Choudhury C.F., Ben-Akiva M., Toledo T. , Lee G. and Rao A. (2006b), Modeling Cooperative Lane Changing and Forced Merging Behaviour, Transportation Research Board, 86th Annual Meeting.

Gipps P.G. (1986), A Model for the Structure of Lane-changing Decisions, Transportation Research, 20B, pp. 403-414.

Lee G., Modelling gap acceptance in freeway merges (2006), M.S thesis, Department of Civil and Environmental Engineering, MIT.

Toledo T. (2003), Integrated Driving Behaviour Modelling, PhD thesis, Department of Civil and Environmental Engineering, MIT.

Wang J., Liu, R., Montgomery F. (2005), A simulation Laboratory for Motorway Merging Behaviour. In: Mahmassani, H (ed.) Transportation and Traffic Theory: Flows, Dynamics and Human Interaction, Elsevier, 5(3), pp.127-140.

Yang Q. and Koutsopoulos H.N. (1996), A microscopic traffic simulator for evaluation of dynamic traffic management systems. Transportation Research 4C, pp. 113-129.

Zhang Y., Owen L.E. and Clark J.E. (1998), A multi-regime approach for microscopic traffic simulation. Transportation Research Board, 77th Annual Meeting.

Transportation and Traffic Theory 2007
Edited by R.E. Allsop, M.G.H. Bell and B.G. Heydecker

31

THE LAGRANGIAN COORDINATES AND WHAT IT MEANS FOR FIRST ORDER TRAFFIC FLOW MODELS

Ludovic Leclercq, Jorge Laval and Estelle Chevallier, Laboratoire d'Ingénierie Circulation Transport (ENTPE/INRETS), Vaulx-en-Velin, France.

INTRODUCTION

Traditionally, first order traffic flow models have been formulated in Eulerian coordinates (x,t) as a scalar conservation law, namely the LWR model of Lighthill and Whitham (1955), and Richards (1956):

$$\partial_t k + \partial_x (kv) = 0 . \tag{1}$$

This model is fully described by vehicle density k. The speed v and the flow q can be derived from a fundamental diagram (FD) $v=V(k)$ or $q=Q(k)=kV(k)$. This model is appealing because of its simplicity, parsimony and its robustness to replicate basic traffic features. Its solution is usually computed with the Godunov scheme (Godunov, 1959), which is based on iterative solutions of Riemann problems (Daganzo, 1994, Lebacque, 1996). Unfortunately, this scheme is known to introduce important numerical viscosity.

Alternatively, Newell (1993) proposed the use of the cumulative count function $N(x,t)$ and conjectured that the LWR solution is the lower envelope of the multiple values of $N(x,t)$ obtained from proper boundary and initial data. Recently, Daganzo (2005) proved Newell's conjecture using variational theory, which reduces the LWR model to the solution of the Hamilton-Jacobi equation:

$$\partial_t N = Q(-\partial_x N) \tag{2}$$

derived from $q=Q(k)$ since $k=-\partial_x N$ and $q=\partial_t N$. This new approach opened the door to powerful numerical methods based on shortest-path algorithms in (x,t) coordinates (Daganzo,2005b).

The aim of this paper is to go further into the use of the N-function by formulating the LWR model in the transformed coordinate system (N,t). These Lagrangian coordinates are fixed to a given fluid particle and move with it in space-time. In this new coordinate system, the purpose is no longer to determine the local density k but the position $X(n,t)$ of vehicle number n. Note that in the continuum, n is not necessarily an integer. In the remainder of the paper, capital N (respectively X) will stand for the $N(x,t)$ (respectively $X(n,t)$) function while n (respectively x) will define a value taken by this function.

This paper is organized as follow: section 2 formulates the LWR model in Lagrangian coordinates as a conservation law and as a variational principle; it also derives relevant numerical schemes. Section 3 analyses the errors introduced by the Godunov scheme in Eulerian coordinates in order to gain some insights about its nature. Section 4 shows how to implement existing and novel extensions using the Lagrangian approach. Finally, section 5 presents a discussion.

THE LWR MODEL IN LAGRANGIAN COORDINATES

This section presents the continuum formulation of the LWR model in Lagrangian coordinates, as a conservation law and as a variational principle. Numerical schemes derived from each formulation will then be proposed. The equivalence between both numerical schemes will be proven under specific assumptions.

Continuum formulation

Lagrangian conservation law

The conservation law in Lagrangian coordinates was first introduced by Courant and Friedrich (1948) in the case of gas dynamics. In traffic flow, equation (1) becomes:

$$\partial_t s + \partial_n v = 0 \tag{3}$$

where the spacing s corresponds to $1/k$. The corresponding fundamental diagram V^* can be expressed as a concave function of s, $v=V^*(s)=V(1/s)$. Therefore, the LWR model in Lagrangian coordinates corresponds to the following hyperbolic equation in s:

$$\partial_t s + \partial_n V^*(s) = 0. \tag{4}$$

Wagner (1987) has proven the equivalence between (4) and (1) for weak solutions, even in vacuum cases where s is not defined; i.e., when k=0.

Lagrangian variational principle

The reader should refer to (Daganzo, 2005; 2005b) for a complete description of this theory in Eulerian coordinates. For simplicity, we will study here homogeneous problems but inhomogeneous ones can be treated similarly as in (Daganzo, 2005). Special inhomogeneous problems will be exemplified in section 4.

What follows shows that the transformation to Lagrangian coordinates preserves the nature of the problem; i.e. that a partial differential equation similar to (2) has to be solved:

$$\partial_t X = V^*(-\partial_n X) \tag{5}$$

derived from $v = V(1/k) = V^*(s)$. To prove the existence of $\partial_t X$ and $\partial_n X$, we only need to show that X exists and is differentiable. To this end, we note that the function $X(n,t)$ can be obtained by inverting $N(x,t)$; i.e., solving for x in $n=N(x,t)$. Two cases may arise:

(a) non-vacuum ($k=\partial_x N \neq 0$): in this case, N is continuous and strictly decreasing in space. Hence, $N(x,t)$ is bijective and the inversion is possible. Thus, $X(n,t)$ exists, is continuous and strictly increasing in n. Furthermore, as N is differentiable except on shockwaves, X verifies the same property.

(b) vaccum ($\partial_x N=0$ and $\partial_n X=+\infty$): this case corresponds to step-jumps in the X-profile with respect to n; i.e., voids in traffic flow. Intuitively, these jumps should not be a problem because a void separates two independent LWR problems: the solution of one does not influence the solution of the other. Reassuringly, Wagner's results (1987) imply that the general problem remains well-posed even when N is not invertible.

Therefore, it is possible to formulate the Lagrangian variational principle analogously to the Eulerian one; i.e., it can be treated similarly as in (Daganzo, 2005). One just has to transpose variables using Table 1.

Table 1: Correspondence between the Eulerian and Lagrangian variational principles

	Eulerian coordinates	**Lagrangian coordinates**
Unknown function	N	X
Main variable	$k=-\partial_x N$	$s=-\partial_n X$
Flux	$q=\partial_t N$	$v=\partial_t X$
FD	$Q(k)$	$V^*(s)$

All the results proven in (Daganzo, 2005 ; 2005b) can thus be applied to the Lagrangian variational formulation of the LWR model. Notably, the value of X at a point P in the (n,t) plane, X_P, can be expressed as a least-cost path problem:

$$X_P = \min\left(B_{\mathcal{P}} + \Delta(\mathcal{P}) : \forall \mathcal{P} \in \mathbf{V} \cap \mathbf{P}_P\right), \text{ where}$$

$\mathbf{V}$: set of all valid paths

$\mathbf{P}_p$: set of all path from the boundary condition to P (6)

$B_{\mathcal{P}}$: X value at the beginning of the path

$\Delta(\mathcal{P})$: cost of path $\mathcal{P}$

Analogously to (Daganzo, 2005), "waves" in (n,t) coordinates are characteristics where s is constant, they have slopes $u=\partial_s V^*(s)$ representing a passing rate. We define two types of passing rates: (a) u is a "possible passing rate" if there exists s such that $u=\partial_s V^*(s)$; (b) $\hat{u}$ is an "allowable passing rate" if $\min(\partial_s V^*(s)) \leq \hat{u} \leq \max(\partial_s V^*(s))$. "Valid paths" are continuous and piecewise differentiable paths $n(t)$ in the (n,t) plane whose slopes $n'(t)$ are allowable passing rates. "Wave paths" are valid paths whose slopes are possible passing rates and are thus composed of a succession of waves.

The cost rate r on a wave path is given by $d_t X$. The scalar r represents the speed of the Eulerian characteristic associated to the passing rate u.

$$r = \mathrm{d}_t X = \partial_t X + \partial_n X \partial_t n = v - su \,. \tag{7}$$

As (5) holds and V^* is concave, one can express r only as a function $R(u)$ using the Legendre transformation as in (Daganzo, 2005):

$$r = R(u) = \sup_s \left\{ V^*(s) - su \right\} . \tag{8}$$

The cost on a Lagrangian valid path $\mathcal{P}$ from B to P is thus:

$$\Delta(\mathcal{P}) = \int_{t_B}^{t_P} R(n'(t))\,dt \,. \tag{9}$$

In the next section, we will show how the Lagrangian variational principle makes it possible to construct a numerical scheme which is exact under few restrictive assumptions.

Numerical resolution

Godunov scheme

In the Godunov scheme, the spacing s is approximated by a constant value, s_i^t, between n and $n+\Delta n$ and is calculated every time step Δt; see Figure 1. Since the flux function V^* for equation (4) is non-decreasing in s, the characteristic speed is always non-negative (traffic anisotropy). The Godunov method reduces in this case to the upwind method:

$$s_i^{t+\Delta t} = s_i^t + \frac{\Delta t}{\Delta n}\left(V^*\left(s_i^t\right) - V^*\left(s_{i-1}^t\right) \right) \tag{10}$$

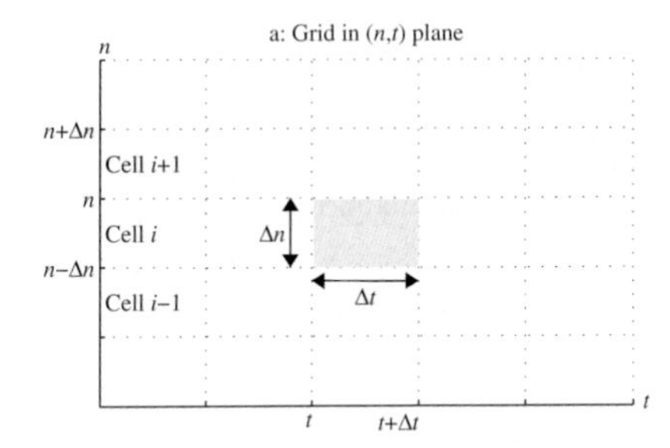

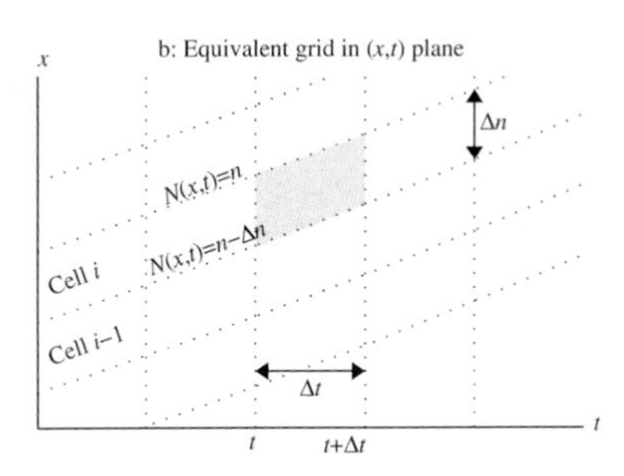

Figure 1: Lagrangian grid

The Courant-Friedrich-Lewy's (CFL) condition (11) defines the stability domain of (10). As the Godunov scheme is consistent and conservative, this also guarantees that it converges (Leveque, 1992).

$$\Delta n \geq \max_s \left| \partial_s \left(V^*(s) \right) \right| \Delta t \tag{11}$$

Notice that non-negative wave speeds imply that Lagrangian rarefaction waves never influence flux values at cell boundaries. The entropy condition is therefore naturally handled in the numerical scheme; i.e., it is not necessary to explicitly include the entropy condition in the numerical solution method.

The Lagrangian Godunov scheme can also be expressed in terms of $X(n,t)$ by noting that the flux $V^*(s_i^t)$ at a boundary n of a cell i is:

$$\frac{X(n,t+\Delta t)-X(n,t)}{\Delta t}=V^*(s_i^t)=V^*\left(-\frac{X(n,t)-X(n-\Delta n,t)}{\Delta n}\right). \tag{12}$$

If we suppose now that Q is triangular as in Figure 2a, V^* can be expressed as:

$$V^*(s)=\min\left(v_m, w\left(k_m s-1\right)\right), \tag{13}$$

where v_m is the free-flow speed, w, the wave speed and k_m, the jam density; see Figure 2b. After simplification (12) becomes:

$$X(n,t+\Delta t)=\min\left(X(n,t)+v_m\Delta t,(1-\alpha)X(n,t)+\alpha X(n-\Delta n,t)-w\Delta t\right), \tag{14}$$

where $\alpha=wk_m\Delta t/\Delta n$. If the CFL condition (11) is satisfied as an equality, then $\Delta n=wk_m\Delta t$ and $\alpha=1$. In this case, equation (14) reduces to:

$$X(n,t+\Delta t)=\min\left(X(n,t)+v_m\Delta t, X(n-\Delta n,t)-w\Delta t\right). \tag{15}$$

When n is an integer and $\Delta n=1$ then $X(n,t)$ corresponds to the position x_n^t of vehicle n at time t and $X(n\text{-}1,t)$ to the position x_{n-1}^t of its leader at the time t. Notice that equation (15) reduces to Newell's simplified car-following model (Newell, 2002) (Daganzo, 2006):

$$x_n^{t+\Delta t} = \min\left(x_n^t + \frac{v_m}{wk_m}, x_{n-1}^t - \frac{1}{k_m} \right). \tag{16}$$

We will show that this scheme is exact using the Lagrangian variational principle.

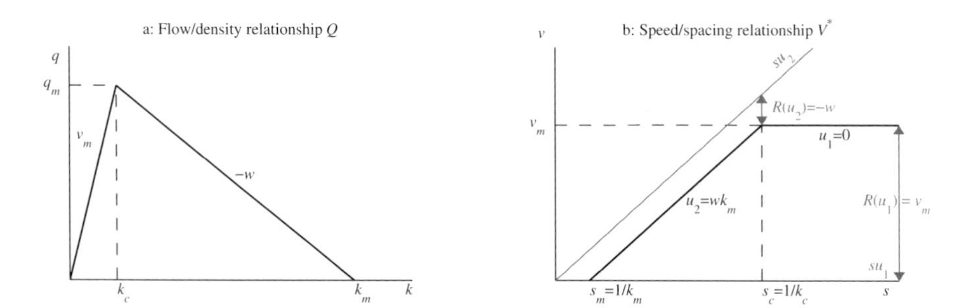

Figure 2: Triangular fundamental diagram

Lagrangian variational principle

Daganzo (2005b) proposed efficient methods to solve the LWR model using the concept of "sufficient networks". A "network" is defined as a directed graph of nodes and arcs in the relevant plane (Eulerian or Lagrangian), where arcs are valid paths. A network is "sufficient" when the least-cost path through the network between every valid pair of nodes is an optimum path. A pair of nodes is said to be "valid" if a valid path exists between them. Notice that this valid path may not necessarily be included in the network. An "optimum path" between a valid pair of nodes is a least-cost valid path between these two points. Notice, again, that this optimum path may not be included in the network and this may introduce errors. In a sufficient network, the solution is exact at every node provided that the initial data is linear between two consecutive initial nodes.

Next, we will apply this method in Lagrangian coordinates to the variational principle (5) supposing that Q is triangular. In this case, waves have only two possible velocities: $u_1=0$ (free-flow) and $u_2=wk_m$ (congestion). The resulting cost rates (8) are: $R(u_1)=v_m$ and $R(u_2)=-w$; see Figure 2b. It can be shown, following the derivation in (Daganzo, 2005b), that any geometric network formed by two families of parallel equidistant lines with slopes u_1 and u_2 and separated by Δn_1 and Δn_2 is sufficient; see Figure 3a. Therefore, with appropriate initial data, the solution at nodes is exact.

Since $u_1=0$ nodes are always lined-up along rows where n values are constant. Furthermore, if one sets $\Delta n_1=\Delta n_2=\Delta n$ nodes also line-up along "time-columns"; see Figure 3b. This defines a rectangular lattice in the (n,t) plane with $\Delta t=\Delta n/wk_m$, which is very practical for computational implementation. Furthermore, with only two incoming arcs per node, the computation of (6) at each node is straightforward; i.e.:

$$\begin{aligned} X(n,t+\Delta t) &= \min\left(X(n,t)+\Delta_{(n,t)\to(n,t+\Delta t)}, X(n-\Delta n,t)+\Delta_{(n-\Delta n,t)\to(n,t+\Delta t)}\right) \\ &= \min\left(X(n,t)+v_m\Delta t, X(n-\Delta n,t)-w\Delta t\right) \end{aligned} \tag{17}$$

where $\Delta_{(n,t)\to(n',t')}$ is the cost of the arc between the two grid-points (n,t) and (n',t').

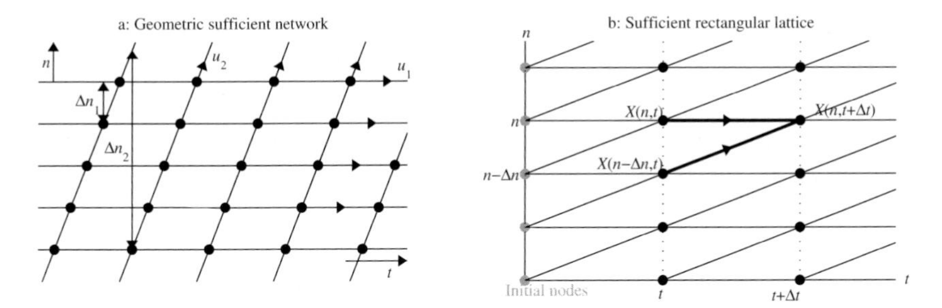

Figure 3: Geometric networks associated to the Lagrangian variational principle

Notice that (17) and (15) are identical. Therefore, the Godunov and the variational schemes are equivalent when Q is triangular and the CFL condition (11) is satisfied as an equality; i.e., when $\Delta n=wk_m\Delta t$. In this case, the Godunov scheme computes the exact solution and no numerical viscosity appears. This will be illustrated in the next section.

NUMERICAL ERRORS OF THE GODUNOV SCHEME IN EULERIAN COORDINATES

This section expresses the Godunov scheme in terms of the N-function in order to obtain insights about the nature of the numerical viscosity introduced by this scheme in Eulerian coordinates. This manipulation will allow us to quantify a bound for the global error. These results enhance the interest of the Lagrangian approach, which induces no numerical errors. Note that Q is supposed triangular in this section.

The Godunov scheme in Eulerian coordinates

In (x,t) coordinates, a highway is partitioned into small sections of length Δx and time into time-steps of duration Δt. The density of cell i at time t is approximated by a constant value, k_i^t. The Godunov scheme expresses the exit flow $q_i^{t\to t+\Delta t}$ of cell i between time t and $t+\Delta t$ as:

$$q_i^{t \to t+\Delta t} = \min\left(\lambda\left(k_i^t\right), \mu\left(k_{i+1}^t\right)\right), \tag{18}$$

where λ and μ correspond to the demand and supply functions; i.e.:

$$\lambda(k) = \min\left(v_m k, q_m\right) \text{ and } \mu(k) = \min\left(w(k_m - k), q_m\right), \tag{19}$$

where q_m is the capacity. This scheme requires the CFL condition to be satisfied:

$$\Delta x \geq v_m \Delta t \,. \tag{20}$$

The density $k_i^{t+\Delta t}$ is updated as usual considering the vehicle conservation law:

$$k_i^{t+\Delta t} = k_i^{t+\Delta t} + \frac{\Delta t}{\Delta x}\left(q_{i-1}^{t \to t+\Delta t} - q_i^{t \to t+\Delta t}\right). \tag{21}$$

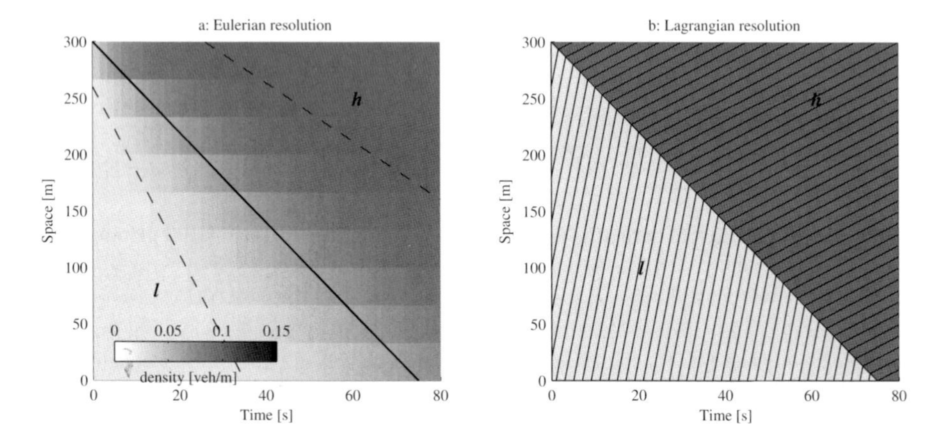

Figure 4: Comparison of the Eulerian and Lagrangian Godunov schemes

It is well known that this scheme induces numerical viscosity especially for shockwave propagating backwards a transition between a low density l and a higher density h. Viscosity appears even if the shockwaves speed w_s is equal to $-w$ and the CFL condition is satisfied as an equality. This is illustrated in Figure 4a where we used v_m=20 m/s; w=4 m/s; k_m=0.15 veh/m; l=0.025 veh/m; h=0.1 veh/m; w_s=-4 m/s; Δt=1.67 s and Δx=33.3 m. Figure 4b shows the same experiment (using Δn=1) solved with the Lagrangian Godunov scheme of section 2, which is exact.

Notice in Figure 4a that the shockwave (shown as a solid slanted line) is increasingly smoothed out as time evolves; dotted lines have been added in the figure to indicate how the time-space region where errors occur grows in time. The exactness of the Lagrangian Godunov scheme is apparent from Figure 4b, as trajectories—represented by solid lines—change speeds precisely on the shockwave; shades of gray have been added to the figure to distinguish both traffic regimes.

Nature of the errors

Even though local truncation errors have been extensively analyzed in the mathematical literature (Leveque, 1992), no analytical expressions for the global errors have been published. The global error is defined here in L_∞ metric for all x and all t. To quantify this error, let us reformulate (18) in terms of $N(x,t)$ using $k=-\partial_x N$ and $q=\partial_t N$:

$$\frac{N(x,t+\Delta t)-N(x,t)}{\Delta t}=\min\left(\lambda\left(-\frac{N(x,t)-N(x-\Delta x,t)}{\Delta x}\right),\mu\left(-\frac{N(x+\Delta x,t)-N(x,t)}{\Delta x}\right)\right). \quad (22)$$

Combining (19) and (22) gives the variational form of the Godunov scheme:

$$N(x,t+\Delta t)=\min\left(N_A, N_B+wk_m\Delta t, N_C+q_m\Delta t\right) \text{ where} \quad (23)$$

$$\begin{vmatrix} N_A=(1-\beta')N(x,t)+\beta' N(x-\Delta x,t) \\ N_B=(1-\beta)N(x,t)+\beta N(x+\Delta x,t) \\ N_C=N(x,t) \\ \beta=w\Delta t/\Delta x \text{ and } \beta'=v_m\Delta t/\Delta x \end{vmatrix} \quad (24)$$

Note that $N(x, t+\Delta t)$ is calculated as the minimum between three wave paths[1], coming from the points A, B and C; see Figure 5a. Their slopes are respectively v_m, $-w$ and 0, and their Eulerian cost rates are 0, wk_m and q_m. The N-values at these points are N_A, N_B and N_C, respectively.

It is clear from Figure 5a that the Godunov scheme is exact only if N_A and N_B are known exactly. However, this is not true in general as A and B are generally not grid-points. Actually, as can be seen from (24), the N-value at these points is a weighted average of the N-values at the endpoints of their associated cell; see Figure 5b. Notice that N_A is known exactly when the CFL condition (20) is satisfied as an equality: i.e., $\Delta x=v_m\Delta t$. This implies that $\beta'=1$ and A is the grid-point $(x-\Delta x,t)$. However, N_B is known exactly only when Q is an isosceles triangle. In this case, $w=v_m$, $\beta=1$ and B is the grid-point $(x+\Delta x,t)$.

With the exception of isosceles FDs – which are not realistic –, numerical errors always arise when N_B is the minimum in (23). This happens for shockwaves propagating backwards. Shockwaves propagating forwards do not induce numerical errors when the CFL condition (20) is satisfied as an equality because N_A or N_C are the minimums in (23).

[1] Notice that horizontal paths are not wave paths in general but when Q is triangular they are optimum paths. Thus, they may be considered as wave paths here.

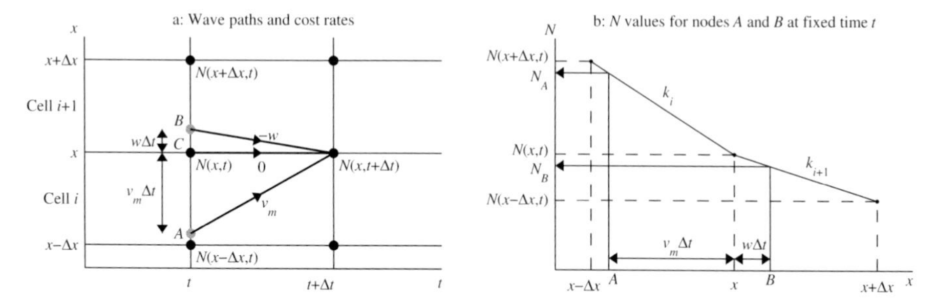

Figure 5: Formulation of the Eulerian Godunov scheme under a variational form

Quantification of the errors

We will now estimate analytically the numerical errors induced by the Godunov scheme, which is possible using equation (23). Let us consider two consecutives cells i and i+1 with densities l and h as shown in Figure 6a. A shockwave appears at time t_0 at the boundary x_1 between these two cells and propagates backwards at a velocity $-w$. For simplicity, we suppose that $\Delta x=v_m\Delta t$ and that the ratio $j=v_m/w$ is an integer[2]. Thus, the shockwave crosses the upstream boundary x_0 of cell i at time t_j after j time-steps; see Figure 6b. At time t_p ($1\leq p \leq j$), the numerical error, e_p, at x_0 between the numerical solution, $\tilde{N}$, and the exact one, N, is given by:

$$e_p = \tilde{N}\left(x_0,t_p\right) - N\left(x_0,t_p\right). \tag{25}$$

In this case, N_B is the minimum in (23). Thus, e_p can be expressed as:

$$\begin{aligned} e_p &= \tilde{N}\left(x_0 + w\Delta t, t_{p-1}\right) - N\left(x_0 + w\Delta t, t_{p-1}\right) \\ &= \tilde{N}\left(x_1, t_{p-1}\right) + l_{p-1}\left(\Delta x - w\Delta t\right) - \left[N\left(x_1,t_{p-1}\right) + \left(p-1\right)hw\Delta t + \left(\Delta x - pw\Delta t\right)l\right] \end{aligned} \tag{26}$$

where l_{p-1} represents the estimated density in cell i at time t_{p-1}. Note that $\tilde{N}(x_1,t_{p-1})$ and $N(x_1,t_{p-1})$ are identical because the flow calculated by the Godunov scheme at the boundary x_1 is always equal to q_h, which is the exact solution. It follows that e_p becomes:

$$e_p = \Delta x\left(1-\alpha\right)\left(l_{p-1} - l\right) - \alpha\left(p-1\right)\Delta x\left(h-l\right), \tag{27}$$

where $\alpha=w\Delta t/\Delta x=w/v_m$.

At time t_p, the estimated density l_p can be expressed by the following recursion:

$$l_p = \left(1-\alpha\right)l_{p-1} + \alpha h = \left(1-\left(1-\alpha\right)^p\right)h + \left(1-\alpha\right)^p l. \tag{28}$$

[2] if it is not the case, it can be shown that the global error computed in this subsection is an upper bound.

By replacing $l_{p\text{-}1}$ in (27), e_p becomes:

$$e_p = \left[-(1-\alpha)^p - \alpha p + 1\right](h-l)\Delta x. \tag{29}$$

Finally, as $j=1/\alpha$, e_j is given by:

$$e_j = (1-\alpha)^{1/\alpha}(l-h)\Delta x. \tag{30}$$

As expected, the error is proportional to Δx and to the density difference l-h. Interestingly, the error is also decreasing in α, which means that the more the congested branch of the FD is flat the more the errors are important. Notice that the error contribution of α can be bound by exp(-1); i.e. $|e_j| \le 0.37|l\text{-}h|\Delta x$. Actually, this bound is rather tight for typical values of α ($4 \le 1/\alpha \le 8$). Notice too, that e_j is an upper bound for the global error (i.e., for all $t \ge t_j$ and all $x \le x_0$) since the Godunov scheme is a contraction mapping and e_j represents the maximum error at x_0 for all t.

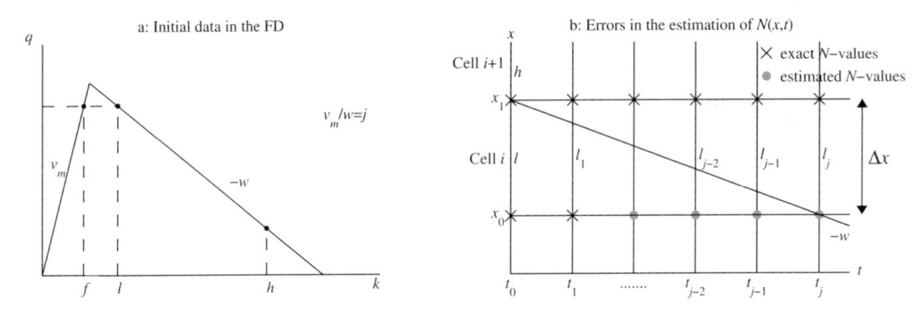

Figure 6: Global error induced by the Godunov scheme for shockwave propagation backwards

Finally, in the more general case of a shockwave, where the upstream density corresponds to a free-flow traffic state (e.g., f in Figure 6a), it can be shown that (30) represents an upper-bound to the global error provided that l corresponds to the congested density associated to the flow $v_m f$; see Figure 6a.

EXTENSIONS USING THE LAGRANGIAN VARIATIONAL PRINCIPLE

This section examines how to incorporate existing extensions to the LWR model into the Lagrangian variational principle. In this section, we will suppose that Q is triangular and that $\Delta n=1$.

Moving bottlenecks

In analogy with the Eulerian case, exogenous moving bottlenecks (see Newell, 1998, Lebacque et *al*, 1998; and Leclercq et *al*, 2004 for a review) can be represented as moving boundary conditions. Thus, they represent shortcuts in the solution network (Daganzo and Menedez, 2005). As described in section 2, the cost rate of each arc of the shortcut becomes the speed of the bottleneck, v_b, while the slope of the arc in (n,t) is the bottleneck's passing rate, r_b; see Figure 7a.

The only caveat is that in multilane/single-pipe highways r_b is not know a priori and therefore the arc cannot be introduced in advance. Note that in single-lane highways (or multilane/multi-pipe) this problem does not exist since r_b is always equal to 0. In fact, this problem only arises in multilane/single-pipe highways when the moving bottleneck is not active. In this case r_b depends on traffic conditions, i.e. $r_b=q-kv_b$; when it is active r_b is set to its maximum value, $\hat{r}_b$.

A general numerical solution method including moving bottlenecks may be described as follows between times t and $t+\Delta t$:

(i) include the shortcut arc (arc a_s in Figure 7b) assuming $r_b = \hat{r}_b$; let B_j be the point of intersections of the shortcut and the network. Notice that this point may intersect either arc a_1 with slope $u_1=0$ (case a in Figure 7b) or arc a_2 with slope $u_2=wk_m$ (case b in Figure 7b).

(ii) calculate the value of X at the point B_j by applying (6) to the two possible paths: $B_{j-1}\rightarrow B_j$ and $(n,t)\rightarrow B_j$ (case a) or $(n-1,t)\rightarrow B_j$ (case b).

(iii) if the optimum path to B_j does not include the shortcut arc then set r_b such that the value of X at B_j is the same on both paths; this condition ensures that r_b is consistent with traffic conditions and that the bottleneck is not active.

(iv) calculate the value of X at the grid-point $(n,t+\Delta t)$ by applying (6) to the two possible paths: $B_j\rightarrow (n,t+\Delta t)$ and $(n-1,t)\rightarrow (n,t+\Delta t)$ (case a) or $(n,t)\rightarrow (n,t+\Delta t)$ (case b).

In the case of a single-lane, the above numerical method simplifies significantly. Only case b in Figure 7b can take place with a horizontal bottleneck trajectory as shown by the dotted line in the figure. Notice that the bottleneck is introduced at a "distance" Δn_b from its leader; this distance is determined when the bottleneck is first introduced, and is constant thereafter.

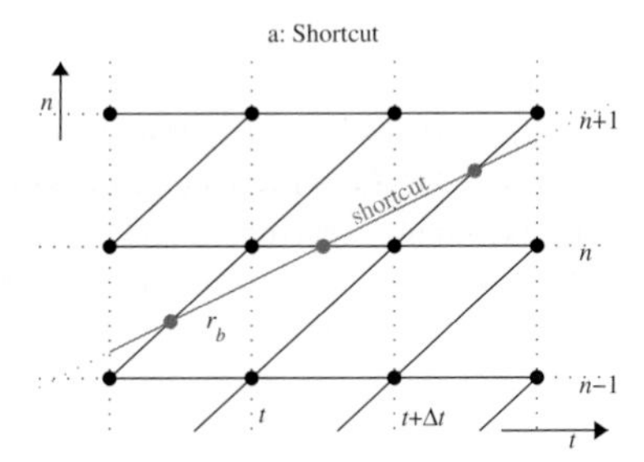

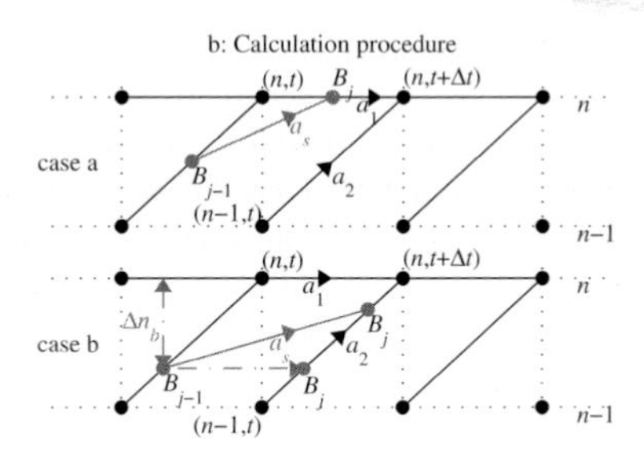

Figure 7: Moving bottlenecks

It can be shown that the position of the bottleneck and its follower are given by:

$$\begin{aligned} X_{B_j} &= \min(X_{B_{j-1}} + v_b \Delta t, X(n-1,t) - \frac{\Delta n_b}{k_m}) \quad \text{(bottleneck)} \\ X(n,t+\Delta t) &= \min(X(n,t) + v_m \Delta t, X_{B_j} - \frac{1-\Delta n_b}{k_m}) \quad \text{(follower)} \end{aligned} \tag{31}$$

Finally, it is worth to note that fixed bottlenecks can be treated in the same way described above by setting v_b=0 and noting that passing rates correspond to the flow crossing the bottleneck. Again, except in cases where r_b can be determined exogenously (e.g., a traffic signal in red phase) the general procedure described above must be utilized.

Self-similar highways

A self-similar highway is composed of successive segments whose FDs are scaled versions of each other; e.g., a highway where the number of lanes varies in space. This means that the set of wave speeds is the same on all segments and that maximum passing rates are multiples of each other. Unfortunately, as opposed to Eulerian coordinates, this problem is harder to treat in Lagrangian coordinates because:

(a) the Lagrangian trajectory of boundaries between segments is not known a priori (for the same reasons than for bottlenecks).

(b) maximum passing rates are different on each segment, which makes it hard to define the network for the whole highway.

(c) if the ratio between different FDs is not a rational number, a sufficient network does not exist and therefore exact solutions can not be found.

Intersections

Existing intersection models split the supply (available capacity) of an outgoing link considering demands on incoming links (see Lebacque and Koshyaran, 2005 for a review). Different models differ in the way they treat demands, but the inputs are the same. Therefore, in the Lagrangian variational principle all one needs to do is computing supplies and demands. This can be accomplished using (19) and properly defining the densities. To this end, we use the method in (Leclercq, 2006) where supplies and demands are computed whenever a vehicle crosses the intersection boundary. It can be shown that in this case the density for computing demands is the inverse of the upstream spacing; the density for computing supplies is the inverse of the downstream spacing. Next, one only has to apply the desired intersection model and incorporate the resulting flow allocation as a fixed boundary condition in space. This boundary corresponds to a bottleneck at the downstream end of each incoming link, and to initial data at the entry of each outgoing link.

The method remains exact as long as the numerical grid-points of every link coincide; i.e., when the FD is the same on all links. Otherwise, one would need to interpolate between grid-points and this would introduce numerical errors that will propagate, and which may or may not grow when passing through neighbouring intersections.

Vehicle characteristics

The Lagrangian framework is the natural environment for introducing vehicle characteristics since it is a coordinate system that "moves" with the flow. Furthermore, if we consider (as in this subsection) a single lane, then $X(n,t)$ represents the trajectory of vehicle n since $\Delta n=1$. In this case, the inclusion of vehicle-specific characteristics becomes straightforward. For example, information such as origin and destination (O/D), vehicle number or drivers' value of time are trivial to incorporate because it does not modify the network (and thus the car-following rule). However, for incorporating characteristics such as maximum desired speed, vehicle acceleration capabilities or driver-specific FDs one must modify the network. Some examples are given next.

Different free-flow speeds

If drivers have different free-flow speeds one may simply change the cost rate of the horizontal arcs of the solution network to the free-flow speed of the particular driver. This is illustrated in Figure 8a, where v_m^n is the desired free-flow speed of vehicle n. Notice that this implementation is straightforward because the network geometry remains unchanged. Only the costs are modified.

Different reaction times

If drivers have different reactions times, they will exhibit different values of w. In this case, the network geometry changes because the slope of the slanted arcs between two consecutive vehicles varies. This is illustrated in Figure 8b where w_n is the wave speed of vehicle n. The time-step implementation of this extension is not straightforward as grid-points are no longer aligned on a regular lattice as shown the by the dotted lines in the figure. It is possible that event programming or approximating w_n by rational numbers may streamline the solution method.

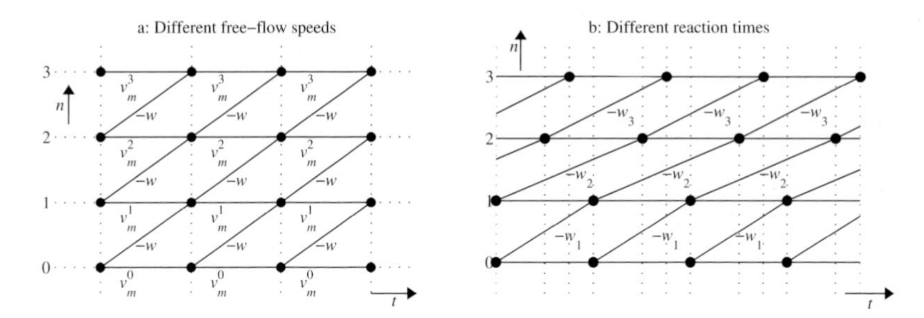

Figure 8: Examples of network for incorporating vehicle characteristics

Bounded vehicle acceleration

In the literature (Lebacque, 2003, Leclercq, 2006) this extension has been conceived for obtaining realistic acceleration profiles when transitioning to a less congested traffic state, as opposed to the infinite accelerations produced by the LWR model.
In Lagrangian coordinates, this can be accomplished similarly as in the case of different free-flow speeds presented above. The only difference is that $v_m^n = v_m^n(t)$; i.e.,

$$v_m^n(t+\Delta t) = v_m^n(t) + a(t)\Delta t\,, \tag{32}$$

where $a(t)$ is the desired acceleration of vehicle n at time t which may depend on vehicle type, roadway geometry, current speed, etc.

Multi-pipe solution method

We refer to a multi-pipe approach when each freeway lane is considered as an independent unit that interacts with adjacent lanes via lane changes. In this case, the solution method for including bottlenecks simplifies significantly compared to multilane/single-pipe highways. This happens because in a single lane bottleneck passing rates are zero, and so are shortcut arc slopes. This implies that one may construct the solution network beforehand without having to modify it due to varying passing rates.

Furthermore, bottlenecks can be used to represent the blockage effect of lane-changing vehicles on the target lane as in Laval and Daganzo (2006). The only complication is the incorporation of forced lane-changing manoeuvres when drivers accept very short non-equilibrium gaps and force their way into the target lane. This phenomenon—commonly observed in the field—can be incorporated using the framework proposed in this paper via the relaxation procedure proposed in Laval and Leclercq (2006). This procedure allows drivers to accept short gaps while gradually attaining safer equilibrium ones.

DISCUSSION

From the authors' point of view, the main advantage of the Lagrangian approach is its exactness when Q is triangular. This is important because a triangular Q is parsimonious while being an accurate representation of reality. Although exactness for triangular FDs is shared by the Eulerian variational numerical methods, further extensions are much easier to incorporate in Lagrangian coordinates provided that a multi-pipe approach is used. This is illustrated in Table 2, which shows the easiness of implementation for the extensions discussed in section 4. Notice that in Lagrangian coordinates, there is only one column because the Godunov and the variational schemes are equivalent when Q is triangular; see section 2. The double entries in the table highlight that a multilane/single-pipe representation makes it difficult to incorporate some extensions in Lagrangian coordinates. Again, this is not the case in the multi-pipe representation because passing rates are known a priori.

Table 2: Extension implementation complexity by approach
(++: very easy; + easy; - not so easy; -- difficult; ? not known as yet)

	Lagrangian approach	**Eulerian approach**	
Extensions	**Godunov/variational schemes**	**Godunov scheme**	**Variational method**
Moving bottlenecks	++ (multi-pipe) -- (multilane / single-pipe)	+ (Daganzo and Laval, 2005a) - (Daganzo and Laval, 2005b)	++
Fixed bottlenecks	++ (multi-pipe) -- (multilane / single-pipe)	++ (Lebacque, 1996)	++
Self-similar highways	++ (multi-pipe) -- (multilane / single-pipe)	++ (Lebacque, 1996)	++
Vehicle characteristics (same FD)	++	- for O/D (Daganzo, 1995)	?
Vehicle characteristics (modified FD)	++ (multi-pipe) ? (multilane / single-pipe)	- for bounded acceleration (Leclercq, 2006); ? otherwise	?
Lane-changing	-	- (Laval and Daganzo, 2006)	?
Intersections	+	+ (Lebacque et Koshyaran, 2005)	+

Table 3 presents a summary of the exactness of the different approaches analysed in this paper. At a glance, the superiority of the Lagrangian approach is apparent as it remains exact in all cases but the Godunov scheme when Q is piecewise linear with more than two wave speeds (PWL). The nature of the error in this case is similar to the Eulerian Godunov case in section 3; i.e., X needs to be interpolated between grid points. We note that when Q is PWL the Eulerian variational numerical method introduces errors because one is forced to introduce horizontal arcs (see Daganzo, 2005b, §3.1) which are not necessarily wave paths. Interestingly, in the Lagrangian counterpart, horizontal arcs are also needed but correspond to wave paths. Therefore, no numerical errors are introduced. Another interesting advantage of the Lagrangian variational method is that it does not require $v_m/w=j$ to be an integer in order to have a rectangular lattice; in the Eulerian counterpart, this is not only necessary but one has to memorize for the j time-steps preceding the current simulation time.

Table 3: Exactness by approach (√:exact; ×: non exact)

	Lagrangian approach		**Eulerian approach**	
	Godunov scheme	**Variational method**	**Godunov scheme**	**Variational method**
***Q* triangular + rectangular lattice**	√	√	×	√ only if v_m/w is an integer and memory is used
***Q* triangular**	√	√	×	√ only for geometric networks
***Q* PWL**	×	√	×	×

With respect to the computational cost for the Godunov scheme, consideration of (15) and (18) reveals that a Lagrangian cell imposes approximately a third of the elementary operations imposed by an Eulerian cell. For the variational approach, the number of elementary operations is roughly equivalent in Lagrangian and in Eulerian coordinates. To have an idea, let us define congestion as a mean density higher than $(k_c+k_m)/2$. For a given Δt, the Lagrangian approach is more efficient than the Eulerian variational method as long as less 30% of the network is congested; it is always more efficient than the Eulerian Godunov method under this definition of congestion.

Finally, further research is needed to ensure that when coupling together different extensions in Lagrangian coordinates no compatibility problems arise. For example, provided that different reaction times imply uneven time-steps, conflicts may arise with bottlenecks or intersections models. The use of memory in Lagrangian networks may provide a solution for such compatibility problems.

ACKNOWLEDGMENT

The authors would like to thank Professor Carlos Daganzo for suggesting the use of variational theory in Lagrangian coordinates. This work has been partially supported by the French ACI-NIM (Nouvelles Interactions des Mathématiques) N°193 (2004).

REFERENCES

Courant, R. and Friedrichs, K.O. (1948). Supersonic Flows and Shock Waves. *Pure Appl. Math*, **1**.

Daganzo, C.F. (2006). In traffic flow, cellular automata = kinematic waves. *Transportation Research*, **40B**(5), 396-403.

Daganzo, C.F. (2005). A variational formulation of kinematic waves: basic theory and complex boundary conditions. *Transportation Research*, **39B**(2), 187-196.

Daganzo, C.F. (2005b). A variational formulation of kinematic waves: Solution methods. *Transportation Research*, **39B**(10), 934-950.

Daganzo, C.F. (1995). The cell transmission model, part II: network traffic. *Transportation Research*, **29B**(2), 79-93.

Daganzo, C.F. (1994). The cell transmission model: A dynamic representation of highway traffic consistent with the hydrodynamic theory. *Transportation Research*, **28B**(4), 269-287.

Daganzo, C.F. and Laval, J.A. (2005). On the numerical treatment of moving bottlenecks. *Transportation Research*, **39B**(1), 31-46.

Daganzo, C.F. and Laval, J.A. (2005b). Moving bottlenecks: A numerical method that converges in flows. *Transportation Research*, **39B**(9), 855-863.

Daganzo, C.F. and Menendez, M. (2005). A variational formulation of kinematic waves: bottlenecks properties and examples. **In**: Mahmassani H.S. (Ed.), *16th ISTTT*, Pergamon, London, 345-364.

Godunov, S.K. (1959). A difference scheme for numerical computation of discontinuous solutions of equations of fluid dynamics. *Mat. Sb.* **47**, 271-290.

Laval, J.A., Daganzo, C.F. (2006). Lane-changing in traffic streams. *Transportation Research*, **40B**(3), 251-264.

Lebacque, J.P. (2003). Two-phase bounded-acceleration traffic flow model: analytical solutions and applications. *Transportation Research Record*, **1852**, 220-230.

Lebacque, J.P. (1996). The Godunov scheme and what it means for first order traffic flow models. **In**: Lesort J.B. (Ed.), *13th ISTTT*, Pergamon, London, 647-678.

Lebacque, J.P., Koshyaran, M.M. (2005). First order macroscopic traffic flow models: intersections modeling, network modeling. **In**: Mahmassani, H.S. (Ed.), *16th ISTTT*, Pergamon, London, 365-386.

Lebacque, J.P., Lesort, J.B., Giorgi, F. (1998). Introducing buses into first-order macroscopic traffic flow models. *Transportation Research Record*, **1644**, 70-79.

Leclercq, L. (2006). Hybrid resolution of the LWR model. *85th Transportation Research Board Annual Meeting* [CDROM]. Washington: Transportation Research Board, 16p. *Submitted for publication.*

Leclercq, L. (2006). Bounded acceleration close to fixed and moving bottlenecks. *Transportation Research B*, accepted for publication.

Leclercq, L., Chanut, S., Lesort, J.B., (2004). Moving bottlenecks in the LWR model: a unified theory. *Transportation Research Record*, **1883**, 3-13.

Leveque, R.J. (1992). Numerical methods for conservation laws. *2nd Edition*. Bâle: Switzerland, Birkhäuser, 214 p.

Lighthill, M.J. and Whitham, J.B. (1955). On kinematic waves II: A theory of traffic flow in long crowded roads. *Proceedings of the Royal Society*, **A229**, 317-345.

Newell, G.F. (2002). A simplified car-following theory: a low-order model. *Transportation Research*, **36B**(3), 195-205.

Newell, G.F. (1998). A moving bottleneck. *Transportation Research*. **32B**(8), 531-537.

Newell, G.F. (1993). A simplified theory of kinematic waves in highway traffic, part I General Theory. *Transportation Research,* **27B**(4), 281-287.

Richards, P.I. (1956). Shockwaves on the highway. *Operations Research*, **4**, 42-51.

Wagner, D. (1987). Equivalence of the Euler and Lagrangian equations of gas dynamics for weak solutions. *J. Diff. Eq.*, **68**, 118-136.

Transportation and Traffic Theory 2007
Edited by R.E. Allsop, M.G.H. Bell and B.G. Heydecker

32

GENERIC SECOND ORDER TRAFFIC FLOW MODELLING

Jean-Patrick Lebacque, Salim Mammar, Habib Haj Salem,
INRETS/GRETIA, Arcueil France.

SUMMARY

The paper presents a generic second order family, which generalizes the ARZ (Aw-Rascle-Zhang) model. This family, called EARZ (Extended Aw-Rascle-Zhang) is characterized by a variable fundamental diagram, by the presence along car trajectories of an invariant related to the fundamental diagram, and by a wide scatter of traffic data in the congested domain. It is shown that the resolution methods developed for the ARZ model can be suitably extended to the EARZ family. A bi-phase traffic flow model introduced recently by Colombo is shown to be close to a model belonging to the EARZ family. The properties of this model are easily deduced from those of the EARZ models.

INTRODUCTION

The Payne-Whitham second order traffic flow model (Payne 1971), a hyperbolic system with relaxation, has been used in many applications. Some of its deficiencies (violation of the anisotropic character of the traffic, Daganzo 1995) have led to the development, by Aw and Rascle 2000, and later but following a different rationale, by Zhang 2002, of a new model, called ARZ in the paper. The ARZ model is also a hyperbolic system of two conservation laws with relaxation.

It can be expressed by the following system:

$$\partial_t \rho + \partial_x (\rho v) = 0 \tag{1a}$$

$$\partial_t (\rho L) + \partial_x (v \rho L) = -\frac{\rho L}{\tau} \tag{1b}$$

The conserved variables of this model are the density $\rho(x,t)$ and relative flow, i.e. the difference between the actual flow and the equilibrium flow:

$$\rho L(x,t) \overset{def}{=} \rho(x,t)\left(v(x,t) - V_e\left(\rho(x,t), x\right)\right) \tag{2}$$

(with: x, t : the position and time, $\rho(x,t)$: the density at time t and location x, v (x,t) the speed at time t and location x, $V_e(\rho, x)$ the *equilibrium speed* (at location x)). L is the relative speed $L(x,t) \overset{def}{=} v(x,t) - V_e(\rho(x,t))$.

It can be shown (Mammar et al 2005) that L is an invariant of the traffic flow in the sense that it is constant along vehicle trajectories resp. exponentially decreasing if the relaxation term is taken into account. It can also be shown (Lebacque et al 2005) that the ARZ model is very close to the classical LWR (Lighthill-Whitham-Richards) first order model, Lighthill-Whitham 1955, Richards 1956. Indeed, one can consider locally that the ARZ model is equivalent to a LWR model with a shifted fundamental diagram, the shift being the product ρL.

The physical interpretation of this invariant is not obvious. Indeed the invariant L of the ARZ model is attached to drivers. For instance a driver with a negative L could be expected to drive eventually at negative speed. It can be shown that this does not occur, which shows that the invariant L is also related to global flow properties. Further, experimental validation has still to be carried out.

The idea of this paper is to introduce a generic family of second order models, the EARZ family, which generalizes the ARZ model in the sense that the ARZ invariant L is replaced by a generic invariant attached to vehicles, and dependent on density and velocity. The purpose of such a generalization is to define a whole family of models, with fundamental diagrams varying as a function of traffic state, in order:

- to find the invariants that fit best experimental data,
- to be able to choose the best model among a complete family of models,
- to develop a comprehensive theory of a large family of second order models based on common analysis methods.

The outline of the paper is the following. After introducing the generic second order traffic flow model, we show that the methods developed for the ARZ model (Lebacque et al 2005, Mammar et al 2005) can be extrapolated with suitable modifications to the generic family. Numerical solution methods based on the Godunov scheme and on a particle discretization are developed for the generic family. Finally, it is shown that the bi-phase model of Colombo 2002, a model difficult to analyze, can be slightly modified in order to fit into the EARZ family. Its properties are then easily deduced. The model reproduces well the scatter of traffic data in the congested domain.

DEFINITION OF THE GENERIC SECOND ORDER MODEL

Definitions

We propose to replace in the ARZ model (1a), (1b) the invariant L by a generic invariant I:

$$I(x,t) \stackrel{def}{=} \mathbf{I}(\rho(x,t), v(x,t)). \tag{3}$$

A generic family of second order models suggests itself, which extends the ARZ model:

$$\partial_t \rho + \partial_x (\rho v) = 0 \tag{4a}$$

$$\partial_t (\rho I) + \partial_x (\rho v I) = -\rho f(I) \tag{4b}$$

with ∂_t , ∂_x the partial derivatives with respect to time and space variables. (4a) means conservation of vehicles, and (4b) means that the quantity $I(x,t)$ is conserved (or relaxed) along trajectories:

$$\dot{I} = \partial_t I + v \partial_x I = -f(I). \tag{5}$$

The functional form of the model's invariant I , and the *relaxation function* $f(I)$, can be chosen so as to best fit field data and driver behavior. The ARZ model is characterized by the identities $\mathbf{I} = v - V_e(\rho)$ and $f(I) = I/\tau$ (Aw and Rascle 2002), that is to say $I \equiv L$.

The system (4) can be rewritten as:

$$\frac{\partial}{\partial t}\begin{pmatrix}\rho \\ \rho I\end{pmatrix} + \frac{\partial}{\partial t}\begin{pmatrix}\rho v \\ \rho I v\end{pmatrix} = \begin{pmatrix}0 \\ -\rho f(I)\end{pmatrix}$$

The system (4) without relaxation is a system of conservation equations with conserved variables $U = \begin{pmatrix}\rho \\ \rho I\end{pmatrix} \stackrel{def}{=} \begin{pmatrix}\rho \\ y\end{pmatrix}$ and flux $F(U) = \begin{pmatrix}\rho v \\ \rho I v\end{pmatrix} = \begin{pmatrix}\rho v \\ y v\end{pmatrix}$. We will call the model (4) without relaxation the *homogeneous EARZ model.* Model (4) with relaxation term will be called *inhomogeneous EARZ model.*

It is necessary to express the speed v as a function of ρ and y. We define

$$I = \mathbf{I}(\rho, v) \Leftrightarrow v = \mathbf{I}_v^{-1}(\rho, I) \stackrel{def}{=} \Im(\rho, I) \tag{6}$$

the *speed function* $\Im$ of model (4). Actually (6) expresses two different but equivalent points of view. From a microscopic perspective, there is a "property" I attached to vehicles and drivers, measuring how far the driver's behaviour is from equilibrium. From a macroscopic perspective the fundamental diagram depends locally on this disequilibrium measure I.

Now the flux function F can be expressed in terms of the components (ρ, y) of U as

$$F(U) = \begin{pmatrix}\rho\, \Im\left(\rho, \frac{y}{\rho}\right) \\ y\, \Im\left(\rho, \frac{y}{\rho}\right)\end{pmatrix} \tag{7}$$

In the case of the ARZ model the speed function is given by: $\Im(\rho, I) = V_e(\rho) + I$, since for this model $I(\rho, v) = v - V_e(\rho)$. The speed function for the ARZ model and for an interpolated model (interpolating the Cremer model (Cremer 1979) with the Aw-Rascle model, Aw and Rascle 2000) is illustrated by the following figure 1 (for various values of I):

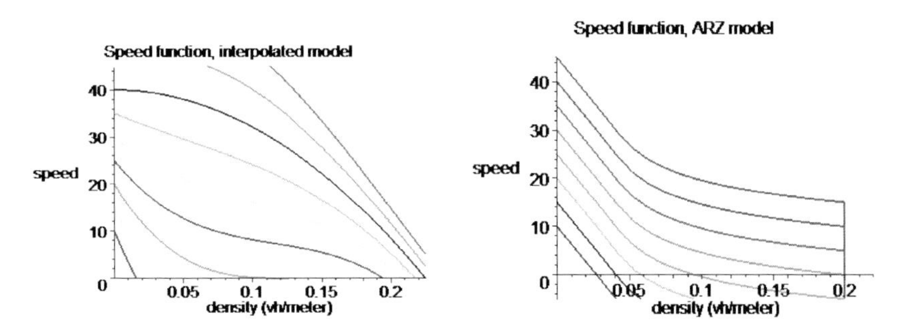

Figure 1: Examples of speed functions (basic units: vehicle, meter, second)

We can now define *the flow function* $\Re$ of the EARZ model (4) as:

$$\Re(\rho, I) \overset{def}{=} \rho \, \Im(\rho, I). \tag{8}$$

In the case of the ARZ model, the flow function is given by

$$\Re(\rho, I) = Q_e(\rho) + I\rho$$

with $Q_e(\rho) \overset{def}{=} \rho V_e(\rho)$ the equilibrium flow-density relationship. In other words the flow function of the ARZ model is the shifted equilibrium flow-density relationship. The flow function for the ARZ model and for the interpolated model (interpolating the Cremer model with the Rascle model) is illustrated by the following figure 2.

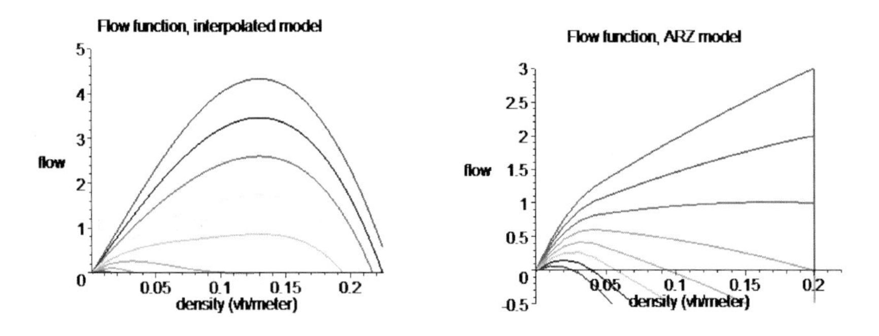

Figure 2: Examples of flow functions (basic units: vehicle, meter, second)

Calculation of eigen-elements

Let us consider in the present sub-section the homogeneous EARZ model. This model can be expressed as:

$$\partial_t U + \partial_x F(U) = 0. \tag{9}$$

The characteristic speeds of (9) constitute the basic tool for describing the propagation of small perturbations, or of large self-similar waves (shock waves or rarefaction waves). Their calculation is standard and has been described in relation to the ARZ model in many publications (Aw Rascle 2000, Aw et al 2002, Zhang 2002, Mammar et al 2005, Lebacque et al 2005). The only problem to be solved is how to overcome the difficulties resulting from the more general setting, i.e. (6), (7). Instead of the variables U we shall use a more practical set of variables:

$$W = \begin{pmatrix} \rho \\ v \end{pmatrix}, \text{ hence} \qquad U = R(W) \overset{def}{=} \begin{pmatrix} \rho \\ \rho\,\mathbf{I}(\rho, v) \end{pmatrix}$$

and the flux function expressed in variables W:

$$F(U) = F(R(W)) \overset{def}{=} G(W) = \begin{pmatrix} \rho v \\ \rho v\,\mathbf{I}(\rho, v) \end{pmatrix}.$$

The differential form of (9):

$$\partial_t U + \nabla_U F(U).\partial_x U = 0$$

is equivalent to

$$\partial_t W + \nabla_W R(W)^{-1}.\nabla_U F(R(W)).\nabla_W R(W).\partial_x W = 0. \tag{10}$$

It follows that the characteristic speeds λ_1, λ_2 of (9) are the eigenvalues of

$$\nabla_W R(W)^{-1}.\nabla_U F(R(W)).\nabla_W R(W) = \nabla_W R(W)^{-1}.\nabla_W G(W)$$

and that the eigenvectors $\mathbf{W}_1, \mathbf{W}_2$ of the matrix $\nabla_W R(W)^{-1}.\nabla_W G(W)$ are related to the eigenvectors $\mathbf{U}_1, \mathbf{U}_2$ of $\nabla_U F(U)$ by the relationship:

$$\mathbf{W}_i = \nabla_W R(W)^{-1}.\mathbf{U}_i(Q(W)) \tag{11}$$

The calculation is straightforward:

$$\nabla_W R(W) = \begin{pmatrix} 1 & 0 \\ I + \rho\partial_\rho \mathbf{I} & \rho\partial_v \mathbf{I} \end{pmatrix} \qquad A = \nabla_W R^{-1}.\nabla_W G(W) = \begin{pmatrix} v & \rho \\ 0 & v - \rho\dfrac{\partial_\rho \mathbf{I}}{\partial_v \mathbf{I}} \end{pmatrix}.$$

The eigenvalues of A are given by

$$\lambda_1(\rho, v) = v - \rho\frac{\partial_\rho \mathbf{I}}{\partial_v \mathbf{I}} \quad \text{and} \quad \lambda_2(\rho, v) = v. \tag{12}$$

It follows from (6) that:

$$\partial_\rho \mathfrak{I} = -\frac{\partial_\rho \mathbf{I}}{\partial_v \mathbf{I}} \tag{13}$$

and therefore the first characteristic speed $\lambda_1(\rho, v)$ can also be expressed as

$$\lambda_1(\rho, v) = v + \rho\partial_\rho \mathfrak{I} = \partial_\rho \mathfrak{R} \tag{14}$$

(by (8) for the expression of $\lambda_1(\rho, v)$ in terms of the flow function $\mathfrak{R}$).

In the rest of the paper we assume that $\Im$ is a decreasing function of the density ρ . This is a natural hypothesis: if the traffic is homogeneous with respect to invariant the *I*, the speed of traffic flow should decrease with density. Thus the velocity of 1-waves cannot exceed the velocity of 2-waves:

$$\lambda_1(W) \leq \lambda_2(W) \quad \forall W .$$

Thus *the generic model respects the anisotropy* of traffic flow.

The eigenvectors $\mathbf{W}_1, \mathbf{W}_2$ are given by:

$$\mathbf{W}_1 = \begin{pmatrix} -\partial_v I \\ \partial_\rho I \end{pmatrix}, \quad \mathbf{W}_2 = \begin{pmatrix} 1 \\ 0 \end{pmatrix}. \tag{15}$$

Expressions (12), (14) and (15) generalize the expressions obtained for the ARZ model (Aw Rascle 2000, Aw et al 2002, Zhang 2002, Mammar et al 2005, Lebacque et al 2005).

1- and 2-waves

In order to check the nature of the 1- and 2-waves associated to the eigen-elements $\lambda_1, \mathbf{U}_1$ and $\lambda_2, \mathbf{U}_2$, it is necessary to calculate the derivatives:

$$\nabla_U \lambda_i . \mathbf{U}_i, \quad i = 1,2 .$$

Formula (11) shows how the eigenvectors change by coordinate change, and it follows that the two eigenvectors constitute two fields, and that the derivatives $\nabla_U \lambda_i . \mathbf{U}_i, \quad i = 1,2$ are the Lie derivatives of the eigenvalues with respect to these fields. Therefore

$$\nabla_U \lambda_i . \mathbf{U}_i = \nabla_W \lambda_i . \mathbf{W}_i, \quad \forall i = 1,2 \tag{16}$$

The derivatives in *W* coordinates are easily obtained:

$$\begin{cases} \nabla_U \lambda_1 . \mathbf{U}_1 = \nabla_W \lambda_1 . \mathbf{W}_1 = -\dfrac{\partial_{\rho\rho}\Re}{\partial_v I} \\ \nabla_U \lambda_2 . \mathbf{U}_2 = \nabla_W \lambda_2 . \mathbf{W}_2 = 0 \end{cases} \tag{17}$$

with: $\partial_{\rho\rho}\Re = 2K + \rho\left(\partial_\rho K + K\partial_v K\right)$ and $\quad K(\rho, v) \overset{def}{=} -\dfrac{\partial_\rho \mathbf{I}}{\partial_v \mathbf{I}}$.

I is constant along 1-field lines, since $\nabla_W I . \mathbf{W}_1 = \left(\partial_\rho \mathbf{I} \;\; \partial_v \mathbf{I}\right) . \begin{pmatrix} -\partial_v \mathbf{I} \\ \partial_\rho \mathbf{I} \end{pmatrix} = 0$.

Thus in the (ρ, v) plane, an integral line of the 1-field is the graph of the speed function $\Im$ for a given value of *I*. Figure 3 below illustrates these results.

Two traffic states U_l and U_r are connected by a 2-wave if they have the same value of the invariant *I*, i.e.

$$I_l = \mathbf{I}(\rho_l, v_l) = I_r = \mathbf{I}(\rho_r, v_r).$$

If the flow function $\Re$ is strictly concave the 1-field is completely non linear, then $\nabla_W \lambda_1 . \mathbf{W}_1 > 0$ and there is a 1-rarefaction wave connecting U_l to U_r if the 1-field is oriented from U_l to U_r. Thus, by (17), under the assumption that the flow function $\Re$ is concave, 1-rarefaction waves are identical to acceleration waves (and 1-shock waves are deceleration waves). Actually the 1-waves of this model are to be interpreted as the kinematical (LWR-like) waves of the model.

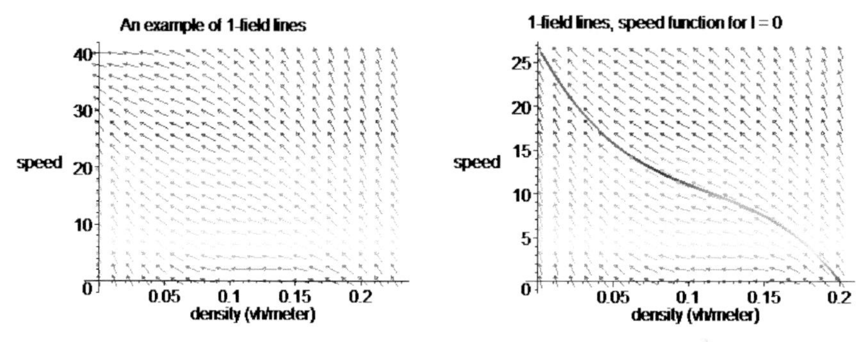

Figure 3: 1-fields, interpolated model (basic units: vehicle, meter, second)

The 2-field is linearly degenerate. 2-field lines are horizontal (constant v), following (15).Two traffic states U_l and U_r are connected by a 2-wave if their speeds v_l and v_r are equal.
The 2-waves of the model can be interpreted as follows: they carry the discontinuities of the invariant I. It should be recalled that 2-waves always travel faster than 1-waves.
All results obtained in this sub-section generalize results obtained previously for the ARZ model.

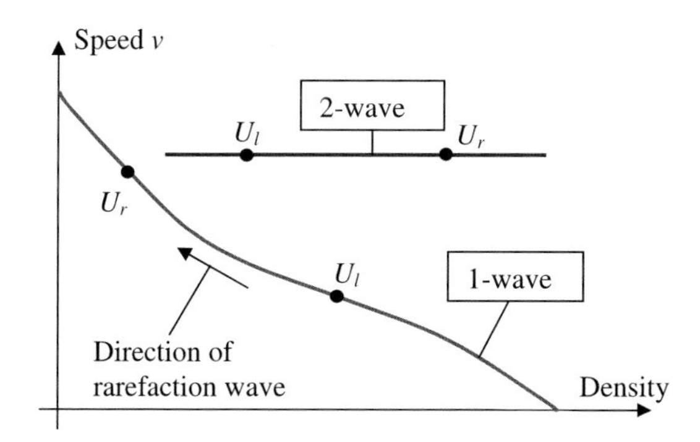

Figure 4: 1- and 2-waves

Shock-waves

Let us consider a shock-wave, with upstream state U_l and downstream state U_r. The Rankine-Hugoniot conservation condition can be expressed as: $F(U_l)-F(U_r)=s(U_l-U_r)$ with s the velocity of the shock. It follows that:

$$0=\det\left|F(U_l)-F(U_r)\;U_l-U_r\right|=\det\begin{vmatrix}\rho_l v_l-\rho_r v_r & \rho_l-\rho_r\\ \rho_l v_l I_l-\rho_r v_r I_r & \rho_l I_l-\rho_r I_r\end{vmatrix}$$

i.e. : $\rho_l\rho_r(v_l-v_r)(I_l-I_r)=0$. The Rankine-Hugoniot conservation condition yields the following three cases:

- ρ_l or $\rho_r=0$ (degenerescence at vanishing density)
- $v_l-v_r=0$ (2-wave, contact discontinuity)
- $I_l-I_r=0$ (1-wave).

The analysis of the Rankine-Hugoniot condition confirms the results obtained in the previous subsection by analyzing the eigen-elements of (4).

Under the assumption that the flow function $\mathfrak{R}$ is concave, there exists a 1-rarefaction wave connecting a traffic state U_l to a state U_r if the 1-field is oriented from U_r to U_l.

Concluding remark: connection between the LWR and the homogeneous EARZ models

Let us assume that there is no relaxation ($f=0$) and that the initial condition of (4) is such that the invariant I is initially piecewise constant. Let $I_1, I_2, \ldots, I_p$ be the values taken initially by I. The invariant I is conserved along vehicle trajectories; it follows that I stays piecewise constant at all times $t>0$.

Thus there are p domains in the (x,t) plane, say $(D_1),(D_2),\ldots,(D_p)$, in which I is uniform:

$$I(x,t)=I_k \quad \forall(x,t)\in(D_k) \quad \forall k=1,\ldots,p.$$

In domain (D_k), the speed is a function of density:

$$v=\mathfrak{I}(\rho,I_k)$$

hence by (8), model (4) reduces to

$$\partial_t\rho+\partial_x(\mathfrak{R}(\rho,I_k))=0\;. \tag{18}$$

Equation (17) expresses that in domain (D_k) the dynamics of traffic flow follow a LWR model, the fundamental diagram of which is $\mathfrak{R}(\rho,I_k)$. This remark applies of course to the special case of the Riemann problem.

The models of the EARZ family can be conceived of as first order models with variable fundamental diagram.

DISCRETIZATION METHODS FOR THE GENERIC EARZ MODEL

Inhomogeneous Riemann problem, shifted supply and demand

The basic properties of the EARZ model generalize with no significant modification the corresponding properties of the ARZ models. Thus using the definition of the *shifted supply and demand* and the solution of the inhomogeneous problem (Lebacque et al 2005), the Godunov scheme can be easily adapted to the EARZ model from the ARZ model. In this section we will therefore simply state the main results and we will only provide proofs inasmuch as they have not been published elsewhere.
The solution of the inhomogeneous Riemann problem is crucial for

- specifying boundary conditions, an absolute prerequisite for network modelling: network entry/exit points, intersections…,
- developing the Godunov discretization scheme.

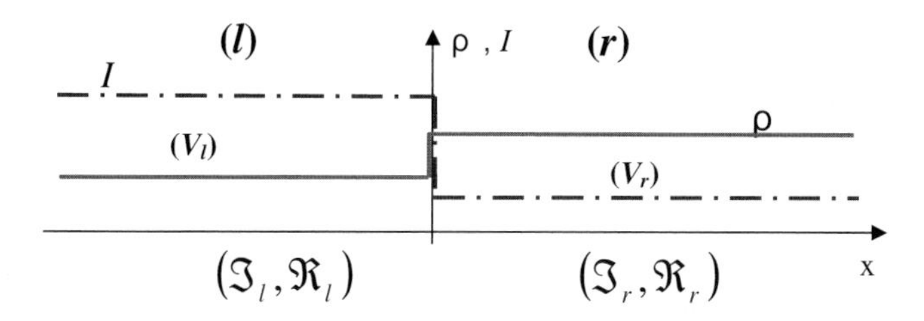

Figure 5: Initial conditions for the inhomogeneous Riemann problem

Let us consider the inhomogeneous Riemann problem, with initial data the traffic states (U_l) and (U_r) on the left- and right-hand side of the origin (see figure 5). Let us recall that *a Riemann problem is called inhomogeneous when the speed and flow functions on the left- and right-hand side of the origin are different*: $(\Im_l, \Re_l) \neq (\Im_r, \Re_r)$. A self-similar solution of (4) with this initial condition is required. It is convenient to describe these states by the values of their density and invariant I: $(V_l) \overset{def}{=} (\rho_l, I_l)$, $(V_r) \overset{def}{=} (\rho_r, I_r)$.

The crucial fact is that the invariant's value I is not changed while traffic crosses the origin. Indeed, the origin may be viewed as a fixed discontinuity and the Rankine-Hugoniot relationship applies, with [.] the jump of any quantity across the discontinuity:

$$0 = \frac{[\rho v]}{[\rho]} = \frac{[\rho v I]}{[\rho I]}.$$

It follows: $[\rho v]=0$, hence $[\rho v I]=\rho v[I]=0$, and $[I]=0$. Another way to interpret this result is to state that I, as an invariant of vehicle trajectories, is unaffected by the presence of fixed discontinuities. Thus the initial discontinuity of I at the origin propagates, and at > 0 time there is no discontinuity of I left at the origin.

The solution to the Riemann problem has only two values of I at any time, $I = I_l$ and $I = I_r$. These values are carried along vehicle trajectories. The discontinuity of this invariant is a 2-wave, which travels at speed v_r . By anisotropy of traffic flow $\lambda_1(W)\leq\lambda_2(W)\quad\forall W$, this wave travels faster than all other waves of the solution. On the right-hand-side of this 2-wave the traffic state is (V_r), and let (V_0) be the traffic state immediately on the left-hand side of this wave.

In order to calculate this state (and others) we need to introduce the following inverses:

$$v=\Im_l(\rho,I)\Leftrightarrow\rho=\Im_{l,\rho}^{-1}(v,I)\quad\text{and}\quad v=\Im_r(\rho,I)\Leftrightarrow\rho=\Im_{r,\rho}^{-1}(v,I). \tag{19}$$

This definition is correct under the assumption that the speed function is a strictly decreasing function of density.

The traffic state (V_0) is characterized by its speed $v_0 = v_r$ and its invariant $I_0 = I_l$. It follows that (V_0) is given by

$$\begin{cases}\rho_0=\Im_{r,\rho}^{-1}(v_r,I_l)\\ v_0=v_r\end{cases} \tag{20}$$

Now in the sector (S) limited on the right by the 2-wave, the invariant I is constant and equal to I_l , and the system (4) reduces to a standard LWR problem:

$$\partial_t\rho+\partial_x\left(\rho\Im(\rho,I_l;x)\right)=0 \tag{21}$$

(note that $\Im(\rho,I;x)=\Im_l(\rho,I)$ if $x<0$, and $\Im(\rho,I;x)=\Im_r(\rho,I)$ if $x>0$).

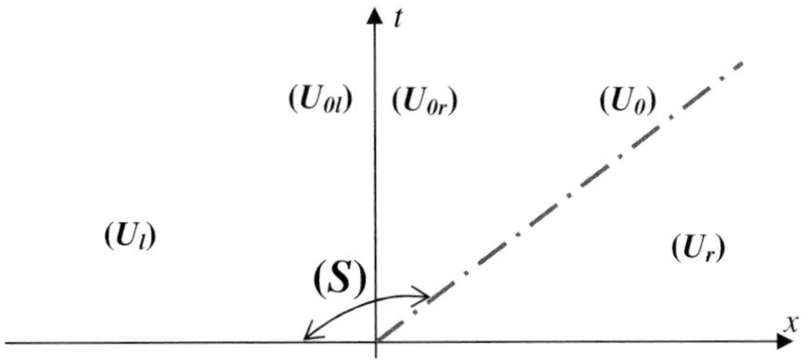

Figure 6: General structure of the Riemann problem solution

In sector (S), the inhomogeneous Riemann problem can be solved for the LWR model (21), with initial conditions (U_l) on the left and (U_0) on the right-hand side of the origin. The methodology is well known (Lebacque 1996, Lebacque et al 2005).

The main elements of the solution to be determined are:

- the flow q_0 at the origin,
- the traffic states immediately on the left- and right-hand side of the origin, (U_{0l}) and (U_{0r}), characterized by their densities ρ_{0l}, ρ_{0r} (the values of their invariant I are: $I_{0l} = I_{0r} = I_l$).

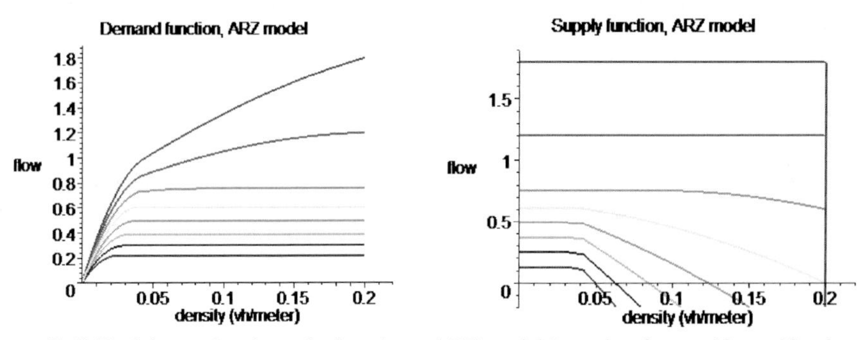

Figure 7: Shifted demand and supply functions: ARZ model I varying from – 12 to +12 m/sec (basic units: vehicle, meter, second)

Let us first defined the *shifted supply and demand* functions (see Figure 7):

$$\begin{cases} \Delta_i(\rho, I) \overset{def}{=} \text{Max}_{0 \le \varsigma \le \rho} \Re_i(\varsigma, I) & \forall i = l, r \\ \Sigma_i(\rho, I) \overset{def}{=} \text{Max}_{\varsigma \ge \rho} \Re_i(\varsigma, I) & \forall i = l, r \end{cases} \tag{22}$$

With $I = I_l$, the shifted supply and demand defined by (22) are the supply and demand associated to model (21) by the usual formulas.

Now we define the upstream demand δ_l and downstream supply σ_r at the origin:

$$\begin{cases} \delta_l \overset{def}{=} \Delta_l(\rho_l, I_l) \\ \sigma_r \overset{def}{=} \Sigma_r(\rho_0, I_l) = \Sigma_r(\Im_{r,\rho}^{-1}(v_r, I_l), I_l) = \Sigma_r(\Im_{r,\rho}^{-1}(\Im_r(\rho_r, I_r), I_l), I_l) \end{cases} \tag{23}$$

the flow at the origin is given by:

$$q_0 = \text{Min}(\delta_l, \sigma_r). \tag{24}$$

The states (U_{0l}) and (U_{0r}) are now given by the following expression

$$(U_{0l}):\begin{cases}\rho_{0l}=\begin{cases}\Delta^{-1}_{l,\rho}(q_0,I_l) & \text{if} \quad q_0=\delta_l\\ \Sigma^{-1}_{l,\rho}(q_0,I_l) & \text{if} \quad q_0=\sigma_r\end{cases}\\ I_{0l}=I_l\end{cases}$$
$$(U_{0r}):\begin{cases}\rho_{0r}=\begin{cases}\Delta^{-1}_{r,\rho}(q_0,I_l) & \text{if} \quad q_0=\delta_l\\ \Sigma^{-1}_{r,\rho}(q_0,I_l) & \text{if} \quad q_0=\sigma_r\end{cases}\\ I_{0r}=I_l\end{cases} \tag{25}$$

with

$$\rho=\Delta^{-1}_{i,\rho}(q,v)\Leftrightarrow q=\Delta_i(\rho,v) \quad \forall i=l,r$$
$$\rho=\Sigma^{-1}_{i,\rho}(q,v)\Leftrightarrow q=\Sigma_i(\rho,v) \quad \forall i=l,r$$

(the standard inverse supply and demand functions).

Finally the solution of the inhomogeneous Riemann problem is completed by calculating the waves $(U_l)\to(U_{0l})$ and $(U_{0r})\to(U_0)$ which can be viewed either as waves of the LWR model (21) or as 1-waves of the EARZ model (4). The $(U_{0r})\to(U_0)$ wave is the 2-wave carrying the I_l-I_r discontinuity and propagating at speed v_r, which bounds the domain (S) on the right, as already mentioned. Figure 11 illustrates (for the 1-phase Colombo model) the resolution of the Riemann problem.

Godunov scheme

The principle of this scheme is well-known. Consider for instance the discretization of traffic flow on a stretch of motorway. Time is divided into time-steps $(t)=[t\Delta t,(t+1)\Delta t]$, the motorway is divided into cells (c) of lengths Δx_c, the traffic state on each cell is approximated by a average cell traffic state $U^t_c=(\rho^t_c,I^t_c)$. The flux between consecutive cells during time-step (t) is obtained by solving an inhomogeneous Riemann problem at the cell boundary, i.e. applying (22), (23), (24).

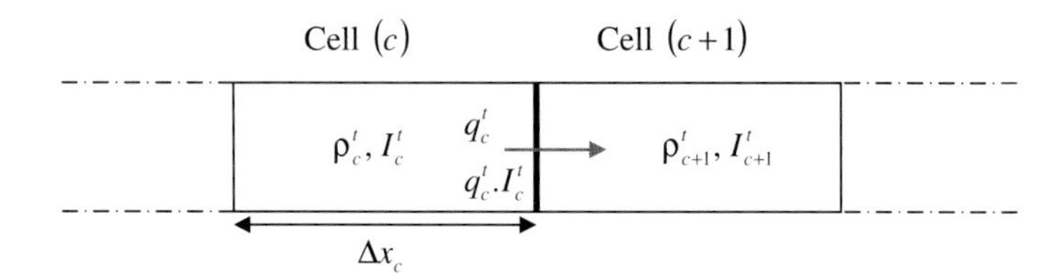

Figure 8: Godunov scheme

It is assumed that there are bounds on I, $I_-\le I\le I_+$, a natural hypothesis: the deviation of traffic from its average behaviour is bounded.

Following Lebacque et al 2005, let us express the scheme for the model without relaxation i.e. $f=0$:

- $\rho_c^{t+1} = \rho_c^t + \frac{\Delta t}{\Delta x_c}\left(q_{c-1}^t - q_c^t\right)$: conservation of vehicles.
- $y_c^{t+1} = y_c^t + \frac{\Delta t}{\Delta x_c}\left(p_{c-1}^t - p_c^t\right)$: conservation of the quantity of invariant I.
- $q_c^t = \mathrm{Min}\left[\Delta_c\left(\rho_c^t, I_c^t\right), \Sigma_{c+1}\left(\Im_{c+1,\rho}^{-1}\left(\Im_{c+1}\left(\rho_{c+1}^t, I_{c+1}^t\right), I_c^t\right), I_c^t\right)\right]$: flow of vehicles, given as the minimum of the shifted supply of cell (c+1) and shifted demand of cell (c), following (23) and (24).
- $p_c^t = q_c^t I_c^t$: the flux of invariant I.
- $I_c^t = \mathbf{P}_{[I_-,I_+]}\left(\frac{y_c^t}{\rho_c^t}\right)$: relationship between the invariant I and the quantity of invariant y.

  $\mathbf{P}_{[I_-,I_+]}$ denotes the projector on the interval $\left[I_-, I_+\right]$.
- Boundary conditions at entry / exit points are given by (23), (24).

The various technical difficulties, the extension to the model with relaxation, and the exact specification of the CFL condition are beyond the scope of this paper. Let us simply mention that under mild hypotheses, the CFL condition can be expressed as:

$$\Delta x_c \geq \Delta t.\,\mathrm{Max}_{\rho\geq 0}\,\Im\left(\rho, I_+\right). \tag{26}$$

Particle discretization

Another approach to the discretization of (4) is the particle (or lagrangian) discretization method, proposed in the context of the ARZ model by Aw et al 2002, Rascle et al 2003.

The idea of this model is to express (4) as:

$$\begin{cases} \dot{x} = \Im(\rho, I) \\ \dot{I} = -f(I) \end{cases} \tag{27}$$

and to consider the cumulative flow function $N(x,t)$. Let $x_n(t)$ denote the trajectory of the n-th vehicle in the system, i.e. the line $N(x,t) = n$. The system (27) can be discretized as

$$\begin{cases} \dot{x}_n(t) = \Im\left(\frac{1}{x_{n-1}(t) - x_n(t)}, I_n(t)\right) \\ \dot{I}_n(t) = -f\left(I_n(t)\right). \end{cases} \tag{28}$$

The quantity $I_n(t)$ denotes the invariant attached to vehicle n. Notice that the local density is approximated as

$$\rho\left(x_n(t)\right) \approx \frac{1}{x_{n-1}(t) - x_n(t)}$$

(the inverse of the inter-vehicular distance). Finally, (28) can be discretized time-wise by the following

$$\begin{cases} x_n(t+\Delta t) = x_n(t) + \Im\left(\dfrac{1}{x_{n-1}(t) - x_n(t)}, I_n(t)\right) \\ I_n(t+\Delta t) = \varphi(I_n(t), \Delta t) \end{cases} \tag{29}$$

with $\varphi(I_0, \tau)$ the solution at time τ of $\dfrac{dI}{d\tau} = -f(I)$, $I_{\tau=0} = I_0$.

In order for vehicles to respect a minimum inter-vehicular distance, a CFL-like condition must be respected. Indeed, (29) can be interpreted as the Godunov scheme for the lagrangian expression of the EARZ model.

Under relatively mild hypotheses, this condition can be shown to express itself as:

$$\Delta t \leq \frac{1}{\rho_{max}(I_+) W_{max}(I_+)} \tag{30}$$

with $\rho_{max}(I)$ the maximum density for which $\Im(\rho, I) \geq 0$ and $W_{max}(I) = -\partial_\rho \Im(\rho, I)_{|\rho=\rho_{max}(I)}$.
I_+ is defined as for the Godunov scheme, as the greatest value taken by the invariant I. This value depends on the initial conditions, the input and the function φ.

The rigorous justification of this scheme, its convergence and the CFL condition (30) are beyond the scope of this paper.

The Godunov scheme and the particle discretization yield similar results for comparable initial and boundary conditions. The Godunov scheme is faster, the particle scheme more flexible (it is possible to assign to each vehicle a value of I).

ANALYSIS OF THE 1-PHASE COLOMBO MODEL

Description of the original 2 phase Colombo model

This model was introduced by Colombo 2002. It is a bi-phase model, designed to explain observations of traffic flow dynamics on Italian motorways (Lombardy), with a phase specific of congested traffic flow, and a phase specific of fluid traffic flow. Each of the phases is associated to a specific density-speed domain.

The model equations are given by

- Fluid phase:

$$\begin{cases} \partial_t \rho + \partial_x \rho v = 0 \\ v = v_f(\rho) . \end{cases} \tag{31}$$

Domain: $\Omega_f = \{\rho / v_f(\rho) \geq \hat{V}\}$.

The notations are those of the original paper of Colombo. $v = v_f(\rho)$ denotes an equilibrium speed density relationship assumed to be linear:

$$v_f(\rho) \overset{def}{=} V - \beta\rho \qquad \text{with}$$

V the free speed and

β a coefficient given by $\beta = \frac{V - \hat{V}}{\hat{R}}$

$\hat{V}$ and $\hat{R}$ some critical speed and density.

It can be noted that (31) is a LWR model with equilibrium relationship $v = v_f(\rho)$.

- Congested phase:

$$\begin{cases} \partial_t \rho + \partial_x \rho v = 0 \\ \partial_t q + \partial_x ((q - q_*) v) = 0. \end{cases} \quad (32)$$

The speed v is given by:

$$v = v_c(\rho, q) \overset{def}{=} \frac{q}{\rho} v_0(\rho) \qquad \text{and} \qquad v_0(\rho) = 1 - \frac{\rho}{R} \text{ (with } R\text{: jam density).}$$

(32) is a second order model, a system of conservation equations with conserved variables $(\rho, q - q_*)$. The notations are again Colombo's original notations. It must be noted that q is **not** the flow (which is given by the product ρv), although q has the dimension of a flow. Indeed $q = \frac{\rho v}{v_0(\rho)}$. Note that q_* is a parameter of the model. The domain of the congested phase (32) is given by

$$\Omega_c = \left\{ (\rho, q) \in [0, R] \times [0, +\infty) / v_c(\rho, q) \leq \hat{V}, \frac{q - q_*}{R} \in \left[\frac{Q_1 - q_*}{R}, \frac{Q_2 - q_*}{R} \right] \right\},$$

with Q_1, Q_2 two parameters such that $0 \leq Q_1 \leq q_* \leq Q_2$.

This model is difficult to analyze, because of the interactions of the two phases. The main difficulties that need be overcome are the analysis of the Riemann problem and the determination of the trajectories of the interfaces between phases.

Trajectory invariant associated to the Colombo model, introduction of a 1-phase model

The second equation of (32) can be rewritten as:

$$\partial_t (q - q_*) + \partial_x (v(q - q_*)) = 0$$

from which it follows that the congested phase admits the following trajectory invariant:

$$\boldsymbol{I}(\rho, v) = \frac{q - q_*}{\rho} = \frac{v}{v_0(\rho)} - \frac{q_*}{\rho}. \quad (33)$$

The speed function associated to this invariant is given by

$$\Im(\rho, I) = \left(I + \frac{q_*}{\rho} \right) v_0(\rho). \quad (34)$$

All known properties of the congested flow (32), as determined for instance in (Colombo 2002), can easily be deduced from the theory elaborated at the beginning of the paper. As an example let us consider the calculation of the smaller characteristic speed $\lambda_1 = v - \rho \dfrac{\partial_\rho \boldsymbol{I}}{\partial_v \boldsymbol{I}} = v + \rho \partial_\rho \Im$ (by (12), (13), (14)).

It follows:

$$\partial_\rho \Im = -\frac{q_*}{\rho^2} v_0(\rho) + v_0'(\rho)\left(I + \frac{q_*}{\rho} \right) \tag{35}$$

in agreement with the expression (3.1): $\lambda_1 = \left(\dfrac{2}{R} - \dfrac{1}{\rho} \right)(q_* - q) - \dfrac{q_*}{R}$, given in Colombo 2002.

Since $v_0(\rho)$ is a positive decreasing function a sufficient condition for $\lambda_1 \le v$ i.e. $\partial_\rho \Im \le 0$ is

$$I \ge -\frac{q_*}{R},$$

which is automatically satisfied in the domain Ω_c in which $I \ge I_1$ and $I + \dfrac{q_*}{R} \ge \dfrac{Q_1}{R}$.

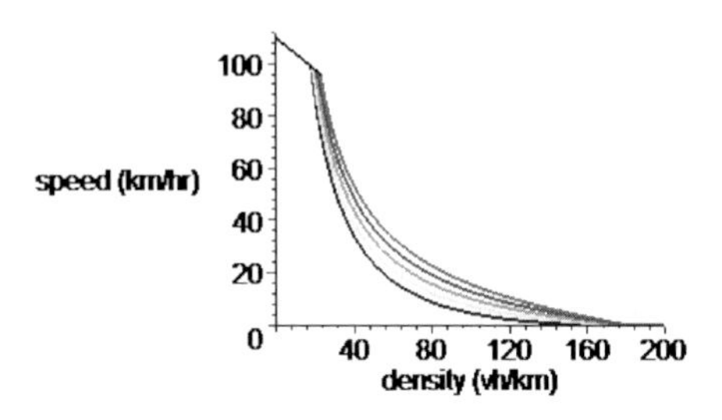

Figure 9: Speed function of the 1-phase Colombo model
(basic units: vehicle, kilometre, hour)

This speed function (34) is only valid in the domain Ω_c. We propose to prolong it to the domain Ω_f by v_f, in order to obtain a 1-phase model behaviourally very close to Colombo's 2-phase model.

Since, depending on the slope β of v_f, the curve $v = v_f$ can intersect the curve $v = \Im$ at one or two points in the interval $[0, R]$, it is necessary to determine the smallest intersection point $\rho \overset{def}{=} \rho_{crit}(I)$ of the two curves by solving $v_f(\rho) = \Im(\rho, I)$. Some basic algebra yields:

$$\rho_{crit}(I) = \frac{1}{2\left(\beta - I/R\right)} \left[V + \frac{q_*}{R} - I - \sqrt{\left(V + \frac{q_*}{R} - I \right)^2 - 4q_*\left(\beta - I/R\right)} \right]. \tag{36}$$

Thus we propose the following speed function

$$v = \Im(\rho, I) \stackrel{def}{=} \begin{cases} v_f(\rho) & \text{if} \quad \rho \le \rho_{crit}(I) \\ \left(I + \dfrac{q_*}{\rho}\right) v_0(\rho) & \text{if} \quad \rho \ge \rho_{crit}(I) \end{cases} \tag{37}$$

This speed function is described by figure 9 above. The resulting model can be written as:

$$\begin{cases} \partial_t \rho + \partial_x (\rho \Im(\rho, I)) = 0 \\ \partial_t (\rho I) + \partial_x (\rho I \, \Im(\rho, I)) = 0 \text{ or } \; - f(I)\rho . \end{cases} \tag{38}$$

In this expression, I is an invariant of trajectories (vehicles) which is a function of position x and time t (and not ρ and v), the conserved variables being ρ and ρI. The traffic speed v is given by $v = \Im(\rho, I)$. I is bounded (definition of domain Ω_f):

$$\frac{Q_1 - q_*}{R} = I_1 \le I \le I_2 = \frac{Q_2 - q_*}{R} .$$

Basic properties of the 1-phase model (38)

The model (37), (38) is of the generalized EARZ type, and its properties can be deduced from the theory outlined in the first sections of this paper. The model is entirely expressed using the flow function:

$$\Re(\rho, I) \stackrel{def}{=} \rho \, \Im(\rho, I) \qquad \text{i.e.:}$$

$$\Re(\rho, I) \stackrel{def}{=} \begin{cases} \rho v_f(\rho) & \text{if} \quad \rho \le \rho_{crit}(I) \\ I \, q_0(\rho) + q_* v_0(\rho) & \text{if} \quad \rho \ge \rho_{crit}(I) \end{cases} \tag{39}$$

$$\text{with} \quad q_0(\rho) \stackrel{def}{=} \rho \, v_0(\rho) = \rho \left(1 - \frac{\rho}{R}\right)$$

The flow function $\Re$ is described by figure 10, left diagram.

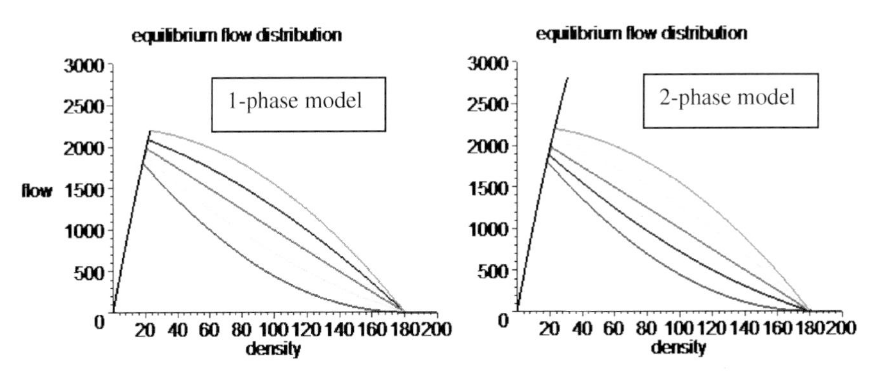

Figure 10: Flow function of the Colombo model, 1-phase and 2-phase (flow in vh/hr, density in vh/km)

The main advantage of the 2-phase model (Figure 10 above, right diagram) over the 1-phase model (Figure 10 above, left diagram) lies in the possibility of extending the fluid phase beyond the density range of the congested phase.

Note that $\rho v_f(\rho)$, $q_0(\rho)$, are concave and $v_0(\rho)$ is linear, thus $\mathfrak{R}$ is concave in the fluid domain, and is concave or convex in the congested domain when I is positive respectively negative. The model predicts a wide scatter of traffic data in the congested domain, due to the range of the dynamic variable I ($I_1 \leq I \leq I_2$).

Thus for negative I values the model predicts acceleration shock-waves, a property which can only be understood by assuming that the distance required for car acceleration is neglected. Such behaviour is very close to the 1-phase approximation of bounded acceleration models (Lebacque 2003).

Remarks

a. The model (37) is not invertible in the sense of (6): it is not possible to define the invariant I as a function **I** of density and speed. Nevertheless, by perturbing slightly the fluid part of model (37), it is possible to construct an invertible model. Such a modification would predict scatter of traffic state points in the fluid domain. It is beyond the scope of this paper.

b. All properties established for the EARZ family apply without change to the 1-phase Colombo model, because they rely on the existence and properties of the function $\mathfrak{I}$ rather than on the function **I**.

c. The 1-phase Colombo model can be modified by using other speed density equilibrium functions than the function $v_0(\rho)$.

Example of resolution of a Riemann problem

The Figure 11 below shows an example of resolution of a Riemann problem with initial conditions.

Using (24), (25), it follows $(U_0) = (U_{0l}) = (U_{0r})$ and $q_0 = \sigma_r = \text{Min}(\delta_l, \sigma_r)$. The solution includes a 1-rarefaction wave and a 2-wave carrying the discontinuity of invariant I.

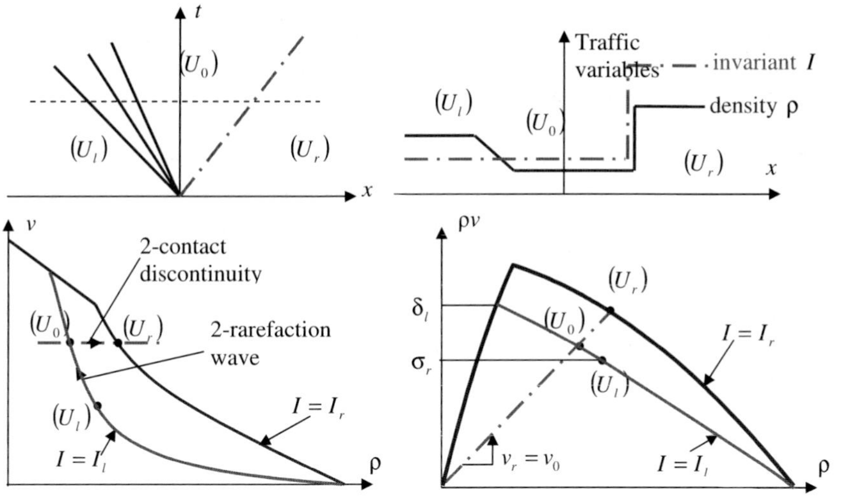

Figure11: Example of Riemann problem for the 1-phase Colombo model

Numerical test results

In order to illustrate the potential of the 1-phase Colombo model, (37), (38), the following numerical experiment was carried out. A stretch of motorway of length 6 km, with two lanes, is considered. The initial data includes some heavy congestion (see figure 12). The entry flow is less than the jam outflow. The simulation is based on the particle discretization (29). The Godunov scheme yields similar results, with reduced scatter. The values of the invariant I are taken randomly in the set $[I_1, I_2]$, with $I_2 = -I_1 = 3.45$ m/s, emulating the distribution of measurement values from A6 south of Paris. Density and flow are measured at a point located inside the jam at a distance 2 km from the entry point. The result (Figure 13) shows a wide scatter of the data points representing the simulated traffic measurements, well in agreement with observed scatter plots.

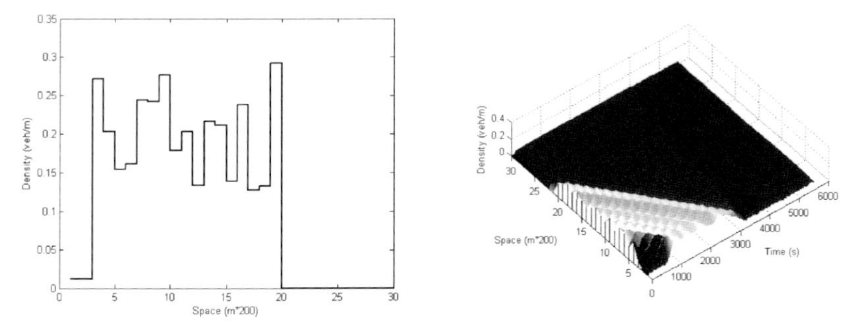

Figure 12: Initial conditions, density dynamics.

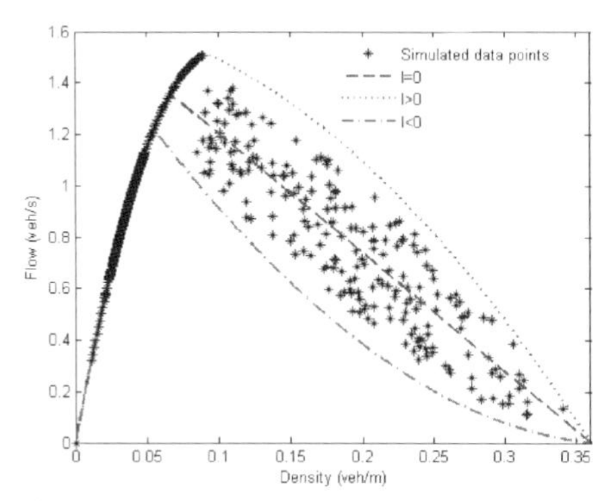

Figure 13: Simulated scatter plot flow vs density

Although the nature of the simulation (particle discretization) and the dispersion of the values of the invariant I in the boundary conditions definitely contribute to this result, the experiment shows the model's potential for reproducing observations. The ARZ model as well as the interpolated model (Figure 2) would obviously do worse, considering their flow function $\Re$.

CONCLUSION

The EARZ model family provides a framework general both in its functional form and in its parameters. The aim of the proposed approach is to obtain, thanks to this flexibility, a better fit with data measurements and better model predictions, while retaining the desirable properties of 2nd order modeling and the solvability of first order LWR type models. The paper provides some essential technical tools: basic model properties, and discretization techniques.

The ARZ family and the 1-phase Colombo model provide two examples of models fitting into the EARZ framework. Other families are conceivable (interpolation for instance). The 1-phase Colombo model likely reproduces best traffic measurements.

By choosing an appropriate family of fundamental diagrams (the flow function $\Re$) it is possible to reproduce any set of traffic measurements, including for instance inverse lambda shapes (Figures 10 and 13). Boundary conditions and simple intersection models can readily be added to the model following for example Lebacque and Khoshyaran 2005.

Nevertheless the EARZ family is at this point purely descriptive in a macroscopic way. Such elements as lane changing, on and off ramps, weaving etc are not accounted for. Actually, the lagrangian discretization as introduced in Aw and Rascle 2000, Rascle et al 2003, and adapted in the present paper, is by essence not designed for modelling refined microscopic behaviour.

Further modelling efforts should target the invariant I. The way EARZ models emulate field measurement scatter plots depends much on the initial and boundary values of I, which are exogenous. The relaxation model for $f(I)$ suggested by Aw and Rascle 2000, Aw et al 2002, is probably incorrect, as it implies a concentration of the scatter plot on the central ($I = 0$) fundamental diagram. Other models should be developed, in order to account for the effects of vehicle manoeuvres, either deterministic, or stochastic. The latter would recapture the element of randomness in the interactions between drivers.

ACKNOWLEDGEMENTS

The authors gratefully acknowledge the support of the ACI (Action concertée incitative) NIM (Nouvelles interfaces des mathématiques) "Modélisation mathématique du trafic automobile" of the French Ministry for Higher Education and Research. The authors thank anonymous referees for helpful comments.

REFERENCES

Aw, A and Rascle, M (2000). Resurrection of second order models of traffic flow. *SIAM Journal of Applied Mathematics*, **60**(3), 916-938.

Aw, A, Klar, A, Materne, T and Rascle M (2002). Derivation of continuum traffic flow models from microscopic follow-the-leader models. *SIAM Journal of applied Mathematics*, **63**, 259-278.

Colombo, R (2002). Hyperbolic phase transitions in traffic flow. *SIAM Journal of Applied Mathematics*, **63**(2), 708-721.

Cremer, M (1979). *Der Verkehrsfluß auf Schnellstraßen*, Springer, Berlin.

Daganzo, C F (1995). Requiem for second order fluid approximation to traffic flow. *Transportation Research*, **29B**(4), 277-286.

Lebacque J P (1996). The Godunov scheme and what it means for first order traffic flow models. *Proceedings of the 13th International Symposium on Transportation and Traffic Theory* (J B Lesort ed), 647-677, Elsevier.

Lebacque J P (2003). Two-phase bounded acceleration traffic flow model: analytical solutions and applications. TRB 82nd annual meeting. *TRR*, **1802**, 220-230.

Lebacque J P and Khoshyaran M M (2005). First order macroscopic traffic flow models: intersection modeling, network modeling. *Proceedings of the 16th International Symposium on Transportation and Traffic Theory* (H S Mahmassani, ed), 365-386.

Lebacque J P, Mammar S and Haj-Salem H (2005). Second order traffic flow modelling: the Riemann problem resolution using supply/demand based approach. *Proceedings of the Euro Working Group on Transportation*. Poznan.

Lighthill M H and Whitham G B (1955). On kinematic waves II: A theory of traffic flow on long crowded roads. *Proceedings of the Royal Society* (London) **A 229**, 317-345.

Mammar S, Lebacque J P and Haj-Salem H (2005). Resolution of the Aw Rascle and Zhang macroscopic traffic flow model. *Proceedings of the 4th IMA Conference on Mathematics in Transport*. London.
Rascle, M, Greenberg J and Klar A (2003). Congestion on Multilane Highways. *SIAM Journal of Applied Mathematics*, **63**(3), 818-833.
Richards P I (1956). Shock-waves on the highway. *Operations Research*, **4**, 42-51.
Payne, H J (1971). Models of freeway traffic and control. *Mathematical Models of Public Systems, Simulation Council Proceedings Series*, **1**(1), 51-61.
Zhang, H M (2002). A non-equilibrium traffic model devoid of gas-like behavior. *Transportation Research*, **36**, 275-290.

Transportation and Traffic Theory 2007
Edited by R.E. Allsop, M.G.H. Bell and B.G. Heydecker

33

UNDERSTANDING TRAFFIC REAKDOWN: A STOCHASTIC APPROACH

Reinhart Kühne, German Aerospace Center, Transportation Studies Group, Berlin, Germany
Reinhard Mahnke, Julia Hinkel, Rostock University, Germany

SUMMARY

Observations from different freeways indicate that traffic breakdowns do not necessary occur at maximum flow and can occur at flows lower or higher than those traditionally accepted as capacity value. The notion breakdown means a stochastic transition from an uncongested traffic state to a congested state which is comparable with a phase transition in nature between different states of matter.

Suggested by recent studies quoting data from German freeways the paper addresses the probabilistic approach for understanding of traffic breakdown. The phenomenon will be described by a balance equation which models the dynamics of jam formation similar to nucleation by two contributions only, called inflow or adhesion rate mainly depending on the traffic volume of the considered road section and discharge rate depending on the length of the congestion.

With this balance equation it is feasible to calculate the dynamics of traffic pattern formation especially the first passage time distribution for a transition from free flow condition to congested traffic including the influence of the parameters affecting the adhesion and discharge rates.

INTRODUCTION

From a traffic engineering point of view a traffic breakdown is defined in a deterministic sense as a

- speed drop $\Delta v > 15$ km/h
- and a mean velocity after speed drop $v_{final} < 75$ km/h
- while the traffic volume before speed drop is $q > 1000$ veh/h/lane

usually based on five–minutes–interval measurement data. This definition is not very critical with respect to the introduced thresholds as it was shown earlier (Kühne and Mahnke, 2005). But observations by different authors from freeways around the world (e. g. in and near Toronto, Canada, by Daganzo et al. (1999) & Lorenz and Elefteriadou, 2000; near Cologne, Germany, by Brilon and Zurlinden, 2003) indicate that traffic breakdowns do not necessary occur at maximum flow and can occur at flows lower or higher than those traditionally accepted as capacity value. The notion *breakdown* means a stochastic transition from an uncongested traffic state to a congested state which is comparable with a phase transition in nature between different states of matter.

Therefore we understand the traffic breakdown is a statistical event which has to be described by probabilities $W(q)$ of breakdown associated with specific vehicular flows q. The stochastic approach based on dynamical equations (balance equation as well as Fokker–Planck equation) to define precisely and to determine analytically the breakdown probability is related to the stochastic concept of capacity (Brilon, Geistefeldt and Regler, 2005) using product limit technique.

Suggested by recent studies quoting data from German freeways we continue to develop the probabilistic approach for understanding of traffic breakdown. The phenomenon will be described by a master equation which balances the effects of growth of a car cluster with those responsible for shrinking. We do so in modelling the dynamics of jam formation similar to nucleation of a supersaturated vapour by two contributions only, called inflow rate responsible for growth and discharge rate responsible for shrinkage, see schematic Fig. 1.

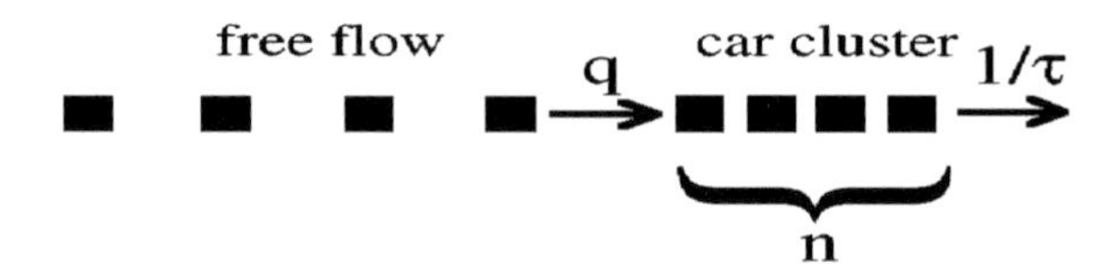

Figure 1. Probabilistic description of traffic pattern formation: q [veh/h] = traffic flow or traffic volume; n = cluster size or spatial queue length (number of congested vehicles) as stochastic variable; τ [$\tau \approx 1.5...2.0$ s/veh] = characteristic time needed for the first car leaving the cluster to become free.

The present paper studies either discrete directed random–walk–dynamics or related continuous drift–diffusion–dynamics together with boundary conditions as a first passage time or escape problem to reach the critical car cluster size n_{esc} for the first time. This critical size summarizes the previous thresholds for speed drop and averaged speed in the definition of a traffic breakdown. The first passage time probability can be calculated from the stochastic differential equation using either forward or backward time development and asking how long the traffic state will remain in free flow (uncongested) conditions. For this aim, special boundary conditions will take into account, (1) $n \geq 0$, i. e. reflecting boundary condition at $n = 0$ in the traffic state space for free flow conditions and (2) $n \leq n_{esc}$, i. e. n_{esc} absorbing boundary condition at $n = n_{esc}$ for reaching critical congested state for the first time which means breaking down of the traffic.

Analytical expressions for the breakdown probability $W(q)$ in comparison with survival probability functions like Weibull distribution are shown together with empirical data points from German freeway A3 near Cologne based on Regler's investigation of 834 breakdown observations on A3 within 5–minutes–intervals (Regler, 2004; Brilon, Geistefeldt and Regler, 2005).

The paper concludes with recommendations and implementations for traffic management and traffic control.

PROBALISTIC DESCRIPTION OF CAR CLUSTER FORMATION

For the following probabilistic description a traffic breakdown is defined as a car cluster formation process. For this we consider a model of traffic flow on a freeway section and study the spontaneous formation of a jam regarded as a large car cluster arising on the road. To get rid of some boundary conditions like entries and exits we can idealise the section by a circular road of length L with N cars moving on it. All the cars are assumed to be identical vehicles and can form two phases. One of them is the set of *freely moving cars* and the other is the congestion called *car cluster*. The cluster is specified by its size n, the number of aggregated cars. Its internal parameters, namely, the headway distance Δx_{clust} and, consequently, the speed of cars in the cluster are treated as fixed values independent of the cluster size n. We note that in the model under consideration there can be only one cluster on the road. The free flow phase is specified also by the corresponding headway distance Δx_{free} that, however, depends strictly speaking on the car cluster size n. The larger the cluster is, the less is the number $(N\text{-}n)$ of the freely moving cars and therefore the larger is the headway distance.

Our model is illustrated in Fig. **2** where we have shown two different regimes of traffic flow, i. e., free traffic flow (left) and congested traffic flow (right). In the congested traffic several jams can exist simultaneously as example two clusters in Fig. 2. Here we consider a simple model where only one car cluster (queue of n cars) is allowed.

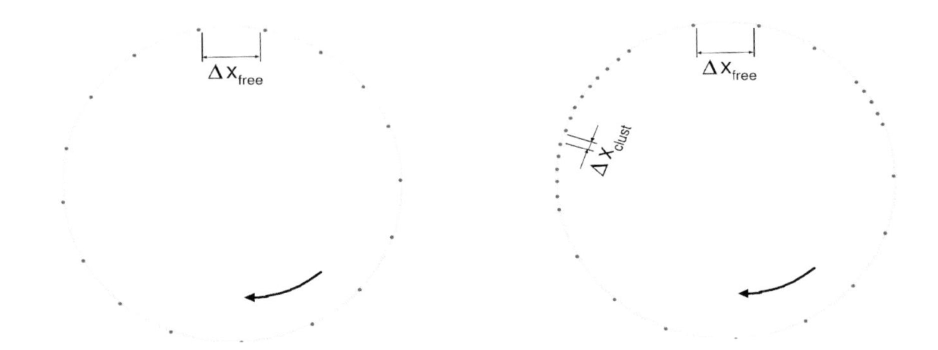

Figure 2. Free traffic flow (left) and congested traffic flow (right) on a one–lane circular road. In the case of congested traffic (right) there are two clusters of different length coexisting with free flow shown as an example. The headway between the cars is Δx_{clust} inside a cluster and Δx_{free} in free flow. The direction of movement is indicated by an arrow.

When a vehicular cluster arises on the road its further growth is due to the attachment of the free cars to its upstream boundary, whereas the cars located near its downstream boundary accelerate to leave it, which decreases the cluster size. These processes are treated as random changes of the cluster size n by ± 1 (see Fig. **3**) and the cluster evolution is described in terms of time variations of the probability function $P(n,t)$ for the cluster to be of size n at time t. Then following (Mahnke, Pieret, 1997; Mahnke et al, 2005) we write the contributions for growth and shrinkage into a balance equation called master equation governing the cluster evolution

$$\frac{\partial P(n,t)}{\partial t} = w_+(n-1)P(n-1,t) + w_-(n+1)P(n+1,t) - [w_+(n)P(n,t) + w_-(n)P(n,t)] \ . \tag{1}$$

The quantity $P(n,t)$ is the probability of finding n vehicles jammed in a cluster at time t.

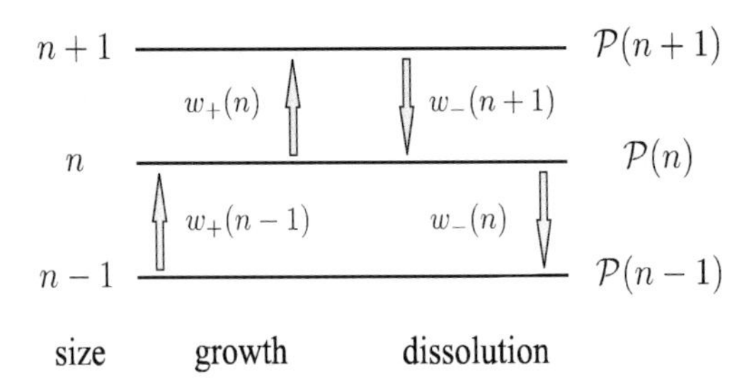

Figure 3. Schematic illustration of cluster transformations.

The growth rate $w_+(n)$ that a car is attached to the cluster takes into account strictly speaking the net time gap for a freely driving car to move up to the cluster. If net time gap and gross time gap are taken equal then this rate is just the traffic flow q (measured in [veh/h])

$$w_+(n) \approx q \tag{2}$$

to which the clustered vehicles as standing or slowly moving cars do not contribute. The attachment rate (2) is thus independent of the cluster size n.
The detachment rate $w_-(n)$ that cars evaporating from the cluster at its downstream front is written as

$$w_-(n) = \frac{1}{\tau(n)} \approx \frac{1}{\tau_\infty} \tag{3}$$

where the value τ_∞ (measured in [sec/veh]) can be interpreted as the characteristic time needed for the first car in the cluster to leave it and to go out from its downstream boundary at a distance about the headway distance in the current free flow state. When the cluster is sufficiently large, it is reasonable to regard the characteristic time $\tau(n)$ as a constant τ_∞. For small clusters the dependence, however, requires special attention, see (Kühne, 2001; Kühne et. al., 2002; Kühne and Mahnke, 2005).

In order to apply well developed techniques of escape problems in the theory of stochastic processes (Gardiner, 2004; Honerkamp, 1994) to the analysis of the traffic breakdown probability we approximate the discrete balance equation (1) the corresponding Langevin equation (4). Our stochastic variable is the size n of a vehicular cluster. The equation of motion which describes the behaviour of a stochastic trajectory $n(t)$ in time t is a stochastic differential equation and for our model under consideration it takes the following form

$$dn(t) = v\,dt + \sqrt{2\,D}\,dW(t) \tag{4}$$

with initial condition $n\,(t{=}0) = n_0 \approx 0$

with a constant driving force (drift velocity $v \approx w_+ - w_-$) as well as a constant fluctuation term (diffusion D). Further on we introduce boundary conditions for the stochastic value $n(t)$. It is naturally to define $n = 0$ as the reflecting border since cluster size $n(t)$ should be always nonnegative. On the other hand, $n = n_{esc}$ is the absorbing or escape value and the breakdown phenomenon appears when cluster size $n(t)$ equals n_{esc}, see Fig. 4 for illustration.

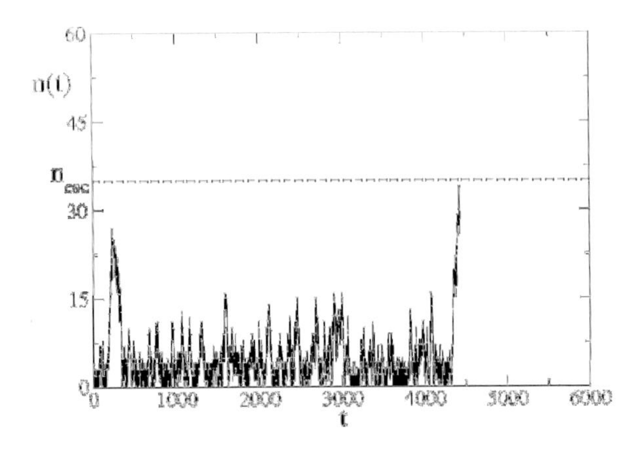

Figure 4. Example of a stochastic trajectory with the initial condition $n(t{=}0){=}0$ showing reflecting boundary at $n = 0$ veh. and absorbing boundary at $n_{esc} = 35$ veh.

Nevertheless, we are not only interested to know the behaviour of one particular stochastic trajectory but we want to study an ensemble as well as. Therefore we approximate the discrete balance equation (1) by the corresponding Fokker–Planck equation (2) called drift–diffusion equation

$$\frac{\partial p(n,t)}{\partial t} = -v\,\frac{\partial p(n,t)}{\partial n} + D\,\frac{\partial^2 p(n,t)}{\partial n^2}\,. \tag{5}$$

Instead of the driving force $v \approx w_+(n) - w_-(n) \approx q - 1/\tau_\infty$ as difference between attachment and detachment rates we would like to consider another important quantity, the so–called potential $U(n)$ defined by $-\,dU/dn = v(n)$ which reads in our case $U(n) = -\,v\,n$ showing a linear function. The functional graph of the potential is presented in Fig. **5** for different scenarios where the system state tends either to $n = 0$ (free traffic flow, $q < 1/\tau_\infty$) or shows undetermined situation ($q \approx 1/\tau_\infty$) or tends to completely congested traffic ($q > 1/\tau_\infty$) with $n = n_{esc}$.

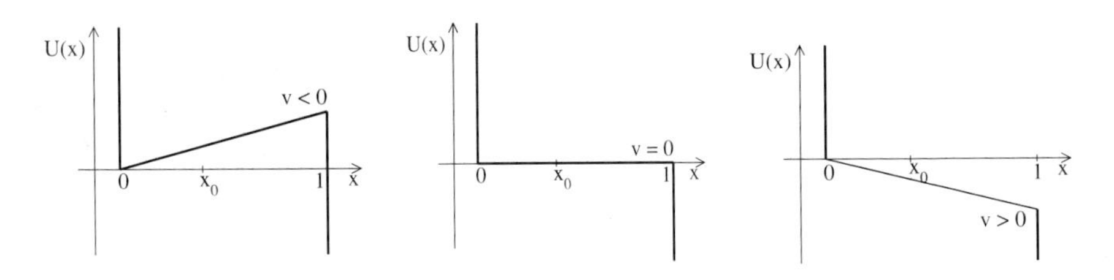

Figure 5. The linear potential $U(x)$ with $x = n/n_{esc}$ for three different scenarios.

Summarizing up to now the following task has been stated. Our aim is to solve the Fokker–Planck equation (5) which can be written as continuity equation

$$\frac{\partial p(n,t)}{\partial t} + \frac{\partial j(n,t)}{\partial n} = 0 \tag{6}$$

with flux

$$j(n,t) = v\, p(n,t) - D\, \frac{\partial p(n,t)}{\partial n} \tag{7}$$

including initial condition as delta function around $n_0 = 0$ given by

$$p(n, t=0) = \delta(n - n_0) \tag{8}$$

together with two boundary conditions:

$n = 0$ – reflecting boundary, i. e. no flux at n_0=0 given by

$$v\, p(n=0,t) \; - \; D\, \left. \frac{\partial p(n,t)}{\partial n} \right|_{n=0} = 0 \tag{9}$$

and
$n = n_{esc}$ – absorbing boundary, given by

$$p(n = n_{esc}, t) = 0 \; . \tag{10}$$

If we know the probability function $p(n,t)$ to be in state n (cluster size) at time t which satisfies the Fokker–Planck equation (5) in agreement with starting condition (8) as well as both boundary conditions (9, 10) we are able to calculate the first passage time distribution function. The exact mathematical results can be found in Hinkel and Mahnke (2006).

BREAKDOWN PROBABILITY

In terms of probabilistic modelling of vehicular traffic the breakdown is an event when the system's state which started at time $t = 0$ with $n = 0$ (free flow) reaches for the first time $n = n_{esc}$ where the escape value n_{esc} is a given cluster size regarded as overcritical.

Following Risken (1996) the distribution function of the first passage times t is given by the outflow probability through the absorbing boundary at $n = n_{esc}$ taking into account the regarded domain $0 \leq n \leq n_{esc}$. The first passage time probability distribution $P(t, n{=}n_{esc})$ or outflow probability at the absorbing boundary n_{esc} is given by

$$\mathcal{P}(t, n = n_{esc}) = -\frac{d}{dt} \int_0^{n_{esc}} p(n,t)dn \tag{11}$$

and has been calculated analytically. The graphical results are shown in Fig. **6**. The system reaches the absorbing boundary $n = n_{esc}$ faster with increasing values of traffic volume q which means increasing drift parameter v.

The first passage time probability density $P(t,n=nesc)$ shown in Fig. **6** can be integrated to obtain the breakdown probability within the time interval $t \in [0, t_{obs}]$ as a cumulative probability

$$W(v, D, t = t_{obs}) = \int_0^{t_{obs}} \mathcal{P}(t, n = n_{esc})\, dt \,. \tag{12}$$

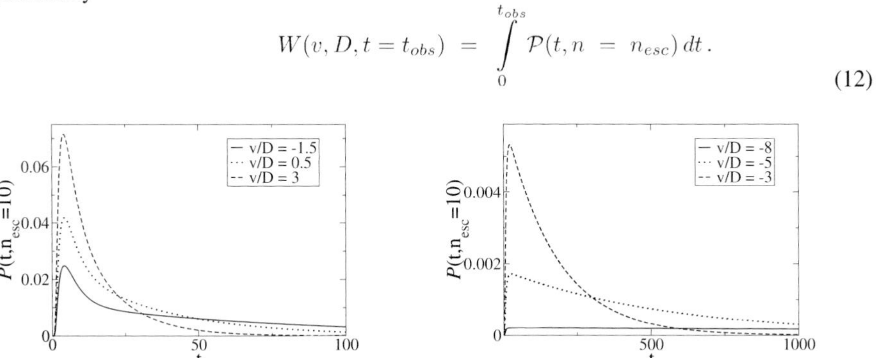

Figure 6. The first passage time probability density distribution $P(t,n=nesc)$ (11) for different values of drift v and diffusion D parameters.

This quantity of practical interest is the probability $W(v,D,t = t_{obs})$ that traffic breakdown takes place within a given observation time interval $t \in [0, t_{obs}]$. It is obtained by integrating the breakdown probability density $P(t,n = n_{esc})$ (11).

The result of integration in three different cases where v/D is larger, equal, or smaller than -2 reads:

$$\frac{v}{D} > -2 \, : \; W(v, D, t_{obs}) \; = \; 2\, e^{\frac{v}{2D}}$$

$$\times \sum_{m=0}^{\infty} \frac{1 - e^{-\left(\tilde{k}_m^2 + (v/2D)^2\right) D\, t_{obs}}}{\tilde{k}_m^2 + (v/2D)^2 + v/2D} \tilde{k}_m \sin\left[\tilde{k}_m\right] \, ; \tag{13}$$

$$\frac{v}{D} = -2 \,:\, W(v, D, t_{obs}) \;=\; e^{-1}\Bigg(3\left(1 - e^{-D\,t_{obs}}\right)$$

$$+\; 2\sum_{m=1}^{\infty} \frac{1 - e^{-\left(\tilde{k}_m^2 + 1\right)D\,t_{obs}}}{\tilde{k}_m} \sin\left[\tilde{k}_m\right]\Bigg) ; \tag{14}$$

$$\frac{v}{D} < -2 \,:\, W(v, D, t_{obs}) \;=\; 2e^{\frac{v}{2D}}\Bigg(-\frac{1 - e^{-\left(-z_0^2 + (v/2D)^2\right)D\,t_{obs}}}{-z_0^2 + (v/2D)^2 + v/2D} z_0 \sinh\left[z_0\right]$$

$$+\; \sum_{m=1}^{\infty} \frac{1 - e^{-\left(\tilde{k}_m^2 + (v/2D)^2\right)D\,t_{obs}}}{\tilde{k}_m^2 + (v/2D)^2 + v/2D} \tilde{k}_m \sin\left[\tilde{k}_m\right]\Bigg) , \tag{15}$$

where the values $\tilde{k}_m$ and z_0 are solutions of transcendental equations

$$\tan \tilde{k}_m \;=\; -\frac{2\,D}{v}\,\tilde{k}_m \tag{16}$$

$$\tanh z_0 \;=\; -\frac{2\,D}{v}\,z_0 \,. \tag{17}$$

We would like to mention that the smallest or ground–state wave vector $\tilde{k}_0$ vanishes when v/D tends to -2 from above, and no continuation of this solution exists on the real axis for $v/D < -2$. A purely imaginary solution $\tilde{k}_0 = iz_0$ appears instead, where z_0 is real (Hinkel and Mahnke, 2006).

Fig. **7** shows the comparison of empirical data with analytical solution (13) – (15). The breakdown probability $W(q)$ as function of flow rate q is presented as analytical solution from traffic flow dynamics (symbols) in agreement with measured capacity distribution function (data provided by Brilon et al., 2005, black dots).

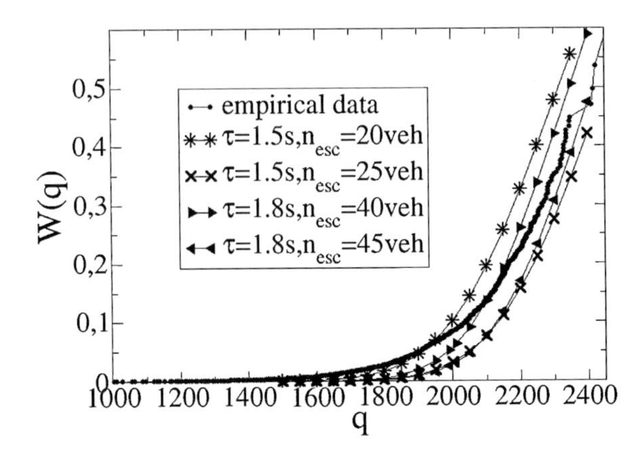

Figure 7. Cumulative breakdown probability $W(q,t = t_{obs})$ for the observation time interval t_{obs}=5 min. Black circles are empirical data from German Autobahn A3 (Regler, 2004). Another points are analytical result (12) for the different values of absorbing barrier $n_{esc} \cdot l_{eff}$ and relaxation time τ. The parameters are l_{eff} = 7 m, n_0 = 0.

The only parameter which allows fitting the empirical data is n_{esc}. All other parameters like τ as inverse discharge rate and the critical density $\rho = 1/l_{eff}$ are fixed within a narrow range due to elementary definitions. The size n_{esc} defines the number of vehicles within a cluster to discriminate congested ($n > n_{esc}$) from non-congested ($n < n_{esc}$) traffic and summarizes therefore the thresholds for a breakdown in comparison with statistical undulations. The value of n_{esc} is about 40 % higher in case of a traffic control system switched on along the regarded road section and shows the effects of traffic control systems by stabilizing traffic flow. It leads to an onset of the breakdown at higher critical traffic volumes in comparison to a situation without a traffic control system. The range n_{esc} is limited to realistic cluster sizes. Very small values like n_{esc} = 2 make no sense as well as too large values n_{esc} = 1000. Such extremes can be omitted.

WEIBULL DISTRIBUTIONS AS FIT FUNCTION

The shape of cumulative breakdown probability $W(q,t = t_{obs})$ in any case reminds to the stochastic distributions used in reliability assessment and support the ideas by Regler (2004) and others. These authors use Weibull distributions as fitting curves with enough parameters to match a broad variety of cumulative distribution functions. The Weibull function is defined as (Muthly, 2004)

$$W(q) = 1 - \exp\left[-\left(\frac{q}{\beta}\right)^{\alpha}\right], \tag{18}$$

where α and β are control parameters of the distribution. We have fitted our calculation for the cumulative (breakdown) probability to Weibull distribution. Fig. **8** confirms the *S*–shaped overall behaviour by fitting of measurements for traffic breakdown probability distribution from field observation for German Autobahn A5 within 5 min measurement intervals.

The following transformations relating the parameters in our equations to the physical observables which have been used

$$v = \left(q - \frac{1}{\tau}\right) l_{eff} , \tag{19}$$

$$D = \frac{1}{2}\left(q + \frac{1}{\tau}\right) l_{eff}^2 , \tag{20}$$

where l_{eff} is the effective length of a car. Here q is the vehicular flow and τ is the characteristic reaction (relaxation) time constant, as introduced earlier.

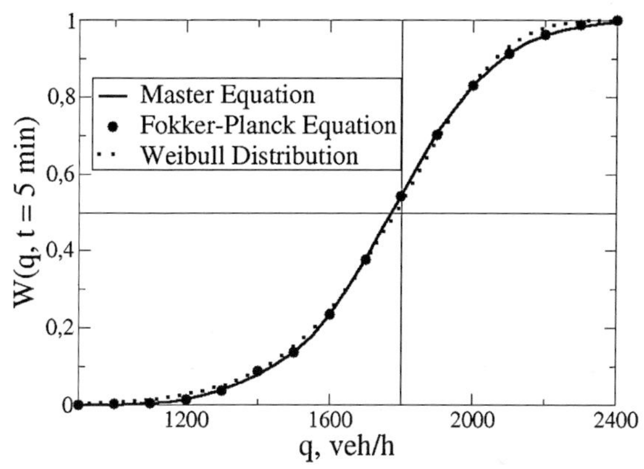

Figure 8. Cumulative breakdown probability $W(q,t = t_{obs})$. The parameters of calculation (Fokker–Planck equation) and simulation (Master equation) are escape cluster size n_{esc}=19 veh., effective length of car l_{eff}=7 m, relaxation time τ=2 s, absorbing boundary $x_{esc} = l_{eff} \cdot n_{esc}$=133 m, observation time t_{obs}=5 min. For comparison Weibull distribution is shown with parameter values α=8.3 and β=1865.

CONCLUSION

We have considered the traffic breakdown phenomenon regarded as a random process developing via nucleation mechanism. The origin of critical jam nuclei proceeds in a metastable phase of traffic flow and seems to be located inside a not too large region on a highway, for example, in the close proximity of a highway bottleneck. The induced complex structure of the congested traffic phase is located upstream of the bottleneck. Keeping these properties in mind, we have applied the probabilistic model regarding the jam emergence as the development of a large car cluster on highway. In these terms the traffic breakdown proceeds through the formation of a certain car cluster of overcritical size in the metastable vehicular flow, which enable us to confine ourselves to the single cluster model.

A method how to calculate the traffic breakdown in this simple physical model has been discussed and developed. A brief summary of the results in a continuum drift–diffusion approximation is presented and the calculated breakdown probability is compared to the Weibull distribution. In the following, after comparing the results with real empirical data one needs to conclude with recommendations for a comprehensive operation improvement and provide necessary steps for a long lasting stabilisation of traffic for a given vehicular flow time series pattern (Schick, 2003; Zurlinden, 2003).

Finally we discuss the influence of controlling measures on breakdown probability. For assessment of the influence of control measures on the breakdown probability the operational definition of a traffic breakdown as a speed drop beyond a certain threshold is used. To quantify the influence of control measures such as traffic actuated speed limits, lane closures and keep in lane recommendations and eventually ramp metering field observations exploring more than 20 000 data points from measurements of freeway sections with and without corridor control systems are analysed.

The investigations were carried out at the autobahn A9 from München to Holledau between July 27 and August 9, 2000. Analysing speed–flow diagrams from one–minute intervals for a three lane section with and without automatic traffic control system we conclude that there is no significant change in the maximum observed traffic flow due to traffic control measures. The stabilising effect of the control system reduces speed drops in the traffic volume range between 3000 and 5400 veh/h/3 lanes and stabilises the final speed above 75 km/h. The beginning of the cumulative distribution is altered by a factor of 3 while the tail of the cumulative distribution is expected to be unchanged as is the maximum observed traffic flow. In conclusion we reflect changes due to automatic traffic control system on the breakdown probability by altering the escaping cluster size n_{esc} and the difference in the discharge rate for small and large clusters.

ACKNOWLEDGMENTS

The authors gratefully acknowledges funding by German Academic Exchange Program (DAAD) to take part at the *summerschool on many–particle physics* in Ekaterinburg (Russia), where part of this work was performed, as well as financial support by German Science Foundation (DFG), grant MA 1508/6 (R. Mahnke).

The authors are indebted to H. Weber (Lulea), J. Kaupuzs (Riga), I. Lubashevsky (Moscow), and P. Wagner (Berlin) for fruitful discussions and support.

REFERENCES

Brilon, W. and H. Zurlinden (2003). Überlastungswahrscheinlichkeiten und Verkehrsleistung als Bemessungskriterium for Straßenverkehrsanlagen (Breakdown Probability and Traffic Efficiency as Design Criteria for Freeways). *Forschung Straßenbau und Straßenverkehrstechnik,* **870**, Bonn.

Brilon, W., J. Geistefeldt and M. Regler (2005). **In**: *Proceedings of the 16th International Symposium on Transportation and Traffic Theory* (H. S. Hahmassani, ed.), 125–144. Elsevier, Oxford.

Daganzo, C. F., M. J. Cassidy and R. L. Bertini (1999). Possible Explanations of Phase Transitions in Highway Traffic, *Transportation Research*, **33A**, 365–379.

Gardiner, C. W. (2004). *Handbook of Stochastic Methods for Physics, Chemistry and the Natural Science.* Springer, Berlin.

Hinkel, J. and R. Mahnke (2006). Outflow Probability for Drift–Diffusion Dynamics, International Journal of Theoretical Physics, *arXiv: cond-mat*/0603579.

Honerkamp, J. (1994). *Stochastic Dynamical Systems. Concepts, Numerical Methods, Data Analysis.* VCH–Publ., New York.

Kühne, R. (2001). Application of Probabilistic Traffic Pattern Analysis to Capacity, Siam Conference, San Diego, July 9 – 19, 2001.

Kühne, R. and R. Mahnke (2005). **In**: *Proceedings of the 16th International Symposium on Transportation and Traffic Theory* (H. S. Hahmassani, ed.), 229–244. Elsevier, Oxford.

Kühne, R., R. Mahnke, I. Lubashevsky and J. Kaupuzs (2002). Probabilistic Description of Traffic Breakdowns, *Physical Review E*, **65**, 066125.

Lorenz, M. and L. Elfteriadou (2000). A Probabilistic Approach to Defining Freeway Capacity and Breakdown. In: *Proceedings 4th Int. Symp. on Highway Capacity,* 84–95, TBR–Circular E-CO18, Transportation Research Board, Washington.

Mahnke, R. and N. Pieret (1997). Stochastic Master–Equation Approach to Aggregation in Freeway Traffic, *Physical Review E*, **56**, 2666–2671.

Mahnke, R., J. Kaupuzs, and I. Lubashevsky (2005). Probabilistic description of traffic flow, *Physics Reports*, **408**(1–2).

Muthly, D. N. P., M. Xie and R. Jiang (2004). *Weibull Models.* Wiley, New Jersey.

Regler, M. (2004). Verkehrsablauf und Kapazität auf Autobahnen (Traffic Flow and Capacity on Freeways), Dissertation, Universität Bochum.

Risken, H. (1996). *The Fokker–Planck Equation. Methods of Solution and Applications.* Springer, Berlin.

Schick, P. (2003). Einfluss von Streckenbeeinflussungsanlagen auf die Kapazität von Autobahnabschnitten sowie die Stabilität des Verkehrsflusses, Dissertation, Universität Stuttgart.

Zurlinden, H. (2003). Überlastungswahrscheinlichkeiten und Verkehrsleistung als Bemessungskriterium für Straßenverkehrsanlagen, FE-Nr.: 03.327/1999/KGB der Bundesanstalt für Straßenwesen.

Transportation and Traffic Theory 2007
Edited by R.E. Allsop, M.G.H. Bell and B.G. Heydecker

34

DEVELOPING A POSITIVE APPROACH TO TRAVEL DEMAND ANALYSIS: SILK THEORY AND BEHAVIORAL USER EQUILIBRIUM

Lei Zhang, Department of Civil, Construction, and Environmental Engineering, Oregon State University, USA

SUMMARY

This paper develops a positive approach to travel demand analysis, which does not assume perfect rationality (i.e. complete information and utility maximization) in travel decision-making. Instead, the process through which travelers learn the characteristics of the transportation system, accumulate spatial knowledge, search for alternatives under imperfect information, and employ subjective beliefs and heuristic rules in decision-making is theorized. The proposed *SILK theory* (for its emphasis on **S**earch, **I**nformation, **L**earning, and **K**nowledge in travel decision-making) is able to produce quantitative models of individual travel behavior in which learning follows Bayes principles and behavioral rules are empirically derived by knowledge-acquisition methods. System-level demand patterns that emerge from individual behaviors are then obtained using agent-based aggregation techniques. This positive approach is demonstrated by the subsequent development of a route choice model for the Twin Cities (Minneapolis-St. Paul, Minnesota) metropolitan area, including the quantification of spatial knowledge and Bayesian learning process, collection of process data for model estimation and validation, the derivation of route search and choice heuristics as production (if-then) rules, and the prediction of aggregate flow patterns. The traffic equilibrium under the adopted positive assumptions is defined as the *Behavioral User Equilibrium* (BUE) at which the subjective search gain is lower than the perceived search cost for all users. Results suggest that normative assumptions, such as perfect information and unlimited human abilities to maximize utility, can produce significant prediction biases. The proposed positive approach, consisting of the SILK theoretical framework, practical data collection techniques, and unconventional travel modeling methods, serves as an alternative framework (to rational behavior theory) for developing travel demand models.

INTRODUCTION

Complexity in travel demand analysis arises from several aspects, including the complexity of the transportation system, the large number of choice dimensions and alternatives, different time scales of travel choices, interaction between travelers, individual difference in terms of preferences, characteristics, and spatial knowledge, and data limitation. These complexities force travel demand modelers to make simplifications. Normative behavioral theory assumes that individuals are rational, have perfect information, and always maximize utility (von Neumann and Morgenstern 1947, Savage, 1954). However, the normative description of travel behavior is obviously unrealistic in complex decision situations. The normative approach also encounters computational feasibility problems in large systems with numerous choice combinations. More recent activity-based microsimulation models tend to make plausible assumptions about the travel decision-making process and focus on model calibration and verification without a consistent micro-behavioral foundation.

This paper addresses these theoretical and modeling issues in travel demand analysis in two steps. First, built upon previous research on spatial behavior (Golledge and Stimson 1997) and search theory (Stigler 1961), a positive (*SILK*) theory of travel behavior is developed, which avoids assumptions of complete information and perfect rationality. The second step demonstrates the positive approach with the development and application of a route choice model in the Twin Cities.

A *positive* theory of travel behavior concerns about how travel decisions are actually made, not how they should be made as in a *normative* theory. A salient problem in travel analysis is not the unawareness of the deficiencies of *normative* behavioral assumptions, but the lack of alternative *positive* theories that are empirically testable and lead to applicable quantitative models. The theoretical contribution of this research lies in the development of a positive theory that can produce behaviorally realistic and operational travel demand models. The proposed SILK theory emphasizes the role of **S**earch, **I**nformation, **L**earning, and **K**nowledge in travel decision-making. Normative assumptions that travelers always have perfect knowledge and maximize utility are removed. The research focus is on how individuals learn about the transportation system and what behavioral rules they actually use to search and choose alternatives. Section 3 summarizes the key assumptions and features of the SILK theory.

Several modeling methods traditionally not used in travel demand analysis are identified appropriate for developing positive travel models. In the positive approach, individuals are no longer assumed to possess perfect knowledge. This requires a quantitative representation of spatial knowledge and a specification of the learning process, for which Bayes rules are adopted. A search process through which travel alternatives are sequentially examined is central to the positive approach, and individual travelers must employ certain search heuristics to deal with various costs associated with searching for alternatives. Travelers must also adopt decision heuristics to choose among known alternatives, which are often poorly described by the utility-maximization principle. These search and decision heuristics are represented by sets of production (if-then) rules in the positive approach. Artificial Intelligence methods, especially those for knowledge acquisition, are selected to empirically derive search and decision rules. Data for the empirical rule-induction process are collected from carefully-designed surveys and field experiments. In the positive approach, each individual traveler has limited and unique knowledge about the transportation system, accumulates knowledge through Bayesian learning, search alternatives using a set of search rules, make and adjust travel choices using a set of decision rules, and interact with each other. This evolutionary description of the individual decision-making process enjoys richer and more realistic representation of travel behavior. The final element of the positive modeling procedure is responsible for aggregating individual behaviors into system demand patterns. Agent-based simulation is found to be a robust and computationally feasible tool for this purpose even on large-scale networks. Section 4 or this paper presents the aforementioned methods for modeling individual travel behavior, followed by a detailed discussion of the agent-based techniques in Section 5.

A demonstration of the proposed SILK theory and positive modeling approach is warranted. The paper documents the development and application of a fully operational traffic assignment model under positive assumptions, while the positive approach also applies to modeling other dimensions of travel choices. The positive traffic assignment model is calibrated, validated, and tested on the Twin Cities (Minneapolis-St. Paul, Minnesota) road network. In the route choice context, network flow patterns emerge from individual network learning, route search, and route changing behavior. User equilibrium principles following assumptions of perfect information and substantive rationality have dominated traffic assignment analysis. Although traffic assignment algorithms under normative assumptions have enjoyed wide acceptance, these normative assumptions hardly correspond to reality. Travelers have limited spatial-temporal knowledge about the transportation network and alternative routes connecting origins and destinations. The dynamic nature of congestion and its extent makes spatial learning a complex task. Instead of optimization principles, simple but effective heuristics may be adopted by travelers to learn and compare alternative routes. In Section 6 of the paper, the traffic flow patterns under traditional normative assumption and under the novel positive assumptions are compared. A novel traffic assignment principle resulting from the positive assumption, Behavioral User Equilibrium (BUE) is also defined in this section, which can replace deterministic or stochastic user equilibrium principles in travel analysis.

The development of the positive travel demand modeling approach in this research is a comprehensive task, which involves theoretical constructs, several modeling methods, experiment design, data collection, and applications. For improved readability, this paper emphasizes the theoretical and methodological contributions. The route choice model is described with just sufficient details to demonstrate the positive approach. As the result of this decision, certain details about the route choice model are excluded, including several surveys and field experiments that provide data for empirical derivation of route search and decision rules. However, readers interested in these topics are referred to appropriate references.

The remainder of this paper is organized as follows. Section 2 briefly reviews previous studies on travel behavior theories and demand models most relevant to our topic. Section 3 summarizes the positive theoretical framework. Section 4 documents the first stage of model development –acquiring individual behavioral rules regarding learning, search, and decision-making. Section 5 outlines the agent-based simulation procedure that links individual behaviors and interactions to aggregate demand. Section 6 presents the application of the positive route choice model on the Twin Cities network. Conclusions are offered in Section 7.

LITERATURE REVIEW

Rational behavior theory assumes individuals are capable of identifying all alternatives, measuring all of their attributes, and always selecting the alternative that maximize their utility (Samuelson 1947, Von Neumann and Morgenstern 1947, Savage 1954). Most travel behavior studies adopt a utility-based theoretical framework that assumes rational behavior to varying degrees (Horowitz 1985). Tversky and Kahneman (1974, pp. 1131) argues on the basis of a number of empirically studies that individuals rely on mental shortcuts or heuristics that "are highly economical and usually effective but ... lead to systematic and predictable errors." There exist at least two general theories of human behavior that do not assume substantive rationality: prospect theory (Kahneman and Tversky 1979) and bounded rationality theory (Simon 1955). Prospect theory suggests that values are assigned to gains and losses and that decision makers use subjective and biased weights to replace probabilities. Bounded rationality was proposed based on the recognition that decision-makers must pay time and other costs to gather information, search for alternatives, and make decisions. With these constraints or bounds, individuals tend to exhibit satisficing behavior instead of utility maximization.

The travel decision-making process involves a search for alternatives, sometimes referred to as the choice set generation process. Search theory originally developed in economics (Stigler 1961, Salop 1973, Rothschild 1974) postulates that a cost characterizes information acquisition when an additional alternative is explored in the search process. Despite the obvious relevance of search theory to travel decision-making that occurs in a complex system and often involves a large number of alternatives, few travel behavior studies apply search theory (Timmermans 1980, Richardson 1982, Williams and Ortuzar 1982). Mahmassani and Change (1987) model travel choices as a bounded rational search process, but use the concept of individual bands instead of search gains and costs. A large body of literature on travel analysis (e.g. Mansky 1977, Swait and Ben-Akiva 1987, Ben-Akiva and Boccara 1995, Cascetta *et al.* 2002) discusses and improves the behavioral realism of the choice set generation process since the assumption of a full choice set (one that includes all alternatives) is often unacceptable and shown to cause prediction biases (Williams and Ortuzar 1982). However, the value of search theory in addressing the choice set generation issue has been largely unexploited.

Several decision theories, besides utility maximization, describe how individuals compare and choose among alternatives (Slovic *et al.* 1977, Svenson 1979, Ben-Akiva and Lerman 1985) including dominance, satisfaction, lexicographic rules, and elimination by aspects. Various knowledge representation methods using machine learning and logical programming have also been developed to simulate human decision-making processes (Durkin 1994, Arentze and Timmermans 2000).

It is uncontroversial that an individual's knowledge about a transportation network and alterative routes are incomplete and often biased. The developmental nature of spatial knowledge (Golledge and Stimson 1997) implies that there is a spatial learning process. However, a general theory that allows quantitative modeling of spatial learning and knowledge accumulation has yet to be developed, which is important for incorporating these elements into travel models. Several studies model route perception update and travel time learning based on Bayesian theory in the dynamic route choice scenario without considering the complete spatial-temporal network knowledge (Iida *et al.* 1992, Jha *et al.* 1996). Asakura *et al.* (2001) use a measure of functional hierarchies of roads to represent the topological aspect of network knowledge. Ramming (2002) models network knowledge as a set of latent variables, estimates coefficients using survey data, and uses fitted values of network knowledge as an explicit variable in discrete choice models. But neither study considers the learning process through which network knowledge is obtained. In order to model imperfect knowledge in travel analysis, it is prerequisite to understand the nature and organization of spatial knowledge. Geographers often describe spatial knowledge as a mental or cognitive map characterized by incompleteness and biases (Tolman 1948, Tversky 1981). There is strong evidence that spatial information is organized in a hierarchical structure. Places (Stevens and Coupe 1978, Chase 1983, Hirtle and Jonides 1985, McNamara 1986) and streets (Pailhous 1970, Elliott and Lesk 1982, Streeter and Vittelo 1986, Peruch *et al.* 1989) are ordered according to their importance to an individual and their prominence (i.e. ease of identification) into primary, secondary, and lower-order entities. A common explanation is that the hierarchical structure of spatial knowledge offers economy of storage of information in memory and simplified the retrieval of knowledge when spatial tasks such as wayfinding and location choices arise. Certain assumptions about spatial knowledge and spatial search in this paper arise from the results of these previous studies.

SILK: A POSITIVE THEORY OF TRAVEL DECISION-MAKING

The proposed SILK theory emphasizes the role of **S**earch, **I**nformation, **L**earning, and **K**nowledge in travel decision-making. Figure 1 illustrates the hypothesized travel decision-making process under this theoretical framework. At any given time, an individual has a certain level of knowledge about places, activities, and transportation networks in an urban area. This knowledge can be drawn upon to solve various spatial tasks such as finding destinations and routes. The problem-solving process consists of several procedural steps in the true behavioral sense. An individual first relies on subjective beliefs to determine the expected gain from a search for alternatives. These beliefs stem from the individual's existing knowledge which is learned from previous experiences. Searching for alternatives involves costs associated with information acquisition and other mental efforts, which can be generalized as a perceived search cost. Subjective search gain and perceived search cost result in a tradeoff that determines when a search for alternatives is initiated or stopped. Although the subjective search gain is defined by individual's beliefs and therefore can be quantitatively derived from quantified spatial knowledge, it is much more difficult to theoretically determine the magnitude of perceived search cost which should be individually different. Therefore, the perceived search cost and its relations with other variables need to be empirically derived.

If an individual decides not to search for alternatives, repetitive learned behavior or habitual behavior is executed. For instance, if the stimulus is increased congestion, the congested route is still used in this case. If the individual decides to search, a search method (or heuristics) is employed to identify alternatives, which is a mapping from spatial knowledge to one or more feasible alternatives. A central hypothesis of the SILK theory is that the search method used by travelers is not a random search in which all feasible alternatives are considered equally favorable. Instead, the search method consists of rules that are systematic and favor certain alternatives (e.g. shorter routes, larger shopping centers).

The subsequent decision step chooses an alternative. The decision rules constitute a mapping from *perceived* attributes of alternatives to a choice. Implementation of the SILK theory requires empirical derivation of the decision rules. The outcome of the decision step is a provisional try behavior. The execution of the provisional try behavior provides first-hand experience about the *actual* attributes of the temporarily-chosen alternative at the time of trial. The positive theory recognizes the role of historical dependency in decision-making. After each round of search, an individual employs the decision rules to compare the newly identified alternative with previously learned alternatives, or with the alternative currently being used. All other things equal, the individual may prefer the currently used alternative due to habit, or prefer the new alternative due to the desire for variety. The true preference as represented by the decision rules, which could be individually different, should be empirically derived from data.

Information from various sources can be gathered by an individual and expands the individual's knowledge through a learning process. The updated knowledge alters the aspiration level and changes subjective beliefs. For instance, repetitively encountering congestion after several rounds of search for alternative routes connecting an origin-destination pair could reduce the degree of belief that an uncongested or less congested route exists, which reduces the subjective search gain and may cause the search process to cease. In other cases, an additional round of search for alternatives may be undertaken if the currently used alternative still does not meet the aspiration level. A theoretically sound and carefully

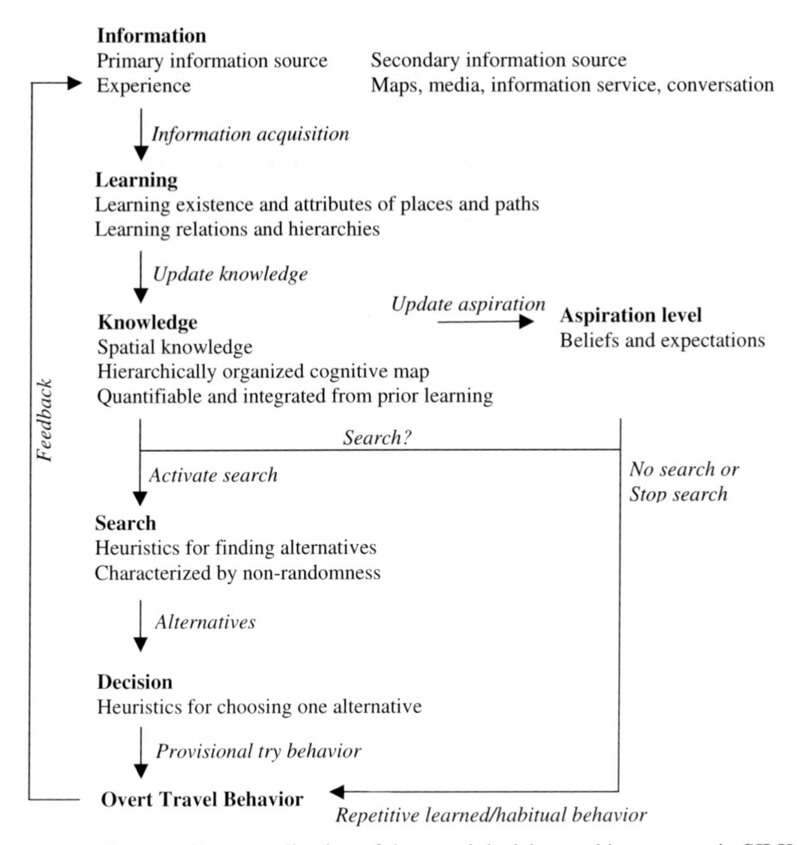

Figure 1. Conceptualization of the travel decision-making process in SILK

specified learning and belief-updating process can effectively describe how individuals adjust their aspiration/expectation levels, and provide guidance for subsequent modeling step. The SILK theory incorporates Bayes rules into the learning and belief updating process and therefore inherits all assumptions underlying the Bayesian theory.

As depicted in Figure 1, the decision-making process in responses to a new spatial task or a new stimulus ends if the search process stops after a certain number of (possibly zero) search rounds. The termination of search is signalized by an individual's subjective search gain lower than the perceived search cost, which has been a standard assumption of search theory in economics. Either a satisfactory alternative has been identified, or the aspiration level has been adjusted downwards sufficiently. Although the conceptualization of subjective search gain, perceived search cost, search initialization, and search stopping conditions in the SILK theory is similar to that in bounded rationality theory, the SILK theory is unique in that by incorporating theoretical constructs from Bayesian learning (belief updating), Artificial Intelligence (rule-based search and decision heuristics), Search theory (the search process), and behavioral science (historical dependency in decision-making) into a coherent framework, it explains the travel decision-making process in its entirety under positive assumptions. Furthermore, the SILK theorization allows consistent quantitative modeling of travel demand without normative assumptions as we shall see in the following sections. Socio-demographic characteristics enter the decision process by influencing the search and decision rules. Various situational, environmental, household, spatial, and temporal constraints limit the search space.

Most assumptions underlying the SILK theory have been justified in previous research supporting Bayesian theory, Search theory, Bounded rationality, and Rule-based decision making (see citations in Section 2). The assumptions regarding the developmental nature of spatial knowledge and historical dependencies in decision making are supported by evidence in behavioral geography (Golledge and Stimson 1997, Ch. 5) and previous empirical studies (Goodwin 1977, Blasé 1979, Williams and Ortuzar 1982, Aarts and Dijksterhuis 2000). A more detailed discussion of the SILK theory with further justifications of the assumptions made is available in Zhang (2006a).

MODELING INDIVIDUAL BEHAVIOR

In order to apply the SILK theory to develop quantitative models, it is necessary to: (1) at the microscopic individual level, quantify and specify spatial knowledge and the learning process (Section 4.1), determine perceived search costs and gains for users with different characteristics (Section 4.2), and empirically derive search rules (Section 4.3) and decision rules (Section 4.4) from data; (2) at the macroscopic system level, derive aggregate demand patterns from individual behaviors and interactions (Section 5). The various modeling methods proposed for the positive approach are elaborated in the development of a positive traffic assignment model. It will become obvious that the methodological framework can be adapted to model other dimensions of travel choices. When applied to the traffic assignment modeling task, the SILK theory suggests that spatial knowledge relevant to route choice can be quantified and improved though a Bayesian learning process. The expected gain from an additional search for alterative routes arises from an individuals' unique knowledge and beliefs. Subjective search gain and perceived search cost jointly determine search starting and stopping conditions. The heuristics individuals develop to search for routes and choose among known routes are represented by production (if-then) rules which should be empirically derived and validated.

Spatial Knowledge and Learning

An individual new to an area has little knowledge about places and roads. However, information regarding the layout of the streets and their hierarchies is usually available through various media (e.g. maps, internet). An important question of spatial cognition and learning is how human beings store the enormous amount of spatial information and retrieve such information effectively. Empirical studies show that road network is hierarchically structured in the cognitive map (Pailhous 1970, Elliott and Lesk 1982, Streeter and Vittelo 1986, Peruch et al. 1989). Individuals in general have better knowledge of the major roads (e.g. freeway network) than secondary (e.g. major arterial streets) and tertiary roads (e.g. connectors, residential streets). It is therefore assumed in this research that individual's *initial* route knowledge includes network connectivity and hierarchy which provide them certain beliefs about travel costs on various levels of roads under uncongested conditions. However, they can only learn the actual travel costs on individual roads after traveling on these roads.

Assume that an individual's perception capabilities allow the separation of a specific route attribute (say travel time) into I categories, and travel time t_i has been observed n_i times between an OD pair in prior experience. The individuals' spatial knowledge about routes connect the OD pair can be quantified as a vector $\boldsymbol{K} = (n_1, \ldots, n_i, \ldots, n_I)$. Bayes rule suggests that when a new alternative route is identified and the travel time observed on that route by the individual falls into category i, the undated knowledge becomes $(n_1, \ldots, n_i+1, \ldots, n_I)$. This updating procedure implies each past observation is weighted the same by the individual. Let vector $\boldsymbol{P}(p_1, \ldots, p_i, \ldots, p_I)$ represent an individual's subjective beliefs, where p_i is the subjective probability that an additional search would produce an alternative route with attribute t_i. In order to establish a quantitative relationship between knowledge $\boldsymbol{K}$ and beliefs $\boldsymbol{P}$, it is assumed that individuals' prior beliefs follow a Dirichlet distribution. Dirichlet is the conjugate prior of the multinomial distribution, and the posterior beliefs will also be a Dirichlet (Rothschild 1974). Let N denote the total number of observations ($N = \Sigma n_i$). This is equivalent to assuming:

$$p_i = n_i / N \tag{1}$$

This assumption is quite general in fact, because it can be shown according to the strong law of large numbers, that as experience accumulates (N becomes large), individuals behave as if their priors are Dirichlet:

$$\lim_{N \to \infty} p_i = n_i / N \tag{2}$$

Search Gain and Search Cost

Under the above assumptions about spatial learning and knowledge, it is possible to drive the subjective search gain. Let an individual's travel time (or another route attribute, or a linear combination of all route attributes) on the route currently used be t. The expected gain (g) in terms of travel time savings per trip from an additional search is:

$$g = \sum_{i(t_i < t)} p_i \cdot (t - t_i) \tag{3}$$

Equation 3 can be further simplified. Since drivers are able to use the best of all searched routes, t is always the minimum of all observed travel times (t_{min}). Since it is assumed that individuals believe there is no congestion until they experience delays in their actual travel, they initially believe it takes t^* to travel between the OD pair, where t^* is the free-flow travel time of the route identified during the first round of search. As the search process proceeds, the subjective probability of finding a route with travel time t^* after N searches is $1/(N+1)$. Therefore, the expression of the subjective search gain becomes:

$$g = (t_{\min} - t^*)/(N+1) \qquad (t^* \le t_{min} \text{ and g is always positive}) \qquad (4)$$

This equation reveals two properties of the subjective search gain: (1) it decreases as the number of searches increases; (2) it decreases if a better route is found (smaller t_{min}).

Although search gain may increase or decrease as the search process proceeds, the search cost a traveler perceives is assumed to be constant for the same traveler throughout the search process. However, perceived search costs for different travelers may be different. If an individual stops searching for alternatives after n rounds of search, the perceived search cost for this individual must be lower than the expected search gain after $(n-1)$ searches such that search n is meaningful, and must be higher than the expected search gain after n searches such that search $(n+1)$ search does not occur. These lower and upper bounds of search cost can be calculated using Equation 4. We can let the average be the estimate of the perceived search cost (c):

$$c_{LOW} = \frac{t_{\min,n} - t^*}{n+1} \qquad (5.1)$$

$$c_{HIGH} = \frac{t_{\min,n-1} - t^*}{n} \qquad (5.2)$$

$$c = \frac{1}{2}(c_{LOW} + c_{HIGH}) \qquad (5.3)$$

In order to empirically derive the distribution of perceived search costs, one needs to observe the search processes of a number of individuals and the order by which alternatives are sequential searched. When direct observations are difficult, carefully designed surveys in which subjects report their previous search processes may also be used. A survey designed for modeling route search behavior and computing route search costs is described in the following section.

A subsequent finding is that the computed search cost (c) is linearly correlated to t^*(correlation coefficient is 0.70) but independent of socio-economic and demographic characteristic such as age, gender, income, and length of residence in the city. This indicates that the distribution of perceive search costs depends on the length of the trip – for any individual, the perceived search cost is higher when the trip is longer. Therefore, it is the distribution of the relative search cost ($c^* = c/t^*$) that is used in the subsequent analysis. The computed cumulative density function of c^* from the survey as well as its log-normal approximation, is plotted in Figure 2. The distribution can be interpreted as follows. If free-flow travel time between an OD pair is 10 minutes according to the initial spatial knowledge, the highest perceived cost of searching for alternative routes connecting this OD pair among all drivers is about equivalent to 3.6 minutes of travel time per trip. About 90 percent of drivers perceive a search cost that is less than 2 minutes per trip (Point A). In other words, 90 percent of drivers will search for new alternative routes for this trip if the subjective search gain is larger than 2 minutes per trip. Only 20 percent of drivers will continue to search for new routes when the subjective search gain is 0.8 minute (Point B).

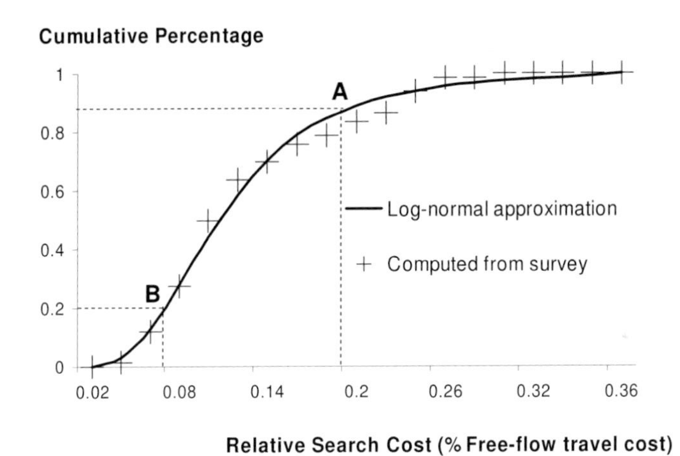

Figure 2. Distribution of relative search costs

Search Rules

Empirical evidence suggests that location search (Humphreys and Whitelaw 1979) is characterized by non-randomness and systematic biases. It should be obvious that route search is not random because travelers tend to first consider routes that appear to be effective. This research adopts production (if-then) rules to represent individuals' route search behavior. In this section, the method for empirically deriving these rules and its application in route search are presented.

Results from several previous empirical studies can facilitate the specification of search rules: (1) route knowledge is hierarchically organized with various levels of roads (Pailhous 1970, Elliott and Lesk 1982, Streeter and Vittelo 1986, Peruch et al. 1989); (2) experienced drivers travel on the basic network (network of roads at the highest level of hierarchy) as further as them can and exit the basic network at the point closest to their destinations. It is therefore hypothesized in this study that the rules individuals use to search for alterative routes can be separated into two categories: (1) how to transfer between different levels of roads; (2) how to travel on sub-networks with each consisting of roads at the same level of hierarchy. It is further assumed that individuals are able to find the shortest path (however defined) within each sub-network. This assumption is, however, much weaker than the assumption that drivers can identify the shortest path between any OD pair, because: (1) each sub-network is much simpler than the complete network; (2) the basic network of the high-order roads (e.g. the freeway sub-network) is known by most drivers; (3) although sub-networks of lower-order roads are more complex, most travel on these lower-order sub-networks is the drive from the origin to an access point on the basic network and from an egress point on the basic network to the destination. Empirical findings supporting this assumption can be found in Bovy and Stern (1990, p.56). They find drivers are well informed about the higher-order roads, and familiar with the lower-order roads only in the vicinity of their major activity locations (home, office etc.). The problem of modeling route search behavior under the above assumptions becomes the derivation of rules drivers use to determine whether or not to use the basic network, and how to access and exit the basic network. This is achieved by collecting route search data from a survey conducted in Fall 2004, and extracting search rules from the data using knowledge-acquisition algorithms.

The data set was collected from 82 drivers in the Twin Cities, Minnesota, who are part time or full time students at the University of Minnesota. During the peak period, about 600,000 vehicle trips are made each hour on this road network that has 7976 nodes, 20194 links, 537 access points (mostly on-ramps) to the basic network (defined as the freeway network), and 530 egress points (mostly off-ramps). During the survey, each subject fills in a questionnaire regarding socio-economic and demographic characteristics, typical travel patterns, and routes considered and chosen for three different origin-destination pairs: home to the university, home to a randomly-selected destination A, and A to another randomly-selected destination B. This design allows for varying degrees of familiarity with the origin-destination pairs. The routes are initially drawn by subjects on an oversized-map, and subsequently coded in GIS as a sequence of network nodes. The subjects are asked to recall the order of the routes they have considered and actually used, which is necessary to empirically derive their perceived search costs. The total number of routes reported is 295 between 165 distinct origin-destination pairs. The average distance of the routes is 14 miles (minimum 0.6 mile, maximum 42 miles) and the average free-flow travel time of the routes (based on the planning network coded by the regional planning agency) is 16 minutes (minimum 1 minute, maximum 54 minutes). In order to focus on the SILK theory and the modeling methodology, the details on how the survey data are further processed for rule induction are discussed elsewhere (Zhang 2006a, b).

If-then rules are selected to represent route search heuristics for several reasons: (1) they are shown to be capable of replicating various types of human heuristics and decision-making processes in previous studies on expert systems; (2) the execution of if-then rules require minimum computational resources which is important for models involving millions of independent decision agents. Various machine learning algorithms (Witten and Frank 2000) are able to derive if-then rules using the collected survey data. From three popular algorithms for deriving if-then classification rules including C4.5 (Quinlan 1986), PRISM (Cendrowska 1987), and RIPPER (Cohen 1995), RIPPER is chosen for its better predictive performance on our dataset. The final route search rule set derived from the survey data consists of 16 rules for selecting access points and 13 rules for selecting egress points on the basic network. The two rule sets are quite similar, and therefore only the rules for selecting access points are shown in disjunctive normal forms, where "Δ" denotes changes or percentage changes (Route B attributes – Route A attributes); "[ ]" constitutes a complete antecedent condition of the if-then rules; "Time" is the total travel time; "Btime" is travel time on the basic network; "Transfer" is number of transfers between different levels of roads.

Choose Route A as the alternative route for consideration, if
[ΔTime = (0.21 ~ infinity)] Rule 1
Or [ΔTime = (0.13 ~ 0.21)
And ΔBtime = (–infinity ~ –0.57)] Rule 2
And ΔBtime = (–0.57 ~ 0.19)
And ΔTransfer = 0 or 1] Rule 3
And Time = (30 minutes ~ infinity)] Rule 4
And ΔBtime = (0.19 ~ infinity)
And ΔTransfer = 1] Rule 5
Or [ΔTime = (0.04 ~ 0.13)
And ΔBtime = (–infinity ~ –0.57)] Rule 6
And ΔBtime = (–0.57 ~ 0.19) And (ΔTransfer = 0 or 1)] Rule 7
And ΔBtime = (–0.57 ~ –0.19) And (Time = (15 ~ 30 minutes)] Rule 8
And ΔBtime = (0.19 ~ 0.57) And (ΔTransfer = 1)] Rule 9
Or [ΔTime = (–0.04 ~ 0.04)
And ΔBtime = (–infinity ~ –0.57)] Rule 10
And ΔBtime = (–0.57 ~ 0.19)
And ΔTransfer = 1] Rule 11
And ΔBtime = (–0.57 ~ –0.19) And (ΔTransfer = 0)] Rule 12
And ΔBtime = (0.19 ~ 0.57) And (ΔTransfer = 1)]
Or [ΔTime = (–0.21 ~ –0.04)
And ΔBtime = (–infinity ~ –0.57)] Rule 13
And ΔBtime = (–0.57 ~ 0.19) And (ΔTransfer = 1)] Rule 14
And ΔBtime = (–0.57 ~ –0.19) And (ΔTransfer = 0)] Rule 15
Otherwise, choose Route B as the alternative route for consideration. Rule 16

For instance, Rule 1 suggests that drivers will identify a specific route in a round of search if its travel time is significantly lower (21%) than other routes. The execution of this rule alone can exclude a large percentage of feasible alternative routes from further considerations. As the travel time difference becomes less apparent (Rule 2 to 15), other factors related to the simplicity of routes, such as the percentage of travel on the basic network, and number of transfers between different sub-networks, also play an important role. Collectively, these rules replicate the heuristics individuals use to identify alternative routes based on their existing spatial knowledge. As knowledge changes (e.g. a congested section in the network is learned), the same rule set can generate different routes (e.g. a new route bypassing the congested section). Repeated executions of these rules in each round of route search produce one alternative route for the subsequent decision step (Section 4.4). Finally, when used to predict behavior, these rules can be executed as deterministic or probabilistic rules (based on their accuracy on the estimation/validation datasets).

Decision Rules

After each round of search, a new alternative is identified. An individual either rejects or changes behavior to use the new alternative. This is determined by a set of decision rules. In the case of route choice, these are rules describing route changing or route switching behavior. Even though during the search process many alternative routes may be considered, the final route choice decision is assumed to be the outcome of a series of route changing decisions. Different from utility maximization, this assumption about the decision step allows for historical dependencies and does not presume unreasonable human information processing and computational capabilities. Similar to the derivation of search rules, an experiment is designed and implemented in Spring 2004 where subjects' actual route changing behaviors can be observed. A machine learning algorithm then extracts decision rules from the collected behavioral data. Again, we briefly summarize the experiment and present the resulting decision rules.

The experiment was designed using both stated preference survey techniques and field observations. Five roughly parallel routes between the University of Minnesota East Bank Campus and Downtown Saint Paul are selected for the experiment. One of the routes is a freeway, and the rest are major arterial streets. Subjects are selected randomly from the University of Minnesota staff list. Each of the 117 subjects is given a pre-test to gather various socio-economic, demographic, and travel pattern data. Their vehicles were then temporarily equipped with a recording GPS unit, which collects vehicle location data at one-second intervals. During the field experiment, each subject was advised to take four of the five selected routes at a given random order. The GPS data were used to confirm that the subjects traveled the correct route, and to calculate the actual route attributes such as total travel time, distance, number of stops, delay time, and speed. At the end of the field experiment, each subject rated the efficiency, easiness, pleasure, and familiarity of the traveled routes on a 1~7 scale (7 being the most efficient, easiest, most pleasant, or most familiar), and ranked the four routes traveled for various trip purposes. This design allows analysts to observe how the subjects change routes as they learn the attributes of alternative routes sequentially in real-world driving scenarios.

The route changing rules for commute trips derived from machine learning algorithms are present below, where Δ denotes changes or percentage changes (new route attributes – current route attributes), and absolute values are attributes of the currently used route.

Change route, if
[ΔTime ≤ -39%] Rule 1
or [ΔTime ≤ -11% and ΔPleasure ≥ -1] Rule 2
or [ΔFamiliarity ≥ 3 and Commute time ≤ 20 min] Rule 3

or [ΔTime ≤ 6% and ΔPleasure ≥ 3] Rule 4
or [ΔTime ≤ 15% and Δ Familiarity ≥ 2 and ΔDelay ≥ -40%] Rule 5
or [Familiarity = 1 and ΔTime ≤ 51% and Commute time ≤ 20 min and Income = 1] Rule 6
or [Delay ≥ 4 min and ΔStops ≤ 0 and Commute distance ≤ 8 miles] Rule 7
or [ΔPleasure ≥ 2 and ΔFamiliarity ≥ 0 and Commute time ≤ 16 min] Rule 8
Otherwise, continue to use the current route. Rule 9

There exists perception threshold in route changing behavior. For instance, Rule 1 implies drivers will change routes as long as travel time can be reduced by more than 39%. Travel time reduction less than 11% is insignificant. Variable "familiarity" is present in several rules, which is evidence of historical dependencies in route choice. A comparison shows that the predictive performance of the route changing rule set is superior to a normative logit model using measures such as hit ratios at the individual level, and route flows at the aggregate level (see Zhang 2006a, b for details).

MODELING SYSTEM DEMAND PATTERNS

In order to estimate System demand patterns, all individual leaning and behavioral rules need to be aggregated to produce macro-level statistics. This is achieved by agent-based simulation. In the case of traffic assignment, the most valuable system statistics are the traffic flows and travel costs on individual links. Zhang and Levinson (2004) describe the agent-based modeling techniques, and discuss its applications in travel analysis.

In the agent-based simulation, each driver agent starts with an initial knowledge about the road network as defined in Section 4.1, decide when to start and then stop the search process based on subjective search gains and perceived search costs computed by equations in Section 4.2, employs search rules developed in Section 4.3 to find alterative routes, and applies decision rules developed in Section 4.4 to select route after each round of search. Currently, a BPR function is used to compute link costs, which may be replaced by a regional microscopic traffic simulator in the future. The aggregate traffic flow pattern from this positive agent-based model is defined as the *Behavioral User Equilibrium (BUE): The BUE is reached on a network when all users with limited spatial knowledge stop searching for alternatives because for each user the perceived search cost exceeds the expected gain from an additional search.* The proposed positive approach tracks the individual decision making process in great details, and still allows the convergence of traffic flows into a well defined equilibrium pattern for planning and engineering applications. Of course, evolutionary factors such as population change could be introduced into the model if demand evolution is more of interest and if an equilibrium solution is not desirable.

The BUE equilibration process can be directly measured by the number of drivers still searching for alterative routes. To prove the existence of BUE, we revisit Equation 4 which defines the subjective search gain (g_i) for traveler i. The traveler's perceived search cost (c_i) is always a positive constant. Therefore, there must exist a positive integer N_i such that after N_i searches $c_i > g_i$, because:

$$\lim_{N_i \to \infty} g_i = \lim_{N_i \to \infty} \frac{t_{\min} - t^*}{N_i + 1} = 0 \tag{6}$$

Let N^* be the maxim of all N_i s. The BUE will surely be reached after N^* search iterations, proves its existence. The BUE exists because users adjust their beliefs and subjective search gains to accommodate unsatisfactory performance of the transportation system. Eventually, an individual stops search either because a good alternative is identified, or because repeated experience with unsatisfactory alternatives leads to decreased expectations. Properties such as the stability and uniqueness properties of the BUE are discussed elsewhere (Zhang 2006a).

DEMONSTRATION OF THE POSITIVE APPROACH

The BUE is solved for the Twin Cities road network that has 7,976 nodes, 20,194 links, and about 600,000 travelers during a typical peak hour. The positive route choice model converges after 43 search iterations (see Figure 3) within 4.2 CPU hours on a 1.7GHz PC. At the equilibrium, the average excess travel time, defined as the percentage difference between travel times on the actually used routes and on the shortest paths, is 15% (0% in deterministic user equilibrium by definition). The perceived search cost and cognitive limitations force users to consider only a small number of routes before ending their search processes. The positive model predicts that more than 80% drivers consider fewer than three alternative routes. The estimated distribution of number of routes actually considered also closely approximates the observed distribution (see Figure 4).

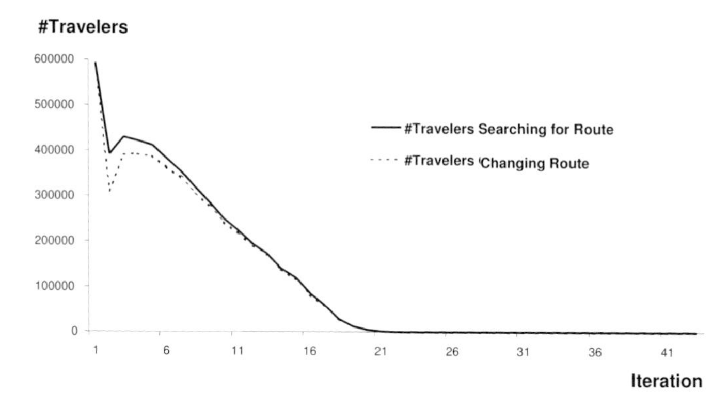

Figure 3. Convergence of BUE on the Twin Cities Road Network

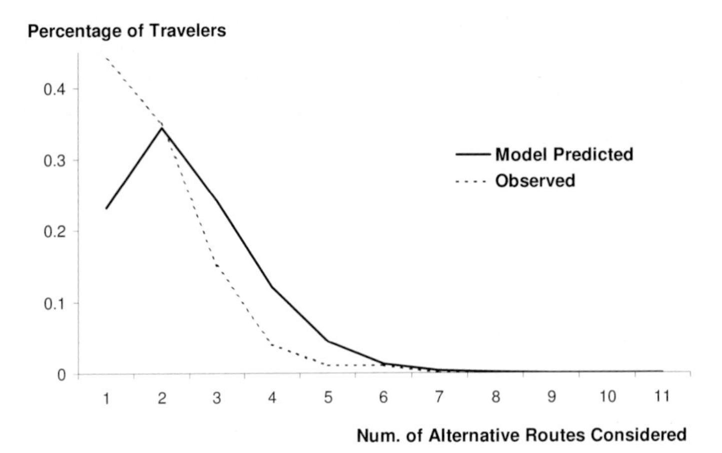

Note: The "Observed" distribution is computed from the route search survey data.

Figure 4. Distribution of the Number of Considered Alternative Routes

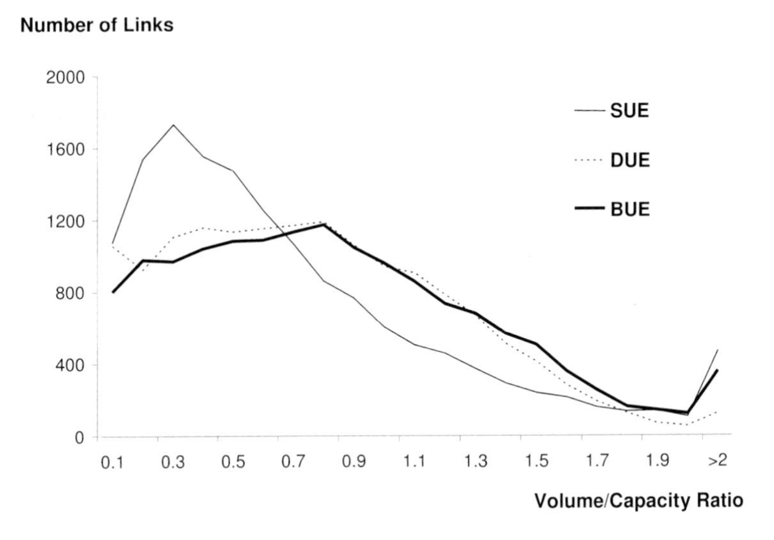

Figure 5. Distributions of Link Volume Capacity Ratios

For comparison purposes, the normative DUE (deterministic user equilibrium, Wardrop 1952) and SUE (stochastic user equilibrium, Daganzo and Sheffi 1977) traffic flows are also computed for the same road network and OD demand. There is notable difference in the estimated total travel time (million vehicle hours per peak hour): 0.21 for DUE, 0.22 for BUE, and 0.25 for SUE. The traffic flows under different behavioral assumptions are clearly different as shown by the correlation coefficients (see Table 1). It is also found that DUE significantly underestimates the level of congestion on the most congested roads in the network compared to BUE (see Figure 5). Comparisons with observed traffic counts show that the BUE (R^2 = 0.53) flows provide slightly better goodness-of-fit than DUE (0.51) and SUE (0.46) flows, thought the quality of the traffic count data is not entirely reliable for common data collection issues. Another application evaluating network performance before and after a policy change demonstrates that different behavioral assumptions (normative vs. positive) can lead to opposite policy recommendations (Zhang 2006c).

Table 1. Correlation Coefficients between Flows at Different Equilibria

Correlation Coefficients	BUE Flow	DUE Flow	SUE Flow
BUE Flow	1	0.92	0.82
DUE Flow		1	0.86
SUE Flow			1

The BUE derived from the positive approach is also different from the Boundedly Rational User Equilibrium (BRUE) proposed by Mahmassani and Chang (1987) who use the notion of indifferent bands to theorize travelers' satisficing behavior under imperfect information. Our approach explicitly theorizes and models the search process and its three important characteristics: search rules, search cost, and search gain. The Bayesian learning process, belief updating and expectation adjustment, and empirical derivation of behavioral rules are also unique in BUE.

CONCLUSIONS

The need for travel forecasting models with improved behavioral realism and capabilities is imperative due to proposals of more sophisticated travel demand management, pricing, and information-related policies. Activity-based models are often criticized for not having a consistent behavioral foundation. Under these circumstances, developing positive theories of travel behavior based on empirical evidence is important, and expected to improve both our understanding of the travel-decision making process and the realism of travel demand models. The SILK theory and the positive modeling approach presented in this paper represent a pilot effort in developing comprehensive travel models that explicitly consider imperfect knowledge, spatial learning, search and decision heuristics, information acquisition and decision cost, and the dynamics of expectations and beliefs in decision making. This research demonstrates that the positive approach for travel demand analysis is feasible, that a general and applicable positive theory can be developed, and that normative and positive behavioral assumptions produce significantly different demand estimates.

One feature of the positive approach may be considered an advantage or a limitation – behavioral rules need to be empirically estimated and validated. It is an advantage because empirically derived rules should better describe the decision-making process than those implied by standard normative assumptions. It is a limitation for its data requirement. Whether or not the increased data collection cost is justified by the improved model realism and performance has always been a controversial question in travel demand forecasting. The total cost of conducting the two surveys and developing BUE on the Twin Cities network is around fifty thousand dollars. The author believes it is worthwhile considering that in many areas flow estimates are used to allocate millions or even billions dollars of future transportation investments.

Many activity-based models use computational process methods, and specify various search heuristics in the generation of choice set. However, the conditions under which search starts and stops in travel decision-making are inadequately explored. It is also uncertain to what extent hypothesized heuristics correspond to reality. This research addresses these issues and makes several contributions. First, the concepts of subjective search gains and costs are developed in this paper, as well as mathematical and empirical methods for deriving these variables that determine the initialization and stopping of search. This theorization of the search process could also be used in other modeling tasks that require choice set generation. Second, this paper distinguishes search heuristics from choice heuristics. The actual rules travelers use to identify alternatives and to make choices should be separately analyzed. Third, machine learning algorithms for extracting production rules have been successfully applied in this study to empirically derive and verify the heuristic rules from survey and experiment data. This should reduce the arbitrariness in specifying heuristics in activity-based and microsimulation models. Finally, the proposed positive route choice model is fully operational as demonstrated on a large real-world network. The Behavioral User Equilibrium (BUE) defined in this study can serve as an alternative traffic assignment principle to normative models such as DUE and SUE. The disaggregate nature of BUE should facilitate the integration of activity-based models (especially the computational process models) and traffic assignment.

Since the positive approach considers imperfect information and search cost, it is therefore more suitable than normative models for studying the demand impact of traveler information services. According to the SILK theory, these information services effectively reduce perceived search costs. Drivers will use information services if the service access cost is lower than their perceived costs of searching for alternatives on their own and lower than their subjective search gains. Information obtained from traveler information services, like information obtained from other sources, can also update spatial knowledge and beliefs and eventually influence behavior. These research opportunities should be pursued in future studies.

REFERENCES

Aarts, H., and Dijksterhuis, A. (2000). The automatic activation of goal-directed behaviour: The case of travel habit. *Journal of Environmental Psychology* 20: 75-82.

Arentze, T., and Timmermans, H.J.P. (2000). *ALBATROSS: A Learning-Based Transportation Oriented Simulation System.* EIRSS, Eindhoven.

Asakura, Y., Yamauchi, T., Hato, E., and Kashiwadani, M. (2001). A route choice model considering topological aspects of a road network. *Journal of the Eastern Asia Society for Transport Studies* 4(3): 55-67.

Ben-Akiva, M., and Boccara, B. (1995). Discrete choice models with latent choice sets. *International Journal in Marketing* 12: 9-24.

Ben-Akiva, M. and Lerman S.R. (1985) *Discrete choice analysis: theory and application to travel demand.* The MIT Press, Cambridge, Massachusetts

Blase, J. (1979). Hysteresis and catastrophe theory: Empirical identification in transport modelling. *Environment and Planning* 11A: 675-688.

Bovy, P.H.L., and Stern, E. (1990). *Route choice: Wayfinding in transport networks.* Kluwer Academic Publishers: The Netherlands.

Cascetta, E., Russo, F., Viola, F.A., and Vitetta, A. (2002). A model of route perception in urban road networks. *Transportation Research* 36B: 577-592.

Cendrowska, J. (1987). PRISM: An algorithm for inducing modular rules. *International Journal of Man-Machine Studies* 27(4): 349-370.

Chase, W.G. (1983). Spatial representations of taxi drivers. In Rogers, D.R. and Sloboda, J.A. (eds.), *Acquisition of symbolic skills*, 391-405, New York, Plenum.

Cohen, W.W. (1995). Fast effective rule induction. In Prieditis, A., and Russell, S., (eds.), *Proceedings of the Twelfth International Conference on Machine Learning*, pp. 115-123, Morgan Kaufmann: San Francisco.

Durkin, J. (1994). *Expert systems: Design and development.* New York: Maxwell

Elliott, R.J. and Lesk, M.E. (1982). Route finding in street maps by computers and people. American Association for Artificial Intelligence-82 Conf., 258-261, Los Altos, CA.

Golledge, R.G., and Stimson, R.J. (1997). *Spatial behavior: A geographic perspective.* The Guilford Express, New York.

Goodwin, P. (1977). Habit and hysteresis in mode choice. *Urban Studies* 14: 95-98.

Hirtle, S.C., and Jonides, J. (1985). Evidence of hierarchies in cognitive maps. *Memory and Cognition* 13(3): 208-217.

Horowitz, J.L. (1985). Travel and location behavior: State of the art and research opportunities. *Transportation Research* 19A: 441-453.

Humphreys, J.S., and Whitelaw, J.S. (1979). Immigrants in an unfamiliar environment: Locational decision-making under constrained circumstances. *Geografiska Annaler* 61B(1): 8-18.

Iida, Y., Akiyama, T., and Uchida, T. (1992). Experimental analysis of dynamic route choice behavior. *Transportation Research* 26B: 17-32.

Jha, M., Madanat, S., and Peeta, S. (1996). Perception updating and day-to-day travel choice dynamics in traffic networks with information provision. *Transportation Research* 6C: 189-212.

Mahmassani, H.S., and Chang, G.L. (1987). On boundedly rational user equilibrium in transportation systems. *Transportation Science* 21(2): 89-99.

Mansky, C.(1977).The structure of random utility models.*Theory and Decision* 8:229-54.

McNamara, T.P. (1986). Mental representations of spatial relations. *Cognitive Psychology* 18: 87-121.

Kahneman, D., and Tversky, A. (1979) Prospect theory: an analysis of decision under risk. *Econometrica* 47(2): 263-291

Pailhous, J. (1970). *La représentation de l'espace urbain*, Paris: Presses Universitaires de France.

Peruch, P., Giraudo, M-D., and Garling, T. (1989). Distance cognition by taxi drivers and the general public. *Journal of Environmental Psychology* 9: 233-239.

Quinlan, J.R. (1986). Induction of decision trees. Machine Learning 1(1): 81-106.

Ramming, M.S. (2002). *Network knowledge and route choice*. Ph.D. Dissertation, Massachusetts Institute of Technology.

Richardson, A. (1982). Search models and choice set generation. *Transportation Research* 16A(5-6): 403-419.

Rothschild, M. (1974). Searching for the lowest price when the distribution of prices is unknown. *Journal of Political Economy*: 689-711

Salop, S.C. (1973). Systematic job search and unemployment. *Review of Economic Studies* 40(2): 191-201.

Samuelson, P.A. (1947) *Foundations of economic analysis*. Harvard University Press.

Savage, L.J. (1954) *The foundations of Statistics*. New York: Wiley.

Simon, H. (1955) A behavioral model of rational choice. *Q. J. of Economics* 69: 99-118

Slovic, P., Fischoff, B., and Lichtenstein, L. (1977). Behavioral decision theory. *Annual Review of Psychology*: 1-39.

Stevens, A., and Coupe, P. (1978). Distortions in judged spatial relations. *Cognitive Psychology* 10: 422-437.

Stigler G.J. (1961) The economics of information. *Journal of Political Economy* 69: 213-225

Streeter, L. A. and Vitello, D. (1986). A profile of drivers' map-reading abilities. *Humans Factors*, 28(2).

Swait, J. and Ben-Akiva, M. (1987). Empirical test of constrained choice discrete model: model choice in San Paulo, Brazil. *Transportation Research* 21B: 103-115.

Swenson, O. (1979). Process description of decision-making. *Organizational Behavior and Human Performance* 23: 86-112.

Timmermnas, H.J.P. (1980). Consumer spatial choice strategies: A comparative study of some alternative behavioral spatial shopping model. *Geoforum* 11: 123-131.

Tolman, E.C. (1948). Cognitive maps in rats and men. *Psychological Review* 55: 189-208.

Tversky, B. (1981). Distortions in memory for maps. *Cognitive Psychology* 13: 407-433.

Tversky, A., and Kahneman, D. (1974) Judgment under uncertainty: Heuristics and biases. *Science*(185): 1124-1131 von Neumann, J. and Morgenstern, O. (1947) *Theory of Games and Economic behavior*. 2nd ed. Princeton University Press

Williams, H. and Ortuzar, J. (1982). Behavior theories of dispersion and misspecification of travel demand models. *Transportation Research* 16B: 169-219.

Witten, I.H., and Frank, E. (2000). *Data mining*. Morgan Kaufmann Publishers: San Francisco.

Zhang, L. (2006a). Search, information, learning and knowledge in travel-decision making. Ph.D. Dissertation, Department of Civil Engineering, University of Minnesota.

Zhang, L. (2006b). An agent-based behavioral model of spatial learning and route choice. In the compendium of papers CD of the 85th Transportation Research Board Annual Meetings, Washington, DC.

Zhang, L. (2006c). Traffic diversion effect of ramp metering. Journal of the Transportation Research Board (Accepted for publication).

Zhang, L., and Levinson, D.M. (2004). Agent-based approach to travel demand modeling: Exploratory analysis. *Transportation Research Record: Journal of the Transportation Research Board* 1898: 28-36.

Transportation and Traffic Theory 2007
Edited by R.E. Allsop, M.G.H. Bell and B.G. Heydecker

35

A NEW MODELING FRAMEWORK FOR TRAVELERS' DAY-TO-DAY ROUTE CHOICE ADJUSTMENT PROCESSES

Fan Yang, ESRI Inc., 380 New York St, Redlands, CA, 92373, USA
Henry X. Liu, Department of Civil Engineering, University of Minnesota, Minneapolis, MN 55455, USA

SUMMARY

This paper presents a new Markov model to study travelers' stochastic behavior in their day-to-day route choice adjustment process. The model is characterized by two components: how often a traveler reconsiders his/her route choice (route-switching rate), and what the probability is to take a certain route (route choice probability). By applying the evolutionary game theory, the conventional perfect information and complete rationality requirements in equilibrium analysis are relaxed. A deterministic mean (expected) route flow dynamic is derived which closely approximates the underlying route flow stochastic process in any finite time span as the travel demand grows large. The mean dynamic is general in that many existing deterministic processes can be considered as its special cases, and more importantly, their meaningful individual behavior explanations are unveiled. It can be shown that with certain reasonable assumptions of behavioral rules of route-switching rate and route choice probability, the Wardrop user equilibrium can be approached by travelers' day-to-day behavior adjustment process, even if an individual traveler only has access to incomplete information and exhibits limited rationality. In addition, the day-to-day mean route flow dynamic may evolve to user equilibrium, the stochastic user equilibrium, system optimal and other disequilibrium states depending on different behavioral rules of route-switching rate and route choice probability, network supply and congestion toll pricing schemes. This model is particularly useful to study the resulting day-to-day disequilibrium traffic evolving pattern when a portion of a network infrastructure is to undergo a scheduled upgrade or when a capacity reduction takes place due to external interventions. We demonstrate this in the case study with three testing scenarios.

INTRODUCTION

Transportation network equilibrium models have become the main thrust of advances in the field of traffic network analysis. The concept of user equilibrium (UE) was first proposed by Wardrop (1952) which stated that at the equilibrium state, no traveler could improve his/her travel cost by unitarily changing routes. The equilibrium flows are considered as the predicted traffic flow in the long run and have been widely used for a variety of transportation planning purposes.

Equilibrium analysis, however, only pays attention to the final "attractor" state while ignoring how travelers dynamically adjust their behavior and traffic flow evolves over "days". With the new advances in the Intelligent Transportation Systems (ITS), travelers may have access to both historical and real-time traffic information; therefore adjust their route choice behavior by their day-to-day learning and information updating processes. Consequently, day-to-day traffic dynamics is of great importance in transportation network analysis, both for a better understanding of the properties of the standard traffic equilibrium model, and for practical reasons related to the monitoring and management of traffic flows. Recent advances in day-to-day dynamic congestion pricing are reported by Friesz et al. (2004) and Yang and Szeto (2006), which demonstrate how transport managers exploit day-to-day time varying pricing schemes to maximize consumer surplus and to realize system optimal, respectively. In addition, it should be noted that, the day-to-day dynamics approach is also highly suited to incorporate within-day dynamics into "doubly dynamics" and can be seen as a more general framework than the equilibrium models (Cascetta and Cantarella, 1991; Friesz et al., 1996; Cantarella et al., 1999; Balijepalli and Watling, 2005).

Existing day-to-day traffic dynamics include continuous-time deterministic processes (e.g., Smith, 1983; Smith, 1984; Friesz et al., 1994; Nagurney and Zhang, 1996; Jin, 2006), and discrete-time stochastic models (e.g. Chang and Mahmassani, 1988; Cascetta, 1989; Davis and Nihan, 1993; Jha et al., 1998; Hazelton and Watling, 2004). Deterministic processes listed above have good mathematical properties in terms of a unique solution trajectory, the equivalence between their fixed points and UE, and the stability property, but ignore random fluctuations of demand and supply, and travelers' behavior under uncertainty given the fact that travelers can only access incomplete traffic information and might not be completely rational. In contrast, stochastic processes address the randomness issue and capture travelers' behavior under uncertainty by means of Markov decision models, however computing the large transition probability matrix in Markov models is analytically difficult and computationally expensive. To overcome this difficulty, Davis and Nihan (1993) was one of the first to use the mean (expected) deterministic dynamic to approximate the stochastic process as the population size grows large, and this approximation was observed in the numerical study in Cantarella and Cascetta (1995). Nevertheless, compared to the deterministic processes, the convergence and stability results for these stochastic models are more complex to analyze. Given the numerous and extensive studies over the classic

Wardropian traffic equilibrium analysis, understanding the linkage between day-to-day traffic dynamics and traffic equilibria is crucial for research.

In this paper, by taking advantages from both deterministic and stochastic approaches, we study the day-to-day traffic dynamics and present a new framework using the evolutionary game theory. The evolutionary game theory applies population dynamical methods into classic game theory, relaxes the requirements for rationality of players and provides insight into the system dynamic evolving behavior (e.g., Weibull, 1995; Hofbauer and Sigmund, 2003). By applying the evolutionary game theory, we first address the stochastic behavior of an individual traveler, i.e., how often one reconsiders his/her route choice (route-switching rate) to capture traveler's habitual behavior or inertia, and what is the probability (route choice probability) that a certain route being chosen to reflect the uncertainty. Since route flow is the sum of all the individual travelers on this route, we next build up the stochastic process of the aggregated route flow. Subsequently we show that when the traffic demand is large, the mean route flow dynamic over finite time spans follows an almost deterministic trajectory. Our goal of this paper is to find some deterministic processes to closely approximate the stochastic process and study the relationship between their fixed points and the traffic equilibria.

Our mean dynamic is general in that many existing deterministic processes can be considered as its special cases (Smith, 1984; Nagurney and Zhang, 1996; Yang, 2005; Jin, 2006). More importantly, more meaningful individual traveler's behavior explanations are unveiled. Besides, it can be concluded from this general framework that if every individual traveler follows certain reasonable rules of route-switching rate and route choice probability, the Wardrop user equilibrium can be approached by travelers' day-to-day behavior adjustment process. Traditionally, the behavior assumption for user equilibrium is that each traveler obtains perfect information and exhibits completely rationally. In contrast, our behavior assumption for an individual is the Markov decision rule, i.e., one makes route choice "today" only depending on the limited road information available from "yesterday", and behaves not completely rationally in that one might choose the non-optimal route with certain probability. Furthermore, the day-to-day mean route flow dynamic may evolve to user equilibrium, the stochastic user equilibrium, system optimal and other disequilibrium states depending on the different behavior rules of revision rate and choice probability, network supply and congestion toll pricing schemes.

The remainder of this paper is organized as follows. We will begin with the formulation of the continuous-time route flow Markov model. The mean route flow dynamic of the underlying Markov model will be presented. Next we will address the relationship between the existing continuous-time deterministic day-to-day processes and the mean dynamic. Computational results will be presented on applying the model to study the resulting traffic pattern under various capacity reduction scenarios. Finally, discussions and conclusion remarks are given.

METHODOLOGY

It is assumed throughout this paper that the travel demand is fixed and day-to-day static, while the elastic demand can be extended by adding a dummy route with zero cost to accommodate the excess demand. Travelers are assumed to behave homogeneously and be "indistinguishable" (Cascetta, 1989).

Formulation of the Continuous-time Route Flow Markov Model

Typically, a transportation network can be considered as a fully-connected directed graph denoted as $G(\boldsymbol{N},\boldsymbol{A})$, consisting of a set of nodes $\boldsymbol{N}$ and a set of links $\boldsymbol{A}$. Let the set of O-D pairs be denoted by W, the traffic demand for O-D pair $w \in W$ by d^w, the set of routes between O-D pair $w \in W$ by P^w, the flow on route $p \in P^w$ by f_p^w, the route travel cost on route p by C_p^w, the flow on link $a \in \boldsymbol{A}$ by x_a, the travel cost on link a by τ_a, the dimensions of the route flow vector $\mathbf{f}$, the link flow vector $\mathbf{x}$, the demand vector $\mathbf{d}$, and the set P^w by n^r, n^a, n^w, and n^p, respectively. Let the matrix A denote the link-route incidence matrix.

The traffic assignment model is to assign O-D demand vector $\mathbf{d} = (d^1, d^2, \cdots, d^w)$ over the transportation network $G(\boldsymbol{N},\boldsymbol{A})$. Recall the definition of the feasible route flow set K is the Cartesian product of each $K^w = \{\mathbf{y} \in R^{n^p} : \mathbf{y} \geq 0, \text{and} \sum_{p \in P^w} y_p^w = d^w\}$; hence, feasible route flows can take any real values in the set K. In the Markov process context, it is more convenient to have the finite state space instead of the real-valued infinite one. In order to make the population large but finite, we will first fix a large number N and multiply it with the demand vector to construct the new demand $(Nd^1, Nd^2, \cdots, Nd^w)$. This rescaled construction is also used in Davis and Nihan (1993) and Balijepalli and Watling (2005). The finite route flow set is denoted by $K^N = \{\mathbf{f} \in K : N\mathbf{f} \in Z^{n^r}\}$ where Z^{n^r} denotes the set of all non-negative integer-valued vectors with dimension n^r. Therefore, $K^N = K \cap \frac{1}{N} Z^{n^r}$ is a fine discrete grid of the continuous real-valued convex set K. The route cost function is $\mathbf{C}: K \rightarrow R_+^{n^r}$ and link cost function $\boldsymbol{\tau}: X \rightarrow R_+^{n^a}$ where X is the feasible link flow set, $X = \{\mathbf{x} : \mathbf{x} = A\mathbf{f}, \mathbf{f} \in K\}$. At each "day" t, Nd^W travelers for O-D pair $X = \{\mathbf{x} : \mathbf{x} = A\mathbf{f}, \mathbf{f} \in K\}$ choose routes in the feasible route set P^w. In this model, the state vector is the discretized route flow random vector $\mathbf{f}_t^N \in K^N$. The route cost random vector $\mathbf{C}_t$ can be expressed by $\mathbf{C}_t = A^T \boldsymbol{\tau}(A\mathbf{f}_t^N)$. The subscript t is dropped from the dynamic variables hereafter to simplify notations if there is no confusion.

Before studying the route flow dynamics, let us first address the individual traveler's stochastic route-choice behavior. The basic two components to model an individual behavior in the evolutionary game theory are the revision rate vector $\boldsymbol{\lambda}^w : K^w \rightarrow R_+^{n^p}$ and choice probability vector $\boldsymbol{\pi}^w : K^w \rightarrow \Delta^w$, where $\Delta^w = \{\mathbf{z}^w \in R^{n^p} : \sum_{p \in P^w} z_p^w = 1, \mathbf{z}^w \geq 0\}$ is the set of probability measure vector on R^{n^p} (Sandholm, 2006). It is assumed that travelers will stay on a route with a non-negligible randomly distributed time interval due to their inertia to switch routes. Route-switching rate describes how fast travelers are revaluating their route choices. Route choice probability defines when travelers reconsider their route choices, the probability that they will choose each route. Route-switching rate and route choice probability both depend on the traffic flow state on the current "day", i.e., the flow distribution. Although route-switching rate and route choice probability are not completely new concepts in traveler's behavior study (e.g., Mahmassani and Chang, 1987; Chang and Mahmassani, 1988), modeling traveler's random inertia is not well understood in the literature. Traveler's inertia is usually modelled as a deterministic and state-independent constant term; for example, either all travelers or a small fixed percentage of travelers will reconsider routes every "day" (Cantarella and Cascetta, 1995; Watling, 1999), while in this paper we model traveler's inertia as a state-dependent random variable. Our approach is based on the following observation: consider the situation where the day-to-day route travel cost varies greatly among routes, it is more logical to assume that travelers will switch routes more frequently, while they will switch routes less frequently if all routes have almost the same travel costs.

To model the individual's inertia, it is assumed that each traveler is associated with a random clock which rings exponentially, that is, the amount of time staying on route p between two consecutive rings is exponentially distributed with the parameter $\lambda_p^w = \lambda_p^w(\mathbf{f})$ for the O-D pair $w \in W$ and the route $p \in P^w$. The route-staying time for every individual traveler is independent of each other and memory-less, i.e., depending only on the current route flow state $\mathbf{f}$ if we assume the continuous-time Markov decision model of individual traveler. If one's clock rings, he/she will reconsider his/her route choice: whether to stay on the same route, or change to another route $q \in P^w$ $(q \neq p)$ with probability $\pi_p^w = \pi_p^w(\mathbf{f})$, as illustrated in Figure 1.

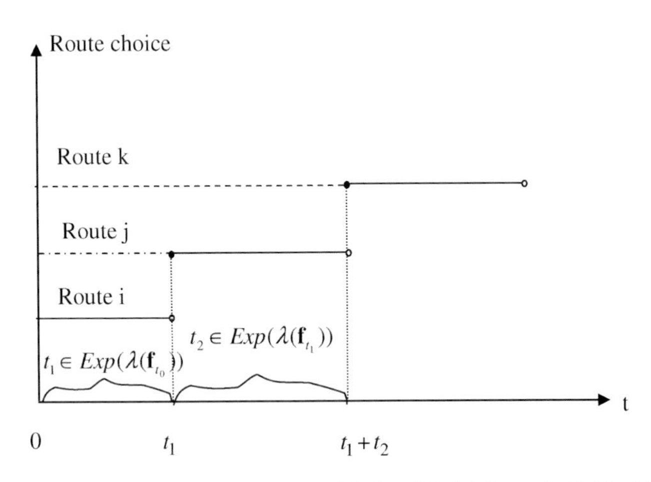

Figure 1 The Continuous-time Markov Model for an Individual Traveler

The statements above construct the homogeneous Markov process for individual traveler's day-to-day route choice behavior. In the following, we will show that the aggregated route flow is a Markov process $\{\mathbf{f}_t^N\}_{t\geq 0}$ within the state space K^N as well.

Let the current time moment be day t and the current route flow state $\mathbf{f} \in K^N$. Therefore, there are $N\mathbf{f}_p^w$ travelers from the O-D pair $w \in W$ on the route $p \in P^w$. All the travelers on route p have the same exponential rate $\lambda_p^w(\mathbf{f})$ while the route-staying time of each traveler is independent of each other. The first clock rings after time t is actually the winner of the underlying independent "exponential race" over all routes. Hence, the route-staying time ℓ before the first traveler's clock rings is exponential distributed with parameter $N\bar{\lambda}$, that is, $\ell \in \exp(N\bar{\lambda})$ where $\bar{\lambda} = \sum_w \sum_p f_p^w \lambda_p^w$ is the weighted average of the exponential rates over all routes. Regardless of the time when the clock alarms, the probability that this clock belongs to the travelers on route p is $\frac{f_p^w \lambda_p^w}{\bar{\lambda}}$. When a traveler on route p reconsiders his/her route choice, he/she will choose route q with probability π_q^w. This choice probability is independent of the history and conditional only upon the current route flow state. Suppose after time ℓ, the number of travelers on route q changes from Nf_q^w to Nf_q^w+1, then it concludes that the first ringing clock belongs to the travelers on route p, and there is only one traveler switching from route p to route q. To sum up all the statements above, we can conclude that the probability that the first ringing clock belonging to a traveler on route p and the traveler switches from route p to route q is

$$\frac{f_p^w \lambda_p^w}{\bar{\lambda}} \pi_q^w \text{, where } \bar{\lambda} = \sum_w \sum_p f_p^w \lambda_p^w \tag{1}$$

Since after time t, the route-staying time before the first route switching $\ell \in \exp(N\bar{\lambda})$, only depends on the route flow vector $\mathbf{f}_t$, and the choice probability π_q^w only depends on $\mathbf{f}_t$ as well regardless of the history, it concludes that this process $\{\mathbf{f}_t^N\}_{t\geq 0}$ is a Markov process.

The Mean Route Flow Dynamic and its Deterministic Approximation

The statements above show that the underlying process $\{\mathbf{f}_t^N\}_{t\geq 0}$ is the continuous-time Markov process. Because of the continuity of the "exponential race", there is only one traveler having the opportunity to reconsider his/her route choice when his/her clock rings first. Hence the actual increment of the route flow vector is $\zeta = \frac{1}{N}(\mathbf{e}_q - \mathbf{e}_p)$, $q \neq p$ if a traveler switches from route p to route q, where $\mathbf{e}_p$ and $\mathbf{e}_q$ are the basis vectors in R^{n^r} with p^{th} and q^{th} element equal to 1 and others 0, respectively. Let t_k denote the time of the kth route-switching opportunity occurring, t_{k+1} for $k+1^{th}$ and ζ^N the random route flow increment at time t_{k+1}. Then the distribution of ζ^N is $P(\zeta^N = \mathbf{z}) = P(\mathbf{f}_{t_{k+1}}^N = \mathbf{f} + \mathbf{z} \mid \mathbf{f}_{t_k}^N = \mathbf{f})$. Generalizing all the statements above, we can write the formulation of the distribution of the increment ζ^N as

$$P(\zeta^N = \frac{1}{N}(\mathbf{e}_q - \mathbf{e}_p)) = \frac{f_p^w \lambda_p^w(\mathbf{f})}{\bar{\lambda}} \pi_q^w(\mathbf{f}) \qquad \text{for some } q \neq p \tag{2}$$

Since the total route-switching rate is $N\bar{\lambda}$, within a small time interval $(t, t+\delta)$, there are $N\bar{\lambda}\delta$ route-switching opportunities. Therefore, it can be seen that even in a very small interval, the switching rate can go to infinite as N grows out of bound. However, for each switching, the upper bound of the expected route flow increment at next revision is of order $\frac{1}{N}$. Hence, the overall expected increment in this interval $(t, t+\delta)$ is of order $\bar{\lambda}\delta$, which is bounded although the population size N approaches very large. Since during $(t, t+\delta)$ there is a large number of switches and each of them generates almost the same expected flow increment, followed by the law of large numbers, the total change of $\{\mathbf{f}_t^N\}$ is mainly determined by its mean dynamic, which can be shown as the deterministic ordinary differential equation (ODE) in the following. The mean dynamic of process $\{\mathbf{f}_t^N\}$ is defined as the deterministic ordinary differential equation whose solution approximates the Markov process $\{\mathbf{f}_t^N\}$. Explicitly, this ODE can be expressed by $\dot{\mathbf{f}} = V^N(\mathbf{f})$, where V^N is a continuous function over the discretized route flow space K^N.

The expected route flow switching increment for every revision opportunity is

$$E\zeta^N = \sum_w \sum_p \sum_{q \neq p} \mathbf{z} P(\zeta^N = \mathbf{z})$$
$$= \sum_w \sum_p \sum_{q \neq p} \frac{1}{N} (\mathbf{e}_q - \mathbf{e}_p) \frac{f_p^w \lambda_p^w(\mathbf{f})}{\bar{\lambda}} \pi_q^w(\mathbf{f}) \tag{3}$$

The expected increment per time unit of the process $\{\mathbf{f}_t^N\}$ can be written as

$$V^N(\mathbf{f}) = (N\bar{\lambda}) E\zeta^N$$
$$= (N\bar{\lambda}) \sum_w \sum_p \sum_{q \neq p} \frac{1}{N} (\mathbf{e}_q - \mathbf{e}_p) \frac{f_p^w \lambda_p^w(\mathbf{f})}{\bar{\lambda}} \pi_q^w(\mathbf{f})$$
$$= \sum_w \sum_p e_p^w \left[\pi_p^w \sum_q f_q^w \lambda_q^w - f_p^w \lambda_p^w \right] \tag{4}$$

Since this vector field $V^N(\mathbf{f})$ is unrelated to the population size N, we can write the mean dynamic for each route flow as

$$V_p^w(\mathbf{f}) = \pi_p^w \sum_q f_q^w \lambda_q^w - f_p^w \lambda_p^w \tag{5}$$

Let the superscript dot denote the derivative with respect to time. Summarizing all the statements above, we can derive the mean dynamic of the route flow stochastic process, given by Theorem 1.

Theorem 1 The mean dynamic of the route flow Markov process $\{\mathbf{f}_t^N\}$ can be expressed by
$\dot{f}_p^w = \pi_p^w \sum_{q \in \boldsymbol{P}^w} f_q^w \lambda_q^w - f_p^w \lambda_p^w$ for all $p \in P^w, w \in W$ *and* $\mathbf{f} \in K$. (6)

The intuitive explanation of the mean dynamic (6) is as follows: the flow change rate on route p can be considered as the difference between the inflow rate into route p and outflow rate from route p. The inflow rate into route p is the multiplication of the total flow switch rate $\sum_{q \in \boldsymbol{P}^w} f_q^w \lambda_q^w$ and the route choice probability π_p^w, and the outflow rate from route p is the multiply of the flow f_p^w and the route-switching rate λ_p^w.

Remark. Some properties of the mean dynamic (6) include the solution trajectory $\mathbf{f}(t)$ non-negativity and flow conservation, i.e., $\mathbf{f}(t) \in K$ given the initial point $\mathbf{f}_0 \in K$. To show the non-negativity property, let us define the boundary of set K is

$K^0 = \{\mathbf{f} \in K : f_i = 0\, and\; f_j > 0\; for\; some\; i\; and\; j \neq i\}$. It is straightforward to show $\dot{f}_i \geq 0$ for $\mathbf{f} \in K^0$ and $f_i = 0$, i.e., at the boundary points, the mean dynamic (6) forces the flow to move in the direction not to decrease f_i any more. The flow conversation property is satisfied after checking $\sum_{q \in P^w} \dot{f}_p^w = 0$.

With the mean dynamic in place, the next question is to see how close this mean dynamic is to the underlying continuous-time Markov process $\{\mathbf{f}_t^N\}$. Intuitively, from the strong law of the large numbers, the stochastic evolution is mainly determined by its mean dynamic. Rigorously, the link between the mean flow dynamic (6) and the finite horizon behavior of Markov process $\{\mathbf{f}_t^N\}$ can be stated in Theorem 2.

Theorem 2 Let $\left\{\{\mathbf{f}_t^N\}_{t \geq 0}\right\}_{N=N_0}^{\infty}$ be a realization of continuous-time route flow Markov processes. Assume that the initial condition $\mathbf{f}_0^N \in K^N$ converges to $\mathbf{f}_0 \in K$. Suppose the mean dynamic is Lipschitz continuous, and $\{\mathbf{f}_t\}_{t \geq 0}$ be the solution to the mean dynamic $\dot{\mathbf{f}} = V(\mathbf{f})$ from initial condition $\mathbf{f}_0$. Then for each $T < \infty$ and $\varepsilon > 0$, one has $\lim_{N \to \infty} P\left[\sup_{\mathbf{f} \in K^N} \left|\mathbf{f}_t^N - \mathbf{f}_t\right| < \varepsilon\right] = 1$.

Proof: see **Appendix 1**.

The Relationship between the Mean dynamic and Existing Day-to-day Dynamics

The arguments above apply the evolutionary game theory to derive the mean route flow dynamic and shows that this mean dynamic is a close approximation to the underlying continuous-time stochastic process. We claim that our model is a general day-to-day dynamics formulation in that many existing deterministic day-to-day processes can be considered as the special cases of our mean dynamic by appropriately setting up the route-switching rate $\boldsymbol{\lambda}$ and the route choice probability $\boldsymbol{\pi}$. By accommodating existing deterministic day-to-day processes into our general frame work, they can be considered as the deterministic approximation of the underlying stochastic processes, and their meaningful individual behavior explanations are recognized.

In the literature, most of the continuous-time deterministic dynamics have the property that there exists unique solution trajectories of the ODEs and the fixed points of the ODEs are the corresponding user equilibrium. They include the proportionally switching dynamical system (Smith, 1984; Smith and Wisten, 1995; Huang and Lam, 2002; Peeta and Yang, 2003), the projected dynamical system (Nagurney and Zhang, 1996; Nagurney and Zhang, 1997), the BNN (Brown-von Neumann-Nash) dynamic (Yang, 2005) and the First-in-First-out (FIFO) dynamical system (Jin, 2006).

First let us address the relationship between the proportional-switch adjustment process model and our mean dynamic (6). The proportional-switch adjustment process (PAP) originates from Smith (1984), although it is not named by Smith (1984). The basic idea of the proportional-switch adjustment process is that travelers on a higher cost route will switch to other lower cost routes in next "day", while the switching rate depends on the cost difference between this route and other routes. The proportionally switching dynamical system can be written as

$$\dot{f}_p^w = \sum_{q \in P^w} f_q^w [C_q^w - C_p^w]_+ - f_p^w \sum_{q \in P^w} [C_p^w - C_q^w]_+ \tag{7}$$

Let the route-switching rate be $\lambda_p^w = \sum_{q \in P^w} [C_p^w - C_q^w]_+$ and route choice probability $\pi_{qp}^w = \dfrac{[C_q^w - C_p^w]_+}{\sum_{s \in P^w} [C_q^w - C_s^w]_+}$ where π_{qp}^w denotes the probability to choose route p when travelers on route q reconsider routes. Then the mean dynamic (6) coincides with the proportionally switching dynamical system (7). The fixed point of (6) under this (λ, π) setting will be user equilibrium.

After investigating the (λ, π) setting for the proportionally switching dynamical system, we can understand better the underlying individual behavior. The behavior explanation for the proportionally switching dynamical system is that, the higher the traveler cost on route p is over other routes for the same O-D pair, the more often a traveler on this route will reconsider switching routes, i.e., he/she has less inertia to change routes. If the travel cost on route p is less than or equal to that of all other routes connecting the same O-D pair, a traveler will stay on this route with probability 1. A traveler will switch from route q to route p with positive probability if the travel cost on route p is lower; otherwise, he/she won't switch at all.

The Projected Dynamical System

The projected dynamical system describes disequilibrium trajectories of traffic dynamics prior to reaching user equilibrium. It has been widely used to solve the variational inequality problem (Nagurney, 1993). It can be expressed by

$$\dot{\mathbf{f}} = \Pi_K(\mathbf{f}, -\mathbf{C}) \tag{8}$$

where the operator Π_K is defined as $\Pi_K(\mathbf{x}, \mathbf{y}) = \lim_{\varepsilon \to 0} \dfrac{P_K(\mathbf{x} + \varepsilon \mathbf{y}) - \mathbf{x}}{\varepsilon}$ and $P_K(\mathbf{x}) = \arg\min_{\mathbf{z} \in K} \|\mathbf{x} - \mathbf{z}\|$ is the projection of vector $\mathbf{x}$ into the feasible set K.

Inspired by Sandholm and Lahkar (2005), we can partition the route index set into three subsets, $Q = \{k : f_k > 0\}$, $Z_G = \{k : f_k = 0, \dot{f}_k > 0\}$ and $Z_H = \{k : f_k = 0, \dot{f}_k = 0\}$. Let $|Z_G|$ and $|Q|$ denote the dimension of set Q and Z_G, respectively. Then we define

$[\tilde{C}_p]_- = \max\left(0, C_p - \frac{\sum_{k\in Z_G} C_k + \sum_{k\in Q} C_k}{|Z_G|+|Q|}\right)$, and $[\tilde{C}_p]_+ = \max\left(0, \frac{\sum_{k\in Z_G} C_k + \sum_{k\in Q} C_k}{|Z_G|+|Q|} - C_p\right)$. The important statement is that we can set up the following (λ, π) rule such that the project dynamical system (8) can be expressed by our mean dynamic (6).

$$\lambda_p(\mathbf{f}) = \begin{cases} \frac{[\tilde{C}_p]_-}{f_p} \; if \; f_p > 0 \\ 0 \; if \; f_p = 0 \end{cases} \text{ and } \pi_p(\mathbf{f}) = \frac{[\tilde{C}_p]_+}{\sum_{k\in Z_G\cup Q} [\tilde{C}_k]_+}$$

This (λ, π) rule indicates the individual behavior explanation for the projected dynamical system is that, the higher the travel cost on route p is over the "weighted average" travel cost $\frac{\sum_{k\in Z_G} C_k + \sum_{k\in Q} C_k}{|Z_G|+|Q|}$, more often travelers on this route would like to change routes, i.e., they exhibit less inertia in their route choices. Travelers will switch to route p with positive probability if the travel cost in route p is lower than the "weighted average" travel cost $\frac{\sum_{k\in Z_G} C_k + \sum_{k\in Q} C_k}{|Z_G|+|Q|}$.

The FIFO Dynamical System

The FIFO dynamical system was developed in Jin (2006) by observing FIFO violation among routes connecting the same O-D pair. Compared with the dynamical systems (7) and (8) above, one distinct feature of FIFO dynamical system is that the right hand side of the FIFO dynamic system (9) is continuously differentiable. The FIFO dynamical system can be written as

$$\dot{f}_p^w = -d^w f_p^w (C_p^w - \overline{C}^w) \tag{9}$$

where $\overline{C}^w = \frac{1}{d^w} \sum_{p\in P^w} C_p^w f_p^w$ is the average travel cost for OD pair w. Let $\lambda_p^w = d^w (C_p^w - \eta^w)$ and $\pi_p^w = {f_p^w}/{d^w}$ where $\eta^w = \min_{p\in P^w} C_p^w$. It is straightforward to check that this (λ, π) rule will make the mean dynamic (6) identical to the FIFO dynamical system (9). The individual behavior explanation for the FIFO dynamical system is that, a traveler on higher cost route reconsiders his/her route choice more often, and the probability for him/her to change to a route p is proportional to how many travelers on the route p.

The BNN Dynamic

The BNN dynamic is a canonical dynamic in microeconomics to model players' dynamical evolving behavior. It was first introduced by Brown and von Neumann (1950) for symmetric zero-sum games and recently studied extensively in Swinkels (1993), Hofbauer (2000), Sandholm (2001) and Hofbauer and Sigmund (2003). An interpretation of BNN dynamic is as follows: during any small time interval, all players in a population are equally likely to switch strategies, and the rate to do so is proportional to the sum of the excess payoffs in the population. Those who switch choose strategies with above average payoffs, choosing each with probability proportional to the strategy's excess payoff (Sandholm, 2001).

The first attempt to model BNN traffic route flow dynamics is in Yang (2005). Let the average route cost be $\bar{C}^w = \frac{1}{d^w} \sum_{p \in \boldsymbol{P}^w} C_p^w f_p^w$ for O-D pair w, and the excess route cost $[\hat{C}_p^w]_+ = \max\{0, -C_p^w + \bar{C}^w\}$ denotes the excess travel cost of route p relative to the average travel cost for O-D pair w. Then the BNN dynamic can be written as

$$\dot{f}_p^w = d^w [\hat{C}_p^w]_+ - f_p^w (\sum_{q \in \boldsymbol{P}^w} [\hat{C}_q^w]_+) \tag{10}$$

There are three important properties of BNN route flow dynamics: first the right hand side of BNN route dynamic is Lipschitz continuous hence it admits a unique solution trajectory. Second, the fixed point f^* of BNN route flow dynamic is equivalent to the Wardrop user equilibrium. Third, the user equilibrium is the state where the excess route cost satisfies $[\hat{C}_p^w]_+ = 0, \forall w, p$ (Friesz and Shah, 2001; Yang, 2005).

Corresponding to the general mean dynamic (6), we have route-switching rate $\lambda_p^w(\mathbf{f}) = \sum_{q \in \boldsymbol{P}^w} [\hat{C}_q^w]_+$ and choice probability $\pi_p^w(\mathbf{f}) = \frac{[\hat{C}_p^w]_+}{\sum_{q \in \boldsymbol{P}^w} [\hat{C}_q^w]_+}$ for BNN dynamic. Indeed, it describes a certain traveler's learning process where the frequency to choose routes with travel cost above average decrease, while the frequency with travel cost below average increase, as long as the route flow is changing. In other words, travelers will choose the routes with positive probability whose travel cost is less than the weighted average travel cost over all feasible routes.

The Logit Dynamic and SUE

In the statements above, we have discussed the cases where the mean flow dynamic converges to user equilibrium. We can find the example of the settings of the appropriate

(λ, π) rules such that the mean dynamic (6) converges to SUE. For example, let $\lambda_p^w \equiv 1$ and $\pi_p^w = \dfrac{\exp(-\theta C_p^w)}{\sum_{q \in P^w} \exp(-\theta C_q^w)}$. Then the mean dynamic (6) becomes

$$\dot{f}_p^w = d^w \frac{\exp(-\theta C_p^w)}{\sum_{p \in P^w} \exp(-\theta C_p^w)} - f_p^w \tag{11}$$

Apparently the fixed point of the dynamic (11) is the logit-based SUE. For (11), the logit discrete choice model is used. Similarly, if the probit model is chosen to define the choice probability, the fixed point of (11) will be the probit-based SUE.

The Day-to-Day Dynamic Toll and System Optimal

Yang and Szeto (2006) show that imposing a day-to-day dynamic congestion toll $\boldsymbol{\beta}(t) = D\boldsymbol{\tau}(t)'\mathbf{x}(t)$ will guide the day-to-day dynamic flow evolving towards system optimal instead of user equilibrium, considering a general drivers' behavior adjustment process (including PAP, PDS, and BNN dynamics). Their day-to-day dynamic toll can be considered as the dynamic version of the classic marginal social cost one, by replacing the system optimal link flow and link travel cost with the ones on the current "day". Moreover, the dynamic total system cost is monotonically decreasing along the day-to-day dynamic flow trajectory until it converges to system optimal flows.

The Disequilibrium Trajectory

Note that the mean dynamic might converge to disequilibrium states as well. The following example demonstrates under certain network supply condition, the mean dynamic converges to a periodic trajectory instead of the equilibrium. Consider the network with one O-D pair and 3 parallel links, and the route travel cost function is $C_1 = 2f_1 + f_2 + 4f_3$, $C_2 = 4f_1 + 2f_2 + f_3$, and $C_3 = f_1 + 4f_2 + 2f_3$. The O-D demand is 1. It is straightforward to show that the only user equilibrium point is evenly distributed route flow $(f_1, f_2, f_3) = (1/3,\ 1/3,\ 1/3)$, which is the center of the triangle in figure 2. However, by solving the BNN dynamic, the solution trajectory turns out to converge to a middle circle regardless of the initial point, rather than the only equilibrium point. Even if the initial point is around the user equilibrium, a small perturbation will cause the dynamic to leave the equilibrium and to head to the middle circle as shown in the phase portrait in figure 2. The solution trajectory also approaches the middle circle when the initial conditions are outside of this circle. In figure 2, the three vertexes are the cases that all the demand is assigned to only one route. The arcs with arrows indicate the solution trajectory of the BNN dynamic given an initial point. The reason why the mean dynamic presents a periodic trajectory lies in the stability property of the equilibrium, which is not the focus of this paper.

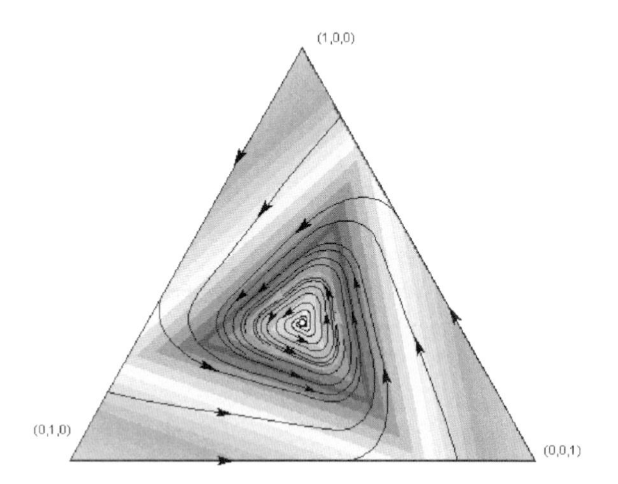

Figure 2 Phase Portrait of the BNN Dynamic for the Three-link Network

EXPERIMENTAL STUDIES

By using the mean dynamic to closely approximate the underlying stochastic route flow changes, we can model the day-to-day disequilibrium dynamic traffic flow evolution from any given initial traffic condition. Therefore, our model is particularly useful to study the resulting dynamic traffic pattern when network supply and demand changes, for example, a portion of a network infrastructure is to undergo a scheduled upgrade or when a capacity reduction takes place due to external interventions. In this section, we will illustrate the evolution of several route flow dynamics (e.g., PAP, BNN and Logit dynamics) undergoing different road conditions.

The simple test network will be the 3×3 grid network with 9 nodes, 12 links and 6 routes from origin 1 to destination 9, as shown in Figure 3. The total O-D flows are 500 units. The route and link correspondence is shown in Table 1.

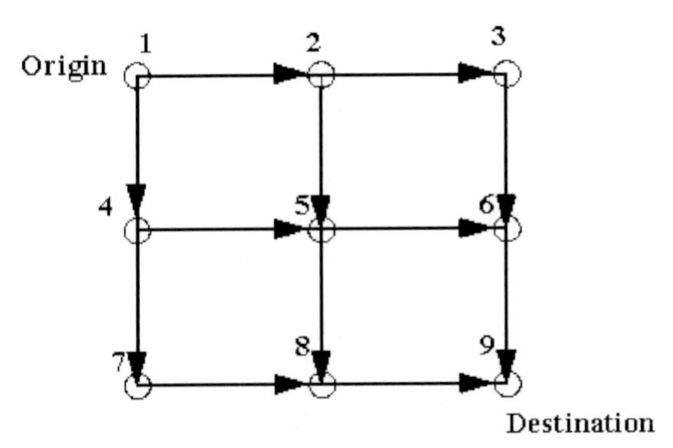

Figure 3 The 3×3 Grid Network

Table 1 Route and Link Correspondence in the 3×3 grid network

Route	Links
1	1-2-3-6-9
2	1-2-5-6-9
3	1-2-5-8-9
4	1-4-7-8-9
5	1-4-5-6-9
6	1-4-5-8-9

The link travel time function is $\tau_a = 1.5[1+0.15(x_a/C_a)^3]$, where x_a is the link flow and C_a is the capacity for link a. All twelve links have the same capacity 2200 vehicles. The initial point of route flow vector is randomly chosen as $\mathbf{f} = (62.5,\ 100,\ 125,\ 75,\ 62.5,\ 75)'$. The unique (stochastic) user equilibrium flows are $\mathbf{x} = [250\ 125\ 250\ 125\ 125\ 125\ 125\ 125\ 125\ 250\ 125\ 250]'$.

Three testing scenarios are listed in Table 2. For scenario 1, there is no capacity reduction of any link. Scenario 2 describes the situation where there is 80% capacity reduction occurred at link 1 (node 1-2) from time t = 3. In scenario 3, the capacity of link 1 returns back to normal from time t = 6.

Table 2 Three Testing Scenarios

Scenario	**Distinctive features**
1	No incident occurred.
2	80% capacity reduction occurred at link 1 (node 1-2) from time t = 3
3	80% capacity reduction at link 1 occurred from time t = 3 and its capacity returned back to normal from time t = 6

The dynamic route flow evolution of PAP, BNN and Logit dynamics are shown below in Figures 4, 5 and 6, corresponding to scenario 1. For both PAP and BNN dynamics, the dynamic route flow pattern smoothly converge to the Wardrop user equilibrium route flow $\mathbf{f}^* = (125,\ 55.55,\ 69.45,\ 125,\ 69.45,\ 55.55)'$, while the Logit dynamic flow converges to SUE route flow $\overline{\mathbf{f}} = (89.6,\ 80.2,\ 80.2,\ 89.6,\ 80.2,\ 80.2)'$. In terms of the convergence rate, the PAP dynamic converges to the UE route flows faster than the BNN dynamic.

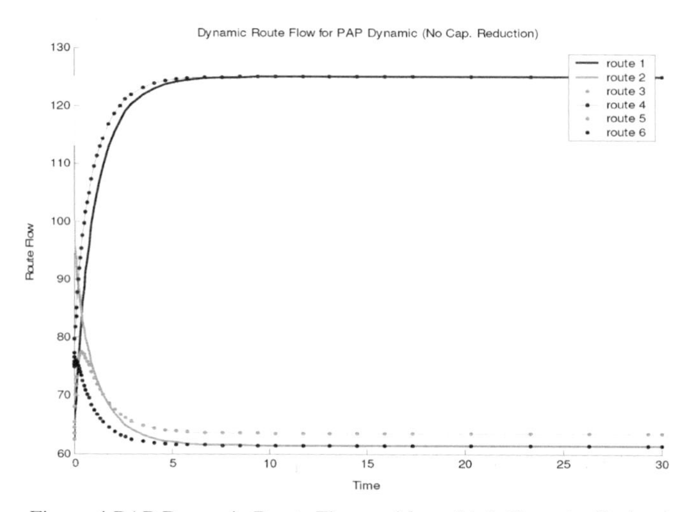

Figure 4 PAP Dynamic Route Flows without Link Capacity Reduction

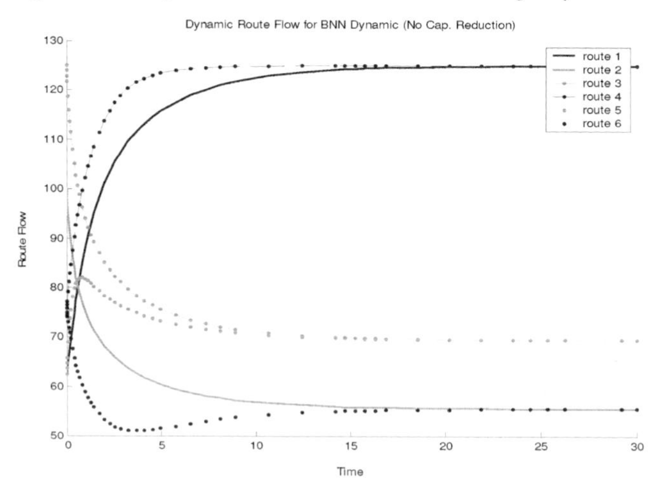

Figure 5 BNN Dynamic Route Flows without Link Capacity Reduction

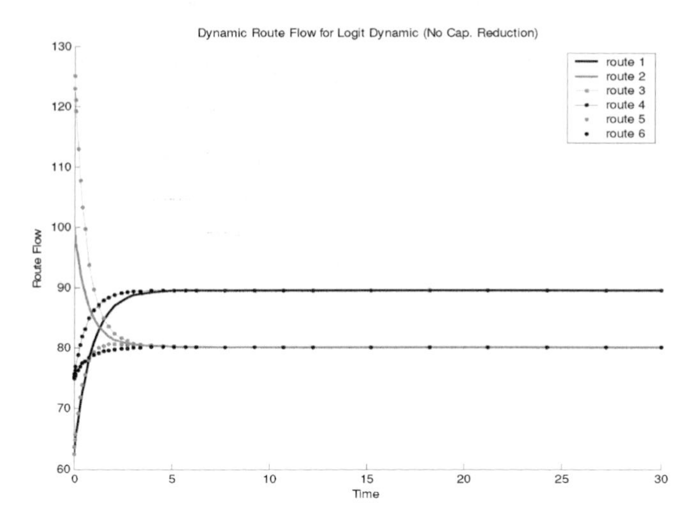

Figure 6 Logit Dynamic Route Flows without Link Capacity Reduction

For scenario 2, there is 80% capacity reduction on link 1 from time $t = 3$. Figures 7, 8 and 9 show the corresponding dynamic route flow evolution for the three dynamics. The route flows are heading their ways to the UE ones before the capacity reduction on link 1. After the external shock to the system, the dynamics depart away from their original trajectory and change dramatically to reflect the drivers' responses to this shock. The route flows on route 1, 2 and 3 decrease fast after the capacity reduction in link 1 because link 1 is part of these routes. On the contrary, the route flows of route 4, 5 and 6 increase dramatically.

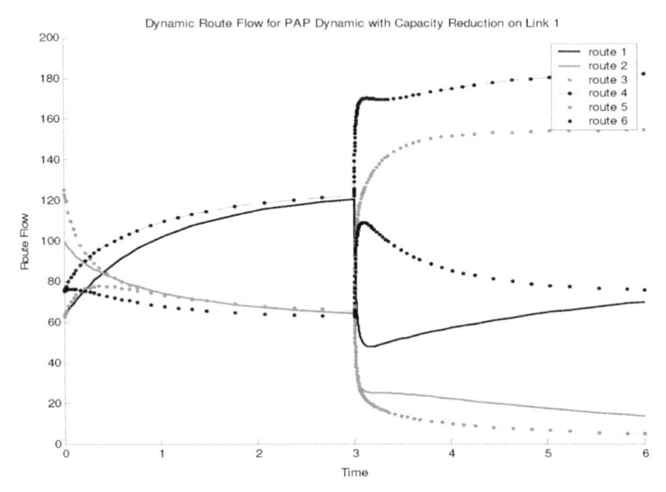

Figure 7 PAP Dynamic Route Flows with Capacity Reduction on Link 1

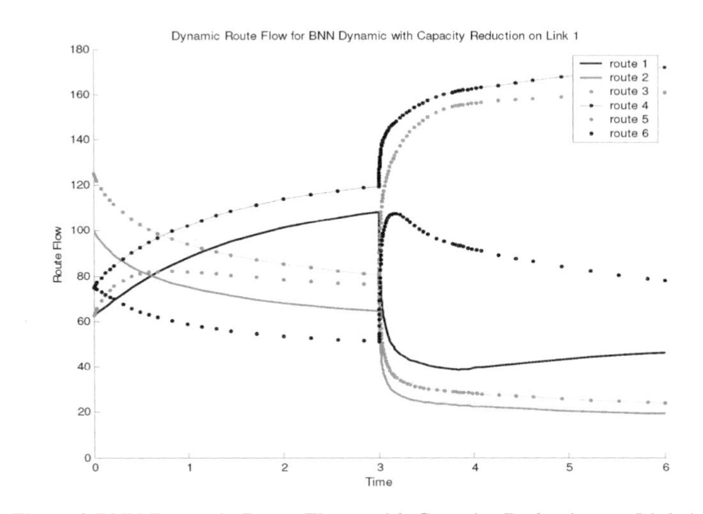

Figure 8 BNN Dynamic Route Flows with Capacity Reduction on Link 1

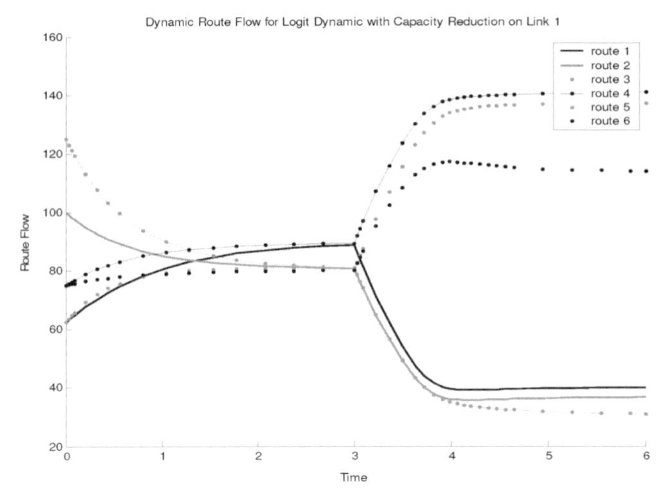

Figure 9 Logit Dynamic Route Flows with Capacity Reduction on Link 1

Scenario 2 is the example of the study on driver's behavior on an unrecoverable external shock (e.g., permanent network change or a long-term construction). Meanwhile, it is also of interest to study the traffic dynamic under some short-term shock, for example, incident, short-term construction, and lane closures. This is the motivation for the study of scenario 3. In this case, there is a severe construction, e.g., 80% of capacity reduction, occurred on link 1 at time t = 3. After three time units, the construction is cleared and the road condition is back to normal at time t = 6. Figures 10, 11 and 12 depict the PAP, BNN, and Logit traffic dynamics for this scenario. The route flows on route 1, 2 and 3 decrease steeply from time t = 3 to t = 6, while route flows on route 4, 5 and 6 increase dramatically in the same period because of the capacity reduction on link 1. After time point t = 6, the capacity on link 1 comes back to normal, the route flows on route 1, 2 and 3 begin to go up very fast, while the route flows on route 4, 5 and 6 start going down. Finally, the PAP and BNN dynamics converge to new UE flow $\mathbf{f}^* = (125, 44, 81, 125, 81, 44)'$, which is different from the original UE flow $\mathbf{f}^* = (125, 55.55, 69.45, 125, 69.45, 55.55)'$ in scenario 1. Indeed, the network condition in scenario 3 after time t = 6 is the same as that in scenario 1. However, the route flow vectors at equilibrium states in these two scenarios are different from each other. This example illustrates an important fact that the route flows at equilibrium may be different because of the impact of some network changes or on-off events, while the equilibrium link flows are the same. The explanation might be that initial route flow conditions for scenario 1 and 3 are different, and that the equilibrium route flows are not unique, thus they eventually converge to different equilibrium route flows. More details will be discussed in a subsequent paper, from the stability point of view. However, as the SUE route flow is unique, the Logit dynamic finally moves to the original SUE route flow as in scenario 1.

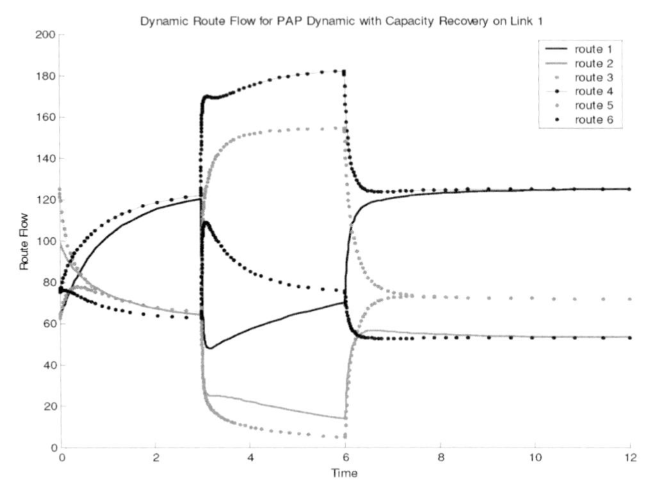

Figure 10 PAP Dynamic Route Flows with Capacity Recovery on Link 1

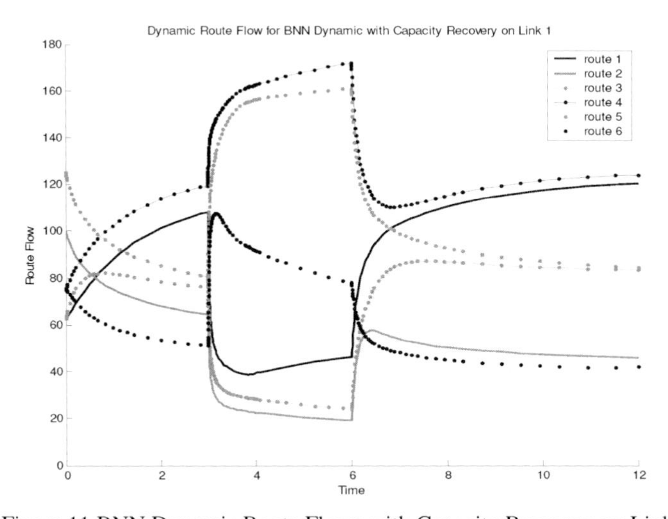

Figure 11 BNN Dynamic Route Flows with Capacity Recovery on Link 1

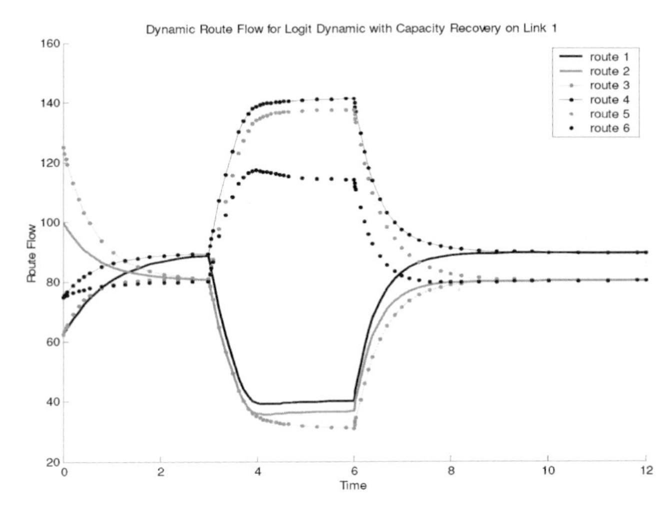

Figure 12 Logit Dynamic Route Flows with Capacity Recovery on Link 1

CONCLUDING REMARKS

In this paper, we present a new framework in modeling travelers' day-to-day route choice behavior adjustment process, which integrates the advantages of both deterministic ODE formulations and discrete-time stochastic processes. Under the assumption that the O-D demand is large, the mean flow dynamic closely approximates the underlying route flow stochastic process. The unique solution trajectory of the mean dynamic in finite-time horizon can provide us the dynamic traffic flow evolution given any initial traffic condition. This mean dynamic is a general day-to-day dynamics formulation in that many existing deterministic processes can be considered as its special cases, and their meaningful individual behavior explanations are recognized.

By constructing some reasonable traveler's stochastic learning schemes, we show the existence of user equilibrium even if one considers traveler's imperfect behavior under uncertainty. User equilibrium can be approached even if the traveler's behavior assumption is relaxed as an individual traveler only has access to limited information and exhibits limited rationality. User equilibrium, as well as the stochastic user equilibrium, might be the fixed points of the mean (deterministic) dynamic, therefore, they can be classified as the day-to-day deterministic equilibria, while the day-to-day stochastic equilibrium may denote the stationary flow distribution of the stochastic process (Watling and Hazelton, 2003).

This study also reveals the important linkage between the fixed points of day-to-day dynamics and the equilibria of traffic assignment problems. Therefore, some higher order Runge-Kutta methods in dynamical system literature may be exploited to solving traffic assignment problems. The case study illustrates an interesting observation that the equilibrium route flows might be changed due to the impact of some network changes or on-off events, while the equilibrium link flows are the same.

One theoretical investigation for future research is to study the local and global stability of the mean dynamic, especially when multiple traffic equilibria exist. For more applications, this research can be used to study detailed representations of temporal and spatial evolution of traffic congestion, thus providing the ability for transportation management authority to utilize the traffic control devices and route guidance systems to direct the traffic flows to more uncongested routes and improve the system performance.

APPENDIX 1

The proof of theorem 2 is mainly based on Kurtz (1970) and Sandholm (2006). For the sake of completeness, we show that the three conditions of theorem in Kurtz (1970) are satisfied.

Lemma 3 (Kurtz 1970) Let V be a Lipschitz continuous vector field. Suppose that for some sequence $\{\delta^N\}_{N=N_0}^{\infty}$ converging to 0, we have

(1) $\lim_{N\to\infty} \sup_{\mathbf{f}\in K^N} \left|V^N(\mathbf{f}) - V(\mathbf{f})\right| = 0$

(2) $\lim_{N\to\infty} \sup_{\mathbf{f}\in K^N} A^N(\mathbf{f}) < \infty$

(3) $\lim_{N\to\infty} \sup_{\mathbf{f}\in K^N} A^N_{\delta^N}(\mathbf{f}) = 0$

and that the initial conditions $\mathbf{f}_0^N \in K^N$ converges to $\mathbf{f}_0 \in K$, where $A^N(\mathbf{f}) = (N\bar{\lambda})E\left(\left|\delta^N\right|\right)$ and $A^N_{\delta^N}(\mathbf{f}) = (N\bar{\lambda})E\left(\left|\delta^N \mathbf{1}_{\delta>\delta^N}\right|\right)$. Let $\{\mathbf{f}_t\}_{t\geq 0}$ be the solution to the mean dynamic $\dot{\mathbf{f}} = V(\mathbf{f})$ from initial condition $\mathbf{f}_0$. Then for each $T < \infty$ and $\varepsilon > 0$, we have $\lim_{N\to\infty} P\left[\sup_{\mathbf{f}\in K^N} \left|\mathbf{f}_t^N - \mathbf{f}_t\right| < \varepsilon\right] = 1$.

Condition (1) of Lemma 3 is straightforward due to the right hand side of the mean dynamic (6) is Lipschitz continuous.

Condition (2) and (3) are also satisfied because the 2-norm of the flow increment ζ^N at any time is always either zero or $\sqrt{2}/N$. The proof is complete if we take the sequence $\delta^N = 2/\sqrt{N}$. □

ACKNOWLEDGEMENT

Comments from two anonymous reviewers are gratefully acknowledged on improving the paper. The authors also would like to thank David Boyce, Ding Zhang, W.Y. Sezto, and Wenlong Jin for their valuable suggestions.

REFERENCES

Balijepalli, N.C. and D.P. Watling (2005). Doubly dynamic equilibrium distribution approximation model for dynamic traffic assignment. *Transportation and Traffic Theory*, H.Mahmassani (Ed), Elsevier, Oxford, 741-760.

Brown, G.W. and J. von Neumann (1950). Solutions of games by differential equations. *Ann. Math. Studies,* **24**, 73-79.

Cantarella, G.E. and E. Cascetta (1995). Dynamic processes and equilibrium in transportation network: towards a unifying theory. *Transportation Science*, **29**, 305-329.

Cantarella, G.E., E. Cascetta, V. Adamo, and V. Astarita (1999). A doubly dynamic traffic assignment model. *Transportation and Traffic Theory*, .A. Ceder (Ed), Elsevier, Jerusalem, Israel, 373-396.

Cascetta, E., 1989. A stochastic process approach to the analysis of temporal dynamics in transportation network. *Transportation Research,* **23B**, 1-17.

Cascetta, E. and G.E. Cantarella (1991). A day-to-day and within-day dynamic stochastic assignment model. *Transportation Research*, **25A**, 277-291.

Chang, G.-L. and H.S. Mahmassani (1988). Travel time prediction and departure time adjustment behavior dynamics in a congested traffic system. *Transportation Research*, **22B**, 217-232.

Davis, G.A. and N.L. Nihan (1993). Large population approximations of a general stochastic traffic assignment model. *Operations Research*, **41**, 169-178.

Friesz, T.L., D.H. Berstein, N.J. Mehta, R.L. Tobin, and S. Ganjalizadeh (1994). Day-to-day dynamic network disequilibrium and idealized traveler information systems. *Operations Research*, **42**, 1120-1136.

Friesz, T.L., D.H. Berstein, and R. Stough (1996). Dynamic systems, variational inequalities, and control theoretic models for predicting time-varying urban network flows. *Transportation Science*, **30**, 14-31.

Friesz, T.L., and S. Shah (2001). An overview of nontraditional formulations of static and dynamic equilibrium network design. *Transportation Research*, **35B**, 5-21.

Friesz, T.L., D. Berstein and N. Kydes (2004). Dynamic congestion pricing in disequilibrium. *Networks and Spatial Economics*, **4**, 181-202.

Hazelton, M.L. and D.P. Watling (2004). Computation of equilibrium distributions of markov traffic assignment models. *Transportation Science*, **38**, 331-342.

Hofbauer, J. (2000). From Nash and Brown to Maynard Smith: Equilibria, Dynamics and ESS, *Selection*, **1**, 81-88.

Hofbauer, J. and K. Sigmund (2003). Evolutionary game dynamics. *BULLETIN (New Series) of the American Mathematical Society*, **40**, 479-519.

Huang, H.J. and W. Lam (2002). Modeling and solving the dynamic user equilibrium route and departure time choice problem in network with queues. *Transportation Research,* **36B**, 253-273.

Jin, W.L. (2006). A Dynamical system model of the traffic assignment problem. *Transportation Research, Part B, Vol. 41(1), pp.32-48.*

Kurtz, T. G. (1970). Solutions of ordinary differential equations as limits of pure jump markov processes. Journal of Applied Probability, **7**, 49-58.

Mahmassani H.S., and G.-L. Chang (1987). On boundedly rational user equilibrium in transportation systems. *Transportation Science*, **21**, 89-99.

Nagurney, A. (1993). *Network Economics: A Variational Inequality Approach.* Kluwer Academic Publishers, Norwell.

Nagurney, A., and D. Zhang (1996). *Projected Dynamical Systems and Variational Inequalities with Applications.* Kluwer Academic Publishers, Boston.

Nagurney, A. and D. Zhang (1997). Projected dynamical systems in the formulation, stability analysis, and computation of fixed-demand traffic network equilibria. *Transportation Science*, **31**, 147-158.

Jha, M., Madanat S., and Peeta, S. (1998). Perception updating and day-to-day travel choice dynamics in traffic networks with information provision. *Transportation Research,* **6C**, 189-212.

Peeta, S. and T.H. Yang (2003). Stability issues for dynamic traffic assignment. *Automatica,* **39**, 21-34.

Sandholm, W. (2001). Potential games with continuous player sets. *Journal of Economic Theory,* **97**, 81-108.

Sandholm, W. and R. Lahkar (2005). The payoff projected dynamic. Working paper, University of Wisconsin-Madison.

Sandholm, W. (2006). *Population Games and Evolutionary Dynamics.* MIT Press, in press.

Smith, M.J. (1984). The stability of a dynamic model of traffic assignment – an application of a method of Lyapunov. *Transportation Science*, **18**, 259-304.

Smith M.J. (1983). The existence and calculation of traffic equilibria. *Transportation Research,* ***17B,*** 291-301.

Smith, M.J. (1984). The stability of a dynamic model of traffic assignment – an application of a method of Lyapunov. *Transportation Science*, **18**, 259-304.

Smith, M. J., and M. B. Wisten (1995). A continuous day-to-day traffic assignment model and the existence of a continuous dynamic user equilibrium. *Annals of Operations Research*, **60**, 59-79.

Swinkels, J. (1993). Adjustment dynamics and rational play in games. *Games Econ. Behav.* **5**, 455-84.

Wardrop, J. G. (1952). Some theoretical aspects of road traffic research. *Proc. Inst. Civil Eng.*, Part II, **1**, 325-378.

Watling, D. (1999). Stability of the stochastic equilibrium assignment problem: a dynamical systems approach. *Transportation Research*, **33B**, 281-312.

Watling, D. and M.L. Hazelton (2003). The dynamics and equilibria of day-to-day assignment models. *Networks and Spatial Economics*, **3**, 349-370.

Weibull, J. (1995). *Evolutionary Game Theory.* The MIT Press, Cambridge, MA.

Yang, F. (2005). An evolutionary game theory approach to the day-to-day traffic dynamics. Ph.D. Dissertation, University of Wisconsin-Madison.

Yang., F. and W.Y., Szeto (2006). Day-to-day dynamic congestion pricing policies towards system optimal. Proceedings of the First International Symposium on Dynamic Traffic Assignment, Leeds, United Kingdom, 266-275.

Transportation and Traffic Theory 2007
Edited by R.E. Allsop, M.G.H. Bell and B.G. Heydecker

36

THE CO-EVOLUTION OF LAND USE AND ROAD NETWORKS

David Levinson, Feng Xie, and Shanjiang Zhu, University of Minnesota, Minneapolis, USA

INTRODUCTION

Transportation and land use are interdependent shapers of urban form. First, changes in land use alter travel demand patterns, which determine traffic flows on transportation infrastructure. Second, changed traffic flows drive the improvement of transportation facilities. Third, new transportation facilities change the accessibility pattern, which drives the re-location of activities and land uses. During this process, both transportation and land use are evolving constantly, leading to salient spatial transformations such as agglomeration and centralization over space and transportation networks. For example, as cities evolved in the first half of the 20th century, we saw a concentration of activities and development at the centers of cities. As freeways were constructed from the 1960s, roads also became more differentiated with regard to their functional designs and running speeds (certainly in the pre-auto era most unpaved streets were equally slow, with paved streets and highways and then freeways, some roads got much faster). Urban agglomeration and differentiated highway networks are referred to as hierarchical systems in this study.

In the context of the co-evolution of land use and road networks, this paper in particular examines the degree to which the dynamics of land use is reinforcing or counteracting hierarchies of road networks. By this we ask will a more hierarchical distribution of activities lead to a more or less hierarchical road network? Observation of historical evidence does not lead to a clear conclusion, as the development of a hierarchy of transit systems during the streetcar/subway era was accompanied with a concentration of development (especially employment) in the center of cities (from an undeveloped state), while the development of a hierarchical road network (from an underdeveloped and largely undifferentiated street system) occurred when those same cities were decentralized from a highly developed state. This paper aims to examine this question in a simulation environment with controlled initial conditions and quantitative measurements of spatial hierarchy.

The remainder of this paper is organized as follows: the next section presents a review of related literature, which is followed by an introduction to the simulation model developed for this study. Then the experiments are outlined, and results are reported, and some sensitivity analyses conducted. The conclusions summarize the findings and suggest future directions for research.

LITERATURE REVIEW

While there have been some investigations of the evolution of transportation infrastructure and that of urban land use separately, few have examined the integrated development of transportation and urban space in an evolutionary way, leaving the co-evolution of transportation and land use still poorly understood.

The investigation of the growth and transformation of transportation infrastructure dates back to 1960s, when a series of studies were conducted by geographers and transportation planners to replicate the changing topology and connectivity of road or rail networks (Garrison and Marble, 1962; Taaffe *et al.*, 1963; Morrill, 1965). The dynamic analysis of transportation networks, however, was based on heuristic and intuitive rules in these studies, due to a lack of understanding on its inherent mechanisms at that time. In recent years, a limited number of attempts have been made to model the dynamics of transportation networks in a more realistic way. Yamins *et al.* (2003) presented a simulation of road growing dynamics on a land use lattice that generates global features observed in urban transportation infrastructure. Van Nes and van der Zijpp (2000) and van Nes (2002) claimed that the emergence of hierarchy in transport networks is a natural phenomenon in maximizing performance while minimizing the resources needed, and also discussed the relation between hierarchy in transport networks and that in spatial structures. Yerra and Levinson (2005) and Levinson and Yerra (2006) incorporated a simplified travel demand model to predict traffic flows on a surface transportation network, and introduced independent agents to invest (disinvest) in individual roads according to the revenue and cost associated with forecasted traffic. They demonstrated that a network could evolve into a hierarchical structure from either a random or a uniform state, even based on completely decentralized decisions of autonomous roads. Previous studies, however, didn't integrate the dynamics of transportation networks with the development of urban space, with land use either ignored or taken as exogenous.

The evolution of urban space has been examined by another steam of studies. The pioneering work by von Thünen (1910) presented a monocentric city surrounded by agricultural land and predicted the rent and land use distribution for competing socio-economic groups. Christaller (1933) introduced central place theory and demonstrated that a hierarchy of central places will emerge on a homogenous plain to serve the surrounding market while minimizing transportation costs. Krugman (1996) explores the phenomenon of self-organization in urban space. He develops an edge city model to demonstrate how interdependent location decisions of businesses within a metropolitan area could lead to a polycentric pattern under the tension between centripetal and centrifugal forces. Based on these theoretical investigations, a host of empirical land use-transport models have been developed to forecast land use development while considering transportation as an important factor. One of the first that gained substantive interest was the Lowry model (Lowry, 1963). Since the 1980s, many integrated land use models have been applied in real cities and some have been developed into commercial packages. Examples include START (Bates *et al.*, 1991), LILT (Mackett, 1983, 1990, 1991), and URBANSIM (Alberi and Waddell, 2000; Waddell, 2002). A comprehensive review of these integrated land use-transport models has been provided by Timmermans (2003). In most of these models, the dynamics of urban space has been played out as the outcome of the location decisions made by residents and businesses, in which both accessibility to employment and accessibility to population play essential roles (Hansen, 1959; Guttenberg, 1960; Huff, 1963).

Although the concept of accessibility connects transportation with land use development, the change of transportation networks has seldom been considered in previous land use-transport models. A possible explanation is that these models are already complicated enough. They usually involve multiple modeling approaches, incorporate numerous constraints and assumptions, and are estimated from empirical data, unavoidably leading to a comprehensive modeling framework including a wide variety of components. These models are so specific and complex that 1) they are difficult to replicate; 2) the relationships between components are entangled and implicit; 3) the emergent large-scale patterns in space and network are difficult to recognize and analyze. Lee (1973) also has an important critique.

In contrast to those complicated and all-encompassing models that do not provide an explicit perspective, this paper models the integrated dynamics of land use and roads in as simple a way as possible that captures salient properties, enabling us to display and analyze the emergent hierarchy and agglomeration patterns of space and network on a large scale, as well as observe the interactions (reinforcement or counteraction) between the dynamics of roads and the development of land uses. The specific simplifications and assumptions made in our model specifications will be discussed later.

Extending Krugman (1996), Levinson and Yerra (2006), and Yerra and Levinson (2005), this paper models the co-evolution of land use and road network as a bottom-up, rather than a top-down process, by which interdependent location decisions of businesses (equivalently referred to as employment or jobs in this paper) and residents (also called population or workers or housing or resident workers) are incorporated, as well as investment decisions of autonomous roads based on predicted traffic on a network. Planners and engineers would argue that while market-based land use may be constrained by zoning and plans, transportation network investments are decisions that are now driven, or coordinated, by centralized organizations such as state departments of transportation or metropolitan planning organizations that make major investment decisions using a forecasting model and planning process to test and evaluate alternative scenarios. Local jurisdictions, of which there are many in some metropolitan areas, make investments on lower level roads. Certainly these organizations do affect new investment, but the decision to build or expand a link is also constrained by many facts on the ground, actual traffic on the link, competing parallel links, and complementary and upstream and downstream links, the costs of expansion, and limited budgets (Levinson and Karamalaputi 2003a,b). According to Krugman (1996), a self-organizing system of urban space and network will evolve into order and pattern, even based on simple, myopic, decentralized decisions of individual businesses and workers. If we can generate convincing collective representations of land use and network structure without any centralized planning or direction, perhaps planning is not as important in shaping urban areas as it is sometimes credited.

MODEL FRAMEWORK

A Simulator of Integrated Growth of Network Growth and Land-use (SIGNAL) is developed in this study to simulate the co-evolution of land use and road networks. An overview and inter-connection of these models is illustrated in Figure 1. The components of the model include *travel demand*, *road investment*, *accessibility*, and *land use*.

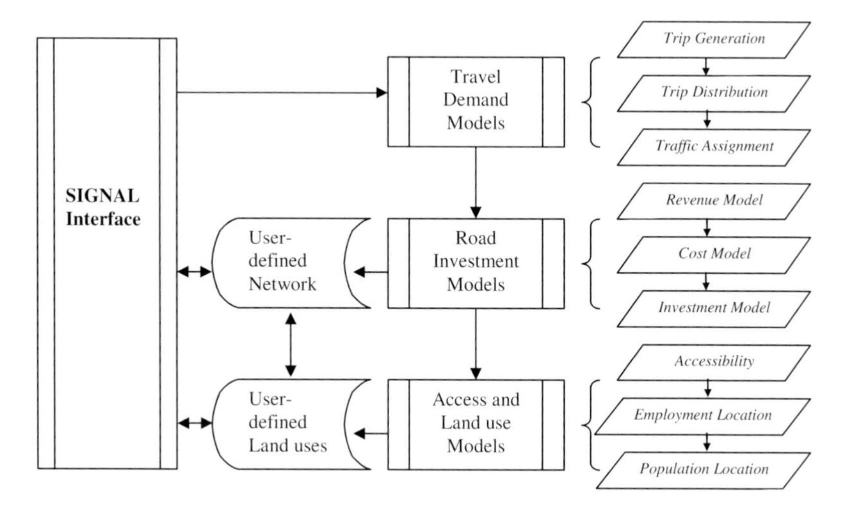

Figure 1. Overview of the SIGNAL model.

Travel demand models

The travel demand model converts population and employment data into traffic using the given network topology and determines the link flows by following the traditional planning steps of trip generation, trip distribution, and traffic assignment (for simplicity, a single mode is assumed) (Ortuzar and Willumsen, 2001).

A simplified *trip generation* model estimates the number of vehicle trips that originate from or are destined to a zone as a linear combination of the quantities of employment and population in this zone, without distinguishing trips by purpose:

$$O_i = \xi_1 E_i + \xi_2 P_i \tag{1}$$

$$D_i = \psi_1 E_i + \psi_2 P_i \tag{2}$$

where O_i and D_i represent the number of trips that originate in or are destined to Zone i respectively, while E_i and P_i are the employment (jobs) and population (resident workers) in this zone.

A doubly constrained gravity-based trip distribution model is adopted to match both trip generation and attraction of locations based on a negative exponential function that assumes the interactions of zones decreases with the travel time between them:

$$T_{ij} = K_i K_j O_i D_j e^{\varepsilon t_{ij}} \tag{3}$$

where T_{ij} is number of trips from zone i to zone j; K_i, K_j are balancing coefficients; O_i is the production of zone i; D_j is the attraction of zone j. The parameters in this trip distribution model have been calibrated using the empirical data in the Twin Cities, (see Levinson *et al.* (2006) for details). The variable t_{ij} is the generalized travel cost from zone i to zone j calculated as:

$$t_{ij} = \begin{cases} \sum_a (\delta^a_{i,j} t_a) + t_{m,i} + t_{m,j} & i \neq j \\ t_{m,i} & i = j \end{cases} \tag{4}$$

where $t_{m,i}$ and $t_{m,j}$ represent the generalized intra-zonal travel time in zones i and j, respectively. The generalized intra-zonal travel time captures a variety of costs incurred on trips that rise with land use intensity. It represents things like higher congestion levels, longer elevator waits in taller buildings, greater difficulty of finding parking, and taking longer to engage in parking that add to local travel time and travel cost in both zones. An intrazonal generalized time penalty acts as a surrogate for all of the above. Assuming a simple quadratic relationship between the generalized intra-zonal travel time and land use density in zone i, $t_{m,i}$ is calculated as:

$$t_{m,i} = t^0_m \left[1 + \left(\frac{G_i}{\bar{G}} \right)^2 \right] \tag{5}$$

where t^0_m is a specified base intra-zonal travel cost for all zones, G_i is the number of activities in zone i, while $\overline{G}$ represents the average number of activities across all zones. In our case,

$$G_i = E_i + P_i \quad (6)$$

where E_i and P_i represent the employment and population in zone i, respectively. They will be discussed later in the land use model.

The inter-zonal travel cost between zone i and zone j is computed as a summation of link travel cost along the shortest path from zone i to zone j, where t_a represents the generalized travel time that a vehicle spends on link a, while $\delta^a_{i,j}$ is a dummy variable equal to 1 if link a belongs to the shortest path from zone i to zone j and 0 otherwise. Dijkstra's Algorithm (Chachra *et al.* 1979) finds the shortest path from each node to all other nodes of the network. The generalized cost of travel time on link a is calculated by incorporating adding monetary cost (toll) (with an appropriate conversion factor) to the actual travel time on this link (tolls are charged by the road agent):

$$t_a = \frac{l_a}{v_a} + \frac{R_a / \eta}{f_a} \quad (7)$$

where l_a, v_a, f_a, and R_a respectively represent the length, average speed, traffic flow, and collected revenue of link a in a given time period. The parameter η represents the average value of time. The calculation of R_a will be discussed later.

A Stochastic User Equilibrium (SUE) is adopted in *traffic assignment* to predict route choices on a network according to perceived travel time, implementing Dial's Algorithm and Method of Successive Average (MSA) (Sheffi, 1985; Davis and Sanderson, 2002). Traffic assignment in a time period starts with the congested travel time resulting from the preceding time period, which makes the convergence in MSA much faster. The convergence rule in MSA specifies a maximal allowable link flow change equal to 0.5 (or a maximum of 100 iterations). A smaller maximal allowable flow change will result in a flow pattern that is closer to the equilibrium, but there is tradeoff between the accuracy and run time. The parameters in the model have also been calibrated by Levinson *et al.* (2006).

Road investment models

Road investment models describe the economic decisions of individual roads as autonomous agents. These decisions in terms of tolling, spending, and investing are abstracted in simple equation forms, also assuming autonomous roads make myopic decisions without considering cooperating with others or saving for the future.

A *revenue* model determines the toll a road collects during a given time period, depending on the traffic that uses this road. To ensure two parallel and opposite one-way links a and b that connect two nodes are always maintained at the same conditions, we assume that a single agent operates both links as a whole. Let f_a and f_b respectively represent the flow traversing link a and link b for a given time period, the total revenue collected on both links by the agent can be calculated as:

$$R_{a+b} = \tau l_a (f_a + f_b) \quad (8)$$

where τ is the regulated toll rate. A regulated toll rate across all the links simulates a distance based tax, which is the most common practice throughout the United States. Both link a and link b have the same length: $l_b = l_a$.

The *cost* to maintain links in their present usable conditions depends on link length, flow and capacity. Suppose link a and link b have the same capacity $C_a = C_b$, the overall spending of the agent operating links a and b is calculated as:

$$S_{a+b} = l_a C_a^{\sigma_2} (f_a^{\sigma_1} + f_b^{\sigma_1}) \quad (9)$$

where the coefficients σ_1 and σ_2 are specified flow and capacity powers in the equation.

An investment model assumes each agent spends all its available revenue at the end of a time period myopically, without saving it for the future. If the revenue exceeds the maintenance cost, remaining revenue will be invested to expand the capacity of subordinate links. In contrast, if the revenue is insufficient to cover the cost, road conditions will deteriorate and link capacity will drop until the link is eventually abandoned. This investment policy adopted by each agent can be expressed in a simplistic form as:

$$C_a^{k+1} = C_a^{k} \left(\frac{R^k_{a+b}}{S^k_{a+b}} \right)^{\rho} \quad (10)$$

where $C_a = C_b$ is the capacity of link a and b, which changes with iteration (k), respectively, while ρ is a specified coefficient that affects the speed of convergence. As implied by Equations (8)-(10) and specified parameters (detailed in Table 1), a network equilibrates when the flow on each link equals road capacity in quantity.

Zhang and Levinson (2005) estimated the relationship between the free flow speed of a link and its capacity in a log-linear model based on the empirical data in the Twin Cities. The log-linear relationship is adopted here to update the free flow speed (v_f) of a link after its capacity is changed:

$$v_{f,a} = \omega_1 + \omega_2 Ln(C_a) \quad (11)$$

where ω_1 and ω_2 are two coefficients in the log linear equation while C_a is the capacity of link a.

The relationship between the free flow speed and congested speed of a link is defined by the BPR function (Bureau of Public Roads, 1964) as:

$$v_{c,a} = v_{f,a}\left[1+\alpha*\left(\frac{f_a}{C_a}\right)^{\beta}\right] \tag{12}$$

where α and β are the coefficients of the function, assumed to equal 0.15 and 4.0, respectively.

Accessibility and land use models

Accessibility reflects the desirability of a place by calculating the opportunities and activities which are available from this place via a road network but are also impeded by the travel cost on the network. Suppose an urban space is divided into J Traffic Analysis Zones (TAZs) or land use cells that contain both employment (jobs) and population (workers). The *accessibility* in each cell (to employment and population) is computed respectively using a negative exponential measure:

$$A_{i,E} = \sum_{j=1}^{J} E_j e^{-\theta t_{ij}} \tag{13}$$

$$A_{i,P} = \sum_{j=1}^{J} P_j e^{-\theta t_{ij}} \tag{14}$$

where $A_{i,E}$ is the accessibility to employment (jobs) from zone i while $A_{i,P}$ is the accessibility to population (workers). The coefficient θ indicates how the accessibility of a zone declines with the increase of travel time to the zone. This coefficient basically represents the same idea with ε in Equation (3), indicating the impedance factor in travel that increases with travel time. Thus θ adopts the same value with ε, though the sensitivity of θ is tested later separately.

A *land use* model is then developed to reflect how the distribution of population and employment respond to the accessibility patterns, while keeping the total population and total employment constant. The land use model is simplified in the sense that accessibility to employment and accessibility to population are the only factors that affect the decision on locations made by businesses and workers. As accessibility is essential in the relationship between transportation and land use, other factors such as land price and administrative policies are excluded to keep this relationship succinct and clear, thus enabling simple accessibility-based rules to which independent location choices can be made. To be representative, our land use model contains both centripetal and centrifugal forces, that is, a force of attraction (e.g. economies of agglomeration) and a force of repulsion (a desire on the resident workers part for spatial separation, keeping all activities from locating at a single point). We assume people want to live near jobs, but far from other people (to maximize available space and to avoid potential competitors for jobs), while businesses (employment) want to be accessible both to other businesses and to people (who are their suppliers of labor and customers). The following stylized models are developed to track the dynamics of population and employment based on independent decisions of businesses with regard to their locations. The first group of equations describes the dynamics of businesses.

$$U_{i,E} = A_{i,E} + \lambda A_{i,P} \tag{15}$$

$$\overline{U_E} = \frac{\sum_{j=1}^{J}(U_{j,E}E_j)}{\sum_{j=1}^{J}E_j} \tag{16}$$

$$\frac{E_i^{k+1} - E_i^k}{E_i^k} = \gamma(U_{i,E}^k - \overline{U}_E^k) \tag{17}$$

The employment utility (desirability) of a zone is estimated as a linear combination of its accessibility to employment and to population in Equation (15). Note that both accessibility to employment and accessibility to population reinforce the employment desirability, indicating a strong centripetal force exists in shaping the pattern of employment (though intrazonal transportation costs do increase with density). The average utility that each business enjoys is calculated in Equation (16), and the influx of businesses to a zone in the next time period (iteration *k+1*) is proportional to the utility above the average that a business can enjoy in the zone as well as the total number of existing businesses, according to Equation (17). It can be easily proven by adding up Equation (17) for all zones that the total employment is ensured to be constant in these equations. The parameters λ and γ are two coefficients in the linear equations.

$$U_{i,P} = A_{i,E} - \mu A_{i,P} \tag{18}$$

$$\overline{U_P} = \frac{\sum_{j=1}^{n}(U_{j,P}P_j)}{\sum_{j=1}^{n}P_j} \tag{19}$$

$$\frac{P_i^{k+1} - P_i^k}{P_i^k} = \gamma(U_{i,P}^k - \overline{U}_P^k) \tag{20}$$

Similarly, the dynamics of population is described in Equations (18)-(20). The only difference lies in Equation (18), in which the residence disutility is determined by a centripetal force and a centrifugal force. A hedonic analysis of home sale prices in the Minneapolis-St. Paul region conducted by El-Geneidy and Levinson (2006) reveals that μ is near 1.0.

Figure 2 illustrates the feedback relationship between the network and land use variables within our system of co-evolution. An arrow with a plus (+) or minus (-) between two boxes shows a positive or negative relationship between the boxes. As can be seen, road expansion increases capacity, which improves free flow speed; the increased capacity increases cost, then forces the capacity back according to the investment rules. The improvement of travel time increases traffic flow, which increases the revenue and facilitates road expansion. The improvement of travel time also increases both accessibility to jobs and accessibility to houses. Employment density is positively associated with both accessibilities while population density is negatively impacted by accessibility to houses. Increased employment or population density increases intrazonal travel time, which offsets the improvement of travel time due to road investment.

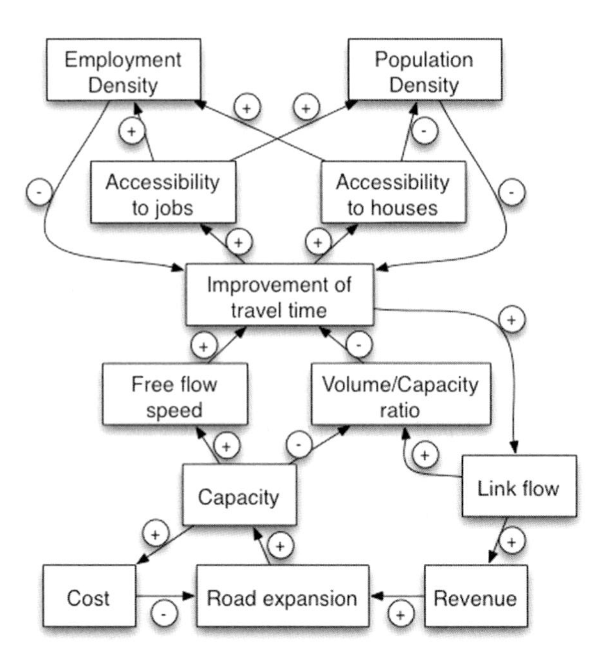

Figure 2. The feedback relationship in the transportation/land use system

After investing (or disinvesting) in each link in the network, computing accessibility, and relocating land uses, the time period is incremented and the whole process is repeated. In this study one time period represents a hypothetical year as the day-to-day traffic on the network is predicted and converted to yearly traffic for road investment models.

HYPOTHESES AND SIMULATION EXPERIMENTS

Two sets of experiments are conducted. The first fixes the land use, and explores how the network evolves in response to those fixed land uses. The second allows both the network and the land use to evolve simultaneously.

The research question is the degree to which hierarchies of road networks are reinforced or counteracted by the dynamics of land use. It is posited that this depends on initial land use and network conditions. Initially flat road networks become more concentrated, and initially concentrated networks become less so when land uses are allowed to vary rather than remain constant. That is, they reinforce to a point, and counteract beyond some point.

Simulation experiments were conducted in a hypothetical metropolitan area where both the population and employment are distributed over a two-dimensional grid. For simplicity, the experiments here are conducted over a square planar surface, stretching 20 km in both dimensions, divided into a 20X20 grid lattice of land use cells (400 zones). Each zone occupies one square kilometer of land. A total of 400,000 people are living in this city, which is equivalent to an average of 1,000 residents in each zone. Total employment equals 400,000 as well (and each resident holds a job). Two-way roads connect the centroids of each pair of adjacent zones, thus forming a 19X19 grid of road network as well, comprising 400 nodes and 1,520 links.

Table 1 lists parameters and their values for our experiments. As explained in Table 1, the toll rate and value of time are adopted from empirical estimates; the coefficients that define the log-linear relationship between link capacity and free flow speed are estimated by Zhang and Levinson (2005) using the empirical data in the Twin Cities. Among those parameters that are arbitrarily specified for the models, some of them were tested in the experiments using sensitivity analysis, which will be discussed later.

Each set of experiments was tested under two different sets of initial conditions. Both sets of initial conditions specify a uniform network in which the same initial conditions are specified for all the links except for their locations: each link is 1 km in length with a free flow speed of 35 km/h, and a capacity of 800 veh/h. The first specifies uniform land uses with both population and employment of each zone equal to 1,000; the second assumes a concentrated distribution of road capacity. The experiments are outlined in Table 2.

Table 1. Model parameters

Parameter	Description	Citation	Value	Source
ξ_1, ξ_2, ψ_1, ψ_2	Coefficients in trip generation and attraction	Eq.(1)	0.5 trips/person, 1.0 trips/person, 1.0 trips/person, 0.5 trips/person	Specified
ε	Trip distribution coefficient	Eq. (3)	0.05/min	Empirical calibrated
t_m^0	Base intra-zonal travel time	Eq. (5)	10 min	Specified
η	Value of time	Eq. (7)	\$10 /h	Empirical estimates
τ	Toll rate	Eq.(8)	\$1.0/ veh-km	Specified
σ_1, σ_2	Coefficients in cost model	Eq.(9)	0, 1	Specified*
ρ	Capacity reduction factor	Eq.(10)	0.25	Specified*
ω_1, ω_2	Coefficients in the *capacity-freeflow speed* loglinear function	Eq.(11)	-30.6 km/hr, 9.8	Empirical estimates
α, β	Coefficients in BPR function	Eq.(12)	0.15, 4.0	Typical values
θ	Impedance factor in accessibility model	Eq.(13), Eq.(14)	0.05/min	Specified*
λ	Coefficient in employment desirability model	Eq.(15)	1.0	Specified*
μ	Coefficient in population desirability model	Eq.(18)	1.0	Empirical estimate
γ	Coefficient in land use model	Eq.(17), Eq.(20)	1.0×10^{-6}	Specified

Note: Analyses were conducted on the sensitivity of asterisked parameters

Table 2. Specification of experiments

No.	Initial conditions			Dynamics	
	Link capacity	Employment	Population	Roads	Land uses
1a	Uniform	Uniform	Uniform	Evolving	Fixed
1b	Concentrated	Uniform	Uniform	Evolving	Fixed
2a	Uniform	Uniform	Uniform	Evolving	Evolving
2b	Concentrated	Uniform	Uniform	Evolving	Evolving

A series of measures of collective properties are developed to track the patterns in the experiments. The Gini index is adopted in this study to indicate the degree of spatial agglomeration for land use and network infrastructure. The Gini index has been widely adopted as a measure of spatial concentration (Krugman, 1991; Chatterjee, 2002). Chatterjee (2003) elaborates the computation of the Gini index based on the Lorenz Curve. The Gini index of land use (employment or population) is computed in this study to reflect how evenly land uses are distributed on the hypothetical space. The index is a number from zero to one, which is equal to zero when employment or population is uniformly located across all zones, while close to one when all employment or population is located in one zone. The more unevenly land use is distributed, the higher value the index is.

Similarly, the Gini index of road capacity is computed to reflect how evenly roads are developed. The index equals zero when all roads have the same capacity while it becomes higher when a larger portion of total capacities are occupied by a smaller number of roads.

In analogy with kinematics, measures of the moment of inertia (I) and the equivalent radius (r) are computed to reflect the spatial clustering patterns of land use and network infrastructure.

The moment of inertia for the spatial distribution of employment is computed as:

$$I = \sum_{j=1}^{n} E_j d_j^2 \qquad (21)$$

where E_j represents the employment of Zone j while d_j is the distance between the centroid of this zone and the center of the hypothetical metropolitan area.

The equivalent radius is then computed as:

$$r = \sqrt{I \Big/ \sum_{j=1}^{n} E_j} \qquad (22)$$

The equivalent radius r essentially reflects how far away employment is distributed from the center of a region. A radius of zero indicates all employment clusters in the center of the region while a larger radius indicates employment is located farther away from the center.

Similarly the equivalent radius can also be computed for the spatial distribution of population as well as road capacity.

RESULTS

Experiments 1(a) and1(b) allow roads to invest in their capacities while fixing the land use, these experiments are similar in nature to those presented in Yerra and Levinson (2005) and Levinson and Yerra (2006), though differing in specific parameters, the route assignment model, and initial conditions. Figure 3 illustrates the fluctuations of average link capacity for 1(a) and 1(b) in the first 50 iterations. As can be seen, with fixed land use road dynamics reaches equilibrium quickly. Whether starting from a uniform state with an average link capacity of 800 veh/h or from a concentrated state with an average capacity of 1426 veh/h, the network adjusts itself in response to the fixed land use pattern and converges to an average capacity of about 983 veh/h. The observation that initially flat network becomes more concentrated while the initially concentrated network becomes less so, and they tend to converge on the same level of average capacity suggests a stable hierarchical distribution of road capacity may emerge from different initial conditions.

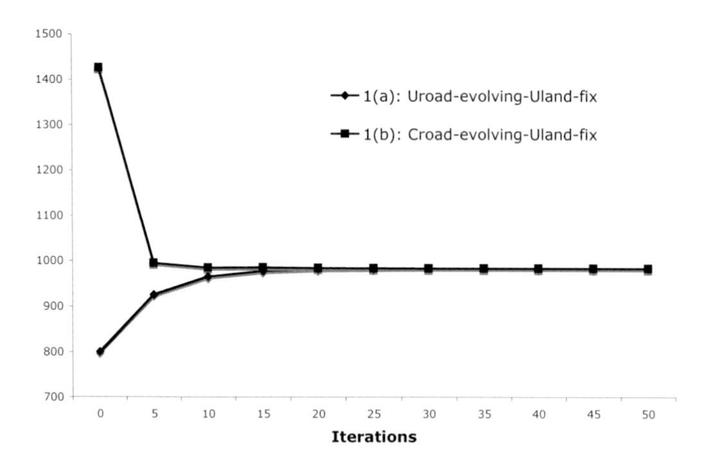

Figure 3. The fluctuations of average capacity with fixed land use

Experiments 2(a)-2(b), on the other hand, allow both land use and network to evolve, thus generating different network structures and land use patterns. The evolving spatial patterns of network and land use were analyzed by plotting the measures of Gini index and equivalent radius for link capacity, employment, and population on a time horizon over 1,000 iterations. Since significant changes in networks occurred during the first 50 or so iterations, the horizontal axis is plotted at a log scale. The plots are summarized in Figures 4 (i)-(iv). Each plot displays four fluctuations from uniform and concentrated initial network with and without land use dynamics, that is, Experiments 1(a), 1(b), 2(a), and 2(b).

Plots 4 (i) and Plot 4(ii) demonstrate how spatial patterns of road capacity distribution evolve over time from the perspectives of agglomeration (reflected by the Gini index) and centralization (reflected by equivalent radius), respectively. As already shown in Figure 3, Experiments 1(a) and 1(b) reached equilibrium with fixed land use within the first 50 iterations and remained unchanged thereafter, both resulting in a Gini index of 0.035 and an equivalent radius of 7.9 km. In Experiments 2(a) and 2(b), on the other hand, the network quickly adjusted its distribution of road capacity to the contemporary traffic pattern from its uniform or concentrated initial state in the first 50 iterations and then gradually changed as the land use evolved. After about 50 iterations, the Gini index in both experiments keeps

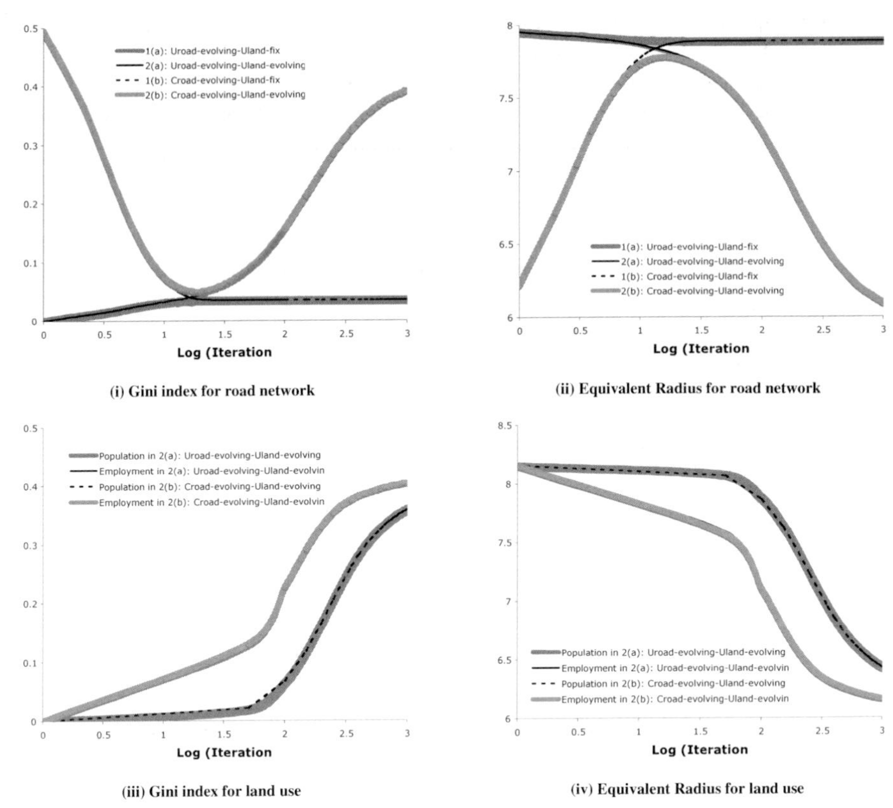

(i) Gini index for road network

(ii) Equivalent Radius for road network

(iii) Gini index for land use

(iv) Equivalent Radius for land use

Figure 4. Measures of spatial patterns

increasing while the radius is dropping, showing a strong trend of agglomeration and centralization of road capacity. More interestingly, whether starting from an uniform or concentrated network, the experiments allowing land use dynamics (2(a) and 2(b)) generate a consistently higher Gini index and lower radius compared to their counterparts with fixed land use, suggesting the evolution of land use distribution reinforces the differentiation of roads.

Plots 4(iii) and 4(iv) illustrate how the spatial patterns of population and employment evolve over time. Starting from a uniform network or a concentrated network, land use patterns display almost the same fluctuation (despite slight differences in the first 100 iterations). Although the Gini index and equivalent radius for both population and employment keep increasing, the distribution of population display a consistently lower Gini index and higher equivalent radius than that of employment, indicating employment has a stronger tendency of agglomeration and centralization, which is consistent with our assumption that employment wants to locate near to each other while people do not like to live together, but want to be near jobs.

Our findings can be further corroborated by the snapshots of emergent network patterns shown in Figure 5. Figure 5 displays two emergent networks over 1,000 iterations in Experiment 1(a) and Experiment 2(a), respectively. Different levels of capacity are displayed in five different levels of greyness in a relative scale. Obviously, the resulting network of Experiment 2(a) with evolving land use is more concentrated than that of 1(a) with fixed land use, suggesting land use dynamics reinforces the hierarchical distribution of road infrastructure in the context of co-evolution of network and land use. Figure 5(i) shows the emergence of beltways that are more important than internal roads, Figure 5(ii) does not have a similar beltway, roads just decline in importance with distance from the centre.

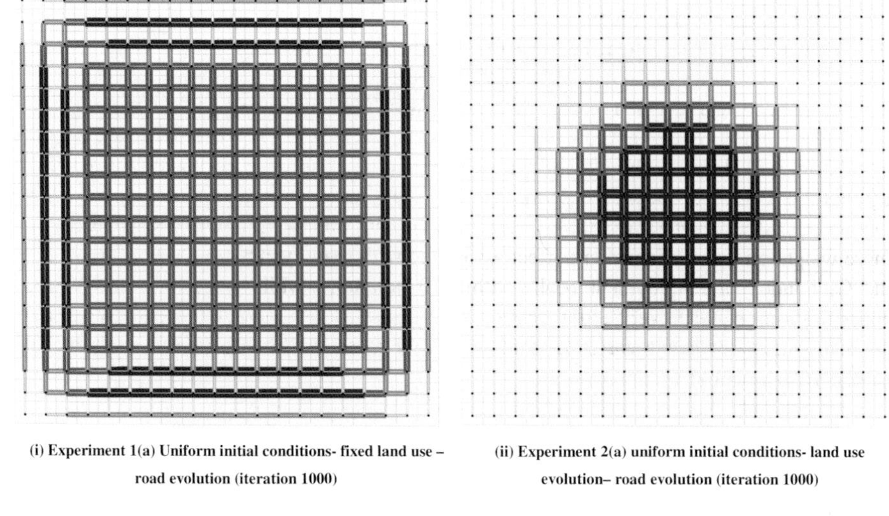

(i) Experiment 1(a) Uniform initial conditions- fixed land use – road evolution (iteration 1000)

(ii) Experiment 2(a) uniform initial conditions- land use evolution– road evolution (iteration 1000)

Figure 5. Emergent network patterns.

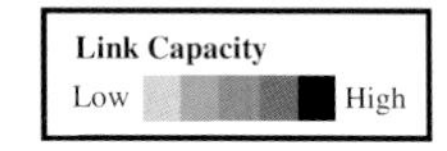

SENSITIVITY

The specified capacity power σ_2 in our road cost model affects the pattern of road infrastructure. A range of σ_2 was tested. Higher values σ_2 (say 1.5) impose a high maintenance cost on roads, and generate a shrinking network infrastructure over time (given initial capacities); lower values for σ_2, on the other hand, set the cost so low that the capacity expanded rapidly. For example, a value of 0.5 expands the average link capacities by 10 times in 20 years; finally the value of 1.0 was chosen for it generated a moderate and reasonable growth of network infrastructure with an increase of average road capacity from 800 veh/h to 1,015 veh/h over 1,000 iterations, which allows us to illustrate other salient points in the model. Another parameter ρ, capacity reduction factor in Equation (10), affects the speed of network convergence without changing the final converged pattern. Taking a higher value of 1.0, for example, the network in Experiment 1(a) converges within only 2 iterations.

Changing the specified values of two parameters λ and θ in the land use model may affect the emerging spatial patterns significantly. The coefficient λ in Equation (15) indicates the importance of accessibility to population in the location choices of employment, relative to accessibility to other employment. When λ equals zero, the location of employment only depends on the accessibility to jobs; while a large λ indicates employment more likely pursues a population-rich location. The accessibility reduction factor θ in Equations (13) and (14) determines how fast the accessibility of a place will decline with the increase of (generalized) travel cost to that place, reflecting the extent to which the change of travel time can affect the location of land use. Different values for the two parameters were tested and the results are summarized in Figure 6 (i)-(iv). To be concise, only the Gini index of land use in the uniform scenario is plotted. This analysis is based off of experiment 1(b).

As can be seen in Plots 6(i) and 6(ii), an increased λ magnifies the concentration of both population and employment. When λ equals 10, employment is rapidly attracted to population-rich places, making these places more attractive to population, and thus forming a positive feedback. As shown in Figure 6(iv), the decrease of θ (the accessibility impedance factor) basically exaggerates and quickens the concentration of employment because it puts more weight on the reinforcement effect associated with road dynamics. When θ equals 0.01, the Gini index of employment peaks within 70 iterations, while when it equals 0.1, the concentrations become very slow. Figure 6(iii) shows that although the concentration of population does not occur as fast as employment, the increase of θ still significantly quickens its concentration process.

CONCLUSIONS

This study models the co-evolution of land use and transportation network as a bottom-up process by which the re-location of activities and expansion of roads are driven by interdependent decisions of individual businesses, workers, and road agents according to simple decision rules. The model was kept simple so that collective spatial patterns of land use distribution can be displayed and analyzed without multiple conflating factors, while the sensitivity of these patterns were also discussed. The Gini index and equivalent radius were adopted to track down the evolution of spatial patterns.

This paper in particular examines the evolution of road networks under the context of the co-evolution of network and land use. Simulation experiments suggest that there may exist an inherently stable hierarchical distribution of road capacity so that flat networks become more

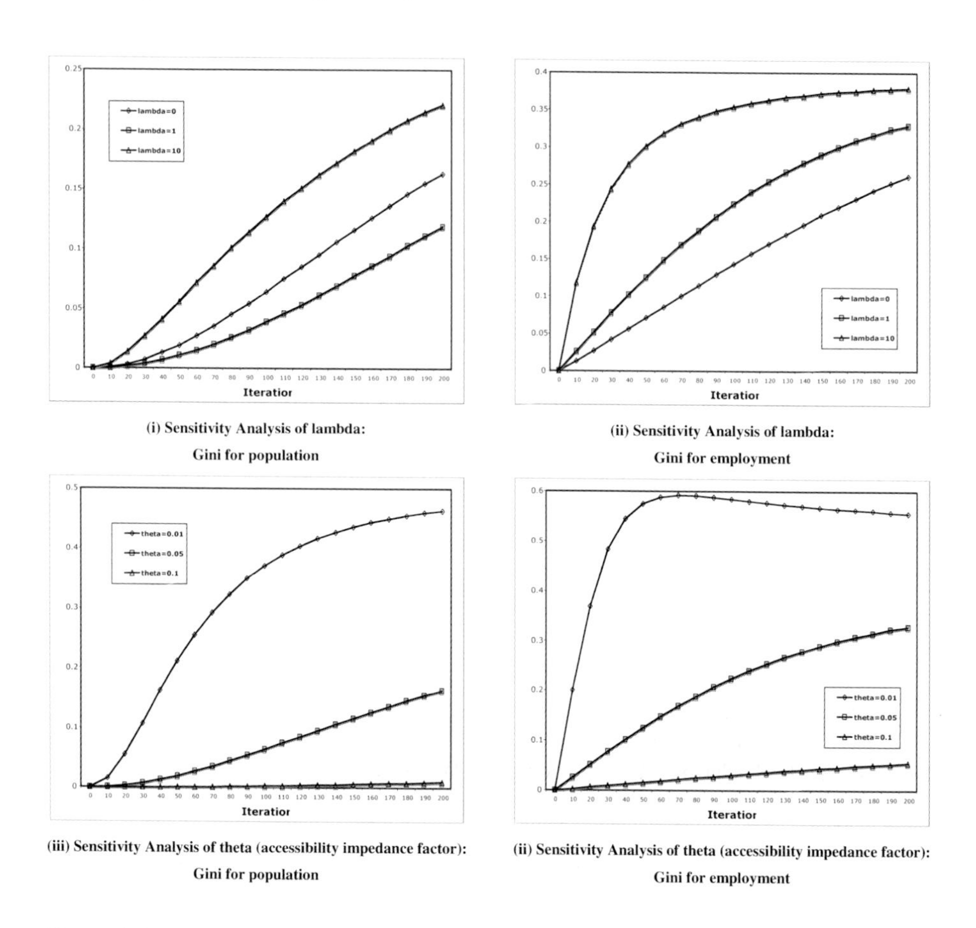

Figure 6. Sensitive analyses for two parameters

concentrated (and concentrated network become less concentrated) given a particular land use pattern. Experimental results also demonstrate that the agglomeration and centralization of road infrastructure is reinforced by the dynamics of employment and population under the tension of pushing and pulling forces. Land use organization and concentration make the road network more concentrated than it otherwise would be. Since it has been replicated in a self-organizing process based on completely decentralized decisions in this study, this reinforcement phenomenon is suggested to be an emergent property of the co-evolution of land use and road network.

As cities have evolved in the 20th century, we have seen a flattening of the density gradient of land use (centers of cities are relatively less important) as more highways are constructed and road networks became more hierarchical (certainly in the pre-auto era most unpaved roads were equally slow, with paved highways and then freeways, some roads got much faster). In a sense the faster roads have enabled decentralization of activities. Our simulation model can be employed in later studies to examine the concentration and flattening of land use that could be reinforced or counteracted by the evolution of road networks.

REFERENCES

Alberti, M. and P. Waddell (2000). An integrated urban development and ecological simulation model. *Integrated Assessment*, **1**, 215-227.

Bates, J., M. Brewer, P. Hanson, D. McDonald, and D.C. Simmonds (1991). Building a strategic model for Edinburgh. *Proceedings of Seminar D, PTRC 19th Summer Annual Meeting*. PTRC, London.

Bureau of Public Roads (1964). *Traffic Assignment Manual*. U.S. Dept. of Commerce, Urban Planning Division, Washington D.C.

Chachra V., P.M. Ghare, and J.M. Moore (1979). *Applications of Graph Theory Algorithms*. North Holland, New York.

Chatterjee, S. (2003). Agglomeration Economies: The Spark That Ignites a City? *Business Review*, **Q4**, 6-13

Chatterjee, S. and G. Carlino (2002). Employment Deconcentration: A New Perspective on America's Postwar Urban Evolution. *Journal of Regional Science*, **42**, 455-475.

Christaller, W. (1933). *Central Places in Southern Germany*. Jena: Gustav Fischer. English translation by C. W. Baskin. Prentice-Hall, London,1966.

Davis, G. and K. Sanderson (2002). Building Our Way Out Of Congestion. *Final Report MN/RC-2002-01*, Minnesota Department of Transportation.

El-Geneidy, A. M. and D. Levinson (2006). Access to Destinations: Development of Accessibility Measures. *Technical Report MN/RC-2006-16*, 31-37, Minnesota Department of Transportation.

Garrison, W.L. and D.F. Marble (1962). The Structure of Transportation Networks. U.S. *Army Transportation Command, Technical Report*, **62-II**, 73-88.

Guttenberg, A.Z. (1960). Urban structure and growth. *Journal of American Inst. Planners*, **26**, 104-110.

Hansen, W.G. (1959). How accessibility shapes land use. *Journal of American Inst. Planners*, **25**, 73-76.

Huff, D.L. (1963). A probabilistic analysis of shopping trade areas. *Land Economics*, **39**, 81-90.

Krugman, P. (1991). *Geography and trade*. MIT Press, Cambridge.

Krugman, P. (1996). *The Self-Organizing Economy*. Malden: Blackwell Publishers Inc., 53-100.

Lee, D., Jr. (1973). Requiem for Large-Scale Models. *Journal of the American Institute of Planners*, **40**, 163-78.

Levinson, D. and R. Karamalaputi (2003). Predicting the Construction of New Highway Links. *Journal of Transportation and Statistics* Vol. 6(2/3) 81–89.

Levinson, D and R. Karamalaputi (2003). Induced Supply: A Model of Highway Network Expansion at the Microscopic Level. *Journal of Transport Economics and Policy*, Volume 37, Part 3, September 2003, 297–318
Levinson, D. and B. Yerra (2006). Self Organization of Surface Transportation Networks. *Transportation Science*, **40(2)**, 179-188
Levinson, D., N. Montes de Oca, and F. Xie (2006). Beyond Business as Usual: Ensuring the Network We Want is the Network We Get. *Technical Report for Minnesota Department of Transportation*. St. Paul, Minnesota.
Lowry, I.S. (1963). Location parameters in the Pittsburgh model, *Papers and proceedings of the Regional Science Association*, **11**, 145-165.
Mackett, R.L. (1983). The Leeds Integrated Transport Model (LILT). *Supplementary Report 805*, Transport and Road Research Laboratory, Crowthorne.
Mackett, R.L. (1990). The systematic application of the LILT model to Dortmund, Leeds and Tokyo, *Transportation Reviews*, **10**, 323-338.
Mackett, R.L. (1991). LILT and MEPLAN: A comparative analysis of land-use and transport policies for Leeds, *Transportation Reviews*, **11**, 131-141.
Morrill, R.L. (1965). Migration and the Growth of Urban Settlement. *Lund Studies in Geography, Series B, Human Geography*, **26**, 65-82.
Ortuzar, Juan de Dios, and L.G.Willumsen (2001). *Modeling Transport*. John Wiley & Songs, LTD, West Sussex.
Sheffi, Y. (1985) *Urban Transportation Networks: equilibrium Analysis with Mathematical Programming Methods*. MIT Press, Cambridge.
Taaffe, E., R.L. Morrill, and P.R. Gould (1963). Transportation Expansion in Underdeveloped Countries: a Comparative Analysis. *Geographical Review*, **53**(4), 503-529.
Timmermans, H. (2003). The Saga of Integrated Land Use-Transport Modeling: How Many More Dreams Before We Wake Up? *Proceeding of 10th International Conference on Travel Behaviour Research*, Lucerne, August 10-15, 2003.
Van Nes, R. (2002). *Design of multimodal transport networks: a hierarchical approach*. Doctoral dissertation Delft University of Technology, published by DUP (Delft University Press) Science, ISBN 90-407-2314-1, 65-79.
Van Nes, R. and N.J. van der Zijpp (2000). Scale-factor 3 for hierarchical road networks: a natural phenomenon? In: *Proceedings TRAIL 6th Annual Congress*, Part 1, 2000, Delft University Press, ISBN 90-407-2135-1, 131-156.
Von Thünen, J. H. (1910). *Isolated State*. Jena: Gustav Fischer. English translation by C. W. Baskin. Pergamon Press, Oxford, New York, 1966.
Yamins, D., S. Rasmussen, and D. Fogel (2003). Growing Urban Roads. *Networks and Spatial Economics*, **3**, 69-85.
Yerra, B. and D. Levinson (2005). The emergence of hierarchy in transportation networks *Annals of Regional Science*, **39(3)**, 541-553.
Zhang, L. and D. Levinson (2005). Road Pricing with Autonomous Links. *Journal of the Transportation Research Board #1932*,147-155.

Previous Symposia and Proceedings

1. General Motors Research Laboratory, Warren, Michigan, 1959: Theory of Traffic Flow (Robert Herman, editor). Elsevier Publishing Company, Amsterdam, The Netherlands, 1961.

2. Road Research Laboratory, London, England, 1963: Proceedings of the Second International Symposium on the Theory of Road Traffic Flow, London 1963 (Joyce Almond, editor). Organisation for Economic Cooperation and Development (OECD), Paris, 1965.

3. Transportation Science Section ORSA, New York, 1965: Vehicular traffic science (L C Edie, R Herman and R Rothery, editors). American Elsevier Publishing Company Inc, New York, 1967.

4. University of Karlsruhe, Germany, 1968: Beiträge zur Theorie des Verkehrsflusses (Wilhelm Leutzbach and Paul Baron, editors). Straßenbau und Straßenverkehrstechnik, Heft 86, 1969.

5. University of California, Berkeley, 1971: Traffic Flow and Transportation (Gordon F Newell, editor). American Elsevier Publishing Company Inc, New York, 1972.

6. University of New South Wales, Sydney, Australia, 1974: Transportation and Traffic Theory (D J Buckley, editor). A H and A W Reed Pty Ltd, Sydney, 1974.

7. The Institute of Systems Science Research, Kyoto, Japan, 1977: Proceedings of the Seventh International Symposium on Transportation and Traffic Theory (Tsuna Sasaki and Takei Yamaoka, editors). The Institute of Systems Science Research, Kyoto, 1977.

8. University of Toronto, Canada, 1981: Proceedings of the Eighth International Symposium on Transportation and Traffic Theory (V F Hurdle, E Hauer and G N Steuart, editors). University of Toronto Press, Toronto, 1983.

9. Delft University of Technology, The Netherlands, 1984: Proceedings of the Ninth International Symposium on Transportation and Traffic Theory (J Volmuller and R Hamerslag, editors). VNU Science Press, Utrecht, 1984.

10. Massachusetts Institute of Technology, Boston, USA, 1987: Transportation and Traffic Theory (Nathan H Gartner and Nigel H M Wilson, editors). Elsevier Science Publishing Company Inc, New York, 1987.

11. University of Tokyo, Japan, 1990: Transportation and Traffic Theory (Masaki Koshi, editor). Elsevier Science Publishing Company Inc, New York, 1990.

12. University of California, Berkeley, USA, 1993: Transportation and Traffic Theory (Carlos F Daganzo, editor). Elsevier Science Publishing Company Inc, New York, 1993.

13. Institut National de Recherche sur les Transports et leur Sécurité, Lyon, France, 1996: Transportation and Traffic Theory (Jean-Baptiste Lesort, editor). Elsevier-Pergamon, Oxford, 1996.

14. Technion - Israel Institute of Technology, Jerusalem, Israel, 1999: Transportation and Traffic Theory (Avishi Ceder, editor). Elsevier-Pergamon, Oxford, 1999.

15. Transport Systems Centre, University of South Australia, Adelaide, Australia, 2002: Transportation and Traffic Theory in the 21st Century (Michael A P Taylor, editor). Pergamon, 2002.

16. University of Maryland, College Park, USA, 2005: Transportation and Traffic Theory: Flow, Dynamics and Human Interaction (Hani S Mahmassani, editor). Elsevier, 2005.

Index